DID YOU KNOW?

The following Technology Manuals are FREE when shrinkwrapped with the Kolman/Hill text:

LINEAR ALGEBRA LABS WITH MATLAB, 3E

David R. Hill and David E. Zitarelli

The order ISBN for Kolman/Hill and Hill/Zitarelli is 0131482734

ATLAST Computer Exercises for Linear Algebra, 2e

Steven Leon, Eugene Herman, and Richard Faulkenberry

The order ISBN for Kolman/Hill and the ATLAST manual is 0131482696

Understanding Linear Algebra using MATLAB

Erwin Kleinfeld and Margaret Kleinfeld

The order ISBN for Kolman/Hill and Kleinfeld/Kleinfeld is 013148270X

VISUALIZING LINEAR ALGEBRA USING MAPLE

Sandra Keith

The order ISBN for Kolman/Hill and Keith is 0131482726

Contact **www.prenhall.com** to order any of these titles

Linear Algebra 2e (0-13-536797-2)
Ken Hoffman and Ray Kunze

Applied Linear Algebra (0-13-085645-2)
Lorenzo Sadun

Applied Linear Algebra 3e (0-13-041260-0)
Ben Noble and Jim Daniel

Introduction to Linear Programming(0-13-035917-3)
Leonid Vaserstein

Introduction to Mathematical Programming (0-13-263765-0)
Russ Walker

Applied Algebra (0-13-067464-8)
Darel Hardy and Carol Walker

The Mathematics of Coding Theory (0-13-101967-8)
Paul Garrett

Introduction to Cryptography with Coding Theory (0-13-061814-4)
Wade Trappe and Larry Washington

Making, Breaking Codes (0-13-030369-0)
Paul Garrett

Invitation to Cryptology (0-13-088976-8)
Tom Barr

An Introduction to Dynamical Systems (0-13-143140-4)
R. Clark Robinson

OTHER PH TEXTS IN INTRODUCTORY LINEAR ALGEBRA

GOODAIRE
Linear Algebra: A First Course Pure and Applied (13-047017-1)

HARDY
Linear Algebra for Engineers and Scientists (0-13-906728-0)

SPENCE/INSEL/FRIEDBERG
Elementary Linear Algebra: A Matrix Approach (0-13-716722-9)

BRETSCHER
Linear Algebra with Applications 3e (0-13-145334-3)

KOLMAN/HILL
Elementary Linear Algebra 8e (0-13-045787-6)

LEON
Linear Algebra with Applications 6e (0-13-033781-1)

UHLIG
Transform Linear Algebra (0-13-041535-9)

HILL/KOLMAN
Modern Matrix Algebra (0-13-948852-9)

EDWARDS/PENNEY
Elementary Linear Algebra (0-13-258260-0)

PH TEXTS IN LINEAR ALGEBRA/DIFFERENTIAL EQUATIONS

EDWARDS/PENNEY
Differential Equations & Linear Algebra 2e (0-13-148146-0)

GOODE
Differential Equations and Linear Algebra 2e (0-13-263757-X)

FARLOW/HALL/MCDILL/WEST
Differential Equations and Linear Algebra (0-13-086250-9)

GREENBERG
Differential Equations and Linear Algebra (0-13-011118-X)

INTRODUCTORY LINEAR ALGEBRA
AN APPLIED FIRST COURSE

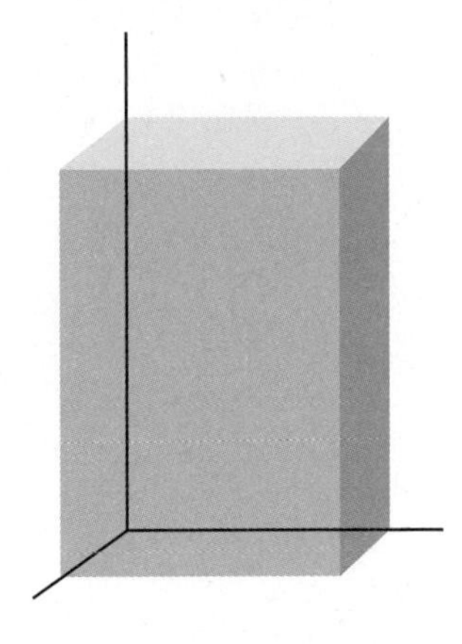

EIGHTH EDITION

INTRODUCTORY LINEAR ALGEBRA

AN APPLIED FIRST COURSE

Bernard Kolman
Drexel University

David R. Hill
Temple University

Upper Saddle River, New Jersey 07458

Library of Congress Cataloging-in-Publication Data

Kolman, Bernard, Hill, David R.

Introductory linear algebra: an applied first course-8th ed./ Bernard Kolman, David R. Hill

p. cm.

Rev. ed. of: Introductory linear algebra with applications. 7th ed. c2001.

Includes bibliographical references and index.

ISBN 0-13-143740-2

1. Algebras, Linear. I. Hill, David R. II. Kolman, Bernard. Introductory linear algebra with applications. III. Title.

QA184.2.K65 2005

512'.5--dc22 2004044755

Executive Acquisitions Editor: *George Lobell*
Editor-in-Chief: *Sally Yagan*
Production Editor: *Jeanne Audino*
Assistant Managing Editor: *Bayani Mendoza de Leon*
Senior Managing Editor: *Linda Mihatov Behrens*
Executive Managing Editor: *Kathleen Schiaparelli*
Vice President/Director of Production and Manufacturing: *David W. Riccardi*
Assistant Manufacturing Manager/Buyer: *Michael Bell*
Manufacturing Manager: *Trudy Pisciotti*
Marketing Manager: *Halee Dinsey*
Marketing Assistant: *Rachel Beckman*
Art Director: *Kenny Beck*
Interior Designer/Cover Designer: *Kristine Carney*
Art Editor: *Thomas Benfatti*
Creative Director: *Carole Anson*
Director of Creative Services: *Paul Belfanti*
Cover Image: *Wassily Kandinsky*, Farbstudien mit Angaben zur Maltechnik, 1913, Städische Galerie im Lenbachhaus, Munich
Cover Image Specialist: *Karen Sanatar*
Art Studio: *Laserwords Private Limited*
Composition: *Dennis Kletzing*

Pearson Prentice Hall
Pearson Education, Inc.
Upper Saddle River, NJ 07458

Printed in the United States of America
10 9 8 7 6 5 4 3

ISBN 0-13-143740-2

Pearson Education Ltd., London
Pearson Education Australia Pty, Limited, Sydney
Pearson Education Singapore, Pte. Ltd.
Pearson Education North Asia Ltd., Hong Kong
Pearson Education Canada, Ltd., Toronto
Pearson Educacion de Mexico, S.A. de C.V.
Pearson Education Japan, Tokyo
Pearson Education Malaysia, Pte. Ltd.

To the memory of Lillie
and to Lisa and Stephen

B. K.

To Suzanne

D. R. H.

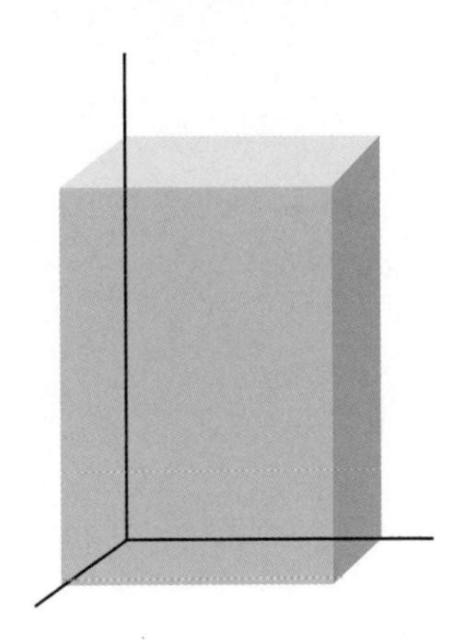

CONTENTS

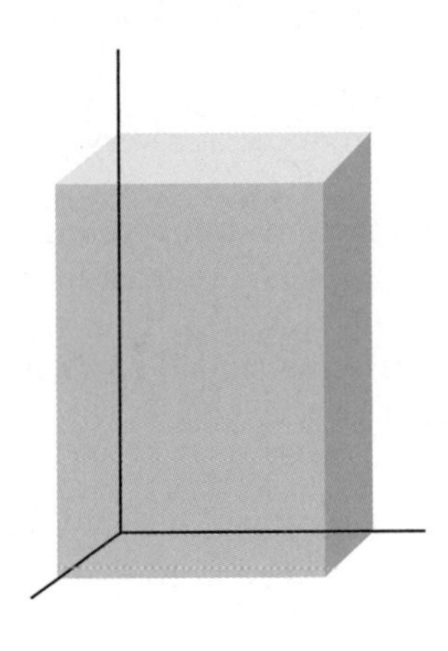

PREFACE

Material Covered

This book presents an introduction to linear algebra and to some of its significant applications. It is designed for a course at the freshman or sophomore level. There is more than enough material for a semester or quarter course. By omitting certain sections, it is possible in a one-semester or quarter course to cover the essentials of linear algebra (including eigenvalues and eigenvectors), to show how the computer is used, and to explore some applications of linear algebra. It is no exaggeration to say that with the many applications of linear algebra in other areas of mathematics, physics, biology, chemistry, engineering, statistics, economics, finance, psychology, and sociology, linear algebra is the undergraduate course that will have the most impact on students' lives. The level and pace of the course can be readily changed by varying the amount of time spent on the theoretical material and on the applications. Calculus is not a prerequisite; examples and exercises using very basic calculus are included and these are labeled "Calculus Required."

The emphasis is on the computational and geometrical aspects of the subject, keeping abstraction to a minimum. Thus we sometimes omit proofs of difficult or less-rewarding theorems while amply illustrating them with examples. The proofs that are included are presented at a level appropriate for the student. We have also devoted our attention to the essential areas of linear algebra; the book does not attempt to cover the subject exhaustively.

What Is New in the Eighth Edition

We have been very pleased by the widespread acceptance of the first seven editions of this book. The reform movement in linear algebra has resulted in a number of techniques for improving the teaching of linear algebra. The **Linear Algebra Curriculum Study Group** and others have made a number of important recommendations for doing this. In preparing the present edition, we have considered these recommendations as well as suggestions from faculty and students. Although many changes have been made in this edition, our objective has remained the same as in the earlier editions:

> **to develop a textbook that will help the instructor to teach and the student to learn the basic ideas of linear algebra and to see some of its applications.**

To achieve this objective, the following features have been developed in this edition:

- New sections have been added as follows:
  - Section 1.5, *Matrix Transformations*, introduces at a very early stage some geometric applications.
  - Section 2.1, *An Introduction to Coding*, along with supporting material on bit matrices throughout the first six chapters, provides an introduction to the basic ideas of coding theory.
  - Section 7.3, *More on Coding*, develops some simple codes and their basic properties related to linear algebra.
- More geometric material has been added.
- New exercises at all levels have been added. Some of these are more open-ended, allowing for exploration and discovery, as well as writing.
- More illustrations have been added.
- MATLAB M-files have been upgraded to more modern versions.
- Key terms have been added at the end of each section, reflecting the increased emphasis in mathematics on communication skills.
- True/false questions now ask the student to justify his or her answer, providing an additional opportunity for exploration and writing.
- Another 25 true/false questions have been added to the cumulative review at the end of the first ten chapters.
- A glossary, new to this edition, has been added.

Exercises

The exercises in this book are grouped into three classes. The first class, *Exercises*, contains routine exercises. The second class, *Theoretical Exercises*, includes exercises that fill in gaps in some of the proofs and amplify material in the text. Some of these call for a verbal solution. In this technological age, it is especially important to be able to write with care and precision; therefore, exercises of this type should help to sharpen such skills. These exercises can also be used to raise the level of the course and to challenge the more capable and interested student. The third class consists of exercises developed by David R. Hill and are labeled by the prefix ML (for MATLAB). These exercises are designed to be solved by an appropriate computer software package.

Answers to all odd-numbered numerical and ML exercises appear in the back of the book. At the end of Chapter 10, there is a cumulative review of the introductory linear algebra material presented thus far, consisting of 100 true/false questions (with answers in the back of the book). The **Instructor's Solutions Manual**, containing answers to all even-numbered exercises and solutions to all theoretical exercises, is available (to instructors only) at no cost from the publisher.

Presentation

We have learned from experience that at the sophomore level, abstract ideas must be introduced quite gradually and must be supported by firm foundations. Thus we begin the study of linear algebra with the treatment of matrices as mere arrays of numbers that arise naturally in the solution of systems of linear equations—a problem already familiar to the student. Much attention has been devoted from one edition to the next to refine and improve the pedagogical aspects of the exposition. The abstract ideas are carefully balanced by the considerable emphasis on the geometrical and computational foundations of the subject.

Material Covered

Chapter 1 deals with matrices and their properties. Section 1.5, *Matrix Transformations*, new to this edition, provides an early introduction to this important topic. This chapter is comprised of two parts: The first part deals with matrices and linear systems and the second part with solutions of linear systems. Chapter 2 (optional) discusses applications of linear equations and matrices to the areas of coding theory, computer graphics, graph theory, electrical circuits, Markov chains, linear economic models, and wavelets. Section 2.1, *An Introduction to Coding*, new to this edition, develops foundations for introducing some basic material in coding theory. To keep this material at a very elementary level, it is necessary to use lengthier technical discussions. Chapter 3 presents the basic properties of determinants rather quickly. Chapter 4 deals with vectors in R^n. In this chapter we also discuss vectors in the plane and give an introduction to linear transformations. Chapter 5 (optional) provides an opportunity to explore some of the many geometric ideas dealing with vectors in R^2 and R^3; we limit our attention to the areas of cross product in R^3 and lines and planes.

In Chapter 6 we come to a more abstract notion, that of a vector space. The abstraction in this chapter is more easily handled after the material covered on vectors in R^n. Chapter 7 (optional) presents three applications of real vector spaces: QR-factorization, least squares, and Section 7.3, *More on Coding*, new to this edition, introducing some simple codes. Chapter 8, on eigenvalues and eigenvectors, the pinnacle of the course, is now presented in three sections to improve pedagogy. The diagonalization of symmetric matrices is carefully developed.

Chapter 9 (optional) deals with a number of diverse applications of eigenvalues and eigenvectors. These include the Fibonacci sequence, differential equations, dynamical systems, quadratic forms, conic sections, and quadric surfaces. Chapter 10 covers linear transformations and matrices. Section 10.4 (optional), *Introduction to Fractals*, deals with an application of a certain nonlinear transformation. Chapter 11 (optional) discusses linear programming, an important application of linear algebra. Section 11.4 presents the basic ideas of the theory of games. Chapter 12, provides a brief introduction to MATLAB (which stands for MATRIX LABORATORY), a very useful software package for linear algebra computation, described below.

Appendix A covers complex numbers and introduces, in a brief but thorough manner, complex numbers and their use in linear algebra. Appendix B presents two more advanced topics in linear algebra: inner product spaces and composite and invertible linear transformations.

Applications

Most of the applications are entirely independent; they can be covered either after completing the entire introductory linear algebra material in the course or they can be taken up as soon as the material required for a particular application has been developed. Brief Previews of most applications are given at appropriate places in the book to indicate how to provide an immediate application of the material just studied. The chart at the end of this Preface, giving the prerequisites for each of the applications, and the Brief Previews will be helpful in deciding which applications to cover and when to cover them.

Some of the sections in Chapters 2, 5, 7, 9, and 11 can also be used as independent student projects. Classroom experience with the latter approach has met with favorable student reaction. Thus the instructor can be quite selective both in the choice of material and in the method of study of these applications.

End of Chapter Material

Every chapter contains a summary of *Key Ideas for Review*, a set of supplementary exercises (answers to all odd-numbered numerical exercises appear in the back of the book), and a chapter test (all answers appear in the back of the book).

MATLAB Software

Although the ML exercises can be solved using a number of software packages, in our judgment MATLAB is the most suitable package for this purpose. MATLAB is a versatile and powerful software package whose cornerstone is its linear algebra capability. MATLAB incorporates professionally developed quality computer routines for linear algebra computation. The code employed by MATLAB is written in the C language and is upgraded as new versions of MATLAB are released. MATLAB is available from The Math Works, Inc., 24 Prime Park Way, Natick, MA 01760, (508) 653-1415; e-mail: **info@mathworks.com** and is not distributed with this book or the instructional routines developed for solving the ML exercises. The Student Edition of MATLAB also includes a version of *Maple*, thereby providing a symbolic computational capability.

Chapter 12 of this edition consists of a brief introduction to MATLAB's capabilities for solving linear algebra problems. Although programs can be written within MATLAB to implement many mathematical algorithms, *it should be noted that the reader of this book is not asked to write programs. The user is merely asked to use* MATLAB *(or any other comparable software package) to solve specific numerical problems.* Approximately 24 instructional M-files have been developed to be used with the ML exercises in this book and are available from the following Prentice Hall Web site: **www.prenhall.com/kolman**. These M-files are designed to transform many of MATLAB's capabilities into courseware. This is done by providing pedagogy that allows the student to interact with MATLAB, thereby letting the student think through all the steps in the solution of a problem and relegating MATLAB to act as a powerful calculator to relieve the drudgery of a tedious computation. Indeed, this is the ideal role for MATLAB (or any other similar package) in a beginning linear algebra course, for in this course, more than in many others, the tedium of lengthy computations makes it almost impossible to solve a modest-size problem. Thus, by introducing pedagogy and reining in the power of MATLAB, these M-files provide a working partnership between the student and the computer. Moreover, the introduction to a powerful tool such as MATLAB early in the student's college career opens the way for other software support in higher-level courses, especially in science and engineering.

Supplements

Student Solutions Manual (0-13-143741-0). Prepared by Dennis Kletzing, Stetson University, and Nina Edelman and Kathy O'Hara, Temple University, contains solutions to all odd-numbered exercises, both numerical and theoretical. It can be purchased from the publisher.

Instructor's Solutions Manual (0-13-143742-9). Contains answers to all even-numbered exercises and solutions to all theoretical exercises—is available (to instructors only) at no cost from the publisher.

Optional combination packages. Provide a computer workbook free of charge when packaged with this book.

- *Linear Algebra Labs with MATLAB*, by David R. Hill and David E. Zitarelli, 3rd edition, ISBN 0-13-124092-7 (supplement and text).
- *Visualizing Linear Algebra with Maple*, by Sandra Z. Keith, ISBN 0-13-124095-1 (supplement and text).
- *ATLAST Computer Exercises for Linear Algebra*, by Steven Leon, Eugene Herman, and Richard Faulkenberry, 2nd edition, ISBN 0-13-124094-3 (supplement and text).
- *Understanding Linear Algebra with MATLAB*, by Erwin and Margaret Kleinfeld, ISBN 0-13-124093-5 (supplement and text).

Prerequisites for Applications

Prerequisites for Applications

Section 2.1	Material on bits in Chapter 1
Section 2.2	Section 1.4
Section 2.3	Section 1.5
Section 2.4	Section 1.6
Section 2.5	Section 1.6
Section 2.6	Section 1.7
Section 2.7	Section 1.7
Section 5.1	Section 4.1 and Chapter 3
Section 5.2	Sections 4.1 and 5.1
Section 7.1	Section 6.8
Section 7.2	Sections 1.6, 1.7, 4.2, 6.9
Section 7.3	Section 2.1
Section 9.1	Section 8.2
Section 9.2	Section 8.2
Section 9.3	Section 9.2
Section 9.4	Section 8.3
Section 9.5	Section 9.4
Section 9.6	Section 9.5
Section 10.4	Section 8.2
Sections 11.1–11.3	Section 1.6
Section 11.4	Sections 11.1–11.3

To Users of Previous Editions:

During the 29-year life of the previous seven editions of this book, the book was primarily used to teach a sophomore-level linear algebra course. This course covered the essentials of linear algebra and used any available extra time to study selected applications of the subject. *In this new edition we have not changed the structural foundation for teaching the essential linear algebra material. Thus, this material can be taught in exactly the same manner as before.* The placement of the applications in a more cohesive and pedagogically unified manner together with the newly added applications and other material should make it easier to teach a richer and more varied course.

Acknowledgments

We are pleased to express our thanks to the following people who thoroughly reviewed the entire manuscript in the first edition: William Arendt, University of Missouri and David Shedler, Virginia Commonwealth University. In the second edition: Gerald E. Bergum, South Dakota State University; James O. Brooks, Villanova University; Frank R. DeMeyer, Colorado State University; Joseph Malkevitch, York College of the City University of New York; Harry W. McLaughlin, Rensselaer Polytechnic Institute; and Lynn Arthur Steen, St. Olaf's College. In the third edition: Jerry Goldman, DePaul University; David R. Hill, Temple University; Allan Krall, The Pennsylvania State University at University Park; Stanley Lukawecki, Clemson University; David Royster, The University of North Carolina; Sandra Welch, Stephen F. Austin State University; and Paul Zweir, Calvin College.

In the fourth edition: William G. Vick, Broome Community College; Carrol G. Wells, Western Kentucky University; Andre L. Yandl, Seattle University; and Lance L. Littlejohn, Utah State University. In the fifth edition: Paul Beem, Indiana University-South Bend; John Broughton, Indiana University of Pennsylvania; Michael Gerahty, University of Iowa; Philippe Loustaunau, George Mason University; Wayne McDaniels, University of Missouri; and Larry Runyan, Shoreline Community College. In the sixth edition: Daniel D. Anderson, University of Iowa; Jürgen Gerlach, Radford University; W. L. Golik, University of Missouri at St. Louis; Charles Heuer, Concordia College; Matt Insall, University of Missouri at Rolla; Irwin Pressman, Carleton University; and James Snodgrass, Xavier University. In the seventh edition: Ali A. Dad-del, University of California-Davis; Herman E. Gollwitzer, Drexel University; John Goulet, Worcester Polytechnic Institute; J. D. Key, Clemson University; John Mitchell, Rensselaer Polytechnic Institute; and Karen Schroeder, Bentley College.

In the eighth edition: Juergen Gerlach, Radford University; Lanita Presson, University of Alabama, Huntsville; Tomaz Pisanski, Colgate University; Mike Daven, Mount Saint Mary College; David Goldberg, Purdue University; Aimee J. Ellington, Virginia Commonwealth University.

We thank Vera Pless, University of Illinois at Chicago, for critically reading the material on coding theory.

We also wish to thank the following for their help with selected portions of the manuscript: Thomas I. Bartlow, Robert E. Beck, and Michael L. Levitan, all of Villanova University; Robert C. Busby, Robin Clark, the late Charles S. Duris, Herman E. Gollwitzer, Milton Schwartz, and the late John H. Staib, all of Drexel University; Avi Vardi; Seymour Lipschutz, Temple University; Oded Kariv, Technion, Israel Institute of Technology; William F. Trench, Trinity University; and Alex Stanoyevitch, the University of Hawaii; and instructors and students from many institutions in the United States and other countries, who shared with us their experiences with the book and offered helpful suggestions.

The numerous suggestions, comments, and criticisms of these people greatly improved the manuscript. To all of them goes a sincere expression of gratitude.

We thank Dennis Kletzing, Stetson University, who typeset the entire manuscript, the *Student Solutions Manual*, and the *Instructor's Manual*. He found a number of errors in the manuscript and cheerfully performed miracles under a very tight schedule. It was a pleasure working with him.

We thank Dennis Kletzing, Stetson University, and Nina Edelman and

Kathy O'Hara, Temple University, for preparing the *Student Solutions Manual*.

We should also like to thank Nina Edelman, Temple University, along with Lilian Brady, for critically reading the page proofs. Thanks also to Blaise deSesa for his help in editing and checking the solutions to the exercises.

Finally, a sincere expression of thanks to Jeanne Audino, Production Editor, who patiently and expertly guided this book from launch to publication; to George Lobell, Executive Editor; and to the entire staff of Prentice Hall for their enthusiasm, interest, and unfailing cooperation during the conception, design, production, and marketing phases of this edition.

Bernard Kolman
bkolman@mcs.drexel.edu

David R. Hill
hill@math.temple.edu

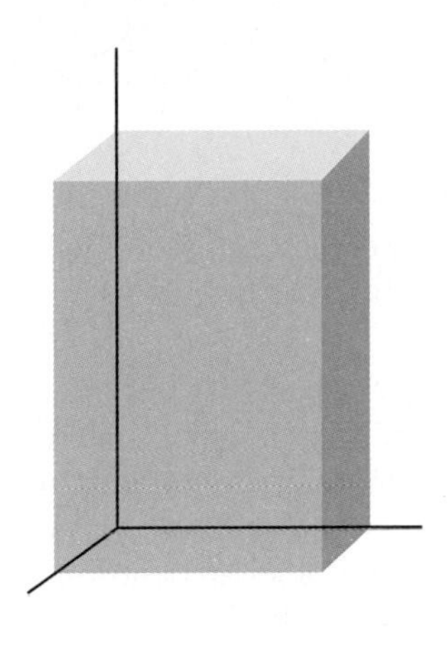

TO THE STUDENT

It is very likely that this course is unlike any other mathematics course that you have studied thus far in at least two important ways. First, it may be your initial introduction to abstraction. Second, it is a mathematics course that may well have the greatest impact on your vocation.

Unlike other mathematics courses, this course will not give you a toolkit of isolated computational techniques for solving certain types of problems. Instead, we will develop a core of material called linear algebra by introducing certain definitions and creating procedures for determining properties and proving theorems. Proving a theorem is a skill that takes time to master, so at first we will only expect you to read and understand the proof of a theorem. As you progress in the course, you will be able to tackle some simple proofs. We introduce you to abstraction slowly, keep it to a minimum, and amply illustrate each abstract idea with concrete numerical examples and applications. Although you will be doing a lot of computations, the goal in most problems is not merely to get the "right" answer, but to understand and explain how to get the answer and then interpret the result.

Linear algebra is used in the everyday world to solve problems in other areas of mathematics, physics, biology, chemistry, engineering, statistics, economics, finance, psychology, and sociology. Applications that use linear algebra include the transmission of information, the development of special effects in film and video, recording of sound, Web search engines on the Internet, and economic analyses. Thus, you can see how profoundly linear algebra affects you. A selected number of applications are included in this book, and if there is enough time, some of these may be covered in this course. Additionally, many of the applications can be used as self-study projects.

There are three different types of exercises in this book. First, there are computational exercises. These exercises and the numbers in them have been carefully chosen so that almost all of them can readily be done by hand. When you use linear algebra in real applications, you will find that the problems are much bigger in size and the numbers that occur in them are not always "nice." This is not a problem because you will almost certainly use powerful software to solve them. A taste of this type of software is provided by the third type of exercises. These are exercises designed to be solved by using a computer and MATLAB, a powerful matrix-based application that is widely used in industry. The second type of exercises are theoretical. Some of these may ask you to prove a result or discuss an idea. In today's world, it is not enough to be able to compute an answer; you often have to prepare a report discussing your solution, justifying the steps in your solution, and interpreting your results.

These types of exercises will give you experience in writing mathematics. Mathematics uses words, not just symbols.

How to Succeed in Linear Algebra

- Read the book slowly with pencil and paper at hand. You might have to read a particular section more than once. Take the time to verify the steps marked "verify" in the text.
- Make sure to do your homework on a timely basis. If you wait until the problems are explained in class, you will miss learning how to solve a problem by yourself. Even if you can't complete a problem, try it anyway, so that when you see it done in class you will understand it more easily. You might find it helpful to work with other students on the material covered in class and on some homework problems.
- Make sure that you ask for help as soon as something is not clear to you. Each abstract idea in this course is based on previously developed ideas—much like laying a foundation and then building a house. If any of the ideas are fuzzy to you or missing, your knowledge of the course will not be sturdy enough for you to grasp succeeding ideas.
- Make use of the pedagogical tools provided in this book. At the end of each section we have a list of key terms; at the end of each chapter we have a list of key ideas for review, supplementary exercises, and a chapter test. At the end of the first ten chapters (completing the core linear algebra material in the course) we have a comprehensive review consisting of 100 true/false questions that ask you to justify your answer. Finally, there is a glossary for linear algebra at the end of the book. Answers to the odd-numbered exercises appear at the end of the book. The *Student Solutions Manual* provides detailed solutions to all odd-numbered exercises, both numerical and theoretical. It can be purchased from the publisher (ISBN 0-13-143742-9).

We assure you that your efforts to learn linear algebra well will be amply rewarded in other courses and in your professional career.

We wish you much success in your study of linear algebra.

Bernard Kolman

David R. Hill

INTRODUCTORY LINEAR ALGEBRA

AN APPLIED FIRST COURSE

CHAPTER 1

LINEAR EQUATIONS AND MATRICES

1.1 LINEAR SYSTEMS

A good many problems in the natural and social sciences as well as in engineering and the physical sciences deal with equations relating two sets of variables. An equation of the type

$$ax = b,$$

expressing the variable b in terms of the variable x and the constant a, is called a **linear equation**. The word *linear* is used here because the graph of the equation above is a straight line. Similarly, the equation

$$a_1x_1 + a_2x_2 + \cdots + a_nx_n = b, \tag{1}$$

expressing b in terms of the variables $x_1, x_2, \ldots, x_n$ and the known constants $a_1, a_2, \ldots, a_n$, is called a **linear equation**. In many applications we are given b and the constants $a_1, a_2, \ldots, a_n$ and must find numbers $x_1, x_2, \ldots, x_n$, called **unknowns**, satisfying (1).

A **solution** to a linear equation (1) is a sequence of n numbers $s_1, s_2, \ldots, s_n$, which has the property that (1) is satisfied when $x_1 = s_1$, $x_2 = s_2$, $\ldots$, $x_n = s_n$ are substituted in (1).

Thus $x_1 = 2$, $x_2 = 3$, and $x_3 = -4$ is a solution to the linear equation

$$6x_1 - 3x_2 + 4x_3 = -13,$$

because

$$6(2) - 3(3) + 4(-4) = -13.$$

This is not the only solution to the given linear equation, since $x_1 = 3$, $x_2 = 1$, and $x_3 = -7$ is another solution.

More generally, a **system of m linear equations in n unknowns $x_1, x_2, \ldots, x_n$**, or simply a **linear system**, is a set of m linear equations each in n unknowns. A linear system can be conveniently denoted by

$$\begin{array}{ccccccccc} a_{11}x_1 & + & a_{12}x_2 & + & \cdots & + & a_{1n}x_n & = & b_1 \\ a_{21}x_1 & + & a_{22}x_2 & + & \cdots & + & a_{2n}x_n & = & b_2 \\ \vdots & & \vdots & & & & \vdots & & \vdots \\ a_{m1}x_1 & + & a_{m2}x_2 & + & \cdots & + & a_{mn}x_n & = & b_m. \end{array} \tag{2}$$

The two subscripts i and j are used as follows. The first subscript i indicates that we are dealing with the ith equation, while the second subscript j is associated with the jth variable x_j. Thus the ith equation is

$$a_{i1}x_1 + a_{i2}x_2 + \cdots + a_{in}x_n = b_i.$$

In (2) the a_{ij} are known constants. Given values of $b_1, b_2, \ldots, b_m$, we want to find values of $x_1, x_2, \ldots, x_n$ that will satisfy each equation in (2).

A **solution** to a linear system (2) is a sequence of n numbers $s_1, s_2, \ldots, s_n$, which has the property that each equation in (2) is satisfied when $x_1 = s_1$, $x_2 = s_2, \ldots, x_n = s_n$ are substituted in (2).

To find solutions to a linear system, we shall use a technique called the **method of elimination**. That is, we eliminate some of the unknowns by adding a multiple of one equation to another equation. Most readers have had some experience with this technique in high school algebra courses. Most likely, the reader has confined his or her earlier work with this method to linear systems in which $m = n$, that is, linear systems having as many equations as unknowns. In this course we shall broaden our outlook by dealing with systems in which we have $m = n$, $m < n$, and $m > n$. Indeed, there are numerous applications in which $m \neq n$. If we deal with two, three, or four unknowns, we shall often write them as x, y, z, and w. In this section we use the method of elimination as it was studied in high school. In Section 1.5 we shall look at this method in a much more systematic manner.

EXAMPLE 1 The director of a trust fund has \$100,000 to invest. The rules of the trust state that both a certificate of deposit (CD) and a long-term bond must be used. The director's goal is to have the trust yield \$7800 on its investments for the year. The CD chosen returns 5% per annum and the bond 9%. The director determines the amount x to invest in the CD and the amount y to invest in the bond as follows:

Since the total investment is \$100,000, we must have $x + y = 100{,}000$. Since the desired return is \$7800, we obtain the equation $0.05x + 0.09y = 7800$. Thus, we have the linear system

$$\begin{aligned} x + y &= 100{,}000 \\ 0.05x + 0.09y &= 7800. \end{aligned} \tag{3}$$

To eliminate x, we add (-0.05) times the first equation to the second, obtaining

$$\begin{aligned} x + y &= 100{,}000 \\ 0.04y &= 2800, \end{aligned}$$

where the second equation has no x term. We have eliminated the unknown x. Then solving for y in the second equation, we have

$$y = 70{,}000,$$

and substituting y into the first equation of (3), we obtain

$$x = 30{,}000.$$

To check that $x = 30{,}000$, $y = 70{,}000$ is a solution to (3), we verify that these values of x and y satisfy *each* of the equations in the given linear system. Thus, the director of the trust should invest \$30,000 in the CD and \$70,000 in the long-term bond. ■

EXAMPLE 2 Consider the linear system

$$\begin{aligned} x - 3y &= -7 \\ 2x - 6y &= 7. \end{aligned} \tag{4}$$

Again, we decide to eliminate x. We add (-2) times the first equation to the second one, obtaining

$$\begin{aligned} x - 3y &= -7 \\ 0x + 0y &= 21 \end{aligned}$$

whose second equation makes no sense. This means that the linear system (4) has no solution. We might have come to the same conclusion from observing that in (4) the left side of the second equation is twice the left side of the first equation, but the right side of the second equation is not twice the right side of the first equation. ■

EXAMPLE 3 Consider the linear system

$$\begin{aligned} x + 2y + 3z &= 6 \\ 2x - 3y + 2z &= 14 \\ 3x + y - z &= -2. \end{aligned} \tag{5}$$

To eliminate x, we add (-2) times the first equation to the second one and (-3) times the first equation to the third one, obtaining

$$\begin{aligned} x + 2y + 3z &= 6 \\ -7y - 4z &= 2 \\ -5y - 10z &= -20. \end{aligned} \tag{6}$$

We next eliminate y from the second equation in (6) as follows. Multiply the third equation of (6) by $\left(-\frac{1}{5}\right)$, obtaining

$$\begin{aligned} x + 2y + 3z &= 6 \\ -7y - 4z &= 2 \\ y + 2z &= 4. \end{aligned}$$

Next we interchange the second and third equations to give

$$\begin{aligned} x + 2y + 3z &= 6 \\ y + 2z &= 4 \\ -7y - 4z &= 2. \end{aligned} \tag{7}$$

We now add 7 times the second equation to the third one, to obtain

$$\begin{aligned} x + 2y + 3z &= 6 \\ y + 2z &= 4 \\ 10z &= 30. \end{aligned}$$

Multiplying the third equation by $\frac{1}{10}$, we have

$$\begin{aligned} x + 2y + 3z &= 6 \\ y + 2z &= 4 \\ z &= 3. \end{aligned} \tag{8}$$

Substituting $z = 3$ into the second equation of (8), we find $y = -2$. Substituting these values of z and y into the first equation of (8), we have $x = 1$. To check that $x = 1$, $y = -2$, $z = 3$ is a solution to (5), we verify that these values of x, y, and z satisfy *each* of the equations in (5). Thus, $x = 1$, $y = -2$, $z = 3$ is a solution to the linear system (5). The importance of the procedure lies in the fact that the linear systems (5) and (8) have exactly the same solutions. System (8) has the advantage that it can be solved quite easily, giving the foregoing values for x, y, and z. ■

EXAMPLE 4 Consider the linear system

$$\begin{aligned} x + 2y - 3z &= -4 \\ 2x + y - 3z &= 4. \end{aligned} \tag{9}$$

Eliminating x, we add (-2) times the first equation to the second one, to obtain

$$\begin{aligned} x + 2y - 3z &= -4 \\ -3y + 3z &= 12. \end{aligned} \tag{10}$$

Solving the second equation in (10) for y, we obtain

$$y = z - 4,$$

where z can be any real number. Then, from the first equation of (10),

$$\begin{aligned} x &= -4 - 2y + 3z \\ &= -4 - 2(z - 4) + 3z \\ &= z + 4. \end{aligned}$$

Thus a solution to the linear system (9) is

$$\begin{aligned} x &= r + 4 \\ y &= r - 4 \\ z &= r, \end{aligned}$$

where r is any real number. This means that the linear system (9) has infinitely many solutions. Every time we assign a value to r, we obtain another solution to (9). Thus, if $r = 1$, then

$$x = 5, \quad y = -3, \quad \text{and} \quad z = 1$$

is a solution, while if $r = -2$, then

$$x = 2, \quad y = -6, \quad \text{and} \quad z = -2$$

is another solution. ■

EXAMPLE 5 Consider the linear system

$$\begin{aligned} x + 2y &= 10 \\ 2x - 2y &= -4 \\ 3x + 5y &= 26. \end{aligned} \tag{11}$$

Eliminating x, we add (-2) times the first equation to the second and (-3) times the first equation to the third one, obtaining

$$\begin{aligned} x + 2y &= 10 \\ -\ 6y &= -24 \\ -y &= -4. \end{aligned}$$

Multiplying the second equation by $\left(-\frac{1}{6}\right)$ and the third one by (-1), we have

$$\begin{aligned} x + 2y &= 10 \\ y &= 4 \\ y &= 4, \end{aligned} \tag{12}$$

which has the same solutions as (11). Substituting $y = 4$ in the first equation of (12), we obtain $x = 2$. Hence $x = 2$, $y = 4$ is a solution to (11). ■

EXAMPLE 6 Consider the linear system

$$\begin{aligned} x + 2y &= 10 \\ 2x - 2y &= -4 \\ 3x + 5y &= 20. \end{aligned} \tag{13}$$

To eliminate x, we add (-2) times the first equation to the second one and (-3) times the first equation to the third one, to obtain

$$\begin{aligned} x + 2y &= 10 \\ -\ 6y &= -24 \\ -y &= -10. \end{aligned}$$

Multiplying the second equation by $\left(-\frac{1}{6}\right)$ and the third one by (-1), we have the system

$$\begin{aligned} x + 2y &= 10 \\ y &= 4 \\ y &= 10, \end{aligned} \tag{14}$$

which has no solution. Since (14) and (13) have the same solutions, we conclude that (13) has no solutions. ■

These examples suggest that a linear system may have one solution (a unique solution), no solution, or infinitely many solutions.

> We have seen that the method of elimination consists of repeatedly performing the following operations:
>
> **1.** Interchange two equations.
>
> **2.** Multiply an equation by a nonzero constant.
>
> **3.** Add a multiple of one equation to another.

It is not difficult to show (Exercises T.1 through T.3) that the method of elimination yields another linear system having exactly the same solutions as the given system. The new linear system can then be solved quite readily.

As you have probably already observed, the method of elimination has been described, so far, in general terms. Thus we have not indicated any rules for selecting the unknowns to be eliminated. Before providing a systematic description of the method of elimination, we introduce, in the next section, the notion of a matrix, which will greatly simplify our notation and will enable us to develop tools to solve many important problems.

Consider now a linear system of two equations in the unknowns x and y:

$$\begin{aligned} a_1x + a_2y &= c_1 \\ b_1x + b_2y &= c_2. \end{aligned} \tag{15}$$

The graph of each of these equations is a straight line, which we denote by l_1 and l_2, respectively. If $x = s_1$, $y = s_2$ is a solution to the linear system (15), then the point (s_1, s_2) lies on both lines l_1 and l_2. Conversely, if the point (s_1, s_2) lies on both lines l_1 and l_2, then $x = s_1$, $y = s_2$ is a solution to the linear system (15). (See Figure 1.1.) Thus we are led geometrically to the same three possibilities mentioned previously.

1. The system has a unique solution; that is, the lines l_1 and l_2 intersect at exactly one point.
2. The system has no solution; that is, the lines l_1 and l_2 do not intersect.
3. The system has infinitely many solutions; that is, the lines l_1 and l_2 coincide.

Figure 1.1 ►

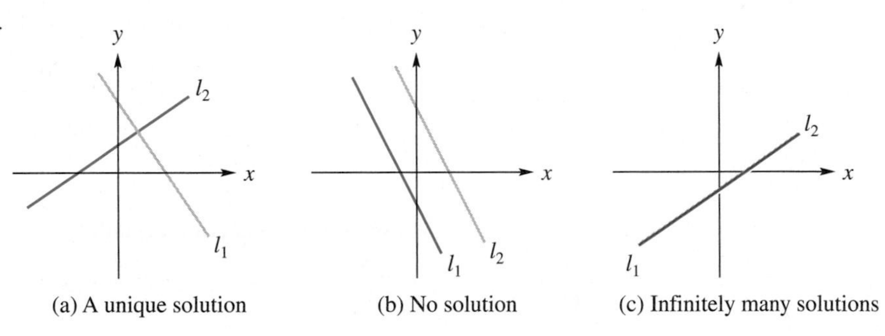

(a) A unique solution (b) No solution (c) Infinitely many solutions

Next, consider a linear system of three equations in the unknowns x, y, and z:

$$\begin{aligned} a_1x + b_1y + c_1z &= d_1 \\ a_2x + b_2y + c_2z &= d_2 \\ a_3x + b_3y + c_3z &= d_3. \end{aligned} \tag{16}$$

The graph of each of these equations is a plane, denoted by P_1, P_2, and P_3, respectively. As in the case of a linear system of two equations in two unknowns, the linear system in (16) can have a unique solution, no solution, or infinitely many solutions. These situations are illustrated in Figure 1.2. For a more concrete illustration of some of the possible cases, the walls (planes) of a room intersect in a unique point, a corner of the room, so the linear system has a unique solution. Next, think of the planes as pages of a book. Three pages of a book (when held open) intersect in a straight line, the spine. Thus, the linear system has infinitely many solutions. On the other hand, when the book is closed, three pages of a book appear to be parallel and do not intersect, so the linear system has no solution.

Figure 1.2 ▶

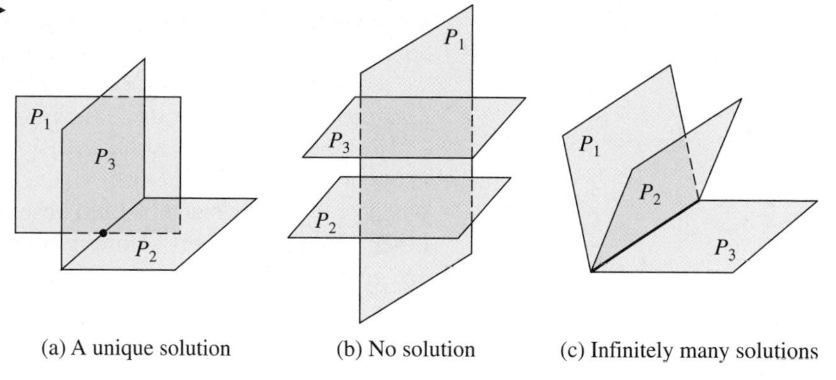

(a) A unique solution (b) No solution (c) Infinitely many solutions

EXAMPLE 7 **(Production Planning)** A manufacturer makes three different types of chemical products: A, B, and C. Each product must go through two processing machines: X and Y. The products require the following times in machines X and Y:

1. One ton of A requires 2 hours in machine X and 2 hours in machine Y.
2. One ton of B requires 3 hours in machine X and 2 hours in machine Y.
3. One ton of C requires 4 hours in machine X and 3 hours in machine Y.

Machine X is available 80 hours per week and machine Y is available 60 hours per week. Since management does not want to keep the expensive machines X and Y idle, it would like to know how many tons of each product to make so that the machines are fully utilized. It is assumed that the manufacturer can sell as much of the products as is made.

To solve this problem, we let x_1, x_2, and x_3 denote the number of tons of products A, B, and C, respectively, to be made. The number of hours that machine X will be used is

$$2x_1 + 3x_2 + 4x_3,$$

which must equal 80. Thus we have

$$2x_1 + 3x_2 + 4x_3 = 80.$$

Similarly, the number of hours that machine Y will be used is 60, so we have

$$2x_1 + 2x_2 + 3x_3 = 60.$$

Mathematically, our problem is to find nonnegative values of x_1, x_2, and x_3 so that

$$\begin{aligned} 2x_1 + 3x_2 + 4x_3 &= 80 \\ 2x_1 + 2x_2 + 3x_3 &= 60. \end{aligned}$$

This linear system has infinitely many solutions. Following the method of Example 4, we see that all solutions are given by

$$\begin{aligned} x_1 &= \frac{20 - x_3}{2} \\ x_2 &= 20 - x_3 \\ x_3 &= \text{any real number such that } 0 \le x_3 \le 20, \end{aligned}$$

since we must have $x_1 \geq 0$, $x_2 \geq 0$, and $x_3 \geq 0$. When $x_3 = 10$, we have

$$x_1 = 5, \qquad x_2 = 10, \qquad x_3 = 10$$

while

$$x_1 = \tfrac{13}{2}, \qquad x_2 = 13, \qquad x_3 = 7$$

when $x_3 = 7$. The reader should observe that one solution is just as good as the other. There is no best solution unless additional information or restrictions are given. ■

Key Terms

Linear equation
Unknowns
Solution to a linear equation
Linear system
Solution to a linear system
Method of elimination
Unique solution
No solution
Infinitely many solutions
Manipulations on a linear system

1.1 Exercises

In Exercises 1 through 14, solve the given linear system by the method of elimination.

1. $\begin{aligned} x + 2y &= 8 \\ 3x - 4y &= 4 \end{aligned}$

2. $\begin{aligned} 2x - 3y + 4z &= -12 \\ x - 2y + z &= -5 \\ 3x + y + 2z &= 1 \end{aligned}$

3. $\begin{aligned} 3x + 2y + z &= 2 \\ 4x + 2y + 2z &= 8 \\ x - y + z &= 4 \end{aligned}$

4. $\begin{aligned} x + y &= 5 \\ 3x + 3y &= 10 \end{aligned}$

5. $\begin{aligned} 2x + 4y + 6z &= -12 \\ 2x - 3y - 4z &= 15 \\ 3x + 4y + 5z &= -8 \end{aligned}$

6. $\begin{aligned} x + y - 2z &= 5 \\ 2x + 3y + 4z &= 2 \end{aligned}$

7. $\begin{aligned} x + 4y - z &= 12 \\ 3x + 8y - 2z &= 4 \end{aligned}$

8. $\begin{aligned} 3x + 4y - z &= 8 \\ 6x + 8y - 2z &= 3 \end{aligned}$

9. $\begin{aligned} x + y + 3z &= 12 \\ 2x + 2y + 6z &= 6 \end{aligned}$

10. $\begin{aligned} x + y &= 1 \\ 2x - y &= 5 \\ 3x + 4y &= 2 \end{aligned}$

11. $\begin{aligned} 2x + 3y &= 13 \\ x - 2y &= 3 \\ 5x + 2y &= 27 \end{aligned}$

12. $\begin{aligned} x - 5y &= 6 \\ 3x + 2y &= 1 \\ 5x + 2y &= 1 \end{aligned}$

13. $\begin{aligned} x + 3y &= -4 \\ 2x + 5y &= -8 \\ x + 3y &= -5 \end{aligned}$

14. $\begin{aligned} 2x + 3y - z &= 6 \\ 2x - y + 2z &= -8 \\ 3x - y + z &= -7 \end{aligned}$

15. Given the linear system

$$\begin{aligned} 2x - y &= 5 \\ 4x - 2y &= t, \end{aligned}$$

(a) determine a value of t so that the system has a solution.

(b) determine a value of t so that the system has no solution.

(c) how many different values of t can be selected in part (b)?

16. Given the linear system

$$\begin{aligned} 2x + 3y - z &= 0 \\ x - 4y + 5z &= 0, \end{aligned}$$

(a) verify that $x_1 = 1$, $y_1 = -1$, $z_1 = -1$ is a solution.

(b) verify that $x_2 = -2$, $y_2 = 2$, $z_2 = 2$ is a solution.

(c) is $x = x_1 + x_2 = -1$, $y = y_1 + y_2 = 1$, and $z = z_1 + z_2 = 1$ a solution to the linear system?

(d) is $3x$, $3y$, $3z$, where x, y, and z are as in part (c), a solution to the linear system?

17. Without using the method of elimination, solve the linear system

$$\begin{aligned} 2x + y - 2z &= -5 \\ 3y + z &= 7 \\ z &= 4. \end{aligned}$$

18. Without using the method of elimination, solve the linear system

$$\begin{aligned} 4x \qquad\qquad &= 8 \\ -2x + 3y \qquad &= -1 \\ 3x + 5y - 2z &= 11. \end{aligned}$$

19. Is there a value of r so that $x = 1$, $y = 2$, $z = r$ is a solution to the following linear system? If there is, find it.

$$\begin{aligned} 2x + 3y - z &= 11 \\ x - y + 2z &= -7 \\ 4x + y - 2z &= 12 \end{aligned}$$

20. Is there a value of r so that $x = r$, $y = 2$, $z = 1$ is a solution to the following linear system? If there is, find it.

$$\begin{aligned} 3x \qquad - 2z &= 4 \\ x - 4y + z &= -5 \\ -2x + 3y + 2z &= 9 \end{aligned}$$

21. Describe the number of points that simultaneously lie in each of the three planes shown in each part of Figure 1.2.

22. Describe the number of points that simultaneously lie in each of the three planes shown in each part of Figure 1.3.

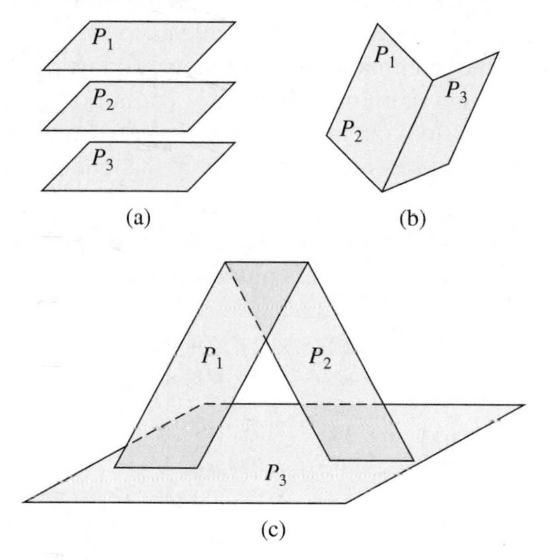

Figure 1.3 ▲

23. An oil refinery produces low-sulfur and high-sulfur fuel. Each ton of low sulfur fuel requires 5 minutes in the blending plant and 4 minutes in the refining plant; each ton of high-sulfur fuel requires 4 minutes in the blending plant and 2 minutes in the refining plant. If the blending plant is available for 3 hours and the refining plant is available for 2 hours, how many tons of each type of fuel should be manufactured so that the plants are fully utilized?

24. A plastics manufacturer makes two types of plastic: regular and special. Each ton of regular plastic requires 2 hours in plant A and 5 hours in plant B; each ton of special plastic requires 2 hours in plant A and 3 hours in plant B. If plant A is available 8 hours per day and plant B is available 15 hours per day, how many tons of each type of plastic can be made daily so that the plants are fully utilized?

25. A dietician is preparing a meal consisting of foods A, B, and C. Each ounce of food A contains 2 units of protein, 3 units of fat, and 4 units of carbohydrate. Each ounce of food B contains 3 units of protein, 2 units of fat, and 1 unit of carbohydrate. Each ounce of food C contains 3 units of protein, 3 units of fat, and 2 units of carbohydrate. If the meal must provide exactly 25 units of protein, 24 units of fat, and 21 units of carbohydrate, how many ounces of each type of food should be used?

26. A manufacturer makes 2-minute, 6-minute, and 9-minute film developers. Each ton of 2-minute developer requires 6 minutes in plant A and 24 minutes in plant B. Each ton of 6-minute developer requires 12 minutes in plant A and 12 minutes in plant B. Each ton of 9-minute developer requires 12 minutes in plant A and 12 minutes in plant B. If plant A is available 10 hours per day and plant B is available 16 hours per day, how many tons of each type of developer can be produced so that the plants are fully utilized?

27. Suppose that the three points $(1, -5)$, $(-1, 1)$, and $(2, 7)$ lie on the parabola $p(x) = ax^2 + bx + c$.

(a) Determine a linear system of three equations in three unknowns that must be solved to find a, b, and c.

(b) Solve the linear system obtained in part (a) for a, b, and c.

28. An inheritance of \$24,000 is to be divided among three trusts, with the second trust receiving twice as much as the first trust. The three trusts pay interest at the rates of 9%, 10%, and 6% annually, respectively, and return a total in interest of \$2210 at the end of the first year. How much was invested in each trust?

Theoretical Exercises

T.1. Show that the linear system obtained by interchanging two equations in (2) has exactly the same solutions as (2).

T.2. Show that the linear system obtained by replacing an equation in (2) by a nonzero constant multiple of the equation has exactly the same solutions as (2).

T.3. Show that the linear system obtained by replacing an equation in (2) by itself plus a multiple of another equation in (2) has exactly the same solutions as (2).

T.4. Does the linear system

$$\begin{aligned} ax + by &= 0 \\ cx + dy &= 0 \end{aligned}$$

always have a solution for any values of a, b, c, and d?

1.2 MATRICES

If we examine the method of elimination described in Section 1.1, we make the following observation. Only the numbers in front of the unknowns x_1, x_2, $\dots$, x_n are being changed as we perform the steps in the method of elimination. Thus we might think of looking for a way of writing a linear system without having to carry along the unknowns. In this section we define an object, a matrix, that enables us to do this—that is, to write linear systems in a compact form that makes it easier to automate the elimination method on a computer in order to obtain a fast and efficient procedure for finding solutions. The use of a matrix is not, however, merely that of a convenient notation. We now develop operations on matrices (plural of matrix) and will work with matrices according to the rules they obey; this will enable us to solve systems of linear equations and solve other computational problems in a fast and efficient manner. Of course, as any good definition should do, the notion of a matrix provides not only a new way of looking at old problems, but also gives rise to a great many new questions, some of which we study in this book.

DEFINITION An $m \times n$ **matrix** A is a rectangular array of mn real (or complex) numbers arranged in m horizontal **rows** and n vertical **columns**:

$$A = \begin{bmatrix} a_{11} & a_{12} & \cdots & \cdots & a_{1j} & \cdots & a_{1n} \\ a_{21} & a_{22} & \cdots & \cdots & a_{2j} & \cdots & a_{2n} \\ \vdots & \vdots & \cdots & \cdots & \vdots & \cdots & \vdots \\ a_{i1} & a_{i2} & \cdots & \cdots & a_{ij} & \cdots & a_{in} \\ \vdots & \vdots & & & \vdots & & \vdots \\ a_{m1} & a_{m2} & \cdots & \cdots & a_{mj} & \cdots & a_{mn} \end{bmatrix} \leftarrow i\text{th row} \tag{1}$$

$\uparrow$ jth column

The ***i*th row** of A is

$$\begin{bmatrix} a_{i1} & a_{i2} & \cdots & a_{in} \end{bmatrix} \qquad (1 \le i \le m);$$

the ***j*th column** of A is

$$\begin{bmatrix} a_{1j} \\ a_{2j} \\ \vdots \\ a_{mj} \end{bmatrix} \qquad (1 \le j \le n).$$

We shall say that A is ***m* by *n*** (written as $\boldsymbol{m \times n}$). If $m = n$, we say that A is a **square matrix of order *n*** and that the numbers $a_{11}, a_{22}, \dots, a_{nn}$ form the **main diagonal** of A. We refer to the number a_{ij}, which is in the ith row and jth column of A, as the ***i*, *j*th element** of A, or the **(*i*, *j*) entry** of A, and we often write (1) as

$$A = \begin{bmatrix} a_{ij} \end{bmatrix}.$$

For the sake of simplicity, we restrict our attention in this book, except for Appendix A, to matrices all of whose entries are real numbers. However, matrices with complex entries are studied and are important in applications.

EXAMPLE 1 Let

$$A = \begin{bmatrix} 1 & 2 & 3 \\ -1 & 0 & 1 \end{bmatrix}, \quad B = \begin{bmatrix} 1 & 4 \\ 2 & -3 \end{bmatrix}, \quad C = \begin{bmatrix} 1 \\ -1 \\ 2 \end{bmatrix},$$

$$D = \begin{bmatrix} 1 & 1 & 0 \\ 2 & 0 & 1 \\ 3 & -1 & 2 \end{bmatrix}, \quad E = \begin{bmatrix} 3 \end{bmatrix}, \quad F = \begin{bmatrix} -1 & 0 & 2 \end{bmatrix}.$$

Then A is a 2×3 matrix with $a_{12} = 2$, $a_{13} = 3$, $a_{22} = 0$, and $a_{23} = 1$; B is a 2×2 matrix with $b_{11} = 1$, $b_{12} = 4$, $b_{21} = 2$, and $b_{22} = -3$; C is a 3×1 matrix with $c_{11} = 1$, $c_{21} = -1$, and $c_{31} = 2$; D is a 3×3 matrix; E is a 1×1 matrix; and F is a 1×3 matrix. In D, the elements $d_{11} = 1$, $d_{22} = 0$, and $d_{33} = 2$ form the main diagonal. ■

For convenience, we focus much of our attention in the illustrative examples and exercises in Chapters 1–7 on matrices and expressions containing only real numbers. Complex numbers will make a brief appearance in Chapters 8 and 9. An introduction to complex numbers, their properties, and examples and exercises showing how complex numbers are used in linear algebra may be found in Appendix A.

A $1 \times n$ or an $n \times 1$ matrix is also called an ***n*-vector** and will be denoted by lowercase boldface letters. When n is understood, we refer to n-vectors merely as **vectors**. In Chapter 4 we discuss vectors at length.

EXAMPLE 2 $\mathbf{u} = \begin{bmatrix} 1 & 2 & -1 & 0 \end{bmatrix}$ is a 4-vector and $\mathbf{v} = \begin{bmatrix} 1 \\ -1 \\ 3 \end{bmatrix}$ is a 3-vector. ■

The n-vector all of whose entries are zero is denoted by $\mathbf{0}$.

Observe that if A is an $n \times n$ matrix, then the rows of A are $1 \times n$ matrices and the columns of A are $n \times 1$ matrices. The set of all n vectors with real entries is denoted by R^n. Similarly, the set of all n-vectors with complex entries is denoted by C^n. As we have already pointed out, in the first seven chapters of this book we will work almost entirely with vectors in R^n.

EXAMPLE 3 **(Tabular Display of Data)** The following matrix gives the airline distances between the indicated cities (in statute miles).

	London	Madrid	New York	Tokyo
London	0	785	3469	5959
Madrid	785	0	3593	6706
New York	3469	3593	0	6757
Tokyo	5959	6706	6757	0

■

EXAMPLE 4 **(Production)** Suppose that a manufacturer has four plants each of which makes three products. If we let a_{ij} denote the number of units of product i made by plant j in one week, then the 4×3 matrix

	Plant 1	Plant 2	Plant 3	Plant 4
Product 1	560	360	380	0
Product 2	340	450	420	80
Product 3	280	270	210	380

gives the manufacturer's production for the week. For example, plant 2 makes 270 units of product 3 in one week. ■

EXAMPLE 5

The wind chill table that follows shows how a combination of air temperature and wind speed makes a body feel colder than the actual temperature. For example, when the temperature is 10°F and the wind is 15 miles per hour, this causes a body heat loss equal to that when the temperature is −18°F with no wind.

	°F					
mph	15	10	5	0	−5	−10
5	12	7	0	−5	−10	−15
10	−3	−9	−15	−22	−27	−34
15	−11	−18	−25	−31	−38	−45
20	−17	−24	−31	−39	−46	−53

This table can be represented as the matrix

$$A = \begin{bmatrix} 5 & 12 & 7 & 0 & -5 & -10 & -15 \\ 10 & -3 & -9 & -15 & -22 & -27 & -34 \\ 15 & -11 & -18 & -25 & -31 & -38 & -45 \\ 20 & -17 & -24 & -31 & -39 & -46 & -53 \end{bmatrix}.$$

■

EXAMPLE 6

With the linear system considered in Example 5 in Section 1.1,

$$\begin{aligned} x + 2y &= 10 \\ 2x - 2y &= -4 \\ 3x + 5y &= 26, \end{aligned}$$

we can associate the following matrices:

$$A = \begin{bmatrix} 1 & 2 \\ 2 & -2 \\ 3 & 5 \end{bmatrix}, \quad \mathbf{x} = \begin{bmatrix} x \\ y \end{bmatrix}, \quad \mathbf{b} = \begin{bmatrix} 10 \\ -4 \\ 26 \end{bmatrix}.$$

In Section 1.3, we shall call A the coefficient matrix of the linear system. ■

DEFINITION

A square matrix $A = \left[a_{ij}\right]$ for which every term off the main diagonal is zero, that is, $a_{ij} = 0$ for $i \neq j$, is called a **diagonal matrix**.

EXAMPLE 7

$$G = \begin{bmatrix} 4 & 0 \\ 0 & -2 \end{bmatrix} \quad \text{and} \quad H = \begin{bmatrix} -3 & 0 & 0 \\ 0 & -2 & 0 \\ 0 & 0 & 4 \end{bmatrix}$$

are diagonal matrices. ■

DEFINITION

A diagonal matrix $A = [a_{ij}]$, for which all terms on the main diagonal are equal, that is, $a_{ij} = c$ for $i = j$ and $a_{ij} = 0$ for $i \neq j$, is called a **scalar matrix**.

EXAMPLE 8

The following are scalar matrices:

$$I_3 = \begin{bmatrix} 1 & 0 & 0 \\ 0 & 1 & 0 \\ 0 & 0 & 1 \end{bmatrix}, \qquad J = \begin{bmatrix} -2 & 0 \\ 0 & -2 \end{bmatrix}.$$

■

The search engines available for information searches and retrieval on the Internet use matrices to keep track of the locations of information, the type of information at a location, keywords that appear in the information, and even the way Web sites link to one another. A large measure of the effectiveness of the search engine Google© is the manner in which matrices are used to determine which sites are referenced by other sites. That is, instead of directly keeping track of the information content of an actual Web page or of an individual search topic, Google's matrix structure focuses on finding Web pages that match the search topic and then presents a list of such pages in the order of their "importance."

Suppose that there are n accessible Web pages during a certain month. A simple way to view a matrix that is part of Google's scheme is to imagine an $n \times n$ matrix A, called the "connectivity matrix," that initially contains all zeros. To build the connections proceed as follows. When it is detected that Web site j links to Web site i, set entry a_{ij} equal to one. Since n is quite large, about 3 billion as of December 2002, most entries of the connectivity matrix A are zero. (Such a matrix is called sparse.) If row i of A contains many ones, then there are many sites linking to site i. Sites that are linked to by many other sites are considered more "important" (or to have a higher rank) by the software driving the Google search engine. Such sites would appear near the top of a list returned by a Google search on topics related to the information on site i. Since Google updates its connectivity matrix about every month, n increases over time and new links and sites are adjoined to the connectivity matrix.

The fundamental technique used by Google© to rank sites uses linear algebra concepts that are somewhat beyond the scope of this course. Further information can be found in the following sources.

1. Berry, Michael W., and Murray Browne. *Understanding Search Engines—Mathematical Modeling and Text Retrieval.* Philadelphia: Siam, 1999.
2. www.google.com/technology/index.html
3. Moler, Cleve. "The World's Largest Matrix Computation: Google's Page Rank Is an Eigenvector of a Matrix of Order 2.7 Billion," MATLAB *News and Notes*, October 2002, pp. 12–13.

Whenever a new object is introduced in mathematics, we must define when two such objects are equal. For example, in the set of all rational numbers, the numbers $\frac{2}{3}$ and $\frac{4}{6}$ are called equal although they are not represented in the same manner. What we have in mind is the definition that $\frac{a}{b}$ equals $\frac{c}{d}$ when $ad = bc$. Accordingly, we now have the following definition.

DEFINITION

Two $m \times n$ matrices $A = [a_{ij}]$ and $B = [b_{ij}]$ are said to be **equal** if $a_{ij} = b_{ij}$, $1 \leq i \leq m$, $1 < j \leq n$, that is, if corresponding elements are equal.

EXAMPLE 9 The matrices

$$A = \begin{bmatrix} 1 & 2 & -1 \\ 2 & -3 & 4 \\ 0 & -4 & 5 \end{bmatrix} \quad \text{and} \quad B = \begin{bmatrix} 1 & 2 & w \\ 2 & x & 4 \\ y & -4 & z \end{bmatrix}$$

are equal if $w = -1$, $x = -3$, $y = 0$, and $z = 5$. ■

We shall now define a number of operations that will produce new matrices out of given matrices. These operations are useful in the applications of matrices.

MATRIX ADDITION

DEFINITION If $A = \begin{bmatrix} a_{ij} \end{bmatrix}$ and $B = \begin{bmatrix} b_{ij} \end{bmatrix}$ are $m \times n$ matrices, then the **sum** of A and B is the $m \times n$ matrix $C = \begin{bmatrix} c_{ij} \end{bmatrix}$, defined by

$$c_{ij} = a_{ij} + b_{ij} \qquad (1 \leq i \leq m, 1 \leq j \leq n).$$

That is, C is obtained by adding corresponding elements of A and B.

EXAMPLE 10 Let

$$A = \begin{bmatrix} 1 & -2 & 4 \\ 2 & -1 & 3 \end{bmatrix} \quad \text{and} \quad B = \begin{bmatrix} 0 & 2 & -4 \\ 1 & 3 & 1 \end{bmatrix}.$$

Then

$$A + B = \begin{bmatrix} 1+0 & -2+2 & 4+(-4) \\ 2+1 & -1+3 & 3+1 \end{bmatrix} = \begin{bmatrix} 1 & 0 & 0 \\ 3 & 2 & 4 \end{bmatrix}.$$ ■

It should be noted that the sum of the matrices A and B is defined only when A and B have the same number of rows and the same number of columns, that is, only when A and B are of the same size.

> We shall now establish the convention that when $A + B$ is formed, both A and B are of the same size.

Thus far, addition of matrices has only been defined for two matrices. Our work with matrices will call for adding more than two matrices. Theorem 1.1 in the next section shows that addition of matrices satisfies the associative property: $A + (B + C) = (A + B) + C$. Additional properties of matrix addition are considered in Section 1.4 and are similar to those satisfied by the real numbers.

EXAMPLE 11 **(Production)** A manufacturer of a certain product makes three models, A, B, and C. Each model is partially made in factory F_1 in Taiwan and then finished in factory F_2 in the United States. The total cost of each product consists of the manufacturing cost and the shipping cost. Then the costs at each factory (in dollars) can be described by the 3×2 matrices F_1 and F_2:

	Manufacturing cost	Shipping cost	
$F_1 =$	32	40	Model A
	50	80	Model B
	70	20	Model C

$$F_2 = \begin{array}{c} \\ \\ \end{array}\begin{matrix} \text{Manufacturing cost} & \text{Shipping cost} \\ \begin{bmatrix} 40 & 60 \\ 50 & 50 \\ 130 & 20 \end{bmatrix} \end{matrix}\begin{matrix} \\ \text{Model A} \\ \text{Model B} \\ \text{Model C} \end{matrix}$$

The matrix $F_1 + F_2$ gives the total manufacturing and shipping costs for each product. Thus the total manufacturing and shipping costs of a model C product are \$200 and \$40, respectively. ■

SCALAR MULTIPLICATION

DEFINITION If $A = \begin{bmatrix} a_{ij} \end{bmatrix}$ is an $m \times n$ matrix and r is a real number, then the **scalar multiple** of A by r, rA, is the $m \times n$ matrix $B = \begin{bmatrix} b_{ij} \end{bmatrix}$, where

$$b_{ij} = ra_{ij} \qquad (1 \le i \le m, 1 \le j \le n).$$

That is, B is obtained by multiplying each element of A by r.

If A and B are $m \times n$ matrices, we write $A + (-1)B$ as $A - B$ and call this the **difference** of A and B.

EXAMPLE 12 Let

$$A = \begin{bmatrix} 2 & 3 & -5 \\ 4 & 2 & 1 \end{bmatrix} \quad \text{and} \quad B = \begin{bmatrix} 2 & -1 & 3 \\ 3 & 5 & -2 \end{bmatrix}.$$

Then

$$A - B = \begin{bmatrix} 2-2 & 3+1 & -5-3 \\ 4-3 & 2-5 & 1+2 \end{bmatrix} = \begin{bmatrix} 0 & 4 & -8 \\ 1 & -3 & 3 \end{bmatrix}.$$

■

EXAMPLE 13 Let $\mathbf{p} = \begin{bmatrix} 18.95 & 14.75 & 8.60 \end{bmatrix}$ be a 3-vector that represents the current prices of three items at a store. Suppose that the store announces a sale so that the price of each item is reduced by 20%.

(a) Determine a 3-vector that gives the price changes for the three items.

(b) Determine a 3-vector that gives the new prices of the items.

Solution (a) Since each item is reduced by 20%, the 3-vector

$$\begin{aligned} 0.20\mathbf{p} &= \begin{bmatrix} (0.20)18.95 & (0.20)14.75 & (0.20)8.60 \end{bmatrix} \\ &= \begin{bmatrix} 3.79 & 2.95 & 1.72 \end{bmatrix} \end{aligned}$$

gives the price reductions for the three items.

(b) The new prices of the items are given by the expression

$$\begin{aligned} \mathbf{p} - 0.20\mathbf{p} &= \begin{bmatrix} 18.95 & 14.75 & 8.60 \end{bmatrix} - \begin{bmatrix} 3.79 & 2.95 & 1.72 \end{bmatrix} \\ &= \begin{bmatrix} 15.16 & 11.80 & 6.88 \end{bmatrix}. \end{aligned}$$

Observe that this expression can also be written as

$$\mathbf{p} - 0.20\mathbf{p} = 0.80\mathbf{p}.$$

■

If $A_1, A_2, \ldots, A_k$ are $m \times n$ matrices and $c_1, c_2, \ldots, c_k$ are real numbers, then an expression of the form

$$c_1A_1 + c_2A_2 + \cdots + c_kA_k \tag{2}$$

is called a **linear combination** of $A_1, A_2, \ldots, A_k$, and $c_1, c_2, \ldots, c_k$ are called **coefficients**.

EXAMPLE 14

(a) If

$$A_1 = \begin{bmatrix} 0 & -3 & 5 \\ 2 & 3 & 4 \\ 1 & -2 & -3 \end{bmatrix} \quad \text{and} \quad A_2 = \begin{bmatrix} 5 & 2 & 3 \\ 6 & 2 & 3 \\ -1 & -2 & 3 \end{bmatrix},$$

then $C = 3A_1 - \frac{1}{2}A_2$ is a linear combination of A_1 and A_2. Using scalar multiplication and matrix addition, we can compute C:

$$C = 3\begin{bmatrix} 0 & -3 & 5 \\ 2 & 3 & 4 \\ 1 & -2 & -3 \end{bmatrix} - \frac{1}{2}\begin{bmatrix} 5 & 2 & 3 \\ 6 & 2 & 3 \\ -1 & -2 & 3 \end{bmatrix}$$

$$= \begin{bmatrix} -\frac{5}{2} & -10 & \frac{27}{2} \\ 3 & 8 & \frac{21}{2} \\ \frac{7}{2} & -5 & -\frac{21}{2} \end{bmatrix}.$$

(b) $2\begin{bmatrix} 3 & -2 \end{bmatrix} - 3\begin{bmatrix} 5 & 0 \end{bmatrix} + 4\begin{bmatrix} -2 & 5 \end{bmatrix}$ is a linear combination of $\begin{bmatrix} 3 & -2 \end{bmatrix}$, $\begin{bmatrix} 5 & 0 \end{bmatrix}$, and $\begin{bmatrix} -2 & 5 \end{bmatrix}$. It can be computed (verify) as $\begin{bmatrix} -17 & 16 \end{bmatrix}$.

(c) $-0.5\begin{bmatrix} 1 \\ -4 \\ -6 \end{bmatrix} + 0.4\begin{bmatrix} 0.1 \\ -4 \\ 0.2 \end{bmatrix}$ is a linear combination of $\begin{bmatrix} 1 \\ -4 \\ -6 \end{bmatrix}$ and $\begin{bmatrix} 0.1 \\ -4 \\ 0.2 \end{bmatrix}$.

It can be computed (verify) as $\begin{bmatrix} -0.46 \\ 0.4 \\ 3.08 \end{bmatrix}$. ■

THE TRANSPOSE OF A MATRIX

DEFINITION If $A = \begin{bmatrix} a_{ij} \end{bmatrix}$ is an $m \times n$ matrix, then the $n \times m$ matrix $A^T = \begin{bmatrix} a_{ij}^T \end{bmatrix}$, where

$$a_{ij}^T = a_{ji} \qquad (1 \le i \le n, 1 \le j \le m)$$

is called the **transpose** of A. Thus, the entries in each row of A^T are the entries in the corresponding column of A.

EXAMPLE 15 Let

$$A = \begin{bmatrix} 4 & -2 & 3 \\ 0 & 5 & -2 \end{bmatrix}, \quad B = \begin{bmatrix} 6 & 2 & -4 \\ 3 & -1 & 2 \\ 0 & 4 & 3 \end{bmatrix}, \quad C = \begin{bmatrix} 5 & 4 \\ -3 & 2 \\ 2 & -3 \end{bmatrix},$$

$$D = \begin{bmatrix} 3 & -5 & 1 \end{bmatrix}, \quad E = \begin{bmatrix} 2 \\ -1 \\ 3 \end{bmatrix}.$$

Then

$$A^T = \begin{bmatrix} 4 & 0 \\ -2 & 5 \\ 3 & -2 \end{bmatrix}, \qquad B^T = \begin{bmatrix} 6 & 3 & 0 \\ 2 & -1 & 4 \\ -4 & 2 & 3 \end{bmatrix},$$

$$C^T = \begin{bmatrix} 5 & -3 & 2 \\ 4 & 2 & -3 \end{bmatrix}, \quad D^T = \begin{bmatrix} 3 \\ -5 \\ 1 \end{bmatrix}, \quad \text{and} \quad E^T = \begin{bmatrix} 2 & -1 & 3 \end{bmatrix}.$$

■

BIT MATRICES (OPTIONAL)

The majority of our work in linear algebra will use matrices and vectors whose entries are real or complex numbers. Hence computations, like linear combinations, are determined using matrix properties and standard arithmetic base 10. However, the continued expansion of computer technology has brought to the forefront the use of binary (base 2) representation of information. In most computer applications like video games, FAX communications, ATM money transfers, satellite communications, DVD videos, or the generation of music CDs, the underlying mathematics is invisible and completely transparent to the viewer or user. Binary coded data is so prevalent and plays such a central role that we will briefly discuss certain features of it in appropriate sections of this book. We begin with an overview of binary addition and multiplication and then introduce a special class of binary matrices that play a prominent role in information and communication theory.

Binary representation of information uses only two symbols 0 and 1. Information is coded in terms of 0 and 1 in a string of **bits**.* For example, the decimal number 5 is represented as the binary string 101, which is interpreted in terms of base 2 as follows:

$$5 = 1(2^2) + 0(2^1) + 1(2^0).$$

The coefficients of the powers of 2 determine the string of bits, 101, which provide the binary representation of 5.

Just as there is arithmetic base 10 when dealing with the real and complex numbers, there is arithmetic using base 2; that is, binary arithmetic. Table 1.1 shows the structure of binary addition and Table 1.2 the structure of binary multiplication.

Table 1.1

$+$	0	1
0	0	1
1	1	0

Table 1.2

$\times$	0	1
0	0	0
1	0	1

The properties of binary arithmetic for combining representations of real numbers given in binary form is often studied in beginning computer science courses or finite mathematics courses. We will not digress to review such topics at this time. However, our focus will be on a particular type of matrix and vector that contain entries that are single binary digits. This class of matrices and vectors are important in the study of information theory and the mathematical field of *error-correcting codes* (also called *coding theory*).

*A bit is a binary digit; that is, either a 0 or 1.

DEFINITION

An $m \times n$ **bit matrix**† is a matrix all of whose entries are (single) bits. That is, each entry is either 0 or 1.

A bit ***n*-vector** (or **vector**) is a $1 \times n$ or $n \times 1$ matrix all of whose entries are bits.

EXAMPLE 16

$A = \begin{bmatrix} 1 & 0 & 0 \\ 1 & 1 & 1 \\ 0 & 1 & 0 \end{bmatrix}$ is a 3×3 bit matrix. ■

EXAMPLE 17

$\mathbf{v} = \begin{bmatrix} 1 \\ 1 \\ 0 \\ 0 \\ 1 \end{bmatrix}$ is a bit 5-vector and $\mathbf{u} = \begin{bmatrix} 0 & 0 & 0 & 0 \end{bmatrix}$ is a bit 4-vector. ■

The definitions of matrix addition and scalar multiplication apply to bit matrices provided we use binary (or base 2) arithmetic for all computations and use the only possible scalars 0 and 1.

EXAMPLE 18

Let $A = \begin{bmatrix} 1 & 0 \\ 1 & 1 \\ 0 & 1 \end{bmatrix}$ and $B = \begin{bmatrix} 1 & 1 \\ 0 & 1 \\ 1 & 0 \end{bmatrix}$. Using the definition of matrix addition and Table 1.1, we have

$$A + B = \begin{bmatrix} 1+1 & 0+1 \\ 1+0 & 1+1 \\ 0+1 & 1+0 \end{bmatrix} = \begin{bmatrix} 0 & 1 \\ 1 & 0 \\ 1 & 1 \end{bmatrix}.$$

■

Linear combinations of bit matrices or bit n-vectors are quite easy to compute using the fact that the only scalars are 0 and 1 together with Tables 1.1 and 1.2.

EXAMPLE 19

Let $c_1 = 1$, $c_2 = 0$, $c_3 = 1$, $\mathbf{u}_1 = \begin{bmatrix} 1 \\ 0 \end{bmatrix}$, $\mathbf{u}_2 = \begin{bmatrix} 0 \\ 1 \end{bmatrix}$, and $\mathbf{u}_3 = \begin{bmatrix} 1 \\ 1 \end{bmatrix}$. Then

$$\begin{aligned} c_1\mathbf{u}_1 + c_2\mathbf{u}_2 + c_3\mathbf{u}_3 &= 1\begin{bmatrix} 1 \\ 0 \end{bmatrix} + 0\begin{bmatrix} 0 \\ 1 \end{bmatrix} + 1\begin{bmatrix} 1 \\ 1 \end{bmatrix} \\ &= \begin{bmatrix} 1 \\ 0 \end{bmatrix} + \begin{bmatrix} 0 \\ 0 \end{bmatrix} + \begin{bmatrix} 1 \\ 1 \end{bmatrix} \\ &= \begin{bmatrix} (1+0)+1 \\ (0+0)+1 \end{bmatrix} \\ &= \begin{bmatrix} 1+1 \\ 0+1 \end{bmatrix} = \begin{bmatrix} 0 \\ 1 \end{bmatrix}. \end{aligned}$$

■

From Table 1.1 we have $0 + 0 = 0$ and $1 + 1 = 0$. Thus the additive inverse of 0 is 0 (as usual) and the additive inverse of 1 is 1. Hence to compute the difference of bit matrices A and B we proceed as follows:

$$A - B = A + (\text{inverse of } 1)\, B = A + 1B = A + B.$$

We see that the difference of bit matrices contributes nothing new to the algebraic relationships among bit matrices.

†A bit matrix is also called a **Boolean matrix**.

Key Terms

Matrix
Rows
Columns
Size of a matrix
Square matrix
Main diagonal of a matrix
Element (or entry) of a matrix
ijth element
(i, j) entry
n-vector (or vector)
Diagonal matrix
Scalar matrix
$\mathbf{0}$, the zero vector
R^n, the set of all n-vectors
Google©
Equal matrices
Matrix addition
Scalar multiplication
Scalar multiple of a matrix
Difference of matrices
Linear combination of matrices
Transpose of a matrix
Bit
Bit (or Boolean) matrix
Upper triangular matrix
Lower triangular matrix

1.2 Exercises

1. Let

$$A = \begin{bmatrix} 2 & -3 & 5 \\ 6 & -5 & 4 \end{bmatrix}, \quad B = \begin{bmatrix} 4 \\ -3 \\ 5 \end{bmatrix},$$

and

$$C = \begin{bmatrix} 7 & 3 & 2 \\ -4 & 3 & 5 \\ 6 & 1 & -1 \end{bmatrix}.$$

(a) What is a_{12}, a_{22}, a_{23}?

(b) What is b_{11}, b_{31}?

(c) What is c_{13}, c_{31}, c_{33}?

2. If

$$\begin{bmatrix} a+b & c+d \\ c-d & a-b \end{bmatrix} = \begin{bmatrix} 4 & 6 \\ 10 & 2 \end{bmatrix},$$

find a, b, c, and d.

3. If

$$\begin{bmatrix} a+2b & 2a-b \\ 2c+d & c-2d \end{bmatrix} = \begin{bmatrix} 4 & -2 \\ 4 & -3 \end{bmatrix},$$

find a, b, c, and d.

In Exercises 4 through 7, let

$$A = \begin{bmatrix} 1 & 2 & 3 \\ 2 & 1 & 4 \end{bmatrix}, \quad B = \begin{bmatrix} 1 & 0 \\ 2 & 1 \\ 3 & 2 \end{bmatrix},$$

$$C = \begin{bmatrix} 3 & -1 & 3 \\ 4 & 1 & 5 \\ 2 & 1 & 3 \end{bmatrix}, \quad D = \begin{bmatrix} 3 & -2 \\ 2 & 4 \end{bmatrix},$$

$$E = \begin{bmatrix} 2 & -4 & 5 \\ 0 & 1 & 4 \\ 3 & 2 & 1 \end{bmatrix}, \quad F = \begin{bmatrix} -4 & 5 \\ 2 & 3 \end{bmatrix},$$

$$\textit{and} \quad O = \begin{bmatrix} 0 & 0 & 0 \\ 0 & 0 & 0 \\ 0 & 0 & 0 \end{bmatrix}.$$

4. If possible, compute the indicated linear combination:

(a) $C + E$ and $E + C$ (b) $A + B$

(c) $D - F$ (d) $-3C + 5O$

(e) $2C - 3E$ (f) $2B + F$

5. If possible, compute the indicated linear combination:

(a) $3D + 2F$

(b) $3(2A)$ and $6A$

(c) $3A + 2A$ and $5A$

(d) $2(D + F)$ and $2D + 2F$

(e) $(2 + 3)D$ and $2D + 3D$

(f) $3(B + D)$

6. If possible, compute:

(a) A^T and $(A^T)^T$

(b) $(C + E)^T$ and $C^T + E^T$

(c) $(2D + 3F)^T$

(d) $D - D^T$

(e) $2A^T + B$

(f) $(3D - 2F)^T$

7. If possible, compute:

(a) $(2A)^T$ (b) $(A - B)^T$

(c) $(3B^T - 2A)^T$

(d) $(3A^T - 5B^T)^T$

(e) $(-A)^T$ and $-(A^T)$

(f) $(C + E + F^T)^T$

8. Is the matrix $\begin{bmatrix} 3 & 0 \\ 0 & 2 \end{bmatrix}$ a linear combination of the matrices $\begin{bmatrix} 1 & 0 \\ 0 & 1 \end{bmatrix}$ and $\begin{bmatrix} 1 & 0 \\ 0 & 0 \end{bmatrix}$? Justify your answer.

9. Is the matrix $\begin{bmatrix} 4 & 1 \\ 0 & -3 \end{bmatrix}$ a linear combination of the matrices $\begin{bmatrix} 1 & 0 \\ 0 & 1 \end{bmatrix}$ and $\begin{bmatrix} 1 & 0 \\ 0 & 0 \end{bmatrix}$? Justify your answer.

10. Let

$$A = \begin{bmatrix} 1 & 2 & 3 \\ 6 & -2 & 3 \\ 5 & 2 & 4 \end{bmatrix} \quad \text{and} \quad I_3 = \begin{bmatrix} 1 & 0 & 0 \\ 0 & 1 & 0 \\ 0 & 0 & 1 \end{bmatrix}.$$

If λ is a real number, compute $\lambda I_3 - A$.

Exercises 11 through 15 involve bit matrices.

11. Let $A = \begin{bmatrix} 1 & 0 & 1 \\ 1 & 1 & 0 \\ 0 & 1 & 1 \end{bmatrix}$, $B = \begin{bmatrix} 0 & 1 & 1 \\ 1 & 0 & 1 \\ 1 & 1 & 0 \end{bmatrix}$, and $C = \begin{bmatrix} 1 & 1 & 0 \\ 0 & 1 & 1 \\ 1 & 0 & 1 \end{bmatrix}$. Compute each of the following.

(a) $A + B$ (b) $B + C$ (c) $A + B + C$

(d) $A + C^T$ (e) $B - C$

12. Let $A = \begin{bmatrix} 1 & 0 \\ 1 & 0 \end{bmatrix}$, $B = \begin{bmatrix} 1 & 0 \\ 0 & 1 \end{bmatrix}$, $C = \begin{bmatrix} 1 & 1 \\ 0 & 0 \end{bmatrix}$, and $D = \begin{bmatrix} 0 & 0 \\ 1 & 0 \end{bmatrix}$. Compute each of the following.

(a) $A + B$ (b) $C + D$ (c) $A + B + (C + D)^T$

(d) $C - B$ (e) $A - B + C - D$

13. Let $A = \begin{bmatrix} 1 & 0 \\ 0 & 0 \end{bmatrix}$.

(a) Find B so that $A + B = \begin{bmatrix} 0 & 0 \\ 0 & 0 \end{bmatrix}$.

(b) Find C so that $A + C = \begin{bmatrix} 1 & 1 \\ 1 & 1 \end{bmatrix}$.

14. Let $\mathbf{u} = \begin{bmatrix} 1 & 1 & 0 & 0 \end{bmatrix}$. Find the bit 4-vector $\mathbf{v}$ so that $\mathbf{u} + \mathbf{v} = \begin{bmatrix} 1 & 1 & 0 & 0 \end{bmatrix}$.

15. Let $\mathbf{u} = \begin{bmatrix} 0 & 1 & 0 & 1 \end{bmatrix}$. Find the bit 4-vector $\mathbf{v}$ so that $\mathbf{u} + \mathbf{v} = \begin{bmatrix} 1 & 1 & 1 & 1 \end{bmatrix}$.

Theoretical Exercises

T.1. Show that the sum and difference of two diagonal matrices is a diagonal matrix.

T.2. Show that the sum and difference of two scalar matrices is a scalar matrix.

T.3. Let

$$A = \begin{bmatrix} a & b & c \\ c & d & e \\ e & e & f \end{bmatrix}.$$

(a) Compute $A - A^T$.

(b) Compute $A + A^T$.

(c) Compute $(A + A^T)^T$.

T.4. Let O be the $n \times n$ matrix all of whose entries are zero. Show that if k is a real number and A is an $n \times n$ matrix such that $kA = O$, then $k = 0$ or $A = O$.

T.5. A matrix $A = \begin{bmatrix} a_{ij} \end{bmatrix}$ is called **upper triangular** if $a_{ij} = 0$ for $i > j$. It is called **lower triangular** if $a_{ij} = 0$ for $i < j$.

$$\begin{bmatrix} a_{11} & a_{12} & \cdots & \cdots & \cdots & a_{1n} \\ 0 & a_{22} & \cdots & \cdots & \cdots & a_{2n} \\ 0 & 0 & a_{33} & \cdots & \cdots & a_{3n} \\ \vdots & \vdots & \vdots & \ddots & & \vdots \\ \vdots & \vdots & \vdots & & \ddots & \vdots \\ 0 & 0 & 0 & \cdots & 0 & a_{nn} \end{bmatrix}$$

Upper triangular matrix
(The elements below the main diagonal are zero.)

$$\begin{bmatrix} a_{11} & 0 & 0 & \cdots & \cdots & 0 \\ a_{21} & a_{22} & 0 & \cdots & \cdots & 0 \\ a_{31} & a_{32} & a_{33} & 0 & \cdots & 0 \\ \vdots & \vdots & \vdots & \ddots & & \vdots \\ \vdots & \vdots & \vdots & & \ddots & 0 \\ a_{n1} & a_{n2} & a_{n3} & \cdots & \cdots & a_{nn} \end{bmatrix}$$

Lower triangular matrix
(The elements above the main diagonal are zero.)

(a) Show that the sum and difference of two upper triangular matrices is upper triangular.

(b) Show that the sum and difference of two lower triangular matrices is lower triangular.

(c) Show that if a matrix is both upper and lower triangular, then it is a diagonal matrix.

T.6. (a) Show that if A is an upper triangular matrix, then A^T is lower triangular.

(b) Show that if A is a lower triangular matrix, then A^T is upper triangular.

T.7. If A is an $n \times n$ matrix, what are the entries on the main diagonal of $A - A^T$? Justify your answer.

T.8. If $\mathbf{x}$ is an n-vector, show that $\mathbf{x} + \mathbf{0} = \mathbf{x}$.

Exercises T.9 through T.18 involve bit matrices.

T.9. Make a list of all possible bit 2-vectors. How many are there?

T.10. Make a list of all possible bit 3-vectors. How many are there?

T.11. Make a list of all possible bit 4-vectors. How many are there?

T.12. How many bit 5-vectors are there? How many bit n-vectors are there?

T.13. Make a list of all possible 2×2 bit matrices. How many are there?

T.14. How many 3×3 bit matrices are there?

T.15. How many $n \times n$ bit matrices are there?

T.16. Let 0 represent OFF and 1 represent ON and

$$A = \begin{bmatrix} \text{ON} & \text{ON} & \text{OFF} \\ \text{OFF} & \text{ON} & \text{OFF} \\ \text{OFF} & \text{ON} & \text{ON} \end{bmatrix}.$$

Find the ON/OFF matrix B so that $A + B$ is a matrix with each entry OFF.

T.17. Let 0 represent OFF and 1 represent ON and

$$A = \begin{bmatrix} \text{ON} & \text{ON} & \text{OFF} \\ \text{OFF} & \text{ON} & \text{OFF} \\ \text{OFF} & \text{ON} & \text{ON} \end{bmatrix}.$$

Find the ON/OFF matrix B so that $A + B$ is a matrix with each entry ON.

T.18. A standard light switch has two positions (or states); either on or off. Let bit matrix

$$A = \begin{bmatrix} 1 & 0 \\ 0 & 1 \\ 1 & 1 \end{bmatrix}$$

represent a bank of light switches where 0 represents OFF and 1 represents ON.

(a) Find a matrix B so that $A + B$ will represent the bank of switches with the state of each switch "reversed."

(b) Let

$$C = \begin{bmatrix} 1 & 1 \\ 0 & 0 \\ 1 & 0 \end{bmatrix}.$$

Will the matrix B from part (a) also "reverse" that state of the bank of switches represented by C? Verify your answer.

(c) If A is any $m \times n$ bit matrix representing a bank of switches, determine an $m \times n$ bit matrix B so that $A + B$ "reverses" all the states of the switches in A. Give reasons why B will "reverse" the states in A.

MATLAB Exercises

In order to use MATLAB *in this section, you should first read Sections 12.1 and 12.2, which give basic information about* MATLAB *and about matrix operations in* MATLAB. *You are urged to do any examples or illustrations of* MATLAB *commands that appear in Sections 12.1 and 12.2 before trying these exercises.*

ML.1. In MATLAB, enter the following matrices.

$$A = \begin{bmatrix} 5 & 1 & 2 \\ -3 & 0 & 1 \\ 2 & 4 & 1 \end{bmatrix},$$

$$B = \begin{bmatrix} 4 * 2 & 2/3 \\ 1/201 & 5 - 8.2 \\ 0.00001 & (9 + 4)/3 \end{bmatrix}.$$

Using MATLAB commands, display the following.

(a) a_{23}, b_{32}, b_{12}

(b) $\text{row}_1(A), \text{col}_3(A), \text{row}_2(B)$

(c) Type MATLAB command **format long** and display matrix B. Compare the elements of B from part (a) with the current display. Note that **format short** displays four decimal places rounded. Reset the format to **format short**.

ML.2. In MATLAB, type the command **H = hilb(5);** (Note that the last character is a semicolon, which suppresses the display of the contents of matrix H. See Section 12.1.) For more information on the **hilb** command, type **help hilb**. Using MATLAB commands, do the following:

(a) Determine the size of H.

(b) Display the contents of H.

(c) Display the contents of H as rational numbers.

(d) Extract as a matrix the first three columns.

(e) Extract as a matrix the last two rows.

Exercises ML.3 through ML.5 use bit matrices and the supplemental instructional commands described in Section 12.9.

ML.3. Use **bingen** to solve Exercises T.10 and T.11.

ML.4. Use **bingen** to solve Exercise T.13. (*Hint*: An $n \times n$ matrix contains the same number of entries as an n^2-vector.)

ML.5. Solve Exercise 11 using **binadd**.

1.3 DOT PRODUCT AND MATRIX MULTIPLICATION

In this section we introduce the operation of matrix multiplication. Unlike matrix addition, matrix multiplication has some properties that distinguish it from multiplication of real numbers.

DEFINITION

The **dot product** or **inner product** of the n-vectors **a** and **b** is the sum of the products of corresponding entries. Thus, if

$$\mathbf{a} = \begin{bmatrix} a_1 \\ a_2 \\ \vdots \\ a_n \end{bmatrix} \quad \text{and} \quad \mathbf{b} = \begin{bmatrix} b_1 \\ b_2 \\ \vdots \\ b_n \end{bmatrix},$$

then

$$\mathbf{a} \cdot \mathbf{b} = a_1b_1 + a_2b_2 + \cdots + a_nb_n = \sum_{i=1}^{n} a_ib_i.^{\dagger} \qquad (1)$$

Similarly, if **a** or **b** (or both) are n-vectors written as a $1 \times n$ matrix, then the dot product $\mathbf{a} \cdot \mathbf{b}$ is given by (1). The dot product of vectors in C^n is defined in Appendix A.2.

The dot product is an important operation that will be used here and in later sections.

EXAMPLE 1

The dot product of

$$\mathbf{u} = \begin{bmatrix} 1 \\ -2 \\ 3 \\ 4 \end{bmatrix} \quad \text{and} \quad \mathbf{v} = \begin{bmatrix} 2 \\ 3 \\ -2 \\ 1 \end{bmatrix}$$

is

$$\mathbf{u} \cdot \mathbf{v} = (1)(2) + (-2)(3) + (3)(-2) + (4)(1) = -6.$$

■

EXAMPLE 2

Let $\mathbf{a} = \begin{bmatrix} x & 2 & 3 \end{bmatrix}$ and $\mathbf{b} = \begin{bmatrix} 4 \\ 1 \\ 2 \end{bmatrix}$. If $\mathbf{a} \cdot \mathbf{b} = -4$, find x.

Solution

We have

$$\begin{aligned} \mathbf{a} \cdot \mathbf{b} = 4x + 2 + 6 &= -4 \\ 4x + 8 &= -4 \\ x &= -3. \end{aligned}$$

■

EXAMPLE 3

(Application: Computing a Course Average) Suppose that an instructor uses four grades to determine a student's course average: quizzes, two hourly exams, and a final exam. These are weighted as 10%, 30%, 30%, and 30%, respectively. If a student's scores are 78, 84, 62, and 85, respectively, we can compute the course average by letting

$$\mathbf{w} = \begin{bmatrix} 0.10 \\ 0.30 \\ 0.30 \\ 0.30 \end{bmatrix} \quad \text{and} \quad \mathbf{g} = \begin{bmatrix} 78 \\ 84 \\ 62 \\ 85 \end{bmatrix}$$

and computing

$$\mathbf{w} \cdot \mathbf{g} = (0.10)(78) + (0.30)(84) + (0.30)(62) + (0.30)(85) = 77.1.$$

Thus, the student's course average is 77.1. ■

† You may already be familiar with this useful notation, the summation notation. It is discussed in detail at the end of this section.

MATRIX MULTIPLICATION

DEFINITION If $A = [a_{ij}]$ is an $m \times p$ matrix and $B = [b_{ij}]$ is a $p \times n$ matrix, then the **product** of A and B, denoted AB, is the $m \times n$ matrix $C = [c_{ij}]$, defined by

$$\begin{aligned} c_{ij} &= a_{i1}b_{1j} + a_{i2}b_{2j} + \cdots + a_{ip}b_{pj} \\ &= \sum_{k=1}^{p} a_{ik}b_{kj} \quad (1 \le i \le m, 1 \le j \le n). \end{aligned} \tag{2}$$

Equation (2) says that the i, jth element in the product matrix is the dot product of the ith row, $\text{row}_i(A)$, and the jth column, $\text{col}_j(B)$, of B; this is shown in Figure 1.4.

Figure 1.4 ►

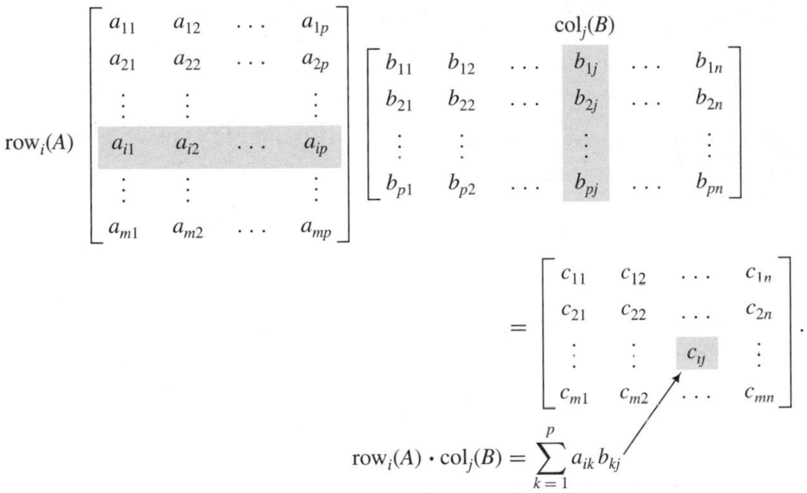

$$\text{row}_i(A)\ \begin{bmatrix} a_{11} & a_{12} & \cdots & a_{1p} \\ a_{21} & a_{22} & \cdots & a_{2p} \\ \vdots & \vdots & & \vdots \\ a_{i1} & a_{i2} & \cdots & a_{ip} \\ \vdots & \vdots & & \vdots \\ a_{m1} & a_{m2} & \cdots & a_{mp} \end{bmatrix} \overset{\text{col}_j(B)}{\begin{bmatrix} b_{11} & b_{12} & \cdots & b_{1j} & \cdots & b_{1n} \\ b_{21} & b_{22} & \cdots & b_{2j} & \cdots & b_{2n} \\ \vdots & \vdots & & \vdots & & \vdots \\ b_{p1} & b_{p2} & \cdots & b_{pj} & \cdots & b_{pn} \end{bmatrix}}$$

$$= \begin{bmatrix} c_{11} & c_{12} & \cdots & c_{1n} \\ c_{21} & c_{22} & \cdots & c_{2n} \\ \vdots & \vdots & c_{ij} & \vdots \\ c_{m1} & c_{m2} & \cdots & c_{mn} \end{bmatrix}.$$

$$\text{row}_i(A) \cdot \text{col}_j(B) = \sum_{k=1}^{p} a_{ik}\, b_{kj}$$

Observe that the product of A and B is defined only when the number of rows of B is exactly the same as the number of columns of A, as is indicated in Figure 1.5.

Figure 1.5 ►

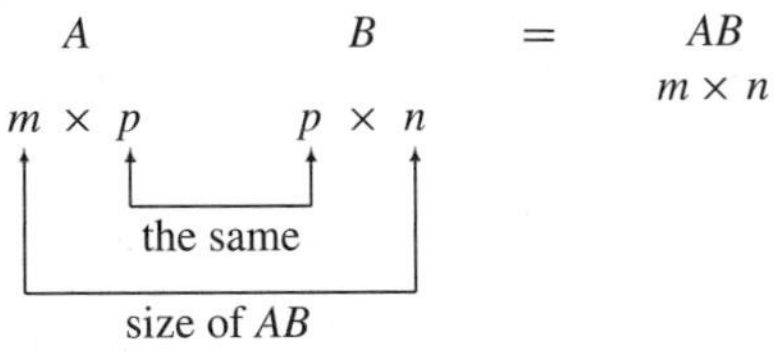

$$\underset{m \times p}{A} \qquad \underset{p \times n}{B} \qquad = \qquad \underset{m \times n}{AB}$$

the same (inner p and p)

size of AB (outer m and n)

EXAMPLE 4 Let

$$A = \begin{bmatrix} 1 & 2 & -1 \\ 3 & 1 & 4 \end{bmatrix} \quad \text{and} \quad B = \begin{bmatrix} -2 & 5 \\ 4 & -3 \\ 2 & 1 \end{bmatrix}.$$

Then

$$\begin{aligned} AB &= \begin{bmatrix} (1)(-2) + (2)(4) + (-1)(2) & (1)(5) + (2)(-3) + (-1)(1) \\ (3)(-2) + (1)(4) + (4)(2) & (3)(5) + (1)(-3) + (4)(1) \end{bmatrix} \\ &= \begin{bmatrix} 4 & -2 \\ 6 & 16 \end{bmatrix}. \end{aligned}$$

■

EXAMPLE 5 Let

$$A = \begin{bmatrix} 1 & -2 & 3 \\ 4 & 2 & 1 \\ 0 & 1 & -2 \end{bmatrix} \quad \text{and} \quad B = \begin{bmatrix} 1 & 4 \\ 3 & -1 \\ -2 & 2 \end{bmatrix}.$$

Compute the (3, 2) entry of AB.

Solution If $AB = C$, then the (3, 2) entry of AB is c_{32}, which is $\text{row}_3(A) \cdot \text{col}_2(B)$. We now have

$$\text{row}_3(A) \cdot \text{col}_2(B) = \begin{bmatrix} 0 & 1 & -2 \end{bmatrix} \cdot \begin{bmatrix} 4 \\ -1 \\ 2 \end{bmatrix} = -5.$$

■

EXAMPLE 6 The linear system

$$\begin{aligned} x + 2y - z &= 2 \\ 3x + 4z &= 5 \end{aligned}$$

can be written (verify) using a matrix product as

$$\begin{bmatrix} 1 & 2 & -1 \\ 3 & 0 & 4 \end{bmatrix} \begin{bmatrix} x \\ y \\ z \end{bmatrix} = \begin{bmatrix} 2 \\ 5 \end{bmatrix}.$$

■

EXAMPLE 7 Let

$$A = \begin{bmatrix} 1 & x & 3 \\ 2 & -1 & 1 \end{bmatrix} \quad \text{and} \quad B = \begin{bmatrix} 2 \\ 4 \\ y \end{bmatrix}.$$

If $AB = \begin{bmatrix} 12 \\ 6 \end{bmatrix}$, find x and y.

Solution We have

$$AB = \begin{bmatrix} 1 & x & 3 \\ 2 & -1 & 1 \end{bmatrix} \begin{bmatrix} 2 \\ 4 \\ y \end{bmatrix} = \begin{bmatrix} 2 + 4x + 3y \\ 4 - 4 + y \end{bmatrix} = \begin{bmatrix} 12 \\ 6 \end{bmatrix}.$$

Then

$$\begin{aligned} 2 + 4x + 3y &= 12 \\ y &= 6, \end{aligned}$$

so $x = -2$ and $y = 6$. ■

The basic properties of matrix multiplication will be considered in the following section. However, multiplication of matrices requires much more care than their addition, since the algebraic properties of matrix multiplication differ from those satisfied by the real numbers. Part of the problem is due to the fact that AB is defined only when the number of columns of A is the same as the number of rows of B. Thus, if A is an $m \times p$ matrix and B is a $p \times n$ matrix, then AB is an $m \times n$ matrix. What about BA? Four different situations may occur:

1. BA may not be defined; this will take place if $n \neq m$.
2. If BA is defined, which means that $m = n$, then BA is $p \times p$ while AB is $m \times m$; thus, if $m \neq p$, AB and BA are of different sizes.

3. If AB and BA are both of the same size, they may be equal.
4. If AB and BA are both of the same size, they may be unequal.

EXAMPLE 8 If A is a 2×3 matrix and B is a 3×4 matrix, then AB is a 2×4 matrix while BA is undefined. ■

EXAMPLE 9 Let A be 2×3 and let B be 3×2. Then AB is 2×2 while BA is 3×3. ■

EXAMPLE 10 Let

$$A = \begin{bmatrix} 1 & 2 \\ -1 & 3 \end{bmatrix} \quad \text{and} \quad B = \begin{bmatrix} 2 & 1 \\ 0 & 1 \end{bmatrix}.$$

Then

$$AB = \begin{bmatrix} 2 & 3 \\ -2 & 2 \end{bmatrix} \quad \text{while} \quad BA = \begin{bmatrix} 1 & 7 \\ -1 & 3 \end{bmatrix}.$$

Thus $AB \neq BA$. ■

One might ask why matrix equality and matrix addition are defined in such a natural way while matrix multiplication appears to be much more complicated. Example 11 provides a motivation for the definition of matrix multiplication.

EXAMPLE 11 **(Ecology)** Pesticides are sprayed on plants to eliminate harmful insects. However, some of the pesticide is absorbed by the plant. The pesticides are absorbed by herbivores when they eat the plants that have been sprayed. To determine the amount of pesticide absorbed by a herbivore, we proceed as follows. Suppose that we have three pesticides and four plants. Let a_{ij} denote the amount of pesticide i (in milligrams) that has been absorbed by plant j. This information can be represented by the matrix

$$A = \begin{array}{c} \\ \\ \\ \end{array}\begin{matrix} \text{Plant 1} & \text{Plant 2} & \text{Plant 3} & \text{Plant 4} \\ \end{matrix}$$

$$A = \left[\begin{array}{cccc} 2 & 3 & 4 & 3 \\ 3 & 2 & 2 & 5 \\ 4 & 1 & 6 & 4 \end{array}\right] \begin{array}{l} \text{Pesticide 1} \\ \text{Pesticide 2} \\ \text{Pesticide 3} \end{array}$$

(columns: Plant 1, Plant 2, Plant 3, Plant 4)

Now suppose that we have three herbivores, and let b_{ij} denote the number of plants of type i that a herbivore of type j eats per month. This information can be represented by the matrix

$$B = \left[\begin{array}{ccc} 20 & 12 & 8 \\ 28 & 15 & 15 \\ 30 & 12 & 10 \\ 40 & 16 & 20 \end{array}\right] \begin{array}{l} \text{Plant 1} \\ \text{Plant 2} \\ \text{Plant 3} \\ \text{Plant 4} \end{array}$$

(columns: Herbivore 1, Herbivore 2, Herbivore 3)

The (i, j) entry in AB gives the amount of pesticide of type i that animal j has absorbed. Thus, if $i = 2$ and $j = 3$, the (2, 3) entry in AB is

$$\begin{aligned} &3(8) + 2(15) + 2(10) + 5(20) \\ &\qquad = 174 \text{ mg of pesticide 2 absorbed by herbivore 3.} \end{aligned}$$

If we now have p carnivores (such as man) who eat the herbivores, we can repeat the analysis to find out how much of each pesticide has been absorbed by each carnivore. ■

It is sometimes useful to be able to find a column in the matrix product AB without having to multiply the two matrices. It can be shown (Exercise T.9) that the jth column of the matrix product AB is equal to the matrix product $A\text{col}_j(B)$.

EXAMPLE 12 Let

$$A = \begin{bmatrix} 1 & 2 \\ 3 & 4 \\ -1 & 5 \end{bmatrix} \quad \text{and} \quad B = \begin{bmatrix} -2 & 3 & 4 \\ 3 & 2 & 1 \end{bmatrix}.$$

Then the second column of AB is

$$A\text{col}_2(B) = \begin{bmatrix} 1 & 2 \\ 3 & 4 \\ -1 & 5 \end{bmatrix} \begin{bmatrix} 3 \\ 2 \end{bmatrix} = \begin{bmatrix} 7 \\ 17 \\ 7 \end{bmatrix}.$$

■

Remark If $\mathbf{u}$ and $\mathbf{v}$ are n-vectors, it can be shown (Exercise T.14) that if we view them as $n \times 1$ matrices, then

$$\mathbf{u} \cdot \mathbf{v} = \mathbf{u}^T\mathbf{v}.$$

This observation will be used in Chapter 4. Similarly, if $\mathbf{u}$ and $\mathbf{v}$ are viewed as $1 \times n$ matrices, then

$$\mathbf{u} \cdot \mathbf{v} = \mathbf{u}\mathbf{v}^T.$$

Finally, if $\mathbf{u}$ is a $1 \times n$ matrix and $\mathbf{v}$ is an $n \times 1$ matrix, then $\mathbf{u} \cdot \mathbf{v} = \mathbf{u}\mathbf{v}$.

EXAMPLE 13 Let $\mathbf{u} = \begin{bmatrix} 1 \\ 2 \\ -3 \end{bmatrix}$ and $\mathbf{v} = \begin{bmatrix} 2 \\ -1 \\ 1 \end{bmatrix}$. Then

$$\mathbf{u} \cdot \mathbf{v} = 1(2) + 2(-1) + (-3)(1) = -3.$$

Moreover,

$$\mathbf{u}^T\mathbf{v} = \begin{bmatrix} 1 & 2 & -3 \end{bmatrix} \begin{bmatrix} 2 \\ -1 \\ 1 \end{bmatrix} = 1(2) + 2(-1) + (-3)(1) = -3.$$

■

THE MATRIX-VECTOR PRODUCT WRITTEN IN TERMS OF COLUMNS

Let

$$A = \begin{bmatrix} a_{11} & a_{12} & \cdots & a_{1n} \\ a_{21} & a_{22} & \cdots & a_{2n} \\ \vdots & \vdots & & \vdots \\ a_{m1} & a_{m2} & \cdots & a_{mn} \end{bmatrix}$$

be an $m \times n$ matrix and let

$$\mathbf{c} = \begin{bmatrix} c_1 \\ c_2 \\ \vdots \\ c_n \end{bmatrix}$$

be an n-vector, that is, an $n \times 1$ matrix. Since A is $m \times n$ and $\mathbf{c}$ is $n \times 1$, the matrix product $A\mathbf{c}$ is the $m \times 1$ matrix

$$A\mathbf{c} = \begin{bmatrix} a_{11} & a_{12} & \cdots & a_{1n} \\ a_{21} & a_{22} & \cdots & a_{2n} \\ \vdots & \vdots & & \vdots \\ a_{m1} & a_{m2} & \cdots & a_{mn} \end{bmatrix} \begin{bmatrix} c_1 \\ c_2 \\ \vdots \\ c_n \end{bmatrix} = \begin{bmatrix} \text{row}_1(A) \cdot \mathbf{c} \\ \text{row}_2(A) \cdot \mathbf{c} \\ \vdots \\ \text{row}_m(A) \cdot \mathbf{c} \end{bmatrix} \tag{3}$$

$$= \begin{bmatrix} a_{11}c_1 + a_{12}c_2 + \cdots + a_{1n}c_n \\ a_{21}c_1 + a_{22}c_2 + \cdots + a_{2n}c_n \\ \vdots \\ a_{m1}c_1 + a_{m2}c_2 + \cdots + a_{mn}c_n \end{bmatrix}.$$

The right side of this expression can be written as

$$c_1 \begin{bmatrix} a_{11} \\ a_{21} \\ \vdots \\ a_{m1} \end{bmatrix} + c_2 \begin{bmatrix} a_{12} \\ a_{22} \\ \vdots \\ a_{m2} \end{bmatrix} + \cdots + c_n \begin{bmatrix} a_{1n} \\ a_{2n} \\ \vdots \\ a_{mn} \end{bmatrix} \tag{4}$$

$$= c_1\text{col}_1(A) + c_2\text{col}_2(A) + \cdots + c_n\text{col}_n(A).$$

Thus the product $A\mathbf{c}$ of an $m \times n$ matrix A and an $n \times 1$ matrix $\mathbf{c}$ can be written as a linear combination of the columns of A, where the coefficients are the entries in $\mathbf{c}$.

EXAMPLE 14 Let

$$A = \begin{bmatrix} 2 & -1 & -3 \\ 4 & 2 & -2 \end{bmatrix} \quad \text{and} \quad \mathbf{c} = \begin{bmatrix} 2 \\ -3 \\ 4 \end{bmatrix}.$$

Then the product $A\mathbf{c}$ written as a linear combination of the columns of A is

$$A\mathbf{c} = \begin{bmatrix} 2 & -1 & -3 \\ 4 & 2 & -2 \end{bmatrix} \begin{bmatrix} 2 \\ -3 \\ 4 \end{bmatrix} = 2\begin{bmatrix} 2 \\ 4 \end{bmatrix} - 3\begin{bmatrix} -1 \\ 2 \end{bmatrix} + 4\begin{bmatrix} -3 \\ -2 \end{bmatrix} = \begin{bmatrix} -5 \\ -6 \end{bmatrix}.$$

■

If A is an $m \times p$ matrix and B is a $p \times n$ matrix, we can then conclude that the jth column of the product AB can be written as a linear combination of the columns of matrix A, where the coefficients are the entries in the jth column of matrix B:

$$\text{col}_j(AB) = A\text{col}_j(B) = b_{1j}\text{col}_1(A) + b_{2j}\text{col}_2(A) + \cdots + b_{pj}\text{col}_p(A).$$

EXAMPLE 15 If A and B are the matrices defined in Example 12, then

$$AB = \begin{bmatrix} 1 & 2 \\ 3 & 4 \\ -1 & 5 \end{bmatrix} \begin{bmatrix} -2 & 3 & 4 \\ 3 & 2 & 1 \end{bmatrix} = \begin{bmatrix} 4 & 7 & 6 \\ 6 & 17 & 16 \\ 17 & 7 & 1 \end{bmatrix}.$$

The columns of AB as linear combinations of the columns of A are given by

$$\text{col}_1(AB) = \begin{bmatrix} 4 \\ 6 \\ 17 \end{bmatrix} = A\text{col}_1(B) = -2\begin{bmatrix} 1 \\ 3 \\ -1 \end{bmatrix} + 3\begin{bmatrix} 2 \\ 4 \\ 5 \end{bmatrix}$$

$$\text{col}_2(AB) = \begin{bmatrix} 7 \\ 17 \\ 7 \end{bmatrix} = A\text{col}_2(B) = 3\begin{bmatrix} 1 \\ 3 \\ -1 \end{bmatrix} + 2\begin{bmatrix} 2 \\ 4 \\ 5 \end{bmatrix}$$

$$\text{col}_3(AB) = \begin{bmatrix} 6 \\ 16 \\ 1 \end{bmatrix} = A\text{col}_3(B) = 4\begin{bmatrix} 1 \\ 3 \\ -1 \end{bmatrix} + 1\begin{bmatrix} 2 \\ 4 \\ 5 \end{bmatrix}.$$

■

LINEAR SYSTEMS

We now generalize Example 6. Consider the linear system of m equations in n unknowns,

$$\begin{array}{ccccccc} a_{11}x_1 & + & a_{12}x_2 & + \cdots + & a_{1n}x_n & = & b_1 \\ a_{21}x_1 & + & a_{22}x_2 & + \cdots + & a_{2n}x_n & = & b_2 \\ \vdots & & \vdots & & \vdots & & \vdots \\ a_{m1}x_1 & + & a_{m2}x_2 & + \cdots + & a_{mn}x_n & = & b_m. \end{array} \tag{5}$$

Now define the following matrices:

$$A = \begin{bmatrix} a_{11} & a_{12} & \cdots & a_{1n} \\ a_{21} & a_{22} & \cdots & a_{2n} \\ \vdots & \vdots & & \vdots \\ a_{m1} & a_{m2} & \cdots & a_{mn} \end{bmatrix}, \qquad \mathbf{x} = \begin{bmatrix} x_1 \\ x_2 \\ \vdots \\ x_n \end{bmatrix}, \qquad \mathbf{b} = \begin{bmatrix} b_1 \\ b_2 \\ \vdots \\ b_m \end{bmatrix}.$$

Then

$$A\mathbf{x} = \begin{bmatrix} a_{11} & a_{12} & \cdots & a_{1n} \\ a_{21} & a_{22} & \cdots & a_{2n} \\ \vdots & \vdots & & \vdots \\ a_{m1} & a_{m2} & \cdots & a_{mn} \end{bmatrix} \begin{bmatrix} x_1 \\ x_2 \\ \vdots \\ x_n \end{bmatrix} = \begin{bmatrix} a_{11}x_1 + a_{12}x_2 + \cdots + a_{1n}x_n \\ a_{21}x_1 + a_{22}x_2 + \cdots + a_{2n}x_n \\ \vdots \qquad \vdots \qquad \vdots \\ a_{m1}x_1 + a_{m2}x_2 + \cdots + a_{mn}x_n \end{bmatrix}.$$

The entries in the product $A\mathbf{x}$ are merely the left sides of the equations in (5). Hence the linear system (5) can be written in matrix form as

$$A\mathbf{x} = \mathbf{b}.$$

The matrix A is called the **coefficient matrix** of the linear system (5), and the matrix

$$\left[\begin{array}{cccc:c} a_{11} & a_{12} & \cdots & a_{1n} & b_1 \\ a_{21} & a_{22} & \cdots & a_{2n} & b_2 \\ \vdots & \vdots & & \vdots & \vdots \\ a_{m1} & a_{m2} & \cdots & a_{mn} & b_m \end{array}\right],$$

obtained by adjoining column $\mathbf{b}$ to A, is called the **augmented matrix** of the linear system (5). The augmented matrix of (5) will be written as $\left[A \,\vdots\, \mathbf{b}\right]$. Conversely, any matrix with more than one column can be thought of as the augmented matrix of a linear system. The coefficient and augmented matrices will play key roles in our method for solving linear systems.

EXAMPLE 16 Consider the linear system

$$\begin{aligned} -2x \quad\;\; + \; z &= 5 \\ 2x + 3y - 4z &= 7 \\ 3x + 2y + 2z &= 3. \end{aligned}$$

Letting

$$A = \begin{bmatrix} -2 & 0 & 1 \\ 2 & 3 & -4 \\ 3 & 2 & 2 \end{bmatrix}, \quad \mathbf{x} = \begin{bmatrix} x \\ y \\ z \end{bmatrix}, \quad \text{and} \quad \mathbf{b} = \begin{bmatrix} 5 \\ 7 \\ 3 \end{bmatrix},$$

we can write the given linear system in matrix form as

$$A\mathbf{x} = \mathbf{b}.$$

The coefficient matrix is A and the augmented matrix is

$$\left[\begin{array}{rrr:r} -2 & 0 & 1 & 5 \\ 2 & 3 & -4 & 7 \\ 3 & 2 & 2 & 3 \end{array}\right].$$

■

EXAMPLE 17 The matrix

$$\left[\begin{array}{rrr:r} 2 & -1 & 3 & 4 \\ 3 & 0 & 2 & 5 \end{array}\right]$$

is the augmented matrix of the linear system

$$\begin{aligned} 2x - y + 3z &= 4 \\ 3x \quad\;\; + 2z &= 5. \end{aligned}$$

■

It follows from our discussion above that the linear system in (5) can be written as a linear combination of the columns of A as

$$x_1 \begin{bmatrix} a_{11} \\ a_{21} \\ \vdots \\ a_{m1} \end{bmatrix} + x_2 \begin{bmatrix} a_{12} \\ a_{22} \\ \vdots \\ a_{m2} \end{bmatrix} + \cdots + x_n \begin{bmatrix} a_{1n} \\ a_{2n} \\ \vdots \\ a_{mn} \end{bmatrix} = \begin{bmatrix} b_1 \\ b_2 \\ \vdots \\ b_m \end{bmatrix}. \qquad (6)$$

Conversely, an equation as in (6) always describes a linear system as in (5).

PARTITIONED MATRICES (OPTIONAL)

If we start out with an $m \times n$ matrix $A = \left[a_{ij}\right]$ and cross out some, but not all, of its rows or columns, we obtain a **submatrix** of A.

EXAMPLE 18 Let

$$A = \begin{bmatrix} 1 & 2 & 3 & 4 \\ -2 & 4 & -3 & 5 \\ 3 & 0 & 5 & -3 \end{bmatrix}.$$

If we cross out the second row and third column, we obtain the submatrix

$$\begin{bmatrix} 1 & 2 & 4 \\ 3 & 0 & -3 \end{bmatrix}.$$

■

A matrix can be partitioned into submatrices by drawing horizontal lines between rows and vertical lines between columns. Of course, the partitioning can be carried out in many different ways.

EXAMPLE 19 The matrix

$$A = \left[\begin{array}{ccc|cc} a_{11} & a_{12} & a_{13} & a_{14} & a_{15} \\ a_{21} & a_{22} & a_{23} & a_{24} & a_{25} \\ \hline a_{31} & a_{32} & a_{33} & a_{34} & a_{35} \\ a_{41} & a_{42} & a_{43} & a_{44} & a_{45} \end{array}\right]$$

is partitioned as

$$A = \left[\begin{array}{c|c} A_{11} & A_{12} \\ \hline A_{21} & A_{22} \end{array}\right].$$

We could also write

$$A = \left[\begin{array}{cc|cc|c} a_{11} & a_{12} & a_{13} & a_{14} & a_{15} \\ a_{21} & a_{22} & a_{23} & a_{24} & a_{25} \\ \hline a_{31} & a_{32} & a_{33} & a_{34} & a_{35} \\ a_{41} & a_{42} & a_{43} & a_{44} & a_{45} \end{array}\right] = \left[\begin{array}{c|c|c} \widehat{A}_{11} & \widehat{A}_{12} & \widehat{A}_{13} \\ \hline \widehat{A}_{21} & \widehat{A}_{22} & \widehat{A}_{23} \end{array}\right], \tag{7}$$

which gives another partitioning of A. We thus speak of **partitioned matrices**. ■

EXAMPLE 20 The augmented matrix of a linear system is a partitioned matrix. Thus, if $A\mathbf{x} = \mathbf{b}$, we can write the augmented matrix of this system as $\left[A \mid \mathbf{b}\right]$. ■

If A and B are both $m \times n$ matrices that are partitioned in the same way, then $A + B$ is obtained simply by adding the corresponding submatrices of A and B. Similarly, if A is a partitioned matrix, then the scalar multiple cA is obtained by forming the scalar multiple of each submatrix.

If A is partitioned as shown in (7) and

$$B = \left[\begin{array}{cc|cc} b_{11} & b_{12} & b_{13} & b_{14} \\ b_{21} & b_{22} & b_{23} & b_{24} \\ \hline b_{31} & b_{32} & b_{33} & b_{34} \\ b_{41} & b_{42} & b_{43} & b_{44} \\ \hline b_{51} & b_{52} & b_{53} & b_{54} \end{array}\right] = \left[\begin{array}{c|c} B_{11} & B_{12} \\ \hline B_{21} & B_{22} \\ \hline B_{31} & B_{32} \end{array}\right],$$

then by straightforward computations we can show that

$$AB = \left[\begin{array}{c|c} (\widehat{A}_{11}B_{11} + \widehat{A}_{12}B_{21} + \widehat{A}_{13}B_{31}) & (\widehat{A}_{11}B_{12} + \widehat{A}_{12}B_{22} + \widehat{A}_{13}B_{32}) \\ \hline (\widehat{A}_{21}B_{11} + \widehat{A}_{22}B_{21} + \widehat{A}_{23}B_{31}) & (\widehat{A}_{21}B_{12} + \widehat{A}_{22}B_{22} + \widehat{A}_{23}B_{32}) \end{array}\right].$$

EXAMPLE 21 Let

$$A = \left[\begin{array}{cc|cc} 1 & 0 & 1 & 0 \\ 0 & 2 & 3 & -1 \\ \hline 2 & 0 & -4 & 0 \\ 0 & 1 & 0 & 3 \end{array}\right] = \left[\begin{array}{c|c} A_{11} & A_{12} \\ \hline A_{21} & A_{22} \end{array}\right]$$

and let

$$B = \left[\begin{array}{ccc|ccc} 2 & 0 & 0 & 1 & 1 & -1 \\ 0 & 1 & 1 & -1 & 2 & 2 \\ \hline 1 & 3 & 0 & 0 & 1 & 0 \\ -3 & -1 & 2 & 1 & 0 & -1 \end{array}\right] = \left[\begin{array}{c|c} B_{11} & B_{12} \\ \hline B_{21} & B_{22} \end{array}\right].$$

Then

$$AB = C = \left[\begin{array}{ccc|ccc} 3 & 3 & 0 & 1 & 2 & -1 \\ 6 & 12 & 0 & -3 & 7 & 5 \\ \hline 0 & -12 & 0 & 2 & -2 & -2 \\ -9 & -2 & 7 & 2 & 2 & -1 \end{array}\right] = \left[\begin{array}{c|c} C_{11} & C_{12} \\ \hline C_{21} & C_{22} \end{array}\right],$$

where C_{11} should be $A_{11}B_{11} + A_{12}B_{21}$. We verify that C_{11} is this expression as follows:

$$\begin{aligned} A_{11}B_{11} + A_{12}B_{21} &= \begin{bmatrix} 1 & 0 \\ 0 & 2 \end{bmatrix}\begin{bmatrix} 2 & 0 & 0 \\ 0 & 1 & 1 \end{bmatrix} + \begin{bmatrix} 1 & 0 \\ 3 & -1 \end{bmatrix}\begin{bmatrix} 1 & 3 & 0 \\ -3 & -1 & 2 \end{bmatrix} \\ &= \begin{bmatrix} 2 & 0 & 0 \\ 0 & 2 & 2 \end{bmatrix} + \begin{bmatrix} 1 & 3 & 0 \\ 6 & 10 & -2 \end{bmatrix} \\ &= \begin{bmatrix} 3 & 3 & 0 \\ 6 & 12 & 0 \end{bmatrix} = C_{11}. \end{aligned}$$

■

This method of multiplying partitioned matrices is also known as **block multiplication**. Partitioned matrices can be used to great advantage in dealing with matrices that exceed the memory capacity of a computer. Thus, in multiplying two partitioned matrices, one can keep the matrices on disk and only bring into memory the submatrices required to form the submatrix products. The latter, of course, can be put out on disk as they are formed. The partitioning must be done so that the products of corresponding submatrices are defined. In contemporary computing technology, parallel-processing computers use partitioned matrices to perform matrix computations more rapidly.

Partitioning of a matrix implies a subdivision of the information into blocks or units. The reverse process is to consider individual matrices as blocks and adjoin them to form a partitioned matrix. The only requirement is that after joining the blocks, all rows have the same number of entries and all columns have the same number of entries.

EXAMPLE 22

Let

$$B = \begin{bmatrix} 2 \\ 3 \end{bmatrix}, \quad C = \begin{bmatrix} 1 & -1 & 0 \end{bmatrix}, \quad \text{and} \quad D = \begin{bmatrix} 9 & 8 & -4 \\ 6 & 7 & 5 \end{bmatrix}.$$

Then we have

$$\left[\begin{array}{c|c} B & D \end{array}\right] = \left[\begin{array}{c|ccc} 2 & 9 & 8 & -4 \\ 3 & 6 & 7 & 5 \end{array}\right], \quad \left[\begin{array}{c} D \\ \hline C \end{array}\right] = \left[\begin{array}{ccc} 9 & 8 & -4 \\ 6 & 7 & 5 \\ \hline 1 & -1 & 0 \end{array}\right],$$

and

$$\left[\begin{array}{c|c} \left[\begin{array}{c} D \\ \hline C \end{array}\right] & C^T \end{array}\right] = \left[\begin{array}{ccc|c} 9 & 8 & -4 & 1 \\ 6 & 7 & 5 & -1 \\ \hline 1 & -1 & 0 & 0 \end{array}\right].$$

■

Adjoining matrix blocks to expand information structures is done regularly in a variety of applications. It is common to keep monthly sales data for a year in a 1×12 matrix and then adjoin such matrices to build a sales history matrix for a period of years. Similarly, results of new laboratory experiments are adjoined to existing data to update a database in a research facility.

We have already noted in Example 20 that the augmented matrix of the linear system $A\mathbf{x} = \mathbf{b}$ is a partitioned matrix. At times we shall need to solve several linear systems in which the coefficient matrix A is the same but the right sides of the systems are different, say $\mathbf{b}$, $\mathbf{c}$, and $\mathbf{d}$. In these cases we shall find it convenient to consider the partitioned matrix $\left[A \mid \mathbf{b} \mid \mathbf{c} \mid \mathbf{d}\right]$. (See Section 6.7.)

SUMMATION NOTATION (OPTIONAL)

We shall occasionally use the **summation notation** and we now review this useful and compact notation, which is widely used in mathematics.

By $\sum_{i=1}^{n} a_i$ we mean

$$a_1 + a_2 + \cdots + a_n.$$

The letter i is called the **index of summation**; it is a dummy variable that can be replaced by another letter. Hence we can write

$$\sum_{i=1}^{n} a_i = \sum_{j=1}^{n} a_j = \sum_{k=1}^{n} a_k.$$

EXAMPLE 23 If

$$a_1 = 3, \quad a_2 = 4, \quad a_3 = 5, \quad \text{and} \quad a_4 = 8,$$

then

$$\sum_{i=1}^{4} a_i = 3 + 4 + 5 + 8 = 20.$$

■

EXAMPLE 24 By $\sum_{i=1}^{n} r_i a_i$ we mean

$$r_1 a_1 + r_2 a_2 + \cdots + r_n a_n.$$

It is not difficult to show (Exercise T.11) that the summation notation satisfies the following properties:

(i) $$\sum_{i=1}^{n} (r_i + s_i) a_i = \sum_{i=1}^{n} r_i a_i + \sum_{i=1}^{n} s_i a_i.$$

(ii) $$\sum_{i=1}^{n} c(r_i a_i) = c\left(\sum_{i=1}^{n} r_i a_i\right).$$

■

EXAMPLE 25 If

$$\mathbf{a} = \begin{bmatrix} a_1 \\ a_2 \\ \vdots \\ a_n \end{bmatrix} \quad \text{and} \quad \mathbf{b} = \begin{bmatrix} b_1 \\ b_2 \\ \vdots \\ b_n \end{bmatrix},$$

then the dot product $\mathbf{a} \cdot \mathbf{b}$ can be expressed using summation notation as

$$\mathbf{a} \cdot \mathbf{b} = a_1 b_1 + a_2 b_2 + \cdots + a_n b_n = \sum_{i=1}^{n} a_i b_i.$$

■

EXAMPLE 26 We can write Equation (2), for the i, jth element in the product of the matrices A and B, in terms of the summation notation as

$$c_{ij} = \sum_{k=1}^{p} a_{ik} b_{kj} \qquad (1 \leq i \leq m,\ 1 \leq j \leq n).$$

■

It is also possible to form double sums. Thus by $\sum_{j=1}^{m} \sum_{i=1}^{n} a_{ij}$ we mean that we first sum on i and then sum the resulting expression on j.

EXAMPLE 27 If $n = 2$ and $m = 3$, we have

$$\begin{aligned}\sum_{j=1}^{3} \sum_{i=1}^{2} a_{ij} &= \sum_{j=1}^{3} (a_{1j} + a_{2j}) \\ &= (a_{11} + a_{21}) + (a_{12} + a_{22}) + (a_{13} + a_{23})\end{aligned} \tag{8}$$

$$\begin{aligned}\sum_{i=1}^{2} \sum_{j=1}^{3} a_{ij} &= \sum_{i=1}^{2} (a_{i1} + a_{i2} + a_{i3}) \\ &= (a_{11} + a_{12} + a_{13}) + (a_{21} + a_{22} + a_{23}) \\ &= \text{right side of (8).}\end{aligned}$$

■

It is not difficult to show, in general (Exercise T.12), that

$$\sum_{i=1}^{n} \sum_{j=1}^{m} a_{ij} = \sum_{j=1}^{m} \sum_{i=1}^{n} a_{ij}. \tag{9}$$

Equation (9) can be interpreted as follows. Let A be the $m \times n$ matrix $\left[a_{ij}\right]$. If we add up the entries in each row of A and then add the resulting numbers, we obtain the same result as when we add up the entries in each column of A and then add the resulting numbers.

EXAMPLES WITH BIT MATRICES (OPTIONAL)

The dot product and the matrix product of bit matrices are computed in the usual manner, but we must recall that the arithmetic involved uses base 2.

EXAMPLE 28 Let $\mathbf{a} = \begin{bmatrix} 1 \\ 0 \\ 1 \end{bmatrix}$ and $\mathbf{b} = \begin{bmatrix} 1 \\ 1 \\ 0 \end{bmatrix}$ be bit vectors. Then

$$\mathbf{a} \cdot \mathbf{b} = (1)(1) + (0)(1) + (1)(0) = 1 + 0 + 0 = 1.$$

■

EXAMPLE 29 Let $A = \begin{bmatrix} 1 & 1 \\ 0 & 1 \end{bmatrix}$ and $B = \begin{bmatrix} 0 & 1 & 0 \\ 1 & 1 & 0 \end{bmatrix}$ be bit matrices. Then

$$AB = \begin{bmatrix} (1)(0) + (1)(1) & (1)(1) + (1)(1) & (1)(0) + (1)(0) \\ (0)(0) + (1)(1) & (0)(1) + (1)(1) & (0)(0) + (1)(0) \end{bmatrix} = \begin{bmatrix} 1 & 0 & 0 \\ 1 & 1 & 0 \end{bmatrix}.$$

■

EXAMPLE 30 Let $A = \begin{bmatrix} 1 & 1 & 1 & x \\ 1 & 1 & 0 & 1 \end{bmatrix}$ and $B = \begin{bmatrix} y \\ 0 \\ 1 \\ 1 \end{bmatrix}$ be bit matrices. If $AB = \begin{bmatrix} 1 \\ 1 \end{bmatrix}$, find x and y.

Solution We have

$$AB = \begin{bmatrix} 1 & 1 & 1 & x \\ 1 & 1 & 0 & 1 \end{bmatrix} \begin{bmatrix} y \\ 0 \\ 1 \\ 1 \end{bmatrix} = \begin{bmatrix} y + 1 + x \\ y + 1 \end{bmatrix} = \begin{bmatrix} 1 \\ 1 \end{bmatrix}.$$

Then $y + 1 + x = 1$ and $y + 1 = 1$. Using base 2 arithmetic, it follows that $y = 0$ and so then $x = 0$. ■

Key Terms

Dot product (inner product)
Product of matrices
Coefficient matrix
Augmented matrix
Submatrix
Partitioned matrix
Block multiplication
Summation notation

1.3 Exercises

In Exercises 1 and 2, compute $\mathbf{a} \cdot \mathbf{b}$.

1. (a) $\mathbf{a} = \begin{bmatrix} 1 & 2 \end{bmatrix}$, $\mathbf{b} = \begin{bmatrix} 4 \\ -1 \end{bmatrix}$

(b) $\mathbf{a} = \begin{bmatrix} -3 & -2 \end{bmatrix}$, $\mathbf{b} = \begin{bmatrix} 1 \\ -2 \end{bmatrix}$

(c) $\mathbf{a} = \begin{bmatrix} 4 & 2 & -1 \end{bmatrix}$, $\mathbf{b} = \begin{bmatrix} 1 \\ 3 \\ 6 \end{bmatrix}$

(d) $\mathbf{a} = \begin{bmatrix} 1 & 1 & 0 \end{bmatrix}$, $\mathbf{b} = \begin{bmatrix} 1 \\ 0 \\ 1 \end{bmatrix}$

2. (a) $\mathbf{a} = \begin{bmatrix} 2 & -1 \end{bmatrix}$, $\mathbf{b} = \begin{bmatrix} 3 \\ 2 \end{bmatrix}$

(b) $\mathbf{a} = \begin{bmatrix} 1 & -1 \end{bmatrix}$, $\mathbf{b} = \begin{bmatrix} 1 \\ 1 \end{bmatrix}$

(c) $\mathbf{a} = \begin{bmatrix} 1 & 2 & 3 \end{bmatrix}$, $\mathbf{b} = \begin{bmatrix} -2 \\ 0 \\ 1 \end{bmatrix}$

(d) $\mathbf{a} = \begin{bmatrix} 1 & 0 & 0 \end{bmatrix}$, $\mathbf{b} = \begin{bmatrix} 1 \\ 0 \\ 0 \end{bmatrix}$

3. Let $\mathbf{a} = \begin{bmatrix} -3 & 2 & x \end{bmatrix}$ and $\mathbf{b} = \begin{bmatrix} -3 \\ 2 \\ x \end{bmatrix}$. If $\mathbf{a} \cdot \mathbf{b} = 17$, find x.

4. Let $\mathbf{w} = \begin{bmatrix} \sin\theta \\ \cos\theta \end{bmatrix}$. Compute $\mathbf{w} \cdot \mathbf{w}$.

5. Find all values of x so that $\mathbf{v} \cdot \mathbf{v} = 1$, where $\mathbf{v} = \begin{bmatrix} \frac{1}{2} \\ -\frac{1}{2} \\ x \end{bmatrix}$.

6. Let $A = \begin{bmatrix} 1 & 2 & x \\ 3 & -1 & 2 \end{bmatrix}$ and $B = \begin{bmatrix} y \\ x \\ 1 \end{bmatrix}$. If $AB = \begin{bmatrix} 6 \\ 8 \end{bmatrix}$, find x and y.

In Exercises 7 and 8, let

$$A = \begin{bmatrix} 1 & 2 & -3 \\ 4 & 0 & -2 \end{bmatrix}, \quad B = \begin{bmatrix} 3 & 1 \\ 2 & 4 \\ -1 & 5 \end{bmatrix},$$

$$C = \begin{bmatrix} 2 & 3 & 1 \\ 3 & -4 & 5 \\ 1 & -1 & -2 \end{bmatrix}, \quad D = \begin{bmatrix} 2 & 3 \\ -1 & -2 \end{bmatrix},$$

$$E = \begin{bmatrix} 1 & 0 & -3 \\ -2 & 1 & 5 \\ 3 & 4 & 2 \end{bmatrix}, \quad \textit{and} \quad F = \begin{bmatrix} 2 & -3 \\ 4 & 1 \end{bmatrix}.$$

7. If possible, compute:

(a) AB (b) BA (c) $CB + D$

(d) $AB + DF$ (e) $BA + FD$

8. If possible, compute:

(a) $A(BD)$ (b) $(AB)D$ (c) $A(C + E)$

(d) $AC + AE$ (e) $(D + F)A$

9. Let $A = \begin{bmatrix} 2 & 3 \\ -1 & 4 \\ 0 & 3 \end{bmatrix}$ and $B = \begin{bmatrix} 3 & -1 & 3 \\ 1 & 2 & 4 \end{bmatrix}$.

Compute the following entries of AB:

(a) The $(1, 2)$ entry (b) The $(2, 3)$ entry

(c) The $(3, 1)$ entry (d) The $(3, 3)$ entry

10. If $I_2 = \begin{bmatrix} 1 & 0 \\ 0 & 1 \end{bmatrix}$ and $D = \begin{bmatrix} 2 & 3 \\ -1 & -2 \end{bmatrix}$, compute DI_2 and I_2D.

11. Let

$$A = \begin{bmatrix} 1 & 2 \\ 3 & 2 \end{bmatrix} \quad \text{and} \quad B = \begin{bmatrix} 2 & -1 \\ -3 & 4 \end{bmatrix}.$$

Show that $AB \neq BA$.

12. If A is the matrix in Example 4 and O is the 3×2 matrix every one of whose entries is zero, compute AO.

In Exercises 13 and 14, let

$$A = \begin{bmatrix} 1 & -1 & 2 \\ 3 & 2 & 4 \\ 4 & -2 & 3 \\ 2 & 1 & 5 \end{bmatrix}$$

and

$$B = \begin{bmatrix} 1 & 0 & -1 & 2 \\ 3 & 3 & -3 & 4 \\ 4 & 2 & 5 & 1 \end{bmatrix}.$$

13. Using the method in Example 12, compute the following columns of AB:

(a) The first column (b) The third column

14. Using the method in Example 12, compute the following columns of AB:

(a) The second column (b) The fourth column

15. Let

$$A = \begin{bmatrix} 2 & -3 & 4 \\ -1 & 2 & 3 \\ 5 & -1 & -2 \end{bmatrix} \quad \text{and} \quad \mathbf{c} = \begin{bmatrix} 2 \\ 1 \\ 4 \end{bmatrix}.$$

Express $A\mathbf{c}$ as a linear combination of the columns of A.

16. Let

$$A = \begin{bmatrix} 1 & -2 & -1 \\ 2 & 4 & 3 \\ 3 & 0 & -2 \end{bmatrix} \quad \text{and} \quad B = \begin{bmatrix} 1 & -1 \\ 3 & 2 \\ 2 & 4 \end{bmatrix}.$$

Express the columns of AB as linear combinations of the columns of A.

17. Let $A = \begin{bmatrix} 2 & -3 & 1 \\ 1 & 2 & 4 \end{bmatrix}$ and $B = \begin{bmatrix} 3 \\ 5 \\ 2 \end{bmatrix}$.

(a) Verify that $AB = 3\mathbf{a}_1 + 5\mathbf{a}_2 + 2\mathbf{a}_3$, where $\mathbf{a}_j$ is the jth column of A for $j = 1, 2, 3$.

(b) Verify that $AB = \begin{bmatrix} (\text{row}_1(A))B \\ (\text{row}_2(A))B \end{bmatrix}$.

18. Write the linear combination

$$3\begin{bmatrix} -2 \\ 3 \end{bmatrix} + 4\begin{bmatrix} 2 \\ 5 \end{bmatrix} + 2\begin{bmatrix} 3 \\ -1 \end{bmatrix}$$

as a product of a 2×3 matrix and a 3-vector.

19. Consider the following linear system:

$$\begin{aligned} 2x + \; w &= 7 \\ 3x + 2y + 3z &= -2 \\ 2x + 3y - 4z &= 3 \\ x + 3z &= 5. \end{aligned}$$

(a) Find the coefficient matrix.

(b) Write the linear system in matrix form.

(c) Find the augmented matrix.

20. Write the linear system with augmented matrix

$$\left[\begin{array}{cccc|c} -2 & -1 & 0 & 4 & 5 \\ -3 & 2 & 7 & 8 & 3 \\ 1 & 0 & 0 & 2 & 4 \\ 3 & 0 & 1 & 3 & 6 \end{array}\right].$$

21. Write the linear system with augmented matrix

$$\left[\begin{array}{ccc|c} 2 & 0 & -4 & 3 \\ 0 & 1 & 2 & 5 \\ 1 & 3 & 4 & -1 \end{array}\right].$$

22. Consider the following linear system:

$$\begin{aligned} 3x - y + 2z &= 4 \\ 2x + y &= 2 \\ y + 3z &= 7 \\ 4x - z &= 4. \end{aligned}$$

(a) Find the coefficient matrix.

(b) Write the linear system in matrix form.

(c) Find the augmented matrix.

23. How are the linear systems whose augmented matrices are

$$\left[\begin{array}{ccc|c} 1 & 2 & 3 & -1 \\ 2 & 3 & 6 & 2 \end{array}\right] \quad \text{and} \quad \left[\begin{array}{ccc|c} 1 & 2 & 3 & -1 \\ 2 & 3 & 6 & 2 \\ 0 & 0 & 0 & 0 \end{array}\right]$$

related?

24. Write each of the following as a linear system in matrix form.

(a) $x\begin{bmatrix} 1 \\ 2 \end{bmatrix} + y\begin{bmatrix} 2 \\ 5 \end{bmatrix} + z\begin{bmatrix} 0 \\ 3 \end{bmatrix} = \begin{bmatrix} 1 \\ 1 \end{bmatrix}$

(b) $x\begin{bmatrix} 1 \\ 1 \\ 2 \end{bmatrix} + y\begin{bmatrix} 2 \\ 1 \\ 0 \end{bmatrix} + z\begin{bmatrix} 1 \\ 2 \\ 2 \end{bmatrix} = \begin{bmatrix} 0 \\ 0 \\ 0 \end{bmatrix}$

25. Write each of the following linear systems as a linear combination of the columns of the coefficient matrix.

(a) $\begin{aligned} x + 2y &= 3 \\ 2x - y &= 5 \end{aligned}$

(b) $\begin{aligned} 2x - 3y + 5z &= -2 \\ x + 4y - z &= 3 \end{aligned}$

26. Let A be an $m \times n$ matrix and B an $n \times p$ matrix. What if anything can you say about the matrix product AB when:

(a) A has a column consisting entirely of zeros?

(b) B has a row consisting entirely of zeros?

27. (a) Find a value of r so that $AB^T = 0$, where

$$A = \begin{bmatrix} r & 1 & -2 \end{bmatrix} \quad \text{and} \quad B = \begin{bmatrix} 1 & 3 & -1 \end{bmatrix}.$$

(b) Give an alternate way to write this product.

28. Find a value of r and a value of s so that $AB^T = 0$, where

$$A = \begin{bmatrix} 1 & r & 1 \end{bmatrix} \quad \text{and} \quad B = \begin{bmatrix} -2 & 2 & s \end{bmatrix}.$$

29. Formulate the method for adding partitioned matrices and verify your method by partitioning the matrices

$$A = \begin{bmatrix} 1 & 3 & -1 \\ 2 & 1 & 0 \\ 2 & -3 & 1 \end{bmatrix} \quad \text{and} \quad B = \begin{bmatrix} 3 & 2 & 1 \\ -2 & 3 & 1 \\ 4 & 1 & 5 \end{bmatrix}$$

in two different ways and finding their sum.

30. Let A and B be the following matrices:

$$A = \begin{bmatrix} 2 & 1 & 3 & 4 & 2 \\ 1 & 2 & 3 & -1 & 4 \\ 2 & 3 & 2 & 1 & 4 \\ 5 & -1 & 3 & 2 & 6 \\ 3 & 1 & 2 & 4 & 6 \\ 2 & -1 & 3 & 5 & 7 \end{bmatrix}$$

and

$$B = \begin{bmatrix} 1 & 2 & 3 & 4 & 1 \\ 2 & 1 & 3 & 2 & -1 \\ 1 & 5 & 4 & 2 & 3 \\ 2 & 1 & 3 & 5 & 7 \\ 3 & 2 & 4 & 6 & 1 \end{bmatrix}.$$

Find AB by partitioning A and B in two different ways.

31. (***Manufacturing Costs***) A furniture manufacturer makes chairs and tables, each of which must go through an assembly process and a finishing process. The times required for these processes are given (in hours) by the matrix

$$A = \begin{array}{c} \begin{array}{cc} \text{Assembly process} & \text{Finishing process} \end{array} \\ \left[\begin{array}{cc} 2 & 2 \\ 3 & 4 \end{array}\right] \end{array} \begin{array}{l} \text{Chair} \\ \text{Table} \end{array}$$

The manufacturer has a plant in Salt Lake City and another in Chicago. The hourly rates for each of the processes are given (in dollars) by the matrix

$$B = \begin{array}{c} \begin{array}{cc} \text{Salt Lake City} & \text{Chicago} \end{array} \\ \left[\begin{array}{cc} 9 & 10 \\ 10 & 12 \end{array}\right] \end{array} \begin{array}{l} \text{Assembly process} \\ \text{Finishing process} \end{array}$$

What do the entries in the matrix product AB tell the manufacturer?

32. (***Ecology—Pollution***) A manufacturer makes two kinds of products, P and Q, at each of two plants, X and Y. In making these products, the pollutants sulfur dioxide, nitric oxide, and particulate matter are produced. The amounts of pollutants produced are given (in kilograms) by the matrix

$$A = \begin{array}{c} \begin{array}{ccc} \text{Sulfur dioxide} & \text{Nitric oxide} & \text{Particulate matter} \end{array} \\ \left[\begin{array}{ccc} 300 & 100 & 150 \\ 200 & 250 & 400 \end{array}\right] \end{array} \begin{array}{l} \text{Product } P \\ \text{Product } Q \end{array}$$

State and federal ordinances require that these pollutants be removed. The daily cost of removing each kilogram of pollutant is given (in dollars) by the matrix

$$B = \begin{array}{c} \begin{array}{cc} \text{Plant } X & \text{Plant } Y \end{array} \\ \left[\begin{array}{cc} 8 & 12 \\ 7 & 9 \\ 15 & 10 \end{array}\right] \end{array} \begin{array}{l} \text{Sulfur dioxide} \\ \text{Nitric oxide} \\ \text{Particulate matter} \end{array}$$

What do the entries in the matrix product AB tell the manufacturer?

33. (***Medicine***) A diet research project consists of adults and children of both sexes. The composition of the participants in the project is given by the matrix

$$A = \begin{array}{c} \begin{array}{cc} \text{Adults} & \text{Children} \end{array} \\ \left[\begin{array}{cc} 80 & 120 \\ 100 & 200 \end{array}\right] \end{array} \begin{array}{l} \text{Male} \\ \text{Female} \end{array}$$

The number of daily grams of protein, fat, and carbohydrate consumed by each child and adult is given by the matrix

$$B = \begin{array}{c} \begin{array}{ccc} \text{Protein} & \text{Fat} & \text{Carbohydrate} \end{array} \\ \left[\begin{array}{ccc} 20 & 20 & 20 \\ 10 & 20 & 30 \end{array}\right] \end{array} \begin{array}{l} \text{Adult} \\ \text{Child} \end{array}$$

(a) How many grams of protein are consumed daily by the males in the project?

(b) How many grams of fat are consumed daily by the females in the project?

34. ***(Business)*** A photography business has a store in each of the following cities: New York, Denver, and Los Angeles. A particular make of camera is available in automatic, semiautomatic, and nonautomatic models. Moreover, each camera has a matched flash unit and a camera is usually sold together with the corresponding flash unit. The selling prices of the cameras and flash units are given (in dollars) by the matrix

	Automatic	Semiautomatic	Nonautomatic	
$A =$	200	150	120	Camera
	50	40	25	Flash unit

The number of sets (camera and flash unit) available at each store is given by the matrix

	New York	Denver	Los Angeles	
$B =$	220	180	100	Automatic
	300	250	120	Semiautomatic
	120	320	250	Nonautomatic

(a) What is the total value of the cameras in New York?

(b) What is the total value of the flash units in Los Angeles?

35. Let $\mathbf{s}_1 = \begin{bmatrix} 18.95 & 14.75 & 8.98 \end{bmatrix}$ and $\mathbf{s}_2 = \begin{bmatrix} 17.80 & 13.50 & 10.79 \end{bmatrix}$ be 3-vectors denoting the current prices of three items at stores A and B, respectively.

(a) Obtain a 2×3 matrix representing the combined information about the prices of the three items at the two stores.

(b) Suppose that each store announces a sale so that the price of each item is reduced by 20%. Obtain a 2×3 matrix representing the sale prices at the two stores.

Exercises 36 through 41 involve bit matrices.

36. For bit vectors **a** and **b** compute $\mathbf{a} \cdot \mathbf{b}$.

(a) $\mathbf{a} = \begin{bmatrix} 1 & 1 & 0 \end{bmatrix}$, $\mathbf{b} = \begin{bmatrix} 0 \\ 1 \\ 1 \end{bmatrix}$

(b) $\mathbf{a} = \begin{bmatrix} 0 & 1 & 1 & 0 \end{bmatrix}$, $\mathbf{b} = \begin{bmatrix} 1 \\ 1 \\ 1 \\ 0 \end{bmatrix}$

37. For bit vectors **a** and **b** compute $\mathbf{a} \cdot \mathbf{b}$.

(a) $\mathbf{a} = \begin{bmatrix} 1 & 1 & 0 \end{bmatrix}$, $\mathbf{b} = \begin{bmatrix} 1 \\ 0 \\ 1 \end{bmatrix}$

(b) $\mathbf{a} = \begin{bmatrix} 1 & 1 \end{bmatrix}$, $\mathbf{b} = \begin{bmatrix} 1 \\ 1 \end{bmatrix}$

38. Let $\mathbf{a} = \begin{bmatrix} 1 & x & 0 \end{bmatrix}$ and $\mathbf{b} = \begin{bmatrix} x \\ 1 \\ 1 \end{bmatrix}$ be bit vectors. If $\mathbf{a} \cdot \mathbf{b} = 0$, find all possible values of x.

39. Let $A = \begin{bmatrix} 1 & 1 & x \\ 0 & y & 1 \end{bmatrix}$ and $B = \begin{bmatrix} 1 \\ 1 \\ 1 \end{bmatrix}$ be bit matrices. If $AB = \begin{bmatrix} 0 \\ 0 \end{bmatrix}$, find x and y.

40. For bit matrices

$$A = \begin{bmatrix} 1 & 1 & 0 \\ 0 & 1 & 0 \\ 0 & 0 & 1 \end{bmatrix} \quad \text{and} \quad B = \begin{bmatrix} 0 & 1 & 0 \\ 1 & 1 & 0 \\ 1 & 0 & 1 \end{bmatrix}$$

compute AB and BA.

41. For bit matrix $A = \begin{bmatrix} 1 & 1 \\ 0 & 1 \end{bmatrix}$, determine a 2×2 bit matrix B so that $AB = \begin{bmatrix} 1 & 0 \\ 0 & 1 \end{bmatrix}$.

Theoretical Exercises

T.1. Let $\mathbf{x}$ be an n-vector.

(a) Is it possible for $\mathbf{x} \cdot \mathbf{x}$ to be negative? Explain.

(b) If $\mathbf{x} \cdot \mathbf{x} = 0$, what is $\mathbf{x}$?

T.2. Let **a**, **b**, and **c** be n-vectors and let k be a real number.

(a) Show that $\mathbf{a} \cdot \mathbf{b} = \mathbf{b} \cdot \mathbf{a}$.

(b) Show that $(\mathbf{a} + \mathbf{b}) \cdot \mathbf{c} = \mathbf{a} \cdot \mathbf{c} + \mathbf{b} \cdot \mathbf{c}$.

(c) Show that $(k\mathbf{a}) \cdot \mathbf{b} = \mathbf{a} \cdot (k\mathbf{b}) = k(\mathbf{a} \cdot \mathbf{b})$.

T.3. (a) Show that if A has a row of zeros, then AB has a row of zeros.

(b) Show that if B has a column of zeros, then AB has a column of zeros.

T.4. Show that the product of two diagonal matrices is a diagonal matrix.

T.5. Show that the product of two scalar matrices is a scalar matrix.

T.6. (a) Show that the product of two upper triangular matrices is upper triangular.

(b) Show that the product of two lower triangular matrices is lower triangular.

T.7. Let A and B be $n \times n$ diagonal matrices. Is $AB = BA$? Justify your answer.

T.8. (a) Let $\mathbf{a}$ be a $1 \times n$ matrix and B an $n \times p$ matrix. Show that the matrix product $\mathbf{a}B$ can be written as a linear combination of the rows of B, where the coefficients are the entries of $\mathbf{a}$.

(b) Let $\mathbf{a} = \begin{bmatrix} 1 & -2 & 3 \end{bmatrix}$ and

$$B = \begin{bmatrix} 2 & 1 & -4 \\ -3 & -2 & 3 \\ 4 & 5 & -2 \end{bmatrix}.$$

Write $\mathbf{a}B$ as a linear combination of the rows of B.

T.9. (a) Show that the jth column of the matrix product AB is equal to the matrix product $A\text{col}_j(B)$.

(b) Show that the ith row of the matrix product AB is equal to the matrix product $\text{row}_i(A)B$.

T.10. Let A be an $m \times n$ matrix whose entries are real numbers. Show that if $AA^T = O$ (the $m \times m$ matrix all of whose entries are zero), then $A = O$.

Exercises T.11 through T.13 depend on material marked optional.

T.11. Show that the summation notation satisfies the following properties:

(a) $\displaystyle\sum_{i=1}^{n}(r_i + s_i)a_i = \sum_{i=1}^{n} r_i a_i + \sum_{i=1}^{n} s_i a_i.$

(b) $\displaystyle\sum_{i=1}^{n} c(r_i a_i) = c\left(\sum_{i=1}^{n} r_i a_i\right).$

T.12. Show that $\displaystyle\sum_{i=1}^{n}\sum_{j=1}^{m} a_{ij} = \sum_{j=1}^{m}\sum_{i=1}^{n} a_{ij}.$

T.13. Answer the following as true or false. If true, prove the result; if false, give a counterexample.

(a) $\displaystyle\sum_{i=1}^{n}(a_i + 1) = \left(\sum_{i=1}^{n} a_i\right) + n$

(b) $\displaystyle\sum_{i=1}^{n}\sum_{j=1}^{m} 1 = mn$

(c) $\displaystyle\sum_{j=1}^{m}\sum_{i=1}^{n} a_i b_j = \left[\sum_{i=1}^{n} a_i\right]\left[\sum_{j=1}^{m} b_j\right]$

T.14. Let $\mathbf{u}$ and $\mathbf{v}$ be n-vectors.

(a) If $\mathbf{u}$ and $\mathbf{v}$ are viewed as $n \times 1$ matrices, show that $\mathbf{u} \cdot \mathbf{v} = \mathbf{u}^T\mathbf{v}$.

(b) If $\mathbf{u}$ and $\mathbf{v}$ are viewed as $1 \times n$ matrices, show that $\mathbf{u} \cdot \mathbf{v} = \mathbf{u}\mathbf{v}^T$.

(c) If $\mathbf{u}$ is viewed as a $1 \times n$ matrix and $\mathbf{v}$ as an $n \times 1$ matrix, show that $\mathbf{u} \cdot \mathbf{v} = \mathbf{u}\mathbf{v}$.

MATLAB Exercises

ML.1. In MATLAB, type the command **clear**, then enter the following matrices:

$$A = \begin{bmatrix} 1 & \frac{1}{2} \\ \frac{1}{3} & \frac{1}{4} \\ \frac{1}{5} & \frac{1}{6} \end{bmatrix}, \quad B = \begin{bmatrix} 5 & -2 \end{bmatrix}, \quad C = \begin{bmatrix} 4 & \frac{5}{4} & \frac{9}{4} \\ 1 & 2 & 3 \end{bmatrix}.$$

Using MATLAB commands, compute each of the following, if possible. Recall that a prime in MATLAB indicates transpose.

(a) $A * C$ (b) $A * B$

(c) $A + C'$ (d) $B * A - C' * A$

(e) $(2 * C - 6 * A') * B'$ (f) $A * C - C * A$

(g) $A * A' + C' * C$

ML.2. Enter the coefficient matrix of the system

$$\begin{aligned} 2x + 4y + 6z &= -12 \\ 2x - 3y - 4z &= 15 \\ 3x + 4y + 5z &= -8 \end{aligned}$$

into MATLAB and call it A. Enter the right-hand side of the system and call it $\mathbf{b}$. Form the augmented matrix associated with this linear system using the MATLAB command **[A b]**. To give the augmented matrix a name, such as **aug**, use the command **aug = [A b]**. (Do not type the period!) Note that no bar appears between the coefficient matrix and the right-hand side in the MATLAB display.

ML.3. Repeat the preceding exercise with the following linear system:

$$\begin{aligned} 4x - 3y + 2z - w &= -5 \\ 2x + y - 3z \phantom{{}+2w} &= 7 \\ -x + 4y + z + 2w &= 8. \end{aligned}$$

ML.4. Enter matrices

$$A = \begin{bmatrix} 1 & -1 & 2 \\ 3 & 2 & 4 \\ 4 & -2 & 3 \\ 2 & 1 & 5 \end{bmatrix}$$

and

$$B = \begin{bmatrix} 1 & 0 & -1 & 2 \\ 3 & 3 & -3 & 4 \\ 4 & 2 & 5 & 1 \end{bmatrix}$$

into MATLAB.

(a) Using MATLAB commands, assign $\text{row}_2(A)$ to $\mathbf{R}$ and $\text{col}_3(B)$ to $\mathbf{C}$. Let $\mathbf{V} = \mathbf{R} * \mathbf{C}$. What is $\mathbf{V}$ in terms of the entries of the product $\mathbf{A} * \mathbf{B}$?

(b) Using MATLAB commands, assign $\text{col}_2(B)$ to $\mathbf{C}$, then compute $\mathbf{V} = \mathbf{A} * \mathbf{C}$. What is $\mathbf{V}$ in terms of the entries of the product $\mathbf{A} * \mathbf{B}$?

(c) Using MATLAB commands, assign $\text{row}_3(A)$ to $\mathbf{R}$, then compute $\mathbf{V} = \mathbf{R} * \mathbf{B}$. What is $\mathbf{V}$ in terms of the entries of the product $\mathbf{A} * \mathbf{B}$?

ML.5. Use the MATLAB command **diag** to form each of the following diagonal matrices. Using **diag** we can form diagonal matrices without typing in all the entries. (To refresh your memory about command **diag**, use MATLAB's help feature.)

(a) The 4×4 diagonal matrix with main diagonal $\begin{bmatrix} 1 & 2 & 3 & 4 \end{bmatrix}$.

(b) The 5×5 diagonal matrix with main diagonal $\begin{bmatrix} 0 & 1 & \frac{1}{2} & \frac{1}{3} & \frac{1}{4} \end{bmatrix}$.

(c) The 5×5 scalar matrix with all 5's on the diagonal.

ML.6. In MATLAB the dot product of a pair of vectors can be computed using the **dot** command. If the vectors **v** and **w** have been entered into MATLAB as either rows or columns, their dot product is computed from the MATLAB command **dot(v, w)**. If the vectors do not have the same number of elements, an error message is displayed.

(a) Use **dot** to compute the dot product of each of the following vectors.

(i) $\mathbf{v} = \begin{bmatrix} 1 & 4 & -1 \end{bmatrix}, \mathbf{w} = \begin{bmatrix} 7 & 2 & 0 \end{bmatrix}$

(ii) $\mathbf{v} = \begin{bmatrix} 2 \\ -1 \\ 0 \\ 6 \end{bmatrix}, \mathbf{w} = \begin{bmatrix} 4 \\ 2 \\ 3 \\ -1 \end{bmatrix}$

(b) Let $\mathbf{a} = \begin{bmatrix} 3 & -2 & 1 \end{bmatrix}$. Find a value for k so that the dot product of $\mathbf{a}$ with $\mathbf{b} = \begin{bmatrix} k & 1 & 4 \end{bmatrix}$ is zero. Verify your results in MATLAB.

(c) For each of the following vectors **v**, compute **dot(v,v)** in MATLAB.

(i) $\mathbf{v} = \begin{bmatrix} 4 & 2 & -3 \end{bmatrix}$

(ii) $\mathbf{v} = \begin{bmatrix} -9 & 3 & 1 & 0 & 6 \end{bmatrix}$

(iii) $\mathbf{v} = \begin{bmatrix} 1 \\ 2 \\ -5 \\ -3 \end{bmatrix}$

What sign is each of these dot products? Explain why this is true for almost all vectors **v**. When is it not true?

Exercises ML.7 through ML.11 use bit matrices and the supplemental instructional commands described in Section 12.9.

ML.7. Use **binprod** to solve Exercise 40.

ML.8. Given the bit vectors $\mathbf{a} = \begin{bmatrix} 1 \\ 1 \\ 0 \\ 1 \end{bmatrix}$ and $\mathbf{b} = \begin{bmatrix} 1 \\ 0 \\ 0 \\ 1 \end{bmatrix}$, use **binprod** to compute $\mathbf{a} \cdot \mathbf{b}$.

ML.9. (a) Use **bingen** to generate a matrix B whose columns are all possible bit 3-vectors.

(b) Define **A = ones(3)** and compute AB using **binprod**.

(c) Describe why AB contains only columns of all zeros or all ones. (*Hint*: Look for a pattern based on the columns of B.)

ML.10. Repeat Exercise ML.9 with 4-vectors and **A = ones(4)**.

ML.11. Let B be the $n \times n$ matrix of all ones. Compute BB for $n = 2, 3, 4$, and 5. What is BB for $n = k$, where k is any positive integer?

1.4 PROPERTIES OF MATRIX OPERATIONS

In this section we consider the algebraic properties of the matrix operations just defined. Many of these properties are similar to familiar properties of the real numbers. However, there will be striking differences between the set of real numbers and the set of matrices in their algebraic behavior under certain operations, for example, under multiplication (as seen in Section 1.3). Most of the properties will be stated as theorems, whose proofs will be left as exercises.

THEOREM 1.1 **(Properties of Matrix Addition)** *Let A, B, C, and D be $m \times n$ matrices.*

(a) $A + B = B + A$.

(b) $A + (B + C) = (A + B) + C$.

(c) *There is a unique $m \times n$ matrix O such that*

$$A + O = A \tag{1}$$

for any $m \times n$ matrix A. The matrix O is called the $m \times n$ **additive identity** *or* **zero matrix**.

(d) *For each $m \times n$ matrix A, there is a unique $m \times n$ matrix D such that*

$$A + D = O. \tag{2}$$

We shall write D as $(-A)$, so that (2) can be written as

$$A + (-A) = O.$$

The matrix $(-A)$ is called the **additive inverse** *or* **negative** *of A.*

Proof (a) To establish (a), we must prove that the i, jth element of $A + B$ equals the i, jth element of $B + A$. The i, jth element of $A + B$ is $a_{ij} + b_{ij}$; the i, jth element of $B + A$ is $b_{ij} + a_{ij}$. Since the elements a_{ij} are real (or complex) numbers,

$$a_{ij} + b_{ij} = b_{ij} + a_{ij} \qquad (1 \leq i \leq m, 1 \leq j \leq n),$$

the result follows.

(b) Exercise T.1.

(c) Let $U = \left[u_{ij}\right]$. Then

$$A + U = A$$

if and only if*

$$a_{ij} + u_{ij} = a_{ij},$$

which holds if and only if $u_{ij} = 0$. Thus U is the $m \times n$ matrix all of whose entries are zero; U is denoted by O.

(d) Exercise T.1. ■

EXAMPLE 1 To illustrate (c) of Theorem 1.1, we note that the 2×2 zero matrix is

$$\begin{bmatrix} 0 & 0 \\ 0 & 0 \end{bmatrix}.$$

If

$$A = \begin{bmatrix} 4 & -1 \\ 2 & 3 \end{bmatrix},$$

we have

$$\begin{bmatrix} 4 & -1 \\ 2 & 3 \end{bmatrix} + \begin{bmatrix} 0 & 0 \\ 0 & 0 \end{bmatrix} = \begin{bmatrix} 4+0 & -1+0 \\ 2+0 & 3+0 \end{bmatrix} = \begin{bmatrix} 4 & -1 \\ 2 & 3 \end{bmatrix}.$$ ■

The 2×3 zero matrix is

$$\begin{bmatrix} 0 & 0 & 0 \\ 0 & 0 & 0 \end{bmatrix}.$$

*The connector "if and only if" means that both statements are true or both statements are false. Thus (1) if $A + U = A$, then $a_{ij} + u_{ij} = a_{ij}$ and (2) if $a_{ij} + u_{ij} = a_{ij}$, then $A + U = A$.

EXAMPLE 2 To illustrate (d) of Theorem 1.1, let

$$A = \begin{bmatrix} 2 & 3 & 4 \\ -4 & 5 & -2 \end{bmatrix}.$$

Then

$$-A = \begin{bmatrix} -2 & -3 & -4 \\ 4 & -5 & 2 \end{bmatrix}.$$

We now have $A + (-A) = O$. ■

EXAMPLE 3 Let

$$A = \begin{bmatrix} 3 & -2 & 5 \\ -1 & 2 & 3 \end{bmatrix} \quad \text{and} \quad B = \begin{bmatrix} 2 & 3 & 2 \\ -3 & 4 & 6 \end{bmatrix}.$$

Then

$$A - B = \begin{bmatrix} 3-2 & -2-3 & 5-2 \\ -1+3 & 2-4 & 3-6 \end{bmatrix} = \begin{bmatrix} 1 & -5 & 3 \\ 2 & -2 & -3 \end{bmatrix}.$$ ■

THEOREM 1.2 **(Properties of Matrix Multiplication)**

(a) *If A, B, and C are of the appropriate sizes, then*

$$A(BC) = (AB)C.$$

(b) *If A, B, and C are of the appropriate sizes, then*

$$A(B + C) = AB + AC.$$

(c) *If A, B, and C are of the appropriate sizes, then*

$$(A + B)C = AC + BC.$$

Proof (a) We omit a general proof here. Exercise T.2 asks the reader to prove the result for a specific case.

(b) Exercise T.3.

(c) Exercise T.3. ■

EXAMPLE 4 Let

$$A = \begin{bmatrix} 5 & 2 & 3 \\ 2 & -3 & 4 \end{bmatrix}, \quad B = \begin{bmatrix} 2 & -1 & 1 & 0 \\ 0 & 2 & 2 & 2 \\ 3 & 0 & -1 & 3 \end{bmatrix},$$

and

$$C = \begin{bmatrix} 1 & 0 & 2 \\ 2 & -3 & 0 \\ 0 & 0 & 3 \\ 2 & 1 & 0 \end{bmatrix}.$$

Then

$$A(BC) = \begin{bmatrix} 5 & 2 & 3 \\ 2 & -3 & 4 \end{bmatrix} \begin{bmatrix} 0 & 3 & 7 \\ 8 & -4 & 6 \\ 9 & 3 & 3 \end{bmatrix} = \begin{bmatrix} 43 & 16 & 56 \\ 12 & 30 & 8 \end{bmatrix}$$

and

$$(AB)C = \begin{bmatrix} 19 & -1 & 6 & 13 \\ 16 & -8 & -8 & 6 \end{bmatrix} \begin{bmatrix} 1 & 0 & 2 \\ 2 & -3 & 0 \\ 0 & 0 & 3 \\ 2 & 1 & 0 \end{bmatrix} = \begin{bmatrix} 43 & 16 & 56 \\ 12 & 30 & 8 \end{bmatrix}.$$ ■

EXAMPLE 5 Let

$$A = \begin{bmatrix} 2 & 2 & 3 \\ 3 & -1 & 2 \end{bmatrix}, \quad B = \begin{bmatrix} 1 & 0 \\ 2 & 2 \\ 3 & -1 \end{bmatrix}, \quad \text{and} \quad C = \begin{bmatrix} -1 & 2 \\ 1 & 0 \\ 2 & -2 \end{bmatrix}.$$

Then

$$A(B + C) = \begin{bmatrix} 2 & 2 & 3 \\ 3 & -1 & 2 \end{bmatrix} \begin{bmatrix} 0 & 2 \\ 3 & 2 \\ 5 & -3 \end{bmatrix} = \begin{bmatrix} 21 & -1 \\ 7 & -2 \end{bmatrix}$$

and

$$AB + AC = \begin{bmatrix} 15 & 1 \\ 7 & -4 \end{bmatrix} + \begin{bmatrix} 6 & -2 \\ 0 & 2 \end{bmatrix} = \begin{bmatrix} 21 & -1 \\ 7 & -2 \end{bmatrix}.$$

■

DEFINITION The $n \times n$ scalar matrix

$$I_n = \begin{bmatrix} 1 & 0 & \cdots & 0 \\ 0 & 1 & \cdots & 0 \\ \vdots & \vdots & & \vdots \\ 0 & 0 & \cdots & 1 \end{bmatrix},$$

all of whose diagonal entries are 1, is called the **identity matrix of order *n*.**

If A is an $m \times n$ matrix, then it is easy to verify (Exercise T.4) that

$$I_m A = A I_n = A.$$

It is also easy to see that every $n \times n$ scalar matrix can be written as rI_n for some r.

EXAMPLE 6 The identity matrix I_2 of order 2 is

$$I_2 = \begin{bmatrix} 1 & 0 \\ 0 & 1 \end{bmatrix}.$$

If

$$A = \begin{bmatrix} 4 & -2 & 3 \\ 5 & 0 & 2 \end{bmatrix},$$

then

$$I_2 A = A.$$

The identity matrix I_3 of order 3 is

$$I_3 = \begin{bmatrix} 1 & 0 & 0 \\ 0 & 1 & 0 \\ 0 & 0 & 1 \end{bmatrix}.$$

Hence

$$AI_3 = A.$$

■

Suppose that A is a square matrix. If p is a positive integer, then we define the **powers of a matrix** as follows:

$$A^p = \underbrace{A \cdot A \cdots A}_{p\ factors}.$$

If A is $n \times n$, we also define

$$A^0 = I_n.$$

For nonnegative integers p and q, some of the familiar laws of exponents for the real numbers can also be proved for matrix multiplication of a square matrix A (Exercise T.5):

$$A^p A^q = A^{p+q}$$

and

$$(A^p)^q = A^{pq}.$$

It should be noted that

$$(AB)^p \neq A^p B^p$$

for square matrices in general. However, if $AB = BA$, then this rule does hold (Exercise T.6).

We now note two other peculiarities of matrix multiplication. If a and b are real numbers, then $ab = 0$ can hold only if a or b is zero. However, this is not true for matrices.

EXAMPLE 7 If

$$A = \begin{bmatrix} 1 & 2 \\ 2 & 4 \end{bmatrix} \quad \text{and} \quad B = \begin{bmatrix} 4 & -6 \\ -2 & 3 \end{bmatrix},$$

then neither A nor B is the zero matrix, but

$$AB = \begin{bmatrix} 0 & 0 \\ 0 & 0 \end{bmatrix}.$$

■

If a, b, and c are real numbers for which $ab = ac$ and $a \neq 0$, it then follows that $b = c$. That is, we can cancel a out. However, the cancellation law does not hold for matrices, as the following example shows.

EXAMPLE 8 If

$$A = \begin{bmatrix} 1 & 2 \\ 2 & 4 \end{bmatrix}, \quad B = \begin{bmatrix} 2 & 1 \\ 3 & 2 \end{bmatrix}, \quad \text{and} \quad C = \begin{bmatrix} -2 & 7 \\ 5 & -1 \end{bmatrix},$$

then

$$AB = AC = \begin{bmatrix} 8 & 5 \\ 16 & 10 \end{bmatrix},$$

but $B \neq C$. ■

Remark In Section 1.7, we investigate a special class of matrices A for which $AB = AC$ does imply that $B = C$.

EXAMPLE 9 **(Business)** Suppose that only two rival companies, R and S, manufacture a certain product. Each year, company R keeps $\frac{1}{4}$ of its customers while $\frac{3}{4}$

switch to S. Each year, S keeps $\frac{2}{3}$ of its customers while $\frac{1}{3}$ switch to R. This information can be displayed in matrix form as

$$A = \begin{array}{c} \begin{matrix} R & S \end{matrix} \\ \begin{bmatrix} \frac{1}{4} & \frac{1}{3} \\ \frac{3}{4} & \frac{2}{3} \end{bmatrix} \end{array} \begin{matrix} R \\ S \end{matrix}$$

When manufacture of the product first starts, R has $\frac{3}{5}$ of the market (the market is the total number of customers) while S has $\frac{2}{5}$ of the market. We denote the initial distribution of the market by

$$\mathbf{x}_0 = \begin{bmatrix} \frac{3}{5} \\ \frac{2}{5} \end{bmatrix}.$$

One year later, the distribution of the market is

$$\mathbf{x}_1 = A\mathbf{x}_0 = \begin{bmatrix} \frac{1}{4} & \frac{1}{3} \\ \frac{3}{4} & \frac{2}{3} \end{bmatrix} \begin{bmatrix} \frac{3}{5} \\ \frac{2}{5} \end{bmatrix} = \begin{bmatrix} \frac{1}{4}\left(\frac{3}{5}\right) + \frac{1}{3}\left(\frac{2}{5}\right) \\ \frac{3}{4}\left(\frac{3}{5}\right) + \frac{2}{3}\left(\frac{2}{5}\right) \end{bmatrix} = \begin{bmatrix} \frac{17}{60} \\ \frac{43}{60} \end{bmatrix}.$$

This can be readily seen as follows. Suppose that the initial market consists of k people, say $k = 12{,}000$, and no change in this number occurs with time. Then, initially, R has $\frac{3}{5}k$ customers, and S has $\frac{2}{5}k$ customers. At the end of the first year, R keeps $\frac{1}{4}$ of its customers and gains $\frac{1}{3}$ of S's customers. Thus R has

$$\tfrac{1}{4}\left(\tfrac{3}{5}k\right) + \tfrac{1}{3}\left(\tfrac{2}{5}k\right) = \left[\tfrac{1}{4}\left(\tfrac{3}{5}\right) + \tfrac{1}{3}\left(\tfrac{2}{5}\right)\right]k = \tfrac{17}{60}k \text{ customers.}$$

When $k = 12{,}000$, R has $\frac{17}{60}(12{,}000) = 3400$ customers. Similarly, at the end of the first year, S keeps $\frac{2}{3}$ of its customers and gains $\frac{3}{4}$ of R's customers. Thus S has

$$\tfrac{3}{4}\left(\tfrac{3}{5}k\right) + \tfrac{2}{3}\left(\tfrac{2}{5}k\right) = \left[\tfrac{3}{4}\left(\tfrac{3}{5}\right) + \tfrac{2}{3}\left(\tfrac{2}{5}\right)\right]k = \tfrac{43}{60}k \text{ customers.}$$

When $k = 12{,}000$, S has $\frac{43}{60}(12{,}000) = 8600$ customers. Similarly, at the end of 2 years, the distribution of the market will be given by

$$\mathbf{x}_2 = A\mathbf{x}_1 = A(A\mathbf{x}_0) = A^2\mathbf{x}_0.$$

If

$$\mathbf{x}_0 = \begin{bmatrix} a \\ b \end{bmatrix},$$

can we determine a and b so that the distribution will be the same from year to year? When this happens, the distribution of the market is said to be **stable**. We proceed as follows. Since R and S control the entire market, we must have

$$a + b = 1. \tag{3}$$

We also want the distribution after 1 year to be unchanged. Hence

$$A\mathbf{x}_0 = \mathbf{x}_0$$

or

$$\begin{bmatrix} \frac{1}{4} & \frac{1}{3} \\ \frac{3}{4} & \frac{2}{3} \end{bmatrix} \begin{bmatrix} a \\ b \end{bmatrix} = \begin{bmatrix} a \\ b \end{bmatrix}.$$

Then

$$\begin{aligned}\tfrac{1}{4}a + \tfrac{1}{3}b &= a\\ \tfrac{3}{4}a + \tfrac{2}{3}b &= b\end{aligned}$$

or

$$\begin{aligned}-\tfrac{3}{4}a + \tfrac{1}{3}b &= 0\\ \tfrac{3}{4}a - \tfrac{1}{3}b &= 0.\end{aligned} \tag{4}$$

Observe that the two equations in (4) are the same. Using Equation (3) and one of the equations in (4), we find (verify) that

$$a = \tfrac{4}{13} \quad \text{and} \quad b = \tfrac{9}{13}.$$

■

The problem described is an example of a **Markov chain**. We shall return to this topic in Section 2.5.

THEOREM 1.3 **(Properties of Scalar Multiplication)** *If r and s are real numbers and A and B are matrices, then*

(a) $r(sA) = (rs)A$
(b) $(r + s)A = rA + sA$
(c) $r(A + B) = rA + rB$
(d) $A(rB) = r(AB) = (rA)B$

Proof Exercise T.12. ■

EXAMPLE 10 Let $r = -2$,

$$A = \begin{bmatrix} 1 & 2 & 3 \\ -2 & 0 & 1 \end{bmatrix}, \quad \text{and} \quad B = \begin{bmatrix} 2 & -1 \\ 1 & 4 \\ 0 & -2 \end{bmatrix}.$$

Then

$$A(rB) = \begin{bmatrix} 1 & 2 & 3 \\ -2 & 0 & 1 \end{bmatrix} \begin{bmatrix} -4 & 2 \\ -2 & -8 \\ 0 & 4 \end{bmatrix} = \begin{bmatrix} -8 & -2 \\ 8 & 0 \end{bmatrix}$$

and

$$r(AB) = (-2) \begin{bmatrix} 4 & 1 \\ -4 & 0 \end{bmatrix} = \begin{bmatrix} -8 & -2 \\ 8 & 0 \end{bmatrix},$$

which illustrates (d) of Theorem 1.3. ■

It is easy to show that $(-1)A = -A$ (Exercise T.13).

THEOREM 1.4 **(Properties of Transpose)** *If r is a scalar and A and B are matrices, then*

(a) $(A^T)^T = A$
(b) $(A + B)^T = A^T + B^T$
(c) $(AB)^T = B^T A^T$
(d) $(rA)^T = rA^T$

Proof We leave the proofs of (a), (b), and (d) as an exercise (Exercise T.14) and prove only (c) here. Thus let $A = \left[a_{ij}\right]$ be $m \times p$ and let $B = \left[b_{ij}\right]$ be $p \times n$. The i, jth element of $(AB)^T$ is c_{ij}^T. Now

$$\begin{aligned} c_{ij}^T = c_{ji} &= \text{row}_j(A) \cdot \text{col}_i(B) \\ &= a_{j1}b_{1i} + a_{j2}b_{2i} + \cdots + a_{jp}b_{pi} \\ &= a_{1j}^T b_{i1}^T + a_{2j}^T b_{i2}^T + \cdots + a_{pj}^T b_{ip}^T \\ &= b_{i1}^T a_{1j}^T + b_{i2}^T a_{2j}^T + \cdots + b_{ip}^T a_{pj}^T \\ &= \text{row}_i(B^T) \cdot \text{col}_j(A^T), \end{aligned}$$

which is the i, jth element of $B^T A^T$. ■

EXAMPLE 11 Let

$$A = \begin{bmatrix} 1 & 3 & 2 \\ 2 & -1 & 3 \end{bmatrix} \quad \text{and} \quad B = \begin{bmatrix} 0 & 1 \\ 2 & 2 \\ 3 & -1 \end{bmatrix}.$$

Then

$$(AB)^T = \begin{bmatrix} 12 & 7 \\ 5 & -3 \end{bmatrix}$$

and

$$B^T A^T = \begin{bmatrix} 0 & 2 & 3 \\ 1 & 2 & -1 \end{bmatrix} \begin{bmatrix} 1 & 2 \\ 3 & -1 \\ 2 & 3 \end{bmatrix} = \begin{bmatrix} 12 & 7 \\ 5 & -3 \end{bmatrix}.$$

■

DEFINITION A matrix $A = \left[a_{ij}\right]$ with real entries is called **symmetric** if

$$A^T = A.$$

That is, A is symmetric if it is a square matrix for which

$$a_{ij} = a_{ji} \qquad \text{(Exercise T.17).}$$

If matrix A is symmetric, then the elements of A are symmetric with respect to the main diagonal of A.

EXAMPLE 12 The matrices

$$A = \begin{bmatrix} 1 & 2 & 3 \\ 2 & 4 & 5 \\ 3 & 5 & 6 \end{bmatrix} \quad \text{and} \quad I_3 = \begin{bmatrix} 1 & 0 & 0 \\ 0 & 1 & 0 \\ 0 & 0 & 1 \end{bmatrix}$$

are symmetric. ■

EXAMPLES WITH BIT MATRICES (OPTIONAL)

All of the matrix operations discussed in this section are valid on bit matrices provided we use arithmetic base 2. Hence the scalars available are only 0 and 1.

EXAMPLE 13 Let $A = \begin{bmatrix} 1 & 0 \\ 1 & 1 \\ 0 & 1 \end{bmatrix}$ be a bit matrix. Find the additive inverse of A.

Solution Let $-A = \begin{bmatrix} a & b \\ c & d \\ e & f \end{bmatrix}$ (the additive inverse of A). Then $A + (-A) = O$. We have

$$\begin{aligned} 1 + a &= 0 \qquad & 0 + b &= 0 \\ 1 + c &= 0 \qquad & 1 + d &= 0 \\ 0 + e &= 0 \qquad & 1 + f &= 0 \end{aligned}$$

so $a = 1$, $b = 0$, $c = 1$, $d = 1$, $e = 0$, and $f = 1$. Hence $-A = A$. (See also Exercise T.38.) ■

EXAMPLE 14 For the bit matrix $A = \begin{bmatrix} 1 & 0 \\ 1 & 0 \end{bmatrix}$ determine a 2×2 bit matrix $B \neq O$ so that $AB = O$.

Solution Let $B = \begin{bmatrix} a & b \\ c & d \end{bmatrix}$. Then

$$AB = \begin{bmatrix} 1 & 0 \\ 1 & 0 \end{bmatrix} \begin{bmatrix} a & b \\ c & d \end{bmatrix} = \begin{bmatrix} a & b \\ a & b \end{bmatrix} = \begin{bmatrix} 0 & 0 \\ 0 & 0 \end{bmatrix}$$

provided $a = b = 0$, $c = 0$ or 1, and $d = 0$ or 1. Thus there are four such matrices,

$$\begin{bmatrix} 0 & 0 \\ 0 & 0 \end{bmatrix}, \quad \begin{bmatrix} 0 & 0 \\ 0 & 1 \end{bmatrix}, \quad \begin{bmatrix} 0 & 0 \\ 1 & 0 \end{bmatrix}, \quad \text{and} \quad \begin{bmatrix} 0 & 0 \\ 1 & 1 \end{bmatrix}.$$

■

Section 2.2, Graph Theory, which can be covered at this time, uses material from this section.

Preview of an Application

Graph Theory (Section 2.2)

In recent years, the need to solve problems dealing with communication among individuals, computers, and organizations has grown at an unprecedented rate. As an example, note the explosive growth of the Internet and the promises of using it to interact with all types of media. Graph theory is an area of applied mathematics that deals with problems such as this one:

Consider a local area network consisting of six users denoted by P_1, P_2, ..., P_6. We say that P_i has "access" to P_j if P_i can directly send a message to P_j. On the other hand, P_i may not be able to send a message directly to P_k, but can send it to P_j, who will then send it to P_k. In this way we say that P_i has "2-stage access" to P_k. In a similar way, we speak of "r-stage access." We may describe the access relation in the network shown in Figure 1.6 by defining the 6×6 matrix $A = [a_{ij}]$, where $a_{ij} = 1$ if P_i has access to P_j and 0 otherwise. Thus A may be

Figure 1.6 ►

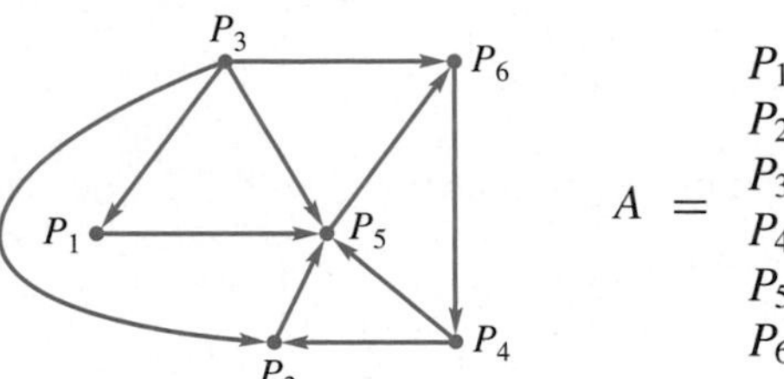

$$A = \begin{array}{c} \\ P_1 \\ P_2 \\ P_3 \\ P_4 \\ P_5 \\ P_6 \end{array} \begin{array}{c} \begin{array}{cccccc} P_1 & P_2 & P_3 & P_4 & P_5 & P_6 \end{array} \\ \begin{bmatrix} 0 & 0 & 0 & 0 & 1 & 0 \\ 0 & 0 & 0 & 0 & 1 & 0 \\ 1 & 1 & 0 & 0 & 1 & 1 \\ 0 & 1 & 0 & 0 & 1 & 0 \\ 0 & 0 & 0 & 0 & 0 & 1 \\ 0 & 0 & 0 & 1 & 0 & 0 \end{bmatrix} \end{array}.$$

Using the matrix A and the techniques from graph theory discussed in Section 2.2, we can determine the number of ways that P_i has r-stage access to P_k, where $r = 1, 2, \ldots$. Many other problems involving communications can be solved using graph theory.

The matrix A above is indeed a bit matrix, but in this situation A is best considered as a matrix in base 10, as will be shown in Section 2.2.

Key Terms

Properties of matrix addition
Additive identity or zero matrix
Additive inverse or negative of a matrix
Properties of matrix multiplication
Identity matrix
Powers of a matrix
Properties of transpose
Symmetric matrix
Skew symmetric matrix

1.4 Exercises

1. Verify Theorem 1.1 for
$$A = \begin{bmatrix} 1 & 2 & -2 \\ 3 & 4 & 5 \end{bmatrix}, \quad B = \begin{bmatrix} 2 & 0 & 1 \\ 3 & -2 & 5 \end{bmatrix},$$
and
$$C = \begin{bmatrix} -4 & -6 & 1 \\ 2 & 3 & 0 \end{bmatrix}.$$

2. Verify (a) of Theorem 1.2 for
$$A = \begin{bmatrix} 1 & 3 \\ 2 & -1 \end{bmatrix}, \quad B = \begin{bmatrix} -1 & 3 & 2 \\ 1 & -3 & 4 \end{bmatrix},$$
and
$$C = \begin{bmatrix} 1 & 0 \\ 3 & -1 \\ 1 & 2 \end{bmatrix}.$$

3. Verify (b) of Theorem 1.2 for
$$A = \begin{bmatrix} 1 & -3 \\ -3 & 4 \end{bmatrix}, \quad B = \begin{bmatrix} 2 & -3 & 2 \\ 3 & -1 & -2 \end{bmatrix},$$
and
$$C = \begin{bmatrix} 0 & 1 & 2 \\ 1 & 3 & -2 \end{bmatrix}.$$

4. Verify (a), (b), and (c) of Theorem 1.3 for $r = 6$, $s = -2$, and
$$A = \begin{bmatrix} 4 & 2 \\ 1 & -3 \end{bmatrix}, \quad B = \begin{bmatrix} 0 & 2 \\ -4 & 3 \end{bmatrix}.$$

5. Verify (d) of Theorem 1.3 for $r = -3$ and
$$A = \begin{bmatrix} 1 & 3 \\ 2 & -1 \end{bmatrix}, \quad B = \begin{bmatrix} -1 & 3 & 2 \\ 1 & -3 & 4 \end{bmatrix}.$$

6. Verify (b) and (d) of Theorem 1.4 for $r = -4$ and
$$A = \begin{bmatrix} 1 & 3 & 2 \\ 2 & 1 & -3 \end{bmatrix}, \quad B = \begin{bmatrix} 4 & 2 & -1 \\ -2 & 1 & 5 \end{bmatrix}.$$

7. Verify (c) of Theorem 1.4 for
$$A = \begin{bmatrix} 1 & 3 & 2 \\ 2 & 1 & -3 \end{bmatrix}, \quad B = \begin{bmatrix} 3 & -1 \\ 2 & 4 \\ 1 & 2 \end{bmatrix}.$$

In Exercises 8 and 9, let
$$A = \begin{bmatrix} 2 & 1 & -2 \\ 3 & 2 & 5 \end{bmatrix}, \quad B = \begin{bmatrix} 2 & -1 \\ 3 & 4 \\ 1 & -2 \end{bmatrix},$$
$$C = \begin{bmatrix} 2 & 1 & 3 \\ -1 & 2 & 4 \\ 3 & 1 & 0 \end{bmatrix}, \quad D = \begin{bmatrix} 2 & -1 \\ -3 & 2 \end{bmatrix},$$
$$E = \begin{bmatrix} 1 & 1 & 2 \\ 2 & -1 & 3 \\ -3 & 2 & -1 \end{bmatrix}, \quad \textit{and} \quad F = \begin{bmatrix} 1 & 0 \\ 2 & -3 \end{bmatrix}.$$

8. If possible, compute:

(a) $(AB)^T$ (b) B^TA^T (c) A^TB^T
(d) BB^T (e) B^TB

9. If possible, compute:

(a) $(3C - 2E)^TB$ (b) $A^T(D + F)$
(c) $B^TC + A$ (d) $(2E)A^T$
(e) $(B^T + A)C$

10. If
$$A = \begin{bmatrix} -2 & 3 \\ 2 & -3 \end{bmatrix} \quad \text{and} \quad B = \begin{bmatrix} 3 & 6 \\ 2 & 4 \end{bmatrix},$$
show that $AB = O$.

11. If
$$A = \begin{bmatrix} -2 & 3 \\ 2 & -3 \end{bmatrix}, \quad B = \begin{bmatrix} -1 & 3 \\ 2 & 0 \end{bmatrix},$$
and
$$C = \begin{bmatrix} -4 & -3 \\ 0 & -4 \end{bmatrix},$$
show that $AB = AC$.

12. If $A = \begin{bmatrix} 0 & 1 \\ 1 & 0 \end{bmatrix}$, show that $A^2 = I_2$.

13. Let $A = \begin{bmatrix} 4 & 2 \\ 1 & 3 \end{bmatrix}$. Find

(a) $A^2 + 3A$
(b) $2A^3 + 3A^2 + 4A + 5I_2$

14. Let $A = \begin{bmatrix} 1 & -1 \\ 2 & 3 \end{bmatrix}$. Find

(a) $A^2 - 2A$
(b) $3A^3 - 2A^2 + 5A - 4I_2$

15. Determine a scalar r such that $A\mathbf{x} = r\mathbf{x}$, where
$$A = \begin{bmatrix} 2 & 1 \\ 1 & 2 \end{bmatrix} \quad \text{and} \quad \mathbf{x} = \begin{bmatrix} 1 \\ 1 \end{bmatrix}.$$

16. Determine a constant k such that $(kA)^T(kA) = 1$, where
$$A = \begin{bmatrix} -2 \\ 1 \\ -1 \end{bmatrix}.$$
Is there more than one value of k that could be used?

17. Let

$$A = \begin{bmatrix} -3 & 2 & 1 \\ 4 & 5 & 0 \end{bmatrix}$$

and $\mathbf{a}_j = \text{col}_j(A)$, $j = 1, 2, 3$. Verify that

$$A^T A = \begin{bmatrix} \mathbf{a}_1 \cdot \mathbf{a}_1 & \mathbf{a}_1 \cdot \mathbf{a}_2 & \mathbf{a}_1 \cdot \mathbf{a}_3 \\ \mathbf{a}_2 \cdot \mathbf{a}_1 & \mathbf{a}_2 \cdot \mathbf{a}_2 & \mathbf{a}_2 \cdot \mathbf{a}_3 \\ \mathbf{a}_3 \cdot \mathbf{a}_1 & \mathbf{a}_3 \cdot \mathbf{a}_2 & \mathbf{a}_3 \cdot \mathbf{a}_3 \end{bmatrix} = \begin{bmatrix} \mathbf{a}_1^T \mathbf{a}_1 & \mathbf{a}_1^T \mathbf{a}_2 & \mathbf{a}_1^T \mathbf{a}_3 \\ \mathbf{a}_2^T \mathbf{a}_1 & \mathbf{a}_2^T \mathbf{a}_2 & \mathbf{a}_2^T \mathbf{a}_3 \\ \mathbf{a}_3^T \mathbf{a}_1 & \mathbf{a}_3^T \mathbf{a}_2 & \mathbf{a}_3^T \mathbf{a}_3 \end{bmatrix}.$$

Exercises 18 through 21 deal with Markov chains, an area that will be studied in greater detail in Section 2.5.

18. Suppose that the matrix A in Example 9 is

$$A = \begin{bmatrix} \frac{1}{3} & \frac{2}{5} \\ \frac{2}{3} & \frac{3}{5} \end{bmatrix} \quad \text{and} \quad \mathbf{x}_0 = \begin{bmatrix} \frac{2}{3} \\ \frac{1}{3} \end{bmatrix}.$$

(a) Find the distribution of the market after 1 year.

(b) Find the stable distribution of the market.

19. Consider two quick food companies, M and N. Each year, company M keeps $\frac{1}{3}$ of its customers, while $\frac{2}{3}$ switch to N. Each year, N keeps $\frac{1}{2}$ of its customers, while $\frac{1}{2}$ switch to M. Suppose that the initial distribution of the market is given by

$$\mathbf{x}_0 = \begin{bmatrix} \frac{1}{3} \\ \frac{2}{3} \end{bmatrix}.$$

(a) Find the distribution of the market after 1 year.

(b) Find the stable distribution of the market.

20. Suppose that in Example 9 there were three rival companies R, S, and T so that the pattern of customer retention and switching is given by the information in the matrix A where

$$A = \begin{array}{c} \begin{matrix} R & S & T \end{matrix} \\ \begin{bmatrix} \frac{1}{3} & \frac{1}{2} & \frac{1}{4} \\ \frac{2}{3} & \frac{1}{4} & \frac{1}{2} \\ 0 & \frac{1}{4} & \frac{1}{4} \end{bmatrix} \begin{matrix} R \\ S \\ T \end{matrix} \end{array}$$

(a) If the initial market distribution is given by

$$\mathbf{x}_0 = \begin{bmatrix} \frac{1}{3} \\ \frac{1}{3} \\ \frac{1}{3} \end{bmatrix},$$

then determine the market distribution after 1 year; after 2 years.

(b) Show that the stable market distribution is given by

$$\mathbf{x} = \begin{bmatrix} \frac{21}{53} \\ \frac{24}{53} \\ \frac{8}{53} \end{bmatrix}.$$

(c) Which company R, S, or T will gain the most market share over a long period of time (assuming that the retention and switching patterns remain the same)? Approximately what percent of the market was gained by this company?

21. Suppose that in Exercise 20 the matrix A was given by

$$A = \begin{array}{c} \begin{matrix} R & S & T \end{matrix} \\ \begin{bmatrix} 0.4 & 0 & 0.4 \\ 0 & 0.5 & 0.4 \\ 0.6 & 0.5 & 0.2 \end{bmatrix} \begin{matrix} R \\ S \\ T \end{matrix} \end{array}$$

(a) If the initial market distribution is given by

$$\mathbf{x}_0 = \begin{bmatrix} \frac{1}{3} \\ \frac{1}{3} \\ \frac{1}{3} \end{bmatrix},$$

then determine the market distribution after 1 year; after 2 years.

(b) Show that the stable market distribution is given by

$$\mathbf{x} = \begin{bmatrix} \frac{10}{37} \\ \frac{12}{37} \\ \frac{15}{37} \end{bmatrix}.$$

(c) Which company R, S, or T will gain the most market share over a long period of time (assuming that the retention and switching patterns remain the same)? Approximately what percent of the market was gained by this company?

Exercises 22 through 25 involve bit matrices.

22. If bit matrix $A = \begin{bmatrix} 1 & 1 \\ 1 & 1 \end{bmatrix}$, show that $A^2 = O$.

23. If bit matrix $A = \begin{bmatrix} 1 & 1 \\ 0 & 1 \end{bmatrix}$, show that $A^2 = I_2$.

24. Let $A = \begin{bmatrix} 0 & 1 \\ 0 & 1 \end{bmatrix}$ be a bit matrix. Find

(a) $A^2 - A$ (b) $A^3 + A^2 + A$

25. Let $A = \begin{bmatrix} 0 & 0 \\ 1 & 1 \end{bmatrix}$ be a bit matrix. Find

(a) $A^2 + A$ (b) $A^4 + A^3 + A^2$

Theoretical Exercises

T.1. Prove properties (b) and (d) of Theorem 1.1.

T.2. If $A = [a_{ij}]$ is a 2×3 matrix, $B = [b_{ij}]$ is a 3×4 matrix, and $C = [c_{ij}]$ is a 4×3 matrix, show that $A(BC) = (AB)C$.

T.3. Prove properties (b) and (c) of Theorem 1.2.

T.4. If A is an $m \times n$ matrix, show that

$$I_m A = A I_n = A.$$

T.5. Let p and q be nonnegative integers and let A be a square matrix. Show that

$$A^p A^q = A^{p+q} \quad \text{and} \quad (A^p)^q = A^{pq}.$$

T.6. If $AB = BA$, and p is a nonnegative integer, show that

$$(AB)^p = A^p B^p.$$

T.7. Show that if A and B are $n \times n$ diagonal matrices, then $AB = BA$.

T.8. Find a 2×2 matrix $B \neq O$ and $B \neq I_2$ such that $AB = BA$, where

$$A = \begin{bmatrix} 1 & 2 \\ 2 & 1 \end{bmatrix}.$$

How many such matrices B are there?

T.9. Find a 2×2 matrix $B \neq O$ and $B \neq I_2$ such that $AB = BA$, where

$$A = \begin{bmatrix} 1 & 2 \\ 0 & 1 \end{bmatrix}.$$

How many such matrices B are there?

T.10. Let $A = \begin{bmatrix} \cos\theta & \sin\theta \\ -\sin\theta & \cos\theta \end{bmatrix}$.

(a) Determine a simple expression for A^2.

(b) Determine a simple expression for A^3.

(c) Conjecture the form of a simple expression for A^k, k a positive integer.

(d) Prove or disprove your conjecture in part (c).

T.11. If p is a nonnegative integer and c is a scalar, show that

$$(cA)^p = c^p A^p.$$

T.12. Prove Theorem 1.3.

T.13. Show that $(-1)A = -A$.

T.14. Complete the proof of Theorem 1.4.

T.15. Show that $(A - B)^T = A^T - B^T$.

T.16. (a) Show that $(A^2)^T = (A^T)^2$.

(b) Show that $(A^3)^T = (A^T)^3$.

(c) Prove or disprove that, for $k = 4, 5, \dots$, $(A^k)^T = (A^T)^k$.

T.17. Show that a square matrix A is symmetric if and only if $a_{ij} = a_{ji}$ for all i, j.

T.18. Show that if A is symmetric, then A^T is symmetric.

T.19. Let A be an $n \times n$ matrix. Show that if $A\mathbf{x} = \mathbf{0}$ for all $n \times 1$ matrices $\mathbf{x}$, then $A = O$.

T.20. Let A be an $n \times n$ matrix. Show that if $A\mathbf{x} = \mathbf{x}$ for all $n \times 1$ matrices $\mathbf{x}$, then $A = I_n$.

T.21. Show that if $AA^T = O$, then $A = O$.

T.22. Show that if A is a symmetric matrix, then A^k, $k = 2, 3, \dots$, is symmetric.

T.23. Let A and B be symmetric matrices.

(a) Show that $A + B$ is symmetric.

(b) Show that AB is symmetric if and only if $AB = BA$.

T.24. A matrix $A = [a_{ij}]$ is called **skew symmetric** if $A^T = -A$. Show that A is skew symmetric if and only if $a_{ij} = -a_{ji}$ for all i, j.

T.25. Describe all skew symmetric scalar matrices. (See Section 1.2 for the definition of scalar matrix.)

T.26. If A is an $n \times n$ matrix, show that AA^T and $A^T A$ are symmetric.

T.27. If A is an $n \times n$ matrix, show that

(a) $A + A^T$ is symmetric.

(b) $A - A^T$ is skew symmetric.

T.28. Show that if A is an $n \times n$ matrix, then A can be written uniquely as $A = S + K$, where S is symmetric and K is skew symmetric.

T.29. Show that if A is an $n \times n$ scalar matrix, then $A = rI_n$ for some real number r.

T.30. Show that $I_n^T = I_n$.

T.31. Let A be an $m \times n$ matrix. Show that if $rA = O$, then $r = 0$ or $A = O$.

T.32. Show that if $A\mathbf{x} = \mathbf{b}$ is a linear system that has more than one solution, then it has infinitely many solutions. (*Hint*: If $\mathbf{u}_1$ and $\mathbf{u}_2$ are solutions, consider $\mathbf{w} = r\mathbf{u}_1 + s\mathbf{u}_2$, where $r + s = 1$.)

T.33. Determine all 2×2 matrices A such that $AB = BA$ for any 2×2 matrix B.

T.34. If A is a skew symmetric matrix, what type of matrix is A^T? Justify your answer.

T.35. What type of matrix is a linear combination of symmetric matrices? (See Section 1.3.) Justify your answer.

T.36. What type of matrix is a linear combination of scalar matrices? (See Section 1.3.) Justify your answer.

T.37. Let $A = [a_{ij}]$ be the $n \times n$ matrix defined by $a_{ii} = r$ and $a_{ij} = 0$ if $i \neq j$. Show that if B is any $n \times n$ matrix, then $AB = rB$.

T.38. If A is any $m \times n$ bit matrix, show that $-A = A$.

T.39. Determine all 2×2 bit matrices A so that $A^2 = O$.

T.40. Determine all 2×2 bit matrices A so that $A^2 = I_2$.

MATLAB Exercises

In order to use MATLAB in this section, you should first have read Chapter 12 through Section 12.3.

ML.1. Use MATLAB to find the smallest positive integer k in each of the following cases. (See also Exercise 12.)

(a) $A^k = I_3$ for $A = \begin{bmatrix} 0 & 0 & 1 \\ 1 & 0 & 0 \\ 0 & 1 & 0 \end{bmatrix}$

(b) $A^k = A$ for $A = \begin{bmatrix} 0 & 1 & 0 & 0 \\ -1 & 0 & 0 & 0 \\ 0 & 0 & 0 & 1 \\ 0 & 0 & 1 & 0 \end{bmatrix}$

ML.2. Use MATLAB to display the matrix A in each of the following cases. Find the smallest value of k such that A^k is a zero matrix. Here **tril**, **ones**, **triu**, **fix**, and **rand** are MATLAB commands. (To see a description, use **help**.)

(a) $A = \textbf{tril(ones(5)}, \mathbf{-1)}$

(b) $A = \textbf{triu(fix(10} * \textbf{rand(7)), 2)}$

ML.3. Let $A = \begin{bmatrix} 1 & -1 & 0 \\ 0 & 1 & -1 \\ -1 & 0 & 1 \end{bmatrix}$. Using command **polyvalm** in MATLAB, compute the following matrix polynomials:

(a) $A^4 - A^3 + A^2 + 2I_3$ (b) $A^3 - 3A^2 + 3A$

ML.4. Let $A = \begin{bmatrix} 0.1 & 0.3 & 0.6 \\ 0.2 & 0.2 & 0.6 \\ 0.3 & 0.3 & 0.4 \end{bmatrix}$. Using MATLAB, compute each of the following matrix expressions:

(a) $(A^2 - 7A)(A + 3I_3)$.

(b) $(A - I_3)^2 + (A^3 + A)$.

(c) Look at the sequence $A, A^2, A^3, \dots, A^8, \dots$. Does it appear to be converging to a matrix? If so, to what matrix?

ML.5. Let $A = \begin{bmatrix} 1 & \frac{1}{2} \\ 0 & \frac{1}{3} \end{bmatrix}$. Use MATLAB to compute members of the sequence $A, A^2, A^3, \dots, A^k, \dots$. Write a description of the behavior of this matrix sequence.

ML.6. Let $A = \begin{bmatrix} \frac{1}{2} & \frac{1}{3} \\ 0 & -\frac{1}{5} \end{bmatrix}$. Repeat Exercise ML.5.

ML.7. Let $A = \begin{bmatrix} 1 & -2 & 1 \\ -1 & 1 & 2 \\ 0 & 2 & 1 \end{bmatrix}$. Use MATLAB to do the following:

(a) Compute $A^T A$ and AA^T. Are they equal?

(b) Compute $B = A + A^T$ and $C = A - A^T$. Show that B is symmetric and C is skew symmetric. (See Exercise T.24.)

(c) Determine a relationship between $B + C$ and A.

Exercises ML.8 through ML.11 use bit matrices and the supplemental instructional commands described in Section 12.9.

ML.8. (a) Use **binrand** to generate a 3×3 bit matrix B.

(b) Use **binadd** to compute $B + B$ and $B + B + B$.

(c) If B were added to itself n times, what would be the result? Explain your answer.

ML.9. Let $B = \textbf{triu(ones(3))}$. Determine k so that $B^k = I_3$.

ML.10. Let $B = \textbf{triu(ones(4))}$. Determine k so that $B^k = I_4$.

ML.11. Let $B = \textbf{triu(ones(5))}$. Determine k so that $B^k = I_5$.

1.5 MATRIX TRANSFORMATIONS

In Section 1.2 we introduced the notation R^n for the set of all n-vectors with real entries. Thus, R^2 denotes the set of all 2-vectors and R^3 denotes the set of all 3-vectors. It is convenient to represent the elements of R^2 and R^3 geometrically as directed line segments in a rectangular coordinate system.[‡] Our approach in this section is intuitive and will enable us to present some interesting geometric applications in the next section (at this early stage of the course). We return in Section 3.1 to a careful and precise study of 2-vectors and 3-vectors.

[‡]You have undoubtedly seen rectangular coordinate systems in your precalculus or calculus courses.

The vector

$$\mathbf{x} = \begin{bmatrix} x \\ y \end{bmatrix}$$

in R^2 is represented by the directed line segment shown in Figure 1.7. The vector

$$\mathbf{x} = \begin{bmatrix} x \\ y \\ z \end{bmatrix}$$

in R^3 is represented by the directed line segment shown in Figure 1.8.

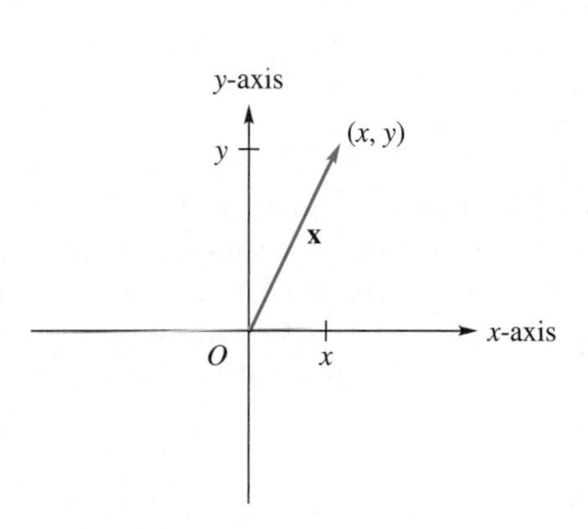

Figure 1.7 ▲

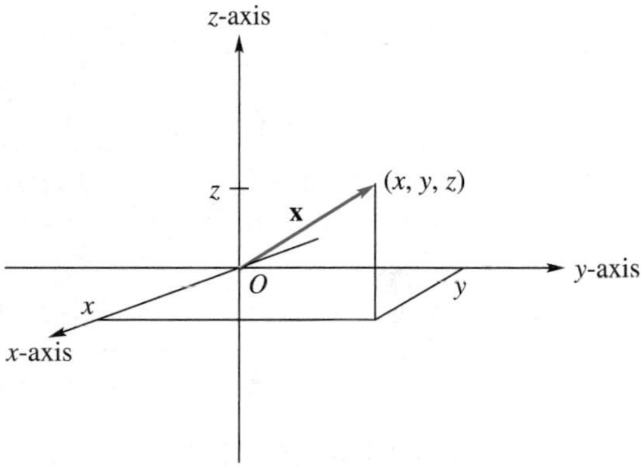

Figure 1.8 ▲

EXAMPLE 1 Figure 1.9 shows geometric representations of the 2-vectors

$$\mathbf{u}_1 = \begin{bmatrix} 1 \\ 2 \end{bmatrix}, \quad \mathbf{u}_2 = \begin{bmatrix} -2 \\ 1 \end{bmatrix}, \quad \text{and} \quad \mathbf{u}_3 = \begin{bmatrix} 0 \\ 1 \end{bmatrix}$$

in a two-dimensional rectangular coordinate system. Figure 1.10 shows geometric representations of the 3-vectors

$$\mathbf{v}_1 = \begin{bmatrix} 1 \\ 2 \\ 3 \end{bmatrix}, \quad \mathbf{v}_2 = \begin{bmatrix} -1 \\ 2 \\ -2 \end{bmatrix}, \quad \text{and} \quad \mathbf{v}_3 = \begin{bmatrix} 0 \\ 0 \\ 1 \end{bmatrix}$$

in a three-dimensional rectangular coordinate system. ■

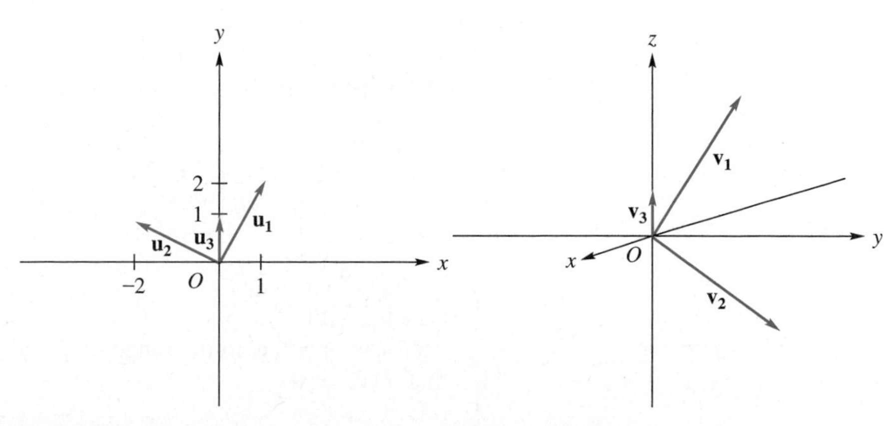

Figure 1.9 ▲

Figure 1.10 ▲

Functions occur in almost every application of mathematics. In this section we give a brief introduction from a geometric point of view to certain functions mapping R^n into R^m. Since we wish to picture these functions, called matrix transformations, we limit most of our discussion in this section to the situation where m and n have the values 2 or 3. In the next section we give an application of these functions to computer graphics in the plane, that is, for m and n equal to 2. In Chapter 4 we consider in more detail a more general function, called a linear transformation mapping R^n into R^m. Since every matrix transformation is a linear transformation, we then learn more about the properties of matrix transformations.

Linear transformations play an important role in many areas of mathematics, as well as in numerous applied problems in the physical sciences, the social sciences, and economics.

If A is an $m \times n$ matrix and $\mathbf{u}$ is an n-vector, then the matrix product $A\mathbf{u}$ is an m-vector. A function f mapping R^n into R^m is denoted by $f: R^n \to R^m$.[§] A **matrix transformation** is a function $f: R^n \to R^m$ defined by $f(\mathbf{u}) = A\mathbf{u}$. The vector $f(\mathbf{u})$ in R^m is called the **image** of $\mathbf{u}$, and the set of all images of the vectors in R^n is called the **range** of f. Although we are limiting ourselves in this section to matrices and vectors with only real entries, an entirely similar discussion can be developed for matrices and vectors with complex entries. (See Appendix A.2.)

EXAMPLE 2 (a) Let f be the matrix transformation defined by

$$f(\mathbf{u}) = \begin{bmatrix} 2 & 4 \\ 3 & 1 \end{bmatrix} \mathbf{u}.$$

The image of $\mathbf{u} = \begin{bmatrix} 2 \\ -1 \end{bmatrix}$ is

$$f(\mathbf{u}) = \begin{bmatrix} 2 & 4 \\ 3 & 1 \end{bmatrix} \begin{bmatrix} 2 \\ -1 \end{bmatrix} = \begin{bmatrix} 0 \\ 5 \end{bmatrix}$$

and the image of $\begin{bmatrix} 1 \\ 2 \end{bmatrix}$ is $\begin{bmatrix} 10 \\ 5 \end{bmatrix}$ (verify).

(b) Let $A = \begin{bmatrix} 1 & 2 & 0 \\ 1 & -1 & 1 \end{bmatrix}$, and consider the matrix transformation defined by

$$f(\mathbf{u}) = A\mathbf{u}.$$

Then the image of $\begin{bmatrix} 1 \\ 0 \\ 1 \end{bmatrix}$ is $\begin{bmatrix} 1 \\ 2 \end{bmatrix}$, the image of $\begin{bmatrix} 0 \\ 1 \\ 3 \end{bmatrix}$ is $\begin{bmatrix} 2 \\ 2 \end{bmatrix}$, and the image of $\begin{bmatrix} -2 \\ 1 \\ 3 \end{bmatrix}$ is $\begin{bmatrix} 0 \\ 0 \end{bmatrix}$ (verify). ■

Observe that if A is an $m \times n$ matrix and $f: R^n \to R^m$ is a matrix transformation mapping R^n into R^m that is defined by $f(\mathbf{u}) = A\mathbf{u}$, then a vector $\mathbf{w}$ in R^m is in the range of f only if we can find a vector $\mathbf{v}$ in R^n such that $f(\mathbf{v}) = \mathbf{w}$.

[§] Appendix A, dealing with sets and functions, may be consulted as needed.

EXAMPLE 3 Let $A = \begin{bmatrix} 1 & 2 \\ -2 & 3 \end{bmatrix}$ and consider the matrix transformation defined by $f(\mathbf{u}) = A\mathbf{u}$. Determine if the vector $\mathbf{w} = \begin{bmatrix} 4 \\ -1 \end{bmatrix}$ is in the range of f.

Solution The question is equivalent to asking if there is a vector $\mathbf{v} = \begin{bmatrix} v_1 \\ v_2 \end{bmatrix}$ such that $f(\mathbf{v}) = \mathbf{w}$. We have

$$A\mathbf{v} = \begin{bmatrix} v_1 + 2v_2 \\ -2v_1 + 3v_2 \end{bmatrix} = \mathbf{w} = \begin{bmatrix} 4 \\ -1 \end{bmatrix}$$

or

$$\begin{aligned} v_1 + 2v_2 &= 4 \\ -2v_1 + 3v_2 &= -1. \end{aligned}$$

Solving this linear system of equations by the familiar method of elimination we get $v_1 = 2$ and $v_2 = 1$ (verify). Thus $\mathbf{w}$ is in the range of f. In particular, if $\mathbf{v} = \begin{bmatrix} 2 \\ 1 \end{bmatrix}$, then $f(\mathbf{v}) = \mathbf{w}$. ■

EXAMPLE 4 **(Production)** A book publisher publishes a book in three different editions: trade, book club, and deluxe. Each book requires a certain amount of paper and canvas (for the cover). The requirements are given (in grams) by the matrix

	Trade	Book Club	Deluxe	
$A =$	300	500	800	Paper
	40	50	60	Canvas

Let

$$\mathbf{x} = \begin{bmatrix} x_1 \\ x_2 \\ x_3 \end{bmatrix}$$

denote the production vector, where x_1, x_2, and x_3 are the number of trade, book club, and deluxe books, respectively, that are published. The matrix transformation $f: R^3 \to R^2$ defined by $f(\mathbf{x}) = A\mathbf{x}$ gives the vector

$$\mathbf{y} = \begin{bmatrix} y_1 \\ y_2 \end{bmatrix},$$

where y_1 is the total amount of paper required and y_2 is the total amount of canvas required. ■

For matrix transformations where m and n are 2 or 3, we can draw pictures showing the effect of the matrix transformation. This will be illustrated in the examples that follow.

EXAMPLE 5 Let $f: R^2 \to R^2$ be the matrix transformation defined by

$$f(\mathbf{u}) = \begin{bmatrix} 1 & 0 \\ 0 & -1 \end{bmatrix}\mathbf{u}.$$

Thus, if $\mathbf{u} = \begin{bmatrix} x \\ y \end{bmatrix}$, then

$$f(\mathbf{u}) = f\left(\begin{bmatrix} x \\ y \end{bmatrix}\right) = \begin{bmatrix} x \\ -y \end{bmatrix}.$$

The effect of the matrix transformation f, called **reflection with respect to the x-axis in R^2**, is shown in Figure 1.11. In Exercise 2 we consider reflection with respect to the y-axis. ■

Figure 1.11 ► Reflection with respect to the x-axis

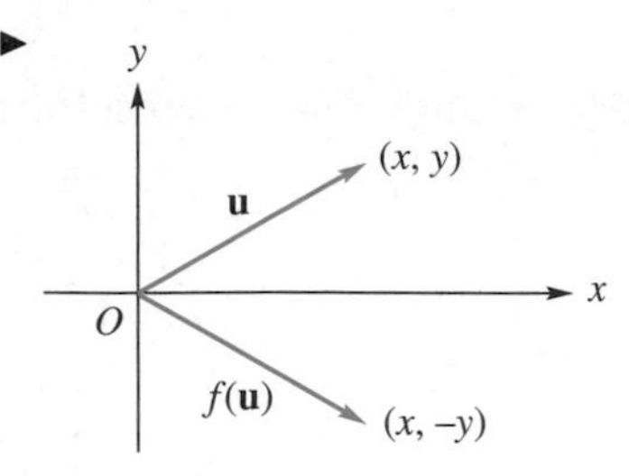

EXAMPLE 6 Let $f\colon R^3 \to R^2$ be the matrix transformation defined by

$$f(\mathbf{u}) = f\left(\begin{bmatrix} x \\ y \\ z \end{bmatrix}\right) = \begin{bmatrix} 1 & 0 & 0 \\ 0 & 1 & 0 \end{bmatrix}\begin{bmatrix} x \\ y \\ z \end{bmatrix}.$$

Then

$$f(\mathbf{u}) = f\left(\begin{bmatrix} x \\ y \\ z \end{bmatrix}\right) = \begin{bmatrix} x \\ y \end{bmatrix}.$$

Figure 1.12 shows the effect of this matrix transformation. (Warning: Carefully note the axes in Figure 1.12.)

Figure 1.12 ►

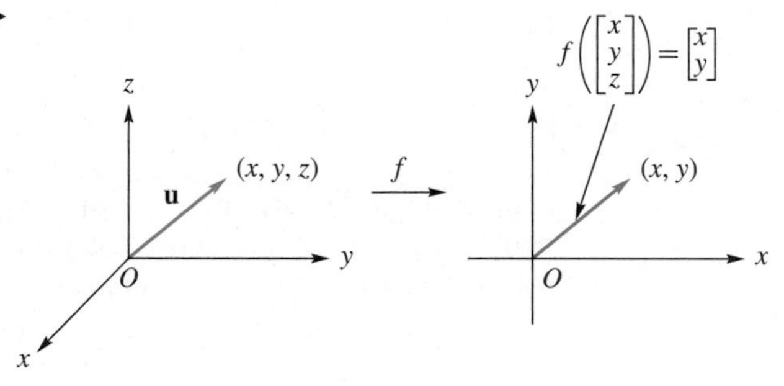

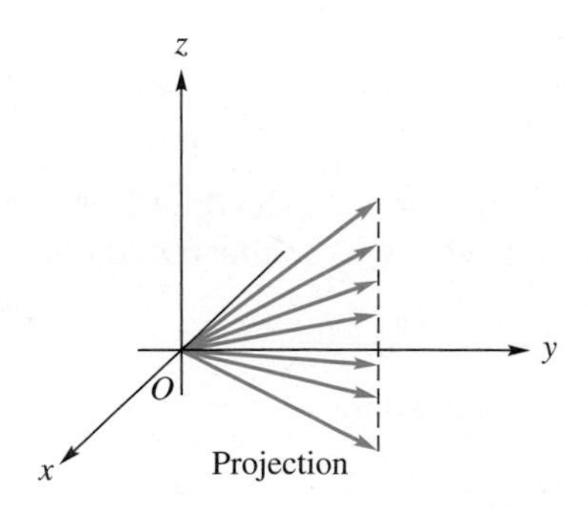

Figure 1.13 ▲

Observe that if

$$\mathbf{v} = \begin{bmatrix} x \\ y \\ s \end{bmatrix},$$

where s is any scalar, then

$$f(\mathbf{v}) = \begin{bmatrix} x \\ y \end{bmatrix} = f(\mathbf{u}).$$

Hence, infinitely many 3-vectors have the same image vector. See Figure 1.13. The matrix transformation f is an example of a type of matrix transformation called **projection**. In this case f is a projection of R^3 into the xy-plane. Note that the image of the 3-vector $\mathbf{v} = \begin{bmatrix} x \\ y \\ z \end{bmatrix}$ under the matrix transformation

$f : R^3 \to R^3$ defined by

$$f(\mathbf{v}) = \begin{bmatrix} 1 & 0 & 0 \\ 0 & 1 & 0 \\ 0 & 0 & 0 \end{bmatrix} \mathbf{v}$$

is $\begin{bmatrix} x \\ y \\ 0 \end{bmatrix}$. The effect of this matrix transformation is shown in Figure 1.14. The picture is almost the same as Figure 1.12, where the image is a 2-vector that lies in the xy-plane, whereas in Figure 1.14 the image is a 3-vector that lies in the xy-plane. Observe that $f(\mathbf{v})$ appears to be the shadow cast by $\mathbf{v}$ onto the xy-planc. ■

Figure 1.14 ►

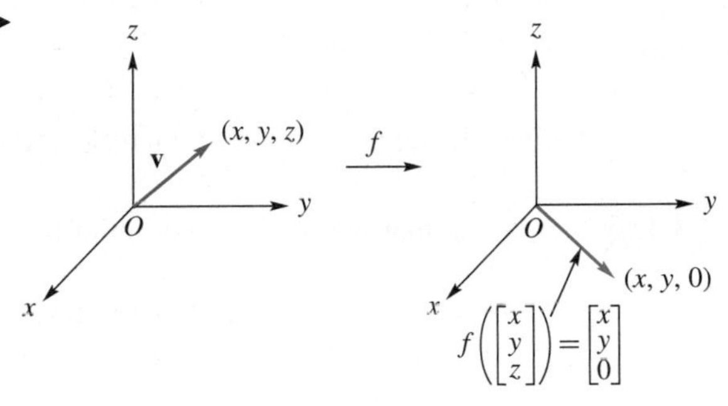

EXAMPLE 7 Let $f : R^3 \to R^3$ be the matrix transformation defined by

$$f(\mathbf{u}) = \begin{bmatrix} r & 0 & 0 \\ 0 & r & 0 \\ 0 & 0 & r \end{bmatrix} \mathbf{u},$$

where r is a real number. It is easily seen that $f(\mathbf{u}) = r\mathbf{u}$. If $r > 1$, f is called **dilation**; if $0 < r < 1$, f is called **contraction**. In Figure 1.15(a) we show the vector $f_1(\mathbf{u}) = 2\mathbf{u}$ and in Figure 1.15(b) the vector $f_2(\mathbf{u}) = \frac{1}{2}\mathbf{u}$. Thus dilation stretches a vector, and contraction shrinks it. Similarly, we can define the matrix transformation $g : R^2 \to R^2$ by

$$g(\mathbf{u}) = \begin{bmatrix} r & 0 \\ 0 & r \end{bmatrix} \mathbf{u}.$$

We also have $g(\mathbf{u}) = r\mathbf{u}$, so again if $r > 1$, g is called **dilation**; if $0 < r < 1$, g is called **contraction**. ■

Figure 1.15 ►

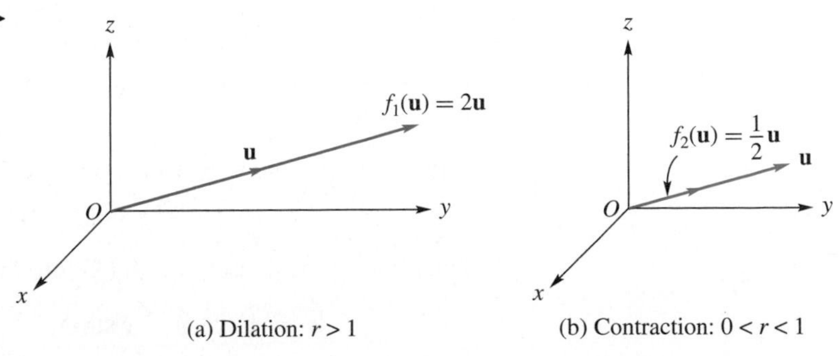

(a) Dilation: $r > 1$

(b) Contraction: $0 < r < 1$

EXAMPLE 8 **(Production)** We return to the book publisher discussed in Example 4. The requirements are given by the production vector

$$\mathbf{x} = \begin{bmatrix} x_1 \\ x_2 \\ x_3 \end{bmatrix},$$

where x_1, x_2, and x_3 are the number of trade, book club, and deluxe books, respectively. The vector

$$\mathbf{y} = A\mathbf{x} = \begin{bmatrix} y_1 \\ y_2 \end{bmatrix}$$

gives y_1, the total amunt of paper required, y_2, the total amount of canvas required. Let c_1 denote the cost per pound of paper and let c_2 denote the cost per pound of canvas. The matrix transformation $g: R^2 \to R^1$ defined by $g(\mathbf{y}) = B\mathbf{y}$, where

$$B = \begin{bmatrix} c_1 & c_2 \end{bmatrix}$$

gives the total cost to manufacture all the books. ■

EXAMPLE 9 Suppose that we rotate every point in R^2 counterclockwise through an angle ϕ about the origin of a rectangular coordinate system. Thus, if the point P has coordinates (x, y), then after rotating, we get the point P' with coordinates (x', y'). To obtain a relationship between the coordinates of P' and those of P, we let $\mathbf{u}$ be the vector $\begin{bmatrix} x \\ y \end{bmatrix}$, which is represented by the directed line segment from the origin to $P(x, y)$. See Figure 1.16(a). Also, let θ be the angle made by $\mathbf{u}$ with the positive x-axis.

Figure 1.16 ►

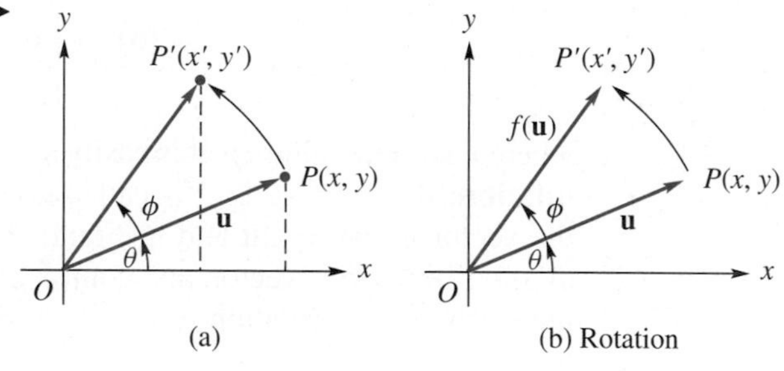

Letting r denote the length of the directed line segment from O to P, we see from Figure 1.16(a) that

$$x = r\cos\theta, \qquad y = r\sin\theta \tag{1}$$

and

$$x' = r\cos(\theta + \phi), \qquad y' = r\sin(\theta + \phi). \tag{2}$$

Using the formulas for the sine and cosine of a sum of angles, the equations in (2) become

$$\begin{aligned} x' &= r\cos\theta\cos\phi - r\sin\theta\sin\phi \\ y' &= r\sin\theta\cos\phi + r\cos\theta\sin\phi. \end{aligned}$$

Substituting the expression in (1) into the last pair of equations, we obtain

$$x' = x\cos\phi - y\sin\phi, \qquad y' = x\sin\phi + y\cos\phi. \tag{3}$$

Solving (3) for x and y, we have

$$x = x' \cos\phi + y' \sin\phi \quad \text{and} \quad y = -x' \sin\phi + y' \cos\phi. \tag{4}$$

Equation (3) gives the coordinates of P' in terms of those of P and (4) expresses the coordinates of P in terms of those of P'. This type of rotation is used to simplify the general equation of second degree

$$ax^2 + bxy + cy^2 + dx + ey + f = 0.$$

Substituting for x and y in terms of x' and y', we obtain

$$a'x'^2 + b'x'y' + c'y'^2 + d'x' + e'y' + f' = 0.$$

The key point is to choose ϕ so that $b' = 0$. Once this is done (we might now have to perform a translation of coordinates), we identify the general equation of second degree as a circle, ellipse, hyperbola, parabola, or a degenerate form of these. This topic will be treated from a linear algebra point of view in Section 9.5.

We may also perform this change of coordinates by considering the matrix transformation $f: R^2 \to R^2$ defined by

$$f\left(\begin{bmatrix} x \\ y \end{bmatrix}\right) = \begin{bmatrix} \cos\phi & -\sin\phi \\ \sin\phi & \cos\phi \end{bmatrix} \begin{bmatrix} x \\ y \end{bmatrix}. \tag{5}$$

Then (5) can be written, using (3), as

$$f(\mathbf{u}) = \begin{bmatrix} x\cos\phi - y\sin\phi \\ x\sin\phi + y\cos\phi \end{bmatrix} = \begin{bmatrix} x' \\ y' \end{bmatrix}.$$

It then follows that the vector $f(\mathbf{u})$ is represented by the directed line segment from O to the point P'. Thus, rotation counterclockwise through an angle ϕ is a matrix transformation. ■

Key Terms

Matrix transformation
Mapping (function)
Image
Range
Reflection
Projection
Dilation
Contraction
Rotation

Preview of an Application

Computer Graphics (Section 2.3)

We are all being dazzled daily by the extensive and expanding use of computer graphics in the areas of video games, special effects in the film and television industries, and in computer-aided design (CAD). In a typical CAD application, a computer model of a product is created and then it is exhaustively tested on a computer to find faults in the design and thereby improve the product.

Matrix transformations play a key role in computer graphics. In Section 2.3 we briefly discuss four simple matrix transformations. Two of these are as follows:

$$f\left(\begin{bmatrix} a_1 \\ a_2 \end{bmatrix}\right) = \begin{bmatrix} a_1 \\ -a_2 \end{bmatrix},$$

which maps

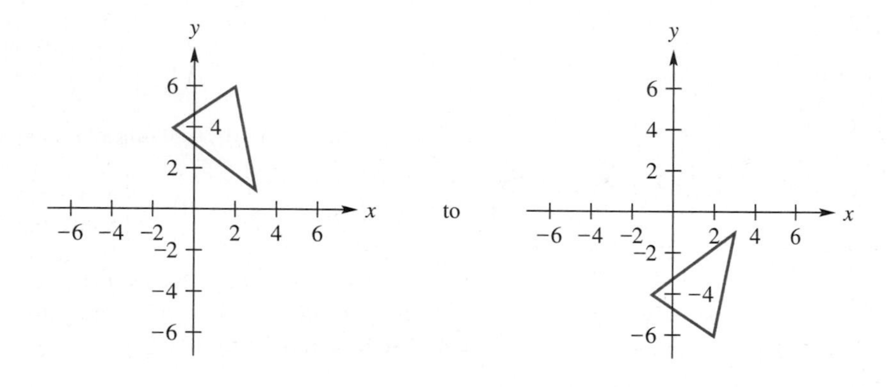

and

$$g\left(\begin{bmatrix} a_1 \\ a_2 \end{bmatrix}\right) = \begin{bmatrix} \cos \frac{5}{18}\pi & -\sin \frac{5}{18}\pi \\ \sin \frac{5}{18}\pi & \cos \frac{5}{18}\pi \end{bmatrix} \begin{bmatrix} a_1 \\ a_2 \end{bmatrix},$$

which maps

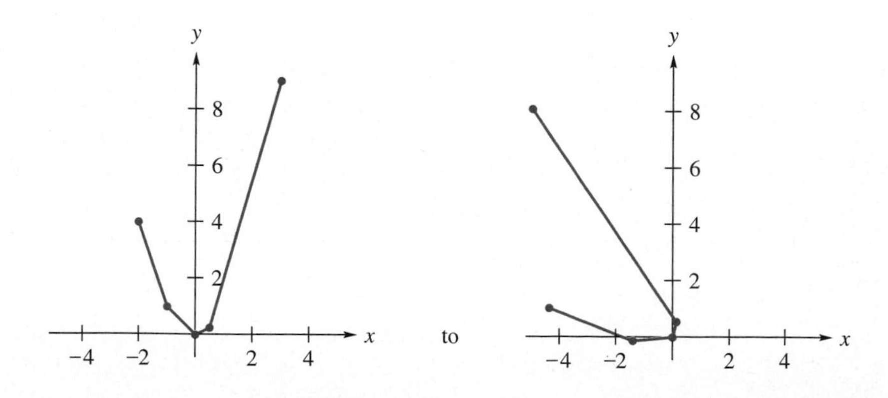

1.5 Exercises

In Exercises 1 through 8, sketch **u** *and its image under the given matrix transformation* f.

1. $f: R^2 \to R^2$ defined by
$$f\left(\begin{bmatrix} x \\ y \end{bmatrix}\right) = \begin{bmatrix} 1 & 0 \\ 0 & -1 \end{bmatrix}\begin{bmatrix} x \\ y \end{bmatrix}; \quad \mathbf{u} = \begin{bmatrix} 2 \\ 3 \end{bmatrix}$$

2. $f: R^2 \to R^2$ **(reflection with respect to the** y**-axis)** defined by
$$f\left(\begin{bmatrix} x \\ y \end{bmatrix}\right) = \begin{bmatrix} -1 & 0 \\ 0 & 1 \end{bmatrix}\begin{bmatrix} x \\ y \end{bmatrix}; \quad \mathbf{u} = \begin{bmatrix} 1 \\ -2 \end{bmatrix}$$

3. $f: R^2 \to R^2$ is a counterclockwise rotation through $30°$; $\mathbf{u} = \begin{bmatrix} -1 \\ 3 \end{bmatrix}$

4. $f: R^2 \to R^2$ is a counterclockwise rotation through $\frac{2}{3}\pi$ radians; $\mathbf{u} = \begin{bmatrix} -2 \\ -3 \end{bmatrix}$

5. $f: R^2 \to R^2$ defined by
$$f\left(\begin{bmatrix} x \\ y \end{bmatrix}\right) = \begin{bmatrix} -1 & 0 \\ 0 & -1 \end{bmatrix}\begin{bmatrix} x \\ y \end{bmatrix}; \quad \mathbf{u} = \begin{bmatrix} 3 \\ 2 \end{bmatrix}$$

6. $f: R^2 \to R^2$ defined by
$$f\left(\begin{bmatrix} x \\ y \end{bmatrix}\right) = \begin{bmatrix} 2 & 0 \\ 0 & 2 \end{bmatrix}\begin{bmatrix} x \\ y \end{bmatrix}; \quad \mathbf{u} = \begin{bmatrix} -3 \\ 3 \end{bmatrix}$$

7. $f: R^3 \to R^3$ defined by
$$f\left(\begin{bmatrix} x \\ y \\ z \end{bmatrix}\right) = \begin{bmatrix} 1 & 0 & 0 \\ 1 & -1 & 0 \\ 0 & 0 & 0 \end{bmatrix}\begin{bmatrix} x \\ y \\ z \end{bmatrix}; \quad \mathbf{u} = \begin{bmatrix} 2 \\ -1 \\ 3 \end{bmatrix}$$

8. $f: R^3 \to R^3$ defined by
$$f\left(\begin{bmatrix} x \\ y \\ z \end{bmatrix}\right) = \begin{bmatrix} 1 & 0 & 1 \\ -1 & 1 & 0 \\ 0 & 0 & 1 \end{bmatrix}\begin{bmatrix} x \\ y \\ z \end{bmatrix}; \quad \mathbf{u} = \begin{bmatrix} 0 \\ -2 \\ 4 \end{bmatrix}$$

In Exercises 9 through 11, let $f: R^2 \to R^2$ *be the matrix transformation defined by* $f(\mathbf{x}) = A\mathbf{x}$, *where*
$$A = \begin{bmatrix} 1 & 3 \\ -1 & 2 \end{bmatrix}.$$

Determine whether the given vector **w** *is in the range of* f.

9. $\mathbf{w} = \begin{bmatrix} 7 \\ 3 \end{bmatrix}$ **10.** $\mathbf{w} = \begin{bmatrix} 4 \\ 1 \end{bmatrix}$ **11.** $\mathbf{w} = \begin{bmatrix} -1 \\ -9 \end{bmatrix}$

In Exercises 12 through 14, let $f: R^2 \to R^3$ *be the matrix transformation defined by* $f(\mathbf{x}) = A\mathbf{x}$, *where*
$$A = \begin{bmatrix} 1 & 2 \\ 0 & 1 \\ 1 & 1 \end{bmatrix}.$$

Determine whether the given vector **w** *is in the range of* f.

12. $\mathbf{w} = \begin{bmatrix} 1 \\ -1 \\ 2 \end{bmatrix}$ **13.** $\mathbf{w} = \begin{bmatrix} 1 \\ 1 \\ 1 \end{bmatrix}$ **14.** $\mathbf{w} = \begin{bmatrix} 0 \\ 0 \\ 0 \end{bmatrix}$

In Exercises 15 through 17, give a geometric description of the matrix transformation $f: R^2 \to R^2$ *defined by* $f(\mathbf{u}) = A\mathbf{u}$ *for the given matrix* A.

15. (a) $A = \begin{bmatrix} -1 & 0 \\ 0 & 1 \end{bmatrix}$ (b) $A = \begin{bmatrix} 0 & -1 \\ 1 & 0 \end{bmatrix}$

16. (a) $A = \begin{bmatrix} 0 & 1 \\ 1 & 0 \end{bmatrix}$ (b) $A = \begin{bmatrix} 0 & -1 \\ -1 & 0 \end{bmatrix}$

17. (a) $A = \begin{bmatrix} 1 & 0 \\ 0 & 0 \end{bmatrix}$ (b) $A = \begin{bmatrix} 0 & 0 \\ 0 & 1 \end{bmatrix}$

18. Some matrix transformations f have the property that $f(\mathbf{u}) - f(\mathbf{v})$, when $\mathbf{u} \neq \mathbf{v}$. That is, the images of different vectors can be the same. For each of the following matrix transformations $f: R^3 \to R^2$ defined by $f(\mathbf{u}) = A\mathbf{u}$, find two different vectors **u** and **v** such that $f(\mathbf{u}) = f(\mathbf{v}) = \mathbf{w}$ for the given vector **w**.

(a) $A = \begin{bmatrix} 1 & 2 & 0 \\ 0 & 1 & -1 \end{bmatrix}$, $\mathbf{w} = \begin{bmatrix} 0 \\ -1 \end{bmatrix}$

(b) $A = \begin{bmatrix} 2 & 1 & 0 \\ 0 & 2 & -1 \end{bmatrix}$, $\mathbf{w} = \begin{bmatrix} 4 \\ 4 \end{bmatrix}$

19. Let $f: R^2 \to R^2$ be the matrix transformation defined by $f(\mathbf{u}) = A\mathbf{u}$, where
$$A = \begin{bmatrix} \cos\phi & -\sin\phi \\ \sin\phi & \cos\phi \end{bmatrix}.$$

For $\phi = 30°$, f defines a counterclockwise rotation by an angle of $30°$.

(a) If $g(\mathbf{u}) = A^2\mathbf{u}$, describe the action of g on **u**.

(b) If $h(\mathbf{u}) = A^3\mathbf{u}$, describe the action of h on **u**.

(c) What is the smallest positive value of k for which $p(\mathbf{u}) = A^k\mathbf{u} = \mathbf{u}$?

Theoretical Exercises

T.1. Let $f: R^n \to R^m$ be a matrix transformation defined by $f(\mathbf{u}) = A\mathbf{u}$, where A is an $m \times n$ matrix.

(a) Show that $f(\mathbf{u} + \mathbf{v}) = f(\mathbf{u}) + f(\mathbf{v})$ for any **u** and **v** in R^n.

(b) Show that $f(c\mathbf{u}) = cf(\mathbf{u})$ for any **u** in R^n and any real number c.

(c) Show that $f(c\mathbf{u} + d\mathbf{v}) = cf(\mathbf{u}) + df(\mathbf{v})$ for any **u** and **v** in R^n and any real numbers c and d.

T.2. Let $f: R^n \to R^m$ be a matrix transformation defined by $f(\mathbf{u}) = A\mathbf{u}$, where A is an $m \times n$ matrix. Show that if $\mathbf{u}$ and $\mathbf{v}$ are vectors in R^n such that $f(\mathbf{u}) = \mathbf{0}$ and $f(\mathbf{v}) = \mathbf{0}$, where

$$\mathbf{0} = \begin{bmatrix} 0 \\ 0 \\ \vdots \\ 0 \\ 0 \end{bmatrix},$$

then $f(c\mathbf{u} + d\mathbf{v}) = \mathbf{0}$ for any real numbers c and d.

T.3. (a) Let $O: R^n \to R^m$ be the matrix transformation defined by $O(\mathbf{u}) = O\mathbf{u}$, where O is the $m \times n$ zero matrix. Show that $O(\mathbf{u}) = \mathbf{0}$ for all $\mathbf{u}$ in R^n.

(b) Let $I: R^n \to R^n$ be the matrix transformation defined by $I(\mathbf{u}) = I_n\mathbf{u}$, where I_n is the identity matrix (see Section 1.4). Show that $I(\mathbf{u}) = \mathbf{u}$ for all $\mathbf{u}$ in R^n.

1.6 SOLUTIONS OF LINEAR SYSTEMS OF EQUATIONS

In this section we shall systematize the familiar method of elimination of unknowns (discussed in Section 1.1) for the solution of linear systems and thus obtain a useful method for solving such systems. This method starts with the augmented matrix of the given linear system and obtains a matrix of a certain form. This new matrix represents a linear system that has exactly the same solutions as the given system but is easier to solve. For example, if

$$\left[\begin{array}{cccc|c} 1 & 0 & 0 & 2 & 4 \\ 0 & 1 & 0 & -1 & -5 \\ 0 & 0 & 1 & 3 & 6 \end{array}\right]$$

represents the augmented matrix of a linear system, then the solution is easily found from the corresponding equations

$$\begin{aligned} x_1 \qquad\qquad + 2x_4 &= 4 \\ x_2 \qquad - x_4 &= -5 \\ x_3 + 3x_4 &= 6. \end{aligned}$$

The task of this section is to manipulate the augmented matrix representing a given linear system into a form from which the solution can easily be found.

DEFINITION

An $m \times n$ matrix A is said to be in **reduced row echelon form** if it satisfies the following properties:

(a) All zero rows, if there are any, appear at the bottom of the matrix.
(b) The first nonzero entry from the left of a nonzero row is a 1. This entry is called a **leading one** of its row.
(c) For each nonzero row, the leading one appears to the right and below any leading one's in preceding rows.
(d) If a column contains a leading one, then all other entries in that column are zero.

A matrix in reduced row echelon form appears as a staircase ("echelon") pattern of leading ones descending from the upper left corner of the matrix.

An $m \times n$ matrix satisfying properties (a), (b), and (c) is said to be in **row echelon form**.

EXAMPLE 1

The following are matrices in reduced row echelon form since they satisfy properties (a), (b), (c), and (d):

$$A = \begin{bmatrix} 1 & 0 & 0 & 0 \\ 0 & 1 & 0 & 0 \\ 0 & 0 & 1 & 0 \\ 0 & 0 & 0 & 1 \end{bmatrix}, \quad B = \begin{bmatrix} 1 & 0 & 0 & 0 & -2 & 4 \\ 0 & 1 & 0 & 0 & 4 & 8 \\ 0 & 0 & 0 & 1 & 7 & -2 \\ 0 & 0 & 0 & 0 & 0 & 0 \\ 0 & 0 & 0 & 0 & 0 & 0 \end{bmatrix},$$

and

$$C = \begin{bmatrix} 1 & 2 & 0 & 0 & 1 \\ 0 & 0 & 1 & 2 & 3 \\ 0 & 0 & 0 & 0 & 0 \end{bmatrix}.$$

The following matrices are not in reduced row echelon form. (Why not?)

$$D = \begin{bmatrix} 1 & 2 & 0 & 4 \\ 0 & 0 & 0 & 0 \\ 0 & 0 & 1 & -3 \end{bmatrix}, \quad E = \begin{bmatrix} 1 & 0 & 3 & 4 \\ 0 & 2 & -2 & 5 \\ 0 & 0 & 1 & 2 \end{bmatrix},$$

$$F = \begin{bmatrix} 1 & 0 & 3 & 4 \\ 0 & 1 & -2 & 5 \\ 0 & 1 & 2 & 2 \\ 0 & 0 & 0 & 0 \end{bmatrix}, \quad G = \begin{bmatrix} 1 & 2 & 3 & 4 \\ 0 & 1 & -2 & 5 \\ 0 & 0 & 1 & 2 \\ 0 & 0 & 0 & 0 \end{bmatrix}.$$

■

EXAMPLE 2 The following are matrices in row echelon form:

$$H = \begin{bmatrix} 1 & 5 & 0 & 2 & -2 & 4 \\ 0 & 1 & 0 & 3 & 4 & 8 \\ 0 & 0 & 0 & 1 & 7 & -2 \\ 0 & 0 & 0 & 0 & 0 & 0 \\ 0 & 0 & 0 & 0 & 0 & 0 \end{bmatrix}, \quad I = \begin{bmatrix} 1 & 0 & 0 & 0 \\ 0 & 1 & 0 & 0 \\ 0 & 0 & 1 & 0 \\ 0 & 0 & 0 & 1 \end{bmatrix},$$

and

$$J = \begin{bmatrix} 0 & 0 & 1 & 3 & 5 & 7 & 9 \\ 0 & 0 & 0 & 0 & 1 & -2 & 3 \\ 0 & 0 & 0 & 0 & 0 & 1 & 2 \\ 0 & 0 & 0 & 0 & 0 & 0 & 1 \\ 0 & 0 & 0 & 0 & 0 & 0 & 0 \end{bmatrix}.$$

■

A useful property of matrices in reduced row echelon form (see Exercise T.9) is that if A is an $n \times n$ matrix in reduced row echelon form not equal to I_n, then A has a row consisting entirely of zeros.

We shall now turn to the discussion of how to transform a given matrix to a matrix in reduced row echelon form.

DEFINITION An **elementary row operation** on an $m \times n$ matrix $A = [a_{ij}]$ is any of the following operations:

(a) Interchange rows r and s of A. That is, replace $a_{r1}, a_{r2}, \dots, a_{rn}$ by $a_{s1}, a_{s2}, \dots, a_{sn}$ and $a_{s1}, a_{s2}, \dots, a_{sn}$ by $a_{r1}, a_{r2}, \dots, a_{rn}$.

(b) Multiply row r of A by $c \neq 0$. That is, replace $a_{r1}, a_{r2}, \dots, a_{rn}$ by $ca_{r1}, ca_{r2}, \dots, ca_{rn}$.

(c) Add d times row r of A to row s of A, $r \neq s$. That is, replace $a_{s1}, a_{s2}, \dots, a_{sn}$ by $a_{s1} + da_{r1}, a_{s2} + da_{r2}, \dots, a_{sn} + da_{rn}$.

Observe that when a matrix is viewed as the augmented matrix of a linear system, the elementary row operations are equivalent, respectively, to interchanging two equations, multiplying an equation by a nonzero constant, and adding a multiple of one equation to another equation.

EXAMPLE 3 Let

$$A = \begin{bmatrix} 0 & 0 & 1 & 2 \\ 2 & 3 & 0 & -2 \\ 3 & 3 & 6 & -9 \end{bmatrix}.$$

Interchanging rows 1 and 3 of A, we obtain

$$B = \begin{bmatrix} 3 & 3 & 6 & -9 \\ 2 & 3 & 0 & -2 \\ 0 & 0 & 1 & 2 \end{bmatrix}.$$

Multiplying the third row of A by $\frac{1}{3}$, we obtain

$$C = \begin{bmatrix} 0 & 0 & 1 & 2 \\ 2 & 3 & 0 & -2 \\ 1 & 1 & 2 & -3 \end{bmatrix}.$$

Adding (-2) times row 2 of A to row 3 of A, we obtain

$$D = \begin{bmatrix} 0 & 0 & 1 & 2 \\ 2 & 3 & 0 & -2 \\ -1 & -3 & 6 & -5 \end{bmatrix}.$$

Observe that in obtaining D from A, row 2 of A *did not change.* ■

DEFINITION An $m \times n$ matrix A is said to be **row equivalent** to an $m \times n$ matrix B if B can be obtained by applying a finite sequence of elementary row operations to the matrix A.

EXAMPLE 4 Let

$$A = \begin{bmatrix} 1 & 2 & 4 & 3 \\ 2 & 1 & 3 & 2 \\ 1 & -2 & 2 & 3 \end{bmatrix}.$$

If we add 2 times row 3 of A to its second row, we obtain

$$B = \begin{bmatrix} 1 & 2 & 4 & 3 \\ 4 & -3 & 7 & 8 \\ 1 & -2 & 2 & 3 \end{bmatrix},$$

so B is row equivalent to A.

Interchanging rows 2 and 3 of B, we obtain

$$C = \begin{bmatrix} 1 & 2 & 4 & 3 \\ 1 & -2 & 2 & 3 \\ 4 & -3 & 7 & 8 \end{bmatrix},$$

so C is row equivalent to B and also row equivalent to A.

Multiplying row 1 of C by 2, we obtain

$$D = \begin{bmatrix} 2 & 4 & 8 & 6 \\ 1 & -2 & 2 & 3 \\ 4 & -3 & 7 & 8 \end{bmatrix},$$

so D is row equivalent to C. It then follows that D is row equivalent to A, since we obtained D by applying three successive elementary row operations to A. ■

It is not difficult to show (Exercise T.2) that

1. every matrix is row equivalent to itself;
2. if A is row equivalent to B, then B is row equivalent to A; and
3. if A is row equivalent to B and B is row equivalent to C, then A is row equivalent to C.

In view of 2, both statements, "A is row equivalent to B" and "B is row equivalent to A," can be replaced by "A and B are row equivalent."

THEOREM 1.5 *Every $m \times n$ matrix is row equivalent to a matrix in row echelon form.* ■

We shall illustrate the proof of the theorem by giving the steps that must be carried out on a specific matrix A to obtain a matrix in row echelon form that is row equivalent to A. We use the following example to illustrate the steps involved.

EXAMPLE 5 Let

$$A = \begin{bmatrix} 0 & 2 & 3 & -4 & 1 \\ 0 & 0 & 2 & 3 & 4 \\ 2 & 2 & -5 & 2 & 4 \\ 2 & 0 & -6 & 9 & 7 \end{bmatrix}.$$

The procedure for transforming a matrix to row echelon form follows.

Procedure

Step 1. Find the first (counting from left to right) column in A not all of whose entries are zero. This column is called the **pivotal column**.

Example

$$A = \begin{bmatrix} 0 & 2 & 3 & -4 & 1 \\ 0 & 0 & 2 & 3 & 4 \\ 2 & 2 & -5 & 2 & 4 \\ 2 & 0 & -6 & 9 & 7 \end{bmatrix}$$

↑ **Pivotal column** of A

Step 2. Identify the first (counting from top to bottom) nonzero entry in the pivotal column. This element is called the **pivot**, which we circle in A.

$$A = \begin{bmatrix} 0 & 2 & 3 & -4 & 1 \\ 0 & 0 & 2 & 3 & 4 \\ \textcircled{2} & 2 & -5 & 2 & 4 \\ 2 & 0 & -6 & 9 & 7 \end{bmatrix}$$

Pivot

Step 3. Interchange, if necessary, the first row with the row where the pivot occurs so that the pivot is now in the first row. Call the new matrix A_1.

$$A_1 = \begin{bmatrix} \textcircled{2} & 2 & -5 & 2 & 4 \\ 0 & 0 & 2 & 3 & 4 \\ 0 & 2 & 3 & -4 & 1 \\ 2 & 0 & -6 & 9 & 7 \end{bmatrix}$$

The first and third rows of A were interchanged.

Step 4. Multiply the first row of A_1 by the reciprocal of the pivot. Thus the entry in the first row and pivotal column (where the pivot was located) is now a 1. Call the new matrix A_2.

$$A_2 = \begin{bmatrix} 1 & 1 & -\frac{5}{2} & 1 & 2 \\ 0 & 0 & 2 & 3 & 4 \\ 0 & 2 & 3 & -4 & 1 \\ 2 & 0 & -6 & 9 & 7 \end{bmatrix}$$

The first row of A_1 was multiplied by $\frac{1}{2}$.

Step 5. Add appropriate multiples of the first row of A_2 to all other rows to make all entries in the pivotal column, except the entry where the pivot was located, equal to zero. Thus all entries in the pivotal column and rows 2, 3, ..., m are zero. Call the new matrix A_3.

$$A_3 = \begin{bmatrix} 1 & 1 & -\frac{5}{2} & 1 & 2 \\ 0 & 0 & 2 & 3 & 4 \\ 0 & 2 & 3 & -4 & 1 \\ 0 & -2 & -1 & 7 & 3 \end{bmatrix}$$

(-2) times the first row of A_2 was added to its fourth row.

Step 6. Identify B as the $(m-1) \times n$ submatrix of A_3 obtained by ignoring or covering the first row of A_3. Repeat Steps 1 through 5 on B.

$$\begin{matrix} 1 & 1 & -\frac{5}{2} & 1 & 2 \end{matrix}$$

$$B = \begin{bmatrix} 0 & 0 & 2 & 3 & 4 \\ 0 & \textcircled{2} & 3 & -4 & 1 \\ 0 & -2 & -1 & 7 & 3 \end{bmatrix}$$

Pivotal column of B **Pivot**

$$\begin{matrix} 1 & 1 & -\frac{5}{2} & 1 & 2 \end{matrix}$$

$$B_1 = \begin{bmatrix} 0 & \textcircled{2} & 3 & -4 & 1 \\ 0 & 0 & 2 & 3 & 4 \\ 0 & -2 & -1 & 7 & 3 \end{bmatrix}$$

The first and second rows of B were interchanged.

$$\begin{matrix} 1 & 1 & -\frac{5}{2} & 1 & 2 \end{matrix}$$

$$B_2 = \begin{bmatrix} 0 & 1 & \frac{3}{2} & -2 & \frac{1}{2} \\ 0 & 0 & 2 & 3 & 4 \\ 0 & -2 & -1 & 7 & 3 \end{bmatrix}$$

The first row of B_1 was multiplied by $\frac{1}{2}$.

$$\begin{matrix} 1 & 1 & -\frac{5}{2} & 1 & 2 \end{matrix}$$

$$B_3 = \begin{bmatrix} 0 & 1 & \frac{3}{2} & -2 & \frac{1}{2} \\ 0 & 0 & 2 & 3 & 4 \\ 0 & 0 & 2 & 3 & 4 \end{bmatrix}$$

2 times the first row of B_2 was added to its third row.

Step 7. Identify C as the $(m-2) \times n$ submatrix of B_3 obtained by ignoring or covering the first row of B_3. Repeat Steps 1 through 5 on C.

$$\begin{matrix} 1 & 1 & -\frac{5}{2} & 1 & 2 \\ 0 & 1 & \frac{3}{2} & -2 & \frac{1}{2} \end{matrix}$$

$$C = \begin{bmatrix} 0 & 0 & \textcircled{2} & 3 & 4 \\ 0 & 0 & 2 & 3 & 4 \end{bmatrix}$$

Pivotal column of C — **Pivot**

$$\begin{matrix} 1 & 1 & -\frac{5}{2} & 1 & 2 \\ 0 & 1 & \frac{3}{2} & -2 & \frac{1}{2} \end{matrix}$$

$$C_1 = C_2 = \begin{bmatrix} 0 & 0 & 1 & \frac{3}{2} & 2 \\ 0 & 0 & 2 & 3 & 4 \end{bmatrix}$$

No rows of C had to be interchanged. The first row of C was multiplied by $\frac{1}{2}$.

$$\begin{matrix} 1 & 1 & -\frac{5}{2} & 1 & 2 \\ 0 & 1 & \frac{3}{2} & -2 & \frac{1}{2} \end{matrix}$$

$$C_3 = \begin{bmatrix} 0 & 0 & 1 & \frac{3}{2} & 2 \\ 0 & 0 & 0 & 0 & 0 \end{bmatrix}$$

(-2) times the first row of C_2 was added to its second row.

Step 8. Identify D as the $(m-3) \times n$ submatrix of C_3 obtained by ignoring or covering the first row of C_3. We now try to repeat Steps 1–5 on D. However, because there is no pivotal row in D, we are finished. The matrix, denoted by H, consisting of D and the shaded rows above D is in row echelon form.

$$\begin{matrix} 1 & 1 & -\frac{5}{2} & 1 & 2 \\ 0 & 1 & \frac{3}{2} & -2 & \frac{1}{2} \\ 0 & 0 & 1 & \frac{3}{2} & 2 \end{matrix}$$

$$D = \begin{bmatrix} 0 & 0 & 0 & 0 & 0 \end{bmatrix}$$

$$H = \begin{bmatrix} 1 & 1 & -\frac{5}{2} & 1 & 2 \\ 0 & 1 & \frac{3}{2} & -2 & \frac{1}{2} \\ 0 & 0 & 1 & \frac{3}{2} & 2 \\ 0 & 0 & 0 & 0 & 0 \end{bmatrix}$$

■

Remark When doing hand computations, it is sometimes possible to avoid fractions by suitably modifying the steps in the procedure.

EXAMPLE 6 Let

$$A = \begin{bmatrix} 2 & 3 \\ 3 & 1 \end{bmatrix}.$$

To find a matrix in row echelon form that is row equivalent to A, we modify the foregoing procedure to avoid fractions and proceed as follows.

Add (-1) times row 1 to row 2 of A to obtain

$$A_1 = \begin{bmatrix} 2 & 3 \\ 1 & -2 \end{bmatrix}.$$

Interchange rows 1 and 2 of A_1 to obtain

$$A_2 = \begin{bmatrix} 1 & -2 \\ 2 & 3 \end{bmatrix}.$$

Add (-2) times row 1 to row 2 to obtain

$$A_3 = \begin{bmatrix} 1 & -2 \\ 0 & 7 \end{bmatrix},$$

a matrix that is in row echelon form and is row equivalent to A. ■

Remark There may be more than one matrix in row echelon form that is row equivalent to a given matrix A. For example, if we perform the following operation on the matrix H from Example 5—add (-1) times the second row of H to its first row—then we obtain the matrix

$$W = \begin{bmatrix} 1 & 0 & -4 & 3 & \frac{3}{2} \\ 0 & 1 & \frac{3}{2} & -2 & \frac{1}{2} \\ 0 & 0 & 1 & \frac{3}{2} & 2 \\ 0 & 0 & 0 & 0 & 0 \end{bmatrix},$$

which is in row echelon form and is row equivalent to A. Thus, H and W are matrices that are in row echelon form and each is row equivalent to A.

In general, if A is a given matrix, then a matrix in row echelon form that is row equivalent to A is called a **row echelon form of** A.

THEOREM 1.6 *Every $m \times n$ matrix is row equivalent to a unique matrix in reduced row echelon form.* ■

The matrix in Theorem 1.6 is called the **reduced row echelon form** of A.

We illustrate the proof of this theorem by also giving the steps that must be carried out on a specific matrix A to obtain a matrix in reduced row echelon form that is row equivalent to A. We omit the proof that the matrix thus obtained is unique. The following example will be used to illustrate the steps involved.

EXAMPLE 7 Find the reduced row echelon form of the matrix A of Example 5.

Solution We start with the row echelon form H of A obtained in Example 5. We add suitable multiples of each nonzero row of H to zero out all entries above a leading 1. Thus, we start by adding $\left(-\frac{3}{2}\right)$ times the third row of H to its second row:

$$J_1 = \begin{bmatrix} 1 & 1 & -\frac{5}{2} & 1 & 2 \\ 0 & 1 & 0 & -\frac{17}{4} & -\frac{5}{2} \\ 0 & 0 & 1 & \frac{3}{2} & 2 \\ 0 & 0 & 0 & 0 & 0 \end{bmatrix}.$$

Next, we add $\frac{5}{2}$ times the third row of J_1 to its first row:

$$J_2 = \begin{bmatrix} 1 & 1 & 0 & \frac{19}{4} & 7 \\ 0 & 1 & 0 & -\frac{17}{4} & -\frac{5}{2} \\ 0 & 0 & 1 & \frac{3}{2} & 2 \\ 0 & 0 & 0 & 0 & 0 \end{bmatrix}.$$

Finally, we add (-1) times the second row of J_2 to its first row:

$$K = \begin{bmatrix} 1 & 0 & 0 & 9 & \frac{19}{2} \\ 0 & 1 & 0 & -\frac{17}{4} & -\frac{5}{2} \\ 0 & 0 & 1 & \frac{3}{2} & 2 \\ 0 & 0 & 0 & 0 & 0 \end{bmatrix},$$

which is in reduced row echelon form and is row equivalent to A.

Note that in this example we started with the bottom nonzero row and worked upward to zero out all entries above leading 1s. ■

Remark The procedure given here for finding the reduced row echelon form of a given matrix is not the only one possible. As an alternate procedure, we could first zero out the entries below a leading 1 and then immediately zero out the entries above the leading 1. This procedure is not as efficient as the one described previously. In actual practice, we do not take the time to identify the matrices $A_1, A_2, \ldots, B_1, B_2, \ldots, C_1, C_2, \ldots$, and so on. We merely start with the given matrix and transform it to reduced row echelon form.

SOLVING LINEAR SYSTEMS

We now apply these results to the solution of linear systems.

THEOREM 1.7 *Let $A\mathbf{x} = \mathbf{b}$ and $C\mathbf{x} = \mathbf{d}$ be two linear systems each of m equations in n unknowns. If the augmented matrices $\left[A \mid \mathbf{b}\right]$ and $\left[C \mid \mathbf{d}\right]$ of these systems are row equivalent, then both linear systems have the same solutions.*

Proof This follows from the definition of row equivalence and from the fact that the three elementary row operations on the augmented matrix turn out to be the three manipulations on a linear system, discussed in Section 1.1, yielding a linear system having the same solutions as the given system. We also note that if one system has no solution, then the other system has no solution. ■

COROLLARY 1.1 *If A and C are row equivalent $m \times n$ matrices, then the linear system $A\mathbf{x} = \mathbf{0}$ and $C\mathbf{x} = \mathbf{0}$ have exactly the same solutions.*

Proof Exercise T.3. ■

The results established thus far provide us with two methods for solving linear systems. The key idea is to start with the linear system $A\mathbf{x} = \mathbf{b}$, then obtain a partitioned matrix $\left[C \mid \mathbf{d}\right]$ in either reduced row echelon form or row echelon form that is row equivalent to the augmented matrix $\left[A \mid \mathbf{b}\right]$. Now $\left[C \mid \mathbf{d}\right]$ represents the linear system $C\mathbf{x} = \mathbf{d}$, which is easier to solve because of the simpler structure of $\left[C \mid \mathbf{d}\right]$, and the set of all solutions to this system gives precisely the set of all solutions to the given system $A\mathbf{x} = \mathbf{b}$. The method where $\left[C \mid \mathbf{d}\right]$ is in reduced row echelon form is called **Gauss***–

*Carl Friedrich Gauss (1777–1855) was born into a poor working-class family in Brunswick, Germany, and died in Göttingen, Germany, the most famous mathematician in the world. He was a child prodigy with a genius that did not impress his father, who called him a "stargazer." However, his teachers were impressed enough to arrange for the Duke of Brunswick to provide a scholarship for Gauss at the local secondary school. As a teenager there, he made original discoveries in number theory and began to speculate about non-Euclidean geometry. His scientific

Jordan[**] **reduction**; the method where $\left[C \mid \mathbf{d}\right]$ is in row echelon form is called **Gaussian elimination**. Strictly speaking, the alternate Gauss–Jordan reduction method described in the Remark above is not as efficient as that used in Examples 5 and 6. In actual practice, neither Gauss–Jordan reduction nor Gaussian elimination is used as much as the method involving the LU-factorization of A that is presented in Section 1.8. However, Gauss–Jordan reduction and Gaussian elimination are fine for small problems, and we use the former frequently in this book.

> The Gauss–Jordan reduction procedure for solving the linear system $A\mathbf{x} = \mathbf{b}$ is as follows.
>
> ***Step 1.*** Form the augmented matrix $\left[A \mid \mathbf{b}\right]$.
>
> ***Step 2.*** Obtain the reduced row echelon form $\left[C \mid \mathbf{d}\right]$ of the augmented matrix $\left[A \mid \mathbf{b}\right]$ by using elementary row operations.
>
> ***Step 3.*** For each nonzero row of the matrix $\left[C \mid \mathbf{d}\right]$, solve the corresponding equation for the unknown associated with the leading one in that row. The rows consisting entirely of zeros can be ignored, because the corresponding equation will be satisfied for any values of the unknowns.

> The Gaussian elimination procedure for solving the linear system $A\mathbf{x} = \mathbf{b}$ is as follows.
>
> ***Step 1.*** Form the augmented matrix $\left[A \mid \mathbf{b}\right]$.
>
> ***Step 2.*** Obtain a row echelon form $\left[C \mid \mathbf{d}\right]$ of the augmented matrix $\left[A \mid \mathbf{b}\right]$ by using elementary row operations.
>
> ***Step 3.*** Solve the linear system corresponding to $\left[C \mid \mathbf{d}\right]$ by **back substitution** (illustrated in Example 11). The rows consisting entirely of zeros can be ignored, because the corresponding equation will be satisfied for any values of the unknowns.

The following examples illustrate the Gauss–Jordan reduction procedure.

---

publications include important contributions in number theory, mathematical astronomy, mathematical geography, statistics, differential geometry, and magnetism. His diaries and private notes contain many other discoveries that he never published.

An austere, conservative man who had few friends and whose private life was generally unhappy, he was very concerned that proper credit be given for scientific discoveries. When he relied on the results of others, he was careful to acknowledge them; and when others independently discovered results in his private notes, he was quick to claim priority.

In his research he used a method of calculation that later generations generalized to row reduction of matrices and named in his honor although the method was used in China almost 2000 years earlier.

**Wilhelm Jordan (1842–1899) was born in southern Germany. He attended college in Stuttgart and in 1868 became full professor of geodesy at the technical college in Karlsruhe, Germany. He participated in surveying several regions of Germany. Jordan was a prolific writer whose major work, *Handbuch der Vermessungskunde* (*Handbook of Geodesy*), was translated into French, Italian, and Russian. He was considered a superb writer and an excellent teacher. Unfortunately, the Gauss–Jordan reduction method has been widely attributed to Camille Jordan (1838–1922), a well-known French mathematician. Moreover, it seems that the method was also discovered independently at the same time by B. I. Clasen, a priest who lived in Luxembourg. This biographical sketch is based on an excellent article: S. C. Althoen and R. McLaughlin, "Gauss–Jordan reduction: A Brief History," *MAA Monthly*, 94 (1987), 130–142.

EXAMPLE 8 Solve the linear system

$$\begin{aligned} x + 2y + 3z &= 9 \\ 2x - y + z &= 8 \\ 3x \phantom{{}+2y} - z &= 3 \end{aligned} \tag{1}$$

by Gauss–Jordan reduction.

Solution ***Step 1.*** The augmented matrix of this linear system is

$$\left[\begin{array}{rrr|r} 1 & 2 & 3 & 9 \\ 2 & -1 & 1 & 8 \\ 3 & 0 & -1 & 3 \end{array}\right].$$

Step 2. We now transform the matrix in Step 1 to reduced row echelon form as follows:

$$\left[\begin{array}{rrr|r} 1 & 2 & 3 & 9 \\ 2 & -1 & 1 & 8 \\ 3 & 0 & -1 & 3 \end{array}\right]$$

$$\left[\begin{array}{rrr|r} 1 & 2 & 3 & 9 \\ 0 & -5 & -5 & -10 \\ 0 & -6 & -10 & -24 \end{array}\right]$$ (-2) times the first row was added to its second row. (-3) times the first row was added to its third row.

$$\left[\begin{array}{rrr|r} 1 & 2 & 3 & 9 \\ 0 & 1 & 1 & 2 \\ 0 & -6 & -10 & -24 \end{array}\right]$$ The second row was multiplied by $\left(-\frac{1}{5}\right)$.

$$\left[\begin{array}{rrr|r} 1 & 2 & 3 & 9 \\ 0 & 1 & 1 & 2 \\ 0 & 0 & -4 & -12 \end{array}\right]$$ 6 times the second row was added to its third row.

$$\left[\begin{array}{rrr|r} 1 & 2 & 3 & 9 \\ 0 & 1 & 1 & 2 \\ 0 & 0 & 1 & 3 \end{array}\right]$$ The third row was multiplied by $\left(-\frac{1}{4}\right)$.

$$\left[\begin{array}{rrr|r} 1 & 2 & 3 & 9 \\ 0 & 1 & 0 & -1 \\ 0 & 0 & 1 & 3 \end{array}\right]$$ (-1) times the third row was added to its first row.

$$\left[\begin{array}{rrr|r} 1 & 2 & 0 & 0 \\ 0 & 1 & 0 & -1 \\ 0 & 0 & 1 & 3 \end{array}\right]$$ (-3) times the third row was added to its first row.

$$\left[\begin{array}{rrr|r} 1 & 0 & 0 & 2 \\ 0 & 1 & 0 & -1 \\ 0 & 0 & 1 & 3 \end{array}\right]$$ (-2) times the second row was added to its first row.

Thus, the augmented matrix is row equivalent to the matrix

$$\left[\begin{array}{rrr|r} 1 & 0 & 0 & 2 \\ 0 & 1 & 0 & -1 \\ 0 & 0 & 1 & 3 \end{array}\right] \tag{2}$$

in reduced row echelon form.

Step 3. The linear system represented by (2) is

$$\begin{aligned} x \qquad &= 2 \\ y \quad &= -1 \\ z &= 3 \end{aligned}$$

so that the unique solution to the given linear system (1) is

$$\begin{aligned} x &= 2 \\ y &= -1 \\ z &= 3. \end{aligned}$$

■

EXAMPLE 9 Solve the linear system

$$\begin{aligned} x + y + 2z - 5w &= 3 \\ 2x + 5y - z - 9w &= -3 \\ 2x + y - z + 3w &= -11 \\ x - 3y + 2z + 7w &= -5 \end{aligned} \tag{3}$$

by Gauss–Jordan reduction.

Solution ***Step 1.*** The augmented matrix of this linear system is

$$\left[\begin{array}{rrrr|r} 1 & 1 & 2 & -5 & 3 \\ 2 & 5 & -1 & -9 & -3 \\ 2 & 1 & -1 & 3 & -11 \\ 1 & -3 & 2 & 7 & -5 \end{array}\right].$$

Step 2. The augmented matrix is row equivalent to the matrix (verify)

$$\left[\begin{array}{rrrr|r} 1 & 0 & 0 & 2 & -5 \\ 0 & 1 & 0 & -3 & 2 \\ 0 & 0 & 1 & -2 & 3 \\ 0 & 0 & 0 & 0 & 0 \end{array}\right], \tag{4}$$

which is in reduced row echelon form.

Step 3. The linear system represented by (4) is

$$\begin{aligned} x \qquad\quad + 2w &= -5 \\ y \quad - 3w &= 2 \\ z - 2w &= 3. \end{aligned}$$

The row in (4) consisting entirely of zeros has been ignored.

Solving each equation for the unknown that corresponds to the leading entry in each row of (4), we obtain

$$\begin{aligned} x &= -5 - 2w \\ y &= 2 + 3w \\ z &= 3 + 2w. \end{aligned}$$

Thus, if we let $w = r$, any real number, then a solution to the linear system (3) is

$$\begin{aligned} x &= -5 - 2r \\ y &= 2 + 3r \\ z &= 3 + 2r \\ w &= r. \end{aligned} \tag{5}$$

Because r can be assigned any real number in (5), the given linear system (3) has infinitely many solutions. ■

EXAMPLE 10 Solve the linear system

$$\begin{aligned} x_1 + 2x_2 - 3x_4 + x_5 &= 2 \\ x_1 + 2x_2 + x_3 - 3x_4 + x_5 + 2x_6 &= 3 \\ x_1 + 2x_2 - 3x_4 + 2x_5 + x_6 &= 4 \\ 3x_1 + 6x_2 + x_3 - 9x_4 + 4x_5 + 3x_6 &= 9 \end{aligned} \tag{6}$$

by Gauss–Jordan reduction.

Solution ***Step 1.*** The augmented matrix of this linear system is

$$\left[\begin{array}{cccccc:c} 1 & 2 & 0 & -3 & 1 & 0 & 2 \\ 1 & 2 & 1 & -3 & 1 & 2 & 3 \\ 1 & 2 & 0 & -3 & 2 & 1 & 4 \\ 3 & 6 & 1 & -9 & 4 & 3 & 9 \end{array}\right].$$

Step 2. The augmented matrix is row equivalent to the matrix (verify)

$$\left[\begin{array}{cccccc:c} 1 & 2 & 0 & -3 & 0 & -1 & 0 \\ 0 & 0 & 1 & 0 & 0 & 2 & 1 \\ 0 & 0 & 0 & 0 & 1 & 1 & 2 \\ 0 & 0 & 0 & 0 & 0 & 0 & 0 \end{array}\right]. \tag{7}$$

Step 3. The linear system represented by (7) is

$$\begin{aligned} x_1 + 2x_2 - 3x_4 - x_6 &= 0 \\ x_3 + 2x_6 &= 1 \\ x_5 + x_6 &= 2. \end{aligned}$$

Solving each equation for the unknown that corresponds to the leading entry in each row of (7), we obtain

$$\begin{aligned} x_1 &= x_6 + 3x_4 - 2x_2 \\ x_3 &= 1 - 2x_6 \\ x_5 &= 2 - x_6. \end{aligned}$$

Letting $x_6 = r$, $x_4 = s$, and $x_2 = t$, a solution to the linear system (6) is

$$\begin{aligned} x_1 &= r + 3s - 2t \\ x_2 &= t \\ x_3 &= 1 - 2r \\ x_4 &= s \\ x_5 &= 2 - r \\ x_6 &= r, \end{aligned} \tag{8}$$

where r, s, and t are any real numbers. Thus (8) is the solution to the given linear system (6). Since r, s, and t can be assigned any real numbers, the given linear system (6) has infinitely many solutions. ■

The following example illustrates the Gaussian elimination procedure and back substitution.

EXAMPLE 11 Solve the linear system given in Example 8 by Gaussian elimination.

Solution *Step 1.* The augmented matrix of this linear system is

$$\left[\begin{array}{ccc|c} 1 & 2 & 3 & 9 \\ 2 & -1 & 1 & 8 \\ 3 & 0 & -1 & 3 \end{array}\right].$$

Step 2. A row echelon form of the augmented matrix is (verify)

$$\left[\begin{array}{ccc|c} 1 & 2 & 3 & 9 \\ 0 & 1 & 1 & 2 \\ 0 & 0 & 1 & 3 \end{array}\right].$$

This augmented matrix corresponds to the equivalent linear system

$$\begin{aligned} x + 2y + 3z &= 9 \\ y + z &= 2 \\ z &= 3. \end{aligned}$$

Step 3. The process of back substitution starts with the equation $z = 3$. We next substitute this value of z in the preceding equation $y + z = 2$ and solve for y, obtaining $y = 2 - z = 2 - 3 = -1$. Finally, we substitute the values just obtained for y and z in the first equation $x + 2y + 3z = 9$ and solve for x, obtaining $x = 9 - 2y - 3z = 9 + 2 - 9 = 2$. Thus, the solution is $x = 2$, $y = -1$, and $z = 3$. ■

EXAMPLE 12 Solve the linear system

$$\begin{aligned} x + 2y + 3z + 4w &= 5 \\ x + 3y + 5z + 7w &= 11 \\ x - z - 2w &= -6 \end{aligned} \tag{9}$$

by Gauss–Jordan reduction.

Solution *Step 1.* The augmented matrix of this linear system is

$$\left[\begin{array}{cccc|c} 1 & 2 & 3 & 4 & 5 \\ 1 & 3 & 5 & 7 & 11 \\ 1 & 0 & -1 & -2 & -6 \end{array}\right].$$

Step 2. The augmented matrix is row equivalent to the matrix (verify)

$$\left[\begin{array}{cccc|c} 1 & 0 & -1 & -2 & 0 \\ 0 & 1 & 2 & 3 & 0 \\ 0 & 0 & 0 & 0 & 1 \end{array}\right]. \tag{10}$$

Step 3. The last equation of the linear system represented by (10) is

$$0x + 0y + 0z + 0w = 1,$$

which has no solution for any x, y, z, and w. Consequently, the given linear system (9) has no solution. ■

The last example is characteristic of the way in which we recognize that a linear system has no solution. That is, a linear system $A\mathbf{x} = \mathbf{b}$ in n unknowns has no solution if and only if its augmented matrix is row equivalent to a matrix in reduced row echelon form or row echelon form, which has a row whose first n elements are zero and whose $(n + 1)$st element is 1 (Exercise T.4).

The linear systems of Examples 8, 9, and 10 each had at least one solution, while the system in Example 12 had no solution. Linear systems with at least one solution are called **consistent**, and linear systems with no solutions are called **inconsistent**. Every inconsistent linear system when in reduced row echelon form or row echelon form results in the situation illustrated in Example 12.

Remarks

1. As we perform elementary row operations, we may encounter a row of the augmented matrix being transformed to reduced row echelon form whose first n entries are zero and whose $(n + 1)$st entry is not zero. In this case, we can stop our computations and conclude that the given linear system is inconsistent.
2. Sometimes we need to solve k linear systems

$$A\mathbf{x} = \mathbf{b}_1, \quad A\mathbf{x} = \mathbf{b}_2, \ldots, \quad A\mathbf{x} = \mathbf{b}_k,$$

all having the same $m \times n$ coefficient matrix A. Instead of solving each linear system separately, we proceed as follows. Form the $m \times (n + k)$ augmented matrix

$$\left[A \mid \mathbf{b}_1 \quad \mathbf{b}_2 \quad \cdots \quad \mathbf{b}_k\right].$$

The reduced row echelon form

$$\left[C \mid \mathbf{d}_1 \quad \mathbf{d}_2 \quad \cdots \quad \mathbf{d}_k\right]$$

of this matrix corresponds to the linear systems

$$C\mathbf{x} = \mathbf{d}_1, \quad C\mathbf{x} = \mathbf{d}_2, \ldots, \quad C\mathbf{x} = \mathbf{d}_k,$$

which have the same solutions as the corresponding given linear systems. This approach will be useful in Section 6.7. Exercises 35 and 36 ask you to explore this technique.

HOMOGENEOUS SYSTEMS

A linear system of the form

$$\begin{aligned}
a_{11}x_1 + a_{12}x_2 + \cdots + a_{1n}x_n &= 0 \\
a_{21}x_1 + a_{22}x_2 + \cdots + a_{2n}x_n &= 0 \\
\vdots \qquad \vdots \qquad\quad\; \vdots \quad &\;\; \vdots \\
a_{m1}x_1 + a_{m2}x_2 + \cdots + a_{mn}x_n &= 0
\end{aligned} \tag{11}$$

is called a **homogeneous system**. We can also write (11) in matrix form as

$$A\mathbf{x} = \mathbf{0}. \tag{12}$$

The solution

$$x_1 = x_2 = \cdots = x_n = 0$$

to the homogeneous system (12) is called the **trivial solution**. A solution $x_1, x_2, \ldots, x_n$ to a homogeneous system in which not all the x_i are zero is called a **nontrivial solution**. We see that a homogeneous system is always consistent, since it always has the trivial solution.

EXAMPLE 13 Consider the homogeneous system

$$\begin{aligned} x + 2y + 3z &= 0 \\ -x + 3y + 2z &= 0 \\ 2x + y - 2z &= 0. \end{aligned} \tag{13}$$

The augmented matrix of this system,

$$\left[\begin{array}{rrr|r} 1 & 2 & 3 & 0 \\ -1 & 3 & 2 & 0 \\ 2 & 1 & -2 & 0 \end{array}\right],$$

is row equivalent to (verify)

$$\left[\begin{array}{rrr|r} 1 & 0 & 0 & 0 \\ 0 & 1 & 0 & 0 \\ 0 & 0 & 1 & 0 \end{array}\right],$$

which is in reduced row echelon form. Hence the solution to (13) is

$$x = y = z = 0,$$

which means that the given homogeneous system (13) has only the trivial solution. ■

EXAMPLE 14 Consider the homogeneous system

$$\begin{aligned} x + y + z + w &= 0 \\ x \qquad\quad + w &= 0 \\ x + 2y + z \qquad &= 0. \end{aligned} \tag{14}$$

The augmented matrix of this system,

$$\left[\begin{array}{rrrr|r} 1 & 1 & 1 & 1 & 0 \\ 1 & 0 & 0 & 1 & 0 \\ 1 & 2 & 1 & 0 & 0 \end{array}\right],$$

is row equivalent to (verify)

$$\left[\begin{array}{rrrr|r} 1 & 0 & 0 & 1 & 0 \\ 0 & 1 & 0 & -1 & 0 \\ 0 & 0 & 1 & 1 & 0 \end{array}\right],$$

which is in reduced row echelon form. Hence the solution to (14) is

$$\begin{aligned} x &= -r \\ y &= r \\ z &= -r \\ w &= r, \end{aligned}$$

where r is any real number. For example, if we let $r = 2$, then

$$x = -2, \quad y = 2, \quad z = -2, \quad w = 2$$

is a nontrivial solution to this homogeneous system. That is,

$$\begin{bmatrix} 1 & 1 & 1 & 1 \\ 1 & 0 & 0 & 1 \\ 1 & 2 & 1 & 0 \end{bmatrix} \begin{bmatrix} -2 \\ 2 \\ -2 \\ 2 \end{bmatrix} = \begin{bmatrix} 0 \\ 0 \\ 0 \end{bmatrix}.$$

(Verify by computing the matrix product on the left side.) Hence, this linear system has infinitely many solutions. ■

Example 14 shows that a homogeneous system may have a nontrivial solution. The following theorem tells of one case when this occurs.

THEOREM 1.8 *A homogeneous system of m equations in n unknowns always has a nontrivial solution if $m < n$, that is, if the number of unknowns exceeds the number of equations.*

Proof Let C be the reduced row echelon form of A. Then the homogeneous systems $A\mathbf{x} = \mathbf{0}$ and $C\mathbf{x} = \mathbf{0}$ are equivalent. If we let r be the number of nonzero rows of C, then $r \leq m$. If $m < n$, we conclude that $r < n$. We are then solving r equations in n unknowns and can solve for r unknowns in terms of the remaining $n - r$ unknowns, the latter being free to take on any real numbers. Thus, by letting one of these $n - r$ unknowns be nonzero, we obtain a nontrivial solution to $C\mathbf{x} = \mathbf{0}$ and thus to $A\mathbf{x} = \mathbf{0}$. ■

We shall also use Theorem 1.8 in the following equivalent form: If A is $m \times n$ and $A\mathbf{x} = \mathbf{0}$ has only the trivial solution, then $m \geq n$.

The following result is important in the study of differential equations. (See Section 9.2.)

Let $A\mathbf{x} = \mathbf{b}$, $\mathbf{b} \neq \mathbf{0}$, be a consistent linear system. If $\mathbf{x}_p$ is a particular solution to the given nonhomogeneous system and $\mathbf{x}_h$ is a solution to the associated homogeneous system $A\mathbf{x} = \mathbf{0}$, then $\mathbf{x}_p + \mathbf{x}_h$ is a solution to the given system $A\mathbf{x} = \mathbf{b}$. Moreover, every solution $\mathbf{x}$ to the nonhomogeneous linear system $A\mathbf{x} = \mathbf{b}$ can be written as $\mathbf{x}_p + \mathbf{x}_h$, where $\mathbf{x}_p$ is a particular solution to the given nonhomogeneous system and $\mathbf{x}_h$ is a solution to the associated homogeneous system $A\mathbf{x} = \mathbf{0}$. For a proof, see Exercise T.13.

EXAMPLE 15 Consider the linear system given in Example 9. A solution to this linear system was given by

$$\begin{aligned} x &= -5 - 2r \\ y &= 2 + 3r \\ z &= 3 + 2r \\ w &= r, \end{aligned}$$

where r is any real number. If we let

$$\mathbf{x} = \begin{bmatrix} x \\ y \\ z \\ w \end{bmatrix},$$

then the solution can be expressed as

$$\mathbf{x} = \begin{bmatrix} -5 - 2r \\ 2 + 3r \\ 3 + 2r \\ r \end{bmatrix} = \begin{bmatrix} -5 \\ 2 \\ 3 \\ 0 \end{bmatrix} + \begin{bmatrix} -2r \\ 3r \\ 2r \\ r \end{bmatrix}.$$

Let

$$\mathbf{x}_p = \begin{bmatrix} -5 \\ 2 \\ 3 \\ 0 \end{bmatrix} \quad \text{and} \quad \mathbf{x}_h = \begin{bmatrix} -2r \\ 3r \\ 2r \\ r \end{bmatrix}.$$

Then $\mathbf{x} = \mathbf{x}_p + \mathbf{x}_h$. Moreover, $\mathbf{x}_p$ is a particular solution to the given system and $\mathbf{x}_h$ is a solution to the associated homogeneous system [verify that $A\mathbf{x}_p = \mathbf{b}$ and $A\mathbf{x}_h = \mathbf{0}$, where A is the coefficient matrix in Example 9 and $\mathbf{b}$ is the right side of Equation (3)]. ■

Remark Homogeneous systems are special and they will play a key role in later chapters in this book.

POLYNOMIAL INTERPOLATION

Suppose we are given the n distinct points $(x_1, y_1), (x_2, y_2), \ldots, (x_n, y_n)$. Can we find a polynomial of degree $n - 1$ or less that "interpolates" the data, that is, passes through the n points? Thus, the polynomial we seek has the form

$$y = a_{n-1}x^{n-1} + a_{n-2}x^{n-2} + \cdots + a_1x + a_0.$$

The n given points can be used to obtain an $n \times n$ linear system whose unknowns are $a_0, a_1, \ldots, a_{n-1}$. It can be shown that this linear system has a unique solution. Thus, there is a unique interpolating polynomial.

We consider the case where $n = 3$ in detail. Here we are given the points $(x_1, y_1), (x_2, y_2), (x_3, y_3)$, where $x_1 \neq x_2$, $x_1 \neq x_3$, and $x_2 \neq x_3$, and seek the polynomial

$$y = a_2x^2 + a_1x + a_0. \tag{15}$$

Substituting the given points in (15), we obtain the linear system

$$\begin{aligned} a_2x_1^2 + a_1x_1 + a_0 &= y_1 \\ a_2x_2^2 + a_1x_2 + a_0 &= y_2 \\ a_2x_3^2 + a_1x_3 + a_0 &= y_3. \end{aligned} \tag{16}$$

We show in Section 3.2 that the linear system (16) has a unique solution. Thus, there is a unique interpolating quadratic polynomial. In general, there is a unique interpolating polynomial of degree $n - 1$ passing through n given points.

EXAMPLE 16 Find the quadratic polynomial that interpolates the points $(1, 3)$, $(2, 4)$, $(3, 7)$.

Solution Setting up the linear system (16) we have

$$\begin{aligned} a_2 + a_1 + a_0 &= 3 \\ 4a_2 + 2a_1 + a_0 &= 4 \\ 9a_2 + 3a_1 + a_0 &= 7 \end{aligned}$$

whose solution is (verify)

$$a_2 = 1, \qquad a_1 = -2, \qquad a_0 = 4.$$

Hence, the quadratic interpolating polynomial is

$$y = x^2 - 2x + 4.$$

Its graph, shown in Figure 1.17, passes through the three given points. ■

Section 2.4, Electrical Circuits, Section 2.5, Markov Chains, and Chapter 11, Linear Programming, which can be studied at this time, use material from this section.

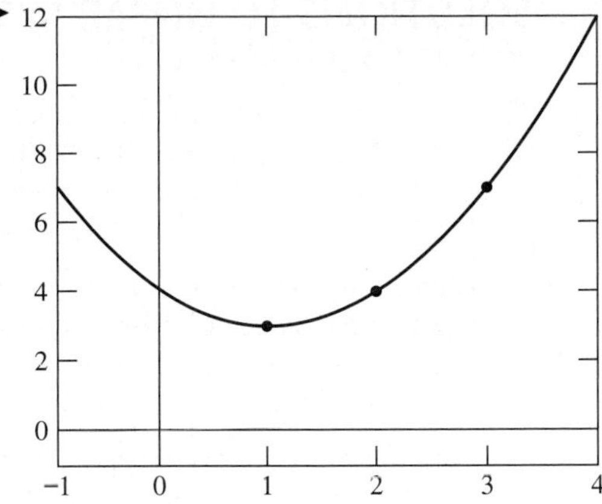

Figure 1.17 ►

TEMPERATURE DISTRIBUTION

A simple model for estimating the temperature distribution on a square plate gives rise to a linear system of equations. To construct the appropriate linear system, we use the following information. The square plate is perfectly insulated on its top and bottom so that the only heat flow is through the plate itself. The four edges are held at various temperatures. To estimate the temperature at an interior point on the plate, we use the rule that it is the average of the temperatures at its four compass point neighbors, to the west, north, east, and south.

EXAMPLE 17 Estimate the temperatures T_i, $i = 1, 2, 3, 4$, at the four equispaced interior points on the plate shown in Figure 1.18.

Solution We now construct the linear system to estimate the temperatures. The points at which we need the temperatures of the plate for this model are indicated in Figure 1.18 by dots. Using our averaging rule, we obtain the equations

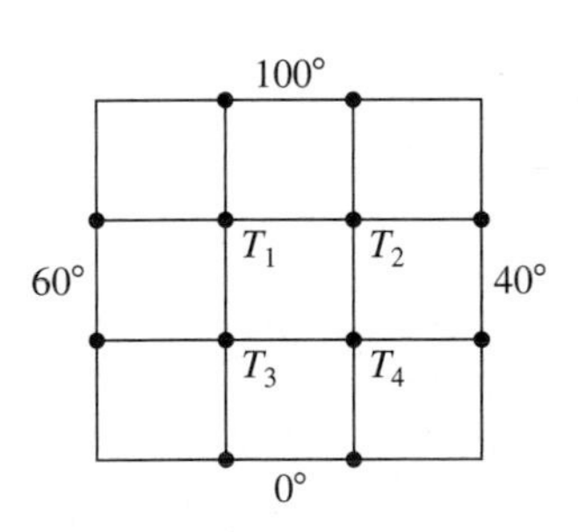

Figure 1.18 ▲

$$
\begin{aligned}
T_1 &= \frac{60 + 100 + T_2 + T_3}{4} & \text{or} \quad & 4T_1 - T_2 - T_3 = 160 \\
T_2 &= \frac{T_1 + 100 + 40 + T_4}{4} & \text{or} \quad & -T_1 + 4T_2 - T_4 = 140 \\
T_3 &= \frac{60 + T_1 + T_4 + 0}{4} & \text{or} \quad & -T_1 + 4T_3 - T_4 = 60 \\
T_4 &= \frac{T_3 + T_2 + 40 + 0}{4} & \text{or} \quad & -T_2 - T_3 + 4T_4 = 40.
\end{aligned}
$$

The augmented matrix for this linear system is (verify)

$$
\left[A \,\vdots\, \mathbf{b}\right] = \left[\begin{array}{rrrr:r} 4 & -1 & -1 & 0 & 160 \\ -1 & 4 & 0 & -1 & 140 \\ -1 & 0 & 4 & -1 & 60 \\ 0 & -1 & -1 & 4 & 40 \end{array}\right].
$$

Using Gaussian elimination or Gauss–Jordan reduction, we obtain the unique solution (verify)

$$
T_1 = 65^\circ, \quad T_2 = 60^\circ, \quad T_3 = 40^\circ, \quad \text{and} \quad T_4 = 35^\circ. \qquad \blacksquare
$$

SOLUTIONS TO LINEAR SYSTEMS INVOLVING BIT MATRICES (OPTIONAL)

The definitions and theorems developed in this section are all valid for systems with bit matrices. Examples 18–21 illustrate the concepts of this section for such systems. We shall refer to such systems as **bit linear systems**.

For bit linear systems the following interpretations apply.

- Elementary row operations on bit matrices are either an interchange of rows or the addition of one row to another row. This follows from the arithmetic properties of binary arithmetic and linear combinations of bit matrices as discussed previously.
- If as a result of the solution process for a consistent linear system an unknown can be assigned any value, we can assign it either a 0 or a 1. Such systems will have more than one solution, but the total number of possible solutions will depend on the number of unknowns that can be assigned in this way. That is, we cannot say such systems have infinitely many solutions.

EXAMPLE 18 Solve the bit linear system

$$\begin{aligned} x + y &= 1 \\ y &= 1 \end{aligned} \tag{17}$$

by Gauss–Jordan reduction.

Solution ***Step 1.*** The augmented matrix of this linear system is

$$\left[\begin{array}{cc:c} 1 & 1 & 1 \\ 0 & 1 & 1 \end{array}\right].$$

Step 2. We next compute the reduced row echelon form of the matrix in Step 1 as follows:

$$\left[\begin{array}{cc:c} 1 & 1 & 1 \\ 0 & 1 & 1 \end{array}\right]$$

$$\left[\begin{array}{cc:c} 1 & 0 & 0 \\ 0 & 1 & 1 \end{array}\right] \quad \text{The second row was added to the first row.}$$

Thus the reduced row echelon form of the augmented matrix is the matrix

$$\left[\begin{array}{cc:c} 1 & 0 & 0 \\ 0 & 1 & 1 \end{array}\right] \tag{18}$$

Step 3. The linear system represented by (18) is

$$\begin{aligned} x \quad &= 0 \\ y &= 1 \end{aligned}$$

so that the unique solution to the given linear system (17) is $x = 0$, $y = 1$. ■

EXAMPLE 19 Solve the bit linear system

$$\begin{aligned} x \quad + z &= 0 \\ y \quad &= 1 \\ x + y + z &= 1 \end{aligned} \tag{19}$$

by Gauss–Jordan reduction.

Solution *Step 1.* The augmented matrix of this linear system is

$$\left[\begin{array}{ccc:c} 1 & 0 & 1 & 0 \\ 0 & 1 & 0 & 1 \\ 1 & 1 & 1 & 1 \end{array}\right].$$

Step 2. We now compute the reduced row echelon form of the matrix in Step 1 as follows:

$$\left[\begin{array}{ccc:c} 1 & 0 & 1 & 0 \\ 0 & 1 & 0 & 1 \\ 1 & 1 & 1 & 1 \end{array}\right]$$

$$\left[\begin{array}{ccc:c} 1 & 0 & 1 & 0 \\ 0 & 1 & 0 & 1 \\ 0 & 1 & 0 & 1 \end{array}\right] \quad \text{The first row was added to the third row.}$$

$$\left[\begin{array}{ccc:c} 1 & 0 & 1 & 0 \\ 0 & 1 & 0 & 1 \\ 0 & 0 & 0 & 0 \end{array}\right] \quad \text{The second row was added to the third row.}$$

Thus the augmented matrix is row equivalent to the matrix

$$\left[\begin{array}{ccc:c} 1 & 0 & 1 & 0 \\ 0 & 1 & 0 & 1 \\ 0 & 0 & 0 & 0 \end{array}\right] \tag{20}$$

Step 3. The linear system represented by (20) is

$$\begin{aligned} x \quad + z &= 0 \\ y \quad &= 1. \end{aligned}$$

Solving each equation for the unknown that corresponds to the leading entry in each row of (20), we obtain

$$\begin{aligned} x &= \text{“}-z\text{” (the additive inverse of } z) \\ y &= 1. \end{aligned}$$

Thus, if we let $z = b$, either bit 0 or 1, then "$-z$" is likewise either 0 or 1. Hence the set of solutions to the bit linear system (19) is

$$\begin{aligned} x &= b \\ y &= 1 \\ z &= b. \end{aligned} \tag{21}$$

Because b is either 0 or 1 in (20), the given linear system (19) has two solutions, either

$$\begin{bmatrix} 0 \\ 1 \\ 0 \end{bmatrix} \quad \text{or} \quad \begin{bmatrix} 1 \\ 1 \\ 1 \end{bmatrix}.$$

■

EXAMPLE 20 Solve the bit linear system

$$\begin{aligned} x + y \quad &= 0 \\ x + y + z &= 1 \\ x + y \quad &= 1 \end{aligned} \tag{22}$$

by Gaussian elimination.

Solution *Step 1.* The augmented matrix of this linear system is

$$\left[\begin{array}{ccc|c} 1 & 1 & 0 & 0 \\ 1 & 1 & 1 & 1 \\ 1 & 1 & 0 & 1 \end{array}\right].$$

Step 2. A row echelon form of the augmented matrix is (verify)

$$\left[\begin{array}{ccc|c} 1 & 1 & 0 & 0 \\ 0 & 0 & 1 & 1 \\ 0 & 0 & 0 & 1 \end{array}\right]. \tag{23}$$

Step 3. The linear system represented by (23) is

$$\begin{aligned} x + y \quad &= 0 \\ z &= 1 \\ 0 &= 1, \end{aligned}$$

which is clearly inconsistent. Hence (22) has no solution. ■

EXAMPLE 21 Solve the bit homogeneous system whose augmented matrix is

$$\left[\begin{array}{cccc|c} 1 & 0 & 1 & 1 & 0 \\ 1 & 1 & 0 & 0 & 0 \\ 0 & 1 & 1 & 1 & 0 \end{array}\right] \tag{24}$$

by Gauss–Jordan reduction.

Solution *Step 1.* The reduced row echelon form of the augmented matrix is (verify)

$$\left[\begin{array}{cccc|c} 1 & 0 & 1 & 1 & 0 \\ 0 & 1 & 1 & 1 & 0 \\ 0 & 0 & 0 & 0 & 0 \end{array}\right] \tag{25}$$

Step 2. The linear system represented by (25) is

$$\begin{aligned} x \quad + z + w &= 0 \\ y + z + w &= 0. \end{aligned}$$

Solving each equation for the unknown that corresponds to the leading entry in each row of (20), we obtain

$$\begin{aligned} x \quad &= \text{“}-z\text{”} + \text{“}-w\text{”} \quad \text{(the additive inverse of } z \text{ and } w\text{)} \\ y &= \text{“}-z\text{”} + \text{“}-w\text{”} \quad \text{(the additive inverse of } z \text{ and } w\text{).} \end{aligned}$$

Thus, if we let $z = b$, either bit 0 or 1, then "$-z$" is likewise either 0 or 1. Similarly, $w = b$ implies w is either 0 or 1. Hence the set of solutions to the bit linear system (25) is

$$\begin{aligned} x &= b_z + b_w \\ y &= b_z + b_w, \end{aligned} \tag{26}$$

where b_z is the bit chosen for z and b_w is that chosen for w. Since there are two choices for both z and w, there are four possible solutions:

$$\begin{bmatrix} 0 \\ 0 \\ 0 \\ 0 \end{bmatrix}, \quad \begin{bmatrix} 1 \\ 1 \\ 0 \\ 1 \end{bmatrix}, \quad \begin{bmatrix} 1 \\ 1 \\ 1 \\ 0 \end{bmatrix}, \quad \text{or} \quad \begin{bmatrix} 0 \\ 0 \\ 1 \\ 1 \end{bmatrix}.$$

■

Preview of Applications

Electrical Circuits (Section 2.4)

An electrical circuit is a closed connection of batteries, resistors (such as lightbulbs) and wires connecting these. The batteries and resistors are denoted on paper by

Batteries Resistors

An example of an electrical circuit is shown in Figure 1.19.

Figure 1.19 ▶

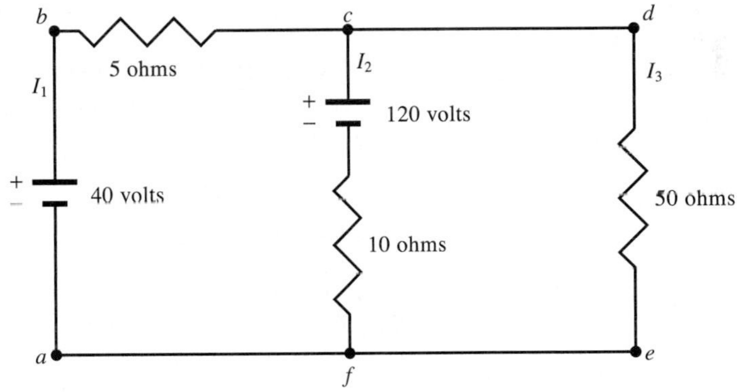

For this circuit, the unknown currents I_1, I_2, and I_3 (in units of amps) are to be determined from the values of resistance (in units of ohms) across each resistor and electrostatic potential (in units of volts) across each battery (as shown in Figure 1.19). Applying two basic laws of physics, to be discussed in Section 2.4, we find that I_1, I_2, and I_3 must satisfy the linear system

$$\begin{bmatrix} 1 & 1 & -1 \\ 1 & -2 & 0 \\ 0 & 1 & 5 \end{bmatrix} \begin{bmatrix} I_1 \\ I_2 \\ I_3 \end{bmatrix} = \begin{bmatrix} 0 \\ -16 \\ 12 \end{bmatrix}.$$

Section 2.4 gives a brief introduction to these types of electrical circuits.

Markov Chains (Section 2.5)

Consider the following problem: A city that has just introduced a new public transportation system has predicted that each year 35% of those presently using public transportation to go to work will go back to driving their car, while 65% will stay with public transportation. They also expect that 45% of those presently driving to work will switch to public transportation, while 55% will continue to drive. Thus the probability that someone presently using public transportation will go back to driving is 0.35. We can display the expected behavior of the commuting population in terms of probabilities by the matrix

$$A = \begin{bmatrix} 0.65 & 0.45 \\ 0.35 & 0.55 \end{bmatrix},$$

which denotes the information in the following display:

		Mode of transportation next year	
		Public transportation	Automobile
Mode of transportation this year	Public transportation	0.65	0.45
	Automobile	0.35	0.55

When the system becomes operational, 15% of the commuters use public transportation, while 85% drive to work. Assuming that the city's population remains constant for a long time, the management of the public transit system would like answers to the following questions:

- What is the percentage of commuters using each mode of transportation after, say, three years?
- What is the percentage of commuters using each mode of transportation in the long run?

This type of problem and the one described in Example 9 of Section 1.4 are Markov chains. The techniques discussed in Section 2.5 enable us to solve these and many other applied problems.

Linear Programming (Chapter 11)

A problem that is typical in a manufacturing process is the following one:

A coffee packer uses Colombian and Kenyan coffee to prepare a regular blend and a deluxe blend. Each pound of regular blend consists of $\frac{1}{2}$ pound of Colombian coffee and $\frac{1}{2}$ pound of Kenyan coffee. Each pound of deluxe blend consists of $\frac{1}{4}$ pound of Colombian coffee and $\frac{3}{4}$ pound of Kenyan coffee. The packer will make a \$2 profit on each pound of regular blend and a \$3 profit on each pound of deluxe blend. If there are 100 pounds of Colombian coffee and 120 pounds of Kenyan coffee, how many pounds of each blend should be packed to make the largest possible profit?

We first translate this problem into mathematical form by letting x denote the number of pounds of regular blend and y the number of pounds of deluxe blend to be packed. Then our problem can be stated as follows:

Find values of x and y that will make the expression

$$z = 2x + 3y$$

as large as possible and satisfy the following restrictions:

$$\begin{aligned} \tfrac{1}{2}x + \tfrac{1}{4}y &\leq 100 \\ \tfrac{1}{2}x + \tfrac{3}{4}y &\leq 120 \\ x &\geq 0 \\ y &\geq 0. \end{aligned}$$

This problem can be readily solved by the techniques of linear programming, a recent area of applied mathematics, which is discussed in Chapter 11.

Key Terms

Reduced row echelon form
Leading one
Row echelon form
Elementary row operation
Row equivalent
Reduced row echelon form of a matrix
Row echelon form of a matrix
Gauss–Jordan reduction
Gaussian elimination
Back substitution
Consistent linear system
Inconsistent linear system
Homogeneous system
Trivial solution
Nontrivial solution
Bit linear systems

1.6 Exercises

In Exercises 1 through 8, determine whether the given matrix is in reduced row echelon form, row echelon form, or neither.

1. $\begin{bmatrix} 1 & 0 & 0 & 0 & -3 \\ 0 & 0 & 1 & 0 & 4 \\ 0 & 0 & 0 & 1 & 2 \end{bmatrix}$

2. $\begin{bmatrix} 0 & 1 & 0 & 0 & 5 \\ 0 & 0 & 1 & 0 & -4 \\ 0 & 0 & 0 & -1 & 3 \end{bmatrix}$

3. $\begin{bmatrix} 1 & 0 & 0 & 0 & 2 \\ 0 & 0 & 1 & 0 & 0 \\ 0 & 0 & 0 & 1 & 3 \\ 0 & 0 & 0 & 0 & 0 \end{bmatrix}$

4. $\begin{bmatrix} 0 & 1 & 0 & 0 & 2 \\ 0 & 0 & 0 & 0 & -1 \\ 0 & 0 & 0 & 1 & 4 \\ 0 & 0 & 0 & 0 & 0 \\ 0 & 0 & 0 & 0 & 1 \end{bmatrix}$

5. $\begin{bmatrix} 1 & 2 & 3 & 1 \\ 0 & 1 & 2 & 3 \\ 0 & 0 & 1 & -4 \\ 0 & 0 & 0 & 0 \end{bmatrix}$

6. $\begin{bmatrix} 1 & 0 & 0 & 1 \\ 0 & 1 & 0 & 2 \\ 0 & 0 & 0 & -1 \\ 0 & 0 & 0 & 0 \end{bmatrix}$

7. $\begin{bmatrix} 0 & 0 & 0 & 0 & 0 \\ 0 & 0 & 1 & 2 & -3 \\ 0 & 0 & 0 & 1 & 0 \\ 0 & 0 & 0 & 0 & 0 \end{bmatrix}$

8. $\begin{bmatrix} 0 & 1 & 0 & 0 & 5 \\ 0 & 0 & 1 & 0 & 4 \\ 0 & 1 & 0 & -2 & 3 \end{bmatrix}$

9. Let

$$A = \begin{bmatrix} 1 & 0 & 3 \\ -3 & 1 & 4 \\ 4 & 2 & 2 \\ 5 & -1 & 5 \end{bmatrix}.$$

Find the matrices obtained by performing the following elementary row operations on A.

(a) Interchanging the second and fourth rows

(b) Multiplying the third row by 3

(c) Adding (-3) times the first row to the fourth row

10. Let

$$A = \begin{bmatrix} 2 & 0 & 4 & 2 \\ 3 & -2 & 5 & 6 \\ -1 & 3 & 1 & 1 \end{bmatrix}.$$

Find the matrices obtained by performing the following elementary row operations on A.

(a) Interchanging the second and third rows

(b) Multiplying the second row by (-4)

(c) Adding 2 times the third row to the first row

11. Find three matrices that are row equivalent to

$$A = \begin{bmatrix} 2 & -1 & 3 & 4 \\ 0 & 1 & 2 & -1 \\ 5 & 2 & -3 & 4 \end{bmatrix}.$$

12. Find three matrices that are row equivalent to

$$\begin{bmatrix} 4 & 3 & 7 & 5 \\ -1 & 2 & -1 & 3 \\ 2 & 0 & 1 & 4 \end{bmatrix}.$$

In Exercises 13 through 16, find a row echelon form of the given matrix.

13. $\begin{bmatrix} 0 & -1 & 2 & 3 \\ 2 & 3 & 4 & 5 \\ 1 & 3 & -1 & 2 \\ 3 & 2 & 4 & 1 \end{bmatrix}$

14. $\begin{bmatrix} 1 & -2 & 0 & 2 \\ 2 & -3 & -1 & 5 \\ 1 & 3 & 2 & 5 \\ 1 & 1 & 0 & 2 \\ 2 & -6 & -2 & 1 \end{bmatrix}$

15. $\begin{bmatrix} 1 & 2 & -3 & 1 \\ -1 & 0 & 3 & 4 \\ 0 & 1 & 2 & -1 \\ 2 & 3 & 0 & -3 \end{bmatrix}$

16. $\begin{bmatrix} 2 & -1 & 0 & 1 & 4 \\ 1 & -2 & 1 & 4 & -3 \\ 5 & -4 & 1 & 6 & 5 \\ -7 & 8 & -3 & -14 & 1 \end{bmatrix}$

17. For each of the matrices in Exercises 13 through 16, find the reduced row echelon form of the given matrix.

18. Let

$$A = \begin{bmatrix} 1 & 2 & 1 \\ -1 & 1 & 2 \\ 2 & 1 & -2 \end{bmatrix}.$$

In each part, determine whether $\mathbf{x}$ is a solution to the linear system $A\mathbf{x} = \mathbf{b}$.

(a) $\mathbf{x} = \begin{bmatrix} 1 \\ 2 \\ 3 \end{bmatrix}$; $\mathbf{b} = \mathbf{0}$ (b) $\mathbf{x} = \begin{bmatrix} 0 \\ 0 \\ 0 \end{bmatrix}$; $\mathbf{b} = \mathbf{0}$

(c) $\mathbf{x} = \begin{bmatrix} -1 \\ 1 \\ 2 \end{bmatrix}$; $\mathbf{b} = \begin{bmatrix} 3 \\ 6 \\ -5 \end{bmatrix}$

(d) $\mathbf{x} = \begin{bmatrix} 1 \\ 2 \\ -3 \end{bmatrix}$; $\mathbf{b} = \begin{bmatrix} 3 \\ 1 \\ 1 \end{bmatrix}$

19. Let

$$A = \begin{bmatrix} 1 & 2 & -1 & 3 \\ 1 & 3 & 0 & 2 \\ -1 & 2 & 1 & 3 \end{bmatrix}.$$

In each part, determine whether $\mathbf{x}$ is a solution to the homogeneous system $A\mathbf{x} = \mathbf{0}$.

(a) $\mathbf{x} = \begin{bmatrix} 5 \\ -3 \\ 5 \\ 2 \end{bmatrix}$ (b) $\mathbf{x} = \begin{bmatrix} 1 \\ 2 \\ 3 \\ 4 \end{bmatrix}$

(c) $\mathbf{x} = \begin{bmatrix} 1 \\ -\frac{3}{5} \\ 1 \\ \frac{2}{5} \end{bmatrix}$ (d) $\mathbf{x} = \begin{bmatrix} 1 \\ 0 \\ 0 \\ -1 \end{bmatrix}$

In Exercises 20 through 22, find all solutions to the given linear system.

20. (a)
$$\begin{aligned} x + y + 2z &= -1 \\ x - 2y + z &= -5 \\ 3x + y + z &= 3 \end{aligned}$$

(b)
$$\begin{aligned} x + y + 3z + 2w &= 7 \\ 2x - y \qquad + 4w &= 8 \\ 3y + 6z \qquad &= 8 \end{aligned}$$

(c)
$$\begin{aligned} x + 2y - 4z &= 3 \\ x - 2y + 3z &= -1 \\ 2x + 3y - z &= 5 \\ 4x + 3y - 2z &= 7 \\ 5x + 2y - 6z &= 7 \end{aligned}$$

(d)
$$\begin{aligned} x + y + z &= 0 \\ x \qquad + z &= 0 \\ 2x + y - 2z &= 0 \\ x + 5y + 5z &= 0 \end{aligned}$$

21. (a)
$$\begin{aligned} x + y + 2z + 3w &= 13 \\ x - 2y + z + w &= 8 \\ 3x + y + z - w &= 1 \end{aligned}$$

(b)
$$\begin{aligned} x + y + z &= 1 \\ x + y - 2z &= 3 \\ 2x + y + z &= 2 \end{aligned}$$

(c)
$$\begin{aligned} 2x + y + z - 2w &= 1 \\ 3x - 2y + z - 6w &= -2 \\ x + y - z - w &= -1 \\ 6x \qquad + z - 9w &= -2 \\ 5x - y + 2z - 8w &= 3 \end{aligned}$$

(d)
$$\begin{aligned} x + 2y + 3z - w &= 0 \\ 2x + y - z + w &= 3 \\ x - y \qquad + w &= -2 \end{aligned}$$

22. (a)
$$\begin{aligned} 2x - y + z &= 3 \\ x - 3y + z &= 4 \\ -5x \qquad - 2z &= -5 \end{aligned}$$

(b)
$$\begin{aligned} x + y + z + w &= 6 \\ 2x + y - z \qquad &= 3 \\ 3x + y \qquad + 2w &= 6 \end{aligned}$$

(c)
$$\begin{aligned} 2x - y + z &= 3 \\ 3x + y - 2z &= -2 \\ x - y + z &= 7 \\ x + 5y + 7z &= 13 \\ x - 7y - 5z &= 12 \end{aligned}$$

(d)
$$\begin{aligned} x + 2y - z &= 0 \\ 2x + y + z &= 0 \\ 5x + 7y + z &= 0 \end{aligned}$$

In Exercises 23 through 26, find all values of a for which the resulting linear system has (a) no solution, (b) a unique solution, and (c) infinitely many solutions.

23.
$$\begin{aligned} x + y - \qquad z &= 2 \\ x + 2y + \qquad z &= 3 \\ x + y + (a^2 - 5)z &= a \end{aligned}$$

24.
$$\begin{aligned} x + y + \qquad z &= 2 \\ 2x + 3y + \qquad 2z &= 5 \\ 2x + 3y + (a^2 - 1)z &= a + 1 \end{aligned}$$

25.
$$\begin{aligned} x + y + \qquad z &= 2 \\ x + 2y + \qquad z &= 3 \\ x + y + (a^2 - 5)z &= a \end{aligned}$$

26.
$$\begin{aligned} x + \qquad y &= 3 \\ x + (a^2 - 8)y &= a \end{aligned}$$

In Exercises 27 through 30, solve the linear system with the given augmented matrix.

27. (a) $\left[\begin{array}{ccc|c} 1 & 1 & 1 & 0 \\ 1 & 1 & 0 & 3 \\ 0 & 1 & 1 & 1 \end{array}\right]$

(b) $\left[\begin{array}{ccc|c} 1 & 2 & 3 & 0 \\ 1 & 1 & 1 & 0 \\ 1 & 1 & 2 & 0 \\ 1 & 3 & 3 & 0 \end{array}\right]$

28. (a) $\left[\begin{array}{ccc|c} 1 & 2 & 3 & 0 \\ 1 & 1 & 1 & 0 \\ 5 & 7 & 9 & 0 \end{array}\right]$

(b) $\left[\begin{array}{ccc|c} 1 & 2 & 1 & 7 \\ 2 & 0 & 1 & 4 \\ 1 & 0 & 2 & 5 \\ 1 & 2 & 3 & 11 \\ 2 & 1 & 4 & 12 \end{array}\right]$

29. (a) $\left[\begin{array}{cccc|c} 1 & 2 & 3 & 1 & 8 \\ 1 & 3 & 0 & 1 & 7 \\ 1 & 0 & 2 & 1 & 3 \end{array}\right]$

(b) $\left[\begin{array}{ccc|c} 1 & -2 & 3 & 4 \\ 2 & -1 & -3 & 5 \\ 3 & 0 & 1 & 2 \\ 3 & -3 & 0 & 7 \end{array}\right]$

30. (a) $\left[\begin{array}{ccc|c} 4 & 2 & -1 & 5 \\ 3 & 3 & 6 & 1 \\ 5 & 1 & -8 & 8 \end{array}\right]$

(b) $\left[\begin{array}{cccc|c} 1 & 1 & 3 & -3 & 0 \\ 0 & 2 & 1 & -3 & 3 \\ 1 & 0 & 2 & -1 & -1 \end{array}\right]$

31. Let $f: R^3 \to R^3$ be the matrix transformation defined by

$$f\left(\begin{bmatrix} x \\ y \\ z \end{bmatrix}\right) = \begin{bmatrix} 4 & 1 & 3 \\ 2 & -1 & 3 \\ 2 & 2 & 0 \end{bmatrix}\begin{bmatrix} x \\ y \\ z \end{bmatrix}.$$

Find x, y, z so that $f\left(\begin{bmatrix} x \\ y \\ z \end{bmatrix}\right) = \begin{bmatrix} 4 \\ 5 \\ -1 \end{bmatrix}$.

32. Let $f: R^3 \to R^3$ be the matrix transformation defined by

$$f\left(\begin{bmatrix} x \\ y \\ z \end{bmatrix}\right) = \begin{bmatrix} 1 & 2 & 3 \\ -3 & -2 & -1 \\ -2 & 0 & 2 \end{bmatrix}\begin{bmatrix} x \\ y \\ z \end{bmatrix}.$$

Find x, y, z so that $f\left(\begin{bmatrix} x \\ y \\ z \end{bmatrix}\right) = \begin{bmatrix} 2 \\ 2 \\ 4 \end{bmatrix}$.

33. Let $f: R^3 \to R^3$ be the matrix transformation defined by

$$f\left(\begin{bmatrix} x \\ y \\ z \end{bmatrix}\right) = \begin{bmatrix} 4 & 1 & 3 \\ 2 & -1 & 3 \\ 2 & 2 & 0 \end{bmatrix}\begin{bmatrix} x \\ y \\ z \end{bmatrix}.$$

Find an equation relating a, b, and c so that we can always compute values of x, y, and z for which

$$f\left(\begin{bmatrix} x \\ y \\ z \end{bmatrix}\right) = \begin{bmatrix} a \\ b \\ c \end{bmatrix}.$$

34. Let $f: R^3 \to R^3$ be the matrix transformation defined by

$$f\left(\begin{bmatrix} x \\ y \\ z \end{bmatrix}\right) = \begin{bmatrix} 1 & 2 & 3 \\ -3 & -2 & -1 \\ -2 & 0 & 2 \end{bmatrix}\begin{bmatrix} x \\ y \\ z \end{bmatrix}.$$

Find an equation relating a, b, and c so that we can always compute values of x, y, and z for which

$$f\left(\begin{bmatrix} x \\ y \\ z \end{bmatrix}\right) = \begin{bmatrix} a \\ b \\ c \end{bmatrix}.$$

In Exercises 35 and 36, solve the linear systems $A\mathbf{x} = \mathbf{b}_1$ and $A\mathbf{x} = \mathbf{b}_2$ separately and then by obtaining the reduced row echelon form of the augmented matrix $\left[A \mid \mathbf{b}_1 \;\; \mathbf{b}_2\right]$. Compare your answers.

35. $A = \begin{bmatrix} 1 & -1 \\ 2 & 3 \end{bmatrix}$, $\mathbf{b}_1 = \begin{bmatrix} 1 \\ -8 \end{bmatrix}$, $\mathbf{b}_2 = \begin{bmatrix} 5 \\ -5 \end{bmatrix}$

36. $A = \begin{bmatrix} 1 & -2 & 0 \\ -3 & 2 & -1 \\ 4 & -2 & 3 \end{bmatrix}$, $\mathbf{b}_1 = \begin{bmatrix} 3 \\ 7 \\ 12 \end{bmatrix}$, $\mathbf{b}_2 = \begin{bmatrix} -4 \\ 6 \\ -10 \end{bmatrix}$

In Exercises 37 and 38, let

$$A = \begin{bmatrix} 1 & 0 & 5 \\ 1 & 1 & 1 \\ 0 & 1 & -4 \end{bmatrix}.$$

37. Find a nontrivial solution to the homogeneous system $(-4I_3 - A)\mathbf{x} = \mathbf{0}$.*

38. Find a nontrivial solution to the homogeneous system $(2I_3 - A)\mathbf{x} = \mathbf{0}$.*

39. Find an equation relating a, b, and c so that the linear system

$$\begin{aligned} x + 2y - 3z &= a \\ 2x + 3y + 3z &= b \\ 5x + 9y - 6z &= c \end{aligned}$$

is consistent for any values of a, b, and c that satisfy that equation.

40. Find an equation relating a, b, and c so that the linear system

$$\begin{aligned} 2x + 2y + 3z &= a \\ 3x - y + 5z &= b \\ x - 3y + 2z &= c \end{aligned}$$

is consistent for any values of a, b, and c that satisfy that equation.

*This type of problem will play a key role in Chapter 8.

41. Find a 2×1 matrix $\mathbf{x}$ with entries not all zero such that $A\mathbf{x} = 4\mathbf{x}$, where

$$A = \begin{bmatrix} 4 & 1 \\ 0 & 2 \end{bmatrix}.$$

[*Hint*: Rewrite the matrix equation $A\mathbf{x} = 4\mathbf{x}$ as $4\mathbf{x} - A\mathbf{x} = (4I_2 - A)\mathbf{x} = \mathbf{0}$ and solve the homogeneous system.]

42. Find a 2×1 matrix $\mathbf{x}$ with entries not all zero such that $A\mathbf{x} = 3\mathbf{x}$, where

$$A = \begin{bmatrix} 2 & 1 \\ 1 & 2 \end{bmatrix}.^*$$

43. Find a 3×1 matrix with entries not all zero such that $A\mathbf{x} = 3\mathbf{x}$, where

$$A = \begin{bmatrix} 1 & 2 & -1 \\ 1 & 0 & 1 \\ 4 & -4 & 5 \end{bmatrix}.^*$$

44. Find a 3×1 matrix $\mathbf{x}$ with entries not all zero such that $A\mathbf{x} = 1\mathbf{x}$, where

$$A = \begin{bmatrix} 1 & 2 & -1 \\ 1 & 0 & 1 \\ 4 & -4 & 5 \end{bmatrix}.$$

In Exercises 45 and 46, solve the given linear system and write the solution $\mathbf{x}$ *as* $\mathbf{x} = \mathbf{x}_p + \mathbf{x}_h$, *where* $\mathbf{x}_p$ *is a particular solution to the given system and* $\mathbf{x}_h$ *is a solution to the associated homogeneous system.*

45.
$$\begin{aligned} x + 2y - z - 2w &= 2 \\ 2x + y - 2z + 3w &= 2 \\ x + 2y + 3z + 4w &= 5 \\ 4x + 5y - 4z - w &= 6 \end{aligned}$$

46.
$$\begin{aligned} x - y - 2z + 3w &= 4 \\ 3x + 2y - z + 2w &= 5 \\ - y - 7z + 9w &= -2 \end{aligned}$$

In Exercises 47 and 48, find the quadratic polynomial that interpolates the given points.

47. $(1, 2)$, $(3, 3)$, $(5, 8)$

48. $(1, 5)$, $(2, 12)$, $(3, 44)$

In Exercises 49 and 50, find the cubic polynomial that interpolates the given points.

49. $(-1, -6)$, $(1, 0)$, $(2, 8)$, $(3, 34)$

50. $(-2, 2)$, $(-1, 2)$, $(1, 2)$, $(2, 10)$

51. A furniture manufacturer makes chairs, coffee tables, and dining-room tables. Each chair requires 10 minutes of sanding, 6 minutes of staining, and 12 minutes of varnishing. Each coffee table requires 12 minutes of sanding, 8 minutes of staining, and 12 minutes of varnishing. Each dining-room table requires 15 minutes of sanding, 12 minutes of staining, and 18 minutes of varnishing. The sanding bench is available 16 hours per week, the staining bench 11 hours per week, and the varnishing bench 18 hours per week. How many (per week) of each type of furniture should be made so that the benches are fully utilized?

52. A book publisher publishes a potential best seller in three different bindings: paperback, book club, and deluxe. Each paperback book requires 1 minute for sewing and 2 minutes for gluing. Each book club book requires 2 minutes for sewing and 4 minutes for gluing. Each deluxe book requires 3 minutes for sewing and 5 minutes for gluing. If the sewing plant is available 6 hours per day and the gluing plant is available 11 hours per day, how many books of each type can be produced per day so that the plants are fully utilized?

53. ***(Calculus Required)*** Construct a linear system of equations to determine a quadratic polynomial

$$p(x) = ax^2 + bx + c$$

that satisfies the conditions $p(0) = f(0)$, $p'(0) = f'(0)$, and $p''(0) = f''(0)$, where $f(x) = e^{2x}$.

54. ***(Calculus Required)*** Construct a linear system of equations to determine a quadratic polynomial

$$p(x) = ax^2 + bx + c$$

that satisfies the conditions $p(1) = f(1)$, $p'(1) = f'(1)$, and $p''(1) = f''(1)$, where $f(x) = xe^{x-1}$.

55. Determine the temperatures at the interior points T_i, $i = 1, 2, 3, 4$ for the plate shown in the figure. (See Example 17.)

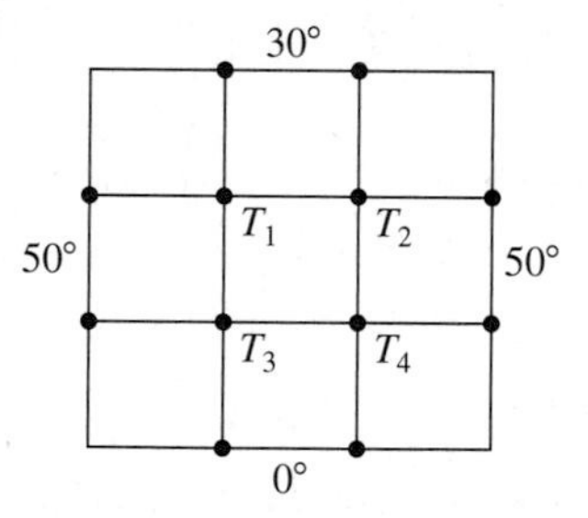

In Exercises 56 through 59, solve the bit linear systems.

56. (a)
$$\begin{aligned} x + y + z &= 0 \\ y + z &= 1 \\ x + y &= 1 \end{aligned}$$
(b)
$$\begin{aligned} x + y + z &= 1 \\ x + z &= 0 \\ y + z &= 1 \end{aligned}$$

57. (a)
$$\begin{aligned} x + y + w &= 0 \\ x + z + w &= 1 \\ y + z + w &= 1 \end{aligned}$$
(b)
$$\begin{aligned} x + y &= 0 \\ x + y + z &= 1 \\ x + y + z + w &= 0 \end{aligned}$$

58. (a)
$$\begin{aligned} x + y + z &= 1 \\ y + z + w &= 1 \\ x + w &= 1 \end{aligned}$$
(b)
$$\begin{aligned} x + y + z &= 0 \\ y + z + w &= 0 \\ x + w &= 0 \end{aligned}$$

*This type of problem will play a key role in Chapter 8.

59. Solve the bit linear system $A\mathbf{x} = \mathbf{c}$, where

(a) $A = \begin{bmatrix} 1 & 1 & 0 \\ 0 & 1 & 0 \\ 1 & 1 & 1 \end{bmatrix}$, $\mathbf{c} = \begin{bmatrix} 0 \\ 1 \\ 0 \end{bmatrix}$

(b) $A = \begin{bmatrix} 1 & 1 & 0 & 1 \\ 1 & 0 & 1 & 1 \\ 0 & 0 & 1 & 1 \\ 0 & 1 & 1 & 0 \end{bmatrix}$, $\mathbf{c} = \begin{bmatrix} 1 \\ 0 \\ 0 \\ 0 \end{bmatrix}$

Theoretical Exercises

T.1. Show that properties (a), (b), and (c) alone [excluding (d)] of the definition of the reduced row echelon form of a matrix A imply that if a column of A contains a leading entry of some row, then all other entries in that column *below the leading entry* are zero.

T.2. Show that

(a) Every matrix is row equivalent to itself.

(b) If A is row equivalent to B, then B is row equivalent to A.

(c) If A is row equivalent to B and B is row equivalent to C, then A is row equivalent to C.

T.3. Prove Corollary 1.1.

T.4. Show that the linear system $A\mathbf{x} = \mathbf{b}$, where A is $n \times n$, has no solution if and only if the reduced row echelon form of the augmented matrix has a row whose first n elements are zero and whose $(n + 1)$st element is 1.

T.5. Let

$$A = \begin{bmatrix} a & b \\ c & d \end{bmatrix}.$$

Show that A is row equivalent to I_2 if and only if $ad - bc \neq 0$.

T.6. (a) Let

$$A = \begin{bmatrix} a & b \\ ka & kb \end{bmatrix}.$$

Use Exercise T.5 to determine if A is row equivalent to I_2.

(b) Let A be a 2×2 matrix with a row consisting entirely of zeros. Use Exercise T.5 to determine if A is row equivalent to I_2.

T.7. Determine the reduced row echelon form of the matrix

$$\begin{bmatrix} \cos\theta & \sin\theta \\ -\sin\theta & \cos\theta \end{bmatrix}.$$

T.8. Let

$$A = \begin{bmatrix} a & b \\ c & d \end{bmatrix}.$$

Show that the homogeneous system $A\mathbf{x} = \mathbf{0}$ has only the trivial solution if and only if $ad - bc \neq 0$.

T.9. Let A be an $n \times n$ matrix in reduced row echelon form. Show that if A is not equal to I_n, then A has a row consisting entirely of zeros.

T.10. Show that the values of λ for which the homogeneous system

$$\begin{aligned} (a - \lambda)x + \qquad by &= 0 \\ cx + (d - \lambda)y &= 0 \end{aligned}$$

has a nontrivial solution satisfy the equation $(a - \lambda)(d - \lambda) - bc = 0$. (*Hint*: See Exercise T.8.)

T.11. Let $\mathbf{u}$ and $\mathbf{v}$ be solutions to the homogeneous linear system $A\mathbf{x} = \mathbf{0}$.

(a) Show that $\mathbf{u} + \mathbf{v}$ is a solution.

(b) Show that $\mathbf{u} - \mathbf{v}$ is a solution.

(c) For any scalar r, show that $r\mathbf{u}$ is a solution.

(d) For any scalars r and s, show that $r\mathbf{u} + s\mathbf{v}$ is a solution.

T.12. Show that if $\mathbf{u}$ and $\mathbf{v}$ are solutions to the linear system $A\mathbf{x} = \mathbf{b}$, then $\mathbf{u} - \mathbf{v}$ is a solution to the associated homogeneous system $A\mathbf{x} = \mathbf{0}$.

T.13. Let $A\mathbf{x} = \mathbf{b}$, $\mathbf{b} \neq \mathbf{0}$, be a consistent linear system.

(a) Show that if $\mathbf{x}_p$ is a particular solution to the given nonhomogeneous system and $\mathbf{x}_h$ is a solution to the associated homogeneous system $A\mathbf{x} = \mathbf{0}$, then $\mathbf{x}_p + \mathbf{x}_h$ is a solution to the given system $A\mathbf{x} = \mathbf{b}$.

(b) Show that every solution $\mathbf{x}$ to the nonhomogeneous linear system $A\mathbf{x} = \mathbf{b}$ can be written as $\mathbf{x}_p + \mathbf{x}_h$, where $\mathbf{x}_p$ is a particular solution to the given nonhomogeneous system and $\mathbf{x}_h$ is a solution to the associated homogeneous system $A\mathbf{x} = \mathbf{0}$. [*Hint*: Let $\mathbf{x} = \mathbf{x}_p + (\mathbf{x} - \mathbf{x}_p)$.]

T.14. Justify the second remark following Example 12.

MATLAB Exercises

In order to use MATLAB *in this section, you should first have read Chapter 12 through Section 12.4.*

ML.1. Let

$$A = \begin{bmatrix} 4 & 2 & 2 \\ -3 & 1 & 4 \\ 1 & 0 & 3 \\ 5 & -1 & 5 \end{bmatrix}.$$

Find the matrices obtained by performing the following row operations in succession on matrix A. Do the row operations directly using the colon operator.

(a) Multiply row 1 by $\frac{1}{4}$.

(b) Add 3 times row 1 to row 2.

(c) Add (-1) times row 1 to row 3.

(d) Add (-5) times row 1 to row 4.

(e) Interchange rows 2 and 4.

ML.2. Let

$$A = \begin{bmatrix} \frac{1}{2} & \frac{1}{3} & \frac{1}{4} & \frac{1}{5} \\ \frac{1}{3} & \frac{1}{4} & \frac{1}{5} & \frac{1}{6} \\ 1 & \frac{1}{2} & \frac{1}{3} & \frac{1}{4} \end{bmatrix}.$$

Find the matrices obtained by performing the following row operations in succession on matrix A. Do the row operations directly using the colon operator.

(a) Multiply row 1 by 2.

(b) Add $\left(-\frac{1}{3}\right)$ times row 1 to row 2.

(c) Add (-1) times row 1 to row 3.

(d) Interchange rows 2 and 3.

ML.3. Use **reduce** to find the reduced row echelon form of matrix A in Exercise ML.1.

ML.4. Use **reduce** to find the reduced row echelon form of matrix A in Exercise ML.2.

ML.5. Use **reduce** to find all solutions to the linear system in Exercise 21(a).

ML.6. Use **reduce** to find all solutions to the linear system in Exercise 20(b).

ML.7. Use **reduce** to find all solutions to the linear system in Exercise 27(b).

ML.8. Use **reduce** to find all solutions to the linear system in Exercise 28(a).

ML.9. Let

$$A = \begin{bmatrix} 1 & 2 \\ 2 & 4 \end{bmatrix}.$$

Use **reduce** to find a nontrivial solution to the homogeneous system

$$(5I_2 - A)\mathbf{x} = \mathbf{0}.$$

[*Hint*: In MATLAB, enter matrix A, then use the command **reduce(5 ∗ eye(size(A)) − A)**.]

ML.10. Let

$$A = \begin{bmatrix} 1 & 5 \\ 5 & 1 \end{bmatrix}.$$

Use **reduce** to find a nontrivial solution to the homogeneous system

$$(-4I_2 - A)\mathbf{x} = \mathbf{0}.$$

[*Hint*: In MATLAB, enter matrix A, then use the command **reduce(− 4 ∗ eye(size(A)) − A)**.]

ML.11. Use **rref** in MATLAB to solve the linear systems in Exercises 27 and 28.

ML.12. MATLAB has an immediate command for solving square linear systems $A\mathbf{x} = \mathbf{b}$. Once the coefficient matrix A and right-hand side $\mathbf{b}$ are entered into MATLAB, command

$$\mathbf{x = A\backslash b}$$

displays the solution as long as A is considered nonsingular. (See the Definition at the beginning of Section 1.7.) The backslash command, \, does not use reduced row echelon form but does initiate numerical methods that are usually discussed in a course in Numerical Analysis. For more details on the command, see D. R. Hill, *Experiments in Computational Matrix Algebra*, New York: Random House, 1988.

(a) Use \ to solve Exercise 27(a).

(b) Use \ to solve Exercise 21(b).

ML.13. The \ command behaves differently than **rref**. Use both \ and **rref** to solve $A\mathbf{x} = \mathbf{b}$, where

$$A = \begin{bmatrix} 1 & 2 & 3 \\ 4 & 5 & 6 \\ 7 & 8 & 9 \end{bmatrix}, \quad \mathbf{b} = \begin{bmatrix} 1 \\ 0 \\ 0 \end{bmatrix}.$$

Exercises ML.14 through ML.16 use bit matrices and the supplemental instructional commands described in Section 12.9.

ML.14. Solve each of the built-in demos in routine **binreduce**. (Enter the command **binreduce** and then choose option **< 1 >** to select a demo.)

ML.15. Use **binreduce** to obtain the reduced row echelon form of the bit augmented matrices of Exercises 56 through 59 and then determine the solution to the corresponding linear system.

ML.16. Use **binreduce** to obtain the reduced row echelon form of the bit augmented matrix

$$\left[\begin{array}{ccccc:c} 1 & 1 & 0 & 0 & 1 & 1 \\ 1 & 0 & 1 & 0 & 0 & 1 \\ 1 & 1 & 1 & 1 & 1 & 0 \end{array}\right]$$

and then determine the solution to the corresponding linear system.

1.7 THE INVERSE OF A MATRIX

In this section we restrict our attention to square matrices and formulate the notion corresponding to the reciprocal of a nonzero number.

DEFINITION An $n \times n$ matrix A is called **nonsingular** (or **invertible**) if there exists an $n \times n$ matrix B such that

$$AB = BA = I_n.$$

The matrix B is called an **inverse** of A. If there exists no such matrix B, then A is called **singular** (or **noninvertible**).

Remark From the preceding definition, it follows that if $AB = BA = I_n$, then A is also an inverse of B.

EXAMPLE 1 Let

$$A = \begin{bmatrix} 2 & 3 \\ 2 & 2 \end{bmatrix} \quad \text{and} \quad B = \begin{bmatrix} -1 & \frac{3}{2} \\ 1 & -1 \end{bmatrix}.$$

Since

$$AB = BA = I_2,$$

we conclude that B is an inverse of A and that A is nonsingular. ■

THEOREM 1.9 *An inverse of a matrix, if it exists, is unique.*

Proof Let B and C be inverses of A. Then $BA = AC = I_n$. Therefore,

$$B = BI_n = B(AC) = (BA)C = I_nC = C,$$

which completes the proof. ■

We shall now write the inverse of A, if it exists, as A^{-1}. Thus

$$AA^{-1} = A^{-1}A = I_n.$$

EXAMPLE 2 Let

$$A = \begin{bmatrix} 1 & 2 \\ 3 & 4 \end{bmatrix}.$$

To find A^{-1}, we let

$$A^{-1} = \begin{bmatrix} a & b \\ c & d \end{bmatrix}.$$

Then we must have

$$AA^{-1} = \begin{bmatrix} 1 & 2 \\ 3 & 4 \end{bmatrix} \begin{bmatrix} a & b \\ c & d \end{bmatrix} = I_2 = \begin{bmatrix} 1 & 0 \\ 0 & 1 \end{bmatrix}$$

so that

$$\begin{bmatrix} a + 2c & b + 2d \\ 3a + 4c & 3b + 4d \end{bmatrix} = \begin{bmatrix} 1 & 0 \\ 0 & 1 \end{bmatrix}.$$

Equating corresponding entries of these two matrices, we obtain the linear systems

$$\begin{aligned} a + 2c &= 1 \\ 3a + 4c &= 0 \end{aligned} \quad \text{and} \quad \begin{aligned} b + 2d &= 0 \\ 3b + 4d &= 1. \end{aligned}$$

The solutions are (verify) $a = -2$, $c = \frac{3}{2}$, $b = 1$, and $d = -\frac{1}{2}$. Moreover, since the matrix

$$\begin{bmatrix} a & b \\ c & d \end{bmatrix} = \begin{bmatrix} -2 & 1 \\ \frac{3}{2} & -\frac{1}{2} \end{bmatrix}$$

also satisfies the property that

$$\begin{bmatrix} -2 & 1 \\ \frac{3}{2} & -\frac{1}{2} \end{bmatrix} \begin{bmatrix} 1 & 2 \\ 3 & 4 \end{bmatrix} = \begin{bmatrix} 1 & 0 \\ 0 & 1 \end{bmatrix},$$

we conclude that A is nonsingular and that

$$A^{-1} = \begin{bmatrix} -2 & 1 \\ \frac{3}{2} & -\frac{1}{2} \end{bmatrix}.$$

■

Remark Not every matrix has an inverse. For instance, consider the following example.

EXAMPLE 3 Let

$$A = \begin{bmatrix} 1 & 2 \\ 2 & 4 \end{bmatrix}.$$

To find A^{-1}, we let

$$A^{-1} = \begin{bmatrix} a & b \\ c & d \end{bmatrix}.$$

Then we must have

$$AA^{-1} = \begin{bmatrix} 1 & 2 \\ 2 & 4 \end{bmatrix} \begin{bmatrix} a & b \\ c & d \end{bmatrix} = I_2 = \begin{bmatrix} 1 & 0 \\ 0 & 1 \end{bmatrix}$$

so that

$$\begin{bmatrix} a + 2c & b + 2d \\ 2a + 4c & 2b + 4d \end{bmatrix} = \begin{bmatrix} 1 & 0 \\ 0 & 1 \end{bmatrix}.$$

Equating corresponding entries of these two matrices, we obtain the linear systems

$$\begin{aligned} a + 2c &= 1 \\ 2a + 4c &= 0 \end{aligned} \quad \text{and} \quad \begin{aligned} b + 2d &= 0 \\ 2b + 4d &= 1. \end{aligned}$$

These linear systems have no solutions, so A has no inverse. Hence A is a singular matrix. ■

The method used in Example 2 to find the inverse of a matrix is not a very efficient one. We shall soon modify it and thereby obtain a much faster method. We first establish several properties of nonsingular matrices.

THEOREM 1.10 **(Properties of the Inverse)**

(a) *If A is a nonsingular matrix, then A^{-1} is nonsingular and*

$$(A^{-1})^{-1} = A.$$

(b) *If A and B are nonsingular matrices, then AB is nonsingular and*

$$(AB)^{-1} = B^{-1}A^{-1}.$$

(c) *If A is a nonsingular matrix, then*

$$(A^T)^{-1} = (A^{-1})^T.$$

Proof (a) A^{-1} is nonsingular if we can find a matrix B such that

$$A^{-1}B = BA^{-1} = I_n.$$

Since A is nonsingular,

$$A^{-1}A = AA^{-1} = I_n.$$

Thus $B = A$ is an inverse of A^{-1}, and since inverses are unique, we conclude that

$$(A^{-1})^{-1} = A.$$

Thus, the inverse of the inverse of the nonsingular matrix A is A.

(b) We have

$$(AB)(B^{-1}A^{-1}) = A(BB^{-1})A^{-1} = AI_nA^{-1} = AA^{-1} = I_n$$

and

$$(B^{-1}A^{-1})(AB) = B^{-1}(A^{-1}A)B = B^{-1}I_nB = B^{-1}B = I_n.$$

Therefore, AB is nonsingular. Since the inverse of a matrix is unique, we conclude that

$$(AB)^{-1} = B^{-1}A^{-1}.$$

Thus, the inverse of a product of two nonsingular matrices is the product of their inverses in reverse order.

(c) We have

$$AA^{-1} = I_n \quad \text{and} \quad A^{-1}A = I_n.$$

Taking transposes, we obtain

$$(AA^{-1})^T = I_n^T = I_n \quad \text{and} \quad (A^{-1}A)^T = I_n^T = I_n.$$

Then

$$(A^{-1})^T A^T = I_n \quad \text{and} \quad A^T(A^{-1})^T = I_n.$$

These equations imply that

$$(A^T)^{-1} = (A^{-1})^T.$$

Thus, the inverse of the transpose of a nonsingular matrix is the transpose of its inverse. ■

EXAMPLE 4 If $A = \begin{bmatrix} 1 & 2 \\ 3 & 4 \end{bmatrix}$, then from Example 2

$$A^{-1} = \begin{bmatrix} -2 & 1 \\ \frac{3}{2} & -\frac{1}{2} \end{bmatrix} \quad \text{and} \quad (A^{-1})^T = \begin{bmatrix} -2 & \frac{3}{2} \\ 1 & -\frac{1}{2} \end{bmatrix}.$$

Also (verify)

$$A^T = \begin{bmatrix} 1 & 3 \\ 2 & 4 \end{bmatrix} \quad \text{and} \quad (A^T)^{-1} = \begin{bmatrix} -2 & \frac{3}{2} \\ 1 & -\frac{1}{2} \end{bmatrix}.$$

■

COROLLARY 1.2 *If $A_1, A_2, \ldots, A_r$ are $n \times n$ nonsingular matrices, then $A_1 A_2 \cdots A_r$ is nonsingular and*

$$(A_1 A_2 \cdots A_r)^{-1} = A_r^{-1} A_{r-1}^{-1} \cdots A_1^{-1}.$$

Proof Exercise T.2. ■

Earlier, we defined a matrix B to be the inverse of A if $AB = BA = I_n$. The following theorem, whose proof we omit, shows that one of these equations follows from the other.

THEOREM 1.11 *Suppose that A and B are $n \times n$ matrices.*

(a) *If $AB = I_n$, then $BA = I_n$.*
(b) *If $BA = I_n$, then $AB = I_n$.* ■

A PRACTICAL METHOD FOR FINDING A^{-1}

We shall now develop a practical method for finding A^{-1}. If A is a given $n \times n$ matrix, we are looking for an $n \times n$ matrix $B = \left[b_{ij}\right]$ such that

$$AB = BA = I_n.$$

Let the columns of B be denoted by the $n \times 1$ matrices $\mathbf{x}_1, \mathbf{x}_2, \ldots, \mathbf{x}_n$, where

$$\mathbf{x}_j = \begin{bmatrix} b_{1j} \\ b_{2j} \\ \vdots \\ b_{ij} \\ \vdots \\ b_{nj} \end{bmatrix} \qquad (1 \le j \le n).$$

Let the columns of I_n be denoted by the $n \times 1$ matrices $\mathbf{e}_1, \mathbf{e}_2, \ldots, \mathbf{e}_n$. Thus

$$\mathbf{e}_j = \begin{bmatrix} 0 \\ 0 \\ \vdots \\ 1 \\ 0 \\ \vdots \\ 0 \end{bmatrix} \longleftarrow j\text{th row.}$$

By Exercise T.9(a) of Section 1.3, the jth column of AB is the $n \times 1$ matrix $A\mathbf{x}_j$. Since equal matrices must agree column by column, it follows that the problem of finding an $n \times n$ matrix $B = A^{-1}$ such that

$$AB = I_n \tag{1}$$

is equivalent to the problem of finding n matrices (each $n \times 1$) $\mathbf{x}_1, \mathbf{x}_2, \ldots, \mathbf{x}_n$ such that

$$A\mathbf{x}_j = \mathbf{e}_j \qquad (1 \le j \le n). \tag{2}$$

Thus finding B is equivalent to solving n linear systems (each is n equations in n unknowns). This is precisely what we did in Example 2. Each of these systems can be solved by the Gauss–Jordan reduction method. To solve the first linear system, we form the augmented matrix $\left[A \mid \mathbf{e}_1\right]$ and compute its reduced row echelon form. We do the same with

$$\left[A \mid \mathbf{e}_2\right], \dots, \left[A \mid \mathbf{e}_n\right].$$

However, if we observe that the coefficient matrix of each of these n linear systems is always A, we can solve all these systems simultaneously. We form the $n \times 2n$ matrix

$$\left[A \mid \mathbf{e}_1 \quad \mathbf{e}_2 \quad \cdots \quad \mathbf{e}_n\right] = \left[A \mid I_n\right]$$

and compute its reduced row echelon form $\left[C \mid D\right]$. The $n \times n$ matrix C is the reduced row echelon form of A. Let $\mathbf{d}_1, \mathbf{d}_2, \dots, \mathbf{d}_n$ be the n columns of D. Then the matrix $\left[C \mid D\right]$ gives rise to the n linear systems

$$C\mathbf{x}_j = \mathbf{d}_j \qquad (1 \le j \le n) \tag{3}$$

or to the matrix equation

$$CB = D. \tag{4}$$

There are now two possible cases.

Case 1. $C = I_n$. Then Equation (3) becomes

$$I_n\mathbf{x}_j = \mathbf{x}_j = \mathbf{d}_j,$$

and $B = D$, so we have obtained A^{-1}.

Case 2. $C \ne I_n$. It then follows from Exercise T.9 in Section 1.6 that C has a row consisting entirely of zeros. From Exercise T.3 in Section 1.3, we observe that the product CB in Equation (4) has a row of zeros. The matrix D in (4) arose from I_n by a sequence of elementary row operations, and it is intuitively clear that D cannot have a row of zeros. The statement that D cannot have a row of zeros can be rigorously established at this point, but we shall ask the reader to accept the argument now. In Section 3.2, an argument using determinants will show the validity of the result. Thus one of the equations $C\mathbf{x}_j = \mathbf{d}_j$ has no solution, so $A\mathbf{x}_j = \mathbf{e}_j$ has no solution and A is singular in this case.

The practical procedure for computing the inverse of matrix A is as follows.

Step 1. Form the $n \times 2n$ matrix $\left[A \mid I_n\right]$ obtained by adjoining the identity matrix I_n to the given matrix A.

Step 2. Compute the reduced row echelon form of the matrix obtained in Step 1 by using elementary row operations. Remember that whatever we do to a row of A we also do to the corresponding row of I_n.

Step 3. Suppose that Step 2 has produced the matrix $\left[C \mid D\right]$ in reduced row echelon form.

(a) If $C = I_n$, then $D = A^{-1}$.

(b) If $C \ne I_n$, then C has a row of zeros. In this case A is singular and A^{-1} does not exist.

EXAMPLE 5 Find the inverse of the matrix

$$A = \begin{bmatrix} 1 & 1 & 1 \\ 0 & 2 & 3 \\ 5 & 5 & 1 \end{bmatrix}.$$

Solution ***Step 1.*** The 3×6 matrix $\left[A \mid I_3\right]$ is

$$\left[A \mid I_3\right] = \left[\begin{array}{ccc|ccc} \multicolumn{3}{c}{A} & \multicolumn{3}{c}{I_3} \\ 1 & 1 & 1 & 1 & 0 & 0 \\ 0 & 2 & 3 & 0 & 1 & 0 \\ 5 & 5 & 1 & 0 & 0 & 1 \end{array}\right].$$

Step 2. We now compute the reduced row echelon form of the matrix obtained in Step 1:

$$\left[\begin{array}{ccc|ccc} \multicolumn{3}{c}{A} & \multicolumn{3}{c}{I_3} \\ 1 & 1 & 1 & 1 & 0 & 0 \\ 0 & 2 & 3 & 0 & 1 & 0 \\ 5 & 5 & 1 & 0 & 0 & 1 \end{array}\right]$$

$$\left[\begin{array}{ccc|ccc} 1 & 1 & 1 & 1 & 0 & 0 \\ 0 & 2 & 3 & 0 & 1 & 0 \\ 0 & 0 & -4 & -5 & 0 & 1 \end{array}\right]$$

(-5) times the first row was added to the third row.

$$\left[\begin{array}{ccc|ccc} 1 & 1 & 1 & 1 & 0 & 0 \\ 0 & 1 & \frac{3}{2} & 0 & \frac{1}{2} & 0 \\ 0 & 0 & -4 & -5 & 0 & 1 \end{array}\right]$$

The second row was multiplied by $\frac{1}{2}$.

$$\left[\begin{array}{ccc|ccc} 1 & 1 & 1 & 1 & 0 & 0 \\ 0 & 1 & \frac{3}{2} & 0 & \frac{1}{2} & 0 \\ 0 & 0 & 1 & \frac{5}{4} & 0 & -\frac{1}{4} \end{array}\right]$$

The third row was multiplied by $(-\frac{1}{4})$.

$$\left[\begin{array}{ccc|ccc} 1 & 1 & 0 & -\frac{1}{4} & 0 & \frac{1}{4} \\ 0 & 1 & 0 & -\frac{15}{8} & \frac{1}{2} & \frac{3}{8} \\ 0 & 0 & 1 & \frac{5}{4} & 0 & -\frac{1}{4} \end{array}\right]$$

$(-\frac{3}{2})$ times the third row was added to the second row.
(-1) times the third row was added to the first row.

$$\left[\begin{array}{ccc|ccc} 1 & 0 & 0 & \frac{13}{8} & -\frac{1}{2} & -\frac{1}{8} \\ 0 & 1 & 0 & -\frac{15}{8} & \frac{1}{2} & \frac{3}{8} \\ 0 & 0 & 1 & \frac{5}{4} & 0 & -\frac{1}{4} \end{array}\right]$$

(-1) times the second row was added to the first row.

Step 3. Since $C = I_3$, we conclude that $D = A^{-1}$. Hence

$$A^{-1} = \begin{bmatrix} \frac{13}{8} & -\frac{1}{2} & -\frac{1}{8} \\ -\frac{15}{8} & \frac{1}{2} & \frac{3}{8} \\ \frac{5}{4} & 0 & -\frac{1}{4} \end{bmatrix}.$$

It is easy to verify that $AA^{-1} = A^{-1}A = I_3$. ■

If the reduced row echelon matrix under A has a row of zeros, then A is singular. Since each matrix under A is row equivalent to A, once a matrix under A has a row of zeros, every subsequent matrix that is row equivalent to A will have a row of zeros. Thus we can stop the procedure as soon as we find a matrix F that is row equivalent to A and has a row of zeros. In this case A^{-1} does not exist.

EXAMPLE 6 Find the inverse of the matrix

$$A = \begin{bmatrix} 1 & 2 & -3 \\ 1 & -2 & 1 \\ 5 & -2 & -3 \end{bmatrix} \quad \text{if it exists.}$$

Solution ***Step 1.*** The 3×6 matrix $[A \mid I_3]$ is

$$[A \mid I_3] = \left[\begin{array}{rrr|rrr} 1 & 2 & -3 & 1 & 0 & 0 \\ 1 & -2 & 1 & 0 & 1 & 0 \\ 5 & -2 & -3 & 0 & 0 & 1 \end{array}\right].$$

Step 2. We compute the reduced row echelon form of the matrix obtained in Step 1. To find A^{-1}, we proceed as follows:

$$\left[\begin{array}{rrr|rrr} 1 & 2 & -3 & 1 & 0 & 0 \\ 1 & -2 & 1 & 0 & 1 & 0 \\ 5 & -2 & -3 & 0 & 0 & 1 \end{array}\right]$$

$$\left[\begin{array}{rrr|rrr} 1 & 2 & -3 & 1 & 0 & 0 \\ 0 & -4 & 4 & -1 & 1 & 0 \\ 5 & -2 & -3 & 0 & 0 & 1 \end{array}\right] \quad (-1) \text{ times the first row was added to the second row.}$$

$$\left[\begin{array}{rrr|rrr} 1 & 2 & -3 & 1 & 0 & 0 \\ 0 & -4 & 4 & -1 & 1 & 0 \\ 0 & -12 & 12 & -5 & 0 & 1 \end{array}\right] \quad (-5) \text{ times the first row was added to the third row.}$$

$$\left[\begin{array}{rrr|rrr} 1 & 2 & -3 & 1 & 0 & 0 \\ 0 & -4 & 4 & -1 & 1 & 0 \\ 0 & 0 & 0 & -2 & -3 & 1 \end{array}\right] \quad (-3) \text{ times the second row was added to the third row.}$$

At this point A is row equivalent to

$$F = \begin{bmatrix} 1 & 2 & -3 \\ 0 & -4 & 4 \\ 0 & 0 & 0 \end{bmatrix}.$$

Since F has a row of zeros, we stop and conclude that A is a singular matrix. ■

Observe that to find A^{-1} we do not have to determine, in advance, whether or not it exists. We merely start the procedure given previously and either obtain A^{-1} or find out that A is singular.

The foregoing discussion for the practical method of obtaining A^{-1} has actually established the following theorem.

THEOREM 1.12 *An $n \times n$ matrix is nonsingular if and only if it is row equivalent to I_n.* ■

LINEAR SYSTEMS AND INVERSES

If A is an $n \times n$ matrix, then the linear system $A\mathbf{x} = \mathbf{b}$ is a system of n equations in n unknowns. Suppose that A is nonsingular. Then A^{-1} exists and we can multiply $A\mathbf{x} = \mathbf{b}$ by A^{-1} on both sides, obtaining

$$\begin{aligned} A^{-1}(A\mathbf{x}) &= A^{-1}\mathbf{b} \\ (A^{-1}A)\mathbf{x} &= A^{-1}\mathbf{b} \\ I_n\mathbf{x} &= A^{-1}\mathbf{b} \\ \mathbf{x} &= A^{-1}\mathbf{b}. \end{aligned}$$

Moreover, $\mathbf{x} = A^{-1}\mathbf{b}$ is clearly a solution to the given linear system. Thus, if A is nonsingular, we have a unique solution.

Applications This method is useful in industrial problems. Many physical models are described by linear systems. This means that if n values are used as inputs (which can be arranged as the $n \times 1$ matrix $\mathbf{x}$), then m values are obtained as outputs (which can be arranged as the $m \times 1$ matrix $\mathbf{b}$) by the rule $A\mathbf{x} = \mathbf{b}$. The matrix A is inherently tied to the process. Thus suppose that a chemical process has a certain matrix A associated with it. Any change in the process may result in a new matrix. In fact, we speak of a **black box**, meaning that the internal structure of the process does not interest us. The problem frequently encountered in systems analysis is that of determining the input to be used to obtain a desired output. That is, we want to solve the linear system $A\mathbf{x} = \mathbf{b}$ for $\mathbf{x}$ as we vary $\mathbf{b}$. If A is a nonsingular square matrix, an efficient way of handling this is as follows: Compute A^{-1} once; then whenever we change $\mathbf{b}$, we find the corresponding solution $\mathbf{x}$ by forming $A^{-1}\mathbf{b}$.

EXAMPLE 7

(Industrial Process) Consider an industrial process whose matrix is the matrix A of Example 5. If $\mathbf{b}$ is the output matrix

$$\begin{bmatrix} 8 \\ 24 \\ 8 \end{bmatrix},$$

then the input matrix $\mathbf{x}$ is the solution to the linear system $A\mathbf{x} = \mathbf{b}$. Then

$$\mathbf{x} = A^{-1}\mathbf{b} = \begin{bmatrix} \frac{13}{8} & -\frac{1}{2} & -\frac{1}{8} \\ -\frac{15}{8} & \frac{1}{2} & \frac{3}{8} \\ \frac{5}{4} & 0 & -\frac{1}{4} \end{bmatrix} \begin{bmatrix} 8 \\ 24 \\ 8 \end{bmatrix} = \begin{bmatrix} 0 \\ 0 \\ 8 \end{bmatrix}.$$

On the other hand, if $\mathbf{b}$ is the output matrix

$$\begin{bmatrix} 4 \\ 7 \\ 16 \end{bmatrix},$$

then (verify)

$$\mathbf{x} = A^{-1}\begin{bmatrix} 4 \\ 7 \\ 16 \end{bmatrix} = \begin{bmatrix} 1 \\ 2 \\ 1 \end{bmatrix}.$$

■

THEOREM 1.13 *If A is an $n \times n$ matrix, the homogeneous system*

$$A\mathbf{x} = \mathbf{0} \tag{5}$$

has a nontrivial solution if and only if A is singular.

Proof Suppose that A is nonsingular. Then A^{-1} exists, and multiplying both sides of (5) by A^{-1}, we have

$$\begin{aligned} A^{-1}(A\mathbf{x}) &= A^{-1}\mathbf{0} \\ (A^{-1}A)\mathbf{x} &= \mathbf{0} \\ I_n\mathbf{x} &= \mathbf{0} \\ \mathbf{x} &= \mathbf{0}. \end{aligned}$$

Hence the only solution to (5) is $\mathbf{x} = \mathbf{0}$.

We leave the proof of the converse—if A is singular, then (5) has a nontrivial solution—as an exercise (Exercise T.3). ■

EXAMPLE 8 Consider the homogeneous system $A\mathbf{x} = \mathbf{0}$, where A is the matrix of Example 5. Since A is nonsingular,

$$\mathbf{x} = A^{-1}\mathbf{0} = \mathbf{0}.$$

We could also solve the given system by Gauss–Jordan reduction. In this case we find that the matrix in reduced row echelon form that is row equivalent to the augmented matrix of the given system,

$$\left[\begin{array}{ccc|c} 1 & 1 & 1 & 0 \\ 0 & 2 & 3 & 0 \\ 5 & 5 & 1 & 0 \end{array}\right],$$

is

$$\left[\begin{array}{ccc|c} 1 & 0 & 0 & 0 \\ 0 & 1 & 0 & 0 \\ 0 & 0 & 1 & 0 \end{array}\right],$$

which again shows that the solution is

$$\mathbf{x} = \mathbf{0}.$$ ■

EXAMPLE 9 Consider the homogeneous system $A\mathbf{x} = \mathbf{0}$, where A is the singular matrix of Example 6. In this case the matrix in reduced row echelon form that is row equivalent to the augmented matrix of the given system,

$$\left[\begin{array}{ccc|c} 1 & 2 & -3 & 0 \\ 1 & -2 & 1 & 0 \\ 5 & -2 & -3 & 0 \end{array}\right],$$

is (verify)

$$\left[\begin{array}{ccc|c} 1 & 0 & -1 & 0 \\ 0 & 1 & -1 & 0 \\ 0 & 0 & 0 & 0 \end{array}\right],$$

which implies that

$$\begin{aligned} x &= r \\ y &= r \\ z &= r, \end{aligned}$$

where r is any real number. Thus the given system has a nontrivial solution. ■

The proof of the following theorem is left as an exercise (Supplementary Exercise T.18).

THEOREM 1.14 *If A is an $n \times n$ matrix, then A is nonsingular if and only if the linear system $A\mathbf{x} = \mathbf{b}$ has a unique solution for every $n \times 1$ matrix $\mathbf{b}$.* ■

We may summarize our results on homogeneous systems and nonsingular matrices in the following list of nonsingular equivalences.

List of Nonsingular Equivalences

The following statements are equivalent.

1. A is nonsingular.
2. $\mathbf{x} = \mathbf{0}$ is the only solution to $A\mathbf{x} = \mathbf{0}$.
3. A is row equivalent to I_n.
4. The linear system $A\mathbf{x} = \mathbf{b}$ has a unique solution for every $n \times 1$ matrix $\mathbf{b}$.

This means that in solving a given problem we can use any of the preceding four statements; they are interchangeable. As you will see throughout this course, a given problem can often be solved in several alternative ways, and sometimes one solution procedure is easier to apply than another. This list of nonsingular equivalences will grow as we progress through the book. By the end of Appendix B it will have expanded to 12 equivalent statements.

INVERSE OF BIT MATRICES (OPTIONAL)

The definitions and theorems developed in this section are all valid for bit matrices. Examples 10 and 11 illustrate the computational procedures developed in this section for bit matrices where, of course, we use arithmetic base 2.

EXAMPLE 10 Find the inverse of the bit matrix

$$A = \begin{bmatrix} 0 & 1 & 1 \\ 1 & 0 & 1 \\ 1 & 1 & 1 \end{bmatrix}.$$

Solution ***Step 1.*** The 3×6 matrix $\left[A \mid I_3\right]$ is

$$\left[A \mid I_3\right] = \left[\begin{array}{ccc|ccc} 0 & 1 & 1 & 1 & 0 & 0 \\ 1 & 0 & 1 & 0 & 1 & 0 \\ 1 & 1 & 1 & 0 & 0 & 1 \end{array}\right].$$

Step 2. We now compute the reduced row echelon form of the matrix obtained in Step 1. To find A^{-1}, we proceed as follows:

$$\begin{array}{c} \qquad A \qquad\qquad\qquad I_3 \\ \left[\begin{array}{ccc|ccc} 0 & 1 & 1 & 1 & 0 & 0 \\ 1 & 0 & 1 & 0 & 1 & 0 \\ 1 & 1 & 1 & 0 & 0 & 1 \end{array}\right] \end{array}$$

$$\left[\begin{array}{ccc|ccc} 1 & 0 & 1 & 0 & 1 & 0 \\ 0 & 1 & 1 & 1 & 0 & 0 \\ 1 & 1 & 1 & 0 & 0 & 1 \end{array}\right]$$ The first and second rows were interchanged.

$$\left[\begin{array}{ccc:ccc} 1 & 0 & 1 & 0 & 1 & 0 \\ 0 & 1 & 1 & 1 & 0 & 0 \\ 0 & 1 & 0 & 0 & 1 & 1 \end{array}\right]$$ The first row was added to the third row.

$$\left[\begin{array}{ccc:ccc} 1 & 0 & 1 & 0 & 1 & 0 \\ 0 & 1 & 0 & 0 & 1 & 1 \\ 0 & 1 & 1 & 1 & 0 & 0 \end{array}\right]$$ The second and third rows were interchanged.

$$\left[\begin{array}{ccc:ccc} 1 & 0 & 1 & 0 & 1 & 0 \\ 0 & 1 & 0 & 0 & 1 & 1 \\ 0 & 0 & 1 & 1 & 1 & 1 \end{array}\right]$$ The second row was added to the third row.

$$\left[\begin{array}{ccc:ccc} 1 & 0 & 0 & 1 & 0 & 1 \\ 0 & 1 & 0 & 0 & 1 & 1 \\ 0 & 0 & 1 & 1 & 1 & 1 \end{array}\right]$$ The third row was added to the first row.

At this point A is row equivalent to I_3, so A is nonsingular and we conclude that

$$A^{-1} = \begin{bmatrix} 1 & 0 & 1 \\ 0 & 1 & 1 \\ 1 & 1 & 1 \end{bmatrix}.$$

■

EXAMPLE 11 Find the inverse of the bit matrix

$$A = \begin{bmatrix} 1 & 1 & 0 \\ 1 & 1 & 1 \\ 0 & 0 & 1 \end{bmatrix}.$$

Solution ***Step 1.*** The 3×6 matrix $\left[A \,\vdots\, I_3\right]$ is

$$\left[A \,\vdots\, I_3\right] = \left[\begin{array}{ccc:ccc} 1 & 1 & 0 & 1 & 0 & 0 \\ 1 & 1 & 1 & 0 & 1 & 0 \\ 0 & 0 & 1 & 0 & 0 & 1 \end{array}\right].$$

Step 2. We now compute the reduced row echelon form of the matrix obtained in Step 1. To find A^{-1}, we proceed as follows:

$$\begin{array}{c} \quad A \qquad\qquad I_3 \\ \left[\begin{array}{ccc:ccc} 1 & 1 & 0 & 1 & 0 & 0 \\ 1 & 1 & 1 & 0 & 1 & 0 \\ 0 & 0 & 1 & 0 & 0 & 1 \end{array}\right] \end{array}$$

$$\left[\begin{array}{ccc:ccc} 1 & 1 & 0 & 1 & 0 & 0 \\ 0 & 0 & 1 & 1 & 1 & 0 \\ 0 & 0 & 1 & 0 & 0 & 1 \end{array}\right]$$ The first row was added to the second row.

$$\left[\begin{array}{ccc:ccc} 1 & 1 & 0 & 1 & 0 & 0 \\ 0 & 0 & 1 & 1 & 1 & 0 \\ 0 & 0 & 0 & 1 & 1 & 1 \end{array}\right].$$ The second row was added to the third row.

At this point we see that A cannot be row equivalent to I_3, since the third row in the coefficient matrix part of the augmented matrix is all zeros. Hence A is singular.

■

EXAMPLE 12 Solve the bit linear system $A\mathbf{x} = \mathbf{b}$, where

$$A = \begin{bmatrix} 0 & 1 & 1 \\ 1 & 0 & 1 \\ 1 & 1 & 1 \end{bmatrix} \quad \text{and} \quad \mathbf{b} = \begin{bmatrix} 0 \\ 1 \\ 1 \end{bmatrix}.$$

Solution From Example 10 we have that A is nonsingular and

$$A^{-1} = \begin{bmatrix} 1 & 0 & 1 \\ 0 & 1 & 1 \\ 1 & 1 & 1 \end{bmatrix}.$$

Thus $A\mathbf{x} = \mathbf{b}$ has a unique solution given by

$$\mathbf{x} = A^{-1}\mathbf{b} = \begin{bmatrix} 1 & 0 & 1 \\ 0 & 1 & 1 \\ 1 & 1 & 1 \end{bmatrix} \begin{bmatrix} 0 \\ 1 \\ 1 \end{bmatrix} = \begin{bmatrix} 1 \\ 0 \\ 0 \end{bmatrix}.$$

■

Section 2.6, Linear Economic Models, Section 2.7, Introduction to Wavelets, and the second half of Section 7.2 (pp. 380–387), Least Squares, which can be studied at this time, use material from this section.

Preview of Applications

Linear Economic Models (Section 2.6)

Economic analysis and forecasting has become increasingly important in our complex modern society. Suppose we have a simple society consisting of only three individuals: a farmer who produces all the food and nothing else, a carpenter who builds all the houses and makes nothing else, and a tailor who makes all the clothes and nothing else. We select units so that each person produces one unit of the commodity that he or she makes during the year. Suppose also that the portion of each commodity that is consumed by each person is given in Table 1.3.

Table 1.3

Goods consumed by:	Goods produced by:		
	Farmer	Carpenter	Tailor
Farmer	$\frac{7}{16}$	$\frac{1}{2}$	$\frac{3}{16}$
Carpenter	$\frac{5}{16}$	$\frac{1}{6}$	$\frac{5}{16}$
Tailor	$\frac{1}{4}$	$\frac{1}{3}$	$\frac{1}{2}$

Thus the farmer consumes $\frac{7}{16}$ of the food produced, $\frac{1}{2}$ of the shelter built by the carpenter, and $\frac{3}{16}$ of the clothes made by the tailor, and so on.

The economist has to determine relative prices p_1, p_2, and p_3 per unit of food, housing, and clothing, respectively, so that no one makes money or loses money. When this situation occurs we say that we have a state of equilibrium. Letting

$$\mathbf{p} = \begin{bmatrix} p_1 \\ p_2 \\ p_3 \end{bmatrix},$$

we find that $\mathbf{p}$ can be obtained by solving the linear system

$$A\mathbf{p} = \mathbf{p}.$$

Section 2.6 discusses this and several other economic models.

Introduction to Wavelets (Section 2.7)

One of the defining characteristics of the twentieth century, which has continued with even greater force into the twenty-first century, is the capability to quickly transmit vast amounts of data. The data to be transmitted includes fingerprint files for law enforcement applications, signal processing for restoring recordings, radio signals from outer space, seismology studies, x-ray results being sent from one medical facility to another, graphical images, and many others. Over the years, a number of schemes have been developed that transform the original data, compress it, transmit it, and recover approximations to the information contained in the original data. Examples of such schemes include Morse code, ciphers of many kinds, and signals used in radio, television, and transmission microwave.

A scheme which is less than 20 years old, known as the method of wavelets, has received a great deal of attention because it can be used successfully in a wide variety of applications in medicine, science, and engineering. The method of wavelets transforms the original data into a form that is equivalent to the given data but is more easily compressed, hence the amount of data to be transmitted has been reduced. Once the compressed data has been transmitted, the next step is to build an approximation to the original data, a wavelet. In Section 2.7, we provide a very elementary introduction to the method of wavelets for small discrete data sets, employing only basic linear algebra techniques.

Least Squares Line Fit (Section 7.2)

The gathering and analyzing of data is a problem that is widely encountered in the sciences, engineering, economics, and the social sciences. When some data are plotted we might obtain Figure 1.20. The least squares problem is that of drawing the straight line that "best fits" the given data. This line is shown in Figure 1.21. The technique for solving this problem is called the method of least squares, which is covered in Section 7.2.

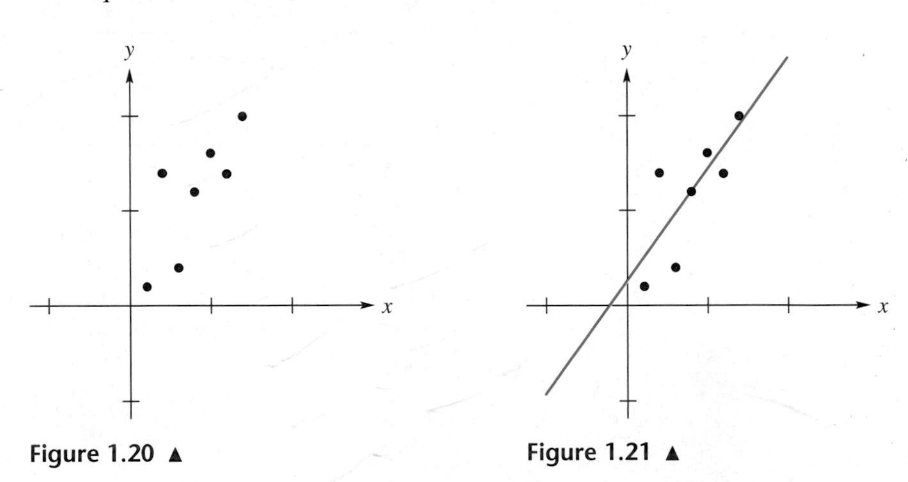

Figure 1.20 ▲ **Figure 1.21** ▲

The procedure for finding the line that best fits the given data is presented in Section 7.2. The justification for this procedure is given in the first part of this section and uses material from Sections 4.2 and 6.9.

Key Terms

Inverse
Nonsingular (or invertible) matrix
Singular (or noninvertible) matrix

1.7 Exercises

In Exercises 1 through 4, use the method of Examples 2 and 3.

1. Show that $\begin{bmatrix} 2 & 1 \\ -2 & 3 \end{bmatrix}$ is nonsingular.

2. Show that $\begin{bmatrix} 2 & 1 \\ -4 & -2 \end{bmatrix}$ is singular.

3. Is the matrix

$$\begin{bmatrix} 1 & 1 \\ 3 & 4 \end{bmatrix}$$

singular or nonsingular? If it is nonsingular, find its inverse.

4. Is the matrix

$$\begin{bmatrix} 1 & 2 & -1 \\ 3 & 2 & 3 \\ 2 & 2 & 1 \end{bmatrix}$$

singular or nonsingular? If it is nonsingular, find its inverse.

In Exercises 5 through 10, find the inverses of the given matrices, if possible.

5. (a) $\begin{bmatrix} 1 & 3 \\ -2 & 6 \end{bmatrix}$ (b) $\begin{bmatrix} 1 & 2 & 3 \\ 1 & 1 & 2 \\ 0 & 1 & 2 \end{bmatrix}$

(c) $\begin{bmatrix} 1 & 1 & 1 & 1 \\ 1 & 2 & -1 & 2 \\ 1 & -1 & 2 & 1 \\ 1 & 3 & 3 & 2 \end{bmatrix}$

6. (a) $\begin{bmatrix} 1 & 3 \\ 2 & 6 \end{bmatrix}$ (b) $\begin{bmatrix} 1 & 2 & 3 \\ 0 & 2 & 3 \\ 1 & 2 & 4 \end{bmatrix}$

(c) $\begin{bmatrix} 1 & 1 & 2 & 1 \\ 0 & -2 & 0 & 0 \\ 0 & 3 & 2 & 1 \\ 1 & 2 & 1 & -2 \end{bmatrix}$

7. (a) $\begin{bmatrix} 1 & 3 \\ 2 & 4 \end{bmatrix}$ (b) $\begin{bmatrix} 1 & 1 & 1 & 1 \\ 1 & 3 & 1 & 2 \\ 1 & 2 & -1 & 1 \\ 5 & 9 & 1 & 6 \end{bmatrix}$

(c) $\begin{bmatrix} 1 & 2 & 1 \\ 1 & 3 & 2 \\ 1 & 0 & 1 \end{bmatrix}$

8. (a) $\begin{bmatrix} 1 & 1 & 1 \\ 1 & 2 & 3 \\ 0 & 1 & 1 \end{bmatrix}$ (b) $\begin{bmatrix} 1 & 2 & 2 \\ 1 & 3 & 1 \\ 1 & 3 & 2 \end{bmatrix}$

(c) $\begin{bmatrix} 1 & 2 & 3 \\ 1 & 1 & 2 \\ 0 & 1 & 1 \end{bmatrix}$

9. (a) $\begin{bmatrix} 1 & 2 & -3 & 1 \\ -1 & 3 & -3 & -2 \\ 2 & 0 & 1 & 5 \\ 3 & 1 & -2 & 5 \end{bmatrix}$

(b) $\begin{bmatrix} 3 & 1 & 2 \\ 2 & 1 & 2 \\ 1 & 2 & 2 \end{bmatrix}$ (c) $\begin{bmatrix} 1 & 2 & 3 \\ 1 & 1 & 2 \\ 1 & 1 & 0 \end{bmatrix}$

10. (a) $\begin{bmatrix} 2 & 1 & 3 \\ 0 & 1 & 2 \\ 1 & 0 & 3 \end{bmatrix}$ (b) $\begin{bmatrix} 1 & -1 & 2 & 3 \\ 4 & 1 & 2 & 0 \\ 2 & -1 & 3 & 1 \\ 4 & 2 & 1 & -5 \end{bmatrix}$

(c) $\begin{bmatrix} 2 & 1 & -2 \\ 3 & 4 & 6 \\ 7 & 6 & 2 \end{bmatrix}$

11. Which of the following linear systems have a nontrivial solution?

(a) $\begin{aligned} x + 2y + 3z &= 0 \\ 2y + 2z &= 0 \\ x + 2y + 3z &= 0 \end{aligned}$ (b) $\begin{aligned} 2x + y - z &= 0 \\ x - 2y - 3z &= 0 \\ -3x - y + 2z &= 0 \end{aligned}$

12. Which of the following linear systems have a nontrivial solution?

(a) $\begin{aligned} x + y + 2z &= 0 \\ 2x + y + z &= 0 \\ 3x - y + z &= 0 \end{aligned}$ (b) $\begin{aligned} x - y + z &= 0 \\ 2x + y &= 0 \\ 2x - 2y + 2z &= 0 \end{aligned}$

(c) $\begin{aligned} 2x - y + 5z &= 0 \\ 3x + 2y - 3z &= 0 \\ x - y + 4z &= 0 \end{aligned}$

13. If $A^{-1} = \begin{bmatrix} 2 & 3 \\ 1 & 4 \end{bmatrix}$, find A.

14. If $A^{-1} = \begin{bmatrix} 3 & 4 \\ -1 & -1 \end{bmatrix}$, find A.

15. Show that a matrix that has a row or column consisting entirely of zeros must be singular.

16. Find all values of a for which the inverse of

$$A = \begin{bmatrix} 1 & 1 & 0 \\ 1 & 0 & 0 \\ 1 & 2 & a \end{bmatrix}$$

exists. What is A^{-1}?

17. Consider an industrial process whose matrix is

$$A = \begin{bmatrix} 2 & 1 & 3 \\ 3 & 2 & -1 \\ 2 & 1 & 1 \end{bmatrix}.$$

Find the input matrix for each of the following output matrices:

(a) $\begin{bmatrix} 30 \\ 20 \\ 10 \end{bmatrix}$ (b) $\begin{bmatrix} 12 \\ 8 \\ 14 \end{bmatrix}$

18. Suppose that $A = \begin{bmatrix} 1 & 3 \\ 2 & 7 \end{bmatrix}$.

(a) Find A^{-1}.

(b) Find $(A^T)^{-1}$. How do $(A^T)^{-1}$ and A^{-1} compare?

19. Is the inverse of a nonsingular symmetric matrix always symmetric? Explain.

20. (a) Is $(A+B)^{-1} = A^{-1} + B^{-1}$ for all A and B?

(b) Is $(cA)^{-1} = \frac{1}{c}A^{-1}$, for $c \neq 0$?

21. For what values of λ does the homogeneous system

$$\begin{aligned} (\lambda - 1)x + \quad\quad 2y &= 0 \\ 2x + (\lambda - 1)y &= 0 \end{aligned}$$

have a nontrivial solution?

22. If A and B are nonsingular, are $A + B$, $A - B$, and $-A$ nonsingular? Explain.

23. If $D = \begin{bmatrix} 4 & 0 & 0 \\ 0 & -2 & 0 \\ 0 & 0 & 3 \end{bmatrix}$, find D^{-1}.

24. If $A^{-1} = \begin{bmatrix} 3 & 2 \\ 1 & 3 \end{bmatrix}$ and $B^{-1} = \begin{bmatrix} 2 & 5 \\ 3 & -2 \end{bmatrix}$, find $(AB)^{-1}$.

25. Solve $A\mathbf{x} = \mathbf{b}$ for $\mathbf{x}$ if

$$A^{-1} = \begin{bmatrix} 2 & 3 \\ 4 & 1 \end{bmatrix} \quad \text{and} \quad \mathbf{b} = \begin{bmatrix} 5 \\ 3 \end{bmatrix}.$$

26. Let A be a 3×3 matrix. Suppose that $\mathbf{x} = \begin{bmatrix} 1 \\ 2 \\ -3 \end{bmatrix}$ is a solution to the homogeneous system $A\mathbf{x} = \mathbf{0}$. Is A singular or nonsingular? Justify your answer.

In Exercises 27 and 28, find the inverse of the given partitioned matrix A and express A^{-1} as a partitioned matrix.

27. $\left[\begin{array}{cc|c} 5 & 2 & 0 \\ 3 & 1 & 0 \\ \hline 0 & 0 & -4 \end{array}\right]$ **28.** $\left[\begin{array}{cc|cc} 1 & 1 & 0 & 0 \\ 2 & 3 & 0 & 0 \\ \hline 0 & 0 & 6 & 7 \\ 0 & 0 & 1 & 1 \end{array}\right]$

In Exercises 29 and 30, find the inverse of the given bit matrices, if possible.

29. (a) $\begin{bmatrix} 1 & 1 & 1 \\ 1 & 1 & 0 \\ 1 & 0 & 0 \end{bmatrix}$ (b) $\begin{bmatrix} 1 & 0 & 1 \\ 1 & 1 & 1 \\ 0 & 1 & 0 \end{bmatrix}$

(c) $\begin{bmatrix} 1 & 1 & 0 & 0 \\ 1 & 1 & 1 & 0 \\ 0 & 1 & 1 & 1 \\ 0 & 0 & 1 & 1 \end{bmatrix}$

30. (a) $\begin{bmatrix} 0 & 1 & 1 \\ 1 & 0 & 1 \\ 1 & 1 & 0 \end{bmatrix}$ (b) $\begin{bmatrix} 1 & 0 & 0 \\ 1 & 1 & 1 \\ 1 & 0 & 1 \end{bmatrix}$

(c) $\begin{bmatrix} 0 & 1 & 1 & 0 \\ 0 & 0 & 1 & 1 \\ 1 & 0 & 0 & 1 \\ 1 & 1 & 0 & 0 \end{bmatrix}$

In Exercises 31 and 32, determine which bit linear systems have a nontrivial solution.

31. (a) $\begin{aligned} x + y + z &= 0 \\ x \quad + z &= 0 \\ y \quad &= 0 \end{aligned}$ (b) $\begin{aligned} x \quad\quad &= 0 \\ x + y + z &= 0 \\ x \quad + z &= 0 \end{aligned}$

32. (a) $\begin{aligned} x + y \quad &= 0 \\ x + y + z &= 0 \\ y + z &= 0 \end{aligned}$ (b) $\begin{aligned} y + z &= 0 \\ x \quad + z &= 0 \\ x + y \quad &= 0 \end{aligned}$

Theoretical Exercises

T.1. Suppose that A and B are square matrices and $AB = O$. If B is nonsingular, find A.

T.2. Prove Corollary 1.2.

T.3. Let A be an $n \times n$ matrix. Show that if A is singular, then the homogeneous system $A\mathbf{x} = \mathbf{0}$ has a nontrivial solution. (*Hint*: Use Theorem 1.12.)

T.4. Show that the matrix

$$A = \begin{bmatrix} a & b \\ c & d \end{bmatrix}$$

is nonsingular if and only if $ad - bc \neq 0$. If this condition holds, show that

$$A^{-1} = \begin{bmatrix} \dfrac{d}{ad-bc} & \dfrac{-b}{ad-bc} \\ \dfrac{-c}{ad-bc} & \dfrac{a}{ad-bc} \end{bmatrix}.$$

T.5. Show that the matrix

$$\begin{bmatrix} \cos\theta & \sin\theta \\ -\sin\theta & \cos\theta \end{bmatrix}$$

is nonsingular, and compute its inverse.

T.6. Show that the inverse of a nonsingular upper (lower) triangular matrix is upper (lower) triangular.

T.7. Show that if A is singular and $A\mathbf{x} = \mathbf{b}$, $\mathbf{b} \neq \mathbf{0}$ has one solution, then it has infinitely many. (*Hint*: Use Exercise T.13 in Section 1.6.)

T.8. Show that if A is a nonsingular symmetric matrix, then A^{-1} is symmetric.

T.9. Let A be a diagonal matrix with nonzero diagonal entries $a_{11}, a_{22}, \dots, a_{nn}$. Show that A^{-1} is nonsingular and that A^{-1} is a diagonal matrix with diagonal entries $1/a_{11}, 1/a_{22}, \dots, 1/a_{nn}$.

T.10. If $B = PAP^{-1}$, express $B^2, B^3, \dots, B^k$, where k is a positive integer, in terms of A, P, and P^{-1}.

T.11. Make a list of all possible 2×2 bit matrices and then determine which are nonsingular. (See Exercise T.13 in Section 1.2.)

T.12. If A and B are nonsingular 3×3 bit matrices, is it possible that $AB = O$? Explain.

T.13. Determine which 2×2 bit matrices A have the property that $A^2 = O$. (See Exercise T.13 in Section 1.2.)

MATLAB Exercises

In order to use MATLAB *in this section, you should first have read Chapter 12 through Section 12.5.*

ML.1. Using MATLAB, determine which of the following matrices are nonsingular. Use command **rref**.

(a) $A = \begin{bmatrix} 1 & 2 \\ -2 & 1 \end{bmatrix}$

(b) $A = \begin{bmatrix} 1 & 2 & 3 \\ 4 & 5 & 6 \\ 7 & 8 & 9 \end{bmatrix}$

(c) $A = \begin{bmatrix} 1 & 2 & 3 \\ 4 & 5 & 6 \\ 7 & 8 & 0 \end{bmatrix}$

ML.2. Using MATLAB, determine which of the following matrices are nonsingular. Use command **rref**.

(a) $A = \begin{bmatrix} 1 & 2 \\ 2 & 4 \end{bmatrix}$ (b) $A = \begin{bmatrix} 1 & 0 & 0 \\ 0 & 1 & 0 \\ 1 & 1 & 1 \end{bmatrix}$

(c) $A = \begin{bmatrix} 1 & 2 & 1 \\ 0 & 1 & 2 \\ 1 & 0 & 0 \end{bmatrix}$

ML.3. Using MATLAB, determine the inverse of each of the following matrices. Use command **rref([A eye(size(A))])**.

(a) $A = \begin{bmatrix} 1 & 3 \\ 1 & 2 \end{bmatrix}$ (b) $A = \begin{bmatrix} 1 & 1 & 2 \\ 2 & 1 & 1 \\ 1 & 2 & 1 \end{bmatrix}$

ML.4. Using MATLAB, determine the inverse of each of the following matrices. Use command **rref([A eye(size(A))])**.

(a) $A = \begin{bmatrix} 2 & 1 \\ 2 & 3 \end{bmatrix}$ (b) $A = \begin{bmatrix} 1 & -1 & 2 \\ 0 & 2 & 1 \\ 1 & 0 & 0 \end{bmatrix}$

ML.5. Using MATLAB, determine a positive integer t so that $(tI - A)$ is singular.

(a) $A = \begin{bmatrix} 1 & 3 \\ 3 & 1 \end{bmatrix}$ (b) $A = \begin{bmatrix} 4 & 1 & 2 \\ 1 & 4 & 1 \\ 0 & 0 & -4 \end{bmatrix}$

Exercises ML.6 through ML.9 use bit matrices and the supplemental instructional commands described in Section 12.9.

ML.6. Determine which of the bit matrices in Exercises 29 and 30 have an inverse using **binreduce**.

ML.7. Determine which of the bit linear systems in Exercises 31 and 32 have a nontrivial solution using **binreduce**.

ML.8. Determine which of the following matrices has an inverse using **binreduce**.

(a) $\begin{bmatrix} 1 & 1 & 0 & 0 \\ 0 & 1 & 1 & 0 \\ 1 & 1 & 1 & 1 \\ 0 & 1 & 0 & 0 \end{bmatrix}$ (b) $\begin{bmatrix} 1 & 1 & 0 & 0 & 1 \\ 0 & 1 & 1 & 1 & 0 \\ 0 & 0 & 0 & 1 & 0 \\ 0 & 1 & 1 & 1 & 1 \\ 1 & 1 & 0 & 1 & 1 \end{bmatrix}$

ML.9. Let $B = \mathbf{bingen(1, 7, 3)}$; that is, the matrix whose columns are the binary representations of the integers 1 through 7 using three bits. Determine two 3×3 submatrices that have an inverse and two that do not.

1.8 LU-FACTORIZATION (OPTIONAL)

In this section we discuss a variant of Gaussian elimination (presented in Section 1.6) that decomposes a matrix as a product of a lower triangular matrix and an upper triangular matrix. This decomposition leads to an algorithm for solving a linear system $A\mathbf{x} = \mathbf{b}$ that is the most widely used method on computers for solving a linear system. A main reason for the popularity of this method is that it provides the cheapest way of solving a linear system for which we repeatedly have to change the right side. This type of situation occurs often in applied problems. For example, an electric utility company must determine the inputs (the unknowns) needed to produce some required outputs (the right sides). The inputs and outputs might be related by a linear system, whose coefficient matrix is fixed, while the right side changes from day to day, or even hour to hour. The decomposition discussed in this section is also useful in solving other problems in linear algebra.

When U is an upper triangular matrix all of whose diagonal entries are different from zero, then the linear system $U\mathbf{x} = \mathbf{b}$ can be solved without transforming the augmented matrix $\left[U \;\vdots\; \mathbf{b}\right]$ to reduced row echelon form or to row echelon form. The augmented matrix of such a system is given by

$$\left[\begin{array}{cccccc:c} u_{11} & u_{12} & u_{13} & \cdots & u_{1n} & b_1 \\ 0 & u_{22} & u_{23} & \cdots & u_{2n} & b_2 \\ 0 & 0 & u_{33} & \cdots & u_{3n} & b_3 \\ \vdots & \vdots & \vdots & \cdots & \vdots & \vdots \\ 0 & 0 & 0 & \cdots & u_{nn} & b_n \end{array}\right].$$

The solution is obtained by the following algorithm:

$$x_n = \frac{b_n}{u_{nn}}$$

$$x_{n-1} = \frac{b_{n-1} - u_{n-1\,n}x_n}{u_{n-1\,n-1}}$$

$$\vdots$$

$$x_j = \frac{b_j - \sum_{k=n}^{j-1} u_{jk}x_k}{u_{jj}}, \qquad j = n, n-1, \ldots, 2, 1.$$

This procedure is merely **back substitution**, which we used in conjunction with Gaussian elimination in Section 1.6, where it was additionally required that the diagonal entries be 1.

In a similar manner, if L is a lower triangular matrix all of whose diagonal entries are different from zero, then the linear system $L\mathbf{x} = \mathbf{b}$ can be solved by **forward substitution**, which consists of the following procedure: The augmented matrix has the form

$$\left[\begin{array}{cccccc} \ell_{11} & 0 & 0 & \cdots & 0 & b_1 \\ \ell_{21} & \ell_{22} & 0 & \cdots & 0 & b_2 \\ \ell_{31} & \ell_{32} & \ell_{33} & \cdots & 0 & b_3 \\ \vdots & \vdots & \vdots & \cdots & \vdots & \vdots \\ \ell_{n1} & \ell_{n2} & \ell_{n3} & \cdots & \ell_{nn} & b_n \end{array}\right]$$

and the solution is given by

$$x_1 = \frac{b_1}{\ell_{11}}$$

$$x_2 = \frac{b_2 - \ell_{21}x_1}{\ell_{22}}$$

$$\vdots$$

$$x_j = \frac{b_j - \sum_{k=1}^{j-1} \ell_{jk}x_k}{\ell_{jj}}, \qquad j = 2, \ldots, n.$$

That is, we proceed from the first equation downward, solving for one unknown from each equation.

We illustrate forward substitution in the following example.

EXAMPLE 1 To solve the linear system

$$\begin{aligned} 5x_1 \qquad\qquad &= 10 \\ 4x_1 - 2x_2 \qquad &= 28 \\ 2x_1 + 3x_2 + 4x_3 &= 26 \end{aligned}$$

we use forward substitution. Hence we obtain from the previous algorithm

$$x_1 = \frac{10}{5} = 2$$

$$x_2 = \frac{28 - 4x_1}{-2} = -10$$

$$x_3 = \frac{26 - 2x_1 - 3x_2}{4} = 13,$$

which implies that the solution to the given lower triangular system of equations is

$$\mathbf{x} = \begin{bmatrix} 2 \\ -10 \\ 13 \end{bmatrix}.$$

■

As illustrated previously, the ease with which systems of equations with upper or lower triangular coefficient matrices can be solved is quite attractive. The forward substitution and back substitution algorithms are fast and simple to use. These are used in another important numerical procedure for solving linear systems of equations, which we develop next.

Suppose that an $n \times n$ matrix A can be written as a product of a matrix L in lower triangular form and a matrix U in upper triangular form; that is,

$$A = LU.$$

In this case we say that A has an **LU-factorization** or an **LU-decomposition**. The LU-factorization of a matrix A can be used to efficiently solve a linear system $A\mathbf{x} = \mathbf{b}$. Substituting LU for A, we have

$$(LU)\mathbf{x} = \mathbf{b}$$

or by (a) of Theorem 1.2 in Section 1.4,

$$L(U\mathbf{x}) = \mathbf{b}.$$

Letting $U\mathbf{x} = \mathbf{z}$, this matrix equation becomes

$$L\mathbf{z} = \mathbf{b}.$$

Since L is in lower triangular form, we solve directly for $\mathbf{z}$ by forward substitution. Once we determine $\mathbf{z}$, since U is in upper triangular form, we solve $U\mathbf{x} = \mathbf{z}$ by back substitution. In summary, if an $n \times n$ matrix A has an LU-factorization, then the solution of $A\mathbf{x} = \mathbf{b}$ can be determined by a forward substitution followed by a back substitution. We illustrate this procedure in the next example.

EXAMPLE 2 Consider the linear system

$$\begin{aligned} 6x_1 - 2x_2 - 4x_3 + 4x_4 &= 2 \\ 3x_1 - 3x_2 - 6x_3 + x_4 &= -4 \\ -12x_1 + 8x_2 + 21x_3 - 8x_4 &= 8 \\ -6x_1 \qquad - 10x_3 + 7x_4 &= -43 \end{aligned}$$

whose coefficient matrix

$$A = \begin{bmatrix} 6 & -2 & -4 & 4 \\ 3 & -3 & -6 & 1 \\ -12 & 8 & 21 & -8 \\ -6 & 0 & -10 & 7 \end{bmatrix}$$

has an LU-factorization where

$$L = \begin{bmatrix} 1 & 0 & 0 & 0 \\ \frac{1}{2} & 1 & 0 & 0 \\ -2 & -2 & 1 & 0 \\ -1 & 1 & -2 & 1 \end{bmatrix} \quad \text{and} \quad U = \begin{bmatrix} 6 & -2 & -4 & 4 \\ 0 & -2 & -4 & -1 \\ 0 & 0 & 5 & -2 \\ 0 & 0 & 0 & 8 \end{bmatrix}$$

(verify). To solve the given system using this LU-factorization, we proceed as follows. Let

$$\mathbf{b} = \begin{bmatrix} 2 \\ -4 \\ 8 \\ -43 \end{bmatrix}.$$

Then we solve $A\mathbf{x} = \mathbf{b}$ by writing it as $LU\mathbf{x} = \mathbf{b}$. First, let $U\mathbf{x} = \mathbf{z}$ and solve $L\mathbf{z} = \mathbf{b}$:

$$\begin{bmatrix} 1 & 0 & 0 & 0 \\ \frac{1}{2} & 1 & 0 & 0 \\ -2 & -2 & 1 & 0 \\ -1 & 1 & -2 & 1 \end{bmatrix} \begin{bmatrix} z_1 \\ z_2 \\ z_3 \\ z_4 \end{bmatrix} = \begin{bmatrix} 2 \\ -4 \\ 8 \\ -43 \end{bmatrix}$$

by forward substitution. We obtain

$$\begin{aligned} z_1 &= 2 \\ z_2 &= -4 - \tfrac{1}{2}z_1 = -5 \\ z_3 &= 8 + 2z_1 + 2z_2 = 2 \\ z_4 &= -43 + z_1 - z_2 + 2z_3 = -32. \end{aligned}$$

Next we solve $U\mathbf{x} = \mathbf{z}$,

$$\begin{bmatrix} 6 & -2 & -4 & 4 \\ 0 & -2 & -4 & -1 \\ 0 & 0 & 5 & -2 \\ 0 & 0 & 0 & 8 \end{bmatrix} \begin{bmatrix} x_1 \\ x_2 \\ x_3 \\ x_4 \end{bmatrix} = \begin{bmatrix} 2 \\ -5 \\ 2 \\ -32 \end{bmatrix},$$

by back substitution. We obtain

$$\begin{aligned} x_4 &= \frac{-32}{8} = -4 \\ x_3 &= \frac{2 + 2x_4}{5} = -1.2 \\ x_2 &= \frac{-5 + 4x_3 + x_4}{-2} = 6.9 \\ x_1 &= \frac{2 + 2x_2 + 4x_3 - 4x_4}{6} = 4.5. \end{aligned}$$

Thus the solution to the given linear system is

$$\mathbf{x} = \begin{bmatrix} 4.5 \\ 6.9 \\ -1.2 \\ -4 \end{bmatrix}.$$

■

Next, we show how to obtain an LU-factorization of a matrix by modifying the Gaussian elimination procedure from Section 1.6. No row interchanges will be permitted and we do not require that the diagonal entries have value 1. At the end of this section we provide a reference that indicates how to enhance the LU-factorization scheme presented to deal with matrices where row interchanges are necessary. We observe that the only elementary row operation permitted is the one that adds a multiple of one row to a different row.

To describe the LU-factorization, we present a step-by-step procedure in the next example.

EXAMPLE 3 Let A be the coefficient matrix of the linear system of Example 2.

$$A = \begin{bmatrix} 6 & -2 & -4 & 4 \\ 3 & -3 & -6 & 1 \\ -12 & 8 & 21 & -8 \\ -6 & 0 & -10 & 7 \end{bmatrix}.$$

We proceed to "zero out" entries below the diagonal entries using only the row operation that adds a multiple of one row to a different row.

Procedure	Matrices Used
Step 1. "Zero out" below the first diagonal entry of A. Add $\left(-\frac{1}{2}\right)$ times the first row of A to the second row of A. Add 2 times the first row of A to the third row of A. Add 1 times the first row of A to the fourth row of A. Call the new resulting matrix U_1.	$U_1 = \begin{bmatrix} 6 & -2 & -4 & 4 \\ 0 & -2 & -4 & -1 \\ 0 & 4 & 13 & 0 \\ 0 & -2 & -14 & 11 \end{bmatrix}$
We begin building a lower triangular matrix, L_1, with 1s on the main diagonal, to record the row operations. Enter the *negatives of the multipliers* used in the row operations in the first column of L_1, below the first diagonal entry of L_1.	$L_1 = \begin{bmatrix} 1 & 0 & 0 & 0 \\ \frac{1}{2} & 1 & 0 & 0 \\ -2 & * & 1 & 0 \\ -1 & * & * & 1 \end{bmatrix}$
Step 2. "Zero out" below the second diagonal entry of U_1. Add 2 times the second row of U_1 to the third row of U_1. Add (-1) times the second row of U_1 to the fourth row of U_1. Call the new resulting matrix U_2.	$U_2 = \begin{bmatrix} 6 & -2 & -4 & 4 \\ 0 & -2 & -4 & -1 \\ 0 & 0 & 5 & -2 \\ 0 & 0 & -10 & 12 \end{bmatrix}$
Enter the negatives of the multipliers from the row operations below the second diagonal entry of L_1. Call the new matrix L_2.	$L_2 = \begin{bmatrix} 1 & 0 & 0 & 0 \\ \frac{1}{2} & 1 & 0 & 0 \\ -2 & -2 & 1 & 0 \\ -1 & 1 & * & 1 \end{bmatrix}$

Step 3. "Zero out" below the third diagonal entry of U_2. Add 2 times the third row of U_2 to the fourth row of U_2. Call the new resulting matrix U_3.

$$U_3 = \begin{bmatrix} 6 & -2 & -4 & 4 \\ 0 & -2 & -4 & -1 \\ 0 & 0 & 5 & -2 \\ 0 & 0 & 0 & 8 \end{bmatrix}$$

Enter the negative of the multiplier below the third diagonal entry of L_2. Call the new matrix L_3.

$$L_3 = \begin{bmatrix} 1 & 0 & 0 & 0 \\ \frac{1}{2} & 1 & 0 & 0 \\ -2 & -2 & 1 & 0 \\ -1 & 1 & -2 & 1 \end{bmatrix}$$

Let $L = L_3$ and $U = U_3$. Then the product LU gives the original matrix A (verify). This linear system of equations was solved in Example 2 using the LU-factorization just obtained. ■

Remark In general, a given matrix may have more than one LU-factorization. For example, if A is the coefficient matrix considered in Example 2, then another LU-factorization is LU, where

$$L = \begin{bmatrix} 2 & 0 & 0 & 0 \\ 1 & -1 & 0 & 0 \\ -4 & 2 & 1 & 0 \\ -2 & -1 & -2 & 2 \end{bmatrix} \quad \text{and} \quad U = \begin{bmatrix} 3 & -1 & -2 & 2 \\ 0 & 2 & 4 & 1 \\ 0 & 0 & 5 & -2 \\ 0 & 0 & 0 & 4 \end{bmatrix}.$$

There are many methods for obtaining an LU-factorization of a matrix, besides the scheme for **storage of multipliers** described in Example 3. It is important to note that if $a_{11} = 0$, then the procedure used in Example 3 fails. Moreover, if the second diagonal entry of U_1 is zero or if the third diagonal entry of U_2 is zero, then the procedure also fails. In such cases we can try rearranging the equations of the system and beginning again or using one of the other methods for LU-factorization. Most computer programs for LU-factorization incorporate row interchanges into the storage of multipliers scheme and use additional strategies to help control roundoff error. If row interchanges are required, then the product of L and U is not necessarily A—it is a matrix that is a permutation of the rows of A. For example, if row interchanges occur when using the **lu** command in MATLAB in the form **[L,U] = lu(A)**, then MATLAB responds as follows: The matrix that it yields as L is not lower triangular, U is upper triangular, and LU is A. The book *Experiments in Computational Matrix Algebra*, by David R. Hill (New York: Random House, 1988, distributed by McGraw-Hill) explores such a modification of the procedure for LU-factorization.

Key Terms

Back substitution
Forward substitution
LU-factorization (or LU-decomposition)

1.8 Exercises

In Exercises 1 through 4, solve the linear system $A\mathbf{x} = \mathbf{b}$ with the given LU-factorization of the coefficient matrix A. Solve the linear system using a forward substitution followed by a back substitution.

1. $A = \begin{bmatrix} 2 & 8 & 0 \\ 2 & 2 & -3 \\ 1 & 2 & 7 \end{bmatrix}, \quad \mathbf{b} = \begin{bmatrix} 18 \\ 3 \\ 12 \end{bmatrix},$

$L = \begin{bmatrix} 2 & 0 & 0 \\ 2 & -3 & 0 \\ 1 & -1 & 4 \end{bmatrix}, \quad U = \begin{bmatrix} 1 & 4 & 0 \\ 0 & 2 & 1 \\ 0 & 0 & 2 \end{bmatrix}$

2. $A = \begin{bmatrix} 8 & 12 & -4 \\ 6 & 5 & 7 \\ 2 & 1 & 6 \end{bmatrix}, \quad \mathbf{b} = \begin{bmatrix} -36 \\ 11 \\ 16 \end{bmatrix},$

$L = \begin{bmatrix} 4 & 0 & 0 \\ 3 & 2 & 0 \\ 1 & 1 & 1 \end{bmatrix}, \quad U = \begin{bmatrix} 2 & 3 & -1 \\ 0 & -2 & 5 \\ 0 & 0 & 2 \end{bmatrix}$

3. $A = \begin{bmatrix} 2 & 3 & 0 & 1 \\ 4 & 5 & 3 & 3 \\ -2 & -6 & 7 & 7 \\ 8 & 9 & 5 & 21 \end{bmatrix}, \quad \mathbf{b} = \begin{bmatrix} -2 \\ -2 \\ -16 \\ -66 \end{bmatrix},$

$L = \begin{bmatrix} 1 & 0 & 0 & 0 \\ 2 & 1 & 0 & 0 \\ -1 & 3 & 1 & 0 \\ 4 & 3 & 2 & 1 \end{bmatrix},$

$U = \begin{bmatrix} 2 & 3 & 0 & 1 \\ 0 & -1 & 3 & 1 \\ 0 & 0 & -2 & 5 \\ 0 & 0 & 0 & 4 \end{bmatrix}$

4. $A = \begin{bmatrix} 4 & 2 & 1 & 0 \\ -4 & -6 & 1 & 3 \\ 8 & 16 & -3 & -4 \\ 20 & 10 & 4 & -3 \end{bmatrix}, \quad \mathbf{b} = \begin{bmatrix} 6 \\ 13 \\ -20 \\ 15 \end{bmatrix},$

$L = \begin{bmatrix} 1 & 0 & 0 & 0 \\ -1 & 1 & 0 & 0 \\ 2 & -3 & 1 & 0 \\ 5 & 0 & -1 & 1 \end{bmatrix},$

$U = \begin{bmatrix} 4 & 2 & 1 & 0 \\ 0 & -4 & 2 & 3 \\ 0 & 0 & 1 & 5 \\ 0 & 0 & 0 & 2 \end{bmatrix}$

In Exercises 5 through 10, find an LU-factorization of the coefficient matrix of the given linear system $A\mathbf{x} = \mathbf{b}$. Solve the linear system using a forward substitution followed by a back substitution.

5. $A = \begin{bmatrix} 2 & 3 & 4 \\ 4 & 5 & 10 \\ 4 & 8 & 2 \end{bmatrix}, \quad \mathbf{b} = \begin{bmatrix} 6 \\ 16 \\ 2 \end{bmatrix}.$

6. $A = \begin{bmatrix} -3 & 1 & -2 \\ -12 & 10 & -6 \\ 15 & 13 & 12 \end{bmatrix}, \quad \mathbf{b} = \begin{bmatrix} 15 \\ 82 \\ -5 \end{bmatrix}$

7. $A = \begin{bmatrix} 4 & 2 & 3 \\ 2 & 0 & 5 \\ 1 & 2 & 1 \end{bmatrix}, \quad \mathbf{b} = \begin{bmatrix} 1 \\ -1 \\ -3 \end{bmatrix}$

8. $A = \begin{bmatrix} -5 & 4 & 0 & 1 \\ -30 & 27 & 2 & 7 \\ 5 & 2 & 0 & 2 \\ 10 & 1 & -2 & 1 \end{bmatrix}, \quad \mathbf{b} = \begin{bmatrix} -17 \\ -102 \\ -7 \\ -6 \end{bmatrix}$

9. $A = \begin{bmatrix} 2 & 1 & 0 & -4 \\ 1 & 0 & 0.25 & -1 \\ -2 & -1.1 & 0.25 & 6.2 \\ 4 & 2.2 & 0.3 & -2.4 \end{bmatrix},$

$\mathbf{b} = \begin{bmatrix} -3 \\ -1.5 \\ 5.6 \\ 2.2 \end{bmatrix}$

10. $A = \begin{bmatrix} 4 & 1 & 0.25 & -0.5 \\ 0.8 & 0.6 & 1.25 & -2.6 \\ -1.6 & -0.08 & 0.01 & 0.2 \\ 8 & 1.52 & -0.6 & -1.3 \end{bmatrix},$

$\mathbf{b} = \begin{bmatrix} -0.15 \\ 9.77 \\ 1.69 \\ -4.576 \end{bmatrix}$

MATLAB Exercises

Routine **lupr** *provides a step-by-step procedure in* MATLAB *for obtaining the LU-factorization discussed in this section. Once we have the LU-factorization, routines* **forsub** *and* **bksub** *can be used to perform the forward and back substitution, respectively. Use* **help** *for further information on these routines.*

ML.1. Use **lupr** in MATLAB to find an LU-factorization of

$$A = \begin{bmatrix} 2 & 8 & 0 \\ 2 & 2 & -3 \\ 1 & 2 & 7 \end{bmatrix}.$$

ML.2. Use **lupr** in MATLAB to find an LU-factorization of

$$A = \begin{bmatrix} 8 & -1 & 2 \\ 3 & 7 & 2 \\ 1 & 1 & 5 \end{bmatrix}.$$

ML.3. Solve the linear system in Example 2 using **lupr**, **forsub**, and **bksub** in MATLAB. Check your LU-factorization using Example 3.

ML.4. Solve Exercises 7 and 8 using **lupr**, **forsub**, and **bksub** in MATLAB.

Key Ideas for Review

- **Method of elimination.** To solve a linear system, repeatedly perform the following operations.

  1. Interchange two equations.
  2. Multiply an equation by a nonzero constant.
  3. Add a multiple of one equation to another equation.

- **Matrix operations. Addition** (see page 14); **scalar multiplication** (see page 15); **transpose** (see page 16); **multiplication** (see page 23).
- **Theorem 1.1.** Properties of matrix addition. See page 39.
- **Theorem 1.2.** Properties of matrix multiplication. See page 41.
- **Theorem 1.3.** Properties of scalar multiplication. See page 45.
- **Theorem 1.4.** Properties of transpose. See page 45.
- **Reduced row echelon form.** See page 62.
- **Procedure for transforming a matrix to reduced row echelon form.** See pages 65–66.
- **Gauss–Jordan reduction procedure** (for solving the linear system $A\mathbf{x} = \mathbf{b}$). See page 70.
- **Gaussian elimination procedure** (for solving the linear system $A\mathbf{x} = \mathbf{b}$). See page 70.
- **Theorem 1.8.** A homogeneous system of m equations in n unknowns always has a nontrivial solution if $m < n$.
- **Theorem 1.10.** Properties of the inverse. See pages 92–93.
- **Practical method for finding A^{-1}.** See pages 94–95.
- **Theorem 1.12.** An $n \times n$ matrix is nonsingular if and only if it is row equivalent to I_n.
- **Theorem 1.13.** If A is an $n \times n$ matrix, the homogeneous system $A\mathbf{x} = \mathbf{0}$ has a nontrivial solution if and only if A is singular.
- **List of Nonsingular Equivalences.** The following statements are equivalent:

  1. A is nonsingular.
  2. $\mathbf{x} = \mathbf{0}$ is the only solution to $A\mathbf{x} = \mathbf{0}$.
  3. A is row equivalent to I_n.
  4. The linear system $A\mathbf{x} = \mathbf{b}$ has a unique solution for every $n \times 1$ matrix $\mathbf{b}$.

- **LU-factorization** (for writing an $n \times n$ matrix A as LU, where L is in lower triangular form and U is in upper triangular form). See Example 3, pages 111–112.

Supplementary Exercises

In Exercises 1 through 3, let

$$A = \begin{bmatrix} 1 & 2 \\ 3 & -2 \end{bmatrix}, \quad B = \begin{bmatrix} 3 & -5 \\ 2 & 4 \end{bmatrix}, \quad \textit{and} \quad C = \begin{bmatrix} 4 & 1 \\ 3 & 2 \end{bmatrix}.$$

1. Compute $2A + BC$ if possible.

2. Compute $A^2 - 2A + 3I_2$ if possible.

3. Compute $A^T + B^T C$ if possible.

4. (a) If A and B are $n \times n$ matrices, when is
$$(A + B)(A - B) = A^2 - B^2?$$
(b) Let A, B, and C be $n \times n$ matrices such that $AC = CA$ and $BC = CB$. Verify that $(AB)C = C(AB)$.

5. (a) Write the augmented matrix of the linear system
$$\begin{aligned} x_1 + 2x_2 - x_3 + \ x_4 &= 7 \\ 2x_1 - \ x_2 \qquad + 2x_4 &= -8. \end{aligned}$$
(b) Write the linear system whose augmented matrix is
$$\left[\begin{array}{cc|c} 3 & 2 & -4 \\ 5 & 1 & 2 \\ 3 & 2 & 6 \end{array}\right].$$

6. Let $f\colon R^2 \to R^2$ be the matrix transformation defined by $f(\mathbf{x}) = A\mathbf{x}$, where
$$A = \begin{bmatrix} 2 & 1 & 2 \\ 1 & 0 & -1 \\ 3 & 1 & k \end{bmatrix}.$$
Determine k so that $\mathbf{w} = \begin{bmatrix} 1 \\ 1 \\ 1 \end{bmatrix}$ is not in the range of f.

7. Let $f\colon R^4 \to R^3$ be the matrix transformation defined by $f(\mathbf{x}) = A\mathbf{x}$, where
$$A = \begin{bmatrix} 0 & 2 & 1 & 0 \\ 1 & 0 & 2 & 1 \\ 1 & 1 & k & t \end{bmatrix}.$$
Determine all values of k and t so that $\mathbf{w} = \begin{bmatrix} 4 \\ 2 \\ 6 \end{bmatrix}$ is in the range of f.

8. If
$$A = \begin{bmatrix} 1 & 3 & -2 & 5 \\ 2 & 1 & 3 & 2 \\ 4 & 7 & -1 & -8 \\ -3 & 1 & -8 & 1 \end{bmatrix},$$
find a matrix C in reduced row echelon form that is row equivalent to A.

9. Find all solutions to the linear system
$$\begin{aligned} x + y - \ z &= 5 \\ 2x + y + \ z &= 2 \\ x - y - 2z &= 3. \end{aligned}$$

10. Find all solutions to the linear system

$$\begin{aligned} x + y - z + w &= 3 \\ 2x - y - z + 2w &= 4 \\ -3y + z &= -2 \\ -3x + 3y + z - 3w &= -5. \end{aligned}$$

11. Find all values of a for which the resulting system has (a) no solution, (b) a unique solution, and (c) infinitely many solutions.

$$\begin{aligned} x + y - z &= 3 \\ x - y + 3z &= 4 \\ x + y + (a^2 - 10)z &= a. \end{aligned}$$

12. Find an equation relating b_1, b_2, and b_3 so that the linear system with augmented matrix

$$\left[\begin{array}{rrr|r} 1 & 1 & -2 & b_1 \\ 2 & -1 & -1 & b_2 \\ 4 & 1 & -5 & b_3 \end{array}\right]$$

has a solution.

13. If

$$A = \begin{bmatrix} 5 & 3 & 1 \\ 0 & 4 & 2 \\ 0 & 0 & 4 \end{bmatrix}$$

and $\lambda = 4$, find all solutions to the homogeneous system $(\lambda I_3 - A)\mathbf{x} = \mathbf{0}$.

14. For what values of a is the linear system

$$\begin{aligned} x_1 + x_3 &= a^2 \\ 2x_1 + x_2 + 3x_3 &= -3a \\ 3x_1 + x_2 + 4x_3 &= -2 \end{aligned}$$

consistent?

15. If possible, find the inverse of the matrix

$$\begin{bmatrix} 1 & 2 & 3 \\ 2 & 5 & 3 \\ 1 & 0 & 8 \end{bmatrix}.$$

16. If possible, find the inverse of the matrix

$$\begin{bmatrix} -1 & 2 & 1 \\ 1 & 0 & 1 \\ -3 & 2 & 3 \end{bmatrix}.$$

17. Does the homogeneous system with coefficient matrix

$$\begin{bmatrix} 1 & -1 & 3 \\ 1 & 2 & -3 \\ 2 & 1 & 0 \end{bmatrix}$$

have a nontrivial solution?

18. If $A^{-1} = \begin{bmatrix} 1 & 2 & -1 \\ 3 & 4 & 2 \\ 0 & 1 & -2 \end{bmatrix}$ and

$$B^{-1} = \begin{bmatrix} 0 & 1 & 1 \\ 1 & 0 & 1 \\ -2 & 3 & 2 \end{bmatrix},$$

compute $(AB)^{-1}$.

19. Solve $A\mathbf{x} = \mathbf{b}$ for $\mathbf{x}$ if

$$A^{-1} = \begin{bmatrix} 1 & 2 & 0 \\ 0 & 1 & 0 \\ 3 & 1 & -1 \end{bmatrix} \quad \text{and} \quad \mathbf{b} = \begin{bmatrix} 2 \\ 1 \\ 3 \end{bmatrix}.$$

20. For what value(s) of λ does the homogeneous system

$$\begin{aligned} (\lambda - 2)x + 2y &= 0 \\ 2x + (\lambda - 2)y &= 0 \end{aligned}$$

have a nontrivial solution?

21. If A is an $n \times n$ matrix and $A^4 = O$, verify that $(I_n - A)^{-1} = I_n + A + A^2 + A^3$.

22. If A is a nonsingular $n \times n$ matrix and c is a nonzero scalar, what is $(cA)^{-1}$?

23. For what values of a does the linear system

$$\begin{aligned} x + y &= 3 \\ 5x + 5y &= a \end{aligned}$$

have (a) no solution; (b) exactly one solution; (c) infinitely many solutions?

24. Find all values of a for which the following linear systems have solutions.

(a) $$\begin{aligned} x + 2y + z &= a^2 \\ x + y + 3z &= a \\ 3x + 4y + 7z &= 8 \end{aligned}$$

(b) $$\begin{aligned} x + 2y + z &= a^2 \\ x + y + 3z &= a \\ 3x + 4y + 8z &= 8 \end{aligned}$$

25. Find all values of a for which the following homogeneous system has nontrivial solutions.

$$\begin{aligned} (1 - a)x + z &= 0 \\ -ay + z &= 0 \\ y - az &= 0. \end{aligned}$$

26. Determine the number of entries on or above the main diagonal of a $k \times k$ matrix when (a) $k = 2$; (b) $k = 3$; (c) $k = 4$; (d) $k = n$.

27. Let

$$A = \begin{bmatrix} 0 & 2 \\ 0 & 5 \end{bmatrix}.$$

(a) Find a $2 \times k$ matrix $B \neq O$ such that $AB = O$ for $k = 1, 2, 3, 4$.

(b) Are your answers to part (a) unique? Explain.

28. Find all 2×2 matrices with real entries of the form

$$A = \begin{bmatrix} a & b \\ 0 & c \end{bmatrix}$$

such that $A^2 = I_2$.

29. An $n \times n$ matrix A (with real entries) is called a **square root** of the $n \times n$ matrix B (with real entries) if $A^2 = B$.

(a) Find a square root of

$$B = \begin{bmatrix} 1 & 1 \\ 0 & 1 \end{bmatrix}.$$

(b) Find a square root of

$$B = \begin{bmatrix} 1 & 0 & 0 \\ 0 & 0 & 0 \\ 0 & 0 & 0 \end{bmatrix}.$$

(c) Find a square root of $B = I_4$.

(d) Show that there is no square root of

$$B = \begin{bmatrix} 0 & 1 \\ 0 & 0 \end{bmatrix}.$$

30. Compute the trace (see Supplementary Exercise T.1) of each of the following matrices.

(a) $\begin{bmatrix} 1 & 0 \\ 2 & 3 \end{bmatrix}$ (b) $\begin{bmatrix} 2 & 2 & 3 \\ 2 & 4 & 4 \\ 3 & -2 & -5 \end{bmatrix}$

(c) $\begin{bmatrix} 1 & 0 & 0 \\ 0 & 1 & 0 \\ 0 & 0 & 1 \end{bmatrix}$

31. Develop a simple expression for the entries of A^n, where n is a positive integer and

$$A = \begin{bmatrix} 1 & \frac{1}{2} \\ 0 & \frac{1}{2} \end{bmatrix}.$$

32. As part of a project, two students must determine the inverse of a given 10×10 matrix A. Each performs the required calculation and they return their results A_1 and A_2, respectively, to the instructor.

(a) What must be true about the two results? Why?

(b) How does the instructor check their work without repeating the calculations?

33. Compute the vector $\mathbf{w}$ for each of the following expressions without computing the inverse of any matrix given that

$$A = \begin{bmatrix} 1 & 0 & -2 \\ 1 & 1 & 0 \\ 0 & 1 & 1 \end{bmatrix}, \quad C = \begin{bmatrix} 1 & 1 & 1 \\ 2 & 3 & 1 \\ 1 & 2 & 1 \end{bmatrix},$$

$$F = \begin{bmatrix} 2 & 1 & 0 \\ -3 & 0 & 2 \\ -1 & 1 & 2 \end{bmatrix}, \quad \mathbf{v} = \begin{bmatrix} 6 \\ 7 \\ -3 \end{bmatrix}.$$

(a) $\mathbf{w} = A^{-1}(C + F)\mathbf{v}$ (b) $\mathbf{w} = (F + 2A)C^{-1}\mathbf{v}$

In Exercises 34 and 35, find an LU-factorization of the coefficient matrix of the given linear system $A\mathbf{x} = \mathbf{b}$. Solve the linear system using a forward substitution followed by a back substitution.

34. $A = \begin{bmatrix} 2 & 2 & 3 \\ 6 & 5 & 7 \\ -6 & -8 & -10 \end{bmatrix}, \quad \mathbf{b} = \begin{bmatrix} -6 \\ -13 \\ 22 \end{bmatrix}$

35. $A = \begin{bmatrix} -2 & 1 & -2 \\ 6 & 1 & 9 \\ -4 & 18 & 5 \end{bmatrix}, \quad \mathbf{b} = \begin{bmatrix} -6 \\ 19 \\ -17 \end{bmatrix}$

Theoretical Exercises

T.1. If $A = \left[a_{ij}\right]$ is an $n \times n$ matrix, then the **trace** of A, $\text{Tr}(A)$, is defined as the sum of all elements on the main diagonal of A,

$$\text{Tr}(A) = a_{11} + a_{22} + \cdots + a_{nn}.$$

Show that

(a) $\text{Tr}(cA) = c\,\text{Tr}(A)$, where c is a real number

(b) $\text{Tr}(A + B) = \text{Tr}(A) + \text{Tr}(B)$

(c) $\text{Tr}(AB) = \text{Tr}(BA)$

(d) $\text{Tr}(A^T) = \text{Tr}(A)$

(e) $\text{Tr}(A^T A) \geq 0$

T.2. Show that if $\text{Tr}(AA^T) = 0$, then $A = O$. (See Exercise T.1.)

T.3. Let A and B be $n \times n$ matrices. Show that if $A\mathbf{x} = B\mathbf{x}$ for all $n \times 1$ matrices $\mathbf{x}$, then $A = B$.

T.4. Show that there are no 2×2 matrices A and B such that

$$AB - BA = \begin{bmatrix} 1 & 0 \\ 0 & 1 \end{bmatrix}.$$

T.5. Show that if A is skew symmetric (see Exercise T.24 in Section 1.4), then A^k is skew symmetric for any positive odd integer k.

T.6. Let A be an $n \times n$ skew symmetric matrix and $\mathbf{x}$ an n-vector. Show that $\mathbf{x}^T A\mathbf{x} = 0$ for all $\mathbf{x}$ in R^n.

T.7. Show that every symmetric upper (or lower) triangular matrix (see Exercise T.5 in Section 1.2) is diagonal.

T.8. Let A be an upper triangular matrix. Show that A is nonsingular if and only if all the entries on the main diagonal of A are nonzero.

T.9. Let A be an $m \times n$ matrix. Show that A is row equivalent to O if and only if $A = O$.

T.10. Let A and B be row equivalent $n \times n$ matrices. Show that A is nonsingular if and only if B is nonsingular.

T.11. Show that if AB is nonsingular, then A and B are nonsingular. (*Hint*: First show that B is nonsingular by considering the homogeneous system $B\mathbf{x} = \mathbf{0}$ and using Theorem 1.13 in Section 1.7.)

T.12. Show that if A is skew symmetric (see Exercise T.24 in Section 1.4), then the elements on the main diagonal of A are all zero.

T.13. Show that if A is skew symmetric and nonsingular, then A^{-1} is skew symmetric.

T.14. If A is an $n \times n$ matrix, then A is called **idempotent** if $A^2 = A$.

(a) Verify that I_n and O are idempotent.

(b) Find an idempotent matrix that is not I_n or O.

(c) Show that the only $n \times n$ nonsingular idempotent matrix is I_n.

T.15. If A is an $n \times n$ matrix, then A is called **nilpotent** if $A^k = O$ for some positive integer k.

(a) Show that every nilpotent matrix is singular.

(b) Verify that

$$A = \begin{bmatrix} 0 & 1 & 1 \\ 0 & 0 & 1 \\ 0 & 0 & 0 \end{bmatrix}$$

is nilpotent.

(c) If A is nilpotent, show that $I_n - A$ is nonsingular. [*Hint*: Find $(I_n - A)^{-1}$ in the cases $A^k = O$, $k = 1, 2, \ldots,$ and look for a pattern.]

T.16. Let A and B be $n \times n$ idempotent matrices. (See Exercise T.14.)

(a) Show that AB is idempotent if $AB = BA$.

(b) Show that if A is idempotent, then A^T is idempotent.

(c) Is $A + B$ idempotent? Justify your answer.

(d) Find all values of k for which kA is also idempotent.

T.17. Show that the product of two 2×2 skew symmetric matrices is diagonal. Is this true for $n \times n$ skew symmetric matrices with $n > 2$?

T.18. Prove Theorem 1.14.

In Exercises T.19 through T.22, let **x** *and* **y** *be column matrices with n elements. The* ***outer product*** *of* **x** *and* **y** *is the matrix product* $\mathbf{x}\mathbf{y}^T$ *that gives the n × n matrix*

$$\begin{bmatrix} x_1y_1 & x_1y_2 & \cdots & x_1y_n \\ x_2y_1 & x_2y_2 & \cdots & x_2y_n \\ \vdots & \vdots & & \vdots \\ x_ny_1 & x_ny_2 & \cdots & x_ny_n \end{bmatrix}.$$

T.19. (a) Form the outer product of **x** and **y**, where

$$\mathbf{x} = \begin{bmatrix} 1 \\ 2 \\ 3 \end{bmatrix} \quad \text{and} \quad \mathbf{y} = \begin{bmatrix} 4 \\ 5 \\ 6 \end{bmatrix}.$$

(b) Form the outer product of **x** and **y**, where

$$\mathbf{x} = \begin{bmatrix} 1 \\ 2 \\ 1 \\ 2 \end{bmatrix} \quad \text{and} \quad \mathbf{y} = \begin{bmatrix} -1 \\ 0 \\ 3 \\ 5 \end{bmatrix}.$$

T.20. Prove or disprove: The outer product of **x** and **y** equals the outer product of **y** and **x**.

T.21. Show that $\text{Tr}(\mathbf{x}\mathbf{y}^T) = \mathbf{x}^T\mathbf{y}$. (See Exercise T.1.)

T.22. Show that the outer product of **x** and **y** is row equivalent to either O or to a matrix with $n - 1$ rows of zeros.

T.23. Let **w** be an $n \times 1$ matrix such that $\mathbf{w}^T\mathbf{w} = 1$. The $n \times n$ matrix

$$H = I_n - 2\mathbf{w}\mathbf{w}^T$$

is called a **Householder matrix**. (Note that a Householder matrix is the identity matrix plus a scalar multiple of an outer product.)

(a) Show that H is symmetric.

(b) Show that $H^{-1} = H^T$.

T.24. Let

$$A = \begin{bmatrix} 2 & 0 \\ -1 & 1 \end{bmatrix}.$$

Show that all 2×2 matrices B such that $AB = BA$ are of the form

$$\begin{bmatrix} r & 0 \\ s - r & s \end{bmatrix},$$

where r and s are any real numbers.

Chapter Test

1. Let $f \colon R^4 \to R^4$ be the matrix transformation defined by $f(\mathbf{x}) = A\mathbf{x}$, where

$$A = \begin{bmatrix} 2 & 1 & 0 & 2 \\ 1 & 0 & 0 & 1 \\ 1 & -1 & 1 & 0 \\ 4 & -1 & 2 & 2 \end{bmatrix}.$$

Determine all vectors $\mathbf{w} = \begin{bmatrix} a \\ b \\ c \\ d \end{bmatrix}$ that are not in the range of f.

2. Find all solutions to the linear system

$$\begin{aligned} x_1 + x_2 + x_3 - 2x_4 &= 3 \\ 2x_1 + x_2 + 3x_3 + 2x_4 &= 5 \\ - x_2 + x_3 + 6x_4 &= 3. \end{aligned}$$

3. Find all values of a for which the resulting linear system has (a) no solution, (b) a unique solution, and (c) infinitely many solutions.

$$\begin{aligned} x + z &= 4 \\ 2x + y + 3z &= 5 \\ -3x - 3y + (a^2 - 5a)z &= a - 8. \end{aligned}$$

4. If possible, find the inverse of the following matrix:

$$\begin{bmatrix} 1 & 2 & -1 \\ 0 & 1 & 1 \\ 1 & 0 & -1 \end{bmatrix}.$$

5. If

$$A = \begin{bmatrix} -1 & -2 \\ -2 & 2 \end{bmatrix},$$

find all values of λ for which the homogeneous system $(\lambda I_2 - A)\mathbf{x} = \mathbf{0}$ has a nontrivial solution.

6. (a) If

$$A^{-1} = \begin{bmatrix} 1 & 3 & 0 \\ 0 & 1 & 1 \\ 1 & -1 & 4 \end{bmatrix}$$

and

$$B^{-1} = \begin{bmatrix} 2 & 1 & 1 \\ 0 & 0 & -2 \\ 1 & 1 & -1 \end{bmatrix},$$

compute $(AB)^{-1}$.

(b) Solve $A\mathbf{x} = \mathbf{b}$ for $\mathbf{x}$ if

$$A^{-1} = \begin{bmatrix} 1 & 0 & -2 \\ 2 & 1 & 3 \\ 4 & 2 & 5 \end{bmatrix} \quad \text{and} \quad \mathbf{b} = \begin{bmatrix} 2 \\ 1 \\ 3 \end{bmatrix}.$$

7. Find the LU-factorization of the coefficient matrix of the linear system $A\mathbf{x} = \mathbf{b}$. Solve the linear system by using a forward substitution followed by a back substitution.

$$A = \begin{bmatrix} 2 & 2 & -1 \\ -8 & -11 & 5 \\ 4 & 13 & -7 \end{bmatrix}, \quad \mathbf{b} = \begin{bmatrix} 3 \\ -14 \\ -5 \end{bmatrix}.$$

8. Answer each of the following as true or false. Justify your answer.

(a) If A and B are $n \times n$ matrices, then

$$(A + B)(A + B) = A^2 + 2AB + B^2.$$

(b) If $\mathbf{u}_1$ and $\mathbf{u}_2$ are solutions to the linear system $A\mathbf{x} = \mathbf{b}$, then $\mathbf{w} = \frac{1}{4}\mathbf{u}_1 + \frac{3}{4}\mathbf{u}_2$ is also a solution to $A\mathbf{x} = \mathbf{b}$.

(c) If A is a nonsingular matrix, then the homogeneous system $A\mathbf{x} = \mathbf{0}$ has a nontrivial solution.

(d) A homogeneous system of three equations in four unknowns has a nontrivial solution.

(e) If A, B, and C are $n \times n$ nonsingular matrices, then $(ABC)^{-1} = C^{-1}A^{-1}B^{-1}$.

CHAPTER 2

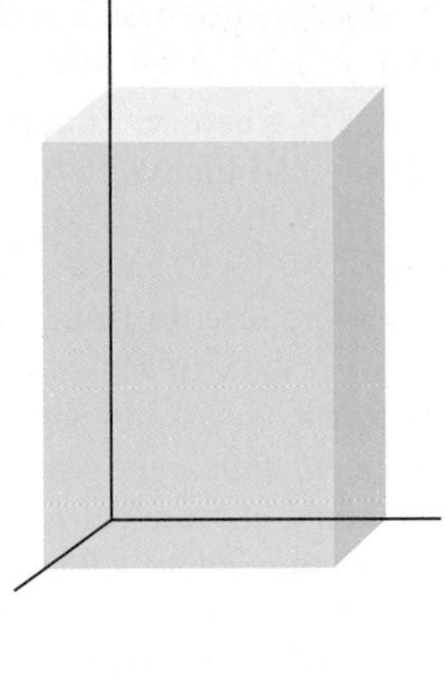

APPLICATIONS OF LINEAR EQUATIONS AND MATRICES (OPTIONAL)

2.1 AN INTRODUCTION TO CODING

Prerequisite. The material on bits in Chapter 1.

In today's global society communication abounds in business, government, research, and education. Data items are transmitted from point to point or recorded in various forms to represent video images, sounds, or combinations of these. Regardless of the distance that the transmission must traverse, the basic process is the same. The data must be sent and received, with the possibility of distortion occurring. The received data must be checked in some way to (hopefully) detect transmission errors.

Coding is a branch of information and communication theory that has developed techniques that help in detecting, and sometimes correcting, errors. It draws extensively on a variety of fields of mathematics, including linear and abstract algebra, number theory, probability and statistics, and combinatorics. We present a brief introduction to coding using linear algebra.

The central aspects of data transmission are doing it fast and doing it cheaply. With this in mind it is reasonable to consider a "shorthand" approach (for example, omitting certain letters from words). Unfortunately, any savings in time with such procedures are offset by an increased likelihood that received data will be misread. Most coding techniques are a sort of reverse shorthand. That is, we send more to aid in detecting transmission errors. A careful selection of how and what more to include is the essence of coding theory.

CODING OF BINARY INFORMATION AND ERROR DETECTION

A **message** is a finite sequence of characters from an alphabet. We shall choose as our alphabet the set $B = \{0, 1\}$. Every character, number, or symbol that we want to transmit will now be represented as a bit m-vector. That is, every character, number, or symbol is to be represented in binary form. Thus a message consists of a set of **words** each of which is a bit m-vector. The set of all bit m-vectors is denoted by B^m.

As we saw in Chapter 1, bit vectors and matrices share the same properties as real (base 10) vectors and matrices, except that we use base 2 arithmetic for computation. (See Tables 1.1 and 1.2 in Section 1.2.) A bit m-vector has the form $\begin{bmatrix} b_1 & b_2 & \cdots & b_m \end{bmatrix}$ or $\begin{bmatrix} b_1 & b_2 & \cdots & b_m \end{bmatrix}^T$, where b_j is either a 0 or a 1. In coding, the matrix notation is often omitted so that we write the bit m-vector as a string of bits in the form $b_1 b_2 \cdots b_m$. When matrix algebra is used we write expressions in standard matrix form.

Figure 2.1 shows the basic process of sending a word from one point to another point over a transmission channel. A vector $\mathbf{x}$ in B^m is sent through a transmission channel and is received as vector $\mathbf{x}_t$ in B^m. In actual practice, the transmission channel may suffer disturbances, which are generally called **noise**. Noise may be due to electrical problems, static electricity, weather interference, and so on. Any of these conditions may cause a 0 to be received as a 1, or vice versa. This erroneous transmission of bits in a word being sent gives rise to the situation where the word received is different from the word that was sent; that is, $\mathbf{x} \neq \mathbf{x}_t$. If errors do occur, then $\mathbf{x}_t$ could be any vector in B^m.

Figure 2.1 ►

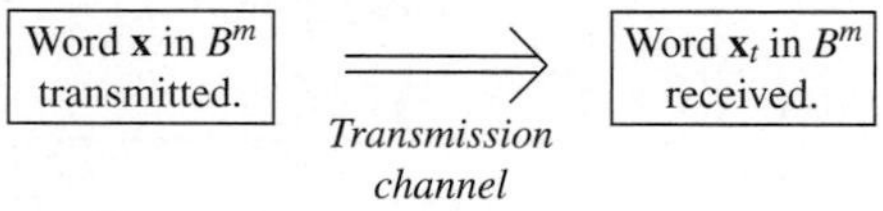

Word $\mathbf{x}$ in B^m transmitted. $\Longrightarrow$ *Transmission channel* $\Longrightarrow$ Word $\mathbf{x}_t$ in B^m received.

The basic task in the transmission of information is to reduce the likelihood of receiving a word that differs from the word that was sent. This can be done as follows. We first choose an integer $n > m$ and a function e from B^m to B^n and further require that e be a one-to-one function; that is, for all $\mathbf{x}$ and $\mathbf{y}$ in B^m, $\mathbf{x} \neq \mathbf{y}$ implies that $e(\mathbf{x}) \neq e(\mathbf{y})$. Thus different words in B^m correspond to different n-vectors in B^n. The function e is called a **coding function**.

EXAMPLE 1

Let $m = 2$, $n = 3$, and $e(b_1 b_2) = b_1 b_2 b_3$, where b_3 is defined to be 0 ($b_3 \equiv 0$). We have

$$e(00) = 000, \quad e(01) = 010, \quad e(10) = 100, \quad e(11) = 110,$$

and it follows that the function e is one-to-one. The function e can be computed using a multiplication by the matrix

$$A = \begin{bmatrix} 1 & 0 \\ 0 & 1 \\ 0 & 0 \end{bmatrix}.$$

Thus e is a matrix transformation from B^2 to B^3 given by

$$e(b_1 b_2) = \begin{bmatrix} 1 & 0 \\ 0 & 1 \\ 0 & 0 \end{bmatrix} \begin{bmatrix} b_1 \\ b_2 \end{bmatrix} = \begin{bmatrix} b_1 \\ b_2 \\ 0 \end{bmatrix}.$$

■

EXAMPLE 2

Let $m = 2$, $n = 3$, and $e(b_1 b_2) = b_1 b_2 b_3$, where b_3 is defined to be $b_1 + b_2$ ($b_3 \equiv b_1 + b_2$). We have

$$e(00) = 000, \quad e(01) = 011, \quad e(10) = 101, \quad e(11) = 110,$$

and it follows that the function e is one-to-one. The function e can be computed using multiplication by the matrix

$$A = \begin{bmatrix} 1 & 0 \\ 0 & 1 \\ 1 & 1 \end{bmatrix}.$$

Thus e is a matrix transformation from B^2 to B^3 given by

$$e(b_1b_2) = \begin{bmatrix} 1 & 0 \\ 0 & 1 \\ 1 & 1 \end{bmatrix} \begin{bmatrix} b_1 \\ b_2 \end{bmatrix} = \begin{bmatrix} b_1 \\ b_2 \\ b_1 + b_2 \end{bmatrix}.$$

■

EXAMPLE 3 Let $m = 2$, $n = 3$, and $e(b_1b_2) = b_100$. We have

$$e(00) = 000, \quad e(01) = 000, \quad e(10) = 100, \quad e(11) = 100,$$

and it follows that the function e is not one-to-one. This function e is a matrix transformation since

$$e(b_1b_2) = \begin{bmatrix} 1 & 0 \\ 0 & 0 \\ 0 & 0 \end{bmatrix} \begin{bmatrix} b_1 \\ b_2 \end{bmatrix} = \begin{bmatrix} b_1 \\ 0 \\ 0 \end{bmatrix}.$$

■

A one-to-one function e from B^m to B^n, $n > m$, is called an (m, n) **encoding function**, and we view it as a means of representing every word in B^m as a unique word in B^n. For a word $\mathbf{b}$ in B^m, $e(\mathbf{b})$ is called the **code word** representing $\mathbf{b}$. The $n-m$ additional bits in the code word can be used to detect transmission errors and, even more amazingly, help correct such errors. Figure 2.2 illustrates the two steps used for transmission: first encode the original word with the function e and then transmit the code word. If the transmission channel is noiseless, then $\mathbf{x}_t = \mathbf{x}$ for all $\mathbf{x}$ in B^n. Since the encoding function e is known, the original word $\mathbf{b}$ can be determined.

Figure 2.2 ►

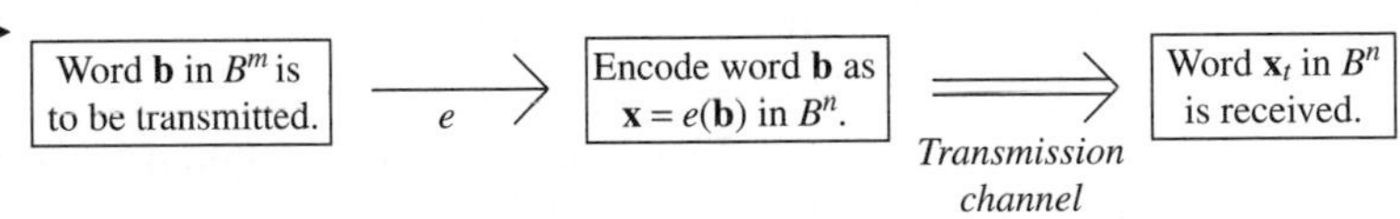

In general, errors in transmission do occur. We will say that the code word $\mathbf{x} = e(\mathbf{b})$ has been transmitted **with k or fewer errors** if $\mathbf{x}$ and $\mathbf{x}_t$ differ in at least 1 but no more than k bits.

Let e be an (m, n) encoding function. We say that e **detects k or fewer errors** if whenever $\mathbf{x} = e(\mathbf{b})$ is transmitted with k or fewer errors, then $\mathbf{x}_t$ is not a code word (thus $\mathbf{x}_t$ could not be $\mathbf{x}$ and therefore could not have been transmitted correctly).

Remark Even if the encoding function e is designed to incorporate error detection capability, errors can still occur.

EXAMPLE 4 Suppose we are interested in transmitting a single bit. That is, we are to transmit either a 0 or a 1. One way to guard against transmission errors is to transmit the message more than once. For example, we could use the (1, 3) encoding function e so that 0 is encoded as 000 and 1 is encoded as 111. In terms of a matrix transformation we have for the single bit b,

$$e(b) = \begin{bmatrix} 1 \\ 1 \\ 1 \end{bmatrix} \begin{bmatrix} b \end{bmatrix} = \begin{bmatrix} b \\ b \\ b \end{bmatrix}.$$

It follows that there are only two valid code words, 000 and 111. If $\mathbf{x} = bbb$ is transmitted so the received word is $\mathbf{x}_t = 001$, then at least one error has occurred. Also, if we receive 001, 110, or 010, we have detected transmission errors since these are invalid code words. The only time errors can go undetected is if $\mathbf{x} = 000$ but $\mathbf{x}_t = 111$, or vice versa. Since the encoding function detects 2 or fewer errors, e is called a **double-error detecting** encoding function.

Suppose we want to perform error correction when errors are detected. If $\mathbf{x}_t = 010$, we know that a transmission error has occurred. But we do not know if there was a single or double error. If $\mathbf{x} = 000$, then one error occurred, but if $\mathbf{x} = 111$, then two errors occurred. One correction strategy in this case is to assume that one error is more likely than two. Thus $\mathbf{x}_t = 010$ is "corrected" to be 000. Figure 2.3 illustrates this correction procedure. If $\mathbf{x}_t = b_1b_2b_3$, then it is decoded as 000 if we can move from vertex $b_1b_2b_3$ in Figure 2.3 to 000 along a single edge; otherwise it is decoded as 111. Thus, with this strategy we have a **single-error correcting code**. But note that if $\mathbf{x} = 000$ and $\mathbf{x}_t = 011$, then with this strategy we would incorrectly decode $\mathbf{x}_t$ as 111. ■

Figure 2.3 ►

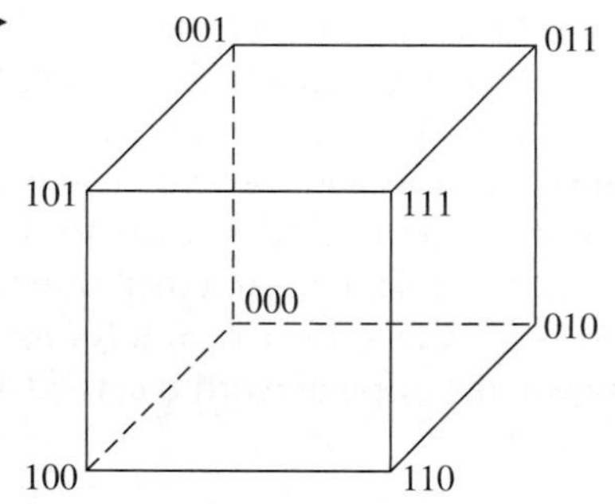

The procedure in Example 4 is often called a three-fold repetition code. See also Exercise T.4.

DEFINITION For a bit n-vector $\mathbf{x}$, the number of 1s in $\mathbf{x}$ is called the weight of $\mathbf{x}$ and is denoted by $|\mathbf{x}|$.

EXAMPLE 5 Find the weight of each of the following words in B^6.

(a) $\mathbf{x} = 011000$; $|\mathbf{x}| = 2$
(b) $\mathbf{x} = 000001$; $|\mathbf{x}| = 1$
(c) $\mathbf{x} = 000000$; $|\mathbf{x}| = 0$
(d) $\mathbf{x} = 101010$; $|\mathbf{x}| = 3$ ■

DEFINITION The encoding function e from B^m to B^{m+1} given by

$$e(\mathbf{b}) = e(b_1b_2 \cdots b_m) = b_1b_2 \cdots b_mb_{m+1} = \mathbf{b}_t,$$

where

$$b_{m+1} = \begin{cases} 0, & \text{if } |\mathbf{b}| \text{ is even} \\ 1, & \text{if } |\mathbf{b}| \text{ is odd} \end{cases}$$

is called the **parity $(m, m+1)$ encoding function** or **the parity $(m, m+1)$ check code**. If $b_{m+1} = 1$, we say $\mathbf{b}_t$ has odd parity and if $b_{m+1} = 0$, we say $\mathbf{b}_t$ has even parity.

EXAMPLE 6 If $m = 3$, then the parity $(3, 4)$ check code produces the code words

$$e(000) = 0000, \quad e(001) = 0011, \quad e(010) = 0101, \quad e(100) = 1001,$$
$$e(101) = 1010, \quad e(110) = 1100, \quad e(011) = 0110, \quad e(111) = 1111.$$

If the transmission channel transmits $\mathbf{x} = 101$ from B^3 as $\mathbf{x}_t = 1011$, then the weight of $\mathbf{x}_t$ is $|\mathbf{x}_t| = 3$. However, since $|101| = 2$ and $\mathbf{x}_t$ has odd parity, we know that an odd number of errors (at least one) has occurred. If the received code word had been $\mathbf{x}_t = 1010$, then $|\mathbf{x}_t| = 2$ and $\mathbf{x}_t$ has even parity. In this case, we cannot conclude that the code word is error free. ■

Remark The parity $(m, m+1)$ check code detects an odd number of errors but does not detect an even number of errors. Despite this limitation, the parity check code is widely used.

Key Terms

Message
Words
Noise
Coding function
(m, n) encoding function
Code word
Detects k or fewer errors
Double-error detecting encoding function
Single-error detecting encoding function
Parity $(m, m+1)$ encoding function [or parity $(m, m+1)$ check code]

2.1 Exercises

All arithmetic operations are to be performed using binary arithmetic.

1. Let e be the function from B^3 to B^4 given by

$$e(b_1b_2b_3) = b_1b_2b_3b_4,$$

where $b_4 = b_1 + b_3$.

(a) Is e one-to-one? If not, determine two different vectors $\mathbf{b}$ and $\mathbf{c}$ in B^3 such that $e(\mathbf{b}) = e(\mathbf{c})$.

(b) Determine the matrix A so that e can be written as a matrix transformation in the form

$$e(b_1b_2b_3) = A\begin{bmatrix} b_1 \\ b_2 \\ b_3 \end{bmatrix} = \begin{bmatrix} b_1 \\ b_2 \\ b_3 \\ b_4 \end{bmatrix}.$$

2. Let e be the function from B^3 to B^4 given by

$$e(b_1b_2b_3) = b_1b_2b_3b_4,$$

where $b_4 = 0$.

(a) Is e one-to-one? If not, determine two different vectors $\mathbf{b}$ and $\mathbf{c}$ in B^3 such that $e(\mathbf{b}) = e(\mathbf{c})$.

(b) Determine the matrix A so that e can be written as a matrix transformation in the form

$$e(b_1b_2b_3) = A\begin{bmatrix} b_1 \\ b_2 \\ b_3 \end{bmatrix} = \begin{bmatrix} b_1 \\ b_2 \\ b_3 \\ 0 \end{bmatrix}.$$

3. Let e be the function from B^3 to B^2 given by

$$e(b_1b_2b_3) = b_1b_2.$$

(a) Is e one-to-one? If not, determine two different vectors $\mathbf{b}$ and $\mathbf{c}$ in B^3 such that $e(\mathbf{b}) = e(\mathbf{c})$.

(b) Determine the matrix A so that e can be written as a matrix transformation in the form

$$e(b_1b_2b_3) = A\begin{bmatrix} b_1 \\ b_2 \\ b_3 \end{bmatrix} = \begin{bmatrix} b_1 \\ b_2 \end{bmatrix}.$$

4. Let e be the function from B^2 to B^3 given by

$$e(b_1b_2) = b_1b_2b_3,$$

where $b_3 = b_1 \times b_2$.

(a) Is e one-to-one? If not, determine two different vectors $\mathbf{b}$ and $\mathbf{c}$ in B^3 such that $e(\mathbf{b}) = e(\mathbf{c})$.

(b) Determine, if possible, the matrix A so that e can be written as a matrix transformation in the form

$$e(b_1b_2) = A\begin{bmatrix} b_1 \\ b_2 \end{bmatrix} = \begin{bmatrix} b_1 \\ b_2 \\ b_1 \times b_2 \end{bmatrix}.$$

5. Determine the weights of the given words.

(a) 01110 (b) 10101 (c) 11000 (d) 00010

6. Determine the weights of the given words.

(a) 101 (b) 111 (c) 011 (d) 010

7. Determine the parity of each of the following words in B^4.

(a) 1101 (b) 0011 (c) 0100 (d) 0000

8. Determine the parity of each of the following words in B^5.

(a) 01101 (b) 00011 (c) 00010 (d) 11111

9. A parity (4, 5) check code is used and the following words are received. Determine whether an error will be detected.

(a) 10100 (b) 01101 (c) 11110 (d) 10000

10. A parity (5, 6) check code is used and the following words are received. Determine whether an error will be detected.

(a) 001101 (b) 101110 (c) 110000 (d) 111010

11. (a) Determine the code words for the parity (2, 3) check code.

(b) Determine whether an error will be detected if the following words are received.

(i) 011 (ii) 111 (iii) 010 (iv) 001

Theoretical Exercises

T.1. Determine the number of words with weight zero in B^2; with weight one; with weight two.

T.2. Determine the number of words with weight zero in B^3; with weight one; with weight two; with weight three.

T.3. Determine the number of words with weight one in B^n; with weight two.

T.4. Let e be an (m, n) encoding function that detects k or fewer errors. We say e produces an **error correcting code**. An error correcting code is called **linear** if the sum (or difference) of any two code words is also a code word.

(a) Show that the error correcting code in Example 4 is linear.

(b) Show that the error correcting code in Exercise 11 is linear.

T.5. Let e be the function from B^2 to B^4 given by the following matrix transformation:

$$e(b_1b_2) = \begin{bmatrix} 1 & 0 \\ 0 & 1 \\ 1 & 0 \\ 0 & 1 \end{bmatrix}\begin{bmatrix} b_1 \\ b_2 \end{bmatrix}.$$

(a) Determine all the code words.

(b) Is this a linear code?

(c) Since all the code words have the same parity, if we used a parity check on a received word would this check detect all possible errors? Explain your answer.

MATLAB Exercises

Exercise ML.1 uses bit matrices and the supplemental instructional commands described in Section 12.9.

ML.1. Develop the code words for the parity (4, 5) check code using the following.

(a) Use command $M = \mathbf{bingen(0, 15, 4)}$ to generate a matrix whose columns are all the vectors in B^4.

(b) Use the command $s = \mathbf{sum(M)}$ to compute a vector whose entries are the weights of the columns of the matrix M.

(c) Construct a 1×16 bit vector $\mathbf{w}$ whose entries are the parity of the columns of the matrix M.

(d) Construct the code words of the parity (4, 5) check code by displaying the matrix $C = [\mathbf{M}; \mathbf{w}]$.

2.2 GRAPH THEORY

Prerequisite. Section 1.4, Properties of Matrix Operations.

Graph theory is a relatively new area of mathematics that is being widely used in formulating models in many problems in business, the social sciences, and the physical sciences. These applications include communications problems and the study of organizations and social structures. In this section we present a very brief introduction to the subject as it relates to matrices and show how these elementary and simple notions can be used in formulating models of some important problems.

DIGRAPHS

DEFINITION

A **directed graph**, or **digraph**, is a finite set of points $P_1, P_2, \ldots, P_n$, called **vertices** or **nodes**, together with a finite set of **directed edges**, each of which joins an ordered pair of distinct vertices. Thus, the directed edge P_iP_j is different from the directed edge P_jP_i. Observe that in a digraph there may be no directed edge from a vertex P_i to any other vertex and there may not be any directed edge from any vertex to the vertex P_i. Also, in a digraph no vertex can be reached from itself by a single directed edge. It can be reached from itself through other vertices. We express this property by saying that there are no loops. Moreover, we shall assume that there are no multiple directed edges joining any two vertices.

EXAMPLE 1

In Figure 2.4 four examples of digraphs are shown. The digraph in Figure 2.4(a) has vertices P_1, P_2, and P_3 and directed edges P_1P_2 and P_2P_3; the digraph in Figure 2.4(b) has vertices P_1, P_2, P_3, and P_4 and directed edges P_1P_2 and P_1P_3; the digraph in Figure 2.4(c) has vertices P_1, P_2, and P_3 and directed edges P_1P_2, P_1P_3, and P_3P_1; the digraph in Figure 2.4(d) has vertices P_1, P_2, and P_3 and directed edges P_2P_1, P_2P_3, P_1P_3, and P_3P_1. A pair of directed edges like P_1P_3 and P_3P_1 are indicated by a curved or straight segment with a double arrow, $P_1 \longleftrightarrow P_3$. ■

Figure 2.4 ►

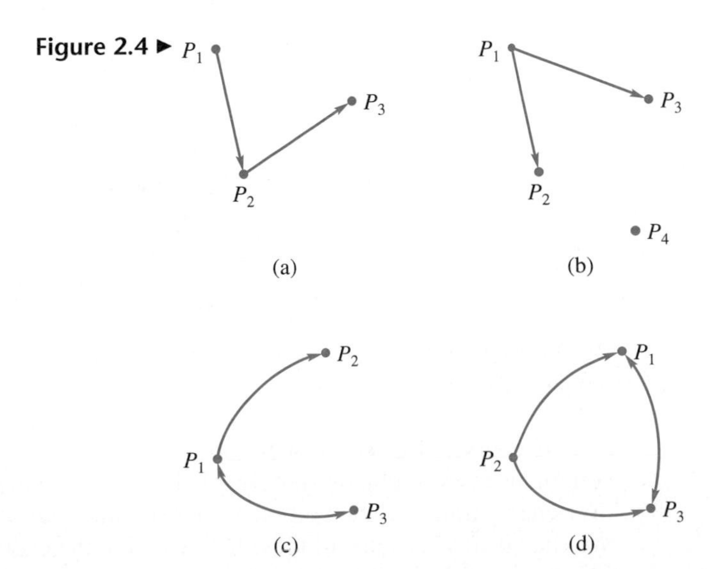

DEFINITION If G is a digraph that has n vertices, then the $n \times n$ matrix $A(G)$, whose i, jth element is 1 if there is a directed edge from P_i to P_j and zero otherwise, is called the **adjacency matrix** of G. Note that $A(G)$ need not be a symmetric matrix.

EXAMPLE 2 The matrices A, B, C, and D are the adjacency matrices of the digraphs in Figure 2.4, respectively (verify).

$$A = \begin{array}{c} \\ P_1 \\ P_2 \\ P_3 \end{array} \begin{array}{c} \begin{array}{ccc} P_1 & P_2 & P_3 \end{array} \\ \begin{bmatrix} 0 & 1 & 0 \\ 0 & 0 & 1 \\ 0 & 0 & 0 \end{bmatrix} \end{array}, \qquad B = \begin{array}{c} \\ P_1 \\ P_2 \\ P_3 \\ P_4 \end{array} \begin{array}{c} \begin{array}{cccc} P_1 & P_2 & P_3 & P_4 \end{array} \\ \begin{bmatrix} 0 & 1 & 1 & 0 \\ 0 & 0 & 0 & 0 \\ 0 & 0 & 0 & 0 \\ 0 & 0 & 0 & 0 \end{bmatrix} \end{array},$$

$$C = \begin{array}{c} \\ P_1 \\ P_2 \\ P_3 \end{array} \begin{array}{c} \begin{array}{ccc} P_1 & P_2 & P_3 \end{array} \\ \begin{bmatrix} 0 & 1 & 1 \\ 0 & 0 & 0 \\ 1 & 0 & 0 \end{bmatrix} \end{array}, \qquad D = \begin{array}{c} \\ P_1 \\ P_2 \\ P_3 \end{array} \begin{array}{c} \begin{array}{ccc} P_1 & P_2 & P_3 \end{array} \\ \begin{bmatrix} 0 & 1 & 1 \\ 0 & 0 & 1 \\ 1 & 0 & 0 \end{bmatrix} \end{array}$$ ■

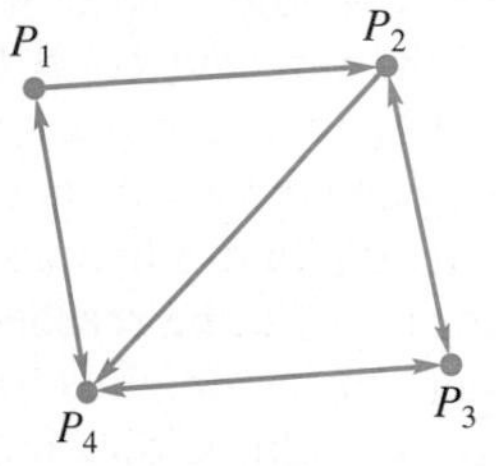

Of course, from a given matrix whose entries are zeros and ones (with diagonal entries equal to zero), we can obtain a digraph whose adjacency matrix is the given matrix.

Figure 2.5 ▲

EXAMPLE 3 The matrix

$$A(G) = \begin{array}{c} \\ P_1 \\ P_2 \\ P_3 \\ P_4 \end{array} \begin{array}{c} \begin{array}{cccc} P_1 & P_2 & P_3 & P_4 \end{array} \\ \begin{bmatrix} 0 & 1 & 0 & 1 \\ 0 & 0 & 1 & 1 \\ 0 & 1 & 0 & 1 \\ 1 & 0 & 1 & 0 \end{bmatrix} \end{array}$$

is the adjacency matrix of the digraph shown in Figure 2.5. ■

EXAMPLE 4 A bowling league consists of seven teams: $T_1, T_2, \ldots, T_7$. Suppose that after a number of games have been played we have the following situation:

T_1 has defeated T_2 and T_5 and lost to T_3;
T_2 has defeated T_5 and lost to T_1 and T_3;
T_3 has defeated T_1 and T_2 and lost to T_4;
T_4 has defeated T_3 and lost to T_7;
T_5 has lost to T_1 and T_2;
T_6 has not played anyone;
T_7 has defeated T_4.

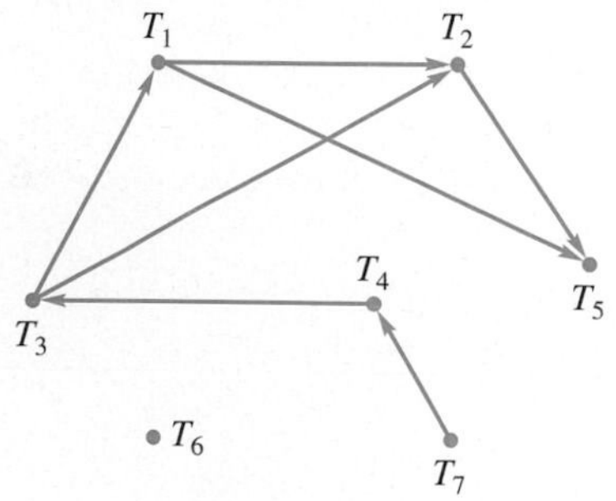

Figure 2.6 ▲

We now obtain the directed graph in Figure 2.6, where $T_i \rightarrow T_j$ means that T_i defeated T_j. ■

Of course, digraphs can be used in a great many situations including communications problems, family relationships, social structures, street maps, flow charts, transportation problems, electrical circuits, and ecological chains. We shall deal with some of these below and in the exercises.

MODELS IN SOCIOLOGY AND IN COMMUNICATIONS

Suppose that we have n individuals $P_1, P_2, \ldots, P_n$, some of whom are associated with each other. We assume that no one is associated with himself. Examples of such relations are as follows:

1. P_i has access to P_j. In this case it may or may not be the case that if P_i has access to P_j, then P_j has access to P_i. For example, many emergency telephones on turnpikes allow a distressed traveler to contact a nearby emergency station but make no provision for the station to contact the traveler. This model can thus be represented by a digraph G as follows. Let $P_1, P_2, \ldots, P_n$ be the vertices of G and draw a directed edge from P_i to P_j if P_i has access to P_j. It is important to observe that this relation need not be transitive. That is, P_i may have access to P_j and P_j may have access to P_k, but P_i need not have access to P_k.
2. P_i influences P_j. This situation is identical to that in 1: if P_i influences P_j, then it may or may not happen that P_j influences P_i.
3. For every pair of individuals, P_i, P_j, either P_i dominates P_j or P_j dominates P_i, but not both. The graph representing this situation is the complete directed graph with n vertices. Such graphs are often called **dominance digraphs**.

EXAMPLE 5

Suppose that six individuals have been meeting in group therapy for a long time and their leader, who is not part of the group, has drawn the digraph G in Figure 2.7 to describe the influence relations among the various individuals. The adjacency matrix of G is

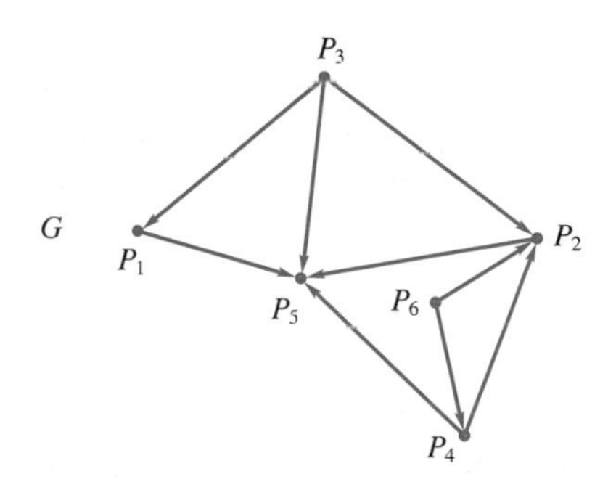

Figure 2.7 ▲

$$A(G) = \begin{array}{c} \\ P_1 \\ P_2 \\ P_3 \\ P_4 \\ P_5 \\ P_6 \end{array} \begin{array}{c} \begin{array}{cccccc} P_1 & P_2 & P_3 & P_4 & P_5 & P_6 \end{array} \\ \begin{bmatrix} 0 & 0 & 0 & 0 & 1 & 0 \\ 0 & 0 & 0 & 0 & 1 & 0 \\ 1 & 1 & 0 & 0 & 1 & 0 \\ 0 & 1 & 0 & 0 & 1 & 0 \\ 0 & 0 & 0 & 0 & 0 & 0 \\ 0 & 1 & 0 & 1 & 0 & 0 \end{bmatrix} \end{array}.$$

Looking at the rows of $A(G)$, we see that P_3 has three 1s in its row so that P_3 influences three people—more than any other individual. Thus P_3 would be declared the leader of the group. On the other hand, P_5 influences no one. ■

EXAMPLE 6

Consider a communication network whose digraph G is shown in Figure 2.8. The adjacency matrix of G is

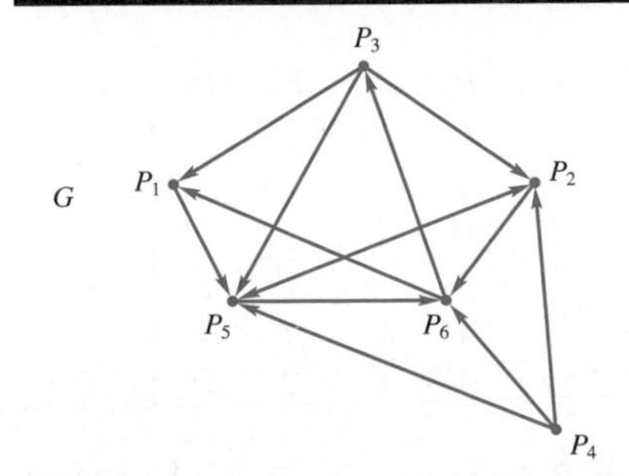

Figure 2.8 ▲

$$A(G) = \begin{array}{c} \\ P_1 \\ P_2 \\ P_3 \\ P_4 \\ P_5 \\ P_6 \end{array} \begin{array}{c} \begin{array}{cccccc} P_1 & P_2 & P_3 & P_4 & P_5 & P_6 \end{array} \\ \begin{bmatrix} 0 & 0 & 0 & 0 & 1 & 0 \\ 0 & 0 & 0 & 0 & 1 & 1 \\ 1 & 1 & 0 & 0 & 1 & 0 \\ 0 & 1 & 0 & 0 & 1 & 1 \\ 0 & 1 & 0 & 0 & 0 & 1 \\ 1 & 0 & 1 & 0 & 0 & 0 \end{bmatrix} \end{array}.$$

■

Although the relation "P_i has access to P_j" need not be transitive, we can speak of two-stage access. We say that P_i has **two-stage access** to P_k if we can find an individual P_j such that P_i has access to P_j and P_j has access to P_k. Similarly, P_i has r-stage access to P_k if we can find $r - 1$ individuals $P_{j_1}, \ldots, P_{j_{r-1}}$, such that P_i has access to P_{j_1}, P_{j_1} has access to $P_{j_2}, \ldots, P_{j_{r-2}}$ has access to $P_{j_{r-1}}$, and $P_{j_{r-1}}$ has access to P_k. Some of the $r + 1$ individuals $P_i, P_{j_1}, \ldots, P_{j_{r-1}}, P_k$ may be the same.

The following theorem, whose proof we omit, can be established.

THEOREM 2.1

Let $A(G)$ be the adjacency matrix of a digraph G and let the rth power of $A(G)$ be B_r:

$$[A(G)]^r = B_r = [\,b_{ij}^{(r)}\,].$$

Then the i, jth element in B_r, $b_{ij}^{(r)}$, is the number of ways in which P_i has access to P_j in r stages. ■

The sum of the elements in the jth column of $[A(G)]^r$ gives the number of ways in which P_j is reached by all the other individuals in r stages.

If we let

$$A(G) + [A(G)]^2 + \cdots + [A(G)]^r = C = \left[c_{ij}\right], \tag{1}$$

then c_{ij} is the number of ways in which P_i has access to P_j in one, two, ..., or r stages.

Similarly, we speak of r-stage dominance, r-stage influence, and so forth. We can also use this model to study the spread of a rumor. Thus c_{ij} in (1) is the number of ways in which P_i has spread the rumor to P_j in one, two, ..., or r stages. In the influence relation, r-stage influence shows the effect of indirect influence.

EXAMPLE 7

If G is the digraph in Figure 2.8, then we find that

$$[A(G)]^2 = \begin{array}{c} \\ P_1 \\ P_2 \\ P_3 \\ P_4 \\ P_5 \\ P_6 \end{array} \!\!\begin{array}{c} \begin{array}{cccccc} P_1 & P_2 & P_3 & P_4 & P_5 & P_6 \end{array} \\ \begin{bmatrix} 0 & 1 & 0 & 0 & 0 & 1 \\ 1 & 1 & 1 & 0 & 0 & 1 \\ 0 & 1 & 0 & 0 & 2 & 2 \\ 1 & 1 & 1 & 0 & 1 & 2 \\ 1 & 0 & 1 & 0 & 1 & 1 \\ 1 & 1 & 0 & 0 & 2 & 0 \end{bmatrix} \end{array}$$

and

$$A(G) + [A(G)]^2 = C = \begin{array}{c} \\ P_1 \\ P_2 \\ P_3 \\ P_4 \\ P_5 \\ P_6 \end{array} \!\!\begin{array}{c} \begin{array}{cccccc} P_1 & P_2 & P_3 & P_4 & P_5 & P_6 \end{array} \\ \begin{bmatrix} 0 & 1 & 0 & 0 & 1 & 1 \\ 1 & 1 & 1 & 0 & 1 & 2 \\ 1 & 2 & 0 & 0 & 3 & 2 \\ 1 & 2 & 1 & 0 & 2 & 3 \\ 1 & 1 & 1 & 0 & 1 & 2 \\ 2 & 1 & 1 & 0 & 2 & 0 \end{bmatrix} \end{array}.$$

Since $c_{35} = 3$, there are three ways in which P_3 has access to P_5 in one or two stages: $P_3 \to P_5$, $P_3 \to P_2 \to P_5$, and $P_3 \to P_1 \to P_5$. ■

In studying organizational structures, we often find subsets of people in which any pair of individuals is related. This is an example of a clique, which we now define.

DEFINITION

A **clique** in a digraph is a subset S of the vertices satisfying the following properties:

(a) S contains three or more vertices.
(b) If P_i and P_j are in S, then there is a directed edge from P_i to P_j and a directed edge from P_j to P_i.
(c) There is no larger subset T of the vertices that satisfies property (b) and contains S [that is, S is a maximal subset satisfying (b)].

EXAMPLE 8

Consider the digraph in Figure 2.9. The set $\{P_1, P_2, P_3\}$ satisfies conditions (a) and (b) for a clique, but it is not a clique, since it fails to satisfy condition (c). That is, $\{P_1, P_2, P_3\}$ is contained in $\{P_1, P_2, P_3, P_4\}$, which satisfies conditions (a), (b), and (c). Thus the only clique in this digraph is $\{P_1, P_2, P_3, P_4\}$. ■

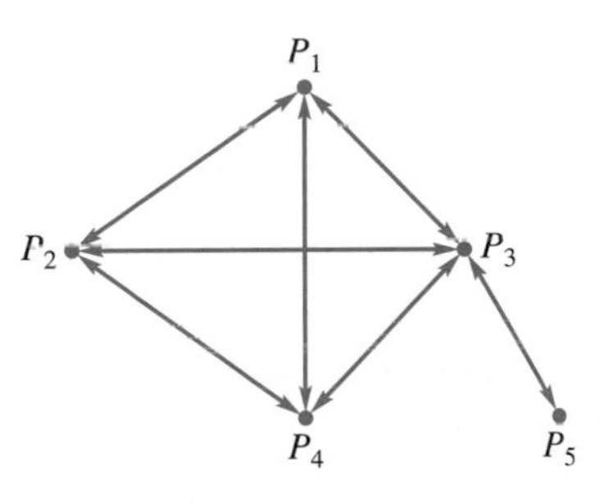

Figure 2.9 ▲

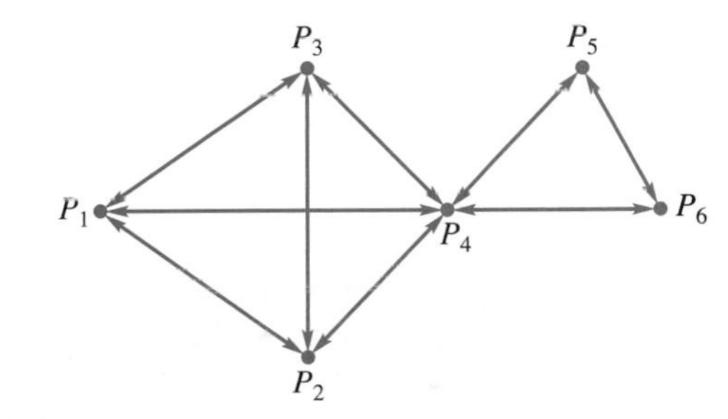

Figure 2.10 ▲

EXAMPLE 9

Consider the digraph in Figure 2.10. In this case we have two cliques:

$$\{P_1, P_2, P_3, P_4\} \quad \text{and} \quad \{P_4, P_5, P_6\}.$$

Moreover, P_4 belongs to both cliques. ■

For large digraphs it is difficult to determine cliques. The following approach provides a useful method for detecting cliques that can easily be implemented on a computer. If $A(G) = [a_{ij}]$ is the given adjacency matrix of a digraph, form a new matrix $S = [s_{ij}]$:

$$s_{ij} = s_{ji} = 1 \quad \text{if} \quad a_{ij} = a_{ji} = 1;$$

otherwise, let $s_{ij} = s_{ji} = 0$. Thus $s_{ij} = 1$ if P_i and P_j have access to each other; otherwise, $s_{ij} = 0$. It should be noted that S is a symmetric matrix ($S = S^T$).

EXAMPLE 10 Consider a digraph with adjacency matrix

$$A(G) = \begin{array}{c} \\ P_1 \\ P_2 \\ P_3 \\ P_4 \\ P_5 \\ P_6 \end{array} \begin{array}{c} \begin{array}{cccccc} P_1 & P_2 & P_3 & P_4 & P_5 & P_6 \end{array} \\ \begin{bmatrix} 0 & 0 & 1 & 1 & 1 & 0 \\ 1 & 0 & 1 & 1 & 1 & 1 \\ 0 & 1 & 0 & 1 & 1 & 1 \\ 1 & 0 & 1 & 0 & 0 & 1 \\ 1 & 1 & 0 & 1 & 0 & 1 \\ 0 & 1 & 1 & 1 & 1 & 0 \end{bmatrix} \end{array}.$$

Then

$$S = \begin{array}{c} \\ P_1 \\ P_2 \\ P_3 \\ P_4 \\ P_5 \\ P_6 \end{array} \begin{array}{c} \begin{array}{cccccc} P_1 & P_2 & P_3 & P_4 & P_5 & P_6 \end{array} \\ \begin{bmatrix} 0 & 0 & 0 & 1 & 1 & 0 \\ 0 & 0 & 1 & 0 & 1 & 1 \\ 0 & 1 & 0 & 1 & 0 & 1 \\ 1 & 0 & 1 & 0 & 0 & 1 \\ 1 & 1 & 0 & 0 & 0 & 1 \\ 0 & 1 & 1 & 1 & 1 & 0 \end{bmatrix} \end{array}.$$

■

The following theorem can be proved.

THEOREM 2.2 *Let $A(G)$ be the adjacency matrix of a digraph and $S = \left[s_{ij}\right]$ be the symmetric matrix defined above, with $S^3 = [\, s_{ij}^{(3)} \,]$, where $s_{ij}^{(3)}$ is the i, jth element in S^3. Then P_i belongs to a clique if and only if the diagonal entry $s_{ii}^{(3)}$ is positive.* ■

Let us briefly consider why we examine the diagonal entries of S^3 in Theorem 2.2. First, note that the diagonal entry $s_{ii}^{(3)}$ of S^3 gives the number of ways in which P_i has access to itself in three stages. If $s_{ii}^{(3)} > 0$, then there is at least one way in which P_i has access to itself. Since a digraph has no loops, this access must occur through two individuals: $P_i \to P_j \to P_k \to P_i$. Thus $s_{ij} \neq 0$. But $s_{ij} \neq 0$ implies that $s_{ji} \neq 0$; so $P_j \to P_i$. Similarly, any two of the individuals in $\{P_i, P_j, P_k\}$ have access to each other. This means that P_i, P_j, and P_k all belong to the same clique. The opposite direction (if P_i is in a clique, then $s_{ii}^{(3)} > 0$) is left as an exercise.

The procedure for determining a clique in a digraph is as follows.

Step 1. If $A = \left[a_{ij}\right]$ is the adjacency matrix of the given digraph, compute the symmetric matrix $S = \left[s_{ij}\right]$, where

$$s_{ij} = s_{ji} = 1 \quad \text{if} \quad a_{ij} = a_{ji} = 1;$$

otherwise, $s_{ij} = 0$.

Step 2. Compute $S^3 = [\, s_{ij}^{(3)} \,]$.

Step 3. P_i belongs to a clique if and only if $s_{ii}^{(3)}$ is positive.

EXAMPLE 11

Consider the digraph in Figure 2.11, whose adjacency matrix is

$$A(G) = \begin{array}{c} \\ P_1 \\ P_2 \\ P_3 \\ P_4 \\ P_5 \end{array} \begin{array}{c} \begin{array}{ccccc} P_1 & P_2 & P_3 & P_4 & P_5 \end{array} \\ \begin{bmatrix} 0 & 1 & 1 & 0 & 1 \\ 1 & 0 & 0 & 0 & 1 \\ 1 & 0 & 0 & 1 & 0 \\ 0 & 1 & 1 & 0 & 0 \\ 1 & 0 & 0 & 0 & 0 \end{bmatrix} \end{array}.$$

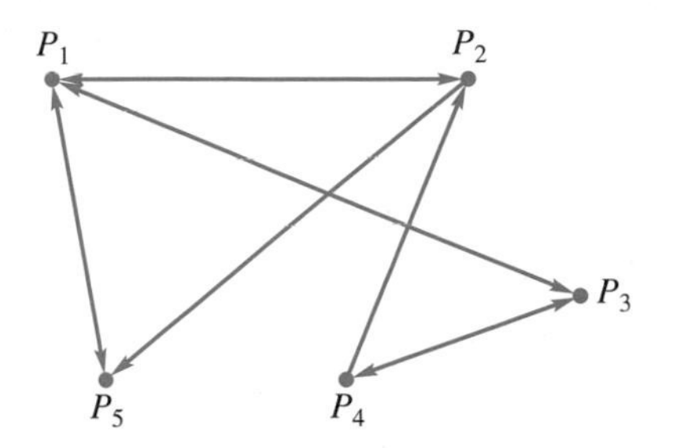

Figure 2.11 ▲

Then (verify)

$$S = \begin{array}{c} \\ P_1 \\ P_2 \\ P_3 \\ P_4 \\ P_5 \end{array} \begin{array}{c} \begin{array}{ccccc} P_1 & P_2 & P_3 & P_4 & P_5 \end{array} \\ \begin{bmatrix} 0 & 1 & 1 & 0 & 1 \\ 1 & 0 & 0 & 0 & 0 \\ 1 & 0 & 0 & 1 & 0 \\ 0 & 0 & 1 & 0 & 0 \\ 1 & 0 & 0 & 0 & 0 \end{bmatrix} \end{array}$$

and

$$S^3 = \begin{array}{c} \\ P_1 \\ P_2 \\ P_3 \\ P_4 \\ P_5 \end{array} \begin{array}{c} \begin{array}{ccccc} P_1 & P_2 & P_3 & P_4 & P_5 \end{array} \\ \begin{bmatrix} 0 & 3 & 4 & 0 & 3 \\ 3 & 0 & 0 & 1 & 0 \\ 4 & 0 & 0 & 2 & 0 \\ 0 & 1 & 2 & 0 & 1 \\ 3 & 0 & 0 & 1 & 0 \end{bmatrix} \end{array}.$$

Since every diagonal entry in S^3 is zero, we conclude that there are no cliques. ■

EXAMPLE 12

Consider the digraph in Figure 2.12, whose adjacency matrix is

$$A(G) = \begin{array}{c} \\ P_1 \\ P_2 \\ P_3 \\ P_4 \\ P_5 \end{array} \begin{array}{c} \begin{array}{ccccc} P_1 & P_2 & P_3 & P_4 & P_5 \end{array} \\ \begin{bmatrix} 0 & 0 & 1 & 1 & 1 \\ 0 & 0 & 1 & 0 & 1 \\ 1 & 0 & 0 & 0 & 1 \\ 0 & 0 & 1 & 0 & 1 \\ 1 & 1 & 1 & 0 & 0 \end{bmatrix} \end{array}.$$

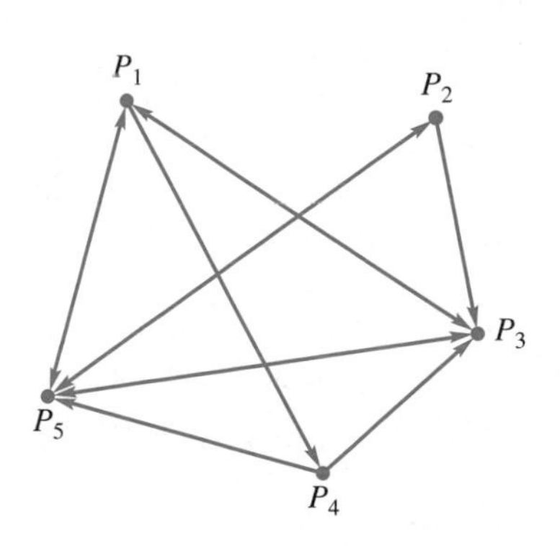

Figure 2.12 ▲

Then (verify)

$$S = \begin{array}{c} \\ P_1 \\ P_2 \\ P_3 \\ P_4 \\ P_5 \end{array} \begin{array}{c} \begin{array}{ccccc} P_1 & P_2 & P_3 & P_4 & P_5 \end{array} \\ \begin{bmatrix} 0 & 0 & 1 & 0 & 1 \\ 0 & 0 & 0 & 0 & 1 \\ 1 & 0 & 0 & 0 & 1 \\ 0 & 0 & 0 & 0 & 0 \\ 1 & 1 & 1 & 0 & 0 \end{bmatrix} \end{array}$$

and

$$S^3 = \begin{array}{c} \\ P_1 \\ P_2 \\ P_3 \\ P_4 \\ P_5 \end{array} \begin{array}{c} \begin{array}{ccccc} P_1 & P_2 & P_3 & P_4 & P_5 \end{array} \\ \begin{bmatrix} 2 & 1 & 3 & 0 & 4 \\ 1 & 0 & 1 & 0 & 3 \\ 3 & 1 & 2 & 0 & 4 \\ 0 & 0 & 0 & 0 & 0 \\ 4 & 3 & 4 & 0 & 2 \end{bmatrix} \end{array}.$$

Since s_{11}, s_{33}, and s_{55} are positive, we conclude that P_1, P_3, and P_5 belong to cliques and in fact they form the only clique in this digraph. ■

We now consider the notion of a strongly connected digraph.

DEFINITION

A **path** joining two individuals P_i and P_k in a digraph is a sequence of distinct vertices $P_i, P_a, P_b, P_c, \ldots, P_r, P_k$ and directed edges $P_i P_a, P_a P_b, \ldots, P_r P_k$.

EXAMPLE 13

Consider the digraph in Figure 2.13. The sequence

$$P_2 \rightarrow P_3 \rightarrow P_4 \rightarrow P_5$$

is a path. The sequence

$$P_2 \rightarrow P_3 \rightarrow P_4 \rightarrow P_2 \rightarrow P_5$$

is not a path, since the vertex P_2 is repeated. ■

Figure 2.13 ►

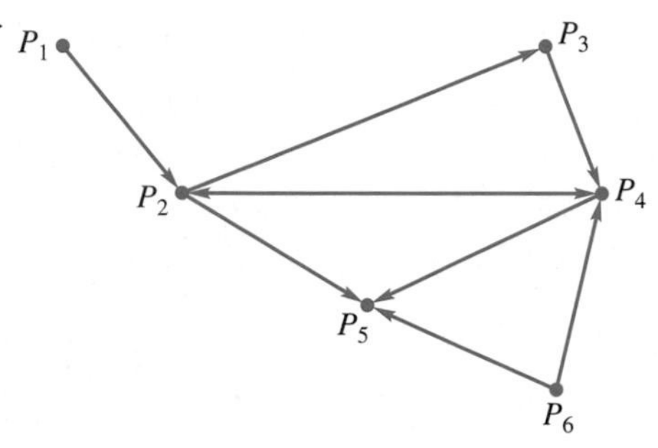

DEFINITION

The digraph G is said to be **strongly connected** if for every two distinct vertices P_i and P_j there is a path from P_i to P_j and a path from P_j to P_i. Otherwise, G is said to be **not strongly connected**.

An example of a strongly connected digraph is provided by the streets of a city.

For many digraphs it is a tedious task to determine whether or not they are strongly connected. First, observe that if our digraph has n vertices, then the number of edges in a path from P_i to P_j cannot exceed $n - 1$, since all the vertices of a path are distinct and we cannot go through more than the n vertices in the digraph. If $[A(G)]^r = [b_{ij}^{(r)}]$, then $b_{ij}^{(r)}$ is the number of ways of getting from P_i to P_j in r stages. A way of getting from P_i to P_j in r stages need not be a path, since it may contain repeated vertices. If these repeated vertices and all edges between repeated vertices are deleted, we do obtain a path between P_i and P_j with at most r edges. For example, if we have $P_1 \rightarrow P_2 \rightarrow P_4 \rightarrow P_3 \rightarrow P_2 \rightarrow P_5$, we can eliminate the second P_2 and all edges between the P_2's and obtain the path $P_1 \rightarrow P_2 \rightarrow P_5$. Hence, if the i, jth element in

$$[A(G)] + [A(G)]^2 + \cdots + [A(G)]^{n-1}$$

is zero, then there is no path from P_i to P_j. Thus the following theorem, half of whose proof we have sketched here, provides a test for strongly connected digraphs.

THEOREM 2.3

A digraph with n vertices is strongly connected if and only if its adjacency matrix $A(G)$ has the property that

$$[A(G)] + [A(G)]^2 + \cdots + [A(G)]^{n-1} = E$$

has no zero entries. ■

> The procedure for determining if a digraph G with n vertices is strongly connected is as follows.
>
> ***Step 1.*** If $A(G)$ is the adjacency matrix of the digraph, compute
>
> $$[A(G)] + [A(G)]^2 + \cdots + [A(G)]^{n-1} = E.$$
>
> ***Step 2.*** G is strongly connected if and only if E has no zero entries.

EXAMPLE 14 Consider the digraph in Figure 2.14. The adjacency matrix is

$$A(G) = \begin{array}{c} \\ P_1 \\ P_2 \\ P_3 \\ P_4 \\ P_5 \end{array} \begin{array}{c} \begin{array}{ccccc} P_1 & P_2 & P_3 & P_4 & P_5 \end{array} \\ \begin{bmatrix} 0 & 1 & 0 & 1 & 0 \\ 0 & 0 & 1 & 1 & 0 \\ 0 & 0 & 0 & 1 & 1 \\ 1 & 0 & 1 & 0 & 0 \\ 0 & 1 & 0 & 1 & 0 \end{bmatrix} \end{array}.$$

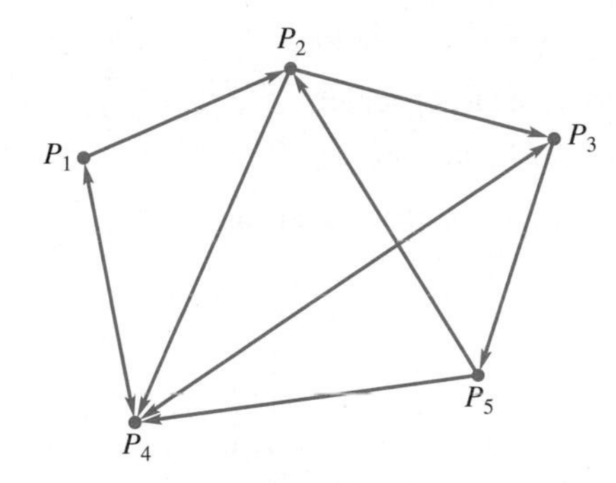

Figure 2.14 ▲

Then (verify)

$$A(G)+[A(G)]^2+[A(G)]^3+[A(G)]^4 = E = \begin{array}{c} \\ P_1 \\ P_2 \\ P_3 \\ P_4 \\ P_5 \end{array} \begin{array}{c} \begin{array}{ccccc} P_1 & P_2 & P_3 & P_4 & P_5 \end{array} \\ \begin{bmatrix} 5 & 5 & 7 & 10 & 3 \\ 5 & 4 & 8 & 10 & 3 \\ 5 & 4 & 7 & 10 & 4 \\ 5 & 4 & 7 & 10 & 4 \\ 5 & 5 & 7 & 10 & 3 \end{bmatrix} \end{array}.$$

Since all the entries in E are positive, the given digraph is strongly connected. ■

This approach can be used to trace the spread of a contaminant in a group of individuals; if there is a path from P_i to P_j, then the contaminant can spread from P_i to P_j.

Further Readings

BERGE, C. *The Theory of Graphs and Its Applications.* New York: Dover Publications, 2001.

BUSACKER, R. G., and T. L. SAATY. *Finite Graphs and Networks: An Introduction with Applications.* New York: McGraw-Hill Book Company, 1965.

CHARTRAND, GARY, and LINDA LESNIAK. *Graphs and Digraphs*, 4th ed. Boca Raton: CRC Press, 2004.

CHARTRAND, GARY, and P. ZHANG. *Introduction to Graph Theory.* New York: McGraw-Hill, 2004.

JOHNSTON, J. B., G. PRICE, and F. S. VAN VLECK. *Linear Equations and Matrices.* Reading, Mass.: Addison-Wesley Publishing Company, Inc., 1966.

KOLMAN, B., R. C. BUSBY, and S. ROSS. *Discrete Mathematical Structures*, 5th ed. Upper Saddle River, N.J.: Prentice Hall, Inc., 2004.

ORE, O. *Graphs and Their Uses*, 2nd ed., revised. Washington, D.C.: Mathematical Association of America, 1996.

TRUDEAU, R. J. *Introduction to Graph Theory.* New York: Dover, 2002.

TUCKER, ALAN. *Applied Combinatorics*, 4th ed. New York: Wiley, 2002.

Key Terms

Directed graph (or digraph)
Vertices (or nodes)
Directed edges
Adjacency matrix
Dominance digraphs
Two-stage access
Clique
Strongly connected

2.2 Exercises

1. Draw a digraph determined by the given matrix.

(a) $\begin{bmatrix} 0 & 1 & 0 & 0 & 0 \\ 0 & 0 & 1 & 1 & 0 \\ 1 & 0 & 0 & 1 & 1 \\ 1 & 1 & 0 & 0 & 1 \\ 1 & 0 & 1 & 1 & 0 \end{bmatrix}$

(b) $\begin{bmatrix} 0 & 1 & 1 & 1 \\ 1 & 0 & 0 & 1 \\ 1 & 1 & 0 & 0 \\ 0 & 1 & 0 & 0 \end{bmatrix}$

2. Write the adjacency matrix of each given digraph.

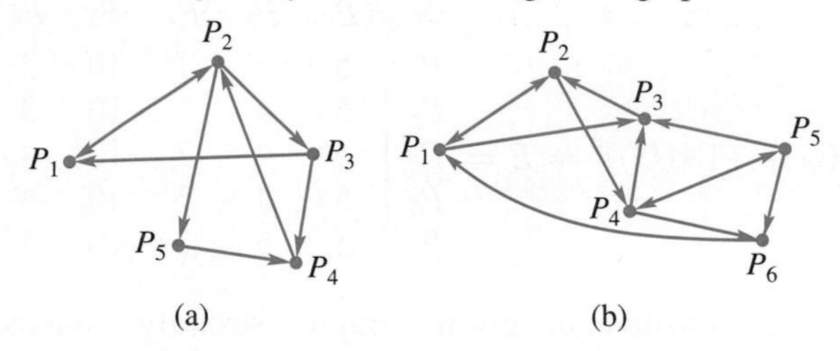

(a) (b)

3. Consider a group of five people, P_1, P_2, P_3, P_4, and P_5, who have been stationed on a remote island to operate a weather station. The following social interactions have been observed:

P_1 gets along with P_2, P_3, and P_4.
P_2 gets along with P_1 and P_5.
P_3 gets along with P_1, P_2, and P_4.
P_4 gets along with P_2, P_3, and P_5.
P_5 gets along with P_4.

Draw a digraph G describing the situation. Write the adjacency matrix representing G.

4. Which of the following matrices can be adjacency matrices for a dominance digraph?

(a) $\begin{bmatrix} 0 & 1 & 1 & 0 \\ 0 & 0 & 1 & 1 \\ 0 & 0 & 0 & 0 \\ 0 & 0 & 1 & 0 \end{bmatrix}$

(b) $\begin{bmatrix} 0 & 0 & 1 & 0 & 0 \\ 1 & 0 & 1 & 0 & 1 \\ 0 & 0 & 0 & 0 & 1 \\ 1 & 1 & 1 & 0 & 1 \\ 1 & 0 & 0 & 0 & 0 \end{bmatrix}$

5. The following data have been obtained by studying a group of six individuals in the course of a sociological study:

Carter influences Smith and Gordon.
Gordon influences Jones.
Smith is influenced by Peters.
Russell is influenced by Carter, Smith, and Gordon.
Peters is influenced by Russell.
Jones influences Carter and Russell.
Smith influences Jones.
Carter is influenced by Peters.
Peters influences Jones and Gordon.

(a) Who influences the most people?
(b) Who is influenced by the most people?

6. Consider a communications network among five individuals with adjacency matrix

$$\begin{array}{c} \\ P_1 \\ P_2 \\ P_3 \\ P_4 \\ P_5 \end{array} \begin{array}{c} \begin{array}{ccccc} P_1 & P_2 & P_3 & P_4 & P_5 \end{array} \\ \begin{bmatrix} 0 & 1 & 1 & 1 & 1 \\ 0 & 0 & 0 & 1 & 1 \\ 0 & 1 & 0 & 1 & 0 \\ 0 & 0 & 0 & 0 & 1 \\ 1 & 0 & 1 & 0 & 0 \end{bmatrix} \end{array}.$$

(a) In how many ways does P_2 have access to P_1 through one individual?
(b) What is the smallest number of individuals through which P_2 has access to himself?

7. Consider the following influence relation among five individuals.

P_1 influences P_2, P_4, and P_5.
P_2 influences P_3 and P_4.
P_3 influences P_1 and P_4.
P_4 influences P_5.
P_5 influences P_2 and P_3.

(a) Can P_4 influence P_1 in at most two stages?
(b) In how many ways does P_1 influence P_4 in exactly three stages?
(c) In how many ways does P_1 influence P_4 in one, two, or three stages?

In Exercises 8 through 10, determine a clique, if there is one, for the digraph with given adjacency matrix.

8. $\begin{bmatrix} 0 & 0 & 0 & 0 & 0 \\ 1 & 0 & 1 & 1 & 1 \\ 0 & 1 & 0 & 1 & 0 \\ 1 & 1 & 1 & 0 & 0 \\ 0 & 0 & 1 & 1 & 0 \end{bmatrix}$

9. $\begin{bmatrix} 0 & 1 & 1 & 0 & 1 & 1 \\ 1 & 0 & 0 & 0 & 0 & 0 \\ 0 & 0 & 0 & 1 & 0 & 0 \\ 1 & 0 & 0 & 0 & 1 & 1 \\ 1 & 0 & 1 & 1 & 0 & 1 \\ 1 & 0 & 1 & 1 & 1 & 0 \end{bmatrix}$

10. $\begin{bmatrix} 0 & 1 & 1 & 0 & 1 \\ 1 & 0 & 1 & 1 & 1 \\ 0 & 0 & 0 & 1 & 0 \\ 1 & 0 & 0 & 0 & 1 \\ 0 & 1 & 0 & 1 & 0 \end{bmatrix}$

11. Consider a communications network among five individuals with adjacency matrix

$$\begin{bmatrix} 0 & 1 & 0 & 0 & 0 \\ 0 & 0 & 1 & 0 & 1 \\ 0 & 0 & 0 & 1 & 0 \\ 1 & 1 & 0 & 0 & 0 \\ 0 & 0 & 1 & 1 & 0 \end{bmatrix}.$$

(a) Can P_3 get a message to P_5 in at most two stages?

(b) What is the minimum number of stages that will guarantee that every person can get a message to any other (different) person?

(c) What is the minimum number of stages that will guarantee that every person can get a message to any person (including himself)?

12. Determine whether the digraph with given adjacency matrix is strongly connected.

(a) $\begin{bmatrix} 0 & 1 & 1 & 1 \\ 0 & 0 & 1 & 1 \\ 1 & 0 & 0 & 1 \\ 0 & 0 & 1 & 0 \end{bmatrix}$

(b) $\begin{bmatrix} 0 & 1 & 0 & 0 & 0 \\ 0 & 0 & 1 & 0 & 0 \\ 1 & 0 & 0 & 0 & 0 \\ 0 & 0 & 0 & 0 & 1 \\ 0 & 0 & 1 & 1 & 0 \end{bmatrix}$

13. Determine whether the digraph with given adjacency matrix is strongly connected.

(a) $\begin{bmatrix} 0 & 0 & 1 & 1 & 1 \\ 1 & 0 & 1 & 1 & 0 \\ 0 & 1 & 0 & 0 & 0 \\ 0 & 1 & 0 & 0 & 1 \\ 1 & 1 & 0 & 0 & 0 \end{bmatrix}$

(b) $\begin{bmatrix} 0 & 0 & 0 & 0 & 1 \\ 0 & 0 & 1 & 1 & 0 \\ 0 & 1 & 0 & 0 & 1 \\ 1 & 0 & 0 & 0 & 0 \\ 0 & 0 & 0 & 1 & 0 \end{bmatrix}$

14. A group of five acrobats performs a pyramiding act in which there must be a supporting path between any two different persons or the pyramid collapses. In the following adjacency matrix, $a_{ij} = 1$ means that P_i supports P_j:

$$\begin{bmatrix} 0 & 0 & 1 & 1 & 0 \\ 1 & 0 & 0 & 0 & 1 \\ 1 & 0 & 0 & 1 & 0 \\ 1 & 0 & 0 & 0 & 1 \\ 0 & 1 & 1 & 0 & 0 \end{bmatrix}.$$

Who can be left out of the pyramid without having a collapse occur?

Theoretical Exercises

T.1. Show that a digraph which has the directed edges P_iP_j and P_jP_i cannot be a dominance graph.

T.2. Prove Theorem 2.1.

T.3. Prove Theorem 2.2.

MATLAB Exercises

MATLAB operators + and ^ can be used to compute sums and powers of a matrix. Hence the computations in Theorems 2.2 and 2.3 are easily performed in MATLAB.

ML.1. Solve Exercise 8 using MATLAB.

ML.2. Determine a clique, if there is one, for the digraph with the following adjacency matrix:

$$\begin{bmatrix} 0 & 1 & 1 & 0 & 1 \\ 1 & 0 & 0 & 1 & 0 \\ 0 & 1 & 0 & 0 & 1 \\ 0 & 1 & 1 & 0 & 1 \\ 1 & 0 & 0 & 1 & 0 \end{bmatrix}.$$

ML.3. Solve Exercise 13 using MATLAB.

2.3 COMPUTER GRAPHICS

Prerequisite. Section 1.5, Matrix Transformatrions.

We are all familiar with the astounding results being developed with computer graphics in the areas of video games and special effects in the film industry. Computer graphics also play a major role in the manufacturing world. *Computer-aided design* (CAD) is used to design a computer model of a product and then, by subjecting the computer model to a variety of tests (carried out on the computer), changes to the current design can be implemented to obtain an improved design. One of the notable successes of this approach has been in the automobile industry, where the computer model can be viewed from different angles to obtain a most pleasing and popular style and can be tested for strength of components, for roadability, for seating comfort, and for safety in a crash.

In this section we give illustrations of matrix transformations $f: R^2 \to R^2$ that are useful in two-dimensional graphics.

EXAMPLE 1 Let $f: R^2 \to R^2$ be the matrix transformation that performs a reflection with respect to the x-axis. (See Example 5 in Section 1.5.) Then f is defined by $f(\mathbf{v}) = A\mathbf{v}$, where

$$A = \begin{bmatrix} 1 & 0 \\ 0 & -1 \end{bmatrix}.$$

Thus we have

$$f(\mathbf{v}) = A\mathbf{v} = \begin{bmatrix} 1 & 0 \\ 0 & -1 \end{bmatrix} \begin{bmatrix} x \\ y \end{bmatrix} = \begin{bmatrix} x \\ -y \end{bmatrix}.$$

To illustrate a reflection with respect to the x-axis in computer graphics, let the triangle T in Figure 2.15(a) have vertices

$$(-1, 4), \quad (3, 1), \quad \text{and} \quad (2, 6).$$

To reflect T with respect to the x-axis, we let

$$\mathbf{v}_1 = \begin{bmatrix} -1 \\ 4 \end{bmatrix}, \quad \mathbf{v}_2 = \begin{bmatrix} 3 \\ 1 \end{bmatrix}, \quad \mathbf{v}_3 = \begin{bmatrix} 2 \\ 6 \end{bmatrix}$$

and compute the images $f(\mathbf{v}_1)$, $f(\mathbf{v}_2)$, and $f(\mathbf{v}_3)$ by forming the products

$$A\mathbf{v}_1 = \begin{bmatrix} 1 & 0 \\ 0 & -1 \end{bmatrix} \begin{bmatrix} -1 \\ 4 \end{bmatrix} = \begin{bmatrix} -1 \\ -4 \end{bmatrix},$$

$$A\mathbf{v}_2 = \begin{bmatrix} 1 & 0 \\ 0 & -1 \end{bmatrix} \begin{bmatrix} 3 \\ 1 \end{bmatrix} = \begin{bmatrix} 3 \\ -1 \end{bmatrix},$$

$$A\mathbf{v}_3 = \begin{bmatrix} 1 & 0 \\ 0 & -1 \end{bmatrix} \begin{bmatrix} 2 \\ 6 \end{bmatrix} = \begin{bmatrix} 2 \\ -6 \end{bmatrix}.$$

These three products can be written in terms of partitioned matrices as

$$A \begin{bmatrix} \mathbf{v}_1 & \mathbf{v}_2 & \mathbf{v}_3 \end{bmatrix} = \begin{bmatrix} -1 & 3 & 2 \\ -4 & -1 & -6 \end{bmatrix}.$$

Thus the image of T has vertices

$$(-1, -4), \quad (3, -1), \quad \text{and} \quad (2, -6)$$

and is displayed in Figure 2.15(b). ■

Figure 2.15 ►

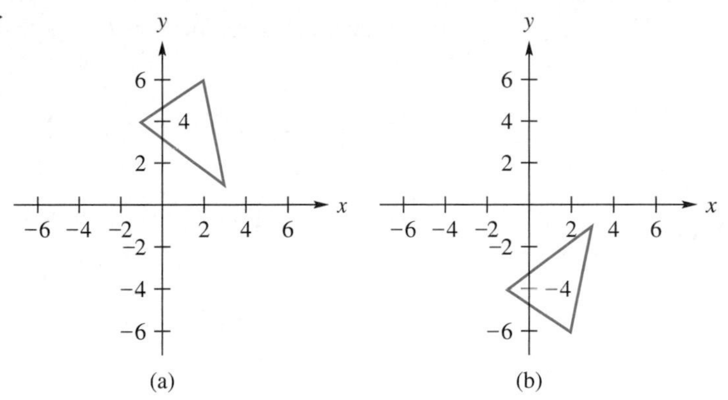

EXAMPLE 2

The matrix transformation $f: R^2 \to R^2$ that performs a reflection with respect to the line $y = -x$ is defined by

$$f(\mathbf{v}) = B\mathbf{v},$$

where

$$B = \begin{bmatrix} 0 & -1 \\ -1 & 0 \end{bmatrix}.$$

To illustrate reflection with respect to the line $y = -x$, we use the triangle T as defined in Example 1 and compute the products

$$B\begin{bmatrix} \mathbf{v}_1 & \mathbf{v}_2 & \mathbf{v}_3 \end{bmatrix} = \begin{bmatrix} 0 & -1 \\ -1 & 0 \end{bmatrix}\begin{bmatrix} -1 & 3 & 2 \\ 4 & 1 & 6 \end{bmatrix} = \begin{bmatrix} -4 & -1 & -6 \\ 1 & -3 & -2 \end{bmatrix}.$$

Thus the image of T has vertices

$$(-4, 1), \quad (-1, -3), \quad \text{and} \quad (-6, -2)$$

and is displayed in Figure 2.16. ■

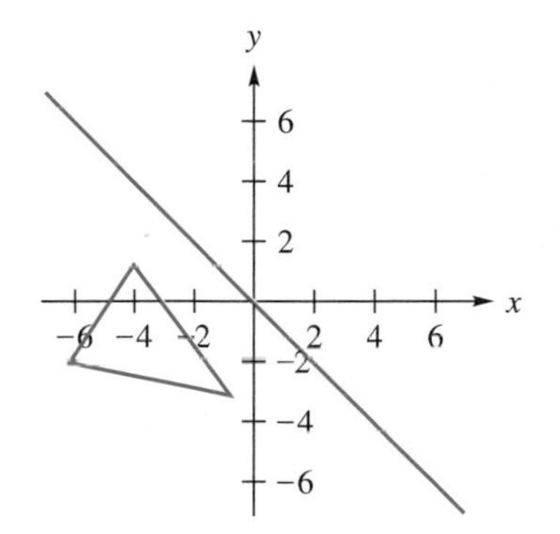

Figure 2.16 ▲

To perform a reflection with respect to the x-axis on the triangle T of Example 1 followed by a reflection with respect to the line $y = -x$, we compute

$$B(A\mathbf{v}_1), \quad B(A\mathbf{v}_2), \quad \text{and} \quad B(A\mathbf{v}_3).$$

It is not difficult to show that reversing the order of these matrix transformations produces a different image (verify). Thus the order in which graphics transformations are performed is important. This is not surprising, since matrix multiplication, unlike multiplication of real numbers, does not satisfy the commutative property.

EXAMPLE 3

Rotations in a plane have been defined in Example 9 of Section 1.5. A plane figure is rotated counterclockwise through an angle ϕ by using the matrix transformation $f: R^2 \to R^2$ defined by $f(\mathbf{v}) = A\mathbf{v}$, where

$$A = \begin{bmatrix} \cos\phi & -\sin\phi \\ \sin\phi & \cos\phi \end{bmatrix}.$$

Now suppose that we wish to rotate the parabola $y = x^2$ counterclockwise through 50°. We start by choosing a sample of points from the parabola, say,

$$(-2, 4), \quad (-1, 1), \quad (0, 0), \quad \left(\tfrac{1}{2}, \tfrac{1}{4}\right), \quad \text{and} \quad (3, 9)$$

[see Figure 2.17(a)]. We then compute the images of these points. Thus letting

$$\mathbf{v}_1 = \begin{bmatrix} -2 \\ 4 \end{bmatrix}, \quad \mathbf{v}_2 = \begin{bmatrix} -1 \\ 1 \end{bmatrix}, \quad \mathbf{v}_3 = \begin{bmatrix} 0 \\ 0 \end{bmatrix}, \quad \mathbf{v}_4 = \begin{bmatrix} \frac{1}{2} \\ \frac{1}{4} \end{bmatrix}, \quad \mathbf{v}_5 = \begin{bmatrix} 3 \\ 9 \end{bmatrix},$$

we compute the products (to four decimal places) (verify)

$$A \begin{bmatrix} \mathbf{v}_1 & \mathbf{v}_2 & \mathbf{v}_3 & \mathbf{v}_4 & \mathbf{v}_5 \end{bmatrix} = \begin{bmatrix} -4.3498 & -1.4088 & 0 & 0.1299 & -4.9660 \\ 1.0391 & -0.1233 & 0 & 0.5437 & 8.0832 \end{bmatrix}.$$

The image points

$$(-4.3498, 1.0391), \quad (-1.4088, -0.1233), \quad (0, 0), \quad (0.1299, 0.5437), \quad \text{and} \quad (-4.9660, 8.0832)$$

are plotted, as shown in Figure 2.17(b), and successive points are connected showing the approximate image of the parabola. ■

Figure 2.17 ►

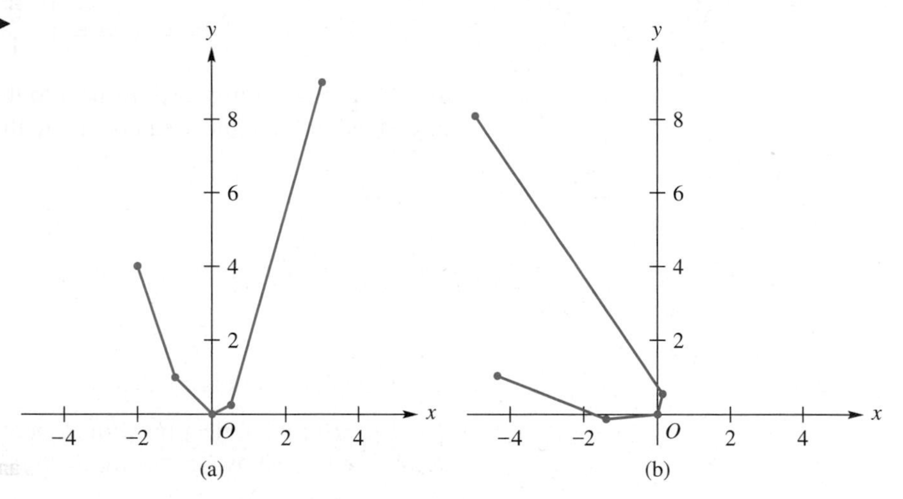

Rotations are particularly useful in achieving the sophisticated effects seen in arcade games and animated computer demonstrations. For example, to show a wheel spinning we can rotate the spokes through an angle θ_1 followed by a second rotation through an angle θ_2 and so on. Let the 2-vector $\mathbf{u} = \begin{bmatrix} a_1 \\ a_2 \end{bmatrix}$ represent a spoke of the wheel; let $f\colon R^2 \to R^2$ be the matrix transformation defined by $f(\mathbf{v}) = A\mathbf{v}$, where

$$A = \begin{bmatrix} \cos\theta_1 & -\sin\theta_1 \\ \sin\theta_1 & \cos\theta_1 \end{bmatrix};$$

and let $g\colon R^2 \to R^2$ be the matrix transformation defined by $g(\mathbf{v}) = B\mathbf{v}$, where

$$B = \begin{bmatrix} \cos\theta_2 & -\sin\theta_2 \\ \sin\theta_2 & \cos\theta_2 \end{bmatrix}.$$

We represent the succession of rotations of the spoke $\mathbf{u}$ by

$$g(f(\mathbf{u})) = g(A\mathbf{u}) = B(A\mathbf{u}).$$

The product $A\mathbf{u}$ is performed first and generates a rotation of $\mathbf{u}$ through the angle θ_1; then the product $B(A\mathbf{u})$ generates the second rotation. We have

$$B(A\mathbf{u}) = B(a_1\text{col}_1(A) + a_2\text{col}_2(A)) = a_1 B\text{col}_1(A) + a_2 B\text{col}_2(A)$$

and the final expression is a linear combination of column vectors $B\text{col}_1(A)$ and $B\text{col}_2(A)$, which we can write as the product

$$\begin{bmatrix} B\text{col}_1(A) & B\text{col}_2(A) \end{bmatrix} \begin{bmatrix} a_1 \\ a_2 \end{bmatrix}.$$

From the definition of matrix multiplication, $\begin{bmatrix} B\text{col}_1(A) & B\text{col}_2(A) \end{bmatrix} = BA$, so we have

$$B(A\mathbf{u}) = (BA)\mathbf{u},$$

which says that instead of applying the transformations in succession, f followed by g, we can achieve the same result by forming the matrix product BA and using it to define a matrix transformation on the spokes of the wheel.

EXAMPLE 4 A **shear in the x-direction** is the matrix transformation defined by

$$f(\mathbf{v}) = \begin{bmatrix} 1 & k \\ 0 & 1 \end{bmatrix} \mathbf{v},$$

where k is a scalar. A shear in the x-direction takes the point (x, y) to the point $(x + ky, y)$. That is, the point (x, y) is moved parallel to the x-axis by the amount ky.

Consider now the rectangle R, shown in Figure 2.18(a), with vertices

$$(0, 0), \quad (0, 2), \quad (4, 0), \quad \text{and} \quad (4, 2).$$

If we apply the shear in the x-direction with $k = 2$, then the image of R is the parallelogram with vertices

$$(0, 0), \quad (4, 2), \quad (4, 0), \quad \text{and} \quad (8, 2),$$

shown in Figure 2.18(b). If we apply the shear in the x-direction with $k = -3$, then the image of R is the parallelogram with vertices

$$(0, 0), \quad (-6, 2), \quad (4, 0), \quad \text{and} \quad (-2, 2),$$

shown in Figure 2.18(c).

In Exercise 3 we consider shears in the y-direction. ■

Figure 2.18 ►

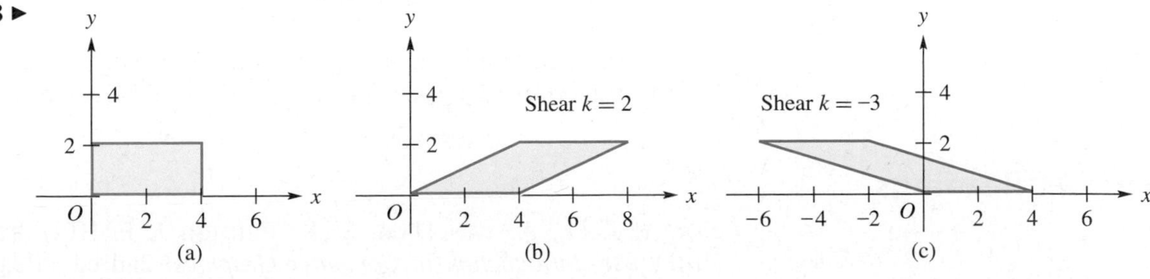

Other matrix transformations used in two-dimensional computer graphics are considered in the exercises at the end of this section. For a detailed discussion of computer graphics, the reader is referred to the books listed in the Further Readings at the end of this section.

In Examples 1 and 2, we applied a matrix transformation to a triangle, a figure that can be specified by giving its three vertices. In Example 3, the figure being transformed was a parabola, which cannot be specified by a finite number of points. In this case we chose a number of points on the parabola to approximate its shape and computed the images of these approximating points, which when joined gave an approximate shape of the image figure.

EXAMPLE 5 Let $f\colon R^2 \to R^2$ be the matrix transformation defined by $f(\mathbf{v}) = A\mathbf{v}$, where

$$A = \begin{bmatrix} h & 0 \\ 0 & k \end{bmatrix}$$

with h and k both nonzero. Suppose that we now wish to apply this matrix transformation to a circle of radius 1 that is centered at the origin (the unit circle). Unfortunately, a circle cannot be specified by a finite number of points. However, each point on the unit circle is described by an ordered pair $(\cos\theta, \sin\theta)$, where the angle θ takes on all values from 0 to 2π radians. Thus, we now represent an arbitrary point on the unit circle by the vector $\mathbf{u} = \begin{bmatrix} \cos\theta \\ \sin\theta \end{bmatrix}$. Hence, the images of the unit circle that are obtained by applying the matrix transformation f are given by

$$f(\mathbf{u}) = A\mathbf{u} = \begin{bmatrix} h & 0 \\ 0 & k \end{bmatrix} \begin{bmatrix} \cos\theta \\ \sin\theta \end{bmatrix} = \begin{bmatrix} h\cos\theta \\ k\sin\theta \end{bmatrix} = \begin{bmatrix} x' \\ y' \end{bmatrix}.$$

We recall that a circle of radius 1 centered at the origin is described by the equation

$$x^2 + y^2 = 1.$$

By Pythagoras' identity, $\sin^2\theta + \cos^2\theta = 1$. Thus, the points $(\cos\theta, \sin\theta)$ lie on the circumference of the unit circle. We now want to obtain an equation describing the image of the unit circle. We have

$$x' = h\cos\theta \quad \text{and} \quad y' = k\sin\theta$$

so

$$\frac{x'}{h} = \cos\theta, \quad \frac{y'}{k} = \sin\theta.$$

It then follows that

$$\left(\frac{x'}{h}\right)^2 + \left(\frac{y'}{k}\right)^2 = 1,$$

which is the equation of an ellipse. Thus the image of the unit circle by the matrix transformation f is an ellipse centered at the origin. See Figure 2.19. ■

Further Readings

FOLEY, J. D., A. VAN DAM, S. K. FEINER, J. F. HUGHES, and R. L. PHILLIPS. *Introduction to Computer Graphics*, 2nd ed. Reading, Mass.: Addison-Wesley, 1996.

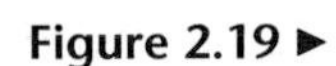

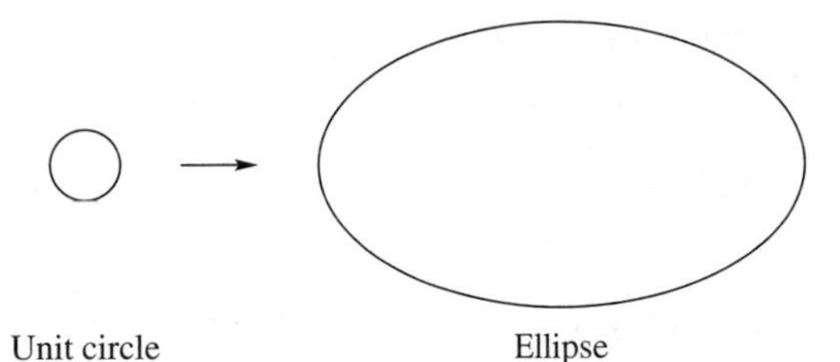

MORTENSON, M. E. *Mathematics for Computer Graphics Applications*, 2nd ed. New York: Industrial Press, Inc., 1999.

ROGERS, D. F., and J. A. ADAMS. *Mathematical Elements for Computer Graphics*, 2nd ed. New York: McGraw-Hill, 1989.

Key Terms

Computer graphics
Computer-aided design
Image
Reflection
Rotation
Shear
Dilation
Contraction

2.3 Exercises

1. Let $f: R^2 \to R^2$ be the matrix transformation defined by $f(\mathbf{v}) = A\mathbf{v}$, where

$$A = \begin{bmatrix} -1 & 0 \\ 0 & 1 \end{bmatrix},$$

that is, f is a reflection with respect to the y-axis. Find and sketch the image of the rectangle R with vertices $(1, 1)$, $(2, 1)$, $(1, 3)$, and $(2, 3)$.

2. Let R be the rectangle with vertices $(1, 1)$, $(1, 4)$, $(3, 1)$, and $(3, 4)$. Let f be the shear in the x-direction with $k = 3$. Find and sketch the image of R.

3. A **shear in the y-direction** is the matrix transformation $f: R^2 \to R^2$ defined by $f(\mathbf{v}) = A\mathbf{v}$, and

$$A = \begin{bmatrix} 1 & 0 \\ k & 1 \end{bmatrix},$$

and k is a scalar. Let R be the rectangle defined in Exercise 2 and let f be the shear in the y-direction with $k = -2$. Find and sketch the image of R.

4. The matrix transformation $f: R^2 \to R^2$ defined by $f(\mathbf{v}) = A\mathbf{v}$, where

$$A = \begin{bmatrix} k & 0 \\ 0 & k \end{bmatrix},$$

and k is a real number, is called **dilation** if $k > 1$ and a **contraction** if $0 < k < 1$. Thus, dilation stretches a vector, whereas contraction shrinks it. If R is the rectangle defined in Exercise 2, find and sketch the image of R for

(a) $k = 4$ (b) $k = \frac{1}{4}$

5. The matrix transformation $f: R^2 \to R^2$ defined by $f(\mathbf{v}) = A\mathbf{v}$, where

$$A = \begin{bmatrix} k & 0 \\ 0 & 1 \end{bmatrix},$$

and k is a real number, is called **dilation in the x-direction** if $k > 1$ and a **contraction in the x-direction** if $0 < k < 1$. If R is the unit square and f is dilation in the x-direction with $k = 2$, find and sketch the image of R.

6. The matrix transformation $f: R^2 \to R^2$ defined by $f(\mathbf{v}) = A\mathbf{v}$, where

$$A = \begin{bmatrix} 1 & 0 \\ 0 & k \end{bmatrix},$$

where k is a real number, is called **dilation in the y-direction** if $k > 1$ and **contraction in the y-direction** if $0 < k < 1$. If R is the unit square and f is the contraction in the y-direction with $k = \frac{1}{2}$, find and sketch the image of R.

7. Let T be the triangle with vertices $(5, 0)$, $(0, 3)$, and $(2, -1)$. Find the coordinates of the vertices of the image of T under the matrix transformation f defined by

$$f(\mathbf{v}) = \begin{bmatrix} -2 & 1 \\ 3 & 4 \end{bmatrix}\mathbf{v}.$$

8. Let T be the triangle with vertices $(1, 1)$, $(-3, -3)$, and $(2, -1)$. Find the coordinates of the vertices of the image of T under the matrix transformation defined by

$$f(\mathbf{v}) = \begin{bmatrix} 4 & -3 \\ -4 & 2 \end{bmatrix}\mathbf{v}.$$

9. Let f be the counterclockwise rotation through 60°. If T is the triangle defined in Exercise 8, find and sketch the image of T under f.

10. Let f_1 be reflection with respect to the y-axis and let f_2 be counterclockwise rotation through $\pi/2$ radians. Show that the result of first performing f_2 and then f_1 is not the same as first performing f_1 and then performing f_2.

11. Let A be the singular matrix $\begin{bmatrix} 1 & 2 \\ 2 & 4 \end{bmatrix}$ and let T be the triangle defined in Exercise 8. Describe the image of T under the matrix transformation $f: R^2 \to R^2$ defined by $f(\mathbf{v}) = A\mathbf{v}$.

12. Let f be the matrix transformation defined in Example 5. Find and sketch the image of the rectangle with vertices $(0, 0)$, $(1, 0)$, $(1, 1)$, and $(0, 1)$ for $h = 2$ and $k = 3$.

13. Let $f: R^2 \to R^2$ be the matrix transformation defined by $f(\mathbf{v}) = A\mathbf{v}$, where

$$A = \begin{bmatrix} 1 & -1 \\ 2 & 3 \end{bmatrix}.$$

Find and sketch the image of the rectangle defined in Exercise 12.

In Exercises 14 and 15, let f_1, f_2, f_3, and f_4 be the following matrix transformations:

f_1: *counterclockwise rotation through the angle* ϕ

f_2: *reflection with respect to the x-axis*

f_3: *reflection with respect to the y-axis*

f_4: *reflection with respect to the line* $y = x$

14. Let S denote the unit square.

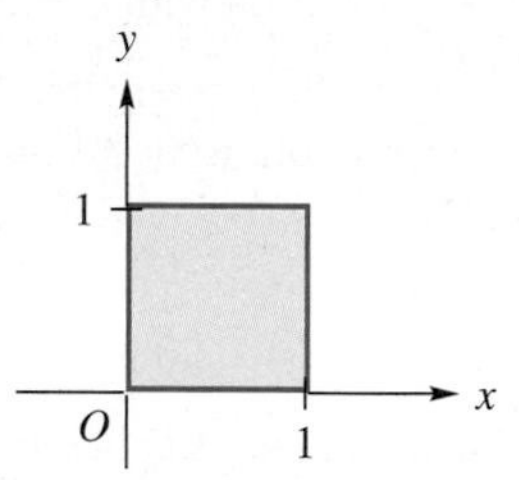

Determine two distinct ways to use the matrix transformations defined on S to obtain the given image. You may apply more than one matrix transformation in succession.

(a)

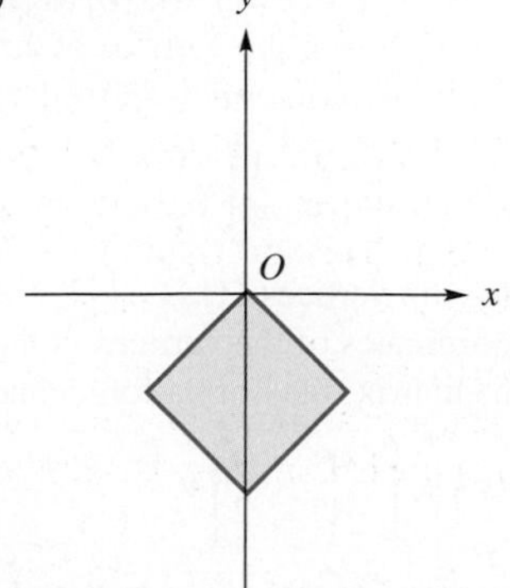

(b)

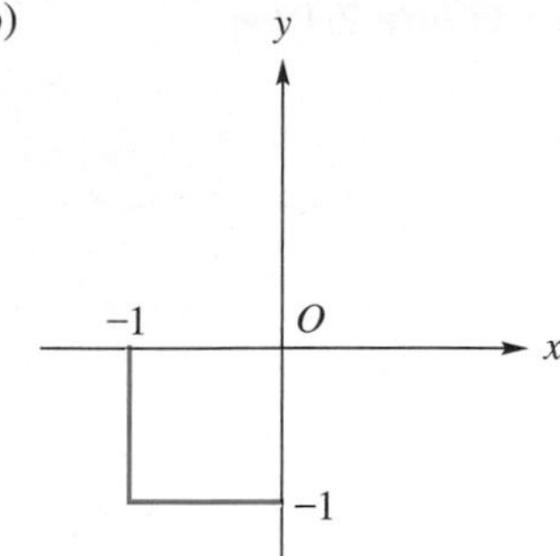

15. Let S denote the triangle shown in the figure.

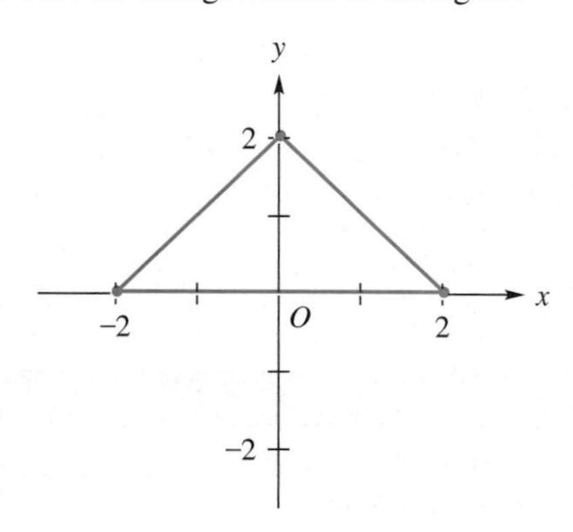

Determine two distinct ways to use the matrix transformations defined on S to obtain the given image. You may apply more than one matrix transformation in succession.

(a)

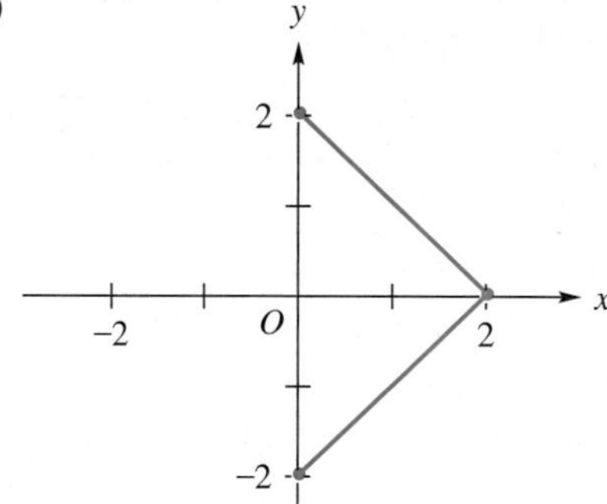

(b)

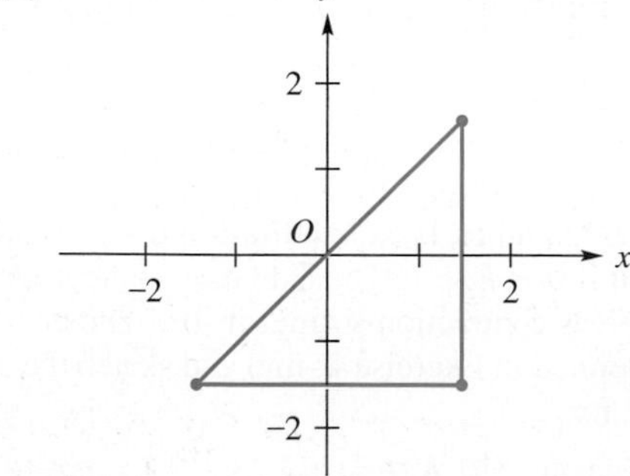

MATLAB Exercises

This section introduced matrix transformations, which are functions whose input and output are vectors that are related by a matrix multiplication: $f(\mathbf{c}) = A\mathbf{c}$. The input $\mathbf{c}$ can be a single vector or collection of vectors that represent a figure or object. (Note that we can view vectors as points and vice versa.) A matrix transformation from R^m to R^n is often called a ***mapping*** *or* ***map****.*

The geometric examples in this section took A to be a 2×2 matrix so that we could easily display the output, the image. In the following exercises we continue this practice and use MATLAB *to construct and display images. The* MATLAB *routines in these exercises provide an opportunity for you to gain experience with the visualization of matrix transformations.*

Exercises ML.1 through ML.4. Use MATLAB *routine* ***matrixtrans*** *to provide further illustrations and experimental capabilities. In* MATLAB *type* ***help matrixtrans*** *and read the brief description. To start this routine type* ***matrixtrans****. At the lower left is the Comment Window that will provide directions for steps to use this routine.*

ML.1. Select the 'Object' Circle by clicking the mouse on the word. Look at the Comment Window. Next click on the View button. The unit circle will appear on the left set of axes.

(a) Click on the MATRIX button, then enter the matrix

$$A = \begin{bmatrix} 0.5 & 0 \\ 0 & 1 \end{bmatrix}$$

by typing **[0.5 0;0 1]** followed by Enter. Click on the MAP IT button to see the image of the unit circle determined by the matrix A.

(b) Click on the Composite button and then the MATRIX button that appears. This time reenter the matrix

$$\begin{bmatrix} 0.5 & 0 \\ 0 & 1 \end{bmatrix}$$

(or use the up arrow key to access the command stack to find this matrix), then click MAP IT.

(c) Write a short description of the actions taken by the matrix transformations in parts (a) and (b).

(d) If we applied this same matrix transformation a third time, where will the image lie in relation to the current figure?

ML.2. (Start **matrixtrans**; or if you are currently using it, click Restart.) Select the 'Object' Square by clicking the mouse on the word. Look at the Comment Window. Next click on the View button. The unit square will appear on the left set of axes.

(a) Now click on the MATRIX button, then enter the matrix $A = \begin{bmatrix} 2 & 0 \\ 0 & 4 \end{bmatrix}$ by typing **[2 0;0 4]** followed by Enter. Click on the MAP IT button to see the image of the unit square determined by the matrix A. What is the area of the image?

(b) Click on the composite button and then on the MATRIX button that appears. This time enter the matrix

$$\begin{bmatrix} \frac{1}{2} & 0 \\ 0 & \frac{1}{4} \end{bmatrix},$$

then click on MAP IT. What is the area of the composite image?

(c) If $A = \begin{bmatrix} 2 & 0 \\ 0 & 4 \end{bmatrix}$ and $B = \begin{bmatrix} \frac{1}{2} & 0 \\ 0 & \frac{1}{4} \end{bmatrix}$, we saw that

$$\begin{aligned} f(\text{unit square}) &= A(\text{unit square}) \\ &= \text{first image} \end{aligned}$$

and

$$\begin{aligned} g(\text{first image}) &= B(\text{first image}) \\ &= \text{composite image.} \end{aligned}$$

Thus we have

$$\begin{aligned} g(f(\text{unit square})) &= B(A(\text{unit square})) \\ &= \text{composite image.} \end{aligned}$$

Compute the matrix BA and explain how the result of this composition is related to the unit square.

ML.3. (Start **matrixtrans**; or if you are currently using it, click Restart.) Select the 'Object' House by clicking the mouse on the word. Look at the Comment Window. Next click on the View button. The house will appear on the left set of axes.

(a) Click the Grid On button. Compute the area of the house. Use a matrix transformation that is a shear in the x-direction where $k = 1$ (see Example 4) and display the image. What is the area of the image? How are the areas related?

(b) Click the Restart button. Select the house again. Use a matrix transformation that is a shear in the x-direction where $k = 0.5$ (see Example 4) and display the image. What is the area of the image?

(c) Click the Restart button. Select the house again. Use a matrix transformation that is a shear in the x-direction where $k = 2$ (see Example 4) and display the image. What is the area of the image? (Inspect the figure carefully.)

ML.4. Use **matrixtrans** to perform each of the following.

(a) Select the Arrow object. Determine a matrix A so that the image is an arrow pointed in the opposite direction. Display the image.

(b) Select the Arrow object. Determine a matrix A so that the image is an arrow pointed in the same direction but only half as long. Display the image.

(c) Select the Arrow object and use the MATLAB matrix

$$\begin{bmatrix} \cos(\text{pi}/4) & \sin(\text{pi}/4) \\ -\sin(\text{pi}/4) & \cos(\text{pi}/4) \end{bmatrix}.$$

Describe the resulting image. What angle does it make with the positive horizontal axis? To help answer this question, use the grid button and then inspect the grid generated on the mapped arrow.

(d) Using part (c), determine the coordinates of the top end of the arrow.

*In Exercises ML.5 through ML.6, use the routine **planelt**. Matrix transformations from R^2 to R^2 are sometimes called plane linear transformations. The MATLAB routine **planelt** lets us experiment with such transformations by choosing the geometric operation we want to perform on a figure. The routine then uses the appropriate matrix to compute the image, displays the matrix, and keeps a graphical record of the original, the previous image, and the current image. Routine **planelt** is quite versatile since you can enter your own figure and/or matrix. To start this routine type **planelt**.*

ML.5. Start routine **planelt**. Read the descriptions that appear and follow the on-screen directions until you get to FIGURE CHOICES. There select the triangle, choose to 'See the Triangle,' and then the option 'Use this Figure. Go to select transformations.'

(a) Select the rotation and use a 45° angle. After the figures are displayed, press Enter. You will see the Plane Linear Transformation menu of options again. If you choose a transformation at this point, it will be used 'compositely' with the transformation just performed. Try this by choosing to reflect the current figure about the y-axis. The figures displayed show the original triangle, this triangle rotated through 45°, and then this image reflected about the y-axis. Record a sketch of the composite figure.

(b) Reverse the order of the transformations in part (a). Record a sketch of the figure obtained from this composition. Compare this sketch with that from part (a). If A is the matrix of the 45° rotation and B is the matrix of the reflection about the y-axis, then explain how we know that $BA \neq AB$.

ML.6. Use **planelt** with the parallelogram. Choose a composition of transformations so that the final figure is that shown in the accompanying figure.

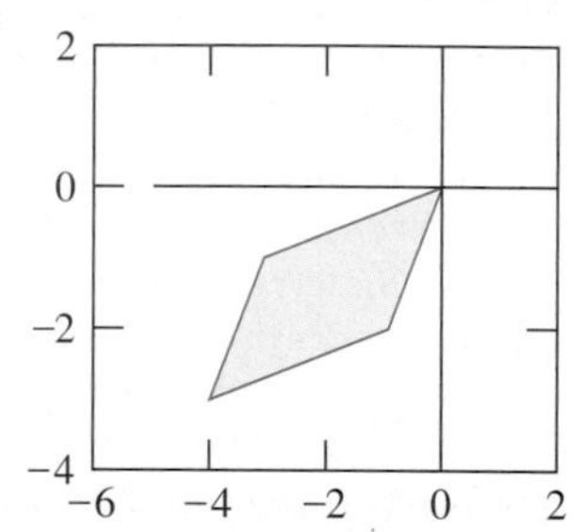

ML.7. Orthogonal projections will play a fundamental role in a variety of situations for us later. While this section developed the algebra for computing projections, MATLAB can help present the geometric aspects. In MATLAB type **help project** and read the description. To start this routine, type **project** and then choose the demo for a first look at this routine. Use **project** to determine $\text{proj}_{\mathbf{w}}\mathbf{u}$; that is, the projection of **u** onto **w**, for each of the following pairs of vectors. Determine whether the projection is longer than the vector **w** and whether it is in the opposite direction.

(a) $\mathbf{u} = \begin{bmatrix} 5 \\ 4 \end{bmatrix}$, $\mathbf{w} = \begin{bmatrix} 3 \\ 1 \end{bmatrix}$

(b) $\mathbf{u} = \begin{bmatrix} 1 \\ -4 \end{bmatrix}$, $\mathbf{w} = \begin{bmatrix} 3 \\ 7 \end{bmatrix}$

(c) $\mathbf{u} = \begin{bmatrix} 5 \\ 3 \\ 1 \end{bmatrix}$, $\mathbf{w} = \begin{bmatrix} 3 \\ 1 \\ -4 \end{bmatrix}$

(d) $\mathbf{u} = \begin{bmatrix} 4 \\ 6 \\ 0 \end{bmatrix}$, $\mathbf{w} = \begin{bmatrix} 2 \\ 3 \\ 8 \end{bmatrix}$

2.4 ELECTRICAL CIRCUITS

Prerequisite. Section 1.6, Solutions of Linear Systems of Equations.

In this section we introduce the basic laws of electrical-circuit analysis and then use these to analyze electrical circuits consisting of batteries, resistors, and wires.

A **battery** is a source of direct current (or voltage) in the circuit, a **resistor** is a device, such as a lightbulb, that reduces the current in a circuit by converting electrical energy into thermal energy, and a **wire** is a conductor that allows a free flow of electrical current. A simple electrical circuit is a closed connection of resistors, batteries, and wires. When circuits are represented by diagrams, batteries, resistors, and wires are depicted as follows:

Batteries   Resistors   Wires

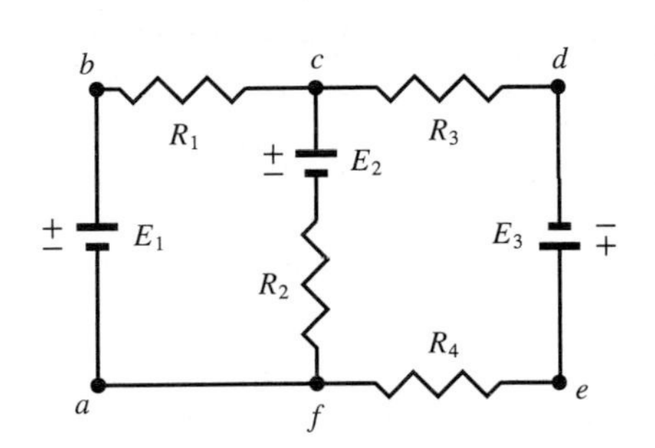

Figure 2.20 ▲

Figure 2.20 shows a simple electrical circuit consisting of three batteries and four resistors connected by wires.

The physical quantities used when discussing electrical circuits are current, resistance, and electrical potential difference across a battery. Electrical potential difference is denoted by E and is measured in volts (V). Current is denoted by I and is measured in amperes (A). Resistance is denoted by R and is measured in ohms (Ω). These units are all related by the equation

$$\text{One Volt} = (\text{One Ampere}) \times (\text{One Ohm}).$$

The electrical potential difference of a battery is taken as positive when measured from the negative terminal ($-$) to the positive terminal ($+$), and negative when measured from the positive terminal ($+$) to the negative terminal ($-$). The electrical potential difference across a resistor (denoted by V) depends on the current flowing through the resistor and its resistance, and is given by Ohm's law:

$$V = \pm IR.$$

The negative ( ) sign is used when the difference is measured across the resistor in the direction of the current flow, while the positive ($+$) sign is used when the difference is measured across the resistor in the direction opposite the current flow.

All electrical circuits consist of voltage loops and current nodes. A **voltage loop** is a closed connection within the circuit. For example, Figure 2.20 contains the three voltage loops

$$a \to b \to c \to f \to a,$$
$$c \to d \to e \to f \to c,$$

and

$$a \to b \to c \to d \to e \to f \to a.$$

A **current node** is a point where three or more segments of wire meet. For example, Figure 2.20 contains two current nodes at points

$$c \quad \text{and} \quad f.$$

Points a, b, d, and e are not current nodes since only two wire segments meet at these points.

The physical laws that govern the flow of current in an electrical circuit are conservation of energy and conservation of charge.

- The *conservation of energy* appears in what is known as **Kirchhoff's voltage law**: Around any voltage loop, the total electrical potential difference is zero.
- The *conservation of charge* appears in what is known as **Kirchhoff's current law**: At any current node, the flow of all currents into the node equals the flow of all currents out of the node. This guarantees that no charge builds up or is depleted at a node, so the current flow is steady through the node.

We are now ready to apply these ideas and the methods for solving linear systems to solving problems involving electrical circuits, which have the following general format. In a circuit containing batteries, resistors, and wires, determine all *unknown* values of electrical potential difference across the batteries, resistance in the resistors, and currents flowing through the wires, given enough *known* values of these same quantities. The following example illustrates this for the standard situation when the unknowns are the currents.

EXAMPLE 1 Figure 2.21 shows the circuit from Figure 2.20 with the batteries having the indicated electrical potentials, measured from the negative terminal to the positive one, and the resistors having the indicated resistances. The problem is to determine the currents that flow through each segment of the circuit.

Figure 2.21 ►

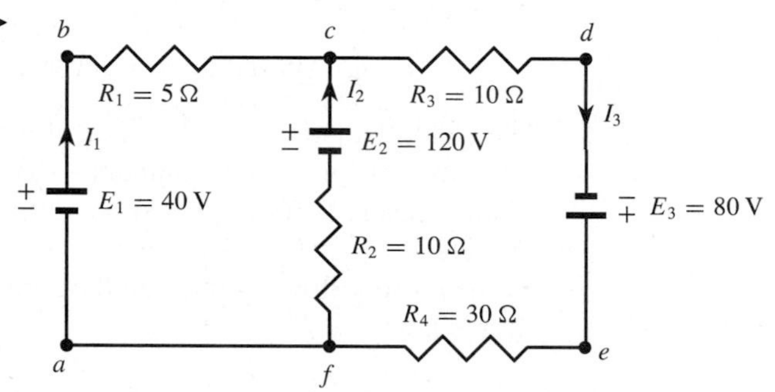

We begin by assigning currents to each segment of the circuit that begins at some node point and ends at some other node point (with no other node points in between). For example, in Figure 2.21, we assign I_1 to the segment $f \to a \to b \to c$, I_2 to the segment $f \to c$, and I_3 to the segment $c \to d \to e \to f$. In addition, we arbitrarily assign directions to these currents as indicated by the arrows in Figure 2.21. If the assigned direction is correct, the computed value of the current will be positive, while if the assigned direction is incorrect, the computed value of the current will be negative. The latter situation then indicates that the actual direction of current flow is just opposite to the assigned one. Using Kirchhoff's current law (the sum of the incoming currents = the sum of the outgoing currents) at points c and f, we have

$$I_1 + I_2 = I_3$$

and

$$I_3 = I_1 + I_2, \tag{1}$$

respectively. Since these two equations contain the same information, only one of them is needed. In general, if a circuit has n nodes, Kirchhoff's current law will yield $n - 1$ useful equations and one equation that is a linear combination of the other $n - 1$ equations.

Next we use Kirchhoff's voltage law. Starting at point a and moving through the battery from (−) to (+) to point b, the potential is +40 V. Moving from point b to point c through the 5-Ω resistor results in a potential difference of $-5I_1$. Moving from point c to point f through the 120-V battery and 10-Ω resistor results in a potential difference of −120 V (across the battery) and a potential difference of $+10I_2$ (across the resistor). Finally, moving along the wire from point f to point a results in no potential difference. To summarize, applying Kirchhoff's voltage law around the closed loop $a \to b \to c \to f \to a$ leads to

$$(+E_1) + (-R_1 I_1) + (-E_2) + (+R_2 I_2) = 0$$

or

$$(+40) + (-5I_1) + (-120) + (10I_2) = 0,$$

which simplifies to

$$I_1 - 2I_2 = -16. \tag{2}$$

In a similar way, applying Kirchhoff's voltage law around the closed loop $c \to d \to e \to f \to c$ leads to

$$(-R_3 I_3) + (+E_3) + (-R_4 I_3) + (-R_2 I_2) + (+E_2) = 0$$

or

$$(-20I_3) + (+80) + (-30I_3) + (-10I_2) + (+120) = 0.$$

This simplifies to $10I_2 + 50I_3 = 200$, or

$$I_2 + 5I_3 = 20. \tag{3}$$

Note that the equation resulting from the voltage loop, $a \to b \to c \to d \to e \to f \to a$, becomes

$$(+E_1) + (-R_1 I_1) + (-R_3 I_3) + (+E_3) + (-R_4 I_3) = 0$$

or

$$(+40) + (-5I_1) + (-20I_3) + (+80) + (-30I_3) = 0,$$

which simplifies to

$$I_1 + 10I_3 = 24.$$

But this is just the linear combination Equation (2) + 2 Equation (3), and it therefore provides no new information. Thus this equation is redundant and can be omitted. In general, a larger outer loop like $a \to b \to c \to d \to e \to f \to a$ provides no new information if all its inner loops, like $a \to b \to c \to f \to a$ and $c \to d \to e \to f \to c$, have already been included.

Equations (1), (2), and (3) lead to the linear system

$$\begin{bmatrix} 1 & 1 & -1 \\ 1 & -2 & 0 \\ 0 & 1 & 5 \end{bmatrix} \begin{bmatrix} I_1 \\ I_2 \\ I_3 \end{bmatrix} = \begin{bmatrix} 0 \\ -16 \\ 20 \end{bmatrix}.$$

Solving this for I_1, I_2, and I_3 yields (verify)

$$I_1 = -3.5\text{ A}, \quad I_2 = 6.25\text{ A}, \quad \text{and} \quad I_3 = 2.75\text{ A}.$$

The *negative* value of I_1 indicates that its true direction is opposite that assigned in Figure 2.21. ■

In general, for an electrical circuit consisting of batteries, resistors, and wires and having n different current assignments, Kirchhoff's voltage and current laws will always lead to n linear equations that have a unique solution.

Key Terms

Battery
Resistor
Wire
Voltage loop
Current node
Kirchhoff's voltage law
Kirchhoff's current law

2.4 Exercises

In Exercises 1 through 4, determine the unknown currents in the given circuit.

1.

b 3 Ω c 4 Ω I_4 d
I_3
10 V
20 V
1 Ω
2 Ω
I_2
I_5
a I_1 + − e
30 V

2.

b 2 Ω I_2 c I_4 d
I_3
5 Ω 6 Ω
40 V
20 V
I_1 I_5
a f 1 Ω e

3.

b I_2 c 20 Ω I_3 d
I_4
10 Ω 100 V
40 Ω
I_1
a f + − e
200 V I_5

4.

a I_1 b I_4 c
+
150 V 5 Ω 4 Ω
300 V −
I_2 I_5
f e d
3 Ω I_3

In Exercises 5 through 8, determine the unknowns in the given circuit.

5.

a 3 Ω b 2 Ω c
I_1 I_2
60 V 80 V E
13 A
f e d
1 Ω 3 Ω

6.

a 5 Ω b 2 Ω I_2 c 10 A d
I_1
10 V E_2 E_3 R
3 A I_3
h g 5 A f e

7.

c
2 Ω R
I_2 28 A
b E d
42 A I_4
4 Ω 3 Ω
I_3 e
a f
I_1

8.

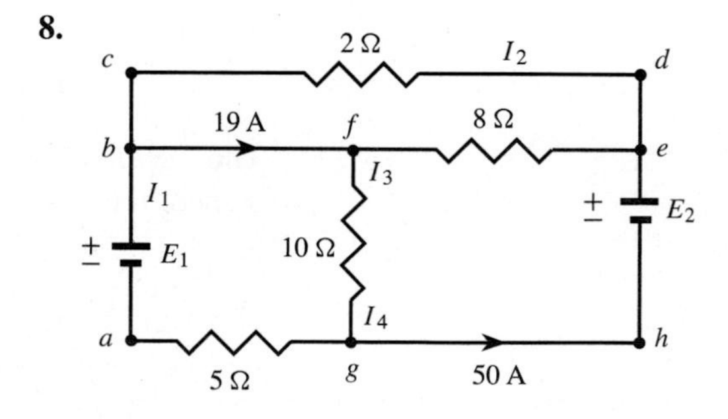

Theoretical Exercises

T.1. For the following circuit, show that

$$I_1 = \left(\frac{R_2}{R_1 + R_2}\right) I = \left(\frac{R}{R_1}\right) I$$

and

$$I_2 = \left(\frac{R_1}{R_1 + R_2}\right) I = \left(\frac{R}{R_2}\right) I,$$

where

$$\frac{1}{R} = \frac{1}{R_1} + \frac{1}{R_2}.$$

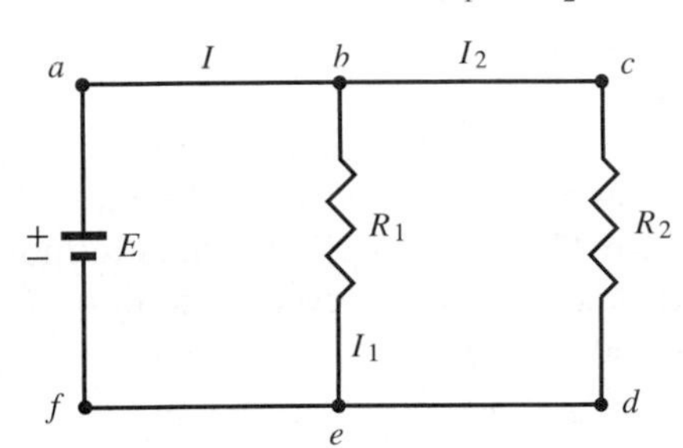

T.2. For the following circuit, show that

$$I_1 = \left(\frac{R}{R_1}\right) I, \qquad I_2 = \left(\frac{R}{R_2}\right) I,$$

and

$$I_3 = \left(\frac{R}{R_3}\right) I,$$

where

$$\frac{1}{R} = \frac{1}{R_1} + \frac{1}{R_2} + \frac{1}{R_3}.$$

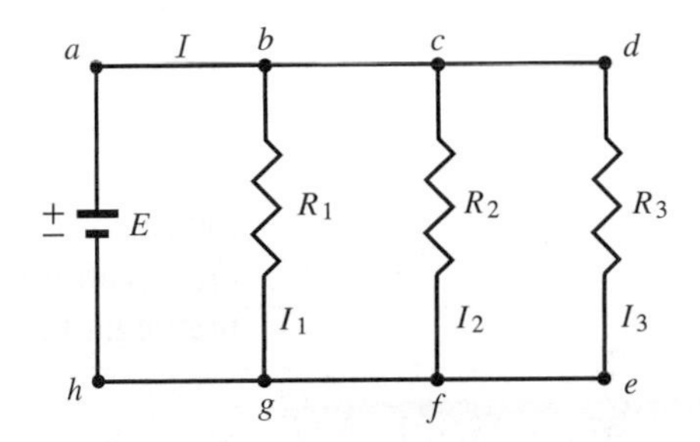

2.5 MARKOV* CHAINS

Prerequisites. Section 1.6, Solutions of Linear Systems of Equations. Basic Ideas of Probability. The Notion of a Limit.

Consider a system that is, at any one time, in one and only one of a finite number of states. For example, the weather in a certain area is either rainy or dry; a person is either a smoker or a nonsmoker; we either go or do not go to college; we live in an urban, suburban, or rural area; we are in the lower-, middle-, or upper-income bracket; we buy a Chevrolet, Ford, or other make of car. As time goes on, the system may move from one state to another, and we assume that the state of the system is observed at fixed time periods (for example, each day, each hour, and so on). In many applications we know the present state of the system and wish to predict the state at the next observation period, or at some future observation period. We often can predict the probability of the system being in a particular state at a future observation period from its past history. The applications mentioned above are of this type.

DEFINITION A **Markov chain** or **Markov process** is a process in which the probability of the system being in a particular state at a given observation period depends only on its state at the immediately preceding observation period.

Suppose that the system has n possible states. For each $i = 1, 2, \ldots, n$, $j = 1, 2, \ldots, n$ let t_{ij} be the probability that if the system is in state j at a

*Andrei Andreevitch Markov (1856–1922) lived most of his life in St. Petersburg, where his father was an official of the Russian forestry department. He was first a student and then a professor at St. Petersburg University. A political liberal, he participated in the protests against the Czarist rule in the first decade of the twentieth century. Although he was interested in many aspects of mathematical analysis, his most important work was in helping to lay the foundations of modern probability theory. His ideas on Markov processes were motivated by a desire to give a strictly rigorous proof of the law of large numbers and to extend the applicability of this law; these ideas appeared in a series of papers between 1906 and 1912.

certain observation period, it will be in state i at the next observation period; t_{ij} is called a **transition probability**. Moreover, t_{ij} applies to every time period; that is, it does not change with time.

Since t_{ij} is a probability, we must have

$$0 \leq t_{ij} \leq 1 \qquad (1 \leq i, j \leq n).$$

Also, if the system is in state j at a certain observation period, then it must be in one of the n states (it may remain in state j) at the next observation period. Thus we have

$$t_{1j} + t_{2j} + \cdots + t_{nj} = 1. \tag{1}$$

It is convenient to arrange the transition probabilities as the $n \times n$ matrix $T = \begin{bmatrix} t_{ij} \end{bmatrix}$, which is called the **transition matrix** of the Markov chain. Other names for a transition matrix are **Markov matrix**, **stochastic matrix**, and **probability matrix**. We see that the entries in each column of T are nonnegative and, from Equation (1), add up to 1.

EXAMPLE 1

Suppose that the weather in a certain city is either rainy or dry. As a result of extensive record keeping it has been determined that the probability of a rainy day following a dry day is $\frac{1}{3}$, and the probability of a rainy day following a rainy day is $\frac{1}{2}$. Let state D be a dry day and state R be a rainy day. Then the transition matrix of this Markov chain is

$$\begin{array}{cc} & \begin{array}{cc} \text{D} & \text{R} \end{array} \\ T = & \begin{bmatrix} \frac{2}{3} & \frac{1}{2} \\ \frac{1}{3} & \frac{1}{2} \end{bmatrix} \begin{array}{c} \text{D} \\ \text{R} \end{array} \end{array}$$

■

Example 9 in Section 1.4 portrays a situation that is similar to that in Example 2, which follows.

EXAMPLE 2

A market research organization is studying a large group of coffee buyers who buy a can of coffee each week. It is found that 50% of those presently using brand A will again buy brand A next week, 25% will switch to brand B, and 25% will switch to another brand. Of those using brand B now, 30% will again buy brand B next week, 60% will switch to brand A, and 10% will switch to another brand. Of those using another brand now, 30% will again buy another brand next week, 40% will switch to brand A, and 30% will switch to brand B. Let states A, B, and D denote brand A, brand B, and another brand, respectively. The probability that a person presently using brand A will switch to brand B is 0.25, the probability that a person presently using brand B will again buy brand B is 0.3, and so on. Thus the transition matrix of this Markov chain is

$$\begin{array}{cc} & \begin{array}{ccc} \text{A} & \text{B} & \text{D} \end{array} \\ T = & \begin{bmatrix} 0.50 & 0.60 & 0.40 \\ 0.25 & 0.30 & 0.30 \\ 0.25 & 0.10 & 0.30 \end{bmatrix} \begin{array}{c} \text{A} \\ \text{B} \\ \text{D} \end{array} \end{array}$$

■

We shall now use the transition matrix of the Markov process to determine the probability of the system being in any of the n states at future times.

Let

$$\mathbf{x}^{(k)} = \begin{bmatrix} p_1^{(k)} \\ p_2^{(k)} \\ \vdots \\ p_n^{(k)} \end{bmatrix} \qquad (k \geq 0)$$

denote the **state vector** of the Markov process at the observation period k, where $p_j^{(k)}$ is the probability that the system is in state j at the observation period k. The state vector $\mathbf{x}^{(0)}$, at the observation period 0, is called the **initial state vector**.

The following theorem, whose proof we omit, is proved by using the basic ideas of probability theory.

THEOREM 2.4

If T is the transition matrix of a Markov process, then the state vector $\mathbf{x}^{(k+1)}$, at the $(k+1)$th observation period, can be determined from the state vector $\mathbf{x}^{(k)}$, at the kth observation period, as

$$\mathbf{x}^{(k+1)} = T\mathbf{x}^{(k)}. \qquad (2) \ ■$$

From (2) we have

$$\begin{aligned} \mathbf{x}^{(1)} &= T\mathbf{x}^{(0)} \\ \mathbf{x}^{(2)} &= T\mathbf{x}^{(1)} = T(T\mathbf{x}^{(0)}) = T^2\mathbf{x}^{(0)} \\ \mathbf{x}^{(3)} &= T\mathbf{x}^{(2)} = T(T^2\mathbf{x}^{(0)}) = T^3\mathbf{x}^{(0)}, \end{aligned}$$

and, in general,

$$\mathbf{x}^{(n)} = T^n\mathbf{x}^{(0)}.$$

That is, to obtain the next state vector, we multiply the current state vector on the left by the matrix T. Thus the transition matrix and the initial state vector completely determine every other state vector.

EXAMPLE 3

Consider Example 1 again. Suppose that when we begin our observations (day 0), it is dry, so the initial state vector is

$$\mathbf{x}^{(0)} = \begin{bmatrix} 1 \\ 0 \end{bmatrix}.$$

Then the state vector on day 1 (the day after we begin our observation) is

$$\mathbf{x}^{(1)} = T\mathbf{x}^{(0)} = \begin{bmatrix} 0.67 & 0.5 \\ 0.33 & 0.5 \end{bmatrix} \begin{bmatrix} 1 \\ 0 \end{bmatrix} = \begin{bmatrix} 0.67 \\ 0.33 \end{bmatrix},$$

where the fractions have been approximated to two decimal places. Thus the probability of no rain on day 1 is 0.67 and the probability of rain on that day is 0.33. Similarly,

$$\begin{aligned} \mathbf{x}^{(2)} &= T\mathbf{x}^{(1)} = \begin{bmatrix} 0.67 & 0.5 \\ 0.33 & 0.5 \end{bmatrix} \begin{bmatrix} 0.67 \\ 0.33 \end{bmatrix} = \begin{bmatrix} 0.614 \\ 0.386 \end{bmatrix} \\ \mathbf{x}^{(3)} &= T\mathbf{x}^{(2)} = \begin{bmatrix} 0.67 & 0.5 \\ 0.33 & 0.5 \end{bmatrix} \begin{bmatrix} 0.614 \\ 0.386 \end{bmatrix} = \begin{bmatrix} 0.604 \\ 0.396 \end{bmatrix} \\ \mathbf{x}^{(4)} &= T\mathbf{x}^{(3)} = \begin{bmatrix} 0.67 & 0.5 \\ 0.33 & 0.5 \end{bmatrix} \begin{bmatrix} 0.604 \\ 0.396 \end{bmatrix} = \begin{bmatrix} 0.603 \\ 0.397 \end{bmatrix} \\ \mathbf{x}^{(5)} &= T\mathbf{x}^{(4)} = \begin{bmatrix} 0.67 & 0.5 \\ 0.33 & 0.5 \end{bmatrix} \begin{bmatrix} 0.603 \\ 0.397 \end{bmatrix} = \begin{bmatrix} 0.603 \\ 0.397 \end{bmatrix}. \end{aligned}$$

From the fourth day on, the state vector of the system is always the same,

$$\begin{bmatrix} 0.603 \\ 0.397 \end{bmatrix}.$$

This means that from the fourth day on, it is dry about 60% of the time, and it rains about 40% of the time. ■

EXAMPLE 4 Consider Example 2 again. Suppose that when our survey begins, we find that brand A has 20% of the market, brand B has 20% of the market, and the other brands have 60% of the market. Thus

$$\mathbf{x}^{(0)} = \begin{bmatrix} 0.2 \\ 0.2 \\ 0.6 \end{bmatrix}.$$

The state vector after the first week is

$$\mathbf{x}^{(1)} = T\mathbf{x}^{(0)} = \begin{bmatrix} 0.50 & 0.60 & 0.40 \\ 0.25 & 0.30 & 0.30 \\ 0.25 & 0.10 & 0.30 \end{bmatrix} \begin{bmatrix} 0.2 \\ 0.2 \\ 0.6 \end{bmatrix} = \begin{bmatrix} 0.4600 \\ 0.2900 \\ 0.2500 \end{bmatrix}.$$

Similarly,

$$\mathbf{x}^{(2)} = T\mathbf{x}^{(1)} = \begin{bmatrix} 0.50 & 0.60 & 0.40 \\ 0.25 & 0.30 & 0.30 \\ 0.25 & 0.10 & 0.30 \end{bmatrix} \begin{bmatrix} 0.4600 \\ 0.2900 \\ 0.2500 \end{bmatrix} = \begin{bmatrix} 0.5040 \\ 0.2770 \\ 0.2190 \end{bmatrix}$$

$$\mathbf{x}^{(3)} = T\mathbf{x}^{(2)} = \begin{bmatrix} 0.50 & 0.60 & 0.40 \\ 0.25 & 0.30 & 0.30 \\ 0.25 & 0.10 & 0.30 \end{bmatrix} \begin{bmatrix} 0.5040 \\ 0.2770 \\ 0.2190 \end{bmatrix} = \begin{bmatrix} 0.5058 \\ 0.2748 \\ 0.2194 \end{bmatrix}$$

$$\mathbf{x}^{(4)} = T\mathbf{x}^{(3)} = \begin{bmatrix} 0.50 & 0.60 & 0.40 \\ 0.25 & 0.30 & 0.30 \\ 0.25 & 0.10 & 0.30 \end{bmatrix} \begin{bmatrix} 0.5058 \\ 0.2748 \\ 0.2194 \end{bmatrix} = \begin{bmatrix} 0.5055 \\ 0.2747 \\ 0.2198 \end{bmatrix}$$

$$\mathbf{x}^{(5)} = T\mathbf{x}^{(4)} = \begin{bmatrix} 0.50 & 0.60 & 0.40 \\ 0.25 & 0.30 & 0.30 \\ 0.25 & 0.10 & 0.30 \end{bmatrix} \begin{bmatrix} 0.5055 \\ 0.2747 \\ 0.2198 \end{bmatrix} = \begin{bmatrix} 0.5055 \\ 0.2747 \\ 0.2198 \end{bmatrix}.$$

Thus as n increases, the state vectors approach the fixed vector

$$\begin{bmatrix} 0.5055 \\ 0.2747 \\ 0.2198 \end{bmatrix}.$$

This means that in the long run, brand A will command about 51% of the market, brand B will retain about 27%, and the other brands will command about 22%. ■

In the last two examples we have seen that as the number of observation periods increases, the state vectors converge to a fixed vector. In this case we say that the Markov process has reached equilibrium. The fixed vector is called the **steady-state vector**. Markov processes are generally used to determine the behavior of a system in the long run; for example, the share of the market that a certain manufacturer can expect to retain on a somewhat permanent basis. Thus the question of whether or not a Markov process reaches **equilibrium** is of paramount importance. The following example shows that not every Markov process reaches equilibrium.

EXAMPLE 5 Let

$$T = \begin{bmatrix} 0 & 1 \\ 1 & 0 \end{bmatrix} \quad \text{and} \quad \mathbf{x}^{(0)} = \begin{bmatrix} \frac{1}{3} \\ \frac{2}{3} \end{bmatrix}.$$

Then

$$\mathbf{x}^{(1)} = \begin{bmatrix} \frac{2}{3} \\ \frac{1}{3} \end{bmatrix}, \quad \mathbf{x}^{(2)} = \begin{bmatrix} \frac{1}{3} \\ \frac{2}{3} \end{bmatrix}, \quad \mathbf{x}^{(3)} = \begin{bmatrix} \frac{2}{3} \\ \frac{1}{3} \end{bmatrix}, \quad \mathbf{x}^{(4)} = \begin{bmatrix} \frac{1}{3} \\ \frac{2}{3} \end{bmatrix}, \ldots.$$

Thus the state vectors oscillate between the vectors

$$\begin{bmatrix} \frac{2}{3} \\ \frac{1}{3} \end{bmatrix} \quad \text{and} \quad \begin{bmatrix} \frac{1}{3} \\ \frac{2}{3} \end{bmatrix}$$

and do not converge to a fixed vector. ■

However, if we demand that the transition matrix of a Markov process satisfy a rather reasonable property, we obtain a large class of Markov processes, many of which arise in practical applications, that *do* reach equilibrium. We now proceed to formulate these ideas.

DEFINITION The vector

$$\mathbf{u} = \begin{bmatrix} u_1 \\ u_2 \\ \vdots \\ u_n \end{bmatrix}$$

is called a **probability vector** if $u_i \geq 0$ $(1 \leq i \leq n)$ and

$$u_1 + u_2 + \cdots + u_n = 1.$$

EXAMPLE 6 The vectors

$$\begin{bmatrix} \frac{1}{2} \\ \frac{1}{4} \\ \frac{1}{4} \end{bmatrix} \quad \text{and} \quad \begin{bmatrix} \frac{1}{3} \\ \frac{2}{3} \\ 0 \end{bmatrix}$$

are probability vectors; the vectors

$$\begin{bmatrix} \frac{1}{5} \\ \frac{1}{5} \\ \frac{2}{5} \end{bmatrix} \quad \text{and} \quad \begin{bmatrix} \frac{1}{3} \\ \frac{1}{2} \\ \frac{1}{2} \end{bmatrix}$$

are not probability vectors. (Why not?) ■

DEFINITION A transition matrix T of a Markov process is called **regular** if all the entries in some power of T are positive. A Markov process is called **regular** if its transition matrix is regular.

The Markov processes in Examples 1 and 2 are regular, since all the entries in the transition matrices themselves are positive.

EXAMPLE 7 The transition matrix

$$T = \begin{bmatrix} 0.2 & 1 \\ 0.8 & 0 \end{bmatrix}$$

is regular, since

$$T^2 = \begin{bmatrix} 0.84 & 0.2 \\ 0.16 & 0.8 \end{bmatrix}.$$

■

We now state the following fundamental theorem of regular Markov processes; the proof, which we omit, can be found in the book by Kemeny and Snell given in Further Readings at the end of the section.

THEOREM 2.5 *If T is the transition matrix of a regular Markov process, then*

(a) *As $n \to \infty$, T^n approaches a matrix*

$$A = \begin{bmatrix} u_1 & u_1 & \cdots & u_1 \\ u_2 & u_2 & \cdots & u_2 \\ \vdots & \vdots & & \vdots \\ u_n & u_n & \cdots & u_n \end{bmatrix},$$

all of whose columns are identical.

(b) *Every column*

$$\mathbf{u} = \begin{bmatrix} u_1 \\ u_2 \\ \vdots \\ u_n \end{bmatrix}$$

of A is a probability vector all of whose components are positive. That is, $u_i > 0$ $(1 \leq i \leq n)$ and

$$u_1 + u_2 + \cdots + u_n = 1.$$

■

We next establish the following result.

THEOREM 2.6 *If T is a regular transition matrix and A and $\mathbf{u}$ are as in Theorem 2.5, then:*

(a) *For any probability vector $\mathbf{x}$, $T^n\mathbf{x} \to \mathbf{u}$ as $n \to \infty$, so that $\mathbf{u}$ is a steady-state vector.*

(b) *The steady-state vector $\mathbf{u}$ is the unique probability vector satisfying the matrix equation $T\mathbf{u} = \mathbf{u}$.*

Proof (a) Let

$$\mathbf{x} = \begin{bmatrix} x_1 \\ x_2 \\ \vdots \\ x_n \end{bmatrix}$$

be a probability vector. Since $T^n \to A$ as $n \to \infty$, we have

$$T^n\mathbf{x} \to A\mathbf{x}.$$

Now

$$A\mathbf{x} = \begin{bmatrix} u_1 & u_1 & \cdots & u_1 \\ u_2 & u_2 & \cdots & u_2 \\ \vdots & \vdots & & \vdots \\ u_n & u_n & \cdots & u_n \end{bmatrix} \begin{bmatrix} x_1 \\ x_2 \\ \vdots \\ x_n \end{bmatrix} = \begin{bmatrix} u_1x_1 + u_1x_2 + \cdots + u_1x_n \\ u_2x_1 + u_2x_2 + \cdots + u_2x_n \\ \vdots \\ u_nx_1 + u_nx_2 + \cdots + u_nx_n \end{bmatrix}$$

$$= \begin{bmatrix} u_1(x_1 + x_2 + \cdots + x_n) \\ u_2(x_1 + x_2 + \cdots + x_n) \\ \vdots \\ u_n(x_1 + x_2 + \cdots + x_n) \end{bmatrix} = \begin{bmatrix} u_1 \\ u_2 \\ \vdots \\ u_n \end{bmatrix},$$

since $x_1 + x_2 + \cdots + x_n = 1$. Hence $T^n\mathbf{x} \to \mathbf{u}$.

(b) Since $T^n \to A$, we also have $T^{n+1} \to A$. However,

$$T^{n+1} = TT^n,$$

so $T^{n+1} \to TA$. Hence $TA = A$. Equating corresponding columns of this equation (using Exercise T.9 of Section 1.3), we have $T\mathbf{u} = \mathbf{u}$. To show that $\mathbf{u}$ is unique, we let $\mathbf{v}$ be another probability vector such that $T\mathbf{v} = \mathbf{v}$. From (a), $T^n\mathbf{v} \to \mathbf{u}$, and since $T\mathbf{v} = \mathbf{v}$, we have $T^n\mathbf{v} = \mathbf{v}$ for all n. Hence $\mathbf{v} = \mathbf{u}$. ■

In Examples 3 and 4 we calculated steady-state vectors by computing the powers $T^n\mathbf{x}$. An alternative way of computing the steady-state vector of a regular transition matrix is as follows. From (b) of Theorem 2.6, we write

$$T\mathbf{u} = \mathbf{u}$$

as

$$T\mathbf{u} = I_n\mathbf{u}$$

or

$$(I_n - T)\mathbf{u} = \mathbf{0}. \tag{3}$$

We have shown that the homogeneous system (3) has a unique solution $\mathbf{u}$ that is a probability vector, so that

$$u_1 + u_2 + \cdots + u_n = 1. \tag{4}$$

The first procedure for computing the steady-state vector $\mathbf{u}$ of a regular transition matrix T is as follows.

Step 1. Compute the powers $T^n\mathbf{x}$, where $\mathbf{x}$ is any probability vector.

Step 2. $\mathbf{u}$ is the limit of the powers $T^n\mathbf{x}$.

The second procedure for computing the steady-state vector $\mathbf{u}$ of a regular transition matrix T is as follows.

Step 1. Solve the homogeneous system

$$(I_n - T)\mathbf{u} = \mathbf{0}.^*$$

Step 2. From the infinitely many solutions obtained in Step 1, determine a unique solution $\mathbf{u}$ by requiring that its components satisfy Equation (4).

*This type of problem will be studied in more detail in Chapter 8.

EXAMPLE 8 Consider the matrix of Example 2. The homogeneous system (3) is (verify)

$$\begin{bmatrix} 0.50 & -0.60 & -0.40 \\ -0.25 & 0.70 & -0.30 \\ -0.25 & -0.10 & 0.70 \end{bmatrix} \begin{bmatrix} u_1 \\ u_2 \\ u_3 \end{bmatrix} = \begin{bmatrix} 0 \\ 0 \\ 0 \end{bmatrix}.$$

The reduced row echelon form of the augmented matrix is (verify)

$$\left[\begin{array}{ccc|c} 1 & 0 & -2.30 & 0 \\ 0 & 1 & -1.25 & 0 \\ 0 & 0 & 0.00 & 0 \end{array}\right].$$

Hence a solution is

$$\begin{aligned} u_1 &= 2.3r \\ u_2 &= 1.25r \\ u_3 &= r, \end{aligned}$$

where r is any real number. From (4), we have

$$2.3r + 1.25r + r = 1,$$

or

$$r = \frac{1}{4.55} \approx 0.2198.$$

Hence

$$\begin{aligned} u_1 &= 0.5055 \\ u_2 &= 0.2747 \\ u_3 &= 0.2198. \end{aligned}$$

These results agree with those obtained in Example 4. ■

Further Readings

KEMENY, JOHN G., and J. LAURIE SNELL. *Finite Markov Chains.* New York: Springer-Verlag, 1976.

MAKI, D. P., and M. THOMPSON. *Mathematical Models and Applications: With Emphasis on the Social, Life, and Management Sciences.* Upper Saddle River, N.J.: Prentice Hall, Inc., 1973.

ROBERTS, FRED S. *Discrete Mathematical Models with Applications to Social, Biological, and Environmental Problems.* Upper Saddle River, N.J.: Prentice Hall, Inc., 1997.

Key Terms

Markov chain (or Markov process)
Transition probability
Transition matrix (or Markov matrix, or stochastic matrix, or probability matrix)
State vector
Initial state vector
Steady-state vector
Equilibrium

2.5 Exercises

1. Which of the following can be transition matrices of a Markov process?

(a) $\begin{bmatrix} 0.3 & 0.7 \\ 0.4 & 0.6 \end{bmatrix}$ (b) $\begin{bmatrix} 0.2 & 0.3 & 0.1 \\ 0.8 & 0.5 & 0.7 \\ 0.0 & 0.2 & 0.2 \end{bmatrix}$

(c) $\begin{bmatrix} 0.55 & 0.33 \\ 0.45 & 0.67 \end{bmatrix}$ (d) $\begin{bmatrix} 0.3 & 0.4 & 0.2 \\ 0.2 & 0.0 & 0.8 \\ 0.1 & 0.3 & 0.6 \end{bmatrix}$

2. Which of the following are probability vectors?

(a) $\begin{bmatrix} \frac{1}{2} \\ \frac{1}{3} \\ \frac{2}{3} \end{bmatrix}$ (b) $\begin{bmatrix} 0 \\ 1 \\ 0 \end{bmatrix}$ (c) $\begin{bmatrix} \frac{1}{4} \\ \frac{1}{6} \\ \frac{1}{3} \\ \frac{1}{4} \end{bmatrix}$ (d) $\begin{bmatrix} \frac{1}{5} \\ \frac{2}{5} \\ \frac{1}{10} \\ \frac{2}{10} \end{bmatrix}$

In Exercises 3 and 4, determine a value of each missing entry, denoted by □, so that the matrix will be a transition matrix of a Markov chain. In some cases there may be more than one correct answer.

3. $\begin{bmatrix} \square & 0.4 & 0.3 \\ 0.3 & \square & 0.5 \\ \square & 0.2 & \square \end{bmatrix}$ **4.** $\begin{bmatrix} 0.2 & 0.1 & 0.3 \\ 0.3 & \square & 0.5 \\ \square & \square & \square \end{bmatrix}$

5. Consider the transition matrix

$$T = \begin{bmatrix} 0.7 & 0.4 \\ 0.3 & 0.6 \end{bmatrix}.$$

(a) If $\mathbf{x}^{(0)} = \begin{bmatrix} 1 \\ 0 \end{bmatrix}$, compute $\mathbf{x}^{(1)}$, $\mathbf{x}^{(2)}$, and $\mathbf{x}^{(3)}$ to three decimal places.

(b) Show that T is regular and find its steady-state vector.

6. Consider the transition matrix

$$T = \begin{bmatrix} 0 & 0.2 & 0.0 \\ 0 & 0.3 & 0.3 \\ 1 & 0.5 & 0.7 \end{bmatrix}.$$

(a) If

$$\mathbf{x}^{(0)} = \begin{bmatrix} 0 \\ 1 \\ 0 \end{bmatrix},$$

compute $\mathbf{x}^{(1)}$, $\mathbf{x}^{(2)}$, $\mathbf{x}^{(3)}$, and $\mathbf{x}^{(4)}$ to three decimal places.

(b) Show that T is regular and find its steady-state vector.

7. Which of the following transition matrices are regular?

(a) $\begin{bmatrix} 0 & \frac{1}{2} \\ 1 & \frac{1}{2} \end{bmatrix}$ (b) $\begin{bmatrix} \frac{1}{2} & 0 & 0 \\ 0 & 1 & \frac{1}{2} \\ \frac{1}{2} & 0 & \frac{1}{2} \end{bmatrix}$

(c) $\begin{bmatrix} 1 & \frac{1}{3} & 0 \\ 0 & \frac{1}{3} & 1 \\ 0 & \frac{1}{3} & 0 \end{bmatrix}$ (d) $\begin{bmatrix} \frac{1}{4} & \frac{3}{5} & \frac{1}{2} \\ \frac{1}{2} & 0 & 0 \\ \frac{1}{4} & \frac{2}{5} & \frac{1}{2} \end{bmatrix}$

8. Show that each of the following transition matrices reaches a state of equilibrium.

(a) $\begin{bmatrix} \frac{1}{2} & 1 \\ \frac{1}{2} & 0 \end{bmatrix}$ (b) $\begin{bmatrix} 0.4 & 0.2 \\ 0.6 & 0.8 \end{bmatrix}$

(c) $\begin{bmatrix} \frac{1}{3} & 1 & \frac{1}{2} \\ \frac{1}{3} & 0 & \frac{1}{4} \\ \frac{1}{3} & 0 & \frac{1}{4} \end{bmatrix}$ (d) $\begin{bmatrix} 0.3 & 0.1 & 0.4 \\ 0.2 & 0.4 & 0.0 \\ 0.5 & 0.5 & 0.6 \end{bmatrix}$

9. Let

$$T = \begin{bmatrix} \frac{1}{2} & 0 \\ \frac{1}{2} & 1 \end{bmatrix}.$$

(a) Show that T is not regular.

(b) Show that $T^n\mathbf{x} \to \begin{bmatrix} 0 \\ 1 \end{bmatrix}$ for any probability vector $\mathbf{x}$.

Thus a Markov chain may have a unique steady-state vector even though its transition matrix is not regular.

10. Find the steady-state vector of each of the following regular matrices.

(a) $\begin{bmatrix} \frac{1}{3} & \frac{1}{2} \\ \frac{2}{3} & \frac{1}{2} \end{bmatrix}$ (b) $\begin{bmatrix} 0.3 & 0.1 \\ 0.7 & 0.9 \end{bmatrix}$

(c) $\begin{bmatrix} \frac{1}{4} & \frac{1}{2} & \frac{1}{3} \\ 0 & \frac{1}{2} & \frac{2}{3} \\ \frac{3}{4} & 0 & 0 \end{bmatrix}$ (d) $\begin{bmatrix} 0.4 & 0.0 & 0.1 \\ 0.2 & 0.5 & 0.3 \\ 0.4 & 0.5 & 0.6 \end{bmatrix}$

11. ***(Psychology)*** A behavioral psychologist places a rat each day in a cage with two doors, A and B. The rat can go through door A, where it receives an electric shock, or through door B, where it receives some food. A record is made of the door through which the rat passes. At the start of the experiment, on a Monday, the rat is equally likely to go through door A as through door B. After going through door A, and receiving a shock, the probability of going through the same door on the next day is 0.3. After going through door B, and receiving food, the probability of going through the same door on the next day is 0.6.

(a) Write the transition matrix for the Markov process.

(b) What is the probability of the rat going through door A on Thursday (the third day after starting the experiment)?

(c) What is the steady-state vector?

12. (***Business***) The subscription department of a magazine sends out a letter to a large mailing list inviting subscriptions for the magazine. Some of the people receiving this letter already subscribe to the magazine, while others do not. From this mailing list, 60% of those who already subscribe will subscribe again, while 25% of those who do not now subscribe will subscribe.

(a) Write the transition matrix for this Markov process.

(b) On the last letter it was found that 40% of those receiving it, ordered a subscription. What percentage of those receiving the current letter can be expected to order a subscription?

13. (***Sociology***) A study has determined that the occupation of a boy, as an adult, depends upon the occupation of his father and is given by the following transition matrix, where P = professional, F = farmer, and L = laborer.

$$\text{Son's occupation}\quad \begin{matrix} & \text{Father's occupation} \\ & \begin{matrix} \text{P} & \text{F} & \text{L} \end{matrix} \\ \begin{matrix} \text{P} \\ \text{F} \\ \text{L} \end{matrix} & \begin{bmatrix} 0.8 & 0.3 & 0.2 \\ 0.1 & 0.5 & 0.2 \\ 0.1 & 0.2 & 0.6 \end{bmatrix} \end{matrix}$$

Thus the probability that the son of a professional will also be a professional is 0.8, and so on.

(a) What is the probability that the grandchild of a professional will also be a professional?

(b) In the long run, what proportion of the population will be farmers?

14. (***Genetics***) Consider a plant that can have red flowers (R), pink flowers (P), or white flowers (W), depending upon the genotypes RR, RW, and WW. When we cross each of these genotypes with a genotype RW, we obtain the transition matrix

$$\text{Flowers of offspring plant}\quad \begin{matrix} & \text{Flowers of parent plant} \\ & \begin{matrix} \text{R} & \text{P} & \text{W} \end{matrix} \\ \begin{matrix} \text{R} \\ \text{P} \\ \text{W} \end{matrix} & \begin{bmatrix} 0.5 & 0.25 & 0.0 \\ 0.5 & 0.50 & 0.5 \\ 0.0 & 0.25 & 0.5 \end{bmatrix} \end{matrix}.$$

Suppose that each successive generation is produced by crossing only with plants of RW genotype. When the process reaches equilibrium, what percentage of the plants will have red, pink, or white flowers?

15. (***Mass Transit***) A new mass transit system has just gone into operation. The transit authority has made studies that predict the percentage of commuters who will change to mass transit (M) or continue driving their automobile (A). The following transition matrix has been obtained:

$$\text{Next year}\quad \begin{matrix} & \text{This year} \\ & \begin{matrix} \text{M} & \text{A} \end{matrix} \\ \begin{matrix} \text{M} \\ \text{A} \end{matrix} & \begin{bmatrix} 0.7 & 0.2 \\ 0.3 & 0.8 \end{bmatrix} \end{matrix}.$$

Suppose that the population of the area remains constant, and that initially 30% of the commuters use mass transit and 70% use their automobiles.

(a) What percentage of the commuters will be using the mass transit system after 1 year? After 2 years?

(b) What percentage of the commuters will be using the mass transit system in the long run?

Theoretical Exercise

T.1. Is the transpose of a transition matrix of a Markov chain also a transition matrix of a Markov chain? Explain.

MATLAB Exercises

The computation of the sequence of vectors $\mathbf{x}^{(1)}, \mathbf{x}^{(2)}, \ldots$ *as in Examples 3 and 4 can easily be done using* MATLAB *commands. Once the transition matrix* T *and initial state vector* $\mathbf{x}^{(0)}$ *are entered into* MATLAB, *the state vector for the* kth *observation period is obtained from the* MATLAB *command*

```
T^k * x
```

ML.1. Use MATLAB to verify the computations of state vectors in Example 3 for periods 1 through 5.

ML.2. In Example 4, if the initial state is changed to

$$\begin{bmatrix} 0.1 \\ 0.3 \\ 0.6 \end{bmatrix},$$

determine $\mathbf{x}^{(5)}$.

ML.3. In MATLAB, enter **help sum** and determine the action of command **sum** on an $m \times n$ matrix. Use command **sum** to determine which of the following are Markov matrices.

(a) $\begin{bmatrix} \frac{2}{3} & \frac{1}{3} & \frac{1}{2} \\ \frac{1}{3} & \frac{1}{3} & \frac{1}{4} \\ 0 & \frac{1}{3} & \frac{1}{4} \end{bmatrix}$

(b) $\begin{bmatrix} 0.5 & 0.6 & 0.7 \\ 0.3 & 0.2 & 0.3 \\ 0.1 & 0.2 & 0.0 \end{bmatrix}$

(c) $\begin{bmatrix} 0.66 & 0.25 & 0.125 \\ 0.33 & 0.25 & 0.625 \\ 0.00 & 0.50 & 0.250 \end{bmatrix}$

2.6 LINEAR ECONOMIC MODELS

Prerequisite. Section 1.7, The Inverse of a Matrix.

As society has grown more complex, increasing attention has been paid to the analysis of economic behavior. Problems of economic behavior are more difficult to handle than problems in the physical sciences for many reasons. For example, we may not know all the factors or variables that must be considered, what data need to be gathered, when we have enough data, or the resulting mathematical problem may be too difficult to solve.

In the 1930s Wassily Leontief, a Harvard University economics professor, developed a pioneering approach to the mathematical analysis of economic behavior. He was awarded the 1973 Nobel Prize in Economics for his work. In this section we give a very brief introduction to the application of linear algebra to economics.

Our approach leans heavily on the presentations in the books by Gale and by Johnston, Price, and van Vleck given in Further Readings, which may be consulted by the reader for a more extensive treatment of this material.

THE LEONTIEF CLOSED MODEL

EXAMPLE 1* Consider a simple society consisting of a farmer, a carpenter, and a tailor. Each produces one commodity: the farmer grows the food, the carpenter builds the homes, and the tailor makes the clothes. For convenience, we may select our units so that each individual produces one unit of each commodity during the year. Suppose that during the year the portion of each commodity that is consumed by each individual is given in Table 2.1.

Table 2.1

Goods Consumed by:	Goods Produced by:		
	Farmer	Carpenter	Tailor
Farmer	$\frac{7}{16}$	$\frac{1}{2}$	$\frac{3}{16}$
Carpenter	$\frac{5}{16}$	$\frac{1}{6}$	$\frac{5}{16}$
Tailor	$\frac{1}{4}$	$\frac{1}{3}$	$\frac{1}{2}$

Thus the farmer consumes $\frac{7}{16}$ of his own produce while the carpenter consumes $\frac{5}{16}$ of the farmer's produce, the carpenter consumes $\frac{5}{16}$ of the clothes made by the tailor, and so on. Let p_1 be the price per unit of food, p_2 the price per unit of housing, and p_3 the price per unit of clothes. We assume that everyone pays the same price for a commodity. Thus the farmer pays the same price for his food as the tailor and carpenter, although he grew the food himself. We are interested in determining the prices p_1, p_2, and p_3 so that we have a state of equilibrium, which is defined as follows: *No one makes money or loses money.*

The farmer's expenditures are

$$\tfrac{7}{16}\,p_1 + \tfrac{1}{2}\,p_2 + \tfrac{3}{16}\,p_3,$$

*This example is due to Johnston, Price, and van Vleck and is described in their book given in Further Readings. The general model for this example is also presented by Gale.

while his income is p_1 because he produces one unit of food. Since expenditures must equal income, we have

$$\tfrac{7}{16}p_1 + \tfrac{1}{2}p_2 + \tfrac{3}{16}p_3 = p_1. \tag{1}$$

Similarly, for the carpenter we obtain

$$\tfrac{5}{16}p_1 + \tfrac{1}{6}p_2 + \tfrac{5}{16}p_3 = p_2, \tag{2}$$

and for the tailor we have

$$\tfrac{1}{4}p_1 + \tfrac{1}{3}p_2 + \tfrac{1}{2}p_3 = p_3. \tag{3}$$

Equations (1), (2), and (3) can be written in matrix notation as

$$A\mathbf{p} = \mathbf{p}, \tag{4}$$

where

$$A = \begin{bmatrix} \frac{7}{16} & \frac{1}{2} & \frac{3}{16} \\ \frac{5}{16} & \frac{1}{6} & \frac{5}{16} \\ \frac{1}{4} & \frac{1}{3} & \frac{1}{2} \end{bmatrix}, \qquad \mathbf{p} = \begin{bmatrix} p_1 \\ p_2 \\ p_3 \end{bmatrix}.$$

Equation (4) can be rewritten as

$$(I_n - A)\mathbf{p} = \mathbf{0}, \tag{5}$$

which is a homogeneous system.

Our problem is that of finding a solution $\mathbf{p}$ to (5) whose components p_i will be nonnegative with at least one positive p_i, since $\mathbf{p} = \mathbf{0}$ means that all the prices are zero, which makes no sense.

Solving (5), we obtain (verify)

$$\mathbf{p} = r\begin{bmatrix} 4 \\ 3 \\ 4 \end{bmatrix},$$

where r is any real number. Letting r be a positive number, we determine the *relative* prices of the commodities. For example, letting $r = 1000$, we find that food costs \$4000 per unit, housing costs \$3000 per unit, and clothes cost \$4000 per unit. ■

EXAMPLE 2 **(The Exchange Model)** Consider now the general problem where we have n manufacturers $M_1, M_2, \ldots, M_n$, and n goods $G_1, G_2, \ldots, G_n$, with M_i making only G_i. Consider a fixed interval of time, say one year, and suppose that M_i only makes one unit of G_i during the year.

In producing good G_i, manufacturer M_i may consume amounts of goods $G_1, G_2, \ldots, G_i, \ldots, G_n$. For example, iron, along with many other ingredients, is used in the manufacture of iron. Let a_{ij} be the amount of good G_j consumed by manufacturer M_i. Then

$$0 \leq a_{ij} \leq 1.$$

Suppose that the model is **closed**; that is, no goods leave or enter the system. This means that the total consumption of each good must equal its total production. Since the total production of G_j is 1, we have

$$a_{1j} + a_{2j} + \cdots + a_{nj} = 1 \qquad (1 \leq j \leq n).$$

If the price per unit of G_k is p_k, then manufacturer M_i pays

$$a_{i1}p_1 + a_{i2}p_2 + \cdots + a_{in}p_n \tag{6}$$

for the goods he uses.

Our problem is that of determining the prices $p_1, p_2, \ldots, p_n$ so that no manufacturer makes money or loses money. That is, so that each manufacturer's income will equal his expenses. Since M_i manufactures only one unit, his income is p_i. Thus from (6) we have

$$\begin{aligned}
a_{11}p_1 + a_{12}p_2 + \cdots + a_{1n}p_n &= p_1 \\
a_{21}p_1 + a_{22}p_2 + \cdots + a_{2n}p_n &= p_2 \\
\vdots \qquad \vdots \qquad\qquad \vdots \quad &\quad \vdots \\
a_{n1}p_1 + a_{n2}p_2 + \cdots + a_{nn}p_n &= p_n,
\end{aligned}$$

which can be written in matrix form as

$$A\mathbf{p} = \mathbf{p}, \tag{7}$$

where

$$A = \begin{bmatrix} a_{ij} \end{bmatrix} \quad \text{and} \quad \mathbf{p} = \begin{bmatrix} p_1 \\ p_2 \\ \vdots \\ p_n \end{bmatrix}.$$

Equation (7) can be rewritten as

$$(I_n - A)\mathbf{p} = \mathbf{0}. \tag{8}$$

Thus our problem is that of finding a vector

$$\mathbf{p} \geq \mathbf{0},$$

with at least one component that is positive and satisfies Equation (8). ■

DEFINITION

An $n \times n$ matrix $A = \begin{bmatrix} a_{ij} \end{bmatrix}$ is called an **exchange matrix** if it satisfies the following two properties:

(a) $a_{ij} \geq 0$ (each entry is nonnegative).

(b) $a_{1j} + a_{2j} + \cdots + a_{nj} = 1$, for $j = 1, 2, \ldots, n$ (the entries in each column add up to 1).

EXAMPLE 3

Matrix A of Example 1 is an exchange matrix, as is the matrix A of Example 2. ■

Our general problem can now be stated as follows: Given an exchange matrix A, find a vector $\mathbf{p} \geq \mathbf{0}$ with at least one positive component, satisfying Equation (8). It can be shown that this problem always has a solution (see the book by Gale, p. 264, given in Further Readings).

In our general problem we required that each manufacturer's income equal his expenses. Instead, we could have required that each manufacturer's expenses not exceed his income. This would have led to

$$A\mathbf{p} \leq \mathbf{p} \tag{9}$$

instead of $A\mathbf{p} = \mathbf{p}$. However, it can be shown (Exercise T.1) that if Equation (9) holds, then Equation (7) will hold. Thus, if no manufacturer spends more than he earns, then everyone's income equals his expenses. An economic interpretation of this statement is that in the Leontief closed model if some manufacturer is making a profit, then at least one manufacturer is taking a loss.

AN INTERNATIONAL TRADE MODEL

EXAMPLE 4

Suppose that n countries $C_1, C_2, \ldots, C_n$ are engaged in trading with each other and that a common currency is in use. We assume that prices are fixed throughout the discussion and that C_j's income y_j comes entirely from selling its goods either internally or to other countries. We also assume that the fraction of C_j's income that is spent on imports from C_i is a fixed number a_{ij} which does not depend on C_j's income y_j. Since the a_{ij}'s are fractions of y_j, we have

$$a_{ij} \geq 0$$
$$a_{1j} + a_{2j} + \cdots + a_{nj} = 1,$$

so that $A = \begin{bmatrix} a_{ij} \end{bmatrix}$ is an exchange matrix. We now wish to determine the total income y_i for each country C_i. Since the value of C_i's exports to C_j is $a_{ij}y_j$, the total income of C_i is

$$a_{i1}y_1 + a_{i2}y_2 + \cdots + a_{in}y_n.$$

Hence, we must have

$$a_{i1}y_1 + a_{i2}y_2 + \cdots + a_{in}y_n = y_i.$$

In matrix notation we must find

$$\mathbf{p} = \begin{bmatrix} y_1 \\ y_2 \\ \vdots \\ y_n \end{bmatrix} \geq \mathbf{0},$$

with at least one $y_i > 0$, so that

$$A\mathbf{p} = \mathbf{p},$$

which is our earlier problem. ■

THE LEONTIEF OPEN MODEL

Suppose that we have n goods $G_1, G_2, \ldots, G_n$ and n activities $M_1, M_2, \ldots, M_n$. Assume that each activity M_i produces only one good G_i and that G_i is produced only by M_i. Let $c_{ij} \geq 0$ be the dollar value of G_i that has to be consumed to produce 1 dollar's worth of G_j. The matrix $C = \begin{bmatrix} c_{ij} \end{bmatrix}$ is called the **consumption matrix**. Observe that c_{ii} may be positive, which means that we may require some amount of G_i to make 1 dollar's worth of G_i.

Let x_i be the dollar value of G_i produced in a fixed period of time, say, 1 year. The vector

$$\mathbf{x} = \begin{bmatrix} x_1 \\ x_2 \\ \vdots \\ x_n \end{bmatrix} \qquad (x_i \geq 0) \tag{10}$$

is called the **production vector**. The expression

$$c_{i1}x_1 + c_{i2}x_2 + \cdots + c_{in}x_n$$

is the total value of the product G_i that is consumed, as determined by the production vector to make x_1 dollar's worth of G_1, x_2 dollar's worth of G_2, and so on. Observe that the expression given by Equation (10) is the ith entry of the matrix product $C\mathbf{x}$. The difference between the dollar value of G_i that is produced and the total dollar value of G_i that is consumed,

$$x_i - (c_{i1}x_1 + c_{i2}x_2 + \cdots + c_{in}x_n), \tag{11}$$

is called the **net production**.

Observe that the expression in Equation (11) is the ith entry in

$$\mathbf{x} - C\mathbf{x} = (I_n - C)\mathbf{x}.$$

Now let d_i represent the dollar value of the outside demand for G_i and let

$$\mathbf{d} = \begin{bmatrix} d_1 \\ d_2 \\ \vdots \\ d_n \end{bmatrix} \qquad (d_i \geq 0)$$

be the **demand vector**.

Our problem can now be stated: Given a demand vector $\mathbf{d} \geq \mathbf{0}$, can we find a production vector $\mathbf{x}$ such that the outside demand $\mathbf{d}$ is met without any surplus? That is, can we find a vector $\mathbf{x} \geq \mathbf{0}$ so that the following equation is satisfied?

$$(I_n - C)\mathbf{x} = \mathbf{d} \tag{12}$$

EXAMPLE 5 Let

$$C = \begin{bmatrix} \frac{1}{4} & \frac{1}{2} \\ \frac{2}{3} & \frac{1}{3} \end{bmatrix}$$

be a consumption matrix. Then

$$I_2 - C = \begin{bmatrix} 1 & 0 \\ 0 & 1 \end{bmatrix} - \begin{bmatrix} \frac{1}{4} & \frac{1}{2} \\ \frac{2}{3} & \frac{1}{3} \end{bmatrix} = \begin{bmatrix} \frac{3}{4} & -\frac{1}{2} \\ -\frac{2}{3} & \frac{2}{3} \end{bmatrix}.$$

Equation (12) becomes

$$\begin{bmatrix} \frac{3}{4} & -\frac{1}{2} \\ -\frac{2}{3} & \frac{2}{3} \end{bmatrix} \begin{bmatrix} x_1 \\ x_2 \end{bmatrix} = \begin{bmatrix} d_1 \\ d_2 \end{bmatrix},$$

so

$$\begin{bmatrix} x_1 \\ x_2 \end{bmatrix} = \begin{bmatrix} \frac{3}{4} & -\frac{1}{2} \\ -\frac{2}{3} & \frac{2}{3} \end{bmatrix}^{-1} \begin{bmatrix} d_1 \\ d_2 \end{bmatrix} = \begin{bmatrix} 4 & 3 \\ 4 & \frac{9}{2} \end{bmatrix} \begin{bmatrix} d_1 \\ d_2 \end{bmatrix} \geq \mathbf{0},$$

since $d_1 \geq 0$ and $d_2 \geq 0$. Thus we can obtain a production vector for any given demand vector. ■

In general, if $(I_n - C)^{-1}$ exists and is $\geq \mathbf{0}$, then $\mathbf{x} = (I_n - C)^{-1}\mathbf{d} \geq \mathbf{0}$ is a production vector for any given demand vector. However, for a given consumption matrix there may be no solution to Equation (12).

EXAMPLE 6 Consider the consumption matrix

$$C = \begin{bmatrix} \frac{1}{2} & \frac{1}{2} \\ \frac{1}{2} & \frac{3}{4} \end{bmatrix}.$$

Then

$$I_2 - C = \begin{bmatrix} \frac{1}{2} & -\frac{1}{2} \\ -\frac{1}{2} & \frac{1}{4} \end{bmatrix}$$

and

$$(I_2 - C)^{-1} = \begin{bmatrix} -2 & -4 \\ -4 & -4 \end{bmatrix},$$

so that

$$\mathbf{x} = (I_2 - C)^{-1}\mathbf{d}$$

is not a production vector for $\mathbf{d} \neq \mathbf{0}$, since all of its components are negative. Thus the problem has no solution. If $\mathbf{d} = \mathbf{0}$, we do have a solution, namely, $\mathbf{x} = \mathbf{0}$, which means that if there is no outside demand, then nothing is produced. ■

DEFINITION An $n \times n$ consumption matrix C is called **productive** if $(I_n - C)^{-1}$ exists and $(I_n - C)^{-1} \geq \mathbf{0}$. That is, C is productive if $(I_n - C)$ is nonsingular and every entry of $(I_n - C)^{-1}$ is nonnegative. In this case, the model is also called **productive**.

It follows that if C is productive, then for any demand vector $\mathbf{d} \geq \mathbf{0}$, the equation

$$(I_n - C)\mathbf{x} = \mathbf{d}$$

has a unique solution $\mathbf{x} \geq \mathbf{0}$.

EXAMPLE 7 Consider the consumption matrix

$$C = \begin{bmatrix} \frac{1}{2} & \frac{1}{3} \\ \frac{1}{4} & \frac{1}{3} \end{bmatrix}.$$

Then

$$(I_2 - C) = \begin{bmatrix} \frac{1}{2} & -\frac{1}{3} \\ -\frac{1}{4} & \frac{2}{3} \end{bmatrix},$$

and

$$(I_2 - C)^{-1} = 4\begin{bmatrix} \frac{2}{3} & \frac{1}{3} \\ \frac{1}{4} & \frac{1}{2} \end{bmatrix}.$$

Thus C is productive. If $\mathbf{d} \geq \mathbf{0}$ is a demand vector, then the equation $(I_n - C)\mathbf{x} = \mathbf{d}$ has the unique solution $\mathbf{x} = (I_n - C)^{-1}\mathbf{d} \geq \mathbf{0}$. ■

Results are given in more advanced books (see the book by Johnston, p. 251, given in Further Readings) for telling when a given consumption matrix is productive.

Further Readings

GALE, DAVID. *The Theory of Linear Economic Models.* New York: McGraw-Hill Book Company, 1960.

JOHNSTON, B., G. PRICE, and F. S. VAN VLECK. *Linear Equations and Matrices.* Reading, Mass.: Addison-Wesley Publishing Co., Inc., 1966.

Key Terms

Leontief closed model
Exchange model
Exchange matrix
International trade model
Leontief open model
Consumption matrix
Production vector
Net production
Demand vector
Productive matrix
Productive model

2.6 Exercises

1. Which of the following matrices are exchange matrices?

(a) $\begin{bmatrix} \frac{1}{3} & 0 & 1 \\ \frac{2}{3} & 1 & 0 \\ \frac{1}{2} & -\frac{1}{2} & 0 \end{bmatrix}$ (b) $\begin{bmatrix} \frac{1}{2} & \frac{1}{3} & \frac{3}{4} \\ \frac{1}{2} & \frac{1}{3} & \frac{1}{4} \\ 0 & \frac{1}{3} & 0 \end{bmatrix}$

(c) $\begin{bmatrix} \frac{1}{3} & -\frac{2}{3} & \frac{1}{2} \\ \frac{2}{3} & \frac{2}{3} & \frac{1}{2} \\ 0 & 1 & 0 \end{bmatrix}$ (d) $\begin{bmatrix} 1 & \frac{1}{4} & \frac{5}{6} \\ 0 & \frac{1}{4} & \frac{1}{6} \\ 0 & \frac{1}{2} & 0 \end{bmatrix}$

In Exercises 2 through 4, find a vector $\mathbf{p} \geq \mathbf{0}$, *with at least one positive component, satisfying Equation (8) for the given exchange matrix.*

2. $\begin{bmatrix} \frac{1}{3} & \frac{2}{3} & 0 \\ \frac{1}{3} & 0 & \frac{1}{4} \\ \frac{1}{3} & \frac{1}{3} & \frac{3}{4} \end{bmatrix}$ **3.** $\begin{bmatrix} \frac{1}{2} & 1 & \frac{2}{3} \\ 0 & 0 & 0 \\ \frac{1}{2} & 0 & \frac{1}{3} \end{bmatrix}$

4. $\begin{bmatrix} 0 & \frac{1}{3} & 1 \\ \frac{1}{6} & \frac{1}{6} & 0 \\ \frac{5}{6} & \frac{1}{2} & 0 \end{bmatrix}$

5. Consider the simple economy of Example 1. Suppose that the farmer consumes $\frac{2}{5}$ of the food, $\frac{1}{3}$ of the shelter, and $\frac{1}{2}$ of the clothes; that the carpenter consumes $\frac{2}{5}$ of the food, $\frac{1}{3}$ of the shelter, and $\frac{1}{2}$ of the clothes; that the tailor consumes $\frac{1}{5}$ of the food, $\frac{1}{3}$ of the shelter, and none of the clothes. Find the exchange matrix A for this problem and a vector $\mathbf{p} \geq \mathbf{0}$ with at least one positive component satisfying Equation (8).

6. Consider the international trade model consisting of three countries, C_1, C_2, and C_3. Suppose that the fraction of C_1's income spent on imports from C_1 is $\frac{1}{4}$, from C_2 is $\frac{1}{2}$, and from C_3 is $\frac{1}{4}$; that the fraction of C_2's income spent on imports from C_1 is $\frac{2}{5}$, from C_2 is $\frac{1}{5}$, and from C_3 is $\frac{2}{5}$; that the fraction of C_3's income spent on imports from C_1 is $\frac{1}{2}$, from C_2 is $\frac{1}{2}$, and from C_3 is 0. Find the income of each country.

In Exercises 7 through 10, determine which matrices are productive.

7. $\begin{bmatrix} \frac{1}{2} & \frac{1}{3} & 0 \\ 0 & \frac{2}{3} & 0 \\ 1 & 0 & 2 \end{bmatrix}$ **8.** $\begin{bmatrix} 0 & \frac{2}{3} & 0 \\ \frac{1}{2} & 0 & 0 \\ 0 & \frac{1}{4} & 0 \end{bmatrix}$

9. $\begin{bmatrix} 0 & \frac{1}{3} & \frac{1}{2} \\ \frac{1}{2} & 0 & \frac{1}{4} \\ \frac{1}{4} & \frac{1}{3} & 0 \end{bmatrix}$ **10.** $\begin{bmatrix} 0 & \frac{1}{3} & \frac{1}{3} \\ \frac{1}{4} & 0 & \frac{1}{6} \\ \frac{1}{3} & \frac{2}{3} & 0 \end{bmatrix}$

11. Suppose that the consumption matrix for the linear production model is

$$\begin{bmatrix} \frac{1}{2} & \frac{1}{2} \\ \frac{1}{2} & \frac{1}{4} \end{bmatrix}.$$

(a) Find the production vector for the demand vector $\begin{bmatrix} 1 \\ 3 \end{bmatrix}$.

(b) Find the production vector for the demand vector $\begin{bmatrix} 2 \\ 0 \end{bmatrix}$.

12. A small town has three primary industries: a copper mine, a railroad, and an electric utility. To produce \$1 of copper, the copper mine uses \$0.20 of copper, \$0.10 of transportation, and \$0.20 of electric power. To provide \$1 of transportation, the railroad uses \$0.10 of copper, \$0.10 of transportation, and \$0.40 of electric power. To provide \$1 of electric power, the electric utility uses \$0.20 of copper, \$0.20 of transportation, and \$0.30 of electric power. Suppose that during the year there is an outside demand of 1.2 million dollars for copper, 0.8 million dollars for transportation, and 1.5 million dollars for electric power. How much should each industry produce to satisfy the demands?

Theoretical Exercise

T.1. In the exchange model (Example 1 or 2), show that $A\mathbf{p} \le \mathbf{p}$ implies that $A\mathbf{p} = \mathbf{p}$.

2.7 INTRODUCTION TO WAVELETS

Prerequisite. Section 1.7, The Inverse of a Matrix.

A major revolution in the nineteenth century was the capability to transmit power. The corresponding revolution of the twentieth century, and continuing into the twenty-first century, is the capability to transmit information. Once the infrastructure was in place to transmit raw data, the burgeoning data needs of government and business required that we figure out how to quickly transmit the essence of the data set so that it could be reconstructed to closely approximate the original information. To do this there have been a multitude of schemes that **transform** the original data set, **compress** it, **transmit** it, and **recover** approximations to the information contained in the original data set. Examples of such schemes include Morse code, ciphers of many kinds (including public key encryption), radio, television, microwave, and methods that employ proprietary digital encoding techniques.

One well-known commercially available digital encoding technique is that developed by the Joint Photographic Experts Group (JPEG) for digital imaging. The JPEG2000 scheme uses **wavelets**, a compression technology that encodes images in a continuous stream. This new technique will create data files that are about 20% smaller, will permit faster downloads, and has the ability to choose an image's size without creating a separate data file. The mathematical topic of wavelets has received wide attention due to its versatility for adaptation to a myriad of applications including compression of data for efficient transmission and accurate approximation of information, image processing (such as FBI fingerprint files), signal processing (like restoration of recordings), seismology, and the numerical solution of partial differential equations.

Thus wavelets have been the object of continued research in the past decade and continue to be adapted for use in an ever-increasing number of scientific and engineering areas.

Our objective is to show how common matrix concepts can be used to reveal the basic nature of wavelets. We shall show how digital information is transformed so that some of the information can be omitted (that is, a compression takes place) and then transmitted so that the received data can be rebuilt to generate a close approximation to the original information. The economy comes about when there is significant reduction in the amount of information that must be transmitted. Thus transformation and compression schemes together with the reconstruction process must be designed with this opportunity for economy in mind. We represent this scenario pictorially in Figure 2.22.

Figure 2.22 ▶

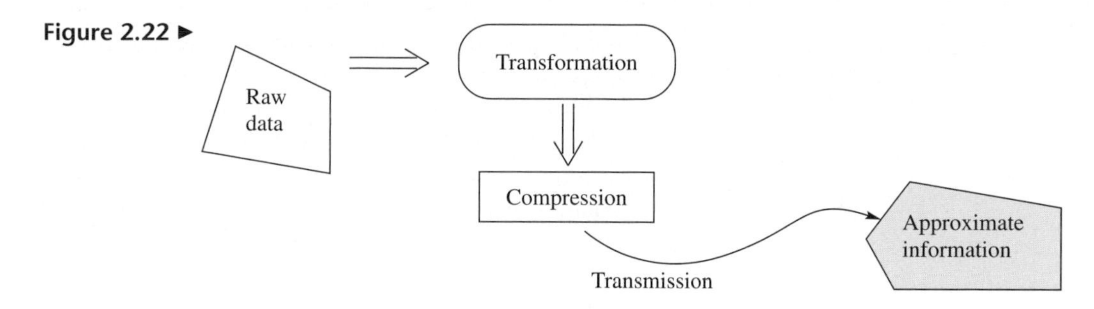

EXAMPLE 1

A dollar's worth of information is displayed by any of the columns in Table 2.2. The entry in a row gives the number of coins of the type corresponding to the row label. The details of the information about any of these five ways to represent a dollar could be transmitted with no loss of information by sending the string of six numbers corresponding to a column. However, several of these representations can be compressed so that fewer than six numbers can be sent and the transmitted data used to exactly reconstruct the original information. For the first column, $\begin{bmatrix} 0 & 0 & 4 & 0 & 0 & 0 \end{bmatrix}^T$, we can compress this information to $\begin{bmatrix} 4 & 3 \end{bmatrix}^T$, where this means four coins of the third type (quarters). Similarly, column two, $\begin{bmatrix} 0 & 0 & 2 & 5 & 0 & 0 \end{bmatrix}^T$, can be compressed to $\begin{bmatrix} 2 & 5 & 3 & 4 \end{bmatrix}^T$, meaning two quarters and five dimes. (Note that the first two entries give the number of coins and the second two entries indicate the position in the list of types of coins.) Although not every column in Table 2.2 can be compressed as simply, for large data sets in which a great many zeros appear (such data are called **sparse**) a compression may be as simple as sending a nonzero digit and its location in the information string. ■

Table 2.2

Dollar coin	0	0	0	0	0
Half dollar	0	0	1	0	1
Quarter	4	2	1	0	1
Dime	0	5	1	0	1
Nickel	0	0	0	1	1
Penny	0	0	15	95	10

Another example in which data compression is useful is imaging, where large amounts of data are often transmitted. The continued advances in calculators and computers have made available easy-to-use graphics for displaying mathematical functions. When the function f is to be graphed, a common procedure is to generate a set of equispaced x values and compute the function values at each of these x values. The graph of f is then displayed by showing the points $(x, f(x))$ connected by straight line segments, or possibly arcs, to present a smooth rendering of the curve. If the spacing between x values is large, then the graph displayed may appear jagged rather than smooth. For high-quality images quite a small spacing may be necessary, hence the original data set $\{(x, f(x))\}$ may be quite large. A graphical image or photograph consists of small dots. A mathematical description of such an image gives the position of each dot, together with a code that denotes a color or shade of gray corresponding to the dot. Such data sets become huge, even for modest size images.

The transmission of large amounts of data for graphical imaging is a real concern and can cause a slowdown on computer networks. One aid to overcoming such difficulties is the use of "transformation + compression" schemes that reduce the amount of data that needs to be transmitted coupled with methods that use the transmitted data to build good approximations to the original image. By **transformation** we mean some process that takes the original, or raw, digital data and produces an equivalent set of data, often by some algebraic manipulation. Ideally, the transformed data should reflect intrinsic qualities of the information within the data. By **compression** we mean a scheme that reduces the overall amount of data that needs to be transmitted so that a good approximation of the original image can be reconstructed. Often the transformation and compression steps are packaged together.

We illustrate the process of transformation and compression using the graph of a function f evaluated at equispaced points. For a string or vector of data of y-coordinates of a set of points on the graph of function f, we develop an equivalent set of data using averaging and differencing: the transformation step. This equivalent data set contains all the information of the original set. A feature of this equivalent form of the data set is that it can be compressed more easily*: the compression step. The compressed data set has lost some of the original information, but in many cases a good approximation can be reconstructed from this smaller set of data: the transmission and reconstruction steps. We show how to use matrix multiplication to perform the transformation step and use algebraic properties to verify that we have an equivalent set of data.

EXAMPLE 2 **(Computing Averages by Matrix Multiplication)**

(a) For the vector $\begin{bmatrix} a \\ b \end{bmatrix}$ we compute the average of the entries using the row-by-column product

$$\begin{bmatrix} \frac{1}{2} & \frac{1}{2} \end{bmatrix} \begin{bmatrix} a \\ b \end{bmatrix} = \begin{bmatrix} \frac{a+b}{2} \end{bmatrix}.$$

(b) The case of four values, where we want to average successive pairs of values (we call this a **pairwise average**), is developed as follows. For the initial data

$$\mathbf{v} = \begin{bmatrix} a \\ b \\ c \\ d \end{bmatrix}$$

we want the result of the matrix multiplication to be

$$\begin{bmatrix} \frac{a+b}{2} \\ \frac{c+d}{2} \end{bmatrix}.$$

*A simple compression scheme is to drop every other data point, but there are techniques that lose less information by taking advantage of regions in which a function does not change much. The use of averaging and differencing has such an advantage.

A matrix A times the 4×1 vector $\mathbf{v}$ is to produce a 2×1 vector, hence the matrix A must be 2×4. Using the result from part (a), it follows that

$$\begin{bmatrix} \frac{1}{2} & \frac{1}{2} & 0 & 0 \\ 0 & 0 & \frac{1}{2} & \frac{1}{2} \end{bmatrix} \begin{bmatrix} a \\ b \\ c \\ d \end{bmatrix} = \begin{bmatrix} \dfrac{a+b}{2} \\ \dfrac{c+d}{2} \end{bmatrix}$$

gives the pairwise average. Hence the matrix that performs the transformation to pairwise averages is

$$A = \begin{bmatrix} \frac{1}{2} & \frac{1}{2} & 0 & 0 \\ 0 & 0 & \frac{1}{2} & \frac{1}{2} \end{bmatrix}.$$

(c) For the case of six values in a vector $\mathbf{v}$, the 3×6 matrix A shown below is used in the matrix product $A\mathbf{v}$ to compute the pairwise average of the entries in $\mathbf{v}$.

$$A = \begin{bmatrix} \frac{1}{2} & \frac{1}{2} & 0 & 0 & 0 & 0 \\ 0 & 0 & \frac{1}{2} & \frac{1}{2} & 0 & 0 \\ 0 & 0 & 0 & 0 & \frac{1}{2} & \frac{1}{2} \end{bmatrix}.$$

(Verify that if $\mathbf{v} = \begin{bmatrix} a & b & c & d & e & f \end{bmatrix}^T$, then $A\mathbf{v}$ gives a 3×1 vector containing the pairwise average of the entries in $\mathbf{v}$.) ■

Our transformation of the data must be such that we can recover the original data from the equivalent alternate form that we produced. To ensure that exact recovery is possible, we develop an equivalent representation of the information in the original vector by pairing another piece of information with an average.

We develop this as follows. We claim that the vector $\mathbf{v} = \begin{bmatrix} a \\ b \end{bmatrix}$ can be replaced by the vector $\mathbf{w} = \begin{bmatrix} c \\ d \end{bmatrix}$, where $c = \dfrac{a+b}{2}$ is the average of the entries in $\mathbf{v}$, and $d = a - c$ is the distance from the first entry of $\mathbf{v}$ to the average. This is obtained from the observation that if we know the values of c and d, then

$$a = c + d \quad \text{and} \quad b = 2c - a,$$

so we have obtained the values of a and b. Next we want a matrix formulation of the transformation of data from

$$\mathbf{v} = \begin{bmatrix} a \\ b \end{bmatrix} \quad \text{to} \quad \mathbf{w} = \begin{bmatrix} c \\ d \end{bmatrix} = \begin{bmatrix} \dfrac{a+b}{2} \\ a - c \end{bmatrix}.$$

Thus we seek a 2×2 matrix A so that

$$A \begin{bmatrix} a \\ b \end{bmatrix} = \begin{bmatrix} c \\ d \end{bmatrix} = \begin{bmatrix} c \\ a - c \end{bmatrix}.$$

From our previous work it follows that the first row of the matrix A should be $\begin{bmatrix} \frac{1}{2} & \frac{1}{2} \end{bmatrix}$ so that c will be the average of a and b. Denote the second row of

A by $\begin{bmatrix} p & q \end{bmatrix}$ and determine these values. It follows by matrix multiplication that

$$\begin{bmatrix} p & q \end{bmatrix}\begin{bmatrix} a \\ b \end{bmatrix} = a - c.$$

Performing the matrix multiplication on the left side, we get $pa + qb = a - c$. Substituting the equivalent expression $2c - a$ for b and collecting terms, we have

$$pa + q(2c - a) = (p - q)a + (2q)c = a - c.$$

Equating coefficients of like terms on the left and right gives the linear system

$$\begin{aligned} p - q &= 1 \\ 2q &= -1, \end{aligned}$$

whose solution is $p = \frac{1}{2}$, $q = -\frac{1}{2}$. Thus the matrix formulation to compute the average and the distance from the first entry to the average is

$$\begin{bmatrix} \frac{1}{2} & \frac{1}{2} \\ \frac{1}{2} & -\frac{1}{2} \end{bmatrix}\begin{bmatrix} a \\ b \end{bmatrix} = \begin{bmatrix} c \\ a - c \end{bmatrix} = \begin{bmatrix} \text{the average of } a \text{ and } b \\ \text{distance from } a \text{ to the average} \end{bmatrix},$$

or if

$$Q = \begin{bmatrix} \frac{1}{2} & \frac{1}{2} \\ \frac{1}{2} & -\frac{1}{2} \end{bmatrix}, \quad \mathbf{v} = \begin{bmatrix} a \\ b \end{bmatrix}, \quad \text{and} \quad \mathbf{w} = \begin{bmatrix} c \\ a - c \end{bmatrix}$$

we have $Q\mathbf{v} = \mathbf{w}$. We call this procedure the **matrix formulation of the average–difference representation**. Note that the matrix Q is nonsingular and that

$$Q^{-1} = \begin{bmatrix} 1 & 1 \\ 1 & -1 \end{bmatrix} \quad \text{(verify)}.$$

Thus, if we know the average of the two values and the distance from the first value to the average, we can recover the original data; that is,

$$\mathbf{v} = Q^{-1}\mathbf{w} = \begin{bmatrix} 1 & 1 \\ 1 & -1 \end{bmatrix}\begin{bmatrix} c \\ a - c \end{bmatrix} = \begin{bmatrix} a \\ b \end{bmatrix}.$$

(This was shown previously using ordinary algebra.)

EXAMPLE 3 We extend the average–difference representation to a vector with more than two entries in order to determine the pairwise averages and corresponding differences.

(a) Let $\mathbf{v} = \begin{bmatrix} a & b & c & d \end{bmatrix}^T$. We now determine the 4×4 matrix A such that

$$A\mathbf{v} = \begin{bmatrix} \text{average of } a \text{ and } b \\ \text{average of } c \text{ and } d \\ \text{distance from } a \text{ to average} \\ \text{distance from } c \text{ to average} \end{bmatrix}.$$

Using Example 2 and following the development immediately above for the case of two data values, a reasonable conjecture for matrix A is

$$\begin{bmatrix} \frac{1}{2} & \frac{1}{2} & 0 & 0 \\ 0 & 0 & \frac{1}{2} & \frac{1}{2} \\ \frac{1}{2} & -\frac{1}{2} & 0 & 0 \\ 0 & 0 & \frac{1}{2} & -\frac{1}{2} \end{bmatrix}.$$

Verify that this conjecture is correct, that A is nonsingular, and find A^{-1}.

(b) For a 6-vector $\mathbf{v} = \begin{bmatrix} a & b & c & d & e & f \end{bmatrix}^T$ make a conjecture for a 6×6 matrix A such that

$$A\mathbf{v} = \begin{bmatrix} \text{average of first data pair} \\ \text{average of second data pair} \\ \text{average of third data pair} \\ \text{distance from } a \text{ to average} \\ \text{distance from } c \text{ to average} \\ \text{distance from } e \text{ to average} \end{bmatrix}.$$

Verify your conjecture and verify that the matrix A is nonsingular. ■

A summary of our developments so far is as follows: Given a vector $\mathbf{v}$ of function values at equispaced points, we have found how to determine a matrix A so that $A\mathbf{v}$ is a vector containing the pairwise averages followed by the distances from the first member of each pair to the average. The next step of our transformation is to apply the same idea to these averages so that in effect we compute averages of averages and their distances from the original averages. However, we must ensure that the distances from the first member of each data pair to its average are preserved. To illustrate that there is technically no new work to be done, merely a reorganization of results we have already determined, we consider the case of four function values designated by $\mathbf{v} = \begin{bmatrix} a & b & c & d \end{bmatrix}^T$. Then let A_1 be the matrix that transforms these data to the averages and differences discussed in part (a) of Example 2. We have

$$A_1\mathbf{v} = \begin{bmatrix} \frac{1}{2} & \frac{1}{2} & 0 & 0 \\ 0 & 0 & \frac{1}{2} & \frac{1}{2} \\ \frac{1}{2} & -\frac{1}{2} & 0 & 0 \\ 0 & 0 & \frac{1}{2} & -\frac{1}{2} \end{bmatrix} \begin{bmatrix} a \\ b \\ c \\ d \end{bmatrix} = \begin{bmatrix} e \\ f \\ a-e \\ c-f \end{bmatrix},$$

where

$$e = \frac{a+b}{2} \quad \text{and} \quad f = \frac{c+d}{2}.$$

Let $\mathbf{w} = \begin{bmatrix} e \\ f \\ a-e \\ c-f \end{bmatrix}$. Next we want to determine a 4×4 matrix A_2 so that

$$\mathbf{u} = A_2\mathbf{w} = \begin{bmatrix} g \\ e-g \\ a-e \\ c-f \end{bmatrix}, \quad \text{where } g = \frac{e+f}{2}.$$

(Note that $e - g$ is the distance from e to the average of e and f.) We see that the last two entries of vector $\mathbf{w}$ and vector

$$\mathbf{u} = \begin{bmatrix} g \\ e-g \\ a-e \\ c-f \end{bmatrix}$$

are to be the same, so we suspect that an identity matrix will be part of A_2. Also, since A_2 is only to transform the first two entries of $\mathbf{w}$, there will be zeros involved. We conjecture that A_2 will be a partitioned matrix that will have the form

$$A_2 = \begin{bmatrix} Q & O_2 \\ O_2 & I_2 \end{bmatrix},$$

where O_2 is a 2×2 matrix of zeros, I_2 is a 2×2 identity matrix, and Q is a 2×2 matrix that must be determined. Inspecting the matrix product

$$\mathbf{u} = A_2\mathbf{w} = \begin{bmatrix} Q & O_2 \\ O_2 & I_2 \end{bmatrix} \begin{bmatrix} e \\ f \\ a-e \\ c-f \end{bmatrix} = \begin{bmatrix} g \\ e-g \\ a-e \\ c-f \end{bmatrix},$$

we find that Q must be chosen so that

$$Q \begin{bmatrix} e \\ f \end{bmatrix} = \begin{bmatrix} g \\ e-g \end{bmatrix}.$$

But this is the same as producing an average and the distance from the first entry to the average. From our previous work it follows that

$$Q = \begin{bmatrix} \frac{1}{2} & \frac{1}{2} \\ \frac{1}{2} & -\frac{1}{2} \end{bmatrix},$$

and thus

$$A_2 = \begin{bmatrix} \frac{1}{2} & \frac{1}{2} & 0 & 0 \\ \frac{1}{2} & -\frac{1}{2} & 0 & 0 \\ 0 & 0 & 1 & 0 \\ 0 & 0 & 0 & 1 \end{bmatrix}.$$

We have the following sequence of data transformations:

$$\mathbf{v} = \begin{bmatrix} a \\ b \\ c \\ d \end{bmatrix} \to A_1\mathbf{v} = \begin{bmatrix} e \\ f \\ a-e \\ c-f \end{bmatrix} = \mathbf{w} \to \mathbf{u} = A_2\mathbf{w} = \begin{bmatrix} g \\ e-g \\ a-e \\ c-f \end{bmatrix}.$$

The preceding set of steps is equivalent to the product $\mathbf{u} = A_2A_1\mathbf{v}$. In the vector

$$\mathbf{w} = \begin{bmatrix} e \\ f \\ a-e \\ c-f \end{bmatrix},$$

entries e and f are called **averages** and $a - e$ and $c - f$ are called the **detail coefficients**. Similarly, in the vector

$$\mathbf{u} = \begin{bmatrix} g \\ e-g \\ a-e \\ c-f \end{bmatrix}$$

entry g is called the **final average** and the last three entries are called the detail coefficients. The intrinsic information of the original data set has been transformed into the detail coefficients and the final average. It can be shown that matrices A_1 and A_2 are nonsingular (see Exercise 6), hence the process is reversible; that is, we have generated an equivalent representation of the original data.

For more than four data values we expect the size of the matrices needed to perform the transformation of the original data to a final average and detail coefficients to grow. Since we are computing averages of pairs of items, it follows that we should always have an even number of items at each level of the averages. Hence, ideally we should use this procedure on sets with $2, 4, 8, 16, \ldots, 2^n$ members. In the event that the size of our data vector is not a power of 2, we can adjoin enough zeros to the end of the vector to fulfill this requirement, and then proceed with the transformation of the data as described above.

EXAMPLE 4

Let $\mathbf{v} = \begin{bmatrix} 37 & 33 & 6 & 16 \end{bmatrix}^T$ represent a sample of a function f at four equispaced points. To determine the average and detail coefficients we transform the data using matrices A_1 and A_2, given previously, as follows:

$$\mathbf{w} = A_2 A_1 \mathbf{v} = A_2(A_1\mathbf{v}) = A_2 \begin{bmatrix} 35 \\ 11 \\ 2 \\ -5 \end{bmatrix} = \begin{bmatrix} 23 \\ 12 \\ 2 \\ -5 \end{bmatrix}. \quad \text{(Verify.)}$$

Thus the final average is 23 and the detail coefficients are 12, 2, and -5. Note that both matrices A_1 and A_2 can be written as partitioned matrices. Let

$$P = \begin{bmatrix} P_1 & P_2 \end{bmatrix} = \begin{bmatrix} \begin{bmatrix} \frac{1}{2} & \frac{1}{2} \\ 0 & 0 \end{bmatrix} & \begin{bmatrix} 0 & 0 \\ \frac{1}{2} & \frac{1}{2} \end{bmatrix} \end{bmatrix}$$

and

$$S = \begin{bmatrix} S_1 & S_2 \end{bmatrix} = \begin{bmatrix} \begin{bmatrix} \frac{1}{2} & -\frac{1}{2} \\ 0 & 0 \end{bmatrix} & \begin{bmatrix} 0 & 0 \\ \frac{1}{2} & -\frac{1}{2} \end{bmatrix} \end{bmatrix}.$$

Then

$$A_1 = \begin{bmatrix} P_1 & P_2 \\ S_1 & S_2 \end{bmatrix}$$

and, from before,

$$A_2 = \begin{bmatrix} Q & O_2 \\ O_2 & I_2 \end{bmatrix}, \quad \text{where } Q = \begin{bmatrix} \frac{1}{2} & \frac{1}{2} \\ \frac{1}{2} & -\frac{1}{2} \end{bmatrix}.$$

This partitioned form for the matrices involved in the transformation is easy to generalize, as shown in Example 5. ■

Once we have the final average and the detail coefficients, we have completed the transformation of the data to the equivalent average–difference representation. Recall that at this stage we can recover all the original data. We then move on to perform a compression. A simple form of compression is to set a detail coefficient equal to zero if its absolute value is below a prescribed **threshold number** ε. Replacing a small detail coefficient by zero in effect replaces a data value with an average of data values. Hence, if the function is not changing rapidly in that region, we obtain a good approximation. (In general, the choice of a threshold value is application dependent and is often determined by experimentation on similar applications.)

Once the compressed data set is determined, we need only transmit the nonzero coefficients and their positions (that is, their column index for our examples). When the data set is large, there can be a dramatic reduction in the amount of data that must be transmitted. Upon receipt of the compressed data we reverse the differencing and averaging transformations to generate an approximation to the original data set. The reversal process uses the inverses of the matrices $A_1, A_2, \ldots, A_n$, thus there are no additional concepts needed at this stage. We merely form the matrix products using the inverses of the matrices that were used in the transformation. The inverses involved are not difficult to compute and have simple partitioned patterns. From Example 4 we have

$$A_1^{-1} = \begin{bmatrix} \begin{bmatrix} \frac{1}{2} & \frac{1}{2} \\ 0 & 0 \end{bmatrix} & \begin{bmatrix} 0 & 0 \\ \frac{1}{2} & \frac{1}{2} \end{bmatrix} \\ \begin{bmatrix} \frac{1}{2} & -\frac{1}{2} \\ 0 & 0 \end{bmatrix} & \begin{bmatrix} 0 & 0 \\ \frac{1}{2} & -\frac{1}{2} \end{bmatrix} \end{bmatrix}^{-1} = \begin{bmatrix} \begin{bmatrix} 1 & 0 \\ 1 & 0 \end{bmatrix} & \begin{bmatrix} 1 & 0 \\ -1 & 0 \end{bmatrix} \\ \begin{bmatrix} 0 & 1 \\ 0 & 1 \end{bmatrix} & \begin{bmatrix} 0 & 1 \\ 0 & -1 \end{bmatrix} \end{bmatrix}$$

$$= \begin{bmatrix} 1 & 0 & 1 & 0 \\ 1 & 0 & -1 & 0 \\ 0 & 1 & 0 & 1 \\ 0 & 1 & 0 & -1 \end{bmatrix}$$

and

$$A_2^{-1} = \begin{bmatrix} \begin{bmatrix} \frac{1}{2} & \frac{1}{2} \\ \frac{1}{2} & -\frac{1}{2} \end{bmatrix} & \begin{bmatrix} 0 & 0 \\ 0 & 0 \end{bmatrix} \\ \begin{bmatrix} 0 & 0 \\ 0 & 0 \end{bmatrix} & \begin{bmatrix} 1 & 0 \\ 0 & 1 \end{bmatrix} \end{bmatrix}^{-1} = \begin{bmatrix} \begin{bmatrix} 1 & 1 \\ 1 & -1 \end{bmatrix} & \begin{bmatrix} 0 & 0 \\ 0 & 0 \end{bmatrix} \\ \begin{bmatrix} 0 & 0 \\ 0 & 0 \end{bmatrix} & \begin{bmatrix} 1 & 0 \\ 0 & 1 \end{bmatrix} \end{bmatrix}$$

$$= \begin{bmatrix} 1 & 1 & 0 & 0 \\ 1 & -1 & 0 & 0 \\ 0 & 0 & 1 & 0 \\ 0 & 0 & 0 & 1 \end{bmatrix}.$$

If $\tilde{\mathbf{w}}$ is the compressed data, then the approximation to the original data is given by $\tilde{\mathbf{y}} = A_1^{-1}A_2^{-1}\tilde{\mathbf{w}}$.

EXAMPLE 5

Suppose that we have the following sample of x and y values from the graph of a function.

x	1	2	3	4	5	6	7	8
y	37	33	6	16	26	28	18	4

Let $\mathbf{v}$ denote the 8×1 vector of y-coordinates, let

$$P = \begin{bmatrix} P_1 & P_2 \end{bmatrix} \quad \text{and} \quad S = \begin{bmatrix} S_1 & S_2 \end{bmatrix},$$

be as in Example 4, and define

$$Z = \begin{bmatrix} 0 & 0 \\ 0 & 0 \end{bmatrix} \quad \text{and} \quad I = \begin{bmatrix} 1 & 0 \\ 0 & 1 \end{bmatrix}.$$

Let A_1 be the 8×8 partitioned matrix given by

$$A_1 = \begin{bmatrix} P_1 & P_2 & Z & Z \\ Z & Z & P_1 & P_2 \\ S_1 & S_2 & Z & Z \\ Z & Z & S_1 & S_2 \end{bmatrix}.$$

Then we have (verify)

$$A_1\mathbf{v} = \begin{bmatrix} 35 & 11 & 27 & 11 & 2 & -5 & -1 & 7 \end{bmatrix}^T,$$

where the first four entries are (pairwise) averages and the last four entries are (first-level) detail coefficients. Next construct the 8×8 matrix A_2 in partitioned form as

$$A_2 = \begin{bmatrix} P_1 & P_2 & Z & Z \\ S_1 & S_2 & Z & Z \\ Z & Z & I & Z \\ Z & Z & Z & I \end{bmatrix}.$$

It follows (verify) that

$$A_2(A_1\mathbf{v}) = \begin{bmatrix} 23 & 19 & 12 & 8 & 2 & -5 & -1 & 7 \end{bmatrix}^T.$$

In this result the first two entries are averages and the last six are (second-level) detail coefficients. Finally, construct the 8×8 matrix A_3 in partitioned form as

$$A_3 = \begin{bmatrix} Q & Z & Z & Z \\ Z & I & Z & Z \\ Z & Z & I & Z \\ Z & Z & Z & I \end{bmatrix},$$

where

$$Q = \begin{bmatrix} \frac{1}{2} & \frac{1}{2} \\ \frac{1}{2} & -\frac{1}{2} \end{bmatrix}$$

as developed previously. It follows (verify) that

$$A_3[A_2(A_1\mathbf{v})] = \mathbf{w} = \begin{bmatrix} 21 & 2 & 12 & 8 & 2 & -5 & -1 & 7 \end{bmatrix}^T,$$

where the first entry is the (final) average of the original eight data points and the remaining seven entries are the (third-level) detail coefficients. To compress the data in this example, set a detail coefficient to zero if its absolute value is less than or equal to 3; that is, use a threshold number $\varepsilon = 3$. The resulting compressed information is the 8×1 vector

$$\tilde{\mathbf{w}} = \begin{bmatrix} 21 & 0 & 12 & 8 & 0 & -5 & 0 & 7 \end{bmatrix}^T.$$

We need only transmit the nonzero entries of this vector and their locations.

Upon receiving the vector $\tilde{\mathbf{w}}$ we can build a model that approximates the original data by computing

$$\tilde{\mathbf{y}} = (A_1)^{-1}(A_2)^{-1}(A_3)^{-1}\tilde{\mathbf{w}} = \begin{bmatrix} 33 & 33 & 4 & 14 & 29 & 29 & 20 & 6 \end{bmatrix}^T.$$

To visually compare the original data with the model developed as a result of the transformation/compression scheme, we plot the original data set and the approximate data set. See Figure 2.23.

x	1	2	3	4	5	6	7	8
$\tilde{y}$	33	33	4	14	29	29	20	6

Figure 2.23 ►

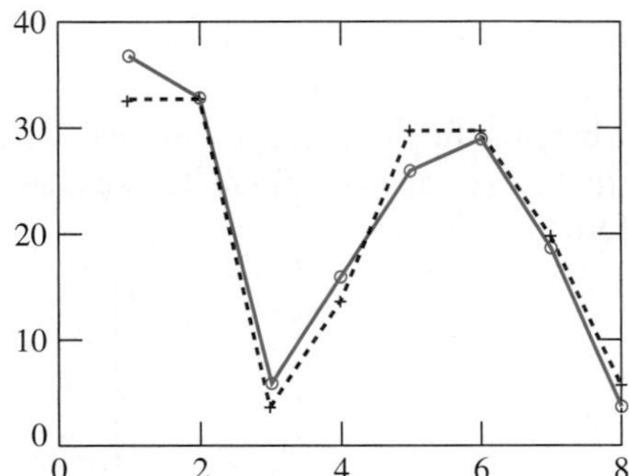

The data points designated by the + signs indicate a model that approximates the original data denoted by the o's. This model is called a **wavelet**, and considering how few data points we used, it is a surprisingly good approximation. ■

To reveal the full potential of wavelets, we must use large data sets. For an initial data set of $512 = 2^9$ points, nine applications of our averaging and differencing technique produce an output of one (final) average and 511 detail coefficients. We can then apply our compression strategy, that is, choose a threshold number that is used to introduce zeros so that we can transmit fewer pieces of data and their locations. Upon reversing the differencing and averaging steps, we obtain a wavelet that often gives us a very good approximation to the original information.

Figures 2.24(a)–(c) show wavelet approximations to discrete samples of $f(x) = e^x \cos(\pi x)$ over the interval $[0, 3]$. The data were chosen at n equispaced points and a threshold number ε was used in the compression. The o's represent the original data sample and the +'s the wavelet data.

Our presentation on wavelets has been limited to equispaced sets of discrete data, ordered pairs. We have illustrated the sequence of operations, that is, transformation of the data to an equivalent representation that can be reversed to obtain the original information, compression of the data by using a threshold number that sets values equal to zero when they are below the chosen threshold number, and the construction of an approximation to the original data, a wavelet.

It is the transmission of the compressed data that presents a significant time savings compared to the transmission of the original data. In addition, the space required to store the compressed data can be significantly smaller than that required to store the original data. We adopted a simplistic view

Figure 2.24 ▶

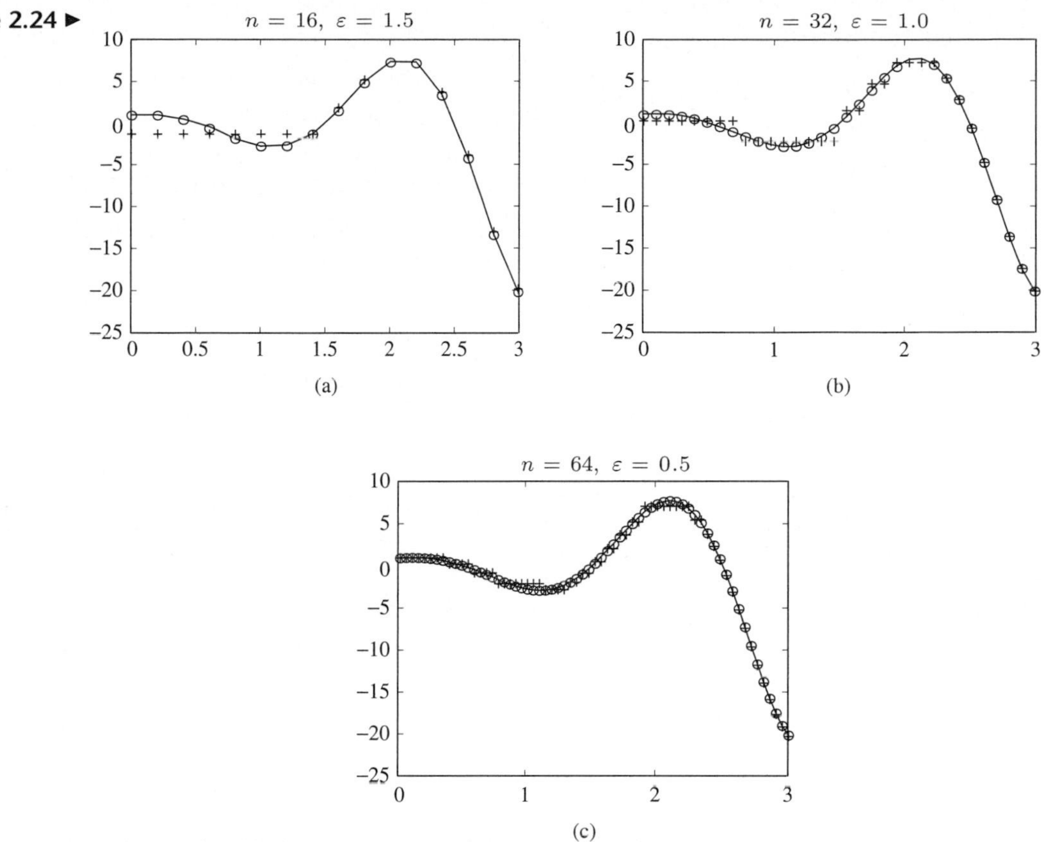

of the transmission, namely, send the nonzero values of the compressed data together with their position in the data string.

There are much better schemes used in applications such as the FBI fingerprint data files that incorporate an encoding procedure that does not include the position of the nonzero compressed data. (For a good description of such a procedure see the book by Aboufadel and Schlicker given in Further Readings at the end of the section.) It is also the case that in many applications long strings of data such as those used in Figures 2.24(a)–(c) are broken into smaller sets to take advantage of regions in which a function does not vary rapidly.

The generalization of wavelets to functions (not just a discrete sample) requires the same steps we have used and also requires further techniques of linear algebra and properties of collections of functions. In succeeding chapters the reader will meet the requisite concepts, in particular *bases*, which are sets used to express matrices or functions in terms of simple building blocks, and *change of bases*, which gives us the mechanism to produce different representations of a matrix or function.

By appropriate choice of bases we can obtain elegant and simple representations of the information contained in the function. Again by judicious choice of bases we get the added benefit of the "zoom-in" property; that is, such representations are designed to deal with fine details that affect only part of the function under study. It is this feature that makes wavelets extremely attractive for applications such as fingerprint encoding. See Further Readings for further details and more general settings for wavelets.

Further Readings

ABOUFADEL, EDWARD, and STEVEN SCHLICKER. *Discovering Wavelets*, New York: Wiley-Interscience, 1999.

CIPRA, BARRY A. "Wavelet Applications Come to the Fore," *SIAM News* (Mathematics That Counts). November 1993.

FRAZIER, MICHAEL W. *An Introduction to Wavelets Through Linear Algebra*. New York: Springer-Verlag, 1999.

HUBBARD, BARBARA BURKE. *The World According to Wavelets: The Story of a Mathematical Technique in the Making*. Cambridge: A.K. Peters, 1995.

MULCAHY, COLM. "Plotting & Scheming with Wavelets." *Mathematics Magazine*, Vol. 69, No. 5, December 1996, pp. 323–343.

NIEVERGELT, YVES. *Wavelets Made Easy*. New York: Springer-Verlag, 1999.

Key Terms

Transform data
Compress data
Transmit data
Recover data
Wavelets
Sparse data
Pairwise average
Matrix formulation of average–difference representation
Detail coefficients
Final average
Threshold number

2.7 Exercises

1. Let $\mathbf{v} = \begin{bmatrix} 87 & 81 & 62 & 64 \end{bmatrix}^T$ represent a sample of a function at four equispaced points. Determine the final average and detail coefficients by computing $A_2 A_1 \mathbf{v}$. Show the result of each stage of the transformation.

2. Let $\mathbf{v} = \begin{bmatrix} 27 & 19 & 5 & 8 \end{bmatrix}^T$ represent a sample of a function at eight equispaced points. Determine the final average and detail coefficients by computing $A_2 A_1 \mathbf{v}$. Show the result of each stage of the transformation. Use a threshold number $\varepsilon = 3$ to determine the compressed data and then compute the wavelet.

3. Let $\mathbf{v} = \begin{bmatrix} 87 & 81 & 62 & 64 & 76 & 78 & 68 & 54 \end{bmatrix}^T$ represent a sample of a function at eight equispaced points. Determine the final average and detail coefficients by computing $A_3 A_2 A_1 \mathbf{v}$. Show the result of each stage of the transformation.

4. Let $\mathbf{v} = \begin{bmatrix} 1 & -6 & -1 & 5 & 2 & -4 & -1 & -3 \end{bmatrix}^T$ represent a sample of a function at eight equispaced points. Determine the final average and detail coefficients by computing $A_3 A_2 A_1 \mathbf{v}$. Show the result of each stage of the transformation. Use a threshold number $\varepsilon = 2$ to determine the compressed data and then compute the wavelet.

5. Explain the difficulty encountered if we try to transform a set of six items in a manner corresponding to the multiplication by matrices like A_1 and A_2 as in Exercise 1. Propose a remedy so that we can complete the procedure.

6. Show that matrices

$$A_1 = \begin{bmatrix} \frac{1}{2} & \frac{1}{2} & 0 & 0 \\ 0 & 0 & \frac{1}{2} & \frac{1}{2} \\ \frac{1}{2} & -\frac{1}{2} & 0 & 0 \\ 0 & 0 & \frac{1}{2} & -\frac{1}{2} \end{bmatrix}$$

and

$$A_2 = \begin{bmatrix} \frac{1}{2} & \frac{1}{2} & 0 & 0 \\ \frac{1}{2} & -\frac{1}{2} & 0 & 0 \\ 0 & 0 & 1 & 0 \\ 0 & 0 & 0 & 1 \end{bmatrix}$$

are nonsingular.

7. Construct the inverse for the partitioned matrix

$$A_3 = \begin{bmatrix} Q & Z & Z & Z \\ Z & I & Z & Z \\ Z & Z & I & Z \\ Z & Z & Z & I \end{bmatrix}$$

of Example 5. (*Hint*: Include Q^{-1} in a partitioned form like that of A_3.)

Key Ideas for Review

- **Encoding function.** See page 121.
- **Theorem 2.1.** Let $A(G)$ be the adjacency matrix of a digraph G and let the rth power of $A(G)$ be B_r:
$$[A(G)]^r = B_r = [b_{ij}^{(r)}].$$
Then the i, jth element in B_r, $b_{ij}^{(r)}$, is the number of ways in which P_i has access to P_j in r stages.
- **Theorem 2.2.** Let $A(G)$ be the adjacency matrix of a digraph and let $S = [s_{ij}]$ be the symmetric matrix defined by $s_{ij} = 1$ if $a_{ij} = 1$ and 0 otherwise, with $S^3 = [s_{ij}^{(3)}]$, where $s_{ij}^{(3)}$ is the i, jth element in S^3. Then P_i belongs to a clique if and only if the diagonal entry $s_{ii}^{(3)}$ is positive.
- **Theorem 2.3.** A digraph with n vertices is strongly connected if and only if its adjacency matrix $A(G)$ has the property that
$$[A(G)] + [A(G)]^2 + \cdots + [A(G)]^{n-1} = E$$
has no zero entries.
- **Theorem 2.4.** If T is the transition matrix of a Markov process, then the state vector $\mathbf{x}^{(k+1)}$, at the $(k+1)$th observation period, can be determined by the state vector $\mathbf{x}^{(k)}$, at the kth observation period, as
$$\mathbf{x}^{(k+1)} = T\mathbf{x}^{(k)}.$$
- **Theorem 2.5.** If T is the transition matrix of a regular Markov process, then:
  (a) As $n \to \infty$, T^n approaches a matrix
$$A = \begin{bmatrix} u_1 & u_1 & \cdots & u_1 \\ u_2 & u_2 & \cdots & u_2 \\ \vdots & \vdots & & \vdots \\ u_n & u_n & \cdots & u_n \end{bmatrix},$$
  all of whose columns are identical.

  (b) Every column
$$\mathbf{u} = \begin{bmatrix} u_1 \\ u_2 \\ \vdots \\ u_n \end{bmatrix}$$
  of A is a probability vector all of whose components are positive. That is, $u_i > 0$ $(1 \le i \le n)$ and
$$u_1 + u_2 + \cdots + u_n = 1.$$
- Shear in the x-direction: $f(\mathbf{v}) = \begin{bmatrix} 1 & k \\ 0 & 1 \end{bmatrix}\mathbf{v}$.
- **Theorem 2.6.** If T is a regular transition matrix and A and $\mathbf{u}$ are as in Theorem 2.5, then:
  (a) For any probability vector $\mathbf{x}$, $T^n\mathbf{x} \to \mathbf{u}$ as $n \to \infty$, so that $\mathbf{u}$ is a steady-state vector.

  (b) The steady-state vector $\mathbf{u}$ is the unique probability vector satisfying the matrix equation $T\mathbf{u} = \mathbf{u}$.
- **Leontief closed model:** Given an exchange matrix A, find a vector $\mathbf{p} \ge \mathbf{0}$ with at least one positive component satisfying $(I_n - A)\mathbf{p} = \mathbf{0}$.
- **Leontief open model.** Given a consumption matrix C, and a demand vector $\mathbf{d}$, find a production vector $\mathbf{x} \ge \mathbf{0}$ satisfying the equation $(I_n - C)\mathbf{x} = \mathbf{d}$.
- **Wavelet applied to a sample of a function:** Given a sample of a function f, we determine a wavelet approximation of f by using averaging and differencing, together with a threshold value to generate a set of approximate detail coefficients.

Supplementary Exercises

1. The parity (4, 5) check code is to be used.

(a) Determine the number of code words.

(b) Determine all the code words that contain exactly two 1 bits.

(c) How many of the code words from part (b) have even parity?

2. Let $f_1 \colon R^2 \to R^2$ be the shear in the x-direction defined by
$$f_1\left(\begin{bmatrix} u_1 \\ u_2 \end{bmatrix}\right) = \begin{bmatrix} u_1 + u_2 \\ u_2 \end{bmatrix},$$
and $f_2 \colon R^2 \to R^2$ the shear in the y-direction defined by
$$f_2\left(\begin{bmatrix} u_1 \\ u_2 \end{bmatrix}\right) = \begin{bmatrix} u_1 \\ 2u_1 + u_2 \end{bmatrix}.$$
The function $f_2 \circ f_1 \colon R^2 \to R^2$ defined by
$$(f_2 \circ f_1)\left(\begin{bmatrix} u_1 \\ u_2 \end{bmatrix}\right) = f_2\left(f_1\left(\begin{bmatrix} u_1 \\ u_2 \end{bmatrix}\right)\right)$$
is a matrix transformation (see Exercise T.1).

(a) Find a matrix associated with $f_2 \circ f_1$.

(b) Let R be the rectangle with vertices (1, 1), (2, 1), (1, 2), and (2, 2). Sketch the image of R under f_1, and under $f_2 \circ f_1$.

3. Consider a communications network among six individuals with adjacency matrix
$$\begin{array}{c} \\ P_1 \\ P_2 \\ P_3 \\ P_4 \\ P_5 \\ P_6 \end{array}\begin{array}{c} \begin{array}{cccccc} P_1 & P_2 & P_3 & P_4 & P_5 & P_6 \end{array} \\ \begin{bmatrix} 0 & 1 & 0 & 0 & 0 & 1 \\ 0 & 0 & 1 & 1 & 0 & 0 \\ 0 & 0 & 0 & 1 & 0 & 0 \\ 1 & 0 & 1 & 0 & 1 & 0 \\ 0 & 0 & 0 & 0 & 0 & 1 \\ 0 & 0 & 0 & 1 & 0 & 0 \end{bmatrix} \end{array}.$$
In how many ways does P_1 have access to P_3 through two individuals?

4. Determine the unknown currents in the following circuit.

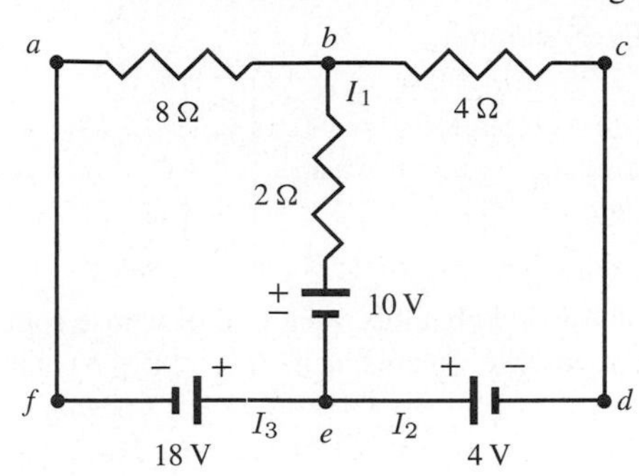

5. A marketing research organization has determined the following behavior of the average student at a certain college. If the student plays a video game on a given day, there is a probability of 0.2 that he or she will play a game on the following day, whereas if the student has not played on a given day, there is a probability of 0.6 that he or she will play on the following day.
   (a) Write the transition matrix for the Markov process.
   (b) If the average student plays a video game on Monday, what is the probability that he or she will play on Friday of the same week?
   (c) In the long run, what is the probability that the average student will play a video game?

6. Consider an isolated village in a remote part of Australia consisting of three households: a rancher (R) who raises all the cattle needed and nothing else; a dairy farmer (D) who produces all the dairy products needed and nothing else; and a vegetable farmer (V) who produces all the necessary vegetables and nothing else. Suppose that the units are selected so that each person produces one unit of each commodity. Suppose that during the year the portion of each commodity that is consumed by each individual is given in Table 2.3. Let p_1, p_2, and p_3 be the prices per unit of cattle, dairy produce, and vegetable produce, respectively. Suppose that everyone pays the same price for a commodity. What prices p_1, p_2, and p_3 should be assigned to the commodities so that we have a state of equilibrium?

Table 2.3

Goods Consumed by:	Goods Produced by:		
	R	D	V
R	$\frac{1}{2}$	$\frac{3}{8}$	$\frac{1}{3}$
D	$\frac{1}{4}$	$\frac{1}{4}$	$\frac{1}{3}$
V	$\frac{1}{4}$	$\frac{3}{8}$	$\frac{1}{3}$

7. Let $\mathbf{v} = \begin{bmatrix} 2 & 4 & 5 & 1 \end{bmatrix}^T$ represent a sample of a function f at the equispaced points $x_1 = -4$, $x_2 = -2$, $x_3 = 0$, $x_4 = 2$.
   (a) Determine the final average and detail coefficients by computing $A_2 A_1 \mathbf{v}$.
   (b) Using a threshold of $\varepsilon = 1$ determine the compressed data and then compute the wavelet.
   (c) On the same set of coordinate axes graph the original data points and the wavelet approximation connecting successive points by straight line segments.

Theoretical Exercise

T.1. Let $f_1 \colon R^2 \to R^2$ and $f_2 \colon R^2 \to R^2$ be matrix transformations. Show that $f_1 \circ f_2 \colon R^2 \to R^2$ defined by $(f_1 \circ f_2)(\mathbf{u}) = f_1(f_2(\mathbf{u}))$ is a matrix transformation.

Chapter Test

1. Let e be the function from B^2 to B^4 given by

$$e(b_1 b_2) = b_1 b_2 b_2 b_1.$$

   (a) Is e one-to-one? If not, determine two different vectors $\mathbf{b}$ and $\mathbf{c}$ in B^2 such that $e(\mathbf{b}) = e(\mathbf{c})$.
   (b) Determine the matrix A so that e can be written as a matrix transformation in the form

$$e(b_1 b_2) = A \begin{bmatrix} b_1 \\ b_2 \end{bmatrix} = \begin{bmatrix} b_1 \\ b_2 \\ b_2 \\ b_1 \end{bmatrix}.$$

   (c) Show that the weight of each code word is even.

2. Let T be the triangle shown in Figure 2.15(a).
   (a) Define the matrix transformation $f \colon R^2 \to R^2$ by $f(\mathbf{u}) = A\mathbf{u}$, where $A = \begin{bmatrix} -1 & 0 \\ 0 & 1 \end{bmatrix}$. Determine the image of T under f.
   (b) Describe the geometric operation that was applied to T under f.
   (c) What is the result of applying f to T twice; that is, $f(f(T))$?

3. Let $f \colon R^2 \to R^2$ be the matrix transformation defined by $f(\mathbf{u}) = A\mathbf{u}$, where $A = \begin{bmatrix} 1 & 2 \\ a & b \end{bmatrix}$. Determine a and b so that the image of the rectangle shown in Figure 2.18(a) is a portion of the straight line $y = x$.

4. Determine a clique, if there is one, for the digraph with the following adjacency matrix:

$$\begin{bmatrix} 0 & 1 & 1 & 0 & 1 & 1 \\ 0 & 0 & 0 & 1 & 0 & 1 \\ 1 & 0 & 0 & 0 & 1 & 1 \\ 0 & 1 & 0 & 0 & 1 & 0 \\ 1 & 0 & 1 & 1 & 0 & 0 \\ 1 & 1 & 1 & 0 & 0 & 0 \end{bmatrix}.$$

5. Determine the unknowns in the following circuit.

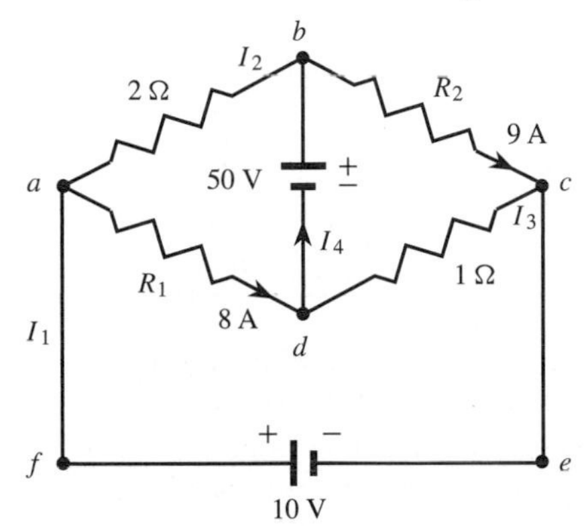

6. A political analyst who has been studying the voting patterns of nonregistered voters in a certain city has come to the following conclusions. If on a given year, a voter votes Republican, then the probability that on the following year he or she will again vote Republican is 0.4. If on a given year, a voter votes Democratic, then the probability that on the following year he or she will again vote Democratic is 0.5.

(a) Write the transition matrix for the Markov process.

(b) If a voter votes Republican in 1996, what is the probability that he or she will vote Republican in 2000?

(c) In the long run, what is the probability of a voter voting Democratic?

7. Consider a city that has three basic industries: a steel plant, a coal mine, and a railroad. To produce \$1 of steel, the steel plant uses \$0.50 of steel, \$0.30 of coal, and \$0.10 of transportation. To mine \$1 of coal, the coal mine uses \$0.10 of steel, \$0.20 of coal, and \$0.30 of transportation. To provide \$1 of transportation, the railroad uses \$0.10 of steel, \$0.40 of coal, and \$0.05 of transportation. Suppose that during the month of December there is an outside demand of 2 million dollars for steel, 1.5 million dollars for coal, and 0.5 million dollars for transportation. How much should each industry produce to satisfy the demands?

8. Let $\mathbf{v} = \begin{bmatrix} 0 & -3 & 0 & 1 \end{bmatrix}^T$ represent a sample of a function f at the equispaced points $x_1 = -6$, $x_2 = -3$, $x_3 = 0$, $x_4 = 3$.

(a) Determine the final average and detail coefficients by computing $A_2 A_1 \mathbf{v}$.

(b) Using a threshold of $\varepsilon = 1$ determine the compressed data and then compute the wavelet.

CHAPTER 3

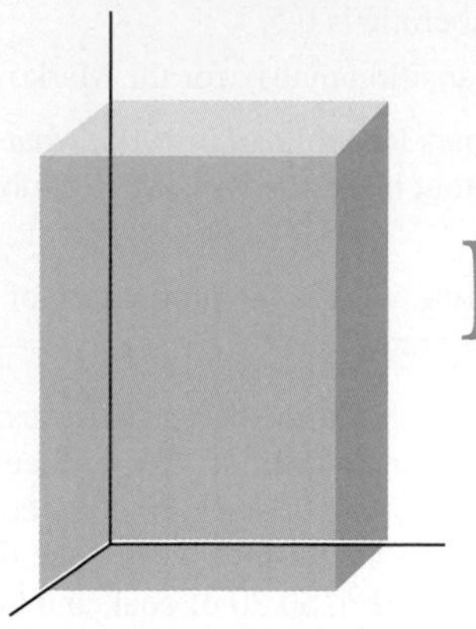

DETERMINANTS

3.1 DEFINITION AND PROPERTIES

In this section we define the notion of a determinant and study some of its properties. Determinants arose first in the solution of linear systems. Although the method given in Chapter 1 for solving such systems is much more efficient than those involving determinants, determinants are useful in other aspects of linear algebra; some of these areas will be considered in Chapter 8. First, we deal rather briefly with permutations, which are used in our definition of determinant. In this chapter, by matrix we mean a square matrix.

DEFINITION Let $S = \{1, 2, \ldots, n\}$ be the set of integers from 1 to n, arranged in ascending order. A rearrangement $j_1 j_2 \cdots j_n$ of the elements of S is called a **permutation** of S.

To illustrate the preceding definition, let $S = \{1, 2, 3, 4\}$. Then 4132 is a permutation of S. It corresponds to the function $f : S \to S$ defined by

$$\begin{aligned} f(1) &= 4 \\ f(2) &= 1 \\ f(3) &= 3 \\ f(4) &= 2. \end{aligned}$$

We can put any one of the n elements of S in first position, any one of the remaining $n - 1$ elements in second position, any one of the remaining $n - 2$ elements in third position, and so on, until the nth position can only be filled by the last remaining element. Thus there are

$$n(n-1)(n-2)\cdots 2 \cdot 1 \tag{1}$$

permutations of S; we denote the set of all permutations of S by S_n.

The expression in Equation (1) is denoted

$$n!, \quad \textbf{n factorial}.$$

We have

$$
\begin{aligned}
1! &= 1 \\
2! &= 2 \cdot 1 = 2 \\
3! &= 3 \cdot 2 \cdot 1 = 6 \\
4! &= 4 \cdot 3 \cdot 2 \cdot 1 = 24 \\
5! &= 5 \cdot 4 \cdot 3 \cdot 2 \cdot 1 = 120 \\
6! &= 6 \cdot 5 \cdot 4 \cdot 3 \cdot 2 \cdot 1 = 720 \\
7! &= 7 \cdot 6 \cdot 5 \cdot 4 \cdot 3 \cdot 2 \cdot 1 = 5040 \\
8! &= 8 \cdot 7 \cdot 6 \cdot 5 \cdot 4 \cdot 3 \cdot 2 \cdot 1 = 40{,}320 \\
9! &= 9 \cdot 8 \cdot 7 \cdot 6 \cdot 5 \cdot 4 \cdot 3 \cdot 2 \cdot 1 = 362{,}880.
\end{aligned}
$$

EXAMPLE 1

S_1 consists of only $1! = 1$ permutation of the set $\{1\}$, namely, 1; S_2 consists of $2! = 2 \cdot 1 = 2$ permutations of the set $\{1, 2\}$, namely, 12 and 21; S_3 consists of $3! = 3 \cdot 2 \cdot 1 = 6$ permutations of the set $\{1, 2, 3\}$, namely, 123, 231, 312, 132, 213, and 321. ■

A permutation $j_1 j_2 \cdots j_n$ of $S = \{1, 2, \ldots, n\}$ is said to have an **inversion** if a larger integer j_r precedes a smaller one j_s. A permutation is called **even** or **odd** according to whether the total number of inversions in it is even or odd. Thus, the permutation 4132 of $S = \{1, 2, 3, 4\}$ has four inversions: 4 before 1, 4 before 3, 4 before 2, and 3 before 2. It is then an even permutation.

If $n \geq 2$, it can be shown that S_n has $n!/2$ even permutations and an equal number of odd permutations.

EXAMPLE 2

In S_2, the permutation 12 is even, since it has no inversions; the permutation 21 is odd, since it has one inversion. ■

EXAMPLE 3

The even permutations in S_3 are 123 (no inversions), 231 (two inversions: 21 and 31); and 312 (two inversions: 31 and 32). The odd permutations in S_3 are 132 (one inversion: 32); 213 (one inversion: 21); and 321 (three inversions: 32, 31, and 21). ■

DEFINITION

Let $A = \left[a_{ij}\right]$ be an $n \times n$ matrix. We define the **determinant** of A (written $\det(A)$ or $|A|$) by

$$\det(A) = |A| = \sum (\pm) a_{1j_1} a_{2j_2} \cdots a_{nj_n}, \tag{2}$$

where the summation ranges over all permutations $j_1 j_2 \cdots j_n$ of the set $S = \{1, 2, \ldots, n\}$. The sign is taken as $+$ or $-$ according to whether the permutation $j_1 j_2 \cdots j_n$ is even or odd.

In each term $(\pm) a_{1j_1} a_{2j_2} \cdots a_{nj_n}$ of $\det(A)$, the row subscripts are in their natural order, whereas the column subscripts are in the order $j_1 j_2 \cdots j_n$. Since the permutation $j_1 j_2 \cdots j_n$ is merely a rearrangement of the numbers from 1 to n, it has no repeats. Thus each term in $\det(A)$ is a product of n elements of A each with its appropriate sign, with exactly one element from each row and exactly one element from each column. Since we sum over all the permutations of the set, $S = \{1, 2, \ldots, n\}$, $\det(A)$ has $n!$ terms in the sum.

EXAMPLE 4 If $A = \begin{bmatrix} a_{11} \end{bmatrix}$ is a 1×1 matrix, then S_1 has only one permutation in it, the identity permutation 1, which is even. Thus $\det(A) = a_{11}$. ■

EXAMPLE 5 If

$$A = \begin{bmatrix} a_{11} & a_{12} \\ a_{21} & a_{22} \end{bmatrix}$$

is a 2×2 matrix, then to obtain $\det(A)$ we write down the terms

$$a_{1\_}a_{2\_} \quad \text{and} \quad a_{1\_}a_{2\_},$$

and fill in the blanks with all possible elements of S_2; the subscripts become 12 and 21. Since 12 is an even permutation, the term $a_{11}a_{22}$ has a $+$ sign associated with it; since 21 is an odd permutation, the term $a_{12}a_{21}$ has a $-$ sign associated with it. Hence

$$\det(A) = a_{11}a_{22} - a_{12}a_{21}.$$

We can also obtain $\det(A)$ by forming the product of the entries on the line from left to right in the following diagram and subtracting from this number the product of the entries on the line from right to left:

$$\begin{matrix} a_{11} & & a_{12} \\ & \times & \\ a_{21} & & a_{22} \end{matrix}.$$

Thus, if

$$A = \begin{bmatrix} 2 & -3 \\ 4 & 5 \end{bmatrix},$$

then $\det(A) = (2)(5) - (-3)(4) = 22$. ■

EXAMPLE 6 If

$$A = \begin{bmatrix} a_{11} & a_{12} & a_{13} \\ a_{21} & a_{22} & a_{23} \\ a_{31} & a_{32} & a_{33} \end{bmatrix},$$

then, to compute $\det(A)$, we write down the six terms

$$a_{1\_}a_{2\_}a_{3\_}, \quad a_{1\_}a_{2\_}a_{3\_}, \quad a_{1\_}a_{2\_}a_{3\_}, \quad a_{1\_}a_{2\_}a_{3\_},$$
$$a_{1\_}a_{2\_}a_{3\_}, \quad \text{and} \quad a_{1\_}a_{2\_}a_{3\_}.$$

All the elements of S_3 are used to fill in the blanks, and if we prefix each term by $+$ or by $-$ according to whether the permutation used is even or odd, we find that

$$\begin{aligned} \det(A) = a_{11}a_{22}a_{33} + a_{12}a_{23}a_{31} + a_{13}a_{21}a_{32} - a_{11}a_{23}a_{32} \\ - a_{12}a_{21}a_{33} - a_{13}a_{22}a_{31}. \end{aligned} \tag{3}$$

We can also obtain $\det(A)$ as follows. Repeat the first and second columns of A as shown below. Form the sum of the products of the entries on the lines from left to right, and subtract from this number the products of the entries on the lines from right to left (verify):

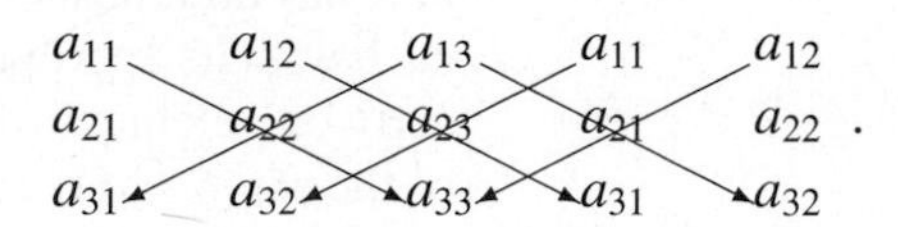

. ■

Warning It should be emphasized that the methods given for evaluating $\det(A)$ in Examples 5 and 6 do not apply for $n \geq 4$.

EXAMPLE 7 Let

$$A = \begin{bmatrix} 1 & 2 & 3 \\ 2 & 1 & 3 \\ 3 & 1 & 2 \end{bmatrix}.$$

Evaluate $\det(A)$.

Solution Substituting in (3), we find that

$$\det(A) = (1)(1)(2) + (2)(3)(3) + (3)(2)(1) - (1)(3)(1) - (2)(2)(2) - (3)(1)(3) = 6.$$

We could obtain the same result by using the easy method given previously (verify). ■

It may already have struck the reader that this is an extremely tedious way of computing determinants for a sizable value of n. In fact, $10! = 3.6288 \times 10^6$ and $20! = 2.4329 \times 10^{18}$ are enormous numbers. We shall soon develop a number of properties satisfied by determinants, which will greatly reduce the computational effort.

Permutations are studied to some depth in abstract algebra courses and in courses dealing with group theory. We shall not make use of permutations in our methods for computing determinants. We require the following property of permutations. If we interchange two numbers in the permutation $j_1 j_2 \cdots j_n$, then the number of inversions is either increased or decreased by an odd number (Exercise T.1).

EXAMPLE 8 The number of inversions in the permutation 54132 is 8. The number of inversions in the permutation 52134 is 5. The permutation 52134 was obtained from 54132 by interchanging 2 and 4. The number of inversions differs by 3, an odd number. ■

PROPERTIES OF DETERMINANTS

THEOREM 3.1 *The determinants of a matrix and its transpose are equal, that is,* $\det(A^T) = \det(A)$.

Proof Let $A = \left[a_{ij}\right]$ and let $A^T = \left[b_{ij}\right]$, where $b_{ij} = a_{ji}$ $(1 \leq i \leq n,\ 1 \leq j \leq n)$. Then from (2) we have

$$\det(A^T) = \sum (\pm) b_{1j_1} b_{2j_2} \cdots b_{nj_n} = \sum (\pm) a_{j_1 1} a_{j_2 2} \cdots a_{j_n n}. \tag{4}$$

We can now rearrange the factors in the term $a_{j_1 1} a_{j_2 2} \cdots a_{j_n n}$ so that the row indices are in their natural order. Thus

$$b_{1j_1} b_{2j_2} \cdots b_{nj_n} = a_{j_1 1} a_{j_2 2} \cdots a_{j_n n} = a_{1k_1} a_{2k_2} \cdots a_{nk_n}.$$

It can be shown, by the properties of permutations discussed in an abstract algebra course,* that the permutation $k_1 k_2 \cdots k_n$, which determines the sign

*See J. Fraleigh, *A First Course in Abstract Algebra*, 7th ed., Reading, Mass.: Addison-Wesley Publishing Company, Inc., 2003; and J. Gallian, *Contemporary Abstract Algebra*, 5th ed., Mass.: Houghton Mifflin, 2002.

associated with $a_{1k_1}a_{2k_2}\cdots a_{nk_n}$, and the permutation $j_1j_2\cdots j_n$, which determines the sign associated with $a_{1j_1}a_{2j_2}\cdots a_{nj_n}$, are both even or both odd. As an example,

$$b_{13}b_{24}b_{35}b_{41}b_{52} = a_{31}a_{42}a_{53}a_{14}a_{25} = a_{14}a_{25}a_{31}a_{42}a_{53};$$

the number of inversions in the permutation 45123 is 6 and the number of inversions in the permutation 34512 is also 6. Since the terms and corresponding signs in (2) and (4) agree, we conclude that $\det(A) = \det(A^T)$. ■

EXAMPLE 9 Let A be the matrix of Example 7. Then

$$A^T = \begin{bmatrix} 1 & 2 & 3 \\ 2 & 1 & 1 \\ 3 & 3 & 2 \end{bmatrix}.$$

Substituting in (3), we find that

$$\begin{aligned}\det(A^T) &= (1)(1)(2) + (2)(1)(3) + (3)(2)(3) \\ &\quad - (1)(1)(3) - (2)(2)(2) - (3)(1)(3) \\ &= 6 = \det(A).\end{aligned}$$

■

Theorem 3.1 will enable us to replace "row" by "column" in many of the additional properties of determinants; we see how to do this in the following theorem.

THEOREM 3.2 *If matrix B results from matrix A by interchanging two rows (columns) of A, then* $\det(B) = -\det(A)$.

Proof Suppose that B arises from A by interchanging rows r and s of A, say $r < s$. Then we have $b_{rj} = a_{sj}$, $b_{sj} = a_{rj}$, and $b_{ij} = a_{ij}$ for $i \neq r$, $i \neq s$. Now

$$\begin{aligned}\det(B) &= \sum(\pm)b_{1j_1}b_{2j_2}\cdots b_{rj_r}\cdots b_{sj_s}\cdots b_{nj_n} \\ &= \sum(\pm)a_{1j_1}a_{2j_2}\cdots a_{sj_r}\cdots a_{rj_s}\cdots a_{nj_n} \\ &= \sum(\pm)a_{1j_1}a_{2j_2}\cdots a_{rj_s}\cdots a_{sj_r}\cdots a_{nj_n}.\end{aligned}$$

The permutation $j_1j_2\cdots j_s\cdots j_r\cdots j_n$ results from the permutation $j_1j_2\cdots j_r\cdots j_s\cdots j_n$ by an interchange of two numbers; the number of inversions in the former differs by an odd number from the number of inversions in the latter (see Exercise T.1). This means that the sign of each term in $\det(B)$ is the negative of the sign of the corresponding term in $\det(A)$. Hence $\det(B) = -\det(A)$.

Now suppose that B is obtained from A by interchanging two columns of A. Then B^T is obtained from A^T by interchanging two rows of A^T. So $\det(B^T) = -\det(A^T)$, but $\det(B^T) = \det(B)$ and $\det(A^T) = \det(A)$. Hence $\det(B) = -\det(A)$. ■

In the results to follow, proofs will be given only for the rows of A; the proofs of the corresponding column case proceed as at the end of the proof of Theorem 3.2.

EXAMPLE 10 We have

$$\begin{vmatrix} 2 & -1 \\ 3 & 2 \end{vmatrix} = 7 \quad\text{and}\quad \begin{vmatrix} 3 & 2 \\ 2 & -1 \end{vmatrix} = -7.$$

■

THEOREM 3.3 *If two rows (columns) of A are equal, then* $\det(A) = 0$.

Proof Suppose that rows r and s of A are equal. Interchange rows r and s of A to obtain a matrix B. Then $\det(B) = -\det(A)$. On the other hand, $B = A$, so $\det(B) = \det(A)$. Thus $\det(A) = -\det(A)$, so $\det(A) = 0$. ■

EXAMPLE 11 Using Theorem 3.3, it follows that

$$\begin{vmatrix} 1 & 2 & 3 \\ -1 & 0 & 7 \\ 1 & 2 & 3 \end{vmatrix} = 0.$$

■

THEOREM 3.4 *If a row (column) of A consists entirely of zeros, then* $\det(A) = 0$.

Proof Let the rth row of A consist entirely of zeros. Since each term in the definition for the determinant of A contains a factor from the rth row, each term in $\det(A)$ is zero. Hence $\det(A) = 0$. ■

EXAMPLE 12 Using Theorem 3.4, it follows that

$$\begin{vmatrix} 1 & 2 & 3 \\ 4 & 5 & 6 \\ 0 & 0 & 0 \end{vmatrix} = 0.$$

■

THEOREM 3.5 *If B is obtained from A by multiplying a row (column) of A by a real number c, then* $\det(B) = c\det(A)$.

Proof Suppose that the rth row of $A = \left[a_{ij}\right]$ is multiplied by c to obtain $B = \left[b_{ij}\right]$. Then $b_{ij} = a_{ij}$ if $i \neq r$ and $b_{rj} = ca_{rj}$. We obtain $\det(B)$ from Equation (2) as

$$\begin{aligned} \det(B) &= \sum (\pm) b_{1j_1} b_{2j_2} \cdots b_{rj_r} \cdots b_{nj_n} \\ &= \sum (\pm) a_{1j_1} a_{2j_2} \cdots (ca_{rj_r}) \cdots a_{nj_n} \\ &= c\left(\sum (\pm) a_{1j_1} a_{2j_2} \cdots a_{rj_r} \cdots a_{nj_n}\right) = c\det(A). \end{aligned}$$

■

We can now use Theorem 3.5 to simplify the computation of $\det(A)$ by factoring out common factors from rows and columns of A.

EXAMPLE 13 We have

$$\begin{vmatrix} 2 & 6 \\ 1 & 12 \end{vmatrix} = 2\begin{vmatrix} 1 & 3 \\ 1 & 12 \end{vmatrix} = (2)(3)\begin{vmatrix} 1 & 1 \\ 1 & 4 \end{vmatrix} = 6(4-1) = 18.$$

■

EXAMPLE 14 We have

$$\begin{vmatrix} 1 & 2 & 3 \\ 1 & 5 & 3 \\ 2 & 8 & 6 \end{vmatrix} = 2\begin{vmatrix} 1 & 2 & 3 \\ 1 & 5 & 3 \\ 1 & 4 & 3 \end{vmatrix} = (2)(3)\begin{vmatrix} 1 & 2 & 1 \\ 1 & 5 & 1 \\ 1 & 4 & 1 \end{vmatrix} = (2)(3)(0) = 0.$$

Here we first factored out 2 from the third row, then 3 from the third column, and then used Theorem 3.3, since the first and third columns are equal. ■

THEOREM 3.6 *If $B = [b_{ij}]$ is obtained from $A = [a_{ij}]$ by adding to each element of the* rth *row (column) of A a constant c times the corresponding element of the* sth *row (column)* $r \neq s$ *of A, then* $\det(B) = \det(A)$.

Proof We prove the theorem for rows. We have $b_{ij} = a_{ij}$ for $i \neq r$, and $b_{rj} = a_{rj} + ca_{sj}$, $r \neq s$, say $r < s$. Then

$$\begin{aligned}\det(B) &= \sum(\pm)b_{1j_1}b_{2j_2}\cdots b_{rj_r}\cdots b_{nj_n}\\ &= \sum(\pm)a_{1j_1}a_{2j_2}\cdots(a_{rj_r} + ca_{sj_r})\cdots a_{sj_s}\cdots a_{nj_n}\\ &= \sum(\pm)a_{1j_1}a_{2j_2}\cdots a_{rj_r}\cdots a_{sj_s}\cdots a_{nj_n}\\ &\quad + \sum(\pm)a_{1j_1}a_{2j_2}\cdots(ca_{sj_r})\cdots a_{sj_s}\cdots a_{nj_n}.\end{aligned}$$

The first sum in this last expression is $\det(A)$; the second sum can be written as

$$c\left[\sum(\pm)a_{1j_1}a_{2j_2}\cdots a_{sj_r}\cdots a_{sj_s}\cdots a_{nj_n}\right].$$

Note that

$$\begin{aligned}&\sum(\pm)a_{1j_1}a_{2j_2}\cdots a_{sj_r}\cdots a_{sj_s}\cdots a_{nj_n}\\ &= \begin{vmatrix} a_{11} & a_{12} & \cdots & a_{1n}\\ a_{21} & a_{22} & \cdots & a_{2n}\\ \vdots & \vdots & & \vdots\\ a_{s1} & a_{s2} & \cdots & a_{sn}\\ \vdots & \vdots & & \vdots\\ a_{s1} & a_{s2} & \cdots & a_{sn}\\ \vdots & \vdots & & \vdots\\ a_{n1} & a_{n2} & \cdots & a_{nn}\end{vmatrix} \begin{matrix} \\ \\ \\ \leftarrow r\text{th row}\\ \\ \leftarrow s\text{th row}\\ \\ \\ \end{matrix}\\ &= 0,\end{aligned}$$

because there are two equal rows. Hence $\det(B) = \det(A) + 0 = \det(A)$. ■

EXAMPLE 15 We have

$$\begin{vmatrix} 1 & 2 & 3\\ 2 & -1 & 3\\ 1 & 0 & 1\end{vmatrix} = \begin{vmatrix} 5 & 0 & 9\\ 2 & -1 & 3\\ 1 & 0 & 1\end{vmatrix},$$

obtained by adding twice the second row to its first row. By applying the definition of determinant to the second determinant, we can see that both have the value 4. ■

THEOREM 3.7 *If a matrix $A = [a_{ij}]$ is upper (lower) triangular (see Exercise T.5, Section 1.2), then*

$$\det(A) = a_{11}a_{22}\cdots a_{nn};$$

that is, the determinant of a triangular matrix is the product of the elements on the main diagonal.

Proof Let $A = \left[a_{ij}\right]$ be upper triangular (that is, $a_{ij} = 0$ for $i > j$). Then a term $a_{1j_1}a_{2j_2}\cdots a_{nj_n}$ in the expression for $\det(A)$ can be nonzero only for $1 \leq j_1$, $2 \leq j_2, \ldots, n \leq j_n$. Now $j_1 j_2 \cdots j_n$ must be a permutation, or rearrangement, of $\{1, 2, \ldots, n\}$. Hence we must have $j_n = n$, $j_{n-1} = n - 1, \ldots, j_2 = 2$, $j_1 = 1$. Thus the only term of $\det(A)$ that can be nonzero is the product of the elements on the main diagonal of A. Since the permutation $12 \cdots n$ has no inversions, the sign associated with it is $+$. Therefore, $\det(A) = a_{11}a_{22}\cdots a_{nn}$.

We leave the proof of the lower triangular case to the reader (Exercise T.2). ■

COROLLARY 3.1 *The determinant of a diagonal matrix is the product of the entries on its main diagonal.*

Proof Exercise T.17. ■

EXAMPLE 16 Let

$$A = \begin{bmatrix} 2 & 3 & 4 \\ 0 & -4 & 5 \\ 0 & 0 & 3 \end{bmatrix}, \quad B = \begin{bmatrix} 3 & 0 & 0 \\ 2 & 5 & 0 \\ 6 & -8 & -4 \end{bmatrix}, \quad C = \begin{bmatrix} -5 & 0 & 0 \\ 0 & 4 & 0 \\ 0 & 0 & -6 \end{bmatrix}.$$

Compute $\det(A)$, $\det(B)$, $\det(C)$.

Solution By Theorem 3.7, $\det(A) = -24$, $\det(B) = -60$. By Corollary 3.1, $\det(C) = 120$. ■

We now introduce the following notation for elementary row and elementary column operations on matrices and determinants.

- Interchange rows (columns) i and j:

$$\mathbf{r}_i \leftrightarrow \mathbf{r}_j \quad (\mathbf{c}_i \leftrightarrow \mathbf{c}_j).$$

- Replace row (column) i by a nonzero value k times row (column) i:

$$k\mathbf{r}_i \rightarrow \mathbf{r}_i \quad (k\mathbf{c}_i \rightarrow \mathbf{c}_i).$$

- Replace row (column) j by a nonzero value k times row (column) i + row (column) j:

$$k\mathbf{r}_i + \mathbf{r}_j \rightarrow \mathbf{r}_j \quad (k\mathbf{c}_i + \mathbf{c}_j \rightarrow \mathbf{c}_j).$$

Using this notation, it is easy to keep track of the elementary row and column operations performed on a matrix. For example, we indicate that we have interchanged the ith and jth rows of A as $A_{\mathbf{r}_i \leftrightarrow \mathbf{r}_j}$. We proceed similarly for column operations.

We can now interpret Theorems 3.2, 3.5, and 3.6 in terms of this notation as follows.

$$\begin{aligned} \det(A_{\mathbf{r}_i \leftrightarrow \mathbf{r}_j}) &= -\det(A), \quad i \neq j \\ \det(A_{k\mathbf{r}_i \rightarrow \mathbf{r}_i}) &= k\det(A) \\ \det(A_{k\mathbf{r}_i + \mathbf{r}_j \rightarrow \mathbf{r}_j}) &= \det(A), \quad i \neq j. \end{aligned}$$

It is convenient to rewrite these properties in terms of $\det(A)$:

$$\det(A) = -\det(A_{\mathbf{r}_i \leftrightarrow \mathbf{r}_j}), \qquad i \neq j$$

$$\det(A) = \frac{1}{k}\det(A_{k\mathbf{r}_i \to \mathbf{r}_i}), \qquad k \neq 0$$

$$\det(A) = \det(A_{k\mathbf{r}_i + \mathbf{r}_j \to \mathbf{r}_j}), \qquad i \neq j.$$

We proceed similarly for column operations.

Theorems 3.2, 3.5, 3.6, and 3.7 are very useful in the evaluation of $\det(A)$. What we do is transform A by means of our elementary row operations to a triangular matrix. Of course, we must keep track of how the determinant of the resulting matrices changes as we perform the elementary row operations.

EXAMPLE 17 Let $A = \begin{bmatrix} 4 & 3 & 2 \\ 3 & -2 & 5 \\ 2 & 4 & 6 \end{bmatrix}$. Compute $\det(A)$.

Solution We have

$$\det(A) = 2\det(A_{\frac{1}{2}\mathbf{r}_3 \to \mathbf{r}_3}) \qquad \text{Multiply row 3 by } \frac{1}{2}.$$

$$= 2\det\left(\begin{bmatrix} 4 & 3 & 2 \\ 3 & -2 & 5 \\ 1 & 2 & 3 \end{bmatrix}\right)$$

$$= 2\det\left(\begin{bmatrix} 4 & 3 & 2 \\ 3 & -2 & 5 \\ 1 & 2 & 3 \end{bmatrix}_{\mathbf{r}_1 \leftrightarrow \mathbf{r}_3}\right) \qquad \text{Interchange rows 1 and 3.}$$

$$= (-1)2\det\left(\begin{bmatrix} 1 & 2 & 3 \\ 3 & -2 & 5 \\ 4 & 3 & 2 \end{bmatrix}\right)$$

$$= -2\det\left(\begin{bmatrix} 1 & 2 & 3 \\ 3 & -2 & 5 \\ 4 & 3 & 2 \end{bmatrix}_{\substack{-3\mathbf{r}_1 + \mathbf{r}_2 \to \mathbf{r}_2 \\ -4\mathbf{r}_1 + \mathbf{r}_3 \to \mathbf{r}_3}}\right) \qquad \text{Zero out below the (1, 1) entry.}$$

$$= -2\det\left(\begin{bmatrix} 1 & 2 & 3 \\ 0 & -8 & -4 \\ 0 & -5 & -10 \end{bmatrix}\right)$$

$$= -2\det\left(\begin{bmatrix} 1 & 2 & 3 \\ 0 & -8 & -4 \\ 0 & -5 & -10 \end{bmatrix}_{-\frac{5}{8}\mathbf{r}_2 + \mathbf{r}_3 \to \mathbf{r}_3}\right) \qquad \text{Zero out below the (2, 2) entry.}$$

$$= -2\det\left(\begin{bmatrix} 1 & 2 & 3 \\ 0 & -8 & -4 \\ 0 & 0 & -\frac{30}{4} \end{bmatrix}\right).$$

Next we compute the determinant of the upper triangular matrix.

$$\det(A) = -2(1)(-8)\left(-\frac{30}{4}\right) = -120 \qquad \text{By Theorem 3.7.}$$

The operations chosen are not the most efficient, but we do avoid fractions during the first few steps. ■

Remark The method used to compute a determinant in Example 17 will be referred to as the **computation via reduction to triangular form**.

We shall omit the proof of the following important theorem.

THEOREM 3.8 *The determinant of a product of two matrices is the product of their determinants; that is,*

$$\det(AB) = \det(A)\det(B).$$

■

EXAMPLE 18 Let

$$A = \begin{bmatrix} 1 & 2 \\ 3 & 4 \end{bmatrix} \quad \text{and} \quad B = \begin{bmatrix} 2 & -1 \\ 1 & 2 \end{bmatrix}.$$

Then

$$|A| = -2 \quad \text{and} \quad |B| = 5.$$

Also,

$$AB = \begin{bmatrix} 4 & 3 \\ 10 & 5 \end{bmatrix}$$

and

$$|AB| = -10 = |A||B|.$$

■

Remark In Example 18 we also have (verify)

$$BA = \begin{bmatrix} -1 & 0 \\ 7 & 10 \end{bmatrix},$$

so $AB \neq BA$. However, $|BA| = |B|\,|A| = -10 = |AB|$.

As an immediate consequence of Theorem 3.8, we can readily compute $\det(A^{-1})$ from $\det(A)$, as the following corollary shows.

COROLLARY 3.2 *If A is nonsingular, then* $\det(A) \neq 0$ *and*

$$\det(A^{-1}) = \frac{1}{\det(A)}.$$

Proof Exercise T.4. ■

EXAMPLE 19 Let

$$A = \begin{bmatrix} 1 & 2 \\ 3 & 4 \end{bmatrix}.$$

Then $\det(A) = -2$ and

$$A^{-1} = \begin{bmatrix} -2 & 1 \\ \frac{3}{2} & -\frac{1}{2} \end{bmatrix}.$$

Now,

$$\det(A^{-1}) = -\frac{1}{2} = \frac{1}{\det(A)}.$$

■

THE DETERMINANT OF BIT MATRICES (OPTIONAL)

The properties and techniques for the determinant developed in this section apply to bit matrices, where computations are carried out using binary arithmetic.

EXAMPLE 20 The determinant of the 2×2 bit matrix

$$A = \begin{bmatrix} 1 & 0 \\ 1 & 1 \end{bmatrix}$$

computed by using the technique developed in Example 5 is

$$\det(A) = (1)(1) - (1)(0) = 1.$$ ■

EXAMPLE 21 The determinant of the 3×3 bit matrix

$$A = \begin{bmatrix} 1 & 0 & 1 \\ 1 & 1 & 0 \\ 0 & 1 & 1 \end{bmatrix}$$

computed by using the technique developed in Example 6 is

$$\begin{aligned} \det(A) &= (1)(1)(1) + (0)(0)(0) + (1)(1)(1) \\ &\quad - (1)(0)(1) - (1)(0)(1) - (0)(1)(1) \\ &= 1 + 0 + 1 - 0 - 0 - 0 = 1 + 1 = 0. \end{aligned}$$ ■

EXAMPLE 22 Use the computation via reduction to triangular form to evaluate the determinant of the bit matrix

$$A = \begin{bmatrix} 0 & 1 & 1 \\ 1 & 1 & 0 \\ 1 & 0 & 1 \end{bmatrix}.$$

Solution

$$\begin{vmatrix} 0 & 1 & 1 \\ 1 & 1 & 0 \\ 1 & 0 & 1 \end{vmatrix}_{\mathbf{r}_1 \leftrightarrow \mathbf{r}_2} = (-1) \begin{vmatrix} 1 & 1 & 0 \\ 0 & 1 & 1 \\ 1 & 0 & 1 \end{vmatrix}_{\mathbf{r}_1 + \mathbf{r}_3 \to \mathbf{r}_3} = (-1) \begin{vmatrix} 1 & 1 & 0 \\ 0 & 1 & 1 \\ 0 & 1 & 1 \end{vmatrix}$$

By Theorem 3.3, $\det(A) = 0$. ■

Key Terms

Permutation
n factorial
Inversion
Even permutation
Odd permutation
Determinant
Computation via reduction to triangular form

3.1 Exercises

1. Find the number of inversions in each of the following permutations of $S = \{1, 2, 3, 4, 5\}$.

(a) 52134 (b) 45213 (c) 42135

(d) 13542 (e) 35241 (f) 12345

2. Determine whether each of the following permutations of $S = \{1, 2, 3, 4\}$ is even or odd.

(a) 4213 (b) 1243 (c) 1234

(d) 3214 (e) 1423 (f) 2431

3. Determine the sign associated with each of the following permutations of $S = \{1, 2, 3, 4, 5\}$.
   (a) 25431 (b) 31245 (c) 21345
   (d) 52341 (e) 34125 (f) 41253

4. In each of the following pairs of permutations of $S = \{1, 2, 3, 4, 5, 6\}$, verify that the number of inversions differs by an odd number.
   (a) 436215 and 416235
   (b) 623415 and 523416
   (c) 321564 and 341562
   (d) 123564 and 423561

In Exercises 5 and 6, evaluate the determinants using Equation (2).

5. (a) $\begin{vmatrix} 2 & -1 \\ 3 & 2 \end{vmatrix}$ (b) $\begin{vmatrix} 0 & 3 & 0 \\ 2 & 0 & 0 \\ 0 & 0 & -5 \end{vmatrix}$

   (c) $\begin{vmatrix} 4 & 2 & 0 \\ 0 & -2 & 5 \\ 0 & 0 & 3 \end{vmatrix}$ (d) $\begin{vmatrix} 4 & 2 & 2 & 0 \\ 2 & 0 & 0 & 0 \\ 3 & 0 & 0 & 1 \\ 0 & 0 & 1 & 0 \end{vmatrix}$

6. (a) $\begin{vmatrix} 2 & 1 \\ 4 & 3 \end{vmatrix}$ (b) $\begin{vmatrix} 0 & 0 & -2 \\ 0 & 3 & 0 \\ 4 & 0 & 0 \end{vmatrix}$

   (c) $\begin{vmatrix} 3 & 4 & 2 \\ 2 & 5 & 0 \\ 3 & 0 & 0 \end{vmatrix}$ (d) $\begin{vmatrix} -4 & 2 & 0 & 0 \\ 2 & 3 & 1 & 0 \\ 3 & 1 & 0 & 2 \\ 1 & 3 & 0 & 3 \end{vmatrix}$

7. Let $A = [a_{ij}]$ be a 4×4 matrix. Write the general expression for $\det(A)$ using Equation (2).

8. If

$$|A| = \begin{vmatrix} a_1 & a_2 & a_3 \\ b_1 & b_2 & b_3 \\ c_1 & c_2 & c_3 \end{vmatrix} = -4,$$

find the determinants of the following matrices:

$$B = \begin{bmatrix} a_3 & a_2 & a_1 \\ b_3 & b_2 & b_1 \\ c_3 & c_2 & c_1 \end{bmatrix},$$

$$C = \begin{bmatrix} a_1 & a_2 & a_3 \\ b_1 & b_2 & b_3 \\ 2c_1 & 2c_2 & 2c_3 \end{bmatrix},$$

and

$$D = \begin{bmatrix} a_1 & a_2 & a_3 \\ b_1 + 4c_1 & b_2 + 4c_2 & b_3 + 4c_3 \\ c_1 & c_2 & c_3 \end{bmatrix}.$$

9. If

$$|A| = \begin{vmatrix} a_1 & a_2 & a_3 \\ b_1 & b_2 & b_3 \\ c_1 & c_2 & c_3 \end{vmatrix} = 3,$$

find the determinants of the following matrices:

$B =$

$$\begin{bmatrix} a_1 + 2b_1 - 3c_1 & a_2 + 2b_2 - 3c_2 & a_3 + 2b_3 - 3c_3 \\ b_1 & b_2 & b_3 \\ c_1 & c_2 & c_3 \end{bmatrix},$$

$$C = \begin{bmatrix} a_1 & 3a_2 & a_3 \\ b_1 & 3b_2 & b_3 \\ c_1 & 3c_2 & c_3 \end{bmatrix},$$

and

$$D = \begin{bmatrix} a_1 & a_2 & a_3 \\ c_1 & c_2 & c_3 \\ b_1 & b_2 & b_3 \end{bmatrix}.$$

10. If

$$A = \begin{bmatrix} 1 & -1 & 2 \\ 3 & 4 & 1 \\ 2 & 5 & 1 \end{bmatrix},$$

verify that $\det(A) = \det(A^T)$.

11. Evaluate:
    (a) $\det\left(\begin{bmatrix} \lambda - 1 & 2 \\ 3 & \lambda - 2 \end{bmatrix}\right)$
    (b) $\det(\lambda I_2 - A)$, where $A = \begin{bmatrix} 4 & 2 \\ -1 & 1 \end{bmatrix}$

12. Evaluate:
    (a) $\det\left(\begin{bmatrix} \lambda - 1 & -1 & -2 \\ 0 & \lambda - 2 & 2 \\ 0 & 0 & \lambda - 3 \end{bmatrix}\right)$
    (b) $\det(\lambda I_3 - A)$, where $A = \begin{bmatrix} -1 & 0 & 1 \\ -2 & 0 & -1 \\ 0 & 0 & 1 \end{bmatrix}$

13. For each of the matrices in Exercise 11, find all values of λ for which the determinant is 0.

14. For each of the matrices in Exercise 12, find all values of λ for which the determinant is 0.

In Exercises 15 and 16, compute the indicated determinant.

15. (a) $\begin{vmatrix} 0 & 2 & -5 \\ 0 & 4 & 6 \\ 0 & 0 & -1 \end{vmatrix}$ (b) $\begin{vmatrix} 6 & 6 & 3 & -2 \\ 0 & 4 & 7 & 5 \\ 0 & 0 & -3 & 2 \\ 0 & 0 & 0 & 2 \end{vmatrix}$

    (c) $\begin{vmatrix} 2 & 0 & 0 \\ 0 & 4 & 0 \\ 0 & 0 & 9 \end{vmatrix}$

16. (a) $\begin{vmatrix} 6 & 0 & 0 & 0 \\ 3 & 4 & 0 & 0 \\ 5 & 9 & -3 & 0 \\ 4 & 1 & -3 & 2 \end{vmatrix}$ (b) $\begin{vmatrix} 7 & 0 & 0 \\ 0 & 8 & 0 \\ 0 & 0 & -3 \end{vmatrix}$

    (c) $\begin{vmatrix} 2 & 6 & -5 \\ 0 & 4 & 0 \\ 0 & 0 & 9 \end{vmatrix}$

In Exercises 17 through 20, evaluate the given determinant via reduction to triangular form.

17. (a) $\begin{vmatrix} 4 & -3 & 5 \\ 5 & 2 & 0 \\ 2 & 0 & 4 \end{vmatrix}$ (b) $\begin{vmatrix} 2 & 0 & 1 & 4 \\ 3 & 2 & -4 & -2 \\ 2 & 3 & -1 & 0 \\ 11 & 8 & -4 & 6 \end{vmatrix}$

(c) $\begin{vmatrix} 4 & 1 & 2 \\ 0 & 2 & 3 \\ 0 & 0 & -3 \end{vmatrix}$

18. (a) $\begin{vmatrix} 4 & 0 & 0 & 0 \\ -1 & 2 & 0 & 0 \\ 1 & 2 & -3 & 0 \\ 1 & 5 & 3 & 5 \end{vmatrix}$ (b) $\begin{vmatrix} 4 & 1 & 3 \\ 2 & 3 & 0 \\ 1 & 3 & 2 \end{vmatrix}$

(c) $\begin{vmatrix} 1 & 2 & 3 \\ 2 & 1 & 0 \\ -3 & 1 & 2 \end{vmatrix}$

19. (a) $\begin{vmatrix} 4 & 2 & 3 & -4 \\ 3 & -2 & 1 & 5 \\ -2 & 0 & 1 & -3 \\ 8 & -2 & 6 & 4 \end{vmatrix}$

(b) $\begin{vmatrix} 1 & 3 & -4 \\ -2 & 1 & 2 \\ -9 & 15 & 0 \end{vmatrix}$ (c) $\begin{vmatrix} 1 & 1 & 2 \\ 0 & 2 & -2 \\ 0 & 0 & 3 \end{vmatrix}$

20. (a) $\begin{vmatrix} 1 & 0 & 1 \\ 1 & 1 & 0 \\ 2 & 1 & 0 \end{vmatrix}$ (b) $\begin{vmatrix} 2 & 0 & 0 & 0 \\ -5 & 3 & 0 & 0 \\ 3 & 2 & 4 & 0 \\ 4 & 2 & 1 & -5 \end{vmatrix}$

(c) $\begin{vmatrix} 1 & 2 & -1 \\ 3 & 2 & 0 \\ 1 & 4 & 3 \end{vmatrix}$

21. Verify that $\det(AB) = \det(A)\det(B)$ for the following:

(a) $A = \begin{bmatrix} 1 & -2 & 3 \\ -2 & 3 & 1 \\ 0 & 1 & 0 \end{bmatrix}$, $B = \begin{bmatrix} 1 & 0 & 2 \\ 3 & -2 & 5 \\ 2 & 1 & 3 \end{bmatrix}$

(b) $A = \begin{bmatrix} 2 & 3 & 6 \\ 0 & 3 & 2 \\ 0 & 0 & -4 \end{bmatrix}$, $B = \begin{bmatrix} 3 & 0 & 0 \\ 4 & 5 & 0 \\ 2 & 1 & -2 \end{bmatrix}$

22. If $|A| = -4$, find

(a) $|A^2|$ (b) $|A^4|$ (c) $|A^{-1}|$

23. If A and B are $n \times n$ matrices with $|A| = 2$ and $|B| = -3$, calculate $|A^{-1}B^T|$.

In Exercises 24 and 25, evaluate the given determinant of the bit matrices using techniques developed in Examples 5 and 6.

24. (a) $\begin{vmatrix} 0 & 1 \\ 1 & 0 \end{vmatrix}$ (b) $\begin{vmatrix} 1 & 1 & 0 \\ 1 & 0 & 1 \\ 1 & 1 & 1 \end{vmatrix}$

(c) $\begin{vmatrix} 1 & 0 & 0 \\ 0 & 1 & 1 \\ 1 & 1 & 0 \end{vmatrix}$

25. (a) $\begin{vmatrix} 1 & 1 \\ 1 & 0 \end{vmatrix}$ (b) $\begin{vmatrix} 0 & 1 & 1 \\ 1 & 1 & 1 \\ 0 & 0 & 1 \end{vmatrix}$

(c) $\begin{vmatrix} 1 & 0 & 0 \\ 1 & 1 & 1 \\ 1 & 1 & 0 \end{vmatrix}$

In Exercises 26 and 27, evaluate the given determinant of the bit matrices via reduction to triangular form.

26. (a) $\begin{vmatrix} 0 & 1 & 0 \\ 1 & 1 & 0 \\ 1 & 1 & 1 \end{vmatrix}$ (b) $\begin{vmatrix} 1 & 1 & 0 & 1 \\ 0 & 1 & 1 & 1 \\ 1 & 0 & 0 & 0 \\ 0 & 1 & 0 & 0 \end{vmatrix}$

27. (a) $\begin{vmatrix} 1 & 0 & 1 \\ 0 & 1 & 1 \\ 1 & 1 & 0 \end{vmatrix}$ (b) $\begin{vmatrix} 1 & 1 & 1 & 0 \\ 1 & 1 & 0 & 1 \\ 1 & 0 & 1 & 1 \\ 0 & 0 & 0 & 1 \end{vmatrix}$

Theoretical Exercises

T.1. Show that if we interchange two numbers in the permutation $j_1 j_2 \cdots j_n$, then the number of inversions is either increased or decreased by an odd number. (*Hint*: First show that if two adjacent numbers are interchanged, the number of inversions is either increased or decreased by 1. Then show that an interchange of any two numbers can be achieved by an odd number of successive interchanges of adjacent numbers.)

T.2. Prove Theorem 3.7 for the lower triangular case.

T.3. Show that if c is a real number and A is $n \times n$, then $\det(cA) = c^n \det(A)$.

T.4. Prove Corollary 3.2.

T.5. Show that if $\det(AB) = 0$, then $\det(A) = 0$ or $\det(B) = 0$.

T.6. Is $\det(AB) = \det(BA)$? Justify your answer.

T.7. Show that if A is a matrix such that in each row and in each column one and only one element is $\neq 0$, then $\det(A) \neq 0$.

T.8. Show that if $AB = I_n$, then $\det(A) \neq 0$ and $\det(B) \neq 0$.

T.9. (a) Show that if $A = A^{-1}$, then $\det(A) = \pm 1$.

(b) Show that if $A^T = A^{-1}$, then $\det(A) = \pm 1$.

T.10. Show that if A is a nonsingular matrix such that $A^2 = A$, then $\det(A) = 1$.

T.11. Show that

$$\det(A^T B^T) = \det(A)\det(B^T)$$
$$= \det(A^T)\det(B).$$

T.12. Show that

$$\begin{vmatrix} a^2 & a & 1 \\ b^2 & b & 1 \\ c^2 & c & 1 \end{vmatrix} = (b-a)(c-a)(b-c).$$

This determinant is called a **Vandermonde*** **determinant**.

T.13. Let $A = [a_{ij}]$ be an upper triangular matrix. Show that A is nonsingular if and only if $a_{ii} \neq 0$, $1 \leq i \leq n$.

T.14. Show that if $\det(A) = 0$, then $\det(AB) = 0$.

T.15. Show that if $A^n = O$, for some positive integer n, then $\det(A) = 0$.

T.16. Show that if A is $n \times n$, with A skew symmetric ($A^T = -A$, see Section 1.4, Exercise T.24), and n odd, then $\det(A) = 0$.

T.17. Prove Corollary 3.1.

T.18. When is a diagonal matrix nonsingular? (*Hint*: See Exercise T.7.)

T.19. Using Exercise T.13 in Section 1.2, determine how many 2 × 2 bit matrices have determinant 0 and how many have determinant 1.

MATLAB Exercises

In order to use MATLAB *in this section, you should first have read Chapter 12 through Section 12.5.*

ML.1. Use the routine **reduce** to perform row operations and keep track by hand of the changes in the determinant as in Example 17.

(a) $A = \begin{bmatrix} 2 & 1 & 3 \\ 1 & 3 & 2 \\ 3 & 2 & 1 \end{bmatrix}$

(b) $A = \begin{bmatrix} 0 & 1 & 3 & -2 \\ -2 & 1 & 1 & 1 \\ 2 & 0 & 1 & 2 \\ 1 & 0 & 0 & 1 \end{bmatrix}$

ML.2. Use routine **reduce** to perform row operations and keep track by hand of the changes in the determinant as in Example 17.

(a) $A = \begin{bmatrix} 1 & 0 & 2 \\ 0 & 2 & 1 \\ 2 & 1 & 0 \end{bmatrix}$

(b) $A = \begin{bmatrix} 1 & 2 & 0 & 0 \\ 2 & 1 & 2 & 0 \\ 0 & 2 & 1 & 2 \\ 0 & 0 & 2 & 1 \end{bmatrix}$

ML.3. MATLAB has command **det**, which returns the value of the determinant of a matrix. Just type **det(A)**. Find the determinant of each of the following matrices using **det**.

(a) $A = \begin{bmatrix} 1 & -1 & 1 \\ 1 & 1 & -1 \\ -1 & 1 & 1 \end{bmatrix}$

(b) $A = \begin{bmatrix} 1 & 2 & 3 & 4 \\ 2 & 3 & 4 & 5 \\ 3 & 4 & 5 & 6 \\ 4 & 5 & 6 & 7 \end{bmatrix}$

ML.4. Use **det** (see Exercise ML.3) to compute the determinant of each of the following.

(a) **5 ∗ eye(size(A)) − A**, where

$$A = \begin{bmatrix} 2 & 3 & 0 \\ 4 & 1 & 0 \\ 0 & 0 & 5 \end{bmatrix}.$$

(b) **(3 ∗ eye(size(A)) − A)^2**, where

$$A = \begin{bmatrix} 1 & 1 \\ 5 & 2 \end{bmatrix}.$$

(c) **invert(A) ∗ A**, where

$$A = \begin{bmatrix} 1 & 1 & 0 \\ 0 & 1 & 0 \\ 1 & 0 & 1 \end{bmatrix}.$$

ML.5. Determine a positive integer t so that **det(t ∗ eye(size(A)) − A) = 0**, where

$$A = \begin{bmatrix} 5 & 2 \\ -1 & 2 \end{bmatrix}.$$

---

*Alexandre-Théophile Vandermonde (1735–1796) was born in Paris. His father, a physician, tried to steer him toward a musical career. His published mathematical work consisted of four papers that were presented over a two-year period. He is generally considered the founder of the theory of determinants and also developed formulas for solving general quadratic, cubic, and quartic equations. Vandermonde was a cofounder of the Conservatoire des Arts et Métiers and was its director from 1782. In 1795 he helped to develop a course in political economy. He was an active revolutionary in the French Revolution and was a member of the Commune of Paris and the club of the Jacobins.

3.2 COFACTOR EXPANSION AND APPLICATIONS

So far, we have been evaluating determinants by using Equation (2) of Section 3.1 and the properties established there. We now develop a different method for evaluating the determinant of an $n \times n$ matrix, which reduces the problem to the evaluation of determinants of matrices of order $n - 1$. We can then repeat the process for these $(n-1) \times (n-1)$ matrices until we get to 2×2 matrices.

DEFINITION

Let $A = \left[a_{ij}\right]$ be an $n \times n$ matrix. Let M_{ij} be the $(n-1) \times (n-1)$ submatrix of A obtained by deleting the ith row and jth column of A. The determinant $\det(M_{ij})$ is called the **minor** of a_{ij}. The **cofactor** A_{ij} of a_{ij} is defined as

$$A_{ij} = (-1)^{i+j} \det(M_{ij}).$$

EXAMPLE 1

Let

$$A = \begin{bmatrix} 3 & -1 & 2 \\ 4 & 5 & 6 \\ 7 & 1 & 2 \end{bmatrix}.$$

Then

$$\det(M_{12}) = \begin{vmatrix} 4 & 6 \\ 7 & 2 \end{vmatrix} = 8 - 42 = -34,$$

$$\det(M_{23}) = \begin{vmatrix} 3 & -1 \\ 7 & 1 \end{vmatrix} = 3 + 7 = 10,$$

and

$$\det(M_{31}) = \begin{vmatrix} -1 & 2 \\ 5 & 6 \end{vmatrix} = -6 - 10 = -16.$$

Also,

$$A_{12} = (-1)^{1+2} \det(M_{12}) = (-1)(-34) = 34,$$

$$A_{23} = (-1)^{2+3} \det(M_{23}) = (-1)(10) = -10,$$

and

$$A_{31} = (-1)^{3+1} \det(M_{31}) = (1)(-16) = -16. \quad \blacksquare$$

If we think of the sign $(-1)^{i+j}$ as being located in position (i, j) of an $n \times n$ matrix, then the signs form a checkerboard pattern that has a $+$ in the $(1, 1)$ position. The patterns for $n = 3$ and $n = 4$ are as follows:

$$\begin{matrix} + & - & + \\ - & + & - \\ + & - & + \end{matrix} \qquad\qquad \begin{matrix} + & - & + & - \\ - & + & - & + \\ + & - & + & - \\ - & + & - & + \end{matrix}$$

$$n = 3 \qquad\qquad\qquad n = 4$$

The following theorem gives another method of evaluating determinants that is not as computationally efficient as reduction to triangular form.

THEOREM 3.9 *Let $A = [a_{ij}]$ be an $n \times n$ matrix. Then for each $1 \leq i \leq n$,*

$$\det(A) = a_{i1}A_{i1} + a_{i2}A_{i2} + \cdots + a_{in}A_{in} \quad \text{(expansion of } \det(A) \text{ along the } i\text{th row)}; \tag{1}$$

and for each $1 \leq j \leq n$,

$$\det(A) = a_{1j}A_{1j} + a_{2j}A_{2j} + \cdots + a_{nj}A_{nj} \quad \text{(expansion of } \det(A) \text{ along the } j\text{th column)}. \tag{2}$$

Proof The first formula follows from the second by Theorem 3.1, that is, from the fact that $\det(A^T) = \det(A)$. We omit the general proof and consider the 3×3 matrix $A = [a_{ij}]$. From (3) in Section 3.1,

$$\begin{aligned}\det(A) &= a_{11}a_{22}a_{33} + a_{12}a_{23}a_{31} + a_{13}a_{21}a_{32} \\ &\quad - a_{11}a_{23}a_{32} - a_{12}a_{21}a_{33} - a_{13}a_{22}a_{31}.\end{aligned} \tag{3}$$

We can write this expression as

$$\begin{aligned}\det(A) &= a_{11}(a_{22}a_{33} - a_{23}a_{32}) + a_{12}(a_{23}a_{31} - a_{21}a_{33}) \\ &\quad + a_{13}(a_{21}a_{32} - a_{22}a_{31}).\end{aligned}$$

Now,

$$A_{11} = (-1)^{1+1}\begin{vmatrix} a_{22} & a_{23} \\ a_{32} & a_{33} \end{vmatrix} = (a_{22}a_{33} - a_{23}a_{32}),$$

$$A_{12} = (-1)^{1+2}\begin{vmatrix} a_{21} & a_{23} \\ a_{31} & a_{33} \end{vmatrix} = (a_{23}a_{31} - a_{21}a_{33}),$$

$$A_{13} = (-1)^{1+3}\begin{vmatrix} a_{21} & a_{22} \\ a_{31} & a_{32} \end{vmatrix} = (a_{21}a_{32} - a_{22}a_{31}).$$

Hence

$$\det(A) = a_{11}A_{11} + a_{12}A_{12} + a_{13}A_{13},$$

which is the expansion of $\det(A)$ along the first row.

If we now write (3) as

$$\begin{aligned}\det(A) &= a_{13}(a_{21}a_{32} - a_{22}a_{31}) + a_{23}(a_{12}a_{31} - a_{11}a_{32}) \\ &\quad + a_{33}(a_{11}a_{22} - a_{12}a_{21}),\end{aligned}$$

we can easily verify that

$$\det(A) = a_{13}A_{13} + a_{23}A_{23} + a_{33}A_{33},$$

which is the expansion of $\det(A)$ along the third column. ■

EXAMPLE 2 To evaluate the determinant

$$\begin{vmatrix} 1 & 2 & -3 & 4 \\ -4 & 2 & 1 & 3 \\ 3 & 0 & 0 & -3 \\ 2 & 0 & -2 & 3 \end{vmatrix},$$

we note that it is best to expand along either the second column or the third row because they each have two zeros. Obviously, the optimal course of action is to expand along a row or column having the largest number of zeros because in that case the cofactors A_{ij} of those a_{ij} that are zero do not have to be evaluated, since $a_{ij}A_{ij} = (0)(A_{ij}) = 0$. Thus, expanding along the third row, we have

$$\begin{aligned}
&\begin{vmatrix} 1 & 2 & -3 & 4 \\ -4 & 2 & 1 & 3 \\ 3 & 0 & 0 & -3 \\ 2 & 0 & -2 & 3 \end{vmatrix} \\
&\quad = (-1)^{3+1}(3)\begin{vmatrix} 2 & -3 & 4 \\ 2 & 1 & 3 \\ 0 & -2 & 3 \end{vmatrix} + (-1)^{3+2}(0)\begin{vmatrix} 1 & -3 & 4 \\ -4 & 1 & 3 \\ 2 & -2 & 3 \end{vmatrix} \\
&\quad\quad + (-1)^{3+3}(0)\begin{vmatrix} 1 & 2 & 4 \\ -4 & 2 & 3 \\ 2 & 0 & 3 \end{vmatrix} + (-1)^{3+4}(-3)\begin{vmatrix} 1 & 2 & -3 \\ -4 & 2 & 1 \\ 2 & 0 & -2 \end{vmatrix}.
\end{aligned} \tag{4}$$

We now evaluate

$$\begin{vmatrix} 2 & -3 & 4 \\ 2 & 1 & 3 \\ 0 & -2 & 3 \end{vmatrix}$$

by expanding along the first column, obtaining

$$(-1)^{1+1}(2)\begin{vmatrix} 1 & 3 \\ -2 & 3 \end{vmatrix} + (-1)^{2+1}(2)\begin{vmatrix} -3 & 4 \\ -2 & 3 \end{vmatrix}$$
$$= (1)(2)(9) + (-1)(2)(-1) = 20.$$

Similarly, we evaluate

$$\begin{vmatrix} 1 & 2 & -3 \\ -4 & 2 & 1 \\ 2 & 0 & -2 \end{vmatrix}$$

by expanding along the third row, obtaining

$$(-1)^{3+1}(2)\begin{vmatrix} 2 & -3 \\ 2 & 1 \end{vmatrix} + (-1)^{3+3}(-2)\begin{vmatrix} 1 & 2 \\ -4 & 2 \end{vmatrix}$$
$$= (1)(2)(8) + (1)(-2)(10) = -4.$$

Substituting in Equation (4), we find the value of the given determinant as

$$(+1)(3)(20) + 0 + 0 + (-1)(-3)(-4) = 48.$$

On the other hand, evaluating the given determinant by expanding along the first column, we have

$$\begin{aligned}
&(-1)^{1+1}(1)\begin{vmatrix} 2 & 1 & 3 \\ 0 & 0 & -3 \\ 0 & -2 & 3 \end{vmatrix} + (-1)^{2+1}(-4)\begin{vmatrix} 2 & -3 & 4 \\ 0 & 0 & -3 \\ 0 & -2 & 3 \end{vmatrix} \\
&\quad + (-1)^{3+1}(3)\begin{vmatrix} 2 & -3 & 4 \\ 2 & 1 & 3 \\ 0 & -2 & 3 \end{vmatrix} + (-1)^{4+1}(2)\begin{vmatrix} 2 & -3 & 4 \\ 2 & 1 & 3 \\ 0 & 0 & -3 \end{vmatrix}
\end{aligned}$$

$$= (1)(1)(-12) + (-1)(-4)(-12) + (1)(3)(20) + (-1)(2)(-24) = 48. \quad \blacksquare$$

We can use the properties of Section 3.1 to introduce many zeros in a given row or column and then expand along that row or column. This is illustrated in the following example.

EXAMPLE 3 Consider the determinant of Example 2. We have

$$\begin{vmatrix} 1 & 2 & -3 & 4 \\ -4 & 2 & 1 & 3 \\ 3 & 0 & 0 & -3 \\ 2 & 0 & -2 & 3 \end{vmatrix}_{\mathbf{c}_4+\mathbf{c}_1\to\mathbf{c}_4} = \begin{vmatrix} 1 & 2 & -3 & 5 \\ -4 & 2 & 1 & -1 \\ 3 & 0 & 0 & 0 \\ 2 & 0 & -2 & 5 \end{vmatrix}$$

$$= (-1)^{3+1}(3)\begin{vmatrix} 2 & -3 & 5 \\ 2 & 1 & -1 \\ 0 & -2 & 5 \end{vmatrix}_{\mathbf{r}_1-\mathbf{r}_2\to\mathbf{r}_1}$$

$$= (-1)^4(3)\begin{vmatrix} 0 & -4 & 6 \\ 2 & 1 & -1 \\ 0 & -2 & 5 \end{vmatrix}$$

$$= (-1)^4(3)(-2)(-8) = 48.$$ ■

THE INVERSE OF A MATRIX

It is interesting to ask what $a_{i1}A_{k1}+a_{i2}A_{k2}+\cdots+a_{in}A_{kn}$ is for $i \neq k$ because, as soon as we answer this question, we shall obtain another method for finding the inverse of a nonsingular matrix.

THEOREM 3.10 *If $A = \left[a_{ij}\right]$ is an $n \times n$ matrix, then*

$$a_{i1}A_{k1} + a_{i2}A_{k2} + \cdots + a_{in}A_{kn} = 0 \quad \textit{for } i \neq k; \tag{5}$$
$$a_{1j}A_{1k} + a_{2j}A_{2k} + \cdots + a_{nj}A_{nk} = 0 \quad \textit{for } j \neq k. \tag{6}$$

Proof We prove only the first formula. The second follows from the first one by Theorem 3.1.

Consider the matrix B obtained from A by replacing the kth row of A by its ith row. Thus B is a matrix having two identical rows—the ith and kth rows. Then $\det(B) = 0$. Now expand $\det(B)$ along the kth row. The elements of the kth row of B are $a_{i1}, a_{i2}, \ldots, a_{in}$. The cofactors of the kth row are $A_{k1}, A_{k2}, \ldots, A_{kn}$. Thus from Equation (1) we have

$$0 = \det(B) = a_{i1}A_{k1} + a_{i2}A_{k2} + \cdots + a_{in}A_{kn},$$

which is what we wanted to show. ■

This theorem says that if we sum the products of the elements of any row (column) times the corresponding cofactors of any other row (column), then we obtain zero.

EXAMPLE 4 Let

$$A = \begin{bmatrix} 1 & 2 & 3 \\ -2 & 3 & 1 \\ 4 & 5 & -2 \end{bmatrix}.$$

Then

$$A_{21} = (-1)^{2+1} \begin{vmatrix} 2 & 3 \\ 5 & -2 \end{vmatrix} = 19, \qquad A_{22} = (-1)^{2+2} \begin{vmatrix} 1 & 3 \\ 4 & -2 \end{vmatrix} = -14,$$

$$A_{23} = (-1)^{2+3} \begin{vmatrix} 1 & 2 \\ 4 & 5 \end{vmatrix} = 3.$$

Now

$$a_{31}A_{21} + a_{32}A_{22} + a_{33}A_{23} = (4)(19) + (5)(-14) + (-2)(3) = 0$$

and

$$a_{11}A_{21} + a_{12}A_{22} + a_{13}A_{23} = (1)(19) + (2)(-14) + (3)(3) = 0. \quad \blacksquare$$

We may combine (1) and (5) as

$$\begin{aligned} a_{i1}A_{k1} + a_{i2}A_{k2} + \cdots + a_{in}A_{kn} &= \det(A) \quad \text{if } i = k \\ &= 0 \quad \text{if } i \neq k. \end{aligned} \tag{7}$$

Similarly, we may combine (2) and (6) as

$$\begin{aligned} a_{1j}A_{1k} + a_{2j}A_{2k} + \cdots + a_{nj}A_{nk} &= \det(A) \quad \text{if } j = k \\ &= 0 \quad \text{if } j \neq k. \end{aligned} \tag{8}$$

DEFINITION Let $A = \left[a_{ij}\right]$ be an $n \times n$ matrix. The $n \times n$ matrix adj A, called the **adjoint** of A, is the matrix whose i, jth element is the cofactor A_{ji} of a_{ji}. Thus

$$\operatorname{adj} A = \begin{bmatrix} A_{11} & A_{21} & \cdots & A_{n1} \\ A_{12} & A_{22} & \cdots & A_{n2} \\ \vdots & \vdots & & \vdots \\ A_{1n} & A_{2n} & \cdots & A_{nn} \end{bmatrix}.$$

Remarks

1. The adjoint of A is formed by taking the transpose of the matrix of cofactors of the elements of A.
2. It should be noted that the term *adjoint* has other meanings in linear algebra in addition to its use in the above definition.

EXAMPLE 5 Let

$$A = \begin{bmatrix} 3 & -2 & 1 \\ 5 & 6 & 2 \\ 1 & 0 & -3 \end{bmatrix}.$$

Compute adj A.

Solution The cofactors of A are

$$A_{11} = (-1)^{1+1}\begin{vmatrix} 6 & 2 \\ 0 & -3 \end{vmatrix} = -18; \quad A_{12} = (-1)^{1+2}\begin{vmatrix} 5 & 2 \\ 1 & -3 \end{vmatrix} = 17;$$

$$A_{13} = (-1)^{1+3}\begin{vmatrix} 5 & 6 \\ 1 & 0 \end{vmatrix} = -6;$$

$$A_{21} = (-1)^{2+1}\begin{vmatrix} -2 & 1 \\ 0 & -3 \end{vmatrix} = -6; \quad A_{22} = (-1)^{2+2}\begin{vmatrix} 3 & 1 \\ 1 & -3 \end{vmatrix} = -10;$$

$$A_{23} = (-1)^{2+3}\begin{vmatrix} 3 & -2 \\ 1 & 0 \end{vmatrix} = -2;$$

$$A_{31} = (-1)^{3+1}\begin{vmatrix} -2 & 1 \\ 6 & 2 \end{vmatrix} = -10; \quad A_{32} = (-1)^{3+2}\begin{vmatrix} 3 & 1 \\ 5 & 2 \end{vmatrix} = -1;$$

$$A_{33} = (-1)^{3+3}\begin{vmatrix} 3 & -2 \\ 5 & 6 \end{vmatrix} = 28.$$

Then

$$\operatorname{adj} A = \begin{bmatrix} -18 & -6 & -10 \\ 17 & -10 & -1 \\ -6 & -2 & 28 \end{bmatrix}.$$

■

THEOREM 3.11 *If $A = \left[a_{ij}\right]$ is an $n \times n$ matrix, then*

$$A(\operatorname{adj} A) = (\operatorname{adj} A)A = \det(A)I_n.$$

Proof We have

$$A(\operatorname{adj} A) = \begin{bmatrix} a_{11} & a_{12} & \cdots & a_{1n} \\ a_{21} & a_{22} & \cdots & a_{2n} \\ \vdots & \vdots & & \vdots \\ a_{i1} & a_{i2} & \cdots & a_{in} \\ \vdots & \vdots & & \vdots \\ a_{n1} & a_{n2} & \cdots & a_{nn} \end{bmatrix} \begin{bmatrix} A_{11} & A_{21} & \cdots & A_{j1} & \cdots & A_{n1} \\ A_{12} & A_{22} & \cdots & A_{j2} & \cdots & A_{n2} \\ \vdots & \vdots & & \vdots & & \vdots \\ A_{1n} & A_{2n} & \cdots & A_{jn} & \cdots & A_{nn} \end{bmatrix}.$$

The i, jth element in the product matrix $A(\operatorname{adj} A)$ is, by (7),

$$\begin{aligned} a_{i1}A_{j1} + a_{i2}A_{j2} + \cdots + a_{in}A_{jn} &= \det(A) \quad \text{if } i = j \\ &= 0 \quad \text{if } i \neq j. \end{aligned}$$

This means that

$$A(\operatorname{adj} A) = \begin{bmatrix} \det(A) & 0 & \cdots & 0 \\ 0 & \det(A) & & 0 \\ \vdots & \vdots & \vdots & \vdots \\ 0 & \cdots & 0 & \det(A) \end{bmatrix} = \det(A)I_n.$$

The i, jth element in the product matrix $(\operatorname{adj} A)A$ is, by (8),

$$\begin{aligned} A_{1i}a_{1j} + A_{2i}a_{2j} + \cdots + A_{ni}a_{nj} &= \det(A) \quad \text{if } i = j \\ &= 0 \quad \text{if } i \neq j. \end{aligned}$$

Thus $(\operatorname{adj} A)A = \det(A)I_n$. ■

EXAMPLE 6 Consider the matrix of Example 5. Then

$$\begin{bmatrix} 3 & -2 & 1 \\ 5 & 6 & 2 \\ 1 & 0 & -3 \end{bmatrix} \begin{bmatrix} -18 & -6 & -10 \\ 17 & -10 & -1 \\ -6 & -2 & 28 \end{bmatrix} = \begin{bmatrix} -94 & 0 & 0 \\ 0 & -94 & 0 \\ 0 & 0 & -94 \end{bmatrix}$$
$$= -94 \begin{bmatrix} 1 & 0 & 0 \\ 0 & 1 & 0 \\ 0 & 0 & 1 \end{bmatrix}$$

and

$$\begin{bmatrix} -18 & -6 & -10 \\ 17 & -10 & -1 \\ -6 & -2 & 28 \end{bmatrix} \begin{bmatrix} 3 & -2 & 1 \\ 5 & 6 & 2 \\ 1 & 0 & -3 \end{bmatrix} = -94 \begin{bmatrix} 1 & 0 & 0 \\ 0 & 1 & 0 \\ 0 & 0 & 1 \end{bmatrix}.$$
■

We now have a new method for finding the inverse of a nonsingular matrix, and we state this result as the following corollary.

COROLLARY 3.3 *If A is an $n \times n$ matrix and* $\det(A) \neq 0$, *then*

$$A^{-1} = \frac{1}{\det(A)}(\operatorname{adj} A) = \begin{bmatrix} \frac{A_{11}}{\det(A)} & \frac{A_{21}}{\det(A)} & \cdots & \frac{A_{n1}}{\det(A)} \\ \frac{A_{12}}{\det(A)} & \frac{A_{22}}{\det(A)} & \cdots & \frac{A_{n2}}{\det(A)} \\ \vdots & \vdots & & \vdots \\ \frac{A_{1n}}{\det(A)} & \frac{A_{2n}}{\det(A)} & \cdots & \frac{A_{nn}}{\det(A)} \end{bmatrix}.$$

Proof By Theorem 3.11, $A(\operatorname{adj} A) = \det(A)I_n$, so if $\det(A) \neq 0$, then

$$A\frac{1}{\det(A)}(\operatorname{adj} A) = \frac{1}{\det(A)}\left[A(\operatorname{adj} A)\right] = \frac{1}{\det(A)}(\det(A)I_n) = I_n.$$

Hence

$$A^{-1} = \frac{1}{\det(A)}(\operatorname{adj} A).$$
■

EXAMPLE 7 Again consider the matrix of Example 5. Then $\det(A) = -94$, and

$$A^{-1} = \frac{1}{\det(A)}(\operatorname{adj} A) = \begin{bmatrix} \frac{18}{94} & \frac{6}{94} & \frac{10}{94} \\ -\frac{17}{94} & \frac{10}{94} & \frac{1}{94} \\ \frac{6}{94} & \frac{2}{94} & -\frac{28}{94} \end{bmatrix}.$$
■

THEOREM 3.12 *A matrix A is nonsingular if and only if* $\det(A) \neq 0$.

Proof If $\det(A) \neq 0$, then Corollary 3.2 gives an expression for A^{-1}, so A is nonsingular.

The converse has already been established in Corollary 3.2, whose proof was left to the reader as Exercise T.4 of Section 3.1. We now prove the converse here. Suppose that A is nonsingular. Then

$$AA^{-1} = I_n.$$

From Theorem 3.8 we have

$$\det(AA^{-1}) = \det(A)\det(A^{-1}) = \det(I_n) = 1,$$

which implies that $\det(A) \neq 0$. This completes the proof. ■

COROLLARY 3.4 *For an $n \times n$ matrix A, the homogeneous system $A\mathbf{x} = \mathbf{0}$ has a nontrivial solution if and only if* $\det(A) = 0$.

Proof If $\det(A) \neq 0$, then, by Theorem 3.12, A is nonsingular and thus $A\mathbf{x} = \mathbf{0}$ has only the trivial solution (Theorem 1.13 in Section 1.7).

Conversely, if $\det(A) = 0$, then A is singular (Theorem 3.12). Suppose that A is row equivalent to a matrix B in reduced row echelon form. It then follows from Theorem 1.12 in Section 1.7 and Exercise T.9 in Section 1.6 that B has a row of zeros. The system $B\mathbf{x} = \mathbf{0}$ has the same solutions as the system $A\mathbf{x} = \mathbf{0}$. Let C_1 be the matrix obtained by deleting the zero rows of B. Then the system $B\mathbf{x} = \mathbf{0}$ has the same solutions as the system $C_1\mathbf{x} = \mathbf{0}$. Since the latter is a homogeneous system of at most $n - 1$ equations in n unknowns, it has a nontrivial solution (Theorem 1.8 in Section 1.6). Hence the given system $A\mathbf{x} = \mathbf{0}$ has a nontrivial solution. We might note that the proof of the converse is essentially the proof of Exercise T.3 in Section 1.7. ■

EXAMPLE 8 Let A be a 4×4 matrix with $\det(A) = -2$.

(a) Describe the set of all solutions to the homogeneous system $A\mathbf{x} = \mathbf{0}$.

(b) If A is transformed to reduced row echelon form B, what is B?

(c) Give an expression for a solution to the linear system $A\mathbf{x} = \mathbf{b}$, where

$$\mathbf{b} = \begin{bmatrix} 1 \\ 2 \\ 3 \\ 4 \end{bmatrix}.$$

(d) Can the linear system $A\mathbf{x} = \mathbf{b}$ have more than one solution? Explain.

(e) Does A^{-1} exist?

Solution

(a) Since $\det(A) \neq 0$, by Corollary 3.4, the homogeneous system has only the trivial solution.

(b) Since $\det(A) \neq 0$, by Theorem 3.12, A is a nonsingular matrix and by Theorem 1.12, $B = I_n$.

(c) A solution to the given system is given by $\mathbf{x} = A^{-1}\mathbf{b}$.

(d) No. The solution given in part (c) is the only one.

(e) Yes. ■

In Section 1.7 we developed a practical method for finding A^{-1}. In describing the situation showing that a matrix is singular and has no inverse, we used the fact that if we start with the identity matrix I_n and use only elementary row operations on I_n, we can never obtain a matrix having a row consisting entirely of zeros. We can now justify this statement as follows. If matrix B results from I_n by interchanging two rows of I_n, then $\det(B) = -\det(I_n) = -1$ (Theorem 3.2); if C results from I_n by multiplying a row of I_n by $c \neq 0$, then $\det(C) = c\det(I_n) = c$ (Theorem 3.5); and if D results from I_n by adding a multiple of a row of I_n to another row of I_n, then $\det(D) = \det(I_n) = 1$ (Theorem 3.6). Performing elementary row operations on I_n will thus never yield a matrix with zero determinant. Now suppose that F is obtained from I_n by a sequence of elementary row operations and F has a row of zeros. Then $\det(F) = 0$ (Theorem 3.4). This contradiction justifies the statement used in Section 1.7.

We might note that the method of inverting a nonsingular matrix given in Corollary 3.3 is much less efficient than the method given in Chapter 1. In fact, the computation of A^{-1} using determinants, as given in Corollary 3.3, becomes too expensive for $n > 4$ from a computing point of view. We discuss these matters in Section 3.3, where we deal with determinants from a computational point of view. However, Corollary 3.3 is still a useful result on other grounds.

We may summarize our results on determinants, homogeneous systems, and nonsingular matrices in the following list of nonsingular equivalences.

List of Nonsingular Equivalences

The following statements are equivalent.

1. A is nonsingular.
2. $\mathbf{x} = \mathbf{0}$ is the only solution to $A\mathbf{x} = \mathbf{0}$.
3. A is row equivalent to I_n.
4. The linear system $A\mathbf{x} = \mathbf{b}$ has a unique solution for every $n \times 1$ matrix $\mathbf{b}$.
5. $\det(A) \neq 0$.

Remark The list of nonsingular equivalences applies to bit matrices.

CRAMER'S* RULE

We can use the result of Theorem 3.11 to obtain another method, known as Cramer's rule, for solving a linear system of n equations in n unknowns with a nonsingular coefficient matrix.

---

*Gabriel Cramer (1704–1752) was born in Geneva, Switzerland, and lived there all his life. Remaining single, he traveled extensively, taught at the Académie de Calvin, and participated actively in civic affairs.

The rule for solving systems of linear equations appeared in an appendix to his 1750 book, *Introduction à l'analyse des lignes courbes algébriques*. It was known previously by other mathematicians but was not widely known or very clearly explained until its appearance in Cramer's influential work.

THEOREM 3.13 *(Cramer's Rule) Let*

$$\begin{aligned} a_{11}x_1 + a_{12}x_2 + \cdots + a_{1n}x_n &= b_1 \\ a_{21}x_1 + a_{22}x_2 + \cdots + a_{2n}x_n &= b_2 \\ \vdots \qquad \vdots \qquad \vdots \qquad \vdots \quad &\quad \vdots \\ a_{n1}x_1 + a_{n2}x_2 + \cdots + a_{nn}x_n &= b_n \end{aligned}$$

be a linear system of n equations in n unknowns and let $A = \left[a_{ij}\right]$ *be the coefficient matrix so that we can write the given system as* $A\mathbf{x} = \mathbf{b}$, *where*

$$\mathbf{b} = \begin{bmatrix} b_1 \\ b_2 \\ \vdots \\ b_n \end{bmatrix}.$$

If $\det(A) \neq 0$, *then the system has the unique solution*

$$x_1 = \frac{\det(A_1)}{\det(A)}, \quad x_2 = \frac{\det(A_2)}{\det(A)}, \ldots, \quad x_n = \frac{\det(A_n)}{\det(A)},$$

where A_i *is the matrix obtained from* A *by replacing the* ith *column of* A *by* $\mathbf{b}$.

Proof If $\det(A) \neq 0$, then, by Theorem 3.12, A is nonsingular. Hence

$$\mathbf{x} = \begin{bmatrix} x_1 \\ x_2 \\ \vdots \\ x_n \end{bmatrix} = A^{-1}\mathbf{b} = \begin{bmatrix} \frac{A_{11}}{\det(A)} & \frac{A_{21}}{\det(A)} & \cdots & \frac{A_{n1}}{\det(A)} \\ \frac{A_{12}}{\det(A)} & \frac{A_{22}}{\det(A)} & \cdots & \frac{A_{n2}}{\det(A)} \\ \vdots & \vdots & & \vdots \\ \frac{A_{1i}}{\det(A)} & \frac{A_{2i}}{\det(A)} & \cdots & \frac{A_{ni}}{\det(A)} \\ \vdots & \vdots & & \vdots \\ \frac{A_{1n}}{\det(A)} & \frac{A_{2n}}{\det(A)} & \cdots & \frac{A_{nn}}{\det(A)} \end{bmatrix} \begin{bmatrix} b_1 \\ b_2 \\ \vdots \\ b_n \end{bmatrix}.$$

This means that

$$x_i = \frac{A_{1i}}{\det(A)}b_1 + \frac{A_{2i}}{\det(A)}b_2 + \cdots + \frac{A_{ni}}{\det(A)}b_n \qquad (1 \leq i \leq n).$$

Now let

$$A_i = \begin{bmatrix} a_{11} & a_{12} & \cdots & a_{1\,i-1} & b_1 & a_{1\,i+1} & \cdots & a_{1n} \\ a_{21} & a_{22} & \cdots & a_{2\,i-1} & b_2 & a_{2\,i+1} & \cdots & a_{2n} \\ \vdots & \vdots & & \vdots & \vdots & \vdots & & \vdots \\ a_{n1} & a_{n2} & \cdots & a_{n\,i-1} & b_n & a_{n\,i+1} & \cdots & a_{nn} \end{bmatrix}.$$

If we evaluate $\det(A_i)$ by expanding along the ith column, we find that

$$\det(A_i) = A_{1i}b_1 + A_{2i}b_2 + \cdots + A_{ni}b_n.$$

Hence

$$x_i = \frac{\det(A_i)}{\det(A)}$$

for $i = 1, 2, \ldots, n$. In this expression for x_i, the determinant of A_i, $\det(A_i)$, can be calculated by any method. It was only in the derivation of the expression for x_i that we had to evaluate it by expanding along the ith column. ■

EXAMPLE 9 Consider the following linear system:

$$\begin{aligned} -2x_1 + 3x_2 - x_3 &= 1 \\ x_1 + 2x_2 - x_3 &= 4 \\ -2x_1 - x_2 + x_3 &= -3. \end{aligned}$$

Then

$$|A| = \begin{vmatrix} -2 & 3 & -1 \\ 1 & 2 & -1 \\ -2 & -1 & 1 \end{vmatrix} = -2.$$

Hence

$$x_1 = \frac{\begin{vmatrix} 1 & 3 & -1 \\ 4 & 2 & -1 \\ -3 & -1 & 1 \end{vmatrix}}{|A|} = \frac{-4}{-2} = 2;$$

$$x_2 = \frac{\begin{vmatrix} -2 & 1 & -1 \\ 1 & 4 & -1 \\ -2 & -3 & 1 \end{vmatrix}}{|A|} = \frac{-6}{-2} = 3;$$

$$x_3 = \frac{\begin{vmatrix} -2 & 3 & 1 \\ 1 & 2 & 4 \\ -2 & -1 & -3 \end{vmatrix}}{|A|} = \frac{-8}{-2} = 4.$$

■

Cramer's rule for solving the linear system $A\mathbf{x} = \mathbf{b}$, where A is $n \times n$, is as follows:

Step 1. Compute $\det(A)$. If $\det(A) = 0$, Cramer's rule is not applicable. Use Gauss–Jordan reduction.

Step 2. If $\det(A) \neq 0$, for each i,

$$x_i = \frac{\det(A_i)}{\det(A)},$$

where A_i is the matrix obtained from A by replacing the ith column of A by $\mathbf{b}$.

We note that Cramer's rule is applicable only to the case where we have n equations in n unknowns and where the coefficient matrix A is nonsingular. If we have to solve a linear system of n equations in n unknowns whose coefficient matrix is singular, then we must use the Gauss–Jordan reduction method as discussed in Section 1.6. Cramer's rule becomes computationally inefficient for $n > 4$, and it is better to use the Gauss–Jordan reduction method.

POLYNOMIAL INTERPOLATION REVISITED

At the end of Section 1.7 we discussed the problem of finding a quadratic polynomial that interpolates the points (x_1, y_1), (x_2, y_2), (x_3, y_3), $x_1 \neq x_2$, $x_1 \neq x_3$, and $x_2 \neq x_3$. Thus, the polynomial has the form

$$y = a_2x^2 + a_1x + a_0 \tag{9}$$

[this was Equation (15) in Section 1.6]. Substituting the given points in (9), we obtain the linear system

$$\begin{aligned} a_2x_1^2 + a_1x_1 + a_0 &= y_1 \\ a_2x_2^2 + a_1x_2 + a_0 &= y_2 \\ a_2x_3^2 + a_1x_3 + a_0 &= y_3. \end{aligned} \tag{10}$$

The coefficient matrix of this linear system is

$$\begin{bmatrix} x_1^2 & x_1 & 1 \\ x_2^2 & x_2 & 1 \\ x_3^2 & x_3 & 1 \end{bmatrix}$$

whose determinant is the Vandermonde determinant (see Exercise T.12 in Section 3.1), which has the value

$$(x_2 - x_1)(x_3 - x_1)(x_2 - x_3).$$

Since the three given points are *distinct*, the Vandermonde determinant is not zero. Hence, the coefficient matrix of the linear system in (10) is nonsingular, which implies that the linear system has a unique solution. Thus there is a unique interpolating quadratic polynomial. The general proof for n points is similar.

OTHER APPLICATIONS OF DETERMINANTS

In Section 4.1 we use determinants to compute the area of a triangle and in Section 5.1 to compute the area of a parallelepiped.

Key Terms

Minor
Cofactor
Adjoint

3.2 Exercises

1. Let

$$A = \begin{bmatrix} 1 & 0 & -2 \\ 3 & 1 & 4 \\ 5 & 2 & -3 \end{bmatrix}.$$

Compute all the cofactors.

2. Let

$$A = \begin{bmatrix} 1 & 0 & 3 & 0 \\ 2 & 1 & 4 & -1 \\ 3 & 2 & 4 & 0 \\ 0 & 3 & -1 & 0 \end{bmatrix}.$$

Compute all the cofactors of the elements in the second row and all the cofactors of the elements in the third column.

In Exercises 3 through 6, evaluate the determinants using Theorem 3.9.

3. (a) $\begin{vmatrix} 1 & 2 & 3 \\ -1 & 5 & 2 \\ 3 & 2 & 0 \end{vmatrix}$

(b) $\begin{vmatrix} 4 & -4 & 2 & 1 \\ 1 & 2 & 0 & 3 \\ 2 & 0 & 3 & 4 \\ 0 & -3 & 2 & 1 \end{vmatrix}$

(c) $\begin{vmatrix} 4 & -2 & 0 \\ 0 & 2 & 4 \\ -1 & -1 & -3 \end{vmatrix}$

4. (a) $\begin{vmatrix} 2 & 2 & -3 & 1 \\ 0 & 1 & 2 & -1 \\ 3 & -1 & 4 & 1 \\ 2 & 3 & 0 & 0 \end{vmatrix}$

(b) $\begin{vmatrix} 0 & 1 & -2 \\ -1 & 3 & 1 \\ 2 & -2 & 3 \end{vmatrix}$

(c) $\begin{vmatrix} 2 & 1 & -3 \\ 0 & 1 & 2 \\ -4 & 2 & 1 \end{vmatrix}$

5. (a) $\begin{vmatrix} 3 & 1 & 2 & -1 \\ 2 & 0 & 3 & -7 \\ 1 & 3 & 4 & -5 \\ 0 & -1 & 1 & -5 \end{vmatrix}$

(b) $\begin{vmatrix} 3 & 1 & 0 \\ 3 & 2 & 1 \\ 0 & 1 & -1 \end{vmatrix}$

(c) $\begin{vmatrix} 3 & -3 & 0 \\ 2 & 0 & 2 \\ 2 & 1 & -3 \end{vmatrix}$

6. (a) $\begin{vmatrix} 0 & 0 & -1 & 3 \\ 0 & 1 & 2 & 1 \\ 2 & -2 & 5 & 2 \\ 3 & 3 & 0 & 0 \end{vmatrix}$

(b) $\begin{vmatrix} 4 & 2 & 0 \\ 1 & 1 & 2 \\ -1 & 3 & 4 \end{vmatrix}$

(c) $\begin{vmatrix} -1 & 2 & -1 \\ 3 & 2 & 1 \\ 1 & 4 & 2 \end{vmatrix}$

7. Verify Theorem 3.10 for the matrix

$$A = \begin{bmatrix} -2 & 3 & 0 \\ 4 & 1 & -3 \\ 2 & 0 & 1 \end{bmatrix}$$

by computing $a_{11}A_{12} + a_{21}A_{22} + a_{31}A_{32}$.

8. Let

$$A = \begin{bmatrix} 2 & 1 & 3 \\ -1 & 2 & 0 \\ 3 & -2 & 1 \end{bmatrix}.$$

(a) Find adj A.

(b) Compute $\det(A)$.

(c) Verify Theorem 3.11; that is, show that

$$A(\text{adj } A) = (\text{adj } A)A = \det(A)I_3.$$

9. Let

$$A = \begin{bmatrix} 6 & 2 & 8 \\ -3 & 4 & 1 \\ 4 & -4 & 5 \end{bmatrix}.$$

(a) Find adj A.

(b) Compute $\det(A)$.

(c) Verify Theorem 3.11; that is, show that

$$A(\text{adj } A) = (\text{adj } A)A = \det(A)I_3.$$

In Exercises 10 through 13, compute the inverses of the given matrices, if they exist, using Corollary 3.3.

10. (a) $\begin{bmatrix} 3 & 2 \\ -3 & 4 \end{bmatrix}$ (b) $\begin{bmatrix} 4 & 2 & 2 \\ 0 & 1 & 2 \\ 1 & 0 & 3 \end{bmatrix}$

(c) $\begin{bmatrix} 2 & 0 & -1 \\ 3 & 7 & 2 \\ 1 & 1 & 0 \end{bmatrix}$

11. (a) $\begin{bmatrix} 1 & 2 & -3 \\ -4 & -5 & 2 \\ -1 & 1 & -7 \end{bmatrix}$ (b) $\begin{bmatrix} 2 & 3 \\ -1 & 2 \end{bmatrix}$

(c) $\begin{bmatrix} 4 & 0 & 2 \\ 0 & 3 & 4 \\ 0 & 1 & -2 \end{bmatrix}$

12. (a) $\begin{bmatrix} 2 & 0 & 1 \\ 3 & 2 & -1 \\ 1 & 0 & 1 \end{bmatrix}$ (b) $\begin{bmatrix} 5 & -1 \\ 2 & -1 \end{bmatrix}$

(c) $\begin{bmatrix} 1 & 2 & 4 \\ 1 & -5 & 6 \\ 3 & -1 & 2 \end{bmatrix}$

13. (a) $\begin{bmatrix} -3 & 1 \\ 2 & 0 \end{bmatrix}$ (b) $\begin{bmatrix} 4 & 0 & 0 \\ 0 & -3 & 0 \\ 0 & 0 & 2 \end{bmatrix}$

(c) $\begin{bmatrix} 0 & 2 & 1 & 3 \\ 2 & -1 & 3 & 4 \\ -2 & 1 & 5 & 2 \\ 0 & 1 & 0 & 2 \end{bmatrix}$

14. Use Theorem 3.12 to determine which of the following matrices are nonsingular.

(a) $\begin{bmatrix} 1 & 2 & 3 \\ 0 & 1 & 2 \\ 2 & -3 & 1 \end{bmatrix}$ (b) $\begin{bmatrix} 1 & 2 \\ 3 & 4 \end{bmatrix}$

(c) $\begin{bmatrix} 1 & 3 & 2 \\ 2 & 1 & 4 \\ 1 & -7 & 2 \end{bmatrix}$

(d) $\begin{bmatrix} 1 & 2 & 0 & 5 \\ 3 & 4 & 1 & 7 \\ -2 & 5 & 2 & 0 \\ 0 & 1 & 2 & -7 \end{bmatrix}$

15. Use Theorem 3.12 to determine which of the following matrices are nonsingular.

(a) $\begin{bmatrix} 4 & 3 & -5 \\ -2 & -1 & 3 \\ 4 & 6 & -2 \end{bmatrix}$

(b) $\begin{bmatrix} 1 & 3 & -1 & 2 \\ 2 & -6 & 4 & 1 \\ 3 & 5 & -1 & 3 \\ 4 & -6 & 5 & 2 \end{bmatrix}$

(c) $\begin{bmatrix} 2 & 2 & -4 \\ 1 & 5 & 2 \\ 3 & 7 & -2 \end{bmatrix}$ (d) $\begin{bmatrix} 0 & 1 & 2 \\ 1 & 2 & 0 \\ 1 & 3 & 4 \end{bmatrix}$

16. Find all values of λ for which

(a) $\det\left(\begin{bmatrix} \lambda - 2 & 2 \\ 3 & \lambda - 3 \end{bmatrix}\right) = 0$

(b) $\det(\lambda I_3 - A) = 0$, where $A = \begin{bmatrix} 1 & 0 & -1 \\ 2 & 0 & 1 \\ 0 & 0 & -1 \end{bmatrix}$

17. Find all values of λ for which

(a) $\det\left(\begin{bmatrix} \lambda - 1 & 4 \\ 0 & \lambda - 4 \end{bmatrix}\right) = 0$

(b) $\det(\lambda I_3 - A) = 0$, where

$$A = \begin{bmatrix} -3 & -1 & -3 \\ 0 & 3 & 0 \\ -2 & -1 & -2 \end{bmatrix}$$

18. Use Corollary 3.4 to find whether the following homogeneous systems have nontrivial solutions.

(a) $\begin{aligned} x - 2y + z &= 0 \\ 2x + 3y + z &= 0 \\ 3x + y + 2z &= 0 \end{aligned}$

(b) $\begin{aligned} x + 2y + w &= 0 \\ x + 2y + 3z &= 0 \\ z + 2w &= 0 \\ y + 2z - w &= 0 \end{aligned}$

19. Repeat Exercise 18 for the following homogeneous systems.

(a) $\begin{aligned} x + 2y - z &= 0 \\ 2x + y + 2z &= 0 \\ 3x - y + z &= 0 \end{aligned}$

(b) $\begin{aligned} x + y + 2z + w &= 0 \\ 2x - y + z - w &= 0 \\ 3x + y + 2z + 3w &= 0 \\ 2x - y - z + w &= 0 \end{aligned}$

In Exercises 20 through 23, if possible, solve the given linear system by Cramer's rule.

20. $\begin{aligned} 2x + 4y + 6z &= 2 \\ x + 2z &= 0 \\ 2x + 3y - z &= -5 \end{aligned}$

21. $\begin{aligned} x + y + z - 2w &= -4 \\ 2y + z + 3w &= 4 \\ 2x + y - z + 2w &= 5 \\ x - y + w &= 4 \end{aligned}$

22. $\begin{aligned} 2x + y + z &= 6 \\ 3x + 2y - 2z &= -2 \\ x + y + 2z &= 4 \end{aligned}$

23. $\begin{aligned} 2x + 3y + 7z &= 2 \\ -2x - 4z &= 0 \\ x + 2y + 4z &= 0 \end{aligned}$

In Exercises 24 and 25, determine which of the following bit matrices are nonsingular using any of the techniques in the list of nonsingular equivalences.

24. (a) $\begin{bmatrix} 1 & 1 & 1 \\ 0 & 1 & 1 \\ 1 & 1 & 0 \end{bmatrix}$ (b) $\begin{bmatrix} 1 & 1 & 0 & 0 \\ 1 & 0 & 1 & 0 \\ 1 & 0 & 0 & 1 \\ 0 & 0 & 1 & 1 \end{bmatrix}$

25. (a) $\begin{bmatrix} 1 & 0 & 1 & 1 \\ 1 & 1 & 0 & 1 \\ 1 & 1 & 1 & 0 \\ 0 & 1 & 1 & 1 \end{bmatrix}$ (b) $\begin{bmatrix} 0 & 1 & 1 & 1 & 1 \\ 1 & 0 & 1 & 1 & 1 \\ 1 & 1 & 0 & 1 & 1 \\ 1 & 1 & 1 & 0 & 1 \\ 1 & 1 & 1 & 1 & 0 \end{bmatrix}$

Theoretical Exercises

T.1. Show by a column (row) expansion that if $A = [a_{ij}]$ is upper (lower) triangular, then $\det(A) = a_{11}a_{22}\cdots a_{nn}$.

T.2. If $A = [a_{ij}]$ is a 3×3 matrix, develop the general expression for $\det(A)$ by expanding (a) along the second column, and (b) along the third row. Compare these answers with those obtained for Example 6 in Section 3.1.

T.3. Show that if A is symmetric, then adj A is also symmetric.

T.4. Show that if A is a nonsingular upper triangular matrix, then A^{-1} is also upper triangular.

T.5. Show that

$$A = \begin{bmatrix} a & b \\ c & d \end{bmatrix}$$

is nonsingular if and only if $ad - bc \neq 0$. If this condition is satisfied, use Corollary 3.3 to find A^{-1}.

T.6. Using Corollary 3.3, find the inverse of

$$A = \begin{bmatrix} 1 & a & a^2 \\ 1 & b & b^2 \\ 1 & c & c^2 \end{bmatrix}.$$

[*Hint*: See Exercise T.12 in Section 3.1, where $\det(A)$ was computed.]

T.7. Show that if A is singular, then adj A is singular. [*Hint*: Show that if A is singular, then $A(\text{adj } A) = O$.]

T.8. Show that if A is an $n \times n$ matrix, then $\det(\text{adj } A) = [\det(A)]^{n-1}$.

T.9. Do Exercise T.10 in Section 1.6 using determinants.

T.10. Let $AB = AC$. Show that if $\det(A) \neq 0$, then $B = C$.

T.11. Let A be an $n \times n$ matrix all of whose entries are integers. Show that if $\det(A) = \pm 1$, then all entries of A^{-1} are integers.

T.12. Show that if A is nonsingular, then adj A is nonsingular and

$$(\text{adj } A)^{-1} = \frac{1}{\det(A)} A = \text{adj } (A^{-1}).$$

T.13. Let A be a 2×2 bit matrix such that A is row equivalent to I_2. Determine all possible such matrices A. (*Hint*: See Exercise T.19 in Section 3.1 or Exercise T.11 in Section 1.6.)

MATLAB Exercises

ML.1. In MATLAB there is a routine **cofactor** that computes the (i, j) cofactor of a matrix. For directions on using this routine, type **help cofactor**. Use **cofactor** to check your hand computations for the matrix A in Exercise 1.

ML.2. Use the **cofactor** routine (see Exercise ML.1) to compute the cofactor of the elements in the second row of

$$A = \begin{bmatrix} 1 & 5 & 0 \\ 2 & -1 & 3 \\ 3 & 2 & 1 \end{bmatrix}.$$

ML.3. Use the **cofactor** routine to evaluate the determinant of A using Theorem 3.9.

$$A = \begin{bmatrix} 4 & 0 & -1 \\ -2 & 2 & -1 \\ 0 & 4 & -3 \end{bmatrix}$$

ML.4. Use the **cofactor** routine to evaluate the determinant of A using Theorem 3.9.

$$A = \begin{bmatrix} -1 & 2 & 0 & 0 \\ 2 & -1 & 2 & 0 \\ 0 & 2 & -1 & 2 \\ 0 & 0 & 2 & -1 \end{bmatrix}$$

ML.5. In MATLAB there is a routine **adjoint**, which computes the adjoint of a matrix. For directions on using this routine, type **help adjoint**. Use **adjoint** to aid in computing the inverses of the matrices in Exercise 11.

3.3 DETERMINANTS FROM A COMPUTATIONAL POINT OF VIEW

In this book we have, by now, developed two methods for solving a linear system of n equations in n unknowns: Gauss–Jordan reduction and Cramer's rule. We also have two methods for finding the inverse of a nonsingular matrix: the method involving determinants and the method discussed in Section 1.7. In this section we discuss criteria to be considered when selecting one or another of these methods.

Most sizable problems in linear algebra are solved on computers so that it is natural to compare two methods by estimating their computing time for the same problem. Since addition is so much faster than multiplication, the number of multiplications is often used as a basis of comparison for two numerical procedures.

Consider the linear system $A\mathbf{x} = \mathbf{b}$, where A is 25×25. If we find $\mathbf{x}$ by Cramer's rule, we must first obtain $\det(A)$. We can find $\det(A)$ by cofactor expansion, say $\det(A) = a_{11}A_{11} + a_{21}A_{21} + \cdots + a_{n1}A_{n1}$, where we have expanded $\det(A)$ along the first column. Note that if each cofactor is available, we require 25 multiplications. Now each cofactor A_{ij} is plus or minus the determinant of a 24×24 matrix, which can be expanded along a given row or column, requiring 24 multiplications. Thus the computation of $\det(A)$ requires more than $25 \times 24 \times \cdots \times 2 \times 1 = 25!$ multiplications. Even if we were to use a futuristic (not very far into the future) computer capable of performing 10 trillion (1×10^{12}) multiplications *per second* (3.15×10^{19} *per year*), it would take *about 49,000 years* to evaluate $\det(A)$. However, Gauss–Jordan reduction takes about $25^3/3$ multiplications, and we would find the solution in less than *one second*. Of course, we can compute $\det(A)$ in a much more efficient way by using elementary row operations to reduce A to triangular form and then use Theorem 3.7 (see Example 17 in Section 3.1). When implemented this way for an $n \times n$ matrix, Cramer's rule will require approximately n^4 multiplications, compared to $n^3/3$ multiplications for Gauss–Jordan reduction. Thus Gauss–Jordan reduction is still much faster.

In general, if we are seeking numerical answers, then any method involving determinants can be used for $n \leq 4$. For $n \geq 5$, determinant-dependent methods are much less efficient than Gauss–Jordan reduction or the method of Section 1.7. for inverting a matrix.

The importance of determinants obviously does not lie in their computational use. Note that methods involving determinants enable one to express the inverse of a matrix and the solution to a linear system of n equations in n unknowns by means of expressions or formulas. Gauss–Jordan reduction and the method for finding A^{-1} given in Section 1.6 do not yield a *formula* for the answer; we must proceed numerically to obtain the answer. Sometimes we do not need numerical answers but an expression for the answer because we may wish to further manipulate the answer. Another important reason for studying determinants is that they play a key role in the study of eigenvalues and eigenvectors, which will be undertaken in Chapter 8.

Key Ideas for Review

- **Theorem 3.1.** $\det(A^T) = \det(A)$.
- **Theorem 3.2.** If B results from A by interchanging two rows (columns) of A, then $\det(B) = -\det(A)$.
- **Theorem 3.3.** If two rows (columns) of A are equal, then $\det(A) = 0$.
- **Theorem 3.4.** If a row (column) of A consists entirely of zeros, then $\det(A) = 0$.
- **Theorem 3.5.** If B is obtained from A by multiplying a row (column) of A by a real number c, then $\det(B) = c\det(A)$.
- **Theorem 3.6.** If B is obtained from A by adding a multiple of a row (column) of A to another row (column) of A, then $\det(B) = \det(A)$.
- **Theorem 3.7.** If $A = \begin{bmatrix} a_{ij} \end{bmatrix}$ is upper (lower) triangular, then $\det(A) = a_{11}a_{22}\cdots a_{nn}$.
- **Theorem 3.8.** $\det(AB) = \det(A)\det(B)$.
- **Theorem 3.9 (Cofactor Expansion).** If $A = \begin{bmatrix} a_{ij} \end{bmatrix}$, then

$$\det(A) = a_{i1}A_{i1} + a_{i2}A_{i2} + \cdots + a_{in}A_{in}$$

and

$$\det(A) = a_{1j}A_{1j} + a_{2j}A_{2j} + \cdots + a_{nj}A_{nj}.$$

- **Corollary 3.3.** If $\det(A) \neq 0$, then

$$A^{-1} = \begin{bmatrix} \dfrac{A_{11}}{\det(A)} & \dfrac{A_{21}}{\det(A)} & \cdots & \dfrac{A_{n1}}{\det(A)} \\ \dfrac{A_{12}}{\det(A)} & \dfrac{A_{22}}{\det(A)} & \cdots & \dfrac{A_{n2}}{\det(A)} \\ \vdots & \vdots & & \vdots \\ \dfrac{A_{1n}}{\det(A)} & \dfrac{A_{2n}}{\det(A)} & \cdots & \dfrac{A_{nn}}{\det(A)} \end{bmatrix}.$$

- **Theorem 3.12.** A is nonsingular if and only if $\det(A) \neq 0$.
- **Corollary 3.4.** If A is an $n \times n$ matrix, then the homogeneous system $A\mathbf{x} = \mathbf{0}$ has a nontrivial solution if and only if $\det(A) = 0$.
- **Theorem 3.13 (Cramer's Rule).** Let $A\mathbf{x} = \mathbf{b}$ be a linear system of n equations in n unknowns. If $\det(A) \neq 0$, then the system has the unique solution

$$x_1 = \frac{\det(A_1)}{\det(A)}, \quad x_2 = \frac{\det(A_2)}{\det(A)}, \ldots, \quad x_n = \frac{\det(A_n)}{\det(A)},$$

where A_i is the matrix obtained from A by replacing the ith column of A by $\mathbf{b}$.

■ **List of Nonsingular Equivalences.** The following statements are equivalent:

1. A is nonsingular.
2. $\mathbf{x} = \mathbf{0}$ is the only solution to $A\mathbf{x} = \mathbf{0}$.
3. A is row equivalent to I_n.
4. The linear system $A\mathbf{x} = \mathbf{b}$ has a unique solution for every $n \times 1$ matrix $\mathbf{b}$.
5. $\det(A) \neq 0$.

Supplementary Exercises

1. Evaluate the following determinants using Equation (2) of Section 3.1.

(a) $\begin{vmatrix} 0 & 2 & 0 \\ 0 & 0 & -3 \\ 4 & 0 & 0 \end{vmatrix}$ (b) $\begin{vmatrix} 3 & 0 & 0 & 0 \\ 0 & -2 & 0 & 0 \\ 0 & 4 & 1 & 0 \\ 3 & 2 & -1 & -4 \end{vmatrix}$

2. If

$$\begin{vmatrix} a_1 & a_2 & a_3 \\ b_1 & b_2 & b_3 \\ c_1 & c_2 & c_3 \end{vmatrix} = 5,$$

find the determinants of the following matrices:

(a) $B = \begin{bmatrix} \frac{1}{2}a_1 & \frac{1}{2}a_2 & \frac{1}{2}a_3 \\ b_1 & b_2 & b_3 \\ c_1 & c_2 & c_3 \end{bmatrix}$

(b) $C = \begin{bmatrix} a_1 - b_1 & a_2 - b_2 & a_3 - b_3 \\ 3b_1 & 3b_2 & 3b_3 \\ 2c_1 & 2c_2 & 2c_3 \end{bmatrix}$

3. Let A be 4×4 and suppose that $|A| = 5$. Compute
(a) $|A^{-1}|$ (b) $|2A|$ (c) $|2A^{-1}|$
(d) $|(2A)^{-1}|$

4. Let $|A| = 3$ and $|B| = 4$. Compute
(a) $|AB|$ (b) $|ABA^T|$ (c) $|B^{-1}AB|$

5. Find all values of λ for which

$$\det\left(\begin{bmatrix} \lambda+2 & -1 & 3 \\ 2 & \lambda-1 & 2 \\ 0 & 0 & \lambda+4 \end{bmatrix}\right) = 0.$$

6. Evaluate

$$\begin{vmatrix} 2 & -1 & 3 \\ 4 & 1 & 5 \\ -2 & -3 & -2 \end{vmatrix}.$$

7. Evaluate

$$\begin{vmatrix} 3 & 2 & -1 & 1 \\ 4 & 1 & 1 & 0 \\ -1 & 2 & 3 & 4 \\ -2 & 3 & 5 & 1 \end{vmatrix}.$$

8. Compute all the cofactors of

$$A = \begin{bmatrix} 2 & -1 & 3 \\ 1 & 4 & 5 \\ -3 & -4 & 6 \end{bmatrix}.$$

9. Evaluate

$$\begin{vmatrix} 3 & 2 & -1 & 0 \\ -1 & 0 & 3 & 2 \\ 4 & 1 & 5 & -2 \\ 1 & 3 & 2 & -3 \end{vmatrix}$$

by cofactor expansion.

10. Let

$$A = \begin{bmatrix} 3 & -1 & 2 \\ 0 & 4 & 5 \\ 1 & 3 & 2 \end{bmatrix}.$$

(a) Find adj A.

(b) Compute $\det(A)$.

(c) Show that $A(\text{adj } A) = \det(A)I_3$.

11. Compute the inverse of the following matrix, if it exists, using Corollary 3.3:

$$\begin{bmatrix} 2 & -1 & 3 \\ 0 & 1 & 2 \\ -1 & 1 & 2 \end{bmatrix}.$$

12. Find all values of λ for which

$$\begin{bmatrix} \lambda-3 & 0 & 3 \\ 0 & \lambda+2 & 0 \\ -5 & 0 & \lambda+5 \end{bmatrix}$$

is singular.

13. If

$$A = \begin{bmatrix} \lambda & 0 & 1 \\ 1 & \lambda-1 & 0 \\ 0 & 0 & \lambda+1 \end{bmatrix},$$

find all values of λ for which the homogeneous system $A\mathbf{x} = \mathbf{0}$ has only the trivial solution.

14. If possible, solve the following linear system by Cramer's rule:

$$\begin{aligned} 3x + 2y - z &= -1 \\ x - y - z &= 0 \\ 2x + y - 2z &= 3. \end{aligned}$$

15. Using only elementary row or elementary column operations and Theorems 3.2, 3.5, and 3.6 (do not expand the determinants), verify the following.

(a) $\begin{vmatrix} a-b & 1 & a \\ b-c & 1 & b \\ c-a & 1 & c \end{vmatrix} = \begin{vmatrix} a & 1 & b \\ b & 1 & c \\ c & 1 & a \end{vmatrix}$

(b) $\begin{vmatrix} 1 & a & bc \\ 1 & b & ca \\ 1 & c & ab \end{vmatrix} = \begin{vmatrix} 1 & a & a^2 \\ 1 & b & b^2 \\ 1 & c & c^2 \end{vmatrix}$

16. Find all values of a for which the linear system

$$\begin{aligned} 2x + ay &= 0 \\ ax + 2y &= 0 \end{aligned}$$

has (a) a unique solution; (b) infinitely many solutions.

17. Find all values of a for which the matrix

$$\begin{bmatrix} a-2 & 2 \\ a-2 & a+2 \end{bmatrix}$$

is nonsingular.

18. Use Cramer's rule to find all values of a for which the linear system

$$\begin{aligned} x - 2y + 2z &= 9 \\ 2x + y \qquad &= a \\ 3x - y - z &= -10 \end{aligned}$$

has the solution in which $y = 1$.

Theoretical Exercises

T.1. Show that if two rows (columns) of the $n \times n$ matrix A are proportional, then $\det(A) = 0$.

T.2. Show that if A is an $n \times n$ matrix, then $\det(AA^T) \geq 0$.

T.3. Let Q be an $n \times n$ matrix in which each entry is 1. Show that $\det(Q - nI_n) = 0$.

T.4. Let P be an invertible matrix. Show that if $B = PAP^{-1}$, then $\det(B) = \det(A)$.

T.5. Show that if A is a singular $n \times n$ matrix, then $A(\text{adj } A) = O$. (*Hint*: See Theorem 3.11.)

T.6. Show that if A is a singular $n \times n$ matrix, then AB is singular for any $n \times n$ matrix B.

T.7. Show that if A and B are square matrices, then

$$\det\left(\begin{bmatrix} A & O \\ O & B \end{bmatrix}\right) = \det(A)\det(B).$$

T.8. Show that if A, B, and C are square matrices, then

$$\det\left(\begin{bmatrix} A & O \\ C & B \end{bmatrix}\right) = \det(A)\det(B).$$

T.9. Let A be an $n \times n$ matrix with integer entries and $\det(A) = \pm 1$. Show that if $\mathbf{b}$ has all integer entries, then every solution to $A\mathbf{x} = \mathbf{b}$ consists of integers.

Chapter Test

1. Evaluate

$$\begin{vmatrix} 1 & 1 & 2 & -1 \\ 0 & 1 & 0 & 3 \\ -1 & 2 & -3 & 4 \\ 0 & 5 & 0 & -2 \end{vmatrix}.$$

2. Let A be 3×3 and suppose that $|A| = 2$. Compute
(a) $|3A|$ (b) $|3A^{-1}|$ (c) $|(3A)^{-1}|$

3. For what value of a is

$$\begin{vmatrix} 2 & 1 & 0 \\ 0 & -1 & 3 \\ 0 & 1 & a \end{vmatrix} + \begin{vmatrix} 0 & a & 1 \\ 1 & 3a & 0 \\ -2 & a & 2 \end{vmatrix} = 14?$$

4. Find all values of a for which the matrix

$$\begin{bmatrix} a^2 & 0 & 3 \\ 5 & a & 2 \\ 3 & 0 & 1 \end{bmatrix}$$

is singular.

5. Solve the following linear system by Cramer's rule.

$$\begin{aligned} x - y + z &= -1 \\ 2x + y - 3z &= 8 \\ x - 2y + 3z &= -5. \end{aligned}$$

6. Answer each of the following as true or false. Justify your answer.

(a) $\det(AA^T) = \det(A^2)$.

(b) $\det(-A) = -\det(A)$.

(c) If $A^T = A^{-1}$, then $\det(A) = 1$.

(d) If $\det(A) = 0$, then $A = O$.

(e) If $\det(A) = 7$, then $A\mathbf{x} = \mathbf{0}$ has only the trivial solution.

(f) The sign of the term $a_{15}a_{23}a_{31}a_{42}a_{54}$ in the expansion of the determinant of a 5×5 matrix is $+$.

(g) If $\det(A) = 0$, then $\det(\text{adj } A) = 0$.

(h) If $B = PAP^{-1}$, and P is nonsingular, then $\det(B) = \det(A)$.

(i) If $A^4 = I_n$, then $\det(A) = 1$.

(j) If $A^2 = A$ and $A \neq I_n$, then $\det(A) = 0$.

CHAPTER 4

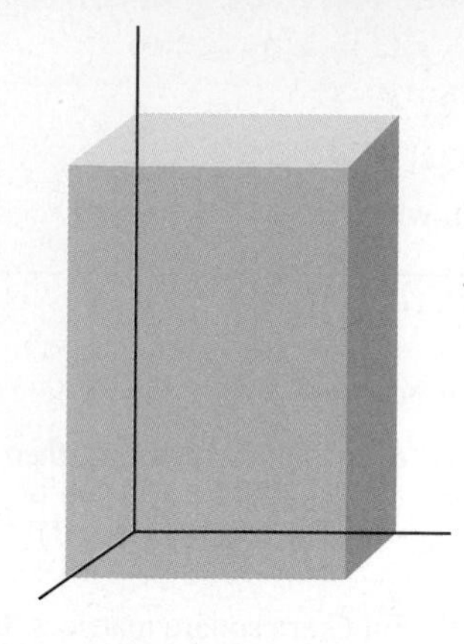

VECTORS IN R^n

4.1 VECTORS IN THE PLANE

COORDINATE SYSTEMS

In many applications we deal with measurable quantities, such as pressure, mass, and speed, which can be described completely by giving their magnitude. There are also many other measurable quantities, such as velocity, force, and acceleration, that require for their description not only magnitude, but also a direction. These are called **vectors**, and their study comprises this chapter. Vectors will be denoted by lowercase boldface letters such as **u**, **v**, **w**, **x**, **y**, and **z**. The real numbers will be called **scalars** and will be denoted by lowercase italic letters.

We recall that the real number system may be visualized as a straight line L, which is usually taken in a horizontal position. A point O, called the **origin**, is chosen on L; O corresponds to the number 0. A point A is chosen to the right of O, thereby fixing the length of OA as 1 and specifying a positive direction. Thus the positive real numbers lie to the right of O; the negative real numbers lie to the left of O (Figure 4.1).

Figure 4.1 ►

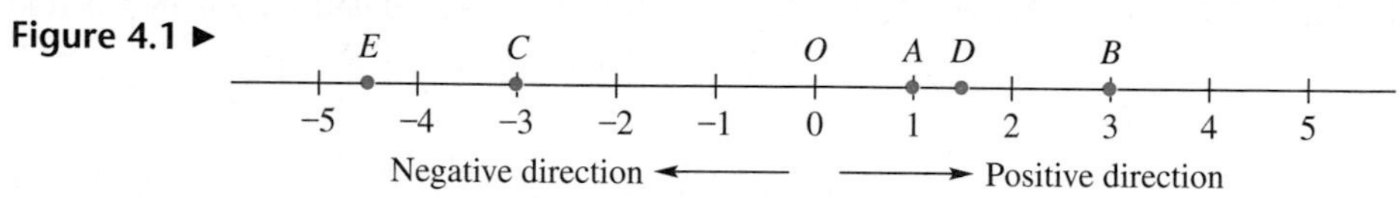

The **absolute value** $|x|$ of the real number x is defined by

$$|x| = \begin{cases} x & \text{if } x \geq 0 \\ -x & \text{if } x < 0. \end{cases}$$

Thus $|3| = 3$, $|-2| = 2$, $|0| = 0$, $\left|-\frac{2}{3}\right| = \frac{2}{3}$, and $|-1.82| = 1.82$.

The real number x corresponding to the point P is called the **coordinate** of P and the point P whose coordinate is x is denoted by $P(x)$. The line L is called a **coordinate axis**. If P is to the right of O, then its coordinate is the length of the segment OP. If Q is to the left of O, then its coordinate is the negative of the length of the segment OQ. The distance between the points P and Q with respective coordinates a and b is $|b - a|$.

EXAMPLE 1

Referring to Figure 4.1, we see that the coordinates of the points B, C, D, and E are, respectively, 3, -3, 1.5, and -4.5. The distance between B and C is $|-3-3| = 6$. The distance between A and B is $|3-1| = 2$. The distance between C and E is $|-4.5-(-3)| = 1.5$. ■

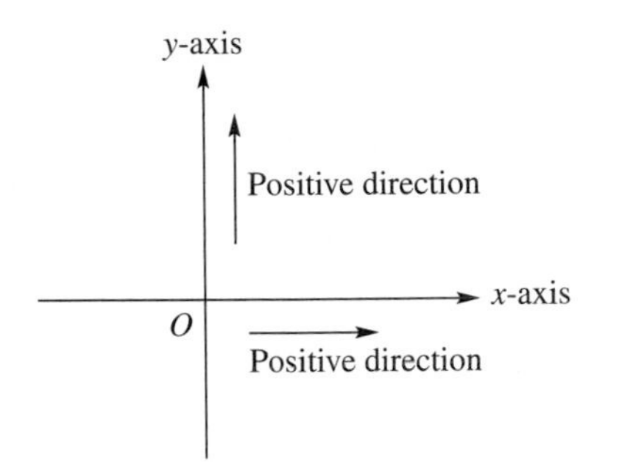

Figure 4.2 ▲

We shall now turn to the analogous situation for the plane. We draw a pair of perpendicular lines intersecting at a point O, called the **origin**. One of the lines, the ***x*-axis**, is usually taken in a horizontal position. The other line, the ***y*-axis**, is then taken in a vertical position. We now choose a point on the x-axis to the right of O and a point on the y-axis above O to fix the units of length and positive directions on the x- and y-axes. Frequently, but not always, these points are chosen so that they are both equidistant from O, that is, so that the same unit of length is used for both axes. The x- and y-axes together are called **coordinate axes** (Figure 4.2). The **orthogonal projection** of a point P in the plane on a line L is the point Q obtained by intersecting L with the line L' passing through P and perpendicular to L [Figures 4.3(a) and (b)].

Figure 4.3 ►

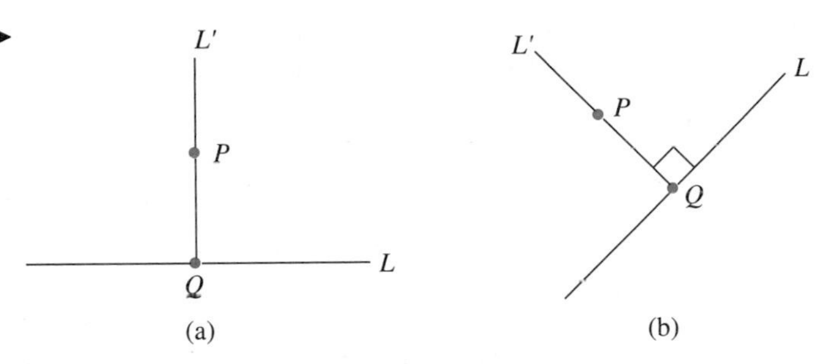

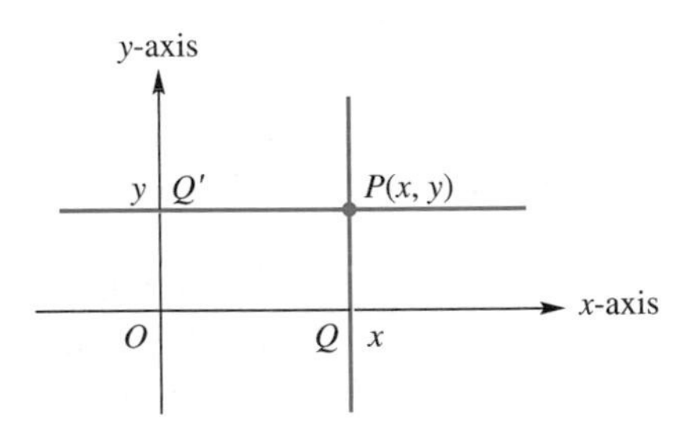

Figure 4.4 ▲

Let P be a point in the plane and let Q be its projection onto the x-axis. The coordinate of Q on the x-axis is called the ***x*-coordinate** of P. Similarly, let Q' be the projection of P onto the y-axis. The coordinate of Q' on the y-axis is called the ***y*-coordinate** of P. Thus with every point in the plane we associate an ordered pair (x, y) of real numbers, its coordinates. The point P with coordinates x and y is denoted by $P(x, y)$. Conversely, it is easy to see (Exercise T.1) how we can associate a point in the plane with each ordered pair (x, y) of real numbers (Figure 4.4). The correspondence given above between points in the plane and ordered pairs of real numbers is called a **rectangular coordinate system** or a **Cartesian coordinate system** (after René Descartes*). The set of points in the plane is denoted by R^2. It is also called **2-space**.

---

*René Descartes (1596–1650) was one of the best-known scientists and philosophers of his day; he was considered by some to be the founder of modern philosophy. After completing a university degree in law, he turned to the private study of mathematics, simultaneously pursuing interests in Parisian night life and in the military, volunteering for brief periods in the Dutch, Bavarian, and French armies. The most productive period of his life was 1628–1648, when he lived in Holland. In 1649 he accepted an invitation from Queen Christina of Sweden to be her private tutor and to establish an Academy of Sciences there. Unfortunately, he had no time for this project, for he died of pneumonia in 1650.

In 1619 Descartes had a dream in which he realized that the method of mathematics is the best way for obtaining truth. However, his only mathematical publication was *La Géométrie*, which appeared as an appendix to his major philosophical work *Discours de la méthode pour bien conduire sa raison, et chercher la vérité dans les sciences* (*Discourse on the Method of Reasoning Well and Seeking Truth in the Sciences*). In *La Géométrie* he proposes the radical idea of doing geometry algebraically. To express a curve algebraically, one chooses any convenient line of reference and, on the line, a point of reference. If y represents the distance from any point of the curve to the reference line and x represents the distance along the line to the reference point, there is an equation relating x and y that represents the curve. The systematic use of

EXAMPLE 2 In Figure 4.5 we show a number of points and their coordinates.

Figure 4.5 ▶

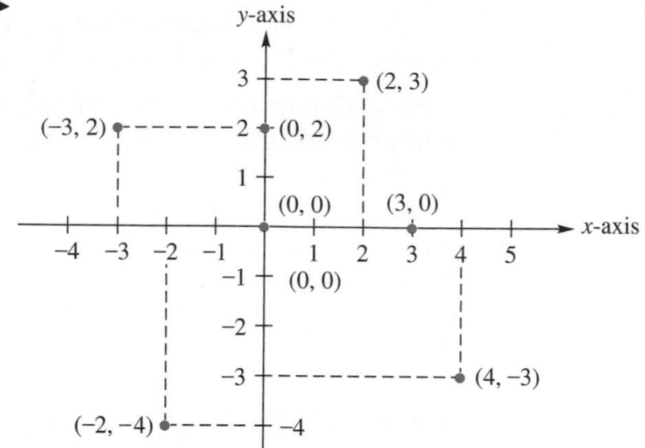

The coordinates of the origin are $(0, 0)$. The coordinates of the projection of the point $P(x, y)$ on the x-axis are $(x, 0)$ and the coordinates of its projection on the y-axis are $(0, y)$. ■

VECTORS

Recall that in Section 1.2 we defined an n-vector and at the beginning of Section 1.3 we briefly introduced vectors algebraically to better understand matrix multiplication. In this section we view 2-vectors geometrically and in the next section we do the same with n-vectors.

Consider the 2-vector

$$\mathbf{u} = \begin{bmatrix} x_1 \\ y_1 \end{bmatrix},$$

where x_1 and y_1 are real numbers. With $\mathbf{u}$ we associate the directed line segment with initial point at the origin $O(0, 0)$ and terminal point at $P(x_1, y_1)$. The directed line segment from O to P is denoted by $\overrightarrow{OP}$; O is called its **tail** and P its **head**. We distinguish tail and head by placing an arrow at the head (Figure 4.6).

Figure 4.6 ▶

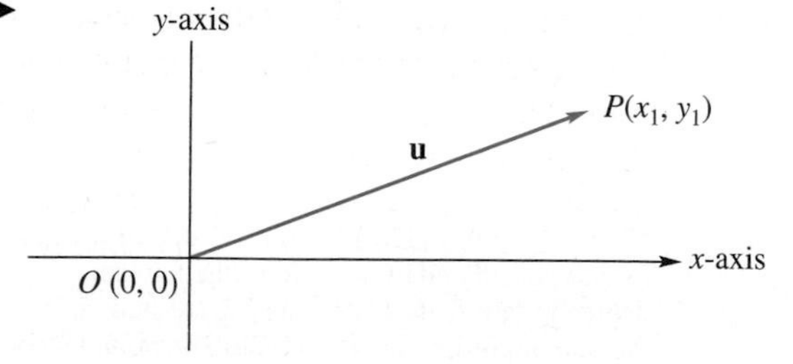

A directed line segment has a **direction**, which is the angle made with the positive x-axis, indicated by the arrow at its head. The **magnitude** of a directed line segment is its length.

EXAMPLE 3 Let

$$\mathbf{u} = \begin{bmatrix} 2 \\ 3 \end{bmatrix}.$$

"Cartesian" coordinates described previously was introduced later in the seventeenth century by authors following up on Descartes's work.

With $\mathbf{u}$ we can associate the directed line segment with tail $O(0, 0)$ and head $P(2, 3)$, shown in Figure 4.7. ■

Figure 4.7 ►

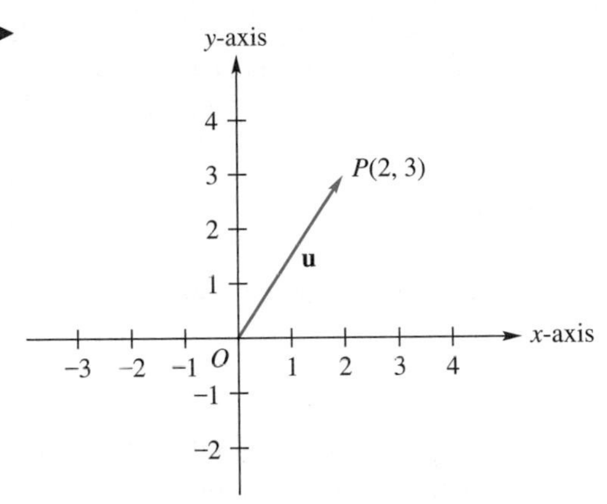

Conversely, with a directed line segment $\overrightarrow{OP}$ with tail $O(0, 0)$ and head $P(x_1, y_1)$, we can associate the 2-vector

$$\begin{bmatrix} x_1 \\ y_1 \end{bmatrix}.$$

EXAMPLE 4 With the directed line segment $\overrightarrow{OP}$ with head $P(4, 5)$, we can associate the 2-vector

$$\begin{bmatrix} 4 \\ 5 \end{bmatrix}.$$

■

DEFINITION A **vector in the plane** is a 2-vector

$$\mathbf{u} = \begin{bmatrix} x_1 \\ y_1 \end{bmatrix},$$

where x_1 and y_1 are real numbers, called the **components of u**. We refer to a vector in the plane merely as a **vector**.

Thus with every vector we can associate a directed line segment, and conversely, with every directed line segment emanating from the origin we can associate a vector. As we have seen, a coordinate system is needed to set up this correspondence. The magnitude and direction of a vector are the magnitude and direction of its associated directed line segment. Frequently, the notions of directed line segment and vector are used interchangeably and a directed line segment is called a **vector**.

Since a vector is a matrix, the vectors

$$\mathbf{u} = \begin{bmatrix} x_1 \\ y_1 \end{bmatrix} \quad \text{and} \quad \mathbf{v} = \begin{bmatrix} x_2 \\ y_2 \end{bmatrix}$$

are said to be **equal** if $x_1 = x_2$ and $y_1 = y_2$. That is, two vectors are equal if their respective components are equal.

EXAMPLE 5 The vectors

$$\begin{bmatrix} 1 \\ 0 \end{bmatrix} \quad \text{and} \quad \begin{bmatrix} 1 \\ -2 \end{bmatrix}$$

are not equal, since their respective components are not equal. ■

With each vector

$$\mathbf{u} = \begin{bmatrix} x_1 \\ y_1 \end{bmatrix}$$

we can also associate in a unique manner the point $P(x_1, y_1)$; conversely, with each point $P(x_1, y_1)$ we can associate in a unique manner the vector

$$\begin{bmatrix} x_1 \\ y_1 \end{bmatrix}.$$

Thus we also write the vector **u** as

$$\mathbf{u} = (x_1, y_1).$$

Of course, this association is obtained by means of the directed line segment $\overrightarrow{OP}$, where O is the origin (Figure 4.6).

Thus the plane may be viewed both as the set of all points or as the set of all vectors. For this reason and, depending upon the context, we sometimes take R^2 as the set of all ordered pairs (x_1, y_1) and sometimes as the set of all 2-vectors

$$\begin{bmatrix} x_1 \\ y_1 \end{bmatrix}.$$

Frequently, in physical applications it is necessary to deal with a directed line segment $\overrightarrow{PQ}$, from the point $P(x_1, y_1)$ (not the origin) to the point $Q(x_2, y_2)$, as shown in Figure 4.8(a). Such a directed line segment will also be called a **vector in the plane**, or simply a **vector** with **tail** $P(x_1, y_1)$ and **head** $Q(x_2, y_2)$. The **components** of such a vector are $x_2 - x_1$ and $y_2 - y_1$. Thus the vector PQ in Figure 4.8(a) can also be represented by the vector $(x_2 - x_1, y_2 - y_1)$ with tail O and head $P''(x_2 - x_1, y_2 - y_1)$. Two such vectors in the plane will be called **equal** if their components are equal. Consider the vectors $\overrightarrow{P_1Q_1}$, $\overrightarrow{P_2Q_2}$, and $\overrightarrow{P_3Q_3}$ joining the points $P_1(3, 2)$ and $Q_1(5, 5)$, $P_2(0, 0)$ and $Q_2(2, 3)$, $P_3(-3, 1)$ and $Q_3(-1, 4)$, respectively, as shown in Figure 4.8(b). Since they all have the same components, they are equal.

To find the head $Q_4(a, b)$ of the vector

$$\overrightarrow{P_4Q_4} = \begin{bmatrix} 2 \\ 3 \end{bmatrix} = \overrightarrow{P_2Q_2},$$

Figure 4.8 ►

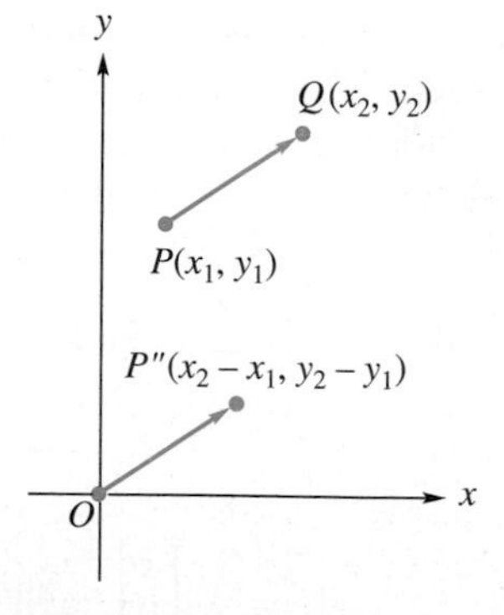

(a) Different directed line segments representing the same vector.

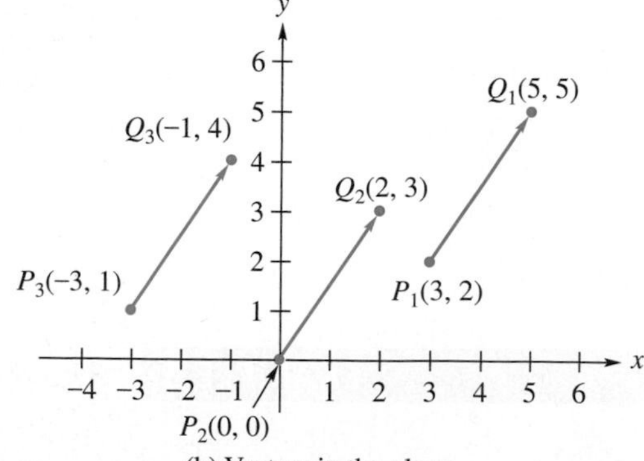

(b) Vectors in the plane.

with tail $P_4(-5, 2)$, we proceed as follows. We must have $a - (-5) = 2$ and $b - 2 = 3$ so that $a = 2 - 5 = -3$ and $b = 3 + 2 = 5$, so the coordinates of Q_4 are $(-3, 5)$. Similarly, to find the tail $P_5(c, d)$ of the vector

$$\overrightarrow{P_5Q_5} = \begin{bmatrix} 2 \\ 3 \end{bmatrix},$$

with head $Q_5(8, 6)$, we must have $8 - c = 2$ and $6 - d = 3$ so that $c = 8 - 2 = 6$ and $d = 6 - 3 = 3$. Hence, the coordinates of P_5 are $(6, 3)$.

LENGTH

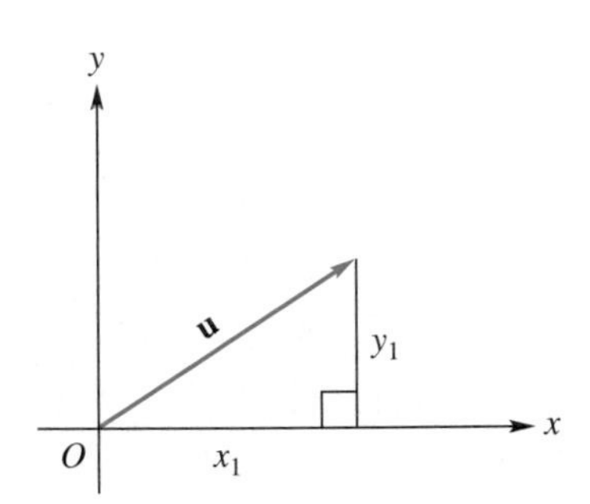

Figure 4.9 ▲

By the Pythagorean theorem (Figure 4.9), the **length** or **magnitude** of the vector $\mathbf{u} = (x_1, y_1)$ is

$$\|\mathbf{u}\| = \sqrt{x_1^2 + y_1^2}. \tag{1}$$

It also follows, by the Pythagorean theorem, that the length of the directed line segment with initial point $P_1(x_1, y_1)$ and terminal point $P_2(x_2, y_2)$ is (Figure 4.10)

$$\|\overrightarrow{P_1P_2}\| = \sqrt{(x_2 - x_1)^2 + (y_2 - y_1)^2}. \tag{2}$$

Equation (2) also gives the distance between the points P_1 and P_2.

Figure 4.10 ►

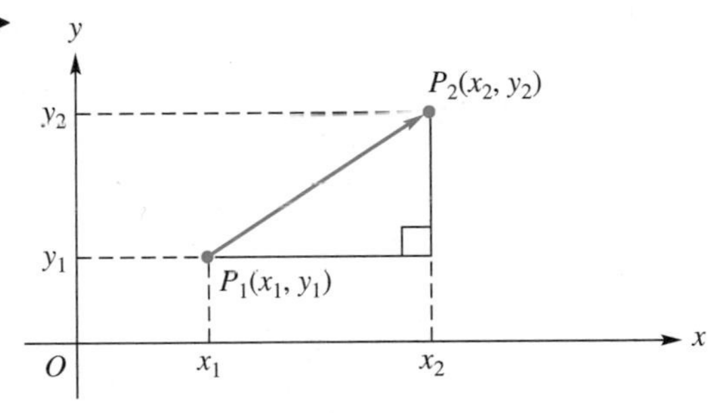

EXAMPLE 6 If $\mathbf{u} = (2, -5)$, then, by Equation (1),

$$\|\mathbf{u}\| = \sqrt{(2)^2 + (-5)^2} = \sqrt{4 + 25} = \sqrt{29}. \quad \blacksquare$$

EXAMPLE 7 The distance between $P(3, 2)$ and $Q(-1, 5)$, or the length of the directed line segment $\overrightarrow{PQ}$, is, by Equation (2),

$$\|\overrightarrow{PQ}\| = \sqrt{(-1 - 3)^2 + (5 - 2)^2} = \sqrt{(-4)^2 + 3^2} = \sqrt{25} = 5. \quad \blacksquare$$

The length of each vector (directed line segment) $\overrightarrow{P_1Q_1}$, $\overrightarrow{P_2Q_2}$, and $\overrightarrow{P_3Q_3}$ in Figure 4.8(b) is $\sqrt{13}$ (verify).

Two nonzero vectors

$$\mathbf{u} = \begin{bmatrix} x_1 \\ y_1 \end{bmatrix} \quad \text{and} \quad \mathbf{v} = \begin{bmatrix} x_2 \\ y_2 \end{bmatrix}$$

are said to be **parallel** if one is a multiple of the other. They are parallel if the lines on which they lie are both vertical or have the same slopes. Thus the vectors (directed line segments) $\overrightarrow{P_1Q_1}$, $\overrightarrow{P_2Q_2}$, and $\overrightarrow{P_3Q_3}$ in Figure 4.8(b) are parallel.

USING DETERMINANTS TO COMPUTE AREA

Consider the triangle with vertices (x_1, y_1), (x_2, y_2), and (x_3, y_3), as shown in Figure 4.11.

Figure 4.11 ▶

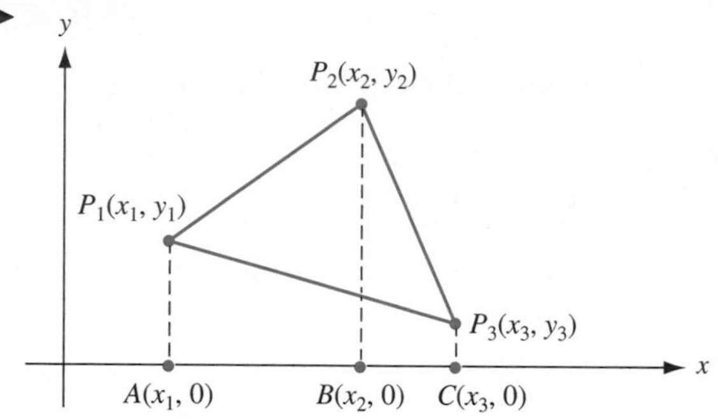

We may compute the area of this triangle as

$$\text{area of trapezoid } AP_1P_2B + \text{area of trapezoid } BP_2P_3C \\ - \text{area of trapezoid } AP_1P_3C.$$

Now recall that the area of a trapezoid is $\frac{1}{2}$ the distance between the parallel sides of the trapezoid times the sum of the lengths of the parallel sides. Thus

$$\begin{aligned}&\text{area of triangle } P_1P_2P_3 \\ &\quad = \tfrac{1}{2}(x_2 - x_1)(y_1 + y_2) + \tfrac{1}{2}(x_3 - x_2)(y_2 + y_3) - \tfrac{1}{2}(x_3 - x_1)(y_1 + y_3) \\ &\quad = \tfrac{1}{2}x_2y_1 - \tfrac{1}{2}x_1y_2 + \tfrac{1}{2}x_3y_2 - \tfrac{1}{2}x_2y_3 - \tfrac{1}{2}x_3y_1 + \tfrac{1}{2}x_1y_3.\end{aligned}$$

It turns out that this expression is

$$\tfrac{1}{2}\det\left(\begin{bmatrix} x_1 & y_1 & 1 \\ x_2 & y_2 & 1 \\ x_3 & y_3 & 1 \end{bmatrix}\right).$$

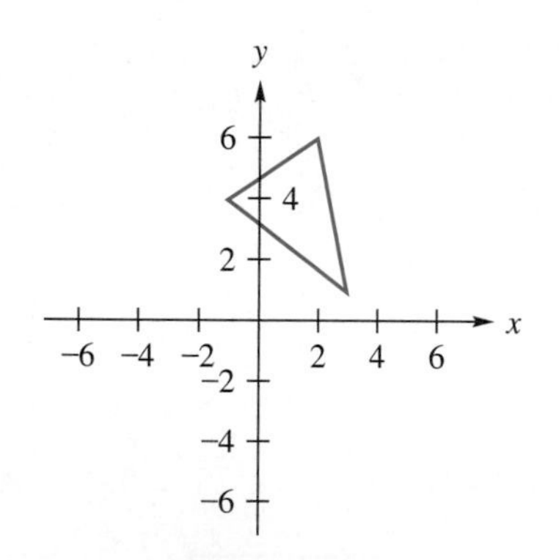

Figure 4.12 ▲

When the points are in the other quadrants or the points are labeled in a different order, the formula just obtained will yield the negative of the area of the triangle. Thus, for a triangle with vertices (x_1, y_1), (x_2, y_2), and (x_3, y_3), we have

$$\text{area of triangle} = \tfrac{1}{2}\left|\det\left(\begin{bmatrix} x_1 & y_1 & 1 \\ x_2 & y_2 & 1 \\ x_3 & y_3 & 1 \end{bmatrix}\right)\right| \tag{3}$$

(the area is $\frac{1}{2}$ the absolute value of the determinant).

EXAMPLE 8 Compute the area of the triangle T shown in Figure 4.12 with vertices $(-1, 4)$, $(3, 1)$, and $(2, 6)$.

Solution By Equation (3), the area of T is

$$\tfrac{1}{2}\left|\det\left(\begin{bmatrix} -1 & 4 & 1 \\ 3 & 1 & 1 \\ 2 & 6 & 1 \end{bmatrix}\right)\right| = \tfrac{1}{2}|17| = 8.5.$$

■

Suppose we now have the parallelogram shown in Figure 4.13. Since a diagonal divides the parallelogram into two equal triangles, it follows from Equation (3) that

$$\text{area of parallelogram} = \left|\det\left(\begin{bmatrix} x_1 & y_1 & 1 \\ x_2 & y_2 & 1 \\ x_3 & y_3 & 1 \end{bmatrix}\right)\right|.$$

Figure 4.13 ►

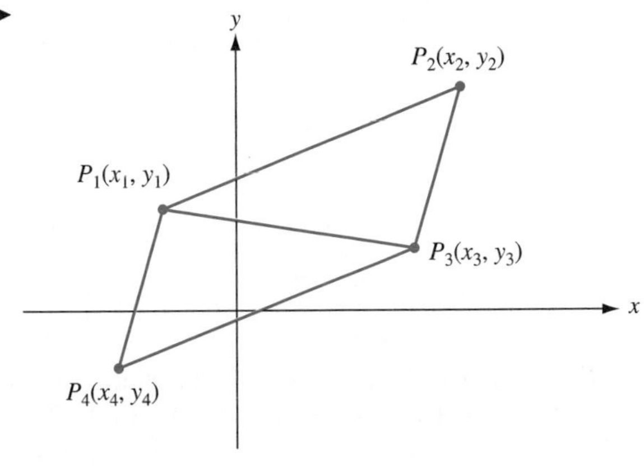

VECTOR OPERATIONS

DEFINITION Let $\mathbf{u} = (x_1, y_1)$* and $\mathbf{v} = (x_2, y_2)$ be two vectors in the plane. The **sum** of the vectors **u** and **v** is the vector

$$(x_1 + x_2, y_1 + y_2)$$

and is denoted by $\mathbf{u} + \mathbf{v}$. Thus vectors are added by adding their components.

EXAMPLE 9 Let $\mathbf{u} = (1, 2)$ and $\mathbf{v} = (3, -4)$. Then

$$\mathbf{u} + \mathbf{v} = (1 + 3, 2 + (-4)) = (4, -2).$$ ■

We can interpret vector addition geometrically as follows. In Figure 4.14, the vector from (x_1, y_1) to $(x_1 + x_2, y_1 + y_2)$ is also **v**. Thus the vector with tail O and head $(x_1 + x_2, y_1 + y_2)$ is $\mathbf{u} + \mathbf{v}$.

Figure 4.14 ►
Vector addition

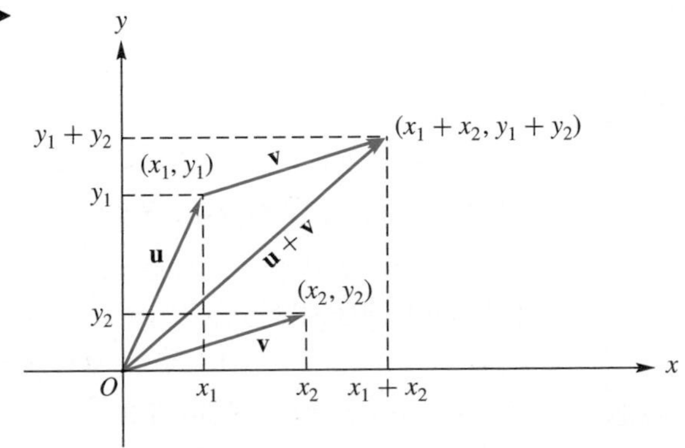

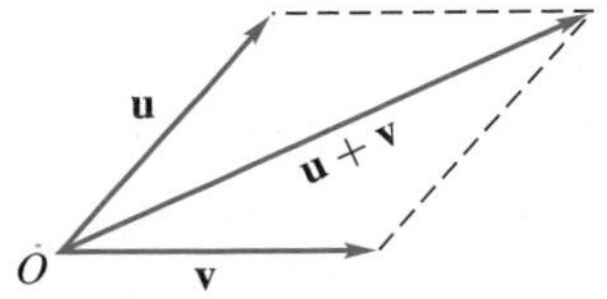

Figure 4.15 ▲
Vector addition

We can also describe $\mathbf{u} + \mathbf{v}$ as the diagonal of the parallelogram defined by **u** and **v**, as shown in Figure 4.15.

Finally, observe that vector addition is a special case of matrix addition.

*Recall that the vector $\mathbf{u} = \begin{bmatrix} x_1 \\ y_1 \end{bmatrix}$ can also be written as (x_1, y_1).

EXAMPLE 10 If $\mathbf{u}$ and $\mathbf{v}$ are as in Example 9, then $\mathbf{u} + \mathbf{v}$ is as shown in Figure 4.16. ■

Figure 4.16 ►

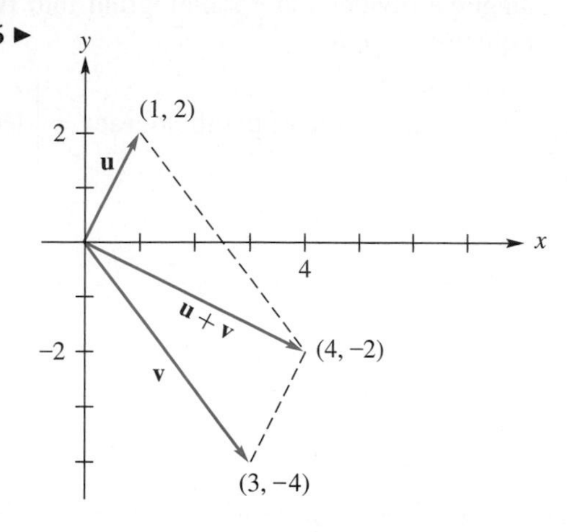

DEFINITION If $\mathbf{u} = (x_1, y_1)$ and c is a scalar (a real number), then the **scalar multiple** $c\mathbf{u}$ of $\mathbf{u}$ by c is the vector (cx_1, cy_1). Thus the scalar multiple $c\mathbf{u}$ of $\mathbf{u}$ by c is obtained by multiplying each component of $\mathbf{u}$ by c.

If $c > 0$, then $c\mathbf{u}$ is in the same direction as $\mathbf{u}$, whereas if $d < 0$, then $d\mathbf{u}$ is in the opposite direction (Figure 4.17).

Figure 4.17 ►
Scalar multiplication

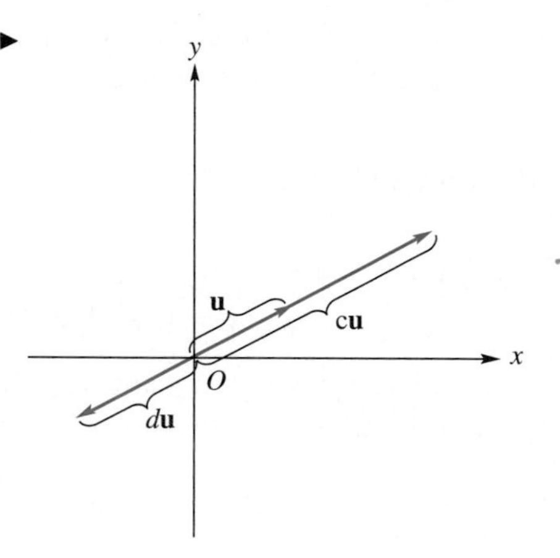

EXAMPLE 11 If $c = 2$, $d = -3$, and $\mathbf{u} = (1, -2)$, then

$$c\mathbf{u} = 2(1, -2) = (2, -4) \quad \text{and} \quad d\mathbf{u} = -3(1, -2) = (-3, 6),$$

which are shown in Figure 4.18. ■

The vector $(0, 0)$ is called the **zero vector** and is denoted by $\mathbf{0}$. If $\mathbf{u}$ is any vector, it follows that (Exercise T.2)

$$\mathbf{u} + \mathbf{0} = \mathbf{u}. \tag{4}$$

We can also show (Exercise T.3) that

$$\mathbf{u} + (-1)\mathbf{u} = \mathbf{0}, \tag{5}$$

and we write $(-1)\mathbf{u}$ as $-\mathbf{u}$ and call it the **negative** of $\mathbf{u}$. Moreover, we write $\mathbf{u} + (-1)\mathbf{v}$ as $\mathbf{u} - \mathbf{v}$ and call it the **difference** of $\mathbf{u}$ and $\mathbf{v}$. The vector $\mathbf{u} - \mathbf{v}$ is shown in Figure 4.19(a).

Observe that while vector addition gives one diagonal of a parallelogram, vector subtraction gives the other diagonal. See Figure 4.19(b).

Figure 4.18 ►

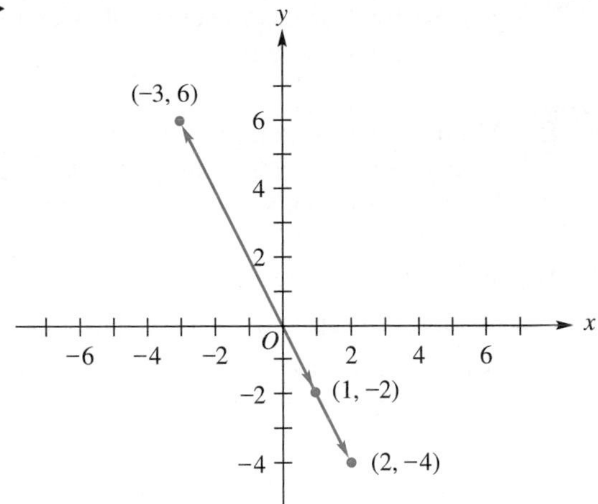

Figure 4.19 ►

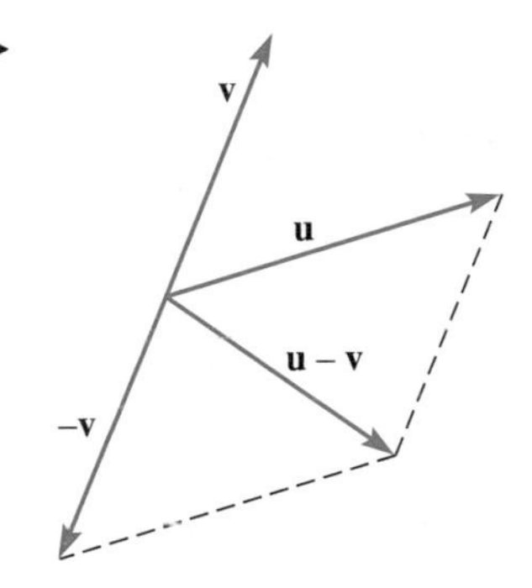

(a) Difference between vectors.

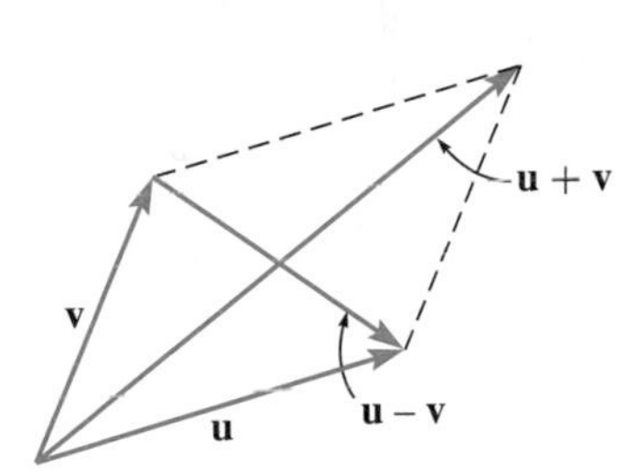

(b) Vector sum and vector difference.

Application (Vectors in Physics) When several forces act on a body, we can find a single force, called the **resultant force**, having the equivalent effect. The resultant force can be determined using vectors. The following example illustrates the method.

EXAMPLE 12 Suppose that a force of 12 pounds is applied to an object along the negative x-axis and a force of 5 pounds is applied to the object along the positive y-axis. Find the magnitude and direction of the resultant force.

Solution In Figure 4.20 we have represented the force along the negative x-axis by the vector $\overrightarrow{OA}$ and the force along the positive y-axis by the vector $\overrightarrow{OB}$. The resultant force is the vector $\overrightarrow{OC} = \overrightarrow{OA} + \overrightarrow{OB}$. Thus the magnitude of the resultant force is 13 pounds and its direction is as indicated in the figure. ■

Figure 4.20 ►

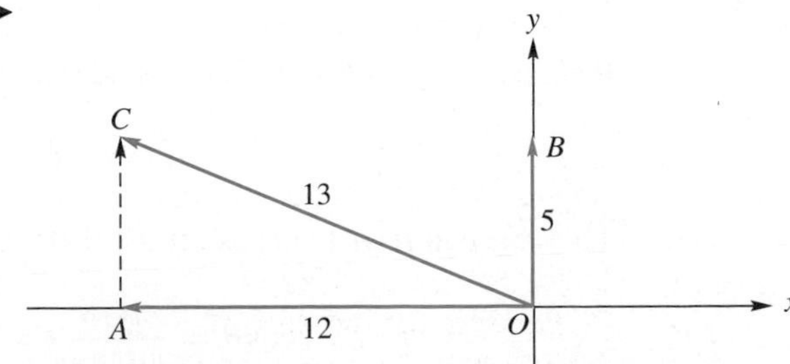

Vectors are also used in physics to deal with velocity problems, as the following example illustrates.

EXAMPLE 13 Suppose that a boat is traveling east across a river at the rate of 4 miles per hour while the river's current is flowing south at a rate of 3 miles per hour. Find the resultant velocity of the boat.

Solution In Figure 4.21, we have represented the velocity of the boat by the vector $\overrightarrow{OA}$ and the velocity of the river's current by the vector $\overrightarrow{OB}$. The resultant velocity is the vector $\overrightarrow{OC} = \overrightarrow{OA} + \overrightarrow{OB}$. Thus the magnitude of the resultant velocity is 5 miles per hour and its direction is as indicated in the figure. ■

Figure 4.21 ►

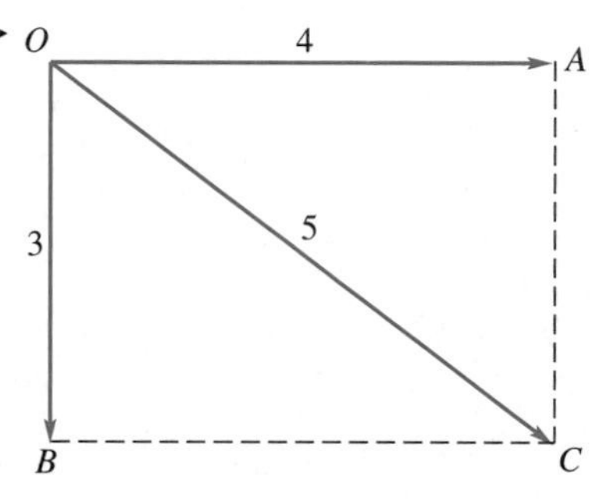

ANGLE BETWEEN TWO VECTORS

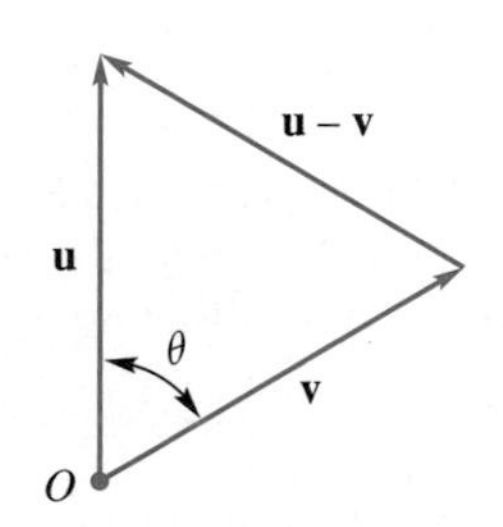

Figure 4.22 ▲

The angle between the nonzero vectors $\mathbf{u} = (x_1, y_1)$ and $\mathbf{v} = (x_2, y_2)$ is the angle θ, $0 \leq \theta \leq \pi$, shown in Figure 4.22. Applying the law of cosines to the triangle in that figure, we obtain

$$\|\mathbf{u} - \mathbf{v}\|^2 = \|\mathbf{u}\|^2 + \|\mathbf{v}\|^2 - 2\|\mathbf{u}\|\|\mathbf{v}\| \cos\theta. \tag{6}$$

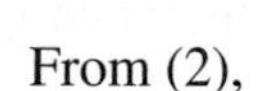

From (2),

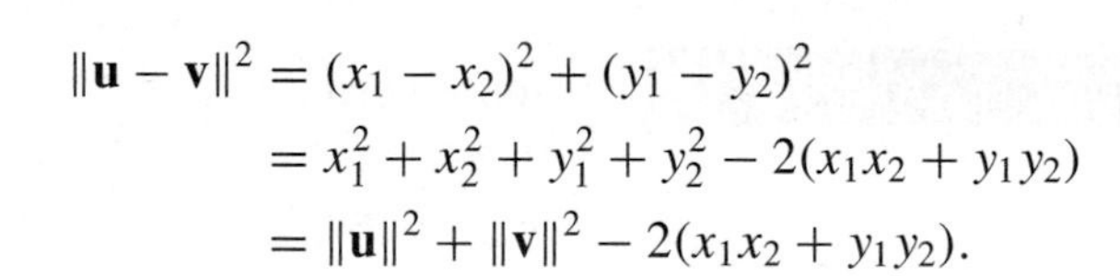

$$\begin{aligned}\|\mathbf{u} - \mathbf{v}\|^2 &= (x_1 - x_2)^2 + (y_1 - y_2)^2 \\ &= x_1^2 + x_2^2 + y_1^2 + y_2^2 - 2(x_1x_2 + y_1y_2) \\ &= \|\mathbf{u}\|^2 + \|\mathbf{v}\|^2 - 2(x_1x_2 + y_1y_2).\end{aligned}$$

If we substitute this expression in (6) and solve for $\cos\theta$ (recall that since $\mathbf{u}$ and $\mathbf{v}$ are nonzero vectors, then $\|\mathbf{u}\| \neq 0$ and $\|\mathbf{v}\| \neq 0$), we obtain

$$\cos\theta = \frac{x_1x_2 + y_1y_2}{\|\mathbf{u}\|\|\mathbf{v}\|}. \tag{7}$$

Recall from the first part of Section 1.3 that the **dot product** of the vectors $\mathbf{u} = (x_1, y_1)$ and $\mathbf{v} = (x_2, y_2)$ is defined to be

$$\mathbf{u} \cdot \mathbf{v} = x_1x_2 + y_1y_2.$$

Thus we can rewrite (7) as

$$\cos\theta = \frac{\mathbf{u} \cdot \mathbf{v}}{\|\mathbf{u}\|\|\mathbf{v}\|} \qquad (0 \leq \theta \leq \pi). \tag{8}$$

EXAMPLE 14 If $\mathbf{u} = (2, 4)$ and $\mathbf{v} = (-1, 2)$, then

$$\mathbf{u} \cdot \mathbf{v} = (2)(-1) + (4)(2) = 6.$$

Also,

$$\|\mathbf{u}\| = \sqrt{2^2 + 4^2} = \sqrt{20}$$

and

$$\|\mathbf{v}\| = \sqrt{(-1)^2 + 2^2} = \sqrt{5}.$$

Hence

$$\cos\theta = \frac{6}{\sqrt{20}\sqrt{5}} = 0.6.$$

We can obtain the approximate angle by using a calculator or a table of cosines; we find that θ is approximately $53°8'$ or 0.93 radian. ■

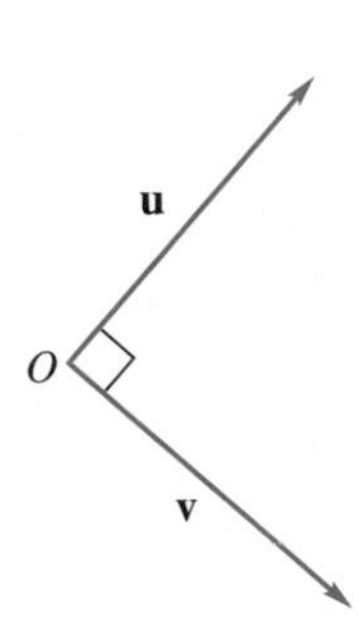

Figure 4.23 ▲ Orthogonal vectors

If $\mathbf{u}$ is a vector in R^2, then we can use the definition of dot product to write

$$\|\mathbf{u}\| = \sqrt{\mathbf{u} \cdot \mathbf{u}}.$$

If the nonzero vectors $\mathbf{u}$ and $\mathbf{v}$ are at right angles (Figure 4.23), then the cosine of the angle θ between them is zero. Hence, from (8), we have $\mathbf{u} \cdot \mathbf{v} = 0$. Conversely, if $\mathbf{u} \cdot \mathbf{v} = 0$, then $\cos\theta = 0$ and the vectors are at right angles. Thus the nonzero vectors $\mathbf{u}$ and $\mathbf{v}$ are **perpendicular** or **orthogonal** if and only if $\mathbf{u} \cdot \mathbf{v} = 0$. We shall also say that two vectors are orthogonal if at least one of them is zero. Therefore, the zero vector is orthogonal to every vector. Thus, we can now say that two vectors $\mathbf{u}$ and $\mathbf{v}$ are orthogonal if and only if $\mathbf{u} \cdot \mathbf{v} = 0$.

EXAMPLE 15 Thc vectors $\mathbf{u} = (2, -4)$ and $\mathbf{v} = (4, 2)$ are orthogonal, since

$$\mathbf{u} \cdot \mathbf{v} = (2)(4) + (-4)(2) = 0$$

(See Figure 4.24). ■

Figure 4.24 ► Orthogonal vectors

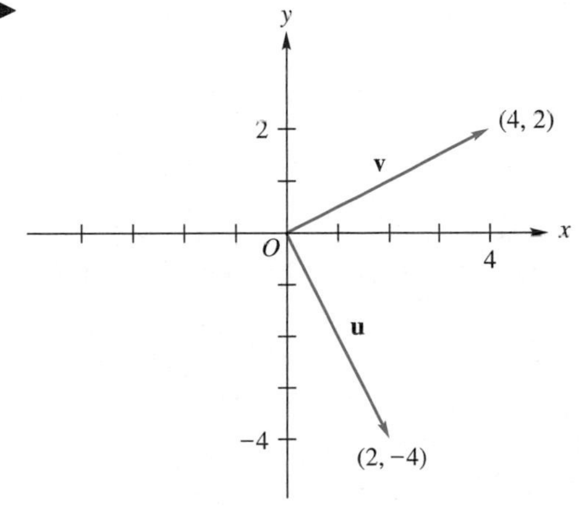

THEOREM 4.1 **(Properties of Dot Product)** *If* $\mathbf{u}$, $\mathbf{v}$, *and* $\mathbf{w}$ *are vectors and* c *is a scalar, then*:

(a) $\mathbf{u} \cdot \mathbf{u} \geq 0$; $\mathbf{u} \cdot \mathbf{u} = 0$ *if and only if* $\mathbf{u} = \mathbf{0}$

(b) $\mathbf{u} \cdot \mathbf{v} = \mathbf{v} \cdot \mathbf{u}$

(c) $(\mathbf{u}+\mathbf{v})\cdot\mathbf{w} = \mathbf{u}\cdot\mathbf{w}+\mathbf{v}\cdot\mathbf{w}$

(d) $(c\mathbf{u})\cdot\mathbf{v} = \mathbf{u}\cdot(c\mathbf{v}) = c(\mathbf{u}\cdot\mathbf{v})$

Proof Exercise T.7. ■

UNIT VECTORS

A **unit vector** is a vector whose length is 1. If $\mathbf{x}$ is any nonzero vector, then the vector

$$\mathbf{u} = \frac{1}{\|\mathbf{x}\|}\mathbf{x}$$

is a unit vector in the direction of $\mathbf{x}$ (Exercise T.5).

EXAMPLE 16 Let $\mathbf{x} = (-3, 4)$. Then

$$\|\mathbf{x}\| = \sqrt{(-3)^2 + 4^2} = 5.$$

Hence the vector $\mathbf{u} = \frac{1}{5}(-3, 4) = \left(-\frac{3}{5}, \frac{4}{5}\right)$ is a unit vector, since

$$\|\mathbf{u}\| = \sqrt{\left(-\tfrac{3}{5}\right)^2 + \left(\tfrac{4}{5}\right)^2} = \sqrt{\tfrac{9+16}{25}} = 1.$$

Also, $\mathbf{u}$ lies in the direction of $\mathbf{x}$ (Figure 4.25). ■

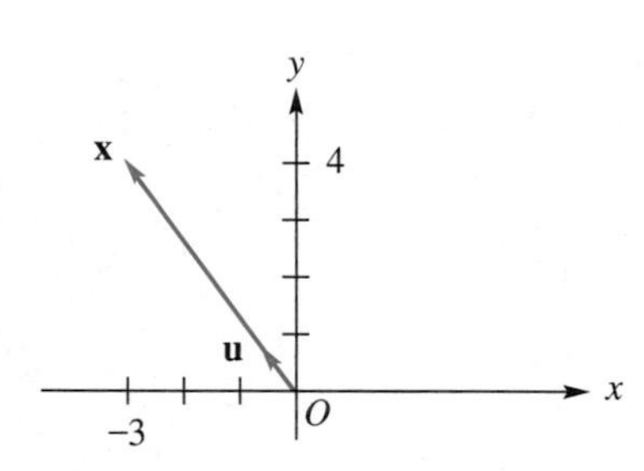

Figure 4.25 ▲

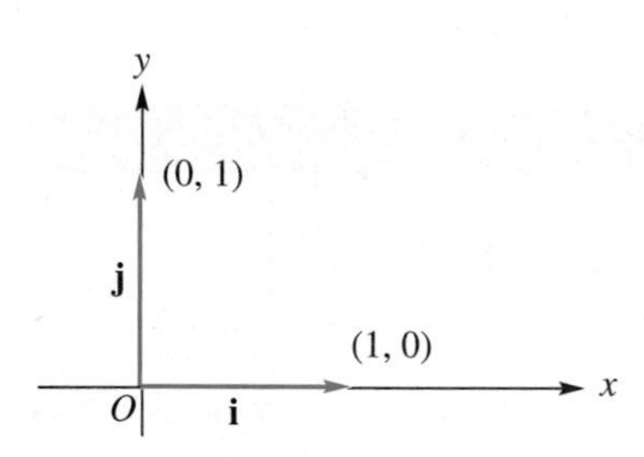

Figure 4.26 ▲
Unit vectors

There are two unit vectors in R^2 that are of special importance. They are $\mathbf{i} = (1, 0)$ and $\mathbf{j} = (0, 1)$, the unit vectors along the positive x- and y-axes, respectively (Figure 4.26).

If $\mathbf{u} = (x_1, y_1)$ is any vector in R^2, then we can write $\mathbf{u}$ in terms of $\mathbf{i}$ and $\mathbf{j}$ as

$$\mathbf{u} = x_1\mathbf{i} + y_1\mathbf{j}.$$

EXAMPLE 17 If

$$\mathbf{u} = (4, -5),$$

then

$$\mathbf{u} = 4\mathbf{i} - 5\mathbf{j}.$$ ■

EXAMPLE 18 The vectors $\mathbf{i}$ and $\mathbf{j}$ are orthogonal [see Exercise 23(b)]. ■

Key Terms

Origin
Absolute value
Coordinate
Coordinate axis
x-axis
y-axis
Rectangular (or Cartesian) coordinate system
2-space
Direction of a directed line segment
Magnitude of a directed line segment
Vector
Components of a vector
Head of a vector
Tail of a vector
Equal vetors
Length (or magnitude) of a vector
Parallel vectors
Sum of vectors
Scalar multiple of vectors
Zero vector
Negative of a vector
Difference of vectors
Resultant force
Dot product
Perpendicular (or orthogonal) vectors
Unit vector

4.1 Exercises

1. Plot the following points in R^2.
(a) $(2, -1)$ (b) $(-1, 2)$ (c) $(3, 4)$
(d) $(-3, -2)$ (e) $(0, 2)$ (f) $(0, -3)$

2. Sketch a directed line segment in R^2 representing each of the following vectors.
(a) $\mathbf{u}_1 = \begin{bmatrix} -2 \\ 3 \end{bmatrix}$ (b) $\mathbf{u}_2 = \begin{bmatrix} 3 \\ 4 \end{bmatrix}$
(c) $\mathbf{u}_3 = \begin{bmatrix} -3 \\ -3 \end{bmatrix}$ (d) $\mathbf{u}_4 = \begin{bmatrix} 0 \\ -3 \end{bmatrix}$

3. Determine the head of the vector $\begin{bmatrix} -2 \\ 5 \end{bmatrix}$ whose tail is at $(3, 2)$. Make a sketch.

4. Determine the head of the vector $\begin{bmatrix} 2 \\ 5 \end{bmatrix}$ whose tail is at $(1, 2)$. Make a sketch.

5. Find $\mathbf{u} + \mathbf{v}$, $\mathbf{u} - \mathbf{v}$, $2\mathbf{u}$, and $3\mathbf{u} - 2\mathbf{v}$ if
(a) $\mathbf{u} = (2, 3)$, $\mathbf{v} = (-2, 5)$
(b) $\mathbf{u} = (0, 3)$, $\mathbf{v} = (3, 2)$
(c) $\mathbf{u} = (2, 6)$, $\mathbf{v} = (3, 2)$

6. Repeat Exercise 5 for
(a) $\mathbf{u} = (-1, 3)$, $\mathbf{v} = (2, 4)$
(b) $\mathbf{u} = (-4, -3)$, $\mathbf{v} = (5, 2)$
(c) $\mathbf{u} = (3, 2)$, $\mathbf{v} = (-2, 0)$

7. Let $\mathbf{u} = (1, 2)$, $\mathbf{v} = (-3, 4)$, $\mathbf{w} = (w_1, 4)$, and $\mathbf{x} = (-2, x_2)$. Find w_1 and x_2 so that
(a) $\mathbf{w} = 2\mathbf{u}$ (b) $\frac{3}{2}\mathbf{x} = \mathbf{v}$ (c) $\mathbf{w} + \mathbf{x} = \mathbf{u}$

8. Let $\mathbf{u} = (-4, 3)$, $\mathbf{v} = (2, -5)$, and $\mathbf{w} = (w_1, w_2)$. Find w_1 and w_2 so that
(a) $\mathbf{w} = 2\mathbf{u} + 3\mathbf{v}$ (b) $\mathbf{u} + \mathbf{w} = 2\mathbf{u} - \mathbf{v}$
(c) $\mathbf{w} = \frac{5}{2}\mathbf{v}$

9. Find the length of the following vectors.
(a) $(1, 2)$ (b) $(-3, -4)$
(c) $(0, 2)$ (d) $(-4, 3)$

10. Find the length of the following vectors.
(a) $(-2, 3)$ (b) $(3, 0)$
(c) $(-4, -5)$ (d) $(3, 2)$

11. Find the distance between the following pairs of points.
(a) $(2, 3)$, $(3, 4)$ (b) $(0, 0)$, $(3, 4)$
(c) $(-3, 2)$, $(0, 1)$ (d) $(0, 3)$, $(2, 0)$

12. Find the distance between the following pairs of points.
(a) $(4, 2)$, $(1, 2)$ (b) $(-2, -3)$, $(0, 1)$
(c) $(2, 4)$, $(-1, 1)$ (d) $(2, 0)$, $(3, 2)$

13. Is it possible to write the vector $(-5, 6)$ as a linear combination (defined before Example 14 in Section 1.3) of the vectors $(1, 2)$ and $(3, 4)$?

14. If possible, find scalars c_1 and c_2, not both zero, so that

$$c_1 \begin{bmatrix} 1 \\ 2 \end{bmatrix} + c_2 \begin{bmatrix} 3 \\ 4 \end{bmatrix} = \begin{bmatrix} 0 \\ 0 \end{bmatrix}.$$

15. Find the area of the triangle with vertices $(3, 3)$, $(-1, -1)$, $(4, 1)$.

16. Find the area of the right triangle with vertices $(0, 0)$, $(0, 3)$, $(4, 0)$. Verify by using the formula $A = \frac{1}{2}$(base)(height).

17. Find the area of the parallelogram with vertices $(2, 3)$, $(5, 3)$, $(4, 5)$, $(7, 5)$.

18. Let Q be the quadrilateral with vertices $(-2, 3)$, $(1, 4)$, $(3, 0)$, and $(-1, -3)$. Find the area of Q.

19. Find a unit vector in the direction of $\mathbf{x}$.
(a) $\mathbf{x} = (3, 4)$ (b) $\mathbf{x} = (-2, -3)$ (c) $\mathbf{x} = (5, 0)$

20. Find a unit vector in the direction of $\mathbf{x}$.
(a) $\mathbf{x} = (2, 4)$ (b) $\mathbf{x} = (0, -2)$ (c) $\mathbf{x} = (-1, -3)$

21. Find the cosine of the angle between each pair of vectors $\mathbf{u}$ and $\mathbf{v}$.
(a) $\mathbf{u} = (1, 2)$, $\mathbf{v} = (2, -3)$
(b) $\mathbf{u} = (1, 0)$, $\mathbf{v} = (0, 1)$
(c) $\mathbf{u} = (-3, -4)$, $\mathbf{v} = (4, -3)$
(d) $\mathbf{u} = (2, 1)$, $\mathbf{v} = (-2, -1)$

22. Find the cosine of the angle between each pair of vectors **u** and **v**.

(a) $\mathbf{u} = (0, -1)$, $\mathbf{v} = (1, 0)$

(b) $\mathbf{u} = (2, 2)$, $\mathbf{v} = (4, -5)$

(c) $\mathbf{u} = (2, -1)$, $\mathbf{v} = (-3, -2)$

(d) $\mathbf{u} = (0, 2)$, $\mathbf{v} = (3, -3)$

23. Show that

(a) $\mathbf{i} \cdot \mathbf{i} = \mathbf{j} \cdot \mathbf{j} = 1$ (b) $\mathbf{i} \cdot \mathbf{j} = 0$

24. Which of the vectors $\mathbf{u}_1 = (1, 2)$, $\mathbf{u}_2 = (0, 1)$, $\mathbf{u}_3 = (-2, -4)$, $\mathbf{u}_4 = (-2, 1)$, $\mathbf{u}_5 = (2, 4)$, $\mathbf{u}_6 = (-6, 3)$ are

(a) orthogonal?

(b) in the same direction?

(c) in opposite directions?

25. Find all constants a such that the vectors $(a, 4)$ and $(2, 5)$ are parallel.

26. Find all constants a such that the vectors $(a, 2)$ and $(a, -2)$ are orthogonal.

27. Write each of the following vectors in terms of **i** and **j**.

(a) $(1, 3)$ (b) $(-2, -3)$

(c) $(-2, 0)$ (d) $(0, 3)$

28. Write each of the following vectors as a 2×1 matrix.

(a) $3\mathbf{i} - 2\mathbf{j}$ (b) $2\mathbf{i}$ (c) $-2\mathbf{i} - 3\mathbf{j}$

29. A ship is being pushed by a tugboat with a force of 300 pounds along the negative y-axis while another tugboat is pushing along the negative x-axis with a force of 400 pounds. Find the magnitude and sketch the direction of the resultant force.

30. Suppose that an airplane is flying with an airspeed of 260 kilometers per hour while a wind is blowing to the west at 100 kilometers per hour. Indicate on a figure the approximate direction that the plane must follow to result in a flight directly south. What will be the resultant speed?

Theoretical Exercises

T.1. Show how we can associate a point in the plane with each ordered pair (x, y) of real numbers.

T.2. Show that $\mathbf{u} + \mathbf{0} = \mathbf{u}$.

T.3. Show that $\mathbf{u} + (-1)\mathbf{u} = \mathbf{0}$.

T.4. Show that if c is a scalar, then $\|c\mathbf{u}\| = |c|\|\mathbf{u}\|$.

T.5. Show that if $\mathbf{x}$ is a nonzero vector, then

$$\mathbf{u} = \frac{1}{\|\mathbf{x}\|}\mathbf{x}$$

is a unit vector in the direction of $\mathbf{x}$.

T.6. Show that

(a) $1\mathbf{u} = \mathbf{u}$

(b) $(rs)\mathbf{u} = r(s\mathbf{u})$, where r and s are scalars

T.7. Prove Theorem 4.1.

T.8. Show that if $\mathbf{w}$ is orthogonal to $\mathbf{u}$ and $\mathbf{v}$, then $\mathbf{w}$ is orthogonal to $r\mathbf{u} + s\mathbf{v}$, where r and s are scalars.

T.9. Let θ be the angle between the nonzero vectors $\mathbf{u} = (x_1, y_1)$ and $\mathbf{v} = (x_2, y_2)$ in the plane. Show that if $\mathbf{u}$ and $\mathbf{v}$ are parallel, then $\cos\theta = \pm 1$.

MATLAB Exercises

The following exercises use the routine **vec2demo**, *which provides a graphical display of vectors in the plane. For a pair of vectors* $\mathbf{u} = (x_1, y_1)$ *and* $\mathbf{v} = (x_2, y_2)$, *routine* **vec2demo** *graphs* **u** *and* **v**, **u** + **v**, **u** − **v**, *and a scalar multiple. Once the vectors* **u** *and* **v** *are entered into* MATLAB, *type*

vec2demo(u, v)

For further information, use **help vec2demo**.

ML.1. Use the routine **vec2demo** with each of the following pairs of vectors. (Square brackets are used in MATLAB.)

(a) $\mathbf{u} = [2 \quad 0]$, $\mathbf{v} = [0 \quad 3]$

(b) $\mathbf{u} = [-3 \quad 1]$, $\mathbf{v} = [2 \quad 2]$

(c) $\mathbf{u} = [5 \quad 2]$, $\mathbf{v} = [-3 \quad 3]$

ML.2. Use the routine **vec2demo** with each of the following pairs of vectors. (Square brackets are used in MATLAB.)

(a) $\mathbf{u} = [2 \quad -2]$, $\mathbf{v} = [1 \quad 3]$

(b) $\mathbf{u} = [0 \quad 3]$, $\mathbf{v} = [-2 \quad 0]$

(c) $\mathbf{u} = [4 \quad -1]$, $\mathbf{v} = [-3 \quad 5]$

ML.3. Choose pairs of vectors **u** and **v** to use with **vec2demo**.

4.2 n-VECTORS

In this section we look at n-vectors from a geometric point of view by generalizing the notions discussed in the preceding section. The case of $n = 3$ will be of special interest, and we shall discuss it in some detail.

As we have already seen in the first part of Section 1.3, an n-vector is an $n \times 1$ matrix

$$\mathbf{u} = \begin{bmatrix} u_1 \\ u_2 \\ \vdots \\ u_n \end{bmatrix},$$

where $u_1, u_2, \ldots, u_n$ are real numbers, which are called the **components** of $\mathbf{u}$. Since an n-vector is an $n \times 1$ matrix, the n-vectors

$$\mathbf{u} = \begin{bmatrix} u_1 \\ u_2 \\ \vdots \\ u_n \end{bmatrix} \quad \text{and} \quad \mathbf{v} = \begin{bmatrix} v_1 \\ v_2 \\ \vdots \\ v_n \end{bmatrix}$$

are said to be **equal** if $u_i = v_i$ $(1 \leq i \leq n)$.

EXAMPLE 1 The 4-vectors $\begin{bmatrix} 1 \\ -2 \\ 3 \\ 4 \end{bmatrix}$ and $\begin{bmatrix} 1 \\ -2 \\ 3 \\ -4 \end{bmatrix}$ are not equal, since their fourth components are not the same. ■

The set of all n vectors is denoted by R^n and is called **n-space**. When the actual value of n need not be specified, we refer to n-vectors simply as **vectors**. The real numbers are called **scalars**. The components of a vector are real numbers and hence the components of a vector are scalars.

VECTOR OPERATIONS

DEFINITION Let

$$\mathbf{u} = \begin{bmatrix} u_1 \\ u_2 \\ \vdots \\ u_n \end{bmatrix} \quad \text{and} \quad \mathbf{v} = \begin{bmatrix} v_1 \\ v_2 \\ \vdots \\ v_n \end{bmatrix}$$

be two vectors in R^n. The **sum** of the vectors $\mathbf{u}$ and $\mathbf{v}$ is the vector

$$\begin{bmatrix} u_1 + v_1 \\ u_2 + v_2 \\ \vdots \\ u_n + v_n \end{bmatrix},$$

and it is denoted by $\mathbf{u} + \mathbf{v}$.

EXAMPLE 2 If $\mathbf{u} = \begin{bmatrix} 1 \\ -2 \\ 3 \end{bmatrix}$ and $\begin{bmatrix} 2 \\ 3 \\ -3 \end{bmatrix}$ are vectors in R^3, then

$$\mathbf{u} + \mathbf{v} = \begin{bmatrix} 1+2 \\ -2+3 \\ 3+(-3) \end{bmatrix} = \begin{bmatrix} 3 \\ 1 \\ 0 \end{bmatrix}.$$

■

DEFINITION If

$$\mathbf{u} = \begin{bmatrix} u_1 \\ u_2 \\ \vdots \\ u_n \end{bmatrix}$$

is a vector in R^n and c is a scalar, then the **scalar multiple** $c\mathbf{u}$ of $\mathbf{u}$ by c is the vector

$$\begin{bmatrix} cu_1 \\ cu_2 \\ \vdots \\ cu_n \end{bmatrix}.$$

EXAMPLE 3 If $\mathbf{u} = \begin{bmatrix} 2 \\ 3 \\ -1 \\ 2 \end{bmatrix}$ is a vector in R^4 and $c = -2$, then

$$c\mathbf{u} = (-2)\begin{bmatrix} 2 \\ 3 \\ -1 \\ 2 \end{bmatrix} = \begin{bmatrix} -4 \\ -6 \\ 2 \\ -4 \end{bmatrix}.$$

■

The operations of vector addition and scalar multiplication satisfy the following properties.

THEOREM 4.2 *Let* $\mathbf{u}$, $\mathbf{v}$, *and* $\mathbf{w}$ *be any vectors in* R^n; *let* c *and* d *be any scalars. Then*

(α) $\mathbf{u} + \mathbf{v}$ *is a vector in* R^n *(i.e.,* R^n *is closed under the operation of vector addition).*

- (a) $\mathbf{u} + \mathbf{v} = \mathbf{v} + \mathbf{u}$
- (b) $\mathbf{u} + (\mathbf{v} + \mathbf{w}) = (\mathbf{u} + \mathbf{v}) + \mathbf{w}$
- (c) *There is a vector* $\mathbf{0}$ *in* R^n *such that* $\mathbf{u} + \mathbf{0} = \mathbf{0} + \mathbf{u} = \mathbf{u}$ *for all* $\mathbf{u}$ *in* R^n.
- (d) *For each vector* $\mathbf{u}$ *in* R^n, *there is a vector* $-\mathbf{u}$ *in* R^n *such that* $\mathbf{u} + (-\mathbf{u}) = \mathbf{0}$.

(β) $c\mathbf{u}$ *is a vector in* R^n *(i.e.,* R^n *is closed under the operation of scalar multiplication).*

- (e) $c(\mathbf{u} + \mathbf{v}) = c\mathbf{u} + c\mathbf{v}$
- (f) $(c + d)\mathbf{u} = c\mathbf{u} + d\mathbf{u}$
- (g) $c(d\mathbf{u}) = (cd)\mathbf{u}$
- (h) $1\mathbf{u} = \mathbf{u}$

Proof (α) and (β) are immediate from the definitions for vector sum and scalar multiple. We verify (f) here and leave the rest of the proof to the reader (Exercise T.1). Thus

$$(c+d)\mathbf{u} = (c+d)\begin{bmatrix} u_1 \\ u_2 \\ \vdots \\ u_n \end{bmatrix} = \begin{bmatrix} (c+d)u_1 \\ (c+d)u_2 \\ \vdots \\ (c+d)u_n \end{bmatrix} = \begin{bmatrix} cu_1 + du_1 \\ cu_2 + du_2 \\ \vdots \\ cu_n + du_n \end{bmatrix}$$

$$= \begin{bmatrix} cu_1 \\ cu_2 \\ \vdots \\ cu_n \end{bmatrix} + \begin{bmatrix} du_1 \\ du_2 \\ \vdots \\ du_n \end{bmatrix} = c\begin{bmatrix} u_1 \\ u_2 \\ \vdots \\ u_n \end{bmatrix} + d\begin{bmatrix} u_1 \\ u_2 \\ \vdots \\ u_n \end{bmatrix} = c\mathbf{u} + d\mathbf{u}.$$

■

It is easy to show that the vectors $\mathbf{0}$ and $-\mathbf{u}$ in Properties (c) and (d) are unique. Moreover,

$$\mathbf{0} = \begin{bmatrix} 0 \\ 0 \\ \vdots \\ 0 \end{bmatrix}$$

and if $\mathbf{u} = \begin{bmatrix} u_1 \\ u_2 \\ \vdots \\ u_n \end{bmatrix}$, then $-\mathbf{u} = \begin{bmatrix} -u_1 \\ -u_2 \\ \vdots \\ -u_n \end{bmatrix}$. The vector $\mathbf{0}$ is called the **zero vector** and $-\mathbf{u}$ is called the **negative** of $\mathbf{u}$. It is easy to verify (Exercise T.2) that

$$-\mathbf{u} = (-1)\mathbf{u}.$$

We shall also write $\mathbf{u} + (-\mathbf{v})$ as $\mathbf{u} - \mathbf{v}$ and call it the **difference** of $\mathbf{u}$ and $\mathbf{v}$.

EXAMPLE 4 If $\mathbf{u}$ and $\mathbf{v}$ are as in Example 2, then

$$\mathbf{u} - \mathbf{v} = \begin{bmatrix} 1 - 2 \\ -2 - 3 \\ 3 - (-3) \end{bmatrix} = \begin{bmatrix} -1 \\ -5 \\ 6 \end{bmatrix}.$$

■

As in the case of R^2, we shall identify the vector

$$\begin{bmatrix} u_1 \\ u_2 \\ \vdots \\ u_n \end{bmatrix}$$

with the point $(u_1, u_2, \ldots, u_n)$ so that points and vectors can be used interchangeably. Thus we may view R^n as consisting of vectors or of points, and we also write

$$\mathbf{u} = (u_1, u_2, \ldots, u_n).$$

Moreover, an n-vector is an $n \times 1$ matrix, vector addition is matrix addition, and scalar multiplication is merely the operation of multiplication of a matrix by a real number. Thus R^n can be viewed as the set of all $n \times 1$ matrices with the operations of matrix addition and scalar multiplication. *The important point here is that no matter how we view R^n—as n-vectors, points, or $n \times 1$ matrices—the algebraic behavior is always the same.*

Application Vectors in R^n can be used to handle large amounts of data. Indeed, a number of computer software products, notably MATLAB, make extensive use of vectors. The following example illustrates these ideas.

EXAMPLE 5 **(Application: Inventory Control)** Suppose that a store handles 100 different items. The inventory on hand at the beginning of the week can be described by the inventory vector $\mathbf{u}$ in R^{100}. The number of items sold at the end of the week can be described by the vector $\mathbf{v}$, and the vector

$$\mathbf{u} - \mathbf{v}$$

represents the inventory at the end of the week. If the store receives a new shipment of goods, represented by the vector $\mathbf{w}$, then its new inventory would be

$$\mathbf{u} - \mathbf{v} + \mathbf{w}.$$ ■

VISUALIZING R^3

We cannot draw pictures of R^n for $n > 3$. However, since R^3 is the world we live in, we can visualize it in a manner similar to that used for R^2.

We first fix a **coordinate system** by choosing a point, called the **origin**, and three lines, called the **coordinate axes**, each passing through the origin, so that each line is perpendicular to the other two. These lines are individually called the ***x*-, *y*-, and *z*-axes**. On each of these coordinate axes we choose a point fixing the units of length and positive directions. Frequently, but not always, the same unit of length is used for all the coordinate axes. In Figures 4.27(a) and (b) we show two of the many possible coordinate systems. The coordinate system shown in Figure 4.27(a) is called a **right-handed coordinate system**; the one shown in Figure 4.27(b) is called **left-handed**. A right-handed system is characterized by the following property. If we curl the fingers of the right hand in the direction of a 90° rotation from the positive x-axis to the positive y-axis, then the thumb will point in the direction of the positive z-axis (Figure 4.28). In this book we use a right-handed coordinate system.

Figure 4.27 ►

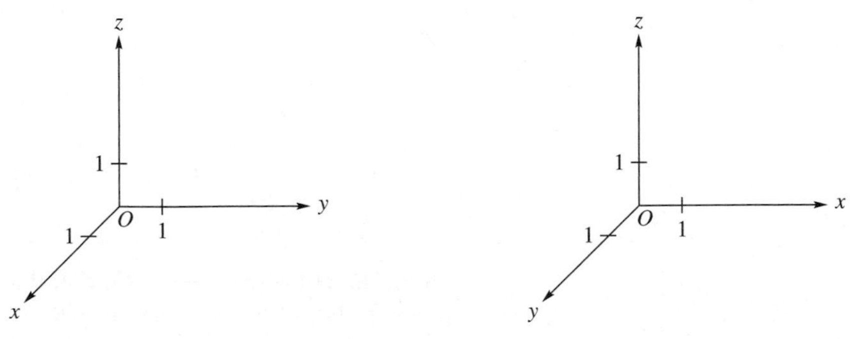

(a) Right-handed coordinate system. (b) Left-handed coordinate system.

The projection of a point P in space on a line L is the point Q obtained by intersecting L with the line L' passing through P and perpendicular to L (Figure 4.29).

The ***x*-coordinate** of the point P is the number associated with the projection of P on the x-axis; similarly for the ***y*- and *z*-coordinates**. These three numbers are called the **coordinates** of P. Thus with each point in space we

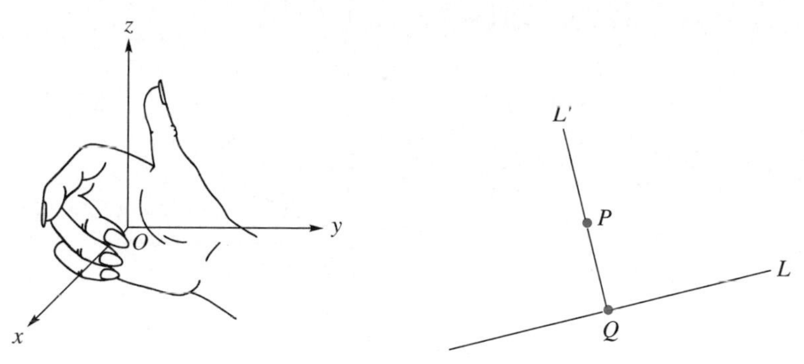

Figure 4.28 ▲

Figure 4.29 ▲

associate an ordered triple (x, y, z) of real numbers, and conversely, with each ordered triple of real numbers, we associate a point in space. This correspondence is called a **rectangular coordinate system**. We write $P(x, y, z)$, or simply (x, y, z) to denote a point in space.

EXAMPLE 6 In Figures 4.30(a) and (b) we show two points and their coordinates. ■

Figure 4.30 ►

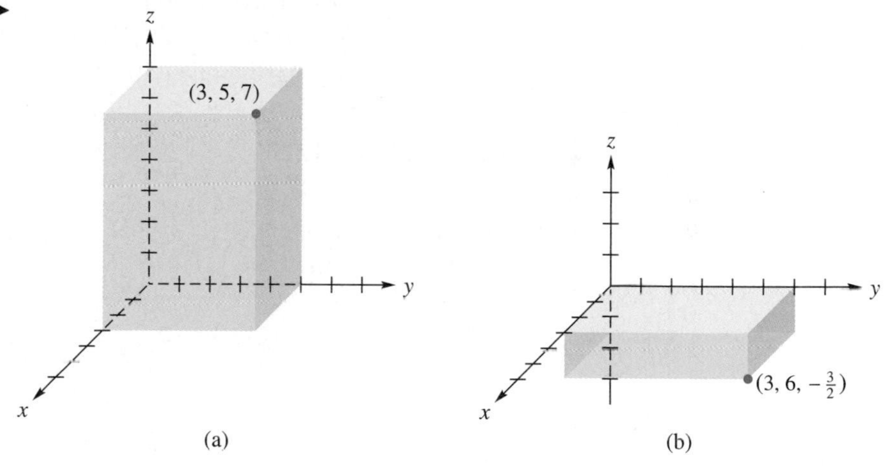

The ***xy*-plane** is the plane determined by the x- and y-axes. Similarly, we have the ***xz*- and *yz*-planes**.

In R^3, the components of a vector $\mathbf{u}$ are denoted by x_1, y_1, and z_1. Thus $\mathbf{u} = (x_1, y_1, z_1)$.

As in the plane, with the vector $\mathbf{u} = (x_1, y_1, z_1)$ we associate the directed line segment $\overrightarrow{OP}$, whose tail is $O(0, 0, 0)$ and whose head is $P(x_1, y_1, z_1)$ [Figure 4.31(a)]. Again as in the plane, in physical applications we often deal with a directed line segment $\overrightarrow{PQ}$, from the point $P(x_1, y_1, z_1)$ (not the origin) to the point $Q(x_2, y_2, z_2)$, as shown in Figure 4.31(b). Such a directed line segment will also be called a **vector in R^3**, or simply a **vector** with tail $P(x_1, y_1, z_1)$ and head $Q(x_2, y_2, z_2)$. The components of such a vector are $x_2 - x_1$, $y_2 - y_1$, and $z_2 - z_1$. Two such vectors in R^3 will be called **equal** if their components are equal. Thus the vector $\overrightarrow{PQ}$ in Figure 4.31(b) can also be represented by the vector $(x_2 - x_1, y_2 - y_1, z_2 - z_1)$ with tail O and head $P''(x_2 - x_1, y_2 - y_1, z_2 - z_1)$.

Figure 4.31 ►

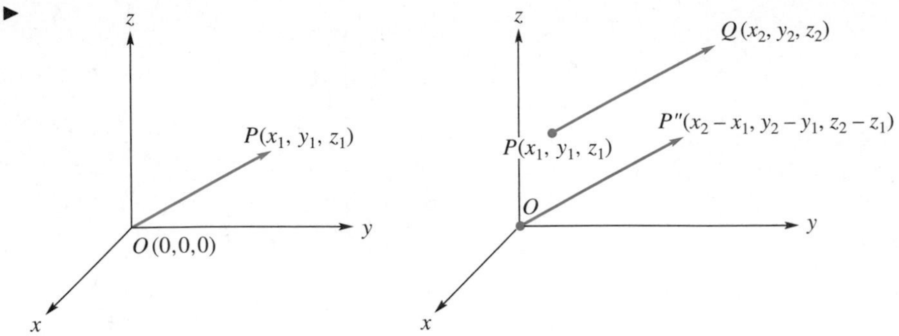

(a) A vector in R^3

(b) Different directed line segments representing the same vector

We shall soon define the length of a vector in R^n and the angle between two nonzero vectors in R^n. Once this has been done, it can be shown that two vectors in R^3 are equal if and only if any directed line segments representing them are parallel, have the same direction, and are of the same length.

The sum $\mathbf{u}+\mathbf{v}$ of the vectors $\mathbf{u} = (x_1, y_1, z_1)$ and $\mathbf{v} = (x_2, y_2, z_2)$ in R^3 is the diagonal of the parallelogram determined by $\mathbf{u}$ and $\mathbf{v}$, as shown in Figure 4.32.

The reader will note that Figure 4.32 looks very much like Figure 4.14 in Section 4.1 in that in both R^2 and R^3 the vector $\mathbf{u} + \mathbf{v}$ is the diagonal of the parallelogram determined by $\mathbf{u}$ and $\mathbf{v}$.

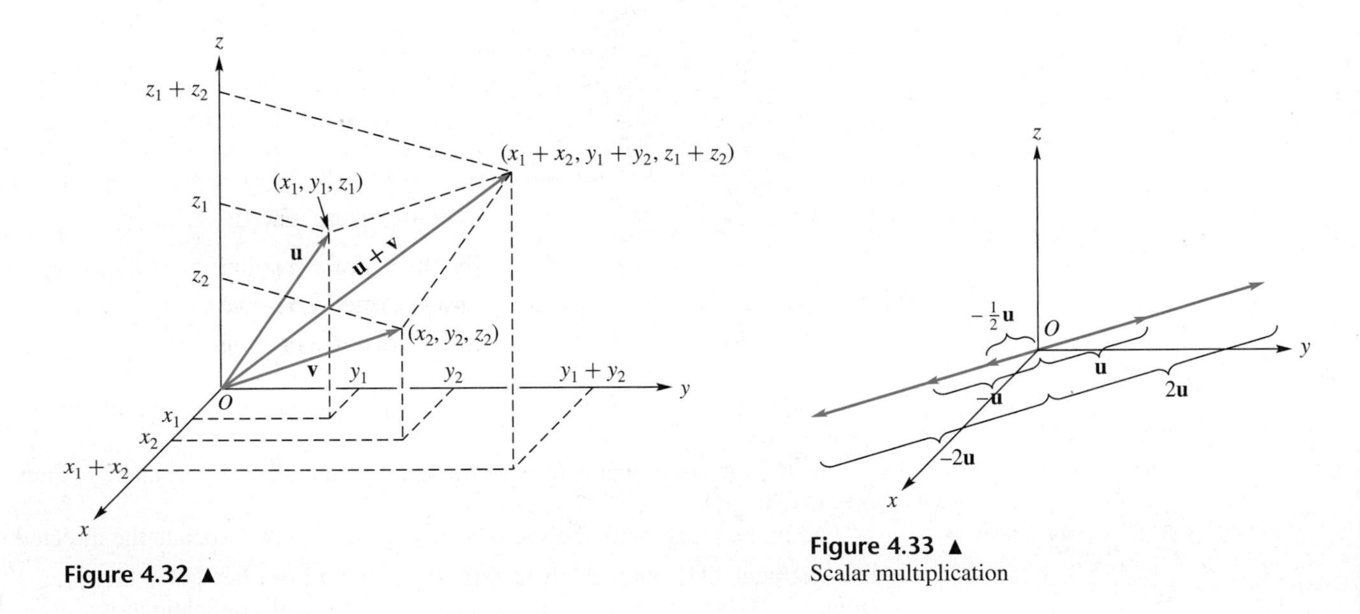

Figure 4.32 ▲

Figure 4.33 ▲
Scalar multiplication

Scalar multiplication in R^3 is shown in Figure 4.33, which resembles Figure 4.17 in Section 4.1.

DOT PRODUCT ON R^n

We shall now define the notion of the length of a vector in R^n by generalizing the corresponding idea for R^2.

DEFINITION

The **length** (also called **magnitude** or **norm**) of the vector $\mathbf{u} = (u_1, u_2, \ldots, u_n)$ in R^n is

$$\|\mathbf{u}\| = \sqrt{u_1^2 + u_2^2 + \cdots + u_n^2}. \tag{1}$$

We also define the distance from the point $(u_1, u_2, \ldots, u_n)$ to the origin by (1). The **distance** between the points $(u_1, u_2, \ldots, u_n)$ and $(v_1, v_2, \ldots, v_n)$ is then defined as the length of the vector $\mathbf{u} - \mathbf{v}$, where

$$\mathbf{u} = (u_1, u_2, \ldots, u_n) \quad \text{and} \quad \mathbf{v} = (v_1, v_2, \ldots, v_n).$$

Thus this distance is given by

$$\|\mathbf{u} - \mathbf{v}\| = \sqrt{(u_1 - v_1)^2 + (u_2 - v_2)^2 + \cdots + (u_n - v_n)^2}. \tag{2}$$

EXAMPLE 7

Let $\mathbf{u} = (2, 3, 2, -1)$ and $\mathbf{v} = (4, 2, 1, 3)$. Then

$$\|\mathbf{u}\| = \sqrt{2^2 + 3^2 + 2^2 + (-1)^2} = \sqrt{18},$$
$$\|\mathbf{v}\| = \sqrt{4^2 + 2^2 + 1^2 + 3^2} = \sqrt{30}.$$

The distance between the points $(2, 3, 2, -1)$ and $(4, 2, 1, 3)$ is the length of the vector $\mathbf{u} - \mathbf{v}$. Thus, from Equation (2),

$$\|\mathbf{u} - \mathbf{v}\| = \sqrt{(2-4)^2 + (3-2)^2 + (2-1)^2 + (-1-3)^2} = \sqrt{22}. \quad \blacksquare$$

It should be noted that in R^3, Equations (1) and (2), for the length of a vector and the distance between two points, do not have to be defined. They can easily be established by means of two applications of the Pythagorean theorem (Exercise T.3).

We shall define the cosine of the angle between two vectors in R^n by generalizing the corresponding formula in R^2. However, we first recall the notion of dot product in R^n as defined in the first part of Section 1.3.

If $\mathbf{u} = (u_1, u_2, \ldots, u_n)$ and $\mathbf{v} = (v_1, v_2, \ldots, v_n)$ are vectors in R^n, then their **dot product** is defined by

$$\mathbf{u} \cdot \mathbf{v} = u_1 v_1 + u_2 v_2 + \cdots + u_n v_n.$$

This is exactly how the dot product was defined in R^2. The dot product in R^n is also called the **standard inner product**.

EXAMPLE 8

If $\mathbf{u}$ and $\mathbf{v}$ are as in Example 7, then

$$\mathbf{u} \cdot \mathbf{v} = (2)(4) + (3)(2) + (2)(1) + (-1)(3) = 13. \quad \blacksquare$$

EXAMPLE 9

(Application: Revenue Monitoring) Consider the store in Example 5. If the vector $\mathbf{p}$ denotes the price of each of the 100 items, then the dot product

$$\mathbf{v} \cdot \mathbf{p}$$

gives the total revenue received at the end of the week. $\blacksquare$

If $\mathbf{u}$ is a vector in R^n, then we can use the definition of dot product in R^n to write

$$\|\mathbf{u}\| = \sqrt{\mathbf{u} \cdot \mathbf{u}}.$$

The dot product in R^n satisfies the same properties as in R^2. We state these properties as the following theorem, which reads just like Theorem 4.1.

THEOREM 4.3 **(Properties of Dot Product)** *If* $\mathbf{u}$, $\mathbf{v}$, *and* $\mathbf{w}$ *are vectors in* R^n *and* c *is a scalar, then*:

(a) $\mathbf{u}\cdot\mathbf{u} \geq 0$; $\mathbf{u}\cdot\mathbf{u} = 0$ *if and only if* $\mathbf{u} = \mathbf{0}$

(b) $\mathbf{u}\cdot\mathbf{v} = \mathbf{v}\cdot\mathbf{u}$

(c) $(\mathbf{u}+\mathbf{v})\cdot\mathbf{w} = \mathbf{u}\cdot\mathbf{w} + \mathbf{v}\cdot\mathbf{w}$

(d) $(c\mathbf{u})\cdot\mathbf{v} = \mathbf{u}\cdot(c\mathbf{v}) = c(\mathbf{u}\cdot\mathbf{v})$

Proof Exercise T.4. ■

We shall now prove a result that will enable us to give a worthwhile definition for the cosine of an angle between two nonzero vectors. This result, called the **Cauchy*–Schwarz** inequality**, has many important applications in mathematics. The proof of this result, although not difficult, is one that is not too natural and does call for a clever start.

THEOREM 4.4 **(Cauchy–Schwarz Inequality)** *If* $\mathbf{u}$ *and* $\mathbf{v}$ *are vectors in* R^n, *then*

$$|\mathbf{u}\cdot\mathbf{v}| \leq \|\mathbf{u}\|\,\|\mathbf{v}\|. \tag{3}$$

(Observe that $|\quad|$ *on the left stands for the absolute value of a real number;* $\|\quad\|$ *on the right denotes the length of a vector.)*

Proof If $\mathbf{u} = 0$, then $\|\mathbf{u}\| = 0$ and $\mathbf{u}\cdot\mathbf{v} = 0$, so (3) holds. Now suppose that $\mathbf{u}$ is nonzero. Let r be a scalar and consider the vector $r\mathbf{u}+\mathbf{v}$. By Theorem 4.3,

$$0 \leq (r\mathbf{u}+\mathbf{v})\cdot(r\mathbf{u}+\mathbf{v}) = r^2\mathbf{u}\cdot\mathbf{u} + 2r\mathbf{u}\cdot\mathbf{v} + \mathbf{v}\cdot\mathbf{v}$$
$$= ar^2 + 2br + c,$$

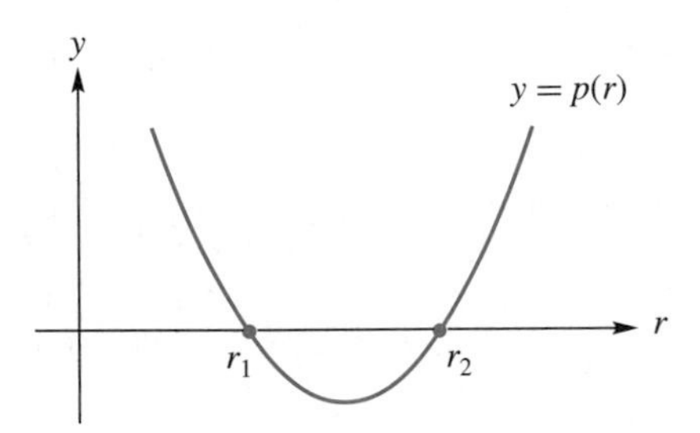

Figure 4.34 ▲

where

$$a = \mathbf{u}\cdot\mathbf{u}, \quad b = \mathbf{u}\cdot\mathbf{v}, \quad \text{and} \quad c = \mathbf{v}\cdot\mathbf{v}.$$

Now $p(r) = ar^2 + 2br + c$ is a quadratic polynomial in r (whose graph is a parabola, opening upward, since $a > 0$) that is nonnegative for all values of r. This means that either this polynomial has no real roots, or if it has real roots, then both roots are equal. [If $p(r)$ had two distinct roots r_1 and r_2, then it would be negative for some value of r between r_1 and r_2, as seen in Figure 4.34.]

Recall that the roots of $p(r)$ are given by the quadratic formula as

$$\frac{-2b+\sqrt{4b^2-4ac}}{2a} \quad \text{and} \quad \frac{-2b-\sqrt{4b^2-4ac}}{2a}$$

---

*Augustin-Louis Cauchy (1789–1857) grew up in a suburb of Paris as a neighbor of several leading mathematicians of the day, attended the École Polytechnique and the École des Ponts et Chaussées, and was for a time a practicing engineer. He was a devout Roman Catholic, with an abiding interest in Catholic charities. He was also strongly devoted to royalty, especially to the Bourbon kings who ruled France after Napoleon's defeat. When Charles X was deposed in 1830, Cauchy voluntarily followed him into exile in Prague.

Cauchy wrote 7 books and more than 700 papers of varying quality, touching all branches of mathematics. He made important contributions to the early theory of determinants, the theory of eigenvalues, the study of ordinary and partial differential equations, the theory of permutation groups, and the foundations of calculus, and he founded the theory of functions of a complex variable.

**Karl Hermann Amandus Schwarz (1843–1921) was born in Poland but was educated and taught in Germany. He was a protégé of Karl Weierstrass and Ernst Eduard Kummer, whose daughter he married. His main contributions to mathematics were in the geometric aspects of analysis, such as conformal mappings and minimal surfaces. In connection with the latter he sought certain numbers associated with differential equations, numbers that have since come to be called eigenvalues. The inequality given previously was used in the search for these numbers.

($a \neq 0$ since $\mathbf{u} \neq \mathbf{0}$). Both roots are equal or there are no real roots when

$$4b^2 - 4ac \leq 0,$$

which means that

$$b^2 \leq ac.$$

Upon taking square roots of both sides and observing that $\sqrt{a} = \sqrt{\mathbf{u} \cdot \mathbf{u}} = \|\mathbf{u}\|$, $\sqrt{c} = \sqrt{\mathbf{v} \cdot \mathbf{v}} = \|\mathbf{v}\|$, we obtain (3). ■

Remark The result widely known as the Cauchy–Schwarz inequality (Theorem 4.4) provides a good example of how nationalistic feelings make their way into science. In Russia this result is generally known as Bunyakovsky's* inequality. In France it is often referred to as *Cauchy's inequality* and in Germany it is frequently called *Schwarz's inequality*. In an attempt to distribute credit for the result among all three contenders, a minority of authors refer to the result as the *CBS inequality*.

EXAMPLE 10 If $\mathbf{u}$ and $\mathbf{v}$ are as in Example 7, then from Example 8, $\mathbf{u} \cdot \mathbf{v} = 13$. Hence

$$|\mathbf{u} \cdot \mathbf{v}| = 13 \leq \|\mathbf{u}\| \|\mathbf{v}\| = \sqrt{18}\sqrt{30}.$$ ■

We shall now use the Cauchy–Schwarz inequality to define the angle between two nonzero vectors in R^n.

If $\mathbf{u}$ and $\mathbf{v}$ are nonzero vectors, then it follows from the Cauchy–Schwarz inequality that

$$\left| \frac{\mathbf{u} \cdot \mathbf{v}}{\|\mathbf{u}\| \|\mathbf{v}\|} \right| \leq 1$$

or

$$-1 \leq \frac{\mathbf{u} \cdot \mathbf{v}}{\|\mathbf{u}\| \|\mathbf{v}\|} \leq 1.$$

Examining the portion of the graph of $y = \cos\theta$ (see Figure 4.35) for $0 \leq \theta \leq \pi$, we see that for any number r in the interval $[-1, 1]$, there is a unique real number θ such that $\cos\theta = r$. This implies that there is a unique real number θ such that

$$\cos\theta = \frac{\mathbf{u} \cdot \mathbf{v}}{\|\mathbf{u}\| \|\mathbf{v}\|}, \qquad 0 \leq \theta \leq \pi. \tag{4}$$

The angle θ is called the **angle between u** and **v**.

In the case of R^3, we can establish by the law of cosines, as was done in R^2, that the cosine of the angle between $\mathbf{u}$ and $\mathbf{v}$ is given by (4). However, for R^n, $n > 3$, we have to define it by (4).

EXAMPLE 11 Let $\mathbf{u} = (1, 0, 0, 1)$ and $\mathbf{v} = (0, 1, 0, 1)$. Then

$$\|\mathbf{u}\| = \sqrt{2}, \quad \|\mathbf{v}\| = \sqrt{2}, \quad \text{and} \quad \mathbf{u} \cdot \mathbf{v} = 1.$$

Thus

$$\cos\theta = \tfrac{1}{2}$$

and $\theta = 60°$ or $\frac{\pi}{3}$ radians. ■

---

*Viktor Yakovlevich Bunyakovsky (1804–1889) was born in Bar, Ukraine. He received a doctorate in Paris in 1825. He carried out additional studies in St. Petersburg and then had a long career there as a professor. Bunyakovsky made important contributions in number theory and also worked in geometry, applied mechanics, and hydrostatics. His proof of the Cauchy–Schwarz inequality appeared in one of his monographs in 1859, 25 years before Schwarz published his proof. He died in St. Petersburg.

Figure 4.35 ►

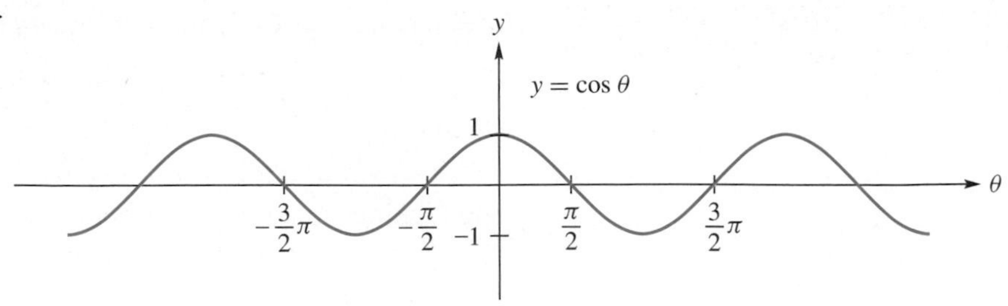

It is very useful to talk about orthogonality and parallelism in R^n, and we accordingly formulate the following definitions.

DEFINITION

Two nonzero vectors $\mathbf{u}$ and $\mathbf{v}$ in R^n are said to be **orthogonal** if $\mathbf{u} \cdot \mathbf{v} = 0$. If one of the vectors is the zero vector, we agree to say that the vectors are orthogonal. They are said to be **parallel** if $|\mathbf{u} \cdot \mathbf{v}| = \|\mathbf{u}\|\|\mathbf{v}\|$. They are in the **same direction** if $\mathbf{u} \cdot \mathbf{v} = \|\mathbf{u}\|\|\mathbf{v}\|$. That is, they are orthogonal if $\cos\theta = 0$, parallel if $\cos\theta = \pm 1$, and in the same direction if $\cos\theta = 1$.

EXAMPLE 12

Consider the vectors $\mathbf{u} = (1, 0, 0, 1)$, $\mathbf{v} = (0, 1, 1, 0)$, and $\mathbf{w} = (3, 0, 0, 3)$. Then

$$\mathbf{u} \cdot \mathbf{v} = 0 \quad \text{and} \quad \mathbf{v} \cdot \mathbf{w} = 0,$$

which implies that $\mathbf{u}$ and $\mathbf{v}$ are orthogonal and $\mathbf{v}$ and $\mathbf{w}$ are orthogonal. We have

$$\mathbf{u} \cdot \mathbf{w} = 6, \quad \|\mathbf{u}\| = \sqrt{2}, \quad \|\mathbf{w}\| = \sqrt{18}, \quad \mathbf{u} \cdot \mathbf{w} = \|\mathbf{u}\|\|\mathbf{w}\|,$$

and the cosine of the angle between $\mathbf{u}$ and $\mathbf{w}$ is 1 (verify). Hence, we conclude that $\mathbf{u}$ and $\mathbf{w}$ are in the same direction. ■

An easy consequence of the Cauchy–Schwarz inequality is the triangle inequality, which we prove next.

THEOREM 4.5

(Triangle Inequality) *If* $\mathbf{u}$ *and* $\mathbf{v}$ *are vectors in* R^n, *then*

$$\|\mathbf{u} + \mathbf{v}\| \leq \|\mathbf{u}\| + \|\mathbf{v}\|.$$

Proof We have, from the definition of the length of a vector,

$$\begin{aligned} \|\mathbf{u} + \mathbf{v}\|^2 &= (\mathbf{u} + \mathbf{v}) \cdot (\mathbf{u} + \mathbf{v}) \\ &= \mathbf{u} \cdot \mathbf{u} + 2(\mathbf{u} \cdot \mathbf{v}) + \mathbf{v} \cdot \mathbf{v} \\ &= \|\mathbf{u}\|^2 + 2(\mathbf{u} \cdot \mathbf{v}) + \|\mathbf{v}\|^2. \end{aligned}$$

By the Cauchy–Schwarz inequality we have

$$\begin{aligned} \|\mathbf{u}\|^2 + 2(\mathbf{u} \cdot \mathbf{v}) + \|\mathbf{v}\|^2 &\leq \|\mathbf{u}\|^2 + 2\|\mathbf{u}\|\|\mathbf{v}\| + \|\mathbf{v}\|^2 \\ &= (\|\mathbf{u}\| + \|\mathbf{v}\|)^2. \end{aligned}$$

Taking square roots, we obtain the desired result. ■

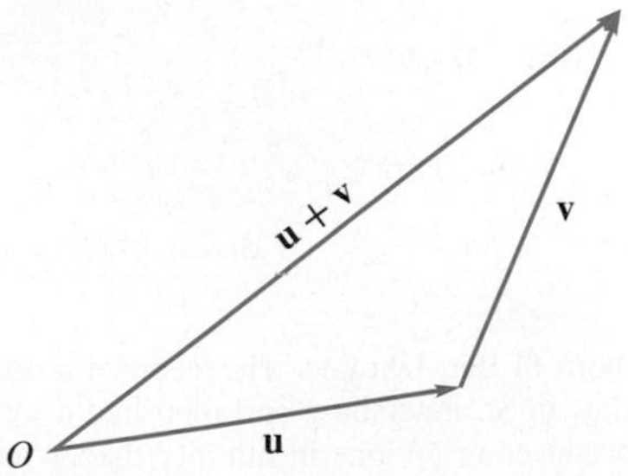

Figure 4.36 ▲
The triangle inequality

The triangle inequality in R^2 and R^3 merely states that the length of a side of a triangle does not exceed the sum of the lengths of the other two sides (Figure 4.36).

EXAMPLE 13

For $\mathbf{u}$ and $\mathbf{v}$ as in Example 12, we have

$$\|\mathbf{u}+\mathbf{v}\| = \sqrt{4} = 2 < \sqrt{2}+\sqrt{2} = \|\mathbf{u}\| + \|\mathbf{v}\|. \qquad \blacksquare$$

Another useful result is the Pythagorean theorem in R^n: If $\mathbf{u}$ and $\mathbf{v}$ are vectors in R^n, then

$$\|\mathbf{u}+\mathbf{v}\|^2 = \|\mathbf{u}\|^2 + \|\mathbf{v}\|^2$$

if and only if $\mathbf{u}$ and $\mathbf{v}$ are orthogonal (Exercise T.10).

DEFINITION

A **unit vector u** in R^n is a vector of length 1. If $\mathbf{x}$ is a nonzero vector, then the vector

$$\mathbf{u} = \frac{1}{\|\mathbf{x}\|}\mathbf{x}$$

is a unit vector in the direction of $\mathbf{x}$.

EXAMPLE 14

If $\mathbf{x} = (1, 0, 0, 1)$, then since $\|\mathbf{x}\| = \sqrt{2}$, the vector $\mathbf{u} = \frac{1}{\sqrt{2}}(1, 0, 0, 1)$ is a unit vector in the direction of $\mathbf{x}$. ■

In the case of R^3 the unit vectors in the positive directions of the x-, y-, and z-axes are denoted by $\mathbf{i} = (1, 0, 0)$, $\mathbf{j} = (0, 1, 0)$, and $\mathbf{k} = (0, 0, 1)$ and are shown in Figure 4.37. If $\mathbf{u} = (x_1, y_1, z_1)$ is any vector in R^3, then we can write $\mathbf{u}$ in terms of $\mathbf{i}$, $\mathbf{j}$, and $\mathbf{k}$ as

$$\mathbf{u} = x_1\mathbf{i} + y_1\mathbf{j} + z_1\mathbf{k}.$$

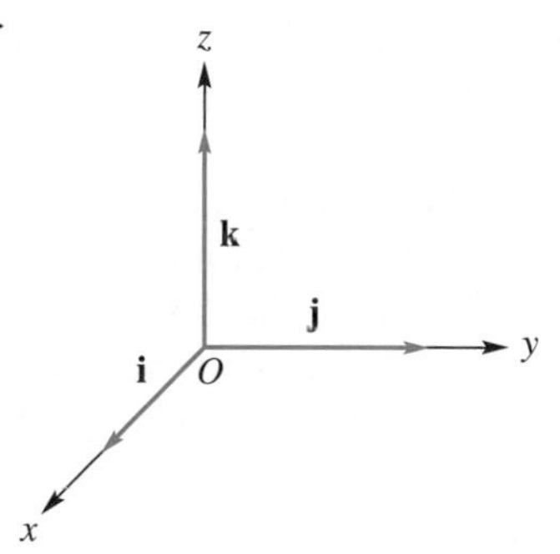

Figure 4.37 ► Unit vectors in R^3

EXAMPLE 15

If $\mathbf{u} = (2, -1, 3)$, then $\mathbf{u} = 2\mathbf{i} - \mathbf{j} + 3\mathbf{k}$. ■

The n-vectors

$$\mathbf{e}_1 = (1, 0, \dots, 0), \quad \mathbf{e}_2 = (0, 1, \dots, 0), \dots, \quad \mathbf{e}_n = (0, 0, \dots, 1)$$

are unit vectors in R^n that are mutually orthogonal. If $\mathbf{u} = (u_1, u_2, \dots, u_n)$ is any vector in R^n, then $\mathbf{u}$ can be written as a linear combination of $\mathbf{e}_1, \mathbf{e}_2, \dots, \mathbf{e}_n$ as

$$\mathbf{u} = u_1\mathbf{e}_1 + u_2\mathbf{e}_2 + \cdots + u_n\mathbf{e}_n.$$

The vector $\mathbf{e}_i$, $1 \le i \le n$, can be viewed as the ith column of the identity matrix I_n. Thus we see that the columns of I_n form a set of n vectors that are mutually orthogonal. We shall discuss such sets of vectors in more detail in Section 6.8.

Summary of Notations for Unit Vectors in R^2 and R^3

In R^2,

$$\mathbf{i} = \mathbf{e}_1 \quad \text{is denoted by} \quad (1, 0) \quad \text{or} \quad \begin{bmatrix} 1 \\ 0 \end{bmatrix}$$

$$\mathbf{j} = \mathbf{e}_2 \quad \text{is denoted by} \quad (0, 1) \quad \text{or} \quad \begin{bmatrix} 0 \\ 1 \end{bmatrix}.$$

In R^3,

$$\mathbf{i} = \mathbf{e}_1 \quad \text{is denoted by} \quad (1, 0, 0) \quad \text{or} \quad \begin{bmatrix} 1 \\ 0 \\ 0 \end{bmatrix}$$

$$\mathbf{j} = \mathbf{e}_2 \quad \text{is denoted by} \quad (0, 1, 0) \quad \text{or} \quad \begin{bmatrix} 0 \\ 1 \\ 0 \end{bmatrix}$$

$$\mathbf{k} = \mathbf{e}_3 \quad \text{is denoted by} \quad (0, 0, 1) \quad \text{or} \quad \begin{bmatrix} 0 \\ 0 \\ 1 \end{bmatrix}.$$

Section 5.1, Cross Product in R^3, and Section 5.2, Lines and Planes, which can be covered at this time, use material from this section.

BIT *n*-VECTORS (OPTIONAL)

In Section 1.2 we defined a bit n-vector as a $1 \times n$ or $n \times 1$ matrix all of whose entries are either 0 or 1. Let B^n be the set of all bit n-vectors. The vector operations of sum and scalar multiple are valid for vectors in B^n provided we use binary arithmetic. In addition, Theorem 4.2 can be verified for B^n, where only the scalars 0 and 1 are permitted. Also, the dot product of bit n-vectors is well defined, as we illustrated in Section 1.3.

In this section a visual model for R^3 was presented. There is a similar, but more restricted, model for B^3. The unit cube as shown in Figure 4.38 provides a visualization of the vectors in B^3. The coordinates of the vertices of the cube correspond to the set of bit 3-vectors. (See Exercise T.10 in Section 1.2.) These vectors can be represented geometrically as directed line segments starting at the origin and terminating at a vertex of the cube. The nonzero vectors of B^3 are shown in Figure 4.39. Note that B^3 has a finite number of vectors while R^3 has infinitely many.

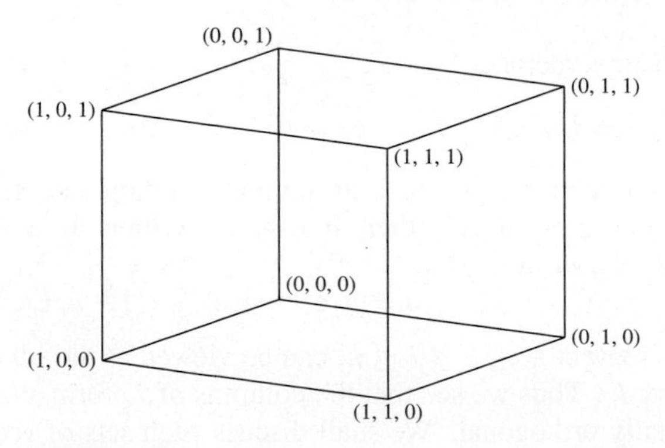

Figure 4.38 ▲

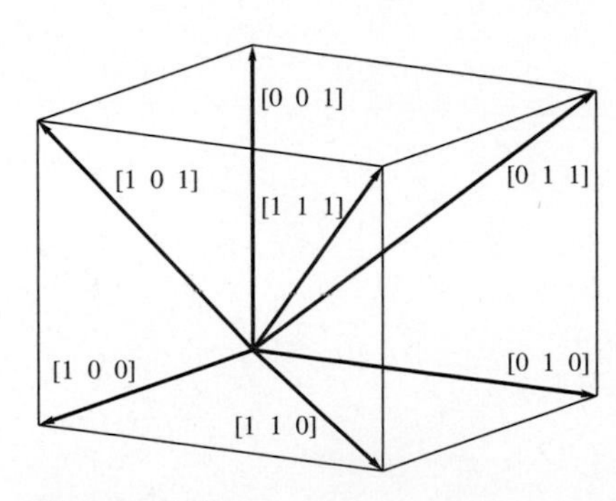

Figure 4.39 ▲

Unfortunately, there is no convenient geometric analog for vector operations on bit 3-vectors that corresponds to the constructions shown in Figures 4.32 and 4.33 for R^3.

The properties of dot product, Theorem 4.3, are not all valid for B^n. See Example 16.

EXAMPLE 16 Let $\mathbf{u} = (1, 1, 0)$ be in B^3. Then

$$\mathbf{u} \cdot \mathbf{u} = (1)(1) + (1)(1) + (0)(0) = 1 + 1 + 0 = 0.$$

Thus there exists a nonzero vector in B^3 such that $\|\mathbf{u}\| = \sqrt{\mathbf{u} \cdot \mathbf{u}} = 0$. ■

This deficiency means that the concept of length of vectors as defined for R^n does not generalize to vectors in B^n. Hence it follows that the Cauchy–Schwarz inequality, angle between vectors, orthogonal vectors, parallel vectors, and the triangle inequality are not applicable to B^n. (See Exercises 38, 39, T.20, and T.21.) However, the bit n-vectors

$$\mathbf{e}_1 = (1, 0, \ldots, 0), \quad \mathbf{e}_2 = (0, 1, \ldots, 0), \ldots, \quad \mathbf{e}_n = (0, 0, \ldots, 1)$$

can be used to express any vector $\mathbf{u}$ in B^n as

$$\mathbf{u} - u_1\mathbf{e}_1 + u_2\mathbf{e}_2 + \cdots + u_n\mathbf{e}_n = \sum_{j=1}^{n} u_j\mathbf{e}_j.$$

Preview of Applications

Cross Product in R^3 (Section 5.1)

All operations on vectors discussed in this text so far have started out with the operation being defined on vectors in R^2, generalized to vectors in R^3, and then further generalized to vectors in R^n. An operation on vectors that is only valid in R^3 is the cross product of two vectors. If $\mathbf{u} = u_1\mathbf{i} + u_2\mathbf{j} + u_3\mathbf{k}$ and $\mathbf{v} = v_1\mathbf{i} + v_2\mathbf{j} + v_3\mathbf{k}$ are vectors in R^3, we define their cross product as

$$\mathbf{u} \times \mathbf{v} = (u_2v_3 - u_3v_2)\mathbf{i} + (u_3v_1 - u_1v_3)\mathbf{j} + (u_1v_2 - u_2v_1)\mathbf{k},$$

a vector perpendicular to the plane determined by $\mathbf{u}$ and $\mathbf{v}$. See Figure A.

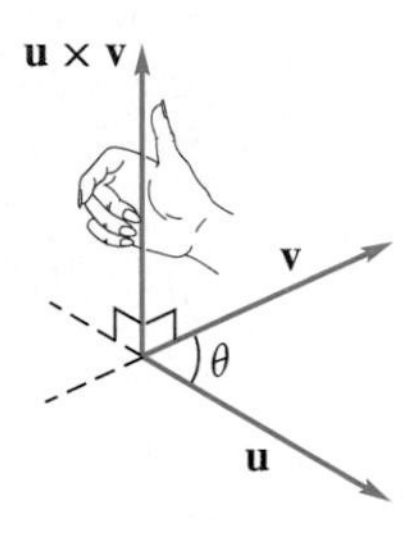

Figure A

The cross product operation is useful in the study of planes and in other applications in mathematics.

Section 5.1 gives the basic properties of the cross product operation.

Lines and Planes (Section 5.2)

The simplest curve in R^2 is a straight line. It is also one of the most useful curves, since it enables us to approximate any curve.

For example, when a computer or graphing calculator sketches a curve, it does it by piecing together a great many line segments, which approximate the curve. In Figure B we show $y = \sin x$ and then its approximations with 5 and 15 equally spaced line segments, respectively. In general, the shorter the line segments, the better the approximation.

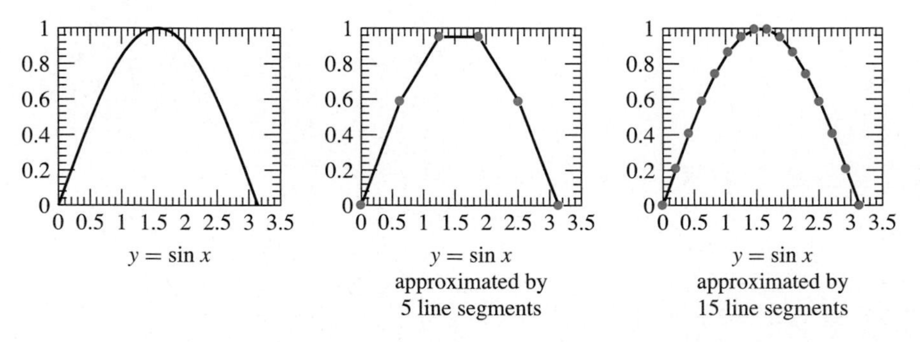

Figure B

The line L in Figure C is described by the vector equation

$$\mathbf{x} = \mathbf{w}_0 + t\mathbf{u} \qquad -\infty < t < \infty.$$

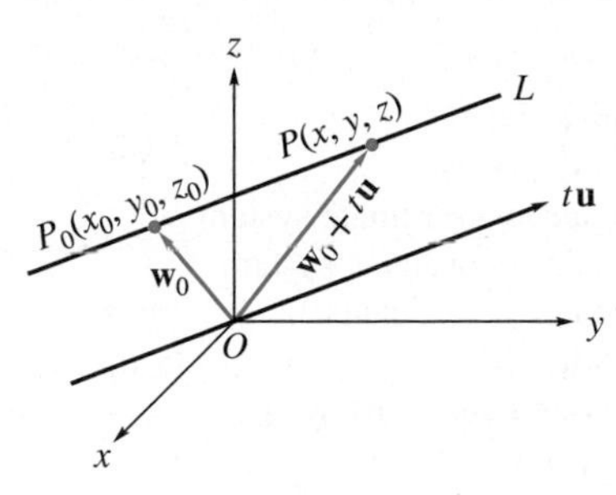

Figure C

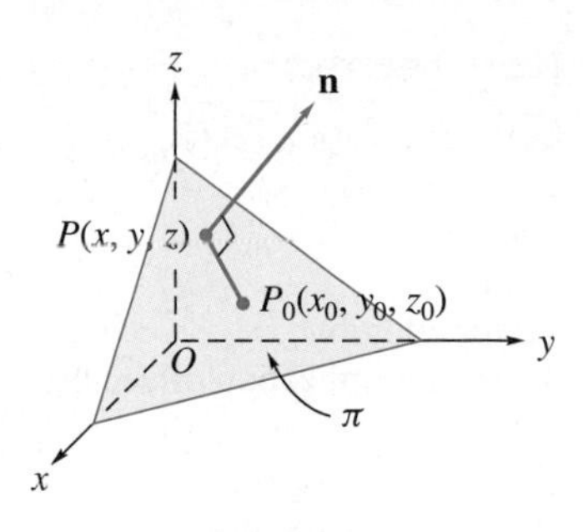

Figure D

Similarly, the simplest surface in R^3 is a plane. Planes can be used to approximate a more complicated surface. The plane π shown in Figure D is described by the set of points $P(x, y, z)$ that satisfy the equation

$$\mathbf{n} \cdot \overrightarrow{P_0P} = 0.$$

Section 5.2 gives a brief introduction to lines and planes from a linear algebra point of view.

Key Terms

Components of a vector
Equal vectors
n-space
Scalars
Vector addition
Scalar multiplication of vectors
Zero vector
Negative of a vector
Difference of vectors
Coordinate system
Coordinate axes
x-, y-, and z-axes
Right-handed coordinate system
Left-handed coordinate system
x-, y-, and z coordinates
coordinates
Rectangular coordinate system
xy-, xz-, and yz-planes
Vector
Equal vectors
Length (or magnitude, or norm) of a vector
Distance between points (or vectors)
Standard inner product
Cauchy–Schwarz inequality
Angle between vectors
Triangle inequality
Unit vector
Parallelogram law

4.2 Exercises

1. Find $\mathbf{u}+\mathbf{v}$, $\mathbf{u}-\mathbf{v}$, $2\mathbf{u}$, and $3\mathbf{u}-2\mathbf{v}$ if

(a) $\mathbf{u}=(1,2,-3)$, $\mathbf{v}=(0,1,-2)$

(b) $\mathbf{u}=(4,-2,1,3)$, $\mathbf{v}=(-1,2,5,-4)$

2. Repeat Exercise 1 for

(a) $\mathbf{u}=\begin{bmatrix}2\\0\\-4\end{bmatrix}$, $\mathbf{v}=\begin{bmatrix}3\\2\\1\end{bmatrix}$

(b) $\mathbf{u}=\begin{bmatrix}-3\\5\\-3\\0\end{bmatrix}$, $\mathbf{v}=\begin{bmatrix}2\\1\\5\\-2\end{bmatrix}$

3. Let

$$\mathbf{u}=\begin{bmatrix}1\\-2\\3\end{bmatrix},\qquad \mathbf{v}=\begin{bmatrix}-3\\-1\\3\end{bmatrix},$$
$$\mathbf{w}=\begin{bmatrix}a\\-1\\b\end{bmatrix},\qquad \mathbf{x}=\begin{bmatrix}3\\c\\2\end{bmatrix}.$$

Find a, b, and c so that

(a) $\mathbf{w}=\frac{1}{2}\mathbf{u}$ (b) $\mathbf{w}+\mathbf{v}=\mathbf{u}$ (c) $\mathbf{w}+\mathbf{x}=\mathbf{v}$

4. Let $\mathbf{u}=(4,-1,-2,3)$, $\mathbf{v}=(3,-2,-4,1)$, $\mathbf{w}=(a,-3,-6,b)$, and $\mathbf{x}=(2,c,d,4)$. Find a, b, c, and d so that

(a) $\mathbf{w}=3\mathbf{u}$ (b) $\mathbf{w}+\mathbf{x}=\mathbf{u}$ (c) $\mathbf{w}-\mathbf{u}=\mathbf{v}$

5. Let $\mathbf{u}=(4,5,-2,3)$, $\mathbf{v}=(3,-2,0,1)$, $\mathbf{w}=(-3,2,-5,3)$, $c=2$, and $d=3$. Verify properties (a) through (h) in Theorem 4.2.

6. Plot the following points in R^3.

(a) $(3,-1,2)$ (b) $(1,0,2)$

(c) $(0,0,-4)$ (d) $(1,0,0)$

(e) $(0,-2,0)$

7. Sketch a directed line segment in R^3 representing each of the following vectors.

(a) $\mathbf{u}_1=(2,-3,-1)$ (b) $\mathbf{u}_2=(0,1,4)$

(c) $\mathbf{u}_3=(0,0,-1)$

8. For each of the following pairs of points in R^3, determine the vector that is associated with the directed line segment whose tail is the first point and whose head is the second point.

(a) $(2,3,-1)$, $(0,0,2)$

(b) $(1,1,0)$, $(0,1,1)$

(c) $(-1,-2,-3)$, $(3,4,5)$

(d) $(1,1,3)$, $(0,0,1)$

9. Determine the head of the vector $(3,4,-1)$ whose tail is $(1,-2,3)$.

10. Find the length of the following vectors.

(a) $(1,2,-3)$ (b) $(2,3,-1,4)$

(c) $(1,0,3)$ (d) $(0,0,3,4)$

11. Find the length of the following vectors.

(a) $(2,3,4)$ (b) $(0,-1,2,3)$

(c) $(-1,-2,0)$ (d) $(1,2,-3,-4)$

12. Find the distance between the following pairs of points.

(a) $(1,-1,2)$, $(3,0,2)$

(b) $(4,2,-1,5)$, $(2,3,-1,4)$

(c) $(0,0,2)$, $(-3,0,0)$

(d) $(1,0,0,2)$, $(3,-1,5,2)$

13. Find the distance between the following pairs of points.

(a) $(1,1,0)$, $(2,-3,1)$

(b) $(4,2,-1,6)$, $(4,3,1,5)$

(c) $(0,2,3)$, $(1,2,-4)$

(d) $(3,4,0,1)$, $(2,2,1,-1)$

14. Is the vector $(2,-2,3)$ a linear combination of the vectors $(1,2,-3)$, $(-1,1,1)$, and $(-1,4,-1)$?

15. If possible, find scalars c_1, c_2, and c_3, not all zero, so that

$$c_1\begin{bmatrix}1\\2\\-1\end{bmatrix}+c_2\begin{bmatrix}1\\3\\-2\end{bmatrix}+c_3\begin{bmatrix}3\\7\\-4\end{bmatrix}=\begin{bmatrix}0\\0\\0\end{bmatrix}.$$

16. Find all constants a such that $\|(1, a, -3, 2)\| = 5$.

17. Find all constants a such that $\mathbf{u} \cdot \mathbf{v} = 0$, where $\mathbf{u} = (a, 2, 1, a)$ and $\mathbf{v} = (a, -1, -2, -3)$.

18. Verify Theorem 4.3 for $c = 3$ and $\mathbf{u} = (1, 2, 3)$, $\mathbf{v} = (1, 2, -4)$, and $\mathbf{w} = (1, 0, 2)$.

19. Verify Theorem 4.4 for $\mathbf{u}$ and $\mathbf{v}$ as in Exercise 18.

20. Find the cosine of the angle between each pair of vectors $\mathbf{u}$ and $\mathbf{v}$.

(a) $\mathbf{u} = (1, 2, 3)$, $\mathbf{v} = (-4, 4, 5)$

(b) $\mathbf{u} = (0, 2, 3, 1)$, $\mathbf{v} = (-3, 1, -2, 0)$

(c) $\mathbf{u} = (0, 0, 1)$, $\mathbf{v} = (2, 2, 0)$

(d) $\mathbf{u} = (2, 0, -1, 3)$, $\mathbf{v} = (-3, -5, 2, -1)$

21. Find the cosine of the angle between each pair of vectors $\mathbf{u}$ and $\mathbf{v}$.

(a) $\mathbf{u} = (2, 3, 1)$, $\mathbf{v} = (3, -2, 0)$

(b) $\mathbf{u} = (1, 2, -1, 3)$, $\mathbf{v} = (0, 0, -1, -2)$

(c) $\mathbf{u} = (2, 0, 1)$, $\mathbf{v} = (2, 2, -1)$

(d) $\mathbf{u} = (0, 4, 2, 3)$, $\mathbf{v} = (0, -1, 2, 0)$

22. Show that

(a) $\mathbf{i} \cdot \mathbf{i} = \mathbf{j} \cdot \mathbf{j} = \mathbf{k} \cdot \mathbf{k} = 1$

(b) $\mathbf{i} \cdot \mathbf{j} = \mathbf{i} \cdot \mathbf{k} = \mathbf{j} \cdot \mathbf{k} = 0$

23. Which of the vectors

$$\mathbf{u}_1 = (4, 2, 6, -8), \quad \mathbf{u}_2 = (-2, 3, -1, -1),$$
$$\mathbf{u}_3 = (-2, -1, -3, 4), \quad \mathbf{u}_4 = (1, 0, 0, 2),$$
$$\mathbf{u}_5 = (1, 2, 3, -4), \quad \text{and } \mathbf{u}_6 = (0, -3, 1, 0)$$

are

(a) orthogonal? (b) parallel?

(c) in the same direction?

24. Find c so that the vector $\mathbf{v} = (2, c, 3)$ is orthogonal to $\mathbf{w} = (1, -2, 1)$.

25. If possible, find a, b, and c not all zero so that $\mathbf{v} = (a, b, c)$ is orthogonal to both

$$\mathbf{w} = (1, 2, 1) \quad \text{and} \quad \mathbf{x} = (1, -1, 1).$$

26. Verify the triangle inequality for $\mathbf{u} = (1, 2, 3, -1)$ and $\mathbf{v} = (1, 0, -2, 3)$.

27. Find a unit vector in the direction of $\mathbf{x}$.

(a) $\mathbf{x} = (2, -1, 3)$ (b) $\mathbf{x} = (1, 2, 3, 4)$

(c) $\mathbf{x} = (0, 1, -1)$ (d) $\mathbf{x} = (0, -1, 2, -1)$

28. Find a unit vector in the direction of $\mathbf{x}$.

(a) $\mathbf{x} = (1, 2, -1)$ (b) $\mathbf{x} = (0, 0, 2, 0)$

(c) $\mathbf{x} = (-1, 0, -2)$ (d) $\mathbf{x} = (0, 0, 3, 4)$

29. Write each of the following vectors in R^3 in terms of $\mathbf{i}$, $\mathbf{j}$, and $\mathbf{k}$.

(a) $(1, 2, -3)$ (b) $(2, 3, -1)$

(c) $(0, 1, 2)$ (d) $(0, 0, -2)$

30. Write each of the following vectors in R^3 as a 3×1 matrix.

(a) $2\mathbf{i} + 3\mathbf{j} - 4\mathbf{k}$ (b) $\mathbf{i} + 2\mathbf{j}$

(c) $-3\mathbf{i}$ (d) $3\mathbf{i} - 2\mathbf{k}$

31. Verify that the triangle with vertices $P_1(2, 3, -4)$, $P_2(3, 1, 2)$, and $P_3(-3, 0, 4)$ is isosceles.

32. Verify that the triangle with vertices $P_1(2, 3, -4)$, $P_2(3, 1, 2)$, and $P_3(7, 0, 1)$ is a right triangle.

33. A large steel manufacturer, who has 2000 employees, lists each employee's salary as a component of a vector $\mathbf{u}$ in R^{2000}. If an 8% across-the-board salary increase has been approved, find an expression involving $\mathbf{u}$ giving all the new salaries.

34. The vector $\mathbf{u} = (20, 30, 80, 10)$ gives the number of receivers, CD players, speakers, and cassette recorders that are on hand in a stereo shop. The vector $\mathbf{v} = (200, 120, 80, 70)$ gives the price (in dollars) of each receiver, CD player, speaker, and cassette recorder, respectively. What does the dot product $\mathbf{u} \cdot \mathbf{v}$ tell the shop owner?

35. A brokerage firm records the high and low values of the price of IBM stock each day. The information for a given week is presented in two vectors, $\mathbf{t}$ and $\mathbf{b}$, in R^5, giving the high and low values, respectively. What expression gives the average daily values of the price of IBM stock for the entire 5-day week?

Exercises 36 through 39 involve bit matrices.

36. Let $\mathbf{u} = (1, 1, 0, 0)$. Find a vector $\mathbf{v}$ in B^4 so that $\mathbf{u} + \mathbf{v} = \mathbf{0}$. Is there more than one such vector $\mathbf{v}$? Explain.

37. Let $\mathbf{u} = (0, 1, 0, 1)$. Find a vector $\mathbf{v}$ in B^4 so that $\mathbf{u} + \mathbf{v} = \mathbf{0}$. Is there more than one such vector $\mathbf{v}$? Explain.

38. Let $\mathbf{u} = (1, 1, 0, 0)$. Find all vectors $\mathbf{v}$ in B^4 so that $\mathbf{u} \cdot \mathbf{v} = 0$.

39. Let $\mathbf{u} = (1, 0, 1)$. Find all vectors $\mathbf{v}$ in B^3 so that $\mathbf{u} \cdot \mathbf{v} = 0$.

Theoretical Exercises

T.1. Prove the rest of Theorem 4.2.

T.2. Show that $-\mathbf{u} = (-1)\mathbf{u}$.

T.3. Establish Equations (1) and (2) in R^3, for the length of a vector and the distance between two points, by using the Pythagorean theorem.

T.4. Prove Theorem 4.3.

T.5. Suppose that $\mathbf{u}$ is orthogonal to both $\mathbf{v}$ and $\mathbf{w}$. Show that $\mathbf{u}$ is orthogonal to any vector of the form $r\mathbf{v} + s\mathbf{w}$, where r and s are scalars.

T.6. Show that if $\mathbf{u} \cdot \mathbf{v} = 0$ for all vectors $\mathbf{v}$, then $\mathbf{u} = \mathbf{0}$.

T.7. Show that $\mathbf{u} \cdot (\mathbf{v} + \mathbf{w}) = \mathbf{u} \cdot \mathbf{v} + \mathbf{u} \cdot \mathbf{w}$.

T.8. Show that if $\mathbf{u} \cdot \mathbf{v} = \mathbf{u} \cdot \mathbf{w}$ for all $\mathbf{u}$, then $\mathbf{v} = \mathbf{w}$.

T.9. Show that if c is a scalar, then $\|c\mathbf{u}\| = |c|\|\mathbf{u}\|$, where $|c|$ is the absolute value of c.

T.10. ***(Pythagorean Theorem in R^n)*** Show that $\|\mathbf{u} + \mathbf{v}\|^2 = \|\mathbf{u}\|^2 + \|\mathbf{v}\|^2$ if and only if $\mathbf{u} \cdot \mathbf{v} = 0$.

T.11. Let A be an $n \times n$ matrix and let $\mathbf{x}$ and $\mathbf{y}$ be vectors in R^n. Show that $A\mathbf{x} \cdot \mathbf{y} = \mathbf{x} \cdot A^T\mathbf{y}$.

T.12. Define the **distance** between two vectors $\mathbf{u}$ and $\mathbf{v}$ in R^n as $d(\mathbf{u}, \mathbf{v}) = \|\mathbf{u} - \mathbf{v}\|$. Show that

(a) $d(\mathbf{u}, \mathbf{v}) \geq 0$

(b) $d(\mathbf{u}, \mathbf{v}) = 0$ if and only if $\mathbf{u} = \mathbf{v}$

(c) $d(\mathbf{u}, \mathbf{v}) = d(\mathbf{v}, \mathbf{u})$

(d) $d(\mathbf{u}, \mathbf{w}) \leq d(\mathbf{u}, \mathbf{v}) + d(\mathbf{v}, \mathbf{w})$

T.13. Prove the **parallelogram law**:

$$\|\mathbf{u} + \mathbf{v}\|^2 + \|\mathbf{u} - \mathbf{v}\|^2 = 2\|\mathbf{u}\|^2 + 2\|\mathbf{v}\|^2.$$

T.14. If $\mathbf{x}$ is a nonzero vector, show that

$$\mathbf{u} = \frac{1}{\|\mathbf{x}\|}\mathbf{x}$$

is a unit vector in the direction of $\mathbf{x}$.

T.15. Show that $\mathbf{u} \cdot \mathbf{v} = \frac{1}{4}\|\mathbf{u} + \mathbf{v}\|^2 - \frac{1}{4}\|\mathbf{u} - \mathbf{v}\|^2$.

T.16. Let $\mathbf{u}$, $\mathbf{v}$, and $\mathbf{w}$ be the vectors defined in Example 12. Let $\mathbf{z} = (0, 2, -2, 0)$. Show that $\mathbf{z}$ is orthogonal to $\mathbf{v}$ but is not parallel to $\mathbf{u}$ or $\mathbf{w}$.

Exercises T.17 through T.21 involve bit matrices.

T.17. Determine the "negative" of each vector in B^3.

T.18. Determine all vectors $\mathbf{v}$ in B^3 such that $\mathbf{v} \cdot \mathbf{v} = 0$.

T.19. Determine the "negative" of each vector in B^4.

T.20. Let $\mathbf{u} = (1, 1, 1)$.

(a) Find $V_{\mathbf{u}}$, the set of all vectors $\mathbf{v}$ in B^3 such that $\mathbf{u} \cdot \mathbf{v} = 0$.

(b) Find $\tilde{V}_{\mathbf{u}}$, the set of all vectors $\mathbf{v}$ in B^3 such that $\mathbf{u} \cdot \mathbf{v} \neq 0$.

(c) Does $V_{\mathbf{u}}$ together with $\tilde{V}_{\mathbf{u}}$ contain all the vectors in B^3? Explain.

T.21. Let $\mathbf{u} = (1, 1, 1, 1)$.

(a) Find $V_{\mathbf{u}}$, the set of all vectors $\mathbf{v}$ in B^4 such that $\mathbf{u} \cdot \mathbf{v} = 0$.

(b) Find $\tilde{V}_{\mathbf{u}}$, the set of all vectors $\mathbf{v}$ in B^4 such that $\mathbf{u} \cdot \mathbf{v} \neq 0$.

(c) Does $V_{\mathbf{u}}$ together with $\tilde{V}_{\mathbf{u}}$ contain all the vectors in B^4? Explain.

MATLAB Exercises

In order to use MATLAB *in this section, you should first have read Section 12.6.*

ML.1. As aid for visualizing vector operations in R^3, we have **vec3demo**. This routine provides a graphical display of vectors in 3-space. For a pair of vectors $\mathbf{u}$ and $\mathbf{v}$, routine **vec3demo** graphs $\mathbf{u}$ and $\mathbf{v}$, $\mathbf{u} + \mathbf{v}$, $\mathbf{u} - \mathbf{v}$, and a scalar multiple. Once the pair of vectors from R^3 are entered into MATLAB, type

vec3demo(u, v)

Use **vec3demo** on each of the following pairs from R^3.

(a) $\mathbf{u} = (2, 6, 4)$, $\mathbf{v} = (6, 2, -5)$

(b) $\mathbf{u} = (3, -5, 4)$, $\mathbf{v} = (7, -1, -2)$

(c) $\mathbf{u} = (4, 0, -5)$, $\mathbf{v} = (0, 6, 3)$

ML.2. Determine the norm or length of each of the following vectors using MATLAB.

(a) $\mathbf{u} = \begin{bmatrix} 2 \\ 2 \\ -1 \end{bmatrix}$ (b) $\mathbf{v} = \begin{bmatrix} 0 \\ 4 \\ -3 \\ 0 \end{bmatrix}$ (c) $\mathbf{w} = \begin{bmatrix} 1 \\ 0 \\ 1 \\ 0 \\ 3 \end{bmatrix}$

ML.3. Determine the distance between each of the following pairs of vectors using MATLAB.

(a) $\mathbf{u} = \begin{bmatrix} 2 \\ 0 \\ 3 \end{bmatrix}$, $\mathbf{v} = \begin{bmatrix} 2 \\ -1 \\ 1 \end{bmatrix}$

(b) $\mathbf{u} = (2, 0, 0, 1)$, $\mathbf{v} = (2, 5, -1, 3)$

(c) $\mathbf{u} = (1, 0, 4, 3)$, $\mathbf{v} = (-1, 1, 2, 2)$

ML.4. Determine the lengths of the sides of the triangle ABC, which has vertices in R^3, given by $A(1, 3, -2)$, $B(4, -1, 0)$, $C(1, 1, 2)$. (*Hint*: Determine a vector for each side and compute its length.)

ML.5. Determine the dot product of each one of the following pairs of vectors using MATLAB.

(a) $\mathbf{u} = (5, 4, -4)$, $\mathbf{v} = (3, 2, 1)$

(b) $\mathbf{u} = (3, -1, 0, 2)$, $\mathbf{v} = (-1, 2, -5, -3)$

(c) $\mathbf{u} = (1, 2, 3, 4, 5)$, $\mathbf{v} = -\mathbf{u}$

ML.6. The norm or length of a vector can be computed using dot products as follows:

$$\|\mathbf{u}\| = \sqrt{\mathbf{u} \cdot \mathbf{u}}.$$

In MATLAB, the right side of the preceding expression is computed as

sqrt(dot(u, u))

Verify this alternative procedure on the vectors in Exercise ML.2.

ML.7. In MATLAB, if the n-vectors $\mathbf{u}$ and $\mathbf{v}$ are entered as columns, then

u′ ∗ v or **v′ ∗ u**

gives the dot product of vectors $\mathbf{u}$ and $\mathbf{v}$. Verify this using the vectors in Exercise ML.5.

ML.8. Use MATLAB to find the angle between each of the following pairs of vectors. (To convert the angle from radians to degrees, multiply by 180/pi.)

(a) $\mathbf{u} = (3, 2, 4, 0)$, $\mathbf{v} = (0, 2, -1, 0)$

(b) $\mathbf{u} = (2, 2, -1)$, $\mathbf{v} = (2, 0, 1)$

(c) $\mathbf{u} = (1, 0, 0, 2)$, $\mathbf{v} = (0, 3, -4, 0)$

ML.9. Use MATLAB to find a unit vector in the direction of the vectors in Exercise ML.2.

4.3 LINEAR TRANSFORMATIONS

In Section 1.5, we introduced matrix transformations, functions that map R^n into R^m. In this section, we present an alternative approach to matrix transformations. We shall now denote a function mapping R^n into R^m by L. In Chapter 10 we consider linear transformations from a much more general point of view and we study their properties in some detail.

DEFINITION A **linear transformation** L of R^n into R^m is a function assigning a unique vector $L(\mathbf{u})$ in R^m to each $\mathbf{u}$ in R^n such that

(a) $L(\mathbf{u} + \mathbf{v}) = L(\mathbf{u}) + L(\mathbf{v})$, for every $\mathbf{u}$ and $\mathbf{v}$ in R^n.

(b) $L(k\mathbf{u}) = kL(\mathbf{u})$, for every $\mathbf{u}$ in R^n and every scalar k.

A function T of R^n into R^m is said to be **nonlinear** if it is not a linear transformation.

The vector $L(\mathbf{u})$ in R^m is called the **image** of $\mathbf{u}$. The set of all images in R^m of the vectors in R^n is called the **range** of L. Since R^n can be viewed as consisting of points or vectors, $L(\mathbf{u})$, for $\mathbf{u}$ in R^n, can be considered as a point or a vector in R^m.

We shall write the fact that L maps R^n into R^m, even if it is not a linear transformation, as

$$L\colon R^n \to R^m.$$

If $n = m$, a linear transformation $L\colon R^n \to R^n$ is also called a **linear operator** on R^n.

Let A be an $m \times n$ matrix. In Section 1.5, we defined a matrix transformation as a function $L\colon R^n \to R^m$ defined by $L(\mathbf{u}) = A\mathbf{u}$. We now show that every matrix transformation is a linear transformation by verifying that properties (a) and (b) in the preceding Definition hold.

If $\mathbf{u}$ and $\mathbf{v}$ are vectors in R^n, then

$$L(\mathbf{u} + \mathbf{v}) = A(\mathbf{u} + \mathbf{v}) = A\mathbf{u} + A\mathbf{v} = L(\mathbf{u}) + L(\mathbf{v}).$$

Moreover, if c is a scalar, then

$$L(c\mathbf{u}) = A(c\mathbf{u}) = c(A\mathbf{u}) = cL(\mathbf{u}).$$

Hence, every matrix transformation is a linear transformation.

For convenience we now summarize the matrix transformations that have already been presented in Section 1.5.

Reflection with respect to the x-axis: $L\colon R^2 \to R^2$ is defined by

$$L\left(\begin{bmatrix} u_1 \\ u_2 \end{bmatrix}\right) = \begin{bmatrix} u_1 \\ -u_2 \end{bmatrix}.$$

Projection into the xy-plane: $L\colon R^3 \to R^2$ is defined by

$$L\left(\begin{bmatrix} u_1 \\ u_2 \\ u_3 \end{bmatrix}\right) = \begin{bmatrix} u_1 \\ u_2 \end{bmatrix}.$$

Dilation: $L\colon R^3 \to R^3$ is defined by $L(\mathbf{u}) = r\mathbf{u}$ for $r > 1$.

Contraction: $L\colon R^3 \to R^3$ is defined by $L(\mathbf{u}) = r\mathbf{u}$ for $0 < r < 1$.

Rotation counterclockwise through an angle ϕ: $L\colon R^2 \to R^2$ is defined by

$$L(\mathbf{u}) = \begin{bmatrix} \cos\phi & -\sin\phi \\ \sin\phi & \cos\phi \end{bmatrix}\mathbf{u}.$$

Shear in the x-direction: $L\colon R^2 \to R^2$ is defined by

$$L(\mathbf{u}) = \begin{bmatrix} 1 & k \\ 0 & 1 \end{bmatrix}\mathbf{u},$$

where k is a scalar.

Shear in the y-direction: $L\colon R^2 \to R^2$ is defined by

$$L(\mathbf{u}) = \begin{bmatrix} 1 & 0 \\ k & 1 \end{bmatrix}\mathbf{u},$$

where k is a scalar.

EXAMPLE 1

Let $L\colon R^3 \to R^2$ be defined by

$$L\left(\begin{bmatrix} u_1 \\ u_2 \\ u_3 \end{bmatrix}\right) = \begin{bmatrix} u_1 + 1 \\ u_2 - u_3 \end{bmatrix}.$$

To determine whether L is a linear transformation, let

$$\mathbf{u} = \begin{bmatrix} u_1 \\ u_2 \\ u_3 \end{bmatrix} \quad \text{and} \quad \mathbf{v} = \begin{bmatrix} v_1 \\ v_2 \\ v_3 \end{bmatrix}.$$

Then

$$L(\mathbf{u}+\mathbf{v}) = L\left(\begin{bmatrix} u_1 \\ u_2 \\ u_3 \end{bmatrix} + \begin{bmatrix} v_1 \\ v_2 \\ v_3 \end{bmatrix}\right) = L\left(\begin{bmatrix} u_1 + v_1 \\ u_2 + v_2 \\ u_3 + v_3 \end{bmatrix}\right)$$
$$= \begin{bmatrix} (u_1 + v_1) + 1 \\ (u_2 + v_2) - (u_3 + v_3) \end{bmatrix}.$$

On the other hand,

$$L(\mathbf{u}) + L(\mathbf{v}) = \begin{bmatrix} u_1 + 1 \\ u_2 - u_3 \end{bmatrix} + \begin{bmatrix} v_1 + 1 \\ v_2 - v_3 \end{bmatrix} = \begin{bmatrix} (u_1 + v_1) + 2 \\ (u_2 - u_3) + (v_2 - v_3) \end{bmatrix}.$$

Since the first coordinates of $L(\mathbf{u} + \mathbf{v})$ and $L(\mathbf{u}) + L(\mathbf{v})$ are different, $L(\mathbf{u} + \mathbf{v}) \neq L(\mathbf{u}) + L(\mathbf{v})$, so we conclude that the function L is not a linear transformation; that is, L is nonlinear. ■

It can be shown that $L\colon R^n \to R^m$ is a linear transformation if and only if $L(a\mathbf{u}+b\mathbf{v}) = aL(\mathbf{u})+bL(\mathbf{v})$ for any real numbers a and b and any vectors $\mathbf{u}$, $\mathbf{v}$ in R^n (see Exercise T.4 in the Supplementary Exercises).

The following two theorems give some additional basic properties of linear transformations from R^n to R^m. The proofs will be left as exercises. Moreover, a more general version of the second theorem below will be proved in Section 10.1.

THEOREM 4.6 *If $L\colon R^n \to R^m$ is a linear transformation, then*

$$L(c_1\mathbf{u}_1 + c_2\mathbf{u}_2 + \cdots + c_k\mathbf{u}_k) = c_1L(\mathbf{u}_1) + c_2L(\mathbf{u}_2) + \cdots + c_kL(\mathbf{u}_k)$$

for any vectors $\mathbf{u}_1, \mathbf{u}_2, \ldots, \mathbf{u}_k$ in R^n and any scalars $c_1, c_2, \ldots, c_k$.

Proof Exercise T.1. ■

THEOREM 4.7 *Let $L\colon R^n \to R^m$ be a linear transformation. Then:*

(a) $L(\mathbf{0}_{R^n}) = \mathbf{0}_{R^m}$

(b) $L(\mathbf{u}-\mathbf{v}) = L(\mathbf{u}) - L(\mathbf{v})$, *for* $\mathbf{u}$ *and* $\mathbf{v}$ *in* R^n

Proof Exercise T.2. ■

COROLLARY 4.1 *Let $T\colon R^n \to R^m$ be a function. If $T(\mathbf{0}_{R^n}) \neq \mathbf{0}_{R^m}$, then T is a nonlinear transformation.*

Proof Let $T(\mathbf{0}_{R^n}) = \mathbf{w} \neq \mathbf{0}_{R^m}$. Then

$$T(\mathbf{0}_{R^n}) = T(\mathbf{0}_{R^n} + \mathbf{0}_{R^n}) = T(\mathbf{0}_{R^n}) + T(\mathbf{0}_{R^n}).$$

However,

$$T(\mathbf{0}_{R^n}) + T(\mathbf{0}_{R^n}) = \mathbf{w}+\mathbf{w} = 2\mathbf{w}.$$

Since $\mathbf{w} \neq 2\mathbf{w}$, T is nonlinear. ■

Remark Example 1 could be solved more easily by using Corollary 4.1 as follows: Since

$$L\left(\begin{bmatrix}0\\0\\0\end{bmatrix}\right) = \begin{bmatrix}1\\0\end{bmatrix} \neq \begin{bmatrix}0\\0\end{bmatrix},$$

L is nonlinear.

These theorems can be used to compute the image of a vector $\mathbf{u}$ in R^2 or R^3 under a linear transformation $L\colon R^2 \to R^n$ once we know $L(\mathbf{i})$ and $L(\mathbf{j})$, where $\mathbf{i} = (1,0)$ and $\mathbf{j} = (0,1)$. Similarly, we can compute $L(\mathbf{u})$ for $\mathbf{u}$ in R^3 under the linear transformation $L\colon R^3 \to R^n$ if we know $L(\mathbf{i})$, $L(\mathbf{j})$, and $L(\mathbf{k})$, where $\mathbf{i} = (1,0,0)$, $\mathbf{j} = (0,1,0)$, and $\mathbf{k} = (0,0,1)$. It follows from the observations in Sections 4.1 and 4.2 that if $\mathbf{v} = (v_1, v_2)$ is any vector in R^2 and $\mathbf{u} = (u_1, u_2, u_3)$ is any vector in R^3, then

$$\mathbf{v} = v_1\mathbf{i} + v_2\mathbf{j} \quad \text{and} \quad \mathbf{u} = u_1\mathbf{i} + u_2\mathbf{j} + u_3\mathbf{k}.$$

EXAMPLE 2 Let $L\colon R^3 \to R^2$ be a linear transformation for which we know that

$$L(1,0,0) = (2,-1), \quad L(0,1,0) = (3,1), \quad \text{and} \quad L(0,0,1) = (-1,2).$$

Find $L(-3,4,2)$.

Solution Since

$$(-3, 4, 2) = -3\mathbf{i} + 4\mathbf{j} + 2\mathbf{k},$$

we have

$$\begin{aligned} L(-3, 4, 2) &= L(-3\mathbf{i} + 4\mathbf{j} + 2\mathbf{k}) \\ &= -3L(\mathbf{i}) + 4L(\mathbf{j}) + 2L(\mathbf{k}) \\ &= -3(2, -1) + 4(3, 1) + 2(-1, 2) \\ &= (4, 11). \end{aligned}$$ ■

More generally, we pointed out after Example 15 in Section 4.2, that if $\mathbf{u} = (u_1, u_2, \ldots, u_n)$ is any vector in R^n, then

$$\mathbf{u} = u_1\mathbf{e}_1 + u_2\mathbf{e}_2 + \cdots + u_n\mathbf{e}_n,$$

where

$$\mathbf{e}_1 = (1, 0, \ldots, 0), \quad \mathbf{e}_2 = (0, 1, \ldots, 0), \ldots, \quad \mathbf{e}_n = (0, 0, \ldots, 1).$$

This implies that if $L: R^n \to R^m$ is a linear transformation for which we know $L(\mathbf{e}_1)$, $L(\mathbf{e}_2)$, ..., $L(\mathbf{e}_n)$, then we can compute $L(\mathbf{u})$. Thus, we can easily compute the image of any vector $\mathbf{u}$ in R^n. (See Theorem 10.3 in Section 10.1.)

EXAMPLE 3 Let $L: R^2 \to R^3$ be defined by

$$L\left(\begin{bmatrix} x \\ y \end{bmatrix}\right) = \begin{bmatrix} 1 & 1 \\ 0 & 1 \\ 1 & -2 \end{bmatrix} \begin{bmatrix} x \\ y \end{bmatrix}.$$

Then L is a linear transformation (verify). Observe that the vector

$$\mathbf{v} = \begin{bmatrix} 2 \\ 3 \\ -7 \end{bmatrix}$$

lies in range L because

$$L\left(\begin{bmatrix} -1 \\ 3 \end{bmatrix}\right) = \begin{bmatrix} 1 & 1 \\ 0 & 1 \\ 1 & -2 \end{bmatrix} \begin{bmatrix} -1 \\ 3 \end{bmatrix} = \begin{bmatrix} 2 \\ 3 \\ -1 - 6 \end{bmatrix} = \begin{bmatrix} 2 \\ 3 \\ -7 \end{bmatrix} = \mathbf{v}.$$

To find out whether the vector

$$\mathbf{w} = \begin{bmatrix} 3 \\ 5 \\ 2 \end{bmatrix}$$

lies in range L we need to determine whether there is a vector

$$\mathbf{u} = \begin{bmatrix} x \\ y \end{bmatrix}$$

so that $L(\mathbf{u}) = \mathbf{w}$. We have

$$L(\mathbf{u}) = L\left(\begin{bmatrix} x \\ y \end{bmatrix}\right) = \begin{bmatrix} 1 & 1 \\ 0 & 1 \\ 1 & -2 \end{bmatrix} \begin{bmatrix} x \\ y \end{bmatrix} = \begin{bmatrix} x + y \\ y \\ x - 2y \end{bmatrix}.$$

Since we want $L(\mathbf{u}) = \mathbf{w}$, we have

$$\begin{bmatrix} x+y \\ y \\ x-2y \end{bmatrix} = \begin{bmatrix} 3 \\ 5 \\ 2 \end{bmatrix}.$$

This means that

$$\begin{aligned} x + y &= 3 \\ y &= 5 \\ x - 2y &= 2. \end{aligned}$$

This linear system of three equations in two unknowns has no solution (verify). Hence, $\mathbf{w}$ is not in range L. ■

EXAMPLE 4 (Cryptology) Cryptology is the technique of coding and decoding messages; it goes back to the time of the ancient Greeks. A simple code is constructed by associating a different number with every letter in the alphabet. For example,

$$\begin{matrix} A & B & C & D & \cdots & X & Y & Z \\ \updownarrow & \updownarrow & \updownarrow & \updownarrow & & \updownarrow & \updownarrow & \updownarrow \\ 1 & 2 & 3 & 4 & \cdots & 24 & 25 & 26 \end{matrix}$$

Suppose that Mark S. and Susan J. are two undercover agents who want to communicate with each other by using a code because they suspect that their phone is being tapped and their mail is being intercepted. In particular, Mark wants to send Susan the message

MEET TOMORROW

Using the substitution scheme given previously, Mark sends the message

13 5 5 20 20 15 13 15 18 18 15 23

A code of this type could be cracked without too much difficulty by a number of techniques, including the analysis of frequency of letters. To make it difficult to crack the code, the agents proceed as follows. First, when they undertook the mission they agreed on a 3×3 nonsingular matrix such as

$$A = \begin{bmatrix} 1 & 2 & 3 \\ 1 & 1 & 2 \\ 0 & 1 & 2 \end{bmatrix}.$$

Mark then breaks the message into four vectors in R^3 (if this cannot be done, we can add extra letters). Thus, we have the vectors

$$\begin{bmatrix} 13 \\ 5 \\ 5 \end{bmatrix}, \quad \begin{bmatrix} 20 \\ 20 \\ 15 \end{bmatrix}, \quad \begin{bmatrix} 13 \\ 15 \\ 18 \end{bmatrix}, \quad \begin{bmatrix} 18 \\ 15 \\ 23 \end{bmatrix}.$$

Mark now defines the linear transformation $L: R^3 \to R^3$ by $L(\mathbf{x}) = A\mathbf{x}$, so the message becomes

$$A \begin{bmatrix} 13 \\ 5 \\ 5 \end{bmatrix} = \begin{bmatrix} 38 \\ 28 \\ 15 \end{bmatrix}, \qquad A \begin{bmatrix} 20 \\ 20 \\ 15 \end{bmatrix} = \begin{bmatrix} 105 \\ 70 \\ 50 \end{bmatrix},$$

$$A \begin{bmatrix} 13 \\ 15 \\ 18 \end{bmatrix} = \begin{bmatrix} 97 \\ 64 \\ 51 \end{bmatrix}, \qquad A \begin{bmatrix} 18 \\ 15 \\ 23 \end{bmatrix} = \begin{bmatrix} 117 \\ 79 \\ 61 \end{bmatrix}.$$

Thus Mark transmits the message

38 28 15 105 70 50 97 64 51 117 79 61

Suppose now that Mark receives the following message from Susan,

77 54 38 71 49 29 68 51 33 76 48 40 86 53 52

which he wants to decode with the same key matrix A as previously. To decode it, Mark breaks the message into five vectors in R^3:

$$\begin{bmatrix} 77 \\ 54 \\ 38 \end{bmatrix}, \quad \begin{bmatrix} 71 \\ 49 \\ 29 \end{bmatrix}, \quad \begin{bmatrix} 68 \\ 51 \\ 33 \end{bmatrix}, \quad \begin{bmatrix} 76 \\ 48 \\ 40 \end{bmatrix}, \quad \begin{bmatrix} 86 \\ 53 \\ 52 \end{bmatrix}$$

and solves the equation

$$L(\mathbf{x}_1) = \begin{bmatrix} 77 \\ 54 \\ 38 \end{bmatrix} = A\mathbf{x}_1$$

for $\mathbf{x}_1$. Since A is nonsingular,

$$\mathbf{x}_1 = A^{-1}\begin{bmatrix} 77 \\ 54 \\ 38 \end{bmatrix} = \begin{bmatrix} 0 & 1 & -1 \\ 2 & -2 & -1 \\ -1 & 1 & 1 \end{bmatrix}\begin{bmatrix} 77 \\ 54 \\ 38 \end{bmatrix} = \begin{bmatrix} 16 \\ 8 \\ 15 \end{bmatrix}.$$

Similarly,

$$\mathbf{x}_2 = A^{-1}\begin{bmatrix} 71 \\ 49 \\ 29 \end{bmatrix} = \begin{bmatrix} 20 \\ 15 \\ 7 \end{bmatrix}, \qquad \mathbf{x}_3 = A^{-1}\begin{bmatrix} 68 \\ 51 \\ 33 \end{bmatrix} = \begin{bmatrix} 18 \\ 1 \\ 16 \end{bmatrix},$$

$$\mathbf{x}_4 = A^{-1}\begin{bmatrix} 76 \\ 48 \\ 40 \end{bmatrix} = \begin{bmatrix} 8 \\ 16 \\ 12 \end{bmatrix}, \qquad \mathbf{x}_5 = A^{-1}\begin{bmatrix} 86 \\ 53 \\ 52 \end{bmatrix} = \begin{bmatrix} 1 \\ 14 \\ 19 \end{bmatrix}.$$

Using our correspondence between letters and numbers, Mark has received the message

PHOTOGRAPH PLANS

Additional material on cryptology may be found in the references given in Further Readings at the end of this section. ■

We have already seen, in general, that if A is an $m \times n$ matrix, then the matrix transformation $L: R^n \to R^m$ defined by $L(\mathbf{x}) = A\mathbf{x}$ for $\mathbf{x}$ in R^n is a linear transformation. In the following theorem, we show that if $L: R^n \to R^m$ is a linear transformation, then L must be a matrix transformation.

THEOREM 4.8 *Let $L: R^n \to R^m$ be a linear transformation. Then there exists a unique $m \times n$ matrix A such that*

$$L(\mathbf{x}) = A\mathbf{x} \tag{1}$$

for $\mathbf{x}$ in R^n.

Proof If

$$\mathbf{x} = \begin{bmatrix} c_1 \\ c_2 \\ \vdots \\ c_n \end{bmatrix}$$

is any vector in R^n, then

$$\mathbf{x} = c_1\mathbf{e}_1 + c_2\mathbf{e}_2 + \cdots + c_n\mathbf{e}_n,$$

so by Theorem 4.6,

$$L(\mathbf{x}) = c_1L(\mathbf{e}_1) + c_2L(\mathbf{e}_2) + \cdots + c_nL(\mathbf{e}_n). \tag{2}$$

If we let A be the $m \times n$ matrix whose jth column is $L(\mathbf{e}_j)$ and

$$\mathbf{x} = \begin{bmatrix} c_1 \\ c_2 \\ \vdots \\ c_n \end{bmatrix},$$

then Equation (2) can be written as

$$L(\mathbf{x}) = A\mathbf{x}.$$

We now show that the matrix A is unique. Suppose that we also have

$$L(\mathbf{x}) = B\mathbf{x} \quad \text{for } \mathbf{x} \text{ in } R^n.$$

Letting $\mathbf{x} = \mathbf{e}_j$, $j = 1, \ldots, n$, we obtain

$$L(\mathbf{e}_j) = A\mathbf{e}_j = \text{col}_j(A)$$

and

$$L(\mathbf{e}_j) = B\mathbf{e}_j = \text{col}_j(B).$$

Thus the columns of A and B agree, so $A = B$. ■

The matrix $A = \begin{bmatrix} L(\mathbf{e}_1) & L(\mathbf{e}_2) & \cdots & L(\mathbf{e}_n) \end{bmatrix}$ in Equation (1) is called the **standard matrix** representing L.

EXAMPLE 5 Let $L\colon R^3 \to R^3$ be the linear operator defined by

$$L\left(\begin{bmatrix} x \\ y \\ z \end{bmatrix}\right) = \begin{bmatrix} x + y \\ y - z \\ x + z \end{bmatrix}.$$

Find the standard matrix representing L and verify Equation (1).

Solution The standard matrix A representing L is the 3×3 matrix whose columns are $L(\mathbf{e}_1)$, $L(\mathbf{e}_2)$, and $L(\mathbf{e}_3)$, respectively. Thus

$$L(\mathbf{e}_1) = L\left(\begin{bmatrix} 1 \\ 0 \\ 0 \end{bmatrix}\right) = \begin{bmatrix} 1+0 \\ 0-0 \\ 1+0 \end{bmatrix} = \begin{bmatrix} 1 \\ 0 \\ 1 \end{bmatrix} = \text{col}_1(A)$$

$$L(\mathbf{e}_2) = L\left(\begin{bmatrix} 0 \\ 1 \\ 0 \end{bmatrix}\right) = \begin{bmatrix} 0+1 \\ 1-0 \\ 0+0 \end{bmatrix} = \begin{bmatrix} 1 \\ 1 \\ 0 \end{bmatrix} = \text{col}_2(A)$$

$$L(\mathbf{e}_3) = L\left(\begin{bmatrix} 0 \\ 0 \\ 1 \end{bmatrix}\right) = \begin{bmatrix} 0+0 \\ 0-1 \\ 0+1 \end{bmatrix} = \begin{bmatrix} 0 \\ -1 \\ 1 \end{bmatrix} = \text{col}_3(A).$$

Hence

$$A = \begin{bmatrix} 1 & 1 & 0 \\ 0 & 1 & -1 \\ 1 & 0 & 1 \end{bmatrix}.$$

Thus we have

$$A\mathbf{x} = \begin{bmatrix} 1 & 1 & 0 \\ 0 & 1 & -1 \\ 1 & 0 & 1 \end{bmatrix} \begin{bmatrix} x \\ y \\ z \end{bmatrix} = \begin{bmatrix} x+y \\ y-z \\ x+z \end{bmatrix} = L(\mathbf{x}),$$

so Equation (1) holds. ■

Further Readings in Cryptology

Elementary presentation

KOHN, BERNICE. *Secret Codes and Ciphers.* Upper Saddle River, N.J.: Prentice Hall, Inc., 1968 (63 pages).

Advanced presentation

FISHER, JAMES L. *Applications-Oriented Algebra.* New York: T. Harper & Row, Publishers, 1977 (Chapter 9, "Coding Theory").

GARRETT, PAUL. *Making, Breaking Codes.* Upper Saddle River, N.J.: Prentice Hall, Inc., 2001.

HARDY, DAREL W., and CAROL L. WALKER, *Applied Algebra, Codes, Ciphers and Discrete Algorithms.* Upper Saddle River, N.J.: Prentice Hall, Inc., 2002.

KAHN, DAVID. *The Codebreakers.* New York: The New American Library Inc., 1973.

Key Terms

Linear transformation
Nonlinear function
Image
Range
Cryptology

4.3 Exercises

1. Which of the following are linear transformations?

(a) $L(x, y) = (x + 1, y, x + y)$

(b) $L\left(\begin{bmatrix} x \\ y \\ z \end{bmatrix}\right) = \begin{bmatrix} x + y \\ y \\ x - z \end{bmatrix}$

(c) $L(x, y) = (x^2 + x, y - y^2)$

2. Which of the following are linear transformations?

(a) $L(x, y, z) = (x - y, x^2, 2z)$

(b) $L\left(\begin{bmatrix} x \\ y \\ z \end{bmatrix}\right) = \begin{bmatrix} 2x - 3y \\ 3y - 2z \\ 2z \end{bmatrix}$

(c) $L(x, y) = (x - y, 2x + 2)$

3. Which of the following are linear transformations?

(a) $L(x, y, z) = (x + y, 0, 2x - z)$

(b) $L\left(\begin{bmatrix} x \\ y \end{bmatrix}\right) = \begin{bmatrix} x^2 - y^2 \\ x^2 + y^2 \end{bmatrix}$

(c) $L(x, y) = (x - y, 0, 2x + 3)$

4. Which of the following are linear transformations?

(a) $L\left(\begin{bmatrix} u_1 \\ u_2 \\ u_3 \\ u_4 \end{bmatrix}\right) = \begin{bmatrix} u_1 \\ u_1^2 + u_2 \\ u_1 - u_3 \end{bmatrix}$

(b) $L\left(\begin{bmatrix} x \\ y \\ z \end{bmatrix}\right) = \begin{bmatrix} 1 & 1 & 0 \\ 0 & -1 & 2 \\ 1 & 1 & -1 \end{bmatrix} \begin{bmatrix} x \\ y \\ z \end{bmatrix}$

(c) $L\left(\begin{bmatrix} x \\ y \\ z \end{bmatrix}\right) = \begin{bmatrix} 0 \\ 0 \\ 0 \end{bmatrix}$

In Exercises 5 through 12, sketch the image of the given point P or vector **u** *under the given linear transformation L.*

5. $L: R^2 \to R^2$ is defined by

$$L(x, y) = (x, -y);\ P = (2, 3).$$

6. $L: R^2 \to R^2$ is defined by

$$L\left(\begin{bmatrix} x \\ y \end{bmatrix}\right) = \begin{bmatrix} 1 & -1 \\ 2 & 1 \end{bmatrix} \begin{bmatrix} x \\ y \end{bmatrix};\ \mathbf{u} = (1, -2).$$

7. $L: R^2 \to R^2$ is a counterclockwise rotation through 30°; $P = (-1, 3)$.

8. $L: R^2 \to R^2$ is a counterclockwise rotation through $\frac{2}{3}\pi$ radians; $\mathbf{u} = (-2, -3)$.

9. $L: R^2 \to R^2$ is defined by $L(\mathbf{u}) = -\mathbf{u}$; $\mathbf{u} = (3, 2)$.

10. $L: R^2 \to R^2$ is defined by $L(\mathbf{u}) = 2\mathbf{u}$; $\mathbf{u} = (-3, 3)$.

11. $L: R^3 \to R^2$ is defined by

$$L\left(\begin{bmatrix} x \\ y \\ z \end{bmatrix}\right) = \begin{bmatrix} x \\ x - y \end{bmatrix};\ \mathbf{u} = (2, -1, 3).$$

12. $L: R^3 \to R^3$ is defined by

$$L\left(\begin{bmatrix} x \\ y \\ z \end{bmatrix}\right) = \begin{bmatrix} 1 & 0 & 1 \\ -1 & 1 & 0 \\ 0 & 0 & 1 \end{bmatrix} \begin{bmatrix} x \\ y \\ z \end{bmatrix};\ \mathbf{u} = (0, -2, 4).$$

13. Let $L: R^3 \to R^3$ be the linear transformation defined by

$$L\left(\begin{bmatrix} x \\ y \\ z \end{bmatrix}\right) = \begin{bmatrix} x + z \\ y + z \\ x + 2y + 2z \end{bmatrix}.$$

Is **w** in range L?

(a) $\mathbf{w} = \begin{bmatrix} 1 \\ -1 \\ 0 \end{bmatrix}$ (b) $\mathbf{w} = \begin{bmatrix} 2 \\ -1 \\ 3 \end{bmatrix}$

14. Let $L: R^3 \to R^3$ be the linear transformation defined by

$$L\left(\begin{bmatrix} x \\ y \\ z \end{bmatrix}\right) = \begin{bmatrix} -1 & 2 & 0 \\ 1 & 1 & 1 \\ 2 & -1 & 1 \end{bmatrix} \begin{bmatrix} x \\ y \\ z \end{bmatrix}.$$

Is **w** in range L?

(a) $\mathbf{w} = \begin{bmatrix} 1 \\ 2 \\ -1 \end{bmatrix}$ (b) $\mathbf{w} = \begin{bmatrix} 1 \\ 3 \\ 2 \end{bmatrix}$

15. Let $L: R^3 \to R^3$ be defined by

$$L\left(\begin{bmatrix} x \\ y \\ z \end{bmatrix}\right) = \begin{bmatrix} 4 & 1 & 3 \\ 2 & -1 & 3 \\ 2 & 2 & 0 \end{bmatrix} \begin{bmatrix} x \\ y \\ z \end{bmatrix}.$$

Find an equation relating a, b, and c so that

$$\mathbf{w} = \begin{bmatrix} a \\ b \\ c \end{bmatrix}$$

will lie in range L.

16. Repeat Exercise 15 if $L: R^3 \to R^3$ is defined by

$$L\left(\begin{bmatrix} x \\ y \\ z \end{bmatrix}\right) = \begin{bmatrix} x + 2y + 3z \\ -3x - 2y - z \\ -2x + 2z \end{bmatrix}.$$

17. Let $L: R^2 \to R^2$ be a linear transformation such that

$$L(\mathbf{i}) = \begin{bmatrix} 2 \\ 3 \end{bmatrix} \quad \text{and} \quad L(\mathbf{j}) = \begin{bmatrix} -1 \\ 2 \end{bmatrix}.$$

Find $L\left(\begin{bmatrix} 4 \\ -3 \end{bmatrix}\right)$.

18. Let $L: R^3 \to R^3$ be a linear transformation such that

$$L(\mathbf{i}) = \begin{bmatrix} 1 \\ 2 \\ -1 \end{bmatrix},\ L(\mathbf{j}) = \begin{bmatrix} 1 \\ 0 \\ 2 \end{bmatrix},\ \text{and}\ L(\mathbf{k}) = \begin{bmatrix} 1 \\ 1 \\ 3 \end{bmatrix}.$$

Find $L\left(\begin{bmatrix} 2 \\ -1 \\ 3 \end{bmatrix}\right)$.

19. Let L be the linear transformation defined in Exercise 11. Find all vectors $\mathbf{x}$ in R^3 such that $L(\mathbf{x}) = \mathbf{0}$.

20. Repeat Exercise 19, where L is the linear transformation defined in Exercise 12.

21. Describe the following linear transformations geometrically.

(a) $L(x, y) = (-x, y)$

(b) $L(x, y) = (-x, -y)$

(c) $L(x, y) = (-y, x)$

22. Describe the following linear transformations geometrically.

(a) $L(x, y) = (y, x)$

(b) $L(x, y) = (-y, -x)$

(c) $L(x, y) = (2x, 2y)$

In Exercises 23 and 24, determine whether L is a linear transformation.

23. $L: R^2 \to R^2$ defined by $L(x, y) = (x + y + 1, x - y)$

24. $L: R^2 \to R^1$ defined by $L(x, y) = \sin x + \sin y$

In Exercises 25 through 30, find the standard matrix representing L.

25. $L: R^2 \to R^2$ is reflection with respect to the y-axis.

26. $L: R^2 \to R^2$ is defined by

$$L\left(\begin{bmatrix} x \\ y \end{bmatrix}\right) = \begin{bmatrix} x - y \\ x + y \end{bmatrix}.$$

27. $L: R^2 \to R^2$ is counterclockwise rotation through $\frac{\pi}{4}$ radians.

28. $L: R^2 \to R^2$ is counterclockwise rotation through $\frac{\pi}{3}$ radians.

29. $L: R^3 \to R^3$ is defined by

$$L\left(\begin{bmatrix} x \\ y \\ z \end{bmatrix}\right) = \begin{bmatrix} x - y \\ x + z \\ y - z \end{bmatrix}.$$

30. $L: R^3 \to R^3$ is defined by $L(\mathbf{u}) = -2\mathbf{u}$.

31. Use the substitution and matrix A of Example 4.

(a) Code the message SEND HIM MONEY.

(b) Decode the message 67 44 41 49 39 19 113 76 62 104 69 55.

32. Use the substitution scheme of Example 4 and the matrix

$$A = \begin{bmatrix} 5 & 3 \\ 2 & 1 \end{bmatrix}.$$

(a) Code the message WORK HARD.

(b) Decode the message

93 36 60 21 159 60 110 43

Theoretical Exercises

T.1. Prove Theorem 4.6.

T.2. Prove Theorem 4.7.

T.3. Show that $L: R^n \to R^n$ defined by $L(\mathbf{u}) = r\mathbf{u}$, where r is a scalar, is a linear operator on R^n.

T.4. Let $\mathbf{u}_0 \neq \mathbf{0}$ be a fixed vector in R^n. Let $L: R^n \to R^n$ be defined by $L(\mathbf{u}) = \mathbf{u} + \mathbf{u}_0$. Determine whether L is a linear transformation. Justify your answer.

T.5. Let $L: R^1 \to R^1$ be defined by $L(\mathbf{u}) = a\mathbf{u} + b$, where a and b are real numbers (of course, $\mathbf{u}$ is a vector in R^1, which in this case means that $\mathbf{u}$ is also a real number). Find all values of a and b such that L is a linear transformation.

T.6. Show that the function $O: R^n \to R^m$ defined by $O(\mathbf{u}) = \mathbf{0}_{R^m}$ is a linear transformation, which is called the **zero linear transformation**.

T.7. Let $I: R^n \to R^n$ be defined by $I(\mathbf{u}) = \mathbf{u}$, for $\mathbf{u}$ in R^n. Show that I is a linear transformation, which is called the **identity operator** on R^n.

T.8. Let $L: R^n \to R^m$ be a linear transformation. Show that if $\mathbf{u}$ and $\mathbf{v}$ are vectors in R^n such that $L(\mathbf{u}) = \mathbf{0}$ and $L(\mathbf{v}) = \mathbf{0}$, then $L(a\mathbf{u} + b\mathbf{v}) = \mathbf{0}$ for any scalars a and b.

T.9. Let $L: R^2 \to R^2$ be the linear transformation defined by $L(\mathbf{u}) = A\mathbf{u}$, where

$$A = \begin{bmatrix} \cos\phi & -\sin\phi \\ \sin\phi & \cos\phi \end{bmatrix}.$$

For $\phi = 30°$, L defines a counterclockwise rotation by $30°$.

(a) If $T_1(\mathbf{u}) = A^2\mathbf{u}$, describe the action of T_1 on $\mathbf{u}$.

(b) If $T_2(\mathbf{u}) = A^{-1}\mathbf{u}$, describe the action of T_2 on $\mathbf{u}$.

(c) What is the smallest positive value of k for which $T(\mathbf{u}) = A^k\mathbf{u} = \mathbf{u}$?

T.10. Let $O: R^n \to R^n$ be the zero linear transformation defined by $O(\mathbf{v}) = \mathbf{0}$ for $\mathbf{v}$ in R^n (see Exercise T.6). Find the standard matrix representing O.

T.11. Let $I: R^n \to R^n$ be the identity linear transformation defined by $I(\mathbf{v}) = \mathbf{v}$ for $\mathbf{v}$ in R^n (see Exercise T.7). Find the standard matrix representing I.

MATLAB Exercises

ML.1. Let $L: R^n \to R^1$ be defined by $L(\mathbf{u}) = \|\mathbf{u}\|$.

(a) Find a pair of vectors $\mathbf{u}$ and $\mathbf{v}$ in R^2 such that

$$L(\mathbf{u}+\mathbf{v}) \neq L(\mathbf{u}) + L(\mathbf{v}).$$

Use MATLAB to do the computations. It follows that L is not a linear transformation.

(b) Find a pair of vectors $\mathbf{u}$ and $\mathbf{v}$ in R^3 such that

$$L(\mathbf{u}+\mathbf{v}) \neq L(\mathbf{u}) + L(\mathbf{v}).$$

Use MATLAB to do the computations.

Key Ideas for Review

- **Theorem 4.2.** Properties of vector addition and scalar multiplication in R^n (see page 230).
- **Theorem 4.3 (Properties of Dot Product).** See page 236.
- **Theorem 4.4 (Cauchy–Schwarz Inequality)**

$$|\mathbf{u} \cdot \mathbf{v}| \leq \|\mathbf{u}\| \, \|\mathbf{v}\|.$$

- **Theorem 4.6.** If $L: R^n \to R^m$ is a linear transformation, then

$$L(c_1\mathbf{u}_1 + c_2\mathbf{u}_2 + \cdots + c_k\mathbf{u}_k) = c_1 L(\mathbf{u}_1) + c_2 L(\mathbf{u}_2) + \cdots + c_k L(\mathbf{u}_k).$$

- **Theorem 4.8.** Let $L: R^n \to R^m$ be a linear transformation. Then there exists a unique $m \times n$ matrix A such that $L(\mathbf{x}) = A\mathbf{x}$ for $\mathbf{x}$ in R^n.

Supplementary Exercises

In Exercises 1 through 3, let

$$\mathbf{u} = (2, -1), \quad \mathbf{v} = (1, 3), \quad \textit{and} \quad \mathbf{w} = (4, 1).$$

1. Sketch a diagram showing that $\mathbf{u}+\mathbf{v} = \mathbf{v}+\mathbf{u}$.

2. Sketch a diagram showing that $(\mathbf{u}+\mathbf{v})+\mathbf{w} = \mathbf{u}+(\mathbf{v}+\mathbf{w})$.

3. Sketch a diagram showing that $2(\mathbf{u}+\mathbf{v}) = 2\mathbf{u}+2\mathbf{v}$.

In Exercises 4 and 5, let $\mathbf{u} = (1, 2, -3)$, $\mathbf{v} = (3, 0, 1)$, *and* $\mathbf{w} = (-2, 1, 1)$.

4. Find $\mathbf{x}$ so that $\mathbf{u}+\mathbf{x} = \mathbf{v}-\mathbf{w}$.

5. Find $\mathbf{x}$ so that $2\mathbf{u}+3\mathbf{x} = \mathbf{w}-5\mathbf{x}$.

6. Write the vector $(1, 2)$ as a linear combination of the vectors $(-2, 3)$ and $(1, -1)$.

7. Let $\mathbf{u} = (1, -1, 2, 3)$, and $\mathbf{v} = (2, 3, 1, -2)$. Compute

(a) $\|\mathbf{u}\|$ (b) $\|\mathbf{v}\|$

(c) $\|\mathbf{u}-\mathbf{v}\|$ (d) $\mathbf{u} \cdot \mathbf{v}$

(e) Cosine of the angle between $\mathbf{u}$ and $\mathbf{v}$

8. If $\mathbf{u} = (x, y)$ is any vector in R^2, show that the vector $\mathbf{v} = (-y, x)$ is orthogonal to $\mathbf{u}$.

9. Find all values of c for which $\|c(1, -2, 2, 0)\| = 9$.

10. Let $\mathbf{u} = (a, 2, a)$ and $\mathbf{v} = (4, -3, 2)$. For what values of a are $\mathbf{u}$ and $\mathbf{v}$ orthogonal?

11. Is $L: R^2 \to R^2$ defined by

$$L(x, y) = (x-1, y-x)$$

a linear transformation?

12. Let $\mathbf{u}_0$ be a fixed vector in R^n. Let $L: R^n \to R^1$ be defined by $L(\mathbf{u}) = \mathbf{u} \cdot \mathbf{u}_0$. Show that L is a linear transformation.

13. Find a unit vector parallel to the vector $(-1, 2, 3)$.

14. Show that a parallelogram is a rhombus, a parallelogram with four equal sides, if and only if its diagonals are orthogonal.

15. If possible, find a and b so that

$$\mathbf{v} = \begin{bmatrix} a \\ b \\ 2 \end{bmatrix}$$

is orthogonal to both

$$\mathbf{w} = \begin{bmatrix} 2 \\ 1 \\ 1 \end{bmatrix} \quad \text{and} \quad \mathbf{x} = \begin{bmatrix} 1 \\ 0 \\ 1 \end{bmatrix}.$$

16. Find a unit vector that is orthogonal to the vector $(1, 2)$.

17. Find the area of the quadrilateral with vertices $(-3, 1)$, $(-2, -2)$, $(2, 4)$, and $(5, 0)$.

18. Find all values of a so that the vector $(a, -5, -2)$ is orthogonal to the vector $(a, a, -3)$.

19. Find the standard matrix representing a clockwise rotation of R^2 through $\frac{\pi}{6}$ radians.

20. Let $L: R^2 \to R^1$ be the linear transformation defined by $L(\mathbf{u}) = \mathbf{u} \cdot \mathbf{u}_0$, where $\mathbf{u}_0 = (1, 2)$. (See Exercise 12.) Find the standard matrix representing L.

21. (a) Write the vector $(1, 3, -2)$ as a linear combination of the vectors $(1, 1, 0)$, $(0, 1, 1)$, and $(0, 0, 1)$.

(b) If $L: R^3 \to R^2$ is the linear transformation for which $L(1, 1, 0) = (2, -1)$, $L(0, 1, 1) = (3, 2)$, and $L(0, 0, 1) = (1, -1)$, find $L(1, 3, -2)$.

22. A river flows south at a rate of 2 miles per hour. If a swimmer is trying to swim westward at a rate of 8 miles per hour, draw a figure showing the magnitude and direction of the resultant velocity.

23. If the 5×3 matrix A is the standard matrix representing the linear transformation $L: R^n \to R^m$, what are the values of n and m?

24. Let $L: R^2 \to R^2$ be the linear transformation defined by

$$L\left(\begin{bmatrix} x \\ y \end{bmatrix}\right) = \begin{bmatrix} x - y \\ x + y \end{bmatrix}.$$

Is $\begin{bmatrix} 2 \\ 3 \end{bmatrix}$ in range L?

25. Let $L: R^3 \to R^3$ be the linear transformation defined by $L(\mathbf{x}) = A\mathbf{x}$, where

$$A = \begin{bmatrix} 1 & 2 & 4 \\ 2 & 3 & 5 \\ -1 & -3 & -7 \end{bmatrix}.$$

Find an equation relating a, b, and c so that $\begin{bmatrix} a \\ b \\ c \end{bmatrix}$ will lie in range L.

Theoretical Exercises

T.1. Let $\mathbf{u}$ and $\mathbf{v}$ be vectors in R^n. Show that $\mathbf{u} \cdot \mathbf{v} = 0$ if and only if $\|\mathbf{u} + \mathbf{v}\| = \|\mathbf{u} - \mathbf{v}\|$.

T.2. Show that for any vectors $\mathbf{u}$, $\mathbf{v}$, and $\mathbf{w}$ in R^2 or R^3 and any scalar c, we have:

(a) $(\mathbf{u} + c\mathbf{v}) \cdot \mathbf{w} = \mathbf{u} \cdot \mathbf{w} + c(\mathbf{v} \cdot \mathbf{w})$

(b) $\mathbf{u} \cdot (c\mathbf{v}) = c(\mathbf{u} \cdot \mathbf{v})$

(c) $(\mathbf{u} + \mathbf{v}) \cdot (c\mathbf{w}) = c(\mathbf{u} \cdot \mathbf{w}) + c(\mathbf{v} \cdot \mathbf{w})$

T.3. Show that the only vector $\mathbf{x}$ in R^2 or R^3 that is orthogonal to every other vector is the zero vector.

T.4. Show that $L: R^n \to R^m$ is a linear transformation if and only if

$$L(a\mathbf{u} + b\mathbf{v}) = aL(\mathbf{u}) + bL(\mathbf{v})$$

for any scalars a and b and any vectors $\mathbf{u}$ and $\mathbf{v}$ in R^n.

T.5. Show that if $\|\mathbf{u}\| = 0$ in R^n, then $\mathbf{u} = \mathbf{0}$.

T.6. Give an example in R^4 showing that $\mathbf{u}$ is orthogonal to $\mathbf{v}$ and $\mathbf{v}$ is orthogonal to $\mathbf{w}$, but $\mathbf{u}$ is not orthogonal to $\mathbf{w}$.

Chapter Test

1. Find the cosine of the angle between the vectors $(1, 2, -1, 4)$ and $(3, -2, 4, 1)$.

2. Find a unit vector in the direction of $(2, -1, 1, 3)$.

3. Is the vector $(1, 2, 3)$ a linear combination of the vectors $(1, 3, 2)$, $(2, 2, -1)$, and $(3, 7, 0)$?

4. Let $L: R^3 \to R^3$ be the linear transformation defined by $L(\mathbf{x}) = A\mathbf{x}$, where

$$A = \begin{bmatrix} 1 & 2 & 0 \\ 2 & -1 & 5 \\ 3 & 2 & 4 \end{bmatrix}.$$

Is $\begin{bmatrix} 1 \\ 2 \\ 3 \end{bmatrix}$ in range L?

5. Let $L: R^2 \to R^3$ be defined by $L(x, y) = (2x + 3y, -2x + 3y, x + y)$. Find the standard matrix representing L.

6. Answer each of the following as true or false. Justify your answer.

(a) In R^n, if $\mathbf{u} \cdot \mathbf{v} = 0$, then $\mathbf{u} = \mathbf{0}$ or $\mathbf{v} = \mathbf{0}$.

(b) In R^n, if $\mathbf{u} \cdot \mathbf{v} = \mathbf{u} \cdot \mathbf{w}$, then $\mathbf{v} = \mathbf{w}$.

(c) In R^n, if $c\mathbf{u} = \mathbf{0}$, then $c = 0$ or $\mathbf{u} = \mathbf{0}$.

(d) In R^n, $\|c\mathbf{u}\| = c\|\mathbf{u}\|$.

(e) In R^n, $\|\mathbf{u} + \mathbf{v}\| = \|\mathbf{u}\| + \|\mathbf{v}\|$.

(f) If $L: R^4 \to R^3$ is a linear transformation defined by $L(\mathbf{x}) = A\mathbf{x}$, then A is 3×4.

(g) The vectors $(1, 0, 1)$ and $(-1, 1, 0)$ are orthogonal.

(h) In R^n, if $\|\mathbf{u}\| = 0$, then $\mathbf{u} = \mathbf{0}$.

(i) In R^n, if $\mathbf{u}$ is orthogonal to $\mathbf{v}$ and $\mathbf{w}$, then $\mathbf{u}$ is orthogonal to $2\mathbf{v} + 3\mathbf{w}$.

(j) If $L: R^n \to R^m$ is a linear transformation, then $L(\mathbf{u}) = L(\mathbf{v})$ implies that $\mathbf{u} = \mathbf{v}$.

CHAPTER 5

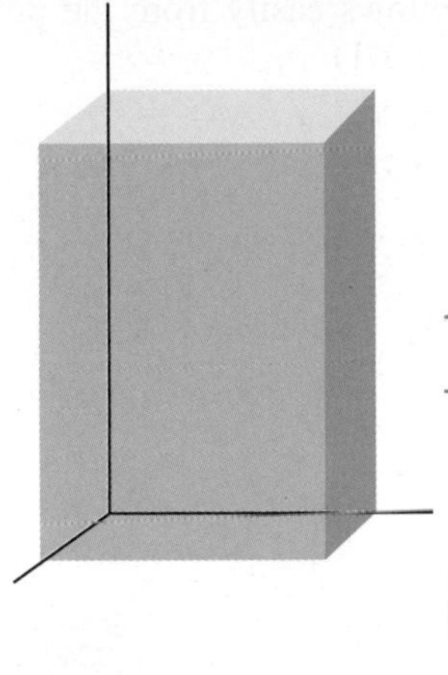

APPLICATIONS OF VECTORS IN R^2 AND R^3 (OPTIONAL)

5.1 CROSS PRODUCT IN R^3

Prerequisites. Section 4.1, Vectors in the Plane. Chapter 3.

In this section we discuss an operation that is meaningful only in R^3. Despite this limitation, it has many important applications in a number of different situations. We shall consider several of these applications in this section.

DEFINITION If

$$\mathbf{u} = u_1\mathbf{i} + u_2\mathbf{j} + u_3\mathbf{k} \quad \text{and} \quad \mathbf{v} = v_1\mathbf{i} + v_2\mathbf{j} + v_3\mathbf{k}$$

are two vectors in R^3, then their **cross product** is the vector $\mathbf{u} \times \mathbf{v}$ defined by

$$\mathbf{u} \times \mathbf{v} = (u_2v_3 - u_3v_2)\mathbf{i} + (u_3v_1 - u_1v_3)\mathbf{j} + (u_1v_2 - u_2v_1)\mathbf{k}. \tag{1}$$

The cross product $\mathbf{u} \times \mathbf{v}$ can also be written as a "determinant,"

$$\mathbf{u} \times \mathbf{v} = \begin{vmatrix} \mathbf{i} & \mathbf{j} & \mathbf{k} \\ u_1 & u_2 & u_3 \\ v_1 & v_2 & v_3 \end{vmatrix}. \tag{2}$$

The right side of (2) is not really a determinant, but it is convenient to think of the computation in this manner. If we expand (2) along the first row, we obtain

$$\mathbf{u} \times \mathbf{v} = \begin{vmatrix} u_2 & u_3 \\ v_2 & v_3 \end{vmatrix}\mathbf{i} - \begin{vmatrix} u_1 & u_3 \\ v_1 & v_3 \end{vmatrix}\mathbf{j} + \begin{vmatrix} u_1 & u_2 \\ v_1 & v_2 \end{vmatrix}\mathbf{k},$$

which is the right side of (1). Observe that the cross product $\mathbf{u} \times \mathbf{v}$ is a vector while the dot product $\mathbf{u} \cdot \mathbf{v}$ is a number.

EXAMPLE 1 Let $\mathbf{u} = 2\mathbf{i} + \mathbf{j} + 2\mathbf{k}$ and $\mathbf{v} = 3\mathbf{i} - \mathbf{j} - 3\mathbf{k}$. Then expanding along the first row, we have

$$\mathbf{u} \times \mathbf{v} = \begin{vmatrix} \mathbf{i} & \mathbf{j} & \mathbf{k} \\ 2 & 1 & 2 \\ 3 & -1 & -3 \end{vmatrix} = -\mathbf{i} + 12\mathbf{j} - 5\mathbf{k}.$$

■

Some of the algebraic properties of cross product are described in the following theorem. The proof, which follows easily from the properties of determinants, is left to the reader (Exercise T.1).

THEOREM 5.1 **(Properties of Cross Product)** *If* **u**, **v**, *and* **w** *are vectors in* R^3 *and* c *is a scalar, then:*

(a) $\mathbf{u} \times \mathbf{v} = -(\mathbf{v} \times \mathbf{u})$
(b) $\mathbf{u} \times (\mathbf{v} + \mathbf{w}) = \mathbf{u} \times \mathbf{v} + \mathbf{u} \times \mathbf{w}$
(c) $(\mathbf{u} + \mathbf{v}) \times \mathbf{w} = \mathbf{u} \times \mathbf{w} + \mathbf{v} \times \mathbf{w}$
(d) $c(\mathbf{u} \times \mathbf{v}) = (c\mathbf{u}) \times \mathbf{v} = \mathbf{u} \times (c\mathbf{v})$
(e) $\mathbf{u} \times \mathbf{u} = \mathbf{0}$
(f) $\mathbf{0} \times \mathbf{u} = \mathbf{u} \times \mathbf{0} = \mathbf{0}$
(g) $\mathbf{u} \times (\mathbf{v} \times \mathbf{w}) = (\mathbf{u} \cdot \mathbf{w})\mathbf{v} - (\mathbf{u} \cdot \mathbf{v})\mathbf{w}$
(h) $(\mathbf{u} \times \mathbf{v}) \times \mathbf{w} = (\mathbf{w} \cdot \mathbf{u})\mathbf{v} - (\mathbf{w} \cdot \mathbf{v})\mathbf{u}$ ■

EXAMPLE 2 It follows from (1) that

$$\mathbf{i} \times \mathbf{i} = \mathbf{j} \times \mathbf{j} = \mathbf{k} \times \mathbf{k} = \mathbf{0}; \qquad \mathbf{i} \times \mathbf{j} = \mathbf{k}, \quad \mathbf{j} \times \mathbf{k} = \mathbf{i}, \quad \mathbf{k} \times \mathbf{i} = \mathbf{j}.$$

Also,

$$\mathbf{j} \times \mathbf{i} = -\mathbf{k}, \qquad \mathbf{k} \times \mathbf{j} = -\mathbf{i}, \qquad \mathbf{i} \times \mathbf{k} = -\mathbf{j}.$$

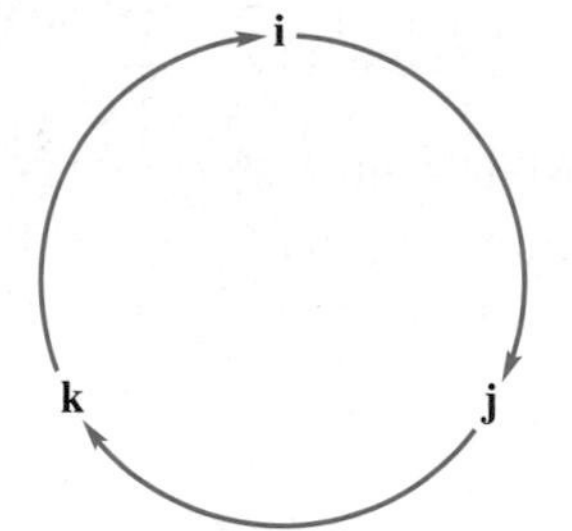

Figure 5.1 ▲

These rules can be remembered by the method indicated in Figure 5.1. Moving around the circle in a clockwise direction, we see that the cross product of two vectors taken in the indicated order is the third vector; moving in a counterclockwise direction, we see that the cross product taken in the indicated order is the negative of the third vector. The cross product of a vector with itself is the zero vector. ■

Although many of the familiar properties of the real numbers hold for the cross product, it should be noted that two important properties do not hold. The commutative law does not hold, since $\mathbf{u} \times \mathbf{v} = -(\mathbf{v} \times \mathbf{u})$. Also, the associative law does not hold, since $\mathbf{i} \times (\mathbf{i} \times \mathbf{j}) = \mathbf{i} \times \mathbf{k} = -\mathbf{j}$ while $(\mathbf{i} \times \mathbf{i}) \times \mathbf{j} = \mathbf{0} \times \mathbf{j} = \mathbf{0}$.

We shall now take a closer look at the geometric properties of the cross product. First, we observe the following additional property of the cross product, whose proof we leave to the reader:

$$(\mathbf{u} \times \mathbf{v}) \cdot \mathbf{w} = \mathbf{u} \cdot (\mathbf{v} \times \mathbf{w}) \qquad \text{Exercise T.2.} \tag{3}$$

It is also easy to show (Exercise T.4) that

$$(\mathbf{u} \times \mathbf{v}) \cdot \mathbf{w} = \begin{vmatrix} u_1 & u_2 & u_3 \\ v_1 & v_2 & v_3 \\ w_1 & w_2 & w_3 \end{vmatrix}. \tag{4}$$

EXAMPLE 3 Let **u** and **v** be as in Example 1, and let $\mathbf{w} = \mathbf{i} + 2\mathbf{j} + 3\mathbf{k}$. Then

$$\mathbf{u} \times \mathbf{v} = -\mathbf{i} + 12\mathbf{j} - 5\mathbf{k} \quad \text{and} \quad (\mathbf{u} \times \mathbf{v}) \cdot \mathbf{w} = 8,$$
$$\mathbf{v} \times \mathbf{w} = 3\mathbf{i} - 12\mathbf{j} + 7\mathbf{k} \quad \text{and} \quad \mathbf{u} \cdot (\mathbf{v} \times \mathbf{w}) = 8,$$

which illustrates Equation (3). ■

From the construction of $\mathbf{u} \times \mathbf{v}$, it follows that $\mathbf{u} \times \mathbf{v}$ is orthogonal to both $\mathbf{u}$ and $\mathbf{v}$; that is,

$$(\mathbf{u} \times \mathbf{v}) \cdot \mathbf{u} = 0, \tag{5}$$

$$(\mathbf{u} \times \mathbf{v}) \cdot \mathbf{v} = 0. \tag{6}$$

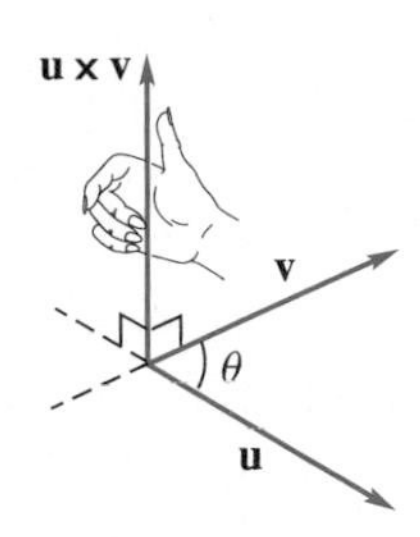

Figure 5.2 ▲

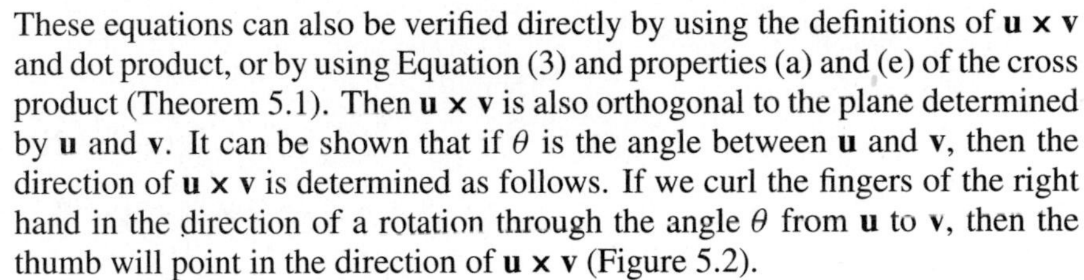

These equations can also be verified directly by using the definitions of $\mathbf{u} \times \mathbf{v}$ and dot product, or by using Equation (3) and properties (a) and (e) of the cross product (Theorem 5.1). Then $\mathbf{u} \times \mathbf{v}$ is also orthogonal to the plane determined by $\mathbf{u}$ and $\mathbf{v}$. It can be shown that if θ is the angle between $\mathbf{u}$ and $\mathbf{v}$, then the direction of $\mathbf{u} \times \mathbf{v}$ is determined as follows. If we curl the fingers of the right hand in the direction of a rotation through the angle θ from $\mathbf{u}$ to $\mathbf{v}$, then the thumb will point in the direction of $\mathbf{u} \times \mathbf{v}$ (Figure 5.2).

The magnitude of $\mathbf{u} \times \mathbf{v}$ can be determined as follows. From the definition of the length of a vector we have

$$\begin{aligned}
\|\mathbf{u} \times \mathbf{v}\|^2 &= (\mathbf{u} \times \mathbf{v}) \cdot (\mathbf{u} \times \mathbf{v}) \\
&= \mathbf{u} \cdot [\mathbf{v} \times (\mathbf{u} \times \mathbf{v})] && \text{by (3)} \\
&= \mathbf{u} \cdot [(\mathbf{v} \cdot \mathbf{v})\mathbf{u} - (\mathbf{v} \cdot \mathbf{u})\mathbf{v}] && \text{by (g) of Theorem 5.1} \\
&= (\mathbf{u} \cdot \mathbf{u})(\mathbf{v} \cdot \mathbf{v}) - (\mathbf{v} \cdot \mathbf{u})(\mathbf{v} \cdot \mathbf{u}) && \text{by (b), (c), and (d) of Theorem 4.3} \\
&= \|\mathbf{u}\|^2 \|\mathbf{v}\|^2 - (\mathbf{u} \cdot \mathbf{v})^2 && \text{by (b) of Theorem 4.3 and the definition of length of a vector.}
\end{aligned}$$

From Equation (4) of Section 4.2 it follows that

$$\mathbf{u} \cdot \mathbf{v} = \|\mathbf{u}\| \|\mathbf{v}\| \cos\theta,$$

where θ is the angle between $\mathbf{u}$ and $\mathbf{v}$. Hence

$$\begin{aligned}
\|\mathbf{u} \times \mathbf{v}\|^2 &= \|\mathbf{u}\|^2 \|\mathbf{v}\|^2 - \|\mathbf{u}\|^2 \|\mathbf{v}\|^2 \cos^2\theta \\
&= \|\mathbf{u}\|^2 \|\mathbf{v}\|^2 (1 - \cos^2\theta) \\
&= \|\mathbf{u}\|^2 \|\mathbf{v}\|^2 \sin^2\theta.
\end{aligned}$$

Taking square roots, we obtain

$$\|\mathbf{u} \times \mathbf{v}\| = \|\mathbf{u}\| \|\mathbf{v}\| \sin\theta. \tag{7}$$

Note that in (7) we do not have to write $|\sin\theta|$, since $\sin\theta$ is nonnegative for $0 \leq \theta \leq \pi$. It follows that vectors $\mathbf{u}$ and $\mathbf{v}$ are parallel if and only if $\mathbf{u} \times \mathbf{v} = \mathbf{0}$ (Exercise T.5).

We now consider several applications of cross product.

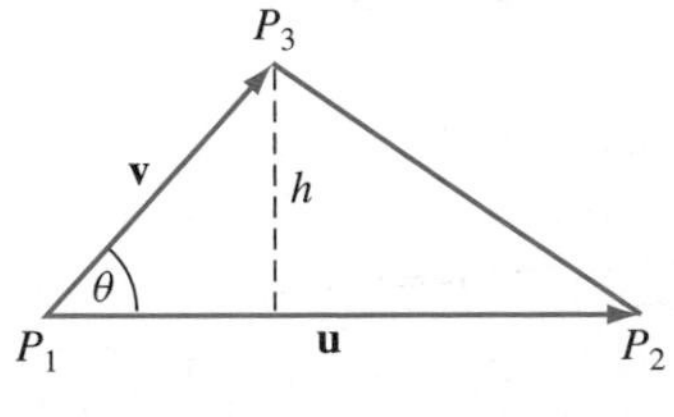

Figure 5.3 ▲

Area of a Triangle Consider the triangle with vertices P_1, P_2, P_3 (Figure 5.3). The area of this triangle is $\frac{1}{2}bh$, where b is the base and h is the height. If we take the segment between P_1 and P_2 to be the base and denote $\overrightarrow{P_1P_2}$ by the vector $\mathbf{u}$, then

$$b = \|\mathbf{u}\|.$$

Letting $\overrightarrow{P_1P_3} = \mathbf{v}$, we find that the height h is given by

$$h = \|\mathbf{v}\| \sin\theta.$$

Hence, by (7), the area A_T of the triangle is

$$A_T = \tfrac{1}{2}\|\mathbf{u}\| \|\mathbf{v}\| \sin\theta = \tfrac{1}{2}\|\mathbf{u} \times \mathbf{v}\|.$$

EXAMPLE 4 Find the area of the triangle with vertices $P_1(2, 2, 4)$, $P_2(-1, 0, 5)$, and $P_3(3, 4, 3)$.

Solution We have

$$\mathbf{u} = \overrightarrow{P_1P_2} = -3\mathbf{i} - 2\mathbf{j} + \mathbf{k}$$
$$\mathbf{v} = \overrightarrow{P_1P_3} = \mathbf{i} + 2\mathbf{j} - \mathbf{k}.$$

Then

$$\begin{aligned} A_T &= \tfrac{1}{2}\|(-3\mathbf{i} - 2\mathbf{j} + \mathbf{k}) \times (\mathbf{i} + 2\mathbf{j} - \mathbf{k})\| \\ &= \tfrac{1}{2}\|-2\mathbf{j} - 4\mathbf{k}\| = \|-\mathbf{j} - 2\mathbf{k}\| = \sqrt{5}. \end{aligned}$$

■

Area of a Parallelogram The area A_P of the parallelogram with adjacent sides $\mathbf{u}$ and $\mathbf{v}$ (Figure 5.4) is $2A_T$, so

$$A_P = \|\mathbf{u} \times \mathbf{v}\|.$$

Figure 5.4 ►

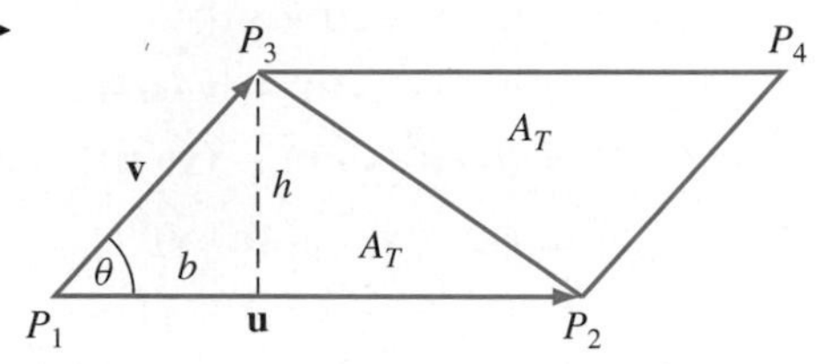

EXAMPLE 5 If P_1, P_2, and P_3 are as in Example 4, then the area of the parallelogram with adjacent sides $\overrightarrow{P_1P_2}$ and $\overrightarrow{P_1P_3}$ is $2\sqrt{5}$. (Verify.) ■

Volume of a Parallelepiped Consider the parallelepiped with a vertex at the origin and edges $\mathbf{u}$, $\mathbf{v}$, and $\mathbf{w}$ (Figure 5.5). The volume V of the parallelepiped is the product of the area of the face containing $\mathbf{v}$ and $\mathbf{w}$ and the distance d from this face to the face parallel to it. Now

$$d = \|\mathbf{u}\|\,|\cos\theta|,$$

where θ is the angle between $\mathbf{u}$ and $\mathbf{v} \times \mathbf{w}$, and the area of the face determined by $\mathbf{v}$ and $\mathbf{w}$ is $\|\mathbf{v} \times \mathbf{w}\|$. Hence

$$V = \|\mathbf{v} \times \mathbf{w}\|\,\|\mathbf{u}\|\,|\cos\theta| = |\mathbf{u} \cdot (\mathbf{v} \times \mathbf{w})|. \tag{8}$$

From Equations (3) and (4), we also have

$$V = \left|\det\left(\begin{bmatrix} u_1 & u_2 & u_3 \\ v_1 & v_2 & v_3 \\ w_1 & w_2 & w_3 \end{bmatrix}\right)\right|. \tag{9}$$

Figure 5.5 ►

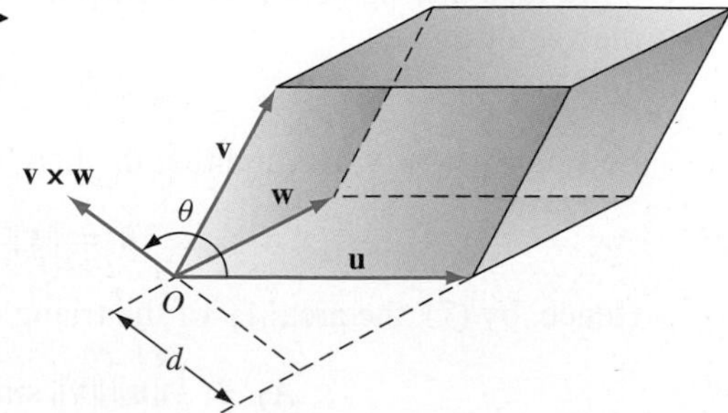

EXAMPLE 6 Consider the parallelepiped with a vertex at the origin and edges $\mathbf{u} = \mathbf{i} - 2\mathbf{j} + 3\mathbf{k}$, $\mathbf{v} = \mathbf{i} + 3\mathbf{j} + \mathbf{k}$, and $\mathbf{w} = 2\mathbf{i} + \mathbf{j} + 2\mathbf{k}$. Then

$$\mathbf{v} \times \mathbf{w} = 5\mathbf{i} - 5\mathbf{k}.$$

Hence $\mathbf{u} \cdot (\mathbf{v} \times \mathbf{w}) = -10$. Thus the volume V is given by (8) as

$$V = |\mathbf{u} \cdot (\mathbf{v} \times \mathbf{w})| = |-10| = 10.$$

We can also compute the volume by Equation (9) as

$$V = \left|\det\left(\begin{bmatrix} 1 & -2 & 3 \\ 1 & 3 & 1 \\ 2 & 1 & 2 \end{bmatrix}\right)\right| = |-10| = 10.$$

■

Key Terms

Cross product
Jacobi identity

5.1 Exercises

In Exercises 1 and 2, compute $\mathbf{u} \times \mathbf{v}$.

1. (a) $\mathbf{u} = 2\mathbf{i} + 3\mathbf{j} + 4\mathbf{k}$, $\mathbf{v} = -\mathbf{i} + 3\mathbf{j} - \mathbf{k}$

(b) $\mathbf{u} = (1, 0, 1)$, $\mathbf{v} = (2, 3, -1)$

(c) $\mathbf{u} = \mathbf{i} - \mathbf{j} + 2\mathbf{k}$, $\mathbf{v} = 3\mathbf{i} - 4\mathbf{j} + \mathbf{k}$

(d) $\mathbf{u} = (2, -1, 1)$, $\mathbf{v} = -2\mathbf{u}$

2. (a) $\mathbf{u} = (1, -1, 2)$, $\mathbf{v} = (3, 1, 2)$

(b) $\mathbf{u} = 2\mathbf{i} + \mathbf{j} - 2\mathbf{k}$, $\mathbf{v} = \mathbf{i} + 3\mathbf{k}$

(c) $\mathbf{u} = 2\mathbf{j} + \mathbf{k}$, $\mathbf{v} = 3\mathbf{u}$

(d) $\mathbf{u} = (4, 0, -2)$, $\mathbf{v} = (0, 2, -1)$

3. Let $\mathbf{u} = \mathbf{i} + 2\mathbf{j} - 3\mathbf{k}$, $\mathbf{v} = 2\mathbf{i} + 3\mathbf{j} + \mathbf{k}$, $\mathbf{w} = 2\mathbf{i} - \mathbf{j} + 2\mathbf{k}$, and $c = -3$. Verify properties (a) through (d) of Theorem 5.1.

4. Let $\mathbf{u} = 2\mathbf{i} - \mathbf{j} + 3\mathbf{k}$, $\mathbf{v} = 3\mathbf{i} + \mathbf{j} - \mathbf{k}$, and $\mathbf{w} = 3\mathbf{i} + \mathbf{j} + 2\mathbf{k}$.

(a) Verify Equation (3).

(b) Verify Equation (4).

5. Let $\mathbf{u} = \mathbf{i} - \mathbf{j} + 2\mathbf{k}$, $\mathbf{v} = 2\mathbf{i} + 2\mathbf{j} - \mathbf{k}$, and $\mathbf{w} = \mathbf{i} + \mathbf{j} - \mathbf{k}$.

(a) Verify Equation (3).

(b) Verify Equation (4).

6. Verify that each of the cross products $\mathbf{u} \times \mathbf{v}$ in Exercise 1 is orthogonal to both $\mathbf{u}$ and $\mathbf{v}$.

7. Verify that each of the cross products $\mathbf{u} \times \mathbf{v}$ in Exercise 2 is orthogonal to both $\mathbf{u}$ and $\mathbf{v}$.

8. Verify Equation (7) for the pairs of vectors in Exercise 1.

9. Find the area of the triangle with vertices $P_1(1, -2, 3)$, $P_2(-3, 1, 4)$, $P_3(0, 4, 3)$.

10. Find the area of the triangle with vertices P_1, P_2, and P_3, where $\overrightarrow{P_1P_2} = 2\mathbf{i} + 3\mathbf{j} - \mathbf{k}$ and $\overrightarrow{P_1P_3} = \mathbf{i} + 2\mathbf{j} + 2\mathbf{k}$.

11. Find the area of the parallelogram with adjacent sides $\mathbf{u} = \mathbf{i} + 3\mathbf{j} - 2\mathbf{k}$ and $\mathbf{v} = 3\mathbf{i} - \mathbf{j} - \mathbf{k}$.

12. Find the volume of the parallelepiped with a vertex at the origin and edges $\mathbf{u} = 2\mathbf{i} - \mathbf{j}$, $\mathbf{v} = \mathbf{i} - 2\mathbf{j} - 2\mathbf{k}$, and $\mathbf{w} = 3\mathbf{i} - \mathbf{j} + \mathbf{k}$.

13. Repeat Exercise 12 for $\mathbf{u} = \mathbf{i} - 2\mathbf{j} + 4\mathbf{k}$, $\mathbf{v} = 3\mathbf{i} + 4\mathbf{j} + \mathbf{k}$, and $\mathbf{w} = -\mathbf{i} + \mathbf{j} + \mathbf{k}$.

Theoretical Exercises

T.1. Prove Theorem 5.1.

T.2. Show that $(\mathbf{u} \times \mathbf{v}) \cdot \mathbf{w} = \mathbf{u} \cdot (\mathbf{v} \times \mathbf{w})$.

T.3. Show that $\mathbf{j} \times \mathbf{i} = -\mathbf{k}$, $\mathbf{k} \times \mathbf{j} = -\mathbf{i}$, $\mathbf{i} \times \mathbf{k} = -\mathbf{j}$.

T.4. Show that

$$(\mathbf{u} \times \mathbf{v}) \cdot \mathbf{w} = \begin{vmatrix} u_1 & u_2 & u_3 \\ v_1 & v_2 & v_3 \\ w_1 & w_2 & w_3 \end{vmatrix}.$$

T.5. Show that $\mathbf{u}$ and $\mathbf{v}$ are parallel if and only if $\mathbf{u} \times \mathbf{v} = \mathbf{0}$.

T.6. Show that $\|\mathbf{u} \times \mathbf{v}\|^2 + (\mathbf{u} \cdot \mathbf{v})^2 = \|\mathbf{u}\|^2 \|\mathbf{v}\|^2$.

T.7. Prove the **Jacobi identity**:

$$(\mathbf{u} \times \mathbf{v}) \times \mathbf{w} + (\mathbf{v} \times \mathbf{w}) \times \mathbf{u} + (\mathbf{w} \times \mathbf{u}) \times \mathbf{v} = \mathbf{0}.$$

MATLAB Exercises

There are two MATLAB *routines that apply to the material in this section. They are* **cross**, *which computes the cross product of a pair of 3-vectors; and* **crossdemo**, *which displays graphically a pair of vectors and their cross product. Using routine* **dot** *with* **cross**, *we can carry out the computations in Example 6. (For directions on the use of* MATLAB *routines, use* **help** *followed by a space and the name of the routine.)*

ML.1. Use **cross** in MATLAB to find the cross product of each of the following pairs of vectors.

(a) $\mathbf{u} = \mathbf{i} - 2\mathbf{j} + 3\mathbf{k}$, $\mathbf{v} = \mathbf{i} + 3\mathbf{j} + \mathbf{k}$

(b) $\mathbf{u} = (1, 0, 3)$, $\mathbf{v} = (1, -1, 2)$

(c) $\mathbf{u} = (1, 2, -3)$, $\mathbf{v} = (2, -1, 2)$

ML.2. Use routine **cross** to find the cross product of each of the following pairs of vectors.

(a) $\mathbf{u} = (2, 3, -1)$, $\mathbf{v} = (2, 3, 1)$

(b) $\mathbf{u} = 3\mathbf{i} - \mathbf{j} + \mathbf{k}$, $\mathbf{v} = 2\mathbf{u}$

(c) $\mathbf{u} = (1, -2, 1)$, $\mathbf{v} = (3, 1, -1)$

ML.3. Use **crossdemo** in MATLAB to display the vectors $\mathbf{u}$, $\mathbf{v}$, and their cross product.

(a) $\mathbf{u} = \mathbf{i} + 2\mathbf{j} + 4\mathbf{k}$, $\mathbf{v} = -2\mathbf{i} + 4\mathbf{j} + 3\mathbf{k}$

(b) $\mathbf{u} = (-2, 4, 5)$, $\mathbf{v} = (0, 1, -3)$

(c) $\mathbf{u} = (2, 2, 2)$, $\mathbf{v} = (3, -3, 3)$

ML.4. Use **cross** in MATLAB to check your answers to Exercises 1 and 2.

ML.5. Use MATLAB to find the volume of the parallelepiped with vertex at the origin and edges $\mathbf{u} = (3, -2, 1)$, $\mathbf{v} = (1, 2, 3)$, and $\mathbf{w} = (2, -1, 2)$.

ML.6. The angle of intersection of two planes in 3-space is the same as the angle of intersection of perpendiculars to the planes. Find the angle of intersection of plane P_1 determined by $\mathbf{x}$ and $\mathbf{y}$ and plane P_2 determined by $\mathbf{v}$, $\mathbf{w}$, where

$$\mathbf{x} = (2, -1, 2), \quad \mathbf{y} = (3, -2, 1),$$
$$\mathbf{v} = (1, 3, 1), \quad \mathbf{w} = (0, 2, -1).$$

5.2 LINES AND PLANES

Prerequisites. Section 4.1, Vectors in the Plane. Section 5.1, Cross Product in R^3.

LINES IN R^2

Any two distinct points $P_1(x_1, y_1)$ and $P_2(x_2, y_2)$ in R^2 (Figure 5.6) determine a straight line whose equation is

$$ax + by + c = 0, \tag{1}$$

where a, b, and c are real numbers, and a and b are not both zero. Since P_1 and P_2 lie on the line, their coordinates satisfy Equation (1):

$$ax_1 + by_1 + c = 0 \tag{2}$$
$$ax_2 + by_2 + c = 0. \tag{3}$$

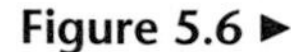
Figure 5.6 ►

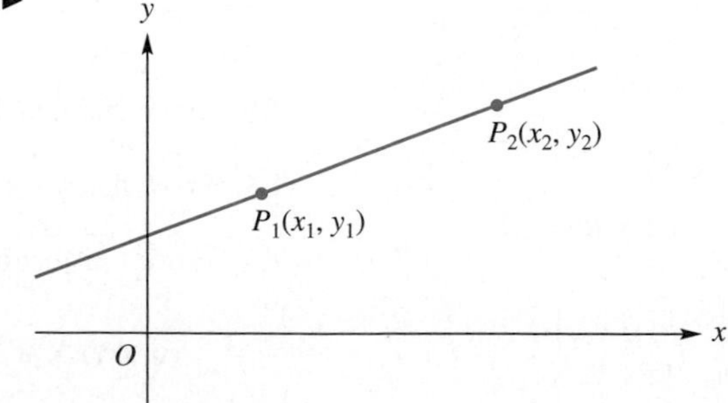

We now write (1), (2), and (3) as a linear system in the unknowns a, b, and c, obtaining

$$\begin{aligned} xa + \ yb + c &= 0 \\ x_1a + y_1b + c &= 0 \\ x_2a + y_2b + c &= 0. \end{aligned} \tag{4}$$

We seek a condition on the values x and y that allow (4) to have a nontrivial solution a, b, and c. Since (4) is a homogeneous system, it has a nontrivial solution if and only if the determinant of the coefficient matrix is zero, that is, if and only if

$$\begin{vmatrix} x & y & 1 \\ x_1 & y_1 & 1 \\ x_2 & y_2 & 1 \end{vmatrix} = 0. \tag{5}$$

Thus every point $P(x, y)$ on the line satisfies (5) and, conversely, a point satisfying (5) lies on the line.

EXAMPLE 1 Find an equation of the line determined by the points $P_1(-1, 3)$ and $P_2(4, 6)$.

Solution Substituting in (5), we obtain

$$\begin{vmatrix} x & y & 1 \\ -1 & 3 & 1 \\ 4 & 6 & 1 \end{vmatrix} = 0.$$

Expanding this determinant in cofactors along the first row, we have (verify)

$$-3x + 5y - 18 = 0.$$ ■

LINES IN R^3

We may recall that in R^2 a line is determined by specifying its slope and one of its points. In R^3 a line is determined by specifying its direction and one of its points. Let $\mathbf{u} = (a, b, c)$ be a nonzero vector in R^3, and let $P_0 = (x_0, y_0, z_0)$ be a point in R^3. Let $\mathbf{w}_0$ be the vector associated with P_0 and let $\mathbf{x}$ be the vector associated with the point $P(x, y, z)$. The line L through P_0 and parallel to $\mathbf{u}$ consists of the points $P(x, y, z)$ (Figure 5.7) such that

$$\mathbf{x} = \mathbf{w}_0 + t\mathbf{u}, \qquad -\infty < t < \infty. \tag{6}$$

Figure 5.7 ►

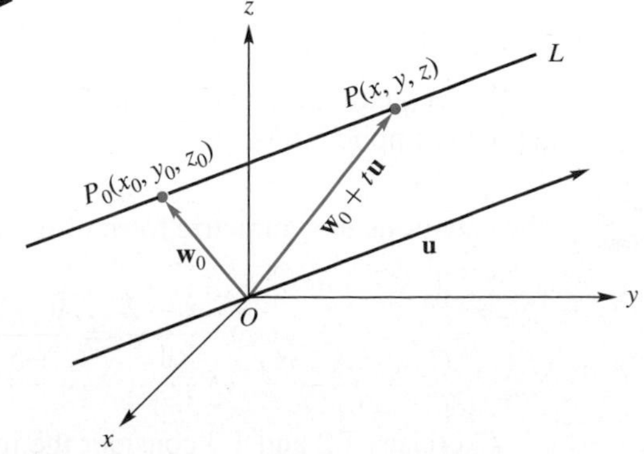

Equation (6) is called a **parametric equation** of L, since it contains the parameter t, which can be assigned any real number. Equation (6) can also be written in terms of the components as

$$\begin{aligned} x &= x_0 + ta \\ y &= y_0 + tb \qquad -\infty < t < \infty \\ z &= z_0 + tc, \end{aligned} \tag{7}$$

which are called **parametric equations** of L.

EXAMPLE 2 Parametric equations of the line through the point $P_0(-3, 2, 1)$, which is parallel to the vector $\mathbf{u} = (2, -3, 4)$, are

$$\begin{aligned} x &= -3 + 2t \\ y &= 2 - 3t \qquad -\infty < t < \infty \\ z &= 1 + 4t. \end{aligned}$$

■

EXAMPLE 3 Find parametric equations of the line L through the points $P_0(2, 3, -4)$ and $P_1(3, -2, 5)$.

Solution The desired line is parallel to the vector $\mathbf{u} = \overrightarrow{P_0P_1}$. Now

$$\mathbf{u} = (3 - 2, -2 - 3, 5 - (-4)) = (1, -5, 9).$$

Since P_0 lies on the line, we can write parametric equations of L as

$$\begin{aligned} x &= 2 + t \\ y &= 3 - 5t \qquad -\infty < t < \infty \\ z &= -4 + 9t. \end{aligned}$$

■

In Example 3 we could have used the point P_2 instead of P_1. In fact, we could use any point on the line in parametric equations of L. Thus a line can be represented in infinitely many ways in parametric form. If a, b, and c are all nonzero in (7), we can solve each equation for t and equate the results to obtain equations in **symmetric form** of the line through P_0 and parallel to $\mathbf{u}$:

$$\frac{x - x_0}{a} = \frac{y - y_0}{b} = \frac{z - z_0}{c}.$$

The equations in symmetric form of a line are useful in some analytic geometry applications.

EXAMPLE 4 The equations in symmetric form of the line in Example 3 are

$$\frac{x - 2}{1} = \frac{y - 3}{-5} = \frac{z + 4}{9}.$$

■

Exercises T.2 and T.3 consider the intersection of two lines in R^3.

PLANES IN R^3

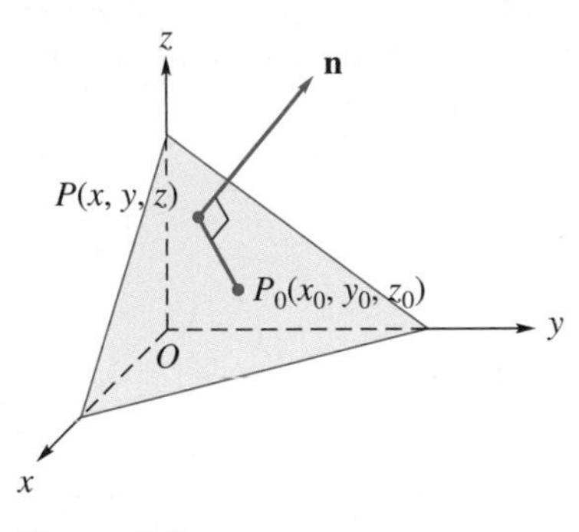

Figure 5.8 ▲

A plane in R^3 can be determined by specifying a point in the plane and a vector perpendicular to the plane. This vector is called a **normal** to the plane.

To obtain an equation of the plane passing through the point $P_0(x_0, y_0, z_0)$ and having the nonzero vector $\mathbf{n} = (a, b, c)$ as a normal, we proceed as follows. A point $P(x, y, z)$ lies in the plane if and only if the vector $\overrightarrow{P_0P}$ is perpendicular to $\mathbf{n}$ (Figure 5.8). Thus $P(x, y, z)$ lies in the plane if and only if

$$\mathbf{n} \cdot \overrightarrow{P_0P} = 0. \tag{8}$$

Since

$$\overrightarrow{P_0P} = (x - x_0, y - y_0, z - z_0),$$

we can write (8) as

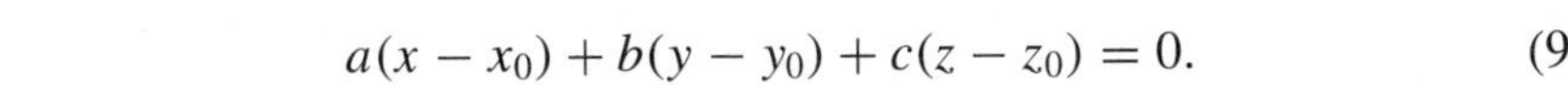

$$a(x - x_0) + b(y - y_0) + c(z - z_0) = 0. \tag{9}$$

EXAMPLE 5 Find an equation of the plane passing through the point $(3, 4, -3)$ and perpendicular to the vector $\mathbf{n} = (5, -2, 4)$.

Solution Substituting in (9), we obtain the equation of the plane as

$$5(x - 3) - 2(y - 4) + 4(z + 3) = 0. \qquad \blacksquare$$

If we multiply out and simplify, (9) can be rewritten as

$$ax + by + cz + d = 0. \tag{10}$$

It is not difficult to show (Exercise T.1) that the graph of an equation of the form given in (10), where a, b, c, and d are constants, is a plane with normal $\mathbf{n} = (a, b, c)$ provided a, b, and c are not all zero.

EXAMPLE 6 Find an equation of the plane passing through the points $P_1(2, -2, 1)$, $P_2(-1, 0, 3)$, and $P_3(5, -3, 4)$.

Solution Let an equation of the desired plane be as given by (10). Since P_1, P_2, and P_3 lie in the plane, their coordinates satisfy (10). Thus we obtain the linear system (verify)

$$\begin{aligned} 2a - 2b + c + d &= 0 \\ -a \phantom{{}- 2b} + 3c + d &= 0 \\ 5a - 3b + 4c + d &= 0. \end{aligned}$$

Solving this system, we have (verify)

$$a = \tfrac{8}{17}r, \qquad b = \tfrac{15}{17}r, \qquad c = -\tfrac{3}{17}r, \qquad d = r,$$

where r is any real number. Letting $r = 17$, we obtain

$$a = 8, \qquad b = 15, \qquad c = -3, \qquad d = 17.$$

Thus, an equation for the desired plane is

$$8x + 15y - 3z + 17 = 0. \tag{11} \quad \blacksquare$$

EXAMPLE 7

A second solution to Example 6 is as follows. Proceeding as in the case of a line in R^2 determined by two distinct points P_1 and P_2, it is not difficult to show (Exercise T.5) that an equation of the plane through the noncollinear points $P_1(x_1, y_1, z_1)$, $P_2(x_2, y_2, z_2)$, and $P_3(x_3, y_3, z_3)$ is

$$\begin{vmatrix} x & y & z & 1 \\ x_1 & y_1 & z_1 & 1 \\ x_2 & y_2 & z_2 & 1 \\ x_3 & y_3 & z_3 & 1 \end{vmatrix} = 0.$$

In our example, the equation of the desired plane is

$$\begin{vmatrix} x & y & z & 1 \\ 2 & -2 & 1 & 1 \\ -1 & 0 & 3 & 1 \\ 5 & -3 & 4 & 1 \end{vmatrix} = 0.$$

Expanding this determinant in cofactors along the first row, we obtain (verify) Equation (11). ■

EXAMPLE 8

A third solution to Example 6 using Section 5.1, Cross Product in R^3, is as follows. The nonparallel vectors $\overrightarrow{P_1P_2} = (-3, 2, 2)$ and $\overrightarrow{P_1P_3} = (3, -1, 3)$ lie in the plane, since the points P_1, P_2, and P_3 lie in the plane. The vector

$$\mathbf{n} = \overrightarrow{P_1P_2} \times \overrightarrow{P_1P_3} = (8, 15, -3)$$

is then perpendicular to both $\overrightarrow{P_1P_2}$ and $\overrightarrow{P_1P_3}$ and is thus a normal to the plane. Using the vector $\mathbf{n}$ and the point $P_1(2, -2, 1)$ in (9), we obtain an equation of the plane as

$$8(x-2) + 15(y+2) - 3(z-1) = 0,$$

which when expanded agrees with Equation (11). ■

The equations of a line in symmetric form can be used to determine two planes whose intersection is the given line.

EXAMPLE 9

Find two planes whose intersection is the line

$$\begin{aligned} x &= -2 + 3t \\ y &= 3 - 2t \qquad -\infty < t < \infty \\ z &= 5 + 4t. \end{aligned}$$

Solution First, find equations of the line in symmetric form as

$$\frac{x+2}{3} = \frac{y-3}{-2} = \frac{z-5}{4}.$$

The given line is then the intersection of the planes

$$\frac{x+2}{3} = \frac{y-3}{-2} \quad \text{and} \quad \frac{x+2}{3} = \frac{z-5}{4}.$$

Thus the given line is the intersection of the planes

$$2x + 3y - 5 = 0 \quad \text{and} \quad 4x - 3z + 23 = 0.$$ ■

Two planes are either parallel or they intersect in a straight line. They are parallel if their normals are parallel. In the following example we determine the line of intersection of two planes.

EXAMPLE 10 Find parametric equations of the line of intersection of the planes

$$\pi_1: 2x + 3y - 2z + 4 = 0 \quad \text{and} \quad \pi_2: x - y + 2z + 3 = 0.$$

Solution Solving the linear system consisting of the equations of π_1 and π_2, we obtain (verify)

$$\begin{aligned} x &= -\tfrac{13}{5} - \tfrac{4}{5}t \\ y &= \tfrac{2}{5} + \tfrac{6}{5}t \qquad -\infty < t < \infty \\ z &= 0 + t \end{aligned}$$

as parametric equations of the line L of intersection of the planes (see Figure 5.9). ■

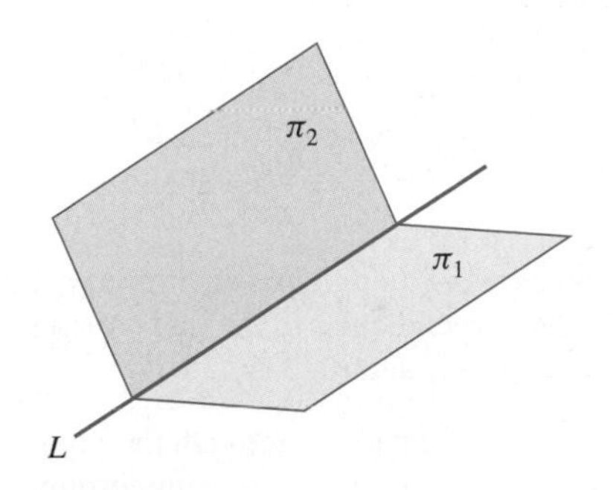

Figure 5.9 ▲

Three planes in R^3 may intersect in a plane, in a line, in a unique point, or have no set of points in common. These possibilities can be detected by solving the linear system consisting of their equations.

Key Terms

Parametric equation(s) of a line
Symmetric form of a line
Normal to a plane
Skew lines

5.2 Exercises

1. In each of the following, find an equation of the line in R^2 determined by the given points.
(a) $P_1(-2, -3)$, $P_2(3, 4)$
(b) $P_1(2, -5)$, $P_2(-3, 4)$
(c) $P_1(0, 0)$, $P_2(-3, 5)$
(d) $P_1(-3, -5)$, $P_2(0, 2)$

2. In each of the following, find the equation of the line in R^2 determined by the given points.
(a) $P_1(1, 1)$, $P_2(2, 2)$
(b) $P_1(1, 2)$, $P_2(1, 3)$
(c) $P_1(2, -4)$, $P_2(-3, -4)$
(d) $P_1(2, -3)$, $P_2(3, -2)$

3. State which of the following points are on the line

$$\begin{aligned} x &= 3 + 2t \\ y &= -2 + 3t \qquad -\infty < t < \infty \\ z &= 4 - 3t. \end{aligned}$$

(a) $(1, 1, 1)$ (b) $(1, -1, 0)$
(c) $(1, 0, -2)$ (d) $\left(4, -\frac{1}{2}, \frac{5}{2}\right)$

4. State which of the following points are on the line

$$\frac{x-4}{-2} = \frac{y+3}{2} = \frac{z-4}{-5}.$$

(a) $(0, 1, -6)$ (b) $(1, 2, 3)$
(c) $(4, -3, 4)$ (d) $(0, 1, -1)$

5. In each of the following, find the parametric equations of the line through the point $P_0(x_0, y_0, z_0)$, which is parallel to the vector $\mathbf{u}$.
(a) $P_0 = (3, 4, -2)$, $\mathbf{u} = (4, -5, 2)$
(b) $P_0 = (3, 2, 4)$, $\mathbf{u} = (-2, 5, 1)$
(c) $P_0 = (0, 0, 0)$, $\mathbf{u} = (2, 2, 2)$
(d) $P_0 = (-2, -3, 1)$, $\mathbf{u} = (2, 3, 4)$

6. In each of the following, find the parametric equations of the line through the given points.
(a) $(2, -3, 1)$, $(4, 2, 5)$ (b) $(-3, -2, -2)$, $(5, 5, 4)$
(c) $(-2, 3, 4)$, $(2, -3, 5)$ (d) $(0, 0, 0)$, $(4, 5, 2)$

7. For each of the lines in Exercise 6, find the equations in symmetric form.

8. State which of the following points are on the plane $3(x - 2) + 2(y + 3) - 4(z - 4) = 0$.
(a) $(0, -2, 3)$ (b) $(1, -2, 3)$
(c) $(1, -1, 3)$ (d) $(0, 0, 4)$

9. In each of the following, find an equation of the plane passing through the given point and perpendicular to the given vector $\mathbf{n}$.

(a) $(0, 2, -3)$, $\mathbf{n} = (3, -2, 4)$

(b) $(-1, 3, 2)$, $\mathbf{n} = (0, 1, -3)$

(c) $(-2, 3, 4)$, $\mathbf{n} = (0, 0, -4)$

(d) $(5, 2, 3)$, $\mathbf{n} = (-1, -2, 4)$

10. In each of the following, find an equation of the plane passing through the given three points.

(a) $(0, 1, 2)$, $(3, -2, 5)$, $(2, 3, 4)$

(b) $(2, 3, 4)$, $(-1, -2, 3)$, $(-5, -4, 2)$

(c) $(1, 2, 3)$, $(0, 0, 0)$, $(-2, 3, 4)$

(d) $(1, 1, 1)$, $(2, 3, 4)$, $(-5, 3, 2)$

11. In each of the following, find parametric equations of the line of intersection of the given planes.

(a) $2x + 3y - 4z + 5 = 0$ and $-3x + 2y + 5z + 6 = 0$

(b) $3x - 2y - 5z + 4 = 0$ and $2x + 3y + 4z + 8 = 0$

(c) $-x + 2y + z = 0$ and $2x - y + 2z + 8 = 0$

12. In each of the following, find a pair of planes whose intersection is the given line.

(a) $\begin{aligned} x &= 2 - 3t \\ y &= 3 + t \\ z &= 2 - 4t \end{aligned}$

(b) $\dfrac{x-2}{-2} = \dfrac{y-3}{4} = \dfrac{z+4}{3}$

(c) $\begin{aligned} x &= 4t \\ y &= 1 + 5t \\ z &= 2 - t \end{aligned}$

13. Are the points $(2, 3, -2)$, $(4, -2, -3)$, and $(0, 8, -1)$ on the same line?

14. Are the points $(-2, 4, 2)$, $(3, 5, 1)$, and $(4, 2, -1)$ on the same line?

15. Find the point of intersection of the lines

$$\begin{aligned} x &= 2 - 3s \\ y &= 3 + 2s \\ z &= 4 + 2s \end{aligned} \quad \text{and} \quad \begin{aligned} x &= 5 + 2t \\ y &= 1 - 3t \\ z &= 2 + t. \end{aligned}$$

16. Which of the following pairs of lines are perpendicular?

(a) $\begin{aligned} x &= 2 + 2t \\ y &= -3 - 3t \\ z &= 4 + 4t \end{aligned}$ and $\begin{aligned} x &= 2 + t \\ y &= 4 - t \\ z &= 5 - t \end{aligned}$

(b) $\begin{aligned} x &= 3 - t \\ y &= 4 + t \\ z &= 2 + 2t \end{aligned}$ and $\begin{aligned} x &= 2t \\ y &= 3 - 2t \\ z &= 4 + 2t \end{aligned}$

17. Show that the following parametric equations define the same line.

$$\begin{aligned} x &= 2 + 3t \\ y &= 3 - 2t \\ z &= -1 + 4t \end{aligned} \quad \text{and} \quad \begin{aligned} x &= -1 - 9t \\ y &= 5 + 6t \\ z &= -5 - 12t \end{aligned}$$

18. Find parametric equations of the line passing through the point $(3, -1, -3)$ and perpendicular to the line passing through the points $(3, -2, 4)$ and $(0, 3, 5)$.

19. Find an equation of the plane passing through the point $(-2, 3, 4)$ and perpendicular to the line passing through the points $(4, -2, 5)$ and $(0, 2, 4)$.

20. Find the point of intersection of the line

$$\begin{aligned} x &= 2 - 3t \\ y &= 4 + 2t \\ z &= 3 - 5t \end{aligned}$$

and the plane $2x + 3y + 4z + 8 = 0$.

21. Find a plane containing the lines

$$\begin{aligned} x &= 3 + 2t \\ y &= 4 - 3t \\ z &= 5 + 4t \end{aligned} \quad \text{and} \quad \begin{aligned} x &= 1 - 2t \\ y &= 7 + 4t \\ z &= 1 - 3t. \end{aligned}$$

22. Find a plane that passes through the point $(2, 4, -3)$ and is parallel to the plane $-2x + 4y - 5z + 6 = 0$.

23. Find a line that passes through the point $(-2, 5, -3)$ and is perpendicular to the plane $2x - 3y + 4z + 7 = 0$.

Theoretical Exercises

T.1. Show that the graph of the equation $ax + by + cz + d = 0$, where a, b, c, and d are constants with a, b, and c not all zero, is a plane with normal $\mathbf{n} = (a, b, c)$.

T.2. Let the lines L_1 and L_2 be given parametrically by

$$L_1\colon \mathbf{x} = \mathbf{w}_0 + s\mathbf{u} \quad \text{and} \quad L_2\colon \mathbf{x} = \mathbf{w}_1 + t\mathbf{v}.$$

Show that

(a) L_1 and L_2 are parallel if and only if $\mathbf{u} = k\mathbf{v}$ for some scalar k.

(b) L_1 and L_2 are identical if and only if $\mathbf{w}_1 - \mathbf{w}_0$ and $\mathbf{u}$ are both parallel to $\mathbf{v}$.

(c) L_1 and L_2 are perpendicular if and only if $\mathbf{u} \cdot \mathbf{v} = 0$.

(d) L_1 and L_2 intersect if and only if $\mathbf{w}_1 - \mathbf{w}_0$ is a linear combination of $\mathbf{u}$ and $\mathbf{v}$.

T.3. The lines L_1 and L_2 in R^3 are said to be **skew** if they are not parallel and do not intersect. Give an example of skew lines L_1 and L_2.

T.4. Consider the planes $a_1x + b_1y + c_1z + d_1 = 0$ and $a_2x + b_2y + c_2z + d_2 = 0$ with normals $\mathbf{n}_1$ and $\mathbf{n}_2$, respectively. Show that if the planes are identical, then $\mathbf{n}_2 = a\mathbf{n}_1$ for some scalar a.

T.5. Show that an equation of the plane through the noncollinear points $P_1(a_1, b_1, c_1)$, $P_2(a_2, b_2, c_2)$, and $P_3(a_3, b_3, c_3)$ is

$$\begin{vmatrix} x & y & z & 1 \\ a_1 & b_1 & c_1 & 1 \\ a_2 & b_2 & c_2 & 1 \\ a_3 & b_3 & c_3 & 1 \end{vmatrix} = 0.$$

Key Ideas for Review

- **Theorem 5.1 (Properties of Cross Product).** See page 260.
- A **parametric equation** of the line through P_0 and parallel to $\mathbf{u}$ is

$$\mathbf{x} = \mathbf{w}_0 + t\mathbf{u}, \qquad -\infty < t < \infty,$$

where $\mathbf{w}_0$ is the vector associated with P_0.

- An equation of the plane with normal $\mathbf{n} = (a, b, c)$ and passing through the point $P_0(x_0, y_0, z_0)$ is

$$a(x - x_0) + b(y - y_0) + c(z - z_0) = 0.$$

Supplementary Exercises

1. Find x and y so that $(x, y, 2) \times (1, 2, 3) = (0, 0, 0)$.

2. Find a vector $\mathbf{u}$ so that $\mathbf{u} \times (3, 2, -1) = (-1, 2, 1)$.

3. Which of the following points are on the line

$$\frac{x-3}{2} = \frac{y+3}{4} = \frac{z+5}{-4}?$$

(a) $(1, 2, 3)$ (b) $(5, 1, -9)$ (c) $(1, -7, -1)$

4. Find parametric equations of the line of intersection of the planes $x - 2y + z + 3 = 0$ and $2x - y + 3z + 4 = 0$.

Chapter Test

1. Find parametric equations of the line through the point $(5, -2, 1)$ that is parallel to the vector $\mathbf{u} = (3, -2, 5)$.

2. Find an equation of the plane passing through the points $(1, 2, -1)$, $(3, 4, 5)$, $(0, 1, 1)$.

3. Answer each of the following as true or false. Justify your answer.

(a) The standard matrix representing the dilation

$$L(\mathbf{u}) = L\left(\begin{bmatrix} u_1 \\ u_2 \end{bmatrix}\right) = \begin{bmatrix} u_1 \\ -2u_2 \end{bmatrix}$$

is $\begin{bmatrix} -1 & 0 \\ 2 & 0 \end{bmatrix}$.

(b) If $\mathbf{u} \times \mathbf{v} = \mathbf{0}$ and $\mathbf{u} \times \mathbf{w} = \mathbf{0}$, then $\mathbf{u} \times (\mathbf{v} + \mathbf{w}) = \mathbf{0}$.

(c) If $\mathbf{v} = -3\mathbf{u}$, then $\mathbf{u} \times \mathbf{v} = \mathbf{0}$.

(d) The point $(2, 3, 4)$ lies in the plane $2x - 3y + z = 5$.

(e) The planes $2x - 3y + 3z = 2$ and $2x + y - z = 4$ are perpendicular.

CHAPTER 6

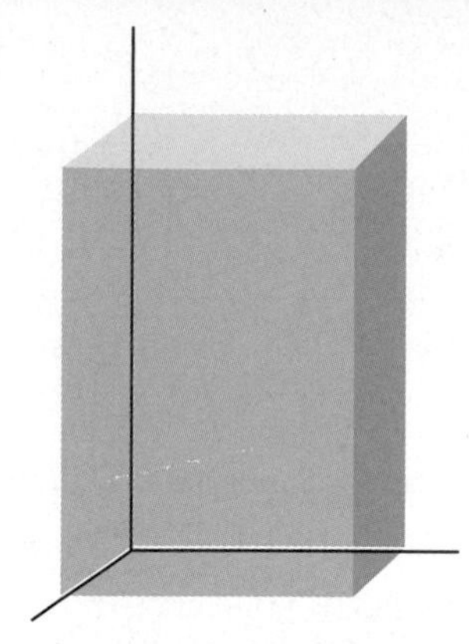

REAL VECTOR SPACES

6.1 VECTOR SPACES

We have already defined R^n and examined some of its basic properties in Theorem 4.2. We must now study the fundamental structure of R^n. In many applications in mathematics, the sciences, and engineering, the notion of a vector space arises. This idea is merely a carefully constructed generalization of R^n. In studying the properties and structure of a vector space, we can study not only R^n, in particular, but many other important vector spaces. In this section we define the notion of a vector space in general and in later sections we study their structure.

DEFINITION 1*

A **real vector space** is a set of elements V together with two operations $\oplus$ and $\odot$ satisfying the following properties:

(α) If $\mathbf{u}$ and $\mathbf{v}$ are any elements of V, then $\mathbf{u} \oplus \mathbf{v}$ is in V (i.e., V is closed under the operation $\oplus$).

(a) $\mathbf{u} \oplus \mathbf{v} = \mathbf{v} \oplus \mathbf{u}$, for $\mathbf{u}$ and $\mathbf{v}$ in V.

(b) $\mathbf{u} \oplus (\mathbf{v} \oplus \mathbf{w}) = (\mathbf{u} \oplus \mathbf{v}) \oplus \mathbf{w}$, for $\mathbf{u}$, $\mathbf{v}$, and $\mathbf{w}$ in V.

(c) There is an element $\mathbf{0}$ in V such that

$$\mathbf{u} \oplus \mathbf{0} = \mathbf{0} \oplus \mathbf{u} = \mathbf{u}, \quad \text{for all } \mathbf{u} \text{ in } V.$$

(d) For each $\mathbf{u}$ in V, there is an element $-\mathbf{u}$ in V such that

$$\mathbf{u} \oplus -\mathbf{u} = \mathbf{0}.$$

(β) If $\mathbf{u}$ is any element of V and c is any real number, then $c \odot \mathbf{u}$ is in V (i.e., V is closed under the operation $\odot$).

(e) $c \odot (\mathbf{u} \oplus \mathbf{v}) = c \odot \mathbf{u} \oplus c \odot \mathbf{v}$, for all real numbers c and all $\mathbf{u}$ and $\mathbf{v}$ in V.

(f) $(c + d) \odot \mathbf{u} = c \odot \mathbf{u} \oplus d \odot \mathbf{u}$, for all real numbers c and d, and all $\mathbf{u}$ in V.

(g) $c \odot (d \odot \mathbf{u}) = (cd) \odot \mathbf{u}$, for all real numbers c and d and all $\mathbf{u}$ in V.

(h) $1 \odot \mathbf{u} = \mathbf{u}$, for all $\mathbf{u}$ in V.

The elements of V are called **vectors**; the real numbers are called **scalars**. The operation $\oplus$ is called **vector addition**; the operation $\odot$ is called **scalar**

*Although the definitions in this book are not numbered, *this* definition is numbered because it will be referred to a number of times in this chapter.

multiplication. The vector $\mathbf{0}$ in property (c) is called a **zero vector**. The vector $-\mathbf{u}$ in property (d) is called a **negative** of $\mathbf{u}$. It can be shown (see Exercises T.5 and T.6) that the vectors $\mathbf{0}$ and $-\mathbf{u}$ are unique.

Property (α) is called the **closure** property for $\oplus$ and property (β) is called the **closure** property for $\odot$. We also say that V is **closed** under the operations of vector addition, $\oplus$, and scalar multiplication, $\odot$.

If we allow the scalars in Definition 1 to be complex numbers, we obtain a **complex vector space**. More generally, the scalars can be members of a field F,† and we obtain a vector space over F. Such spaces are important in many applications in mathematics and the physical sciences. We provide a brief introduction to complex vector spaces in Appendix A. Although most of our attention in this book will be focused on real vector spaces, we now take a brief look at a vector space over the field consisting of the bits 0 and 1, with the operations of binary addition and binary multiplication. In this case, we take the set of vectors V to be B^n, the set of bit n-vectors. With the operations of addition of bit n-vectors using binary addition and scalar multiplication with bit scalars, all the properties listed in Definition 1 are valid. [Note that in Section 1.4 we remarked that Theorem 1.1 and Theorem 1.3 (a)–(c) are valid for bit matrices, hence for B^n; properties (α), (β), and (h) of Definition 1 are also valid.] (See Exercises T.7–T.9.) Hence B^n is a vector space.

EXAMPLE 1

Consider the set R^n together with the operations of vector addition and scalar multiplication as defined in Section 4.2. Theorem 4.2 in Section 4.2 established the fact that R^n is a vector space under the operations of addition and scalar multiplication of n-vectors. ■

EXAMPLE 2

Consider the set V of all ordered triples of real numbers of the form $(x, y, 0)$ and define the operations $\oplus$ and $\odot$ by

$$\begin{aligned}(x, y, 0) \oplus (x', y', 0) &= (x + x', y + y', 0)\\ c \odot (x, y, 0) &= (cx, cy, 0).\end{aligned}$$

It is then not difficult to show (Exercise 7) that V is a vector space, since it satisfies all the properties of Definition 1. ■

EXAMPLE 3

Consider the set V of all ordered triples of real numbers (x, y, z) and define the operations $\oplus$ and $\odot$ by

$$\begin{aligned}(x, y, z) \oplus (x', y', z') &= (x + x', y + y', z + z')\\ c \odot (x, y, z) &= (cx, y, z).\end{aligned}$$

It is then easy to verify (Exercise 8) that properties (α), (β), (a), (b), (c), (d), and (e) of Definition 1 hold. Here $\mathbf{0} = (0, 0, 0)$ and the negative of the vector (x, y, z) is the vector $(-x, -y, -z)$. For example, to verify property (e) we proceed as follows. First,

$$\begin{aligned}c \odot [(x, y, z) \oplus (x', y', z')] &= c \odot (x + x', y + y', z + z')\\ &= (c(x + x'), y + y', z + z').\end{aligned}$$

†A field is an algebraic structure enjoying the arithmetic properties shared by the real, complex, and rational numbers. Fields are studied in detail in an abstract algebra course.

Also,

$$\begin{aligned} c \odot (x, y, z) \oplus c \odot (x', y', z') &= (cx, y, z) \oplus (cx', y', z') \\ &= (cx + cx', y + y', z + z') \\ &= (c(x + x'), y + y', z + z'). \end{aligned}$$

However, we now show that property (f) fails to hold. Thus

$$(c + d) \odot (x, y, z) = ((c + d)x, y, z).$$

On the other hand,

$$\begin{aligned} c \odot (x, y, z) \oplus d \odot (x, y, z) &= (cx, y, z) \oplus (dx, y, z) \\ &= (cx + dx, y + y, z + z) \\ &= ((c + d)x, 2y, 2z). \end{aligned}$$

Thus V is not a vector space under the prescribed operations. Incidentally, properties (g) and (h) *do* hold for this example. ■

EXAMPLE 4 Consider the set M_{23} of all 2×3 matrices under the usual operations of matrix addition and scalar multiplication. In Section 1.4 (Theorems 1.1 and 1.3) we have established that the properties in Definition 1 hold, thereby making M_{23} into a vector space. Similarly, the set of all $m \times n$ matrices under the usual operations of matrix addition and scalar multiplication is a vector space. This vector space will be denoted by M_{mn}. ■

EXAMPLE 5 Let $F[a, b]$ be the set of all real-valued functions that are defined on the interval $[a, b]$. If f and g are in V, we define $f \oplus g$ by

$$(f \oplus g)(t) = f(t) + g(t).$$

If f is in $F[a, b]$ and c is a scalar, we define $c \odot f$ by

$$(c \odot f)(t) = cf(t).$$

Then $F[a, b]$ is a vector space (Exercise 9). Similarly, the set of all real-valued functions defined for all real numbers denoted by $F(-\infty, \infty)$ is a vector space. ■

Another source of examples of vector spaces will be sets of polynomials; therefore, we recall some well-known facts about such functions. A **polynomial** (in t) is a function that is expressible as

$$p(t) = a_n t^n + a_{n-1} t^{n-1} + \cdots + a_1 t + a_0, \tag{1}$$

where n is an integer ≥ 0 and the coefficients $a_0, a_1, \ldots, a_n$ are real numbers.

EXAMPLE 6 The following functions are polynomials:

$$\begin{aligned} p_1(t) &= 3t^4 - 2t^2 + 5t - 1 \\ p_2(t) &= 2t + 1 \\ p_3(t) &= 4. \end{aligned}$$

The following functions are not polynomials (explain why):

$$f_4(t) = 2\sqrt{t} - 6 \quad \text{and} \quad f_5(t) = \frac{1}{t^2} - 2t + 1.$$ ■

The polynomial $p(t)$ in (1) is said to have **degree n** if $a_n \neq 0$. Thus the degree of a polynomial is the highest power having a nonzero coefficient.

EXAMPLE 7

The polynomials defined in Example 6 have the following degrees:

$$p_1(t): \text{degree } 4$$
$$p_2(t): \text{degree } 1$$
$$p_3(t): \text{degree } 0. \qquad \blacksquare$$

The **zero polynomial** is defined as

$$0t^n + 0t^{n-1} + \cdots + 0t + 0.$$

Note that, by definition, the zero polynomial has no degree.

We now let P_n be the set of all polynomials of degree $\leq n$ together with the zero polynomial. Thus $2t^2 - 3t + 5$, $2t + 1$, and 1 are a members of P_2.

EXAMPLE 8

If

$$p(t) = a_n t^n + a_{n-1}t^{n-1} + \cdots + a_1 t + a_0$$

and

$$q(t) = b_n t^n + b_{n-1}t^{n-1} + \cdots + b_1 t + b_0,$$

we define $p(t) \oplus q(t)$ as

$$p(t) \oplus q(t) = (a_n + b_n)t^n + (a_{n-1} + b_{n-1})t^{n-1} + \cdots + (a_1 + b_1)t + (a_0 + b_0)$$

(i.e., add coefficients of like-power terms). If c is a scalar, we also define $c \odot p(t)$ as

$$c \odot p(t) = (ca_n)t^n + (ca_{n-1})t^{n-1} + \cdots + (ca_1)t + (ca_0)$$

(i.e., multiply each coefficient by c). We now show that P_n is a vector space.

Let $p(t)$ and $q(t)$ defined previously be elements of P_n; that is, they are polynomials of degree $\leq n$ or the zero polynomial. Then the preceding definitions of the operations $\oplus$ and $\odot$ show that $p(t) \oplus q(t)$ and $c \odot p(t)$, for any scalar c, are polynomials of degree $\leq n$ or the zero polynomial. That is, $p(t) \oplus q(t)$ and $c \odot p(t)$ are in P_n so that (α) and (β) in Definition 1 hold. To verify property (a), we observe that

$$q(t) \oplus p(t) = (b_n + a_n)t^n + (b_{n-1} + a_{n-1})t^{n-1} + \cdots + (b_1 + a_1)t + (a_0 + b_0),$$

and since $a_i + b_i = b_i + a_i$ holds for the real numbers, we conclude that $p(t) \oplus q(t) = q(t) \oplus p(t)$. Similarly, we verify property (b). The zero polynomial is the element $\mathbf{0}$ needed in property (c). If $p(t)$ is as given previously, then its negative, $-p(t)$, is

$$-a_n t^n - a_{n-1}t^{n-1} - \cdots - a_1 t - a_0.$$

We shall now verify property (f) and leave the verification of the remaining properties to the reader. Thus

$$\begin{aligned}
(c + d) \odot p(t) &= (c + d)a_n t^n + (c + d)a_{n-1}t^{n-1} + \cdots + (c + d)a_1 t \\
&\quad + (c + d)a_0 \\
&= ca_n t^n + da_n t^n + ca_{n-1}t^{n-1} + da_{n-1}t^{n-1} + \cdots + ca_1 t \\
&\quad + da_1 t + ca_0 + da_0 \\
&= c(a_n t^n + a_{n-1}t^{n-1} + \cdots + a_1 t + a_0) \\
&\quad + d(a_n t^n + a_{n-1}t^{n-1} + \cdots + a_1 t + a_0) \\
&= c \odot p(t) \oplus d \odot p(t).
\end{aligned} \qquad \blacksquare$$

EXAMPLE 9 Let V be the set of all real numbers with the operations $\mathbf{u} \oplus \mathbf{v} = \mathbf{u} - \mathbf{v}$ ($\oplus$ is ordinary subtraction) and $c \odot \mathbf{u} = c\mathbf{u}$ ($\odot$ is ordinary multiplication). Is V a vector space? If it is not, which properties in Definition 1 fail to hold?

Solution If $\mathbf{u}$ and $\mathbf{v}$ are in V, and c is a scalar, then $\mathbf{u} \oplus \mathbf{v}$ and $c \odot \mathbf{u}$ are in V so that (α) and (β) in Definition 1 hold. However, property (a) fails to hold, as we can see by taking $\mathbf{u} = 2$ and $\mathbf{v} = 3$:

$$\mathbf{u} \oplus \mathbf{v} = 2 \oplus 3 = -1$$

and

$$\mathbf{v} \oplus \mathbf{u} = 3 \oplus 2 = 1.$$

Also, properties (b), (c), and (d) fail to hold (verify). Properties (e), (g), and (h) hold, but property (f) does not hold, as we can see by taking $c = 2$, $d = 3$, and $\mathbf{u} = 4$:

$$(c + d) \odot \mathbf{u} = (2 + 3) \odot 4 = 5 \odot 4 = 20$$

while

$$c \odot \mathbf{u} \oplus d \odot \mathbf{u} = 2 \odot 4 \oplus 3 \odot 4 = 8 \oplus 12 = -4.$$

Thus V is not a vector space. ■

For each natural number n, we have just defined the vector space P_n of all polynomials of degree $\leq n$ together with the zero polynomial. We could also consider the space P of *all* polynomials (of any degree), together with the zero polynomial. Here P is the mathematical union of all the vector spaces P_n. Two polynomials $p(t)$ of degree n and $g(t)$ of degree m are added in P in the same way as they would be added in P_r, where r is the maximum of the two numbers m and n. Then P is a vector space (Exercise 10).

> To verify that a given set V with two operations $\oplus$ and $\odot$ is a real vector space, we must show that it satisfies all the properties of Definition 1. The first thing to check is whether (α) and (β) hold, for, if either of these fails, we do not have a vector space. If both (α) and (β) hold, it is recommended that (c), the existence of a zero element, be verified next. Naturally, if (c) fails to hold, we do not have a vector space and do not have to check the remaining properties.

We frequently refer to a real vector space simply as a **vector space**. We also write $\mathbf{u} \oplus \mathbf{v}$ simply as $\mathbf{u} + \mathbf{v}$ and $c \odot \mathbf{u}$ simply as $c\mathbf{u}$, being careful to keep the particular operation in mind.

Many other important examples of vector spaces occur in numerous areas of mathematics.

The advantage of Definition 1 is that it is not concerned with the question of what a vector is. For example, is a vector in R^3 a point, a directed line segment, or a 3×1 matrix? Definition 1 deals only with the algebraic behavior of the elements in a vector space. In the case of R^3, whichever point of view we take, the algebraic behavior is the same. The mathematician abstracts those features that all such objects have in common (i.e., those properties that make them all behave alike) and defines a new structure, called a real vector space. We can now talk about properties of all vector spaces without having to refer to any one vector space in particular. Thus a "vector" is now merely an element of a vector space, and it no longer needs to be associated with a directed line segment. The following theorem presents several useful properties common to all vector spaces.

THEOREM 6.1 *If V is a vector space, then:*

(a) $0\mathbf{u} = \mathbf{0}$, *for every* $\mathbf{u}$ *in* V.
(b) $c\mathbf{0} = \mathbf{0}$, *for every scalar* c.
(c) *If* $c\mathbf{u} = \mathbf{0}$, *then* $c = 0$ *or* $\mathbf{u} = \mathbf{0}$.
(d) $(-1)\mathbf{u} = -\mathbf{u}$, *for every* $\mathbf{u}$ *in* V.

Proof (a) We have

$$0\mathbf{u} = (0 + 0)\mathbf{u} = 0\mathbf{u} + 0\mathbf{u}, \tag{2}$$

by (f) of Definition 1. Adding $-0\mathbf{u}$ to both sides of (2), we obtain by (b), (c), and (d) of Definition 1,

$$\begin{aligned} \mathbf{0} = 0\mathbf{u} + (-0\mathbf{u}) &= (0\mathbf{u} + 0\mathbf{u}) + (-0\mathbf{u}) \\ &= 0\mathbf{u} + [0\mathbf{u} + (-0\mathbf{u})] = 0\mathbf{u} + \mathbf{0} = 0\mathbf{u}. \end{aligned}$$

(b) Exercise T.1.

(c) Suppose that $c\mathbf{u} = \mathbf{0}$ and $c \neq 0$. We have

$$\mathbf{u} = 1\mathbf{u} = \left(\frac{1}{c}c\right)\mathbf{u} = \frac{1}{c}(c\mathbf{u}) = \frac{1}{c}\mathbf{0} = \mathbf{0}$$

by (b) of this theorem and (g) and (h) of Definition 1.

(d) $(-1)\mathbf{u} + \mathbf{u} = (-1)\mathbf{u} + (1)\mathbf{u} = (-1 + 1)\mathbf{u} = 0\mathbf{u} = \mathbf{0}$. Since $-\mathbf{u}$ is unique, we conclude that $(-1)\mathbf{u} = -\mathbf{u}$. ■

Notation for Vector Spaces Appearing in This Section

R^n,	the vector space of all n-vectors with real components
M_{mn},	the vector space of all $m \times n$ matrices
$F[a, b]$,	the vector space of all real-valued functions that are defined on the interval $[a, b]$
$F(-\infty, \infty)$,	the vector space of all real-valued functions defined for all real numbers
P_n,	the vector space of all polynomials of degree $\leq n$ together with the zero polynomial
P,	the vector space of all polynomials together with the zero polynomial

Key Terms

Real vector space
Vectors
Scalars
Vector addition
Scalar multiplication
Zero vector
Negative of a vector
Closure properties
Complex vector space
Polynomial
Degree of a polynomial
Zero polynomial

6.1 Exercises

In Exercises 1 through 4, determine whether the given set V is closed under the operations $\oplus$ and $\odot$.

1. V is the set of all ordered pairs of real numbers (x, y), where $x > 0$ and $y > 0$;

$$(x, y) \oplus (x', y') = (x + x', y + y')$$

and

$$c \odot (x, y) = (cx, cy).$$

2. V is the set of all ordered triples of real numbers of the form $(0, y, z)$;

$$(0, y, z) \oplus (0, y', z') = (0, y + y', z + z')$$

and

$$c \odot (0, y, z) = (0, 0, cz).$$

3. V is the set of all polynomials of the form $at^2 + bt + c$, where a, b, and c are real numbers with $b = a + 1$;

$$(a_1t^2 + b_1t + c_1) \oplus (a_2t^2 + b_2t + c_2) = (a_1 + a_2)t^2 + (b_1 + b_2)t + (c_1 + c_2)$$

and

$$r \odot (at^2 + bt + c) = (ra)t^2 + (rb)t + rc.$$

4. V is the set of all 2×2 matrices

$$\begin{bmatrix} a & b \\ c & d \end{bmatrix},$$

where $a = d$; $\oplus$ is matrix addition and $\odot$ is scalar multiplication.

5. Verify in detail that R^2 is a vector space.

6. Verify in detail that R^3 is a vector space.

7. Verify that the set in Example 2 is a vector space.

8. Verify that all the properties of Definition 1, except property (f), hold for the set in Example 3.

9. Show that the set in Example 5 is a vector space.

10. Show that the space P of all polynomials is a vector space.

In Exercises 11 through 17, determine whether the given set together with the given operations is a vector space. If it is not a vector space, list the properties of Definition 1 that fail to hold.

11. The set of all ordered triples of real numbers (x, y, z) with the operations

$$(x, y, z) \oplus (x', y', z') = (x', y + y', z')$$

and

$$c \odot (x, y, z) = (cx, cy, cz)$$

12. The set of all ordered triples of real numbers (x, y, z) with the operations

$$(x, y, z) \oplus (x', y', z') = (x + x', y + y', z + z')$$

and

$$c \odot (x, y, z) = (x, 1, z)$$

13. The set of all ordered triples of real numbers of the form $(0, 0, z)$ with the operations

$$(0, 0, z) \oplus (0, 0, z') = (0, 0, z + z')$$

and

$$c \odot (0, 0, z) = (0, 0, cz)$$

14. The set of all real numbers with the usual operations of addition and multiplication

15. The set of all ordered pairs of real numbers (x, y), where $x \leq 0$, with the usual operations in R^2

16. The set of all ordered pairs of real numbers (x, y) with the operations $(x, y) \oplus (x', y') = (x + x', y + y')$ and $c \odot (x, y) = (0, 0)$

17. The set of all positive real numbers $\mathbf{u}$ with the operations $\mathbf{u} \oplus \mathbf{v} = \mathbf{uv}$ and $c \odot \mathbf{u} = \mathbf{u}^c$

18. Let V be the set of all real numbers; define $\oplus$ by $\mathbf{u} \oplus \mathbf{v} = 2\mathbf{u} - \mathbf{v}$ and $\odot$ by $c \odot \mathbf{u} = c\mathbf{u}$. Is V a vector space?

19. Let V be the set consisting of a single element $\mathbf{0}$. Let $\mathbf{0} \oplus \mathbf{0} = \mathbf{0}$ and $c \odot \mathbf{0} = \mathbf{0}$. Show that V is a vector space.

20. (a) If V is a vector space that has a nonzero vector, how many vectors are in V?

(b) Describe all vector spaces having a finite number of vectors.

Theoretical Exercises

In Exercises T.1 through T.4, establish the indicated result for a real vector space V.

T.1. Show that $c\,\mathbf{0} = \mathbf{0}$ for every scalar c.

T.2. Show that $-(-\mathbf{u}) = \mathbf{u}$.

T.3. Show that if $\mathbf{u} + \mathbf{v} = \mathbf{u} + \mathbf{w}$, then $\mathbf{v} = \mathbf{w}$.

T.4. Show that if $\mathbf{u} \neq \mathbf{0}$ and $a\mathbf{u} = b\mathbf{u}$, then $a = b$.

T.5. Show that a vector space has only one zero vector.

T.6. Show that a vector $\mathbf{u}$ in a vector space has only one negative $-\mathbf{u}$.

T.7. Show that B^n is closed under binary addition of bit n-vectors.

T.8. Show that B^n is closed under scalar multiplication by bits 0 and 1.

T.9. Show that property (h) is valid for all vectors in B^n.

MATLAB Exercises

The concepts discussed in this section are not easily implemented in MATLAB *routines. The items in Definition 1 must hold for* all *vectors. Just because we demonstrate in* MATLAB *that a property of Definition 1 holds for a few vectors it does not imply that it holds for all such vectors. You must guard against such faulty reasoning. However, if, for a particular choice of vectors, we show that a property fails in* MATLAB, *then we have established that the property does not hold in all possible cases. Hence the property is considered to be false. In this way we might be able to show that a set is not a vector space.*

ML.1. Let V be the set of all 2×2 matrices with operations given by the following MATLAB commands:

$$A \oplus B \quad \text{is} \quad A.*B$$
$$k \odot A \quad \text{is} \quad k+A$$

Is V a vector space? (*Hint*: Enter some 2×2 matrices and experiment with the MATLAB commands to understand their behavior before checking the conditions in Definition 1.)

ML.2. Following Example 8, we discuss the vector space P_n of polynomials of degree n or less. Operations on polynomials can be performed in linear algebra software by associating a row matrix of size $n+1$ with polynomial $p(t)$ of P_n. The row matrix consists of the coefficients of $p(t)$ using the association

$$p(t) = a_n t^n + a_{n-1}t^{n-1} + \cdots + a_1 t + a_0$$
$$\rightarrow \begin{bmatrix} a_n & a_{n-1} & \cdots & a_1 & a_0 \end{bmatrix}.$$

If any term of $p(t)$ is explicitly missing, a zero is used for its coefficient. Then the addition of polynomials corresponds to matrix addition and multiplication of a polynomial by a scalar corresponds to scalar multiplication of matrices. Use MATLAB to perform the following operations on polynomials, using the matrix association described above. Let $n = 3$ and

$$p(t) = 2t^3 + 5t^2 + t - 2$$
$$q(t) = t^3 + 3t + 5.$$

(a) $p(t) + q(t)$ (b) $5p(t)$

(c) $3p(t) - 4q(t)$

6.2 SUBSPACES

In this section we begin to analyze the structure of a vector space. First, it is convenient to have a name for a subset of a given vector space that is itself a vector space with respect to the same operations as those in V. Thus we have the following.

DEFINITION Let V be a vector space and W a nonempty subset of V. If W is a vector space with respect to the operations in V, then W is called a **subspace** of V.

EXAMPLE 1 Every vector space has at least two subspaces, itself and the subspace $\{\mathbf{0}\}$ consisting only of the zero vector [recall that $\mathbf{0} \oplus \mathbf{0} = \mathbf{0}$ and $c \odot \mathbf{0} = \mathbf{0}$ in any vector space (see Exercise 19 in Section 6.1)]. The subspace $\{\mathbf{0}\}$ is called the **zero subspace**. ■

EXAMPLE 2 Let W be the subset of R^3 consisting of all vectors of the form $(a, b, 0)$, where a and b are any real numbers, with the usual operations of vector addition and scalar multiplication. To check if W is a subspace of R^3, we first see whether properties (α) and (β) of Definition 1 hold. Thus let $\mathbf{u} = (a_1, b_1, 0)$ and $\mathbf{v} = (a_2, b_2, 0)$ be vectors in W. Then $\mathbf{u} + \mathbf{v} = (a_1, b_1, 0) + (a_2, b_2, 0) = (a_1 + a_2, b_1 + b_2, 0)$ is in W, since the third component is zero. Also, if c is a scalar, then $c\mathbf{u} = c(a_1, b_1, 0) = (ca_1, cb_1, 0)$ is in W. Thus properties (α) and (β) of Definition 1 hold. We can easily verify that properties (a)–(h) hold. Hence W is a subspace of R^3. ■

Before listing other subspaces we pause to develop a labor-saving result. We just noted that to verify that a nonempty subset W of a vector space V is a subspace, we must check that (α), (β), and (a)–(h) of Definition 1 hold.

However, the following theorem says that it is enough to merely check that (α) and (β) hold. That is, we only need to verify that W is closed under the operations $\oplus$ and $\odot$.

THEOREM 6.2 *Let V be a vector space with operations $\oplus$ and $\odot$ and let W be a nonempty subset of V. Then W is a subspace of V if and only if the following conditions hold:*

(α) *If $\mathbf{u}$ and $\mathbf{v}$ are any vectors in W, then $\mathbf{u} \oplus \mathbf{v}$ is in W.*

(β) *If c is any real number and $\mathbf{u}$ is any vector in W, then $c \odot \mathbf{u}$ is in W.*

Proof Exercise T.1. ■

Remarks

1. Observe that the subspace consisting only of the zero vector (see Example 1) is a nonempty subspace.
2. If a subset W of a vector space V does not contain the zero vector, then W is not a subspace of V. (See Exercise T.13.)

EXAMPLE 3 Consider the set W consisting of all 2×3 matrices of the form

$$\begin{bmatrix} a & b & 0 \\ 0 & c & d \end{bmatrix},$$

where a, b, c, and d are arbitrary real numbers. Then W is a subset of the vector space M_{23} defined in Example 4 of Section 6.1. Show that W is a subspace of M_{23}. Note that a 2×3 matrix is in W provided its $(1, 3)$ and $(2, 1)$ entries are zero.

Solution Consider $\mathbf{u} = \begin{bmatrix} a_1 & b_1 & 0 \\ 0 & c_1 & d_1 \end{bmatrix}$ and $\mathbf{v} = \begin{bmatrix} a_2 & b_2 & 0 \\ 0 & c_2 & d_2 \end{bmatrix}$ in W. Then

$$\mathbf{u} + \mathbf{v} = \begin{bmatrix} a_1 + a_2 & b_1 + b_2 & 0 \\ 0 & c_1 + c_2 & d_1 + d_2 \end{bmatrix} \quad \text{is in } W$$

so that (α) of Theorem 6.2 is satisfied. Also, if k is a scalar, then

$$k\mathbf{u} = \begin{bmatrix} ka_1 & kb_1 & 0 \\ 0 & kc_1 & kd_1 \end{bmatrix} \quad \text{is in } W$$

so that (β) of Theorem 6.2 is satisfied. Hence W is a subspace of M_{23}. ■

We can also show that a nonempty subset W of a vector space V is a subspace of V if and only if $a\mathbf{u} + b\mathbf{v}$ is in W for any vectors $\mathbf{u}$ and $\mathbf{v}$ in W and any scalars a and b (Exercise T.2).

EXAMPLE 4 Which of the following subsets of R^2 with the usual operations of vector addition and scalar multiplication are subspaces?

(a) W_1 is the set of all vectors of the form $\begin{bmatrix} x \\ y \end{bmatrix}$, where $x \geq 0$.

(b) W_2 is the set of all vectors of the form $\begin{bmatrix} x \\ y \end{bmatrix}$, where $x \geq 0$, $y \geq 0$.

(c) W_3 is the set of all vectors of the form $\begin{bmatrix} x \\ y \end{bmatrix}$, where $x = 0$.

Solution (a) W_1 is the right half of the xy-plane (see Figure 6.1). It is not a subspace of R^2 because if we take the vector $\begin{bmatrix} 2 \\ 3 \end{bmatrix}$ in W_1, then the scalar multiple

$$-3\begin{bmatrix} 2 \\ 3 \end{bmatrix} = \begin{bmatrix} -6 \\ -9 \end{bmatrix}$$

is not in W_1, so property (β) in Theorem 6.2 does not hold.

Figure 6.1 ►

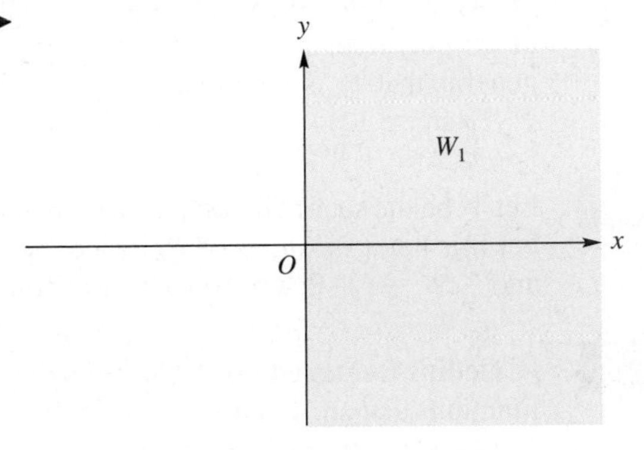

(b) W_2 is the first quadrant of the xy-plane (see Figure 6.2). The same vector and scalar multiple as in part (a) shows that W_2 is not a subspace.

Figure 6.2 ►

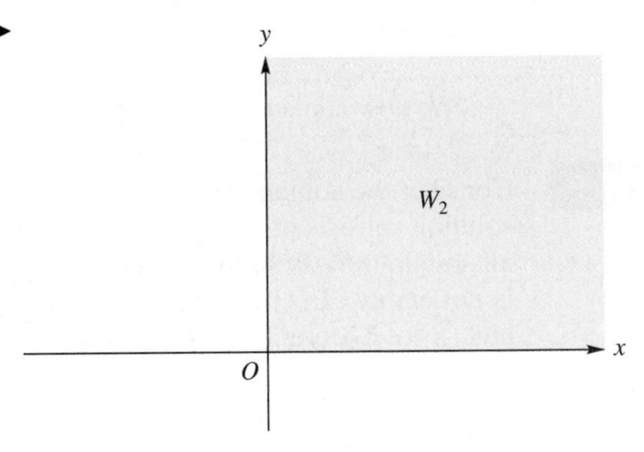

(c) W_3 is the y-axis in the xy-plane (see Figure 6.3). To see if W_3 is a subspace, let

$$\mathbf{u} = \begin{bmatrix} 0 \\ b_1 \end{bmatrix} \quad \text{and} \quad \mathbf{v} = \begin{bmatrix} 0 \\ b_2 \end{bmatrix}$$

be vectors in W_3. Then

$$\mathbf{u} + \mathbf{v} = \begin{bmatrix} 0 \\ b_1 \end{bmatrix} + \begin{bmatrix} 0 \\ b_2 \end{bmatrix} = \begin{bmatrix} 0 \\ b_1 + b_2 \end{bmatrix},$$

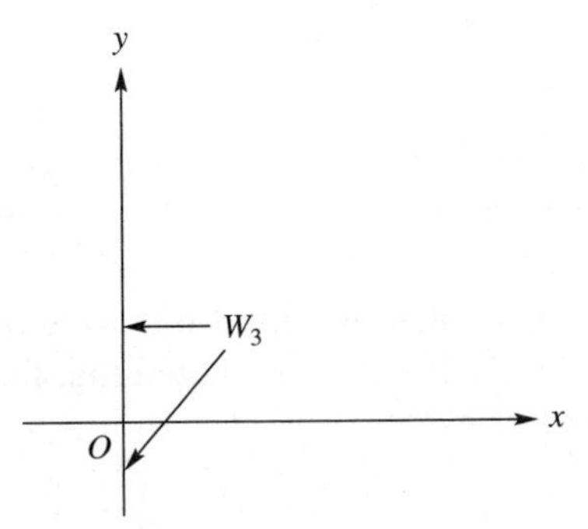

Figure 6.3 ▲

which is in W_3 so property (α) in Theorem 6.2 holds. Moreover, if c is a scalar, then

$$c\mathbf{u} = c\begin{bmatrix} 0 \\ b_1 \end{bmatrix} = \begin{bmatrix} 0 \\ cb_1 \end{bmatrix},$$

which is in W_3 so property (β) in Theorem 6.2 holds. Hence W_3 is a subspace of R^2. ■

EXAMPLE 5 Let W be the subset of R^3 consisting of all vectors of the form $(a, b, 1)$, where a and b are any real numbers. To check whether properties (α) and (β) of Theorem 6.2 hold, we let $\mathbf{u} = (a_1, b_1, 1)$ and $\mathbf{v} = (a_2, b_2, 1)$ be vectors in W. Then $\mathbf{u} + \mathbf{v} = (a_1, b_1, 1) + (a_2, b_2, 1) = (a_1 + a_2, b_1 + b_2, 2)$, which is not in W, since the third component is 2 and not 1. Since (α) of Theorem 6.2 does not hold, W is not a subspace of R^3. ■

EXAMPLE 6 In Section 6.1, we let P_n denote the vector space consisting of all polynomials of degree $\leq n$ and the zero polynomial, and let P denote the vector space of all polynomials. It is easy to verify that P_2 is a subspace of P_3 and, in general, that P_n is a subspace of P_{n+1} (Exercise 11). Also, P_n is a subspace of P (Exercise 12). ■

EXAMPLE 7 Let V be the set of all polynomials of degree exactly $= 2$; V is a *subset* of P_2, but it is not a *subspace* of P_2, since the sum of the polynomials $2t^2 + 3t + 1$ and $-2t^2 + t + 2$, a polynomial of degree 1, is not in V. ■

EXAMPLE 8 **(Calculus Required)** Let $C[a, b]$ denote the set of all real-valued continuous functions that are defined on the interval $[a, b]$. If f and g are in $C[a, b]$, then $f + g$ is in $C[a, b]$, since the sum of two continuous functions is continuous. Similarly, if c is a scalar, then cf is in $C[a, b]$. Hence $C[a, b]$ is a subspace of the vector space of all real-valued functions that are defined on $[a, b]$, which was introduced in Example 5 of Section 6.1. If the functions are defined for all real numbers, the vector space is denoted by $C(-\infty, \infty)$. ■

We now come to a very important example of a subspace.

EXAMPLE 9 Consider the homogeneous system $A\mathbf{x} = \mathbf{0}$, where A is an $m \times n$ matrix. A solution consists of a vector $\mathbf{x}$ in R^n. Let W be the subset of R^n consisting of all solutions to the homogeneous system. Since $A\mathbf{0} = \mathbf{0}$, we conclude that W is not empty. To check that W is a subspace of R^n, we verify properties (α) and (β) of Theorem 6.2. Thus let $\mathbf{x}$ and $\mathbf{y}$ be solutions. Then

$$A\mathbf{x} = \mathbf{0} \quad \text{and} \quad A\mathbf{y} = \mathbf{0}.$$

Now

$$A(\mathbf{x} + \mathbf{y}) = A\mathbf{x} + A\mathbf{y} = \mathbf{0} + \mathbf{0} = \mathbf{0},$$

so $\mathbf{x} + \mathbf{y}$ is a solution. Also, if c is a scalar, then

$$A(c\mathbf{x}) = c(A\mathbf{x}) = c\mathbf{0} = \mathbf{0},$$

so $c\mathbf{x}$ is also a solution. Hence W is a subspace of R^n. ■

The subspace W, discussed in Example 9, is called the **solution space** of the homogeneous system, or the **null space** of A. It should be noted that the set of all solutions to the linear system $A\mathbf{x} = \mathbf{b}$, where A is $m \times n$, is not a subspace of R^n if $\mathbf{b} \neq \mathbf{0}$ (Exercise T.3).

EXAMPLE 10 A simple way of constructing subspaces in a vector space is as follows. Let $\mathbf{v}_1$ and $\mathbf{v}_2$ be fixed vectors in a vector space V and let W be the set of all linear combinations (see Section 1.3) of $\mathbf{v}_1$ and $\mathbf{v}_2$, that is, W consists of all vectors in V of the form $a_1\mathbf{v}_1 + a_2\mathbf{v}_2$, where a_1 and a_2 are any real numbers. To show that W is a subspace of V, we verify properties (α) and (β) of Theorem 6.2.

Thus let

$$\mathbf{w}_1 = a_1\mathbf{v}_1 + a_2\mathbf{v}_2 \quad \text{and} \quad \mathbf{w}_2 = b_1\mathbf{v}_1 + b_2\mathbf{v}_2$$

be vectors in W. Then

$$\mathbf{w}_1 + \mathbf{w}_2 = (a_1\mathbf{v}_1 + a_2\mathbf{v}_2) + (b_1\mathbf{v}_1 + b_2\mathbf{v}_2) = (a_1 + b_1)\mathbf{v}_1 + (a_2 + b_2)\mathbf{v}_2,$$

which is in W. Also, if c is a scalar, then

$$c\mathbf{w}_1 = (ca_1)\mathbf{v}_1 + (ca_2)\mathbf{v}_2$$

is in W. Hence W is a subspace of V. ■

The construction carried out in Example 10 for two vectors can easily be performed for more than two vectors. We now give a formal definition.

DEFINITION Let $\mathbf{v}_1, \mathbf{v}_2, \ldots, \mathbf{v}_k$ be vectors in a vector space V. A vector $\mathbf{v}$ in V is called a **linear combination** of $\mathbf{v}_1, \mathbf{v}_2, \ldots, \mathbf{v}_k$ if

$$\mathbf{v} = c_1\mathbf{v}_1 + c_2\mathbf{v}_2 + \cdots + c_k\mathbf{v}_k$$

for some real numbers $c_1, c_2, \ldots, c_k$. (See also Section 1.3.)

In Figure 6.4 we show the vector $\mathbf{v}$ in R^2 or R^3 as a linear combination of the vectors $\mathbf{v}_1$ and $\mathbf{v}_2$.

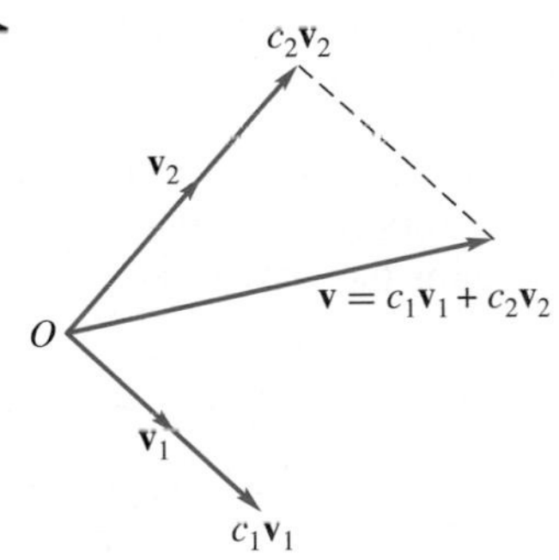

Figure 6.4 ► Linear combination of two vectors

EXAMPLE 11 In R^3 let

$$\mathbf{v}_1 = (1, 2, 1), \quad \mathbf{v}_2 = (1, 0, 2), \quad \text{and} \quad \mathbf{v}_3 = (1, 1, 0).$$

The vector

$$\mathbf{v} = (2, 1, 5)$$

is a linear combination of $\mathbf{v}_1$, $\mathbf{v}_2$, and $\mathbf{v}_3$ if we can find real numbers c_1, c_2, and c_3 so that

$$c_1\mathbf{v}_1 + c_2\mathbf{v}_2 + c_3\mathbf{v}_3 = \mathbf{v}.$$

Substituting for $\mathbf{v}$, $\mathbf{v}_1$, $\mathbf{v}_2$, and $\mathbf{v}_3$, we have

$$c_1(1, 2, 1) + c_2(1, 0, 2) + c_3(1, 1, 0) = (2, 1, 5).$$

Combining terms on the left and equating corresponding entries leads to the linear system (verify)

$$\begin{aligned} c_1 + c_2 + c_3 &= 2 \\ 2c_1 + c_3 &= 1 \\ c_1 + 2c_2 &= 5. \end{aligned}$$

Solving this linear system by the methods of Chapter 1 gives (verify) $c_1 = 1$, $c_2 = 2$, and $c_3 = -1$, which means that $\mathbf{v}$ is a linear combination of $\mathbf{v}_1$, $\mathbf{v}_2$, and $\mathbf{v}_3$. Thus

$$\mathbf{v} = \mathbf{v}_1 + 2\mathbf{v}_2 - \mathbf{v}_3.$$

Figure 6.5 shows $\mathbf{v}$ as a linear combination of $\mathbf{v}_1$, $\mathbf{v}_2$, and $\mathbf{v}_3$. ■

Figure 6.5 ▶

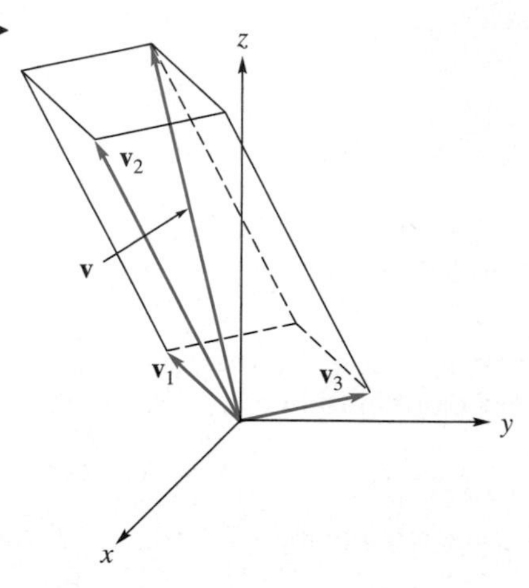

DEFINITION If $S = \{\mathbf{v}_1, \mathbf{v}_2, \ldots, \mathbf{v}_k\}$ is a set of vectors in a vector space V, then the set of all vectors in V that are linear combinations of the vectors in S is denoted by

$$\text{span } S \quad \text{or} \quad \text{span } \{\mathbf{v}_1, \mathbf{v}_2, \ldots, \mathbf{v}_k\}.$$

In Figure 6.6 we show a portion of span $\{\mathbf{v}_1, \mathbf{v}_2\}$, where $\mathbf{v}_1$ and $\mathbf{v}_2$ are noncollinear vectors in R^3; span $\{\mathbf{v}_1, \mathbf{v}_2\}$ is a plane that passes through the origin and contains the vectors $\mathbf{v}_1$ and $\mathbf{v}_2$.

Figure 6.6 ▶

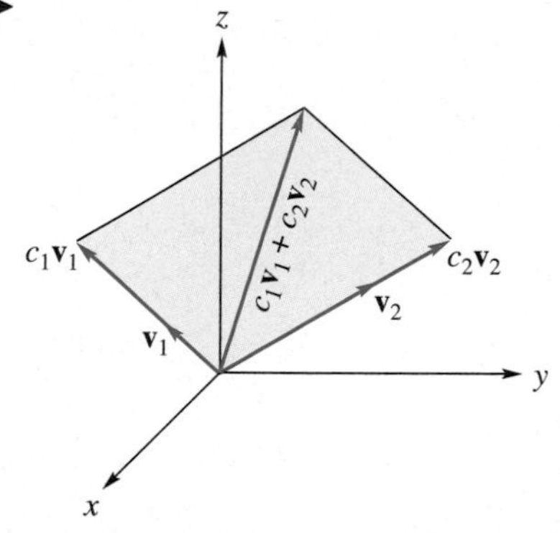

EXAMPLE 12 Consider the set S of 2×3 matrices given by

$$S = \left\{ \begin{bmatrix} 1 & 0 & 0 \\ 0 & 0 & 0 \end{bmatrix}, \begin{bmatrix} 0 & 1 & 0 \\ 0 & 0 & 0 \end{bmatrix}, \begin{bmatrix} 0 & 0 & 0 \\ 0 & 1 & 0 \end{bmatrix}, \begin{bmatrix} 0 & 0 & 0 \\ 0 & 0 & 1 \end{bmatrix} \right\}.$$

Then span S is the set in M_{23} consisting of all vectors of the form

$$a \begin{bmatrix} 1 & 0 & 0 \\ 0 & 0 & 0 \end{bmatrix} + b \begin{bmatrix} 0 & 1 & 0 \\ 0 & 0 & 0 \end{bmatrix} + c \begin{bmatrix} 0 & 0 & 0 \\ 0 & 1 & 0 \end{bmatrix} + d \begin{bmatrix} 0 & 0 & 0 \\ 0 & 0 & 1 \end{bmatrix}$$

$$= \begin{bmatrix} a & b & 0 \\ 0 & c & d \end{bmatrix}, \quad \text{where } a, b, c, \text{ and } d \text{ are real numbers.}$$

That is, span S is the subset of M_{23} consisting of all matrices of the form

$$\begin{bmatrix} a & b & 0 \\ 0 & c & d \end{bmatrix},$$

where a, b, c, and d are real numbers. ■

THEOREM 6.3 *Let $S = \{\mathbf{v}_1, \mathbf{v}_2, \ldots, \mathbf{v}_k\}$ be a set of vectors in a vector space V. Then span S is a subspace of V.*

Proof See Exercise T.4. ■

EXAMPLE 13 In P_2 let

$$\mathbf{v}_1 = 2t^2 + t + 2, \quad \mathbf{v}_2 = t^2 - 2t, \quad \mathbf{v}_3 = 5t^2 - 5t + 2, \quad \mathbf{v}_4 = -t^2 - 3t - 2.$$

Determine if the vector

$$\mathbf{u} = t^2 + t + 2$$

belongs to span $\{\mathbf{v}_1, \mathbf{v}_2, \mathbf{v}_3, \mathbf{v}_4\}$.

Solution If we can find scalars c_1, c_2, c_3, and c_4 so that

$$c_1\mathbf{v}_1 + c_2\mathbf{v}_2 + c_3\mathbf{v}_3 + c_4\mathbf{v}_4 = \mathbf{u},$$

then $\mathbf{u}$ belongs to span $\{\mathbf{v}_1, \mathbf{v}_2, \mathbf{v}_3, \mathbf{v}_4\}$. Substituting for $\mathbf{u}$, $\mathbf{v}_1$, $\mathbf{v}_2$, $\mathbf{v}_3$, and $\mathbf{v}_4$, we have

$$c_1(2t^2 + t + 2) + c_2(t^2 - 2t) + c_3(5t^2 - 5t + 2) + c_4(-t^2 - 3t - 2) \\ = t^2 + t + 2$$

or

$$(2c_1 + c_2 + 5c_3 - c_4)t^2 + (c_1 - 2c_2 - 5c_3 - 3c_4)t + (2c_1 + 2c_3 - 2c_4) \\ = t^2 + t + 2.$$

Now two polynomials agree for all values of t only if the coefficients of respective powers of t agree. Thus we get the linear system

$$\begin{aligned} 2c_1 + c_2 + 5c_3 - c_4 &= 1 \\ c_1 - 2c_2 - 5c_3 - 3c_4 &= 1 \\ 2c_1 \phantom{{}- 2c_2} + 2c_3 - 2c_4 &= 2. \end{aligned}$$

To investigate whether or not this system of linear equations is consistent, we form the augmented matrix and transform it to reduced row echelon form, obtaining (verify)

$$\left[\begin{array}{cccc|c} 1 & 0 & 1 & -1 & 0 \\ 0 & 1 & 3 & 1 & 0 \\ 0 & 0 & 0 & 0 & 1 \end{array}\right],$$

which indicates that the system is inconsistent, that is, it has no solution. Hence $\mathbf{u}$ does not belong to span $\{\mathbf{v}_1, \mathbf{v}_2, \mathbf{v}_3, \mathbf{v}_4\}$. ■

Remark In general, to determine if a specific vector $\mathbf{v}$ belongs to span S, we investigate the consistency of an appropriate linear system.

SUBSPACES IN B^n (OPTIONAL)

The notions of subspace, linear combination, and span are valid for any vector space, hence for B^n. Examples 14–16 provide illustrations of these concepts for B^n.

EXAMPLE 14 Let $V = B^3$ and let $W = \{\mathbf{w}_1, \mathbf{w}_2, \mathbf{w}_3, \mathbf{w}_4\}$, where

$$\mathbf{w}_1 = \begin{bmatrix} 0 \\ 0 \\ 0 \end{bmatrix}, \quad \mathbf{w}_2 = \begin{bmatrix} 1 \\ 0 \\ 0 \end{bmatrix}, \quad \mathbf{w}_3 = \begin{bmatrix} 0 \\ 1 \\ 0 \end{bmatrix}, \quad \text{and} \quad \mathbf{w}_4 = \begin{bmatrix} 1 \\ 1 \\ 0 \end{bmatrix}.$$

Determine if W is a subspace of V.

Solution Apply Theorem 6.2 using bit scalars and binary arithmetic. We have (verify)

$$\begin{array}{llll} \mathbf{w}_1 + \mathbf{w}_1 = \mathbf{w}_1, & \mathbf{w}_1 + \mathbf{w}_2 = \mathbf{w}_2, & \mathbf{w}_1 + \mathbf{w}_3 = \mathbf{w}_3, & \mathbf{w}_1 + \mathbf{w}_4 = \mathbf{w}_4, \\ \mathbf{w}_2 + \mathbf{w}_2 = \mathbf{w}_1, & \mathbf{w}_2 + \mathbf{w}_3 = \mathbf{w}_4, & \mathbf{w}_2 + \mathbf{w}_4 = \mathbf{w}_3, & \mathbf{w}_3 + \mathbf{w}_3 = \mathbf{w}_1, \\ \mathbf{w}_3 + \mathbf{w}_4 = \mathbf{w}_2, & \mathbf{w}_4 + \mathbf{w}_4 = \mathbf{w}_1, & & \end{array}$$

so W is closed under vector addition. Also,

$$0\mathbf{w}_j = \mathbf{w}_1, \quad \text{and} \quad 1\mathbf{w}_j = \mathbf{w}_j \quad \text{for } j = 1, 2, 3, 4,$$

hence W is closed under scalar multiplication. Thus W is a subspace of B^3. ■

EXAMPLE 15 Linear combinations of vectors in B^n can be formed using only the scalars 0 and 1. For the vectors $\mathbf{w}_j$, $j = 1, 2, 3, 4$ in Example 14 and the scalars $c_1 = c_2 = c_3 = c_4 = 1$ we have the linear combination

$$c_1\mathbf{w}_1 + c_2\mathbf{w}_2 + c_3\mathbf{w}_3 + c_4\mathbf{w}_4 = \begin{bmatrix} 0 \\ 0 \\ 0 \end{bmatrix} + \begin{bmatrix} 1 \\ 0 \\ 0 \end{bmatrix} + \begin{bmatrix} 0 \\ 1 \\ 0 \end{bmatrix} + \begin{bmatrix} 1 \\ 1 \\ 0 \end{bmatrix} = \begin{bmatrix} 0 \\ 0 \\ 0 \end{bmatrix} = \mathbf{w}_1.$$

(See also Example 19 in Section 1.2.) ■

EXAMPLE 16 In B^3 let

$$\mathbf{v}_1 = \begin{bmatrix} 1 \\ 1 \\ 0 \end{bmatrix}, \quad \mathbf{v}_2 = \begin{bmatrix} 0 \\ 1 \\ 1 \end{bmatrix}, \quad \text{and} \quad \mathbf{v}_3 = \begin{bmatrix} 1 \\ 1 \\ 1 \end{bmatrix}.$$

Determine if the vector $\mathbf{u} = \begin{bmatrix} 1 \\ 0 \\ 0 \end{bmatrix}$ belongs to span $\{\mathbf{v}_1, \mathbf{v}_2, \mathbf{v}_3\}$.

Solution If we can find (bit) scalars c_1, c_2, and c_3 so that

$$c_1\mathbf{v}_1 + c_2\mathbf{v}_2 + c_3\mathbf{v}_3 = \mathbf{u},$$

then $\mathbf{u}$ belongs to span $\{\mathbf{v}_1, \mathbf{v}_2, \mathbf{v}_3\}$. Substituting for $\mathbf{u}$, $\mathbf{v}_1$, $\mathbf{v}_2$, and $\mathbf{v}_3$, we have

$$c_1 \begin{bmatrix} 1 \\ 1 \\ 0 \end{bmatrix} + c_2 \begin{bmatrix} 0 \\ 1 \\ 1 \end{bmatrix} + c_3 \begin{bmatrix} 1 \\ 1 \\ 1 \end{bmatrix} = \begin{bmatrix} 1 \\ 0 \\ 0 \end{bmatrix}.$$

This linear combination can be written as the matrix product (verify)

$$\begin{bmatrix} 1 & 0 & 1 \\ 1 & 1 & 1 \\ 0 & 1 & 1 \end{bmatrix} \begin{bmatrix} c_1 \\ c_2 \\ c_3 \end{bmatrix} = \begin{bmatrix} 1 \\ 0 \\ 0 \end{bmatrix},$$

which has the form of a linear system. Forming the corresponding augmented matrix and transforming it to reduced row echelon form, we obtain (verify)

$$\left[\begin{array}{ccc|c} 1 & 0 & 0 & 0 \\ 0 & 1 & 0 & 1 \\ 0 & 0 & 1 & 1 \end{array}\right].$$

Hence the linear system is consistent with $c_1 = 0$, $c_2 = 1$, and $c_3 = 1$. Thus $\mathbf{u}$ is in span $\{\mathbf{v}_1, \mathbf{v}_2, \mathbf{v}_3\}$. ■

Key Terms

Subspace
Zero subspace
Closure property
Solution space
Linear combination

6.2 Exercises

1. The set W consisting of all the points in R^2 of the form (x, x) is a straight line. Is W a subspace of R^2? Explain.

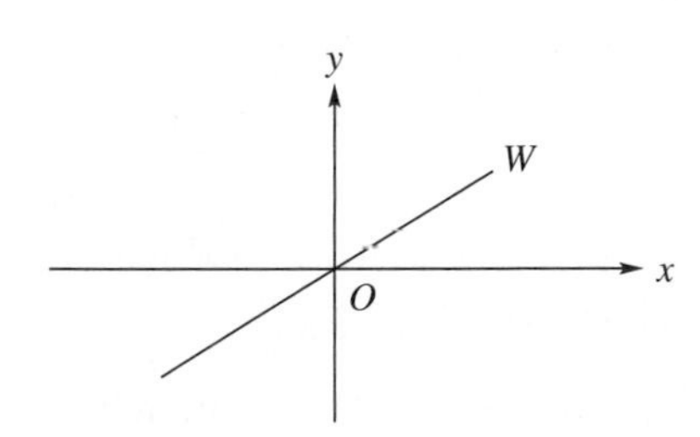

2. Let W be the set of all points in R^3 that lie in the xy-plane. Is W a subspace of R^3? Explain.

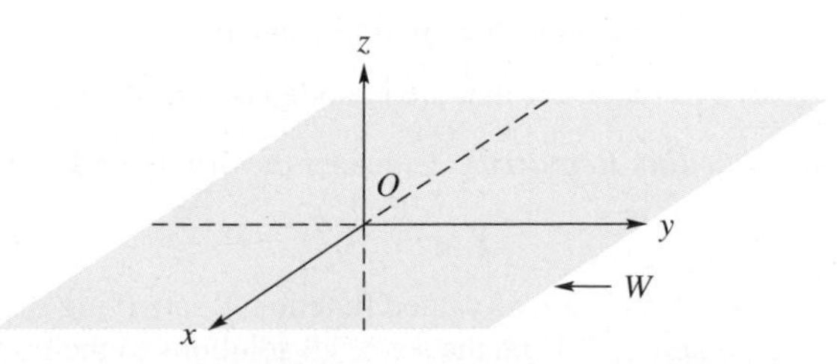

3. Consider the circle in the xy-plane centered at the origin whose equation is $x^2 + y^2 = 1$. Let W be the set of all vectors whose tail is at the origin and whose head is a point inside or on the circle. Is W a subspace of R^2? Explain.

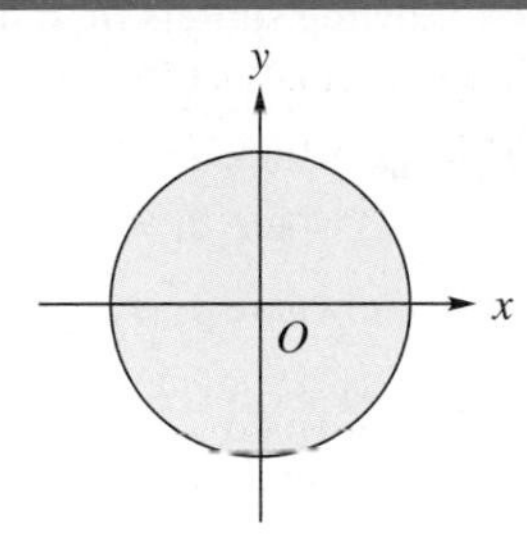

4. Consider the unit square shown in the accompanying figure. Let W be the set of all vectors of the form $\begin{bmatrix} x \\ y \end{bmatrix}$, where $0 \le x \le 1$, $0 \le y \le 1$. That is, W is the set of all vectors whose tail is at the origin and whose head is a point inside or on the square. Is W a subspace of R^2? Explain.

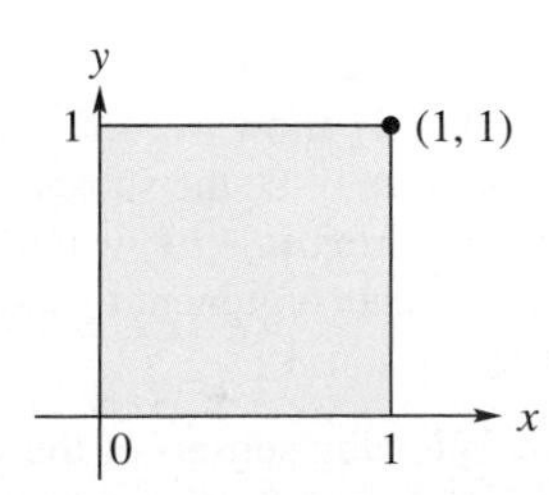

5. Which of the following subsets of R^3 are subspaces of R^3? The set of all vectors of the form

(a) $(a, b, 2)$

(b) (a, b, c), where $c = a + b$

(c) (a, b, c), where $c > 0$

6. Which of the following subsets of R^3 are subspaces of R^3? The set of all vectors of the form

(a) (a, b, c), where $a = c = 0$

(b) (a, b, c), where $a = -c$

(c) (a, b, c), where $b = 2a + 1$

7. Which of the following subsets of R^4 are subspaces of R^4? The set of all vectors of the form

(a) (a, b, c, d), where $a - b = 2$

(b) (a, b, c, d), where $c = a + 2b$ and $d = a - 3b$

(c) (a, b, c, d), where $a = 0$ and $b = -d$

8. Which of the following subsets of R^4 are subspaces of R^4? The set of all vectors of the form

(a) (a, b, c, d), where $a = b = 0$

(b) (a, b, c, d), where $a = 1$, $b = 0$, and $c + d = 1$

(c) (a, b, c, d), where $a > 0$ and $b < 0$

9. Which of the following subsets of P_2 are subspaces? The set of all polynomials of the form

(a) $a_2t^2 + a_1t + a_0$, where $a_0 = 0$

(b) $a_2t^2 + a_1t + a_0$, where $a_0 = 2$

(c) $a_2t^2 + a_1t + a_0$, where $a_2 + a_1 = a_0$

10. Which of the following subsets of P_2 are subspaces? The set of all polynomials of the form

(a) $a_2t^2 + a_1t + a_0$, where $a_1 = 0$ and $a_0 = 0$

(b) $a_2t^2 + a_1t + a_0$, where $a_1 = 2a_0$

(c) $a_2t^2 + a_1t + a_0$, where $a_2 + a_1 + a_0 = 2$

11. (a) Show that P_2 is a subspace of P_3.

(b) Show that P_n is a subspace of P_{n+1}.

12. Show that P_n is a subspace of P.

13. Show that P is a subspace of the vector space defined in Example 5 of Section 6.1.

14. Let $\mathbf{u} = (1, 2, -3)$ and $\mathbf{v} = (-2, 3, 0)$ be two vectors in R^3 and let W be the subset of R^3 consisting of all vectors of the form $a\mathbf{u} + b\mathbf{v}$, where a and b are any real numbers. Give an argument to show that W is a subspace of R^3.

15. Let $\mathbf{u} = (2, 0, 3, -4)$ and $\mathbf{v} = (4, 2, -5, 1)$ be two vectors in R^4 and let W be the subset of R^4 consisting of all vectors of the form $a\mathbf{u} + b\mathbf{v}$, where a and b are any real numbers. Give an argument to show that W is a subspace of R^4.

16. Which of the following subsets of the vector space M_{23} defined in Example 4 of Section 6.1 are subspaces? The set of all matrices of the form

(a) $\begin{bmatrix} a & b & c \\ d & 0 & 0 \end{bmatrix}$, where $b = a + c$

(b) $\begin{bmatrix} a & b & c \\ d & 0 & 0 \end{bmatrix}$, where $c > 0$

(c) $\begin{bmatrix} a & b & c \\ d & e & f \end{bmatrix}$, where $a = -2c$ and $f = 2e + d$

17. Which of the following subsets of the vector space M_{23} defined in Example 4 of Section 6.1 are subspaces? The set of all matrices of the form

(a) $\begin{bmatrix} a & b & c \\ d & e & f \end{bmatrix}$, where $a = 2c + 1$

(b) $\begin{bmatrix} 0 & 1 & a \\ b & c & 0 \end{bmatrix}$

(c) $\begin{bmatrix} a & b & c \\ d & e & f \end{bmatrix}$, where $a + c = 0$ and $b + d + f = 0$

18. Which of the following subsets of the vector space M_{nn} are subspaces?

(a) The set of all $n \times n$ symmetric matrices

(b) The set of all $n \times n$ nonsingular matrices

(c) The set of all $n \times n$ diagonal matrices

19. Which of the following subsets of the vector space M_{nn} are subspaces?

(a) The set of all $n \times n$ singular matrices

(b) The set of all $n \times n$ upper triangular matrices

(c) The set of all $n \times n$ matrices whose determinant is 1

20. ***(Calculus Required)*** Which of the following subsets are subspaces of the vector space $C(-\infty, \infty)$ defined in Example 8?

(a) All nonnegative functions

(b) All constant functions

(c) All functions f such that $f(0) = 0$

(d) All functions f such that $f(0) = 5$

(e) All differentiable functions.

21. ***(Calculus Required)*** Which of the following subsets of the vector space $C(-\infty, \infty)$ defined in Example 8 are subspaces?

(a) All integrable functions

(b) All bounded functions

(c) All functions that are integrable on $[a, b]$

(d) All functions that are bounded on $[a, b]$

22. ***(Calculus Required)*** Consider the differential equation

$$y'' - y' + 2y = 0.$$

A solution is a real-valued function f satisfying the equation. Let V be the set of all solutions to the given differential equation; define $\oplus$ and $\odot$ as in Example 5 in Section 6.1. Show that V is a subspace of the vector space of all real-valued functions defined on $(-\infty, \infty)$. (See also Section 9.2.)

23. Determine which of the following subsets of R^2 are subspaces.

(a)

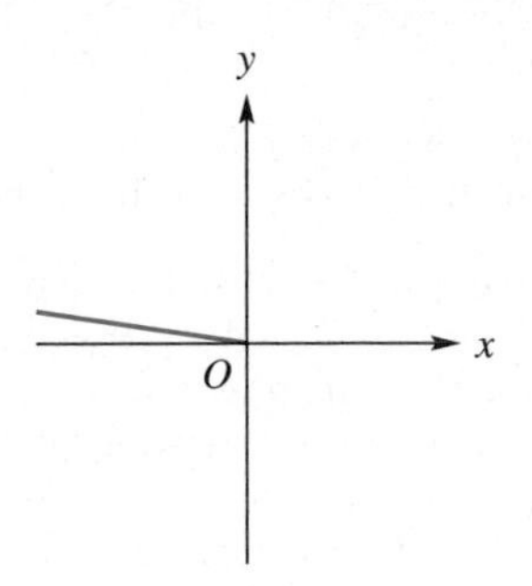

(b)

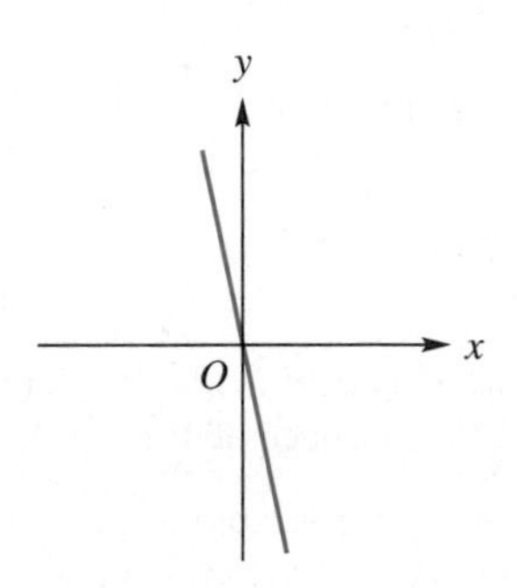

24. Determine which of the following subsets of R^2 are subspaces.

(a)

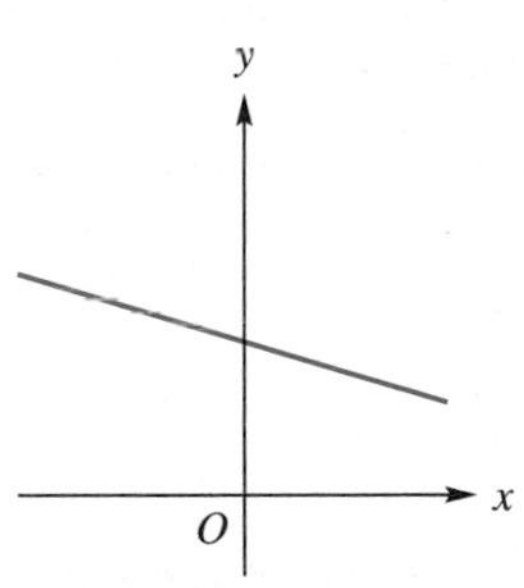

(b)

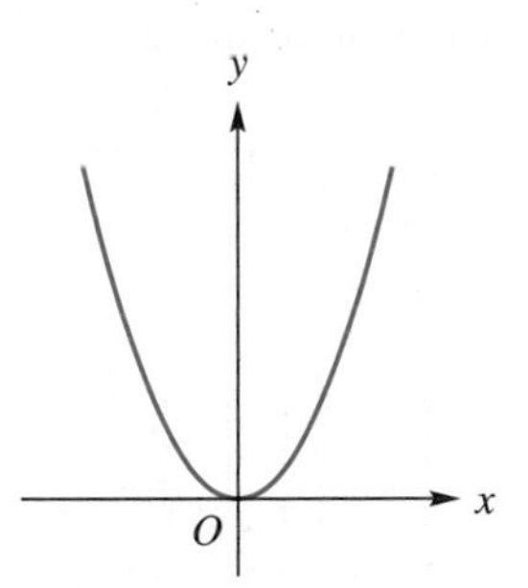

25. In each part, determine whether the given vector $\mathbf{v}$ belongs to span $\{\mathbf{v}_1, \mathbf{v}_2, \mathbf{v}_3\}$, where

$$\mathbf{v}_1 = (1, 0, 0, 1), \qquad \mathbf{v}_2 = (1, -1, 0, 0),$$

and

$$\mathbf{v}_3 = (0, 1, 2, 1).$$

(a) $\mathbf{v} = (-1, 4, 2, 2)$ (b) $\mathbf{v} = (1, 2, 0, 1)$

(c) $\mathbf{v} = (-1, 1, 4, 3)$ (d) $\mathbf{v} = (0, 1, 1, 0)$

26. Which of the following vectors are linear combinations of

$$A_1 = \begin{bmatrix} 1 & -1 \\ 0 & 3 \end{bmatrix}, \quad A_2 = \begin{bmatrix} 1 & 1 \\ 0 & 2 \end{bmatrix}, \quad A_3 = \begin{bmatrix} 2 & 2 \\ -1 & 1 \end{bmatrix}?$$

(a) $\begin{bmatrix} 5 & 1 \\ -1 & 9 \end{bmatrix}$ (b) $\begin{bmatrix} -3 & -1 \\ 3 & 2 \end{bmatrix}$

(c) $\begin{bmatrix} 3 & -2 \\ 3 & 2 \end{bmatrix}$ (d) $\begin{bmatrix} 1 & 0 \\ 2 & 1 \end{bmatrix}$

27. In each part, determine whether the given vector $p(t)$ belongs to span $\{p_1(t), p_2(t), p_3(t)\}$, where

$$p_1(t) = t^2 - t,$$
$$p_2(t) = t^2 - 2t + 1,$$
$$p_3(t) = -t^2 + 1.$$

(a) $p(t) = 3t^2 - 3t + 1$ (b) $p(t) = t^2 - t + 1$

(c) $p(t) = t + 1$ (d) $p(t) = 2t^2 - t - 1$

Exercises 28 through 33 use bit matrices.

28. Let $V = B^3$. Determine if

$$W = \left\{ \begin{bmatrix} 0 \\ 1 \\ 1 \end{bmatrix}, \begin{bmatrix} 0 \\ 1 \\ 0 \end{bmatrix}, \begin{bmatrix} 0 \\ 0 \\ 1 \end{bmatrix} \right\}$$

is a subspace of V.

29. Let $V = B^3$. Determine if

$$W = \left\{ \begin{bmatrix} 1 \\ 0 \\ 0 \end{bmatrix}, \begin{bmatrix} 0 \\ 0 \\ 1 \end{bmatrix}, \begin{bmatrix} 1 \\ 0 \\ 1 \end{bmatrix}, \begin{bmatrix} 0 \\ 0 \\ 0 \end{bmatrix} \right\}$$

is a subspace of V.

30. Let $V = B^4$. Determine if W, the set of all vectors in V with first entry zero, is a subspace of V.

31. Let $V = B^4$. Determine if W, the set of all vectors in V with second entry one, is a subspace of V.

32. Let

$$S = \left\{ \begin{bmatrix} 1 \\ 0 \\ 0 \end{bmatrix}, \begin{bmatrix} 1 \\ 1 \\ 0 \end{bmatrix}, \begin{bmatrix} 1 \\ 1 \\ 1 \end{bmatrix} \right\}.$$

Determine if $\mathbf{u} = \begin{bmatrix} 1 \\ 0 \\ 1 \end{bmatrix}$ belongs to span S.

33. Let

$$S = \left\{ \begin{bmatrix} 0 \\ 0 \\ 1 \end{bmatrix}, \begin{bmatrix} 1 \\ 0 \\ 1 \end{bmatrix}, \begin{bmatrix} 0 \\ 1 \\ 1 \end{bmatrix} \right\}.$$

Determine if $\mathbf{u} = \begin{bmatrix} 1 \\ 1 \\ 1 \end{bmatrix}$ belongs to span S.

Theoretical Exercises

T.1. Prove Theorem 6.2.

T.2. Show that a subset W of a vector space V is a subspace of V if and only if the following condition holds: If $\mathbf{u}$ and $\mathbf{v}$ are any vectors in W and a and b are any scalars, then $a\mathbf{u} + b\mathbf{v}$ is in W.

T.3. Show that the set of all solutions to $A\mathbf{x} = \mathbf{b}$, where A is $m \times n$, is not a subspace of R^n if $\mathbf{b} \neq \mathbf{0}$.

T.4. Prove Theorem 6.3.

T.5. Let $S = \{\mathbf{v}_1, \mathbf{v}_2, \dots, \mathbf{v}_k\}$ be a set of vectors in a vector space V, and let W be a subspace of V containing S. Show that W contains span S.

T.6. If A is a nonsingular matrix, what is the null space of A? Justify your answer.

T.7. Let $\mathbf{x}_0$ be a fixed vector in a vector space V. Show that the set W consisting of all scalar multiples $c\mathbf{x}_0$ of $\mathbf{x}_0$ is a subspace of V.

T.8. Let A be an $m \times n$ matrix. Is the set W of all vectors $\mathbf{x}$ in R^n such that $A\mathbf{x} \neq \mathbf{0}$ a subspace of R^n? Justify your answer.

T.9. Show that the only subspaces of R^1 are $\{\mathbf{0}\}$ and R^1 itself.

T.10. Let W_1 and W_2 be subspaces of a vector space V. Let $W_1 + W_2$ be the set of all vectors $\mathbf{v}$ in V such that $\mathbf{v} = \mathbf{w}_1 + \mathbf{w}_2$, where $\mathbf{w}_1$ is in W_1 and $\mathbf{w}_2$ is in W_2. Show that $W_1 + W_2$ is a subspace of V.

T.11. Let W_1 and W_2 be subspaces of a vector space V with $W_1 \cap W_2 = \{\mathbf{0}\}$. Let $W_1 + W_2$ be as defined in Exercise T.10. Suppose that $V = W_1 + W_2$. Show that every vector in V can be uniquely written as $\mathbf{w}_1 + \mathbf{w}_2$, where $\mathbf{w}_1$ is in W_1 and $\mathbf{w}_2$ is in W_2. In this case we write $V = W_1 \oplus W_2$ and say that V is the **direct sum** of the subspaces W_1 and W_2.

T.12. Show that the set of all points in the plane $ax + by + cz = 0$ is a subspace of R^3.

T.13. Show that if a subset W of a vector space V does not contain the zero vector, then W is not a subspace of V.

T.14. Let $V = B^3$ and $W = \{\mathbf{w}_1\}$, where $\mathbf{w}_1$ is any vector in B^3. Is W a subspace of V?

T.15. Let $V = B^3$. Determine if there is a subspace of V that contains exactly three different vectors.

T.16. In Example 14, W is a subspace of B^3 with exactly four vectors. Determine two other subspaces of B^3 that contain exactly four vectors.

T.17. Determine all the subspaces of B^3 that contain

$$\mathbf{v} = \begin{bmatrix} 1 \\ 1 \\ 1 \end{bmatrix}.$$

MATLAB Exercises

ML.1. Let V be R^3 and let W be the subset of V of vectors of the form $(2, a, b)$, where a and b are any real numbers. Is W a subspace of V? Use the following MATLAB commands to help you determine the answer.

```
a1 = fix(10 * randn);
a2 = fix(10 * randn);
b1 = fix(10 * randn);
b2 = fix(10 * randn);
v = [2 a1 b1]
w = [2 a2 b2]
v + w
3 * v
```

ML.2. Let V be P_2 and let W be the subset of V of vectors of the form $ax^2 + bx + 5$, where a and b are arbitrary real numbers. With each such polynomial in W we associate a vector $(a, b, 5)$ in R^3. Construct commands like those in Exercise ML.1 to show that W is not a subspace of V.

Before solving the following MATLAB *exercises, you should have read Section 12.7.*

ML.3. Use MATLAB to determine if vector $\mathbf{v}$ is a linear combination of the members of set S.

(a) $S = \{\mathbf{v}_1, \mathbf{v}_2, \mathbf{v}_3\}$
$= \{(1, 0, 0, 1), (0, 1, 1, 0), (1, 1, 1, 1)\}$
$\mathbf{v} = (0, 1, 1, 1)$

(b) $S = \{\mathbf{v}_1, \mathbf{v}_2, \mathbf{v}_3\}$

$$= \left\{ \begin{bmatrix} 1 \\ 2 \\ -1 \end{bmatrix}, \begin{bmatrix} 2 \\ -1 \\ 0 \end{bmatrix}, \begin{bmatrix} -1 \\ 8 \\ -3 \end{bmatrix} \right\}$$

$$\mathbf{v} = \begin{bmatrix} 0 \\ 5 \\ -2 \end{bmatrix}$$

ML.4. Use MATLAB to determine if $\mathbf{v}$ is a linear combination of the members of set S. If it is, express $\mathbf{v}$ in terms of the members of S.

(a) $S = \{\mathbf{v}_1, \mathbf{v}_2, \mathbf{v}_3\}$
$= \{(1, 2, 1), (3, 0, 1), (1, 8, 3)\}$
$\mathbf{v} = (-2, 14, 4)$

(b) $S = \{A_1, A_2, A_3\}$

$$= \left\{ \begin{bmatrix} 1 & 2 \\ 1 & 0 \end{bmatrix}, \begin{bmatrix} 2 & -1 \\ 1 & 2 \end{bmatrix}, \begin{bmatrix} -3 & 1 \\ 0 & 1 \end{bmatrix} \right\}$$

$\mathbf{v} = I_2$

ML.5. Use MATLAB to determine if $\mathbf{v}$ is a linear combination of the members of set S. If it is, express $\mathbf{v}$ in terms of the members of S.

(a) $S = \{\mathbf{v}_1, \mathbf{v}_2, \mathbf{v}_3, \mathbf{v}_4\}$

$$= \left\{ \begin{bmatrix} 1 \\ 2 \\ 1 \\ 0 \\ 1 \end{bmatrix}, \begin{bmatrix} 0 \\ 1 \\ 2 \\ -1 \\ 1 \end{bmatrix}, \begin{bmatrix} 2 \\ 1 \\ 0 \\ 0 \\ -1 \end{bmatrix}, \begin{bmatrix} -2 \\ 1 \\ 1 \\ 1 \\ 1 \end{bmatrix} \right\}$$

$$\mathbf{v} = \begin{bmatrix} 0 \\ -1 \\ 1 \\ -2 \\ 1 \end{bmatrix}$$

(b) $S = \{p_1(t), p_2(t), p_3(t)\}$
$= \{2t^2 - t + 1, t^2 - 2, t - 1\}$
$\mathbf{v} = p(t) = 4t^2 + t - 5$

ML.6. In each part, determine whether $\mathbf{v}$ belongs to span S, where

$$S = \{\mathbf{v}_1, \mathbf{v}_2, \mathbf{v}_3\} = \{(1, 1, 0, 1), (1, -1, 0, 1), (0, 1, 2, 1)\}.$$

(a) $\mathbf{v} = (2, 3, 2, 3)$

(b) $\mathbf{v} = (2, -3, -2, 3)$

(c) $\mathbf{v} = (0, 1, 2, 3)$

ML.7. In each part, determine whether $p(t)$ belongs to span S, where

$$S = \{p_1(t), p_2(t), p_3(t)\} = \{t - 1, t + 1, t^2 + t + 1\}.$$

(a) $p(t) = t^2 + 2t + 4$

(b) $p(t) = 2t^2 + t - 2$

(c) $p(t) = -2t^2 + 1$

6.3 LINEAR INDEPENDENCE

Thus far we have defined a mathematical system called a real vector space and noted some of its properties. We further observe that the only real vector space having a finite number of vectors in it is the vector space whose only vector is $\mathbf{0}$, for if $\mathbf{v} \neq \mathbf{0}$ is in a vector space V, then by Exercise T.4 in Section 6.1, $c\mathbf{v} \neq c'\mathbf{v}$, where c and c' are distinct real numbers, and so V has infinitely many vectors in it. However, in this section and the following one we show that most vector spaces V studied here have a set composed of a finite number of vectors that completely describe V; that is, we can write every vector in V as a linear combination of the vectors in this set. It should be noted that, in general, there is more than one such set describing V. We now turn to a formulation of these ideas.

DEFINITION

The vectors $\mathbf{v}_1, \mathbf{v}_2, \ldots, \mathbf{v}_k$ in a vector space V are said to **span** V if every vector in V is a linear combination of $\mathbf{v}_1, \mathbf{v}_2, \ldots, \mathbf{v}_k$. Moreover, if $S = \{\mathbf{v}_1, \mathbf{v}_2, \ldots, \mathbf{v}_k\}$, then we also say that the set S **spans** V, or that $\{\mathbf{v}_1, \mathbf{v}_2, \ldots, \mathbf{v}_k\}$ **spans** V, or that V is **spanned by** S, or in the language of Section 6.2, span $S = V$.

The procedure to check if the vectors $\mathbf{v}_1, \mathbf{v}_2, \ldots, \mathbf{v}_k$ span the vector space V is as follows.

Step 1. Choose an arbitrary vector $\mathbf{v}$ in V.

Step 2. Determine if $\mathbf{v}$ is a linear combination of the given vectors. If it is, then the given vectors span V. If it is not, they do not span V.

Again, in Step 2, we investigate the consistency of a linear system, but this time for a right side that represents an arbitrary vector in a vector space V.

EXAMPLE 1 Let V be the vector space R^3 and let

$$\mathbf{v}_1 = (1, 2, 1), \quad \mathbf{v}_2 = (1, 0, 2), \quad \text{and} \quad \mathbf{v}_3 = (1, 1, 0).$$

Do $\mathbf{v}_1$, $\mathbf{v}_2$, and $\mathbf{v}_3$ span V?

Solution ***Step 1.*** Let $\mathbf{v} = (a, b, c)$ be any vector in R^3, where a, b, and c are arbitrary real numbers.

Step 2. We must find out whether there are constants c_1, c_2, and c_3 such that

$$c_1\mathbf{v}_1 + c_2\mathbf{v}_2 + c_3\mathbf{v}_3 = \mathbf{v}.$$

This leads to the linear system (verify)

$$\begin{aligned} c_1 + c_2 + c_3 &= a \\ 2c_1 \qquad + c_3 &= b \\ c_1 + 2c_2 \qquad &= c. \end{aligned}$$

A solution is (verify)

$$c_1 = \frac{-2a + 2b + c}{3}, \qquad c_2 = \frac{a - b + c}{3}, \qquad c_3 = \frac{4a - b - 2c}{3}.$$

Since we have obtained a solution for every choice of a, b, and c, we conclude that $\mathbf{v}_1$, $\mathbf{v}_2$, $\mathbf{v}_3$ span R^3. This is equivalent to saying that span $\{\mathbf{v}_1, \mathbf{v}_2, \mathbf{v}_3\} = R^3$. ■

EXAMPLE 2 Show that

$$S = \left\{ \begin{bmatrix} 1 & 0 \\ 0 & 0 \end{bmatrix}, \begin{bmatrix} 0 & 1 \\ 1 & 0 \end{bmatrix}, \begin{bmatrix} 0 & 0 \\ 0 & 1 \end{bmatrix} \right\}$$

spans the subspace of M_{22} consisting of all symmetric matrices.

Solution ***Step 1.*** An arbitrary symmetric matrix has the form

$$A = \begin{bmatrix} a & b \\ b & c \end{bmatrix},$$

where a, b, and c are any real numbers.

Step 2. We must find constants d_1, d_2, and d_3 such that

$$d_1 \begin{bmatrix} 1 & 0 \\ 0 & 0 \end{bmatrix} + d_2 \begin{bmatrix} 0 & 1 \\ 1 & 0 \end{bmatrix} + d_3 \begin{bmatrix} 0 & 0 \\ 0 & 1 \end{bmatrix} = A = \begin{bmatrix} a & b \\ b & c \end{bmatrix},$$

which leads to a linear system whose solution is (verify)

$$d_1 = a, \qquad d_2 = b, \qquad d_3 = c.$$

Since we have found a solution for every choice of a, b, and c, we conclude that S spans the given subspace. ■

EXAMPLE 3 Let V be the vector space P_2. Let $S = \{p_1(t), p_2(t)\}$, where $p_1(t) = t^2+2t+1$ and $p_2(t) = t^2 + 2$. Does S span P_2?

Solution ***Step 1.*** Let $p(t) = at^2 + bt + c$ be any polynomial in P_2, where a, b, and c are any real numbers.

Step 2. We must find out whether there are constants c_1 and c_2 such that

$$p(t) = c_1 p_1(t) + c_2 p_2(t)$$

or

$$at^2 + bt + c = c_1(t^2 + 2t + 1) + c_2(t^2 + 2).$$

Thus

$$(c_1 + c_2)t^2 + (2c_1)t + (c_1 + 2c_2) = at^2 + bt + c.$$

Since two polynomials agree for all values of t only if the coefficients of respective powers of t agree, we obtain the linear system

$$\begin{aligned} c_1 + c_2 &= a \\ 2c_1 &= b \\ c_1 + 2c_2 &= c. \end{aligned}$$

Using elementary row operations on the augmented matrix of this linear system, we obtain (verify)

$$\left[\begin{array}{cc:c} 1 & 0 & 2a - c \\ 0 & 1 & c - a \\ 0 & 0 & b - 4a + 2c \end{array}\right].$$

If $b - 4a + 2c \neq 0$, then the system is inconsistent and there is no solution. Hence $S = \{p_1(t), p_2(t)\}$ does not span P_2. For example, the polynomial $3t^2 + 2t - 1$ cannot be written as a linear combination of $p_1(t)$ and $p_2(t)$. ■

EXAMPLE 4

The vectors $\mathbf{e}_1 = \mathbf{i} = (1, 0)$ and $\mathbf{e}_2 = \mathbf{j} = (0, 1)$ span R^2, for as was observed in Sections 4.1 and 4.2, if $\mathbf{u} = (u_1, u_2)$ is any vector in R^2, then $\mathbf{u} = u_1\mathbf{e}_1 + u_2\mathbf{e}_2$. As was noted in Section 4.2, every vector $\mathbf{u}$ in R^3 can be written as a linear combination of the vectors $\mathbf{e}_1 = \mathbf{i} = (1, 0, 0)$, $\mathbf{e}_2 = \mathbf{j} = (0, 1, 0)$, and $\mathbf{e}_3 = \mathbf{k} = (0, 0, 1)$. Thus $\mathbf{e}_1$, $\mathbf{e}_2$, and $\mathbf{e}_3$ span R^3. Similarly, the vectors $\mathbf{e}_1 = (1, 0, \ldots, 0)$, $\mathbf{e}_2 = (0, 1, 0, \ldots, 0)$, $\ldots$, $\mathbf{e}_n = (0, 0, \ldots, 1)$ span R^n, since any vector $\mathbf{u} = (u_1, u_2, \ldots, u_n)$ in R^n can be written as

$$\mathbf{u} = u_1\mathbf{e}_1 + u_2\mathbf{e}_2 + \cdots + u_n\mathbf{e}_n.$$

■

EXAMPLE 5

The set $S = \{t^n, t^{n-1}, \ldots, t, 1\}$ spans P_n, since every polynomial in P_n is of the form

$$a_0 t^n + a_1 t^{n-1} + \cdots + a_{n-1}t + a_n,$$

which is a linear combination of the elements in S. ■

EXAMPLE 6

Consider the homogeneous linear system $A\mathbf{x} = \mathbf{0}$, where

$$A = \begin{bmatrix} 1 & 1 & 0 & 2 \\ -2 & -2 & 1 & -5 \\ 1 & 1 & -1 & 3 \\ 4 & 4 & -1 & 9 \end{bmatrix}.$$

From Example 9 in Section 6.2, the set of all solutions to $A\mathbf{x} = \mathbf{0}$ forms a subspace of R^4. To determine a spanning set for the solution space of this homogeneous system, we find that the reduced row echelon form of the augmented matrix is (verify)

$$\left[\begin{array}{cccc|c} 1 & 1 & 0 & 2 & 0 \\ 0 & 0 & 1 & -1 & 0 \\ 0 & 0 & 0 & 0 & 0 \\ 0 & 0 & 0 & 0 & 0 \end{array}\right].$$

The general solution is then given by

$$\begin{aligned} x_1 &= -r - 2s \\ x_2 &= r \\ x_3 &= s \\ x_4 &= s, \end{aligned}$$

where r and s are any real numbers. In matrix form we have that any member of the solution space is given by

$$\mathbf{x} = r\begin{bmatrix} -1 \\ 1 \\ 0 \\ 0 \end{bmatrix} + s\begin{bmatrix} -2 \\ 0 \\ 1 \\ 1 \end{bmatrix}.$$

Hence the vectors $\begin{bmatrix} -1 \\ 1 \\ 0 \\ 0 \end{bmatrix}$ and $\begin{bmatrix} -2 \\ 0 \\ 1 \\ 1 \end{bmatrix}$ span the solution space. ■

LINEAR INDEPENDENCE

DEFINITION The vectors $\mathbf{v}_1, \mathbf{v}_2, \ldots, \mathbf{v}_k$ in a vector space V are said to be **linearly dependent** if there exist constants $c_1, c_2, \ldots, c_k$, not all zero, such that

$$c_1\mathbf{v}_1 + c_2\mathbf{v}_2 + \cdots + c_k\mathbf{v}_k = \mathbf{0}. \tag{1}$$

Otherwise, $\mathbf{v}_1, \mathbf{v}_2, \ldots, \mathbf{v}_k$ are called **linearly independent**. That is, $\mathbf{v}_1, \mathbf{v}_2, \ldots, \mathbf{v}_k$ are linearly independent if whenever $c_1\mathbf{v}_1 + c_2\mathbf{v}_2 + \cdots + c_k\mathbf{v}_k = \mathbf{0}$, we must have

$$c_1 = c_2 = \cdots = c_k = 0.$$

That is, the *only* linear combination of $\mathbf{v}_1, \mathbf{v}_2, \ldots, \mathbf{v}_k$ that yields the zero vector is that in which all the coefficients are zero. If $S = \{\mathbf{v}_1, \mathbf{v}_2, \ldots, \mathbf{v}_k\}$, then we also say that the set S is **linearly dependent** or **linearly independent** if the vectors have the corresponding property defined previously.

It should be emphasized that for any vectors $\mathbf{v}_1, \mathbf{v}_2, \ldots, \mathbf{v}_k$, Equation (1) always holds if we choose all the scalars $c_1, c_2, \ldots, c_k$ equal to zero. The important point in this definition is whether or not it is possible to satisfy (1) with at least one of the scalars different from zero.

The procedure to determine if the vectors $\mathbf{v}_1, \mathbf{v}_2, \ldots, \mathbf{v}_k$ are linearly dependent or linearly independent is as follows.

Step 1. Form Equation (1), which leads to a homogeneous system.

Step 2. If the homogeneous system obtained in Step 1 has only the trivial solution, then the given vectors are linearly independent; if it has a nontrivial solution, then the vectors are linearly dependent.

EXAMPLE 7 Determine whether the vectors

$$\begin{bmatrix} -1 \\ 1 \\ 0 \\ 0 \end{bmatrix} \quad \text{and} \quad \begin{bmatrix} -2 \\ 0 \\ 1 \\ 1 \end{bmatrix}$$

found in Example 6 as spanning the solution space of $A\mathbf{x} = \mathbf{0}$ are linearly dependent or linearly independent.

Solution Forming Equation (1),

$$c_1 \begin{bmatrix} -1 \\ 1 \\ 0 \\ 0 \end{bmatrix} + c_2 \begin{bmatrix} -2 \\ 0 \\ 1 \\ 1 \end{bmatrix} = \begin{bmatrix} 0 \\ 0 \\ 0 \\ 0 \end{bmatrix},$$

we obtain the homogeneous system

$$\begin{aligned} -c_1 - 2c_2 &= 0 \\ c_1 + 0c_2 &= 0 \\ 0c_1 + c_2 &= 0 \\ 0c_1 + c_2 &= 0, \end{aligned}$$

whose only solution is $c_1 = c_2 = 0$. Hence the given vectors are linearly independent. ■

EXAMPLE 8 Are the vectors $\mathbf{v}_1 = (1, 0, 1, 2)$, $\mathbf{v}_2 = (0, 1, 1, 2)$, and $\mathbf{v}_3 = (1, 1, 1, 3)$ in R^4 linearly dependent or linearly independent?

Solution We form Equation (1),

$$c_1\mathbf{v}_1 + c_2\mathbf{v}_2 + c_3\mathbf{v}_3 = \mathbf{0},$$

and solve for c_1, c_2, and c_3. The resulting homogeneous system is (verify)

$$\begin{aligned} c_1 + c_3 &= 0 \\ c_2 + c_3 &= 0 \\ c_1 + c_2 + c_3 &= 0 \\ 2c_1 + 2c_2 + 3c_3 &= 0, \end{aligned}$$

which has as its only solution $c_1 = c_2 = c_3 = 0$ (verify), showing that the given vectors are linearly independent. ■

EXAMPLE 9 Consider the vectors

$$\mathbf{v}_1 = (1, 2, -1), \qquad \mathbf{v}_2 = (1, -2, 1), \qquad \mathbf{v}_3 = (-3, 2, -1),$$

and

$$\mathbf{v}_4 = (2, 0, 0) \quad \text{in } R^3.$$

Is $S = \{\mathbf{v}_1, \mathbf{v}_2, \mathbf{v}_3, \mathbf{v}_4\}$ linearly dependent or linearly independent?

Solution Setting up Equation (1), we are led to the homogeneous system (verify)

$$\begin{aligned} c_1 + c_2 - 3c_3 + 2c_4 &= 0 \\ 2c_1 - 2c_2 + 2c_3 &= 0 \\ -c_1 + c_2 - c_3 &= 0, \end{aligned}$$

a homogeneous system of three equations in four unknowns. By Theorem 1.8, Section 1.6, we are assured of the existence of a nontrivial solution. Hence, S is linearly dependent. In fact, two of the infinitely many solutions are

$$\begin{aligned} &c_1 = 1, \quad c_2 = 2, \quad c_3 = 1, \quad c_4 = 0; \\ &c_1 = 1, \quad c_2 = 1, \quad c_3 = 0, \quad c_4 = -1. \end{aligned}$$

■

EXAMPLE 10 The vectors $\mathbf{e}_1$ and $\mathbf{e}_2$ in R^2, defined in Example 4, are linearly independent, since

$$c_1(1, 0) + c_2(0, 1) = (0, 0)$$

can hold only if $c_1 = c_2 = 0$. Similarly, the vectors $\mathbf{e}_1$, $\mathbf{e}_2$, and $\mathbf{e}_3$ in R^3, and more generally, the vectors $\mathbf{e}_1, \mathbf{e}_2, \dots, \mathbf{e}_n$ in R^n are linearly independent (Exercise T.1). ■

Corollary 6.4 in Section 6.6, to follow, gives another way of testing whether n given vectors in R^n are linearly dependent or linearly independent. We form the matrix A, whose columns are the given n vectors. Then the given vectors are linearly independent if and only if $\det(A) \neq 0$. Thus, in Example 10,

$$A = \begin{bmatrix} 1 & 0 \\ 0 & 1 \end{bmatrix}$$

and $\det(A) = 1$ so that $\mathbf{e}_1$ and $\mathbf{e}_2$ are linearly independent.

EXAMPLE 11 Consider the vectors

$$p_1(t) = t^2 + t + 2, \qquad p_2(t) = 2t^2 + t, \qquad p_3(t) = 3t^2 + 2t + 2.$$

To find out whether $S = \{p_1(t), p_2(t), p_3(t)\}$ is linearly dependent or linearly independent, we set up Equation (1) and solve for c_1, c_2, and c_3. The resulting homogeneous system is (verify)

$$\begin{aligned} c_1 + 2c_2 + 3c_3 &= 0 \\ c_1 + c_2 + 2c_3 &= 0 \\ 2c_1 + 2c_3 &= 0, \end{aligned}$$

which has infinitely many solutions (verify). A particular solution is $c_1 = 1$, $c_2 = 1$, $c_3 = -1$, so

$$p_1(t) + p_2(t) - p_3(t) = 0.$$

Hence S is linearly dependent. ■

EXAMPLE 12 If $\mathbf{v}_1, \mathbf{v}_2, \dots, \mathbf{v}_k$ are k vectors in any vector space and $\mathbf{v}_i$ is the zero vector, Equation (1) holds by letting $c_i = 1$ and $c_j = 0$ for $j \neq i$. Thus $S = \{\mathbf{v}_1, \mathbf{v}_2, \dots, \mathbf{v}_k\}$ is linearly dependent. ■

Example 12 shows that *every set of vectors containing the zero vector is linearly dependent.*

Let S_1 and S_2 be finite subsets of a vector space and let S_1 be a subset of S_2. Then (a) if S_1 is linearly dependent, so is S_2; and (b) if S_2 is linearly independent, so is S_1 (Exercise T.2).

We consider next the meaning of linear independence in R^2 and R^3. Suppose that $\mathbf{v}_1$ and $\mathbf{v}_2$ are linearly dependent in R^2. Then there exist scalars c_1 and c_2, not both zero, such that

$$c_1\mathbf{v}_1 + c_2\mathbf{v}_2 = \mathbf{0}.$$

If $c_1 \neq 0$, then

$$\mathbf{v}_1 = \left(-\frac{c_2}{c_1}\right)\mathbf{v}_2.$$

If $c_2 \neq 0$, then

$$\mathbf{v}_2 = \left(-\frac{c_1}{c_2}\right)\mathbf{v}_1.$$

Thus one of the vectors is a scalar multiple of the other. Conversely, suppose that $\mathbf{v}_1 = c\mathbf{v}_2$. Then

$$1\mathbf{v}_1 - c\mathbf{v}_2 = \mathbf{0},$$

and since the coefficients of $\mathbf{v}_1$ and $\mathbf{v}_2$ are not both zero, it follows that $\mathbf{v}_1$ and $\mathbf{v}_2$ are linearly dependent. Thus $\mathbf{v}_1$ and $\mathbf{v}_2$ are linearly dependent in R^2 if and only if one of the vectors is a multiple of the other. Geometrically, two vectors in R^2 are linearly dependent if and only if they both lie on the same line passing through the origin [Figure 6.7(a)].

Figure 6.7 ►

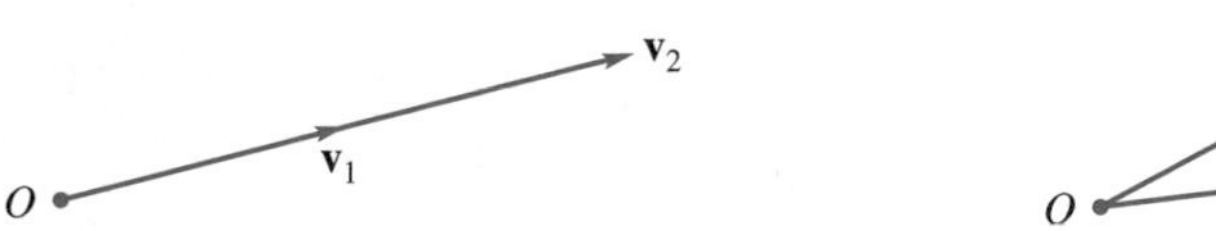

(a) Linearly dependent vectors in R^2 (b) Linearly independent vectors in R^2

Suppose now that $\mathbf{v}_1$, $\mathbf{v}_2$, and $\mathbf{v}_3$ are linearly dependent in R^3. Then we can write

$$c_1\mathbf{v}_1 + c_2\mathbf{v}_2 + c_3\mathbf{v}_3 = \mathbf{0},$$

where c_1, c_2, and c_3 are not all zero, say $c_2 \neq 0$. Then

$$\mathbf{v}_2 = \left(-\frac{c_1}{c_2}\right)\mathbf{v}_1 - \left(\frac{c_3}{c_2}\right)\mathbf{v}_3,$$

which means that $\mathbf{v}_2$ is in the subspace W spanned by $\mathbf{v}_1$ and $\mathbf{v}_3$.

Now W is either a plane through the origin (when $\mathbf{v}_1$ and $\mathbf{v}_3$ are linearly independent), or a line through the origin (when $\mathbf{v}_1$ and $\mathbf{v}_3$ are linearly dependent), or the origin (when $\mathbf{v}_1 = \mathbf{v}_2 = \mathbf{v}_3 = \mathbf{0}$). Since a line through the origin always lies in a plane through the origin, we conclude that $\mathbf{v}_1$, $\mathbf{v}_2$, and $\mathbf{v}_3$ all lie in the same plane through the origin. Conversely, suppose that $\mathbf{v}_1$, $\mathbf{v}_2$, and $\mathbf{v}_3$ all lie in the same plane through the origin. Then either all three vectors are the zero vector, or all three vectors lie on the same line through the origin, or all three vectors lie in a plane through the origin spanned by two vectors, say $\mathbf{v}_1$ and $\mathbf{v}_3$. Thus, in all these cases, $\mathbf{v}_2$ is a linear combination of $\mathbf{v}_1$ and $\mathbf{v}_3$:

$$\mathbf{v}_2 = a_1\mathbf{v}_1 + a_3\mathbf{v}_3.$$

Then

$$a_1\mathbf{v}_1 - 1\mathbf{v}_2 + a_3\mathbf{v}_3 = \mathbf{0},$$

which means that $\mathbf{v}_1$, $\mathbf{v}_2$, and $\mathbf{v}_3$ are linearly dependent. Hence three vectors in R^3 are linearly dependent if and only if they all lie in the same plane passing through the origin [Figure 6.8(a)].

Figure 6.8 ►

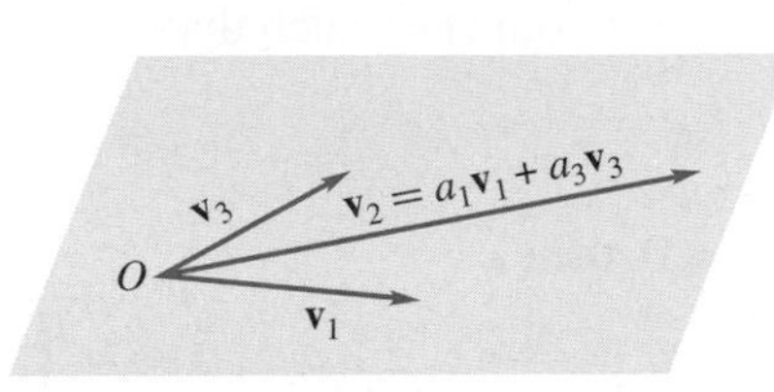

(a) Linearly dependent vectors in R^3

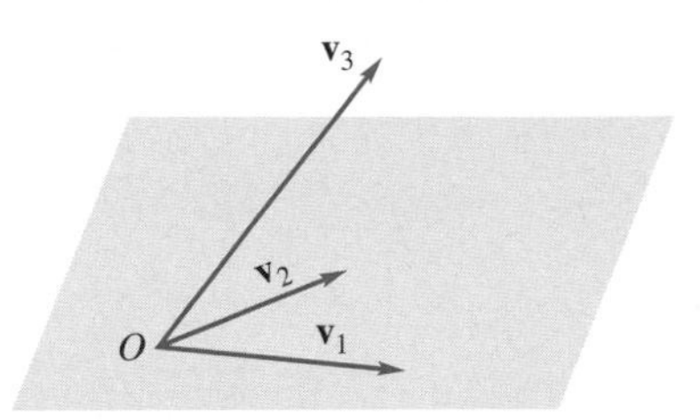

(b) Linearly independent vectors in R^3

More generally, let $\mathbf{u}$ and $\mathbf{v}$ be nonzero vectors in a vector space V. We can show (Exercise T.13) that $\mathbf{u}$ and $\mathbf{v}$ are linearly dependent if and only if there is a scalar k such that $\mathbf{v} = k\mathbf{u}$. Equivalently, $\mathbf{u}$ and $\mathbf{v}$ are linearly independent if and only if neither vector is a multiple of the other. This approach will not work with sets having three or more vectors. Instead, we use the result given by the following theorem.

THEOREM 6.4 *The nonzero vectors $\mathbf{v}_1, \mathbf{v}_2, \ldots, \mathbf{v}_n$ in a vector space V are linearly dependent if and only if one of the vectors $\mathbf{v}_j$, $j \geq 2$, is a linear combination of the preceding vectors $\mathbf{v}_1, \mathbf{v}_2, \ldots, \mathbf{v}_{j-1}$.*

Proof If $\mathbf{v}_j$ is a linear combination of $\mathbf{v}_1, \mathbf{v}_2, \ldots, \mathbf{v}_{j-1}$,

$$\mathbf{v}_j = c_1\mathbf{v}_1 + c_2\mathbf{v}_2 + \cdots + c_{j-1}\mathbf{v}_{j-1},$$

then

$$c_1\mathbf{v}_1 + c_2\mathbf{v}_2 + \cdots + c_{j-1}\mathbf{v}_{j-1} + (-1)\mathbf{v}_j + 0\mathbf{v}_{j+1} + \cdots + 0\mathbf{v}_n = \mathbf{0}.$$

Since at least one coefficient, -1, is nonzero, we conclude that $\mathbf{v}_1, \mathbf{v}_2, \ldots, \mathbf{v}_n$ are linearly dependent.

Conversely, suppose that $\mathbf{v}_1, \mathbf{v}_2, \ldots, \mathbf{v}_n$ are linearly dependent. Then there exist scalars $c_1, c_2, \ldots, c_n$, not all zero, such that

$$c_1\mathbf{v}_1 + c_2\mathbf{v}_2 + \cdots + c_n\mathbf{v}_n = \mathbf{0}.$$

Now let j be the largest subscript for which $c_j \neq 0$. If $j > 1$, then

$$\mathbf{v}_j = -\left(\frac{c_1}{c_j}\right)\mathbf{v}_1 - \left(\frac{c_2}{c_j}\right)\mathbf{v}_2 - \cdots - \left(\frac{c_{j-1}}{c_j}\right)\mathbf{v}_{j-1}.$$

If $j = 1$, then $c_1\mathbf{v}_1 = \mathbf{0}$, which implies that $\mathbf{v}_1 = \mathbf{0}$, a contradiction to the hypothesis that none of the vectors are the zero vector. Thus one of the vectors $\mathbf{v}_j$ is a linear combination of the preceding vectors $\mathbf{v}_1, \mathbf{v}_2, \ldots, \mathbf{v}_{j-1}$. ■

EXAMPLE 13 If $\mathbf{v}_1$, $\mathbf{v}_2$, $\mathbf{v}_3$, and $\mathbf{v}_4$ are as in Example 9, then we find (verify) that

$$\mathbf{v}_1 + \mathbf{v}_2 + 0\mathbf{v}_3 - \mathbf{v}_4 = \mathbf{0},$$

so $\mathbf{v}_1$, $\mathbf{v}_2$, $\mathbf{v}_3$, and $\mathbf{v}_4$ are linearly dependent. We then have

$$\mathbf{v}_4 = \mathbf{v}_1 + \mathbf{v}_2.$$

■

Remarks

1. We observe that Theorem 6.4 does not say that *every* vector $\mathbf{v}$ is a linear combination of the preceding vectors. Thus, in Example 9, we also have the equation $\mathbf{v}_1 + 2\mathbf{v}_2 + \mathbf{v}_3 + 0\mathbf{v}_4 = \mathbf{0}$. We cannot solve, in this equation, for $\mathbf{v}_4$ as a linear combination of $\mathbf{v}_1$, $\mathbf{v}_2$, and $\mathbf{v}_3$, since its coefficient is zero.
2. We can also prove that if $S = \{\mathbf{v}_1, \mathbf{v}_2, \ldots, \mathbf{v}_k\}$ is a set of vectors in a vector space V, then S is linearly dependent if and only if one of the vectors in S is a linear combination of all the other vectors in S (see Exercise T.3). For instance, in Example 13,
$$\mathbf{v}_1 = -\mathbf{v}_2 - 0\mathbf{v}_3 + \mathbf{v}_4 \quad \text{and} \quad \mathbf{v}_2 = -\tfrac{1}{2}\mathbf{v}_1 - \tfrac{1}{2}\mathbf{v}_3 - 0\mathbf{v}_4.$$
3. Observe that if $\mathbf{v}_1, \mathbf{v}_2, \ldots, \mathbf{v}_k$ are linearly independent vectors in a vector space, then they must be distinct and none can be the zero vector.

The following result will be used in Section 6.4 as well as in several other places. Suppose that $S = \{\mathbf{v}_1, \mathbf{v}_2, \ldots, \mathbf{v}_n\}$ spans a vector space V and $\mathbf{v}_j$ is a linear combination of the preceding vectors in S. Then the set

$$S_1 = \{\mathbf{v}_1, \mathbf{v}_2, \ldots, \mathbf{v}_{j-1}, \mathbf{v}_{j+1}, \ldots, \mathbf{v}_n\},$$

consisting of S with $\mathbf{v}_j$ deleted, also spans V. To show this result, observe that if $\mathbf{v}$ is any vector in V, then, since S spans V, we can find scalars $a_1, a_2, \ldots, a_n$ such that

$$\mathbf{v} = a_1\mathbf{v}_1 + a_2\mathbf{v}_2 + \cdots + a_{j-1}\mathbf{v}_{j-1} + a_j\mathbf{v}_j + a_{j+1}\mathbf{v}_{j+1} + \cdots + a_n\mathbf{v}_n.$$

Now if

$$\mathbf{v}_j = b_1\mathbf{v}_1 + b_2\mathbf{v}_2 + \cdots + b_{j-1}\mathbf{v}_{j-1},$$

then

$$\begin{aligned}\mathbf{v} &= a_1\mathbf{v}_1 + a_2\mathbf{v}_2 + \cdots + a_{j-1}\mathbf{v}_{j-1} + a_j(b_1\mathbf{v}_1 + b_2\mathbf{v}_2 + \cdots + b_{j-1}\mathbf{v}_{j-1}) \\ &\quad + a_{j+1}\mathbf{v}_{j+1} + \cdots + a_n\mathbf{v}_n \\ &= c_1\mathbf{v}_1 + c_2\mathbf{v}_2 + \cdots + c_{j-1}\mathbf{v}_{j-1} + c_{j+1}\mathbf{v}_{j+1} + \cdots + c_n\mathbf{v}_n,\end{aligned}$$

which means that span $S_1 = V$.

EXAMPLE 14 Consider the set of vectors $S = \{\mathbf{v}_1, \mathbf{v}_2, \mathbf{v}_3, \mathbf{v}_4\}$ in R^4, where

$$\mathbf{v}_1 = \begin{bmatrix}1\\1\\0\\0\end{bmatrix}, \quad \mathbf{v}_2 = \begin{bmatrix}1\\0\\1\\0\end{bmatrix}, \quad \mathbf{v}_3 = \begin{bmatrix}0\\1\\1\\0\end{bmatrix}, \quad \text{and} \quad \mathbf{v}_4 = \begin{bmatrix}2\\1\\1\\0\end{bmatrix},$$

and let $W = \text{span } S$. Since $\mathbf{v}_4 = \mathbf{v}_1 + \mathbf{v}_2$, we conclude that $W = \text{span } S_1$, where $S_1 = \{\mathbf{v}_1, \mathbf{v}_2, \mathbf{v}_3\}$. ■

SPAN AND LINEAR INDEPENDENCE IN B^n (OPTIONAL)

The concepts of linear independence, linear dependence, and span are valid regardless of the nature of the scalars or vectors of a vector space. For the vector space B^n we recall that only 0 and 1 are permitted as scalars and all the arithmetic operations are performed using binary arithmetic.

EXAMPLE 15 Let V be the vector space B^3 and let

$$\mathbf{v}_1 = \begin{bmatrix} 1 \\ 1 \\ 0 \end{bmatrix}, \quad \mathbf{v}_2 = \begin{bmatrix} 1 \\ 0 \\ 1 \end{bmatrix}, \quad \text{and} \quad \mathbf{v}_3 = \begin{bmatrix} 0 \\ 1 \\ 1 \end{bmatrix}.$$

Do $\mathbf{v}_1$, $\mathbf{v}_2$, and $\mathbf{v}_3$ span V?

Solution Let $\mathbf{v} = \begin{bmatrix} a \\ b \\ c \end{bmatrix}$ be any vector in B^3, where a, b, and c are any of the bits 0 or 1. We must determine if there are scalars c_1, c_2, and c_3 (which are bits 0 or 1) such that

$$c_1\mathbf{v}_1 + c_2\mathbf{v}_2 + c_3\mathbf{v}_3 = \mathbf{v}.$$

This leads to the linear system

$$\begin{aligned} c_1 + c_2 &= a \\ c_1 + c_3 &= b \\ c_2 + c_3 &= c. \end{aligned}$$

We form the augmented matrix and obtain its reduced row echelon form (verify):

$$\left[\begin{array}{ccc|c} 1 & 0 & 1 & a + a + b \\ 0 & 1 & 1 & a + b \\ 0 & 0 & 0 & a + b + c \end{array}\right].$$

The system is inconsistent if the choice of bits for a, b, and c are such that $a + b + c \neq 0$. For example, if $\mathbf{v} = \begin{bmatrix} 1 \\ 1 \\ 1 \end{bmatrix}$, then the system is inconsistent; hence $\mathbf{v}_1$, $\mathbf{v}_2$, and $\mathbf{v}_3$ do not span V. ■

EXAMPLE 16 Are the vectors $\mathbf{v}_1$, $\mathbf{v}_2$, and $\mathbf{v}_3$ in Example 15 linearly independent?

Solution Setting up Equation (1), we are led to the homogeneous linear system

$$\begin{aligned} c_1 + c_2 &= 0 \\ c_1 + c_3 &= 0 \\ c_2 + c_3 &= 0. \end{aligned}$$

The reduced row echelon form of this system is (verify)

$$\left[\begin{array}{ccc|c} 1 & 0 & 1 & 0 \\ 0 & 1 & 1 & 0 \\ 0 & 0 & 0 & 0 \end{array}\right].$$

Thus $c_1 = -c_3$ and $c_2 = -c_3$, where c_3 is either 0 or 1. Choosing $c_3 = 1$, we find a nontrivial solution $c_1 = c_2 = c_3 = 1$ (verify). Thus $\mathbf{v}_1$, $\mathbf{v}_2$, and $\mathbf{v}_3$ are linearly dependent. ■

Key Terms

Span
Linearly dependent
Linearly independent

6.3 Exercises

1. Which of the following vectors span R^2?

(a) $(1, 2), (-1, 1)$

(b) $(0, 0), (1, 1), (-2, -2)$

(c) $(1, 3), (2, -3), (0, 2)$

(d) $(2, 4), (-1, 2)$

2. Which of the following sets of vectors span R^3?

(a) $\{(1, -1, 2), (0, 1, 1)\}$

(b) $\{(1, 2, -1), (6, 3, 0), (4, -1, 2), (2, -5, 4)\}$

(c) $\{(2, 2, 3), (-1, -2, 1), (0, 1, 0)\}$

(d) $\{(1, 0, 0), (0, 1, 0), (0, 0, 1), (1, 1, 1)\}$

3. Which of the following vectors span R^4?

(a) $(1, 0, 0, 1), (0, 1, 0, 0), (1, 1, 1, 1), (1, 1, 1, 0)$

(b) $(1, 2, 1, 0), (1, 1, -1, 0), (0, 0, 0, 1)$

(c) $(6, 4, -2, 4), (2, 0, 0, 1), (3, 2, -1, 2), (5, 6, -3, 2), (0, 4, -2, -1)$

(d) $(1, 1, 0, 0), (1, 2, -1, 1), (0, 0, 1, 1), (2, 1, 2, 1)$

4. Which of the following sets of polynomials span P_2?

(a) $\{t^2 + 1, t^2 + t, t + 1\}$

(b) $\{t^2 + 1, t - 1, t^2 + t\}$

(c) $\{t^2 + 2, 2t^2 - t + 1, t + 2, t^2 + t + 4\}$

(d) $\{t^2 + 2t - 1, t^2 - 1\}$

5. Do the polynomials $t^3 + 2t + 1$, $t^2 - t + 2$, $t^3 + 2$, $-t^3 + t^2 - 5t + 2$ span P_3?

6. Find a set of vectors spanning the solution space of $A\mathbf{x} = \mathbf{0}$, where

$$A = \begin{bmatrix} 1 & 0 & 1 & 0 \\ 1 & 2 & 3 & 1 \\ 2 & 1 & 3 & 1 \\ 1 & 1 & 2 & 1 \end{bmatrix}.$$

7. Find a set of vectors spanning the null space of

$$A = \begin{bmatrix} 1 & 1 & 2 & -1 \\ 2 & 3 & 6 & -2 \\ -2 & 1 & 2 & 2 \\ 0 & -2 & -4 & 0 \end{bmatrix}.$$

8. Let

$$\mathbf{x}_1 = \begin{bmatrix} 2 \\ -1 \\ 1 \end{bmatrix}, \quad \mathbf{x}_2 = \begin{bmatrix} 4 \\ -7 \\ -1 \end{bmatrix}, \quad \mathbf{x}_3 = \begin{bmatrix} 1 \\ 2 \\ 2 \end{bmatrix}$$

belong to the solution space of $A\mathbf{x} = \mathbf{0}$. Is $\{\mathbf{x}_1, \mathbf{x}_2, \mathbf{x}_3\}$ linearly independent?

9. Let

$$\mathbf{x}_1 = \begin{bmatrix} 1 \\ 2 \\ 0 \\ 1 \end{bmatrix}, \quad \mathbf{x}_2 = \begin{bmatrix} 1 \\ 0 \\ -1 \\ 1 \end{bmatrix}, \quad \mathbf{x}_3 = \begin{bmatrix} 1 \\ 6 \\ 2 \\ 0 \end{bmatrix}$$

belong to the null space of A. Is $\{\mathbf{x}_1, \mathbf{x}_2, \mathbf{x}_3\}$ linearly independent?

10. Which of the following sets of vectors in R^3 are linearly dependent? For those that are, express one vector as a linear combination of the rest.

(a) $\{(1, 2, -1), (3, 2, 5)\}$

(b) $\{(4, 2, 1), (2, 6, -5), (1, -2, 3)\}$

(c) $\{(1, 1, 0), (0, 2, 3), (1, 2, 3), (3, 6, 6)\}$

(d) $\{(1, 2, 3), (1, 1, 1), (1, 0, 1)\}$

11. Consider the vector space R^4. Follow the directions of Exercise 10.

(a) $\{(1, 1, 2, 1), (1, 0, 0, 2), (4, 6, 8, 6), (0, 3, 2, 1)\}$

(b) $\{(1, -2, 3, -1), (-2, 4, -6, 2)\}$

(c) $\{(1, 1, 1, 1), (2, 3, 1, 2), (3, 1, 2, 1), (2, 2, 1, 1)\}$

(d) $\{(4, 2, -1, 3), (6, 5, -5, 1), (2, -1, 3, 5)\}$

12. Consider the vector space P_2. Follow the directions of Exercise 10.

(a) $\{t^2 + 1, t - 2, t + 3\}$

(b) $\{2t^2 + 1, t^2 + 3, t\}$

(c) $\{3t + 1, 3t^2 + 1, 2t^2 + t + 1\}$

(d) $\{t^2 - 4, 5t^2 - 5t - 6, 3t^2 - 5t + 2\}$

13. Consider the vector space M_{22}. Follow the directions of Exercise 10.

(a) $\left\{\begin{bmatrix} 1 & 1 \\ 1 & 2 \end{bmatrix}, \begin{bmatrix} 1 & 0 \\ 0 & 2 \end{bmatrix}, \begin{bmatrix} 0 & 3 \\ 1 & 2 \end{bmatrix}, \begin{bmatrix} 2 & 6 \\ 4 & 6 \end{bmatrix}\right\}$

(b) $\left\{\begin{bmatrix} 1 & 1 \\ 1 & 1 \end{bmatrix}, \begin{bmatrix} 1 & 0 \\ 0 & 2 \end{bmatrix}, \begin{bmatrix} 0 & 1 \\ 0 & 2 \end{bmatrix}\right\}$

(c) $\left\{\begin{bmatrix} 1 & 1 \\ 1 & 1 \end{bmatrix}, \begin{bmatrix} 2 & 3 \\ 1 & 2 \end{bmatrix}, \begin{bmatrix} 3 & 1 \\ 2 & 1 \end{bmatrix}, \begin{bmatrix} 2 & 2 \\ 1 & 1 \end{bmatrix}\right\}$

14. Let V be the vector space of all real-valued continuous functions. Follow the directions of Exercise 10.

(a) $\{\cos t, \sin t, e^t\}$ (b) $\{t, e^t, \sin t\}$

(c) $\{t^2, t, e^t\}$ (d) $\{\cos^2 t, \sin^2 t, \cos 2t\}$

15. For what values of c are the vectors $(-1, 0, -1)$, $(2, 1, 2)$, and $(1, 1, c)$ in R^3 linearly dependent?

16. For what values of λ are the vectors $t + 3$ and $2t + \lambda^2 + 2$ in P_1 linearly dependent?

17. Determine if the vectors $\begin{bmatrix} 1 \\ 1 \\ 0 \end{bmatrix}$, $\begin{bmatrix} 1 \\ 0 \\ 1 \end{bmatrix}$, and $\begin{bmatrix} 1 \\ 1 \\ 1 \end{bmatrix}$ span B^3.

18. Determine if the vectors $\begin{bmatrix} 0 \\ 1 \\ 1 \end{bmatrix}$, $\begin{bmatrix} 1 \\ 0 \\ 0 \end{bmatrix}$, and $\begin{bmatrix} 1 \\ 1 \\ 0 \end{bmatrix}$ span B^3.

19. Determine if the vectors $\begin{bmatrix}1\\1\\0\\0\end{bmatrix}, \begin{bmatrix}0\\1\\1\\0\end{bmatrix}, \begin{bmatrix}0\\0\\1\\1\end{bmatrix}$, and $\begin{bmatrix}1\\0\\0\\1\end{bmatrix}$ span B^4.

20. Determine if the vectors in Exercise 17 are linearly independent.

21. Determine if the vectors in Exercise 19 are linearly independent.

22. Show that $\mathbf{v}_1$, $\mathbf{v}_2$, and $\mathbf{v}_3$ in Example 15 are linearly dependent using Theorem 6.4.

Theoretical Exercises

T.1. Show that the vectors $\mathbf{e}_1, \mathbf{e}_2, \dots, \mathbf{e}_n$ in R^n are linearly independent.

T.2. Let S_1 and S_2 be finite subsets of a vector space and let S_1 be a subset of S_2. Show that:

(a) If S_1 is linearly dependent, so is S_2.

(b) If S_2 is linearly independent, so is S_1.

T.3. Let $S = \{\mathbf{v}_1, \mathbf{v}_2, \dots, \mathbf{v}_k\}$ be a set of vectors in a vector space. Show that S is linearly dependent if and only if one of the vectors in S is a linear combination of all the other vectors in S.

T.4. Suppose that $S = \{\mathbf{v}_1, \mathbf{v}_2, \mathbf{v}_3\}$ is a linearly independent set of vectors in a vector space V. Show that $T = \{\mathbf{w}_1, \mathbf{w}_2, \mathbf{w}_3\}$ is also linearly independent, where $\mathbf{w}_1 = \mathbf{v}_1 + \mathbf{v}_2 + \mathbf{v}_3$, $\mathbf{w}_2 = \mathbf{v}_2 + \mathbf{v}_3$, and $\mathbf{w}_3 = \mathbf{v}_3$.

T.5. Suppose that $S = \{\mathbf{v}_1, \mathbf{v}_2, \mathbf{v}_3\}$ is a linearly independent set of vectors in a vector space V. Is $T = \{\mathbf{w}_1, \mathbf{w}_2, \mathbf{w}_3\}$, where $\mathbf{w}_1 = \mathbf{v}_1 + \mathbf{v}_2$, $\mathbf{w}_2 = \mathbf{v}_1 + \mathbf{v}_3$, $\mathbf{w}_3 = \mathbf{v}_2 + \mathbf{v}_3$, linearly dependent or linearly independent? Justify your answer.

T.6. Suppose that $S = \{\mathbf{v}_1, \mathbf{v}_2, \mathbf{v}_3\}$ is a linearly dependent set of vectors in a vector space V. Is $T = \{\mathbf{w}_1, \mathbf{w}_2, \mathbf{w}_3\}$, where $\mathbf{w}_1 = \mathbf{v}_1$, $\mathbf{w}_2 = \mathbf{v}_1 + \mathbf{v}_2$, $\mathbf{w}_3 = \mathbf{v}_1 + \mathbf{v}_2 + \mathbf{v}_3$, linearly dependent or linearly independent? Justify your answer.

T.7. Let $\mathbf{v}_1$, $\mathbf{v}_2$, and $\mathbf{v}_3$ be vectors in a vector space such that $\{\mathbf{v}_1, \mathbf{v}_2\}$ is linearly independent. Show that if $\mathbf{v}_3$ does not belong to span $\{\mathbf{v}_1, \mathbf{v}_2\}$, then $\{\mathbf{v}_1, \mathbf{v}_2, \mathbf{v}_3\}$ is linearly independent.

T.8. Let A be an $m \times n$ matrix in reduced row echelon form. Show that the nonzero rows of A, viewed as vectors in R^n, form a linearly independent set of vectors.

T.9. Let $S = \{\mathbf{u}_1, \mathbf{u}_2, \dots, \mathbf{u}_k\}$ be a set of vectors in a vector space, and let $T = \{\mathbf{v}_1, \mathbf{v}_2, \dots, \mathbf{v}_m\}$, where each $\mathbf{v}_i$, $i = 1, 2, \dots, m$, is a linear combination of the vectors in S. Show that

$$\mathbf{w} = b_1\mathbf{v}_1 + b_2\mathbf{v}_2 + \cdots + b_m\mathbf{v}_m$$

is a linear combination of the vectors in S.

T.10. Suppose that $\{\mathbf{v}_1, \mathbf{v}_2, \dots, \mathbf{v}_n\}$ is a linearly independent set of vectors in R^n. Show that if A is an $n \times n$ nonsingular matrix, then $\{A\mathbf{v}_1, A\mathbf{v}_2, \dots, A\mathbf{v}_n\}$ is linearly independent.

T.11. Let S_1 and S_2 be finite subsets of a vector space V and let S_1 be a subset of S_2. If S_2 is linearly dependent, show by examples that S_1 may be either linearly dependent or linearly independent.

T.12. Let S_1 and S_2 be finite subsets of a vector space and let S_1 be a subset of S_2. If S_1 is linearly independent, show by examples that S_2 may be either linearly dependent or linearly independent.

T.13. Let $\mathbf{u}$ and $\mathbf{v}$ be nonzero vectors in a vector space V. Show that $\{\mathbf{u}, \mathbf{v}\}$ is linearly dependent if and only if there is a scalar k such that $\mathbf{v} = k\mathbf{u}$. Equivalently, $\{\mathbf{u}, \mathbf{v}\}$ is linearly independent if and only if one of the vectors is not a multiple of the other.

T.14. ***(Uses material from Section 5.1)*** Let $\mathbf{u}$ and $\mathbf{v}$ be linearly independent vectors in R^3. Show that $\mathbf{u}$, $\mathbf{v}$, and $\mathbf{u} \times \mathbf{v}$ form a basis for R^3. [*Hint*: Form Equation (1) and take the dot product with $\mathbf{u} \times \mathbf{v}$.]

T.15. Let W be a subspace of V spanned by the vectors $\mathbf{w}_1, \mathbf{w}_2, \dots, \mathbf{w}_k$. Is there any vector $\mathbf{v}$ in V such that the span of $\{\mathbf{w}_1, \mathbf{w}_2, \dots, \mathbf{w}_k, \mathbf{v}\}$ will also be W? Describe all such vectors $\mathbf{v}$.

MATLAB Exercises

ML.1. Determine if S is linearly independent or linearly dependent.

(a) $S = \{(1, 0, 0, 1), (0, 1, 1, 0), (1, 1, 1, 1)\}$

(b) $S = \left\{\begin{bmatrix}1 & 2\\1 & 0\end{bmatrix}, \begin{bmatrix}2 & -1\\1 & 2\end{bmatrix}, \begin{bmatrix}-3 & 1\\0 & 1\end{bmatrix}\right\}$

(c) $S = \left\{\begin{bmatrix}1\\2\\1\\0\\1\end{bmatrix}, \begin{bmatrix}0\\1\\2\\-1\\1\end{bmatrix}, \begin{bmatrix}2\\1\\0\\0\\-1\end{bmatrix}, \begin{bmatrix}-2\\1\\1\\1\\1\end{bmatrix}\right\}$

ML.2. Find a spanning set of the solution space of $A\mathbf{x} = \mathbf{0}$, where

$$A = \begin{bmatrix} 1 & 2 & 0 & 1 \\ 1 & 1 & 1 & 2 \\ 2 & -1 & 5 & 7 \\ 0 & 2 & -2 & -2 \end{bmatrix}.$$

6.4 BASIS AND DIMENSION

In this section we continue our study of the structure of a vector space V by determining a smallest set of vectors in V that completely describes V.

BASIS

DEFINITION The vectors $\mathbf{v}_1, \mathbf{v}_2, \ldots, \mathbf{v}_k$ in a vector space V are said to form a **basis** for V if (a) $\mathbf{v}_1, \mathbf{v}_2, \ldots, \mathbf{v}_k$ span V and (b) $\mathbf{v}_1, \mathbf{v}_2, \ldots, \mathbf{v}_k$ are linearly independent.

Remark If $\mathbf{v}_1, \mathbf{v}_2, \ldots, \mathbf{v}_k$ form a basis for a vector space V, then they must be nonzero (see Example 12 in Section 6.3) and distinct and so we write them as a set $\{\mathbf{v}_1, \mathbf{v}_2, \ldots, \mathbf{v}_k\}$.

EXAMPLE 1 The vectors $\mathbf{e}_1 = (1, 0)$ and $\mathbf{e}_2 = (0, 1)$ form a basis for R^2, the vectors $\mathbf{e}_1$, $\mathbf{e}_2$, and $\mathbf{e}_3$ form a basis for R^3 and, in general, the vectors $\mathbf{e}_1, \mathbf{e}_2, \ldots, \mathbf{e}_n$ form a basis for R^n. Each of these sets of vectors is called the **natural basis** or **standard basis** for R^2, R^3, and R^n, respectively. ■

EXAMPLE 2 Show that the set $S = \{\mathbf{v}_1, \mathbf{v}_2, \mathbf{v}_3, \mathbf{v}_4\}$, where $\mathbf{v}_1 = (1, 0, 1, 0)$, $\mathbf{v}_2 = (0, 1, -1, 2)$, $\mathbf{v}_3 = (0, 2, 2, 1)$, and $\mathbf{v}_4 = (1, 0, 0, 1)$ is a basis for R^4.

Solution To show that S is linearly independent, we form the equation

$$c_1\mathbf{v}_1 + c_2\mathbf{v}_2 + c_3\mathbf{v}_3 + c_4\mathbf{v}_4 = \mathbf{0}$$

and solve for c_1, c_2, c_3, and c_4. Substituting for $\mathbf{v}_1$, $\mathbf{v}_2$, $\mathbf{v}_3$, and $\mathbf{v}_4$, we obtain the linear system (verify)

$$\begin{aligned} c_1 \qquad\qquad\qquad + c_4 &= 0 \\ c_2 + 2c_3 \qquad &= 0 \\ c_1 - c_2 + 2c_3 \qquad &= 0 \\ 2c_2 + c_3 + c_4 &= 0, \end{aligned}$$

which has as its only solution $c_1 = c_2 = c_3 = c_4 = 0$ (verify), showing that S is linearly independent. Observe that the coefficient matrix of the preceding linear system consists of the vectors $\mathbf{v}_1$, $\mathbf{v}_2$, $\mathbf{v}_3$, and $\mathbf{v}_4$ written in column form.

To show that S spans R^4, we let $\mathbf{v} = (a, b, c, d)$ be any vector in R^4. We now seek constants k_1, k_2, k_3, and k_4 such that

$$k_1\mathbf{v}_1 + k_2\mathbf{v}_2 + k_3\mathbf{v}_3 + k_4\mathbf{v}_4 = \mathbf{v}.$$

Substituting for $\mathbf{v}_1$, $\mathbf{v}_2$, $\mathbf{v}_3$, $\mathbf{v}_4$, and $\mathbf{v}$, we find a solution (verify) for k_1, k_2, k_3, and k_4 to the resulting linear system for any a, b, c, and d. Hence S spans R^4 and is a basis for R^4. ■

EXAMPLE 3 Show that the set $S = \{t^2 + 1, t - 1, 2t + 2\}$ is a basis for the vector space P_2.

Solution We must show that S spans V and is linearly independent. To show that it spans V, we take any vector in V, that is, a polynomial $at^2 + bt + c$, and must find constants a_1, a_2, and a_3 such that

$$\begin{aligned} at^2 + bt + c &= a_1(t^2 + 1) + a_2(t - 1) + a_3(2t + 2) \\ &= a_1 t^2 + (a_2 + 2a_3)t + (a_1 - a_2 + 2a_3). \end{aligned}$$

Since two polynomials agree for all values of t only if the coefficients of respective powers of t agree, we get the linear system

$$\begin{aligned} a_1 \qquad\qquad &= a \\ a_2 + 2a_3 &= b \\ a_1 - a_2 + 2a_3 &= c. \end{aligned}$$

Solving, we have

$$a_1 = a, \quad a_2 = \frac{a + b - c}{2}, \quad a_3 = \frac{c + b - a}{4}.$$

Hence S spans V.

To illustrate this result, suppose that we are given the vector $2t^2 + 6t + 13$. Here $a = 2$, $b = 6$, and $c = 13$. Substituting in the foregoing expressions for a_1, a_2, and a_3, we find that

$$a_1 = 2, \quad a_2 = -\tfrac{5}{2}, \quad a_3 = \tfrac{17}{4}.$$

Hence

$$2t^2 + 6t + 13 = 2(t^2 + 1) - \tfrac{5}{2}(t - 1) + \tfrac{17}{4}(2t + 2).$$

To show that S is linearly independent, we form

$$a_1(t^2 + 1) + a_2(t - 1) + a_3(2t + 2) = 0.$$

Then

$$a_1 t^2 + (a_2 + 2a_3)t + (a_1 - a_2 + 2a_3) = 0.$$

Again, this can hold for all values of t only if

$$\begin{aligned} a_1 \qquad\qquad &= 0 \\ a_2 + 2a_3 &= 0 \\ a_1 - a_2 + 2a_3 &= 0. \end{aligned}$$

The only solution to this homogeneous system is $a_1 = a_2 = a_3 = 0$, which implies that S is linearly independent. Thus S is a basis for P_2. ■

The set of vectors $\{t^n, t^{n-1}, \ldots, t, 1\}$ forms a basis for the vector space P_n called the **natural basis** or **standard basis** for P_n. It has already been shown in Example 5 of Section 6.3 that this set of vectors spans P_n. Linear independence is left as an exercise (Exercise T.15).

EXAMPLE 4 Find a basis for the subspace V of P_2, consisting of all vectors of the form $at^2 + bt + c$, where $c = a - b$.

Solution Every vector in V is of the form

$$at^2 + bt + a - b$$

which can be written as

$$a(t^2 + 1) + b(t - 1),$$

so the vectors $t^2 + 1$ and $t - 1$ span V. Moreover, these vectors are linearly independent because neither one is a multiple of the other. This conclusion could also be reached (with more work) by writing the equation

$$a_1(t^2 + 1) + a_2(t - 1) = 0$$

or

$$t^2 a_1 + t a_2 + (a_1 - a_2) = 0.$$

Since this equation is to hold for all values of t, we must have $a_1 = 0$ and $a_2 = 0$. ■

A vector space V is called **finite-dimensional** if there is a finite subset of V that is a basis for V. If there is no such finite subset of V, then V is called **infinite-dimensional**.

Almost all the vector spaces considered in this book are finite-dimensional. However, we point out that there are many infinite-dimensional vector spaces that are extremely important in mathematics and physics; their study lies beyond the scope of this book. The vector space P, consisting of all polynomials, and the vector space $C(-\infty, \infty)$, consisting of all continuous functions $f: R \to R$, are not finite-dimensional.

We now establish some results about finite-dimensional vector spaces that will tell about the number of vectors in a basis, compare two different bases, and give properties of bases.

THEOREM 6.5 *If $S = \{\mathbf{v}_1, \mathbf{v}_2, \ldots, \mathbf{v}_n\}$ is a basis for a vector space V, then every vector in V can be written in one and only one way as a linear combination of the vectors in S.*

Proof First, every vector $\mathbf{v}$ in V can be written as a linear combination of the vectors in S because S spans V. Now let

$$\mathbf{v} = c_1\mathbf{v}_1 + c_2\mathbf{v}_2 + \cdots + c_n\mathbf{v}_n \tag{1}$$

and

$$\mathbf{v} = d_1\mathbf{v}_1 + d_2\mathbf{v}_2 + \cdots + d_n\mathbf{v}_n. \tag{2}$$

Subtracting (2) from (1), we obtain

$$\mathbf{0} = (c_1 - d_1)\mathbf{v}_1 + (c_2 - d_2)\mathbf{v}_2 + \cdots + (c_n - d_n)\mathbf{v}_n.$$

Since S is linearly independent, it follows that $c_i - d_i = 0$, $1 \le i \le n$, so $c_i = d_i$, $1 \le i \le n$. Hence there is only one way to express $\mathbf{v}$ as a linear combination of the vectors in S. ■

We can also show (Exercise T.11) that if $S = \{\mathbf{v}_1, \mathbf{v}_2, \ldots, \mathbf{v}_n\}$ is a set of vectors in a vector space V such that every vector in V can be written in one and only one way as a linear combination of the vectors in S, then S is a basis for V.

THEOREM 6.6 *Let $S = \{\mathbf{v}_1, \mathbf{v}_2, \ldots, \mathbf{v}_n\}$ be a set of nonzero vectors in a vector space V and let $W = \text{span } S$. Then some subset of S is a basis for W.*

Proof **Case I.** If S is linearly independent, then since S already spans W, we conclude that S is a basis for W.

Case II. If S is linearly dependent, then

$$c_1\mathbf{v}_1 + c_2\mathbf{v}_2 + \cdots + c_n\mathbf{v}_n = \mathbf{0}, \tag{3}$$

where $c_1, c_2, \ldots, c_n$ are not all zero. Thus, some $\mathbf{v}_j$ is a linear combination of the preceding vectors in S (Theorem 6.4). We now delete $\mathbf{v}_j$ from S, getting a subset S_1 of S. Then, by the observation made just before Example 14 in Section 6.3, we conclude that $S_1 = \{\mathbf{v}_1, \mathbf{v}_2, \ldots, \mathbf{v}_{j-1}, \mathbf{v}_{j+1}, \ldots, \mathbf{v}_n\}$ also spans W.

If S_1 is linearly independent, then S_1 is a basis. If S_1 is linearly dependent, delete a vector of S_1 that is a linear combination of the preceding vectors of S_1 and get a new set S_2 which spans W. Continuing, since S is a finite set, we will eventually find a subset T of S that is linearly independent and spans W. The set T is a basis for W.

Alternative Constructive Proof When V is R^m, $n \geq m$. We take the vectors in S as $m \times 1$ matrices and form Equation (3). This equation leads to a homogeneous system in the n unknowns $c_1, c_2, \ldots, c_n$; the columns of its $m \times n$ coefficient matrix A are $\mathbf{v}_1, \mathbf{v}_2, \ldots, \mathbf{v}_n$. We now transform A to a matrix B in reduced row echelon form, having r nonzero rows, $1 \leq r \leq m$. Without loss of generality, we may assume that the r leading 1s in the r nonzero rows of B occur in the first r columns. Thus we have

$$B = \begin{bmatrix} 1 & 0 & 0 & \cdots & 0 & b_{1\,r+1} & \cdots & b_{1\,n} \\ 0 & 1 & 0 & \cdots & 0 & b_{2\,r+1} & \cdots & b_{2\,n} \\ 0 & 0 & 1 & \cdots & 0 & b_{3\,r+1} & \cdots & b_{3\,n} \\ \vdots & \vdots & \vdots & \ddots & \vdots & \vdots & & \vdots \\ 0 & 0 & 0 & \cdots & 1 & b_{r\,r+1} & \cdots & b_{r\,n} \\ 0 & 0 & 0 & \cdots & 0 & 0 & \cdots & 0 \\ \vdots & \vdots & \vdots & & \vdots & \vdots & \ddots & \vdots \\ 0 & 0 & 0 & \cdots & 0 & 0 & \cdots & 0 \end{bmatrix}.$$

Solving for the unknowns corresponding to the leading 1s, we see that $c_1, c_2, \ldots, c_r$ can be solved for in terms of the other unknowns $c_{r+1}, c_{r+2}, \ldots, c_n$. Thus

$$\begin{aligned} c_1 &= -b_{1\,r+1}c_{r+1} - b_{1\,r+2}c_{r+2} - \cdots - b_{1\,n}c_n, \\ c_2 &= -b_{2\,r+1}c_{r+1} - b_{2\,r+2}c_{r+2} - \cdots - b_{2\,n}c_n, \\ &\vdots \\ c_r &= -b_{r\,r+1}c_{r+1} - b_{r\,r+2}c_{r+2} - \cdots - b_{r\,n}c_n, \end{aligned} \tag{4}$$

where $c_{r+1}, c_{r+2}, \ldots, c_n$ can be assigned arbitrary real values. Letting

$$c_{r+1} = 1, \quad c_{r+2} = 0, \ldots, \quad c_n = 0$$

in Equation (4) and using these values in Equation (3), we have

$$-b_{1\,r+1}\mathbf{v}_1 - b_{2\,r+1}\mathbf{v}_2 - \cdots - b_{r\,r+1}\mathbf{v}_r + \mathbf{v}_{r+1} = \mathbf{0},$$

which implies that $\mathbf{v}_{r+1}$ is a linear combination of $\mathbf{v}_1, \mathbf{v}_2, \ldots, \mathbf{v}_r$. By the observation made just before Example 14 in Section 6.3, the set of vectors obtained from S by deleting $\mathbf{v}_{r+1}$ spans W. Similarly, letting

$$c_{r+1} = 0, \quad c_{r+2} = 1, \quad c_{r+3} = 0, \ldots, \quad c_n = 0,$$

we find that $\mathbf{v}_{r+2}$ is a linear combination of $\mathbf{v}_1, \mathbf{v}_2, \ldots, \mathbf{v}_r$ and the set of vectors obtained from S by deleting $\mathbf{v}_{r+1}$ and $\mathbf{v}_{r+2}$ spans W. Continuing in this manner, $\mathbf{v}_{r+3}, \mathbf{v}_{r+4}, \ldots, \mathbf{v}_n$ are linear combinations of $\mathbf{v}_1, \mathbf{v}_2, \ldots, \mathbf{v}_r$, so it follows that $\{\mathbf{v}_1, \mathbf{v}_2, \ldots, \mathbf{v}_r\}$ spans W.

We next show that $\{\mathbf{v}_1, \mathbf{v}_2, \ldots, \mathbf{v}_r\}$ is linearly independent. Consider the matrix B_D obtained by deleting from B all columns not containing a leading 1. In this case, B_D consists of the first r columns of B. Thus,

$$B_D = \begin{bmatrix} 1 & 0 & 0 & \cdots & 0 \\ 0 & 1 & 0 & \cdots & 0 \\ 0 & 0 & 1 & & \\ & & & \ddots & \vdots \\ 0 & 0 & & \cdots & 1 \\ 0 & 0 & & \cdots & 0 \\ \vdots & \vdots & & & \vdots \\ 0 & 0 & & \cdots & 0 \end{bmatrix}.$$

Let A_D be the matrix obtained from A by deleting the columns corresponding to the columns that were deleted in B to obtain B_D. In this case, the columns of A_D are $\mathbf{v}_1, \mathbf{v}_2, \ldots, \mathbf{v}_r$, the first r columns of A. Since A and B are row equivalent, so are A_D and B_D. Then the homogeneous systems

$$A_D\mathbf{x} = \mathbf{0} \quad \text{and} \quad B_D\mathbf{x} = \mathbf{0}$$

are equivalent. Recall now that the homogeneous system $B_D\mathbf{x} = \mathbf{0}$ can be written equivalently as

$$x_1\mathbf{w}_1 + x_2\mathbf{w}_2 + \cdots + x_r\mathbf{w}_r = \mathbf{0}, \tag{5}$$

where

$$\mathbf{x} = \begin{bmatrix} x_1 \\ x_2 \\ \vdots \\ x_r \end{bmatrix}$$

and $\mathbf{w}_1, \mathbf{w}_2, \ldots, \mathbf{w}_r$ are the columns of B_D. Since the columns of B_D form a linearly independent set of vectors in R^m, Equation (5) has only the trivial solution. Hence, $A_D\mathbf{x} = \mathbf{0}$ also has only the trivial solution. Thus the columns of A_D are linearly independent. That is, $\{\mathbf{v}_1, \mathbf{v}_2, \ldots, \mathbf{v}_r\}$ is linearly independent. ■

The first proof of Theorem 6.6 leads to a simple procedure for finding a subset T of a set S so that T is a basis for span S.

Let $S = \{\mathbf{v}_1, \mathbf{v}_2, \ldots, \mathbf{v}_n\}$ be a set of nonzero vectors in a vector space V. The procedure for finding a subset of S that is a basis for $W = \text{span } S$ is as follows.

Step 1. Form Equation (3),

$$c_1\mathbf{v}_1 + c_2\mathbf{v}_2 + \cdots + c_n\mathbf{v}_n = \mathbf{0},$$

which we solve for $c_1, c_2, \ldots, c_n$. If these are all zero, then S is linearly independent and is then a basis for W.

Step 2. If $c_1, c_2, \ldots, c_n$ are not all zero, then S is linearly dependent, so one of the vectors in S — say, $\mathbf{v}_j$ — is a linear combination of the preceding vectors in S. Delete $\mathbf{v}_j$ from S, getting the subset S_1, which also spans W.

Step 3. Repeat Step 1, using S_1 instead of S. By repeatedly deleting vectors of S we obtain a subset T of S that spans W and is linearly independent. Thus T is a basis for W.

This procedure can be rather tedious, since *every time* we delete a vector from S, we must solve a linear system. In Section 6.6 we shall present a much more efficient procedure for finding a basis for $W = \text{span } S$, but the basis is *not* guaranteed to be a subset of S. In many cases this is not a cause for concern, since one basis for $W = \text{span } S$ is as good as any other basis. However, there are cases when the vectors in S have some special properties and we want the basis for $W = \text{span } S$ to have the same properties, so we want the basis to be a subset of S. If $V = R^m$, the alternative proof of Theorem 6.6 yields a very efficient procedure (see Example 5 below) for finding a basis for $W = \text{span } S$ consisting of vectors from S.

Let $V = R^m$ and let $S = \{\mathbf{v}_1, \mathbf{v}_2, \ldots, \mathbf{v}_n\}$ be a set of nonzero vectors in V. The procedure for finding a subset of S that is a basis for $W = \text{span } S$ is as follows.

Step 1. Form Equation (3),

$$c_1\mathbf{v}_1 + c_2\mathbf{v}_2 + \cdots + c_n\mathbf{v}_n = \mathbf{0}.$$

Step 2. Construct the augmented matrix associated with the homogeneous system of Equation (3), and transform it to reduced row echelon form.

Step 3. The vectors corresponding to the columns containing the leading 1s form a basis for $W = \text{span } S$.

Recall that in the alternative proof of Theorem 6.6 we assumed without loss of generality that the r leading 1s in the r nonzero rows of B occur in the first r columns. Thus, if $S = \{\mathbf{v}_1, \mathbf{v}_2, \ldots, \mathbf{v}_6\}$ and the leading 1s occur in columns 1, 3, and 4, then $\{\mathbf{v}_1, \mathbf{v}_3, \mathbf{v}_4\}$ is a basis for span S.

Remark In Step 2 of the procedure in the preceding box, it is sufficient to transform the augmented matrix to row echelon form (see Section 1.6).

EXAMPLE 5 Let $S = \{\mathbf{v}_1, \mathbf{v}_2, \mathbf{v}_3, \mathbf{v}_4, \mathbf{v}_5\}$ be a set of vectors in R^4, where

$$\mathbf{v}_1 = (1, 2, -2, 1), \qquad \mathbf{v}_2 = (-3, 0, -4, 3),$$

$$\mathbf{v}_3 = (2, 1, 1, -1), \quad \mathbf{v}_4 = (-3, 3, -9, 6), \quad \text{and} \quad \mathbf{v}_5 = (9, 3, 7, -6).$$

Find a subset of S that is a basis for $W = \text{span } S$.

Solution ***Step 1.*** Form Equation (3),

$$c_1(1, 2, -2, 1) + c_2(-3, 0, -4, 3) + c_3(2, 1, 1, -1) + c_4(-3, 3, -9, 6) + c_5(9, 3, 7, -6) = (0, 0, 0, 0).$$

Step 2. Equating corresponding components, we obtain the homogeneous system

$$\begin{aligned} c_1 - 3c_2 + 2c_3 - 3c_4 + 9c_5 &= 0 \\ 2c_1 \qquad + c_3 + 3c_4 + 3c_5 &= 0 \\ -2c_1 - 4c_2 + c_3 - 9c_4 + 7c_5 &= 0 \\ c_1 + 3c_2 - c_3 + 6c_4 - 6c_5 &= 0. \end{aligned}$$

The reduced row echelon form of the associated augmented matrix is (verify)

$$\left[\begin{array}{ccccc|c} 1 & 0 & \frac{1}{2} & \frac{3}{2} & \frac{3}{2} & 0 \\ 0 & 1 & -\frac{1}{2} & \frac{3}{2} & -\frac{5}{2} & 0 \\ 0 & 0 & 0 & 0 & 0 & 0 \\ 0 & 0 & 0 & 0 & 0 & 0 \end{array}\right].$$

Step 3. The leading 1s appear in columns 1 and 2, so $\{\mathbf{v}_1, \mathbf{v}_2\}$ is a basis for $W = \text{span } S$. ■

Remark In the alternative proof of Theorem 6.6 when $V = R^n$, the order of the vectors in the original spanning set S determines which basis for W is obtained. If, for example, we consider Example 5, where $S = \{\mathbf{w}_1, \mathbf{w}_2, \mathbf{w}_3, \mathbf{w}_4, \mathbf{w}_5\}$ with $\mathbf{w}_1 = \mathbf{v}_4$, $\mathbf{w}_2 = \mathbf{v}_3$, $\mathbf{w}_3 = \mathbf{v}_2$, $\mathbf{w}_4 = \mathbf{v}_1$, and $\mathbf{w}_5 = \mathbf{v}_5$, then the reduced row echelon form of the augmented matrix is (verify)

$$\left[\begin{array}{ccccc|c} 1 & 0 & \frac{1}{3} & \frac{1}{3} & -\frac{1}{3} & 0 \\ 0 & 1 & -1 & 1 & 4 & 0 \\ 0 & 0 & 0 & 0 & 0 & 0 \\ 0 & 0 & 0 & 0 & 0 & 0 \end{array}\right].$$

It now follows that $\{\mathbf{w}_1, \mathbf{w}_2\} = \{\mathbf{v}_4, \mathbf{v}_3\}$ is a basis for $W = \text{span } S$.

We shall soon establish a main result (Corollary 6.1) of this section, which will tell us about the number of vectors in two different bases. First, observe that if $\{\mathbf{v}_1, \mathbf{v}_2, \ldots, \mathbf{v}_n\}$ is a basis for a vector space V, then $\{c\mathbf{v}_1, \mathbf{v}_2, \ldots, \mathbf{v}_n\}$ is also a basis if $c \neq 0$ (Exercise T.9). Thus a nonzero real vector space always has infinitely many bases.

THEOREM 6.7 *If $S = \{\mathbf{v}_1, \mathbf{v}_2, \ldots, \mathbf{v}_n\}$ is a basis for a vector space V and $T = \{\mathbf{w}_1, \mathbf{w}_2, \ldots, \mathbf{w}_r\}$ is a linearly independent set of vectors in V, then $r \leq n$.*

Proof Let $T_1 = \{\mathbf{w}_1, \mathbf{v}_1, \ldots, \mathbf{v}_n\}$. Since S spans V, so does T_1. Since $\mathbf{w}_1$ is a linear combination of the vectors in S, we find that T_1 is linearly dependent. Then, by Theorem 6.4, some $\mathbf{v}_j$ is a linear combination of the preceding vectors in T_1. Delete that particular vector $\mathbf{v}_j$.

Let $S_1 = \{\mathbf{w}_1, \mathbf{v}_1, \ldots, \mathbf{v}_{j-1}, \mathbf{v}_{j+1}, \ldots, \mathbf{v}_n\}$. Note that S_1 spans V. Next, let $T_2 = \{\mathbf{w}_2, \mathbf{w}_1, \mathbf{v}_1, \ldots, \mathbf{v}_{j-1}, \mathbf{v}_{j+1}, \ldots, \mathbf{v}_n\}$. Then T_2 is linearly dependent and some vector in T_2 is a linear combination of the preceding vectors in T_2. Since T is linearly independent, this vector cannot be $\mathbf{w}_1$, so it is $\mathbf{v}_i$, $i \neq j$. Repeat this process over and over. If the $\mathbf{v}$ vectors are all eliminated before we can run out of $\mathbf{w}$ vectors, then the resulting set of $\mathbf{w}$ vectors, a subset of T, is linearly dependent, which implies that T is also linearly dependent. Since we have reached a contradiction, we conclude that the number r of $\mathbf{w}$ vectors must be no greater than the number n of $\mathbf{v}$ vectors. That is, $r \leq n$. ■

COROLLARY 6.1 *If $S = \{\mathbf{v}_1, \mathbf{v}_2, \ldots, \mathbf{v}_n\}$ and $T = \{\mathbf{w}_1, \mathbf{w}_2, \ldots, \mathbf{w}_m\}$ are bases for a vector space, then $n = m$.*

Proof Since T is a linearly independent set of vectors, Theorem 6.7 implies that $m \leq n$. Similarly, $n \leq m$ because S is linearly independent. Hence $n = m$. ■

Thus, although a vector space has many bases, we have just shown that for a particular vector space V, all bases have the same number of vectors. We can then make the following definition.

DIMENSION

DEFINITION The **dimension** of a nonzero vector space V is the number of vectors in a basis for V. We often write **dim** V for the dimension of V. Since the set $\{\mathbf{0}\}$ is linearly dependent, it is natural to say that the vector space $\{\mathbf{0}\}$ has dimension **zero**.

EXAMPLE 6 The dimension of R^2 is 2; the dimension of R^3 is 3; and in general, the dimension of R^n is n. ■

EXAMPLE 7 The dimension of P_2 is 3; the dimension of P_3 is 4; and in general, the dimension of P_n is $n + 1$. ■

It can be shown that all finite-dimensional vector spaces of the same dimension differ only in the nature of their elements; their algebraic properties are identical.

It can also be shown that if V is a finite-dimensional vector space, then every nonzero subspace W of V has a finite basis and dim $W \leq$ dim V (Exercise T.2).

EXAMPLE 8 The subspace W of R^4 considered in Example 5 has dimension 2. ■

We might also consider the subspaces of R^2 [recall that R^2 can be viewed as the (x, y)-plane]. First, we have $\{\mathbf{0}\}$ and R^2, the trivial subspaces of dimension 0 and 2, respectively. Now the subspace V of R^2 spanned by a vector $\mathbf{v} \neq \mathbf{0}$ is a one-dimensional subspace of R^2; V is represented by a line through the origin. Thus the subspaces of R^2 are $\{\mathbf{0}\}$, R^2, and all the lines through the origin. Similarly, we ask you to show (Exercise T.8) that the subspaces of R^3 are $\{\mathbf{0}\}$, R^3, all lines through the origin, and all planes through the origin.

It can be shown that if a vector space V has dimension n, then any set of $n+1$ vectors in V is necessarily linearly dependent (Exercise T.3). Any set of more than n vectors in R^n is linearly dependent. Thus the four vectors in R^3 considered in Example 9 of Section 6.3 were shown to be linearly dependent. Also, if a vector space V is of dimension n, then no set of $n - 1$ vectors in V can span V (Exercise T.4). Thus in Example 3 of Section 6.3, polynomials $p_1(t)$ and $p_2(t)$ do not span P_2, which is of dimension 3.

We now come to a theorem that we shall have occasion to use several times in constructing a basis containing a given set of linearly independent vectors. We shall leave the proof as an exercise (Exercise T.5). The example following the theorem completely imitates the proof.

THEOREM 6.8 *If S is a linearly independent set of vectors in a finite-dimensional vector space V, then there is a basis T for V, which contains S.* ■

Theorem 6.8 says that a linearly independent set of vectors in a vector space V can be extended to a basis for V.

EXAMPLE 9 Suppose that we wish to find a basis for R^4 that contains the vectors $\mathbf{v}_1 = (1, 0, 1, 0)$ and $\mathbf{v}_2 = (-1, 1, -1, 0)$.

We use Theorem 6.8 as follows. First, let $\{\mathbf{e}_1, \mathbf{e}_2, \mathbf{e}_3, \mathbf{e}_4\}$ be the natural basis for R^4, where

$$\mathbf{e}_1 = (1, 0, 0, 0), \qquad \mathbf{e}_2 = (0, 1, 0, 0), \qquad \mathbf{e}_3 = (0, 0, 1, 0),$$

and

$$\mathbf{e}_4 = (0, 0, 0, 1).$$

Form the set $S = \{\mathbf{v}_1, \mathbf{v}_2, \mathbf{e}_1, \mathbf{e}_2, \mathbf{e}_3, \mathbf{e}_4\}$. Since $\{\mathbf{e}_1, \mathbf{e}_2, \mathbf{e}_3, \mathbf{e}_4\}$ spans R^4, so does S. We now use the alternative proof of Theorem 6.6 to find a subset of S that is a basis for R^4. Thus, we form Equation (3),

$$c_1\mathbf{v}_1 + c_2\mathbf{v}_2 + c_3\mathbf{e}_1 + c_4\mathbf{e}_2 + c_5\mathbf{e}_3 + c_6\mathbf{e}_4 = \mathbf{0},$$

which leads to the homogeneous system

$$\begin{aligned} c_1 - c_2 + c_3 \qquad\qquad\qquad &= 0 \\ - c_2 \qquad + c_4 \qquad\quad &= 0 \\ c_1 - c_2 \qquad\qquad + c_5 \quad &= 0 \\ c_6 &= 0. \end{aligned}$$

Transforming the augmented matrix to reduced row echelon form, we obtain (verify)

$$\left[\begin{array}{cccccc|c} 1 & 0 & 0 & 1 & 1 & 0 & 0 \\ 0 & 1 & 0 & 1 & 0 & 0 & 0 \\ 0 & 0 & 1 & 0 & -1 & 0 & 0 \\ 0 & 0 & 0 & 0 & 0 & 1 & 0 \end{array}\right].$$

Since the leading 1s appear in columns 1, 2, 3, and 6, we conclude that $\{\mathbf{v}_1, \mathbf{v}_2, \mathbf{e}_1, \mathbf{e}_4\}$ is a basis for R^4 containing $\mathbf{v}_1$ and $\mathbf{v}_2$. ■

From the definition of a basis, a set of vectors in a vector space V is a basis for V if it spans V and is linearly independent. However, if we are given the additional information that the dimension of V is n, we need only verify one of the two conditions. This is the content of the following theorem.

THEOREM 6.9 *Let V be an n-dimensional vector space, and let $S = \{\mathbf{v}_1, \mathbf{v}_2, \dots, \mathbf{v}_n\}$ be a set of n vectors in V.*

(a) *If S is linearly independent, then it is a basis for V.*

(b) *If S spans V, then it is a basis for V.*

Proof Exercise T.6. ■

As a particular application of Theorem 6.9, we have the following. To determine if a subset S of R^n is a basis for R^n, first count the number of elements in S. If S has n elements, we can use either part (a) or part (b) of Theorem 6.9 to determine whether S is or is not a basis. If S does not have n elements, it is not a basis for R^n. (Why?) The same line of reasoning applies to any vector space or subspace whose *dimension is known.*

EXAMPLE 10 In Example 5, $W = \text{span } S$ is a subspace of R^4, so $\dim W \leq 4$. Since S contains five vectors, we conclude by Corollary 6.1 that S is not a basis for W. In Example 2, since $\dim R^4 = 4$ and the set S contains four vectors, it is possible for S to be a basis for R^4. If S is linearly independent *or* spans R^4, it is a basis; otherwise it is not a basis. Thus, we need only check one of the conditions in Theorem 6.9, not both. ■

BASIS AND DIMENSION IN B^n (OPTIONAL)

The definitions and theorems of this section are valid for general vector spaces, hence for B^n with the proviso that we use binary arithmetic. Examples 11–15 illustrate the concepts of this section for the vector space B^n.

EXAMPLE 11 The vectors

$$\mathbf{b}_1 = \begin{bmatrix} 1 \\ 0 \end{bmatrix} \quad \text{and} \quad \mathbf{b}_2 = \begin{bmatrix} 0 \\ 1 \end{bmatrix}$$

form a basis for B^2; the vectors

$$\mathbf{b}_1 = \begin{bmatrix} 1 \\ 0 \\ 0 \end{bmatrix}, \quad \mathbf{b}_2 = \begin{bmatrix} 0 \\ 1 \\ 0 \end{bmatrix}, \quad \text{and} \quad \mathbf{b}_3 = \begin{bmatrix} 0 \\ 0 \\ 1 \end{bmatrix}$$

form a basis for B^3; and, in general, the columns of I_n form a basis for B^n. Each of these sets of vectors is called the natural basis or standard basis for B^2, B^3, and B^n, respectively. ■

EXAMPLE 12 The dimension of B^2 is 2, the dimension of B^3 is 3, and, in general, the dimension of B^n is n. It follows that B^n is a finite-dimensional vector space. ■

EXAMPLE 13 Show that the set $S = \{\mathbf{v}_1, \mathbf{v}_2, \mathbf{v}_3\}$, where

$$\mathbf{v}_1 = \begin{bmatrix} 1 \\ 1 \\ 0 \end{bmatrix}, \quad \mathbf{v}_2 = \begin{bmatrix} 0 \\ 1 \\ 1 \end{bmatrix}, \quad \text{and} \quad \mathbf{v}_3 = \begin{bmatrix} 0 \\ 1 \\ 0 \end{bmatrix}$$

is a basis for B^3.

Solution To show that S is linearly independent, we form the equation

$$c_1\mathbf{v}_1 + c_2\mathbf{v}_2 + c_3\mathbf{v}_3 = \mathbf{0}$$

and solve for c_1, c_2, and c_3. Substituting for $\mathbf{v}_1$, $\mathbf{v}_2$, and $\mathbf{v}_3$, we obtain the linear system (verify)

$$\begin{aligned} c_1 \qquad\qquad &= 0 \\ c_1 + c_2 + c_3 &= 0 \\ c_2 \qquad &= 0, \end{aligned}$$

which has as its only solution $c_1 = c_2 = c_3 = 0$, showing that S is linearly independent.

It follows from Theorem 6.9(a) that S is a basis for B^3. ■

EXAMPLE 14 In Example 14 of Section 6.2, it was shown that $W = \{\mathbf{w}_1, \mathbf{w}_2, \mathbf{w}_3, \mathbf{w}_4\}$, where

$$\mathbf{w}_1 = \begin{bmatrix} 0 \\ 0 \\ 0 \end{bmatrix}, \quad \mathbf{w}_2 = \begin{bmatrix} 1 \\ 0 \\ 0 \end{bmatrix}, \quad \mathbf{w}_3 = \begin{bmatrix} 0 \\ 1 \\ 0 \end{bmatrix}, \quad \text{and} \quad \mathbf{w}_4 = \begin{bmatrix} 1 \\ 1 \\ 0 \end{bmatrix}$$

is a subspace of B^3. It follows that $\{\mathbf{w}_2, \mathbf{w}_3\}$ is a basis for W (verify) and hence $\dim W = 2$. ■

EXAMPLE 15 Let $S = \{\mathbf{v}_1, \mathbf{v}_2, \mathbf{v}_3, \mathbf{v}_4\}$, where

$$\mathbf{v}_1 = \begin{bmatrix} 1 \\ 0 \\ 0 \\ 1 \end{bmatrix}, \quad \mathbf{v}_2 = \begin{bmatrix} 0 \\ 1 \\ 1 \\ 1 \end{bmatrix}, \quad \mathbf{v}_3 = \begin{bmatrix} 1 \\ 1 \\ 0 \\ 0 \end{bmatrix}, \quad \text{and} \quad \mathbf{v}_4 = \begin{bmatrix} 0 \\ 0 \\ 1 \\ 0 \end{bmatrix}.$$

Determine if S is a basis for B^4.

Solution We must determine if S spans B^4 and is linearly independent. If $\mathbf{w} = \begin{bmatrix} a \\ b \\ c \\ d \end{bmatrix}$ is any vector in B^4, then we must find (bit) constants c_1, c_2, c_3, and c_4 such that

$$c_1\mathbf{v}_1 + c_2\mathbf{v}_2 + c_3\mathbf{v}_3 + c_4\mathbf{v}_4 = \mathbf{w}.$$

Substituting for $\mathbf{v}_1$, $\mathbf{v}_2$, $\mathbf{v}_3$, $\mathbf{v}_4$, and $\mathbf{w}$, we obtain the linear system (verify)

$$\begin{aligned} c_1 \qquad + c_3 \qquad\quad &= a \\ c_2 + c_3 \qquad\quad &= b \\ c_2 \qquad\quad + c_4 &= c \\ c_1 + c_2 \qquad\qquad\quad &= d. \end{aligned}$$

We form the augmented matrix and use row opertions: Add row 1 to row 4, add row 2 to row 3, and add row 2 to row 4, to obtain the equivalent augmented matrix (verify)

$$\left[\begin{array}{cccc|c} 1 & 0 & 1 & 0 & a \\ 0 & 1 & 1 & 0 & b \\ 0 & 0 & 1 & 1 & b+c \\ 0 & 0 & 0 & 0 & a+b+d \end{array}\right].$$

It follows that this system is inconsistent if the choice of bits a, b, and d is such that $a + b + d \neq 0$. For example, if

$$\mathbf{w} = \begin{bmatrix} 0 \\ 1 \\ 0 \\ 0 \end{bmatrix},$$

then the system is inconsistent; hence S does not span B^4 and is not a basis for B^4. ■

Key Terms

Basis
Natural (or standard) basis
Finite-dimensional vector space
Infinite-dimensional vector space
Dimension

6.4 Exercises

1. Which of the following sets of vectors are bases for R^2?

(a) $\{(1, 3), (1, -1)\}$

(b) $\{(0, 0), (1, 2), (2, 4)\}$

(c) $\{(1, 2), (2, -3), (3, 2)\}$

(d) $\{(1, 3), (-2, 6)\}$

2. Which of the following sets of vectors are bases for R^3?

(a) $\{(1, 2, 0), (0, 1, -1)\}$

(b) $\{(1, 1, -1), (2, 3, 4), (4, 1, -1), (0, 1, -1)\}$

(c) $\{(3, 2, 2), (-1, 2, 1), (0, 1, 0)\}$

(d) $\{(1, 0, 0), (0, 2, -1), (3, 4, 1), (0, 1, 0)\}$

3. Which of the following sets of vectors are bases for R^4?

(a) $\{(1, 0, 0, 1), (0, 1, 0, 0), (1, 1, 1, 1), (0, 1, 1, 1)\}$

(b) $\{(1, -1, 0, 2), (3, -1, 2, 1), (1, 0, 0, 1)\}$

(c) $\{(-2, 4, 6, 4), (0, 1, 2, 0), (-1, 2, 3, 2), (-3, 2, 5, 6), (-2, -1, 0, 4)\}$

(d) $\{(0, 0, 1, 1), (-1, 1, 1, 2), (1, 1, 0, 0), (2, 1, 2, 1)\}$

4. Which of the following sets of vectors are bases for P_2?

(a) $\{-t^2 + t + 2, 2t^2 + 2t + 3, 4t^2 - 1\}$

(b) $\{t^2 + 2t - 1, 2t^2 + 3t - 2\}$

(c) $\{t^2 + 1, 3t^2 + 2t, 3t^2 + 2t + 1, 6t^2 + 6t + 3\}$

(d) $\{3t^2 + 2t + 1, t^2 + t + 1, t^2 + 1\}$

5. Which of the following sets of vectors are bases for P_3?

(a) $\{t^3 + 2t^2 + 3t, 2t^3 + 1, 6t^3 + 8t^2 + 6t + 4, t^3 + 2t^2 + t + 1\}$

(b) $\{t^3 + t^2 + 1, t^3 - 1, t^3 + t^2 + t\}$

(c) $\{t^3 + t^2 + t + 1, t^3 + 2t^2 + t + 3, 2t^3 + t^2 + 3t + 2, t^3 + t^2 + 2t + 2\}$

(d) $\{t^3 - t, t^3 + t^2 + 1, t - 1\}$

6. Show that the matrices

$$\begin{bmatrix} 1 & 1 \\ 0 & 0 \end{bmatrix}, \begin{bmatrix} 0 & 0 \\ 1 & 1 \end{bmatrix}, \begin{bmatrix} 1 & 0 \\ 0 & 1 \end{bmatrix}, \begin{bmatrix} 0 & 1 \\ 1 & 1 \end{bmatrix}$$

form a basis for the vector space M_{22}.

In Exercises 7 and 8, determine which of the given subsets forms a basis for R^3. Express the vector (2, 1, 3) as a linear combination of the vectors in each subset that is a basis.

7. (a) $\{(1, 1, 1), (1, 2, 3), (0, 1, 0)\}$

(b) $\{(1, 2, 3), (2, 1, 3), (0, 0, 0)\}$

8. (a) $\{(2, 1, 3), (1, 2, 1), (1, 1, 4), (1, 5, 1)\}$

(b) $\{(1, 1, 2), (2, 2, 0), (3, 4, -1)\}$

In Exercises 9 and 10, determine which of the given subsets forms a basis for P_2. Express $5t^2 - 3t + 8$ as a linear combination of the vectors in each subset that is a basis.

9. (a) $\{t^2 + t, t - 1, t + 1\}$

(b) $\{t^2 + 1, t - 1\}$

10. (a) $\{t^2 + t, t^2, t^2 + 1\}$

(b) $\{t^2 + 1, t^2 - t + 1\}$

11. Let $S = \{\mathbf{v}_1, \mathbf{v}_2, \mathbf{v}_3, \mathbf{v}_4\}$, where

$$\mathbf{v}_1 = (1, 2, 2), \qquad \mathbf{v}_2 = (3, 2, 1),$$
$$\mathbf{v}_3 = (11, 10, 7), \quad \text{and} \quad \mathbf{v}_4 = (7, 6, 4).$$

Find a basis for the subspace $W = \text{span } S$ of R^3. What is $\dim W$?

12. Let $S = \{\mathbf{v}_1, \mathbf{v}_2, \mathbf{v}_3, \mathbf{v}_4, \mathbf{v}_5\}$, where

$$\mathbf{v}_1 = (1, 1, 0, -1), \quad \mathbf{v}_2 = (0, 1, 2, 1),$$
$$\mathbf{v}_3 = (1, 0, 1, -1), \quad \mathbf{v}_4 = (1, 1, -6, -3),$$

and $\mathbf{v}_5 = (-1, -5, 1, 0)$. Find a basis for the subspace $W = \text{span } S$ of R^4. What is $\dim W$?

13. Consider the following subset of P_3:

$$S = \{t^3 + t^2 - 2t + 1, t^2 + 1, t^3 - 2t, 2t^3 + 3t^2 - 4t + 3\}.$$

Find a basis for the subspace $W = \text{span } S$. What is $\dim W$?

14. Let

$$S = \left\{ \begin{bmatrix} 1 & 0 \\ 0 & 1 \end{bmatrix}, \begin{bmatrix} 0 & 1 \\ 1 & 0 \end{bmatrix}, \begin{bmatrix} 1 & 1 \\ 1 & 1 \end{bmatrix}, \begin{bmatrix} -1 & 1 \\ 1 & -1 \end{bmatrix} \right\}.$$

Find a basis for the subspace $W = \text{span } S$ of M_{22}.

15. Find a basis for M_{23}. What is the dimension of M_{23}? Generalize to M_{mn}.

16. Consider the following subset of the vector space of all real-valued functions

$$S = \{\cos^2 t, \sin^2 t, \cos 2t\}.$$

Find a basis for the subspace $W = \text{span } S$. What is $\dim W$?

In Exercises 17 and 18, find a basis for the given subspaces of R^3 and R^4.

17. (a) All vectors of the form (a, b, c), where $b = a + c$

(b) All vectors of the form (a, b, c), where $b = a$

(c) All vectors of the form (a, b, c), where $2a + b - c = 0$

18. (a) All vectors of the form (a, b, c), where $a = 0$

(b) All vectors of the form $(a + c, a - b, b + c, -a + b)$

(c) All vectors of the form (a, b, c), where $a - b + 5c = 0$

In Exercises 19 and 20, find the dimensions of the given subspaces of R^4.

19. (a) All vectors of the form (a, b, c, d), where $d = a + b$

(b) All vectors of the form (a, b, c, d), where $c = a - b$ and $d = a + b$

20. (a) All vectors of the form (a, b, c, d), where $a = b$

(b) All vectors of the form $(a + c, -a + b, -b - c, a + b + 2c)$

21. Find a basis for the subspace of P_2 consisting of all vectors of the form $at^2 + bt + c$, where $c = 2a - 3b$.

22. Find a basis for the subspace of P_3 consisting of all vectors of the form $at^3 + bt^2 + ct + d$, where $a = b$ and $c = d$.

23. Find the dimensions of the subspaces of R^2 spanned by the vectors in Exercise 1.

24. Find the dimensions of the subspaces of R^3 spanned by the vectors in Exercise 2.

25. Find the dimensions of the subspaces of R^4 spanned by the vectors in Exercise 3.

26. Find the dimension of the subspace of P_2 consisting of all vectors of the form $at^2 + bt + c$, where $c = b - 2a$.

27. Find the dimension of the subspace of P_3 consisting of all vectors of the form $at^3 + bt^2 + ct + d$, where $b = 3a - 5d$ and $c = d + 4a$.

28. Find a basis for R^3 that includes the vectors

(a) $(1, 0, 2)$

(b) $(1, 0, 2)$ and $(0, 1, 3)$

29. Find a basis for R^4 that includes the vectors $(1, 0, 1, 0)$ and $(0, 1, -1, 0)$.

30. Find all values of a for which $\{(a^2, 0, 1), (0, a, 2), (1, 0, 1)\}$ is a basis for R^3.

31. Find a basis for the subspace W of M_{33} consisting of all symmetric matrices.

32. Find a basis for the subspace of M_{33} consisting of all diagonal matrices.

33. Give an example of a two-dimensional subspace of R^4.

34. Give an example of a two-dimensional subspace of P_3.

In Exercises 35 and 36, find a basis for the given plane.

35. $2x - 3y + 4z = 0$. **36.** $x + y - 3z = 0$.

37. Determine if the vectors

$$\begin{bmatrix} 1 \\ 1 \\ 1 \end{bmatrix}, \begin{bmatrix} 1 \\ 1 \\ 0 \end{bmatrix}, \begin{bmatrix} 1 \\ 0 \\ 1 \end{bmatrix}$$

are a basis for B^3.

38. Determine if the vectors

$$\begin{bmatrix} 1 \\ 1 \\ 0 \end{bmatrix}, \begin{bmatrix} 1 \\ 0 \\ 0 \end{bmatrix}, \begin{bmatrix} 0 \\ 1 \\ 1 \end{bmatrix}$$

are a basis for B^3.

39. Determine if the vectors

$$\begin{bmatrix} 1 \\ 0 \\ 0 \\ 1 \end{bmatrix}, \begin{bmatrix} 0 \\ 0 \\ 1 \\ 1 \end{bmatrix}, \begin{bmatrix} 0 \\ 1 \\ 1 \\ 0 \end{bmatrix}, \begin{bmatrix} 1 \\ 1 \\ 0 \\ 0 \end{bmatrix}$$

are a basis for B^4.

40. Determine if the vectors

$$\begin{bmatrix} 1 \\ 1 \\ 0 \\ 1 \end{bmatrix}, \begin{bmatrix} 1 \\ 0 \\ 1 \\ 1 \end{bmatrix}, \begin{bmatrix} 0 \\ 1 \\ 1 \\ 0 \end{bmatrix}, \begin{bmatrix} 1 \\ 1 \\ 1 \\ 1 \end{bmatrix}$$

are a basis for B^4.

Theoretical Exercises

T.1. Suppose that in the nonzero vector space V, the largest number of vectors in a linearly independent set is m. Show that any set of m linearly independent vectors in V is a basis for V.

T.2. Show that if V is a finite-dimensional vector space, then every nonzero subspace W of V has a finite basis and $\dim W \leq \dim V$.

T.3. Show that if $\dim V = n$, then any $n + 1$ vectors in V are linearly dependent.

T.4. Show that if $\dim V = n$, then no set of $n - 1$ vectors in V can span V.

T.5. Prove Theorem 6.8.

T.6. Prove Theorem 6.9.

T.7. Show that if W is a subspace of a finite-dimensional vector space V and $\dim W = \dim V$, then $W = V$.

T.8. Show that the subspaces of R^3 are $\{\mathbf{0}\}$, R^3, all lines through the origin, and all planes through the origin.

T.9. Show that if $\{\mathbf{v}_1, \mathbf{v}_2, \ldots, \mathbf{v}_n\}$ is a basis for a vector space V and $c \neq 0$, then $\{c\mathbf{v}_1, \mathbf{v}_2, \ldots, \mathbf{v}_n\}$ is also a basis for V.

T.10. Let $S = \{\mathbf{v}_1, \mathbf{v}_2, \mathbf{v}_3\}$ be a basis for vector space V. Then show that $T = \{\mathbf{w}_1, \mathbf{w}_2, \mathbf{w}_3\}$, where

$$\mathbf{w}_1 = \mathbf{v}_1 + \mathbf{v}_2 + \mathbf{v}_3,$$
$$\mathbf{w}_2 = \mathbf{v}_2 + \mathbf{v}_3,$$

and

$$\mathbf{w}_3 = \mathbf{v}_3,$$

is also a basis for V.

T.11. Let

$$S = \{\mathbf{v}_1, \mathbf{v}_2, \ldots, \mathbf{v}_n\}$$

be a set of nonzero vectors in a vector space V such that every vector in V can be written in one and only one way as a linear combination of the vectors in S. Show that S is a basis for V.

T.12. Suppose that

$$\{\mathbf{v}_1, \mathbf{v}_2, \ldots, \mathbf{v}_n\}$$

is a basis for R^n. Show that if A is an $n \times n$ nonsingular matrix, then

$$\{A\mathbf{v}_1, A\mathbf{v}_2, \ldots, A\mathbf{v}_n\}$$

is also a basis for R^n. (*Hint*: See Exercise T.10 in Section 6.3.)

T.13. Suppose that

$$\{\mathbf{v}_1, \mathbf{v}_2, \ldots, \mathbf{v}_n\}$$

is a linearly independent set of vectors in R^n and let A be a singular matrix. Prove or disprove that

$$\{A\mathbf{v}_1, A\mathbf{v}_2, \ldots, A\mathbf{v}_n\}$$

is linearly independent.

T.14. Show that the vector space P of all polynomials is not finite-dimensional. [*Hint*: Suppose that

$$\{p_1(t), p_2(t), \ldots, p_k(t)\}$$

is a finite basis for P. Let $d_j = \text{degree } p_j(t)$. Establish a contradiction.]

T.15. Show that the set of vectors

$$\{t^n, t^{n-1}, \ldots, t, 1\}$$

in P_n is linearly independent.

T.16. Show that if the sum of the vectors $\mathbf{v}_1, \mathbf{v}_2, \ldots, \mathbf{v}_n$ from B^n is $\mathbf{0}$, then these vectors cannot form a basis for B^n.

T.17. Let $S = \{\mathbf{v}_1, \mathbf{v}_2, \mathbf{v}_3\}$ be a set of vectors in B^3.

(a) Find linearly independent vectors $\mathbf{v}_1, \mathbf{v}_2, \mathbf{v}_3$ such that $\mathbf{v}_1 + \mathbf{v}_2 + \mathbf{v}_3 \neq \mathbf{0}$.

(b) Find linearly dependent vectors $\mathbf{v}_1, \mathbf{v}_2, \mathbf{v}_3$ such that $\mathbf{v}_1 + \mathbf{v}_2 + \mathbf{v}_3 \neq \mathbf{0}$.

MATLAB Exercises

In order to use MATLAB *in this section, you should have read Section 12.7. In the following exercises we relate the theory developed in the section to computational procedures in* MATLAB *that aid in analyzing the situation.*

To determine if a set $S = \{\mathbf{v}_1, \mathbf{v}_2, \ldots, \mathbf{v}_k\}$ *is a basis for a vector space* V*, the definition requires that we show span* $S = V$ *and* S *is linearly independent. However, if we know that* dim $V = k$*, then Theorem 6.9 implies that we need only show that either span* $S = V$ *or* S *is linearly independent. The linear independence, in this special case, is easily analyzed using* MATLAB*'s* **rref** *command. Construct the homogeneous system* $\mathbf{Ax} = \mathbf{0}$ *associated with the linear independence/dependence question. Then* S *is linearly independent if and only if*

$$\mathbf{rref(A)} = \begin{bmatrix} \mathbf{I}_k \\ \mathbf{0} \end{bmatrix}.$$

In Exercises ML.1 through ML.6, if this special case can be applied, do so; otherwise, determine if S *is a basis for* V *in the conventional manner.*

ML.1. $S = \{(1, 2, 1), (2, 1, 1), (2, 2, 1)\}$ in $V = R^3$

ML.2. $S = \{2t - 2, t^2 - 3t + 1, 2t^2 - 8t + 4\}$ in $V = P_2$

ML.3. $S = \{(1, 1, 0, 0), (2, 1, 1, -1), (0, 0, 1, 1), (1, 2, 1, 2)\}$ in $V = R^4$

ML.4. $S = \{(1, 2, 1, 0), (2, 1, 3, 1), (2, -2, 4, 2)\}$ in $V = \text{span } S$

ML.5. $S = \{(1, 2, 1, 0), (2, 1, 3, 1), (2, 2, 1, 2)\}$ in $V = \text{span } S$

ML.6. $V =$ the subspace of R^3 of all vectors of the form (a, b, c), where $b = 2a - c$ and $S = \{(0, 1, -1), (1, 1, 1)\}$.

In Exercises ML.7 through ML.9, use MATLAB*'s* **rref** *command to determine a subset of* S *that is a basis for span* S*. See Example 5.*

ML.7. $S = \{(1, 1, 0, 0), (-2, -2, 0, 0), (1, 0, 2, 1), (2, 1, 2, 1), (0, 1, 1, 1)\}$.

What is dim span S? Does span $S = R^4$?

ML.8. $S = \left\{ \begin{bmatrix} 1 & 2 \\ 1 & 2 \end{bmatrix}, \begin{bmatrix} 1 & 0 \\ 1 & 1 \end{bmatrix}, \begin{bmatrix} 0 & 2 \\ 0 & 1 \end{bmatrix}, \begin{bmatrix} 2 & 4 \\ 2 & 4 \end{bmatrix}, \begin{bmatrix} 1 & 0 \\ 0 & 1 \end{bmatrix} \right\}$.

What is dim span S? Does span $S = M_{22}$?

ML.9. $S = \{t - 2, 2t - 1, 4t - 2, t^2 - t + 1, t^2 + 2t + 1\}$.
What is dim span S? Does span $S = P_2$?

An interpretation of Theorem 6.8 is that any linearly independent subset S *of vector space* V *can be extended to a basis for* V*. Following the ideas in Example 9, use* MATLAB*'s* **rref** *command to extend* S *to a basis for* V *in Exercises ML.10 through ML.12.*

ML.10. $S = \{(1, 1, 0, 0), (1, 0, 1, 0)\}$, $V = R^4$

ML.11. $S = \{t^3 - t + 1, t^3 + 2\}$, $V = P_3$

ML.12. $S = \{(0, 3, 0, 2, -1)\}$,
$V =$ the subspace of R^5 consisting of all vectors of the form (a, b, c, d, e), where $c = a$, $b = 2d + e$

6.5 HOMOGENEOUS SYSTEMS

Homogeneous systems play a central role in linear algebra. This will be seen in Chapter 8, where the foundations of the subject are all integrated to solve one of the major problems occurring in a wide variety of applications. In this section we deal with several problems involving homogeneous systems that will be fundamental in Chapter 8. Here we are able to focus our attention on these problems without being distracted by the additional material in Chapter 8.

Consider the homogeneous system

$$A\mathbf{x} = \mathbf{0},$$

where A is an $m \times n$ matrix. As we have already observed in Example 9 of Section 6.2, the set of all solutions to this homogeneous system is a subspace of R^n. An extremely important problem, which will occur repeatedly in Chapter 8, is that of finding a basis for this solution space. To find such a basis, we use the method of Gauss–Jordan reduction presented in Section 1.6. Thus we transform the augmented matrix $\left[A \mid \mathbf{0}\right]$ of the system to a matrix $\left[B \mid \mathbf{0}\right]$ in reduced row echelon form, where B has r nonzero rows, $1 \le r \le m$. Without loss of generality we may assume that the leading 1s in the r nonzero rows of B occur in the first r columns. If $r = n$, then

$$
\left[B \,\vdots\, \mathbf{0}\right] = \overbrace{\begin{bmatrix} 1 & 0 & \cdots & 0 & 0 \\ 0 & 1 & \cdots & 0 & 0 \\ \vdots & \vdots & \ddots & \vdots & \vdots \\ 0 & 0 & \cdots & 0 & 1 \\ 0 & 0 & \cdots & & 0 \\ \vdots & \vdots & & & \vdots \\ 0 & 0 & \cdots & & 0 \end{bmatrix}}^{n}\begin{matrix} \vdots\, 0 \\ \vdots\, 0 \\ \vdots\, \vdots \\ \vdots\, 0 \\ \vdots\, 0 \\ \vdots\, \vdots \\ \vdots\, 0 \end{matrix} \quad \begin{matrix} \left.\right\} r = n \\ \left.\right\} m \end{matrix}
$$

and the only solution to $A\mathbf{x} = \mathbf{0}$ is the trivial one. The solution space has no basis (sometimes referred to as an **empty basis**) and its dimension is zero.

If $r < n$, then

$$
\left[B \,\vdots\, \mathbf{0}\right] = \overbrace{\begin{bmatrix} 1 & 0 & 0 & \cdots & 0 & b_{1\,r+1} & \cdots & b_{1n} \\ 0 & 1 & 0 & \cdots & 0 & b_{2\,r+1} & \cdots & b_{2n} \\ 0 & 0 & 1 & \cdots & 0 & \vdots & & \vdots \\ \vdots & \vdots & \vdots & \ddots & \vdots & & & \\ 0 & 0 & 0 & \cdots & 1 & b_{r\,r+1} & \cdots & b_{rn} \\ 0 & 0 & 0 & \cdots & 0 & 0 & \cdots & 0 \\ \vdots & \vdots & \vdots & & \vdots & \vdots & & \vdots \\ 0 & 0 & 0 & \cdots & 0 & \cdots & & 0 \end{bmatrix}}^{n}\begin{matrix} \vdots\, 0 \\ \vdots\, 0 \\ \vdots\, \vdots \\ \\ \vdots\, 0 \\ \vdots\, 0 \\ \vdots\, \vdots \\ \vdots\, 0 \end{matrix} \quad \begin{matrix} \left.\right\} r \\ \left.\right\} m \end{matrix}
$$

Solving for the unknowns corresponding to the leading 1s, we have

$$
\begin{aligned}
x_1 &= -b_{1\,r+1}x_{r+1} - b_{1\,r+2}x_{r+2} - \cdots - b_{1n}x_n \\
x_2 &= -b_{2\,r+1}x_{r+1} - b_{2\,r+2}x_{r+2} - \cdots - b_{2n}x_n \\
&\vdots \\
x_r &= -b_{r\,r+1}x_{r+1} - b_{r\,r+2}x_{r+2} - \cdots - b_{rn}x_n,
\end{aligned}
$$

where $x_{r+1}, x_{r+2}, \ldots, x_n$ can be assigned arbitrary real values s_j,

$j = 1, 2, \ldots, p$, and $p = n - r$. Thus

$$\mathbf{x} = \begin{bmatrix} x_1 \\ x_2 \\ \vdots \\ x_r \\ x_{r+1} \\ x_{r+2} \\ \vdots \\ x_n \end{bmatrix} = \begin{bmatrix} -b_{1\,r+1}s_1 - b_{1\,r+2}s_2 - \cdots - b_{1n}s_p \\ -b_{2\,r+1}s_1 - b_{2\,r+2}s_2 - \cdots - b_{2n}s_p \\ \vdots \\ -b_{r\,r+1}s_1 - b_{r\,r+2}s_2 - \cdots - b_{rn}s_p \\ s_1 \\ s_2 \\ \vdots \\ s_p \end{bmatrix}$$

$$= s_1 \begin{bmatrix} -b_{1\,r+1} \\ -b_{2\,r+1} \\ \vdots \\ -b_{r\,r+1} \\ 1 \\ 0 \\ 0 \\ \vdots \\ 0 \\ 0 \end{bmatrix} + s_2 \begin{bmatrix} -b_{1\,r+2} \\ -b_{2\,r+2} \\ \vdots \\ -b_{r\,r+2} \\ 0 \\ 1 \\ 0 \\ \vdots \\ 0 \\ 0 \end{bmatrix} + \cdots + s_p \begin{bmatrix} -b_{1n} \\ -b_{2n} \\ \vdots \\ -b_{rn} \\ 0 \\ 0 \\ 0 \\ \vdots \\ 0 \\ 1 \end{bmatrix}.$$

Since $s_1, s_2, \ldots, s_p$ can be assigned arbitrary real values, we make the following choices for these values:

$$\begin{array}{lllll} s_1 = 1, & s_2 = 0, & \ldots, & & s_p = 0 \\ s_1 = 0, & s_2 = 1, & \ldots, & & s_p = 0 \\ & & \vdots & & \\ s_1 = 0, & s_2 = 0, & \ldots, & s_{p-1} = 0, & s_p = 1, \end{array}$$

obtaining the solutions

$$\mathbf{x}_1 = \begin{bmatrix} -b_{1\,r+1} \\ -b_{2\,r+1} \\ \vdots \\ -b_{r\,r+1} \\ 1 \\ 0 \\ 0 \\ \vdots \\ 0 \\ 0 \end{bmatrix}, \quad \mathbf{x}_2 = \begin{bmatrix} -b_{1\,r+2} \\ -b_{2\,r+2} \\ \vdots \\ -b_{r\,r+2} \\ 0 \\ 1 \\ 0 \\ \vdots \\ 0 \\ 0 \end{bmatrix}, \ldots, \quad \mathbf{x}_p = \begin{bmatrix} -b_{1n} \\ -b_{2n} \\ \vdots \\ -b_{rn} \\ 0 \\ 0 \\ 0 \\ \vdots \\ 0 \\ 1 \end{bmatrix}.$$

Since

$$\mathbf{x} = s_1\mathbf{x}_1 + s_2\mathbf{x}_2 + \cdots + s_p\mathbf{x}_p,$$

we see that $\{\mathbf{x}_1, \mathbf{x}_2, \ldots, \mathbf{x}_p\}$ spans the solution space of $A\mathbf{x} = \mathbf{0}$. Moreover, if we form the equation

$$c_1\mathbf{x}_1 + c_2\mathbf{x}_2 + \cdots + c_p\mathbf{x}_p = \mathbf{0},$$

its coefficient matrix is the matrix whose columns are $\mathbf{x}_1, \mathbf{x}_2, \dots, \mathbf{x}_p$. If we look at rows $r+1, r+2, \dots, n$ of this matrix, we readily see that

$$c_1 = c_2 = \cdots = c_p = 0.$$

Hence $\{\mathbf{x}_1, \mathbf{x}_2, \dots, \mathbf{x}_p\}$ is linearly independent and forms a basis for the solution space of $A\mathbf{x} = \mathbf{0}$, the null space of A.

The procedure for finding a basis for the solution space of a homogeneous system $A\mathbf{x} = \mathbf{0}$, or the null space of A, where A is $m \times n$, is as follows.

Step 1. Solve the given homogeneous system by Gauss–Jordan reduction. If the solution contains no arbitrary constants, then the solution space is $\{\mathbf{0}\}$, which has no basis; the dimension of the solution space is zero.

Step 2. If the solution $\mathbf{x}$ contains arbitrary constants, write $\mathbf{x}$ as a linear combination of vectors $\mathbf{x}_1, \mathbf{x}_2, \dots, \mathbf{x}_p$ with $s_1, s_2, \dots, s_p$ as coefficients:

$$\mathbf{x} = s_1\mathbf{x}_1 + s_2\mathbf{x}_2 + \cdots + s_p\mathbf{x}_p.$$

Step 3. The set of vectors $\{\mathbf{x}_1, \mathbf{x}_2, \dots, \mathbf{x}_p\}$ is a basis for the solution space of $A\mathbf{x} = \mathbf{0}$; the dimension of the solution space is p.

Remark In Step 1, suppose that the matrix in reduced row echelon form to which $\left[A \;\vdots\; \mathbf{0}\right]$ has been transformed has r nonzero rows (also r leading 1s). Then $p = n - r$. That is, the dimension of the solution space is $n - r$. Moreover, a solution $\mathbf{x}$ to $A\mathbf{x} = \mathbf{0}$ has $n - r$ arbitrary constants.

DEFINITION If A is an $m \times n$ matrix, we refer to the dimension of the null space of A as the **nullity** of A, denoted by nullity A.

Remark The nullity of a matrix A is the number of arbitrary constants in the solution to the homogeneous system $A\mathbf{x} = \mathbf{0}$.

EXAMPLE 1 Find a basis for and the dimension of the solution space W of the homogeneous system

$$\begin{bmatrix} 1 & 1 & 4 & 1 & 2 \\ 0 & 1 & 2 & 1 & 1 \\ 0 & 0 & 0 & 1 & 2 \\ 1 & -1 & 0 & 0 & 2 \\ 2 & 1 & 6 & 0 & 1 \end{bmatrix} \begin{bmatrix} x_1 \\ x_2 \\ x_3 \\ x_4 \\ x_5 \end{bmatrix} = \begin{bmatrix} 0 \\ 0 \\ 0 \\ 0 \\ 0 \end{bmatrix}.$$

Solution ***Step 1.*** To solve the given system by the Gauss–Jordan reduction method, we transform the augmented matrix to reduced row echelon form, obtaining (verify)

$$\left[\begin{array}{ccccc:c} 1 & 0 & 2 & 0 & 1 & 0 \\ 0 & 1 & 2 & 0 & -1 & 0 \\ 0 & 0 & 0 & 1 & 2 & 0 \\ 0 & 0 & 0 & 0 & 0 & 0 \\ 0 & 0 & 0 & 0 & 0 & 0 \end{array}\right].$$

Every solution is of the form (verify)

$$\mathbf{x} = \begin{bmatrix} -2s - t \\ -2s + t \\ s \\ -2t \\ t \end{bmatrix}, \tag{1}$$

where s and t are any real numbers.

Step 2. Every vector in W is a solution and is then of the form given by Equation (1). We can then write every vector in W as

$$\mathbf{x} = s\begin{bmatrix} -2 \\ -2 \\ 1 \\ 0 \\ 0 \end{bmatrix} + t\begin{bmatrix} -1 \\ 1 \\ 0 \\ -2 \\ 1 \end{bmatrix}. \tag{2}$$

Since s and t can take on any values, we first let $s = 1$, $t = 0$, and then let $s = 0$, $t = 1$, in Equation (2), obtaining as solutions

$$\mathbf{x}_1 = \begin{bmatrix} -2 \\ -2 \\ 1 \\ 0 \\ 0 \end{bmatrix} \quad \text{and} \quad \mathbf{x}_2 = \begin{bmatrix} -1 \\ 1 \\ 0 \\ -2 \\ 1 \end{bmatrix}.$$

Step 3. The set $\{\mathbf{x}_1, \mathbf{x}_2\}$ is a basis for W. Moreover, $\dim W = 2$. ■

The following example illustrates a type of problem that we will be solving frequently in Chapter 8.

EXAMPLE 2 Find a basis for the solution space of the homogeneous system $(\lambda I_3 - A)\mathbf{x} = \mathbf{0}$ for $\lambda = -2$ and

$$A = \begin{bmatrix} -3 & 0 & 1 \\ 2 & 1 & 0 \\ 0 & 0 & -2 \end{bmatrix}.$$

Solution We form $-2I_3 - A$:

$$-2\begin{bmatrix} 1 & 0 & 0 \\ 0 & 1 & 0 \\ 0 & 0 & 1 \end{bmatrix} - \begin{bmatrix} -3 & 0 & -1 \\ 2 & 1 & 0 \\ 0 & 0 & -2 \end{bmatrix} = \begin{bmatrix} 1 & 0 & 1 \\ -2 & -3 & 0 \\ 0 & 0 & 0 \end{bmatrix}.$$

This last matrix is the coefficient matrix of the homogeneous system, so we transform the augmented matrix

$$\left[\begin{array}{ccc|c} 1 & 0 & 1 & 0 \\ -2 & -3 & 0 & 0 \\ 0 & 0 & 0 & 0 \end{array}\right]$$

to reduced row echelon form, obtaining (verify)

$$\left[\begin{array}{ccc|c} 1 & 0 & 1 & 0 \\ 0 & 1 & -\frac{2}{3} & 0 \\ 0 & 0 & 0 & 0 \end{array}\right].$$

Every solution is then of the form (verify)

$$\mathbf{x} = \begin{bmatrix} -s \\ \frac{2}{3}s \\ s \end{bmatrix},$$

where s is any real number. Then every vector in the solution can be written as

$$\mathbf{x} = s\begin{bmatrix} -1 \\ \frac{2}{3} \\ 1 \end{bmatrix},$$

so

$$\left\{\begin{bmatrix} -1 \\ \frac{2}{3} \\ 1 \end{bmatrix}\right\}$$

is a basis for the solution space. ■

Another important problem that we have to solve often in Chapter 8 is illustrated in the following example.

EXAMPLE 3 Find all real numbers λ such that the homogeneous system $(\lambda I_2 - A)\mathbf{x} = \mathbf{0}$ has a nontrivial solution for

$$A = \begin{bmatrix} 1 & 5 \\ 3 & -1 \end{bmatrix}.$$

Solution We form $\lambda I_2 - A$:

$$\lambda\begin{bmatrix} 1 & 0 \\ 0 & 1 \end{bmatrix} - \begin{bmatrix} 1 & 5 \\ 3 & -1 \end{bmatrix} = \begin{bmatrix} \lambda - 1 & -5 \\ -3 & \lambda + 1 \end{bmatrix}.$$

The homogeneous system $(\lambda I_2 - A)\mathbf{x} = \mathbf{0}$ is then

$$\begin{bmatrix} \lambda - 1 & -5 \\ -3 & \lambda + 1 \end{bmatrix}\begin{bmatrix} x_1 \\ x_2 \end{bmatrix} = \begin{bmatrix} 0 \\ 0 \end{bmatrix}.$$

By Corollary 3.4, this homogeneous system has a nontrivial solution if and only if the determinant of the coefficient matrix is zero; that is, if and only if

$$\det\left(\begin{bmatrix} \lambda - 1 & -5 \\ -3 & \lambda + 1 \end{bmatrix}\right) = 0,$$

or if and only if

$$(\lambda - 1)(\lambda + 1) - 15 = 0$$
$$\lambda^2 - 16 = 0$$
$$\lambda = 4 \quad \text{or} \quad \lambda = -4.$$

Thus, when $\lambda = 4$ or -4, the homogeneous system $(\lambda I_2 - A)\mathbf{x} = \mathbf{0}$ for the given matrix A has a nontrivial solution. ■

RELATIONSHIP BETWEEN NONHOMOGENEOUS LINEAR SYSTEMS AND HOMOGENEOUS SYSTEMS

We have already noted in Section 6.2 (see the comment after Example 9) that if A is $m \times n$, then the set of all solutions to the linear system $A\mathbf{x} = \mathbf{b}$, $\mathbf{b} \neq \mathbf{0}$, is not a subspace of R^n. The following example illustrates a geometric relationship between the set of all solutions to the nonhomogeneous system $A\mathbf{x} = \mathbf{b}$, $\mathbf{b} \neq \mathbf{0}$, and the associated homogeneous system $A\mathbf{x} = \mathbf{0}$.

EXAMPLE 4 Consider the linear system

$$\begin{bmatrix} 1 & 2 & -3 \\ 2 & 4 & -6 \\ 3 & 6 & -9 \end{bmatrix} \begin{bmatrix} x_1 \\ x_2 \\ x_3 \end{bmatrix} = \begin{bmatrix} 2 \\ 4 \\ 6 \end{bmatrix}.$$

The set of all solutions to this linear system consists of all vectors of the form

$$\mathbf{x} = \begin{bmatrix} 2 - 2r + 3s \\ r \\ s \end{bmatrix}$$

(verify), which can be written as

$$\mathbf{x} = \begin{bmatrix} 2 \\ 0 \\ 0 \end{bmatrix} + r \begin{bmatrix} -2 \\ 1 \\ 0 \end{bmatrix} + s \begin{bmatrix} 3 \\ 0 \\ 1 \end{bmatrix}.$$

The set of all solutions to the associated homogeneous system is the two-dimensional subspace of R^3 consisting of all vectors of the form

$$\mathbf{x} = r \begin{bmatrix} -2 \\ 1 \\ 0 \end{bmatrix} + s \begin{bmatrix} 3 \\ 0 \\ 1 \end{bmatrix}.$$

This subspace is a plane Π_1 passing through the origin; the set of all solutions to the given nonhomogeneous system is a plane Π_2 that does not pass through the origin and is obtained by shifting Π_1 parallel to itself. This situation is illustrated in Figure 6.9. ■

Figure 6.9 ►

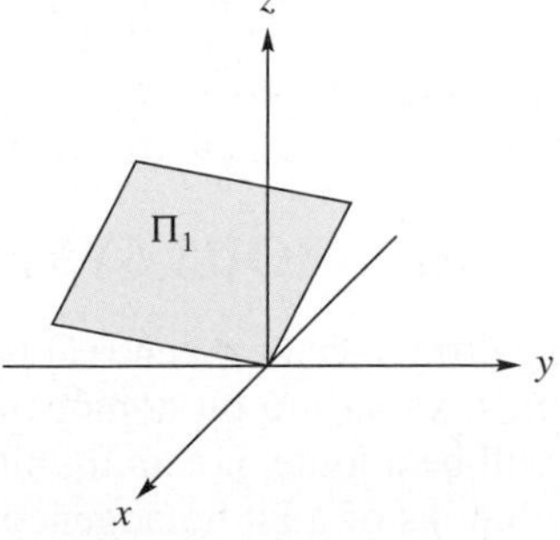

Π_1 is the solution space to $A\mathbf{x} = \mathbf{0}$.

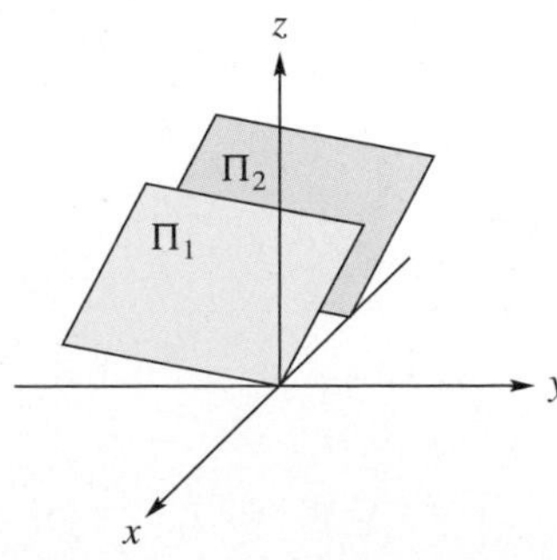

Π_2 is the set of all solutions to $A\mathbf{x} = \mathbf{b}$.

The following result, which is important in the study of differential equations, was presented in Section 1.6 and its proof left to Exercise T.13 of that section.

If $\mathbf{x}_p$ is a particular solution to the nonhomogeneous system $A\mathbf{x} = \mathbf{b}$, $\mathbf{b} \neq \mathbf{0}$, and $\mathbf{x}_h$ is a solution to the associated homogeneous system $A\mathbf{x} = \mathbf{0}$, then $\mathbf{x}_p + \mathbf{x}_h$ is a solution to the given system $A\mathbf{x} = \mathbf{b}$. Moreover, every solution $\mathbf{x}$ to the nonhomogeneous linear system $A\mathbf{x} = \mathbf{b}$ can be written as $\mathbf{x}_p + \mathbf{x}_h$, where $\mathbf{x}_p$ is a particular solution to the given nonhomogeneous system and $\mathbf{x}_h$ is a solution to the associated homogeneous system $A\mathbf{x} = \mathbf{0}$. Thus, in Example 4,

$$\mathbf{x}_p = \begin{bmatrix} 2 \\ 0 \\ 0 \end{bmatrix} \quad \text{and} \quad \mathbf{x}_h = r \begin{bmatrix} -2 \\ 1 \\ 0 \end{bmatrix} + s \begin{bmatrix} 3 \\ 0 \\ 1 \end{bmatrix},$$

where r and s are any real numbers.

EXAMPLE 5 Consider the linear system

$$\begin{aligned} x_1 + 2x_2 \phantom{{}+x_3} - 3x_4 + x_5 \phantom{{}+2x_6} &= 2 \\ x_1 + 2x_2 + x_3 - 3x_4 + x_5 + 2x_6 &= 3 \\ x_1 + 2x_2 \phantom{{}+x_3} - 3x_4 + 2x_5 + x_6 &= 4 \\ 3x_1 + 6x_2 + x_3 - 9x_4 + 4x_5 + 3x_6 &= 9, \end{aligned}$$

which was defined in Example 10 in Section 1.6. There we found that a solution to the given linear system is given by

$$\mathbf{x} = \begin{bmatrix} r + 3s - 2t \\ t \\ 1 - 2r \\ s \\ 2 - r \\ r \end{bmatrix} = \begin{bmatrix} 0 \\ 0 \\ 1 \\ 0 \\ 2 \\ 0 \end{bmatrix} + \begin{bmatrix} r + 3s - 2t \\ t \\ -2r \\ s \\ -r \\ r \end{bmatrix},$$

where r, s, and t are any real numbers. Let

$$\mathbf{x}_p = \begin{bmatrix} 0 \\ 0 \\ 1 \\ 0 \\ 2 \\ 0 \end{bmatrix} \quad \text{and} \quad \mathbf{x}_h = \begin{bmatrix} r + 3s - 2t \\ t \\ -2r \\ s \\ -r \\ r \end{bmatrix}.$$

Then $\mathbf{x} = \mathbf{x}_p + \mathbf{x}_h$. Moreover, $\mathbf{x}_p$ is a particular solution to the given system (verify) and $\mathbf{x}_h$ is a solution to the associated homogeneous system. ■

BIT HOMOGENEOUS SYSTEMS (OPTIONAL)

The primary difference between real homogeneous systems and bit homogeneous systems is that if a bit homogeneous system has a nontrivial solution, then there will be a finite, not an infinite, number of nontrivial solutions. The set of all solutions of a bit homogeneous system is a subspace, and that subspace contains a finite number of vectors.

EXAMPLE 6 Find a basis for the solution space W of the bit homogeneous system

$$\begin{aligned} x + y \quad &= 0 \\ y + z &= 0 \\ x \quad + z &= 0. \end{aligned}$$

Solution We form the augmented matrix and determine its reduced row echelon form to obtain (verify)

$$\left[\begin{array}{ccc|c} 1 & 0 & 1 & 0 \\ 0 & 1 & 1 & 0 \\ 0 & 0 & 0 & 0 \end{array}\right].$$

Thus we have $x = -z$ and $y = -z$, where z is either 0 or 1. (Recall that the negative of a bit is itself.) Hence the set of solutions consists of all vectors in B^3 of the form $\begin{bmatrix} z \\ z \\ z \end{bmatrix}$ and so there are exactly two solutions $\begin{bmatrix} 0 \\ 0 \\ 0 \end{bmatrix}$ and $\begin{bmatrix} 1 \\ 1 \\ 1 \end{bmatrix}$. A basis for the solution space is $\left\{\begin{bmatrix} 1 \\ 1 \\ 1 \end{bmatrix}\right\}$. (Verify.) ■

EXAMPLE 7 In Example 21 of Section 1.6 we showed that the bit homogeneous system corresponding to the augmented matrix

$$\left[\begin{array}{cccc|c} 1 & 0 & 1 & 1 & 0 \\ 1 & 1 & 0 & 0 & 0 \\ 0 & 1 & 1 & 1 & 0 \end{array}\right]$$

has reduced row echelon form

$$\left[\begin{array}{cccc|c} 1 & 0 & 1 & 1 & 0 \\ 0 & 1 & 1 & 1 & 0 \\ 0 & 0 & 0 & 0 & 0 \end{array}\right].$$

Hence there are two arbitrary bits b_1 and b_2 in the solution (verify) given by

$$\mathbf{x} = \begin{bmatrix} b_1 + b_2 \\ b_1 + b_2 \\ b_1 \\ b_2 \end{bmatrix}. \tag{3}$$

From Equation (3) we have that

$$\mathbf{x} = b_1 \begin{bmatrix} 1 \\ 1 \\ 1 \\ 0 \end{bmatrix} + b_2 \begin{bmatrix} 1 \\ 1 \\ 0 \\ 1 \end{bmatrix}$$

so

$$\left\{\begin{bmatrix} 1 \\ 1 \\ 1 \\ 0 \end{bmatrix}, \begin{bmatrix} 1 \\ 1 \\ 0 \\ 1 \end{bmatrix}\right\}$$

is a basis for the solution space W. (Verify.) There are exactly three nontrivial solutions: one for each of the choices $b_1 = 0$, $b_2 = 1$; $b_1 = 1$, $b_2 = 0$; and $b_1 = b_2 = 1$. Hence

$$W = \left\{ \begin{bmatrix} 0 \\ 0 \\ 0 \\ 0 \end{bmatrix}, \begin{bmatrix} 1 \\ 1 \\ 0 \\ 1 \end{bmatrix}, \begin{bmatrix} 1 \\ 1 \\ 1 \\ 0 \end{bmatrix}, \begin{bmatrix} 0 \\ 0 \\ 1 \\ 1 \end{bmatrix} \right\}.$$

■

EXAMPLE 8 Let $A\mathbf{x} = \mathbf{b}$ be a bit system such that the reduced row echelon form of the augmented matrix $\left[A \mid \mathbf{b}\right]$ is given by

$$\left[\begin{array}{cccc|c} 1 & 0 & 1 & 0 & 0 \\ 0 & 1 & 1 & 1 & 1 \\ 0 & 0 & 0 & 0 & 0 \end{array}\right].$$

Hence we have a nonhomogeneous system of two equations in four unknowns that can be expressed in the form

$$\begin{aligned} x_1 \quad + x_3 \qquad &= 0 \\ x_2 + x_3 + x_4 &= 1. \end{aligned}$$

It follows that there are two arbitrary bits since

$$\begin{aligned} x_1 &= -x_3 \\ x_2 &= -x_3 - x_4 + 1. \end{aligned}$$

Let $x_3 = b_1$ and $x_4 = b_2$ (recall that the negative of a bit is itself). Then we have the solution

$$\mathbf{x} = \begin{bmatrix} b_1 \\ b_1 + b_2 + 1 \\ b_1 \\ b_2 \end{bmatrix}.$$

The particular solution is

$$\mathbf{x}_p = \begin{bmatrix} 0 \\ 1 \\ 0 \\ 0 \end{bmatrix}$$

and the solution of the corresponding homogeneous system is

$$\mathbf{x}_h = b_1 \begin{bmatrix} 1 \\ 1 \\ 1 \\ 0 \end{bmatrix} + b_2 \begin{bmatrix} 0 \\ 1 \\ 0 \\ 1 \end{bmatrix}.$$

■

Key Terms

Nullity

6.5 Exercises

1. Let

$$A = \begin{bmatrix} 2 & -1 & -2 \\ -4 & 2 & 4 \\ -8 & 4 & 8 \end{bmatrix}.$$

(a) Find the set of all solutions to $A\mathbf{x} = \mathbf{0}$.

(b) Express each solution as a linear combination of two vectors in R^3.

(c) Sketch these vectors in a three-dimensional coordinate system to show that the solution space is a plane through the origin.

2. Let

$$A = \begin{bmatrix} 1 & 1 & -2 \\ -2 & -2 & 4 \\ -1 & -1 & 2 \end{bmatrix}.$$

(a) Find the set of all solutions to $A\mathbf{x} = \mathbf{0}$.

(b) Express each solution as a linear combination of two vectors in R^3.

(c) Sketch these vectors in a three-dimensional coordinate system to show that the solution space is a plane through the origin.

In Exercises 3 through 10, find a basis for and the dimension of the solution space of the given homogeneous system.

3. $$\begin{aligned} x_1 + x_2 + x_3 + x_4 &= 0 \\ 2x_1 + x_2 - x_3 + x_4 &= 0 \end{aligned}$$

4. $$\begin{bmatrix} 1 & -1 & 1 & -2 & 1 \\ 3 & -3 & 2 & 0 & 2 \end{bmatrix} \begin{bmatrix} x_1 \\ x_2 \\ x_3 \\ x_4 \\ x_5 \end{bmatrix} = \begin{bmatrix} 0 \\ 0 \end{bmatrix}$$

5. $$\begin{aligned} x_1 + 2x_2 - x_3 + 3x_4 &= 0 \\ 2x_1 + 2x_2 - x_3 + 2x_4 &= 0 \\ x_1 \qquad + 3x_3 + 3x_4 &= 0 \end{aligned}$$

6. $$\begin{aligned} x_1 - x_2 + 2x_3 + 3x_4 + 4x_5 &= 0 \\ -x_1 + 2x_2 + 3x_3 + 4x_4 + 5x_5 &= 0 \\ x_1 - x_2 + 3x_3 + 5x_4 + 6x_5 &= 0 \\ 3x_1 - 4x_2 + x_3 + 2x_4 + 3x_5 &= 0 \end{aligned}$$

7. $$\begin{bmatrix} 1 & 2 & 1 & 2 & 1 \\ 1 & 2 & 2 & 1 & 2 \\ 2 & 4 & 3 & 3 & 3 \\ 0 & 0 & 1 & -1 & -1 \end{bmatrix} \begin{bmatrix} x_1 \\ x_2 \\ x_3 \\ x_4 \\ x_5 \end{bmatrix} = \begin{bmatrix} 0 \\ 0 \\ 0 \\ 0 \end{bmatrix}$$

8. $$\begin{bmatrix} 1 & 0 & 2 \\ 2 & 1 & 3 \\ 3 & 1 & 2 \end{bmatrix} \begin{bmatrix} x_1 \\ x_2 \\ x_3 \end{bmatrix} = \begin{bmatrix} 0 \\ 0 \\ 0 \end{bmatrix}$$

9. $$\begin{bmatrix} 1 & 2 & 2 & -1 & 1 \\ 0 & 2 & 2 & -2 & -1 \\ 2 & 6 & 2 & -4 & 1 \\ 1 & 4 & 0 & -3 & 0 \end{bmatrix} \begin{bmatrix} x_1 \\ x_2 \\ x_3 \\ x_4 \\ x_5 \end{bmatrix} = \begin{bmatrix} 0 \\ 0 \\ 0 \\ 0 \end{bmatrix}$$

10. $$\begin{bmatrix} 1 & 2 & -3 & 2 & 1 & 3 \\ 1 & 2 & -4 & 3 & 3 & 4 \\ -2 & -4 & 6 & 4 & -3 & 2 \\ 0 & 0 & -1 & 5 & 1 & 9 \\ 1 & 2 & -3 & -2 & 0 & 7 \end{bmatrix} \begin{bmatrix} x_1 \\ x_2 \\ x_3 \\ x_4 \\ x_5 \\ x_6 \end{bmatrix} = \begin{bmatrix} 0 \\ 0 \\ 0 \\ 0 \\ 0 \end{bmatrix}$$

In Exercises 11 and 12, find a basis for the null space of the given matrix A.

11. $$A = \begin{bmatrix} 1 & 2 & 3 & -1 \\ 2 & 3 & 2 & 0 \\ 3 & 4 & 1 & 1 \\ 1 & 1 & -1 & 1 \end{bmatrix}$$

12. $$A = \begin{bmatrix} 1 & -1 & 2 & 1 & 0 \\ 2 & 0 & 1 & -1 & 3 \\ 5 & -1 & 3 & 0 & 3 \\ 4 & -2 & 5 & 1 & 3 \\ 1 & 3 & -4 & -5 & 6 \end{bmatrix}$$

In Exercises 13 through 16, find a basis for the solution space of the homogeneous system $(\lambda I_n - A)\mathbf{x} = \mathbf{0}$ for the given scalar λ and given matrix A.

13. $\lambda = 1$, $A = \begin{bmatrix} 3 & 2 \\ 1 & 2 \end{bmatrix}$

14. $\lambda = -3$, $A = \begin{bmatrix} -4 & -3 \\ 2 & 3 \end{bmatrix}$

15. $\lambda = 1$, $A = \begin{bmatrix} 0 & 0 & 1 \\ 1 & 0 & -3 \\ 0 & 1 & 3 \end{bmatrix}$

16. $\lambda = 3$, $A = \begin{bmatrix} 1 & 1 & -2 \\ -1 & 2 & 1 \\ 0 & 1 & -1 \end{bmatrix}$

In Exercises 17 through 20, find all real numbers λ such that the homogeneous system $(\lambda I_n - A)\mathbf{x} = \mathbf{0}$ has a nontrivial solution.

17. $A = \begin{bmatrix} 2 & 3 \\ 2 & -3 \end{bmatrix}$

18. $A = \begin{bmatrix} 3 & 0 \\ 2 & -2 \end{bmatrix}$

19. $A = \begin{bmatrix} 0 & 0 & 0 \\ 0 & 1 & -1 \\ 1 & 0 & 0 \end{bmatrix}$

20. $A = \begin{bmatrix} -2 & 0 & 0 \\ 0 & -2 & -3 \\ 0 & 4 & 5 \end{bmatrix}$

In Exercises 21 and 22, solve the given linear system and write the solution $\mathbf{x}$ as $\mathbf{x} = \mathbf{x}_p + \mathbf{x}_h$, where $\mathbf{x}_p$ is a particular solution to the given system and $\mathbf{x}_h$ is a solution to the associated homogeneous system.

21. $$\begin{aligned} x + 2y - z - w &= 3 \\ x + y + 3z + 2w &= -2 \\ 2x - y + 4z + 3w &= 1 \\ 2x - 2y + 8z + 6w &= -4 \end{aligned}$$

22. $$\begin{aligned} x - y + 2z + 2w &= 1 \\ -x + 2y + 3z + 2w &= 0 \\ 2x + 2y + z \qquad &= 4 \end{aligned}$$

In Exercises 23 through 26, find a basis for and the dimension of the solution space of the given bit homogeneous system.

23. $\begin{aligned} x_1 + x_2 + x_4 &= 0 \\ x_1 + x_3 + x_4 &= 0 \\ x_1 + x_2 + x_3 &= 0 \end{aligned}$

24. $$\begin{bmatrix} 0 & 0 & 1 & 1 \\ 1 & 0 & 0 & 1 \\ 1 & 0 & 1 & 0 \end{bmatrix} \begin{bmatrix} x_1 \\ x_2 \\ x_3 \\ x_4 \end{bmatrix} = \begin{bmatrix} 0 \\ 0 \\ 0 \end{bmatrix}$$

25. $$\begin{bmatrix} 1 & 1 & 0 & 0 & 1 \\ 1 & 0 & 1 & 1 & 1 \\ 0 & 1 & 1 & 0 & 0 \end{bmatrix} \begin{bmatrix} x_1 \\ x_2 \\ x_3 \\ x_4 \\ x_5 \end{bmatrix} = \begin{bmatrix} 0 \\ 0 \\ 0 \end{bmatrix}$$

26. $$\begin{bmatrix} 1 & 1 & 0 & 0 & 0 \\ 1 & 1 & 1 & 0 & 0 \\ 0 & 1 & 1 & 1 & 0 \\ 0 & 0 & 1 & 1 & 1 \\ 0 & 0 & 0 & 1 & 1 \end{bmatrix} \begin{bmatrix} x_1 \\ x_2 \\ x_3 \\ x_4 \\ x_5 \end{bmatrix} = \begin{bmatrix} 0 \\ 0 \\ 0 \\ 0 \\ 0 \end{bmatrix}$$

In Exercises 27 and 28, solve the given bit system and write the solution $\mathbf{x}$ *as* $\mathbf{x}_p + \mathbf{x}_h$.

27. $$\begin{bmatrix} 1 & 1 & 0 & 1 \\ 1 & 0 & 1 & 1 \\ 0 & 1 & 1 & 0 \\ 0 & 0 & 1 & 1 \end{bmatrix} \begin{bmatrix} x_1 \\ x_2 \\ x_3 \\ x_4 \end{bmatrix} = \begin{bmatrix} 1 \\ 1 \\ 0 \\ 0 \end{bmatrix}$$

28. $$\begin{bmatrix} 1 & 0 & 0 & 0 & 1 \\ 1 & 1 & 0 & 0 & 0 \\ 1 & 1 & 1 & 0 & 1 \\ 1 & 0 & 1 & 0 & 0 \end{bmatrix} \begin{bmatrix} x_1 \\ x_2 \\ x_3 \\ x_4 \\ x_5 \end{bmatrix} = \begin{bmatrix} 1 \\ 1 \\ 1 \\ 1 \end{bmatrix}$$

Theoretical Exercises

T.1. Let $S = \{\mathbf{x}_1, \mathbf{x}_2, \dots, \mathbf{x}_k\}$ be a set of solutions to a homogeneous system $A\mathbf{x} = \mathbf{0}$. Show that every vector in span S is a solution to $A\mathbf{x} = \mathbf{0}$.

T.2. Show that if the $n \times n$ coefficient matrix A of the homogeneous system $A\mathbf{x} = \mathbf{0}$ has a row or column of zeros, then $A\mathbf{x} = \mathbf{0}$ has a nontrivial solution.

T.3. (a) Show that the zero matrix is the only 3×3 matrix whose null space has dimension 3.

(b) Let A be a nonzero 3×3 matrix and suppose that $A\mathbf{x} = \mathbf{0}$ has a nontrivial solution. Show that the dimension of the null space of A is either 1 or 2.

T.4. Matrices A and B are $m \times n$ and their reduced row echelon forms are the same. What is the relationship between the null space of A and the null space of B?

MATLAB Exercises

In Exercises ML.1 through ML.3, use MATLAB*'s* **rref** *command to aid in finding a basis for the null space of A. You may also use routine* **homsoln**. *For directions, use* **help**.

ML.1. $A = \begin{bmatrix} 1 & 1 & 2 & 2 & 1 \\ 2 & 0 & 4 & 2 & 4 \\ 1 & 1 & 2 & 2 & 1 \end{bmatrix}$

ML.2. $A = \begin{bmatrix} 2 & 2 & 2 \\ 1 & 2 & 1 \\ 3 & 1 & 0 \\ 0 & 0 & 1 \\ 1 & 0 & 0 \end{bmatrix}$

ML.3. $A = \begin{bmatrix} 1 & 4 & 7 & 0 \\ 2 & 5 & 8 & -1 \\ 3 & 6 & 9 & -2 \end{bmatrix}$

ML.4. For the matrix

$$A = \begin{bmatrix} 1 & 2 \\ 2 & 1 \end{bmatrix}$$

and $\lambda = 3$, the homogeneous system $(\lambda I_2 - A)\mathbf{x} = \mathbf{0}$ has a nontrivial solution. Find such a solution using MATLAB commands.

ML.5. For the matrix

$$A = \begin{bmatrix} 1 & 2 & 3 \\ 3 & 2 & 1 \\ 2 & 1 & 3 \end{bmatrix}$$

and $\lambda = 6$, the homogeneous linear system $(\lambda I_3 - A)\mathbf{x} = \mathbf{0}$ has a nontrivial solution. Find such a solution using MATLAB commands.

6.6 THE RANK OF A MATRIX AND APPLICATIONS

In this section we obtain another effective method for finding a basis for a vector space V spanned by a given set of vectors $S = \{\mathbf{v}_1, \mathbf{v}_2, \dots, \mathbf{v}_n\}$. In the proof of Theorem 6.6 we developed a technique for choosing a basis for V that is a *subset* of S. The method to be developed in this section produces a basis for V that is *not* guaranteed to be a subset of S. We shall also attach a unique number to a matrix A that we later show gives us information about the dimension of the solution space of a homogeneous system with coefficient matrix A.

DEFINITION Let

$$A = \begin{bmatrix} a_{11} & a_{12} & \cdots & a_{1n} \\ a_{21} & a_{22} & \cdots & a_{2n} \\ \vdots & \vdots & & \vdots \\ a_{m1} & a_{m2} & \cdots & a_{mn} \end{bmatrix}$$

be an $m \times n$ matrix. The rows of A,

$$\begin{aligned} \mathbf{v}_1 &= (a_{11}, a_{12}, \ldots, a_{1n}) \\ \mathbf{v}_2 &= (a_{21}, a_{22}, \ldots, a_{2n}) \\ &\vdots \\ \mathbf{v}_m &= (a_{m1}, a_{m2}, \ldots, a_{mn}), \end{aligned}$$

considered as vectors in R^n, span a subspace of R^n, called the **row space** of A. Similarly, the columns of A,

$$\mathbf{w}_1 = \begin{bmatrix} a_{11} \\ a_{21} \\ \vdots \\ a_{m1} \end{bmatrix}, \quad \mathbf{w}_2 = \begin{bmatrix} a_{12} \\ a_{22} \\ \vdots \\ a_{m2} \end{bmatrix}, \ldots, \quad \mathbf{w}_n = \begin{bmatrix} a_{1n} \\ a_{2n} \\ \vdots \\ a_{mn} \end{bmatrix},$$

considered as vectors in R^m, span a subspace of R^m, called the **column space** of A.

THEOREM 6.10 *If A and B are two $m \times n$ row equivalent matrices, then the row spaces of A and B are equal.*

Proof If A and B are row equivalent, then the rows of B are obtained from those of A by a finite number of the three elementary row operations. Thus each row of B is a linear combination of the rows of A. Hence the row space of B is contained in the row space of A. Similarly, A can be obtained from B by a finite number of the three elementary row operations, so the row space of A is contained in the row space of B. Hence the row spaces of A and B are equal. ■

Remark For a related result see Theorem 1.7 in Section 1.6, where we showed that if two augmented matrices are row equivalent, then their corresponding linear systems have the same solutions.

It follows from Theorem 6.10 that if we take a given matrix A and find its reduced row echelon form B, then the row spaces of A and B are equal. Furthermore, recall from Exercise T.8 in Section 6.3 that the nonzero rows of a matrix that is in reduced row echelon form are linearly independent and thus form a basis for its row space. We can use this method to find a basis for a vector space spanned by a given set of vectors in R^n, as illustrated in Example 1.

EXAMPLE 1 Let $S = \{\mathbf{v}_1, \mathbf{v}_2, \mathbf{v}_3, \mathbf{v}_4\}$, where

$$\mathbf{v}_1 = (1, -2, 0, 3, -4), \qquad \mathbf{v}_2 = (3, 2, 8, 1, 4),$$
$$\mathbf{v}_3 = (2, 3, 7, 2, 3), \qquad \mathbf{v}_4 = (-1, 2, 0, 4, -3),$$

and let V be the subspace of R^5 given by $V = \text{span } S$. Find a basis for V.

Solution Note that V is the row space of the matrix A whose rows are the given vectors

$$A = \begin{bmatrix} 1 & -2 & 0 & 3 & -4 \\ 3 & 2 & 8 & 1 & 4 \\ 2 & 3 & 7 & 2 & 3 \\ -1 & 2 & 0 & 4 & -3 \end{bmatrix}.$$

Using elementary row operations, we find that A is row equivalent to the matrix (verify)

$$B = \begin{bmatrix} 1 & 0 & 2 & 0 & 1 \\ 0 & 1 & 1 & 0 & 1 \\ 0 & 0 & 0 & 1 & -1 \\ 0 & 0 & 0 & 0 & 0 \end{bmatrix},$$

which is in reduced row echelon form. The row spaces of A and B are identical, and a basis for the row space of B consists of the nonzero rows of B. Hence

$$\mathbf{w}_1 = (1, 0, 2, 0, 1), \quad \mathbf{w}_2 = (0, 1, 1, 0, 1), \quad \text{and} \quad \mathbf{w}_3 = (0, 0, 0, 1, -1)$$

form a basis for V. ■

Remark Observe in Example 1 that the rows of the matrices A and B are different, but their row spaces are the same.

We may now summarize the method used in Example 1 to find a basis for the subspace V of R^n given by $V = \text{span } S$, where S is a set of vectors in R^n.

> The procedure for finding a basis for the subspace V of R^n given by $V = \text{span } S$, where $S = \{\mathbf{v}_1, \mathbf{v}_2, \dots, \mathbf{v}_k\}$ is a set of vectors in R^n that are given in row form, is as follows.
>
> ***Step 1.*** Form the matrix
>
> $$A = \begin{bmatrix} \mathbf{v}_1 \\ \mathbf{v}_2 \\ \vdots \\ \mathbf{v}_k \end{bmatrix}$$
>
> whose rows are the given vectors in S.
>
> ***Step 2.*** Transform A to reduced row echelon form, obtaining the matrix B.
>
> ***Step 3.*** The nonzero rows of B form a basis for V.

Of course, the basis that has been obtained by the procedure used in Example 1 produced a basis that is not a subset of the spanning set. However, the basis for a subspace V of R^n that is obtained in this manner is analogous, in its simplicity, to the standard basis for R^n. Thus if $\mathbf{v} = (a_1, a_2, \dots, a_n)$ is a vector in V and $\{\mathbf{v}_1, \mathbf{v}_2, \dots, \mathbf{v}_k\}$ is a basis for V obtained by the method of Example 1 where the leading 1s occur in columns $j_1, j_2, \dots, j_k$, then it can be shown that

$$\mathbf{v} = a_{j_1}\mathbf{v}_1 + a_{j_2}\mathbf{v}_2 + \cdots + a_{j_k}\mathbf{v}_k.$$

EXAMPLE 2 Let V be the subspace of Example 1. Given that the vector

$$\mathbf{v} = (5, 4, 14, 6, 3)$$

is in V, write $\mathbf{v}$ as a linear combination of the basis determined in Example 1.

Solution We have $j_1 = 1$, $j_2 = 2$, and $j_3 = 4$, so

$$\mathbf{v} = 5\mathbf{w}_1 + 4\mathbf{w}_2 + 6\mathbf{w}_3. \qquad \blacksquare$$

Remark The solution to Example 1 also yields a basis for the row space of matrix A in that example. Observe that the vectors in this basis are not rows of matrix A.

DEFINITION The dimension of the row space of A is called the **row rank** of A, and the dimension of the column space of A is called the **column rank** of A.

EXAMPLE 3 Find a basis for the row space of the matrix A defined in the solution of Example 1 that contains only row vectors from A. Also compute the row rank of the matrix A.

Solution Using the procedure in the alternative proof of Theorem 6.6, we form the equation

$$c_1(1, -2, 0, 3, -4) + c_2(3, 2, 8, 1, 4) \\ + c_3(2, 3, 7, 2, 3) + c_4(-1, 2, 0, 4, -3) = (0, 0, 0, 0, 0),$$

whose augmented matrix is

$$\left[\begin{array}{rrrr|r} 1 & 3 & 2 & -1 & 0 \\ -2 & 2 & 3 & 2 & 0 \\ 0 & 8 & 7 & 0 & 0 \\ 3 & 1 & 2 & 4 & 0 \\ -4 & 4 & 3 & -3 & 0 \end{array}\right] = \left[A^T \;\vdots\; \mathbf{0}\right]; \qquad (1)$$

that is, the coefficient matrix is A^T. Transforming the augmented matrix $\left[A^T \;\vdots\; \mathbf{0}\right]$ in (1) to reduced row echelon form, we obtain (verify)

$$\left[\begin{array}{rrrr|r} 1 & 0 & 0 & \frac{11}{24} & 0 \\ 0 & 1 & 0 & -\frac{49}{24} & 0 \\ 0 & 0 & 1 & \frac{7}{3} & 0 \\ 0 & 0 & 0 & 0 & 0 \\ 0 & 0 & 0 & 0 & 0 \end{array}\right]. \qquad (2)$$

Since the leading 1s in (2) occur in columns 1, 2, and 3, we conclude that the first three rows of A form a basis for the row space of A. That is,

$$\{(1, -2, 0, 3, -4), (3, 2, 8, 1, 4), (2, 3, 7, 2, 3)\}$$

is a basis for the row space of A. The row rank of A is 3. $\blacksquare$

EXAMPLE 4 Find a basis for the column space of the matrix A defined in the solution of Example 1 and compute the column rank of A.

Solution 1 Writing the columns of A as row vectors, we obtain the matrix A^T, which when transformed to reduced row echelon form is (as we saw in Example 3)

$$\begin{bmatrix} 1 & 0 & 0 & \frac{11}{24} \\ 0 & 1 & 0 & -\frac{49}{24} \\ 0 & 0 & 1 & \frac{7}{3} \\ 0 & 0 & 0 & 0 \\ 0 & 0 & 0 & 0 \end{bmatrix}.$$

Thus the vectors $\left(1, 0, 0, \frac{11}{24}\right)$, $\left(0, 1, 0, -\frac{49}{24}\right)$, and $\left(0, 0, 1, \frac{7}{3}\right)$ form a basis for the row space of A^T. Hence the vectors

$$\begin{bmatrix} 1 \\ 0 \\ 0 \\ \frac{11}{24} \end{bmatrix}, \quad \begin{bmatrix} 0 \\ 1 \\ 0 \\ -\frac{49}{24} \end{bmatrix}, \quad \text{and} \quad \begin{bmatrix} 0 \\ 0 \\ 1 \\ \frac{7}{3} \end{bmatrix}$$

form a basis for the column space of A and we conclude that the column rank of A is 3.

Solution 2 If we want to find a basis for the column space of A that contains only the column vectors from A, we follow the procedure developed in the proof of Theorem 6.6, forming the equation

$$c_1 \begin{bmatrix} 1 \\ 3 \\ 2 \\ -1 \end{bmatrix} + c_2 \begin{bmatrix} -2 \\ 2 \\ 3 \\ 2 \end{bmatrix} + c_3 \begin{bmatrix} 0 \\ 8 \\ 7 \\ 0 \end{bmatrix} + c_4 \begin{bmatrix} 3 \\ 1 \\ 2 \\ 4 \end{bmatrix} + c_5 \begin{bmatrix} -4 \\ 4 \\ 3 \\ -3 \end{bmatrix} = \begin{bmatrix} 0 \\ 0 \\ 0 \\ 0 \end{bmatrix}$$

whose augmented matrix is $\left[A \,\vdots\, \mathbf{0}\right]$. Transforming this matrix to reduced row echelon form, we obtain (as in Example 1)

$$\left[\begin{array}{ccccc:c} 1 & 0 & 2 & 0 & 1 & 0 \\ 0 & 1 & 1 & 0 & 1 & 0 \\ 0 & 0 & 0 & 1 & -1 & 0 \\ 0 & 0 & 0 & 0 & 0 & 0 \end{array}\right].$$

Since the leading 1s occur in columns 1, 2, and 4, we conclude that the first, second, and fourth columns of A form a basis for the column space of A. That is,

$$\left\{ \begin{bmatrix} 1 \\ 3 \\ 2 \\ -1 \end{bmatrix}, \begin{bmatrix} -2 \\ 2 \\ 3 \\ 2 \end{bmatrix}, \begin{bmatrix} 3 \\ 1 \\ 2 \\ 4 \end{bmatrix} \right\}$$

is a basis for the column space of A. The column rank of A is 3. ■

From Examples 3 and 4 we observe that the row and column ranks of A are equal. This is always true, and is a very important result in linear algebra. We now turn to the proof of this theorem.

THEOREM 6.11 *The row rank and column rank of the $m \times n$ matrix $A = \left[a_{ij}\right]$ are equal.*

Proof Let $\mathbf{w}_1, \mathbf{w}_2, \ldots, \mathbf{w}_n$ denote the columns of A. To determine the dimension of the column space of A, we use the procedure in the alternative proof of Theorem 6.6. Thus we consider the equation

$$c_1\mathbf{w}_1 + c_2\mathbf{w}_2 + \cdots + c_n\mathbf{w}_n = \mathbf{0}.$$

We now transform the augmented matrix, $\left[A \;\vdots\; \mathbf{0}\right]$, of this homogeneous system to reduced row echelon form. The vectors corresponding to the columns containing the leading 1s form a basis for the column space of A. Thus the column rank of A is the number of leading 1s. But this number is also the number of nonzero rows in the reduced row echelon form matrix that is row equivalent to A, so it is the row rank of A. Thus row rank A = column rank A. ■

DEFINITION Since row rank A = column rank A, we now merely refer to the **rank** of an $m \times n$ matrix and write rank A.

We next summarize the procedure for computing the rank of a matrix.

> The procedure for computing the rank of the matrix A is as follows.
>
> ***Step 1.*** Using elementary row operations, transform A to a matrix B in reduced row echelon form.
>
> ***Step 2.*** Rank A = the number of nonzero rows of B.

If A is an $m \times n$ matrix, we have defined (see Section 6.5) the nullity of A as the dimension of the null space of A, that is, the dimension of the solution space of $A\mathbf{x} = \mathbf{0}$. If A is transformed to a matrix B in reduced row echelon form having r nonzero rows, then we know that the dimension of the solution space of $A\mathbf{x} = \mathbf{0}$ is $n - r$. Since r is also the rank of A, we have obtained a fundamental relationship between the rank and nullity of A, which we state in the following theorem.

THEOREM 6.12 *If A is an $m \times n$ matrix, then rank A + nullity $A = n$.* ■

EXAMPLE 5 Let

$$A = \begin{bmatrix} 1 & 1 & 4 & 1 & 2 \\ 0 & 1 & 2 & 1 & 1 \\ 0 & 0 & 0 & 1 & 2 \\ 1 & -1 & 0 & 0 & 2 \\ 2 & 1 & 6 & 0 & 1 \end{bmatrix},$$

which was defined in Example 1 of Section 6.5. When A is transformed to reduced row echelon form, we obtain

$$\begin{bmatrix} 1 & 0 & 2 & 0 & 1 \\ 0 & 1 & 2 & 0 & -1 \\ 0 & 0 & 0 & 1 & 2 \\ 0 & 0 & 0 & 0 & 0 \\ 0 & 0 & 0 & 0 & 0 \end{bmatrix}.$$

Then rank $A = 3$ and nullity $A = 2$. This agrees with the result obtained in the solution of Example 1 of Section 6.5, where we found that the dimension of the solution space of $A\mathbf{x} = \mathbf{0}$ is 2. ■

The following example will be used to illustrate geometrically some of the ideas discussed previously.

EXAMPLE 6 Let

$$A = \begin{bmatrix} 3 & -1 & 2 \\ 2 & 1 & 3 \\ 7 & 1 & 8 \end{bmatrix}.$$

Transforming A to reduced row echelon form we obtain (verify)

$$\begin{bmatrix} 1 & 0 & 1 \\ 0 & 1 & 1 \\ 0 & 0 & 0 \end{bmatrix},$$

so we conclude the following:

- rank $A = 2$.
- dimension of row space of $A = 2$, so the row space of A is a two-dimensional subspace of R^3, that is, a plane passing through the origin.

From the reduced row echelon form of A, we see that every solution to the homogeneous system $A\mathbf{x} = \mathbf{0}$ is of the form

$$\mathbf{x} = \begin{bmatrix} -r \\ -r \\ r \end{bmatrix},$$

where r is an arbitrary constant (verify), so the solution space of this homogeneous system, or the null space of A, is a line passing through the origin. Moreover, the dimension of the null space of A, or the nullity of A, is 1. Thus, Theorem 6.12 has been verified.

Of course, we already know that the dimension of the column space of A is also 2. We could also obtain this result by finding a basis consisting of two vectors for the column space of A. Thus, the column space of A is also a two-dimensional subspace of R^3, that is, a plane passing through the origin. These results are illustrated in Figure 6.10. ■

Figure 6.10 ►

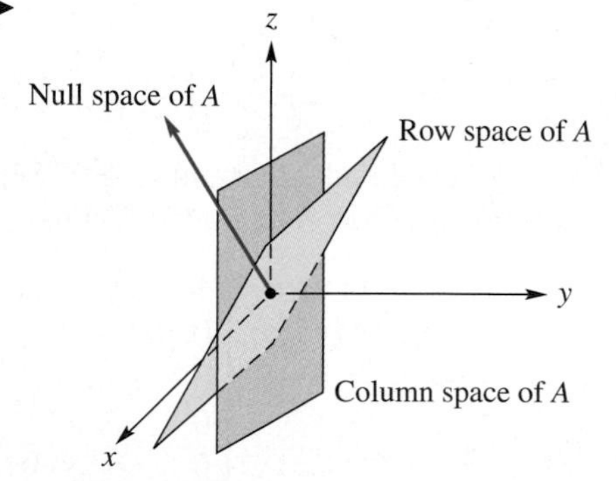

Remark The concepts of rank and nullity are also applied to linear transformations as discussed in Section 10.2.

RANK AND SINGULARITY

The rank of a square matrix can be used to determine whether the matrix is singular or nonsingular, as the following theorem shows.

THEOREM 6.13 *An $n \times n$ matrix A is nonsingular if and only if rank $A = n$.*

Proof Suppose that A is nonsingular. Then A is row equivalent to I_n (Theorem 1.12, Section 1.7), and so rank $A = n$.

Conversely, let rank $A = n$. Suppose that the reduced row echelon form of A is B. Then rank $B = n$. Hence B has no zero rows and since it is in reduced row echelon form, it must be I_n. Thus A is row equivalent to I_n and so A is nonsingular (Theorem 1.12, Section 1.7). ■

An immediate consequence of Theorem 6.13 is the following corollary, which is a criterion for the rank of an $n \times n$ matrix to be n.

COROLLARY 6.2 *If A is an $n \times n$ matrix, then rank $A = n$ if and only if $\det(A) \neq 0$.*

Proof Exercise T.1. ■

Another result easily obtained from Theorem 6.13 is contained in the following corollary.

COROLLARY 6.3 *Let A be an $n \times n$ matrix. The linear system $A\mathbf{x} = \mathbf{b}$ has a unique solution for every $n \times 1$ matrix $\mathbf{b}$ if and only if rank $A = n$.*

Proof Exercise T.2. ■

The following corollary gives another method of testing whether n given vectors in R^n are linearly dependent or linearly independent.

COROLLARY 6.4 *Let $S = \{\mathbf{v}_1, \mathbf{v}_2, \ldots, \mathbf{v}_n\}$ be a set of n vectors in R^n and let A be the matrix whose rows (columns) are the vectors in S. Then S is linearly independent if and only if $\det(A) \neq 0$.*

Proof Exercise T.3. ■

For $n > 4$, the method of Corollary 6.4 for testing linear dependence is not as efficient as the direct method of Section 6.3, calling for the solution of a homogeneous system.

COROLLARY 6.5 *The homogeneous system $A\mathbf{x} = \mathbf{0}$ of n linear equations in n unknowns has a nontrivial solution if and only if rank $A < n$.*

Proof This follows from Corollary 6.2 and from the fact that $A\mathbf{x} = \mathbf{0}$ has a nontrivial solution if and only if A is singular (Theorem 1.13, Section 1.7). ■

EXAMPLE 7 Let

$$A = \begin{bmatrix} 1 & 2 & 0 \\ 0 & 1 & 3 \\ 2 & 1 & 3 \end{bmatrix}.$$

The reduced row echelon form of A is I_3 (verify). Thus rank $A = 3$ and matrix

A is nonsingular. Moreover, the homogeneous system $A\mathbf{x} = \mathbf{0}$ has only the trivial solution. ■

EXAMPLE 8 Let

$$A = \begin{bmatrix} 1 & 2 & 0 \\ 1 & 1 & -3 \\ 1 & 3 & 3 \end{bmatrix}.$$

Then A is row equivalent to

$$\begin{bmatrix} 1 & 0 & -6 \\ 0 & 1 & 3 \\ 0 & 0 & 0 \end{bmatrix},$$

a matrix in reduced row echelon form. Hence rank $A < 3$, and A is singular. Moreover, $A\mathbf{x} = \mathbf{0}$ has a nontrivial solution. ■

APPLICATIONS OF RANK TO THE LINEAR SYSTEM $A\mathbf{x} = \mathbf{b}$

In Corollary 6.5 we have seen that the rank of A provides us with information about the existence of a nontrivial solution to the homogeneous system $A\mathbf{x} = \mathbf{0}$. We shall now obtain some results that use the rank of A to provide information about the solutions to the linear system $A\mathbf{x} = \mathbf{b}$, where $\mathbf{b}$ is an arbitrary $n \times 1$ matrix. When $\mathbf{b} \neq \mathbf{0}$, the linear system is said to be **nonhomogeneous**.

THEOREM 6.14 *The linear system $A\mathbf{x} = \mathbf{b}$ has a solution if and only if rank A = rank $\left[A \,\vdots\, \mathbf{b}\right]$; that is, if and only if the ranks of the coefficient and augmented matrices are equal.*

Proof First, observe that if $A = \left[a_{ij}\right]$ is $m \times n$, then the given linear system may be written as

$$x_1 \begin{bmatrix} a_{11} \\ a_{21} \\ \vdots \\ a_{m1} \end{bmatrix} + x_2 \begin{bmatrix} a_{12} \\ a_{22} \\ \vdots \\ a_{m2} \end{bmatrix} + \cdots + x_n \begin{bmatrix} a_{1n} \\ a_{2n} \\ \vdots \\ a_{mn} \end{bmatrix} = \begin{bmatrix} b_1 \\ b_2 \\ \vdots \\ b_m \end{bmatrix}. \tag{3}$$

Suppose now that $A\mathbf{x} = \mathbf{b}$ has a solution. Then there exist values of $x_1, x_2, \ldots, x_n$ that satisfy Equation (3). Thus, $\mathbf{b}$ is a linear combination of the columns of A and so belongs to the column space of A. Hence the dimensions of the column spaces of A and $\left[A \,\vdots\, \mathbf{b}\right]$ are equal, so rank A = rank $\left[A \,\vdots\, \mathbf{b}\right]$.

Conversely, suppose that rank A = rank $\left[A \,\vdots\, \mathbf{b}\right]$. Then $\mathbf{b}$ is in the column space of A, which means that we can find values of $x_1, x_2, \ldots, x_n$ that satisfy Equation (3). Hence $A\mathbf{x} = \mathbf{b}$ has a solution. ■

Remarks

1. Theorem 6.14 implies that $A\mathbf{x} = \mathbf{b}$ is inconsistent if and only if $\mathbf{b}$ is not in the column space of A.
2. The result given in Theorem 6.14, while of interest, is not of great computational value, since we usually are interested in finding a solution rather than in knowing whether or not a solution exists.

EXAMPLE 9 Consider the linear system

$$\begin{bmatrix} 2 & 1 & 3 \\ 1 & -2 & 2 \\ 0 & 1 & 3 \end{bmatrix} \begin{bmatrix} x_1 \\ x_2 \\ x_3 \end{bmatrix} = \begin{bmatrix} 1 \\ 2 \\ 3 \end{bmatrix}.$$

Since rank $A = \text{rank}\left[A \,\vdots\, \mathbf{b}\right] = 3$, the linear system has a solution. ■

EXAMPLE 10 The linear system

$$\begin{bmatrix} 1 & 2 & 3 \\ 1 & -3 & 4 \\ 2 & -1 & 7 \end{bmatrix} \begin{bmatrix} x_1 \\ x_2 \\ x_3 \end{bmatrix} = \begin{bmatrix} 4 \\ 5 \\ 6 \end{bmatrix}$$

has no solution because rank $A = 2$ and rank $\left[A \,\vdots\, \mathbf{b}\right] = 3$ (verify). ■

We now extend our list of nonsingular equivalences.

List of Nonsingular Equivalences

The following statements are equivalent for an $n \times n$ matrix A.

1. A is nonsingular.
2. $\mathbf{x} = \mathbf{0}$ is the only solution to $A\mathbf{x} = \mathbf{0}$.
3. A is row equivalent to I_n.
4. The linear system $A\mathbf{x} = \mathbf{b}$ has a unique solution for every $n \times 1$ matrix $\mathbf{b}$.
5. $\det(A) \neq 0$.
6. A has rank n.
7. A has nullity 0.
8. The rows of A form a linearly independent set of n vectors in R^n.
9. The columns of A form a linearly independent set of n vectors in R^n.

Key Terms

Row space
Column space
Row rank
Column rank
Rank
Nonhomogeneous system

6.6 Exercises

1. Let $S = \{\mathbf{v}_1, \mathbf{v}_2, \mathbf{v}_3, \mathbf{v}_4, \mathbf{v}_5\}$, where

$$\mathbf{v}_1 = (1, 2, 3), \quad \mathbf{v}_2 = (2, 1, 4),$$
$$\mathbf{v}_3 = (-1, -1, 2), \quad \mathbf{v}_4 = (0, 1, 2),$$

and $\mathbf{v}_5 = (1, 1, 1)$. Find a basis for the subspace $V = \text{span } S$ of R^3.

2. Let $S = \{\mathbf{v}_1, \mathbf{v}_2, \mathbf{v}_3, \mathbf{v}_4, \mathbf{v}_5\}$, where

$$\mathbf{v}_1 = (1, 1, 2, 1), \quad \mathbf{v}_2 = (1, 0, -3, 1),$$
$$\mathbf{v}_3 = (0, 1, 1, 2), \quad \mathbf{v}_4 = (0, 0, 1, 1),$$

and $\mathbf{v}_5 = (1, 0, 0, 1)$. Find a basis for the subspace $V = \text{span } S$ of R^4.

3. Let $S = \{\mathbf{v}_1, \mathbf{v}_2, \mathbf{v}_3, \mathbf{v}_4, \mathbf{v}_5\}$, where

$$\mathbf{v}_1 = \begin{bmatrix} 1 \\ 2 \\ 1 \\ 2 \end{bmatrix}, \quad \mathbf{v}_2 = \begin{bmatrix} 2 \\ 1 \\ 2 \\ 1 \end{bmatrix},$$

$$\mathbf{v}_3 = \begin{bmatrix} 3 \\ 2 \\ 3 \\ 2 \end{bmatrix}, \quad \mathbf{v}_4 = \begin{bmatrix} 3 \\ 3 \\ 3 \\ 3 \end{bmatrix}, \quad \text{and} \quad \mathbf{v}_5 = \begin{bmatrix} 5 \\ 3 \\ 5 \\ 3 \end{bmatrix}.$$

Find a basis for the subspace $V = \text{span } S$ of R^4.

4. Let $S = \{\mathbf{v}_1, \mathbf{v}_2, \mathbf{v}_3, \mathbf{v}_4, \mathbf{v}_5\}$, where

$$\mathbf{v}_1 = \begin{bmatrix} 1 \\ 2 \\ 1 \\ 1 \end{bmatrix}, \quad \mathbf{v}_2 = \begin{bmatrix} 2 \\ 1 \\ 3 \\ 1 \end{bmatrix},$$

$$\mathbf{v}_3 = \begin{bmatrix} 0 \\ 2 \\ 1 \\ 2 \end{bmatrix}, \quad \mathbf{v}_4 = \begin{bmatrix} 3 \\ 2 \\ 1 \\ 4 \end{bmatrix}, \quad \text{and} \quad \mathbf{v}_5 = \begin{bmatrix} 5 \\ 0 \\ 0 \\ -1 \end{bmatrix}.$$

Find a basis for the subspace $V = \text{span } S$ of R^4.

In Exercises 5 and 6, find a basis for the row space of A
(a) *consisting of vectors that are* not *row vectors of A;*
(b) *consisting of vectors that* are *row vectors of A.*

5. $A = \begin{bmatrix} 1 & 2 & -1 \\ 1 & 9 & -1 \\ -3 & 8 & 3 \\ -2 & 3 & 2 \end{bmatrix}$

6. $A = \begin{bmatrix} 1 & 2 & -1 & 3 \\ 3 & 5 & 2 & 0 \\ 0 & 1 & 2 & 1 \\ -1 & 0 & -2 & 7 \end{bmatrix}$

In Exercises 7 and 8, find a basis for the column space of A
(a) *consisting of vectors that are* not *column vectors of A;*
(b) *consisting of vectors that* are *column vectors of A.*

7. $A = \begin{bmatrix} 1 & -2 & 7 & 0 \\ 1 & -1 & 4 & 0 \\ 3 & 2 & -3 & 5 \\ 2 & 1 & -1 & 3 \end{bmatrix}$

8. $A = \begin{bmatrix} -2 & 2 & 3 & 7 & 1 \\ -2 & 2 & 4 & 8 & 0 \\ -3 & 3 & 2 & 8 & 4 \\ 4 & -2 & 1 & -5 & -7 \end{bmatrix}$

In Exercises 9 and 10, compute a basis for the row space of A, the column space of A, the row space of A^T, and the column space of A^T. Write a short description giving the relationships among these bases.

9. $A = \begin{bmatrix} 1 & -2 & 5 \\ 2 & 3 & 2 \\ 0 & -7 & 8 \end{bmatrix}$

10. $A = \begin{bmatrix} 2 & -3 & -7 & 11 \\ 3 & -1 & -7 & 13 \\ 1 & 2 & 0 & 2 \end{bmatrix}$

In Exercises 11 and 12, compute the row and column ranks of A, verifying Theorem 6.11.

11. $A = \begin{bmatrix} 1 & 2 & 3 & 2 & 1 \\ 3 & 1 & -5 & -2 & 1 \\ 7 & 8 & -1 & 2 & 5 \end{bmatrix}$

12. $A = \begin{bmatrix} 1 & 3 & 2 & 0 & 0 & 1 \\ 2 & 1 & -5 & 1 & 2 & 0 \\ 3 & 2 & 5 & 1 & -2 & 1 \\ 5 & 8 & 9 & 1 & -2 & 2 \\ 9 & 9 & 4 & 2 & 0 & 2 \end{bmatrix}$

In Exercises 13 through 17, compute the rank and nullity of A and verify Theorem 6.12.

13. $A = \begin{bmatrix} 1 & 2 & 1 & 3 \\ 2 & 1 & -4 & -5 \\ 7 & 8 & -5 & -1 \\ 10 & 14 & -2 & 8 \end{bmatrix}$

14. $A = \begin{bmatrix} 1 & 2 & 1 & 3 \\ 2 & 1 & -4 & -5 \\ 1 & 1 & 0 & 0 \\ 0 & 0 & 1 & 1 \end{bmatrix}$

15. $A = \begin{bmatrix} 1 & 2 & 3 \\ -1 & 2 & 1 \\ 3 & 1 & 2 \end{bmatrix}$ **16.** $A = \begin{bmatrix} 1 & -2 & -1 \\ 2 & -1 & 3 \\ 7 & -8 & 3 \end{bmatrix}$

17. $A = \begin{bmatrix} 1 & -2 & -1 \\ 2 & -1 & 3 \\ 7 & -8 & 3 \\ 5 & -7 & 0 \end{bmatrix}$

18. If A is a 3×4 matrix, what is the largest possible value for rank A?

19. If A is a 4×6 matrix, show that the columns of A are linearly dependent.

20. If A is a 5×3 matrix, show that the rows of A are linearly dependent.

In Exercises 21 and 22, let A be a 7×3 matrix whose rank is 3.

21. Are the rows of A linearly dependent or linearly independent? Justify your answer.

22. Are the columns of A linearly dependent or linearly independent? Justify your answer.

In Exercises 23 through 25, use Theorem 6.13 to determine whether each matrix is singular or nonsingular.

23. $\begin{bmatrix} 1 & 2 & -3 \\ -1 & 2 & 3 \\ 0 & 8 & 0 \end{bmatrix}$ **24.** $\begin{bmatrix} 1 & 2 & -3 \\ -1 & 2 & 3 \\ 0 & 1 & 1 \end{bmatrix}$

25. $\begin{bmatrix} 1 & 1 & 4 & -1 \\ 1 & 2 & 3 & 2 \\ -1 & 3 & 2 & 1 \\ -2 & 6 & 12 & -4 \end{bmatrix}$

In Exercises 26 and 27, use Corollary 6.3 to determine whether the linear system $A\mathbf{x} = \mathbf{b}$ has a unique solution for every 3×1 matrix $\mathbf{b}$.

26. $A = \begin{bmatrix} 1 & 2 & -2 \\ 0 & 8 & -7 \\ 3 & -2 & 1 \end{bmatrix}$

27. $A = \begin{bmatrix} 1 & -1 & 2 \\ 3 & 2 & 3 \\ 1 & -2 & 1 \end{bmatrix}$

Use Corollary 6.4 to do Exercises 28 and 29.

28. Is

$$S = \left\{ \begin{bmatrix} 2 \\ 2 \\ 3 \end{bmatrix}, \begin{bmatrix} 1 \\ 0 \\ 2 \end{bmatrix}, \begin{bmatrix} 0 \\ 1 \\ 3 \end{bmatrix} \right\}$$

a linearly independent set of vectors in R^3?

29. Is

$$S = \left\{ \begin{bmatrix} 4 \\ 1 \\ 2 \end{bmatrix}, \begin{bmatrix} 2 \\ 5 \\ -5 \end{bmatrix}, \begin{bmatrix} 2 \\ -1 \\ 3 \end{bmatrix} \right\}$$

a linearly independent set of vectors in R^3?

In Exercises 30 through 32, find which homogeneous systems have a nontrivial solution for the given matrix A by using Corollary 6.5.

30. $A = \begin{bmatrix} 1 & 1 & 2 & -1 \\ 1 & 3 & -1 & 2 \\ 1 & 1 & 1 & 3 \\ 1 & 2 & 1 & 1 \end{bmatrix}$

31. $A = \begin{bmatrix} 1 & 2 & 3 \\ 0 & 1 & 0 \\ 1 & 0 & 3 \end{bmatrix}$

32. $A = \begin{bmatrix} 1 & 2 & -1 \\ 2 & -1 & 3 \\ 5 & -4 & 3 \end{bmatrix}$

In Exercises 33 through 36, determine which of the linear systems have a solution by using Theorem 6.14.

33. $\begin{bmatrix} 1 & 2 & 5 & -2 \\ 2 & 3 & -2 & 4 \\ 5 & 1 & 0 & 2 \end{bmatrix} \begin{bmatrix} x_1 \\ x_2 \\ x_3 \\ x_4 \end{bmatrix} = \begin{bmatrix} 0 \\ 0 \\ 0 \end{bmatrix}$

34. $\begin{bmatrix} 1 & 2 & 5 & -2 \\ 2 & 3 & -2 & 4 \\ 5 & 1 & 0 & 2 \end{bmatrix} \begin{bmatrix} x_1 \\ x_2 \\ x_3 \\ x_4 \end{bmatrix} = \begin{bmatrix} 1 \\ -13 \\ 3 \end{bmatrix}$

35. $\begin{bmatrix} 1 & -2 & -3 & 4 \\ 4 & -1 & -5 & 6 \\ 2 & 3 & 1 & -2 \end{bmatrix} \begin{bmatrix} x_1 \\ x_2 \\ x_3 \\ x_4 \end{bmatrix} = \begin{bmatrix} 1 \\ 2 \\ 2 \end{bmatrix}$

36. $\begin{bmatrix} 1 & 1 & 1 \\ 1 & -1 & 1 \\ 5 & 1 & 5 \end{bmatrix} \begin{bmatrix} x_1 \\ x_2 \\ x_3 \end{bmatrix} = \begin{bmatrix} 6 \\ 2 \\ 5 \end{bmatrix}$

In Exercises 37 through 40, compute the rank of the given bit matrix.

37. $\begin{bmatrix} 1 & 0 & 1 \\ 0 & 1 & 1 \\ 1 & 1 & 0 \end{bmatrix}$

38. $\begin{bmatrix} 1 & 0 & 1 & 0 \\ 0 & 1 & 0 & 1 \\ 0 & 1 & 1 & 0 \\ 1 & 1 & 0 & 0 \end{bmatrix}$

39. $\begin{bmatrix} 1 & 1 & 0 & 0 \\ 1 & 1 & 1 & 0 \\ 0 & 1 & 1 & 1 \\ 0 & 0 & 1 & 1 \end{bmatrix}$

40. $\begin{bmatrix} 1 & 1 & 0 & 1 & 1 \\ 0 & 1 & 1 & 1 & 0 \\ 1 & 0 & 1 & 0 & 1 \\ 1 & 0 & 0 & 0 & 1 \end{bmatrix}$

Theoretical Exercises

T.1. Prove Corollary 6.2.

T.2. Prove Corollary 6.3.

T.3. Prove Corollary 6.4.

T.4. Let A be an $n \times n$ matrix. Show that the homogeneous system $A\mathbf{x} = \mathbf{0}$ has a nontrivial solution if and only if the columns of A are linearly dependent.

T.5. Let A be an $n \times n$ matrix. Show that rank $A = n$ if and only if the columns of A are linearly independent.

T.6. Let A be an $n \times n$ matrix. Show that the rows of A are linearly independent if and only if the columns of A span R^n.

T.7. Let A be an $m \times n$ matrix. Show that the linear system $A\mathbf{x} = \mathbf{b}$ has a solution for every $m \times 1$ matrix $\mathbf{b}$ if and only if rank $A = m$.

T.8. Let A be an $m \times n$ matrix. Show that the columns of A are linearly independent if and only if the homogeneous system $A\mathbf{x} = \mathbf{0}$ has only the trivial solution.

T.9. Let A be an $m \times n$ matrix. Show that the linear system $A\mathbf{x} = \mathbf{b}$ has at most one solution for every $m \times 1$ matrix $\mathbf{b}$ if and only if the associated homogeneous system $A\mathbf{x} = \mathbf{0}$ has only the trivial solution.

T.10. Let A be an $m \times n$ matrix with $m \neq n$. Show that either the rows or the columns of A are linearly dependent.

T.11. Suppose that the linear system $A\mathbf{x} = \mathbf{b}$, where A is $m \times n$, is consistent (has a solution). Show that the solution is unique if and only if rank $A = n$.

T.12. Show that if A is an $m \times n$ matrix such that AA^T is nonsingular, then rank $A = m$.

MATLAB Exercises

Given a matrix A, the nonzero rows of **rref(A)** *form a basis for the row space of A and the nonzero rows of* **rref(A′)** *transformed to columns give a basis for the column space of A.*

ML.1. Solve Exercises 1 through 4 using MATLAB.

To find a basis for the row space of A that consists of rows of A, we compute **rref(A′)**. *The leading 1s point to the original rows of A that give us a basis for the row space. See Example 3.*

ML.2. Determine two bases for the row space of A that have no vectors in common.

(a) $A = \begin{bmatrix} 1 & 3 & 1 \\ 2 & 5 & 0 \\ 4 & 11 & 2 \\ 6 & 9 & 1 \end{bmatrix}$

(b) $A = \begin{bmatrix} 2 & 1 & 2 & 0 \\ 0 & 0 & 0 & 0 \\ 1 & 2 & 2 & 1 \\ 4 & 5 & 6 & 2 \\ 3 & 3 & 4 & 1 \end{bmatrix}$

ML.3. Repeat Exercise ML.2 for column spaces.

To compute the rank of a matrix A in MATLAB, use the command **rank(A)**.

ML.4. Compute the rank and nullity of each of the following matrices.

(a) $\begin{bmatrix} 3 & 2 & 1 \\ 1 & 2 & -1 \\ 2 & 1 & 3 \end{bmatrix}$

(b) $\begin{bmatrix} 1 & 2 & 1 & 2 & 1 \\ 2 & 1 & 0 & 0 & 2 \\ 1 & -1 & -1 & -2 & 1 \\ 3 & 0 & -1 & -2 & 3 \end{bmatrix}$

ML.5. Using only the rank command, determine which of the following linear systems is consistent.

(a) $\begin{bmatrix} 1 & 2 & 4 & -1 \\ 0 & 1 & 2 & 0 \\ 3 & 1 & 1 & -2 \end{bmatrix} \begin{bmatrix} x_1 \\ x_2 \\ x_3 \\ x_4 \end{bmatrix} = \begin{bmatrix} 21 \\ 8 \\ 16 \end{bmatrix}$

(b) $\begin{bmatrix} 1 & 2 & 1 \\ 1 & 1 & 0 \\ 2 & 1 & -1 \end{bmatrix} \begin{bmatrix} x_1 \\ x_2 \\ x_3 \end{bmatrix} = \begin{bmatrix} 3 \\ 3 \\ 3 \end{bmatrix}$

(c) $\begin{bmatrix} 1 & 2 \\ 2 & 0 \\ 2 & 1 \\ -1 & 2 \end{bmatrix} \begin{bmatrix} x_1 \\ x_2 \end{bmatrix} = \begin{bmatrix} 3 \\ 2 \\ 3 \\ 2 \end{bmatrix}$

6.7 COORDINATES AND CHANGE OF BASIS

COORDINATES

If V is an n-dimensional vector space, we know that V has a basis S with n vectors in it; so far we have not paid much attention to the order of the vectors in S. However, in the discussion of this section we speak of an **ordered basis** $S = \{\mathbf{v}_1, \mathbf{v}_2, \dots, \mathbf{v}_n\}$ for V; thus $S_1 = \{\mathbf{v}_2, \mathbf{v}_1, \dots, \mathbf{v}_n\}$ is a different ordered basis for V.

If $S = \{\mathbf{v}_1, \mathbf{v}_2, \dots, \mathbf{v}_n\}$ is an ordered basis for the n-dimensional vector space V, then every vector $\mathbf{v}$ in V can be uniquely expressed in the form

$$\mathbf{v} = c_1\mathbf{v}_1 + c_2\mathbf{v}_2 + \cdots + c_n\mathbf{v}_n,$$

where $c_1, c_2, \dots, c_n$ are real numbers. We shall refer to

$$\left[\mathbf{v}\right]_S = \begin{bmatrix} c_1 \\ c_2 \\ \vdots \\ c_n \end{bmatrix}$$

as the **coordinate vector of v with respect to the ordered basis S**. The entries of $\left[\mathbf{v}\right]_S$ are called the **coordinates of v with respect to S**.

Observe that the coordinate vector $\left[\mathbf{v}\right]_S$ depends on the order in which the vectors in S are listed; a change in the order of this listing may change the coordinates of $\mathbf{v}$ with respect to S. All bases considered in this section are assumed to be ordered bases.

EXAMPLE 1 Let $S = \{\mathbf{v}_1, \mathbf{v}_2, \mathbf{v}_3, \mathbf{v}_4\}$ be a basis for R^4, where

$$\mathbf{v}_1 = (1, 1, 0, 0), \qquad \mathbf{v}_2 = (2, 0, 1, 0),$$
$$\mathbf{v}_3 = (0, 1, 2, -1), \qquad \mathbf{v}_4 = (0, 1, -1, 0).$$

If

$$\mathbf{v} = (1, 2, -6, 2),$$

compute $[\mathbf{v}]_S$.

Solution To find $[\mathbf{v}]_S$ we need to compute constants c_1, c_2, c_3, and c_4 such that

$$c_1\mathbf{v}_1 + c_2\mathbf{v}_2 + c_3\mathbf{v}_3 + c_4\mathbf{v}_4 = \mathbf{v},$$

which is just a linear combination problem. The previous equation leads to the linear system whose augmented matrix is (verify)

$$\left[\begin{array}{cccc|c} 1 & 2 & 0 & 0 & 1 \\ 1 & 0 & 1 & 1 & 2 \\ 0 & 1 & 2 & -1 & -6 \\ 0 & 0 & -1 & 0 & 2 \end{array}\right], \tag{1}$$

or, equivalently,

$$\left[\begin{array}{cccc|c} \mathbf{v}_1^T & \mathbf{v}_2^T & \mathbf{v}_3^T & \mathbf{v}_4^T & \mathbf{v}^T \end{array}\right].$$

Transforming the matrix in (1) to reduced row echelon form, we obtain the solution (verify)

$$c_1 = 3, \qquad c_2 = -1, \qquad c_3 = -2, \qquad c_4 = 1,$$

so the coordinate vector of $\mathbf{v}$ with respect to the basis S is

$$[\mathbf{v}]_S = \begin{bmatrix} 3 \\ -1 \\ -2 \\ 1 \end{bmatrix}.$$

■

EXAMPLE 2 Let $S = \{\mathbf{e}_1, \mathbf{e}_2, \mathbf{e}_3\}$ be the natural basis for R^3 and let

$$\mathbf{v} = (2, -1, 3).$$

Compute $[\mathbf{v}]_S$.

Solution Since S is the natural basis,

$$\mathbf{v} = 2\mathbf{e}_1 - 1\mathbf{e}_2 + 3\mathbf{e}_3,$$

so

$$[\mathbf{v}]_S = \begin{bmatrix} 2 \\ -1 \\ 3 \end{bmatrix}.$$

■

Remark In Example 2 the coordinate vector $[\mathbf{v}]_S$ of $\mathbf{v}$ with respect to S agrees with $\mathbf{v}$ because S is the natural basis for R^3.

EXAMPLE 3 Let V be P_1, the vector space of all polynomials of degree ≤ 1, and let $S = \{\mathbf{v}_1, \mathbf{v}_2\}$ and $T = \{\mathbf{w}_1, \mathbf{w}_2\}$ be bases for P_1, where

$$\mathbf{v}_1 = t, \qquad \mathbf{v}_2 = 1, \qquad \mathbf{w}_1 = t + 1, \qquad \mathbf{w}_2 = t - 1.$$

Let $\mathbf{v} = p(t) = 5t - 2$.

(a) Compute $\left[\mathbf{v}\right]_S$.

(b) Compute $\left[\mathbf{v}\right]_T$.

Solution (a) Since S is the standard or natural basis for P_1, we have

$$5t - 2 = 5t + (-2)(1).$$

Hence

$$\left[\mathbf{v}\right]_S = \begin{bmatrix} 5 \\ -2 \end{bmatrix}.$$

(b) To compute $\left[\mathbf{v}\right]_T$, we must write $\mathbf{v}$ as a linear combination of $\mathbf{w}_1$ and $\mathbf{w}_2$. Thus,

$$5t - 2 = c_1(t + 1) + c_2(t - 1),$$

or

$$5t - 2 = (c_1 + c_2)t + (c_1 - c_2).$$

Equating coefficients of like powers of t, we obtain the linear system

$$\begin{aligned} c_1 + c_2 &= 5 \\ c_1 - c_2 &= -2, \end{aligned}$$

whose solution is (verify)

$$c_1 = \tfrac{3}{2} \quad \text{and} \quad c_2 = \tfrac{7}{2}.$$

Hence

$$\left[\mathbf{v}\right]_T = \begin{bmatrix} \frac{3}{2} \\ \frac{7}{2} \end{bmatrix}.$$

■

In some important ways the coordinate vectors of elements in a vector space behave algebraically in ways that are similar to the way the vectors themselves behave. For example, it is not difficult to show (see Exercise T.2) that if S is a basis for an n-dimensional vector space V, $\mathbf{v}$ and $\mathbf{w}$ are vectors in V, and k is a scalar, then

$$\left[\mathbf{v} + \mathbf{w}\right]_S = \left[\mathbf{v}\right]_S + \left[\mathbf{w}\right]_S \tag{2}$$

and

$$\left[k\mathbf{v}\right]_S = k\left[\mathbf{v}\right]_S. \tag{3}$$

That is, the coordinate vector of a sum of two vectors is the sum of the coordinate vectors, and the coordinate vector of a scalar multiple of a vector is the scalar multiple of the coordinate vector. Moreover, the results in Equations (2) and (3) can be generalized to show that

$$\left[k_1\mathbf{v}_1 + k_2\mathbf{v}_2 + \cdots + k_n\mathbf{v}_n\right]_S = k_1\left[\mathbf{v}_1\right]_S + k_2\left[\mathbf{v}_2\right]_S + \cdots + k_n\left[\mathbf{v}_n\right]_S.$$

That is, the coordinate vector of a linear combination of vectors is the same linear combination of the individual coordinate vectors.

PICTURING A VECTOR SPACE

The choice of an ordered basis and the consequent assignment of a coordinate vector for every $\mathbf{v}$ in V enables us to "picture" the vector space. We illustrate this notion by using Example 3. Choose a fixed point O in the plane R^2 and draw any two arrows $\mathbf{w}_1$ and $\mathbf{w}_2$ from O that depict the basis vectors t and 1 in the ordered basis $S = \{t, 1\}$ for P_1 (see Figure 6.11). The directions of $\mathbf{w}_1$ and $\mathbf{w}_2$ determine two lines, which we call the **x_1- and x_2-axes**, respectively. The positive direction on the x_1-axis is in the direction of $\mathbf{w}_1$; the negative direction on the x_1-axis is along $-\mathbf{w}_1$. Similarly, the positive direction on the x_2-axis is in the direction of $\mathbf{w}_2$; the negative direction on the x_2-axis is along $-\mathbf{w}_2$. The lengths of $\mathbf{w}_1$ and $\mathbf{w}_2$ determine the scales on the x_1- and x_2-axes, respectively. If $\mathbf{v}$ is a vector in P_1, we can write $\mathbf{v}$, uniquely, as $\mathbf{v} = c_1\mathbf{w}_1 + c_2\mathbf{w}_2$. We now mark off a segment of length $|c_1|$ on the x_1-axis (in the positive direction if c_1 is positive and in the negative direction if c_1 is negative) and draw a line through the endpoint of this segment parallel to $\mathbf{w}_2$. Similarly, mark off a segment of length $|c_2|$ on the x_2-axis (in the positive direction if c_2 is positive and in the negative direction if c_2 is negative) and draw a line through the endpoint of this segment parallel to $\mathbf{w}_1$. We draw a directed line segment from O to the point of intersection of these two lines. This directed line segment represents $\mathbf{v}$.

Figure 6.11 ►

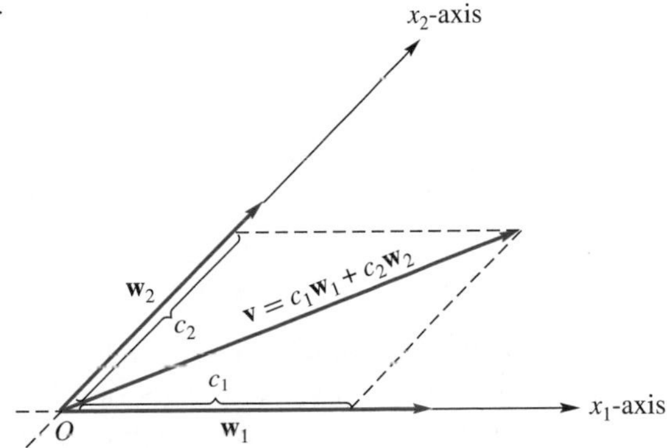

TRANSITION MATRICES

Suppose now that $S = \{\mathbf{v}_1, \mathbf{v}_2, \ldots, \mathbf{v}_n\}$ and $T = \{\mathbf{w}_1, \mathbf{w}_2, \ldots, \mathbf{w}_n\}$ are bases for the n-dimensional vector space V. We shall examine the relationship between the coordinate vectors $\left[\mathbf{v}\right]_S$ and $\left[\mathbf{v}\right]_T$ of the vector $\mathbf{v}$ in V with respect to the bases S and T, respectively.

If $\mathbf{v}$ is any vector in V, then

$$\mathbf{v} = c_1\mathbf{w}_1 + c_2\mathbf{w}_2 + \cdots + c_n\mathbf{w}_n \tag{4}$$

so that

$$\left[\mathbf{v}\right]_T = \begin{bmatrix} c_1 \\ c_2 \\ \vdots \\ c_n \end{bmatrix}.$$

Then

$$\begin{aligned}[\mathbf{v}]_S &= [c_1\mathbf{w}_1 + c_2\mathbf{w}_2 + \cdots + c_n\mathbf{w}_n]_S \\ &= [c_1\mathbf{w}_1]_S + [c_2\mathbf{w}_2]_S + \cdots + [c_n\mathbf{w}_n]_S \\ &= c_1[\mathbf{w}_1]_S + c_2[\mathbf{w}_2]_S + \cdots + c_n[\mathbf{w}_n]_S.\end{aligned}$$

Let the coordinate vector of $\mathbf{w}_j$ with respect to S be denoted by

$$[\mathbf{w}_j]_S = \begin{bmatrix} a_{1j} \\ a_{2j} \\ \vdots \\ a_{nj} \end{bmatrix}.$$

Then

$$[\mathbf{v}]_S = c_1\begin{bmatrix} a_{11} \\ a_{21} \\ \vdots \\ a_{n1} \end{bmatrix} + c_2\begin{bmatrix} a_{12} \\ a_{22} \\ \vdots \\ a_{n2} \end{bmatrix} + \cdots + c_n\begin{bmatrix} a_{1n} \\ a_{2n} \\ \vdots \\ a_{nn} \end{bmatrix} = \begin{bmatrix} a_{11} & a_{12} & \cdots & a_{1n} \\ a_{21} & a_{22} & \cdots & a_{2n} \\ \vdots & \vdots & & \vdots \\ a_{n1} & a_{n2} & \cdots & a_{nn} \end{bmatrix}\begin{bmatrix} c_1 \\ c_2 \\ \vdots \\ c_n \end{bmatrix}$$

or

$$[\mathbf{v}]_S = P_{S\leftarrow T}[\mathbf{v}]_T, \tag{5}$$

where

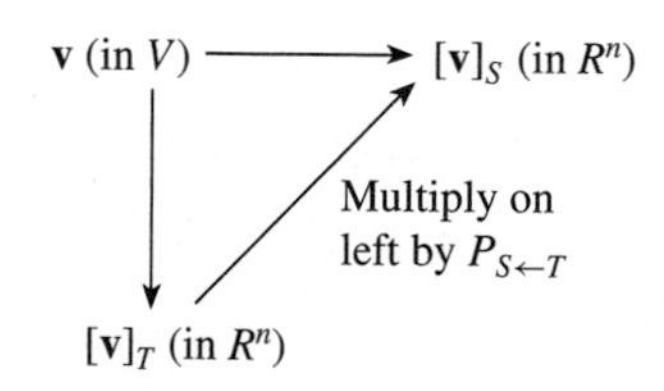

Figure 6.12 ▲

$$P_{S\leftarrow T} = \begin{bmatrix} a_{11} & a_{12} & \cdots & a_{1n} \\ a_{21} & a_{22} & \cdots & a_{2n} \\ \vdots & \vdots & & \vdots \\ a_{n1} & a_{n2} & \cdots & a_{nn} \end{bmatrix} = \begin{bmatrix} [\mathbf{w}_1]_S & [\mathbf{w}_2]_S & \cdots & [\mathbf{w}_n]_S \end{bmatrix}$$

is called the **transition matrix from the *T*-basis to the *S*-basis**. Equation (5) says that the coordinate vector of $\mathbf{v}$ with respect to the basis S is the transition matrix $P_{S\leftarrow T}$ times the coordinate vector of $\mathbf{v}$ with respect to the basis T. Figure 6.12 illustrates Equation (5).

We may now summarize the procedure just developed for computing the transition matrix from the T-basis to the S-basis.

> The procedure for computing the transition matrix $P_{S\leftarrow T}$ from the basis $T = \{\mathbf{w}_1, \mathbf{w}_2, \ldots, \mathbf{w}_n\}$ for V to the basis $S = \{\mathbf{v}_1, \mathbf{v}_2, \ldots, \mathbf{v}_n\}$ for V is as follows.
>
> ***Step 1.*** Compute the coordinate vector of $\mathbf{w}_j$, $j = 1, 2, \ldots, n$, with respect to the basis S. This means that we first have to express $\mathbf{w}_j$ as a linear combination of the vectors in S:
>
> $$a_{1j}\mathbf{v}_1 + a_{2j}\mathbf{v}_2 + \cdots + a_{nj}\mathbf{v}_n = \mathbf{w}_j, \qquad j = 1, 2, \ldots, n.$$
>
> We now solve for $a_{1j}, a_{2j}, \ldots, a_{nj}$ by Gauss–Jordan reduction, transforming the augmented matrix of this linear system to reduced row echelon form.
>
> ***Step 2.*** The .transition matrix $P_{S\leftarrow T}$ from the T-basis to the S-basis is formed by choosing $[\mathbf{w}_j]_S$ as the jth column of $P_{S\leftarrow T}$.

EXAMPLE 4 Let V be R^3 and let $S = \{\mathbf{v}_1, \mathbf{v}_2, \mathbf{v}_3\}$ and $T = \{\mathbf{w}_1, \mathbf{w}_2, \mathbf{w}_3\}$ be bases for R^3, where

$$\mathbf{v}_1 = \begin{bmatrix} 2 \\ 0 \\ 1 \end{bmatrix}, \quad \mathbf{v}_2 = \begin{bmatrix} 1 \\ 2 \\ 0 \end{bmatrix}, \quad \mathbf{v}_3 = \begin{bmatrix} 1 \\ 1 \\ 1 \end{bmatrix}$$

and

$$\mathbf{w}_1 = \begin{bmatrix} 6 \\ 3 \\ 3 \end{bmatrix}, \quad \mathbf{w}_2 = \begin{bmatrix} 4 \\ -1 \\ 3 \end{bmatrix}, \quad \mathbf{w}_3 = \begin{bmatrix} 5 \\ 5 \\ 2 \end{bmatrix}.$$

(a) Compute the transition matrix $P_{S \leftarrow T}$ from the T-basis to the S-basis.

(b) Verify Equation (5) for $\mathbf{v} = \begin{bmatrix} 4 \\ -9 \\ 5 \end{bmatrix}$.

Solution (a) To compute $P_{S \leftarrow T}$, we find a_1, a_2, a_3 such that

$$a_1\mathbf{v}_1 + a_2\mathbf{v}_2 + a_3\mathbf{v}_3 = \mathbf{w}_1.$$

In this case we are led to a linear system of three equations in three unknowns, whose augmented matrix is

$$\begin{bmatrix} \mathbf{v}_1 & \mathbf{v}_2 & \mathbf{v}_3 & \vdots & \mathbf{w}_1 \end{bmatrix}.$$

That is, the augmented matrix is

$$\left[\begin{array}{ccc:c} 2 & 1 & 1 & 6 \\ 0 & 2 & 1 & 3 \\ 1 & 0 & 1 & 3 \end{array}\right].$$

Similarly, to find b_1, b_2, b_3 and c_1, c_2, c_3 such that

$$b_1\mathbf{v}_1 + b_2\mathbf{v}_2 + b_3\mathbf{v}_3 = \mathbf{w}_2$$
$$c_1\mathbf{v}_1 + c_2\mathbf{v}_2 + c_3\mathbf{v}_3 = \mathbf{w}_3,$$

we are led to two linear systems, each of three equations in three unknowns, whose augmented matrices are

$$\begin{bmatrix} \mathbf{v}_1 & \mathbf{v}_2 & \mathbf{v}_3 & \vdots & \mathbf{w}_2 \end{bmatrix} \quad \text{and} \quad \begin{bmatrix} \mathbf{v}_1 & \mathbf{v}_2 & \mathbf{v}_3 & \vdots & \mathbf{w}_3 \end{bmatrix},$$

or, specifically,

$$\left[\begin{array}{ccc:c} 2 & 1 & 1 & 4 \\ 0 & 2 & 1 & -1 \\ 1 & 0 & 1 & 3 \end{array}\right] \quad \text{and} \quad \left[\begin{array}{ccc:c} 2 & 1 & 1 & 5 \\ 0 & 2 & 1 & 5 \\ 1 & 0 & 1 & 2 \end{array}\right].$$

Since the coefficient matrix of all three linear systems is $\begin{bmatrix} \mathbf{v}_1 & \mathbf{v}_2 & \mathbf{v}_3 \end{bmatrix}$, we can transform the three augmented matrices to reduced row echelon form simultaneously by transforming the partitioned matrix

$$\begin{bmatrix} \mathbf{v}_1 & \mathbf{v}_2 & \mathbf{v}_3 & \vdots & \mathbf{w}_1 & \vdots & \mathbf{w}_2 & \vdots & \mathbf{w}_3 \end{bmatrix}$$

to reduced row echelon form.* Thus we transform

$$\left[\begin{array}{ccc:c:c:c} 2 & 1 & 1 & 6 & 4 & 5 \\ 0 & 2 & 1 & 3 & -1 & 5 \\ 1 & 0 & 1 & 3 & 3 & 2 \end{array}\right]$$

*This has been discussed in Remark 2 following Example 12 in Section 1.6.

to reduced row echelon form, obtaining (verify)

$$\left[\begin{array}{ccc:c:c:c} 1 & 0 & 0 & 2 & 2 & 1 \\ 0 & 1 & 0 & 1 & -1 & 2 \\ 0 & 0 & 1 & 1 & 1 & 1 \end{array}\right],$$

which implies that the transition matrix from the T-basis to the S-basis is

$$P_{S\leftarrow T} = \begin{bmatrix} 2 & 2 & 1 \\ 1 & -1 & 2 \\ 1 & 1 & 1 \end{bmatrix}.$$

(b) If

$$\mathbf{v} = \begin{bmatrix} 4 \\ -9 \\ 5 \end{bmatrix},$$

then to express $\mathbf{v}$ in terms of the T-basis, we use Equation (4). From the associated linear system we find that (verify)

$$\mathbf{v} = \begin{bmatrix} 4 \\ -9 \\ 5 \end{bmatrix} = 1\begin{bmatrix} 6 \\ 3 \\ 3 \end{bmatrix} + 2\begin{bmatrix} 4 \\ -1 \\ 3 \end{bmatrix} - 2\begin{bmatrix} 5 \\ 5 \\ 2 \end{bmatrix} = 1\mathbf{w}_1 + 2\mathbf{w}_2 - 2\mathbf{w}_3,$$

so

$$\left[\mathbf{v}\right]_T = \begin{bmatrix} 1 \\ 2 \\ -2 \end{bmatrix}.$$

Then by Equation (5) we find that $\left[\mathbf{v}\right]_S$ is

$$P_{S\leftarrow T}\left[\mathbf{v}\right]_T = \begin{bmatrix} 2 & 2 & 1 \\ 1 & -1 & 2 \\ 1 & 1 & 1 \end{bmatrix}\begin{bmatrix} 1 \\ 2 \\ -2 \end{bmatrix} = \begin{bmatrix} 4 \\ -5 \\ 1 \end{bmatrix}.$$

If we compute $\left[\mathbf{v}\right]_S$ directly by setting up and solving the associated linear system, we find that (verify)

$$\mathbf{v} = \begin{bmatrix} 4 \\ -9 \\ 5 \end{bmatrix} = 4\begin{bmatrix} 2 \\ 0 \\ 1 \end{bmatrix} - 5\begin{bmatrix} 1 \\ 2 \\ 0 \end{bmatrix} + 1\begin{bmatrix} 1 \\ 1 \\ 1 \end{bmatrix} = 4\mathbf{v}_1 - 5\mathbf{v}_2 + 1\mathbf{v}_3,$$

so

$$\left[\mathbf{v}\right]_S = \begin{bmatrix} 4 \\ -5 \\ 1 \end{bmatrix}.$$

Hence

$$\left[\mathbf{v}\right]_S = P_{S\leftarrow T}\left[\mathbf{v}\right]_T. \quad \blacksquare$$

We next show that the transition matrix $P_{S\leftarrow T}$ from the T-basis to the S-basis is nonsingular and that $P_{S\leftarrow T}^{-1}$ is the transition matrix from the S-basis to the T-basis.

THEOREM 6.15 *Let $S = \{\mathbf{v}_1, \mathbf{v}_2, \ldots, \mathbf{v}_n\}$ and $T = \{\mathbf{w}_1, \mathbf{w}_2, \ldots, \mathbf{w}_n\}$ be bases for the n-dimensional vector space V. Let $P_{S\leftarrow T}$ be the transition matrix from the T-basis to the S-basis. Then $P_{S\leftarrow T}$ is nonsingular and $P_{S\leftarrow T}^{-1}$ is the transition matrix from the S-basis to the T-basis.*

Proof We proceed by showing that the null space of $P_{S \leftarrow T}$ contains only the zero vector. Suppose that $P_{S \leftarrow T} \left[\mathbf{v}\right]_T = \mathbf{0}_{R^n}$ for some $\mathbf{v}$ in V. From Equation (5) we have

$$P_{S \leftarrow T} \left[\mathbf{v}\right]_T = \left[\mathbf{v}\right]_S = \mathbf{0}_{R^n}.$$

If $\mathbf{v} = b_1\mathbf{v}_1 + b_2\mathbf{v}_2 + \cdots + b_n\mathbf{v}_n$, then

$$\begin{bmatrix} b_1 \\ b_2 \\ \vdots \\ b_n \end{bmatrix} = \left[\mathbf{v}\right]_S = \mathbf{0}_{R^n} = \begin{bmatrix} 0 \\ 0 \\ \vdots \\ 0 \end{bmatrix},$$

so

$$\mathbf{v} = 0\mathbf{v}_1 + 0\mathbf{v}_2 + \cdots + 0\mathbf{v}_n = \mathbf{0}_V.$$

Hence $\left[\mathbf{v}\right]_T = \mathbf{0}_{R^n}$. Thus the homogeneous system $P_{S \leftarrow T}\mathbf{x} = \mathbf{0}$ has only the trivial solution; it then follows from Theorem 1.13 that $P_{S \leftarrow T}$ is nonsingular. Multiplying both sides of Equation (5) on the left by $P_{S \leftarrow T}^{-1}$, we have

$$\left[\mathbf{v}\right]_T = P_{S \leftarrow T}^{-1} \left[\mathbf{v}\right]_S.$$

That is, $P_{S \leftarrow T}^{-1}$ is then the transition matrix from the S-basis to the T-basis; the jth column of $P_{S \leftarrow T}^{-1}$ is $\left[\mathbf{v}_j\right]_T$. ■

Remark In Exercises T.5 through T.7 we ask you to show that if S and T are bases for the vector space R^n, then

$$P_{S \leftarrow T} = M_S^{-1} M_T,$$

where M_S is the $n \times n$ matrix whose jth column is $\mathbf{v}_j$ and M_T is the $n \times n$ matrix whose jth column is $\mathbf{w}_j$. This formula implies that $P_{S \leftarrow T}$ is nonsingular and it is helpful in solving some of the exercises in this section.

EXAMPLE 5 Let S and T be the bases for R^3 defined in Example 4. Compute the transition matrix $Q_{T \leftarrow S}$ from the S-basis to the T-basis directly and show that $Q_{T \leftarrow S} = P_{S \leftarrow T}^{-1}$.

Solution $Q_{T \leftarrow S}$ is the matrix whose columns are the solution vectors to the linear system obtained from the vector equations

$$\begin{aligned} a_1\mathbf{w}_1 + a_2\mathbf{w}_2 + a_3\mathbf{w}_3 &= \mathbf{v}_1 \\ b_1\mathbf{w}_1 + b_2\mathbf{w}_2 + b_3\mathbf{w}_3 &= \mathbf{v}_2 \\ c_1\mathbf{w}_1 + c_2\mathbf{w}_2 + c_3\mathbf{w}_3 &= \mathbf{v}_3. \end{aligned}$$

As in Example 4, we can solve these linear systems simultaneously by transforming the partitioned matrix

$$\left[\mathbf{w}_1 \quad \mathbf{w}_2 \quad \mathbf{w}_3 \;\vdots\; \mathbf{v}_1 \;\vdots\; \mathbf{v}_2 \;\vdots\; \mathbf{v}_3\right]$$

to reduced row echelon form. That is, we transform

$$\left[\begin{array}{ccc|c|c|c} 6 & 4 & 5 & 2 & 1 & 1 \\ 3 & -1 & 5 & 0 & 2 & 1 \\ 3 & 3 & 2 & 1 & 0 & 1 \end{array}\right]$$

to reduced row echelon form, obtaining (verify)

$$\left[\begin{array}{ccc|c|c|c} 1 & 0 & 0 & \frac{3}{2} & \frac{1}{2} & -\frac{5}{2} \\ 0 & 1 & 0 & -\frac{1}{2} & -\frac{1}{2} & \frac{3}{2} \\ 0 & 0 & 1 & -1 & 0 & 2 \end{array}\right],$$

so

$$Q_{T \leftarrow S} = \begin{bmatrix} \frac{3}{2} & \frac{1}{2} & -\frac{5}{2} \\ -\frac{1}{2} & -\frac{1}{2} & \frac{3}{2} \\ -1 & 0 & 2 \end{bmatrix}.$$

Multiplying $Q_{T \leftarrow S}$ by $P_{S \leftarrow T}$, we find (verify) that $Q_{T \leftarrow S} P_{S \leftarrow T} = I_3$, so we conclude that $Q_{T \leftarrow S} = P_{S \leftarrow T}^{-1}$. ■

EXAMPLE 6 Let V be P_1 and let $S = \{\mathbf{v}_1, \mathbf{v}_2\}$ and $T = \{\mathbf{w}_1, \mathbf{w}_2\}$ be bases for P_1, where

$$\mathbf{v}_1 = t, \qquad \mathbf{v}_2 = t - 3, \qquad \mathbf{w}_1 = t - 1, \qquad \mathbf{w}_2 = t + 1.$$

(a) Compute the transition matrix $P_{S \leftarrow T}$ from the T-basis to the S-basis.

(b) Verify Equation (5) for $\mathbf{v} = 5t + 1$.

(c) Compute the transition matrix $Q_{T \leftarrow S}$ from the S-basis to the T-basis and show that $Q_{T \leftarrow S} = P_{S \leftarrow T}^{-1}$.

Solution (a) To compute $P_{S \leftarrow T}$, we need to solve the vector equations

$$a_1\mathbf{v}_1 + a_2\mathbf{v}_2 = \mathbf{w}_1$$
$$b_1\mathbf{v}_1 + b_2\mathbf{v}_2 = \mathbf{w}_2$$

simultaneously by transforming the resulting partitioned matrix (verify)

$$\left[\begin{array}{cc|c|c} 1 & 1 & 1 & 1 \\ 0 & -3 & -1 & 1 \end{array}\right]$$

to reduced row echelon form. The result is (verify)

$$\left[\begin{array}{cc|c|c} 1 & 0 & \frac{2}{3} & \frac{4}{3} \\ 0 & 1 & \frac{1}{3} & -\frac{1}{3} \end{array}\right],$$

so

$$P_{S \leftarrow T} = \begin{bmatrix} \frac{2}{3} & \frac{4}{3} \\ \frac{1}{3} & -\frac{1}{3} \end{bmatrix}.$$

(b) If $\mathbf{v} = 5t + 1$, then expressing $\mathbf{v}$ in terms of the T-basis, we have (verify)

$$\mathbf{v} = 5t + 1 = 2(t - 1) + 3(t + 1),$$

so

$$\left[\mathbf{v}\right]_T = \begin{bmatrix} 2 \\ 3 \end{bmatrix}.$$

To verify this, set up and solve the associated linear system resulting from the vector equation

$$\mathbf{v} = a_1\mathbf{w}_1 + a_2\mathbf{w}_2.$$

Then

$$[\mathbf{v}]_S = P_{S\leftarrow T}[\mathbf{v}]_T = \begin{bmatrix} \frac{2}{3} & \frac{4}{3} \\ \frac{1}{3} & -\frac{1}{3} \end{bmatrix} \begin{bmatrix} 2 \\ 3 \end{bmatrix} = \begin{bmatrix} \frac{16}{3} \\ -\frac{1}{3} \end{bmatrix}.$$

Computing $[\mathbf{v}]_S$ directly from the associated linear system arising from the vector equation

$$\mathbf{v} = b_1\mathbf{w}_1 + b_2\mathbf{w}_2$$

we find that (verify)

$$\mathbf{v} = 5t + 1 = \tfrac{16}{3}t - \tfrac{1}{3}(t-3),$$

so

$$[\mathbf{v}]_S = \begin{bmatrix} \frac{16}{3} \\ -\frac{1}{3} \end{bmatrix}.$$

Hence

$$[\mathbf{v}]_S = P_{S\leftarrow T}[\mathbf{v}]_T,$$

which is Equation (5).

(c) The transition matrix $Q_{T\leftarrow S}$ from the S-basis to the T-basis is obtained (verify) by transforming the partitioned matrix

$$\left[\begin{array}{rr|r|r} 1 & 1 & 1 & 1 \\ -1 & 1 & 0 & -3 \end{array}\right]$$

to reduced row echelon form, obtaining (verify)

$$\left[\begin{array}{rr|r|r} 1 & 0 & \frac{1}{2} & 2 \\ 0 & 1 & \frac{1}{2} & -1 \end{array}\right].$$

Hence

$$Q_{T\leftarrow S} = \begin{bmatrix} \frac{1}{2} & 2 \\ \frac{1}{2} & -1 \end{bmatrix}.$$

Multiplying $Q_{T\leftarrow S}$ by $P_{S\leftarrow T}$, we find (verify) that $Q_{T\leftarrow S}P_{S\leftarrow T} = I_2$, so $Q_{T\leftarrow S} = P^{-1}_{S\leftarrow T}$. ■

Key Terms

Ordered basis
Coordinate vector
Transition matrix

6.7 Exercises

All bases considered in these exercises are assumed to be ordered bases. In Exercises 1 through 6, compute the coordinate vector of $\mathbf{v}$ *with respect to the given basis S for V.*

1. V is R^2, $S = \left\{\begin{bmatrix} 1 \\ 0 \end{bmatrix}, \begin{bmatrix} 0 \\ 1 \end{bmatrix}\right\}$, $\mathbf{v} = \begin{bmatrix} 3 \\ -2 \end{bmatrix}$.

2. V is R^3, $S = \{(1,-1,0), (0,1,0), (1,0,2)\}$, $\mathbf{v} = (2,-1,-2)$.

3. V is P_1, $S = \{t+1, t-2\}$, $\mathbf{v} = t+4$.

4. V is P_2, $S = \{t^2 - t + 1, t + 1, t^2 + 1\}$, $\mathbf{v} = 4t^2 - 2t + 3$.

5. V is M_{22}, $S = \left\{ \begin{bmatrix} 1 & 0 \\ 0 & 0 \end{bmatrix}, \begin{bmatrix} 0 & 0 \\ 1 & 0 \end{bmatrix}, \begin{bmatrix} 0 & 1 \\ 0 & 0 \end{bmatrix}, \begin{bmatrix} 0 & 0 \\ 0 & 1 \end{bmatrix} \right\}$, $\mathbf{v} = \begin{bmatrix} 1 & 0 \\ -1 & 2 \end{bmatrix}$.

6. V is M_{22}, $S = \left\{ \begin{bmatrix} 1 & -1 \\ 0 & 0 \end{bmatrix}, \begin{bmatrix} 0 & 1 \\ 1 & 0 \end{bmatrix}, \begin{bmatrix} 1 & 0 \\ 0 & -1 \end{bmatrix}, \begin{bmatrix} 1 & 0 \\ -1 & 0 \end{bmatrix} \right\}$, $\mathbf{v} = \begin{bmatrix} 1 & 3 \\ -2 & 2 \end{bmatrix}$.

In Exercises 7 through 12, compute the vector **v** *if the coordinate vector* $[\mathbf{v}]_S$ *is given with respect to the basis S for V.*

7. V is R^2, $S = \left\{ \begin{bmatrix} 2 \\ 1 \end{bmatrix}, \begin{bmatrix} -1 \\ 1 \end{bmatrix} \right\}$, $[\mathbf{v}]_S = \begin{bmatrix} 1 \\ 2 \end{bmatrix}$.

8. V is R^3, $S = \{(0, 1, -1), (1, 0, 0), (1, 1, 1)\}$, $[\mathbf{v}]_S = \begin{bmatrix} -1 \\ 1 \\ 2 \end{bmatrix}$.

9. V is P_1, $S = \{t, 2t - 1\}$, $[\mathbf{v}]_S = \begin{bmatrix} -2 \\ 3 \end{bmatrix}$.

10. V is P_2, $S = \{t^2 + 1, t + 1, t^2 + t\}$, $[\mathbf{v}]_S = \begin{bmatrix} 3 \\ -1 \\ -2 \end{bmatrix}$.

11. V is M_{22}, $S = \left\{ \begin{bmatrix} -1 & 0 \\ 1 & 0 \end{bmatrix}, \begin{bmatrix} 2 & 2 \\ 0 & 1 \end{bmatrix}, \begin{bmatrix} 1 & 2 \\ -1 & 3 \end{bmatrix}, \begin{bmatrix} 0 & 0 \\ 2 & 3 \end{bmatrix} \right\}$, $[\mathbf{v}]_S = \begin{bmatrix} 2 \\ 1 \\ -1 \\ 3 \end{bmatrix}$.

12. V is M_{22}, $S = \left\{ \begin{bmatrix} 1 & -2 \\ 0 & 0 \end{bmatrix}, \begin{bmatrix} -1 & 3 \\ 0 & 1 \end{bmatrix}, \begin{bmatrix} 1 & 0 \\ 0 & 0 \end{bmatrix}, \begin{bmatrix} 0 & -1 \\ 1 & 0 \end{bmatrix} \right\}$, $[\mathbf{v}]_S = \begin{bmatrix} 0 \\ 1 \\ 0 \\ 2 \end{bmatrix}$.

13. Let $S = \{(1, 2), (0, 1)\}$ and $T = \{(1, 1), (2, 3)\}$ be bases for R^2. Let $\mathbf{v} = (1, 5)$ and $\mathbf{w} = (5, 4)$.

(a) Find the coordinate vectors of **v** and **w** with respect to the basis T.

(b) What is the transition matrix $P_{S \leftarrow T}$ from the T- to the S-basis?

(c) Find the coordinate vectors of **v** and **w** with respect to S using $P_{S \leftarrow T}$.

(d) Find the coordinate vectors of **v** and **w** with respect to S directly.

(e) Find the transition matrix $Q_{T \leftarrow S}$ from the S- to the T-basis.

(f) Find the coordinate vectors of **v** and **w** with respect to T using $Q_{T \leftarrow S}$. Compare the answers with those of (a).

14. Let

$$S = \left\{ \begin{bmatrix} 1 \\ 0 \\ 1 \end{bmatrix}, \begin{bmatrix} -1 \\ 0 \\ 0 \end{bmatrix}, \begin{bmatrix} 0 \\ 1 \\ 2 \end{bmatrix} \right\}$$

and

$$T = \left\{ \begin{bmatrix} -1 \\ 1 \\ 0 \end{bmatrix}, \begin{bmatrix} 1 \\ 2 \\ -1 \end{bmatrix}, \begin{bmatrix} 0 \\ 1 \\ 0 \end{bmatrix} \right\}$$

be bases for R^3. Let

$$\mathbf{v} = \begin{bmatrix} 1 \\ 3 \\ 8 \end{bmatrix} \quad \text{and} \quad \mathbf{w} = \begin{bmatrix} -1 \\ 8 \\ -2 \end{bmatrix}.$$

Follow the directions for Exercise 13.

15. Let $S = \{t^2 + 1, t - 2, t + 3\}$ and $T = \{2t^2 + t, t^2 + 3, t\}$ be bases for P_2. Let $\mathbf{v} = 8t^2 - 4t + 6$ and $\mathbf{w} = 7t^2 - t + 9$. Follow the directions for Exercise 13.

16. Let $S = \{t^2 + t + 1, t^2 + 2t + 3, t^2 + 1\}$ and $T = \{t + 1, t^2, t^2 + 1\}$ be bases for P_2. Also let $\mathbf{v} = -t^2 + 4t + 5$ and $\mathbf{w} = 2t^2 - 6$. Follow the directions for Exercise 13.

17. Let

$$S = \left\{ \begin{bmatrix} 1 & 0 \\ 0 & 0 \end{bmatrix}, \begin{bmatrix} 0 & 1 \\ 1 & 0 \end{bmatrix}, \begin{bmatrix} 0 & 2 \\ 0 & 1 \end{bmatrix}, \begin{bmatrix} 0 & 0 \\ 1 & 1 \end{bmatrix} \right\}$$

and

$$T = \left\{ \begin{bmatrix} 1 & 1 \\ 0 & 0 \end{bmatrix}, \begin{bmatrix} 0 & 0 \\ 1 & 0 \end{bmatrix}, \begin{bmatrix} 0 & 0 \\ 0 & 1 \end{bmatrix}, \begin{bmatrix} 1 & 0 \\ 0 & 0 \end{bmatrix} \right\}$$

be bases for M_{22}. Let

$$\mathbf{v} = \begin{bmatrix} 1 & 1 \\ 1 & 1 \end{bmatrix} \quad \text{and} \quad \mathbf{w} = \begin{bmatrix} 1 & 2 \\ -2 & 1 \end{bmatrix}.$$

Follow the directions for Exercise 13.

18. Let

$$S = \left\{ \begin{bmatrix} -1 & -1 \\ 0 & 1 \end{bmatrix}, \begin{bmatrix} 1 & 0 \\ 0 & 1 \end{bmatrix}, \begin{bmatrix} 0 & -1 \\ 0 & 0 \end{bmatrix}, \begin{bmatrix} 1 & 0 \\ 1 & 0 \end{bmatrix} \right\}$$

and

$$T = \left\{ \begin{bmatrix} 1 & 1 \\ 0 & 0 \end{bmatrix}, \begin{bmatrix} 1 & 0 \\ 1 & -1 \end{bmatrix}, \begin{bmatrix} 1 & 0 \\ 0 & 0 \end{bmatrix}, \begin{bmatrix} 0 & 1 \\ 0 & 1 \end{bmatrix} \right\}$$

be bases for M_{22}. Let

$$\mathbf{v} = \begin{bmatrix} 0 & 0 \\ 3 & -1 \end{bmatrix} \quad \text{and} \quad \mathbf{w} = \begin{bmatrix} -2 & 3 \\ -1 & 3 \end{bmatrix}.$$

Follow the directions for Exercise 13.

19. Let $S = \{(1, -1), (2, 1)\}$ and $T = \{(3, 0), (4, -1)\}$ be bases for R^2. If **v** is in R^2 and

$$[\mathbf{v}]_T = \begin{bmatrix} 1 \\ 2 \end{bmatrix},$$

determine $[\mathbf{v}]_S$.

20. Let $S = \{t + 1, t - 2\}$ and $T = \{t - 5, t - 2\}$ be bases for P_1. If $\mathbf{v}$ is in P_1 and

$$\left[\mathbf{v}\right]_T = \begin{bmatrix} -1 \\ 3 \end{bmatrix},$$

determine $\left[\mathbf{v}\right]_S$.

21. Let $S = \{(-1, 2, 1), (0, 1, 1), (-2, 2, 1)\}$ and $T = \{(-1, 1, 0), (0, 1, 0), (0, 1, 1)\}$ be bases for R^3. If $\mathbf{v}$ is in R^3 and

$$\left[\mathbf{v}\right]_S = \begin{bmatrix} 2 \\ 0 \\ 1 \end{bmatrix},$$

determine $\left[\mathbf{v}\right]_T$.

22. Let $S = \{t^2, t - 1, 1\}$ and $T = \{t^2 + t + 1, t + 1, 1\}$ be bases for P_2. If $\mathbf{v}$ is in P_2 and

$$\left[\mathbf{v}\right]_S = \begin{bmatrix} 1 \\ 2 \\ 3 \end{bmatrix},$$

determine $\left[\mathbf{v}\right]_T$.

23. Let $S = \{\mathbf{v}_1, \mathbf{v}_2, \mathbf{v}_3\}$ and $T = \{\mathbf{w}_1, \mathbf{w}_2, \mathbf{w}_3\}$ be bases for R^3, where

$$\mathbf{v}_1 = (1, 0, 1), \quad \mathbf{v}_2 = (1, 1, 0), \quad \mathbf{v}_3 = (0, 0, 1).$$

If the transition matrix from T to S is

$$\begin{bmatrix} 1 & 1 & 2 \\ 2 & 1 & 1 \\ -1 & -1 & 1 \end{bmatrix},$$

determine T.

24. Let $S = \{\mathbf{v}_1, \mathbf{v}_2\}$ and $T = \{\mathbf{w}_1, \mathbf{w}_2\}$ be bases for P_1, where

$$\mathbf{w}_1 = t, \qquad \mathbf{w}_2 = t - 1.$$

If the transition matrix from S to T is

$$\begin{bmatrix} 2 & 3 \\ -1 & 2 \end{bmatrix},$$

determine S.

25. Let $S = \{\mathbf{v}_1, \mathbf{v}_2\}$ and $T = \{\mathbf{w}_1, \mathbf{w}_2\}$ be bases for R^2, where

$$\mathbf{v}_1 = (1, 2), \qquad \mathbf{v}_2 = (0, 1).$$

If the transition matrix from S to T is

$$\begin{bmatrix} 2 & 1 \\ 1 & 1 \end{bmatrix},$$

determine T.

26. Let $S = \{\mathbf{v}_1, \mathbf{v}_2\}$ and $T = \{\mathbf{w}_1, \mathbf{w}_2\}$ be bases for P_1, where

$$\mathbf{w}_1 = t - 1, \qquad \mathbf{w}_2 = t + 1.$$

If the transition matrix from T to S is

$$\begin{bmatrix} 1 & 2 \\ 2 & 3 \end{bmatrix},$$

determine S.

Theoretical Exercises

T.1. Let $S = \{\mathbf{v}_1, \mathbf{v}_2, \dots, \mathbf{v}_n\}$ be a basis for the n-dimensional vector space V, and let $\mathbf{v}$ and $\mathbf{w}$ be two vectors in V. Show that $\mathbf{v} = \mathbf{w}$ if and only if $\left[\mathbf{v}\right]_S = \left[\mathbf{w}\right]_S$.

T.2. Show that if S is a basis for an n-dimensional vector space V, $\mathbf{v}$ and $\mathbf{w}$ are vectors in V, and k is a scalar, then

$$\left[\mathbf{v} + \mathbf{w}\right]_S = \left[\mathbf{v}\right]_S + \left[\mathbf{w}\right]_S$$

and

$$\left[k\mathbf{v}\right]_S = k\left[\mathbf{v}\right]_S.$$

T.3. Let S be a basis for an n-dimensional vector space V. Show that if $\{\mathbf{w}_1, \mathbf{w}_2, \dots, \mathbf{w}_k\}$ is a linearly independent set of vectors in V, then

$$\{\left[\mathbf{w}_1\right]_S, \left[\mathbf{w}_2\right]_S, \dots, \left[\mathbf{w}_k\right]_S\}$$

is a linearly independent set of vectors in R^n.

T.4. Let $S = \{\mathbf{v}_1, \mathbf{v}_2, \dots, \mathbf{v}_n\}$ be a basis for an n-dimensional vector space V. Show that

$$\{\left[\mathbf{v}_1\right]_S, \left[\mathbf{v}_2\right]_S, \dots, \left[\mathbf{v}_n\right]_S\}$$

is a basis for R^n.

In Exercises T.5 through T.7, let $S = \{\mathbf{v}_1, \mathbf{v}_2, \dots, \mathbf{v}_n\}$ and $T = \{\mathbf{w}_1, \mathbf{w}_2, \dots, \mathbf{w}_n\}$ be bases for the vector space R^n.

T.5. Let M_S be the $n \times n$ matrix whose jth column is $\mathbf{v}_j$ and let M_T be the $n \times n$ matrix whose jth column is $\mathbf{w}_j$. Show that M_S and M_T are nonsingular. (*Hint*: Consider the homogeneous systems $M_S\mathbf{x} = \mathbf{0}$ and $M_T\mathbf{x} = \mathbf{0}$.)

T.6. If $\mathbf{v}$ is a vector in V, show that

$$\mathbf{v} = M_S\left[\mathbf{v}\right]_S \quad \text{and} \quad \mathbf{v} = M_T\left[\mathbf{v}\right]_T.$$

T.7. (a) Use Equation (5) and Exercises T.5 and T.6 to show that

$$P_{S \leftarrow T} = M_S^{-1} M_T.$$

(b) Show that $P_{S \leftarrow T}$ is nonsingular.

(c) Verify the result in part (a) for Example 4.

MATLAB Exercises

Finding the coordinates of a vector with respect to a basis is a linear combination problem. Hence, once the corresponding linear system is constructed, we can use MATLAB *routine* **reduce** *or* **rref** *to find its solution. The solution gives us the desired coordinates. (The discussion in Section 12.7 will be helpful as an aid for constructing the necessary linear system.)*

ML.1. Let $V = R^3$ and

$$S = \left\{ \begin{bmatrix} 1 \\ 2 \\ 1 \end{bmatrix}, \begin{bmatrix} 2 \\ 1 \\ 0 \end{bmatrix}, \begin{bmatrix} 1 \\ 0 \\ 2 \end{bmatrix} \right\}.$$

Show that S is a basis for V and find $\left[\mathbf{v}\right]_S$ for each of the following vectors.

(a) $\mathbf{v} = \begin{bmatrix} 8 \\ 4 \\ 7 \end{bmatrix}$ (b) $\mathbf{v} = \begin{bmatrix} 2 \\ 0 \\ -3 \end{bmatrix}$

(c) $\mathbf{v} = \begin{bmatrix} 4 \\ 3 \\ 3 \end{bmatrix}$

ML.2. Let $V = R^4$ and $S = \{(1, 0, 1, 1), (1, 2, 1, 3), (0, 2, 1, 1), (0, 1, 0, 0)\}$. Show that S is a basis for V and find $\left[\mathbf{v}\right]_S$ for each of the following vectors.

(a) $\mathbf{v} = (4, 12, 8, 14)$

(b) $\mathbf{v} = \left(\frac{1}{2}, 0, 0, 0\right)$

(c) $\mathbf{v} = \left(1, 1, 1, \frac{7}{3}\right)$

ML.3. Let V be the vector space of all 2×2 matrices and

$$S = \left\{ \begin{bmatrix} 1 & 2 \\ 1 & 2 \end{bmatrix}, \begin{bmatrix} 0 & 2 \\ 1 & 0 \end{bmatrix}, \begin{bmatrix} 3 & 1 \\ -1 & 0 \end{bmatrix}, \begin{bmatrix} -1 & 0 \\ 0 & 0 \end{bmatrix} \right\}.$$

Show that S is a basis for V and find $\left[\mathbf{v}\right]_S$ for each of the following vectors.

(a) $\mathbf{v} = \begin{bmatrix} 1 & 0 \\ 0 & 1 \end{bmatrix}$ (b) $\mathbf{v} = \begin{bmatrix} 2 & \frac{10}{3} \\ \frac{7}{6} & 2 \end{bmatrix}$

(c) $\mathbf{v} = \begin{bmatrix} 1 & 1 \\ 1 & 1 \end{bmatrix}$

Finding the transition matrix $P_{S \leftarrow T}$ from the T-basis to the S-basis is also a linear combination problem. $P_{S \leftarrow T}$ is the matrix whose columns are the coordinates of the vectors in T with respect to the S-basis. Following the ideas developed in Example 4, we can find matrix $P_{S \leftarrow T}$ using routine **reduce** *or* **rref**. *The idea is to construct a matrix A whose columns correspond to the vectors in S (see Section 12.7) and a matrix B whose columns correspond to the vectors in T. Then* MATLAB *command* **rref([A B])** *gives* **[I $\mathbf{P_{S \leftarrow T}}$]**.

In Exercises ML.4 through ML.6, use the MATLAB *techniques described above to find the transition matrix $P_{S \leftarrow T}$ from the T-basis to the S-basis.*

ML.4. $V = R^3$,

$$S = \left\{ \begin{bmatrix} 1 \\ 1 \\ 0 \end{bmatrix}, \begin{bmatrix} 0 \\ 1 \\ 1 \end{bmatrix}, \begin{bmatrix} 1 \\ 0 \\ 1 \end{bmatrix} \right\},$$

$$T = \left\{ \begin{bmatrix} 2 \\ 1 \\ 1 \end{bmatrix}, \begin{bmatrix} 1 \\ 2 \\ 1 \end{bmatrix}, \begin{bmatrix} 1 \\ 1 \\ 2 \end{bmatrix} \right\}$$

ML.5. $V = P_3$, $S = \{t - 1, t + 1, t^2 + t, t^3 - t\}$, $T = \{t^2, 1 - t, 2 - t^2, t^3 + t^2\}$

ML.6. $V = R^4$, $S = \{(1, 2, 3, 0), (0, 1, 2, 3), (3, 0, 1, 2), (2, 3, 0, 1)\}$, $T =$ natural basis

ML.7. Let $V = R^3$ and suppose that we have bases

$$S = \left\{ \begin{bmatrix} 1 \\ 1 \\ 1 \end{bmatrix}, \begin{bmatrix} 1 \\ 2 \\ 1 \end{bmatrix}, \begin{bmatrix} 0 \\ 1 \\ 1 \end{bmatrix} \right\},$$

$$T = \left\{ \begin{bmatrix} 1 \\ 0 \\ 1 \end{bmatrix}, \begin{bmatrix} 1 \\ 1 \\ 0 \end{bmatrix}, \begin{bmatrix} 0 \\ 1 \\ 2 \end{bmatrix} \right\},$$

and

$$U = \left\{ \begin{bmatrix} 2 \\ 1 \\ 1 \end{bmatrix}, \begin{bmatrix} -1 \\ 2 \\ 1 \end{bmatrix}, \begin{bmatrix} 1 \\ -2 \\ 1 \end{bmatrix} \right\}.$$

(a) Find the transition matrix P from U to T.

(b) Find the transition matrix Q from T to S.

(c) Find the transition matrix Z from U to S.

(d) Does $Z = PQ$ or QP?

6.8 ORTHONORMAL BASES IN R^n

From our work with the natural bases for R^2, R^3, and, in general, for R^n, we know that when these bases are present, the computations are kept to a minimum. A subspace W of R^n need not contain any of the natural basis vectors, but in this section we want to show that it has a basis with the same properties. That is, we want to show that W contains a basis S such that every vector in S is of unit length and every two vectors in S are orthogonal. The method used to obtain such a basis is the Gram–Schmidt process, which is presented in this section.

DEFINITION

A set $S = \{\mathbf{u}_1, \mathbf{u}_2, \ldots, \mathbf{u}_k\}$ in R^n is called **orthogonal** if every pair of distinct vectors in S are orthogonal, that is, if $\mathbf{u}_i \cdot \mathbf{u}_j = 0$ for $i \neq j$. An **orthonormal** set of vectors is an orthogonal set of unit vectors. That is, $S = \{\mathbf{u}_1, \mathbf{u}_2, \ldots, \mathbf{u}_k\}$ is orthonormal if $\mathbf{u}_i \cdot \mathbf{u}_j = 0$ for $i \neq j$, and $\mathbf{u}_i \cdot \mathbf{u}_i = 1$ for $i = 1, 2, \ldots, k$.

EXAMPLE 1

If $\mathbf{x}_1 = (1, 0, 2)$, $\mathbf{x}_2 = (-2, 0, 1)$, and $\mathbf{x}_3 = (0, 1, 0)$, then $\{\mathbf{x}_1, \mathbf{x}_2, \mathbf{x}_3\}$ is an orthogonal set in R^3. The vectors

$$\mathbf{u}_1 = \left(\frac{1}{\sqrt{5}}, 0, \frac{2}{\sqrt{5}}\right) \quad \text{and} \quad \mathbf{u}_2 = \left(-\frac{2}{\sqrt{5}}, 0, \frac{1}{\sqrt{5}}\right)$$

are unit vectors in the directions of $\mathbf{x}_1$ and $\mathbf{x}_2$, respectively. Since $\mathbf{x}_3$ is also of unit length, it follows that $\{\mathbf{u}_1, \mathbf{u}_2, \mathbf{x}_3\}$ is an orthonormal set. Also, span $\{\mathbf{x}_1, \mathbf{x}_2, \mathbf{x}_3\}$ is the same as span $\{\mathbf{u}_1, \mathbf{u}_2, \mathbf{x}_3\}$. ■

EXAMPLE 2

The natural basis

$$\{(1, 0, 0), (0, 1, 0), (0, 0, 1)\}$$

is an orthonormal set in R^3. More generally, the natural basis in R^n is an orthonormal set. ■

An important result about orthogonal sets is the following theorem.

THEOREM 6.16

Let $S = \{\mathbf{u}_1, \mathbf{u}_2, \ldots, \mathbf{u}_k\}$ be an orthogonal set of nonzero vectors in R^n. Then S is linearly independent.

Proof Consider the equation

$$c_1\mathbf{u}_1 + c_2\mathbf{u}_2 + \cdots + c_k\mathbf{u}_k = \mathbf{0}. \tag{1}$$

Taking the dot product of both sides of (1) with $\mathbf{u}_i$, $1 \leq i \leq k$, we have

$$(c_1\mathbf{u}_1 + c_2\mathbf{u}_2 + \cdots + c_k\mathbf{u}_k) \cdot \mathbf{u}_i = \mathbf{0} \cdot \mathbf{u}_i. \tag{2}$$

By properties (c) and (d) of Theorem 4.3, Section 4.2, the left side of (2) is

$$c_1(\mathbf{u}_1 \cdot \mathbf{u}_i) + c_2(\mathbf{u}_2 \cdot \mathbf{u}_i) + \cdots + c_k(\mathbf{u}_k \cdot \mathbf{u}_i),$$

and the right side is 0. Since $\mathbf{u}_j \cdot \mathbf{u}_i = 0$ if $i \neq j$, (2) becomes

$$0 = c_i(\mathbf{u}_i \cdot \mathbf{u}_i) = c_i \|\mathbf{u}_i\|^2. \tag{3}$$

By (a) of Theorem 4.3, Section 4.2, $\|\mathbf{u}_i\| \neq 0$, since $\mathbf{u}_i \neq \mathbf{0}$. Hence (3) implies that $c_i = 0$, $1 \leq i \leq k$, and S is linearly independent. ■

COROLLARY 6.6

An orthonormal set of vectors in R^n is linearly independent.

Proof Exercise T.2. ■

DEFINITION

From Theorem 6.9 of Section 6.4 and Corollary 6.6, it follows that an orthonormal set of n vectors in R^n is a basis for R^n (Exercise T.3). An **orthogonal (orthonormal) basis** for a vector space is a basis that is an orthogonal (orthonormal) set.

We have already seen that the computational effort required to solve a given problem is often reduced when we are dealing with the natural basis for R^n. This reduction in computational effort is due to the fact that we are dealing with an orthonormal basis. For example, if $S = \{\mathbf{u}_1, \mathbf{u}_2, \ldots, \mathbf{u}_n\}$ is a basis for R^n, then if $\mathbf{v}$ is any vector in V, we can write $\mathbf{v}$ as

$$\mathbf{v} = c_1\mathbf{u}_1 + c_2\mathbf{u}_2 + \cdots + c_n\mathbf{u}_n.$$

The coefficients $c_1, c_2, \ldots, c_n$ are obtained by solving a linear system of n equations in n unknowns. (See Section 6.3.)

However, if S is orthonormal, we can obtain the same result with much less work. This is the content of the following theorem.

THEOREM 6.17 *Let $S = \{\mathbf{u}_1, \mathbf{u}_2, \ldots, \mathbf{u}_n\}$ be an orthonormal basis for R^n and $\mathbf{v}$ any vector in R^n. Then*

$$\mathbf{v} = c_1\mathbf{u}_1 + c_2\mathbf{u}_2 + \cdots + c_n\mathbf{u}_n,$$

where

$$c_i = \mathbf{v} \cdot \mathbf{u}_i \qquad 1 \le i \le n.$$

Proof Exercise T.4(a). ■

COROLLARY 6.7 *Let $S = \{\mathbf{u}_1, \mathbf{u}_2, \ldots, \mathbf{u}_n\}$ be an orthogonal basis for R^n and $\mathbf{v}$ any vector in R^n. Then*

$$\mathbf{v} = c_1\mathbf{u}_1 + c_2\mathbf{u}_2 + \cdots + c_n\mathbf{u}_n,$$

where

$$c_i = \frac{\mathbf{v} \cdot \mathbf{u}_i}{\mathbf{u}_i \cdot \mathbf{u}_i} \qquad 1 \le i \le n.$$

Proof Exercise T.4(b). ■

EXAMPLE 3 Let $S = \{\mathbf{u}_1, \mathbf{u}_2, \mathbf{u}_3\}$ be an orthonormal basis for R^3, where

$$\mathbf{u}_1 = \left(\tfrac{2}{3}, -\tfrac{2}{3}, \tfrac{1}{3}\right), \quad \mathbf{u}_2 = \left(\tfrac{2}{3}, \tfrac{1}{3}, -\tfrac{2}{3}\right), \quad \text{and} \quad \mathbf{u}_3 = \left(\tfrac{1}{3}, \tfrac{2}{3}, \tfrac{2}{3}\right).$$

Write the vector $\mathbf{v} = (3, 4, 5)$ as a linear combination of the vectors in S.

Solution We have

$$\mathbf{v} = c_1\mathbf{u}_1 + c_2\mathbf{u}_2 + c_3\mathbf{u}_3.$$

Theorem 6.17 shows that c_1, c_2, and c_3 can be obtained without having to solve a linear system of three equations in three unknowns. Thus

$$c_1 = \mathbf{v} \cdot \mathbf{u}_1 = 1, \qquad c_2 = \mathbf{v} \cdot \mathbf{u}_2 = 0, \qquad c_3 = \mathbf{v} \cdot \mathbf{u}_3 = 7,$$

and $\mathbf{v} = \mathbf{u}_1 + 7\mathbf{u}_3$. ■

THEOREM 6.18 **(Gram*–Schmidt** Process)** *Let W be a nonzero subspace of R^n with basis $S = \{\mathbf{u}_1, \mathbf{u}_2, \ldots, \mathbf{u}_m\}$. Then there exists an orthonormal basis $T = \{\mathbf{w}_1, \mathbf{w}_2, \ldots, \mathbf{w}_m\}$ for W.*

---

*Jörgen Pederson Gram (1850–1916) was a Danish actuary.

**Erhard Schmidt (1876–1959) taught at several leading German Universities and was a student of both Hermann Amandus Schwarz and David Hilbert. He made important contributions to the study of integral equations and partial differential equations and, as part of this study, he introduced the method for finding an orthonormal basis in 1907. In 1908 he wrote a paper on infinitely many linear equations in infinitely many unknowns, in which he founded the theory of Hilbert spaces and in which he again used his method.

Proof The proof is constructive; that is, we develop the desired basis T. However, we first find an orthogonal basis $T^* = \{\mathbf{v}_1, \mathbf{v}_2, \ldots, \mathbf{v}_m\}$ for W.

First, we pick any one of the vectors in S, say $\mathbf{u}_1$, and call it $\mathbf{v}_1$. Thus $\mathbf{v}_1 = \mathbf{u}_1$. We now look for a vector $\mathbf{v}_2$ in the subspace W_1 of W spanned by $\{\mathbf{u}_1, \mathbf{u}_2\}$ that is orthogonal to $\mathbf{v}_1$. Since $\mathbf{v}_1 = \mathbf{u}_1$, W_1 is also the subspace spanned by $\{\mathbf{v}_1, \mathbf{u}_2\}$. Thus

$$\mathbf{v}_2 = c_1\mathbf{v}_1 + c_2\mathbf{u}_2.$$

We try to determine c_1 and c_2 so that $\mathbf{v}_1 \cdot \mathbf{v}_2 = 0$. Now

$$0 = \mathbf{v}_2 \cdot \mathbf{v}_1 = (c_1\mathbf{v}_1 + c_2\mathbf{u}_2) \cdot \mathbf{v}_1 = c_1(\mathbf{v}_1 \cdot \mathbf{v}_1) + c_2(\mathbf{u}_2 \cdot \mathbf{v}_1). \qquad (4)$$

Since $\mathbf{v}_1 \neq \mathbf{0}$ (why?), $\mathbf{v}_1 \cdot \mathbf{v}_1 \neq 0$, and solving for c_1 and c_2 in (4), we have

$$c_1 = -c_2 \frac{\mathbf{u}_2 \cdot \mathbf{v}_1}{\mathbf{v}_1 \cdot \mathbf{v}_1}.$$

We may assign an arbitrary nonzero value to c_2. Thus, letting $c_2 = 1$, we obtain

$$c_1 = -\frac{\mathbf{u}_2 \cdot \mathbf{v}_1}{\mathbf{v}_1 \cdot \mathbf{v}_1}.$$

Hence

$$\mathbf{v}_2 = c_1\mathbf{v}_1 + c_2\mathbf{u}_2 = \mathbf{u}_2 - \left(\frac{\mathbf{u}_2 \cdot \mathbf{v}_1}{\mathbf{v}_1 \cdot \mathbf{v}_1}\right)\mathbf{v}_1.$$

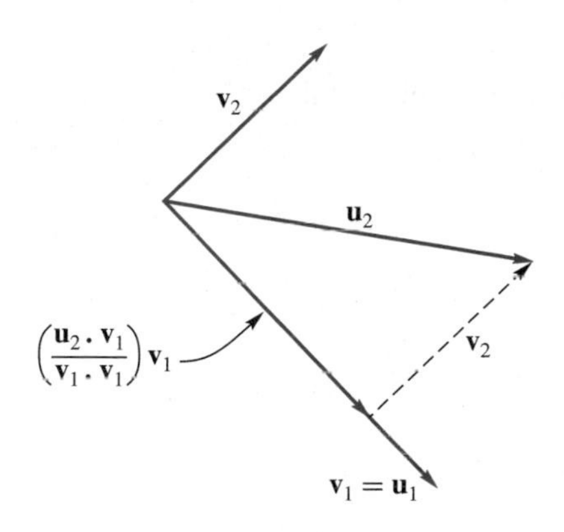

Figure 6.13 ▲

Notice that at this point we have an orthogonal subset $\{\mathbf{v}_1, \mathbf{v}_2\}$ of W (see Figure 6.13).

Next, we look for a vector $\mathbf{v}_3$ in the subspace W_2 of W spanned by $\{\mathbf{u}_1, \mathbf{u}_2, \mathbf{u}_3\}$ that is orthogonal to both $\mathbf{v}_1$ and $\mathbf{v}_2$. Of course, W_2 is also the subspace spanned by $\{\mathbf{v}_1, \mathbf{v}_2, \mathbf{u}_3\}$ (why?). Thus

$$\mathbf{v}_3 = d_1\mathbf{v}_1 + d_2\mathbf{v}_2 + d_3\mathbf{u}_3.$$

We now try to find d_1 and d_2 so that

$$\mathbf{v}_3 \cdot \mathbf{v}_1 = 0 \quad \text{and} \quad \mathbf{v}_3 \cdot \mathbf{v}_2 = 0.$$

Now

$$0 = \mathbf{v}_3 \cdot \mathbf{v}_1 = (d_1\mathbf{v}_1 + d_2\mathbf{v}_2 + d_3\mathbf{u}_3) \cdot \mathbf{v}_1 = d_1(\mathbf{v}_1 \cdot \mathbf{v}_1) + d_3(\mathbf{u}_3 \cdot \mathbf{v}_1), \qquad (5)$$

$$0 = \mathbf{v}_3 \cdot \mathbf{v}_2 = (d_1\mathbf{v}_1 + d_2\mathbf{v}_2 + d_3\mathbf{u}_3) \cdot \mathbf{v}_2 = d_2(\mathbf{v}_2 \cdot \mathbf{v}_2) + d_3(\mathbf{u}_3 \cdot \mathbf{v}_2). \qquad (6)$$

In obtaining the right sides of (5) and (6), we have used the fact that $\mathbf{v}_1 \cdot \mathbf{v}_2 = 0$. Observe that $\mathbf{v}_2 \neq \mathbf{0}$ (why?). Solving (5) and (6) for d_1 and d_2, respectively, we obtain

$$d_1 = -d_3 \frac{\mathbf{u}_3 \cdot \mathbf{v}_1}{\mathbf{v}_1 \cdot \mathbf{v}_1} \quad \text{and} \quad d_2 = -d_3 \frac{\mathbf{u}_3 \cdot \mathbf{v}_2}{\mathbf{v}_2 \cdot \mathbf{v}_2}.$$

We may assign an arbitrary nonzero value to d_3. Thus, letting $d_3 = 1$, we have

$$d_1 = -\frac{\mathbf{u}_3 \cdot \mathbf{v}_1}{\mathbf{v}_1 \cdot \mathbf{v}_1} \quad \text{and} \quad d_2 = -\frac{\mathbf{u}_3 \cdot \mathbf{v}_2}{\mathbf{v}_2 \cdot \mathbf{v}_2}.$$

Hence

$$\mathbf{v}_3 = \mathbf{u}_3 - \left(\frac{\mathbf{u}_3 \cdot \mathbf{v}_1}{\mathbf{v}_1 \cdot \mathbf{v}_1}\right)\mathbf{v}_1 - \left(\frac{\mathbf{u}_3 \cdot \mathbf{v}_2}{\mathbf{v}_2 \cdot \mathbf{v}_2}\right)\mathbf{v}_2.$$

Notice that at this point we have an orthogonal subset $\{\mathbf{v}_1, \mathbf{v}_2, \mathbf{v}_3\}$ of W (see Figure 6.14).

Figure 6.14 ►

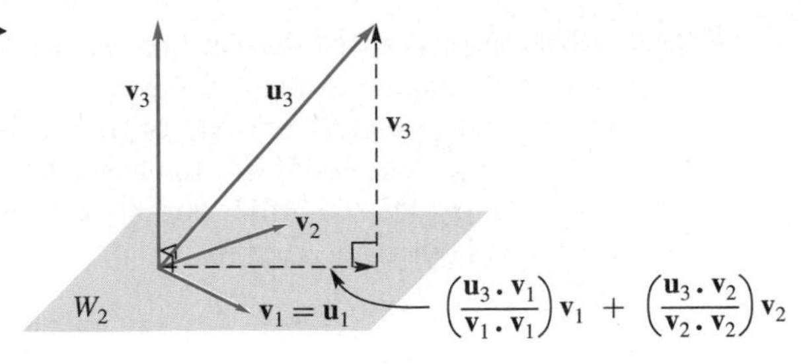

We next seek a vector $\mathbf{v}_4$ in the subspace W_3 of W spanned by the set $\{\mathbf{u}_1, \mathbf{u}_2, \mathbf{u}_3, \mathbf{u}_4\}$, and thus by $\{\mathbf{v}_1, \mathbf{v}_2, \mathbf{v}_3, \mathbf{u}_4\}$, that is orthogonal to $\mathbf{v}_1$, $\mathbf{v}_2$, and $\mathbf{v}_3$. We can then write

$$\mathbf{v}_4 = \mathbf{u}_4 - \left(\frac{\mathbf{u}_4 \cdot \mathbf{v}_1}{\mathbf{v}_1 \cdot \mathbf{v}_1}\right)\mathbf{v}_1 - \left(\frac{\mathbf{u}_4 \cdot \mathbf{v}_2}{\mathbf{v}_2 \cdot \mathbf{v}_2}\right)\mathbf{v}_2 - \left(\frac{\mathbf{u}_4 \cdot \mathbf{v}_3}{\mathbf{v}_3 \cdot \mathbf{v}_3}\right)\mathbf{v}_3.$$

We continue in this manner until we have an orthogonal set $T^* = \{\mathbf{v}_1, \mathbf{v}_2, \ldots, \mathbf{v}_m\}$ of m vectors. It then follows that T^* is a basis for W. If we now normalize the $\mathbf{v}_i$, that is, let

$$\mathbf{w}_i = \frac{1}{\|\mathbf{v}_i\|}\mathbf{v}_i \qquad (1 \le i \le m),$$

then $T = \{\mathbf{w}_1, \mathbf{w}_2, \ldots, \mathbf{w}_m\}$ is an orthonormal basis for W. ■

We now summarize the Gram–Schmidt process.

> The Gram–Schmidt process for computing an orthonormal basis $T = \{\mathbf{w}_1, \mathbf{w}_2, \ldots, \mathbf{w}_m\}$ for a nonzero subspace W of R^n with basis $S = \{\mathbf{u}_1, \mathbf{u}_2, \ldots, \mathbf{u}_m\}$ is as follows.
>
> ***Step 1.*** Let $\mathbf{v}_1 = \mathbf{u}_1$.
>
> ***Step 2.*** Compute the vectors $\mathbf{v}_2, \mathbf{v}_3, \ldots, \mathbf{v}_m$, successively, one at a time, by the formula
>
> $$\mathbf{v}_i = \mathbf{u}_i - \left(\frac{\mathbf{u}_i \cdot \mathbf{v}_1}{\mathbf{v}_1 \cdot \mathbf{v}_1}\right)\mathbf{v}_1 - \left(\frac{\mathbf{u}_i \cdot \mathbf{v}_2}{\mathbf{v}_2 \cdot \mathbf{v}_2}\right)\mathbf{v}_2 - \cdots - \left(\frac{\mathbf{u}_i \cdot \mathbf{v}_{i-1}}{\mathbf{v}_{i-1} \cdot \mathbf{v}_{i-1}}\right)\mathbf{v}_{i-1}.$$
>
> The set of vectors $T^* = \{\mathbf{v}_1, \mathbf{v}_2, \ldots, \mathbf{v}_m\}$ is an orthogonal set.
>
> ***Step 3.*** Let
>
> $$\mathbf{w}_i = \frac{1}{\|\mathbf{v}_i\|}\mathbf{v}_i \qquad (1 \le i \le m).$$
>
> Then $T = \{\mathbf{w}_1, \mathbf{w}_2, \ldots, \mathbf{w}_m\}$ is an orthonormal basis for W.

Remark It is not difficult to show that if $\mathbf{u}$ and $\mathbf{v}$ are vectors in R^n such that $\mathbf{u} \cdot \mathbf{v} = 0$, then $\mathbf{u} \cdot (c\mathbf{v}) = 0$ for any scalar c (Exercise T.7). This result can often be used to simplify hand computations in the Gram–Schmidt process. As soon as a vector $\mathbf{v}_i$ is computed in Step 2, multiply it by a proper scalar to clear any fractions that may be present. We shall use this approach in our computational work with the Gram–Schmidt process.

EXAMPLE 4 Consider the basis $S = \{\mathbf{u}_1, \mathbf{u}_2, \mathbf{u}_3\}$ for R^3, where

$$\mathbf{u}_1 = (1, 1, 1), \quad \mathbf{u}_2 = (-1, 0, -1), \quad \text{and} \quad \mathbf{u}_3 = (-1, 2, 3).$$

Use the Gram–Schmidt process to transform S to an orthonormal basis for R^3.

Solution ***Step 1.*** Let $\mathbf{v}_1 = \mathbf{u}_1 = (1, 1, 1)$.
Step 2. We now compute $\mathbf{v}_2$ and $\mathbf{v}_3$:

$$\mathbf{v}_2 = \mathbf{u}_2 - \left(\frac{\mathbf{u}_2 \cdot \mathbf{v}_1}{\mathbf{v}_1 \cdot \mathbf{v}_1}\right)\mathbf{v}_1 = (-1, 0, -1) - \left(\tfrac{-2}{3}\right)(1, 1, 1) = \left(-\tfrac{1}{3}, \tfrac{2}{3}, -\tfrac{1}{3}\right).$$

Multiplying $\mathbf{v}_2$ by 3 to clear fractions, we obtain $(-1, 2, -1)$, which we now use as $\mathbf{v}_2$. Then

$$\begin{aligned}\mathbf{v}_3 &= \mathbf{u}_3 - \left(\frac{\mathbf{u}_3 \cdot \mathbf{v}_1}{\mathbf{v}_1 \cdot \mathbf{v}_1}\right)\mathbf{v}_1 - \left(\frac{\mathbf{u}_3 \cdot \mathbf{v}_2}{\mathbf{v}_2 \cdot \mathbf{v}_2}\right)\mathbf{v}_2 \\ &= (-1, 2, 3) - \tfrac{4}{3}(1, 1, 1) - \tfrac{2}{6}(-1, 2, -1) = (-2, 0, 2).\end{aligned}$$

Thus

$$T^* = \{\mathbf{v}_1, \mathbf{v}_2, \mathbf{v}_3\} = \{(1, 1, 1), (-1, 2, -1), (-2, 0, 2)\}$$

is an orthogonal basis for R^3.
Step 3. Let

$$\begin{aligned}\mathbf{w}_1 &= \frac{1}{\|\mathbf{v}_1\|}\mathbf{v}_1 = \frac{1}{\sqrt{3}}(1, 1, 1) \\ \mathbf{w}_2 &= \frac{1}{\|\mathbf{v}_2\|}\mathbf{v}_2 = \frac{1}{\sqrt{6}}(-1, 2, -1) \\ \mathbf{w}_3 &= \frac{1}{\|\mathbf{v}_3\|}\mathbf{v}_3 = \frac{1}{\sqrt{8}}(-2, 0, 2) = \left(-\frac{1}{\sqrt{2}}, 0, \frac{1}{\sqrt{2}}\right).\end{aligned}$$

Then

$$\begin{aligned}T &= \{\mathbf{w}_1, \mathbf{w}_2, \mathbf{w}_3\} \\ &= \left\{\left(\frac{1}{\sqrt{3}}, \frac{1}{\sqrt{3}}, \frac{1}{\sqrt{3}}\right), \left(-\frac{1}{\sqrt{6}}, \frac{2}{\sqrt{6}}, -\frac{1}{\sqrt{6}}\right), \left(-\frac{1}{\sqrt{2}}, 0, \frac{1}{\sqrt{2}}\right)\right\}\end{aligned}$$

is an orthonormal basis for R^3. ■

EXAMPLE 5 Let W be the subspace of R^4 with basis $S = \{\mathbf{u}_1, \mathbf{u}_2, \mathbf{u}_3\}$, where

$$\mathbf{u}_1 = (1, -2, 0, 1), \quad \mathbf{u}_2 = (-1, 0, 0, -1), \quad \text{and} \quad \mathbf{u}_3 = (1, 1, 0, 0).$$

Use the Gram–Schmidt process to transform S to an orthonormal basis for W.

Solution ***Step 1.*** Let $\mathbf{v}_1 = \mathbf{u}_1 = (1, -2, 0, 1)$.
Step 2. We now compute $\mathbf{v}_2$ and $\mathbf{v}_3$:

$$\begin{aligned}\mathbf{v}_2 = \mathbf{u}_2 - \left(\frac{\mathbf{u}_2 \cdot \mathbf{v}_1}{\mathbf{v}_1 \cdot \mathbf{v}_1}\right)\mathbf{v}_1 &= (-1, 0, 0, -1) - \left(\tfrac{-2}{6}\right)(1, -2, 0, 1) \\ &= \left(-\tfrac{2}{3}, -\tfrac{2}{3}, 0, -\tfrac{2}{3}\right).\end{aligned}$$

Multiplying $\mathbf{v}_2$ by 3 to clear fractions, we obtain $(-2, -2, 0, -2)$, which we now use as $\mathbf{v}_2$. Then

$$\begin{aligned}\mathbf{v}_3 &= \mathbf{u}_3 - \left(\frac{\mathbf{u}_3 \cdot \mathbf{v}_1}{\mathbf{v}_1 \cdot \mathbf{v}_1}\right)\mathbf{v}_1 - \left(\frac{\mathbf{u}_3 \cdot \mathbf{v}_2}{\mathbf{v}_2 \cdot \mathbf{v}_2}\right)\mathbf{v}_2 \\ &= (1, 1, 0, 0) - \left(\tfrac{-1}{6}\right)(1, -2, 0, 1) - \left(\tfrac{-4}{12}\right)(-2, -2, 0, -2) \\ &= \left(\tfrac{1}{2}, 0, 0, -\tfrac{1}{2}\right).\end{aligned}$$

Multiplying $\mathbf{v}_3$ by 2 to clear fractions, we obtain $(1, 0, 0, -1)$, which we now use as $\mathbf{v}_3$. Thus

$$T^* = \{(1, -2, 0, 1), (-2, -2, 0, -2), (1, 0, 0, -1)\}$$

is an orthogonal basis for W.

Step 3. Let

$$\mathbf{w}_1 = \frac{1}{\|\mathbf{v}_1\|}\mathbf{v}_1 = \frac{1}{\sqrt{6}}(1, -2, 0, 1)$$

$$\mathbf{w}_2 = \frac{1}{\|\mathbf{v}_2\|}\mathbf{v}_2 = \frac{1}{\sqrt{12}}(-2, -2, 0, -2) = \frac{1}{\sqrt{3}}(-1, -1, 0, -1)$$

$$\mathbf{w}_3 = \frac{1}{\|\mathbf{v}_3\|}\mathbf{v}_3 = \frac{1}{\sqrt{2}}(1, 0, 0, -1).$$

Then

$$\begin{aligned}T &= \{\mathbf{w}_1, \mathbf{w}_2, \mathbf{w}_3\} \\ &= \left\{\left(\frac{1}{\sqrt{6}}, -\frac{2}{\sqrt{6}}, 0, \frac{1}{\sqrt{6}}\right), \left(-\frac{1}{\sqrt{3}}, -\frac{1}{\sqrt{3}}, 0, -\frac{1}{\sqrt{3}}\right), \left(\frac{1}{\sqrt{2}}, 0, 0, -\frac{1}{\sqrt{2}}\right)\right\}\end{aligned}$$

is an orthonormal basis for W. ■

Remarks

1. In solving Example 5, as soon as a vector is computed we multiplied it by an appropriate scalar to eliminate any fractions that may be present. This optional step results in simpler computations when working by hand. If this approach is taken, the resulting basis, while orthonormal, may differ from the orthonormal basis obtained by not clearing fractions. Most computer implementations of the Gram–Schmidt process, including those developed with MATLAB, do not clear fractions.
2. We make one final observation with regard to the Gram–Schmidt process. In our proof of Theorem 6.18 we first obtained an orthogonal basis T^* and then normalized all the vectors in T^* to obtain the orthonormal basis T. Of course, an alternative course of action is to normalize each vector as soon as we produce it.

Key Terms

Orthogonal set
Orthonormal set
Orthogonal basis
Orthonormal basis
Gram–Schmidt process

6.8 Exercises

1. Which of the following are orthogonal sets of vectors?
   (a) $\{(1, -1, 2), (0, 2, -1), (-1, 1, 1)\}$
   (b) $\{(1, 2, -1, 1), (0, -1, -2, 0), (1, 0, 0, -1)\}$
   (c) $\{(0, 1, 0, -1), (1, 0, 1, 1), (-1, 1, -1, 2)\}$

2. Which of the following are orthonormal sets of vectors?
   (a) $\left\{\left(\frac{1}{3}, \frac{2}{3}, \frac{2}{3}\right), \left(\frac{2}{3}, \frac{1}{3}, -\frac{2}{3}\right), \left(\frac{2}{3}, -\frac{2}{3}, \frac{1}{3}\right)\right\}$
   (b) $\left\{\left(\frac{1}{\sqrt{2}}, 0, -\frac{1}{\sqrt{2}}\right), \left(\frac{1}{\sqrt{3}}, \frac{1}{\sqrt{3}}, \frac{1}{\sqrt{3}}\right), (0, 1, 0)\right\}$
   (c) $\{(0, 2, 2, 1), (1, 1, -2, 2), (0, -2, 1, 2)\}$

In Exercises 3 and 4, let $V = R^3$.

3. Let $\mathbf{u} = (1, 1, -2)$ and $\mathbf{v} = (a, -1, 2)$. For what values of a are $\mathbf{u}$ and $\mathbf{v}$ orthogonal?

4. Let $\mathbf{u} = \left(\frac{1}{\sqrt{2}}, 0, \frac{1}{\sqrt{2}}\right)$, and $\mathbf{v} = \left(a, \frac{1}{\sqrt{2}}, -b\right)$. For what values of a and b is $\{\mathbf{u}, \mathbf{v}\}$ an orthonormal set?

5. Use the Gram–Schmidt process to find an orthonormal basis for the subspace of R^3 with basis $\{(1, -1, 0), (2, 0, 1)\}$.

6. Use the Gram–Schmidt process to find an orthonormal basis for the subspace of R^3 with basis $\{(1, 0, 2), (-1, 1, 0)\}$.

7. Use the Gram–Schmidt process to find an orthonormal basis for the subspace of R^4 with basis $\{(1, -1, 0, 1), (2, 0, 0, -1), (0, 0, 1, 0)\}$.

8. Use the Gram–Schmidt process to find an orthonormal basis for the subspace of R^4 with basis $\{(1, 1, -1, 0), (0, 2, 0, 1), (-1, 0, 0, 1)\}$.

9. Use the Gram–Schmidt process to transform the basis $\{(1, 2), (-3, 4)\}$ for R^2 into (a) an orthogonal basis; (b) an orthonormal basis.

10. (a) Use the Gram–Schmidt process to transform the basis $\{(1, 1, 1), (0, 1, 1), (1, 2, 3)\}$ for R^3 into an orthonormal basis for R^3.
    (b) Use Theorem 6.17 to write $(2, 3, 1)$ as a linear combination of the basis obtained in part (a).

11. Find an orthonormal basis for R^3 containing the vectors $\left(\frac{2}{3}, -\frac{2}{3}, \frac{1}{3}\right)$ and $\left(\frac{2}{3}, \frac{1}{3}, -\frac{2}{3}\right)$.

12. Use the Gram–Schmidt process to construct an orthonormal basis for the subspace W of R^3 *spanned* by $\{(1, 1, 1), (2, 2, 2), (0, 0, 1), (1, 2, 3)\}$.

13. Use the Gram–Schmidt process to construct an orthonormal basis for the subspace W of R^4 *spanned* by $\{(1, 1, 0, 0), (2, -1, 0, 1), (3, -3, 0, -2), (1, -2, 0, -3)\}$.

14. Find an orthonormal basis for the subspace of R^3 consisting of all vectors of the form $(a, a + b, b)$.

15. Find an orthonormal basis for the subspace of R^4 consisting of all vectors of the form
$$(a, a + b, c, b + c).$$

16. Find an orthonormal basis for the subspace of R^3 consisting of all vectors (a, b, c) such that
$$a + b + c = 0.$$

17. Find an orthonormal basis for the subspace of R^4 consisting of all vectors (a, b, c, d) such that
$$a - b - 2c + d = 0.$$

18. Find an orthonormal basis for the solution space of the homogeneous system
$$\begin{aligned} x_1 + x_2 - x_3 &= 0 \\ 2x_1 + x_2 + 2x_3 &= 0. \end{aligned}$$

19. Find an orthonormal basis for the solution space of the homogeneous system
$$\begin{bmatrix} 1 & 1 & -1 \\ 2 & 1 & 3 \\ 1 & 2 & -6 \end{bmatrix} \begin{bmatrix} x_1 \\ x_2 \\ x_3 \end{bmatrix} = \begin{bmatrix} 0 \\ 0 \\ 0 \end{bmatrix}.$$

20. Consider the orthonormal basis
$$S = \left\{\left(\frac{1}{\sqrt{2}}, \frac{1}{\sqrt{2}}\right), \left(-\frac{1}{\sqrt{2}}, \frac{1}{\sqrt{2}}\right)\right\}$$
for R^2. Using Theorem 6.17 write the vector $(2, 3)$ as a linear combination of the vectors in S.

21. Consider the orthonormal basis
$$S = \left\{\left(\frac{1}{\sqrt{5}}, 0, \frac{2}{\sqrt{5}}\right), \left(-\frac{2}{\sqrt{5}}, 0, \frac{1}{\sqrt{5}}\right), (0, 1, 0)\right\}$$
for R^3. Using Theorem 6.17 write the vector $(2, -3, 1)$ as a linear combination of the vectors in S.

Theoretical Exercises

T.1. Verify that the natural basis for R^n is an orthonormal set.

T.2. Prove Corollary 6.6.

T.3. Show that an orthonormal set of n vectors in R^n is a basis for R^n.

T.4. (a) Prove Theorem 6.17.
(b) Prove Corollary 6.7.

T.5. Let $\mathbf{u}, \mathbf{v}_1, \mathbf{v}_2, \ldots, \mathbf{v}_n$ be vectors in R^n. Show that if $\mathbf{u}$ is orthogonal to $\mathbf{v}_1, \mathbf{v}_2, \ldots, \mathbf{v}_n$, then $\mathbf{u}$ is orthogonal to every vector in span $\{\mathbf{v}_1, \mathbf{v}_2, \ldots, \mathbf{v}_n\}$.

T.6. Let $\mathbf{u}$ be a fixed vector in R^n. Show that the set of all vectors in R^n that are orthogonal to $\mathbf{u}$ is a subspace of R^n.

T.7. Let $\mathbf{u}$ and $\mathbf{v}$ be vectors in R^n. Show that if $\mathbf{u} \cdot \mathbf{v} = 0$, then $\mathbf{u} \cdot (c\mathbf{v}) = 0$ for any scalar c.

T.8. Suppose that $\{\mathbf{v}_1, \mathbf{v}_2, \ldots, \mathbf{v}_n\}$ is an orthonormal set in R^n. Let the matrix A be given by $A = \begin{bmatrix} \mathbf{v}_1 & \mathbf{v}_2 & \cdots & \mathbf{v}_n \end{bmatrix}$. Show that A is nonsingular and compute its inverse. Give three different examples of such a matrix in R^2 or R^3.

T.9. Suppose that $\{\mathbf{v}_1, \mathbf{v}_2, \ldots, \mathbf{v}_n\}$ is an orthogonal set in R^n. Let A be the matrix whose jth column is $\mathbf{v}_j$, $j = 1, 2, \ldots, n$. Prove or disprove: A is nonsingular.

T.10. Let $S = \{\mathbf{u}_1, \mathbf{u}_2, \ldots, \mathbf{u}_k\}$ be an orthonormal basis for a subspace W of R^n, where $n > k$. Discuss how to construct an orthonormal basis for V that includes S.

T.11. Let $\{\mathbf{u}_1, \ldots, \mathbf{u}_k, \mathbf{u}_{k+1}, \ldots, \mathbf{u}_n\}$ be an orthonormal basis for R^n, $S = \text{span}\,\{\mathbf{u}_1, \ldots, \mathbf{u}_k\}$, and $T = \text{span}\,\{\mathbf{u}_{k+1}, \ldots, \mathbf{u}_n\}$. For any $\mathbf{x}$ in S and any $\mathbf{y}$ in T, show that $\mathbf{x} \cdot \mathbf{y} = 0$.

MATLAB Exercises

The Gram–Schmidt process takes a basis S for a subspace W of R^n and produces an orthonormal basis T for W. The algorithm to produce the orthonormal basis T that is given in this section is implemented in MATLAB *in routine* **gschmidt**. *Type* **help gschmidt** *for directions.*

ML.1. Use **gschmidt** to produce an orthonormal basis for R^3 from the basis

$$S = \left\{ \begin{bmatrix} 1 \\ 1 \\ 0 \end{bmatrix}, \begin{bmatrix} 1 \\ 0 \\ 0 \end{bmatrix}, \begin{bmatrix} 0 \\ 1 \\ 1 \end{bmatrix} \right\}.$$

Your answer will be in decimal form; rewrite it in terms of $\sqrt{2}$.

ML.2. Use **gschmidt** to produce an orthonormal basis for R^4 from the basis $S = \{(1, 0, 1, 1), (1, 2, 1, 3), (0, 2, 1, 1), (0, 1, 0, 0)\}$.

ML.3. In R^3, $S = \{(0, -1, 1), (0, 1, 1), (1, 1, 1)\}$ is a basis. Find an orthonormal basis T from S and then find $\begin{bmatrix}\mathbf{v}\end{bmatrix}_T$ for each of the following vectors.

(a) $\mathbf{v} = (1, 2, 0)$ (b) $\mathbf{v} = (1, 1, 1)$

(c) $\mathbf{v} = (-1, 0, 1)$

ML.4. Find an orthonormal basis for the subspace of R^4 consisting of all vectors of the form

$$(a, 0, a + b, b + c),$$

where a, b, and c are any real numbers.

6.9 ORTHOGONAL COMPLEMENTS

Let W_1 and W_2 be subspaces of a vector space V. Let $W_1 + W_2$ be the set of all vectors $\mathbf{v}$ in V such that $\mathbf{v} = \mathbf{w}_1 + \mathbf{w}_2$, where $\mathbf{w}_1$ is in W_1 and $\mathbf{w}_2$ is in W_2. In Exercise T.10 in Section 6.2, we asked you to show that $W_1 + W_2$ is a subspace of V. In Exercise T.11 in Section 6.2, we asked you to show that if $V = W_1 + W_2$ and $W_1 \cap W_2 = \{\mathbf{0}\}$, then V is the direct sum of W_1 and W_2 and we write $V = W_1 \oplus W_2$. Moreover, in this case every vector in V can be uniquely written as $\mathbf{w}_1 + \mathbf{w}_2$, where $\mathbf{w}_1$ is in W_1 and $\mathbf{w}_2$ is in W_2. (See Exercise T.11 in Section 6.2.) In this section we show that if W is a subspace of R^n, then R^n can be written as a direct sum of W and another subspace of R^n. This subspace will be used to examine a basic relationship between four vector spaces associated with a matrix.

DEFINITION

Let W be a subspace of R^n. A vector $\mathbf{u}$ in R^n is said to be **orthogonal** to W if it is orthogonal to every vector in W. The set of all vectors in R^n that are orthogonal to all the vectors in W is called the **orthogonal complement** of W in R^n and is denoted by $W^\perp$ (read "W perp").

EXAMPLE 1

Let W be the subspace of R^3 consisting of all multiples of the vector

$$\mathbf{w} = (2, -3, 4).$$

Thus $W = \text{span}\,\{\mathbf{w}\}$, so W is a one-dimensional subspace of W. Then a vector $\mathbf{u}$ in R^3 belongs to $W^\perp$ if and only if $\mathbf{u}$ is orthogonal to $c\mathbf{w}$, for any scalar c. It can be shown that $W^\perp$ is the plane with normal $\mathbf{w}$. ■

Observe that if W is a subspace of R^n, then the zero vector always belongs to $W^\perp$ (Exercise T.1). Moreover, the orthogonal complement of R^n is the zero subspace and the orthogonal complement of the zero subspace is R^n itself (Exercise T.2).

THEOREM 6.19 *Let W be a subspace of R^n. Then:*

(a) $W^\perp$ *is a subspace of* R^n.

(b) $W \cap W^\perp = \{\mathbf{0}\}$.

Proof (a) Let $\mathbf{u}_1$ and $\mathbf{u}_2$ be in $W^\perp$. Then $\mathbf{u}_1$ and $\mathbf{u}_2$ are orthogonal to each vector $\mathbf{w}$ in W. We now have

$$(\mathbf{u}_1 + \mathbf{u}_2) \cdot \mathbf{w} = \mathbf{u}_1 \cdot \mathbf{w} + \mathbf{u}_2 \cdot \mathbf{w} = 0 + 0 = 0,$$

so $\mathbf{u}_1 + \mathbf{u}_2$ is in $W^\perp$. Also, let $\mathbf{u}$ be in $W^\perp$ and let c be a real scalar. Then for any vector $\mathbf{w}$ in W, we have

$$(c\mathbf{u}) \cdot \mathbf{w} = c(\mathbf{u} \cdot \mathbf{w}) = c0 = 0,$$

so $c\mathbf{u}$ is in W, which implies that $W^\perp$ is closed under vector addition and scalar multiplication and hence is a subspace of R^n.

(b) Let $\mathbf{u}$ be a vector in $W \cap W^\perp$. Then $\mathbf{u}$ is in both W and $W^\perp$, so $\mathbf{u} \cdot \mathbf{u} = 0$. From Theorem 4.3 in Section 4.2 it follows that $\mathbf{u} = \mathbf{0}$. ■

In Exercise T.3 we ask you to show that if W is a subspace of R^n that is spanned by a set of vectors S, then a vector $\mathbf{u}$ in R^n belongs to $W^\perp$ if and only if $\mathbf{u}$ is orthogonal to every vector in S. This result can be helpful in finding $W^\perp$, as shown in the next example.

EXAMPLE 2 Let W be the subspace of R^4 with basis $\{\mathbf{w}_1, \mathbf{w}_2\}$, where

$$\mathbf{w}_1 = (1, 1, 0, 1) \quad \text{and} \quad \mathbf{w}_2 = (0, -1, 1, 1).$$

Find a basis for $W^\perp$.

Solution Let $\mathbf{u} = (a, b, c, d)$ be a vector in $W^\perp$. Then $\mathbf{u} \cdot \mathbf{w}_1 = 0$ and $\mathbf{u} \cdot \mathbf{w}_2 = 0$. Thus we have

$$\mathbf{u} \cdot \mathbf{w}_1 = a + b + d = 0 \quad \text{and} \quad \mathbf{u} \cdot \mathbf{w}_2 = -b + c + d = 0.$$

Solving the homogeneous system

$$\begin{aligned} a + b \phantom{{}+c} + d &= 0 \\ -b + c + d &= 0, \end{aligned}$$

we obtain (verify)

$$a = -r - 2s, \qquad b = r + s, \qquad c = r, \qquad d = s.$$

Then

$$\mathbf{u} = (-r - 2s, r + s, r, s) = r(-1, 1, 1, 0) + s(-2, 1, 0, 1).$$

Hence the vectors $(-1, 1, 1, 0)$ and $(-2, 1, 0, 1)$ span $W^\perp$. Since they are not multiples of each other, they are linearly independent and thus form a basis for $W^\perp$. ■

THEOREM 6.20 *Let W be a subspace of R^n. Then*

$$R^n = W \oplus W^{\perp}.$$

Proof Let $\dim W = m$. Then W has a basis consisting of m vectors. By the Gram–Schmidt process we can transform this basis to an orthonormal basis. Thus let $S = \{\mathbf{w}_1, \mathbf{w}_2, \ldots, \mathbf{w}_m\}$ be an orthonormal basis for W. If $\mathbf{v}$ is a vector in R^n, let

$$\mathbf{w} = (\mathbf{v} \cdot \mathbf{w}_1)\mathbf{w}_1 + (\mathbf{v} \cdot \mathbf{w}_2)\mathbf{w}_2 + \cdots + (\mathbf{v} \cdot \mathbf{w}_m)\mathbf{w}_m \tag{1}$$

and

$$\mathbf{u} = \mathbf{v} - \mathbf{w}. \tag{2}$$

Since $\mathbf{w}$ is a linear combination of vectors in S, $\mathbf{w}$ belongs to W. We next show that $\mathbf{u}$ lies in $W^{\perp}$ by showing that $\mathbf{u}$ is orthogonal to every vector in S, a basis for W. For each $\mathbf{w}_i$ in S, we have

$$\begin{aligned} \mathbf{u} \cdot \mathbf{w}_i &= (\mathbf{v} - \mathbf{w}) \cdot \mathbf{w}_i = \mathbf{v} \cdot \mathbf{w}_i - \mathbf{w} \cdot \mathbf{w}_i \\ &= \mathbf{v} \cdot \mathbf{w}_i - [(\mathbf{v} \cdot \mathbf{w}_1)\mathbf{w}_1 + (\mathbf{v} \cdot \mathbf{w}_2)\mathbf{w}_2 + \cdots + (\mathbf{v} \cdot \mathbf{w}_m)\mathbf{w}_m] \cdot \mathbf{w}_i \\ &= \mathbf{v} \cdot \mathbf{w}_i - (\mathbf{v} \cdot \mathbf{w}_i)(\mathbf{w}_i \cdot \mathbf{w}_i) \\ &= 0 \end{aligned}$$

since $\mathbf{w}_i \cdot \mathbf{w}_j = 0$ for $i \neq j$ and $\mathbf{w}_i \cdot \mathbf{w}_i = 1$, $1 \leq i \leq m$. Thus $\mathbf{u}$ is orthogonal to every vector in W and so lies in $W^{\perp}$. Hence

$$\mathbf{v} = \mathbf{w} + \mathbf{u},$$

which means that $R^n = W + W^{\perp}$. From part (b) of Theorem 6.19, it follows that

$$R^n = W \oplus W^{\perp}. \qquad \blacksquare$$

Remark As pointed out at the beginning of this section, we also conclude that the vectors $\mathbf{w}$ and $\mathbf{u}$ defined by Equations (1) and (2) are unique.

EXAMPLE 3 In Example 2, the subspace W had the basis

$$\{\mathbf{w}_1, \mathbf{w}_2\} = \{(1, 1, 0, 1), (0, -1, 1, 1)\}$$

and we determined that $W^{\perp}$ had the basis

$$\{\mathbf{w}_3, \mathbf{w}_4\} = \{(-1, 1, 1, 0), (-2, 1, 0, 1)\}.$$

If $\mathbf{v} = (-1, 1, 4, 3)$, find a vector $\mathbf{w}$ in W and a vector $\mathbf{u}$ in $W^{\perp}$ so that $\mathbf{v} = \mathbf{w} + \mathbf{u}$.

Solution Following the procedure used in the proof of Theorem 6.20, we use the Gram–Schmidt process to determine an orthonormal basis for W, as

$$\mathbf{u}_1 = \frac{1}{\sqrt{3}}(1, 1, 0, 1) \quad \text{and} \quad \mathbf{u}_2 = \frac{1}{\sqrt{3}}(0, -1, 1, 1)$$

(verify). Let

$$\begin{aligned}\mathbf{w} &= (\mathbf{v}\cdot\mathbf{u}_1)\mathbf{u}_1 + (\mathbf{v}\cdot\mathbf{u}_2)\mathbf{u}_2\\ &= \sqrt{3}\mathbf{u}_1 + 2\sqrt{3}\mathbf{u}_2\\ &= (1, 1, 0, 1) + 2(0, -1, 1, 1)\\ &= (1, -1, 2, 3).\end{aligned}$$

Then we compute

$$\mathbf{u} = \mathbf{v} - \mathbf{w} = (-1, 1, 4, 3) - (1, -1, 2, 3) = (-2, 2, 2, 0).$$

(Verify that $\mathbf{u}$ is in $W^\perp$.) It follows that $\mathbf{v} = \mathbf{w} + \mathbf{u}$, for $\mathbf{w}$ in W and $\mathbf{u}$ in $W^\perp$. ■

THEOREM 6.21 *If W is a subspace of R^n, then*

$$(W^\perp)^\perp = W.$$

Proof First, if $\mathbf{w}$ is any vector in W, then $\mathbf{w}$ is orthogonal to every vector $\mathbf{u}$ in $W^\perp$, so $\mathbf{w}$ is in $(W^\perp)^\perp$. Hence W is a subspace of $(W^\perp)^\perp$. Conversely, let $\mathbf{v}$ be an arbitrary vector in $(W^\perp)^\perp$. Then, by Theorem 6.20, $\mathbf{v}$ can be written as

$$\mathbf{v} = \mathbf{w} + \mathbf{u},$$

where $\mathbf{w}$ is in W and $\mathbf{u}$ is in $W^\perp$. Since $\mathbf{u}$ is in $W^\perp$, it is orthogonal to $\mathbf{v}$ and $\mathbf{w}$. Thus

$$0 = \mathbf{u}\cdot\mathbf{v} = \mathbf{u}\cdot(\mathbf{w} + \mathbf{u}) = \mathbf{u}\cdot\mathbf{w} + \mathbf{u}\cdot\mathbf{u} = \mathbf{u}\cdot\mathbf{u}$$

or

$$\mathbf{u}\cdot\mathbf{u} = 0,$$

which implies that $\mathbf{u} = \mathbf{0}$. Then $\mathbf{v} = \mathbf{w}$, so $\mathbf{v}$ belongs to W. Hence it follows that $(W^\perp)^\perp = W$. ■

Remark Since W is the orthogonal complement of $W^\perp$ and $W^\perp$ is also the orthogonal complement of W, we say that W and $W^\perp$ are **orthogonal complements**.

RELATIONS AMONG THE FUNDAMENTAL VECTOR SPACES ASSOCIATED WITH A MATRIX

If A is a given $m \times n$ matrix, we associate the following four fundamental vector spaces with A: the null space of A (a subspace of R^n), the row space of A (a subspace of R^m), the null space of A^T (a subspace of R^m), and the column space of A (a subspace of R^m). The following theorem shows that pairs of these four vector spaces are orthogonal complements.

THEOREM 6.22 *If A is a given $m \times n$ matrix, then:*

(a) *The null space of A is the orthogonal complement of the row space of A.*

(b) *The null space of A^T is the orthogonal complement of the column space of A.*

Proof (a) Before proving the result, let us verify that the two vector spaces that we wish to show are the same have equal dimensions. If r is the rank of A, then the dimension of the null space of A is $n-r$ (Theorem 6.12 in Section 6.6). Since the dimension of the row space of A is r, then by Theorem 6.20, the dimension of its orthogonal complement is also $n - r$. Let the vector $\mathbf{x}$ in R^n be in the null space of A. Then $A\mathbf{x} = \mathbf{0}$. Let the vectors $\mathbf{v}_1, \mathbf{v}_2, \ldots, \mathbf{v}_m$ in R^n denote the rows of A. Then the entries in the $m \times 1$ matrix $A\mathbf{x}$ are $\mathbf{v}_1\mathbf{x}, \mathbf{v}_2\mathbf{x}, \ldots, \mathbf{v}_m\mathbf{x}$. Thus we have

$$\mathbf{v}_1\mathbf{x} = 0, \quad \mathbf{v}_2\mathbf{x} = 0, \ldots, \quad \mathbf{v}_m\mathbf{x} = 0. \tag{3}$$

Hence $\mathbf{x}$ is orthogonal to the vectors $\mathbf{v}_1, \mathbf{v}_2, \ldots, \mathbf{v}_m$, which span the row space of A. It then follows that $\mathbf{x}$ is orthogonal to every vector in the row space of A, so $\mathbf{x}$ lies in the orthogonal complement of the row space of A. Hence the null space of A is contained in the orthogonal complement of the row space of A. It then follows from Exercise T.7 in Section 6.4 that the null space of A equals the orthogonal complement of the row space of A.

(b) To establish the result, replace A by A^T in part (a) to conclude that the null space of A^T is the orthogonal complement of the row space of A^T. Since the row space of A^T is the column space of A, we have established part (b). ■

EXAMPLE 4 Let

$$A = \begin{bmatrix} 1 & -2 & 1 & 0 & 2 \\ 1 & -1 & 4 & 1 & 3 \\ -1 & 3 & 2 & 1 & -1 \\ 2 & -3 & 5 & 1 & 5 \end{bmatrix}.$$

Compute the four fundamental vector spaces associated with A and verify Theorem 6.22.

Solution We first transform A to reduced row echelon form, obtaining (verify)

$$B = \begin{bmatrix} 1 & 0 & 7 & 2 & 4 \\ 0 & 1 & 3 & 1 & 1 \\ 0 & 0 & 0 & 0 & 0 \\ 0 & 0 & 0 & 0 & 0 \end{bmatrix}.$$

Solving the linear system $B\mathbf{x} = \mathbf{0}$, we find (verify) that

$$S = \left\{ \begin{bmatrix} -7 \\ -3 \\ 1 \\ 0 \\ 0 \end{bmatrix}, \begin{bmatrix} -2 \\ -1 \\ 0 \\ 1 \\ 0 \end{bmatrix}, \begin{bmatrix} -4 \\ -1 \\ 0 \\ 0 \\ 1 \end{bmatrix} \right\}$$

is a basis for the null space of A. Moreover,

$$T = \{(1, 0, 7, 2, 4), (0, 1, 3, 1, 1)\}$$

is a basis for the row space of A. Since the vectors in S and T are orthogonal, it follows that S is a basis for the orthogonal complement of the row space of A, where we take the vectors in S in horizontal form. Next,

$$A^T = \begin{bmatrix} 1 & 1 & -1 & 2 \\ -2 & -1 & 3 & -3 \\ 1 & 4 & 2 & 5 \\ 0 & 1 & 1 & 1 \\ 2 & 3 & -1 & 5 \end{bmatrix}.$$

Transforming A^T to reduced row echelon form, we obtain (verify)

$$C = \begin{bmatrix} 1 & 0 & -2 & 1 \\ 0 & 1 & 1 & 1 \\ 0 & 0 & 0 & 0 \\ 0 & 0 & 0 & 0 \\ 0 & 0 & 0 & 0 \end{bmatrix}.$$

Solving the linear system $A^T\mathbf{x} = \mathbf{0}$, we find (verify) that

$$S' = \left\{ \begin{bmatrix} 2 \\ -1 \\ 1 \\ 0 \end{bmatrix}, \begin{bmatrix} -1 \\ -1 \\ 0 \\ 1 \end{bmatrix} \right\}$$

is a basis for the null space of A^T. Moreover, the nonzero rows of C read vertically yield the following basis for the column space of A:

$$T' = \left\{ \begin{bmatrix} 1 \\ 0 \\ -2 \\ 1 \end{bmatrix}, \begin{bmatrix} 0 \\ 1 \\ 1 \\ 1 \end{bmatrix} \right\}.$$

Since the vectors in S' and T' are orthogonal, it follows that S' is a basis for the orthogonal complement of the column space of A. ■

EXAMPLE 5 Find a basis for the orthogonal complement of the subspace W of R^5 spanned by the vectors

$$\mathbf{w}_1 = (2, -1, 0, 1, 2), \quad \mathbf{w}_2 = (1, 3, 1, -2, -4),$$
$$\mathbf{w}_3 = (3, 2, 1, -1, -2), \quad \mathbf{w}_4 = (7, 7, 3, -4, -8),$$
$$\mathbf{w}_5 = (1, -4, -1, -1, -2).$$

Solution 1 Let $\mathbf{u} = (a, b, c, d, e)$ be an arbitrary vector in $W^{\perp}$. Since $\mathbf{u}$ is orthogonal to each of the given vectors spanning W, we have a linear system of five equations in five unknowns, whose coefficient matrix is (verify)

$$A = \begin{bmatrix} 2 & -1 & 0 & 1 & 2 \\ 1 & 3 & 1 & -2 & -4 \\ 3 & 2 & 1 & -1 & -2 \\ 7 & 7 & 3 & -4 & -8 \\ 1 & -4 & -1 & -1 & -2 \end{bmatrix}.$$

Solving the homogeneous system $A\mathbf{x} = \mathbf{0}$, we obtain the following basis for the solution space (verify):

$$S = \left\{ \begin{bmatrix} -\frac{1}{7} \\ -\frac{2}{7} \\ 1 \\ 0 \\ 0 \end{bmatrix}, \begin{bmatrix} 0 \\ 0 \\ 0 \\ -2 \\ 1 \end{bmatrix} \right\}.$$

These vectors taken in horizontal form provide a basis for $W^{\perp}$.

Solution 2 Form the matrix whose rows are the given vectors. This matrix is A as shown in Solution 1, so the row space of A is W. By Theorem 6.22, $W^{\perp}$ is the null space of A. Thus we obtain the same basis for $W^{\perp}$ as in Solution 1. ■

The following example will be used to geometrically illustrate Theorem 6.22.

EXAMPLE 6 Let

$$A = \begin{bmatrix} 3 & -1 & 2 \\ 2 & 1 & 3 \\ 7 & 1 & 8 \end{bmatrix}.$$

Transforming A to reduced row echelon form, we obtain (verify)

$$\begin{bmatrix} 1 & 0 & 1 \\ 0 & 1 & 1 \\ 0 & 0 & 0 \end{bmatrix},$$

so the row space of A is a two-dimensional subspace of R^3, that is, a plane passing through the origin, with basis $\{(1, 0, 1), (0, 1, 1)\}$. The null space of A is a one-dimensional subspace of R^3 with basis

$$\left\{ \begin{bmatrix} -1 \\ -1 \\ 1 \end{bmatrix} \right\},$$

(verify). Since this basis vector is orthogonal to the two basis vectors for the row space of A given previously, the null space of A is orthogonal to the row space of A; that is, the null space of A is the orthogonal complement of the row space of A.

Next, transforming A^T to reduced row echelon form, we have

$$\begin{bmatrix} 1 & 0 & 1 \\ 0 & 1 & 2 \\ 0 & 0 & 0 \end{bmatrix},$$

(verify). It follows that

$$\left\{ \begin{bmatrix} -1 \\ -2 \\ 1 \end{bmatrix} \right\}$$

is a basis for the null space of A^T (verify). Hence the null space of A^T is a line through the origin. Moreover,

$$\left\{\begin{bmatrix}1\\0\\1\end{bmatrix}, \begin{bmatrix}0\\1\\2\end{bmatrix}\right\}$$

is a basis for the column space of A^T (verify), so the column space of A^T is a plane through the origin. Since every basis vector for the null space of A^T is orthogonal to every basis vector for the column space of A^T, we conclude that the null space of A^T is the orthogonal complement of the column space of A^T. These results are illustrated in Figure 6.15. ■

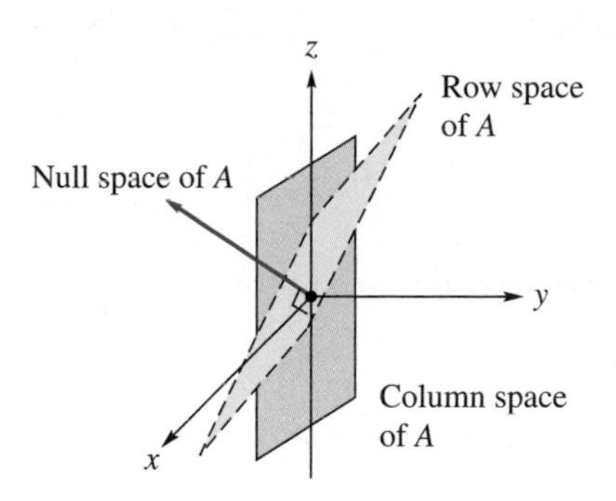

Figure 6.15 ▲

PROJECTIONS AND APPLICATIONS

In Theorem 6.20 and in the Remark following the theorem, we have shown that if W is a subspace of R^n with orthonormal basis $\{\mathbf{w}_1, \mathbf{w}_2, \ldots, \mathbf{w}_m\}$ and $\mathbf{v}$ is any vector in R^n, then there exist unique vectors $\mathbf{w}$ in W and $\mathbf{u}$ in $W^\perp$ such that

$$\mathbf{v} = \mathbf{w} + \mathbf{u}.$$

Moreover, as we saw in Equation (1),

$$\mathbf{w} = (\mathbf{v} \cdot \mathbf{w}_1)\mathbf{w}_1 + (\mathbf{v} \cdot \mathbf{w}_2)\mathbf{w}_2 + \cdots + (\mathbf{v} \cdot \mathbf{w}_m)\mathbf{w}_m,$$

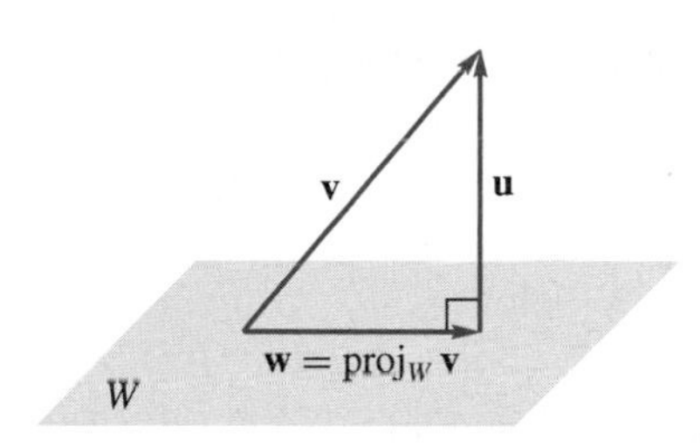

Figure 6.16 ▲

which is called the **orthogonal projection** of $\mathbf{v}$ on W and is denoted by $\text{proj}_W\mathbf{v}$. In Figure 6.16, we illustrate Theorem 6.20 when W is a two-dimensional subspace of R^3 (a plane through the origin).

Often an orthonormal basis has many fractions, so it is helpful to also have a formula giving $\text{proj}_W\mathbf{v}$ when W has an *orthogonal* basis. In Exercise T.6, we ask you to show that if $\{\mathbf{w}_1, \mathbf{w}_2, \ldots, \mathbf{w}_m\}$ is an orthogonal basis for W, then

$$\text{proj}_W\mathbf{v} = \frac{\mathbf{v} \cdot \mathbf{w}_1}{\mathbf{w}_1 \cdot \mathbf{w}_1}\mathbf{w}_1 + \frac{\mathbf{v} \cdot \mathbf{w}_2}{\mathbf{w}_2 \cdot \mathbf{w}_2}\mathbf{w}_2 + \cdots + \frac{\mathbf{v} \cdot \mathbf{w}_m}{\mathbf{w}_m \cdot \mathbf{w}_m}\mathbf{w}_m.$$

Remark The Gram–Schmidt process described in Theorem 6.18 can be rephrased in terms of projections at each step. Thus, Figure 6.13 (the first step in the Gram–Schmidt process) can be relabeled as follows:

Figure 6.13 (relabeled) ►

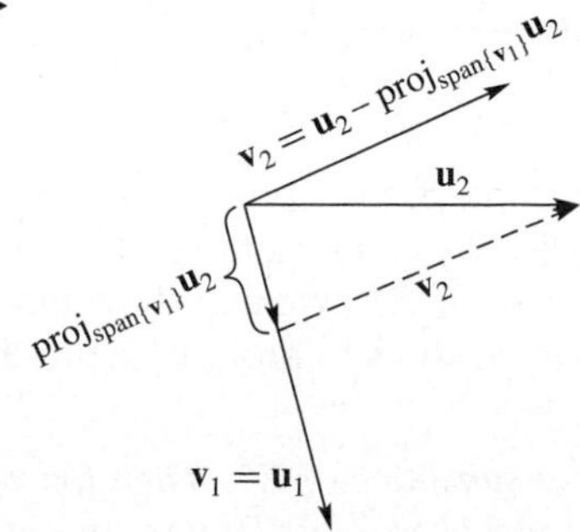

EXAMPLE 7 Let W be the two-dimensional subspace of R^3 with orthonormal basis $\{\mathbf{w}_1, \mathbf{w}_2\}$, where

$$\mathbf{w}_1 = \left(\frac{2}{3}, -\frac{1}{3}, -\frac{2}{3}\right) \quad \text{and} \quad \mathbf{w}_2 = \left(\frac{1}{\sqrt{2}}, 0, \frac{1}{\sqrt{2}}\right).$$

Using the standard inner product on R^3, find the orthogonal projection of

$$\mathbf{v} = (2, 1, 3)$$

on W and the vector $\mathbf{u}$ that is orthogonal to every vector in W.

Solution From Equation (1) we have

$$\mathbf{w} = \text{proj}_W \mathbf{v} = (\mathbf{v} \cdot \mathbf{w}_1)\mathbf{w}_1 + (\mathbf{v} \cdot \mathbf{w}_2)\mathbf{w}_2 = -1\,\mathbf{w}_1 + \frac{5}{\sqrt{2}}\mathbf{w}_2 = \left(\frac{11}{6}, \frac{1}{3}, \frac{19}{6}\right)$$

and

$$\mathbf{u} = \mathbf{v} - \mathbf{w} = \left(\frac{1}{6}, \frac{2}{3}, -\frac{1}{6}\right).$$

■

It is clear from Figure 6.16 that the distance from $\mathbf{v}$ to the plane W is given by the length of the vector $\mathbf{u} = \mathbf{v} - \mathbf{w}$, that is, by

$$\left\|\mathbf{v} - \text{proj}_W \mathbf{v}\right\|.$$

We prove this result in general in Theorem 6.23 below.

EXAMPLE 8 Let W be the subspace of R^3 defined in Example 7 and let $\mathbf{v} = (1, 1, 0)$. Find the distance from $\mathbf{v}$ to W.

Solution We first compute

$$\text{proj}_W \mathbf{v} = (\mathbf{v} \cdot \mathbf{w}_1)\mathbf{w}_1 + (\mathbf{v} \cdot \mathbf{w}_2)\mathbf{w}_2 = \frac{1}{3}\mathbf{w}_1 + \frac{1}{\sqrt{2}}\mathbf{w}_2 = \left(\frac{13}{18}, -\frac{1}{9}, \frac{5}{18}\right).$$

Then

$$\mathbf{v} - \text{proj}_W \mathbf{v} = (1, 1, 0) - \left(\frac{13}{18}, -\frac{1}{9}, \frac{5}{18}\right) = \left(\frac{5}{18}, \frac{10}{9}, -\frac{5}{18}\right)$$

and

$$\left\|\mathbf{v} - \text{proj}_W \mathbf{v}\right\| = \sqrt{\frac{25}{324} + \frac{100}{81} + \frac{25}{324}} = \frac{5\sqrt{2}}{6},$$

so the distance from $\mathbf{v}$ to W is $\dfrac{5\sqrt{2}}{6}$. ■

In Example 8, $\|\mathbf{v} - \text{proj}_W \mathbf{v}\|$ represented the distance in 3-space from $\mathbf{v}$ to the plane W. We can generalize this notion of distance from a vector in R^n to a subspace W of R^n. We can show that the vector in W that is closest to $\mathbf{v}$ is in fact $\text{proj}_W \mathbf{v}$, so $\|\mathbf{v} - \text{proj}_W \mathbf{v}\|$ represents the distance from $\mathbf{v}$ to W.

THEOREM 6.23 *Let W be a subspace of R^n. Then for vector $\mathbf{v}$ belonging to R^n, the vector in W closest to $\mathbf{v}$ is $\text{proj}_W \mathbf{v}$. That is, $\|\mathbf{v} - \mathbf{w}\|$, for $\mathbf{w}$ belonging to W, is minimized when $\mathbf{w} = \text{proj}_W \mathbf{v}$.*

Proof Let $\mathbf{w}$ be any vector in W. Then

$$\mathbf{v} - \mathbf{w} = \left(\mathbf{v} - \mathrm{proj}_W\mathbf{v}\right) + \left(\mathrm{proj}_W\mathbf{v} - \mathbf{w}\right).$$

Since $\mathbf{w}$ and $\mathrm{proj}_W\mathbf{v}$ are both in W, $\mathrm{proj}_W\mathbf{v} - \mathbf{w}$ is in W. By Theorem 6.20, $\mathbf{v} - \mathrm{proj}_W\mathbf{v}$ is orthogonal to every vector in W, so

$$\begin{aligned}\|\mathbf{v}-\mathbf{w}\|^2 &= (\mathbf{v} - \mathbf{w}) \cdot (\mathbf{v} - \mathbf{w}) \\ &= \left((\mathbf{v} - \mathrm{proj}_W\mathbf{v}) + (\mathrm{proj}_W\mathbf{v} - \mathbf{w})\right) \cdot \left((\mathbf{v} - \mathrm{proj}_W\mathbf{v}) + (\mathrm{proj}_W\mathbf{v} - \mathbf{w})\right) \\ &= \left\|\mathbf{v} - \mathrm{proj}_W\mathbf{v}\right\|^2 + \left\|\mathrm{proj}_W\mathbf{v} - \mathbf{w}\right\|^2.\end{aligned}$$

If $\mathbf{w} \neq \mathrm{proj}_W\mathbf{v}$, then $\|\mathrm{proj}_W\mathbf{v} - \mathbf{w}\|^2$ is positive and

$$\left\|\mathbf{v} - \mathbf{w}\right\|^2 > \left\|\mathbf{v} - \mathrm{proj}_W\mathbf{v}\right\|^2.$$

Thus it follows that $\mathrm{proj}_W\mathbf{v}$ is the vector in W that minimizes $\|\mathbf{v} - \mathbf{w}\|^2$ and hence minimizes $\|\mathbf{v} - \mathbf{w}\|$. ■

In Example 7,

$$\mathbf{w} = \mathrm{proj}_W\mathbf{v} = \left(\tfrac{11}{6}, \tfrac{1}{3}, \tfrac{19}{6}\right)$$

is the vector in $W = \mathrm{span}\,\{\mathbf{w}_1, \mathbf{w}_2\}$ that is closest to $\mathbf{v} = (2, 1, 3)$.

Key Terms

Orthogonal vectors
Orthogonal complement(s)
Fundamental vector spaces associated with a matrix
Orthogonal projection

6.9 Exercises

1. Let W be a subspace of R^3 with basis $\{\mathbf{w}_1, \mathbf{w}_2\}$, where $\mathbf{w}_1 = (1, 0, 1)$ and $\mathbf{w}_2 = (1, 1, 1)$. Express $\mathbf{v} = (2, 2, 0)$ as a sum of a vector $\mathbf{w}$ in W and a vector $\mathbf{u}$ in $W^\perp$.

2. Let W be a subspace of R^4 with basis $\{\mathbf{w}_1, \mathbf{w}_2\}$, where $\mathbf{w}_1 = (1, 1, 1, 0)$ and $\mathbf{w}_2 = (1, 0, 0, 0)$. Express $\mathbf{v} = (1, 1, 0, 0)$ as a sum of a vector $\mathbf{w}$ in W and a vector $\mathbf{u}$ in $W^\perp$.

3. Let W be the subspace of R^3 spanned by the vector $\mathbf{w} = (2, -3, 1)$.

(a) Find a basis for $W^\perp$.

(b) Describe $W^\perp$ geometrically. (You may use a verbal or pictorial description.)

4. Let $W = \mathrm{span}\,\{(1, 2, -1), (-1, 3, 2)\}$.

(a) Find a basis for $W^\perp$.

(b) Describe $W^\perp$ geometrically. (You may use a verbal or pictorial description.)

5. Let W be the subspace of R^5 spanned by the vectors $\mathbf{w}_1$, $\mathbf{w}_2$, $\mathbf{w}_3$, $\mathbf{w}_4$, $\mathbf{w}_5$, where

$\mathbf{w}_1 = (2, -1, 1, 3, 0)$, $\mathbf{w}_2 = (1, 2, 0, 1, -2)$,

$\mathbf{w}_3 = (4, 3, 1, 5, -4)$, $\mathbf{w}_4 = (3, 1, 2, -1, 1)$,

$\mathbf{w}_5 = (2, -1, 2, -2, 3)$.

Find a basis for $W^\perp$.

In Exercises 6 and 7, compute the four fundamental vector spaces associated with A and verify Theorem 6.22.

6. $A = \begin{bmatrix} 1 & 5 & 3 & 7 \\ 2 & 0 & -4 & -6 \\ 4 & 7 & -1 & 2 \end{bmatrix}$

7. $A = \begin{bmatrix} 2 & -1 & 3 & 4 \\ 0 & -3 & 7 & -2 \\ 1 & 1 & -2 & 3 \\ 1 & 4 & -9 & 5 \end{bmatrix}$

In Exercises 8 and 9, find $\mathrm{proj}_W\mathbf{v}$ *for the given vector* $\mathbf{v}$ *and subspace W.*

8. Let W be the subspace of R^3 with basis

$$\left(\frac{1}{\sqrt{5}}, 0, \frac{2}{\sqrt{5}}\right), \quad \left(-\frac{2}{\sqrt{5}}, 0, \frac{1}{\sqrt{5}}\right).$$

(a) $\mathbf{v} = (3, 4, -1)$ (b) $\mathbf{v} = (2, 1, 3)$

(c) $\mathbf{v} = (-5, 0, 1)$

9. Let W be the subspace of R^4 with basis $(1, 1, 0, 1), (0, 1, 1, 0), (-1, 0, 0, 1)$.

(a) $\mathbf{v} = (2, 1, 3, 0)$ (b) $\mathbf{v} = (0, -1, 1, 0)$

(c) $\mathbf{v} = (0, 2, 0, 3)$

10. Let W be the subspace of R^3 with orthonormal basis $\{\mathbf{w}_1, \mathbf{w}_2\}$, where

$$\mathbf{w}_1 = (0, 1, 0) \quad \text{and} \quad \mathbf{w}_2 = \left(\frac{1}{\sqrt{5}}, 0, \frac{2}{\sqrt{5}}\right).$$

Write the vector $\mathbf{v} = (1, 2, -1)$ as $\mathbf{w} + \mathbf{u}$, with $\mathbf{w}$ in W and $\mathbf{u}$ in $W^\perp$.

11. Let W be the subspace of R^4 with orthonormal basis $\{\mathbf{w}_1, \mathbf{w}_2, \mathbf{w}_3\}$, where

$$\mathbf{w}_1 = \left(\frac{1}{\sqrt{2}}, 0, 0, -\frac{1}{\sqrt{2}}\right), \quad \mathbf{w}_2 = (0, 0, 1, 0),$$

$$\text{and} \quad \mathbf{w}_3 = \left(\frac{1}{\sqrt{2}}, 0, 0, \frac{1}{\sqrt{2}}\right).$$

Write the vector $\mathbf{v} = (1, 0, 2, 3)$ as $\mathbf{w} + \mathbf{u}$, with $\mathbf{w}$ in W and $\mathbf{u}$ in $W^\perp$.

12. Let W be the subspace of R^3 defined in Exercise 10, and let $\mathbf{v} = (-1, 0, 1)$. Find the distance from $\mathbf{v}$ to W.

13. Let W be the subspace of R^4 defined in Exercise 11, and let $\mathbf{v} = (1, 2, -1, 0)$. Find the distance from $\mathbf{v}$ to W.

14. Find the distance from the point $(2, 3, -1)$ to the plane $3x - 2y + z = 0$. (*Hint*: First find an orthonormal basis for the plane.)

Theoretical Exercises

T.1. Show that if W is a subspace of R^n, then the zero vector of R^n belongs to $W^\perp$.

T.2. Show that the orthogonal complement of R^n is the zero subspace and the orthogonal complement of the zero subspace is R^n itself.

T.3. Show that if W is a subspace of R^n that is spanned by a set of vectors S, then a vector $\mathbf{u}$ in R^n belongs to $W^\perp$ if and only if $\mathbf{u}$ is orthogonal to every vector in S.

T.4. Let A be an $m \times n$ matrix. Show that every vector $\mathbf{v}$ in R^n can be written uniquely as $\mathbf{w} + \mathbf{u}$, where $\mathbf{w}$ is in the null space of A and $\mathbf{u}$ is in the column space of A^T.

T.5. Let W be a subspace of R^n. Show that if $\mathbf{w}_1, \mathbf{w}_2, \ldots, \mathbf{w}_r$ is a basis for W and $\mathbf{u}_1, \mathbf{u}_2, \ldots, \mathbf{u}_s$ is a basis for $W^\perp$, then $\mathbf{w}_1, \mathbf{w}_2, \ldots, \mathbf{w}_r, \mathbf{u}_1, \mathbf{u}_2, \ldots, \mathbf{u}_s$ is a basis for R^n, and $n = \dim W + \dim W^\perp$.

T.6. Let W be a subspace of R^n and let $\{\mathbf{w}_1, \mathbf{w}_2, \ldots, \mathbf{w}_m\}$ be an orthogonal basis for W. Show that if $\mathbf{v}$ is any vector in R^n, then

$$\text{proj}_W \mathbf{v} = \frac{\mathbf{v} \cdot \mathbf{w}_1}{\mathbf{w}_1 \cdot \mathbf{w}_1} \mathbf{w}_1 + \frac{\mathbf{v} \cdot \mathbf{w}_2}{\mathbf{w}_2 \cdot \mathbf{w}_2} \mathbf{w}_2 + \cdots + \frac{\mathbf{v} \cdot \mathbf{w}_m}{\mathbf{w}_m \cdot \mathbf{w}_m} \mathbf{w}_m.$$

T.7. Let W be the row space of the $m \times n$ matrix A. Show that if $A\mathbf{x} = \mathbf{0}$, then $\mathbf{x}$ is in $W^\perp$.

MATLAB Exercises

ML.1. Find the projection of $\mathbf{v}$ onto $\mathbf{w}$. (Recall that we have routines **dot** and **norm** available in MATLAB.)

(a) $\mathbf{v} = \begin{bmatrix} 1 \\ 5 \\ -1 \\ 2 \end{bmatrix}, \quad \mathbf{w} = \begin{bmatrix} 0 \\ 1 \\ 2 \\ 1 \end{bmatrix}$

(b) $\mathbf{v} = \begin{bmatrix} 1 \\ -2 \\ 3 \\ 0 \\ 1 \end{bmatrix}, \quad \mathbf{w} = \begin{bmatrix} 1 \\ 1 \\ 1 \\ 1 \\ 1 \end{bmatrix}$

ML.2. Let $S = \{\mathbf{w1}, \mathbf{w2}\}$, where

$$\mathbf{w1} = \begin{bmatrix} 1 \\ 0 \\ 1 \\ 1 \end{bmatrix} \quad \text{and} \quad \mathbf{w2} = \begin{bmatrix} 1 \\ 1 \\ -1 \\ 0 \end{bmatrix},$$

and let $W = \text{span } S$.

(a) Show that S is an orthogonal basis for W.

(b) Let

$$\mathbf{v} = \begin{bmatrix} 2 \\ 1 \\ 2 \\ 1 \end{bmatrix}.$$

Compute $\text{proj}_{\mathbf{w1}} \mathbf{v}$.

(c) For vector $\mathbf{v}$ in part (b), compute $\text{proj}_W \mathbf{v}$.

ML.3. Plane P in R^3 has orthogonal basis $\{\mathbf{w1}, \mathbf{w2}\}$, where

$$\mathbf{w1} = \begin{bmatrix} 1 \\ 2 \\ 3 \end{bmatrix} \quad \text{and} \quad \mathbf{w2} = \begin{bmatrix} 0 \\ -3 \\ 2 \end{bmatrix}.$$

(a) Find the projection of

$$\mathbf{v} = \begin{bmatrix} 2 \\ 4 \\ 8 \end{bmatrix}$$

onto P.

(b) Find the distance from $\mathbf{v}$ to P.

ML.4. Let W be the subspace of R^4 with basis

$$S = \left\{ \begin{bmatrix} 1 \\ 1 \\ 0 \\ 1 \end{bmatrix}, \begin{bmatrix} 2 \\ -1 \\ 0 \\ 0 \end{bmatrix}, \begin{bmatrix} 0 \\ 1 \\ 0 \\ 1 \end{bmatrix} \right\} \quad \text{and} \quad \mathbf{v} = \begin{bmatrix} 0 \\ 0 \\ 1 \\ 1 \end{bmatrix}.$$

(a) Find $\text{proj}_W \mathbf{v}$.

(b) Find the distance from $\mathbf{v}$ to W.

ML.5. Let

$$T = \begin{bmatrix} 1 & 0 \\ 0 & 1 \\ 1 & 1 \\ 1 & 0 \\ 1 & 0 \end{bmatrix} \quad \text{and} \quad \mathbf{b} = \begin{bmatrix} 1 \\ 1 \\ 1 \\ 1 \\ 1 \end{bmatrix}.$$

(a) Show that system $T\mathbf{x} = \mathbf{b}$ is inconsistent.

(b) Since $T\mathbf{x} = \mathbf{b}$ is inconsistent, $\mathbf{b}$ is not in the column space of T. One approach to find an approximate solution is to find a vector $\mathbf{y}$ in the column space of T so that $T\mathbf{y}$ is as close as possible to $\mathbf{b}$. We can do this by finding the projection $\mathbf{p}$ of $\mathbf{b}$ onto the column space of T. Find this projection $\mathbf{p}$ (which will be $T\mathbf{y}$).

Key Ideas for Review

- **Vector space.** See page 272.
- **Theorem 6.2.** Let V be a vector space with operations $\oplus$ and $\odot$ and let W be a nonempty subset of V. Then W is a subspace of V if and only if
  (α) If $\mathbf{u}$ and $\mathbf{v}$ are any vectors in W, then $\mathbf{u} \oplus \mathbf{v}$ is in W.
  (β) If $\mathbf{u}$ is in W and c is a scalar, then $c \odot \mathbf{u}$ is in W.
- $S = \{\mathbf{v}_1, \mathbf{v}_2, \ldots, \mathbf{v}_k\}$ is linearly dependent if and only if one of the vectors in S is a linear combination of all the other vectors in S.
- **Theorem 6.5.** If $S = \{\mathbf{v}_1, \mathbf{v}_2, \ldots, \mathbf{v}_n\}$ is a basis for a vector space V, then every vector in V can be written in one and only one way as a linear combination of the vectors in S.
- **Theorem 6.7.** If $S = \{\mathbf{v}_1, \mathbf{v}_2, \ldots, \mathbf{v}_n\}$ is a basis for a vector space V and $T = \{\mathbf{w}_1, \mathbf{w}_2, \ldots, \mathbf{w}_r\}$ is a linearly independent set of vectors in V, then $r \leq n$.
- **Corollary 6.1.** All bases for a vector space V must have the same number of vectors.
- **Theorem 6.9.** Let V be an n-dimensional vector space, and let $S = \{\mathbf{v}_1, \mathbf{v}_2, \ldots, \mathbf{v}_n\}$ be a set of n vectors in V.
  (a) If S is linearly independent, then it is a basis for V.
  (b) If S spans V, then it is a basis for V.
- **Basis for solution space of $A\mathbf{x} = \mathbf{0}$.** See page 320.
- **Theorem 6.11.** Row rank A = column rank A.
- **Theorem 6.12.** If A is an $m \times n$ matrix, then rank A + nullity $A = n$.
- **Theorem 6.13.** An $n \times n$ matrix A is nonsingular if and only if rank $A = n$.
- **Corollary 6.2.** If A is an $n \times n$ matrix, then rank $A = n$ if and only if $\det(A) \neq 0$.
- **Corollary 6.5.** If A is $n \times n$, then $A\mathbf{x} = \mathbf{0}$ has a nontrivial solution if and only if rank $A < n$.
- **Theorem 6.14.** The linear system $A\mathbf{x} = \mathbf{b}$ has a solution if and only if rank A = rank $\left[A \;\vdots\; \mathbf{b}\right]$.
- **List of Nonsingular Equivalences.** The following statements are equivalent for an $n \times n$ matrix A.
  1. A is nonsingular.
  2. $\mathbf{x} = \mathbf{0}$ is the only solution to $A\mathbf{x} = \mathbf{0}$.
  3. A is row equivalent to I_n.
  4. The linear system $A\mathbf{x} = \mathbf{b}$ has a unique solution for every $n \times 1$ matrix $\mathbf{b}$.
  5. $\det(A) \neq 0$.
  6. A has rank n.
  7. A has nullity 0.
  8. The rows of A form a linearly independent set of n vectors in R^n.
  9. The columns of A form a linearly independent set of n vectors in R^n.
- **Theorem 6.15.** Let $S = \{\mathbf{v}_1, \mathbf{v}_2, \ldots, \mathbf{v}_n\}$ and $T = \{\mathbf{w}_1, \mathbf{w}_2, \ldots, \mathbf{w}_n\}$ be bases for the n-dimensional vector space V. Let $P_{S \leftarrow T}$ be the transition matrix from T to S. Then $P_{S \leftarrow T}$ is nonsingular and $P_{S \leftarrow T}^{-1}$ is the transition matrix from S to T.
- **Theorem 6.16.** An orthogonal set of nonzero vectors in R^n is linearly independent.
- **Theorem 6.17.** If $\{\mathbf{u}_1, \mathbf{u}_2, \ldots, \mathbf{u}_n\}$ is an orthonormal basis for R^n and $\mathbf{v}$ is a vector in R^n, then $\mathbf{v} = (\mathbf{v} \cdot \mathbf{u}_1)\mathbf{u}_1 + (\mathbf{v} \cdot \mathbf{u}_2)\mathbf{u}_2 + \cdots + (\mathbf{v} \cdot \mathbf{u}_n)\mathbf{u}_n$.
- **Theorem 6.18 (Gram–Schmidt Process).** See pages 354–355.
- **Theorem 6.20.** Let W be a subspace of R^n. Then $R^n = W \oplus W^\perp$.
- **Theorem 6.22.** If A is a given $m \times n$ matrix, then
  (a) The null space of A is the orthogonal complement of the row space of A.
  (b) The null space of A^T is the orthogonal complement of the column space of A.

Supplementary Exercises

1. Determine whether the set of all ordered pairs of real numbers with the operations

$$(x, y) \oplus (x', y') = (x - x', y + y')$$
$$c \odot (x, y) = (cx, cy)$$

is a vector space.

2. Consider the set W of all vectors in R^4 of the form (a, b, c, d), where $a = b + c$ and $d = a + 1$. Is W a subspace of R^4?

3. Is the vector $(4, 2, 1)$ a linear combination of the vectors $(1, 2, 1)$, $(-1, 1, 2)$, and $(-3, -3, 0)$?

4. Do the vectors $(-1, 2, 1)$, $(2, 1, -1)$, and $(0, 5, -1)$ span R^3?

5. Is the set of vectors

$$\{t^2 + 2t + 2, 2t^2 + 3t + 1, -t - 3\}$$

linearly dependent or linearly independent? If it is linearly dependent, write one of the vectors as a linear combination of the other two.

6. Does the set of vectors

$$\{t - 1, t^2 + 2t + 1, t^2 + t - 2\}$$

form a basis for P_2?

7. Find a basis for the subspace of R^4 consisting of all vectors of the form

$$(a + b, b + c, a - b - 2c, b + c).$$

What is the dimension of this subspace?

8. Find the dimension of the subspace of P_2 consisting of all polynomials $a_0t^2 + a_1t + a_2$, where $a_2 = 0$.

9. Find a basis for the solution space of the homogeneous system

$$\begin{aligned} x_1 + 2x_2 - x_3 + x_4 + 2x_5 &= 0 \\ x_1 + x_2 + 2x_3 - 3x_4 + x_5 &= 0. \end{aligned}$$

What is the dimension of the solution space? Find a basis for $V = \text{span } S$.

10. Let

$$S = \{(1, 2, -1, 2), (0, -1, 3, -6), (2, 3, 1, -2), (3, 2, 1, -4), (1, -1, 0, -2)\}.$$

Find a basis for $V = \text{span } S$.

11. For what value(s) of λ is the set

$$\{t + 3, 2t + \lambda^2 + 2\}$$

linearly independent?

12. Let

$$A = \begin{bmatrix} \lambda^2 & 0 & 1 \\ -1 & \lambda & -2 \\ -1 & 0 & -1 \end{bmatrix}.$$

For what value(s) of λ does the homogeneous system $A\mathbf{x} = \mathbf{0}$ have only the trivial solution?

13. For what values of a is the vector $(a^2, a, 1)$ in span $\{(1, 2, 3), (1, 1, 1), (0, 1, 2)\}$?

14. For what values of a is the vector $(a^2, -3a, -2)$ in span $\{(1, 2, 3), (0, 1, 1), (1, 3, 4)\}$?

15. Let $\mathbf{v}_1, \mathbf{v}_2, \mathbf{v}_3$ be vectors in a vector space. Verify that

$$\text{span } \{\mathbf{v}_1, \mathbf{v}_2, \mathbf{v}_3\} = \text{span } \{\mathbf{v}_1 + \mathbf{v}_2, \mathbf{v}_1 - \mathbf{v}_3, \mathbf{v}_1 + \mathbf{v}_3\}.$$

16. For what values of k will the set S form a basis for R^6?

$$S = \{(1, 2, -2, 1, 1, 1), (0, 2, 0, 0, 0, 0), (0, 5, k, 1, 0, 0), (0, 2, 1, k, 0, 0), (0, 1, -2, 4, -3, 1), (0, 3, 1, 1, 1, 0)\}.$$

17. Consider the vector space R^2.

(a) For what values of m and b will all vectors of the form $(x, mx + b)$ be a subspace of R^2?

(b) For what values of r will the set of all vectors of the form (x, rx^2) be a subspace of R^2?

18. Let A be a fixed $n \times n$ matrix and let the set of all $n \times n$ matrices B such that $AB = BA$ be denoted by $C(A)$. Is $C(A)$ a subspace of the vector space of all $n \times n$ matrices?

19. Show that a subspace W of R^3 coincides with R^3 if and only if it contains the vectors $(1, 0, 0)$, $(0, 1, 0)$, and $(0, 0, 1)$.

20. Let A be a fixed $m \times n$ matrix and define W to be the subset of all $m \times 1$ matrices $\mathbf{b}$ in R^m for which the linear system $A\mathbf{x} = \mathbf{b}$ has a solution.

(a) Is W a subspace of R^m?

(b) What is the relationship between W and the column space of A?

21. ***(Calculus Required)*** Let $C[a, b]$ denote the set of all real-valued continuous functions defined on $[a, b]$. If f and g are in $C[a, b]$, we define $f \oplus g$ by $(f \oplus g)(t) = f(t) + g(t)$, for t in $[a, b]$. If f is in $C[a, b]$ and c is a scalar, we define $c \odot f$ by $(c \odot f)(t) = cf(t)$, for t in $[a, b]$.

(a) Show that $C[a, b]$ is a real vector space.

(b) Let $W(k)$ be the set of all functions in $C[a, b]$ with $f(a) = k$. For what values of k will $W(k)$ be a subspace of $C[a, b]$?

(c) Let $t_1, t_2, \ldots, t_n$ be a fixed set of points in $[a, b]$. Show that the subset of all functions f in $C[a, b]$ that have roots at $t_1, t_2, \ldots, t_n$, that is, $f(t_i) = 0$ for $i = 1, 2, \ldots, n$, forms a subspace.

22. Let $V = \text{span } \{\mathbf{v}_1, \mathbf{v}_2\}$, where

$$\mathbf{v}_1 = (1, 0, 2) \quad \text{and} \quad \mathbf{v}_2 = (1, 1, 1).$$

Find a basis S for R^3 that includes $\mathbf{v}_1$ and $\mathbf{v}_2$. (*Hint*: Use Theorem 6.8; see Example 9 in Section 6.4.)

23. Find the rank of the matrix

$$\begin{bmatrix} 1 & 2 & -1 & 3 & 1 \\ 0 & 1 & -3 & 2 & 3 \\ 2 & 3 & 1 & 4 & -1 \\ -1 & 2 & 2 & 2 & -5 \\ 3 & 1 & -1 & 2 & 4 \end{bmatrix}.$$

24. Let A be an 8×5 matrix. How large can the dimension of the row space be? Explain.

25. What can you say about the dimension of the solution space of a homogeneous system of 8 equations in 10 unknowns?

26. Is it possible that all nontrivial solutions of a homogeneous system of 5 equations in 7 unknowns be multiples of each other? Explain.

27. Let $S = \{(1, 0, -1), (0, 1, 1), (0, 0, 1)\}$ and $T = \{(1, 0, 0), (0, 1, -1), (1, -1, 2)\}$ be bases for R^3. Let $\mathbf{v} = (2, 3, 5)$.

(a) Find the coordinate vector of $\mathbf{v}$ with respect to T directly.

(b) Find the coordinate vector of $\mathbf{v}$ with respect to S directly.

(c) Find the transition matrix $P_{S \leftarrow T}$ from T to S.

(d) Find the coordinate vector of $\mathbf{v}$ with respect to S using $P_{S \leftarrow T}$ and the answer to part (a).

(e) Find the transition matrix $Q_{T \leftarrow S}$ from S to T.

(f) Find the coordinate vector of $\mathbf{v}$ with respect to T using $Q_{T \leftarrow S}$ and the answer to part (b).

28. Let $S = \{\mathbf{v}_1, \mathbf{v}_2\}$ and $T = \{\mathbf{w}_1, \mathbf{w}_2\}$ be bases for P_1, where

$$\mathbf{v}_1 = t + 2, \qquad \mathbf{v}_2 = 2t - 1.$$

If the transition matrix from T to S is

$$\begin{bmatrix} 3 & -1 \\ 2 & 1 \end{bmatrix},$$

determine T.

29. Let

$$\mathbf{u} = \left(\frac{1}{\sqrt{2}}, 0, -\frac{1}{\sqrt{2}}\right) \quad \text{and} \quad \mathbf{v} = (a, -1, -b).$$

For what values of a and b is $\{\mathbf{u}, \mathbf{v}\}$ an orthonormal set?

30. Exercise T.6 of Section 6.8 shows that the set of all vectors in R^n that are orthogonal to a fixed vector $\mathbf{u}$ is a subspace of R^n. For $\mathbf{u} = (1, -2, 1)$ find an orthogonal basis for the subspace of vectors in R^3 that are orthogonal to $\mathbf{u}$. [*Hint*: Solve the linear system $\mathbf{u} \cdot \mathbf{v} = 0$, when $\mathbf{v} = (y_1, y_2, y_3)$.]

31. Find an orthonormal basis for the set of vectors $\mathbf{x}$ such that $A\mathbf{x} = \mathbf{0}$ when

(a) $A = \begin{bmatrix} 1 & 0 & 5 & -2 \\ 0 & 1 & -2 & -5 \end{bmatrix}$

(b) $A = \begin{bmatrix} 1 & 0 & 5 & -2 \\ 0 & 1 & -2 & 4 \end{bmatrix}$

32. Given the orthonormal basis

$$S = \left\{\left(\frac{1}{\sqrt{2}}, 0, -\frac{1}{\sqrt{2}}\right), (0, 1, 0), \left(\frac{1}{\sqrt{2}}, 0, \frac{1}{\sqrt{2}}\right)\right\}$$

for R^3, write the vector $(1, 2, 3)$ as a linear combination of the vectors in S.

33. Use the Gram–Schmidt process to find an orthonormal basis for the subspace of R^4 with basis $\{(1, 0, 0, -1), (1, -1, 0, 0), (0, 1, 0, 1)\}$.

34. Let W be the subspace of R^4 spanned by the vectors $\mathbf{w}_1, \mathbf{w}_2, \mathbf{w}_3, \mathbf{w}_4$, where

$$\mathbf{w}_1 = (2, 0, -1, 3), \qquad \mathbf{w}_2 = (1, 2, 2, -5),$$
$$\mathbf{w}_3 = (3, 2, 1, -2), \qquad \mathbf{w}_4 = (7, 2, -1, 4).$$

Find a basis for $W^\perp$.

35. Let $W = \text{span}\,\{(1, 0, 1), (0, 1, 0)\}$ in R^3.

(a) Find a basis for $W^\perp$.

(b) Show that vectors $(1, 0, 1)$, $(0, 1, 0)$ and the basis for $W^\perp$ from part (a) form a basis for R^3.

(c) Write the vector $\mathbf{v}$ as $\mathbf{w} + \mathbf{u}$ with $\mathbf{w}$ in W and $\mathbf{u}$ in $W^\perp$.

(i) $\mathbf{v} = (1, 0, 0)$ (ii) $\mathbf{v} = (1, 2, 3)$

36. Let W be the subspace of R^3 defined in Exercise 35. Find the distance from $\mathbf{v} = (2, 1, -2)$ to W.

37. Compute the four fundamental vector spaces associated with

$$A = \begin{bmatrix} 2 & 3 & -1 & 2 \\ 1 & 1 & -2 & 3 \\ 2 & 1 & 4 & 2 \end{bmatrix}.$$

Theoretical Exercises

T.1. Let A be an $n \times n$ matrix. Show that A is nonsingular if and only if the dimension of the solution space of the homogeneous system $A\mathbf{x} = \mathbf{0}$ is zero.

T.2. Let $\{\mathbf{v}_1, \mathbf{v}_2, \ldots, \mathbf{v}_n\}$ be a basis for an n-dimensional vector space V. Show that if A is a nonsingular $n \times n$ matrix, then $\{A\mathbf{v}_1, A\mathbf{v}_2, \ldots, A\mathbf{v}_n\}$ is also a basis for V.

T.3. Let $S = \{\mathbf{v}_1, \mathbf{v}_2, \ldots, \mathbf{v}_n\}$ be a basis for R^n. Show that if $\mathbf{u}$ is a vector in R^n that is orthogonal to each vector in S, then $\mathbf{u} = \mathbf{0}$.

T.4. Show that rank A = rank A^T, for any $m \times n$ matrix A.

T.5. Let A and B be $m \times n$ matrices that are row equivalent.

(a) Show that rank A = rank B.

(b) Show that for $\mathbf{x}$ in R^n, $A\mathbf{x} = \mathbf{0}$ if and only if $B\mathbf{x} = \mathbf{0}$.

T.6. Let A be $m \times n$ and B be $n \times k$ matrices.

(a) Show that

$$\text{rank}\,(AB) \leq \min\{\text{rank}\,A, \text{rank}\,B\}.$$

(b) Find A and B such that

$$\text{rank}\,(AB) < \min\{\text{rank}\,A, \text{rank}\,B\}.$$

(c) If $k = n$ and B is nonsingular, show that

$$\text{rank}\,(AB) = \text{rank}\,A.$$

(d) If $m = n$ and A is nonsingular, show that

$$\text{rank}\,(AB) = \text{rank}\,B.$$

(e) For nonsingular matrices P and Q, what is rank (PAQ)?

T.7. Let $S = \{\mathbf{v}_1, \mathbf{v}_2, \dots, \mathbf{v}_n\}$ be an orthonormal basis for R^n and let $\{a_1, \dots, a_n\}$ be any set of scalars none of which is zero. Show that $T = \{a_1\mathbf{v}_1, a_2\mathbf{v}_2, \dots, a_n\mathbf{v}_n\}$ is an orthogonal basis for V. How should the scalars $a_1, \dots, a_n$ be chosen so that T is an orthonormal basis for R^n?

T.8. Let B be an $m \times n$ matrix with orthonormal columns $\mathbf{b}_1, \mathbf{b}_2, \dots, \mathbf{b}_n$.

(a) Show that $m \geq n$.

(b) Show that $B^T B = I_n$.

Chapter Test

1. Consider the set W of all vectors in R^3 of the form (a, b, c), where $a + b + c = 0$. Is W a subspace of R^3?

2. Find a basis for the solution space of the homogeneous system

$$\begin{aligned} x_1 + 3x_2 + 3x_3 - x_4 + 2x_5 &= 0 \\ x_1 + 2x_2 + 2x_3 - 2x_4 + 2x_5 &= 0 \\ x_1 + x_2 + x_3 - 3x_4 + 2x_5 &= 0. \end{aligned}$$

3. Does the set of vectors $\{(1, -1, 1), (1, -3, 1), (1, -2, 2)\}$ form a basis for R^3?

4. For what value(s) of λ is the set of vectors $\{(\lambda^2 - 5, 1, 0), (2, -2, 3), (2, 3, -3)\}$ linearly dependent?

5. Use the Gram–Schmidt process to find an orthonormal basis for the subspace of R^4 with basis $\{(1, 0, -1, 0), (1, -1, 0, 0), (3, 1, 0, 0)\}$.

6. Answer each of the following as true or false. Justify your answer.

(a) All vectors of the form $(a, 0, -a)$ form a subspace of R^3.

(b) In R^n, $\|c\mathbf{x}\| = c\|\mathbf{x}\|$.

(c) Every set of vectors in R^3 containing two vectors is linearly independent.

(d) The solution space of the homogeneous system $A\mathbf{x} = \mathbf{0}$ is spanned by the columns of A.

(e) If the columns of an $n \times n$ matrix form a basis for R^n, so do the rows.

(f) If A is an 8×8 matrix such that the homogeneous system $A\mathbf{x} = \mathbf{0}$ has only the trivial solution, then rank $A < 8$.

(g) Every orthonormal set of five vectors in R^5 is a basis for R^5.

(h) Every linearly independent set of vectors in R^3 contains three vectors.

(i) If A is an $n \times n$ symmetric matrix, then rank $A = n$.

(j) Every set of vectors spanning R^3 contains at least three vectors.

CHAPTER 7

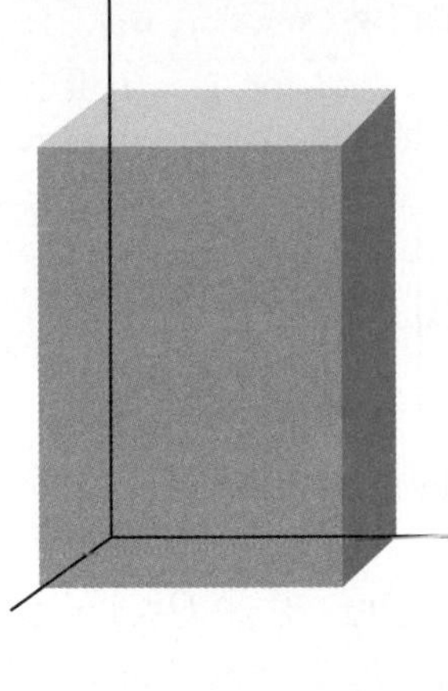

APPLICATIONS OF REAL VECTOR SPACES (OPTIONAL)

7.1 QR-FACTORIZATION

Prerequisite. Section 6.8, Orthonormal Bases in R^n.

In Section 1.8 we discussed the LU-factorization of a matrix and showed how it leads to a very efficient method for solving a linear system. We now discuss another factorization of a matrix A, called the **QR-factorization** of A. This type of factorization is widely used in computer codes to find the eigenvalues of a matrix (see Chapter 8), to solve linear systems, and to find least squares approximations (see Section 7.2 for a discussion of least squares).

THEOREM 7.1 *If A is an $m \times n$ matrix with linearly independent columns, then A can be factored as $A = QR$, where Q is an $m \times n$ matrix whose columns form an orthonormal basis for the column space of A and R is an $n \times n$ nonsingular upper triangular matrix.*

Proof Let $\mathbf{u}_1, \mathbf{u}_2, \dots, \mathbf{u}_n$ denote the linearly independent columns of A, which form a basis for the column space of A. By using the Gram–Schmidt process (see Theorem 6.18 in Section 6.8), we can obtain an orthonormal basis $\mathbf{w}_1, \mathbf{w}_2, \dots, \mathbf{w}_n$ for the column space of A. Recall how this orthonormal basis was obtained. We first constructed an orthogonal basis $\mathbf{v}_1, \mathbf{v}_2, \dots, \mathbf{v}_n$ as follows: $\mathbf{v}_1 = \mathbf{u}_1$ and then for $i = 2, 3, \dots, n$ we have

$$\mathbf{v}_i = \mathbf{u}_i - \frac{\mathbf{u}_i \cdot \mathbf{v}_1}{\mathbf{v}_1 \cdot \mathbf{v}_1}\mathbf{v}_1 - \frac{\mathbf{u}_i \cdot \mathbf{v}_2}{\mathbf{v}_2 \cdot \mathbf{v}_2}\mathbf{v}_2 - \cdots - \frac{\mathbf{u}_i \cdot \mathbf{v}_{i-1}}{\mathbf{v}_{i-1} \cdot \mathbf{v}_{i-1}}\mathbf{v}_{i-1}. \tag{1}$$

Finally, $\mathbf{w}_i = \dfrac{1}{\|\mathbf{v}_i\|}\mathbf{v}_i$ for $i = 1, 2, 3, \dots, n$. Now each of the vectors $\mathbf{u}_i$ can be written as a linear combination of the $\mathbf{w}$-vectors:

$$\begin{aligned}
\mathbf{u}_1 &= r_{11}\mathbf{w}_1 + r_{21}\mathbf{w}_2 + \cdots + r_{n1}\mathbf{w}_n \\
\mathbf{u}_2 &= r_{12}\mathbf{w}_1 + r_{22}\mathbf{w}_2 + \cdots + r_{n2}\mathbf{w}_n \\
&\ \ \vdots \\
\mathbf{u}_n &= r_{1n}\mathbf{w}_1 + r_{2n}\mathbf{w}_2 + \cdots + r_{nn}\mathbf{w}_n.
\end{aligned} \tag{2}$$

From Theorem 6.17 we have

$$r_{ji} = \mathbf{u}_i \cdot \mathbf{w}_j.$$

Moreover, from Equation (1), we see that $\mathbf{u}_i$ lies in

$$\text{span}\,\{\mathbf{v}_1, \mathbf{v}_2, \ldots, \mathbf{v}_i\} = \text{span}\,\{\mathbf{w}_1, \mathbf{w}_2, \ldots, \mathbf{w}_i\}.$$

Since $\mathbf{w}_j$ is orthogonal to span $\{\mathbf{w}_1, \mathbf{w}_2, \ldots, \mathbf{w}_i\}$ for $j > i$, it is orthogonal to $\mathbf{u}_i$. Hence $r_{ji} = 0$ for $j > i$. Let Q be the matrix whose columns are $\mathbf{w}_1, \mathbf{w}_2, \ldots, \mathbf{w}_j$. Let

$$\mathbf{r}_j = \begin{bmatrix} r_{1j} \\ r_{2j} \\ \vdots \\ r_{nj} \end{bmatrix}.$$

Then the equations in (2) can be written in matrix form (see Exercise T.9 in Section 1.3) as

$$A = \begin{bmatrix} \mathbf{u}_1 & \mathbf{u}_2 & \cdots & \mathbf{u}_n \end{bmatrix} = \begin{bmatrix} Q\mathbf{r}_1 & Q\mathbf{r}_2 & \cdots & Q\mathbf{r}_n \end{bmatrix} = QR,$$

where R is the matrix whose columns are $\mathbf{r}_1, \mathbf{r}_2, \ldots, \mathbf{r}_n$. Thus

$$R = \begin{bmatrix} r_{11} & r_{12} & \cdots & r_{1n} \\ 0 & r_{22} & \cdots & r_{2n} \\ 0 & 0 & \cdots & \\ \vdots & \vdots & & \vdots \\ 0 & 0 & \cdots & r_{nn} \end{bmatrix}.$$

We now show that R is nonsingular. Let $\mathbf{x}$ be a solution to the linear system $R\mathbf{x} = \mathbf{0}$. Multiplying this equation by Q on the left, we have

$$Q(R\mathbf{x}) = (QR)\mathbf{x} = A\mathbf{x} = Q\mathbf{0} = \mathbf{0}.$$

As we know from Equation (3) in Section 1.3, the homogeneous system $A\mathbf{x} = \mathbf{0}$ can be written as

$$x_1\mathbf{u}_1 + x_2\mathbf{u}_2 + \cdots + x_n\mathbf{u}_n = \mathbf{0},$$

where $x_1, x_2, \ldots, x_n$ are the components of the vector $\mathbf{x}$. Since the columns of A are linearly independent,

$$x_1 = x_2 = \cdots = x_n = 0,$$

so $\mathbf{x}$ must be the zero vector. Then Theorem 1.13 implies that R is nonsingular. In Exercise T.1 we ask you to show that the diagonal entries r_{ii} of R are nonzero by first expressing $\mathbf{u}_i$ as a linear combination of $\mathbf{v}_1, \mathbf{v}_2, \ldots, \mathbf{v}_i$ and then computing $r_{ii} = \mathbf{u}_i \cdot \mathbf{w}_i$. This provides another proof of the nonsingularity of R. ■

The procedure for finding the QR-factorization of an $m \times n$ matrix A with linearly independent columns is as follows.

Step 1. Let the columns of A be denoted by $\mathbf{u}_1, \mathbf{u}_2, \ldots, \mathbf{u}_n$ and let W be the subspace of R^n with these vectors as basis.

Step 2. Transform the basis $\{\mathbf{u}_1, \mathbf{u}_2, \ldots, \mathbf{u}_n\}$ for W by using the Gram–Schmidt process to an orthonormal basis $\{\mathbf{w}_1, \mathbf{w}_2, \ldots, \mathbf{w}_n\}$. Let

$$Q = \begin{bmatrix} \mathbf{w}_1 & \mathbf{w}_2 & \cdots & \mathbf{w}_n \end{bmatrix}$$

be the matrix whose columns are $\mathbf{w}_1, \mathbf{w}_2, \ldots, \mathbf{w}_n$.

Step 3. Compute $R = \begin{bmatrix} r_{ij} \end{bmatrix}$, where

$$r_{ji} = \mathbf{u}_i \cdot \mathbf{w}_j.$$

EXAMPLE 1 Find the QR-factorization of

$$A = \begin{bmatrix} 1 & -1 & -1 \\ 1 & 0 & 0 \\ 1 & -1 & 0 \\ 0 & 1 & -1 \end{bmatrix}.$$

Solution Letting the columns of A be denoted by $\mathbf{u}_1$, $\mathbf{u}_2$, and $\mathbf{u}_3$, we let W be the subspace of R^4 with $\mathbf{u}_1, \mathbf{u}_2, \mathbf{u}_3$ as a basis. Applying the Gram–Schmidt process to this basis, starting with $\mathbf{u}_1$, we find (verify) the following orthonormal basis for the column space of A:

$$\mathbf{w}_1 = \begin{bmatrix} \frac{1}{\sqrt{3}} \\ \frac{1}{\sqrt{3}} \\ \frac{1}{\sqrt{3}} \\ 0 \end{bmatrix}, \quad \mathbf{w}_2 = \begin{bmatrix} -\frac{1}{\sqrt{15}} \\ \frac{2}{\sqrt{15}} \\ -\frac{1}{\sqrt{15}} \\ \frac{3}{\sqrt{15}} \end{bmatrix}, \quad \mathbf{w}_3 = \begin{bmatrix} -\frac{4}{\sqrt{35}} \\ \frac{3}{\sqrt{35}} \\ \frac{1}{\sqrt{35}} \\ -\frac{3}{\sqrt{35}} \end{bmatrix}.$$

Then

$$Q = \begin{bmatrix} \frac{1}{\sqrt{3}} & -\frac{1}{\sqrt{15}} & -\frac{4}{\sqrt{35}} \\ \frac{1}{\sqrt{3}} & \frac{2}{\sqrt{15}} & \frac{3}{\sqrt{35}} \\ \frac{1}{\sqrt{3}} & -\frac{1}{\sqrt{15}} & \frac{1}{\sqrt{35}} \\ 0 & \frac{3}{\sqrt{15}} & -\frac{3}{\sqrt{35}} \end{bmatrix} \approx \begin{bmatrix} 0.5774 & -0.2582 & -0.6761 \\ 0.5774 & 0.5164 & 0.5071 \\ 0.5774 & -0.2582 & 0.1690 \\ 0 & 0.7746 & -0.5071 \end{bmatrix}$$

and

$$R = \begin{bmatrix} r_{11} & r_{12} & r_{13} \\ 0 & r_{22} & r_{23} \\ 0 & 0 & r_{33} \end{bmatrix},$$

where $r_{ji} = \mathbf{u}_i \cdot \mathbf{w}_j$. Thus

$$R = \begin{bmatrix} \frac{3}{\sqrt{3}} & -\frac{2}{\sqrt{3}} & -\frac{1}{\sqrt{3}} \\ 0 & \frac{5}{\sqrt{15}} & -\frac{2}{\sqrt{15}} \\ 0 & 0 & \frac{7}{\sqrt{35}} \end{bmatrix} \approx \begin{bmatrix} 1.7321 & -1.1547 & -0.5774 \\ 0 & 1.2910 & -0.5164 \\ 0 & 0 & 1.1832 \end{bmatrix}.$$

As you can verify, $A = QR$. ■

Remark State-of-the-art computer implementations (such as in MATLAB) yield an alternative QR-factorization of an $m \times n$ matrix A as the product of an $m \times m$ matrix Q and an $m \times n$ matrix $R = [r_{ij}]$, where $r_{ij} = 0$ if $i > j$. Thus, if A is 5×3, then

$$R = \begin{bmatrix} * & * & * \\ 0 & * & * \\ 0 & 0 & * \\ 0 & 0 & 0 \\ 0 & 0 & 0 \end{bmatrix}.$$

Key Terms

QR-factorization

7.1 Exercises

In Exercises 1 through 6, compute the QR-factorization of A.

1. $A = \begin{bmatrix} 1 & 2 \\ -1 & 3 \end{bmatrix}$

2. $A = \begin{bmatrix} 1 & 2 \\ -1 & -2 \\ 1 & 1 \end{bmatrix}$

3. $A = \begin{bmatrix} 1 & 0 & -1 \\ 2 & -3 & 3 \\ -1 & 2 & 4 \end{bmatrix}$

4. $A = \begin{bmatrix} 2 & -1 \\ -1 & 3 \\ 0 & 1 \end{bmatrix}$

5. $A = \begin{bmatrix} 1 & 0 & 2 \\ -1 & 2 & 0 \\ -1 & -2 & 2 \end{bmatrix}$

6. $A = \begin{bmatrix} 2 & -1 & 1 \\ 1 & 2 & -2 \\ 0 & 1 & -2 \end{bmatrix}$

Theoretical Exercises

T.1. In the proof of Theorem 7.1, show that r_{ii} is nonzero by first expressing $\mathbf{u}_i$ as a linear combination of $\mathbf{v}_1, \mathbf{v}_2, \ldots, \mathbf{v}_i$ and then computing $r_{ii} = \mathbf{u}_i \cdot \mathbf{w}_i$.

T.2. Show that every nonsingular matrix has a QR-factorization.

7.2 LEAST SQUARES

Prerequisites. Section 1.6, Solutions of Linear Systems of Equations, Section 1.7, The Inverse of a Matrix, Section 4.2, n-Vectors, Section 6.9, Orthogonal Complements.

From Chapter 1 we recall that an $m \times n$ linear system $A\mathbf{x} = \mathbf{b}$ is inconsistent if it has no solution. In the proof of Theorem 6.14 in Section 6.6 we show that $A\mathbf{x} = \mathbf{b}$ is consistent if and only if $\mathbf{b}$ belongs to the column space of A. Equivalently, $A\mathbf{x} = \mathbf{b}$ is inconsistent if and only if $\mathbf{b}$ is *not* in the column space of A. Inconsistent systems do indeed arise in many situations and we must determine how to deal with them. Our approach is to change the problem so that we do not require that the matrix equation $A\mathbf{x} = \mathbf{b}$ be satisfied. Instead, we seek a vector $\widehat{\mathbf{x}}$ in R^n such that $A\widehat{\mathbf{x}}$ is as close to $\mathbf{b}$ as possible. If W is the column space of A, then from Theorem 6.23 in Section 6.9, it follows that the vector in W that is closest to $\mathbf{b}$ is $\text{proj}_W\mathbf{b}$. That is, $\|\mathbf{b} - \mathbf{w}\|$, for $\mathbf{w}$ in W, is minimized when $\mathbf{w} = \text{proj}_W\mathbf{b}$. Thus, if we find $\widehat{\mathbf{x}}$ such that $A\widehat{\mathbf{x}} = \text{proj}_W\mathbf{b}$, then we are assured that $\|\mathbf{b} - A\widehat{\mathbf{x}}\|$ will be as small as possible. As shown in the proof of Theorem 6.23, $\mathbf{b} - \text{proj}_W\mathbf{b} = \mathbf{b} - A\widehat{\mathbf{x}}$ is orthogonal to every vector in W. (See Figure 7.1.) It then follows that $\mathbf{b} - A\widehat{\mathbf{x}}$ is orthogonal to each column of A. In terms of a matrix equation, we have

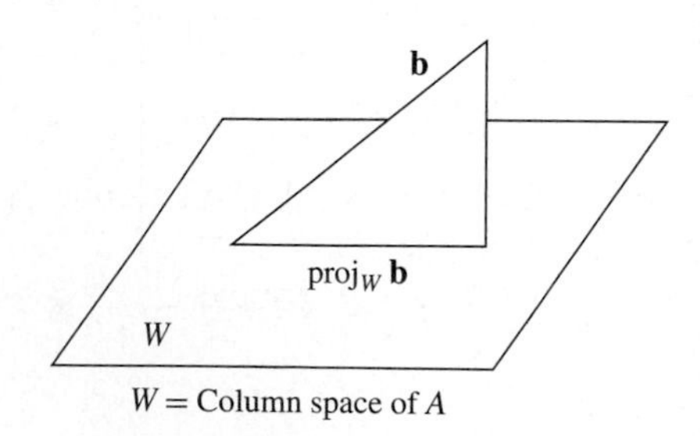

Figure 7.1 ▲

$$A^T(A\widehat{\mathbf{x}} - \mathbf{b}) = \mathbf{0}$$

or, equivalently,

$$A^TA\widehat{\mathbf{x}} = A^T\mathbf{b}.$$

Thus, $\widehat{\mathbf{x}}$ is a solution to

$$A^TA\mathbf{x} = A^T\mathbf{b}. \tag{1}$$

Any solution to (1) is called a **least squares solution** to the linear system $A\mathbf{x} = \mathbf{b}$. (**Warning:** In general, $A\widehat{\mathbf{x}} \neq \mathbf{b}$.) Equation (1) is called the **normal system** of equations associated with $A\mathbf{x} = \mathbf{b}$, or just the normal system. Observe that if $A\mathbf{x} = \mathbf{b}$ is consistent, then a solution to this system is a least squares solution. In particular, if A is nonsingular, a least squares solution to $A\mathbf{x} = \mathbf{b}$ is just the usual solution $\mathbf{x} = A^{-1}\mathbf{b}$ (see Exercise T.2).

To compute a least squares solution $\widehat{\mathbf{x}}$ to the linear system $A\mathbf{x} = \mathbf{b}$, we can proceed as follows. Let $\{\mathbf{w}_1, \mathbf{w}_2, \ldots, \mathbf{w}_m\}$ be an orthonormal basis for the column space W of A. Then Equation (1) of Section 6.9 yields

$$\text{proj}_W\mathbf{b} = (\mathbf{b} \cdot \mathbf{w}_1)\mathbf{w}_1 + (\mathbf{b} \cdot \mathbf{w}_2)\mathbf{w}_2 + \cdots + (\mathbf{b} \cdot \mathbf{w}_m)\mathbf{w}_m.$$

Recall that one way to find an orthonormal basis for W consisted of first finding a basis for W by transforming A^T to reduced row echelon form and then taking the transposes of the nonzero rows as a basis for W. Next apply the Gram–Schmidt process to this basis to find an orthonormal basis for W. The procedure just outlined is theoretically valid when we assume that exact arithmetic is used. However, even small numerical errors, due to, say, roundoff, may adversely affect the results. Thus more sophisticated algorithms are required for numerical applications. [See D. Hill, *Experiments in Computational Matrix Algebra* (New York: Random House, 1988), distributed by McGraw-Hill.] We shall not pursue the general case here, but turn our attention to an important special case.

Remark An alternative method for finding $\text{proj}_W\mathbf{b}$ is as follows. Solve Equation (1) for $\widehat{\mathbf{x}}$, the least squares solution to the linear system $A\mathbf{x} = \mathbf{b}$. Then $A\widehat{\mathbf{x}}$ will be $\text{proj}_W\mathbf{b}$.

THEOREM 7.2 *If A is an $m \times n$ matrix with* rank $A = n$, *then A^TA is nonsingular and the linear system $A\mathbf{x} = \mathbf{b}$ has a unique least squares solution given by $\widehat{\mathbf{x}} = (A^TA)^{-1}A^T\mathbf{b}$. That is, the normal system of equations has a unique solution.*

Proof Since A has rank n, the columns of A are linearly independent. The matrix A^TA is nonsingular provided the linear system $A^TA\mathbf{x} = \mathbf{0}$ has only the zero solution. Multiplying both sides of $A^TA\mathbf{x} = \mathbf{0}$ by $\mathbf{x}^T$ gives

$$\mathbf{0} = \mathbf{x}^TA^TA\mathbf{x} = (A\mathbf{x})^T(A\mathbf{x}) = (A\mathbf{x}) \cdot (A\mathbf{x}).$$

It then follows from Theorem 4.3 in Section 4.2 that $A\mathbf{x} = \mathbf{0}$. This implies that we have a linear combination of the linearly independent columns of A that is zero. Therefore, $\mathbf{x} = \mathbf{0}$. Thus A^TA is nonsingular and Equation (1) has the unique solution $\widehat{\mathbf{x}} = (A^TA)^{-1}A^T\mathbf{b}$. ■

The procedure for finding the least squares solution $\widehat{\mathbf{x}}$ to $A\mathbf{x} = \mathbf{b}$ is as follows.

Step 1. Form A^TA and $A^T\mathbf{b}$.

Step 2. Solve the normal system

$$A^TA\mathbf{x} = A^T\mathbf{b}$$

for $\mathbf{x}$ by Gauss–Jordan reduction.

EXAMPLE 1 Determine a least squares solution to $A\mathbf{x} = \mathbf{b}$, where

$$A = \begin{bmatrix} 1 & 2 & -1 & 3 \\ 2 & 1 & 1 & 2 \\ -2 & 3 & 4 & 1 \\ 4 & 2 & 1 & 0 \\ 0 & 2 & 1 & 3 \\ 1 & -1 & 2 & 0 \end{bmatrix}, \qquad \mathbf{b} = \begin{bmatrix} 1 \\ 5 \\ -2 \\ 1 \\ 3 \\ 5 \end{bmatrix}.$$

Solution Using row reduction, we can show that rank $A = 4$ (verify). We now form the normal system $A^T A\mathbf{x} = A^T\mathbf{b}$ (verify),

$$\begin{bmatrix} 26 & 5 & -1 & 5 \\ 5 & 23 & 13 & 17 \\ -1 & 13 & 24 & 6 \\ 5 & 17 & 6 & 23 \end{bmatrix} \mathbf{x} = \begin{bmatrix} 24 \\ 4 \\ 10 \\ 20 \end{bmatrix}.$$

By Theorem 7.2, the normal system has a unique solution. Applying Gauss–Jordan reduction, we have the unique least squares solution (verify)

$$\widehat{\mathbf{x}} \approx \begin{bmatrix} 0.9990 \\ -2.0643 \\ 1.1039 \\ 1.8902 \end{bmatrix}.$$

If W is the column space of A, then (verify)

$$\operatorname{proj}_W \mathbf{b} = A\widehat{\mathbf{x}} \approx \begin{bmatrix} 1.4371 \\ 4.8181 \\ -1.8852 \\ 0.9713 \\ 2.6459 \\ 5.2712 \end{bmatrix},$$

which is the vector in W such that $\|\mathbf{b} - \mathbf{w}\|$, $\mathbf{w}$ in W, is minimized. That is,

$$\min_{\mathbf{w} \text{ in } W} \|\mathbf{b} - \mathbf{w}\| = \|\mathbf{b} - A\widehat{\mathbf{x}}\|. \qquad \blacksquare$$

When A is an $m \times n$ matrix whose rank is n, it is computationally more efficient to solve Equation (1) by Gauss–Jordan reduction than to determine $(A^T A)^{-1}$ and then form the product $(A^T A)^{-1} A^T\mathbf{b}$. An even better approach is to use the QR-factorization of A, which is discussed next.

LEAST SQUARES USING QR-FACTORIZATION

Let $A = QR$ be a QR-factorization of A. Substituting this expression for A into Equation (1), we obtain

$$(QR)^T (QR)\mathbf{x} = (QR)^T \mathbf{b}$$

or

$$R^T (Q^T Q) R\mathbf{x} = R^T Q^T \mathbf{b}.$$

Since the columns of Q form an orthonormal set, we have $Q^T Q = I_m$, so

$$R^T R\mathbf{x} = R^T Q^T \mathbf{b}.$$

Since R^T is a nonsingular matrix, we obtain

$$R\mathbf{x} = Q^T\mathbf{b}.$$

Using the fact that R is upper triangular, we easily solve this linear system by back substitution to obtain $\widehat{\mathbf{x}}$.

EXAMPLE 2 Solve Example 1 using the QR-factorization of A.

Solution We use the Gram–Schmidt process, carrying out all computations in MATLAB. We find that Q is given by (verify)

$$Q = \begin{bmatrix} -0.1961 & -0.3851 & 0.5099 & 0.3409 \\ -0.3922 & -0.1311 & -0.1768 & 0.4244 \\ 0.3922 & -0.7210 & -0.4733 & -0.2177 \\ -0.7845 & -0.2622 & -0.1041 & -0.5076 \\ 0 & -0.4260 & 0.0492 & 0.4839 \\ -0.1961 & 0.2540 & -0.6867 & 0.4055 \end{bmatrix}$$

and R is given by (verify)

$$R = \begin{bmatrix} -5.0990 & -0.9806 & 0.1961 & -0.9806 \\ 0 & -4.6945 & -2.8102 & -3.4164 \\ 0 & 0 & -4.0081 & 0.8504 \\ 0 & 0 & 0 & 3.1054 \end{bmatrix}.$$

Then

$$Q^T\mathbf{b} = \begin{bmatrix} 4.7068 \\ -0.1311 \\ 2.8172 \\ 5.8699 \end{bmatrix}.$$

Finally, solving

$$R\mathbf{x} = Q^T\mathbf{b}$$

we find (verify)

$$\mathbf{x} = \begin{bmatrix} 0.9990 \\ -2.0643 \\ 1.1039 \\ 1.8902 \end{bmatrix},$$

which agrees with $\widehat{\mathbf{x}}$ as obtained in Example 1. ■

LEAST SQUARES LINE FIT

The problem of gathering and analyzing data is one that is present in many facets of human activity. Frequently, we measure a value of y for a given value of x and then plot the points (x, y) on graph paper. From the resulting graph we try to develop a relationship between the variables x and y that can then be used to predict new values of y for given values of x.

EXAMPLE 3 In the manufacture of product XXX, the amount of the compound beta present in the product is controlled by the amount of the ingredient alpha used in the process. In manufacturing a gallon of XXX, the amount of alpha used and the amount of beta present are recorded. The following data were obtained:

Alpha Used (ounces/gallon)	3	4	5	6	7	8	9	10	11	12
Beta Present (ounces/gallon)	4.5	5.5	5.7	6.6	7.0	7.7	8.5	8.7	9.5	9.7

The points in this table are plotted in Figure 7.2. ■

Figure 7.2 ►

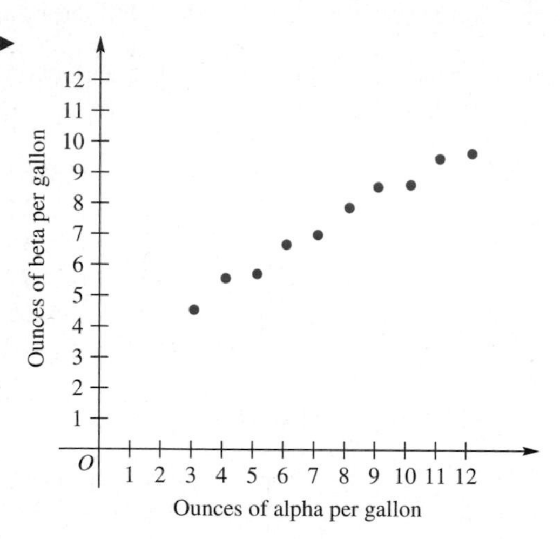

Suppose it is assumed that the relationship between the amount of alpha used and the amount of beta present is given by a linear equation so that the graph is a straight line. It would thus not be reasonable to connect the plotted points by drawing a curve that passes through every point. Moreover, the data are of a *probabilistic* nature. That is, they are not *deterministic* in the sense that if we repeated the experiment, we would expect slightly different values of beta for the same values of alpha, since all measurements are subject to experimental error. Thus the plotted points do not lie exactly on a straight line. We shall now use the method of least squares to obtain the straight line that "best fits" the given data. This line is called the **least squares line**.

Suppose that we are given n points $(x_1, y_1), (x_2, y_2), \ldots, (x_n, y_n)$, where at least two of the x_i are distinct. We are interested in finding the least squares line

$$y = b_1 x + b_0 \tag{2}$$

that "best fits the data." If the points $(x_1, y_1), (x_2, y_2), \ldots, (x_n, y_n)$ were exactly on the least squares line, we would have

$$y_i = b_1 x_i + b_0. \tag{3}$$

Since some of these points may not lie exactly on the line, we have

$$y_i = b_1 x_i + b_0 + d_i, \qquad i = 1, 2, \ldots, n, \tag{4}$$

where d_i is the vertical deviation from the point (x_i, y_i) to the (desired) least squares line. The quantity d_i can be positive, negative, or zero. In Figure 7.3 we show five data points $(x_1, y_1), (x_2, y_2), (x_3, y_3), (x_4, y_4), (x_5, y_5)$ and their corresponding deviations $d_1, d_2, \ldots, d_5$ from the least squares line.

Figure 7.3 ►

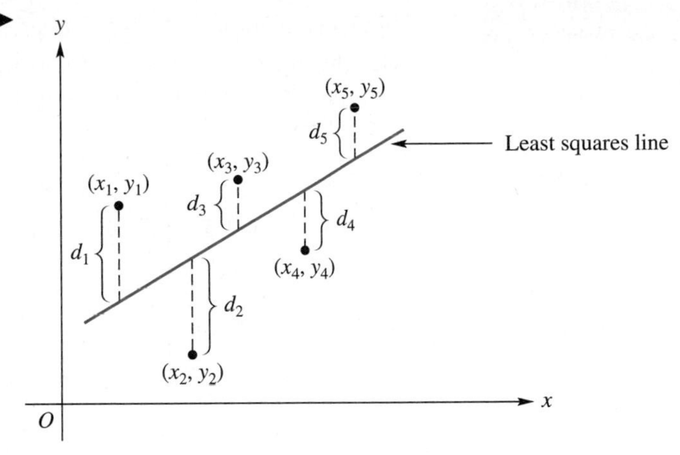

If we let

$$\mathbf{b} = \begin{bmatrix} y_1 \\ y_2 \\ \vdots \\ y_n \end{bmatrix}, \quad A = \begin{bmatrix} x_1 & 1 \\ x_2 & 1 \\ \vdots & \vdots \\ x_n & 1 \end{bmatrix}, \quad \mathbf{x} = \begin{bmatrix} b_1 \\ b_0 \end{bmatrix}, \quad \text{and} \quad \mathbf{d} = \begin{bmatrix} d_1 \\ d_2 \\ \vdots \\ d_n \end{bmatrix},$$

then we may write the n equations in (4) as a single matrix equation

$$\mathbf{b} = A\mathbf{x} + \mathbf{d}.$$

Since the linear system $A\mathbf{x} = \mathbf{b}$ is usually inconsistent, we find a least squares solution $\widehat{\mathbf{x}}$ to $A\mathbf{x} = \mathbf{b}$. Recall that at least two of the x_i are distinct, so rank $A = 2$. Theorem 7.2 then implies that $A\mathbf{x} = \mathbf{b}$ has a unique least squares solution given by $\widehat{\mathbf{x}} = (A^T A)^{-1} A^T \mathbf{b}$. We are then assured that $A\widehat{\mathbf{x}}$ will be as close as possible to $\mathbf{b}$. Since $\mathbf{d} = \mathbf{b} - A\widehat{\mathbf{x}}$, the magnitude of $\mathbf{d}$ will be as small as possible (the magnitude of a vector is discussed in Section 4.2). For each x_i, $i = 1, 2, \ldots, n$, let $\widehat{y_i} = b_1 x_i + b$. Then $(x_i, \widehat{y_i})$ are the points on the least squares line. Using material from Section 4.2, it follows that $\|\mathbf{d}\|$ is minimized. Equivalently,

$$\|\mathbf{d}\|^2 = d_1^2 + d_2^2 + \cdots + d_n^2$$

will be minimized.

The procedure for finding the least squares line $y = b_1 x + b_0$ for the data $(x_1, y_1), (x_2, y_2), \ldots, (x_n, y_n)$, where at least two of the x_i are different, is as follows.

Step 1. Let

$$\mathbf{b} = \begin{bmatrix} y_1 \\ y_2 \\ \vdots \\ y_n \end{bmatrix}, \quad A = \begin{bmatrix} x_1 & 1 \\ x_2 & 1 \\ \vdots & \vdots \\ x_n & 1 \end{bmatrix}, \quad \text{and} \quad \mathbf{x} = \begin{bmatrix} b_1 \\ b_0 \end{bmatrix}.$$

Step 2. Solve the normal system

$$A^T A\mathbf{x} = A^T \mathbf{b}$$

for $\mathbf{x}$ by Gauss–Jordan reduction.

EXAMPLE 4 (a) Find an equation of the least squares line for the data in Example 3.

(b) Use the equation obtained in part (a) to predict the number of ounces of beta present in a gallon of product XXX if 30 ounces of alpha are used per gallon.

Solution (a) We have

$$\mathbf{b} = \begin{bmatrix} 4.5 \\ 5.5 \\ 5.7 \\ 6.6 \\ 7.0 \\ 7.7 \\ 8.5 \\ 8.7 \\ 9.5 \\ 9.7 \end{bmatrix}, \qquad A = \begin{bmatrix} 3 & 1 \\ 4 & 1 \\ 5 & 1 \\ 6 & 1 \\ 7 & 1 \\ 8 & 1 \\ 9 & 1 \\ 10 & 1 \\ 11 & 1 \\ 12 & 1 \end{bmatrix}, \qquad \mathbf{x} = \begin{bmatrix} b_1 \\ b_0 \end{bmatrix}.$$

Then

$$A^T A = \begin{bmatrix} 645 & 75 \\ 75 & 10 \end{bmatrix} \quad \text{and} \quad A^T \mathbf{b} = \begin{bmatrix} 598.6 \\ 73.4 \end{bmatrix}.$$

Solving the normal system $A^T A\mathbf{x} = A^T\mathbf{b}$ by Gauss–Jordan reduction we obtain (verify)

$$\mathbf{x} = \begin{bmatrix} b_1 \\ b_0 \end{bmatrix} = \begin{bmatrix} 0.583 \\ 2.967 \end{bmatrix}.$$

Then

$$b_1 = 0.583 \quad \text{and} \quad b_0 = 2.967.$$

Hence, an equation for the least squares line, shown in Figure 7.4, is

$$y = 0.583x + 2.967, \tag{5}$$

where y is the amount of beta present and x is the amount of alpha used.

(b) If $x = 30$, then substituting in (5), we obtain

$$y = 20.457.$$

Thus there would be 20.457 ounces of beta present in a gallon of XXX. ■

Figure 7.4 ►

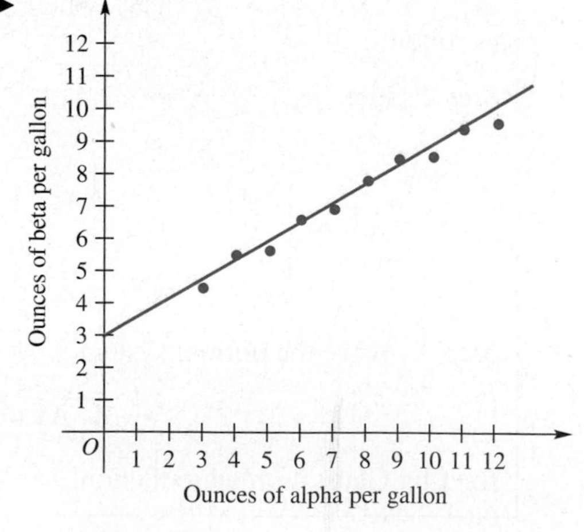

LEAST SQUARES POLYNOMIAL FIT

The method presented for obtaining the least squares line fit for a given set of points can be easily generalized to the problem of finding a polynomial of prescribed degree that "best fits" the given data.

Thus, suppose we are given n data points

$$(x_1, y_1), (x_2, y_2), \ldots, (x_n, y_n),$$

where at least $m+1$ of the x_i are distinct, and wish to construct a mathematical model of the form

$$y = a_m x^m + a_{m-1} x^{m-1} + \cdots + a_1 x + a_0, \qquad m \leq n - 1,$$

that "best fits" these data. As in the least squares line fit, since some of the n data points do not lie exactly on the graph of the least squares polynomial, we have

$$y_i = a_m x_i^m + a_{m-1} x_i^{m-1} + \cdots + a_1 x_i + a_0 + d_i, \qquad i = 1, 2, \ldots, n. \quad (6)$$

Letting

$$\mathbf{b} = \begin{bmatrix} y_1 \\ y_2 \\ \vdots \\ y_n \end{bmatrix}, \qquad A = \begin{bmatrix} x_1^m & x_1^{m-1} & \cdots & x_1^2 & x_1 & 1 \\ x_2^m & x_2^{m-1} & \cdots & x_2^2 & x_2 & 1 \\ \vdots & \vdots & & \vdots & \vdots & \vdots \\ x_n^m & x_n^{m-1} & \cdots & x_n^2 & x_n & 1 \end{bmatrix}, \qquad \mathbf{x} = \begin{bmatrix} a_m \\ a_{m-1} \\ \vdots \\ a_1 \\ a_0 \end{bmatrix},$$

and

$$\mathbf{d} = \begin{bmatrix} d_1 \\ d_2 \\ \vdots \\ d_n \end{bmatrix},$$

we may write the n equations in (6) as the matrix equation

$$\mathbf{b} = A\mathbf{x} + \mathbf{d}.$$

As in the case of the least squares line fit, a solution $\widehat{\mathbf{x}}$ to the normal system

$$A^T A\mathbf{x} = A^T\mathbf{b}$$

is a least squares solution to $A\mathbf{x} = \mathbf{b}$. With this solution we will be assured that $\|\mathbf{d}\| = \|\mathbf{b} - A\widehat{\mathbf{x}}\|$ is minimized.

The procedure for finding the least squares polynomial

$$y = a_m x^m + a_{m-1}x^{m-1} + \cdots + a_1 x + a_0$$

that best fits the data $(x_1, y_1), (x_2, y_2), \ldots, (x_n, y_n)$, where $m \leq n - 1$ and at least $m + 1$ of the x_i are distinct, is as follows.

Step 1. Form

$$\mathbf{b} = \begin{bmatrix} y_1 \\ y_2 \\ \vdots \\ y_n \end{bmatrix}, \quad A = \begin{bmatrix} x_1^m & x_1^{m-1} & \cdots & x_1^2 & x_1 & 1 \\ x_2^m & x_2^{m-1} & \cdots & x_2^2 & x_2 & 1 \\ \vdots & \vdots & & \vdots & \vdots & \vdots \\ x_n^m & x_n^{m-1} & \cdots & x_n^2 & x_n & 1 \end{bmatrix}, \quad \text{and} \quad \mathbf{x} = \begin{bmatrix} a_m \\ a_{m-1} \\ \vdots \\ a_1 \\ a_0 \end{bmatrix}.$$

Step 2. Solve the normal system

$$A^T A\mathbf{x} = A^T\mathbf{b}$$

for $\mathbf{x}$ by Gauss–Jordan reduction.

EXAMPLE 5

The following data show atmospheric pollutants y_i (relative to an EPA standard) at half-hour intervals t_i.

t_i	1	1.5	2	2.5	3	3.5	4	4.5	5
y_i	−0.15	0.24	0.68	1.04	1.21	1.15	0.86	0.41	−0.08

A plot of these data points, as shown in Figure 7.5, suggests that a quadratic polynomial

$$y = a_2 t^2 + a_1 t + a_0$$

may produce a good model for these data.

Figure 7.5 ►

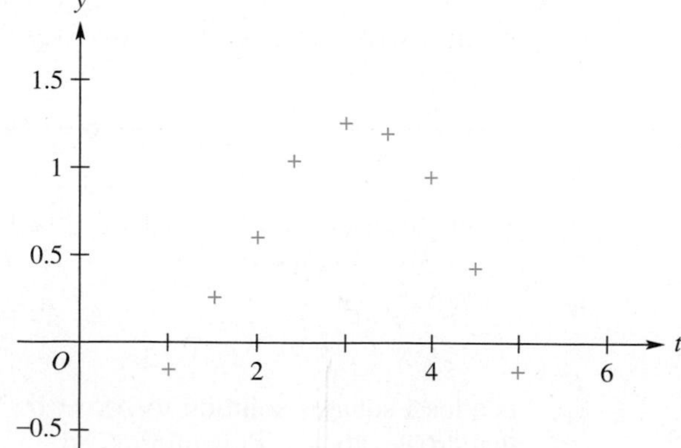

We now have

$$A = \begin{bmatrix} 1 & 1 & 1 \\ 2.25 & 1.5 & 1 \\ 4 & 2 & 1 \\ 6.25 & 2.5 & 1 \\ 9 & 3 & 1 \\ 12.25 & 3.5 & 1 \\ 16 & 4 & 1 \\ 20.25 & 4.5 & 1 \\ 25 & 5 & 1 \end{bmatrix}, \quad \mathbf{x} = \begin{bmatrix} a_2 \\ a_1 \\ a_0 \end{bmatrix}, \quad \mathbf{b} = \begin{bmatrix} -0.15 \\ 0.24 \\ 0.68 \\ 1.04 \\ 1.21 \\ 1.15 \\ 0.86 \\ 0.41 \\ -0.08 \end{bmatrix}.$$

The rank of A is 3 (verify), and the normal system is

$$\begin{bmatrix} 1583.25 & 378 & 96 \\ 378 & 96 & 27 \\ 96 & 27 & 9 \end{bmatrix} \begin{bmatrix} a_2 \\ a_1 \\ a_0 \end{bmatrix} = \begin{bmatrix} 54.65 \\ 16.71 \\ 5.36 \end{bmatrix}.$$

Applying Gauss–Jordan reduction we obtain (verify)

$$\widehat{\mathbf{x}} \approx \begin{bmatrix} -0.3274 \\ 2.0067 \\ -1.9317 \end{bmatrix},$$

so we obtain the quadratic polynomial model

$$y = -0.3274t^2 + 2.0067t - 1.9317.$$

Figure 7.6 shows the data set indicated with $+$ and the graph of y. We see that y is close to each data point but is not required to go through the data. ■

Figure 7.6 ►

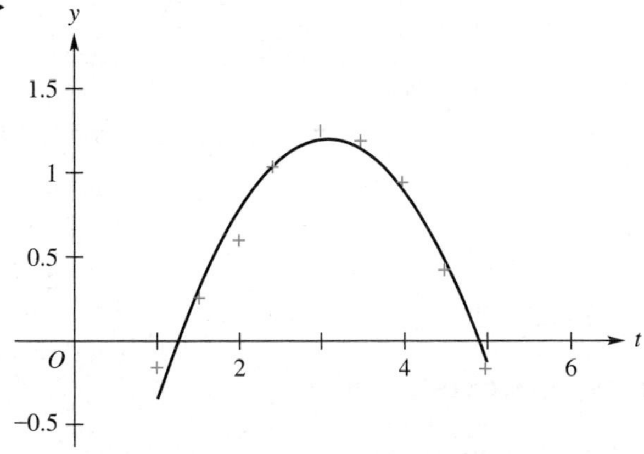

Key Terms

Least squares solution
Normal system
Least squares lines
Least squares polynomial

7.2 Exercises

In Exercises 1 through 4, find a least squares solution to $A\mathbf{x} = \mathbf{b}$.

1. $A = \begin{bmatrix} 2 & 1 \\ 1 & 0 \\ 0 & -1 \\ -1 & 1 \end{bmatrix}$ and $\mathbf{b} = \begin{bmatrix} 3 \\ 1 \\ 2 \\ -1 \end{bmatrix}$

2. $A = \begin{bmatrix} 3 & -2 \\ 2 & -3 \\ 1 & -1 \\ 2 & 3 \\ 3 & 4 \end{bmatrix}$ and $\mathbf{b} = \begin{bmatrix} 2 \\ -1 \\ 0 \\ 1 \\ 0 \end{bmatrix}$

3. $A = \begin{bmatrix} 1 & 2 & 1 \\ 1 & 3 & 2 \\ 2 & 5 & 3 \\ 2 & 0 & 1 \\ 3 & 1 & 1 \end{bmatrix}$ and $\mathbf{b} = \begin{bmatrix} -1 \\ 2 \\ 0 \\ 1 \\ -2 \end{bmatrix}$

4. $A = \begin{bmatrix} -1 & 0 & 0 & 4 \\ 4 & -2 & 0 & 0 \\ 3 & 0 & 1 & 0 \\ 0 & 0 & -1 & 2 \\ 0 & 2 & -2 & 0 \\ 2 & 0 & 0 & 1 \end{bmatrix}$ and $\mathbf{b} = \begin{bmatrix} -1 \\ 1 \\ 2 \\ 0 \\ 0 \\ 1 \end{bmatrix}$

5. Solve Exercise 1 using the QR-factorization of A.

6. Solve Exercise 3 using the QR-factorization of A.

In Exercises 7 through 10, find the least squares line for the given data points.

7. (2, 1), (3, 2), (4, 3), (5, 2)

8. (3, 2), (4, 3), (5, 2), (6, 4), (7, 3)

9. (2, 3), (3, 4), (4, 3), (5, 4), (6, 3), (7, 4)

10. (3, 3), (4, 5), (5, 4), (6, 5), (7, 5), (8, 6), (9, 5), (10, 6)

In Exercises 11 and 12, find a quadratic least squares polynomial for the given data.

11. (0, 3.2), (0.5, 1.6), (1, 2), (2, −0.4), (2.5, −0.8), (3, −1.6), (4, 0.3), (5, 2.2)

12. (0.5, −1.6), (1, 0.4), (1.5, 0.7), (2, 1.8), (2.5, 1.6), (3, 2.2), (3.5, 1.7), (4, 2.2), (4.5, 1.6), (5, 1.5)

13. In an experiment designed to determine the extent of a person's natural orientation, a subject is put in a special room and kept there for a certain length of time. He is then asked to find a way out of a maze and a record is made of the time it takes the subject to accomplish this task. The following data are obtained:

Time in Room (hours)	1	2	3	4	5	6
Time to Find Way Out of Maze (minutes)	0.8	2.1	2.6	2.0	3.1	3.3

Let x denote the number of hours in the room and let y denote the number of minutes that it takes the subject to find his way out.

(a) Find the least squares line relating x and y.

(b) Use the equation obtained in (a) to estimate the time it will take the subject to find his way out of the maze after 10 hours in the room.

14. A steel producer gathers the following data.

Year	1997	1998	1999	2000	2001	2002
Annual Sales (millions of dollars)	1.2	2.3	3.2	3.6	3.8	5.1

Represent the years 1997, . . . , 2002 as 0, 1, 2, 3, 4, 5, respectively, and let x denote the year. Let y denote the annual sales (in millions of dollars).

(a) Find the least squares line relating x and y.

(b) Use the equation obtained in (a) to estimate the annual sales for the year 2006.

15. A sales organization obtains the following data relating the number of salespersons to annual sales.

Number of Salespersons	5	6	7	8	9	10
Annual Sales (millions of dollars)	2.3	3.2	4.1	5.0	6.1	7.2

Let x denote the number of salespersons and let y denote the annual sales (in millions of dollars).

(a) Find the least squares line relating x and y.

(b) Use the equation obtained in (a) to estimate the annual sales when there are 14 salespersons.

16. The distributor of a new car has obtained the following data.

Number of Weeks After Introduction of Car	*Gross Receipts per Week* (millions of dollars)
1	0.8
2	0.5
3	3.2
4	4.3
5	4
6	5.1
7	4.3
8	3.8
9	1.2
10	0.8

Let x denote the gross receipts per week (in millions of dollars) t weeks after the introduction of the car.

(a) Find a least squares quadratic polynomial for the given data.

(b) Use the equation in part (a) to estimate the gross receipts 12 weeks after the introduction of the car.

17. Given $A\mathbf{x} = \mathbf{b}$, where

$$A = \begin{bmatrix} 1 & 3 & -3 \\ 2 & 4 & -2 \\ 0 & -1 & 2 \\ 1 & 2 & -1 \end{bmatrix} \quad \text{and} \quad \mathbf{b} = \begin{bmatrix} 1 \\ 0 \\ 0 \\ 1 \end{bmatrix}.$$

(a) Show that rank $A = 2$.

(b) Since rank $A \neq$ number of columns, Theorem 7.2 cannot be used to determine a least squares solution $\widehat{\mathbf{x}}$. Follow the general procedure as discussed prior to Theorem 7.2 to find a least squares solution. Is the solution unique?

Theoretical Exercises

T.1. Suppose that we wish to find the least squares line for the n data points $(x_1, y_1), (x_2, y_2), \ldots, (x_n, y_n)$, so that $m = 1$ in (6). Show that if at least two x-coordinates are unequal, then the matrix $A^T A$ is nonsingular, where A is the matrix resulting from the n equations in (6).

T.2. Let A be $n \times n$ and nonsingular. From the normal system of equations in (1), show that the least squares solution to $A\mathbf{x} = \mathbf{b}$ is $\widehat{\mathbf{x}} = A^{-1}\mathbf{b}$.

MATLAB Exercises

Routine **lsqline** *in* MATLAB *will compute the least squares line for data you supply and graph both the line and the data points. To use* **lsqline**, *put the x-coordinates of your data into a vector* **x** *and the corresponding y-coordinates into a vector* **y** *and then type* **lsqline(x, y)**. *For more information, use* **help lsqline**.

ML.1. Solve Exercise 9 in MATLAB using **lsqline**.

ML.2. Use **lsqline** to determine the solution to Exercise 13. Then estimate the time it will take the subject to find his way out of the maze after 7 hours; 8 hours; 9 hours.

ML.3. An experiment was conducted on the temperatures of a fluid in a newly designed container. The following data were obtained.

Time (minutes)	0	2	3	5	9
Temperature (°F)	185	170	166	152	110

(a) Determine the least squares line.

(b) Estimate the temperature at $x = 1, 6, 8$ minutes.

(c) Estimate the time at which the temperature of the fluid was 160°F.

ML.4. At time $t = 0$ an object is dropped from a height of 1 meter above a fluid. A recording device registers the height of the object above the surface of the fluid at $\frac{1}{2}$ second intervals, with a negative value indicating the object is below the surface of the fluid. The following table of data is the result.

Time (seconds)	*Depth* (meters)
0	1
0.5	0.88
1	0.54
1.5	0.07
2	−0.42
2.5	−0.80
3	−0.99
3.5	−0.94
4	−0.65
4.5	−0.21

(a) Determine the least squares quadratic polynomial.

(b) Estimate the depth at $t = 5$ and $t = 6$ seconds.

(c) Estimate the time the object breaks through the surface of the fluid the second time.

ML.5. Determine the least squares quadratic polynomial for the following table of data. Use this data model to predict the value of y when $x = 7$.

x	y
-3	0.5
-2.5	0
-2	-1.125
-1.5	-1.875
-1	-1
0	0.9375
0.5	2.8750
1	4.75
1.5	8.25
2	11.5

7.3 MORE ON CODING*

Prerequisite. Section 2.1, An Introduction to Coding

In Section 2.1, An Introduction to Coding, we briefly discussed binary codes and error detection. In this section we show how the techniques of linear algebra can be used to develop error-correcting binary codes. We stress the role that matrices, vector spaces, and associated concepts play in constructing some simple codes and cite several references at the end of the section for a more in-depth study of the fascinating area of coding theory.

The examples and exercises in Section 2.1 contained codes generated by a coding function $e\colon B^m \to B^n$, where $n > m$ and e is one-to-one. In this section we approach the generation of codes from a related point of view. Here we will generate binary codes using a matrix transformation from B^m to B^n, and will choose the matrix of the transformation in such a way that we will be guaranteed that there is a one-to-one correspondence between the message vector from B^m and the code word in B^n. For a message vector $\mathbf{b}$ in B^m, define the matrix transformation

$$e(\mathbf{b}) = C\mathbf{b},$$

where C is the $n \times m$ matrix

$$\begin{bmatrix} I_m \\ D \end{bmatrix} \tag{1}$$

and D is an $(n - m) \times m$ matrix. Each code word in B^n has the form

$$C\mathbf{b} = \begin{bmatrix} I_m \\ D \end{bmatrix} \mathbf{b} = \begin{bmatrix} \mathbf{b} \\ D\mathbf{b} \end{bmatrix},$$

where $D\mathbf{b}$ is an $(n - m) \times 1$ vector. The matrix D will be chosen to aid in error detection and correction.

THEOREM 7.3 *The matrix transformation $e\colon B^m \to B^n$ defined by*

$$e(\mathbf{b}) = C\mathbf{b} = \begin{bmatrix} I_m \\ D \end{bmatrix} \mathbf{b}, \quad \text{is one-to-one,}$$

where $\mathbf{b}$ is an m-vector and D is an $(n - m) \times m$ matrix.

*Throughout this section only binary arithmetic is used.

Proof We have $n > m$ and rank $C = m$ (verify). Thus the columns of C are linearly independent. If $\mathbf{a}$ and $\mathbf{b}$ are m-vectors such that $e(\mathbf{a}) = e(\mathbf{b})$, it follows that

$$e(\mathbf{a}) - e(\mathbf{b}) = C\mathbf{a} - C\mathbf{b} = C(\mathbf{a} - \mathbf{b}) = \mathbf{0}.$$

Hence we have a linear combination of the columns of C, which is the zero vector. But since the columns of C are linearly independent, the only such linear combination for which this is true is the trivial linear combination, thus $\mathbf{a} = \mathbf{b}$, so e is one-to-one. ■

We call the matrix $C = \begin{bmatrix} I_m \\ D \end{bmatrix}$ the **code matrix** or **generator matrix** of the code generated by e.

There are 2^m possible messages in B^m, and since e is one-to-one there are exactly 2^m code words in B^n. The code words are all the possible linear combinations of the columns of C; hence the code words are exactly the column space of the matrix C. Thus the code words form a subspace of B^n, which is then closed with respect to addition and scalar multiplication. This implies that the code generated by the code function e is linear. (See Exercise T.4 in Section 2.1.)

EXAMPLE 1 The (2, 3) code of Example 1 in Section 2.1 has the code matrix

$$C = \begin{bmatrix} 1 & 0 \\ 0 & 1 \\ 0 & 0 \end{bmatrix}.$$

The code words are the subspace of B^3 consisting of all vectors of the form

$$\begin{bmatrix} b_1 \\ b_2 \\ 0 \end{bmatrix}.$$

These vectors form a subspace of B^3 of dimension 2. ■

EXAMPLE 2 The code matrix of the parity (3, 4) code (see Example 6 in Section 2.1) is

$$C = \begin{bmatrix} 1 & 0 & 0 \\ 0 & 1 & 0 \\ 0 & 0 & 1 \\ 1 & 1 & 1 \end{bmatrix}.$$

The code words are all the linear combinations of the columns of C, which is the set of all vectors in B^4 given by

$$C\mathbf{b} = \begin{bmatrix} 1 & 0 & 0 \\ 0 & 1 & 0 \\ 0 & 0 & 1 \\ 1 & 1 & 1 \end{bmatrix} \begin{bmatrix} b_1 \\ b_2 \\ b_3 \end{bmatrix} = \begin{bmatrix} b_1 \\ b_2 \\ b_3 \\ b_1 + b_2 + b_3 \end{bmatrix}.$$

These vectors form a subspace of B^4 of dimension 3. (See Exercise 1.) ■

EXAMPLE 3 The code matrix of the parity $(m, m + 1)$ code (see Section 2.1) is

$$C = \begin{bmatrix} I_m \\ \mathbf{u} \end{bmatrix},$$

where $\mathbf{u}$ is the $1 \times m$ matrix of all 1s. (See Exercise 2.) ■

EXAMPLE 4 Determine the code words of the (3, 5) code with code matrix

$$C = \begin{bmatrix} 1 & 0 & 0 \\ 0 & 1 & 0 \\ 0 & 0 & 1 \\ 1 & 1 & 0 \\ 1 & 0 & 1 \end{bmatrix}.$$

Solution There will be 2^3 code words, which are the images of the vectors $\mathbf{b}$ of B^3 by the matrix transformation $e\colon B^3 \to B^5$ defined by $e(\mathbf{b}) = C\mathbf{b}$. We can display the code words by forming the product of C with the 3×8 matrix whose columns are all the vectors of B^3:

$$\begin{bmatrix} 1 & 0 & 0 \\ 0 & 1 & 0 \\ 0 & 0 & 1 \\ 1 & 1 & 0 \\ 1 & 0 & 1 \end{bmatrix} \begin{bmatrix} 0 & 1 & 0 & 0 & 1 & 1 & 0 & 1 \\ 0 & 0 & 1 & 0 & 1 & 0 & 1 & 1 \\ 0 & 0 & 0 & 1 & 0 & 1 & 1 & 1 \end{bmatrix} = \begin{bmatrix} 0 & 1 & 0 & 0 & 1 & 1 & 0 & 1 \\ 0 & 0 & 1 & 0 & 1 & 0 & 1 & 1 \\ 0 & 0 & 0 & 1 & 0 & 1 & 1 & 1 \\ 0 & 1 & 1 & 0 & 0 & 1 & 1 & 0 \\ 0 & 1 & 0 & 1 & 1 & 0 & 1 & 0 \end{bmatrix}.$$

The columns of the matrix on the right in the preceding display are the code words and they form a subspace of B^5 of dimension 3. ■

Once we have selected our coding function e, or, equivalently, the code matrix C of the form given in (1), we can encode and transmit our message. As we discussed in Section 2.1 and illustrated in Figure 2.2, a message $\mathbf{b}$ in B^m is first encoded as $\mathbf{x} = e(\mathbf{b})$, a vector in B^n, and is then transmitted. The received word $\mathbf{x}_t$ is also a vector in B^n. Upon receipt of $\mathbf{x}_t$ we need to determine if it is a code word, and if it is not, we would like to use an error correction procedure to associate $\mathbf{x}_t$ with a code word. We emphasize that it is never possible to detect all errors. This general scheme is illustrated in Figure 7.7, where the final step is to decode the code word.

Figure 7.7 ►

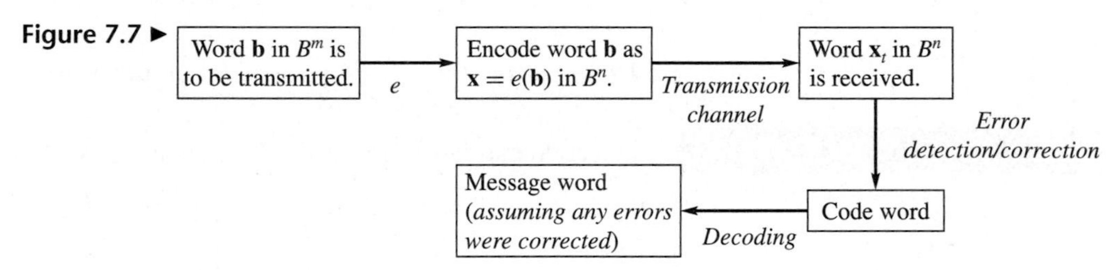

Linear algebra is used to encode the message, to detect and (try) to correct any transmission errors, and to decode the code word.

For our purposes we will assume that no errors have been made in transmission if the received word $\mathbf{x}_t$ is a code word. For a code matrix C of the form in (1) we will show how to determine a **check matrix*** G such that $G\mathbf{x}_t = \mathbf{0}$ when $\mathbf{x}_t$ is a code word and $G\mathbf{x}_t \neq \mathbf{0}$ otherwise. Since $\mathbf{x}_t$ is a code word if it is in the column space of C, one way to proceed is to determine a matrix G such that $GC = O$. For such a matrix G, G "times" any code word must be the zero vector; that is, code words are in the null space of G. Naturally $G \neq O$, so it must in some way rely on the contents of C. The main observation we make is that a bit added to itself is zero. Hence we construct G so that a row of G times a column of C is a sum of a bit with itself. Rather than give an abstract construction for G, we provide several examples, which illustrate the construction and then state the general result in Theorem 7.4.

*Also known as a **parity-check matrix**.

EXAMPLE 5 For the code matrix

$$C = \begin{bmatrix} 1 & 0 & 0 \\ 0 & 1 & 0 \\ 0 & 0 & 1 \\ 1 & 1 & 1 \end{bmatrix}$$

of Example 2, the check matrix is $G = \begin{bmatrix} 1 & 1 & 1 & 1 \end{bmatrix}$. To verify this we can compute GC and show that the result is O, but it is more informative to proceed as follows. Let

$$\mathbf{x}_t = \begin{bmatrix} k_1 \\ k_2 \\ k_3 \\ k_4 \end{bmatrix}$$

be the received word, where k_j, $j = 1, 2, 3, 4$ are bits. From Example 2, a code word must satisfy

$$k_1 + k_2 + k_3 = k_4$$

or, equivalently (recall that the negative of a bit is itself),

$$k_1 + k_2 + k_3 - k_4 = k_1 + k_2 + k_3 + k_4 = 0. \tag{2}$$

We see that if $G\mathbf{x}_t = k_1 + k_2 + k_3 + k_4 = 0$, then $\mathbf{x}_t$ is a code word. If $G\mathbf{x}_t = 1$, then $\mathbf{x}_t$ is not a code word. However, if there is more than one error in $\mathbf{x}_t$, then it may not get detected. For example, for the message $\mathbf{b} = \begin{bmatrix} 0 \\ 1 \\ 1 \end{bmatrix}$ we have

$$e(\mathbf{b}) = C\mathbf{b} = \begin{bmatrix} 0 \\ 1 \\ 1 \\ 0 \end{bmatrix} \quad \text{(verify)}.$$

If the received word, $\mathbf{x}_t = \begin{bmatrix} 1 \\ 1 \\ 1 \\ 1 \end{bmatrix}$, is a code word but not $e(\mathbf{b})$, then $G\mathbf{x}_t = 0$ and so the check matrix would fail to detect the two errors in the first and fourth bits. ■

EXAMPLE 6 For the code matrix

$$C = \begin{bmatrix} 1 & 0 & 0 \\ 0 & 1 & 0 \\ 0 & 0 & 1 \\ 1 & 1 & 0 \\ 1 & 0 & 1 \end{bmatrix}$$

of Example 4, the check matrix is

$$G = \begin{bmatrix} 1 & 1 & 0 & 1 & 0 \\ 1 & 0 & 1 & 0 & 1 \end{bmatrix}.$$

This follows since $GC = O$ (see Exercise 4) and hence every code word is in the null space of G; in addition, if $G\mathbf{y} = 0$, then $\mathbf{y}$ is a code word. The latter statement follows since a basis for the null space of G is

$$S = \left\{ \begin{bmatrix} 1 \\ 1 \\ 1 \\ 0 \\ 0 \end{bmatrix}, \begin{bmatrix} 0 \\ 1 \\ 0 \\ 1 \\ 0 \end{bmatrix}, \begin{bmatrix} 1 \\ 1 \\ 0 \\ 0 \\ 1 \end{bmatrix} \right\} \quad \text{(see Exercise 5)},$$

and each vector in S is a code word as shown in Example 4. ■

EXAMPLE 7 Suppose that we have a (3, 5) code with code matrix

$$C = \begin{bmatrix} 1 & 0 & 0 \\ 0 & 1 & 0 \\ 0 & 0 & 1 \\ b_1 & b_2 & b_3 \\ b_4 & b_5 & b_6 \end{bmatrix},$$

where the b_j, $j = 1, 2, 3, 4, 5, 6$ are bits. The matrix

$$G = \begin{bmatrix} b_1 & b_2 & b_3 & 1 & 0 \\ b_4 & b_5 & b_6 & 0 & 1 \end{bmatrix}$$

satisfies

$$GC = \begin{bmatrix} b_1 + b_1 & b_2 + b_2 & b_3 + b_3 \\ b_4 + b_4 & b_5 + b_5 & b_6 + b_6 \end{bmatrix} = \begin{bmatrix} 0 & 0 & 0 \\ 0 & 0 & 0 \end{bmatrix}.$$

In addition, the dimension of the null space of G is 3 since the rank G is 2 (verify). The three columns of C are linearly independent and are in the null space of G, hence must be a basis for the null space of G. It follows that any member of the null space of G is a code word. Thus we conclude that G is the check matrix for the code generated by C. ■

THEOREM 7.4 *The check matrix for a code matrix of the form*

$$C = \begin{bmatrix} I_m \\ D \end{bmatrix}$$

for D an $(n - m) \times m$ matrix is $G = \begin{bmatrix} D & I_{n-m} \end{bmatrix}$.

Proof See Exercise T.1. ■

Using the check matrix, we can determine if the received word $\mathbf{x}_t$ is a code word or not. If $\mathbf{x}_t$ is a code word, we can proceed to decode it. If $\mathbf{x}_t$ is not a code word, we know that there has been at least one transmission error, so we need a strategy that indicates whether we should try to correct the error and then decode the corrected word or to request that the message be resent.

Suppose $\mathbf{x}_t$ is not a code word and that a single error has occurred. Then $\mathbf{x}_t = \mathbf{x} + \mathbf{e}_i$, where $\mathbf{e}_i$ is the ith column of the $n \times n$ identity matrix (for some unknown value i). We have

$$G\mathbf{x}_t = G\mathbf{x} + G\mathbf{e}_i = \mathbf{0} + \mathbf{col}_i(G) = \mathbf{col}_i(G).$$

(We only know that some error has been made, not a particular value for i.) As long as the columns of G are distinct, we can compare column $G\mathbf{x}_t$ with the columns of G to determine the value of i. Then we can say that the single error occurred in the ith bit, correct it, and hence proceed to correctly decode the message. (If two columns of G are identical, then we cannot determine which bit is in error.)

On the other hand, assume that the check matrix G can be used to correctly detect all received words $\mathbf{x}_t$ in which at most one error has occurred. If the ith column of G was all zeros and if $\mathbf{x} = \mathbf{0}$ was received in the form $\mathbf{x}_t = \mathbf{e}_i$, we would have $G\mathbf{e}_i = \mathbf{0}$. Thus we would incorrectly accept $\mathbf{e}_i$. So it follows that no column of G can be all zeros. Next suppose that the matrix G has two identical columns, say the ith and jth, with $i \neq j$. If an encoded message $\mathbf{x}$ is received as $\mathbf{x}_t = \mathbf{x} + \mathbf{e}_i$, that is, with a single error in the ith position, we would have

$$G\mathbf{x}_t = G\mathbf{x} + G\mathbf{e}_i = \mathbf{0} + \mathbf{col}_i(G) = \mathbf{col}_i(G) = \mathbf{col}_j(G).$$

This implies that we cannot determine which bit, the ith or the jth, is in error; hence we will be unable to correctly detect the single error in the message. This is a contradiction of the assumption that the check matrix G can be used to detect correctly all received words $\mathbf{x}_t$ in which at most one error has occurred.

Theorem 7.5, which we state without proof, summarizes the properties of a check matrix, which can be used to correct single errors.

THEOREM 7.5 *A check matrix G for a code with code matrix C will detect any single error in transmission if and only if no column of G is all zeros and the columns are distinct.*

Remark Theorem 7.5 does not require that G and C have the forms specified in Theorem 7.4.

EXAMPLE 8 For the code matrix

$$C = \begin{bmatrix} 1 & 0 & 0 \\ 0 & 1 & 0 \\ 0 & 0 & 1 \\ 1 & 1 & 0 \\ 1 & 0 & 1 \end{bmatrix}$$

of Example 4, the check matrix is

$$G = \begin{bmatrix} 1 & 1 & 0 & 1 & 0 \\ 1 & 0 & 1 & 0 & 1 \end{bmatrix},$$

as shown in Example 6. The columns of the matrix G are not distinct, so Theorem 7.5 tells us that the check matrix cannot be used to detect single errors. For example, for

$$\mathbf{x}_t = \begin{bmatrix} 0 \\ 1 \\ 1 \\ 1 \\ 0 \end{bmatrix},$$

not a code word (see Example 4), the product $G\mathbf{x}_t = \begin{bmatrix} 0 \\ 1 \end{bmatrix}$. Comparing this vector to the columns of G, we see that either bit 3 or bit 5 is in error. We have no way to determine which bit is in error (in fact, possibly both are in error) and thus no way to implement a correction. ■

EXAMPLE 9 For the coding function $e\colon B^3 \to B^6$ with code matrix

$$C = \begin{bmatrix} 1 & 0 & 0 \\ 0 & 1 & 0 \\ 0 & 0 & 1 \\ 1 & 1 & 0 \\ 0 & 1 & 1 \\ 1 & 0 & 1 \end{bmatrix}$$

the check matrix is

$$G = \begin{bmatrix} 1 & 1 & 0 & 1 & 0 & 0 \\ 0 & 1 & 1 & 0 & 1 & 0 \\ 1 & 0 & 1 & 0 & 0 & 1 \end{bmatrix} \quad \text{(verify).}$$

If the received word is

$$\mathbf{x}_t = \begin{bmatrix} 1 \\ 1 \\ 1 \\ 0 \\ 0 \\ 1 \end{bmatrix},$$

then

$$G\mathbf{x}_t = \begin{bmatrix} 0 \\ 0 \\ 1 \end{bmatrix} \quad \text{(verify).}$$

Thus an error has been detected. Comparing this vector with the columns of matrix G, which satisfies Theorem 7.5, we infer that bit number 6 is in error. We can correct the error to obtain the code word

$$\begin{bmatrix} 1 \\ 1 \\ 1 \\ 0 \\ 0 \\ 0 \end{bmatrix}.$$

We would then proceed to decode this word by stripping off the last three bits. Multiplying the code word by the 2×6 matrix $\begin{bmatrix} I_3 & O \end{bmatrix}$ will perform that task. (Verify.) ■

EXAMPLE 10 For the coding function $e\colon B^2 \to B^5$ with code matrix

$$C = \begin{bmatrix} 1 & 1 \\ 1 & 0 \\ 1 & 0 \\ 0 & 1 \\ 0 & 1 \end{bmatrix}$$

the check matrix is

$$G = \begin{bmatrix} 0 & 0 & 0 & 1 & 1 \\ 0 & 1 & 1 & 0 & 0 \\ 1 & 0 & 1 & 0 & 1 \end{bmatrix} \quad \text{(verify).}$$

If the received word is

$$\mathbf{x}_t = \begin{bmatrix} 1 \\ 1 \\ 1 \\ 1 \\ 1 \end{bmatrix},$$

then

$$G\mathbf{x}_t = \begin{bmatrix} 0 \\ 0 \\ 1 \end{bmatrix} \quad \text{(verify)}.$$

Thus an error has been detected. Comparing this vector with the columns of matrix G, which does satisfy Theorem 7.5, we infer that bit number 1 is in error. We can correct the error to obtain the received code word $\begin{bmatrix} 0 \\ 1 \\ 1 \\ 1 \\ 1 \end{bmatrix}$. We would then proceed to decode this word by stripping off the last three bits. Multiplying the code word by the 2×5 matrix $\begin{bmatrix} I_2 & O_3 \end{bmatrix}$ will perform that task (verify). ■

HAMMING* CODES

The check matrix for a single error-correcting code must satisfy Theorem 7.5. A clever way of constructing such a check matrix is named after Richard Hamming, a pioneer in coding theory. We start our development of Hamming codes by first devising a check matrix that satisfies Theorem 7.5 using the following construction. Let the columns of the check matrix be the binary representation of the integers from 1 to n. (See Exercise T.2.) We call a check matrix in this form a **Hamming (check) matrix** and denote it as $H(n)$. A general Hamming matrix will be denoted simply as H.

EXAMPLE 11 When $n = 7$ the binary representations are

1	2	3	4	5	6	7
001	010	011	100	101	110	111

and hence the corresponding Hamming matrix is

$$H(7) = \begin{bmatrix} 0 & 0 & 0 & 1 & 1 & 1 & 1 \\ 0 & 1 & 1 & 0 & 0 & 1 & 1 \\ 1 & 0 & 1 & 0 & 1 & 0 & 1 \end{bmatrix}. \tag{3}$$

Also note that the matrix G in Example 10 is $H(5)$. ■

---

*Richard W. Hamming (1915–1998) was born in Chicago, Illinois, and died in Monterey, California. He received his B.S. from the University of Chicago in 1937, his M.A. from the University of Nebraska in 1939, and his Ph.D. from the University of Illinois. In 1945 he joined the Manhattan Project in Los Alamos, working on the development of the first atomic bomb. In 1946 he accepted a position at the Bell Telephone Laboratories, where he remained until 1976, when he joined the Computer Science department at the Naval Postgraduate School in Monterey, California. He was best known for his pioneering work on error detection and correction, an area in which he wrote the fundamental paper in 1950. He also did important work in numerical analysis and in differential equations. In the mid-1950s, Hamming's work on the early IBM 650 computer led to the development of a programming language that later became one of the programming languages in current use. He received many awards in his lifetime, and in 1988 the Institute of Electrical and Electronic Engineers (IEEE) created a prestigious award named "the Hamming Medal" in his honor.

The Hamming matrix $H(7)$ in Example 11 is not in the form given in Theorem 7.4. However, we can use the following result to our advantage:

> If matrix Q_p is obtained from matrix Q by a rearrangement of the columns of Q, then the null space of Q_p consists of the vectors of the null space of Q with their entries rearranged in the same manner as used to obtain Q_p from Q. (See Exercise T.3.)

Thus the matrix $H(7)$ of Example 11 has the same null space as

$$H(7)_p = \begin{bmatrix} 0 & 1 & 1 & 1 & 1 & 0 & 0 \\ 1 & 0 & 1 & 1 & 0 & 1 & 0 \\ 1 & 1 & 0 & 1 & 0 & 0 & 1 \end{bmatrix}. \qquad (4)$$

Note that there are other rearrangements of the columns of the matrix $H(7)$ that have the same property. When we perform such a rearrangement, we will do it so that the columns preceding those of the identity matrix are in decreasing order for the integer they represent. Here we have

$$H(7)_p = \begin{bmatrix} 0 & 1 & 1 & 1 & 1 & 0 & 0 \\ 1 & 0 & 1 & 1 & 0 & 1 & 0 \\ 1 & 1 & 0 & 1 & 0 & 0 & 1 \end{bmatrix}.$$
$$\begin{matrix} 3 & 5 & 6 & 7 & 4 & 2 & 1 \end{matrix}$$

Once a Hamming check matrix is rearranged into a form $\begin{bmatrix} D & I_{n-m} \end{bmatrix}$ as discussed previously, we can construct the corresponding code matrix

$$C = \begin{bmatrix} I_m \\ D \end{bmatrix}.$$

The code matrix corresponding to the check matrix in Equation (4) is

$$C = \begin{bmatrix} 1 & 0 & 0 & 0 \\ 0 & 1 & 0 & 0 \\ 0 & 0 & 1 & 0 \\ 0 & 0 & 0 & 1 \\ 0 & 1 & 1 & 1 \\ 1 & 0 & 1 & 1 \\ 1 & 1 & 0 & 1 \end{bmatrix}, \qquad (5)$$

which defines a coding function $e\colon B^4 \to B^7$ as the matrix transformation defined by $e(\mathbf{b}) = C\mathbf{b}$, for a vector $\mathbf{b}$ in B^4.

In practice, Hamming matrices are not rearranged as in the preceding discussion. We wanted to demonstrate that the results developed in this section are applicable to Hamming matrices. A Hamming matrix H has the crucial property that if there is a single error in the ith position of the received word $\mathbf{x}_t$, then $H\mathbf{x}_t$ is equal to the ith column of H, which is exactly the number i in binary form. Hence there is no need to compare the result of the product $H\mathbf{x}_t$ with the columns of the Hamming matrix H. Thus we obtain the location of the bit that is in error directly. In Example 10, $G = H(5)$. The product

$$G\mathbf{x}_t = \begin{bmatrix} 0 \\ 0 \\ 1 \end{bmatrix}$$

gives the binary code for 1 and that is the bit that was in error.

Since Hamming matrices H are so useful in detecting single errors, we need only construct the corresponding encoding matrix C. The columns of C are a basis for the null space of H, and this basis is not difficult to obtain by using row reduction.

EXAMPLE 12 Find the code matrix C for the Hamming matrix $H(5)$.

Solution $H(5) = \begin{bmatrix} 0 & 0 & 0 & 1 & 1 \\ 0 & 1 & 1 & 0 & 0 \\ 1 & 0 & 1 & 0 & 1 \end{bmatrix}$. Construct the augmented matrix

$$\left[\begin{array}{ccccc|c} 0 & 0 & 0 & 1 & 1 & 0 \\ 0 & 1 & 1 & 0 & 0 & 0 \\ 1 & 0 & 1 & 0 & 1 & 0 \end{array}\right],$$

and find its reduced row echelon form. Using row operations with binary arithmetic, we obtain (verify)

$$\left[\begin{array}{ccccc|c} 1 & 0 & 1 & 0 & 1 & 0 \\ 0 & 1 & 1 & 0 & 0 & 0 \\ 0 & 0 & 0 & 1 & 1 & 0 \end{array}\right].$$

We now obtain the general solution to this homogeneous system and find that it is a linear combination of the vectors

$$\begin{bmatrix} 1 \\ 0 \\ 0 \\ 1 \\ 1 \end{bmatrix} \quad \text{and} \quad \begin{bmatrix} 1 \\ 1 \\ 1 \\ 0 \\ 0 \end{bmatrix}.$$

Hence a code matrix is

$$C = \begin{bmatrix} 1 & 1 \\ 0 & 1 \\ 0 & 1 \\ 1 & 0 \\ 1 & 0 \end{bmatrix}.$$

The columns of the code matrix can be interchanged to produce another code. The code words will be the same. ■

A **Hamming code** consists of a (Hamming) check matrix H and its corresponding code matrix C as we have described previously. We designate Hamming codes by the integer n used to construct H and the dimension of the null space of H. As we noted earlier, the dimension of the null space of the check matrix is the dimension of the subspace of code words and is a distinguishing feature of a code. $H(7)$ has rank 3 so the dimension of its null space is 4. We refer to the corresponding code as a $(7, 4)$ Hamming code. Example 10 is a $(5, 2)$ Hamming code, in fact two such codes. The naming conventions for Hamming codes differ; here we have adopted a simple device that stresses the linear algebra foundation.

For a code matrix in the form

$$C = \begin{bmatrix} I_m \\ D \end{bmatrix}$$

with D an $(n-m)\times m$ matrix and check matrix $G = \begin{bmatrix} D & I_{n-m} \end{bmatrix}$ as illustrated in Example 9 for the case $m = 3$, $n = 6$, we gave a procedure to decode a code word. We multiply the code word by a matrix of the form $\begin{bmatrix} I_m & O_{n-m} \end{bmatrix}$, which "strips" off the $n - m$ bottom bits, returning the m top bits. From the form of the code matrix C we can see that we get the correct message that was transmitted. In the case of Hamming codes we know that we can rearrange the columns of H into the form $\begin{bmatrix} D & I_{n-m} \end{bmatrix}$. [See the discussion involving Equations (4) and (5).] Such a rearrangement can be accomplished by multiplying H on the right by a **permutation matrix**; that is, a matrix whose columns are a rearrangement of the columns of the identity matrix. If we call the appropriate permutation matrix P, then $HP = \begin{bmatrix} D & I_{n-m} \end{bmatrix}$. Since P is a rearrangement of the identity matrix, it is nonsingular and P^{-1} exists. If we use a code C matrix that corresponds to a Hamming matrix H and it does not have the form

$$\begin{bmatrix} I_m \\ D \end{bmatrix},$$

then we have to reverse the rearrangement that is built-in to H, so we multiply the code word by $\begin{bmatrix} I_m & O_{n-m} \end{bmatrix} P^{-1}$ to obtain the original message. We illustrate such a decoding in Examples 13 and 14.

EXAMPLE 13 Suppose that we have used the (5, 2) Hamming code with

$$H(5) = \begin{bmatrix} 0 & 0 & 0 & 1 & 1 \\ 0 & 1 & 1 & 0 & 0 \\ 1 & 0 & 1 & 0 & 1 \end{bmatrix}$$

and the code matrix

$$C = \begin{bmatrix} 1 & 1 \\ 0 & 1 \\ 0 & 1 \\ 1 & 0 \\ 1 & 0 \end{bmatrix},$$

not of the form

$$\begin{bmatrix} I_m \\ D \end{bmatrix}.$$

(See Example 12.) Let the original message be $\mathbf{b} = \begin{bmatrix} 1 \\ 1 \end{bmatrix}$. It follows that the corresponding code word is

$$C\mathbf{b} = \mathbf{v} = \begin{bmatrix} 0 \\ 1 \\ 1 \\ 1 \\ 1 \end{bmatrix}.$$

The receiver must decode the contents of $\mathbf{v}$ to obtain the message we sent. Since the receiver knows that we used a (5, 2) Hamming code in the form of the check matrix as given previously, the permutation matrix P and P^{-1} can be constructed ahead of time so they are ready for use. Here we show that construction. To find P so that

$$HP = \begin{bmatrix} D & I_{5-2} \end{bmatrix} = \begin{bmatrix} * & * & 1 & 0 & 0 \\ * & * & 0 & 1 & 0 \\ * & * & 0 & 0 & 1 \end{bmatrix}$$

we note that we can perform the following column shifts (the contents of columns 1 and 2 are the submatrix D of the coding matrix of the form $\begin{bmatrix} I_2 \\ D \end{bmatrix}$):

$$H(5) = \begin{bmatrix} 0 & 0 & 0 & 1 & 1 \\ 0 & 1 & 1 & 0 & 0 \\ 1 & 0 & 1 & 0 & 1 \end{bmatrix}$$

$$\begin{array}{rccccc} & \downarrow & \downarrow & \downarrow & \downarrow & \downarrow \\ \text{goes to column} & 5 & 4 & 1 & 3 & 2\,. \end{array}$$

Define the matrix P to have rows such that

$$\begin{aligned} &\mathbf{row}_1(P) = \mathbf{row}_5(I_5), \quad \mathbf{row}_2(P) = \mathbf{row}_4(I_5), \\ &\mathbf{row}_3(P) = \mathbf{row}_1(I_5), \quad \mathbf{row}_4(P) = \mathbf{row}_3(I_5), \\ \text{and} \quad &\mathbf{row}_5(P) = \mathbf{row}_2(I_5); \end{aligned}$$

that is, the rows of P are the 54132 permutations of the rows of I_5. We get

$$P = \begin{bmatrix} 0 & 0 & 0 & 0 & 1 \\ 0 & 0 & 0 & 1 & 0 \\ 1 & 0 & 0 & 0 & 0 \\ 0 & 0 & 1 & 0 & 0 \\ 0 & 1 & 0 & 0 & 0 \end{bmatrix}$$

and in fact $P^{-1} = P^T$. (Verify.) The matrix P is an example of an orthogonal matrix, which is studied in Chapter 8. Thus to decode the contents of the code word $\mathbf{v}$ we compute

$$\begin{bmatrix} 1 & 0 & 0 & 0 & 0 \\ 0 & 1 & 0 & 0 & 0 \end{bmatrix} P^T \mathbf{v} = \begin{bmatrix} 1 & 0 & 0 & 0 & 0 \\ 0 & 1 & 0 & 0 & 0 \end{bmatrix} \begin{bmatrix} 0 & 0 & 1 & 0 & 0 \\ 0 & 0 & 0 & 0 & 1 \\ 0 & 0 & 0 & 1 & 0 \\ 0 & 1 & 0 & 0 & 0 \\ 1 & 0 & 0 & 0 & 0 \end{bmatrix} \begin{bmatrix} 0 \\ 1 \\ 1 \\ 1 \\ 1 \end{bmatrix}$$

$$= \begin{bmatrix} 1 \\ 1 \end{bmatrix} = \mathbf{b}.$$

The product

$$\begin{bmatrix} 1 & 0 & 0 & 0 & 0 \\ 0 & 1 & 0 & 0 & 0 \end{bmatrix} \begin{bmatrix} 0 & 0 & 1 & 0 & 0 \\ 0 & 0 & 0 & 0 & 1 \\ 0 & 0 & 0 & 1 & 0 \\ 0 & 1 & 0 & 0 & 0 \\ 1 & 0 & 0 & 0 & 0 \end{bmatrix} = \begin{bmatrix} 0 & 0 & 1 & 0 & 0 \\ 0 & 1 & 0 & 0 & 0 \end{bmatrix}$$

can be computed ahead of time and stored for decoding use. ■

EXAMPLE 14 Suppose that we have used the (7, 4) Hamming code with

$$H(7) = \begin{bmatrix} 0 & 0 & 0 & 1 & 1 & 1 & 1 \\ 0 & 1 & 1 & 0 & 0 & 1 & 1 \\ 1 & 0 & 1 & 0 & 1 & 0 & 1 \end{bmatrix}$$

and the code matrix

$$C = \begin{bmatrix} 1 & 1 & 0 & 1 \\ 1 & 0 & 1 & 1 \\ 1 & 0 & 0 & 0 \\ 0 & 1 & 1 & 1 \\ 0 & 1 & 0 & 0 \\ 0 & 0 & 1 & 0 \\ 0 & 0 & 0 & 1 \end{bmatrix}.$$

(See Exercise 21.) Let the original message be

$$\mathbf{b} = \begin{bmatrix} 0 \\ 0 \\ 1 \\ 1 \end{bmatrix}.$$

It follows that the corresponding code word is (verify)

$$\mathbf{v} = \begin{bmatrix} 1 \\ 0 \\ 0 \\ 0 \\ 0 \\ 1 \\ 1 \end{bmatrix}.$$

The receiver must decode the contents of $\mathbf{v}$ to obtain the message we sent. Since the receiver knows that we used a $(7, 4)$ Hamming code in the form of the check matrix as given previously, the permutation matrix P and P^{-1} can be constructed ahead of time so they are ready for use. To find P we follow the procedure used in Example 13 and we see that

$$\begin{array}{rccccccc} H(7) = & \left[\begin{matrix} 0 \\ 0 \\ 1 \end{matrix}\right. & \begin{matrix} 0 \\ 1 \\ 0 \end{matrix} & \begin{matrix} 0 \\ 1 \\ 1 \end{matrix} & \begin{matrix} 1 \\ 0 \\ 0 \end{matrix} & \begin{matrix} 1 \\ 0 \\ 1 \end{matrix} & \begin{matrix} 1 \\ 1 \\ 0 \end{matrix} & \left.\begin{matrix} 1 \\ 1 \\ 1 \end{matrix}\right] \\ & \downarrow & \downarrow & \downarrow & \downarrow & \downarrow & \downarrow & \downarrow \\ \text{goes to column} & 7 & 6 & 1 & 5 & 2 & 3 & 4 \end{array}$$

and it follows that the permutation matrix is (verify)

$$P = \begin{bmatrix} 0 & 0 & 0 & 0 & 0 & 0 & 1 \\ 0 & 0 & 0 & 0 & 0 & 1 & 0 \\ 1 & 0 & 0 & 0 & 0 & 0 & 0 \\ 0 & 0 & 0 & 0 & 1 & 0 & 0 \\ 0 & 1 & 0 & 0 & 0 & 0 & 0 \\ 0 & 0 & 1 & 0 & 0 & 0 & 0 \\ 0 & 0 & 0 & 1 & 0 & 0 & 0 \end{bmatrix}.$$

Then $P^{-1} = P^T$ (verify) and to decode the contents of code word $\mathbf{v}$ we compute

$$\begin{bmatrix} 1 & 0 & 0 & 0 & 0 & 0 & 0 \\ 0 & 1 & 0 & 0 & 0 & 0 & 0 \\ 0 & 0 & 1 & 0 & 0 & 0 & 0 \\ 0 & 0 & 0 & 1 & 0 & 0 & 0 \end{bmatrix} P^T \mathbf{v}$$

$$= \begin{bmatrix} 1 & 0 & 0 & 0 & 0 & 0 & 0 \\ 0 & 1 & 0 & 0 & 0 & 0 & 0 \\ 0 & 0 & 1 & 0 & 0 & 0 & 0 \\ 0 & 0 & 0 & 1 & 0 & 0 & 0 \end{bmatrix} \begin{bmatrix} 0 & 0 & 1 & 0 & 0 & 0 & 0 \\ 0 & 0 & 0 & 0 & 1 & 0 & 0 \\ 0 & 0 & 0 & 0 & 0 & 1 & 0 \\ 0 & 0 & 0 & 0 & 0 & 0 & 1 \\ 0 & 0 & 0 & 1 & 0 & 0 & 0 \\ 0 & 1 & 0 & 0 & 0 & 0 & 0 \\ 1 & 0 & 0 & 0 & 0 & 0 & 0 \end{bmatrix} \begin{bmatrix} 1 \\ 0 \\ 0 \\ 0 \\ 0 \\ 1 \\ 1 \end{bmatrix}$$

$$= \begin{bmatrix} 0 \\ 0 \\ 1 \\ 1 \end{bmatrix} = \mathbf{b}.$$

■

There is a whole family of Hamming codes, which can correct single errors, but no more. A brief discussion of why only single errors can be corrected is given in Exercises T.5–T.10. A more detailed discussion can be found in a number of the references given at the end of this section.

The Hamming codes had a great impact on the theory and development of error-correcting codes—so much that the (7, 4) Hamming check matrix $H(7)$ appears on the Hamming IEEE medal, one of the highest awards bestowed by the IEEE. Subsequent developments have produced codes that correct multiple errors. Such codes are used in today's digital devices, and their development requires significantly more mathematics. For example, a family of codes known as Reed–Solomon codes is used by NASA and for compact disk recording. See the References at the end of the section for further information.

In our examples we have looked at the coding step operating on a single word. In practice a text message is first converted to binary representation as a long string of bits. It is then reshaped into a matrix X with $(n - (k - 1))$ rows and padded with zeros as needed in the final column. Then the coding function e is applied as a matrix transformation using the coding matrix C. The encoded message of the product CX is reshaped into a single long string, it is then transmitted, the error detection/correction process is applied to an appropriately reshaped matrix, and finally the message is decoded.

In this brief look at the linear algebra foundations of coding theory, we have stressed the role that matrices, vector spaces, and associated concepts play in constructing some simple codes. More extensive developments of coding can be found in the References. The Exercises focus on the basic linear algebra ideas we used in this section.

Further Readings

CHILDS, L. N. *Concrete Introduction to Higher Algebra*, 2nd ed. New York: Springer-Verlag, 2000.

CIPRA, B. *The Ubiquitous Reed–Solomon Codes*. SIAM News, Vol. 26, No. 1, January 1993. **http://www.siam.org./siamnews/mtc/mtc193.htm**.

COHEN, S. *Aspects of Coding*, UMAP Module 336. Lexington, Mass.: COMAP, Inc., 1979.

HAMMING, R. W. *Coding and Information Theory*. Upper Saddle River, N.J.: Prentice Hall 1980.

HILL, R. *A First Course in Coding Theory*. New York: Oxford University Press, 1986.

KOLMAN, B., R. C. BUSBY, and S. C. ROSS. *Discrete Mathematical Structures*, 5th ed. Upper Saddle River, N.J.: Pearson Education, 2004.

MORGAN, S. P. *Richard Wesley Hamming, 1915–1998*. Notices of the American Mathematical Society 45 No. 9, 1998, pp. 972–977.

RICE, B. F., and C. O. WILDE. *Error-Correcting Codes I*, UMAP Module 346. Lexington, Mass.: COMAP, Inc., 1979.

SMITH, R. *MATLAB Project Book for Linear Algebra*, Upper Saddle River, N.J.: Prentice Hall, 1997.

Key Terms

Code matrix
Check matrix
Hamming (check) matrix
Hamming codes
Permutation matrix
Reed–Solomon codes
Minimum (Hamming) distance

7.3 Exercises

1. Describe the subspace of code words for the code of Example 3.

2. Show that the parity $(m, m+1)$ code of Section 2.1 has code matrix

$$C = \begin{bmatrix} I_m \\ \mathbf{u} \end{bmatrix},$$

where $\mathbf{u}$ is the $1 \times m$ matrix of all 1s.

3. Determine the code words for the code given in Example 5.

4. Show that the product of the check matrix and the code matrix in Example 6 is the zero matrix.

5. Show that a basis for the null space of the check matrix in Example 6 is the set given in the example.

6. Use Example 7 to determine the check matrix for the code matrix

$$C = \begin{bmatrix} 1 & 0 & 0 \\ 0 & 1 & 0 \\ 0 & 0 & 1 \\ 0 & 1 & 1 \\ 1 & 1 & 0 \end{bmatrix}.$$

7. Use Example 7 to determine the check matrix for the code matrix

$$C = \begin{bmatrix} 1 & 0 & 0 \\ 0 & 1 & 0 \\ 0 & 0 & 1 \\ 1 & 1 & 1 \\ 1 & 0 & 0 \end{bmatrix}.$$

8. Find a basis for the null space of the check matrix

$$G = \begin{bmatrix} 0 & 1 & 1 & 1 & 0 \\ 1 & 0 & 1 & 0 & 1 \end{bmatrix}.$$

9. Find a basis for the null space of the check matrix

$$G = \begin{bmatrix} 1 & 1 & 1 & 0 & 0 \\ 0 & 1 & 0 & 1 & 0 \\ 1 & 1 & 0 & 0 & 1 \end{bmatrix}.$$

10. Find the dimension of the subspace of code words for the code whose check matrix is

$$G = \begin{bmatrix} 0 & 1 & 1 & 0 & 0 \\ 1 & 0 & 0 & 1 & 0 \\ 1 & 1 & 0 & 0 & 1 \end{bmatrix}.$$

11. Find the dimension of the subspace of code words for the code whose check matrix is

$$G = \begin{bmatrix} 1 & 0 & 0 & 1 & 0 & 0 \\ 0 & 0 & 1 & 0 & 1 & 0 \\ 1 & 1 & 1 & 0 & 0 & 1 \end{bmatrix}.$$

12. Apply Theorem 7.5 to determine if the code with check matrix G will detect any single error in transmission for

(a) the check matrix of Exercise 8

(b) the check matrix of Exercise 9

13. Apply Theorem 7.5 to determine if the code with check matrix G will detect any single error in transmission for

(a) the check matrix of Exercise 10

(b) the check matrix of Exercise 11

14. For the code in Example 9, determine which, if any, of the following received words have a single error; if it has an error correct it.

(a) $\mathbf{x}_t = \begin{bmatrix} 0 \\ 1 \\ 0 \\ 1 \\ 1 \\ 0 \end{bmatrix}$ (b) $\mathbf{x}_t = \begin{bmatrix} 1 \\ 0 \\ 1 \\ 1 \\ 1 \\ 1 \end{bmatrix}$ (c) $\mathbf{x}_t = \begin{bmatrix} 0 \\ 0 \\ 1 \\ 0 \\ 1 \\ 1 \end{bmatrix}$

15. For the code in Example 9, determine which, if any, of the following received words have a single error; if it has an error correct it.

(a) $\mathbf{x}_t = \begin{bmatrix} 1 \\ 1 \\ 1 \\ 0 \\ 0 \\ 0 \end{bmatrix}$ (b) $\mathbf{x}_t = \begin{bmatrix} 1 \\ 0 \\ 1 \\ 0 \\ 1 \\ 1 \end{bmatrix}$ (c) $\mathbf{x}_t = \begin{bmatrix} 0 \\ 1 \\ 1 \\ 1 \\ 1 \\ 1 \end{bmatrix}$

Exercises 16 through 22 involve Hamming codes.

16. Construct $H(4)$. (Use 3 bits.)

17. Construct $H(6)$. (Use 3 bits.)

18. Determine a code matrix for $H(4)$.

19. Determine a code matrix for $H(6)$.

20. Use the (5, 2) Hamming code given in Example 12. Determine which, if any, of the following received words have a single error; if it has an error, correct it.

(a) $\mathbf{x}_t = \begin{bmatrix} 1 \\ 1 \\ 0 \\ 0 \\ 0 \end{bmatrix}$ (b) $\mathbf{x}_t = \begin{bmatrix} 1 \\ 1 \\ 0 \\ 1 \\ 1 \end{bmatrix}$ (c) $\mathbf{x}_t = \begin{bmatrix} 0 \\ 1 \\ 1 \\ 1 \\ 1 \end{bmatrix}$

21. Use the (7, 4) Hamming code given in Example 14. Determine which, if any, of the following received words have a single error; if it has an error, correct it.

(a) $\mathbf{x}_t = \begin{bmatrix} 0 \\ 0 \\ 1 \\ 0 \\ 1 \\ 1 \\ 0 \end{bmatrix}$ (b) $\mathbf{x}_t = \begin{bmatrix} 1 \\ 1 \\ 1 \\ 1 \\ 1 \\ 0 \\ 0 \end{bmatrix}$ (c) $\mathbf{x}_t = \begin{bmatrix} 1 \\ 0 \\ 0 \\ 0 \\ 1 \\ 1 \\ 1 \end{bmatrix}$

22. Use the (7, 4) Hamming code given in Example 14. Decode the following received code words.

(a) $\mathbf{x}_t = \begin{bmatrix} 0 \\ 1 \\ 1 \\ 0 \\ 0 \\ 1 \\ 1 \end{bmatrix}$ (b) $\mathbf{x}_t = \begin{bmatrix} 1 \\ 1 \\ 1 \\ 1 \\ 1 \\ 1 \\ 1 \end{bmatrix}$ (c) $\mathbf{x}_t = \begin{bmatrix} 0 \\ 0 \\ 0 \\ 1 \\ 1 \\ 1 \\ 1 \end{bmatrix}$

Theoretical Exercises

T.1. Prove Theorem 7.4.

T.2. Give an argument to show that if the columns of the check matrix G are the binary representation of the integers from 1 to n, then G satisfies Theorem 7.5.

T.3. Show that if the matrix Q_p is obtained from the matrix Q by a rearrangement of the columns of Q, then the null space of Q_p consists of the vectors of the null space of Q with their entries rearranged in the same manner as used to obtain Q_p from Q.

T.4. The **Hamming distance** between two vectors $\mathbf{v}$ and $\mathbf{w}$ in B^n is denoted by $H(\mathbf{v}, \mathbf{w})$ and is defined to be the number of positions in which $\mathbf{v}$ and $\mathbf{w}$ differ. Compute the Hamming distance between the following pairs of vectors.

(a) $\mathbf{v} = \begin{bmatrix} 1 \\ 1 \\ 0 \\ 1 \end{bmatrix}$, $\mathbf{w} = \begin{bmatrix} 1 \\ 0 \\ 0 \\ 0 \end{bmatrix}$

(b) $\mathbf{v} = \begin{bmatrix} 0 \\ 0 \\ 0 \\ 1 \end{bmatrix}$, $\mathbf{w} = \begin{bmatrix} 1 \\ 1 \\ 1 \\ 1 \end{bmatrix}$

(c) $\mathbf{v} = \begin{bmatrix} 1 \\ 1 \\ 0 \\ 0 \\ 1 \\ 1 \end{bmatrix}$, $\mathbf{w} = \begin{bmatrix} 0 \\ 1 \\ 1 \\ 1 \\ 1 \\ 0 \end{bmatrix}$

T.5. The weight of a bit n-vector $\mathbf{x}$ as defined in Section 2.1 is the number of 1s in $\mathbf{x}$ and is denoted by $|\mathbf{x}|$. Show that $H(\mathbf{u}, \mathbf{v}) = |\mathbf{u} - \mathbf{v}| = |\mathbf{u} + \mathbf{v}|$; that is, the Hamming distance between two vectors is the weight of their difference or sum.

T.6. Use Example 14 to form all the code words of the (7, 4) Hamming code. Show that the weight of all the nonzero code words is greater than or equal to 3.

T.7. Use Example 12 to form all the code words of the (5, 2) Hamming code. Show that the weight of all the nonzero code words is greater than or equal to 3.

T.8. Let $\mathbf{u}$, $\mathbf{v}$, and $\mathbf{w}$ be vectors in B^n. Prove the following.

(a) $H(\mathbf{u}, \mathbf{v}) = H(\mathbf{v}, \mathbf{u})$

(b) $H(\mathbf{u}, \mathbf{v}) \geq 0$

(c) $H(\mathbf{u}, \mathbf{v}) = 0$ if and only if $\mathbf{u} = \mathbf{v}$

(d) $H(\mathbf{u}, \mathbf{v}) \leq H(\mathbf{u}, \mathbf{w}) + H(\mathbf{w}, \mathbf{v})$

T.9. The **minimum** (Hamming) **distance** of a code is the minimum of the distances between all distinct code words.

(a) Use Exercise T.6 to determine the minimum distance of the (7, 4) Hamming code.

(b) Use Exercise T.7 to determine the minimum distance of the (5, 2) Hamming code.

T.10. The following theorem can be proved. (See Kolman, Busby, and Ross in the References.)

A code can detect k or fewer errors if and only if its minimum distance is at least $k + 1$.

(a) How many errors can the (7, 4) Hamming code detect?

(b) How many errors can the (5, 2) Hamming code detect?

MATLAB Exercises

The following exercises use the routines **bingen** *and* **binprod**. *See Section 12.9 or use* MATLAB*'s help for a description of these routines.*

ML.1. (a) Use **bingen** to determine $H(8)$.

(b) Determine a code matrix C for $H(8)$. (*Hint*: **binreduce** will be useful.)

(c) Verify that your matrix C from part (b) is correct. (*Hint*: Use **binprod**.)

ML.2. Determine all the code words in the (8, 4) Hamming code that uses the code matrix C from Exercise ML.1. (*Hint*: Use **bingen** and **binprod**.)

ML.3. (a) Use **bingen** to determine $H(15)$.

(b) Determine a code matrix C for $H(15)$. (*Hint*: **binreduce** will be useful.)

(c) Verify that your matrix C from part (b) is correct. (*Hint*: Use **binprod**.)

ML.4. Determine all the code words in the (15, 11) Hamming code that uses the code matrix C from Exercise ML.3. (*Hint*: Use **bingen** and **binprod**.)

Key Ideas for Review

- **QR-factorization** (for writing an $m \times n$ matrix A with linearly independent columns as QR, where Q is an $m \times n$ matrix whose columns form an orthonormal basis for the column space of A and R is an $n \times n$ nonsingular upper triangular matrix). See page 376.
- **Theorem 7.1.** If A is an $m \times n$ matrix with linearly independent columns, then A can be factored as $A = QR$, where Q is an $m \times n$ matrix whose columns form an orthonormal basis for the column space of A and R is an $n \times n$ nonsingular upper triangular matrix.
- **Theorem 7.2.** If A is an $m \times n$ matrix with rank $A = n$, then $A^T A$ is nonsingular and the linear system $A\mathbf{x} = \mathbf{b}$ has a unique least squares solution given by $\widehat{\mathbf{x}} = (A^T A)^{-1} A^T \mathbf{b}$. That is, the normal system of equations has a unique solution.
- **Method of least squares** for finding the straight line $y = b_1 x + b_0$ that best fits the data $(x_1, y_1), (x_2, y_2), \dots, (x_n, y_n)$. Let

$$\mathbf{b} = \begin{bmatrix} y_1 \\ y_2 \\ \vdots \\ y_n \end{bmatrix}, \quad A = \begin{bmatrix} x_1 & 1 \\ x_2 & 1 \\ \vdots & \vdots \\ x_n & 1 \end{bmatrix}, \quad \text{and} \quad \mathbf{x} = \begin{bmatrix} b_1 \\ b_0 \end{bmatrix}.$$

Then b_1 and b_0 can be found by solving the normal system $A^T A\mathbf{x} = A^T \mathbf{b}$ for $\mathbf{x}$ by Gauss–Jordan reduction.

- **Method of least squares** for finding the least squares polynomial

$$y = a_m x^m + a_{m-1} x^{m-1} + \cdots + a_1 x + a_0$$

that best fits the data $(x_1, y_1), (x_2, y_2), \dots, (x_n, y_n)$. Let

$$\mathbf{b} = \begin{bmatrix} y_1 \\ y_2 \\ \vdots \\ y_n \end{bmatrix},$$

$$A = \begin{bmatrix} x_1^m & x_1^{m-1} & \cdots & x_1^2 & x_1 & 1 \\ x_2^m & x_2^{m-1} & \cdots & x_2^2 & x_2 & 1 \\ \vdots & \vdots & & \vdots & \vdots & \vdots \\ x_n^m & x_n^{m-1} & \cdots & x_n^2 & x_n & 1 \end{bmatrix},$$

and

$$\mathbf{x} = \begin{bmatrix} a_m \\ a_{m-1} \\ \vdots \\ a_1 \\ a_0 \end{bmatrix}.$$

Solve the normal system $A^T A\mathbf{x} = A^T \mathbf{b}$ for $\mathbf{x}$ by Gauss–Jordan reduction.

- **Code Matrix (Generator Matrix):** For a coding function $e\colon B^m \to B^n$ where $n > m$ and e is one-to-one we define a matrix transformation $e(\mathbf{b}) = C\mathbf{b}$, where C is an $n \times m$ matrix given in (1) on page 390. C is called the code matrix (generator matrix).
- For a code matrix C we determine an $n - m \times n$ **check matrix** G such that $G\mathbf{x}_t = \mathbf{0}$ when $\mathbf{x}_t$ is a code word and $G\mathbf{x}_t \neq \mathbf{0}$ otherwise. It follows that G is such that $GC = O$ and the code words in R^n are exactly the null space of G.

- **Theorem 7.5** A check matrix G for a code with code matrix C will detect any single error in transmission if and only if no column of G is all zeros and columns are distinct.
- **Hamming (check) matrix** A check matrix whose columns are the binary representations of the integers from 1 to n. A Hamming matrix H has the crucial property that if there is a single error in the ith position of the received word $\mathbf{x}_t$, then $H\mathbf{x}_t$ is equal to the ith column of H, which is exactly the number i in binary form.

Supplementary Exercises

1. Consider the Hamming check matrix

$$H(8) = \begin{bmatrix} 0 & 0 & 0 & 0 & 0 & 0 & 0 & 1 \\ 0 & 0 & 0 & 1 & 1 & 1 & 1 & 0 \\ 0 & 1 & 1 & 0 & 0 & 1 & 1 & 0 \\ 1 & 0 & 1 & 0 & 1 & 0 & 1 & 0 \end{bmatrix}.$$

   (a) Determine a basis for the corresponding subspace of code words.

   (b) Determine if received word $\mathbf{x}_t = \begin{bmatrix} 1 \\ 1 \\ 1 \\ 1 \\ 1 \\ 1 \\ 1 \\ 0 \end{bmatrix}$ is a code word. If it is not, then determine the bit that is in error.

2. Find a QR-factorization of

$$A = \begin{bmatrix} 1 & 0 & 1 \\ -1 & 1 & 2 \\ 2 & 2 & -1 \end{bmatrix}.$$

3. Find the least squares line for the following data points:

$$(0, 1), \quad (3, 2), \quad (5, 4), \quad (8, 10).$$

4. Find the least squares quadratic polynomial for the following data points:

$$(-1.5, 1.3), \quad (-1, 1), \quad (0, 2.8), \quad (0.5, 3.2),$$
$$(1, 3), \quad (1.5, 3.3), \quad (2, 3.6), \quad (3, 2.8).$$

Chapter Test

1. Find a basis for the subspace of code words for the Hamming check matrix

$$H(6) = \begin{bmatrix} 0 & 0 & 0 & 1 & 1 & 1 \\ 0 & 1 & 1 & 0 & 0 & 1 \\ 1 & 0 & 1 & 0 & 1 & 0 \end{bmatrix}.$$

2. Example 10 is a Hamming (5, 2) code. Determine if received word $\mathbf{x}_t = \begin{bmatrix} 1 \\ 1 \\ 0 \\ 1 \\ 1 \end{bmatrix}$ is a code word. If it is not, then determine the bit that is in error.

3. Find a QR-factorization of

$$A = \begin{bmatrix} 2 & 1 \\ -1 & -1 \\ -2 & 3 \end{bmatrix}.$$

4. Find a least squares solution to $A\mathbf{x} = \mathbf{b}$, where

$$A = \begin{bmatrix} -3 & 1 \\ 4 & 2 \\ 3 & 5 \\ 0 & 1 \end{bmatrix} \quad \text{and} \quad \mathbf{b} = \begin{bmatrix} 2 \\ -1 \\ 3 \\ 4 \end{bmatrix}.$$

5. A physical equipment center determines the relationship between the length of time spent on a certain piece of equipment and the number of calories lost. The following data are obtained:

Time on Equipment (minutes)	5	8	10	12	15
Lost Calories (hundreds)	3.2	5.5	6.8	7.8	9.2

   Let x denote the number of minutes on the equipment, and let y denote the number of calories lost (in hundreds).

   (a) Find the least squares line relating x and y.

   (b) Use the equation obtained in part (a) to estimate the number of calories lost after 20 minutes on the equipment.

CHAPTER 8

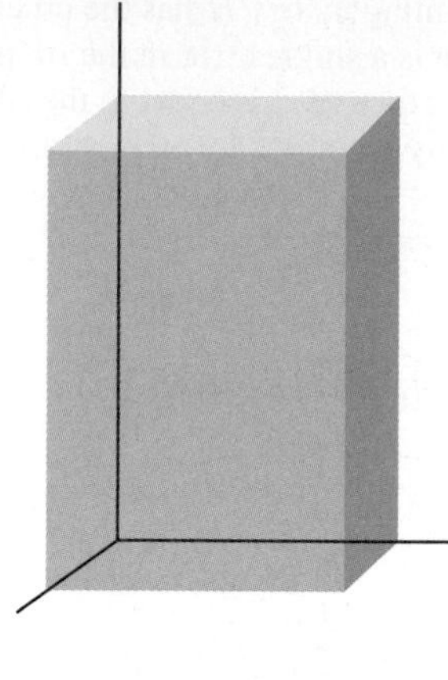

EIGENVALUES, EIGENVECTORS, AND DIAGONALIZATION

In the first seven chapters of this book we have used real numbers as entries of matrices and as scalars. Accordingly, we have only dealt with real vector spaces and with the vector space B^n, where the scalars and entries in a vector are the bits 0 and 1. In this chapter, we deal with matrices that have complex entries and with complex vector spaces. You may consult Appendix A for an introduction to complex numbers and for linear algebra with complex numbers.

8.1 EIGENVALUES AND EIGENVECTORS

In this chapter every matrix considered is a square matrix. Let A be an $n \times n$ matrix. Then, as we have seen in Sections 1.5 and 4.3, the function $L: R^n \to R^n$ defined by $L(\mathbf{x}) = A\mathbf{x}$, for $\mathbf{x}$ in R^n, is a linear transformation. A question of considerable importance in a great many applied problems is the determination of vectors $\mathbf{x}$, if there are any, such that $\mathbf{x}$ and $A\mathbf{x}$ are parallel (see Examples 1 and 2). Such questions arise in all applications involving vibrations; they arise in aerodynamics, elasticity, nuclear physics, mechanics, chemical engineering, biology, differential equations, and others. In this section we shall formulate this problem precisely; we also define some pertinent terminology. In the next section we solve this problem for symmetric matrices and briefly discuss the situation in the general case.

DEFINITION Let A be an $n \times n$ matrix. The number λ is called an **eigenvalue** of A if there exists a *nonzero* vector $\mathbf{x}$ in R^n such that

$$A\mathbf{x} = \lambda\mathbf{x}. \tag{1}$$

Every nonzero vector $\mathbf{x}$ satisfying (1) is called an **eigenvector of A associated with the eigenvalue λ**. The word *eigenvalue* is a hybrid one (*eigen* in German means "proper"). Eigenvalues are also called **proper values**, **characteristic values**, and **latent values**; and eigenvectors are also called **proper vectors**, and so on, accordingly.

Note that $\mathbf{x} = \mathbf{0}$ always satisfies (1), but $\mathbf{0}$ is not an eigenvector, since we insist that an eigenvector be a nonzero vector.

Remark In the preceding definition, the number λ can be real or complex and the vector $\mathbf{x}$ can have real or complex components.

EXAMPLE 1 If A is the identity matrix I_n, then the only eigenvalue is $\lambda = 1$; every nonzero vector in R^n is an eigenvector of A associated with the eigenvalue $\lambda = 1$:

$$I_n\mathbf{x} = 1\mathbf{x}.$$

■

EXAMPLE 2 Let

$$A = \begin{bmatrix} 0 & \frac{1}{2} \\ \frac{1}{2} & 0 \end{bmatrix}.$$

Then

$$A\begin{bmatrix} 1 \\ 1 \end{bmatrix} = \begin{bmatrix} 0 & \frac{1}{2} \\ \frac{1}{2} & 0 \end{bmatrix}\begin{bmatrix} 1 \\ 1 \end{bmatrix} = \begin{bmatrix} \frac{1}{2} \\ \frac{1}{2} \end{bmatrix} = \frac{1}{2}\begin{bmatrix} 1 \\ 1 \end{bmatrix}$$

so that

$$\mathbf{x}_1 = \begin{bmatrix} 1 \\ 1 \end{bmatrix}$$

is an eigenvector of A associated with the eigenvalue $\lambda_1 = \frac{1}{2}$. Also,

$$A\begin{bmatrix} 1 \\ -1 \end{bmatrix} = \begin{bmatrix} 0 & \frac{1}{2} \\ \frac{1}{2} & 0 \end{bmatrix}\begin{bmatrix} 1 \\ -1 \end{bmatrix} = \begin{bmatrix} -\frac{1}{2} \\ \frac{1}{2} \end{bmatrix} = -\frac{1}{2}\begin{bmatrix} 1 \\ -1 \end{bmatrix}$$

so that

$$\mathbf{x}_2 = \begin{bmatrix} 1 \\ -1 \end{bmatrix}$$

is an eigenvector of A associated with the eigenvalue $\lambda_2 = -\frac{1}{2}$. Figure 8.1 shows that $\mathbf{x}_1$ and $A\mathbf{x}_1$ are parallel, and $\mathbf{x}_2$ and $A\mathbf{x}_2$ are parallel also. This illustrates the fact that if $\mathbf{x}$ is an eigenvector of A, then $\mathbf{x}$ and $A\mathbf{x}$ are parallel.

■

Figure 8.1 ►

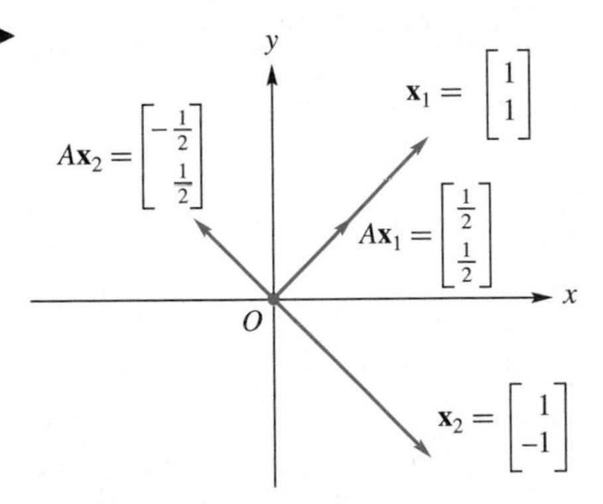

Let λ be an eigenvalue of A with corresponding eigenvector $\mathbf{x}$. In Figure 8.2 we show $\mathbf{x}$ and $A\mathbf{x}$ for the cases $\lambda > 1$, $0 < \lambda < 1$, and $\lambda < 0$.

An eigenvalue λ of A can have associated with it many different eigenvectors. In fact, if $\mathbf{x}$ is an eigenvector of A associated with λ (i.e., $A\mathbf{x} = \lambda\mathbf{x}$) and r is any nonzero real number, then

$$A(r\mathbf{x}) = r(A\mathbf{x}) = r(\lambda\mathbf{x}) = \lambda(r\mathbf{x}).$$

Thus $r\mathbf{x}$ is also an eigenvector of A associated with λ.

Figure 8.2 ►

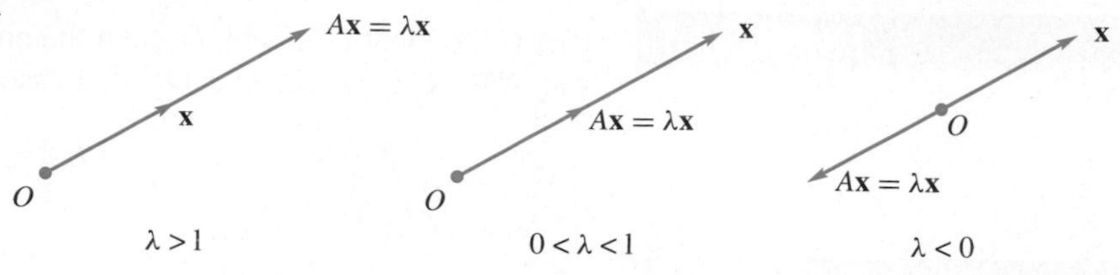

Remark Note that two eigenvectors associated with the same eigenvalue need not be in the same direction. They must only be parallel. Thus, in Example 2, it can be verified easily that $\mathbf{x}_3 = \begin{bmatrix} -1 \\ -1 \end{bmatrix}$ is another eigenvector associated with the eigenvalue $\lambda_1 = \frac{1}{2}$.

EXAMPLE 3 Let

$$A = \begin{bmatrix} 0 & 0 \\ 0 & 1 \end{bmatrix}.$$

Then

$$A\begin{bmatrix} 1 \\ 0 \end{bmatrix} = \begin{bmatrix} 0 & 0 \\ 0 & 1 \end{bmatrix}\begin{bmatrix} 1 \\ 0 \end{bmatrix} = \begin{bmatrix} 0 \\ 0 \end{bmatrix} = 0\begin{bmatrix} 1 \\ 0 \end{bmatrix}$$

so that $\mathbf{x}_1 = \begin{bmatrix} 1 \\ 0 \end{bmatrix}$ is an eigenvector of A associated with the eigenvalue $\lambda_1 = 0$. Also,

$$\mathbf{x}_2 = \begin{bmatrix} 0 \\ 1 \end{bmatrix}$$

is an eigenvector of A associated with the eigenvalue $\lambda_2 = 1$ (verify). ■

Example 3 points out the fact that although the zero vector, by definition, cannot be an eigenvector, the number zero can be an eigenvalue.

COMPUTING EIGENVALUES AND EIGENVECTORS

Thus far we have found the eigenvalues and associated eigenvectors of a given matrix by inspection, geometric arguments, or very simple algebraic approaches. In the following example, we compute the eigenvalues and associated eigenvectors of a matrix by a somewhat more systematic method.

EXAMPLE 4 Let

$$A = \begin{bmatrix} 1 & 1 \\ -2 & 4 \end{bmatrix}.$$

We wish to find the eigenvalues of A and their associated eigenvectors. Thus we wish to find all real numbers λ and all nonzero vectors

$$\mathbf{x} = \begin{bmatrix} x_1 \\ x_2 \end{bmatrix}$$

satisfying (1), that is,

$$\begin{bmatrix} 1 & 1 \\ -2 & 4 \end{bmatrix}\begin{bmatrix} x_1 \\ x_2 \end{bmatrix} = \lambda\begin{bmatrix} x_1 \\ x_2 \end{bmatrix}. \tag{2}$$

Equation (2) becomes

$$\begin{aligned} x_1 + x_2 &= \lambda x_1 \\ -2x_1 + 4x_2 &= \lambda x_2, \end{aligned}$$

or

$$\begin{aligned} (\lambda - 1)x_1 - x_2 &= 0 \\ 2x_1 + (\lambda - 4)x_2 &= 0. \end{aligned} \tag{3}$$

Equation (3) is a homogeneous system of two equations in two unknowns. From Corollary 3.4 of Section 3.2, it follows that the homogeneous system in (3) has a nontrivial solution if and only if the determinant of its coefficient matrix is zero; that is, if and only if

$$\begin{vmatrix} \lambda - 1 & -1 \\ 2 & \lambda - 4 \end{vmatrix} = 0.$$

This means that

$$(\lambda - 1)(\lambda - 4) + 2 = 0,$$

or

$$\lambda^2 - 5\lambda + 6 = 0 = (\lambda - 3)(\lambda - 2).$$

Hence

$$\lambda_1 = 2 \quad \text{and} \quad \lambda_2 = 3$$

are the eigenvalues of A. To find all eigenvectors of A associated with $\lambda_1 = 2$, we form the linear system

$$A\mathbf{x} = 2\mathbf{x},$$

or

$$\begin{bmatrix} 1 & 1 \\ -2 & 4 \end{bmatrix} \begin{bmatrix} x_1 \\ x_2 \end{bmatrix} = 2 \begin{bmatrix} x_1 \\ x_2 \end{bmatrix}.$$

This gives

$$\begin{aligned} x_1 + x_2 &= 2x_1 \\ -2x_1 + 4x_2 &= 2x_2 \end{aligned}$$

or

$$\begin{aligned} (2 - 1)x_1 - x_2 &= 0 \\ 2x_1 + (2 - 4)x_2 &= 0 \end{aligned}$$

or

$$\begin{aligned} x_1 - x_2 &= 0 \\ 2x_1 - 2x_2 &= 0. \end{aligned}$$

Note that we could have obtained this last homogeneous system by merely substituting $\lambda = 2$ in (3). All solutions to this last system are given by

$$\begin{aligned} x_1 &= x_2 \\ x_2 &= \text{any real number } r. \end{aligned}$$

Hence all eigenvectors associated with the eigenvalue $\lambda_1 = 2$ are given by $\begin{bmatrix} r \\ r \end{bmatrix}$, r any nonzero real number. In particular,

$$\mathbf{x}_1 = \begin{bmatrix} 1 \\ 1 \end{bmatrix}$$

is an eigenvector associated with $\lambda_1 = 2$. Similarly, for $\lambda_2 = 3$ we obtain, from (3),

$$\begin{aligned} (3-1)x_1 - \qquad x_2 &= 0 \\ 2x_1 + (3-4)x_2 &= 0 \end{aligned}$$

or

$$\begin{aligned} 2x_1 - x_2 &= 0 \\ 2x_1 - x_2 &= 0. \end{aligned}$$

All solutions to this last homogeneous system are given by

$$\begin{aligned} x_1 &= \tfrac{1}{2}x_2 \\ x_2 &= \text{any real number } r. \end{aligned}$$

Hence all eigenvectors associated with the eigenvalue $\lambda_2 = 3$ are given by $\begin{bmatrix} \frac{1}{2}r \\ r \end{bmatrix}$, r any nonzero real number. In particular,

$$\mathbf{x}_2 = \begin{bmatrix} 1 \\ 2 \end{bmatrix}$$

is an eigenvector associated with the eigenvalue $\lambda_2 = 3$. ■

In Examples 1, 2, and 3 we found eigenvalues and eigenvectors by inspection, whereas in Example 4 we proceeded in a more systematic fashion. We use the procedure of Example 4 as our standard method, as follows.

DEFINITION Let $A = \left[a_{ij}\right]$ be an $n \times n$ matrix. The determinant

$$f(\lambda) = \det(\lambda I_n - A) = \begin{vmatrix} \lambda - a_{11} & -a_{12} & \cdots & -a_{1n} \\ -a_{21} & \lambda - a_{22} & \cdots & -a_{2n} \\ \vdots & \vdots & & \vdots \\ -a_{n1} & -a_{n2} & \cdots & \lambda - a_{nn} \end{vmatrix} \tag{4}$$

is called the **characteristic polynomial of** A. The equation

$$f(\lambda) = \det(\lambda I_n - A) = 0$$

is called the **characteristic equation of** A.

EXAMPLE 5 Let

$$A = \begin{bmatrix} 1 & 2 & -1 \\ 1 & 0 & 1 \\ 4 & -4 & 5 \end{bmatrix}.$$

The characteristic polynomial of A is (verify)

$$\begin{aligned} f(\lambda) = \det(\lambda I_3 - A) &= \begin{vmatrix} \lambda - 1 & -2 & 1 \\ -1 & \lambda - 0 & -1 \\ -4 & 4 & \lambda - 5 \end{vmatrix} \\ &= \lambda^3 - 6\lambda^2 + 11\lambda - 6. \end{aligned}$$

■

Recall from Chapter 3 that each term in the expansion of the determinant of an $n \times n$ matrix is a product of n elements of the matrix, containing exactly one element from each row and exactly one element from each column. Thus, if we expand $f(\lambda) = \det(\lambda I_n - A)$, we obtain a polynomial of degree n. A polynomial of degree n with real coefficients has n roots (counting repeats), some of which may be complex numbers. The expression involving λ^n in the characteristic polynomial of A comes from the product

$$(\lambda - a_{11})(\lambda - a_{22})\cdots(\lambda - a_{nn}),$$

so the coefficient of λ^n is 1. We can then write

$$f(\lambda) = \det(\lambda I_n - A) = \lambda^n + c_1\lambda^{n-1} + c_2\lambda^{n-2} + \cdots + c_{n-1}\lambda + c_n.$$

If we let $\lambda = 0$ in $\det(\lambda I_n - A)$ as well as in the expression on the right, then we get $\det(-A) = c_n$, which shows that the constant term c_n is $(-1)^n \det(A)$. This result can be used to establish the following theorem.

THEOREM 8.1 *An $n \times n$ matrix A is singular if and only if 0 is an eigenvalue of A.*

Proof Exercise T.7(b). ■

We now extend our list of nonsingular equivalences.

List of Nonsingular Equivalences

The following statements are equivalent for an $n \times n$ matrix A.

1. A is nonsingular.
2. $\mathbf{x} = \mathbf{0}$ is the only solution to $A\mathbf{x} = \mathbf{0}$.
3. A is row equivalent to I_n.
4. The linear system $A\mathbf{x} = \mathbf{b}$ has a unique solution for every $n \times 1$ matrix $\mathbf{b}$.
5. $\det(A) \neq 0$.
6. A has rank n.
7. A has nullity 0.
8. The rows of A form a linearly independent set of n vectors in R^n.
9. The columns of A form a linearly independent set of n vectors in R^n.
10. Zero is *not* an eigenvalue of A.

We now connect the characteristic polynomial of a matrix with its eigenvalues in the following theorem.

THEOREM 8.2 *The eigenvalues of A are the roots of the characteristic polynomial of A.*

Proof Let λ be an eigenvalue of A with associated eigenvector $\mathbf{x}$. Then $A\mathbf{x} = \lambda\mathbf{x}$, which can be rewritten as

$$A\mathbf{x} = (\lambda I_n)\mathbf{x}$$

or

$$(\lambda I_n - A)\mathbf{x} = \mathbf{0}, \tag{5}$$

a homogeneous system of n equations in n unknowns. This system has a nontrivial solution if and only if the determinant of its coefficient matrix vanishes (Corollary 3.4, Section 3.2), that is, if and only if $\det(\lambda I_n - A) = 0$.

Conversely, if λ is a root of the characteristic polynomial of A, then $\det(\lambda I_n - A) = 0$, so the homogeneous system (5) has a nontrivial solution $\mathbf{x}$. Hence λ is an eigenvalue of A. ■

Thus, to find the eigenvalues of a given matrix A, we must find the roots of its characteristic polynomial $f(\lambda)$. There are many methods for finding approximations to the roots of a polynomial, some of them more effective than others; indeed, many computer programs are available to find the roots of a polynomial. Two results that are sometimes useful in this connection are as follows: The product of all the roots of the polynomial

$$f(\lambda) = \lambda^n + a_1\lambda^{n-1} + \cdots + a_{n-1}\lambda + a_n$$

is $(-1)^n a_n$, and if $a_1, a_2, \ldots, a_n$ are integers, then $f(\lambda)$ cannot have a rational root that is not already an integer. Thus, as possible rational roots of $f(\lambda)$, one need only try the integer factors of a_n. Of course, $f(\lambda)$ might well have irrational roots or complex roots. However, to minimize the computational effort and as a convenience to the reader, many of the characteristic polynomials to be considered in the rest of this chapter have only integer roots, and each of these roots is a factor of the constant term of the characteristic polynomial of A. The corresponding eigenvectors are obtained by substituting the value of λ in Equation (5) and solving the resulting homogeneous system. The solution to this type of problem has already been studied in Section 6.5.

EXAMPLE 6 Consider the matrix of Example 5. The characteristic polynomial is

$$f(\lambda) = \lambda^3 - 6\lambda^2 + 11\lambda - 6.$$

The possible integer roots of $f(\lambda)$ are ± 1, ± 2, ± 3, and ± 6. By substituting these values in $f(\lambda)$, we find that $f(1) = 0$, so $\lambda = 1$ is a root of $f(\lambda)$. Hence $(\lambda - 1)$ is a factor of $f(\lambda)$. Dividing $f(\lambda)$ by $(\lambda - 1)$, we obtain (verify)

$$f(\lambda) = (\lambda - 1)(\lambda^2 - 5\lambda + 6).$$

Factoring $\lambda^2 - 5\lambda + 6$, we have

$$f(\lambda) = (\lambda - 1)(\lambda - 2)(\lambda - 3).$$

The eigenvalues of A are then

$$\lambda_1 = 1, \qquad \lambda_2 = 2, \qquad \lambda_3 = 3.$$

To find an eigenvector $\mathbf{x}_1$ associated with $\lambda_1 = 1$, we form the linear system

$$(1I_3 - A)\mathbf{x} = \mathbf{0},$$

$$\begin{bmatrix} 1-1 & -2 & 1 \\ -1 & 1 & -1 \\ -4 & 4 & 1-5 \end{bmatrix} \begin{bmatrix} x_1 \\ x_2 \\ x_3 \end{bmatrix} = \begin{bmatrix} 0 \\ 0 \\ 0 \end{bmatrix}$$

or

$$\begin{bmatrix} 0 & -2 & 1 \\ -1 & 1 & -1 \\ -4 & 4 & -4 \end{bmatrix} \begin{bmatrix} x_1 \\ x_2 \\ x_3 \end{bmatrix} = \begin{bmatrix} 0 \\ 0 \\ 0 \end{bmatrix}.$$

A solution is

$$\begin{bmatrix} -\frac{1}{2}r \\ \frac{1}{2}r \\ r \end{bmatrix}$$

for any number r. Thus, for $r = 2$,

$$\mathbf{x}_1 = \begin{bmatrix} -1 \\ 1 \\ 2 \end{bmatrix}$$

is an eigenvector of A associated with $\lambda_1 = 1$.

To find an eigenvector $\mathbf{x}_2$ associated with $\lambda_2 = 2$, we form the linear system

$$(2I_3 - A)\mathbf{x} = \mathbf{0},$$

that is,

$$\begin{bmatrix} 2-1 & -2 & 1 \\ -1 & 2 & -1 \\ -4 & 4 & 2-5 \end{bmatrix} \begin{bmatrix} x_1 \\ x_2 \\ x_3 \end{bmatrix} = \begin{bmatrix} 0 \\ 0 \\ 0 \end{bmatrix}$$

or

$$\begin{bmatrix} 1 & -2 & 1 \\ -1 & 2 & -1 \\ -4 & 4 & -3 \end{bmatrix} \begin{bmatrix} x_1 \\ x_2 \\ x_3 \end{bmatrix} = \begin{bmatrix} 0 \\ 0 \\ 0 \end{bmatrix}.$$

A solution is

$$\begin{bmatrix} -\frac{1}{2}r \\ \frac{1}{4}r \\ r \end{bmatrix}$$

for any number r. Thus, for $r = 4$,

$$\mathbf{x}_2 = \begin{bmatrix} -2 \\ 1 \\ 4 \end{bmatrix}$$

is an eigenvector of A associated with $\lambda_2 = 2$.

To find an eigenvector $\mathbf{x}_3$ associated with $\lambda_3 = 3$, we form the linear system

$$(3I_3 - A)\mathbf{x} = \mathbf{0},$$

and find that a solution is (verify)

$$\begin{bmatrix} -\frac{1}{4}r \\ \frac{1}{4}r \\ r \end{bmatrix}$$

for any number r. Thus, for $r = 4$,

$$\mathbf{x}_3 = \begin{bmatrix} -1 \\ 1 \\ 4 \end{bmatrix}$$

is an eigenvector of A associated with $\lambda_3 = 3$. ■

EXAMPLE 7 Compute the eigenvalues and associated eigenvectors of

$$A = \begin{bmatrix} 0 & 0 & 3 \\ 1 & 0 & -1 \\ 0 & 1 & 3 \end{bmatrix}.$$

Solution The characteristic polynomial of A is

$$p(\lambda) = \det(\lambda I_3 - A) = \begin{vmatrix} \lambda - 0 & 0 & -3 \\ -1 & \lambda - 0 & 1 \\ 0 & -1 & \lambda - 3 \end{vmatrix} = \lambda^3 - 3\lambda^2 + \lambda - 3$$

(verify). We find that $\lambda = 3$ is a root of $p(\lambda)$. Dividing $p(\lambda)$ by $(\lambda - 3)$, we obtain $p(\lambda) = (\lambda - 3)(\lambda^2 + 1)$. The eigenvalues of A are then

$$\lambda_1 = 3, \quad \lambda_2 = i, \quad \lambda_3 = -i.$$

To obtain an eigenvector $\mathbf{x}_1$ associated with $\lambda_1 = 3$, we substitute $\lambda = 3$ in (5), obtaining

$$\begin{bmatrix} 3 - 0 & 0 & -3 \\ -1 & 3 - 0 & 1 \\ 0 & -1 & 3 - 3 \end{bmatrix} \begin{bmatrix} x_1 \\ x_2 \\ x_3 \end{bmatrix} = \begin{bmatrix} 0 \\ 0 \\ 0 \end{bmatrix}.$$

We find that the vector $\begin{bmatrix} r \\ 0 \\ r \end{bmatrix}$ is a solution for any number r (verify). Letting $r = 1$, we conclude that

$$\mathbf{x}_1 = \begin{bmatrix} 1 \\ 0 \\ 1 \end{bmatrix}$$

is an eigenvector of A associated with $\lambda_1 = 3$. To obtain an eigenvector $\mathbf{x}_2$ associated with $\lambda_2 = i$, we substitute $\lambda = i$ in (5), obtaining

$$\begin{bmatrix} i - 0 & 0 & -3 \\ -1 & i - 0 & 1 \\ 0 & -1 & i - 3 \end{bmatrix} \begin{bmatrix} x_1 \\ x_2 \\ x_3 \end{bmatrix} = \begin{bmatrix} 0 \\ 0 \\ 0 \end{bmatrix}.$$

We find that the vector $\begin{bmatrix} (-3i)r \\ (-3 + i)r \\ r \end{bmatrix}$ is a solution for any number r (verify). Letting $r = 1$, we conclude that

$$\mathbf{x}_2 = \begin{bmatrix} -3i \\ -3 + i \\ 1 \end{bmatrix}$$

is an eigenvector of A associated with $\lambda_2 = i$. Similarly, we find that

$$\mathbf{x}_3 = \begin{bmatrix} 3i \\ -3 - i \\ 1 \end{bmatrix}$$

is an eigenvector of A associated with $\lambda_3 = -i$. ■

> The procedure for finding the eigenvalues and associated eigenvectors of a matrix is as follows.
>
> ***Step 1.*** Determine the roots of the characteristic polynomial $f(\lambda) = \det(\lambda I_n - A)$. These are the eigenvalues of A.
>
> ***Step 2.*** For each eigenvalue λ, find all the nontrivial solutions to the homogeneous system $(\lambda I_n - A)\mathbf{x} = \mathbf{0}$. These are the eigenvectors of A associated with the eigenvalue λ.

Of course, the characteristic polynomial of a matrix may have some complex roots and it may even have no real roots. However, in the important case of symmetric matrices, all the roots of the characteristic polynomial are real. We shall prove this in Section 8.3 (Theorem 8.6).

Eigenvalues and eigenvectors satisfy many important and interesting properties. For example, if A is an upper (lower) triangular matrix, or a diagonal matrix, then the eigenvalues of A are the elements on the main diagonal of A (Exercise T.3). The set S consisting of all eigenvectors of A associated with λ_j as well as the zero vector is a subspace of R^n (Exercise T.1) called the **eigenspace associated with** λ_j. Other properties are developed in the exercises for this section.

It must be pointed out that the method for finding the eigenvalues of a linear transformation or matrix by obtaining the roots of the characteristic polynomial is not practical for $n > 4$, since it involves evaluating a determinant. Efficient numerical methods for finding eigenvalues and associated eigenvectors are studied in numerical analysis courses.

Warning When finding the eigenvalues and associated eigenvectors of a matrix A, do not make the common mistake of first transforming A to reduced row echelon form B and then finding the eigenvalues and eigenvectors of B. To see quickly how this approach fails, consider the matrix A defined in Example 4. Its eigenvalues are $\lambda_1 = 2$ and $\lambda_2 = 3$. Since A is a nonsingular matrix, when we transform it to reduced row echelon form B, we have $B = I_2$. The eigenvalues of I_2 are $\lambda_1 = 1$ and $\lambda_2 = 1$.

We now turn to examine briefly three applications of eigenvalues and eigenvectors. The first two of these applications have already been seen in this book; the third one is new. Chapter 9 is devoted entirely to a deeper study of several additional applications of eigenvalues and eigenvectors.

MARKOV CHAINS

We have already discussed Markov processes or chains in Sections 1.4 and 2.5. Let T be a regular transition matrix of a Markov process. In Theorem 2.5 we showed that as $n \to \infty$, T^n approaches a matrix A, all of whose columns are the identical vector $\mathbf{u}$. Also, Theorem 2.6 showed that $\mathbf{u}$ is a steady-state vector, which is the unique probability vector satisfying the matrix equation $T\mathbf{u} = \mathbf{u}$. This means that $\lambda = 1$ is an eigenvalue of T and $\mathbf{u}$ is an associated eigenvector. Finally, since the columns of A add up to 1, it follows from Exercise T.14 that $\lambda = 1$ is an eigenvalue of A.

LINEAR ECONOMIC MODELS

In Section 2.6 we discussed the Leontief closed model, consisting of a society made up of a farmer, a carpenter, and a tailor, where each person produces one unit of each commodity during the year. The exchange matrix A gives the portion of each commodity that is consumed by each individual during the year. The problem facing the economic planner is that of determining the prices p_1, p_2, and p_3 of the three commodities so that no one makes money or loses money, that is, so that we have a state of equilibrium. Let $\mathbf{p}$ denote the price vector. Then the problem is that of finding a solution $\mathbf{p}$ to the linear system $A\mathbf{p} = \mathbf{p}$ whose components p_i will be nonnegative with at least one positive p_i. It follows from Exercise T.14 that $\lambda = 1$ is an eigenvalue of A and $\mathbf{p}$ is an associated eigenvector.

STABLE AGE DISTRIBUTION IN A POPULATION

Consider a population of animals that can live to a maximum age of n years (or any other time unit). Suppose that the number of males in the population is always a fixed percentage of the female population. Thus, in studying the growth of the entire population, we can ignore the male population and concentrate our attention on the female population. We divide the female population into $n + 1$ age groups as follows:

$$\begin{aligned} x_i^{(k)} &= \text{number of females of age } i \text{ who are alive at time } k,\ 0 \leq i \leq n; \\ f_i &= \text{fraction of females of age } i \text{ who will be alive a year later;} \\ b_i &= \text{average number of females born to a female of age } i. \end{aligned}$$

Let

$$\mathbf{x}^{(k)} = \begin{bmatrix} x_0^{(k)} \\ x_1^{(k)} \\ \vdots \\ x_n^{(k)} \end{bmatrix} \qquad (k \geq 0)$$

denote the age distribution vector at time k.

The number of females in the first age group (age zero) at time $k + 1$ is merely the total number of females born from time k to time $k + 1$. There are $x_0^{(k)}$ females in the first age group at time k and each of these females, on the average, produces b_0 female offspring, so the first age group produces a total of $b_0 x_0^{(k)}$ females. Similarly, the $x_1^{(k)}$ females in the second age group (age 1) produce a total of $b_1 x_1^{(k)}$ females. Thus

$$x_0^{(k+1)} = b_0 x_0^{(k)} + b_1 x_1^{(k)} + \cdots + b_n x_n^{(k)}. \tag{6}$$

The number $x_1^{(k+1)}$ of females in the second age group at time $k + 1$ is the number of females from the first age group at time k who are alive a year later. Thus

$$x_1^{(k+1)} = \begin{pmatrix} \text{fraction of females in} \\ \text{first age group who are} \\ \text{alive a year later} \end{pmatrix} \times \begin{pmatrix} \text{number of females in} \\ \text{first age group} \end{pmatrix},$$

or

$$x_1^{(k+1)} = f_0 x_0^{(k)},$$

and, in general,

$$x_j^{(k+1)} = f_{j-1} x_{j-1}^{(k)} \qquad (1 \leq j \leq n). \tag{7}$$

We can write (6) and (7), using matrix notation, as

$$\mathbf{x}^{(k+1)} = A\mathbf{x}^{(k)} \qquad (k \geq 1), \tag{8}$$

where

$$A = \begin{bmatrix} b_0 & b_1 & b_2 & \cdots & b_{n-1} & b_n \\ f_0 & 0 & 0 & \cdots & 0 & 0 \\ 0 & f_1 & 0 & \cdots & 0 & 0 \\ \vdots & \vdots & \vdots & & \vdots & \vdots \\ 0 & 0 & 0 & \cdots & f_{n-1} & 0 \end{bmatrix},$$

and A is called a **Leslie*** **matrix**. We can use Equation (8) to try to determine a distribution of the population by age groups at time $k+1$ so that the number of females in each age group at time $k+1$ will be a fixed multiple of the number in the corresponding age group at time k. That is, if λ is the multiplier, we want

$$\mathbf{x}^{(k+1)} = \lambda\mathbf{x}^{(k)}$$

or

$$A\mathbf{x}^{(k)} = \lambda\mathbf{x}^{(k)}.$$

Thus λ is an eigenvalue of A and $\mathbf{x}^{(k)}$ is a corresponding eigenvector. If $\lambda = 1$, the number of females in each age group will be the same year after year. If we can find an eigenvector $\mathbf{x}^{(k)}$ corresponding to the eigenvalue $\lambda = 1$, we say that we have a **stable age distribution**.

EXAMPLE 8

Consider a beetle that can live to a maximum age of two years and whose population dynamics are represented by the Leslie matrix

$$A = \begin{bmatrix} 0 & 0 & 6 \\ \frac{1}{2} & 0 & 0 \\ 0 & \frac{1}{3} & 0 \end{bmatrix}.$$

We find that $\lambda = 1$ is an eigenvalue of A with corresponding eigenvector

$$\begin{bmatrix} 6 \\ 3 \\ 1 \end{bmatrix}.$$

Thus, if the numbers of females in the three groups are proportional to 6:3:1, we have a stable age distribution. That is, if we have 600 females in the first age group, 300 in the second, and 100 in the third, then year after year the number of females in each age group will remain the same. ■

Population growth problems of the type considered in Example 8 have applications to animal harvesting. For further discussion of elementary applications of eigenvalues and eigenvectors, see D. R. Hill, *Experiments in Computational Matrix Algebra*, New York: Random House, 1988 or D. R. Hill and D. E. Zitarelli, *Linear Algebra Labs with MATLAB*, 3rd ed., Upper Saddle River, N.J.: Prentice Hall, 2004.

The theoretical exercises in this section contain many useful properties of eigenvalues. The reader is encouraged to develop a list of facts about eigenvalues and eigenvectors.

---

*See P. H. Leslie, "On the Use of Matrices in Certain Population Mathematics," *Biometrika* 33, 1945.

Key Terms

Eigenvalue
Eigenvector
Proper value
Characteristic value
Latent value
Characteristic polynomial
Characteristic equation
Roots of the characteristic polynomial
Eigenspace
Leslie matrix
Stable age distribution
Invariant subspace

8.1 Exercises

1. Let $A = \begin{bmatrix} 3 & -1 \\ -2 & 2 \end{bmatrix}$.

(a) Verify that $\lambda_1 = 1$ is an eigenvalue of A and $\mathbf{x}_1 = \begin{bmatrix} r \\ 2r \end{bmatrix}$, $r \neq 0$, is an associated eigenvector.

(b) Verify that $\lambda_1 = 4$ is an eigenvalue of A and $\mathbf{x}_2 = \begin{bmatrix} r \\ -r \end{bmatrix}$, $r \neq 0$, is an associated eigenvector.

2. Let $A = \begin{bmatrix} 2 & 2 & 3 \\ 1 & 2 & 1 \\ 2 & -2 & 1 \end{bmatrix}$.

(a) Verify that $\lambda_1 = -1$ is an eigenvalue of A and $\mathbf{x}_1 = \begin{bmatrix} 1 \\ 0 \\ -1 \end{bmatrix}$ is an associated eigenvector.

(b) Verify that $\lambda_2 = 2$ is an eigenvalue of A and $\mathbf{x}_2 = \begin{bmatrix} -2 \\ -3 \\ 2 \end{bmatrix}$ is an associated eigenvector.

(c) Verify that $\lambda_3 = 4$ is an eigenvalue of A and $\mathbf{x}_3 = \begin{bmatrix} 8 \\ 5 \\ 2 \end{bmatrix}$ is an associated eigenvector.

In Exercises 3 through 7, find the characteristic polynomial of each matrix.

3. $\begin{bmatrix} 1 & 2 & 1 \\ 0 & 1 & 2 \\ -1 & 3 & 2 \end{bmatrix}$ **4.** $\begin{bmatrix} 2 & 1 \\ -1 & 3 \end{bmatrix}$

5. $\begin{bmatrix} 4 & -1 & 3 \\ 0 & 2 & 1 \\ 0 & 0 & 3 \end{bmatrix}$ **6.** $\begin{bmatrix} 4 & 2 \\ 3 & 3 \end{bmatrix}$

7. $\begin{bmatrix} 2 & 1 & 2 \\ 2 & 2 & -2 \\ 3 & 1 & 1 \end{bmatrix}$

In Exercises 8 through 15, find the characteristic polynomial, eigenvalues, and eigenvectors of each matrix.

8. $\begin{bmatrix} 0 & 1 & 2 \\ 0 & 0 & 3 \\ 0 & 0 & 0 \end{bmatrix}$ **9.** $\begin{bmatrix} 1 & 0 & 0 \\ -1 & 3 & 0 \\ 3 & 2 & -2 \end{bmatrix}$

10. $\begin{bmatrix} 1 & 1 \\ 1 & 1 \end{bmatrix}$ **11.** $\begin{bmatrix} 1 & -1 \\ 2 & 4 \end{bmatrix}$

12. $\begin{bmatrix} 2 & -2 & 3 \\ 0 & 3 & -2 \\ 0 & -1 & 2 \end{bmatrix}$ **13.** $\begin{bmatrix} 2 & 2 & 3 \\ 1 & 2 & 1 \\ 2 & -2 & 1 \end{bmatrix}$

14. $\begin{bmatrix} 2 & 0 & 0 \\ 3 & -1 & 0 \\ 0 & 4 & 3 \end{bmatrix}$ **15.** $\begin{bmatrix} 1 & 2 & 3 & 4 \\ 0 & -1 & 3 & 2 \\ 0 & 0 & 3 & 3 \\ 0 & 0 & 0 & 2 \end{bmatrix}$

16. Find the characteristic polynomial, the eigenvalues and associated eigenvectors of each of the following matrices.

(a) $\begin{bmatrix} 0 & 1 \\ -1 & 0 \end{bmatrix}$ (b) $\begin{bmatrix} -2 & -4 & -8 \\ 1 & 0 & 0 \\ 0 & 1 & 0 \end{bmatrix}$

(c) $\begin{bmatrix} 2-i & 2i & 0 \\ 1 & 0 & 0 \\ 0 & 1 & 0 \end{bmatrix}$ (d) $\begin{bmatrix} 5 & 2 \\ -1 & 3 \end{bmatrix}$

17. Find all the eigenvalues and associated eigenvectors of each of the following matrices.

(a) $\begin{bmatrix} -1 & -1+i \\ 1 & 0 \end{bmatrix}$ (b) $\begin{bmatrix} i & 1 & 0 \\ 1 & i & 0 \\ 0 & 0 & 1 \end{bmatrix}$

(c) $\begin{bmatrix} 0 & -1 & 0 \\ 1 & 0 & 0 \\ 0 & 1 & 0 \end{bmatrix}$ (d) $\begin{bmatrix} 0 & 0 & -9 \\ 0 & 1 & 0 \\ 1 & 0 & 0 \end{bmatrix}$

In Exercises 18 and 19, find bases for the eigenspaces (see Exercise T.1) associated with each eigenvalue.

18. $\begin{bmatrix} 2 & 3 & 0 \\ 0 & 1 & 0 \\ 0 & 0 & 2 \end{bmatrix}$ **19.** $\begin{bmatrix} 2 & 2 & 3 & 4 \\ 0 & 2 & 3 & 2 \\ 0 & 0 & 1 & 1 \\ 0 & 0 & 0 & 1 \end{bmatrix}$

In Exercises 20 through 23, find a basis for the eigenspace (see Exercise T.1) associated with λ.

20. $\begin{bmatrix} 0 & 0 & 1 \\ 0 & 1 & 0 \\ 1 & 0 & 0 \end{bmatrix}$, $\lambda = 1$ **21.** $\begin{bmatrix} 2 & 1 & 0 \\ 1 & 2 & 1 \\ 0 & 1 & 2 \end{bmatrix}$, $\lambda = 2$

22. $\begin{bmatrix} 3 & 0 & 0 \\ -2 & 3 & -2 \\ 2 & 0 & 5 \end{bmatrix}$, $\lambda = 3$

23. $\begin{bmatrix} 4 & 2 & 0 & 0 \\ 3 & 3 & 0 & 0 \\ 0 & 0 & 2 & 5 \\ 0 & 0 & 0 & 2 \end{bmatrix}$, $\lambda = 2$

24. Let $A = \begin{bmatrix} 0 & -4 & 0 \\ 1 & 0 & 0 \\ 0 & 1 & 0 \end{bmatrix}$.

(a) Find a basis for the eigenspace associated with the eigenvaluc $\lambda_1 = 2i$.

(b) Find a basis for the eigenspace associated with the eigenvalue $\lambda_2 = -2i$.

25. Let $A = \begin{bmatrix} 3 & 0 & 0 & 0 \\ 0 & 3 & 0 & 0 \\ 4 & 0 & 0 & 3 \\ 0 & 0 & -3 & 0 \end{bmatrix}$.

(a) Find a basis for the eigenspace associated with the eigenvalue $\lambda_1 = 3$.

(b) Find a basis for the eigenspace associated with the eigenvalue $\lambda_2 = 3i$.

26. Let A be the matrix of Exercise 1. Find the eigenvalues and eigenvectors of A^2 and verify Exercise T.5.

27. Consider a living organism that can live to a maximum age of 2 years and whose Leslie matrix is

$$A = \begin{bmatrix} 0 & 0 & 8 \\ \frac{1}{4} & 0 & 0 \\ 0 & \frac{1}{2} & 0 \end{bmatrix}.$$

Find a stable age distribution.

28. Consider a living organism that can live to a maximum age of 2 years and whose Leslie matrix is

$$A = \begin{bmatrix} 0 & 4 & 0 \\ \frac{1}{4} & 0 & 0 \\ 0 & \frac{1}{2} & 0 \end{bmatrix}.$$

Show that a stable age distribution exists and find one.

Theoretical Exercises

T.1. Let λ_j be a particular eigenvalue of the $n \times n$ matrix A. Show that the subset S of R^n consisting of the zero vector and all eigenvectors of A associated with λ_j is a subspace of R^n, called the **eigenspace** associated with the eigenvalue λ_j.

T.2. In Exercise T.1 why do we have to include the zero vector in the subset S?

T.3. Show that if A is an upper (lower) triangular matrix or a diagonal matrix, then the eigenvalues of A are the elements on the main diagonal of A.

T.4. Show that A and A^T have the same eigenvalues. What, if anything, can we say about the associated eigenvectors of A and A^T?

T.5. If λ is an eigenvalue of A with associated eigenvector $\mathbf{x}$, show that λ^k is an eigenvalue of $A^k = A \cdot A \cdots A$ (k factors) with associated eigenvector $\mathbf{x}$, where k is a positive integer.

T.6. An $n \times n$ matrix A is called **nilpotent** if $A^k = O$ for some positive integer k. Show that if A is nilpotent, then the only eigenvalue of A is 0. (*Hint*: Use Exercise T.5.)

T.7. Let A be an $n \times n$ matrix.

(a) Show that $\det(A)$ is the product of all the roots of the characteristic polynomial of A.

(b) Show that A is singular if and only if 0 is an eigenvalue of A.

T.8. Let λ be an eigenvalue of the nonsingular matrix A with associated eigenvector $\mathbf{x}$. Show that $1/\lambda$ is an eigenvalue of A^{-1} with associated eigenvector $\mathbf{x}$.

T.9. Let A be any $n \times n$ real matrix.

(a) Show that the coefficient of λ^{n-1} in the characteristic polynomial of A is given by $-\operatorname{Tr}(A)$, where $\operatorname{Tr}(A)$ denotes the trace of A (see Supplementary Exercise T.1 in Chapter 1).

(b) Show that $\operatorname{Tr}(A)$ is the sum of the eigenvalues of A.

(c) Show that the constant term of the characteristic polynomial of A is $\pm$ times the product of the eigenvalues of A.

(d) Show that $\det(A)$ is the product of the eigenvalues of A.

T.10. Let A be an $n \times n$ matrix with eigenvalues λ_1 and λ_2, where $\lambda_1 \neq \lambda_2$. Let S_1 and S_2 be the eigenspaces associated with λ_1 and λ_2, respectively. Explain why the zero vector is the only vector that is in both S_1 and S_2.

T.11. Let λ be an eigenvalue of A with associated eigenvector $\mathbf{x}$. Show that $\lambda + r$ is an eigenvalue of $A + rI_n$ with associated eigenvector $\mathbf{x}$. Thus, adding a scalar multiple of the identity matrix to A merely shifts the eigenvalues by the scalar multiple.

T.12. Let A be a square matrix.

(a) Suppose that the homogeneous system $A\mathbf{x} = \mathbf{0}$ has a nontrivial solution $\mathbf{x} = \mathbf{u}$. Show that $\mathbf{u}$ is an eigenvector of A.

(b) Suppose that 0 is an eigenvalue of A and $\mathbf{v}$ is an associated eigenvector. Show that the homogeneous system $A\mathbf{x} = \mathbf{0}$ has a nontrivial solution.

T.13. Let A and B be $n \times n$ matrices such that $A\mathbf{x} = \lambda\mathbf{x}$ and $B\mathbf{x} = \mu\mathbf{x}$. Show that:

(a) $(A + B)\mathbf{x} = (\lambda + \mu)\mathbf{x}$

(b) $(AB)\mathbf{x} = (\lambda\mu)\mathbf{x}$

T.14. Show that if A is a matrix all of whose columns add up to 1, then $\lambda = 1$ is an eigenvalue of A. (*Hint*: Consider the product $A^T\mathbf{x}$, where $\mathbf{x}$ is a vector all of whose entries are 1 and use Exercise T.4.)

T.15. Let A be an $n \times n$ matrix and consider the linear operator on R^n defined by $L(\mathbf{u}) = A\mathbf{u}$, for $\mathbf{u}$ in R^n. A subspace W of R^n is called **invariant** under L if for any $\mathbf{w}$ in W, $L(\mathbf{w})$ is also in W. Show that an eigenspace of A is invariant under L (see Exercise T.1).

MATLAB Exercises

MATLAB has a pair of commands that can be used to find the characteristic polynomial and eigenvalues of a matrix. Command **poly(A)** *gives the coefficients of the characteristic polynomial of matrix A, starting with the highest-degree term. If we set* **v** = **poly(A)** *and then use command* **roots(v)**, *we obtain the roots of the characteristic polynomial of A. This process can also find complex eigenvalues, which are discussed in Appendix A.2.*

Once we have an eigenvalue λ of A, we can use **rref** *or* **homsoln** *to find a corresponding eigenvector from the linear system* $(\lambda I - A)\mathbf{x} = \mathbf{0}$.

ML.1. Find the characteristic polynomial of each of the following matrices using MATLAB.

(a) $A = \begin{bmatrix} 1 & 2 \\ 2 & -1 \end{bmatrix}$

(b) $A = \begin{bmatrix} 2 & 4 & 0 \\ 1 & 2 & 1 \\ 0 & 4 & 2 \end{bmatrix}$

(c) $A = \begin{bmatrix} 1 & 0 & 0 & 0 \\ 2 & -2 & 0 & 0 \\ 0 & 0 & 2 & -1 \\ 0 & 0 & -1 & 2 \end{bmatrix}$

ML.2. Use the **poly** and **roots** commands in MATLAB to find the eigenvalues of the following matrices:

(a) $A = \begin{bmatrix} 1 & -3 \\ 3 & -5 \end{bmatrix}$ (b) $A = \begin{bmatrix} 3 & -1 & 4 \\ -1 & 0 & 1 \\ 4 & 1 & 2 \end{bmatrix}$

(c) $A = \begin{bmatrix} 2 & -2 & 0 \\ 1 & -1 & 0 \\ 1 & -1 & 0 \end{bmatrix}$ (d) $A = \begin{bmatrix} 2 & 4 \\ 3 & 6 \end{bmatrix}$

ML.3. In each of the following cases, λ is an eigenvalue of A. Use MATLAB to find a corresponding eigenvector.

(a) $\lambda = 3$, $A = \begin{bmatrix} 1 & 2 \\ -1 & 4 \end{bmatrix}$

(b) $\lambda = -1$, $A = \begin{bmatrix} 4 & 0 & 0 \\ 1 & 3 & 0 \\ 2 & 1 & -1 \end{bmatrix}$

(c) $\lambda = 2$, $A = \begin{bmatrix} 2 & 1 & 2 \\ 2 & 2 & -2 \\ 3 & 1 & 1 \end{bmatrix}$

ML.4. Consider a living organism that can live to a maximum age of two years and whose Leslie matrix is

$$\begin{bmatrix} 0.2 & 0.8 & 0.3 \\ 0.9 & 0 & 0 \\ 0 & 0.7 & 0 \end{bmatrix}.$$

Find a stable age distribution.

8.2 DIAGONALIZATION

In this section we show how to find the eigenvalues and associated eigenvectors of a given matrix A by finding the eigenvalues and eigenvectors of a related matrix B that has the same eigenvalues and eigenvectors as A. The matrix B has the helpful property that its eigenvalues are easily obtained. Thus, we will have found the eigenvalues of A. In Section 8.3, this approach will shed much light on the eigenvalue-eigenvector problem. For convenience, we only work with matrices all of whose entries and eigenvalues are real numbers.

SIMILAR MATRICES

DEFINITION A matrix B is said to be **similar** to a matrix A if there is a nonsingular matrix P such that

$$B = P^{-1}AP.$$

EXAMPLE 1 Let

$$A = \begin{bmatrix} 1 & 1 \\ -2 & 4 \end{bmatrix}$$

be the matrix of Example 4 in Section 8.1. Let

$$P = \begin{bmatrix} 1 & 1 \\ 1 & 2 \end{bmatrix}.$$

Then

$$P^{-1} = \begin{bmatrix} 2 & -1 \\ -1 & 1 \end{bmatrix}$$

and

$$B = P^{-1}AP = \begin{bmatrix} 2 & -1 \\ -1 & 1 \end{bmatrix} \begin{bmatrix} 1 & 1 \\ -2 & 4 \end{bmatrix} \begin{bmatrix} 1 & 1 \\ 1 & 2 \end{bmatrix} = \begin{bmatrix} 2 & 0 \\ 0 & 3 \end{bmatrix}.$$

Thus B is similar to A. ■

We shall let the reader (Exercise T.1) show that the following elementary properties hold for similarity:

1. A is similar to A.
2. If B is similar to A, then A is similar to B.
3. If A is similar to B and B is similar to C, then A is similar to C.

By property 2 we replace the statements "A is similar to B" and "B is similar to A" by "A and B are similar."

DEFINITION

We shall say that the matrix A is **diagonalizable** if it is similar to a diagonal matrix. In this case we also say that A **can be diagonalized**.

EXAMPLE 2

If A and B are as in Example 1, then A is diagonalizable, since it is similar to B. ■

THEOREM 8.3

Similar matrices have the same eigenvalues.

Proof

Let A and B be similar. Then $B = P^{-1}AP$, for some nonsingular matrix P. We prove that A and B have the same characteristic polynomials, $f_A(\lambda)$ and $f_B(\lambda)$, respectively. We have

$$\begin{aligned} f_B(\lambda) &= \det(\lambda I_n - B) = \det(\lambda I_n - P^{-1}AP) \\ &= \det(P^{-1}\lambda I_n P - P^{-1}AP) = \det(P^{-1}(\lambda I_n - A)P) \\ &= \det(P^{-1})\det(\lambda I_n - A)\det(P) \qquad (1) \\ &= \det(P^{-1})\det(P)\det(\lambda I_n - A) \\ &= \det(\lambda I_n - A) = f_A(\lambda). \end{aligned}$$

Since $f_A(\lambda) = f_B(\lambda)$, it follows that A and B have the same eigenvalues. ■

It follows from Exercise T.3 in Section 8.1 that the eigenvalues of a diagonal matrix are the entries on its main diagonal. The following theorem establishes when a matrix is diagonalizable.

THEOREM 8.4

An $n \times n$ matrix A is diagonalizable if and only if it has n linearly independent eigenvectors.

Proof Suppose that A is similar to D. Then

$$P^{-1}AP = D,$$

a diagonal matrix, so

$$AP = PD. \tag{2}$$

Let

$$D = \begin{bmatrix} \lambda_1 & 0 & \cdots & 0 \\ 0 & \lambda_2 & \cdots & 0 \\ \vdots & & & \vdots \\ 0 & \cdots & 0 & \lambda_n \end{bmatrix},$$

and let $\mathbf{x}_j$, $j = 1, 2, \ldots, n$, be the jth column of P. From Exercise T.9 in Section 1.3, it follows that the jth column of the matrix AP is $A\mathbf{x}_j$, and the jth column of PD is $\lambda_j \mathbf{x}_j$.

Thus from (2) we have

$$A\mathbf{x}_j = \lambda_j \mathbf{x}_j. \tag{3}$$

Since P is a nonsingular matrix, by Theorem 6.13 in Section 6.6 its columns are linearly independent and so are all nonzero. Hence λ_j is an eigenvalue of A and $\mathbf{x}_j$ is a corresponding eigenvector.

Conversely, suppose that $\lambda_1, \lambda_2, \ldots, \lambda_n$ are n eigenvalues of A and that the corresponding eigenvectors $\mathbf{x}_1, \mathbf{x}_2, \ldots, \mathbf{x}_n$ are linearly independent. Let $P = \begin{bmatrix} \mathbf{x}_1 & \mathbf{x}_2 & \cdots & \mathbf{x}_n \end{bmatrix}$ be the matrix whose jth column is $\mathbf{x}_j$. Since the columns of P are linearly independent, it follows from Theorem 6.13 in Section 6.6 that P is nonsingular. From (3) we obtain (2), which implies that A is diagonalizable. This completes the proof. ■

Remark If A is a diagonalizable matrix, then $P^{-1}AP = D$, where D is a diagonal matrix. It follows from the proof of Theorem 8.4 that the diagonal elements of D are the eigenvalues of A. Moreover, P is a matrix whose columns are, respectively, n linearly independent eigenvectors of A. Observe also that in Theorem 8.4, the order of the columns of P determines the order of the diagonal entries in D.

EXAMPLE 3 Let A be as in Example 1. The eigenvalues are $\lambda_1 = 2$ and $\lambda_2 = 3$. (See Example 4 in Section 8.1.) The corresponding eigenvectors

$$\mathbf{x}_1 = \begin{bmatrix} 1 \\ 1 \end{bmatrix} \quad \text{and} \quad \mathbf{x}_2 = \begin{bmatrix} 1 \\ 2 \end{bmatrix}$$

are linearly independent. Hence A is diagonalizable. Here

$$P = \begin{bmatrix} 1 & 1 \\ 1 & 2 \end{bmatrix} \quad \text{and} \quad P^{-1} = \begin{bmatrix} 2 & -1 \\ -1 & 1 \end{bmatrix}.$$

Thus, as in Example 1,

$$P^{-1}AP = \begin{bmatrix} 2 & -1 \\ -1 & 1 \end{bmatrix} \begin{bmatrix} 1 & 1 \\ -2 & 4 \end{bmatrix} \begin{bmatrix} 1 & 1 \\ 1 & 2 \end{bmatrix} = \begin{bmatrix} 2 & 0 \\ 0 & 3 \end{bmatrix}.$$

On the other hand, if we let $\lambda_1 = 3$ and $\lambda_2 = 2$, then

$$\mathbf{x}_1 = \begin{bmatrix} 1 \\ 2 \end{bmatrix} \quad \text{and} \quad \mathbf{x}_2 = \begin{bmatrix} 1 \\ 1 \end{bmatrix}.$$

Then

$$P = \begin{bmatrix} 1 & 1 \\ 2 & 1 \end{bmatrix} \quad \text{and} \quad P^{-1} = \begin{bmatrix} -1 & 1 \\ 2 & -1 \end{bmatrix}.$$

Hence

$$P^{-1}AP = \begin{bmatrix} -1 & 1 \\ 2 & -1 \end{bmatrix} \begin{bmatrix} 1 & 1 \\ -2 & 4 \end{bmatrix} \begin{bmatrix} 1 & 1 \\ 2 & 1 \end{bmatrix} = \begin{bmatrix} 3 & 0 \\ 0 & 2 \end{bmatrix}.$$

■

EXAMPLE 4 Let

$$A = \begin{bmatrix} 1 & 1 \\ 0 & 1 \end{bmatrix}.$$

The eigenvalues of A are $\lambda_1 = 1$ and $\lambda_2 = 1$. Eigenvectors associated with λ_1 and λ_2 are vectors of the form

$$\begin{bmatrix} r \\ 0 \end{bmatrix},$$

where r is any nonzero real number. Since A does not have two linearly independent eigenvectors, we conclude that A is not diagonalizable. ■

The following is a useful theorem because it identifies a large class of matrices that can be diagonalized.

THEOREM 8.5 *If the roots of the characteristic polynomial of an $n \times n$ matrix A are all distinct (i.e., different from each other), then A is diagonalizable.*

Proof Let $\lambda_1, \lambda_2, \ldots, \lambda_n$ be the distinct eigenvalues of A and let $S = \{\mathbf{x}_1, \mathbf{x}_2, \ldots, \mathbf{x}_n\}$ be a set of associated eigenvectors. We wish to show that S is linearly independent.

Suppose that S is linearly dependent. Then Theorem 6.4 of Section 6.3 implies that some vector $\mathbf{x}_j$ is a linear combination of the preceding vectors in S. We can assume that $S_1 = \{\mathbf{x}_1, \mathbf{x}_2, \ldots, \mathbf{x}_{j-1}\}$ is linearly independent, for otherwise one of the vectors in S_1 is a linear combination of the preceding ones, and we can choose a new set S_2, and so on. We thus have that S_1 is linearly independent and that

$$\mathbf{x}_j = c_1\mathbf{x}_1 + c_2\mathbf{x}_2 + \cdots + c_{j-1}\mathbf{x}_{j-1}, \tag{4}$$

where $c_1, c_2, \ldots, c_{j-1}$ are scalars. Premultiplying (multiplying on the left) both sides of Equation (4) by A, we obtain

$$\begin{aligned} A\mathbf{x}_j &= A(c_1\mathbf{x}_1 + c_2\mathbf{x}_2 + \cdots + c_{j-1}\mathbf{x}_{j-1}) \\ &= c_1A\mathbf{x}_1 + c_2A\mathbf{x}_2 + \cdots + c_{j-1}A\mathbf{x}_{j-1}. \end{aligned} \tag{5}$$

Since $\lambda_1, \lambda_2, \ldots, \lambda_j$ are eigenvalues of A and $\mathbf{x}_1, \mathbf{x}_2, \ldots, \mathbf{x}_j$, its associated eigenvectors, we know that $A\mathbf{x}_i = \lambda_i\mathbf{x}_i$ for $i = 1, 2, \ldots, j$. Substituting in (5), we have

$$\lambda_j\mathbf{x}_j = c_1\lambda_1\mathbf{x}_1 + c_2\lambda_2\mathbf{x}_2 + \cdots + c_{j-1}\lambda_{j-1}\mathbf{x}_{j-1}. \tag{6}$$

Multiplying (4) by λ_j, we obtain

$$\lambda_j\mathbf{x}_j = \lambda_jc_1\mathbf{x}_1 + \lambda_jc_2\mathbf{x}_2 + \cdots + \lambda_jc_{j-1}\mathbf{x}_{j-1}. \tag{7}$$

Subtracting (7) from (6), we have

$$\begin{aligned} 0 &= \lambda_j \mathbf{x}_j - \lambda_j \mathbf{x}_j \\ &= c_1(\lambda_1 - \lambda_j)\mathbf{x}_1 + c_2(\lambda_2 - \lambda_j)\mathbf{x}_2 + \cdots + c_{j-1}(\lambda_{j-1} - \lambda_j)\mathbf{x}_{j-1}. \end{aligned}$$

Since S_1 is linearly independent, we must have

$$c_1(\lambda_1 - \lambda_j) = 0, \quad c_2(\lambda_2 - \lambda_j) = 0, \ldots, \quad c_{j-1}(\lambda_{j-1} - \lambda_j) = 0.$$

Now

$$\lambda_1 - \lambda_j \neq 0, \quad \lambda_2 - \lambda_j \neq 0, \ldots, \quad \lambda_{j-1} - \lambda_j \neq 0$$

(because the λ's are distinct), which implies that

$$c_1 = c_2 = \cdots = c_{j-1} = 0.$$

From (4) we conclude that $\mathbf{x}_j = \mathbf{0}$, which is impossible if $\mathbf{x}_j$ is an eigenvector. Hence S is linearly independent, and from Theorem 8.4 it follows that A is diagonalizable. ■

Remark In the proof of Theorem 8.5, we have actually established the following somewhat stronger result: Let A be an $n \times n$ matrix and let $\lambda_1, \lambda_2, \ldots, \lambda_k$ be k distinct eigenvalues of A with associated eigenvectors $\mathbf{x}_1, \mathbf{x}_2, \ldots, \mathbf{x}_k$. Then $\mathbf{x}_1, \mathbf{x}_2, \ldots, \mathbf{x}_k$ are linearly independent (Exercise T.11).

If the roots of the characteristic polynomial of A are not all distinct, then A may or may not be diagonalizable. The characteristic polynomial of A can be written as the product of n factors, each of the form $\lambda - \lambda_j$, where λ_j is a root of the characteristic polynomial and the eigenvalues of A are the roots of the characteristic polynomial of A. Thus the characteristic polynomial can be written as

$$(\lambda - \lambda_1)^{k_1}(\lambda - \lambda_2)^{k_2}\cdots(\lambda - \lambda_r)^{k_r},$$

where $\lambda_1, \lambda_2, \ldots, \lambda_r$ are the distinct eigenvalues of A, and $k_1, k_2, \ldots, k_r$ are integers whose sum is n. The integer k_i is called the **multiplicity** of λ_i. Thus in Example 4, $\lambda = 1$ is an eigenvalue of

$$A = \begin{bmatrix} 1 & 1 \\ 0 & 1 \end{bmatrix}$$

of multiplicity 2. It can be shown that A can be diagonalized if and only if for each eigenvalue λ_j of multiplicity k_j we can find k_j linearly independent eigenvectors. This means that the solution space of the linear system $(\lambda_j I_n - A)\mathbf{x} = \mathbf{0}$ has dimension k_j. It can also be shown that if λ_j is an eigenvalue of A of multiplicity k_j, then we can never find more than k_j linearly independent eigenvectors associated with λ_j. We consider the following examples.

EXAMPLE 5 Let

$$A = \begin{bmatrix} 0 & 0 & 1 \\ 0 & 1 & 2 \\ 0 & 0 & 1 \end{bmatrix}.$$

The characteristic polynomial of A is $f(\lambda) = \lambda(\lambda - 1)^2$, so the eigenvalues of A are $\lambda_1 = 0$, $\lambda_2 = 1$, and $\lambda_3 = 1$. Thus $\lambda_2 = 1$ is an eigenvalue of multiplicity 2. We now consider the eigenvectors associated with the eigenvalues $\lambda_2 = \lambda_3 = 1$. They are obtained by solving the linear system $(1I_3 - A)\mathbf{x} = \mathbf{0}$:

$$\begin{bmatrix} 1 & 0 & -1 \\ 0 & 0 & -2 \\ 0 & 0 & 0 \end{bmatrix} \begin{bmatrix} x_1 \\ x_2 \\ x_3 \end{bmatrix} = \begin{bmatrix} 0 \\ 0 \\ 0 \end{bmatrix}.$$

A solution is any vector of the form

$$\begin{bmatrix} 0 \\ r \\ 0 \end{bmatrix},$$

where r is any number, so the dimension of the solution space of the linear system $(1I_3 - A)\mathbf{x} = \mathbf{0}$ is 1. There do not exist two linearly independent eigenvectors associated with $\lambda_2 = 1$. Thus A cannot be diagonalized. ■

EXAMPLE 6 Let

$$A = \begin{bmatrix} 0 & 0 & 0 \\ 0 & 1 & 0 \\ 1 & 0 & 1 \end{bmatrix}.$$

The characteristic polynomial of A is $f(\lambda) = \lambda(\lambda - 1)^2$, so the eigenvalues of A are $\lambda_1 = 0$, $\lambda_2 = 1$, $\lambda_3 = 1$; $\lambda_2 = 1$ is again an eigenvalue of multiplicity 2. Now we consider the solution space of $(1I_3 - A)\mathbf{x} = \mathbf{0}$, that is, of

$$\begin{bmatrix} 1 & 0 & 0 \\ 0 & 0 & 0 \\ -1 & 0 & 0 \end{bmatrix} \begin{bmatrix} x_1 \\ x_2 \\ x_3 \end{bmatrix} = \begin{bmatrix} 0 \\ 0 \\ 0 \end{bmatrix}.$$

A solution is any vector of the form

$$\begin{bmatrix} 0 \\ r \\ s \end{bmatrix}$$

for any numbers r and s. Thus we can take as eigenvectors $\mathbf{x}_2$ and $\mathbf{x}_3$ the vectors

$$\mathbf{x}_2 = \begin{bmatrix} 0 \\ 1 \\ 0 \end{bmatrix} \quad \text{and} \quad \mathbf{x}_3 = \begin{bmatrix} 0 \\ 0 \\ 1 \end{bmatrix}.$$

Now we look for an eigenvector associated with $\lambda_1 = 0$. We have to solve the homogeneous system $(0I_3 - A)\mathbf{x} = \mathbf{0}$, or

$$\begin{bmatrix} 0 & 0 & 0 \\ 0 & -1 & 0 \\ -1 & 0 & -1 \end{bmatrix} \begin{bmatrix} x_1 \\ x_2 \\ x_3 \end{bmatrix} = \begin{bmatrix} 0 \\ 0 \\ 0 \end{bmatrix}.$$

A solution is any vector of the form

$$\begin{bmatrix} t \\ 0 \\ -t \end{bmatrix}$$

for any number t. Thus

$$\mathbf{x}_1 = \begin{bmatrix} 1 \\ 0 \\ -1 \end{bmatrix}$$

is an eigenvector associated with $\lambda_1 = 0$. Since $\mathbf{x}_1$, $\mathbf{x}_2$, and $\mathbf{x}_3$ are linearly independent, A can be diagonalized. ■

Thus an $n \times n$ matrix will fail to be diagonalizable only if it does not have n linearly independent eigenvectors.

The procedure for diagonalizing a matrix A is as follows.

Step 1. Form the characteristic polynomial $f(\lambda) = \det(\lambda I_n - A)$ of A.

Step 2. Find the roots of the characteristic polynomial of A.

Step 3. For each eigenvalue λ_j of A of multiplicity k_j, find a basis for the solution space of $(\lambda_j I_n - A)\mathbf{x} = \mathbf{0}$ (the eigenspace associated with λ_j). If the dimension of the eigenspace is less than k_j, then A is not diagonalizable. We thus determine n linearly independent eigenvectors of A. In Section 6.5 we solved the problem of finding a basis for the solution space of a homogeneous system.

Step 4. Let P be the matrix whose columns are the n linearly independent eigenvectors determined in Step 3. Then $P^{-1}AP = D$, a diagonal matrix whose diagonal elements are the eigenvalues of A that correspond to the columns of P.

Preview of Applications

The Fibonacci Sequence (Section 9.1)

The sequence of numbers

$$1, 1, 2, 3, 5, 8, 13, 21, 34, 55, 89, \ldots$$

is called the *Fibonacci sequence*. The first two numbers in this sequence are 1, 1 and the next number is obtained by adding the two preceding ones. Thus, in general, if $u_0 = 1$ and $u_1 = 1$, then for $n \geq 2$

$$u_n = u_{n-1} + u_{n-2}. \qquad (*)$$

The Fibonacci sequence occurs in a wide variety of applications such as the distribution of leaves on some trees, the arrangement of seeds on sunflowers, search techniques in numerical methods, generating random numbers in statistics, and others.

The preceding equation can be used to *successively* compute values of u_n for any desired value of n, which can be tedious if n is large. In addition to $(*)$ we write

$$u_{n-1} = u_{n-1} \qquad (**)$$

and define

$$\mathbf{w}_k = \begin{bmatrix} u_{k+1} \\ u_k \end{bmatrix} \quad \text{and} \quad A = \begin{bmatrix} 1 & 1 \\ 1 & 0 \end{bmatrix}, \qquad 0 \leq k \leq n-1.$$

Then $(*)$ and $(**)$ can be written in matrix form as

$$\mathbf{w}_{n-1} = A\mathbf{w}_{n-2}.$$

By diagonalizing the matrix A, we obtain the following formula for calculating u_n directly:

$$u_n = \frac{1}{\sqrt{5}}\left[\left(\frac{1+\sqrt{5}}{2}\right)^{n+1} - \left(\frac{1-\sqrt{5}}{2}\right)^{n+1}\right].$$

Section 9.1 provides a brief discussion of the Fibonacci sequence of numbers.

Differential Equations (Section 9.2) (Calculus Required)

A differential equation is an equation that involves an unknown function and its derivatives. Differential equations occur in a wide variety of applications.

As an example, suppose that we have a system consisting of two interconnected tanks each containing a brine solution. Tank R contains $x(t)$ pounds of salt in 200 gallons of brine and tank S contains $y(t)$ pounds of salt in 300 gallons of brine. The mixture in each tank is kept uniform by constant stirring. Starting at $t = 0$, brine is pumped from tank R to tank S at 20 gallons/minute and from tank S to tank R at 20 gallons/minute. (See Figure A.)

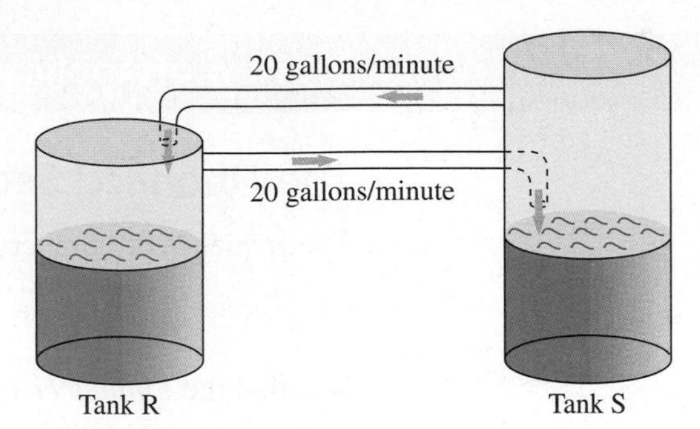

Figure A

When we set up a mathematical model for this problem, we find that $x(t)$ and $y(t)$ must satisfy the following system of differential equations:

$$\begin{aligned} x'(t) &= -\tfrac{1}{10}x(t) + \tfrac{2}{30}y(t) \\ y'(t) &= \tfrac{1}{10}x(t) - \tfrac{2}{30}y(t), \end{aligned}$$

which can be written in matrix form as

$$\begin{bmatrix} x'(t) \\ y'(t) \end{bmatrix} = \begin{bmatrix} -\frac{1}{10} & \frac{2}{30} \\ \frac{1}{10} & -\frac{2}{30} \end{bmatrix} \begin{bmatrix} x(t) \\ y(t) \end{bmatrix}.$$

Section 9.2 gives an introduction to the solution of homogeneous linear systems of differential equations.

Dynamical Systems (Section 9.3) (Calculus Required)

A system of first-order differential equations in which the derivatives do not explicitly depend on the independent variable is referred to as an (autonomous) dynamical system. We focus our attention on the case of linear homogeneous systems of two equations that can be written in the form

$$\frac{dx}{dt} = ax + by$$

$$\frac{dy}{dt} = cx + dy$$

for which initial conditions $x(0) = k_1$, $y(0) = k_2$ are specified. By determining the eigenvalues and eigenvectors of the matrix

$$A = \begin{bmatrix} a & b \\ c & d \end{bmatrix}$$

we can predict the behavior of the curve given by the ordered pairs $(x(t), y(t))$, called a trajectory of the system.

Section 9.3 provides an elementary introduction to dynamical systems.

Key Terms

Similar matrices
Diagonalizable
Diagonalized
Distinct eigenvalues
Multiplicity of an eigenvalue
Defective matrix

8.2 Exercises

In Exercises 1 through 8, determine whether the given matrix is diagonalizable.

1. $\begin{bmatrix} 1 & 4 \\ 1 & -2 \end{bmatrix}$

2. $\begin{bmatrix} 1 & 0 \\ -2 & 1 \end{bmatrix}$

3. $\begin{bmatrix} 1 & 1 & -2 \\ 4 & 0 & 4 \\ 1 & -1 & 4 \end{bmatrix}$

4. $\begin{bmatrix} 1 & 2 & 3 \\ 0 & -1 & 2 \\ 0 & 0 & 2 \end{bmatrix}$

5. $\begin{bmatrix} 3 & 1 & 0 \\ 0 & 3 & 1 \\ 0 & 0 & 3 \end{bmatrix}$

6. $\begin{bmatrix} -2 & 2 \\ 5 & 1 \end{bmatrix}$

7. $\begin{bmatrix} 2 & 0 & 3 \\ 0 & 1 & 0 \\ 0 & 1 & 2 \end{bmatrix}$

8. $\begin{bmatrix} 2 & 3 & 3 & 5 \\ 3 & 2 & 2 & 3 \\ 0 & 0 & 2 & 2 \\ 0 & 0 & 0 & 2 \end{bmatrix}$

9. Find a 2×2 nondiagonal matrix whose eigenvalues are 2 and -3, and associated eigenvectors are

$$\begin{bmatrix} -1 \\ 2 \end{bmatrix} \text{ and } \begin{bmatrix} 1 \\ 1 \end{bmatrix},$$

respectively.

10. Find a 3×3 nondiagonal matrix whose eigenvalues are -2, -2, and 3, and associated eigenvectors are

$$\begin{bmatrix} 1 \\ 0 \\ 1 \end{bmatrix}, \begin{bmatrix} 0 \\ 1 \\ 1 \end{bmatrix}, \text{ and } \begin{bmatrix} 1 \\ 1 \\ 1 \end{bmatrix},$$

respectively.

In Exercises 11 through 22, find, if possible, a nonsingular matrix P such that $P^{-1}AP$ is diagonal.

11. $\begin{bmatrix} 4 & 2 & 3 \\ 2 & 1 & 2 \\ -1 & -2 & 0 \end{bmatrix}$

12. $\begin{bmatrix} 1 & 1 & 2 \\ 0 & 1 & 0 \\ 0 & 1 & 3 \end{bmatrix}$

13. $\begin{bmatrix} 1 & 2 & 3 \\ 0 & 1 & 0 \\ 2 & 1 & 2 \end{bmatrix}$

14. $\begin{bmatrix} 0 & -1 \\ 2 & 3 \end{bmatrix}$

15. $\begin{bmatrix} 8 & 1 & 0 \\ 0 & 8 & 0 \\ 8 & 0 & 0 \end{bmatrix}$

16. $\begin{bmatrix} 3 & 0 & 0 \\ 1 & 3 & 1 \\ 0 & 0 & 1 \end{bmatrix}$

17. $\begin{bmatrix} 3 & -2 & 1 \\ 0 & 2 & 0 \\ 0 & 0 & 0 \end{bmatrix}$

18. $\begin{bmatrix} 2 & 2 & 2 \\ 2 & 2 & 2 \\ 2 & 2 & 2 \end{bmatrix}$

19. $\begin{bmatrix} 3 & 0 & 0 \\ 2 & 3 & 0 \\ 0 & 0 & 3 \end{bmatrix}$

20. $\begin{bmatrix} 1 & 0 & 1 \\ 0 & 1 & 0 \\ 0 & 1 & 2 \end{bmatrix}$

21. $\begin{bmatrix} 4 & 1 & 0 & 0 \\ 0 & 4 & 0 & 0 \\ 0 & 0 & 4 & 0 \\ 1 & 1 & 0 & 0 \end{bmatrix}$

22. $\begin{bmatrix} 3 & 0 & 0 & 0 \\ 2 & 3 & 0 & 0 \\ 0 & 0 & 5 & 1 \\ 0 & 0 & 0 & 6 \end{bmatrix}$

23. Let A be a 2×2 matrix whose eigenvalues are 3 and 4, and associated eigenvectors are

$$\begin{bmatrix} -1 \\ 1 \end{bmatrix} \text{ and } \begin{bmatrix} 2 \\ 1 \end{bmatrix},$$

respectively. Without computation, find a diagonal matrix D that is similar to A and nonsingular matrix P such that $P^{-1}AP = D$.

24. Let A be a 3×3 matrix whose eigenvalues are -3, 4, and 4, and associated eigenvectors are

$$\begin{bmatrix} -1 \\ 0 \\ 1 \end{bmatrix}, \begin{bmatrix} 0 \\ 0 \\ 1 \end{bmatrix}, \text{ and } \begin{bmatrix} 0 \\ 1 \\ 1 \end{bmatrix},$$

respectively. Without computation, find a diagonal matrix D that is similar to A and nonsingular matrix P such that $P^{-1}AP = D$.

In Exercises 25 through 28, find two matrices that are similar to A.

25. $A = \begin{bmatrix} 3 & 4 \\ 0 & 0 \end{bmatrix}$

26. $A = \begin{bmatrix} 1 & 2 \\ 3 & 0 \end{bmatrix}$

27. $A = \begin{bmatrix} 2 & 0 & 0 \\ 1 & 1 & 0 \\ -1 & 0 & 1 \end{bmatrix}$

28. $A = \begin{bmatrix} 2 & 0 & 0 \\ 1 & 2 & 1 \\ 0 & 1 & 2 \end{bmatrix}$

In Exercises 29 through 32, determine whether the given matrix is similar to a diagonal matrix.

29. $\begin{bmatrix} 2 & 3 & 0 \\ 0 & 1 & 0 \\ 0 & 0 & 2 \end{bmatrix}$

30. $\begin{bmatrix} 2 & 3 & 1 \\ 0 & 1 & 0 \\ 0 & 0 & 2 \end{bmatrix}$

31. $\begin{bmatrix} -3 & 0 \\ 1 & 2 \end{bmatrix}$

32. $\begin{bmatrix} 1 & 1 & 0 \\ 2 & 2 & 0 \\ 3 & 3 & 3 \end{bmatrix}$

In Exercises 33 through 36, show that each matrix is diagonalizable and find a diagonal matrix similar to the given matrix.

33. $\begin{bmatrix} 4 & 2 \\ 3 & 3 \end{bmatrix}$

34. $\begin{bmatrix} 3 & 2 \\ 6 & 4 \end{bmatrix}$

35. $\begin{bmatrix} 2 & -2 & 3 \\ 0 & 3 & -2 \\ 0 & -1 & 2 \end{bmatrix}$

36. $\begin{bmatrix} 0 & -2 & 1 \\ 1 & 3 & -1 \\ 0 & 0 & 1 \end{bmatrix}$

In Exercises 37 through 40, show that the given matrix is not diagonalizable.

37. $\begin{bmatrix} 1 & 1 \\ 0 & 1 \end{bmatrix}$

38. $\begin{bmatrix} 2 & 0 & 0 \\ 3 & 2 & 0 \\ 0 & 0 & 5 \end{bmatrix}$

39. $\begin{bmatrix} 10 & 11 & 3 \\ -3 & -4 & -3 \\ -8 & -8 & -1 \end{bmatrix}$

40. $\begin{bmatrix} 2 & 3 & 3 & 5 \\ 3 & 2 & 2 & 3 \\ 0 & 0 & 1 & 1 \\ 0 & 0 & 0 & 1 \end{bmatrix}$

*A matrix A is called **defective** if A has an eigenvalue λ of multiplicity $m > 1$ for which the associated eigenspace has a basis of fewer than m vectors; that is, the dimension of the eigenspace associated with λ is less than m. In Exercises 41 through 44, use the eigenvalues of the given matrix to determine if the matrix is defective.*

41. $\begin{bmatrix} 8 & 7 \\ 0 & 8 \end{bmatrix}$, $\lambda = 8, 8$

42. $\begin{bmatrix} 3 & 0 & 0 \\ -2 & 3 & -2 \\ 2 & 0 & 5 \end{bmatrix}$, $\lambda = 3, 3, 5$

43. $\begin{bmatrix} 3 & 3 & 3 \\ 3 & 3 & 3 \\ -3 & -3 & -3 \end{bmatrix}$, $\lambda = 0, 0, 3$

44. $\begin{bmatrix} 0 & 0 & 1 & 0 \\ 0 & 0 & 0 & -1 \\ 1 & 0 & 0 & 0 \\ 0 & -1 & 0 & 0 \end{bmatrix}$, $\lambda = 1, 1, -1, -1$

45. Let $D = \begin{bmatrix} 2 & 0 \\ 0 & -2 \end{bmatrix}$. Compute D^9.

46. Let $A = \begin{bmatrix} 3 & -5 \\ 1 & -3 \end{bmatrix}$. Compute A^9. (*Hint*: Find a matrix P such that $P^{-1}AP$ is a diagonal matrix D and show that $A^9 = PD^9P^{-1}$.)

Theoretical Exercises

T.1. Show that:

(a) A is similar to A.

(b) If B is similar to A, then A is similar to B.

(c) If A is similar to B and B is similar to C, then A is similar to C.

T.2. Show that if A is nonsingular and diagonalizable, then A^{-1} is diagonalizable.

T.3. Let

$$A = \begin{bmatrix} a & b \\ c & d \end{bmatrix}.$$

Find necessary and sufficient conditions for A to be diagonalizable.

T.4. Let A and B be nonsingular $n \times n$ matrices. Show that AB^{-1} and $B^{-1}A$ have the same eigenvalues.

T.5. Prove or disprove: Every nonsingular matrix is similar to a diagonal matrix.

T.6. If A and B are nonsingular, show that AB and BA are similar.

T.7. Show that if A is diagonalizable, then:

(a) A^T is diagonalizable.

(b) A^k is diagonalizable, where k is a positive integer.

T.8. Show that if A and B are similar matrices, then A^k and B^k, for any nonnegative integer k, are similar.

T.9. Show that if A and B are similar matrices, then $\det(A) = \det(B)$.

T.10. Let A be an $n \times n$ matrix and let $B = P^{-1}AP$ be similar to A. Show that if $\mathbf{x}$ is an eigenvector of A associated with the eigenvalue λ of A, then $P^{-1}\mathbf{x}$ is an eigenvector of B associated with the eigenvalue λ of the matrix B.

T.11. Let $\lambda_1, \lambda_2, \dots, \lambda_k$ be distinct eigenvalues of a matrix A with associated eigenvectors $\mathbf{x}_1, \mathbf{x}_2, \dots, \mathbf{x}_k$. Show that $\mathbf{x}_1, \mathbf{x}_2, \dots, \mathbf{x}_k$ are linearly independent. (*Hint*: See the proof of Theorem 8.5.)

T.12. Show that if A and B are similar matrices, then they have the same characteristic polynomial.

MATLAB Exercises

ML.1. Use MATLAB to determine if A is diagonalizable. If it is, find a nonsingular matrix P so that $P^{-1}AP$ is diagonal.

(a) $A = \begin{bmatrix} 0 & 2 \\ -1 & 3 \end{bmatrix}$

(b) $A = \begin{bmatrix} 1 & -3 \\ 3 & -5 \end{bmatrix}$

(c) $A = \begin{bmatrix} 0 & 0 & 4 \\ 5 & 3 & 6 \\ 6 & 0 & 5 \end{bmatrix}$

ML.2. Use MATLAB and the hint in Exercise 46 to compute A^{30}, where

$$A = \begin{bmatrix} -1 & 1 & -1 \\ -2 & 2 & -1 \\ -2 & 2 & -1 \end{bmatrix}.$$

ML.3. Repeat Exercise ML.2 for

$$A = \begin{bmatrix} -1 & 1.5 & -1.5 \\ -2 & 2.5 & -1.5 \\ -2 & 2.0 & -1.0 \end{bmatrix}.$$

Display your answer in both **format short** and **format long**.

ML.4. Use MATLAB to investigate the sequences

$$A, A^3, A^5, \ldots \quad \text{and} \quad A^2, A^4, A^6, \ldots$$

for matrix A in Exercise ML.2. Write a brief description of the behavior of these sequences. Describe $\lim_{n\to\infty} A^n$.

8.3 DIAGONALIZATION OF SYMMETRIC MATRICES

In this section we consider the diagonalization of symmetric matrices (an $n \times n$ matrix A with real entries such that $A = A^T$). We restrict our attention to this case because it is easier to handle than that of general matrices and also because symmetric matrices arise in many applied problems.

As an example of such a problem, consider the task of identifying the conic represented by the equation

$$2x^2 + 2xy + 2y^2 = 9,$$

which can be written in matrix form as

$$\begin{bmatrix} x & y \end{bmatrix} \begin{bmatrix} 2 & 1 \\ 1 & 2 \end{bmatrix} \begin{bmatrix} x \\ y \end{bmatrix} = 9.$$

Observe that the matrix used here is a symmetric matrix. This problem is discussed in detail in Section 9.5. We shall merely remark here that the solution calls for the determination of the eigenvalues and eigenvectors of the matrix

$$\begin{bmatrix} 2 & 1 \\ 1 & 2 \end{bmatrix}.$$

The x- and y-axes are then rotated to a new set of axes, which lie along the eigenvectors of the matrix. In the new set of axes, the given conic can be identified readily.

We omit the proof of the following important theorem (see D. R. Hill, *Experiments in Computational Matrix Algebra*, New York: Random House, 1988).

THEOREM 8.6 *All the roots of the characteristic polynomial of a symmetric matrix are real numbers.* ■

THEOREM 8.7 *If A is a symmetric matrix, then eigenvectors that are associated with distinct eigenvalues of A are orthogonal.*

Proof First, we shall let the reader verify (Exercise T.1) the property that if $\mathbf{x}$ and $\mathbf{y}$ are vectors in R^n, then

$$(A\mathbf{x}) \cdot \mathbf{y} = \mathbf{x} \cdot (A^T\mathbf{y}). \tag{1}$$

Now let $\mathbf{x}_1$ and $\mathbf{x}_2$ be eigenvectors of A associated with the distinct eigenvalues λ_1 and λ_2 of A. We then have

$$A\mathbf{x}_1 = \lambda_1\mathbf{x}_1 \quad \text{and} \quad A\mathbf{x}_2 = \lambda_2\mathbf{x}_2.$$

Now

$$\begin{aligned}\lambda_1(\mathbf{x}_1 \cdot \mathbf{x}_2) &= (\lambda_1\mathbf{x}_1) \cdot \mathbf{x}_2 = (A\mathbf{x}_1) \cdot \mathbf{x}_2 \\ &= \mathbf{x}_1 \cdot (A^T\mathbf{x}_2) = \mathbf{x}_1 \cdot (A\mathbf{x}_2) \\ &= \mathbf{x}_1 \cdot (\lambda_2\mathbf{x}_2) = \lambda_2(\mathbf{x}_1 \cdot \mathbf{x}_2),\end{aligned} \tag{2}$$

where we have used the fact that $A = A^T$. Thus

$$\lambda_1(\mathbf{x}_1 \cdot \mathbf{x}_2) = \lambda_2(\mathbf{x}_1 \cdot \mathbf{x}_2)$$

and subtracting, we obtain

$$\begin{aligned}0 &= \lambda_1(\mathbf{x}_1 \cdot \mathbf{x}_2) - \lambda_2(\mathbf{x}_1 \cdot \mathbf{x}_2) \\ &= (\lambda_1 - \lambda_2)(\mathbf{x}_1 \cdot \mathbf{x}_2).\end{aligned} \tag{3}$$

Since $\lambda_1 \neq \lambda_2$, we conclude that $\mathbf{x}_1 \cdot \mathbf{x}_2 = 0$, so $\mathbf{x}_1$ and $\mathbf{x}_2$ are orthogonal. ■

EXAMPLE 1 Given the symmetric matrix

$$A = \begin{bmatrix} 0 & 0 & -2 \\ 0 & -2 & 0 \\ -2 & 0 & 3 \end{bmatrix},$$

we find that the characteristic polynomial of A is (verify)

$$f(\lambda) = (\lambda + 2)(\lambda - 4)(\lambda + 1),$$

so the eigenvalues of A are

$$\lambda_1 = -2, \quad \lambda_2 = 4, \quad \text{and} \quad \lambda_3 = -1.$$

Then we can find the associated eigenvectors by solving the homogeneous system $(\lambda_j I_3 - A)\mathbf{x} = \mathbf{0}$, $j = 1, 2, 3$ and obtain the respective eigenvectors (verify)

$$\mathbf{x}_1 = \begin{bmatrix} 0 \\ 1 \\ 0 \end{bmatrix}, \quad \mathbf{x}_2 = \begin{bmatrix} -1 \\ 0 \\ 2 \end{bmatrix}, \quad \text{and} \quad \mathbf{x}_3 = \begin{bmatrix} 2 \\ 0 \\ 1 \end{bmatrix}.$$

It is easy to check that $\{\mathbf{x}_1, \mathbf{x}_2, \mathbf{x}_3\}$ is an orthogonal set of vectors in R^3 (and is thus linearly independent by Theorem 6.16, Section 6.8). Thus A is diagonalizable and is similar to

$$D = \begin{bmatrix} -2 & 0 & 0 \\ 0 & 4 & 0 \\ 0 & 0 & -1 \end{bmatrix}.$$

■

We recall that if A can be diagonalized, then there exists a nonsingular matrix P such that $P^{-1}AP$ is diagonal. Moreover, the columns of P are eigenvectors of A. Now, if the eigenvectors of A form an orthogonal set S, as happens when A is symmetric and the eigenvalues of A are distinct, then since any scalar multiple of an eigenvector of A is also an eigenvector of A, we can normalize S to obtain an orthonormal set $T = \{\mathbf{x}_1, \mathbf{x}_2, \ldots, \mathbf{x}_n\}$ of eigenvectors of A. The jth column of P is the eigenvector $\mathbf{x}_j$ associated with λ_j, and we now examine what type of matrix P must be. We can write P as

$$P = \begin{bmatrix} \mathbf{x}_1 & \mathbf{x}_2 & \cdots & \mathbf{x}_n \end{bmatrix}.$$

Then

$$P^T = \begin{bmatrix} \mathbf{x}_1^T \\ \mathbf{x}_2^T \\ \vdots \\ \mathbf{x}_n^T \end{bmatrix},$$

where $\mathbf{x}_i^T$, $1 \le i \le n$, is the transpose of the $n \times 1$ matrix (or vector) $\mathbf{x}_i$. We find that the i, jth entry in $P^T P$ is $\mathbf{x}_i \cdot \mathbf{x}_j$ (verify). Since

$$\mathbf{x}_i \cdot \mathbf{x}_j = \begin{cases} 1 & \text{if } i = j \\ 0 & \text{if } i \neq j, \end{cases}$$

then $P^T P = I_n$. Thus $P^T = P^{-1}$. Such matrices are important enough to have a special name.

DEFINITION A nonsingular matrix A is called an **orthogonal matrix** if

$$A^{-1} = A^T.$$

We can also say that a nonsingular matrix A is orthogonal if $A^T A = I_n$.

EXAMPLE 2 Let

$$A = \begin{bmatrix} \frac{2}{3} & -\frac{2}{3} & \frac{1}{3} \\ \frac{2}{3} & \frac{1}{3} & -\frac{2}{3} \\ \frac{1}{3} & \frac{2}{3} & \frac{2}{3} \end{bmatrix}.$$

It is easy to check that $A^T A = I_n$. Hence A is an orthogonal matrix. ■

EXAMPLE 3 Let A be the matrix of Example 1. We already know that the set of eigenvectors

$$\left\{ \begin{bmatrix} 0 \\ 1 \\ 0 \end{bmatrix}, \begin{bmatrix} -1 \\ 0 \\ 2 \end{bmatrix}, \begin{bmatrix} 2 \\ 0 \\ 1 \end{bmatrix} \right\}$$

is orthogonal. If we normalize these vectors, we find that

$$T = \left\{ \begin{bmatrix} 0 \\ 1 \\ 0 \end{bmatrix}, \begin{bmatrix} -\frac{1}{\sqrt{5}} \\ 0 \\ \frac{2}{\sqrt{5}} \end{bmatrix}, \begin{bmatrix} \frac{2}{\sqrt{5}} \\ 0 \\ \frac{1}{\sqrt{5}} \end{bmatrix} \right\}$$

is an orthonormal set of vectors. The matrix P such that $P^{-1}AP$ is diagonal is the matrix whose columns are the vectors in T. Thus

$$P = \begin{bmatrix} 0 & -\frac{1}{\sqrt{5}} & \frac{2}{\sqrt{5}} \\ 1 & 0 & 0 \\ 0 & \frac{2}{\sqrt{5}} & \frac{1}{\sqrt{5}} \end{bmatrix}.$$

We leave it to the reader to verify (Exercise 4) that P is an orthogonal matrix and that

$$P^{-1}AP = P^TAP = \begin{bmatrix} -2 & 0 & 0 \\ 0 & 4 & 0 \\ 0 & 0 & -1 \end{bmatrix}.$$

■

The following theorem is not difficult to show, and we leave its proof to the reader (Exercise T.3).

THEOREM 8.8 *The $n \times n$ matrix A is orthogonal if and only if the columns (and rows) of A form an orthonormal set of vectors in R^n.* ■

If A is an orthogonal matrix, then it is easy to show that $\det(A) = \pm 1$ (Exercise T.4).

We now turn to the general situation for a symmetric matrix; even if A has eigenvalues whose multiplicities are greater than one, it turns out that we can still diagonalize A. We omit the proof of the theorem. For a proof, see J. M. Ortega, *Matrix Theory, a Second Course*, New York: Plenum Press, 1987.

THEOREM 8.9 *If A is a symmetric $n \times n$ matrix, then there exists an orthogonal matrix P such that $P^{-1}AP = D$, a diagonal matrix. The eigenvalues of A lie on the main diagonal of D.* ■

Thus, not only is a symmetric matrix always diagonalizable, but it is diagonalizable by means of an orthogonal matrix. In such a case, we say that A is **orthogonally diagonalizable**.

It can be shown that if a symmetric matrix A has an eigenvalue λ_j of multiplicity k_j, then the solution space of the linear system $(\lambda_j I_n - A)\mathbf{x} = \mathbf{0}$ (the eigenspace of λ_j) has dimension k_j. This means that there exist k_j linearly independent eigenvectors of A associated with the eigenvalue λ_j. By the Gram–Schmidt process, we can construct an orthonormal basis for the solution space. Thus we obtain a set of k_j orthonormal eigenvectors associated with the eigenvalue λ_j. Since eigenvectors associated with distinct eigenvalues are orthogonal, if we form the set of all eigenvectors, we get an orthonormal set. Hence the matrix P whose columns are the eigenvectors is orthogonal.

The procedure for diagonalizing a symmetric matrix A by an orthogonal matrix P is as follows.

Step 1. Form the characteristic polynomial $f(\lambda) = \det(\lambda I_n - A)$.

Step 2. Find the roots of the characteristic polynomial of A. These will all be real.

Step 3. For each eigenvalue λ_j of A of multiplicity k_j, find a basis of k_j eigenvectors for the solution space of $(\lambda_j I_n - A)\mathbf{x} = \mathbf{0}$ (the eigenspace of λ_j).

Step 4. For each eigenspace, transform the basis obtained in Step 3 to an orthonormal basis by the Gram–Schmidt process. The totality of all these orthonormal bases determines an orthonormal set of n linearly independent eigenvectors of A.

Step 5. Let P be the matrix whose columns are the n linearly independent eigenvectors determined in Step 4. Then P is an orthogonal matrix and $P^{-1}AP = P^TAP = D$, a diagonal matrix whose diagonal elements are the eigenvalues of A that correspond to the columns of P.

EXAMPLE 4 Let

$$A = \begin{bmatrix} 0 & 2 & 2 \\ 2 & 0 & 2 \\ 2 & 2 & 0 \end{bmatrix}.$$

The characteristic polynomial of A is (verify)

$$f(\lambda) = (\lambda + 2)^2(\lambda - 4),$$

so the eigenvalues are

$$\lambda_1 = -2, \quad \lambda_2 = -2, \quad \text{and} \quad \lambda_3 = 4.$$

That is, -2 is an eigenvalue whose multiplicity is 2. Next, to find the eigenvectors associated with λ_1 and λ_2, we solve the homogeneous linear system $(-2I_3 - A)\mathbf{x} = \mathbf{0}$:

$$\begin{bmatrix} -2 & -2 & -2 \\ -2 & -2 & -2 \\ -2 & -2 & -2 \end{bmatrix} \begin{bmatrix} x_1 \\ x_2 \\ x_3 \end{bmatrix} = \begin{bmatrix} 0 \\ 0 \\ 0 \end{bmatrix}. \tag{4}$$

A basis for the solution space of (4) consists of the eigenvectors (verify)

$$\mathbf{x}_1 = \begin{bmatrix} -1 \\ 1 \\ 0 \end{bmatrix} \quad \text{and} \quad \mathbf{x}_2 = \begin{bmatrix} -1 \\ 0 \\ 1 \end{bmatrix}.$$

Now $\mathbf{x}_1$ and $\mathbf{x}_2$ are not orthogonal, since $\mathbf{x}_1 \cdot \mathbf{x}_2 \neq 0$. We can use the Gram–Schmidt process to obtain an orthonormal basis for the solution space of (4) (the eigenspace of $\lambda_1 = -2$) as follows. Let

$$\mathbf{y}_1 = \mathbf{x}_1 = \begin{bmatrix} -1 \\ 1 \\ 0 \end{bmatrix}$$

and

$$\mathbf{y}_2 = \mathbf{x}_2 - \left(\frac{\mathbf{x}_2 \cdot \mathbf{y}_1}{\mathbf{y}_1 \cdot \mathbf{y}_1}\right)\mathbf{y}_1 = \begin{bmatrix} -\frac{1}{2} \\ -\frac{1}{2} \\ 1 \end{bmatrix}.$$

Let

$$\mathbf{y}_2^* = 2\mathbf{y}_2 = \begin{bmatrix} -1 \\ -1 \\ 2 \end{bmatrix}.$$

The set $\{\mathbf{y}_1, \mathbf{y}_2^*\}$ is an orthogonal set of eigenvectors. Normalizing these eigenvectors, we obtain

$$\mathbf{z}_1 = \frac{1}{\|\mathbf{y}_1\|}\mathbf{y}_1 = \frac{1}{\sqrt{2}}\begin{bmatrix} -1 \\ 1 \\ 0 \end{bmatrix} \quad \text{and} \quad \mathbf{z}_2 = \frac{1}{\|\mathbf{y}_2^*\|}\mathbf{y}_2^* = \frac{1}{\sqrt{6}}\begin{bmatrix} -1 \\ -1 \\ 2 \end{bmatrix}.$$

The set $\{\mathbf{z}_1, \mathbf{z}_2\}$ is an orthonormal basis of eigenvectors of A for the solution space of (4). Now we find a basis for the solution space of $(4I_3 - A)\mathbf{x} = \mathbf{0}$,

$$\begin{bmatrix} 4 & -2 & -2 \\ -2 & 4 & -2 \\ -2 & -2 & 4 \end{bmatrix}\begin{bmatrix} x_1 \\ x_2 \\ x_3 \end{bmatrix} = \begin{bmatrix} 0 \\ 0 \\ 0 \end{bmatrix}, \tag{5}$$

to consist of (verify)

$$\mathbf{x}_3 = \begin{bmatrix} 1 \\ 1 \\ 1 \end{bmatrix}.$$

Normalizing this vector, we have the eigenvector

$$\mathbf{z}_3 = \frac{1}{\sqrt{3}}\begin{bmatrix} 1 \\ 1 \\ 1 \end{bmatrix}$$

as an orthonormal basis for the solution space of (5). Since eigenvectors associated with distinct eigenvalues are orthogonal, we observe that $\mathbf{z}_3$ is orthogonal to both $\mathbf{z}_1$ and $\mathbf{z}_2$. Thus the set $\{\mathbf{z}_1, \mathbf{z}_2, \mathbf{z}_3\}$ is an orthonormal basis for R^3 consisting of eigenvectors of A. The matrix P is the matrix whose jth column is $\mathbf{z}_j$:

$$P = \begin{bmatrix} -\frac{1}{\sqrt{2}} & -\frac{1}{\sqrt{6}} & \frac{1}{\sqrt{3}} \\ \frac{1}{\sqrt{2}} & -\frac{1}{\sqrt{6}} & \frac{1}{\sqrt{3}} \\ 0 & \frac{2}{\sqrt{6}} & \frac{1}{\sqrt{3}} \end{bmatrix}.$$

We leave it to the reader to verify that

$$P^{-1}AP = P^TAP = \begin{bmatrix} -2 & 0 & 0 \\ 0 & -2 & 0 \\ 0 & 0 & 4 \end{bmatrix}.$$

See also Exercises 17 and 18. ■

EXAMPLE 5 Let

$$A = \begin{bmatrix} 1 & 2 & 0 & 0 \\ 2 & 1 & 0 & 0 \\ 0 & 0 & 1 & 2 \\ 0 & 0 & 2 & 1 \end{bmatrix}.$$

The characteristic polynomial of A is (verify)

$$f(\lambda) = (\lambda + 1)^2(\lambda - 3)^2,$$

so the eigenvalues of A are

$$\lambda_1 = -1, \quad \lambda_2 = -1, \quad \lambda_3 = 3, \quad \text{and} \quad \lambda_4 = 3.$$

We find (verify) that a basis for the solution space of

$$(-1I_3 - A)\mathbf{x} = \mathbf{0} \tag{6}$$

consists of the eigenvectors

$$\mathbf{x}_1 = \begin{bmatrix} 1 \\ -1 \\ 0 \\ 0 \end{bmatrix} \quad \text{and} \quad \mathbf{x}_2 = \begin{bmatrix} 0 \\ 0 \\ 1 \\ -1 \end{bmatrix},$$

which are orthogonal. Normalizing these eigenvectors, we obtain

$$\mathbf{z}_1 = \begin{bmatrix} \frac{1}{\sqrt{2}} \\ -\frac{1}{\sqrt{2}} \\ 0 \\ 0 \end{bmatrix} \quad \text{and} \quad \mathbf{z}_2 = \begin{bmatrix} 0 \\ 0 \\ \frac{1}{\sqrt{2}} \\ -\frac{1}{\sqrt{2}} \end{bmatrix}$$

as an orthonormal basis of eigenvectors for the solution space of (6). We also find (verify) that a basis for the solution space of

$$(3I_3 - A)\mathbf{x} = \mathbf{0} \tag{7}$$

consists of the eigenvectors

$$\mathbf{x}_3 = \begin{bmatrix} 1 \\ 1 \\ 0 \\ 0 \end{bmatrix} \quad \text{and} \quad \mathbf{x}_4 = \begin{bmatrix} 0 \\ 0 \\ 1 \\ 1 \end{bmatrix},$$

which are orthogonal. Normalizing these eigenvectors, we obtain

$$\mathbf{z}_3 = \begin{bmatrix} \frac{1}{\sqrt{2}} \\ \frac{1}{\sqrt{2}} \\ 0 \\ 0 \end{bmatrix} \quad \text{and} \quad \mathbf{z}_4 = \begin{bmatrix} 0 \\ 0 \\ \frac{1}{\sqrt{2}} \\ \frac{1}{\sqrt{2}} \end{bmatrix}$$

as an orthonormal basis of eigenvectors for the solution space of (7). Since eigenvectors associated with distinct eigenvalues are orthogonal, we conclude that

$$\{\mathbf{z}_1, \mathbf{z}_2, \mathbf{z}_3, \mathbf{z}_4\}$$

is an orthonormal basis for R^4 consisting of eigenvectors of A. The matrix P is the matrix whose jth column is $\mathbf{z}_j$:

$$P = \begin{bmatrix} \frac{1}{\sqrt{2}} & 0 & \frac{1}{\sqrt{2}} & 0 \\ -\frac{1}{\sqrt{2}} & 0 & \frac{1}{\sqrt{2}} & 0 \\ 0 & \frac{1}{\sqrt{2}} & 0 & \frac{1}{\sqrt{2}} \\ 0 & -\frac{1}{\sqrt{2}} & 0 & \frac{1}{\sqrt{2}} \end{bmatrix}.$$

■

Suppose now that A is an $n \times n$ matrix that is orthogonally diagonalizable. Thus we have an orthogonal matrix P such that $P^{-1}AP$ is a diagonal matrix D. Then $P^{-1}AP = D$, or $A = PDP^{-1}$. Since $P^{-1} = P^T$, we can write $A = PDP^T$. Then

$$A^T = (PDP^T)^T = (P^T)^T D^T P^T = PDP^T = A$$

($D = D^T$ since D is a diagonal matrix). Thus A is symmetric. This result, together with Theorem 8.9, yields the following theorem.

THEOREM 8.10 *An $n \times n$ matrix A is orthogonally diagonalizable if and only if A is symmetric.* ■

Some remarks about nonsymmetric matrices are in order at this point. Theorem 8.5 assures us that A is diagonalizable if all the roots of its characteristic polynomial are distinct. We also saw examples, in Section 8.2, of nonsymmetric matrices with repeated eigenvalues that were diagonalizable and others that were not diagonalizable. There are some striking differences between the symmetric and nonsymmetric cases, which we now summarize. If A is nonsymmetric, then the roots of its characteristic polynomial need not all be real numbers; if an eigenvalue λ_j has multiplicity k_j, then the solution space of $(\lambda_j I_n - A)\mathbf{x} = \mathbf{0}$ may have dimension $< k_j$; if the roots of the characteristic polynomial of A are all real, it is still possible for A not to have n linearly independent eigenvectors (which means that A cannot be diagonalized); eigenvectors associated with distinct eigenvalues need not be orthogonal. Thus, in Example 6 of Section 9.1, the eigenvectors $\mathbf{x}_1$ and $\mathbf{x}_3$ associated with the eigenvalues $\lambda_1 = 0$ and $\lambda_3 = 1$, respectively, are not orthogonal. If a matrix A cannot be diagonalized, then we can often find a matrix B similar to A that is "nearly diagonal." The matrix B is said to be in **Jordan canonical form**; its treatment lies beyond the scope of this book but is studied in advanced books on linear algebra.*

It should be noted that in many applications, we need only find a diagonal matrix D that is similar to the given matrix A; that is, we do not need the orthogonal matrix P such that $P^{-1}AP = D$. Many of the matrices to be diagonalized in applied problems are symmetric. Of course, as previously stated, the methods for finding eigenvalues that have been presented in this chapter are not recommended for matrices larger than 4×4 because of the need to evaluate determinants. Efficient numerical methods for diagonalizing a symmetric matrix are discussed in numerical analysis courses.

---

*See, for example, K. Hoffman and R. Kunze, *Linear Algebra*, 2nd ed., Upper Saddle River, N.J.: Prentice Hall, Inc., 1971, or J. M. Ortega, *Matrix Theory, a Second Course*, New York: Plenum Press, 1987.

Sections 9.4, 9.5, and 9.6, on Quadratic Forms, Conic Sections, and Quadric Surfaces, respectively, which can be taken up at this time, use material from this section.

Preview of an Application

Quadratic Forms (Section 9.4)
Conic Sections (Section 9.5)
Quadric Surfaces (Section 9.6)

The conic sections are the curves obtained by intersecting a right circular cone with a plane. Depending on the position of the plane relative to the cone, the result can be an ellipse, circle, parabola, or hyperbola, or degenerate forms of these such as a point, a line, a pair of lines, or the empty set. These curves occur in a wide variety of applications in everyday life, ranging from the orbit of the planets to navigation devices and the design of headlights in automobiles.

Conic sections are in standard form when they are centered at the origin and their foci lie on a coordinate axis. In this case, they are easily identified by their simple equations. The graph of the equation

$$x^2 + 4xy - 2y^2 = 8$$

shown in Figure A is a hyperbola, which is not in standard form. This equation can be written in matrix form as

$$\begin{bmatrix} x & y \end{bmatrix} \begin{bmatrix} 1 & 2 \\ 2 & -2 \end{bmatrix} \begin{bmatrix} x \\ y \end{bmatrix} = 8.$$

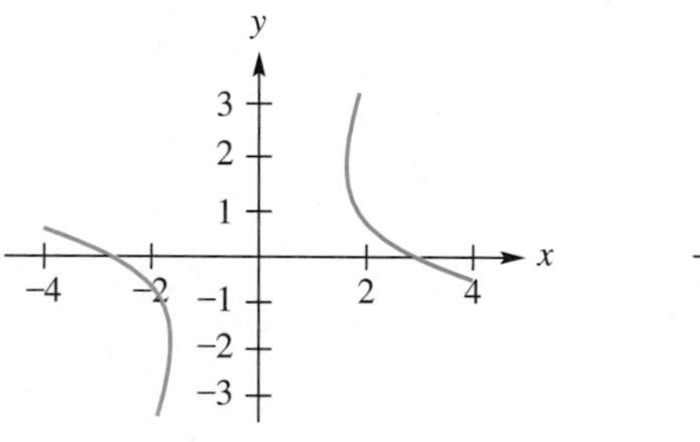

Figure A

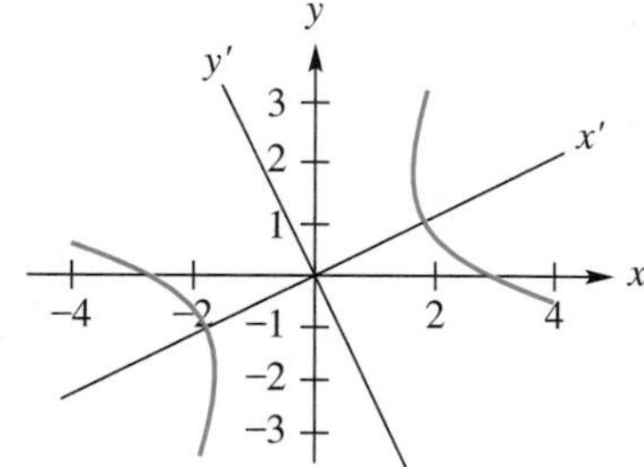

Figure B

The left side of this equation is called a quadratic form and these are discussed in Section 9.4. By diagonalizing the symmetric matrix

$$\begin{bmatrix} 1 & 2 \\ 2 & -2 \end{bmatrix}$$

that occurs in the quadratic form, we can transform the given equation to

$$\frac{x'^2}{4} - \frac{y'^2}{\frac{8}{3}} = 1,$$

whose graph is the hyperbola in standard form shown in Figure B. As you can see, the x- and y-axes have been rotated to the x'- and y'-axes. Section 9.5 discusses the rotation of axes to identify conic sections.

Section 9.6 discusses the analogous problem of identifying quadric surfaces in R^3.

We now provide a summary of properties of eigenvalues and eigenvectors.

SUMMARY OF PROPERTIES OF EIGENVALUES AND EIGENVECTORS

Let A be an $n \times n$ matrix.

- If $\mathbf{x}$ and $\mathbf{y}$ are eigenvectors associated with the eigenvalue λ of A, then if $\mathbf{x} + \mathbf{y} \neq \mathbf{0}$, it follows that $\mathbf{x} + \mathbf{y}$ is also an eigenvector associated with λ.
- If $\mathbf{x}$ is an eigenvector associated with the eigenvalue λ of A, then $k\mathbf{x}$, $k \neq 0$, is also an eigenvector associated with λ.
- If λ is an eigenvalue of A and $\mathbf{x}$ is an associated eigenvector, then for any nonnegative integer k, λ^k is an eigenvalue of A^k and $\mathbf{x}$ is an associated eigenvector.
- If λ and μ are distinct eigenvalues of A with associated eigenvectors $\mathbf{x}$ and $\mathbf{y}$, respectively, then $\mathbf{x}$ and $\mathbf{y}$ are linearly independent. That is, *eigenvectors that are associated with distinct eigenvalues are linearly independent.*
- A and A^T have the same eigenvalues.
- If A is a diagonal, upper triangular, or lower triangular matrix, then its eigenvalues are the entries on its main diagonal.
- The eigenvalues of a symmetric matrix are all real.
- Eigenvectors associated with distinct eigenvalues of a symmetric matrix are orthogonal.
- $\det(A)$ is the product of all the roots of the characteristic polynomial of A. [Equivalently, $\det(A)$ is the product of the eigenvalues of A.]
- A is singular if and only if 0 is an eigenvalue of A.
- Similar matrices have the same eigenvalues.
- A is diagonalizable if and only if A has n linearly independent eigenvectors.
- If A has n distinct eigenvalues, then A is diagonalizable.

Key Terms

Orthogonal matrix
Orthogonally diagonalizable
Jordan canonical form

8.3 Exercises

1. Verify that

$$P = \begin{bmatrix} \frac{2}{3} & -\frac{2}{3} & \frac{1}{3} \\ \frac{2}{3} & \frac{1}{3} & -\frac{2}{3} \\ \frac{1}{3} & \frac{2}{3} & \frac{2}{3} \end{bmatrix}$$

is an orthogonal matrix.

2. Find the inverse of each of the following orthogonal matrices.

(a) $A = \begin{bmatrix} 1 & 0 & 0 \\ 0 & \cos\theta & \sin\theta \\ 0 & -\sin\theta & \cos\theta \end{bmatrix}$

(b) $B = \begin{bmatrix} 1 & 0 & 0 \\ 0 & \frac{1}{\sqrt{2}} & -\frac{1}{\sqrt{2}} \\ 0 & -\frac{1}{\sqrt{2}} & -\frac{1}{\sqrt{2}} \end{bmatrix}$

3. Verify Theorem 8.8 for the matrices in Exercise 2.

4. Verify that the matrix P in Example 3 is an orthogonal matrix.

In Exercises 5 through 10, orthogonally diagonalize each given matrix A, giving the diagonal matrix D and the diagonalizing orthogonal matrix P.

5. $\begin{bmatrix} 2 & 2 \\ 2 & 2 \end{bmatrix}$

6. $\begin{bmatrix} 0 & 0 & 1 \\ 0 & 0 & 0 \\ 1 & 0 & 0 \end{bmatrix}$

7. $\begin{bmatrix} 0 & 0 & 0 \\ 0 & 2 & 2 \\ 0 & 2 & 2 \end{bmatrix}$

8. $\begin{bmatrix} 0 & 0 & 0 & 0 \\ 0 & 0 & 0 & 0 \\ 0 & 0 & 0 & 1 \\ 0 & 0 & 1 & 0 \end{bmatrix}$

9. $\begin{bmatrix} 0 & -1 & -1 \\ -1 & 0 & -1 \\ -1 & -1 & 0 \end{bmatrix}$

10. $\begin{bmatrix} -1 & 2 & 2 \\ 2 & -1 & 2 \\ 2 & 2 & -1 \end{bmatrix}$

In Exercises 11 through 18, orthogonally diagonalize each given matrix.

11. $\begin{bmatrix} 2 & 1 \\ 1 & 2 \end{bmatrix}$

12. $\begin{bmatrix} 2 & 2 & 0 & 0 \\ 2 & 2 & 0 & 0 \\ 0 & 0 & 2 & 2 \\ 0 & 0 & 2 & 2 \end{bmatrix}$

13. $\begin{bmatrix} 1 & 1 & 0 \\ 1 & 1 & 0 \\ 0 & 0 & 1 \end{bmatrix}$

14. $\begin{bmatrix} 1 & 0 & 0 \\ 0 & 3 & -2 \\ 0 & -2 & 3 \end{bmatrix}$

15. $\begin{bmatrix} 1 & 0 & 0 \\ 0 & 1 & 1 \\ 0 & 1 & 1 \end{bmatrix}$

16. $\begin{bmatrix} 0 & 0 & 0 & 1 \\ 0 & 0 & 0 & 0 \\ 0 & 0 & 0 & 0 \\ 1 & 0 & 0 & 0 \end{bmatrix}$

17. $\begin{bmatrix} 2 & 1 & 1 \\ 1 & 2 & 1 \\ 1 & 1 & 2 \end{bmatrix}$

18. $\begin{bmatrix} -3 & 0 & -1 \\ 0 & -2 & 0 \\ -1 & 0 & -3 \end{bmatrix}$

Theoretical Exercises

T.1. Show that if $\mathbf{x}$ and $\mathbf{y}$ are vectors in R^n, then $(A\mathbf{x}) \cdot \mathbf{y} = \mathbf{x} \cdot (A^T\mathbf{y})$.

T.2. Show that if A is an $n \times n$ orthogonal matrix and $\mathbf{x}$ and $\mathbf{y}$ are vectors in R^n, then $(A\mathbf{x}) \cdot (A\mathbf{y}) = \mathbf{x} \cdot \mathbf{y}$.

T.3. Prove Theorem 8.8.

T.4. Show that if A is an orthogonal matrix, then $\det(A) = \pm 1$.

T.5. Prove Theorem 8.9 for the 2×2 case by studying the possible roots of the characteristic polynomial of A.

T.6. Show that if A and B are orthogonal matrices, then AB is an orthogonal matrix.

T.7. Show that if A is an orthogonal matrix, then A^{-1} is also orthogonal.

T.8. (a) Verify that the matrix

$$\begin{bmatrix} \cos\theta & -\sin\theta \\ \sin\theta & \cos\theta \end{bmatrix}$$

is orthogonal.

(b) Show that if A is an orthogonal 2×2 matrix, then there exists a real number θ such that either

$$A = \begin{bmatrix} \cos\theta & -\sin\theta \\ \sin\theta & \cos\theta \end{bmatrix}$$

or

$$A = \begin{bmatrix} \cos\theta & \sin\theta \\ \sin\theta & -\cos\theta \end{bmatrix}.$$

T.9. Show that if $A^T A\mathbf{y} = \mathbf{y}$ for any $\mathbf{y}$ in R^n, then $A^T A = I_n$.

T.10. Show that if A is nonsingular and orthogonally diagonalizable, then A^{-1} is orthogonally diagonalizable.

MATLAB Exercises

The MATLAB *command* **eig** *will produce the eigenvalues and a set of orthonormal eigenvectors for a symmetric matrix A. Use the command in the form*

$$[\mathbf{V}, \mathbf{D}] = \mathbf{eig}(\mathbf{A})$$

The matrix V will contain the orthonormal eigenvectors, and matrix D will be diagonal with the corresponding eigenvalues on the main diagonal.

ML.1. Use **eig** to find the eigenvalues of A and an orthogonal matrix P so that $P^T AP$ is diagonal.

(a) $A = \begin{bmatrix} 6 & 6 \\ 6 & 6 \end{bmatrix}$

(b) $A = \begin{bmatrix} 1 & 2 & 2 \\ 2 & 1 & 2 \\ 2 & 2 & 1 \end{bmatrix}$

(c) $A = \begin{bmatrix} 4 & 1 & 0 \\ 1 & 4 & 1 \\ 0 & 1 & 4 \end{bmatrix}$

ML.2. Command **eig** can be applied to any matrix, but the matrix V of eigenvectors need not be orthogonal. For each of the following, use **eig** to determine which matrices A are such that V is orthogonal. If V is not orthogonal, then discuss briefly whether it can or cannot be replaced by an orthogonal matrix of eigenvectors.

(a) $A = \begin{bmatrix} 1 & 2 \\ -1 & 4 \end{bmatrix}$

(b) $A = \begin{bmatrix} 2 & 1 & 2 \\ 2 & 2 & -2 \\ 3 & 1 & 1 \end{bmatrix}$

(c) $A = \begin{bmatrix} 1 & -3 \\ 3 & -5 \end{bmatrix}$

(d) $A = \begin{bmatrix} 1 & 0 & 0 \\ 0 & 1 & 1 \\ 0 & 1 & 1 \end{bmatrix}$

Key Ideas for Review

- **Theorem 8.2.** The eigenvalues of A are the roots of the characteristic polynomial of A.
- **Theorem 8.3.** Similar matrices have the same eigenvalues.
- **Theorem 8.4.** An $n \times n$ matrix A is diagonalizable if and only if it has n linearly independent eigenvectors. In this case A is similar to a diagonal matrix D, with $D = P^{-1}AP$, whose diagonal elements are the eigenvalues of A, and P is a matrix whose columns are n linearly independent eigenvectors of A.
- **Theorem 8.6.** All the roots of the characteristic polynomial of a symmetric matrix are real numbers.
- **Theorem 8.7.** If A is a symmetric matrix, then eigenvectors that belong to distinct eigenvalues of A are orthogonal.
- **Theorem 8.8.** The $n \times n$ matrix A is orthogonal if and only if the columns of A form an orthonormal set of vectors in R^n.
- **Theorem 8.9.** If A is a symmetric $n \times n$ matrix, then there exists an orthogonal matrix P ($P^{-1} = P^T$) such that $P^{-1}AP = D$, a diagonal matrix. The eigenvalues of A lie on the main diagonal of D.
- **List of Nonsingular Equivalences.** The following statements are equivalent for an $n \times n$ matrix A.
  1. A is nonsingular.
  2. $\mathbf{x} = \mathbf{0}$ is the only solution to $A\mathbf{x} = \mathbf{0}$.
  3. A is row equivalent to I_n.
  4. The linear system $A\mathbf{x} = \mathbf{b}$ has a unique solution for every $n \times 1$ matrix $\mathbf{b}$.
  5. $\det(A) \neq 0$.
  6. A has rank n.
  7. A has nullity 0.
  8. The rows of A form a linearly independent set of n vectors in R^n.
  9. The columns of A form a linearly independent set of n vectors in R^n.
  10. Zero is *not* an eigenvalue of A.

Supplementary Exercises

1. Find the characteristic polynomial, eigenvalues, and eigenvectors of the matrix

$$\begin{bmatrix} -2 & 0 & 0 \\ 3 & 2 & 3 \\ 4 & -1 & 6 \end{bmatrix}.$$

In Exercises 2 and 3, determine whether the given matrix is diagonalizable.

2. $A = \begin{bmatrix} -1 & 0 & 0 \\ 3 & 2 & 0 \\ 4 & -1 & 2 \end{bmatrix}$

3. $A = \begin{bmatrix} 2 & 2 & 0 \\ 5 & -1 & 3 \\ 0 & 0 & 0 \end{bmatrix}$

In Exercises 4 and 5, find, if possible, a nonsingular matrix P and a diagonal matrix D so that A is similar to D.

4. $A = \begin{bmatrix} 1 & 0 & 0 \\ 2 & -1 & -6 \\ 2 & 0 & -2 \end{bmatrix}$

5. $A = \begin{bmatrix} 1 & 0 & 1 \\ 0 & 1 & 0 \\ -1 & 0 & 1 \end{bmatrix}$

6. If possible, find a diagonal matrix D so that

$$A = \begin{bmatrix} 0 & 0 & 0 \\ 0 & 1 & 0 \\ 2 & 2 & 1 \end{bmatrix}$$

is similar to D.

7. Find bases for the eigenspaces associated with each eigenvalue of the matrix

$$A = \begin{bmatrix} 0 & 0 & 1 \\ 0 & 2 & 0 \\ 0 & 0 & 2 \end{bmatrix}.$$

8. Is the matrix

$$\begin{bmatrix} \frac{2}{3} & \frac{1}{\sqrt{5}} & 1 \\ \frac{2}{3} & 0 & 0 \\ \frac{1}{3} & -\frac{2}{\sqrt{5}} & 0 \end{bmatrix}$$

orthogonal?

In Exercises 9 and 10, orthogonally diagonalize the given matrix A, giving the orthogonal matrix P and the diagonal matrix D.

9. $A = \begin{bmatrix} 1 & 1 & 1 \\ 1 & 1 & 1 \\ 1 & 1 & 1 \end{bmatrix}$

10. $A = \begin{bmatrix} -3 & 0 & -4 \\ 0 & 5 & 0 \\ -4 & 0 & 3 \end{bmatrix}$

11. If $A^2 = A$, what are possible values for the eigenvalues of A? Justify your answer.

12. Let $p_1(\lambda)$ be the characteristic polynomial of A_{11} and $p_2(\lambda)$ the characteristic polynomial of A_{22}. What is the characteristic polynomial of each of the following partitioned matrices?

(a) $A = \begin{bmatrix} A_{11} & O \\ O & A_{22} \end{bmatrix}$ (b) $A = \begin{bmatrix} A_{11} & O \\ A_{12} & A_{22} \end{bmatrix}$

(*Hint*: See Supplementary Exercises T.7 and T.8 in Chapter 3.)

Theoretical Exercises

T.1. Show that if a matrix A is similar to a diagonal matrix D, then $\text{Tr}(A) = \text{Tr}(D)$, where $\text{Tr}(A)$ is the trace of A. [*Hint*: See Supplementary Exercise T.1 in Chapter 1, where part (c) establishes $\text{Tr}(AB) = \text{Tr}(BA)$.]

T.2. Show that $\text{Tr}(A)$ is the sum of the eigenvalues of A (see Supplementary Exercise T.1 in Chapter 1).

T.3. Let A and B be similar matrices. Show that:

(a) A^T and B^T are similar.

(b) rank A = rank B.

(c) A is nonsingular if and only if B is nonsingular.

(d) If A and B are nonsingular, then A^{-1} and B^{-1} are similar.

(e) $\text{Tr}(A) = \text{Tr}(B)$.

T.4. Show that if A is an orthogonal matrix, then A^T is also orthogonal.

T.5. Let A be an orthogonal matrix. Show that cA is orthogonal if and only if $c = \pm 1$.

T.6. Let

$$A = \begin{bmatrix} a & b \\ c & d \end{bmatrix}.$$

Show that the characteristic polynomial $f(\lambda)$ of A is given by

$$f(\lambda) = \lambda^2 - \text{Tr}(A)\lambda + \det(A),$$

where $\text{Tr}(A)$ denotes the trace of A (see Supplementary Exercise T.1 in Chapter 1).

T.7. The **Cayley–Hamilton theorem** states that a matrix satisfies its characteristic equation; that is, if A is an $n \times n$ matrix with characteristic polynomial

$$f(\lambda) = \lambda^n + a_1\lambda^{n-1} + \cdots + a_{n-1}\lambda + a_n,$$

then

$$A^n + a_1A^{n-1} + \cdots + a_{n-1}A + a_nI_n = O.$$

The proof and applications of this result, unfortunately, lie beyond the scope of this book. Verify the Cayley–Hamilton theorem for the following matrices.

(a) $\begin{bmatrix} 1 & 2 & 3 \\ 2 & -1 & 5 \\ 3 & 2 & 1 \end{bmatrix}$ (b) $\begin{bmatrix} 1 & 2 & 3 \\ 0 & 2 & 2 \\ 0 & 0 & -3 \end{bmatrix}$

(c) $\begin{bmatrix} 3 & 3 \\ 2 & 4 \end{bmatrix}$

T.8. Let A be an $n \times n$ matrix whose characteristic polynomial is

$$f(\lambda) = \lambda^n + a_1\lambda^{n-1} + \cdots + a_{n-1}\lambda + a_n.$$

If A is nonsingular, show that

$$A^{-1} = -\frac{1}{a_n}(A^{n-1} + a_1A^{n-2} + \cdots + a_{n-2}A + a_{n-1}I_n).$$

[*Hint*: Use the Cayley–Hamilton theorem (Exercise T.7).]

Chapter Test

1. If possible, find a nonsingular matrix P and a diagonal matrix D so that A is similar to D, where

$$A = \begin{bmatrix} 1 & 0 & 0 \\ 5 & 2 & 0 \\ 4 & 3 & 2 \end{bmatrix}.$$

2. Verify that the following matrix is orthogonal:

$$\begin{bmatrix} -\frac{1}{\sqrt{2}} & \frac{1}{\sqrt{3}} & \frac{1}{\sqrt{6}} \\ 0 & \frac{1}{\sqrt{3}} & -\frac{2}{\sqrt{6}} \\ \frac{1}{\sqrt{2}} & \frac{1}{\sqrt{3}} & \frac{1}{\sqrt{6}} \end{bmatrix}.$$

3. Orthogonally diagonalize A, giving the orthogonal matrix P and the diagonal matrix D.

$$A = \begin{bmatrix} -1 & -4 & -8 \\ -4 & -7 & 4 \\ -8 & 4 & -1 \end{bmatrix}.$$

4. Answer each of the following as true or false. Justify your answer.

(a) If A is an $n \times n$ orthogonal matrix, then rank $A < n$.

(b) If A is diagonalizable, then each of its eigenvalues has multiplicity one.

(c) If none of the eigenvalues of A are zero, then $\det(A) \neq 0$.

(d) If A and B are similar, then $\det(A) = \det(B)$.

(e) If $\mathbf{x}$ and $\mathbf{y}$ are eigenvectors of A associated with the distinct eigenvalues λ_1 and λ_2, respectively, then $\mathbf{x} + \mathbf{y}$ is an eigenvector of A associated with the eigenvalue $\lambda_1 + \lambda_2$.

CHAPTER 9

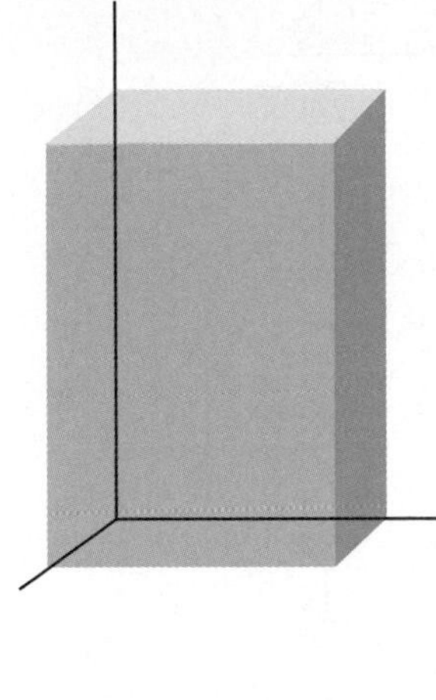

APPLICATIONS OF EIGENVALUES AND EIGENVECTORS (OPTIONAL)

9.1 THE FIBONACCI SEQUENCE

Prerequisite. Section 8.2, Diagonalization.

In 1202, Leonardo of Pisa, also called Fibonacci,* wrote a book on mathematics in which he posed the following problem. A pair of newborn rabbits begins to breed at the age of 1 month, and thereafter produces one pair of offspring per month. Suppose that we start with a pair of newly born rabbits and that none of the rabbits produced from this pair die. How many pairs of rabbits will there be at the beginning of each month?

The breeding pattern of the rabbits is shown in Figure 9.1, where an arrow indicates an offspring pair. At the beginning of month 0, we have the newly born pair of rabbits P_1. At the beginning of month 1 we still have only the original pair of rabbits P_1, which have not yet produced any offspring. At the beginning of month 2 we have the original pair P_1 and its first pair of offspring, P_2. At the beginning of month 3 we have the original pair P_1, its first pair of offspring P_2 born at the beginning of month 2, and its second pair of offspring, P_3. At the beginning of month 4 we have P_1, P_2, and P_3; P_4, the offspring of P_1; and P_5, the offspring of P_2. Let u_n denote the number of pairs of rabbits at the beginning of month n. We see that

$$u_0 = 1, \qquad u_1 = 1, \qquad u_2 = 2, \qquad u_3 = 3, \qquad u_4 = 5, \qquad u_5 = 8.$$

To obtain a formula for u_n, we proceed as follows. The number of pairs of rabbits that are alive at the beginning of month n is u_{n-1}, the number of pairs who were alive the previous month, plus the number of pairs newly born at the beginning of month n. The latter number is u_{n-2}, since a pair of rabbits

*Leonardo Fibonacci of Pisa (about 1170–1250) was born and lived most of his life in Pisa, Italy. When he was about 20, his father was appointed director of Pisan commercial interests in northern Africa, now a part of Algeria. Leonardo accompanied his father to Africa and for several years traveled extensively throughout the Mediterranean area on behalf of his father. During these travels he learned the Hindu–Arabic method of numeration and calculation and decided to promote its use in Italy. This was one purpose of his most famous book, *Liber Abaci*, which appeared in 1202 and contained the rabbit problem stated here.

Figure 9.1 ►

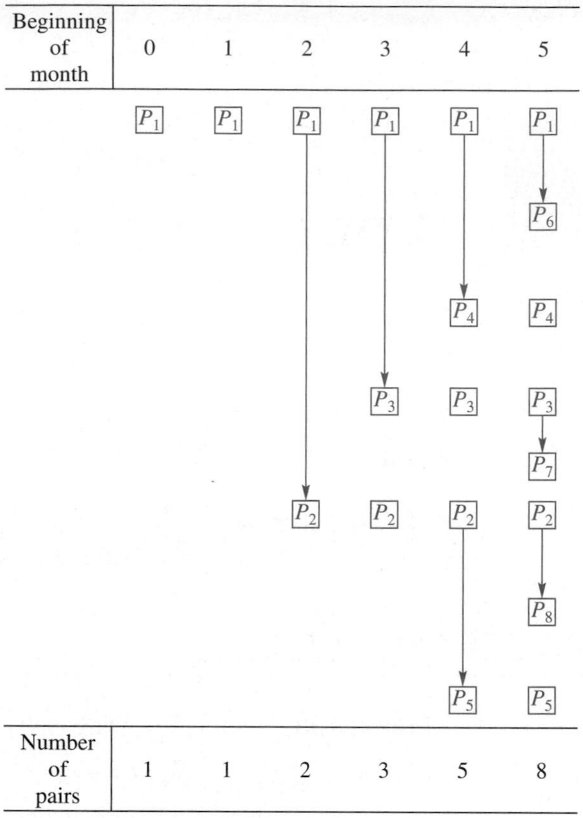

Beginning of month	0	1	2	3	4	5
Number of pairs	1	1	2	3	5	8

produces a pair of offspring starting with its second month of life. Thus

$$u_n = u_{n-1} + u_{n-2}. \tag{1}$$

That is, each number is the sum of its two predecessors. The resulting sequence of numbers, called a **Fibonacci sequence**, occurs in a remarkably varied number of applications, such as the distribution of leaves on certain trees, the arrangements of seeds on sunflowers, search techniques in numerical analysis, generating random numbers in statistics, and others.

To compute u_n by using the **recursion relation** (or difference equation) (1), we have to compute $u_0, u_1, \dots, u_{n-2}, u_{n-1}$. This can be rather tedious for large n. We now develop a formula that will enable us to calculate u_n directly.

In addition to Equation (1) we write

$$u_{n-1} = u_{n-1},$$

so we now have

$$\begin{aligned} u_n &= u_{n-1} + u_{n-2} \\ u_{n-1} &= u_{n-1}, \end{aligned}$$

which can be written in matrix form as

$$\begin{bmatrix} u_n \\ u_{n-1} \end{bmatrix} = \begin{bmatrix} 1 & 1 \\ 1 & 0 \end{bmatrix} \begin{bmatrix} u_{n-1} \\ u_{n-2} \end{bmatrix}. \tag{2}$$

We now define in general

$$\mathbf{w}_k = \begin{bmatrix} u_{k+1} \\ u_k \end{bmatrix} \quad \text{and} \quad A = \begin{bmatrix} 1 & 1 \\ 1 & 0 \end{bmatrix} \qquad (0 \le k \le n-1)$$

so that

$$\mathbf{w}_0 = \begin{bmatrix} u_1 \\ u_0 \end{bmatrix} = \begin{bmatrix} 1 \\ 1 \end{bmatrix},$$

$$\mathbf{w}_1 = \begin{bmatrix} u_2 \\ u_1 \end{bmatrix} = \begin{bmatrix} 2 \\ 1 \end{bmatrix}, \ldots, \quad \mathbf{w}_{n-2} = \begin{bmatrix} u_{n-1} \\ u_{n-2} \end{bmatrix}, \quad \text{and} \quad \mathbf{w}_{n-1} = \begin{bmatrix} u_n \\ u_{n-1} \end{bmatrix}.$$

Then (2) can be written as

$$\mathbf{w}_{n-1} = A\mathbf{w}_{n-2}.$$

Thus

$$\begin{aligned} \mathbf{w}_1 &= A\mathbf{w}_0 \\ \mathbf{w}_2 &= A\mathbf{w}_1 = A(A\mathbf{w}_0) = A^2\mathbf{w}_0 \\ \mathbf{w}_3 &= A\mathbf{w}_2 = A(A^2\mathbf{w}_0) = A^3\mathbf{w}_0 \\ &\vdots \\ \mathbf{w}_{n-1} &= A^{n-1}\mathbf{w}_0. \end{aligned} \tag{3}$$

Hence, to find u_n, we merely have to calculate A^{n-1}, which is still rather tedious if n is large. To avoid this difficulty, we find a diagonal matrix B that is similar to A. The characteristic equation of A is

$$\begin{vmatrix} \lambda - 1 & -1 \\ -1 & \lambda \end{vmatrix} = \lambda^2 - \lambda - 1 = 0.$$

The eigenvalues of A are (verify)

$$\lambda_1 = \frac{1+\sqrt{5}}{2} \quad \text{and} \quad \lambda_2 = \frac{1-\sqrt{5}}{2}$$

so that

$$D = \begin{bmatrix} \frac{1+\sqrt{5}}{2} & 0 \\ 0 & \frac{1-\sqrt{5}}{2} \end{bmatrix}.$$

Corresponding eigenvectors are (verify)

$$\mathbf{x}_1 = \begin{bmatrix} \frac{1+\sqrt{5}}{2} \\ 1 \end{bmatrix} \quad \text{and} \quad \mathbf{x}_2 = \begin{bmatrix} \frac{1-\sqrt{5}}{2} \\ 1 \end{bmatrix}.$$

Thus

$$P = \begin{bmatrix} \frac{1+\sqrt{5}}{2} & \frac{1-\sqrt{5}}{2} \\ 1 & 1 \end{bmatrix}, \qquad P^{-1} = \begin{bmatrix} \frac{1}{\sqrt{5}} & -\frac{1-\sqrt{5}}{2\sqrt{5}} \\ -\frac{1}{\sqrt{5}} & \frac{1+\sqrt{5}}{2\sqrt{5}} \end{bmatrix},$$

and

$$A = PDP^{-1}.$$

Hence (verify) for any nonnegative integer k,

$$A^k = PD^kP^{-1}.$$

Since D is diagonal, D^k is easy to calculate; its entries are the diagonal entries of D raised to the kth power. From (3) we have

$$\mathbf{w}_{n-1} = A^{n-1}\mathbf{w}_0 = PD^{n-1}P^{-1}\mathbf{w}_0$$

$$= \begin{bmatrix} \frac{1+\sqrt{5}}{2} & \frac{1-\sqrt{5}}{2} \\ 1 & 1 \end{bmatrix} \begin{bmatrix} \left(\frac{1+\sqrt{5}}{2}\right)^{n-1} & 0 \\ 0 & \left(\frac{1-\sqrt{5}}{2}\right)^{n-1} \end{bmatrix} \begin{bmatrix} \frac{1}{\sqrt{5}} & -\frac{1-\sqrt{5}}{2\sqrt{5}} \\ -\frac{1}{\sqrt{5}} & \frac{1+\sqrt{5}}{2\sqrt{5}} \end{bmatrix} \begin{bmatrix} 1 \\ 1 \end{bmatrix}.$$

This equation gives the formula (verify)

$$u_n = \frac{1}{\sqrt{5}}\left[\left(\frac{1+\sqrt{5}}{2}\right)^{n+1} - \left(\frac{1-\sqrt{5}}{2}\right)^{n+1}\right]$$

for calculating u_n directly.

Using a handheld calculator, we find that for $n = 50$, u_{50} is approximately 20.365 billion.

Further Readings

A Primer for the Fibonacci Numbers. San Jose State University, San Jose, Calif., 1972.

VOROBYOV, N. N., *The Fibonacci Numbers*. Boston: D. C. Heath and Company, 1963.

Key Terms

Fibonacci sequence
Recursion relation

9.1 Exercises

1. Compute the eigenvalues and eigenvectors of

$$A = \begin{bmatrix} 1 & 1 \\ 1 & 0 \end{bmatrix}$$

and verify that they are as given in the text.

2. Verify that if $A = PBP^{-1}$ and k is a positive integer, then $A^k = PB^kP^{-1}$.

3. Using a handheld calculator or MATLAB, compute

(a) u_8 (b) u_{12} (c) u_{20}

4. Consider the rabbit-breeding problem but suppose that each pair of rabbits now produces two pairs of rabbits starting with its second month of life and continuing every month thereafter.

(a) Formulate a recursion relation for the number u_n of rabbits at the beginning of month n.

(b) Develop a formula for calculating u_n directly.

Theoretical Exercise

T.1. Let $A = \begin{bmatrix} 1 & 1 \\ 1 & 0 \end{bmatrix}$. It can then be shown that

$$A^{n+1} = \begin{bmatrix} u_{n+1} & u_n \\ u_n & u_{n-1} \end{bmatrix}.$$

Use this result to obtain the formula

$$u_{n+1}u_{n-1} - u_n^2 = (-1)^{n+1}.$$

9.2 DIFFERENTIAL EQUATIONS

Prerequisite. Section 8.2, Diagonalization. Calculus required.

A **differential equation** is an equation that involves an unknown function and its derivatives. An important, simple example of a differential equation is

$$\frac{d}{dt}x(t) = rx(t),$$

where r is a constant. The idea here is to find a function $x(t)$ that will satisfy the given differential equation; that is, $x'(t) = rx(t)$. This differential equation is discussed further in this section.

Differential equations occur often in all branches of science and engineering; linear algebra is helpful in the formulation and solution of differential equations. In this section we provide only a brief survey of the approach; books on differential equations deal with the subject in much greater detail, and several suggestions for further reading are given at the end of this section.

HOMOGENEOUS LINEAR SYSTEMS

We consider the **homogeneous linear system** of differential equations

$$\begin{aligned} x_1'(t) &= a_{11}x_1(t) + a_{12}x_2(t) + \cdots + a_{1n}x_n(t) \\ x_2'(t) &= a_{21}x_1(t) + a_{22}x_2(t) + \cdots + a_{2n}x_n(t) \\ &\vdots \\ x_n'(t) &= a_{n1}x_1(t) + a_{n2}x_2(t) + \cdots + a_{nn}x_n(t), \end{aligned} \tag{1}$$

where the a_{ij} are known constants. We seek functions $x_1(t), x_2(t), \ldots, x_n(t)$ defined and differentiable on the real line satisfying (1).

We can write (1) in matrix form by letting

$$\mathbf{x}(t) = \begin{bmatrix} x_1(t) \\ x_2(t) \\ \vdots \\ x_n(t) \end{bmatrix}, \qquad A = \begin{bmatrix} a_{11} & a_{12} & \cdots & a_{1n} \\ a_{21} & a_{22} & \cdots & a_{2n} \\ \vdots & \vdots & & \vdots \\ a_{n1} & a_{n2} & \cdots & a_{nn} \end{bmatrix},$$

and defining

$$\mathbf{x}'(t) = \begin{bmatrix} x_1'(t) \\ x_2'(t) \\ \vdots \\ x_n'(t) \end{bmatrix}.$$

Then (1) can be written as

$$\mathbf{x}'(t) = A\mathbf{x}(t). \tag{2}$$

We shall often write (2) more briefly as

$$\mathbf{x}' = A\mathbf{x}.$$

With this notation, an n-vector function

$$\mathbf{x}(t) = \begin{bmatrix} x_1(t) \\ x_2(t) \\ \vdots \\ x_n(t) \end{bmatrix}$$

satisfying (2) is called a **solution** to the given system. It can be shown (Exercise T.1) that the set of all solutions to the homogeneous linear system of differential equations (1) is a subspace of the vector space of differentiable real-valued n-vector functions.

We leave it to the reader to verify that if $\mathbf{x}^{(1)}(t), \mathbf{x}^{(2)}(t), \dots, \mathbf{x}^{(n)}(t)$ are all solutions to (2), then any linear combination

$$\mathbf{x}(t) = b_1\mathbf{x}^{(1)}(t) + b_2\mathbf{x}^{(2)}(t) + \cdots + b_n\mathbf{x}^{(n)}(t) \tag{3}$$

is also a solution to (2).

A set of vector functions $\{\mathbf{x}^{(1)}(t), \mathbf{x}^{(2)}(t), \dots, \mathbf{x}^{(n)}(t)\}$ is said to be a **fundamental system** for (1) if every solution to (1) can be written in the form (3). In this case, the right side of (3), where $b_1, b_2, \dots, b_n$ are arbitrary constants, is said to be the **general solution** to (2).

It can be shown (see the book by Boyce and DiPrima or the book by Cullen cited in Further Readings) that any system of the form (2) has a fundamental system (in fact, infinitely many).

In general, differential equations arise in the course of solving physical problems. Typically, once a general solution to the differential equation has been obtained, the physical constraints of the problem impose certain definite values on the arbitrary constants in the general solution, giving rise to a **particular solution**. An important particular solution is obtained by finding a solution $\mathbf{x}(t)$ to Equation (2) such that $\mathbf{x}(0) = \mathbf{x}_0$, an **initial condition**, where $\mathbf{x}_0$ is a given vector. This problem is called an **initial value problem**. If the general solution (3) is known, then the initial value problem can be solved by setting $t = 0$ in (3) and determining the constants $b_1, b_2, \dots, b_n$ so that

$$\mathbf{x}_0 = b_1\mathbf{x}^{(1)}(0) + b_2\mathbf{x}^{(2)}(0) + \cdots + b_n\mathbf{x}^{(n)}(0).$$

It is easily seen that this is actually an $n \times n$ linear system with unknowns $b_1, b_2, \dots, b_n$. This linear system can also be written as

$$C\mathbf{b} = \mathbf{x}_0, \tag{4}$$

where

$$\mathbf{b} = \begin{bmatrix} b_1 \\ b_2 \\ \vdots \\ b_n \end{bmatrix}$$

and C is the $n \times n$ matrix whose columns are $\mathbf{x}^{(1)}(0), \mathbf{x}^{(2)}(0), \dots, \mathbf{x}^{(n)}(0)$, respectively. It can be shown (see the book by Boyce and DiPrima or the book by Cullen cited in Further Readings) that if $\mathbf{x}^{(1)}(t), \mathbf{x}^{(2)}(t), \dots, \mathbf{x}^{(n)}(t)$ form a fundamental system for (1), then C is nonsingular, so (4) always has a unique solution.

EXAMPLE 1 The simplest system of the form (1) is the single equation

$$\frac{dx}{dt} = ax, \tag{5}$$

where a is a constant. From calculus, the solutions to this equation are of the form

$$x = be^{at}; \tag{6}$$

that is, this is the general solution to (5). To solve the initial value problem

$$\frac{dx}{dt} = ax, \qquad x(0) = x_0,$$

we set $t = 0$ in (6) and obtain $b = x_0$. Thus the solution to the initial value problem is

$$x = x_0 e^{at}. \qquad \blacksquare$$

The system (2) is said to be **diagonal** if the matrix A is diagonal. Then (1) can be rewritten as

$$\begin{aligned} x_1'(t) &= a_{11}x_1(t) \\ x_2'(t) &= \qquad\quad a_{22}x_2(t) \\ &\vdots \\ x_n'(t) &= \qquad\qquad\qquad a_{nn}x_n(t). \end{aligned} \tag{7}$$

This system is easy to solve, since the equations can be solved separately. Applying the results of Example 1 to each equation in (7), we obtain

$$\begin{aligned} x_1(t) &= b_1 e^{a_{11}t} \\ x_2(t) &= b_2 e^{a_{22}t} \\ \vdots \quad &\quad \vdots \\ x_n(t) &= b_n e^{a_{nn}t}, \end{aligned} \tag{8}$$

where $b_1, b_2, \dots, b_n$ are arbitrary constants. Writing (8) in vector form yields

$$\mathbf{x}(t) = \begin{bmatrix} b_1 e^{a_{11}t} \\ b_2 e^{a_{22}t} \\ \vdots \\ b_n e^{a_{nn}t} \end{bmatrix} = b_1 \begin{bmatrix} 1 \\ 0 \\ 0 \\ \vdots \\ 0 \end{bmatrix} e^{a_{11}t} + b_2 \begin{bmatrix} 0 \\ 1 \\ 0 \\ \vdots \\ 0 \end{bmatrix} e^{a_{22}t} + \cdots + b_n \begin{bmatrix} 0 \\ 0 \\ \vdots \\ 0 \\ 1 \end{bmatrix} e^{a_{nn}t}.$$

This implies that the vector functions

$$\mathbf{x}^{(1)}(t) = \begin{bmatrix} 1 \\ 0 \\ 0 \\ \vdots \\ 0 \end{bmatrix} e^{a_{11}t}, \quad \mathbf{x}^{(2)}(t) = \begin{bmatrix} 0 \\ 1 \\ 0 \\ \vdots \\ 0 \end{bmatrix} e^{a_{22}t}, \dots, \quad \mathbf{x}^{(n)}(t) = \begin{bmatrix} 0 \\ 0 \\ \vdots \\ 0 \\ 1 \end{bmatrix} e^{a_{nn}t}$$

form a fundamental system for the diagonal system (7).

EXAMPLE 2 The diagonal system

$$\begin{bmatrix} x_1' \\ x_2' \\ x_3' \end{bmatrix} = \begin{bmatrix} 3 & 0 & 0 \\ 0 & -2 & 0 \\ 0 & 0 & 4 \end{bmatrix} \begin{bmatrix} x_1 \\ x_2 \\ x_3 \end{bmatrix} \tag{9}$$

can be written as three equations:

$$\begin{aligned} x_1' &= 3x_1 \\ x_2' &= -2x_2 \\ x_3' &= 4x_3. \end{aligned}$$

Solving these equations, we obtain

$$x_1 = b_1 e^{3t}, \qquad x_2 = b_2 e^{-2t}, \qquad x_3 = b_3 e^{4t},$$

where b_1, b_2, and b_3 are arbitrary constants. Thus

$$\mathbf{x}(t) = \begin{bmatrix} b_1 e^{3t} \\ b_2 e^{-2t} \\ b_3 e^{4t} \end{bmatrix} = b_1 \begin{bmatrix} 1 \\ 0 \\ 0 \end{bmatrix} e^{3t} + b_2 \begin{bmatrix} 0 \\ 1 \\ 0 \end{bmatrix} e^{-2t} + b_3 \begin{bmatrix} 0 \\ 0 \\ 1 \end{bmatrix} e^{4t}$$

is the general solution to (9) and the functions

$$\mathbf{x}^{(1)}(t) = \begin{bmatrix} 1 \\ 0 \\ 0 \end{bmatrix} e^{3t}, \qquad \mathbf{x}^{(2)}(t) = \begin{bmatrix} 0 \\ 1 \\ 0 \end{bmatrix} e^{-2t}, \qquad \mathbf{x}^{(3)}(t) = \begin{bmatrix} 0 \\ 0 \\ 1 \end{bmatrix} e^{4t}$$

form a fundamental system for (9). ■

If the system (2) is not diagonal, then it cannot be solved as simply as the system in the preceding example. However, there is an extension of this method that yields the general solution in the case where A is diagonalizable. Suppose that A is diagonalizable and P is a nonsingular matrix such that

$$P^{-1}AP = D, \tag{10}$$

where D is diagonal. Then multiplying the given system

$$\mathbf{x}' = A\mathbf{x}$$

on the left by P^{-1}, we obtain

$$P^{-1}\mathbf{x}' = P^{-1}A\mathbf{x}.$$

Since $P^{-1}P = I_n$, we can rewrite the last equation as

$$P^{-1}\mathbf{x}' = (P^{-1}AP)(P^{-1}\mathbf{x}). \tag{11}$$

Temporarily, let

$$\mathbf{u} = P^{-1}\mathbf{x}. \tag{12}$$

Since P^{-1} is a constant matrix,

$$\mathbf{u}' = P^{-1}\mathbf{x}'. \tag{13}$$

Therefore, substituting (10), (12), and (13) into (11), we obtain

$$\mathbf{u}' = D\mathbf{u}. \tag{14}$$

Equation (14) is a diagonal system and can be solved by the methods just discussed. Before proceeding, however, let us recall from Theorem 8.4 in Section 8.2 it follows that

$$D = \begin{bmatrix} \lambda_1 & 0 & \cdots & 0 \\ 0 & \lambda_2 & \cdots & 0 \\ \vdots & \vdots & & \vdots \\ 0 & 0 & \cdots & \lambda_n \end{bmatrix},$$

where $\lambda_1, \lambda_2, \ldots, \lambda_n$ are the eigenvalues of A, and that the columns of P are linearly independent eigenvectors of A associated, respectively, with $\lambda_1, \lambda_2, \ldots, \lambda_n$. From the discussion just given for diagonal systems, the general solution to (14) is

$$\mathbf{u}(t) = b_1\mathbf{u}^{(1)}(t) + b_2\mathbf{u}^{(2)}(t) + \cdots + b_n\mathbf{u}^{(n)}(t) = \begin{bmatrix} b_1 e^{\lambda_1 t} \\ b_2 e^{\lambda_2 t} \\ \vdots \\ b_n e^{\lambda_n t} \end{bmatrix},$$

where

$$\mathbf{u}^{(1)}(t) = \begin{bmatrix} 1 \\ 0 \\ 0 \\ \vdots \\ 0 \end{bmatrix} e^{\lambda_1 t}, \quad \mathbf{u}^{(2)}(t) = \begin{bmatrix} 0 \\ 1 \\ 0 \\ \vdots \\ 0 \end{bmatrix} e^{\lambda_2 t}, \ldots, \quad \mathbf{u}^{(n)}(t) = \begin{bmatrix} 0 \\ 0 \\ \vdots \\ 0 \\ 1 \end{bmatrix} e^{\lambda_n t} \tag{15}$$

and $b_1, b_2, \ldots, b_n$ are arbitrary constants. From Equation (12), $\mathbf{x} = P\mathbf{u}$, so the general solution to the given system $\mathbf{x}' = A\mathbf{x}$ is

$$\mathbf{x}(t) = P\mathbf{u}(t) = b_1 P\mathbf{u}^{(1)}(t) + b_2 P\mathbf{u}^{(2)}(t) + \cdots + b_n P\mathbf{u}^{(n)}(t). \tag{16}$$

However, since the constant vectors in (15) are the columns of the identity matrix and $PI_n = P$, (16) can be rewritten as

$$\mathbf{x}(t) = b_1\mathbf{p}_1 e^{\lambda_1 t} + b_2\mathbf{p}_2 e^{\lambda_2 t} + \cdots + b_n\mathbf{p}_n e^{\lambda_n t}, \tag{17}$$

where $\mathbf{p}_1, \mathbf{p}_2, \ldots, \mathbf{p}_n$ are the columns of P, and therefore eigenvectors of A associated with $\lambda_1, \lambda_2, \ldots, \lambda_n$, respectively.

We summarize the preceding discussion in the following theorem.

THEOREM 9.1

If the $n \times n$ matrix A has n linearly independent eigenvectors $\mathbf{p}_1, \mathbf{p}_2, \ldots, \mathbf{p}_n$ associated with the eigenvalues $\lambda_1, \lambda_2, \ldots, \lambda_n$, respectively, then the general solution to the homogeneous linear system of differential equations

$$\mathbf{x}' = A\mathbf{x}$$

is given by (17). ■

The procedure for obtaining the general solution to the homogeneous linear system $\mathbf{x}'(t) = A\mathbf{x}(t)$, where A is diagonalizable, is as follows.

Step 1. Compute the eigenvalues $\lambda_1, \lambda_2, \ldots, \lambda_n$ of A.

Step 2. Compute eigenvectors $\mathbf{p}_1, \mathbf{p}_2, \ldots, \mathbf{p}_n$ of A associated, respectively, with $\lambda_1, \lambda_2, \ldots, \lambda_n$.

Step 3. The general solution is given by Equation (17).

EXAMPLE 3 For the system

$$\mathbf{x}' = \begin{bmatrix} 1 & -1 \\ 2 & 4 \end{bmatrix} \mathbf{x},$$

the matrix

$$A = \begin{bmatrix} 1 & -1 \\ 2 & 4 \end{bmatrix}$$

has eigenvalues $\lambda_1 = 2$ and $\lambda_2 = 3$ with associated eigenvectors (verify)

$$\mathbf{p}_1 = \begin{bmatrix} 1 \\ -1 \end{bmatrix} \quad \text{and} \quad \mathbf{p}_2 = \begin{bmatrix} 1 \\ -2 \end{bmatrix}.$$

These eigenvectors are automatically linearly independent, since they are associated with distinct eigenvalues (see the Remark following Theorem 8.5 in Section 8.2). Hence the general solution to the given system is

$$\mathbf{x}(t) = b_1 \begin{bmatrix} 1 \\ -1 \end{bmatrix} e^{2t} + b_2 \begin{bmatrix} 1 \\ -2 \end{bmatrix} e^{3t}.$$

In terms of components, this can be written as

$$\begin{aligned} x_1(t) &= b_1 e^{2t} + b_2 e^{3t} \\ x_2(t) &= -b_1 e^{2t} - 2b_2 e^{3t}. \end{aligned}$$

■

EXAMPLE 4 Consider the following homogeneous linear system of differential equations:

$$\mathbf{x}' = \begin{bmatrix} x_1' \\ x_2' \\ x_3' \end{bmatrix} = \begin{bmatrix} 0 & 1 & 0 \\ 0 & 0 & 1 \\ 8 & -14 & 7 \end{bmatrix} \begin{bmatrix} x_1 \\ x_2 \\ x_3 \end{bmatrix}.$$

The characteristic polynomial of A is (verify)

$$f(\lambda) = \lambda^3 - 7\lambda^2 + 14\lambda - 8$$

or

$$f(\lambda) = (\lambda - 1)(\lambda - 2)(\lambda - 4),$$

so the eigenvalues of A are $\lambda_1 = 1$, $\lambda_2 = 2$, and $\lambda_3 = 4$. Associated eigenvectors are (verify)

$$\begin{bmatrix} 1 \\ 1 \\ 1 \end{bmatrix}, \quad \begin{bmatrix} 1 \\ 2 \\ 4 \end{bmatrix}, \quad \begin{bmatrix} 1 \\ 4 \\ 16 \end{bmatrix},$$

respectively. The general solution is then given by

$$\mathbf{x}(t) = b_1 \begin{bmatrix} 1 \\ 1 \\ 1 \end{bmatrix} e^{t} + b_2 \begin{bmatrix} 1 \\ 2 \\ 4 \end{bmatrix} e^{2t} + b_3 \begin{bmatrix} 1 \\ 4 \\ 16 \end{bmatrix} e^{4t},$$

where b_1, b_2, and b_3 are arbitrary constants. ■

EXAMPLE 5 For the system of Example 4 solve the initial value problem determined by the **initial conditions** $x_1(0) = 4$, $x_2(0) = 6$, and $x_3(0) = 8$.

Solution We write our general solution in the form $\mathbf{x} = P\mathbf{u}$ as

$$\mathbf{x}(t) = \begin{bmatrix} 1 & 1 & 1 \\ 1 & 2 & 4 \\ 1 & 4 & 16 \end{bmatrix} \begin{bmatrix} b_1 e^{t} \\ b_2 e^{2t} \\ b_3 e^{4t} \end{bmatrix}.$$

Now

$$\mathbf{x}(0) = \begin{bmatrix} 4 \\ 6 \\ 8 \end{bmatrix} = \begin{bmatrix} 1 & 1 & 1 \\ 1 & 2 & 4 \\ 1 & 4 & 16 \end{bmatrix} \begin{bmatrix} b_1 e^{0} \\ b_2 e^{0} \\ b_3 e^{0} \end{bmatrix}$$

or

$$\begin{bmatrix} 1 & 1 & 1 \\ 1 & 2 & 4 \\ 1 & 4 & 16 \end{bmatrix} \begin{bmatrix} b_1 \\ b_2 \\ b_3 \end{bmatrix} = \begin{bmatrix} 4 \\ 6 \\ 8 \end{bmatrix}. \tag{18}$$

Solving (18) by Gauss–Jordan reduction, we obtain (verify)

$$b_1 = \tfrac{4}{3}, \qquad b_2 = 3, \qquad b_3 = -\tfrac{1}{3}.$$

Therefore, the solution to the initial value problem is

$$\mathbf{x}(t) = \tfrac{4}{3} \begin{bmatrix} 1 \\ 1 \\ 1 \end{bmatrix} e^{t} + 3 \begin{bmatrix} 1 \\ 2 \\ 4 \end{bmatrix} e^{2t} - \tfrac{1}{3} \begin{bmatrix} 1 \\ 4 \\ 16 \end{bmatrix} e^{4t}.$$

■

We now recall several facts from Chapter 8. If A does not have distinct eigenvalues, then we may or may not be able to diagonalize A. Let λ be an eigenvalue of A of multiplicity k. Then A can be diagonalized if and only if the dimension of the eigenspace associated with λ is k, that is, if and only if the rank of the matrix $(\lambda I_n - A)$ is $n - k$ (verify). If the rank of $(\lambda I_n - A)$ is $n - k$, then we can find k linearly independent eigenvectors of A associated with λ.

EXAMPLE 6 Consider the linear system

$$\mathbf{x}' = A\mathbf{x} = \begin{bmatrix} 1 & 0 & 0 \\ 0 & 3 & -2 \\ 0 & -2 & 3 \end{bmatrix} \mathbf{x}.$$

The eigenvalues of A are $\lambda_1 = \lambda_2 = 1$ and $\lambda_3 = 5$ (verify). The rank of the matrix

$$(1I_3 - A) = \begin{bmatrix} 0 & 0 & 0 \\ 0 & -2 & 2 \\ 0 & 2 & -2 \end{bmatrix}$$

is 1 and the linearly independent eigenvectors

$$\begin{bmatrix} 1 \\ 0 \\ 0 \end{bmatrix} \quad \text{and} \quad \begin{bmatrix} 0 \\ 1 \\ 1 \end{bmatrix}$$

are associated with the eigenvalue 1 (verify). The eigenvector

$$\begin{bmatrix} 0 \\ 1 \\ -1 \end{bmatrix}$$

is associated with the eigenvalue 5 (verify). The general solution to the given system is then

$$\mathbf{x}(t) = b_1 \begin{bmatrix} 1 \\ 0 \\ 0 \end{bmatrix} e^t + b_2 \begin{bmatrix} 0 \\ 1 \\ 1 \end{bmatrix} e^t + b_3 \begin{bmatrix} 0 \\ 1 \\ -1 \end{bmatrix} e^{5t},$$

where b_1, b_2, and b_3 are arbitrary constants. ■

If we cannot diagonalize A as in the examples, we have a considerably more difficult situation. Methods for dealing with such problems are discussed in more advanced books (see Further Readings).

APPLICATION—A DIFFUSION PROCESS

The following example is a modification of an example presented by Derrick and Grossman in *Elementary Differential Equations with Applications* (see Further Readings).

EXAMPLE 7 Consider two adjoining cells separated by a permeable membrane and suppose that a fluid flows from the first cell to the second one at a rate (in milliliters per minute) that is numerically equal to three times the volume (in milliliters) of the fluid in the first cell. It then flows out of the second cell at a rate (in milliliters per minute) that is numerically equal to twice the volume in the second cell. Let $x_1(t)$ and $x_2(t)$ denote the volumes of the fluid in the first and second cells at time t, respectively. Assume that initially the first cell has 40 milliliters of fluid, while the second one has 5 milliliters of fluid. Find the volume of fluid in each cell at time t. (See Figure 9.2.)

Solution The change in volume of the fluid in each cell is the difference between the amount flowing in and the amount flowing out. Since no fluid flows into the first cell, we have

$$\frac{dx_1(t)}{dt} = -3x_1(t),$$

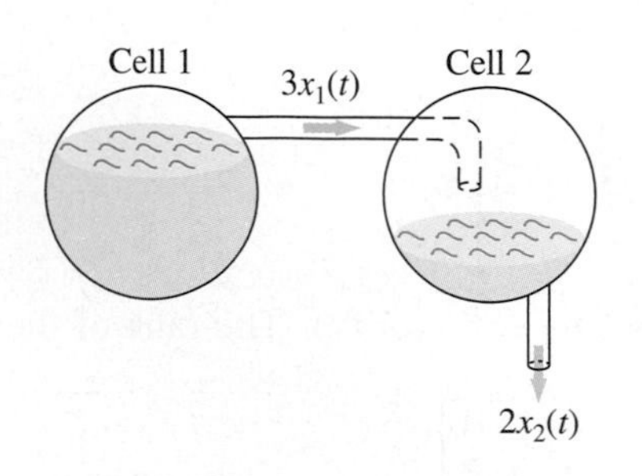

Figure 9.2 ▲

where the minus sign indicates that the fluid is flowing out of the cell. The flow $3x_1(t)$ from the first cell flows into the second cell. The flow out of the second cell is $2x_2(t)$. Thus the change in volume of the fluid in the second cell is given by

$$\frac{dx_2(t)}{dt} = 3x_1(t) - 2x_2(t).$$

We have then obtained the linear system

$$\begin{aligned} \frac{dx_1(t)}{dt} &= -3x_1(t) \\ \frac{dx_2(t)}{dt} &= 3x_1(t) - 2x_2(t), \end{aligned}$$

which can be written in matrix form as

$$\begin{bmatrix} x_1'(t) \\ x_2'(t) \end{bmatrix} = \begin{bmatrix} -3 & 0 \\ 3 & -2 \end{bmatrix} \begin{bmatrix} x_1(t) \\ x_2(t) \end{bmatrix}.$$

The eigenvalues of the matrix

$$A = \begin{bmatrix} -3 & 0 \\ 3 & -2 \end{bmatrix}$$

are (verify)

$$\lambda_1 = -3, \qquad \lambda_2 = -2$$

and corresponding associated eigenvectors are (verify)

$$\begin{bmatrix} 1 \\ -3 \end{bmatrix}, \qquad \begin{bmatrix} 0 \\ 1 \end{bmatrix}.$$

Hence the general solution is given by

$$\mathbf{x}(t) = \begin{bmatrix} x_1(t) \\ x_2(t) \end{bmatrix} = b_1 \begin{bmatrix} 1 \\ -3 \end{bmatrix} e^{-3t} + b_2 \begin{bmatrix} 0 \\ 1 \end{bmatrix} e^{-2t}.$$

Using the initial conditions, we find that (verify)

$$b_1 = 40, \qquad b_2 = 125.$$

Thus the volume of fluid in each cell at time t is given by

$$\begin{aligned} x_1(t) &= \phantom{-120e^{-3t}+{}} 40e^{-3t} \\ x_2(t) &= -120e^{-3t} + 125e^{-2t}. \end{aligned}$$

■

It should also be pointed out that many differential equations cannot be solved in the sense that we can write a formula for the solution. Numerical methods, some of which are studied in numerical analysis, exist for obtaining numerical solutions to differential equations; computer codes for some of these methods are widely available.

Further Readings

BOYCE, W. E., and R. C. DIPRIMA. *Elementary Differential Equations*, 8th ed. New York: John Wiley & Sons, Inc., 2004.

CULLEN, C. G. *Linear Algebra and Differential Equations*, 2nd ed. Boston: PWS-Kent, 1991.

DERRICK, W. R., and S. I. GROSSMAN. *Elementary Differential Equations with Applications*, 4th ed. Reading, Mass.: Addison-Wesley, 1997.

DETTMAN, J. H. *Introduction to Linear Algebra and Differential Equations*. New York: Dover, 1986.

GOODE, S. W. *Differential Equations and Linear Algebra*, 2nd ed. Upper Saddle River, N.J.: Prentice Hall, Inc., 2000.

RABENSTEIN, A. L. *Elementary Differential Equations with Linear Algebra*, 4th ed. Philadelphia: W. B. Saunders, Harcourt Brace Jovanovich, 1992.

Key Terms

Differential equation
Homogeneous linear system
Solution
Fundamental system
General solution
Particular solution
Initial solution
Initial value problem
Diagonal system

9.2 Exercises

1. Consider the homogeneous linear system of differential equations

$$\begin{bmatrix} x_1' \\ x_2' \\ x_3' \end{bmatrix} = \begin{bmatrix} -3 & 0 & 0 \\ 0 & 4 & 0 \\ 0 & 0 & 2 \end{bmatrix} \begin{bmatrix} x_1 \\ x_2 \\ x_3 \end{bmatrix}.$$

(a) Find the general solution.

(b) Find the solution to the initial value problem determined by the initial conditions $x_1(0) = 3$, $x_2(0) = 4$, $x_3(0) = 5$.

2. Consider the homogeneous linear system of differential equations

$$\begin{bmatrix} x_1' \\ x_2' \\ x_3' \end{bmatrix} = \begin{bmatrix} 1 & 0 & 0 \\ 0 & -2 & 1 \\ 0 & 0 & 3 \end{bmatrix} \begin{bmatrix} x_1 \\ x_2 \\ x_3 \end{bmatrix}.$$

(a) Find the general solution.

(b) Find the solution to the initial value problem determined by the initial conditions $x_1(0) = 2$, $x_2(0) = 7$, $x_3(0) = 20$.

3. Find the general solution to the homogeneous linear system of differential equations

$$\begin{bmatrix} x_1' \\ x_2' \\ x_3' \end{bmatrix} = \begin{bmatrix} 4 & 0 & 0 \\ 3 & -5 & 0 \\ 2 & 1 & 2 \end{bmatrix} \begin{bmatrix} x_1 \\ x_2 \\ x_3 \end{bmatrix}.$$

4. Find the general solution to the homogeneous linear system of differential equations

$$\begin{bmatrix} x_1' \\ x_2' \\ x_3' \end{bmatrix} = \begin{bmatrix} 2 & 3 & 0 \\ 0 & 1 & 0 \\ 0 & 0 & 2 \end{bmatrix} \begin{bmatrix} x_1 \\ x_2 \\ x_3 \end{bmatrix}.$$

5. Find the general solution to the homogeneous linear system of differential equations

$$\begin{bmatrix} x_1' \\ x_2' \\ x_3' \end{bmatrix} = \begin{bmatrix} 5 & 0 & 0 \\ 0 & -4 & 3 \\ 0 & 3 & 4 \end{bmatrix} \begin{bmatrix} x_1 \\ x_2 \\ x_3 \end{bmatrix}.$$

6. Find the general solution to the homogeneous linear system of differential equations

$$\begin{bmatrix} x_1' \\ x_2' \end{bmatrix} = \begin{bmatrix} 3 & -2 \\ -2 & 3 \end{bmatrix} \begin{bmatrix} x_1 \\ x_2 \end{bmatrix}.$$

7. Find the general solution to the homogeneous linear system of differential equations

$$\begin{bmatrix} x_1' \\ x_2' \\ x_3' \end{bmatrix} = \begin{bmatrix} 1 & 2 & 3 \\ 0 & 1 & 0 \\ 2 & 1 & 2 \end{bmatrix} \begin{bmatrix} x_1 \\ x_2 \\ x_3 \end{bmatrix}.$$

8. Find the general solution to the homogeneous linear system of differential equations

$$\begin{bmatrix} x_1' \\ x_2' \\ x_3' \end{bmatrix} = \begin{bmatrix} 1 & 1 & 2 \\ 0 & 1 & 0 \\ 0 & 1 & 3 \end{bmatrix} \begin{bmatrix} x_1 \\ x_2 \\ x_3 \end{bmatrix}.$$

9. Consider two competing species that live in the same forest and let $x_1(t)$ and $x_2(t)$ denote the respective populations of the species at time t. Suppose that the initial populations are $x_1(0) = 500$ and $x_2(0) = 200$. If the growth rates of the species are given by

$$\begin{aligned} x_1'(t) &= -3x_1(t) + 6x_2(t) \\ x_2'(t) &= x_1(t) - 2x_2(t), \end{aligned}$$

what is the population of each species at time t?

10. Suppose that we have a system consisting of two interconnected tanks, each containing a brine solution. Tank A contains $x(t)$ pounds of salt in 200 gallons of brine and tank B contains $y(t)$ pounds of salt in 300 gallons of brine. The mixture in each tank is kept uniform by constant stirring. Starting at $t = 0$, brine is pumped from tank A to tank B at 20 gallons/minute and from tank B to tank A at 20 gallons/minute. Find the amount of salt in each tank at time t if $x(0) = 10$ and $y(0) = 40$.

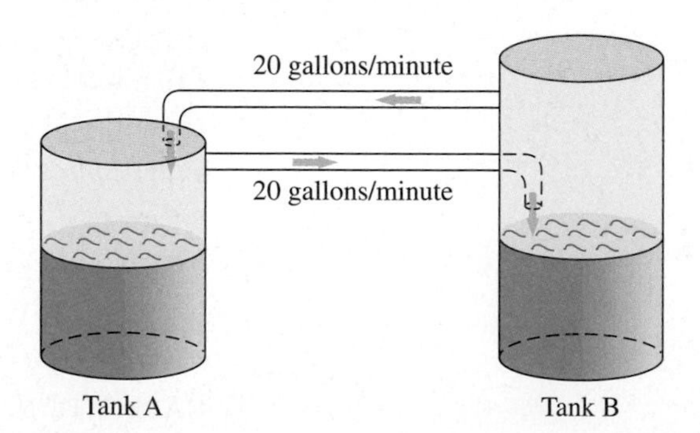

Theoretical Exercise

T.1. Show that the set of all solutions to the homogeneous linear system of differential equations $\mathbf{x}' = A\mathbf{x}$, where A is $n \times n$, is a subspace of the vector space of all differentiable real-valued n-vector functions. This subspace is called the **solution space** of the given linear system.

MATLAB Exercises

For the linear system $\mathbf{x}' = A\mathbf{x}$, this section shows how to construct the general solution $\mathbf{x}(t)$ provided A is diagonalizable. In MATLAB we can determine the eigenvalues and eigenvectors of A as in Section 8.1, or we can use the command **eig**, *which is discussed in Section 8.3. The* **eig** *command may give eigenvectors different from those found by your hand computations, but recall that eigenvectors are not unique.*

ML.1. Solve Example 4 using the **eig** command in MATLAB.

ML.2. Solve Example 6 using the **eig** command in MATLAB.

ML.3. Solve Example 7 using the **eig** command in MATLAB.

9.3 DYNAMICAL SYSTEMS

Prerequisite. Section 9.2, Differential Equations.

In Section 9.2 we studied how to solve homogeneous linear systems of differential equations for which an initial condition had been specified. We called such systems initial value problems and wrote them in the form

$$\mathbf{x}'(t) = A\mathbf{x}(t), \qquad \mathbf{x}(0) = \mathbf{x}_0, \tag{1}$$

where

$$\mathbf{x}(t) = \begin{bmatrix} x_1(t) \\ x_2(t) \\ \vdots \\ x_n(t) \end{bmatrix}, \qquad A = \begin{bmatrix} a_{11} & a_{12} & \cdots & \cdots & a_{1n} \\ a_{21} & a_{22} & \cdots & \cdots & a_{2n} \\ \vdots & \vdots & \cdots & \cdots & \vdots \\ \vdots & \vdots & \cdots & \cdots & \vdots \\ a_{n1} & a_{n2} & \cdots & \cdots & a_{nn} \end{bmatrix},$$

and $\mathbf{x}_0$ is a specified vector of constants. In the case that A was diagonalizable we used the eigenvalues and eigenvectors of A to construct a particular solution to (1).

In this section we focus our attention on the case $n = 2$, and for ease of reference we use x and y instead of x_1 and x_2. Such homogeneous linear systems of differential equations can be written in the form

$$\begin{aligned} \frac{dx}{dt} &= ax + by \\ \frac{dy}{dt} &= cx + dy, \end{aligned} \tag{2}$$

where a, b, c, and d are real constants, or

$$\mathbf{x}'(t) = \frac{d}{dt}\begin{bmatrix} x \\ y \end{bmatrix} = A\begin{bmatrix} x \\ y \end{bmatrix} = A\mathbf{x}(t), \tag{3}$$

where

$$A = \begin{bmatrix} a & b \\ c & d \end{bmatrix}.$$

...or the systems (2) and (3) we try to describe properties of the solution based on the differential equation itself. This area is called the **qualitative theory** of differential equations and was studied extensively by J. H. Poincaré.*

The systems (2) or (3) are called **autonomous** differential equations since the rates of change $\frac{dx}{dt}$ and $\frac{dy}{dt}$ depend explicitly only on the values of x and y, not on the independent variable t. For our purposes we will call the independent variable t **time** and then (2) and (3) are said to be time-independent systems. Using this convention, the systems in (2) and (3) provide a model for the change of x and y as time goes by. Hence such systems are called **dynamical systems**. We will use this terminology throughout our discussion in this section.

A qualitative analysis of dynamical systems is interested in such questions as:

- Are there any constant solutions?
- If there are constant solutions, do nearby solutions move toward or away from the constant solution?
- What is the behavior of solutions as $t \to \pm\infty$?
- Are there any solutions that oscillate?

Each of these questions has a geometric flavor. Hence we introduce a helpful device for studying the behavior of dynamical systems. If we regard t, time, as a parameter, then $x = x(t)$ and $y = y(t)$ will represent a curve in the xy-plane. Such a curve is called a **trajectory** or an **orbit** of the systems (2) and (3). The xy-plane is called the **phase plane** of the dynamical system.

A simple physical model of a dynamical system is the mechanical system known as a mass–spring system. A mass is attached to a spring and set into motion. The position of the mass changes with time. To model this system mathematically we make certain simplifying assumptions that permit us to develop a simple second-order differential equation that describes the dynamical changes of the mass–spring system. The term *idealized system* is often used for what we will describe because of the assumptions we make in order to obtain the equation of motion of the mass.

We start with an unstretched spring of length L attached to a rigid support like a beam; [see Figure 9.3(a)]. We attach a mass m to the spring. The result is that the spring stretches an additional length ΔL. The resulting position of the mass is called its **equilibrium position**; [see Figure 9.3(b)]. Let $x(t)$ measure the vertical displacement of the mass from its equilibrium position at time t; [see Figure 9.3(c)]. We take $x(t)$ to be positive for displacements below the equilibrium position; that is, the positive direction is downward.

---

*Jules Henri Poincaré (1854–1912) was born in Nancy, France, to a well-to-do family, many of whose members played key roles in the French government. As a youngster, he was clumsy and absentminded but showed great talent in mathematics. In 1873 he entered the École Polytechnique, from which he received his doctorate. He then began a university career, finally joining the University of Paris in 1881, where he remained until his death. Poincaré is considered the last of the universalists in mathematics, that is, someone who can work in all branches of mathematics, both pure and applied. His doctoral dissertation dealt with the existence of solutions to differential equations. In applied mathematics, he made contributions to the fields of optics, electricity, elasticity, thermodynamics, quantum mechanics, the theory of relativity, and cosmology. In pure mathematics, he was one of the principal creators of algebraic topology and made numerous contributions to algebraic geometry, analytic functions, and number theory. He was the first person to think of chaos in connection with his work in astronomy. In his later years, he wrote several books popularizing mathematics. In some of these books he dealt with the psychological processes involved in mathematical discovery and with the aesthetic aspects of mathematics.

Figure 9.3 ►

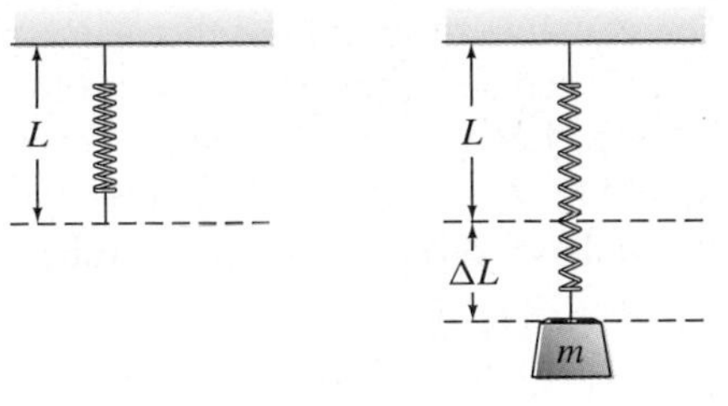

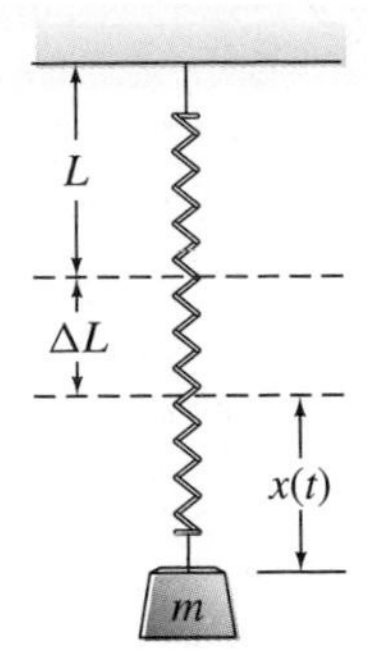

(a) Natural length of the spring (b) Mass at its equilibrium position (c) Mass in motion, displaced a distance $x(t)$ from equilibrium

To determine a differential equation that provides a model for the motion of the mass–spring system we use Newton's second law of motion, which says that the sum of the forces, F, acting on the object is equal to the mass, m, times its acceleration, a; that is, $F = ma$. Here is where we make some assumptions to obtain an idealized model for the motion of the mass. We will assume that there are only two forces acting on the mass; the force of gravity and the restoring force of the spring. (*Note*: We have ignored friction and air resistance and assumed that the mass will move only vertically.) Hence we have the following forces acting on the system:

- Weight, given by mg, where g is the acceleration due to gravity.
- The restoring force of the spring that is oppositely directed to the force of gravity and is given by $-k[\Delta L + x(t)]$, where k is called the *spring constant* (a measure of stiffness of the spring), ΔL is the additional length of the spring due to attaching the mass, and $x(t)$ is the displacement from equilibrium. (This formulation is known as Hooke's law of springs.)

If we denote the acceleration as $x''(t)$ (the second derivative of displacement with respect to time), then Newton's second law of motion gives us the equation

$$mx''(t) = mg - k[\Delta L + x(t)]. \tag{4}$$

We simplify this equation of motion of the system as follows. When the mass is at equilibrium, the restoring force of the spring is equal to the weight of the mass; that is, $k\Delta L = mg$. Thus Equation (4) can be simplified to obtain

$$mx''(t) + kx(t) = 0. \tag{5}$$

Equation (5) is often referred to as the equation that describes **simple harmonic motion**. Intuitively, we expect that when the mass is displaced from equilibrium and released it exhibits an up-and-down motion that repeats itself over and over again. In order to carefully inspect the motion of the system we proceed mathematically to develop the trajectory of the system in the phase plane. We first determine a fundamental system for Equation (5) in Example 1. Then we develop some concepts that will reveal the nature of the trajectories and return to analyze the trajectories of this physical model in Example 6.

EXAMPLE 1 The equation of motion of the mass-spring system developed above in Equation (5) can be written as a system of first-order linear differential equations as follows. Let $x_1(t) = x(t)$ and $x_2(t) = x'(t)$. Next differentiate each of these equations to obtain $x_1'(t) = x'(t) = x_2(t)$ and $x_2'(t) = x''(t)$. From Equation (5) we have $x''(t) = -\frac{k}{m}x(t)$. It follows that

$$x_2'(t) = x''(t) = -\frac{k}{m}x(t) = -\frac{k}{m}x_1(t).$$

We now have the pair of differential equations

$$\begin{aligned} x_1'(t) &= x_2(t) \\ x_2'(t) &= -\frac{k}{m}x_1(t), \end{aligned}$$

which we write in matrix form as

$$\begin{bmatrix} x_1(t) \\ x_2(t) \end{bmatrix}' = \begin{bmatrix} 0 & 1 \\ -\frac{k}{m} & 0 \end{bmatrix} \begin{bmatrix} x_1(t) \\ x_2(t) \end{bmatrix}. \tag{6}$$

Letting

$$\mathbf{x}(t) = \begin{bmatrix} x_1(t) \\ x_2(t) \end{bmatrix} \quad \text{and} \quad A = \begin{bmatrix} 0 & 1 \\ -\frac{k}{m} & 0 \end{bmatrix},$$

we can write Equation (6) as $\mathbf{x}'(t) = A\mathbf{x}(t)$. To find a fundamental system for Equation (6) we compute the eigenvalues and corresponding eigenvectors of matrix A. We find that the eigenvalues λ_1 and λ_2 are complex with

$$\lambda_1 = i\sqrt{\frac{k}{m}} \quad \text{and} \quad \lambda_2 = -i\sqrt{\frac{k}{m}} \quad \text{(verify)}.$$

The associated eigenvectors are

$$\mathbf{p}_1 = \begin{bmatrix} -i\sqrt{\frac{m}{k}} \\ 1 \end{bmatrix} \quad \text{and} \quad \mathbf{p}_2 = \begin{bmatrix} i\sqrt{\frac{m}{k}} \\ 1 \end{bmatrix},$$

respectively (verify). Thus following Theorem 9.1 and Equation (17) in Section 9.2, for arbitrary constants b_1 and b_2, we have

$$\mathbf{x}(t) = b_1\mathbf{p}_1 e^{i\sqrt{\frac{k}{m}}\,t} + b_2\mathbf{p}_2 e^{-i\sqrt{\frac{k}{m}}\,t}. \tag{7}$$

Since the eigenvalues and eigenvectors are complex, we are unable to determine the vector $\mathbf{x}(t)$ or the trajectories of this system at this time. In this section we will develop procedures for determining trajectories of dynamical systems in the case $n = 2$ that will include the case of complex eigenvalues and eigenvectors. We complete the analysis of the mass–spring dynamical system in Example 6. ■

EXAMPLE 2 The system

$$\mathbf{x}'(t) = A\mathbf{x}(t) = \begin{bmatrix} 0 & 1 \\ -1 & 0 \end{bmatrix}\mathbf{x}(t) \quad \text{or} \quad \begin{aligned} \frac{dx}{dt} &= y \\ \frac{dy}{dt} &= -x \end{aligned}$$

has the general solution*

$$\begin{aligned} x &= b_1 \sin(t) + b_2 \cos(t) \\ y &= b_1 \cos(t) - b_2 \sin(t). \end{aligned} \tag{8}$$

It follows that the trajectories satisfy** $x^2 + y^2 = c^2$, where $c^2 = b_1^2 + b_2^2$. Hence the trajectories of this dynamical system are circles in the phase plane centered at the origin. We note that if an initial condition $\mathbf{x}(0) = \begin{bmatrix} k_1 \\ k_2 \end{bmatrix}$ is specified, then setting $t = 0$ in (8) gives the linear system

$$\begin{aligned} b_1 \sin(0) + b_2 \cos(0) &= k_1 \\ b_1 \cos(0) - b_2 \sin(0) &= k_2. \end{aligned}$$

It follows that the solution is $b_2 = k_1$ and $b_1 = k_2$ so that the corresponding particular solution to the initial value problem $\mathbf{x}'(t) = A\mathbf{x}(t)$, $\mathbf{x}_0 = \begin{bmatrix} k_1 \\ k_2 \end{bmatrix}$ determines the trajectory $x^2 + y^2 = k_1^2 + k_2^2$, that is, a circle centered at the origin with radius $\sqrt{k_1^2 + k_2^2}$. ■

A sketch of the trajectories of a dynamical system in the phase plane is called a **phase portrait**. A phase portrait usually contains the sketches of a few trajectories and an indication of the direction in which the curve is traversed. This is done by placing arrowheads on the trajectory to indicate the direction of motion of a point (x, y) as t increases. The direction is indicated by the **velocity vector**

$$\mathbf{v} = \begin{bmatrix} \dfrac{dx}{dt} \\ \dfrac{dy}{dt} \end{bmatrix}.$$

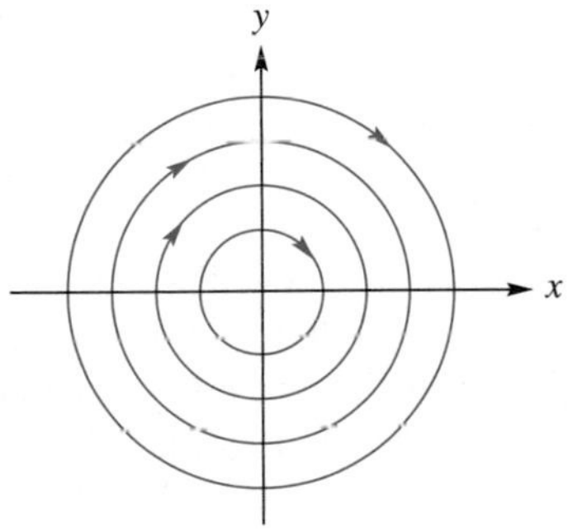

Figure 9.4 ▲

For the dynamical system in Example 2 we have $\mathbf{v} = \begin{bmatrix} y \\ -x \end{bmatrix}$. Thus in the phase plane for $x > 0$ and $y > 0$, the vector $\mathbf{v}$ is oriented downward to the right; hence these trajectories are traversed clockwise as indicated in Figure 9.4. (**Warning:** In other dynamical systems not all trajectories are traversed in the same direction.)

One of the questions posed earlier concerned the existence of constant solutions. For a dynamical system (which we regard as a model depicting the change of x and y as time goes by) to have a constant solution both $\dfrac{dx}{dt}$ and $\dfrac{dy}{dt}$ must be zero. That is, the system does not change. It follows that points in the phase plane that correspond to constant solutions are determined by solving

$$\begin{aligned} \frac{dx}{dt} &= ax + by = 0 \\ \frac{dy}{dt} &= cx + dy = 0, \end{aligned}$$

*We verify this later in this section.

**The trajectories can be obtained directly by noting that we can eliminate t to get $\frac{dy}{dx} = \frac{-x}{y}$. Then separating the variables gives $y\,dy = -x\,dx$ and upon integrating we get $\frac{y^2}{2} = -\frac{x^2}{2} + k^2$ or, equivalently, $x^2 + y^2 = c^2$.

which leads to the homogeneous linear system

$$A\mathbf{x} = \begin{bmatrix} a & b \\ c & d \end{bmatrix} \begin{bmatrix} x \\ y \end{bmatrix} = \begin{bmatrix} 0 \\ 0 \end{bmatrix}.$$

We know that one solution to this homogeneous system is $x = 0$, $y = 0$ and that there exist other solutions if and only if $\det(A) = 0$. In Example 2,

$$A = \begin{bmatrix} 0 & 1 \\ -1 & 0 \end{bmatrix} \quad \text{and} \quad \det(A) = 1.$$

Thus for the dynamical system in Example 2, the only point in the phase plane that corresponds to a constant solution is $x = 0$ and $y = 0$, the origin.

DEFINITION

A point in the phase plane at which both $\dfrac{dx}{dt}$ and $\dfrac{dy}{dt}$ are zero is called an **equilibrium point**, or **fixed point**, of the dynamical system.

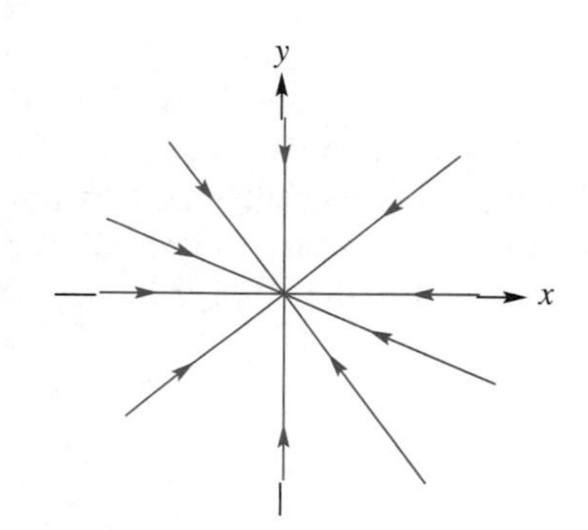

Figure 9.5 ▲
A stable equilibrium point at (0, 0)

The behavior of trajectories near an equilibrium point gives a way to characterize different types of equilibrium points. If trajectories through all points near an equilibrium point converge to the equilibrium point, then we say that the equilibrium point is **stable**, or an **attractor**. This is the case for the origin shown in Figure 9.5, where the trajectories are all straight lines heading into the origin. The dynamical system whose phase portrait is shown in Figure 9.5 is

$$\frac{dx}{dt} = -x$$

$$\frac{dy}{dt} = -y,$$

which we discuss later in Example 4. Another situation is shown in Figure 9.4, where again the only equilibrium point is the origin. In this case, trajectories through points near the equilibrium point stay a small distance away. In such a case, the equilibrium point is called **marginally stable**. At other times, nearby trajectories tend to move away from an equilibrium point. In such cases, we say that the equilibrium point is **unstable**, or a **repelling** point. (See Figure 9.6.) In addition, we can have equilibrium points where nearby trajectories on one side move toward it and on the other side move away from it. Such an equilibrium point is called a **saddle point**. (See Figure 9.7.)

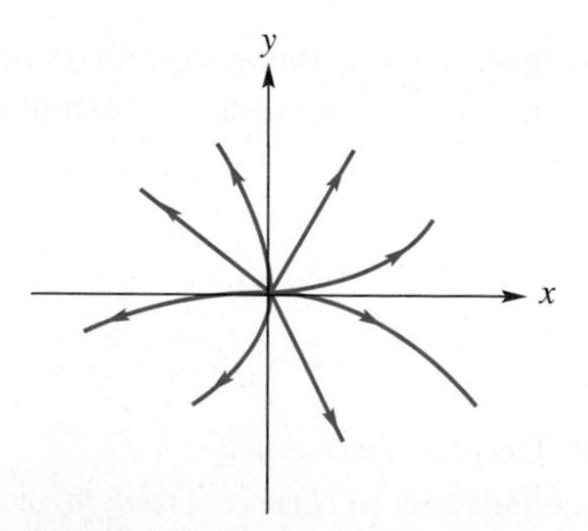

Figure 9.6 ▲
An unstable equilibrium point at (0, 0)

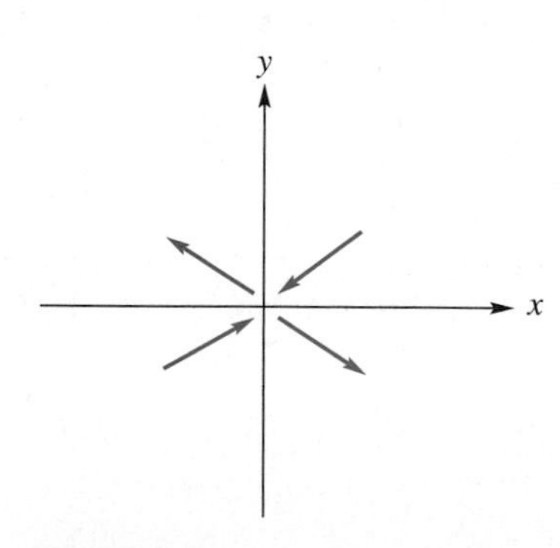

Figure 9.7 ▲
A saddle point at (0, 0)

From the developments in Section 9.2, we expect that the eigenvalues and eigenvectors of the coefficient matrix $A = \begin{bmatrix} a & b \\ c & d \end{bmatrix}$ of the dynamical system will determine features of the phase portrait of the system. From Equation (17) in Section 9.2 we have

$$\mathbf{x}(t) = \begin{bmatrix} x \\ y \end{bmatrix} = b_1\mathbf{p}_1 e^{\lambda_1 t} + b_2\mathbf{p}_2 e^{\lambda_2 t}, \tag{9}$$

where λ_1 and λ_2 are the eigenvalues of A, and $\mathbf{p}_1$ and $\mathbf{p}_2$ are associated eigenvectors. In Sections 8.1 through 8.3, eigenvalues and eigenvectors were to be real numbers and real vectors, respectively. We no longer require them to be real; they can be complex. (See Section A.1 in Appendix A for a discussion of this more general case.) We also require that both λ_1 and λ_2 be nonzero.* Hence A is nonsingular. (Explain why.) And so the only equilibrium point is $x = 0$, $y = 0$, the origin.

To show how we use the eigenvalues and associated eigenvectors of A to determine the phase portrait, we treat the case of complex eigenvalues separately from the case of real eigenvalues.

CASE λ_1 AND λ_2 REAL

For real eigenvalues (and eigenvectors) the phase plane interpretation of Equation (9) is that $\mathbf{x}(t)$ is in span $\{\mathbf{p}_1, \mathbf{p}_2\}$. Hence $\mathbf{p}_1$ and $\mathbf{p}_2$ are trajectories. It follows that the eigenvectors $\mathbf{p}_1$ and $\mathbf{p}_2$ determine lines or rays through the origin in the phase plane, and a phase portrait for this case has the general form shown in Figure 9.8. To complete the portrait we need more than the special trajectories corresponding to the directions of the eigenvectors. These other trajectories depend on the values of λ_1 and λ_2.

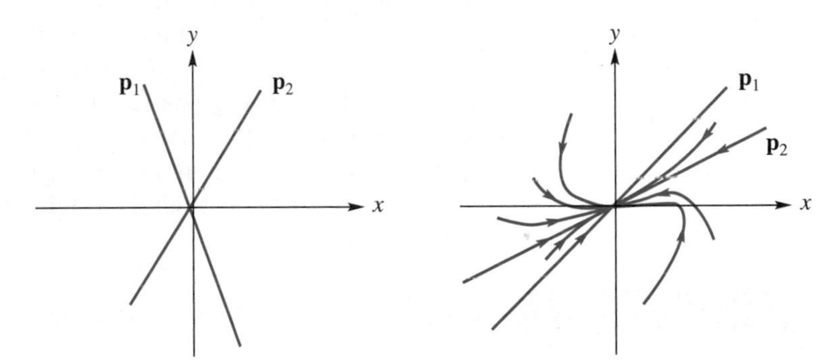

Figure 9.8 ▲ Figure 9.9 ▲

Eigenvalues negative and distinct: $\lambda_1 < \lambda_2 < 0$. From (9), as $t \to \infty$, $\mathbf{x}(t)$ gets small. Hence all the trajectories tend toward the equilibrium point at the origin as $t \to \infty$. See Example 3 and Figure 9.9.

EXAMPLE 3 Determine the phase plane portrait of the dynamical system

$$\mathbf{x}'(t) = A\mathbf{x}(t) = \begin{bmatrix} -2 & -2 \\ 1 & -5 \end{bmatrix}\mathbf{x}(t).$$

---

*It can be shown that if both eigenvalues of A are zero, then all solutions to (2) as given in (5) are either constants or constants and straight lines. In addition, we can show that if one eigenvalue of A is zero and the other nonzero, then there is a line of equilibrium points. See Further Readings at the end of this section.

We begin by finding the eigenvalues and associated eigenvectors of A. We find (verify)

$$\lambda_1 = -4, \quad \lambda_2 = -3 \quad \text{and} \quad \mathbf{p}_1 = \begin{bmatrix} 1 \\ 1 \end{bmatrix}, \quad \mathbf{p}_2 = \begin{bmatrix} 2 \\ 1 \end{bmatrix}.$$

It follows that

$$\mathbf{x}(t) = \begin{bmatrix} x \\ y \end{bmatrix} = b_1\mathbf{p}_1 e^{-4t} + b_2\mathbf{p}_2 e^{-3t}$$

and as $t \to \infty$, $\mathbf{x}(t)$ gets small. It is helpful to rewrite this expression in the form

$$\mathbf{x}(t) = \begin{bmatrix} x \\ y \end{bmatrix} = b_1\mathbf{p}_1 e^{-4t} + b_2\mathbf{p}_2 e^{-3t} = e^{-3t}(b_1\mathbf{p}_1 e^{-t} + b_2\mathbf{p}_2).$$

As long as $b_2 \neq 0$, the term $b_1\mathbf{p}_1 e^{-t}$ is negligible in comparison to $b_2\mathbf{p}_2$. This implies that as $t \to \infty$ all trajectories, except those starting on $\mathbf{p}_1$, will align themselves in the direction of $\mathbf{p}_2$ as they get close to the origin. Hence the phase portrait appears as shown in Figure 9.9. The origin is an attractor. ■

Eigenvalues positive and distinct: $\lambda_1 > \lambda_2 > 0$. From (9), as $t \to \infty$, $\mathbf{x}(t)$ gets large. Hence all the trajectories tend to move away from the equilibrium point at the origin. The phase portrait for such dynamical systems is like that in Figure 9.9, except that all the arrowheads are reversed, indicating motion away from the origin. In this case (0, 0) is called an **unstable** equilibrium point.

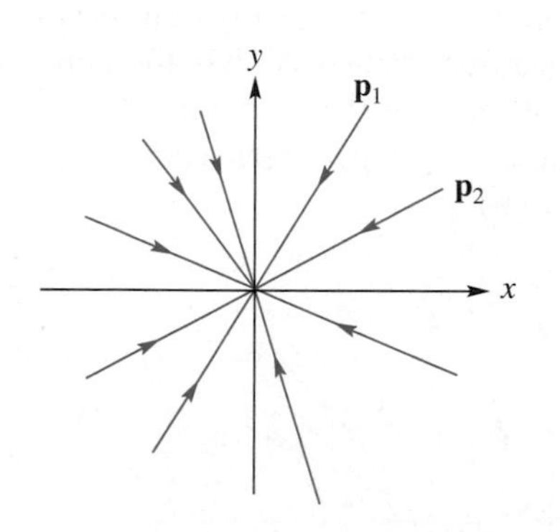

Figure 9.10 ▲

Both eigenvalues negative but equal: $\lambda_1 = \lambda_2 < 0$. All trajectories move to a stable equilibrium point at the origin, but they may bend differently than the trajectories depicted in Figure 9.9. Their behavior depends upon the number of linearly independent eigenvectors of matrix A. If there are two linearly independent eigenvectors, then $\mathbf{x}(t) = e^{\lambda_1 t}(b_1\mathbf{p}_1 + b_2\mathbf{p}_2)$, which is a multiple of the constant vector $b_1\mathbf{p}_1 + b_2\mathbf{p}_2$. Thus it follows that all the trajectories are lines through the origin and, since $\lambda_1 < 0$, motion along them is toward the origin. See Figure 9.10. We illustrate this in Example 4.

EXAMPLE 4 The matrix A of the dynamical system

$$\mathbf{x}'(t) = A\mathbf{x}(t) = \begin{bmatrix} -1 & 0 \\ 0 & -1 \end{bmatrix} \mathbf{x}(t)$$

has eigenvalues $\lambda_1 = \lambda_2 = -1$ and associated eigenvectors

$$\mathbf{p}_1 = \begin{bmatrix} 1 \\ 0 \end{bmatrix} \quad \text{and} \quad \mathbf{p}_2 = \begin{bmatrix} 0 \\ 1 \end{bmatrix} \qquad \text{(verify)}$$

It follows that

$$\mathbf{x}(t) = e^{-t}(b_1\mathbf{p}_1 + b_2\mathbf{p}_2) = e^{-t}\begin{bmatrix} b_1 \\ b_2 \end{bmatrix},$$

so $x = b_1 e^{-t}$ and $y = b_2 e^{-t}$. If $b_1 \neq 0$, then $y = \dfrac{b_2}{b_1}x$. If $b_1 = 0$, then we are on the trajectory in the direction of $\mathbf{p}_2$. It follows that all trajectories are straight lines through the origin as in Figures 9.5 and 9.10. ■

If there is only one linearly independent eigenvector, then it can be shown that all trajectories passing through points not on the eigenvector align themselves to be tangent to the eigenvector at the origin. We will not develop this case, but the phase portrait is similar to Figure 9.11. In the case that the eigenvalues are positive and equal, the phase portraits for these two cases are like Figures 9.10 and 9.11 with the arrowheads reversed.

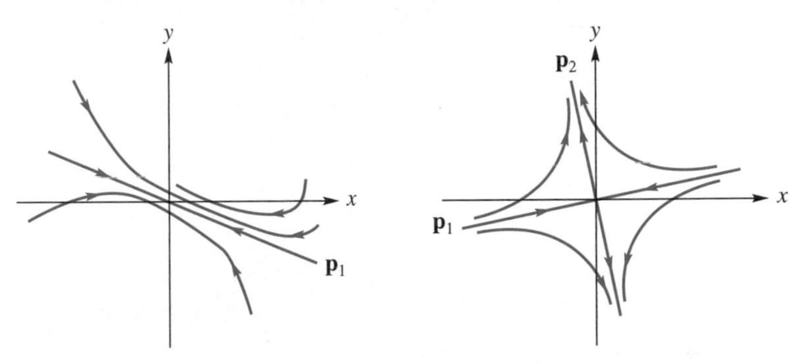

Figure 9.11 ▲ **Figure 9.12** ▲

One positive eigenvalue and one negative eigenvalue: $\lambda_1 < 0 < \lambda_2$. From (9), as $t \to \infty$ one of the terms is increasing while the other term is decreasing. This causes a trajectory, which is not in the direction of an eigenvector, to head toward the origin but bend away as t gets larger. The origin in this case is called a **saddle point**. The phase portrait resembles Figure 9.12.

EXAMPLE 5 Determine the phase plane portrait of the dynamical system

$$\mathbf{x}'(t) = A\mathbf{x}(t) = \begin{bmatrix} 1 & -1 \\ -2 & 0 \end{bmatrix} \mathbf{x}(t).$$

We begin by finding the eigenvalues and associated eigenvectors of A. We find (verify)

$$\lambda_1 = -1, \quad \lambda_2 = 2 \quad \text{and} \quad \mathbf{p}_1 = \begin{bmatrix} 1 \\ 2 \end{bmatrix}, \quad \mathbf{p}_2 = \begin{bmatrix} 1 \\ -1 \end{bmatrix}.$$

It follows that the origin is a saddle point and that we have

$$\mathbf{x}(t) = \begin{bmatrix} x \\ y \end{bmatrix} = b_1\mathbf{p}_1 e^{-t} + b_2\mathbf{p}_2 e^{2t}.$$

We see that if $b_1 \neq 0$ and $b_2 = 0$, then the motion is in the direction of eigenvector $\mathbf{p}_1$ and toward the origin. Similarly, if $b_1 = 0$ and $b_2 \neq 0$, then the motion is in the direction of $\mathbf{p}_2$ but away from the origin. If we look at the components of the original system and eliminate t, we obtain (verify)

$$\frac{dy}{dx} = \frac{2x}{y - x}.$$

This expression tells us the slope along trajectories in the phase plane. Inspecting this expression we see the following: (Explain.)

- All trajectories crossing the y-axis have horizontal tangents.
- As a trajectory crosses the line $y = x$ it has a vertical tangent.

- Whenever a trajectory crosses the x-axis it has slope -2.

Using the general form of a saddle point, shown in Figure 9.12, we can produce quite an accurate phase portrait for this dynamical system, shown in Figure 9.13. ■

Figure 9.13 ►

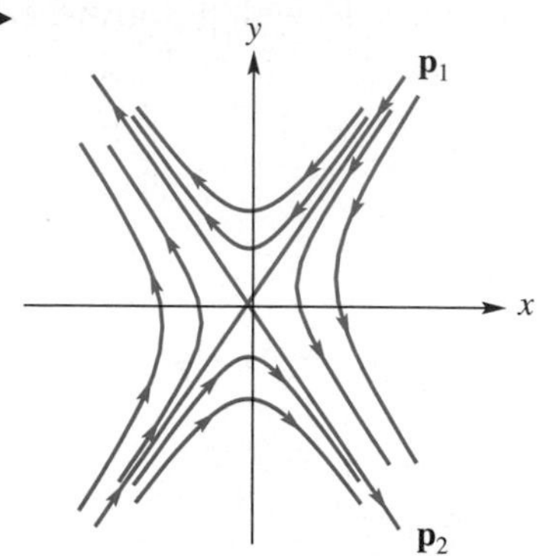

BOTH EIGENVALUES COMPLEX NUMBERS

For real 2×2 matrices A, the characteristic equation $\det(\lambda I_2 - A) = 0$ is a quadratic polynomial. If the roots of this quadratic equation λ_1 and λ_2 are complex numbers, then they are conjugates of one another. (See Section A.2.) If $\lambda_1 = \alpha + \beta i$, where α and β are real numbers with $\beta \neq 0$, then $\lambda_2 = \overline{\lambda_1} = \alpha - \beta i$. Hence in Equation (9) we have the exponential of a complex number:

$$e^{\lambda_1 t} = e^{(\alpha+\beta i)t} = e^{\alpha t} e^{\beta i t}.$$

The term $e^{\alpha t}$ is a standard exponential function, but $e^{\beta i t}$ is quite different since $i = \sqrt{-1}$. Fortunately there is a simple way to express such an exponential function in terms of more manageable functions. We use Euler's identity,

$$e^{i\theta} = \cos(\theta) + i\sin(\theta),$$

which we state without proof. Using Euler's identity we have

$$e^{\lambda_1 t} = e^{(\alpha+\beta i)t} = e^{\alpha t} e^{\beta i t} = e^{\alpha t}(\cos(\beta t) + i\sin(\beta t))$$

and

$$\begin{aligned} e^{\lambda_2 t} = e^{(\alpha-\beta i)t} &= e^{\alpha t} e^{-\beta i t} \\ &= e^{\alpha t}(\cos(-\beta t) + i\sin(-\beta t)) = e^{\alpha t}(\cos(\beta t) - i\sin(\beta t)). \end{aligned}$$

It can be shown that the system given in (9) can be written so that the components $x(t)$ and $y(t)$ are linear combinations of $e^{\alpha t}\cos(\beta t)$ and $e^{\alpha t}\sin(\beta t)$ with real coefficients. The behavior of the trajectories can now be analyzed by considering the sign of α, since $\beta \neq 0$.

Complex eigenvalues: $\lambda_1 = \alpha + \beta i$ and $\lambda_2 = \alpha - \beta i$ with $\alpha = 0$, $\beta \neq 0$. For this case $x(t)$ and $y(t)$ are linear combinations of $\cos(\beta t)$ and $\sin(\beta t)$. It can be shown that the trajectories are ellipses whose major and minor axes are determined by the eigenvectors. (See Example 2 for a particular case.) The motion is periodic since the trajectories are closed curves. The origin is a marginally stable equilibrium point, since the trajectories through points near the origin do not move very far away. (See Figure 9.4.)

Complex eigenvalues: $\lambda_1 = \alpha + \beta i$ **and** $\lambda_2 = \alpha - \beta i$ **with** $\alpha \neq 0$, $\beta \neq 0$. For this case $x(t)$ and $y(t)$ are linear combinations of $e^{\alpha t}\cos(\beta t)$ and $e^{\alpha t}\sin(\beta t)$. It can be shown that the trajectories are spirals. If $\alpha > 0$, then the spiral moves outward, away from the origin. Thus the origin is an unstable equilibrium point. If $\alpha < 0$, then the spiral moves inward, toward the origin, so the origin is a stable equilibrium point. The phase portrait is a collection of spirals as shown in Figure 9.14, with arrowheads appropriately affixed.

Figure 9.14 ►

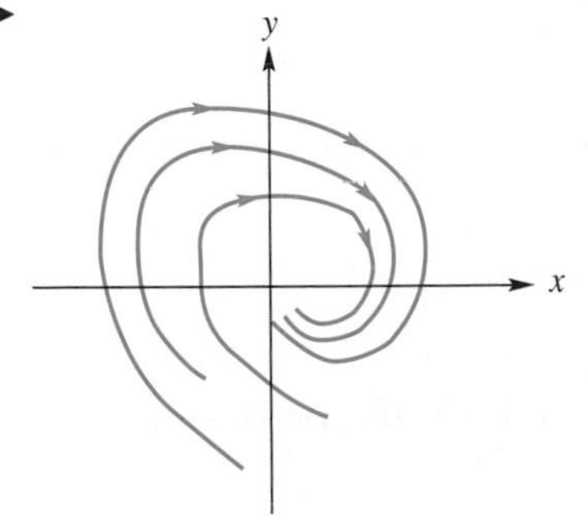

EXAMPLE 6

We can use the preceding concepts for complex eigenvalues and eigenvectors to determine the trajectories of the mass–spring dynamical system discussed in Example 1. Recall that the eigenvalues of the matrix A are

$$\lambda_1 = i\sqrt{\frac{k}{m}} \quad \text{and} \quad \lambda_2 = -i\sqrt{\frac{k}{m}}$$

with associated eigenvectors

$$\mathbf{p}_1 = \begin{bmatrix} -i\sqrt{\frac{m}{k}} \\ 1 \end{bmatrix} \quad \text{and} \quad \mathbf{p}_2 = \begin{bmatrix} i\sqrt{\frac{m}{k}} \\ 1 \end{bmatrix},$$

respectively. From our preceding discussion it follows that the trajectories are ellipses whose major and minor axes are determined by the eigenvectors of A. Using Euler's identity, the solution of the dynamical system $\mathbf{x}'(t) = A\mathbf{x}(t)$ from Example 1 can be expressed as follows:

$$\begin{aligned}
\mathbf{x}(t) &= b_1\mathbf{p}_1 e^{i\sqrt{\frac{k}{m}}\,t} + b_2\mathbf{p}_2 e^{-i\sqrt{\frac{k}{m}}\,t} \\
&= b_1\begin{bmatrix} -i\sqrt{\frac{m}{k}} \\ 1 \end{bmatrix}\left[\cos\left(\sqrt{\frac{k}{m}}\,t\right) + i\sin\left(\sqrt{\frac{k}{m}}\,t\right)\right] \\
&\quad + b_2\begin{bmatrix} i\sqrt{\frac{m}{k}} \\ 1 \end{bmatrix}\left[\cos\left(\sqrt{\frac{k}{m}}\,t\right) - i\sin\left(\sqrt{\frac{k}{m}}\,t\right)\right] \\
&= \cos\left(\sqrt{\frac{k}{m}}\,t\right)\left(b_1\begin{bmatrix} -i\sqrt{\frac{m}{k}} \\ 1 \end{bmatrix} + b_2\begin{bmatrix} i\sqrt{\frac{m}{k}} \\ 1 \end{bmatrix}\right) \\
&\quad + i\sin\left(\sqrt{\frac{k}{m}}\,t\right)\left(b_1\begin{bmatrix} -i\sqrt{\frac{m}{k}} \\ 1 \end{bmatrix} - b_2\begin{bmatrix} i\sqrt{\frac{m}{k}} \\ 1 \end{bmatrix}\right)
\end{aligned}$$

$$= \cos\left(\sqrt{\frac{k}{m}}\,t\right)\begin{bmatrix} i\sqrt{\frac{m}{k}}\,(-b_1+b_2) \\ b_1+b_2 \end{bmatrix} + i\sin\left(\sqrt{\frac{k}{m}}\,t\right)\begin{bmatrix} -i\sqrt{\frac{m}{k}}\,(b_1+b_2) \\ b_1-b_2 \end{bmatrix}$$

$$= \cos\left(\sqrt{\frac{k}{m}}\,t\right)\begin{bmatrix} i\sqrt{\frac{m}{k}}(-b_1+b_2) \\ b_1+b_2 \end{bmatrix} + \sin\left(\sqrt{\frac{k}{m}}\,t\right)\begin{bmatrix} \sqrt{\frac{m}{k}}(b_1+b_2) \\ i(b_1-b_2) \end{bmatrix}.$$

Thus

$$x_1(t) = i\sqrt{\frac{m}{k}}\,(-b_1+b_2)\cos\left(\sqrt{\frac{k}{m}}\,t\right) + \sqrt{\frac{m}{k}}\,(b_1+b_2)\sin\left(\sqrt{\frac{k}{m}}\,t\right)$$

and

$$x_2(t) = (b_1+b_2)\cos\left(\sqrt{\frac{k}{m}}\,t\right) + i(b_1-b_2)\sin\left(\sqrt{\frac{k}{m}}\,t\right).$$

Letting

$$c_1 = i(-b_1+b_2)\sqrt{\frac{m}{k}}, \qquad c_2 = (b_1+b_2)\sqrt{\frac{m}{k}},$$
$$c_3 = b_1+b_2, \qquad c_4 = i(b_1-b_2),$$

we have

$$x_1(t) = c_1\cos\left(\sqrt{\frac{k}{m}}\,t\right) + c_2\sin\left(\sqrt{\frac{k}{m}}\,t\right) \tag{10}$$

and

$$x_2(t) = c_3\cos\left(\sqrt{\frac{k}{m}}\,t\right) + c_4\sin\left(\sqrt{\frac{k}{m}}\,t\right).$$

Since $x_1'(t) = x_2(t)$, it follows that $c_3 = c_2\sqrt{\frac{k}{m}}$ and $c_4 = -c_1\sqrt{\frac{k}{m}}$ (verify); hence

$$x_2(t) = c_2\sqrt{\frac{k}{m}}\cos\left(\sqrt{\frac{k}{m}}\,t\right) - c_1\sqrt{\frac{k}{m}}\sin\left(\sqrt{\frac{k}{m}}\,t\right). \tag{11}$$

To see that the trajectories in the phase plane are indeed ellipses, we use Equations (10) and (11) as follows. Compute

$$[x_1(t)]^2 + \left(\frac{x_2(t)}{\sqrt{\frac{k}{m}}}\right)^2$$

and show that the result is a constant. We have

$$
\begin{aligned}
[x_1(t)]^2 + \left(\frac{x_2(t)}{\sqrt{\frac{k}{m}}}\right)^2 &= \left[c_1 \cos\left(\sqrt{\frac{k}{m}}\,t\right) + c_2 \sin\left(\sqrt{\frac{k}{m}}\,t\right)\right]^2 \\
&\quad + \left[c_2 \cos\left(\sqrt{\frac{k}{m}}\,t\right) - c_1 \sin\left(\sqrt{\frac{k}{m}}\,t\right)\right]^2 \\
&= c_1^2 \cos^2\left(\sqrt{\frac{k}{m}}\,t\right) + 2c_1c_2 \cos\left(\sqrt{\frac{k}{m}}\,t\right) \sin\left(\sqrt{\frac{k}{m}}\,t\right) \\
&\quad + c_2^2 \sin^2\left(\sqrt{\frac{k}{m}}\,t\right) + c_2^2 \cos^2\left(\sqrt{\frac{k}{m}}\,t\right) \\
&\quad - 2c_1c_2 \cos\left(\sqrt{\frac{k}{m}}\,t\right) \sin\left(\sqrt{\frac{k}{m}}\,t\right) + c_1^2 \sin^2\left(\sqrt{\frac{k}{m}}\,t\right) \\
&= (c_1^2 + c_2^2) \cos^2\left(\sqrt{\frac{k}{m}}\,t\right) + (c_1^2 + c_2^2) \sin^2\left(\sqrt{\frac{k}{m}}\,t\right) \\
&= c_1^2 + c_2^2.
\end{aligned}
$$

Hence the displacement of the mass from equilibrium, which is given by $x_1(t)$, varies in a repetitive pattern as time changes because the trajectory follows the path of an ellipse in the x_1x_2-plane, the phase plane. ■

The dynamical system in (2) may appear quite special. However, the experience gained by the qualitative analysis we have presented is the key to understanding the behavior of nonlinear dynamical systems of the form

$$
\begin{aligned}
\frac{dx}{dt} &= f(x, y) \\
\frac{dy}{dt} &= g(x, y).
\end{aligned}
$$

A simple example of a nonlinear dynamical system is obtained from a predator–prey model in which a species x, the predator, can only exhibit a population growth by killing another species y, the prey. Such models are often given in terms of fox and rabbit populations, where the rabbits are the primary food for the foxes. In such a situation one population influences the births and deaths in the other population. In general, the interactions of the predator and the prey result in a cycle of population changes for each species. Using field studies to estimate birth, death, and kill rates, ecologists develop a dynamical system to model the populations $x(t)$ and $y(t)$ of the two species. One such system of differential equations that is commonly determined has the form

$$
\begin{aligned}
x'(t) &= -ay(t) - ax(t)y(t) \\
y'(t) &= bx(t) + bx(t)y(t)
\end{aligned}
$$

where $a > 0$ and $b > 0$. This nonlinear dynamical system is approximated by the linear dynamical system

$$\begin{aligned} x'(t) &= -ay(t) \\ y'(t) &= bx(t) \end{aligned}$$

or, equivalently, in matrix form

$$\mathbf{x}'(t) = \begin{bmatrix} x'(t) \\ y'(t) \end{bmatrix} = \begin{bmatrix} 0 & -a \\ b & 0 \end{bmatrix} \begin{bmatrix} x(t) \\ y(t) \end{bmatrix} = A\mathbf{x}(t).$$

The eigenvalues of the matrix A are complex (verify), hence the trajectories of the linear dynamical systems are ellipses and the populations of foxes and rabbits behave in a cyclical fashion for the linear model. We suspect that the trajectories of the nonlinear dynamical system also have a cyclical pattern, but not necessarily elliptical.

The extension of the phase plane and phase portraits to such nonlinear systems is beyond the scope of this brief introduction. Such topics are part of courses on differential equations and can be found in the books listed in Further Readings.

Further Readings

BOYCE, W. E., AND R. C. DIPRIMA. *Elementary Differential Equations*, 8th ed., New York: John Wiley & Sons, Inc., 2004.

CAMPBELL, S. L. *An Introduction to Differential Equations and Their Applications*, 2nd ed. Belmont, California: Wadsworth Publishing Co., 1990.

FARLOW, S. J. *An Introduction to Differential Equations and Their Applications*. New York: McGraw-Hill, Inc., 1994.

Key Terms

Qualitative theory of differential equations
Autonomous differential equations
Time
Dynamical systems
Trajectory (or orbit)
Phase plane
Equilibrium position
Simple harmonic motion
Phase portrait
Velocity vector
Equilibrium point (or fixed point)
Stable point (or attractor)
Marginally stable point
Unstable (or repelling) point
Saddle point

9.3 Exercises

For each of the dynamical systems in Exercises 1 through 10, determine the nature of the equilibrium point at the origin, and describe the phase portrait.

1. $\mathbf{x}'(t) = \begin{bmatrix} -1 & 0 \\ 0 & -3 \end{bmatrix} \mathbf{x}(t)$

2. $\mathbf{x}'(t) = \begin{bmatrix} 2 & 0 \\ 0 & 1 \end{bmatrix} \mathbf{x}(t)$

3. $\mathbf{x}'(t) = \begin{bmatrix} -1 & 2 \\ 0 & -1 \end{bmatrix} \mathbf{x}(t)$

4. $\mathbf{x}'(t) = \begin{bmatrix} -2 & 0 \\ 3 & 1 \end{bmatrix} \mathbf{x}(t)$

5. $\mathbf{x}'(t) = \begin{bmatrix} 1 & 1 \\ 3 & -1 \end{bmatrix} \mathbf{x}(t)$

6. $\mathbf{x}'(t) = \begin{bmatrix} -1 & -1 \\ 1 & -1 \end{bmatrix} \mathbf{x}(t)$

7. $\mathbf{x}'(t) = \begin{bmatrix} -2 & 1 \\ 2 & -3 \end{bmatrix} \mathbf{x}(t)$

8. $\mathbf{x}'(t) = \begin{bmatrix} -2 & 1 \\ -1 & -2 \end{bmatrix} \mathbf{x}(t)$

9. $\mathbf{x}'(t) = \begin{bmatrix} 3 & -13 \\ 1 & -3 \end{bmatrix} \mathbf{x}(t)$

10. $\mathbf{x}'(t) = \begin{bmatrix} 3 & 2 \\ 2 & 3 \end{bmatrix} \mathbf{x}(t)$

Theoretical Exercise

T.1. A variation of the mass–spring system discussed in Example 1 is used in the design of a shock absorber. In this device a mass is suspended on a coiled spring that is anchored at its lower end. The mass is also attached to a piston that moves in a fluid-filled cylinder and resists the motion of the mass. (See Figure 9.15.) Using an analysis similar to that in Example 1 we can show that a linear second-order differential equation that models the displacement from equilibrium, $x(t)$, of the mass m is

$$x''(t) + 2rx'(t) + \frac{k}{m}x(t) = 0, \qquad (12)$$

where k is the spring constant and $r \geq 0$ is a parameter involved in the resistive force exerted by the piston. (Again we have imposed simplifying assumptions within the modeling process to obtain this equation.)

Shock absorber

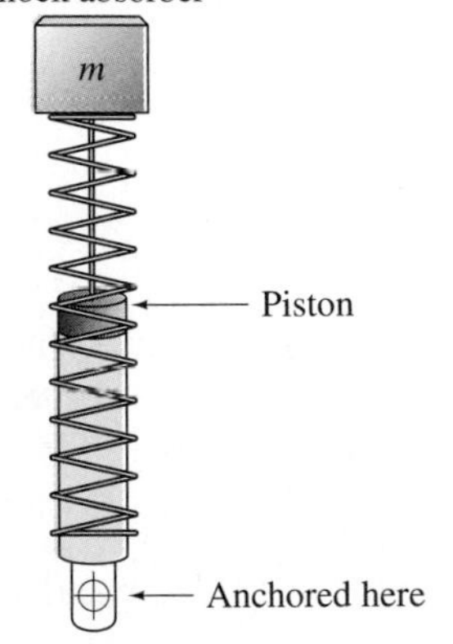

Figure 9.15 ▲

(a) Show that Equation (12) can be formulated as a dynamical system $\mathbf{x}'(t) = A\mathbf{x}(t)$, where

$$\mathbf{x}(t) = \begin{bmatrix} x_1(t) \\ x_2(t) \end{bmatrix} = \begin{bmatrix} x(t) \\ x'(t) \end{bmatrix}$$

and

$$A = \begin{bmatrix} 0 & 1 \\ -\frac{k}{m} & -2r \end{bmatrix}.$$

(b) Let m and r each be fixed at 1 unit. Determine the eigenvalues of A and describe the trajectories of the dynamical system for each of the following values of the spring constant k.

(i) $k = 0.75$ (ii) $k = 1$

(iii) $k = 2$ (iv) $k = 10$

(c) For $m = 1$ and $r = 1$, determine the characteristic polynomial of A and an expression for the eigenvalues of A in terms of the spring constant k. Beyond what value of k will the eigenvalues of A be guaranteed to be complex?

(d) Let m and k each be fixed at 1 unit. Determine the eigenvalues of A and describe the trajectories of the dynamical system for each of the following values of the parameter r.

(i) $r = 0$ (ii) $r = \frac{1}{2}$

(iii) $r = 1$ (iv) $r = \sqrt{2}$

(e) In part (d) for $r = 0$, describe the motion of a rider on a motorcycle if such a shock absorber were installed on the bike and the bike hit a bump in the road.

(f) For $m = 1$ and $k = 1$, determine the characteristic polynomial of A and an expression for the eigenvalues of A in terms of the parameter r. Beyond what value of r will the eigenvalues of A be guaranteed to be real?

9.4 QUADRATIC FORMS

Prerequisite. Section 8.3, Diagonalization of Symmetric Matrices.

In your precalculus and calculus courses you have seen that the graph of the equation

$$ax^2 + 2bxy + cy^2 = d, \qquad (1)$$

where a, b, c, and d are real numbers, is a **conic section** centered at the origin of a rectangular Cartesian coordinate system in two-dimensional space. Similarly, the graph of the equation

$$ax^2 + 2dxy + 2exz + by^2 + 2fyz + cz^2 = g, \qquad (2)$$

where a, b, c, d, e, f, and g are real numbers, is a **quadric surface** centered at the origin of a rectangular Cartesian coordinate system in three-dimensional space. If a conic section or quadric surface is not centered at the origin, its equations are more complicated than those given in (1) and (2).

The identification of the conic section or quadric surface that is the graph of a given equation often requires the rotation and translation of the coordinate axes. These methods can best be understood as an application of eigenvalues and eigenvectors of matrices and will be discussed in Sections 9.5 and 9.6.

The expressions on the left sides of Equations (1) and (2) are examples of quadratic forms. Quadratic forms arise in statistics, mechanics, and in other problems in physics; in quadratic programming; in the study of maxima and minima of functions of several variables; and in other applied problems. In this section we use our results on eigenvalues and eigenvectors of matrices to give a brief treatment of real quadratic forms in n variables. In Section 9.5 we apply these results to the classification of the conic sections, and in Section 9.6 to the classification of the quadric surfaces.

DEFINITION If A is a symmetric matrix, then the function $g\colon R^n \to R^1$ (a real-valued function on R^n) defined by

$$g(\mathbf{x}) = \mathbf{x}^T A\mathbf{x},$$

where

$$\mathbf{x} = \begin{bmatrix} x_1 \\ x_2 \\ \vdots \\ x_n \end{bmatrix},$$

is called a **real quadratic form in the variables $x_1, x_2, \dots, x_n$**. The matrix A is called the **matrix of the quadratic form g**. We shall also denote the quadratic form by $g(\mathbf{x})$.

EXAMPLE 1 Write the left side of (1) as the quadratic form in the variables x and y.

Solution Let

$$\mathbf{x} = \begin{bmatrix} x \\ y \end{bmatrix} \quad \text{and} \quad A = \begin{bmatrix} a & b \\ b & c \end{bmatrix}.$$

Then the left side of (1) is the quadratic form

$$g(\mathbf{x}) = \mathbf{x}^T A\mathbf{x}.$$

■

EXAMPLE 2 Write the left side of (2) as the quadratic form.

Solution Let

$$\mathbf{x} = \begin{bmatrix} x \\ y \\ z \end{bmatrix} \quad \text{and} \quad A = \begin{bmatrix} a & d & e \\ d & b & f \\ e & f & c \end{bmatrix}.$$

Then the left side of (2) is the quadratic form

$$g(\mathbf{x}) = \mathbf{x}^T A\mathbf{x}.$$

■

EXAMPLE 3 The following expressions are quadratic forms:

(a) $3x^2 - 5xy - 7y^2 = \begin{bmatrix} x & y \end{bmatrix} \begin{bmatrix} 3 & -\frac{5}{2} \\ -\frac{5}{2} & -7 \end{bmatrix} \begin{bmatrix} x \\ y \end{bmatrix}$

(b) $3x^2 - 7xy + 5xz + 4y^2 - 4yz - 3z^2 = \begin{bmatrix} x & y & z \end{bmatrix} \begin{bmatrix} 3 & -\frac{7}{2} & \frac{5}{2} \\ -\frac{7}{2} & 4 & -2 \\ \frac{5}{2} & -2 & -3 \end{bmatrix} \begin{bmatrix} x \\ y \\ z \end{bmatrix}$ ■

Suppose now that $g(\mathbf{x}) = \mathbf{x}^T A\mathbf{x}$ is a quadratic form. To simplify the quadratic form, we change from the variables $x_1, x_2, \ldots, x_n$ to the variables $y_1, y_2, \ldots, y_n$, where we assume that the old variables are related to the new variables by $\mathbf{x} = P\mathbf{y}$ for some orthogonal matrix P. Then

$$g(\mathbf{x}) = \mathbf{x}^T A\mathbf{x} = (P\mathbf{y})^T A(P\mathbf{y}) = \mathbf{y}^T (P^T AP)\mathbf{y} = \mathbf{y}^T B\mathbf{y},$$

where $B = P^T AP$. We shall let you verify that if A is a symmetric matrix, then $P^T AP$ is also symmetric (Exercise T.1). Thus

$$h(\mathbf{y}) = \mathbf{y}^T B\mathbf{y}$$

is another quadratic form and $g(\mathbf{x}) = h(\mathbf{y})$.

This situation is important enough to formulate the following definitions.

DEFINITION

If A and B are $n \times n$ matrices, we say that B is **congruent** to A if $B = P^T AP$ for a nonsingular matrix P.

In light of Exercise T.2 we do not distinguish between the statements "A is congruent to B," and "B is congruent to A." Each of these statements can be replaced by "A and B are congruent."

DEFINITION

Two quadratic forms g and h with matrices A and B, respectively, are said to be **equivalent** if A and B are congruent.

The congruence of matrices and equivalence of forms are more general concepts than similarity of symmetric matrices by an orthogonal matrix P, since P is required only to be nonsingular. We shall consider here the more restrictive situation with P orthogonal.

EXAMPLE 4

Consider the quadratic form in the variables x and y defined by

$$g(\mathbf{x}) = 2x^2 + 2xy + 2y^2 = \begin{bmatrix} x & y \end{bmatrix} \begin{bmatrix} 2 & 1 \\ 1 & 2 \end{bmatrix} \begin{bmatrix} x \\ y \end{bmatrix}. \tag{3}$$

We now change from the variables x and y to the variables x' and y'. Suppose that the old variables are related to the new variables by the equations

$$x = \frac{1}{\sqrt{2}}x' - \frac{1}{\sqrt{2}}y' \quad \text{and} \quad y = \frac{1}{\sqrt{2}}x' + \frac{1}{\sqrt{2}}y', \tag{4}$$

which can be written in matrix form as

$$\mathbf{x} = \begin{bmatrix} x \\ y \end{bmatrix} = \begin{bmatrix} \frac{1}{\sqrt{2}} & -\frac{1}{\sqrt{2}} \\ \frac{1}{\sqrt{2}} & \frac{1}{\sqrt{2}} \end{bmatrix} \begin{bmatrix} x' \\ y' \end{bmatrix} = P\mathbf{y},$$

where the orthogonal (hence nonsingular) matrix

$$P = \begin{bmatrix} \frac{1}{\sqrt{2}} & -\frac{1}{\sqrt{2}} \\ \frac{1}{\sqrt{2}} & \frac{1}{\sqrt{2}} \end{bmatrix} \quad \text{and} \quad \mathbf{y} = \begin{bmatrix} x' \\ y' \end{bmatrix}.$$

We shall soon see why and how this particular matrix P was selected. Substituting in (3), we obtain

$$\begin{aligned} g(\mathbf{x}) &= \mathbf{x}^T A\mathbf{x} = (P\mathbf{y})^T A(P\mathbf{y}) = \mathbf{y}^T P^T A P\mathbf{y} \\ &= \begin{bmatrix} x' & y' \end{bmatrix} \begin{bmatrix} \frac{1}{\sqrt{2}} & -\frac{1}{\sqrt{2}} \\ \frac{1}{\sqrt{2}} & \frac{1}{\sqrt{2}} \end{bmatrix}^T \begin{bmatrix} 2 & 1 \\ 1 & 2 \end{bmatrix} \begin{bmatrix} \frac{1}{\sqrt{2}} & -\frac{1}{\sqrt{2}} \\ \frac{1}{\sqrt{2}} & \frac{1}{\sqrt{2}} \end{bmatrix} \begin{bmatrix} x' \\ y' \end{bmatrix} \\ &= \begin{bmatrix} x' & y' \end{bmatrix} \begin{bmatrix} 3 & 0 \\ 0 & 1 \end{bmatrix} \begin{bmatrix} x' \\ y' \end{bmatrix} = h(\mathbf{y}) \\ &= 3x'^2 + y'^2. \end{aligned}$$

Thus the matrices

$$\begin{bmatrix} 2 & 1 \\ 1 & 2 \end{bmatrix} \quad \text{and} \quad \begin{bmatrix} 3 & 0 \\ 0 & 1 \end{bmatrix}$$

are congruent and the quadratic forms g and h are equivalent. ■

We now turn to the question of how to select the matrix P.

THEOREM 9.2 **(Principal Axes Theorem)** *Any quadratic form in n variables $g(\mathbf{x}) = \mathbf{x}^T A\mathbf{x}$ is equivalent by means of an orthogonal matrix P to a quadratic form, $h(\mathbf{y}) = \lambda_1 y_1^2 + \lambda_2 y_2^2 + \cdots + \lambda_n y_n^2$, where*

$$\mathbf{y} = \begin{bmatrix} y_1 \\ y_2 \\ \vdots \\ y_n \end{bmatrix}$$

and $\lambda_1, \lambda_2, \ldots, \lambda_n$ are the eigenvalues of the matrix A of g.

Proof If A is the matrix of g, then, since A is symmetric, we know, by Theorem 8.9 in Section 8.3, that A can be diagonalized by an orthogonal matrix. This means that there exists an orthogonal matrix P such that $D = P^{-1}AP$ is a diagonal matrix. Since P is orthogonal, $P^{-1} = P^T$, so $D = P^T AP$. Moreover, the elements on the main diagonal of D are the eigenvalues, $\lambda_1, \lambda_2, \ldots, \lambda_n$ of A, which are real numbers. The quadratic form h with matrix D is given by

$$h(\mathbf{y}) = \lambda_1 y_1^2 + \lambda_2 y_2^2 + \cdots + \lambda_n y_n^2;$$

g and h are equivalent. ■

EXAMPLE 5 Consider the quadratic form g in the variables x, y, and z defined by

$$g(\mathbf{x}) = 2x^2 + 4y^2 + 6yz - 4z^2.$$

Determine a quadratic form h of the form in Theorem 9.2 to which g is equivalent.

Solution The matrix of g is

$$A = \begin{bmatrix} 2 & 0 & 0 \\ 0 & 4 & 3 \\ 0 & 3 & -4 \end{bmatrix},$$

and the eigenvalues of A are

$$\lambda_1 = 2, \quad \lambda_2 = 5, \quad \text{and} \quad \lambda_3 = -5 \qquad \text{(verify)}.$$

Let h be the quadratic form in the variables x', y', and z' defined by

$$h(\mathbf{y}) = 2x'^2 + 5y'^2 - 5z'^2.$$

Then g and h are equivalent by means of some orthogonal matrix. Note that $\widehat{h}(\mathbf{y}) = -5x'^2 + 2y'^2 + 5z'^2$ is also equivalent to g. ■

Observe that to apply Theorem 9.2 to diagonalize a given quadratic form, as shown in Example 5, we do not need to know the eigenvectors of A (nor the matrix P); we only require the eigenvalues of A.

To understand the significance of Theorem 9.2, we consider quadratic forms in two and three variables. As we have already observed at the beginning of this section, the graph of the equation

$$g(\mathbf{x}) = \mathbf{x}^T A\mathbf{x} = 1,$$

where $\mathbf{x}$ is a vector in R^2 and A is a symmetric 2×2 matrix, is a conic section centered at the origin of the xy-plane. From Theorem 9.2 it follows that there is a Cartesian coordinate system in the xy-plane with respect to which the equation of this conic section is

$$ax'^2 + by'^2 = 1,$$

where a and b are real numbers. Similarly, the graph of the equation

$$g(\mathbf{x}) = \mathbf{x}^T A\mathbf{x} = 1,$$

where $\mathbf{x}$ is a vector in R^3 and A is a symmetric 3×3 matrix, is a quadric surface centered at the origin of the xyz Cartesian coordinate system. From Theorem 9.2 it follows that there is a Cartesian coordinate system in 3-space with respect to which the equation of the quadric surface is

$$ax'^2 + by'^2 + cz'^2 = 1,$$

where a, b, and c are real numbers. The principal axes of the conic section or quadric surface lie along the new coordinate axes, and this is the reason for calling Theorem 9.2 the **principal axes theorem**.

EXAMPLE 6 Consider the conic section whose equation is

$$g(\mathbf{x}) = 2x^2 + 2xy + 2y^2 = 9.$$

From Example 4 it follows that this conic section can also be described by the equation

$$h(\mathbf{y}) = 3x'^2 + y'^2 = 9,$$

which can be rewritten as

$$\frac{x'^2}{3} + \frac{y'^2}{9} = 1.$$

The graph of this equation is an ellipse (Figure 9.16) whose axis is along the y'-axis. The major axis is of length 6; the minor axis is of length $2\sqrt{3}$. We now note that there is a very close connection between the eigenvectors of the matrix of (3) and the location of the x'- and y'-axes.

Since $\mathbf{x} = P\mathbf{y}$, we have $\mathbf{y} = P^{-1}\mathbf{x} = P^T\mathbf{x} = P\mathbf{x}$ (P is orthogonal and, in this example, also symmetric). Thus

$$x' = \frac{1}{\sqrt{2}}x + \frac{1}{\sqrt{2}}y \quad \text{and} \quad y' = -\frac{1}{\sqrt{2}}x + \frac{1}{\sqrt{2}}y.$$

This means that, in terms of the x- and y-axes, the x'-axis lies along the vector

$$\mathbf{x}_1 = \begin{bmatrix} \frac{1}{\sqrt{2}} \\ \frac{1}{\sqrt{2}} \end{bmatrix}$$

and the y'-axis lies along the vector

$$\mathbf{x}_2 = \begin{bmatrix} -\frac{1}{\sqrt{2}} \\ \frac{1}{\sqrt{2}} \end{bmatrix}.$$

Now $\mathbf{x}_1$ and $\mathbf{x}_2$ are the columns of the matrix

$$P = \begin{bmatrix} \frac{1}{\sqrt{2}} & -\frac{1}{\sqrt{2}} \\ \frac{1}{\sqrt{2}} & \frac{1}{\sqrt{2}} \end{bmatrix},$$

which in turn are eigenvectors of the matrix of (3). Thus the x'- and y'-axes lie along the eigenvectors of the matrix of (3) (see Figure 9.16). ■

Figure 9.16 ►

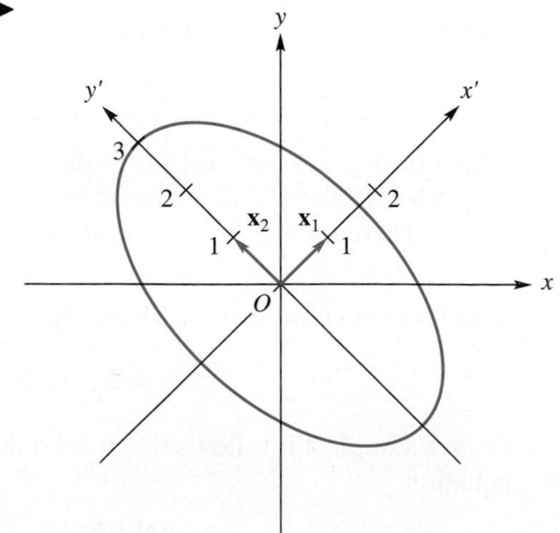

The situation described in Example 6 is true in general. That is, the principal axes of a conic section or quadric surface lie along the eigenvectors of the matrix of the quadratic form.

Let $g(\mathbf{x}) = \mathbf{x}^T A\mathbf{x}$ be a quadratic form in n variables. Then we know that g is equivalent to the quadratic form

$$h(\mathbf{y}) = \lambda_1 y_1^2 + \lambda_2 y_2^2 + \cdots + \lambda_n y_n^2,$$

where $\lambda_1, \lambda_2, \ldots, \lambda_n$ are eigenvalues of the symmetric matrix A of g, and hence are all real. We can label the eigenvalues so that all the positive eigenvalues of A, if any, are listed first, followed by all the negative eigenvalues, if any, followed by the zero eigenvalues, if any. Thus let $\lambda_1, \lambda_2, \ldots, \lambda_p$ be positive, $\lambda_{p+1}, \lambda_{p+2}, \ldots, \lambda_r$ be negative, and $\lambda_{r+1}, \lambda_{r+2}, \ldots, \lambda_n$ be zero. We now define the diagonal matrix H whose entries on the main diagonal are

$$\frac{1}{\sqrt{\lambda_1}}, \frac{1}{\sqrt{\lambda_2}}, \ldots, \frac{1}{\sqrt{\lambda_p}}, \frac{1}{\sqrt{-\lambda_{p+1}}}, \frac{1}{\sqrt{-\lambda_{p+2}}}, \ldots, \frac{1}{\sqrt{-\lambda_r}}, 1, 1, \ldots, 1,$$

with $n-r$ ones. Let D be the diagonal matrix whose entries on the main diagonal are $\lambda_1, \lambda_2, \ldots, \lambda_p, \lambda_{p+1}, \ldots, \lambda_r, \lambda_{r+1}, \ldots, \lambda_n$; A and D are congruent. Let $D_1 = H^T DH$ be the matrix whose diagonal elements are $1, 1, \ldots, 1, -1, \ldots, -1, 0, 0, \ldots, 0$ (p ones, $r-p$ negative ones, and $n-r$ zeros); D and D_1 are then congruent. From Exercise T.2 it follows that A and D_1 are congruent. In terms of quadratic forms, we have established Theorem 9.3.

THEOREM 9.3 *A quadratic form $g(\mathbf{x}) = \mathbf{x}^T A\mathbf{x}$ in n variables is equivalent to a quadratic form*

$$h(\mathbf{y}) = y_1^2 + y_2^2 + \cdots + y_p^2 - y_{p+1}^2 - y_{p+2}^2 - \cdots - y_r^2. \qquad \blacksquare$$

It is clear that the rank of the matrix D_1 is r, the number of nonzero entries on its main diagonal. Now it can be shown that congruent matrices have equal ranks. Since the rank of D_1 is r, the rank of A is also r. We also refer to r as the **rank** of the quadratic form g whose matrix is A. It can be shown that the number p of positive terms in the quadratic form h of Theorem 9.3 is unique; that is, no matter how we simplify the given quadratic form g to obtain an equivalent quadratic form, the latter will always have p positive terms. Hence the quadratic form h in Theorem 9.3 is unique; it is often called the **canonical form** of a quadratic form in n variables. The difference between the number of positive eigenvalues and the number of negative eigenvalues is $s = p - (r - p) = 2p - r$ and is called the **signature** of the quadratic form. Thus, if g and h are equivalent quadratic forms, then they have equal ranks and signatures. However, it can also be shown that if g and h have equal ranks and signatures, then they are equivalent.

EXAMPLE 7 Consider the quadratic form in x_1, x_2, x_3, given by

$$g(\mathbf{x}) = 3x_2^2 + 8x_2x_3 - 3x_3^2 = \mathbf{x}^T A\mathbf{x} = \begin{bmatrix} x_1 & x_2 & x_3 \end{bmatrix} \begin{bmatrix} 0 & 0 & 0 \\ 0 & 3 & 4 \\ 0 & 4 & -3 \end{bmatrix} \begin{bmatrix} x_1 \\ x_2 \\ x_3 \end{bmatrix}.$$

The eigenvalues of A are (verify)

$$\lambda_1 = 5, \quad \lambda_2 = -5, \quad \text{and} \quad \lambda_3 = 0.$$

In this case A is congruent to

$$D = \begin{bmatrix} 5 & 0 & 0 \\ 0 & -5 & 0 \\ 0 & 0 & 0 \end{bmatrix}.$$

If we let

$$H = \begin{bmatrix} \frac{1}{\sqrt{5}} & 0 & 0 \\ 0 & \frac{1}{\sqrt{5}} & 0 \\ 0 & 0 & 1 \end{bmatrix},$$

then

$$D_1 = H^T D H = \begin{bmatrix} 1 & 0 & 0 \\ 0 & -1 & 0 \\ 0 & 0 & 0 \end{bmatrix}$$

and A are congruent, and the given quadratic form is equivalent to the canonical form

$$h(\mathbf{y}) = y_1^2 - y_2^2.$$

The rank of g is 2, and since $p = 1$, the signature $s = 2p - r = 0$. ■

As a final application of quadratic forms we consider positive definite symmetric matrices.

DEFINITION

A symmetric $n \times n$ matrix A is called **positive definite** if $\mathbf{x}^T A\mathbf{x} > 0$ for every nonzero vector $\mathbf{x}$ in R^n.

If A is a symmetric matrix, then $\mathbf{x}^T A\mathbf{x}$ is a quadratic form $g(\mathbf{x}) = \mathbf{x}^T A\mathbf{x}$ and, by Theorem 9.2, g is equivalent to h, where

$$h(\mathbf{y}) = \lambda_1 y_1^2 + \lambda_2 y_2^2 + \cdots + \lambda_p y_p^2 + \lambda_{p+1} y_{p+1}^2 + \lambda_{p+2} y_{p+2}^2 + \cdots + \lambda_r y_r^2.$$

Now A is positive definite if and only if $h(\mathbf{y}) > 0$ for each $\mathbf{y} \neq \mathbf{0}$. However, this can happen if and only if all summands in $h(\mathbf{y})$ are positive and $r = n$. These remarks have established the following theorem.

THEOREM 9.4

A symmetric matrix A is positive definite if and only if all the eigenvalues of A are positive. ■

A quadratic form is then called **positive definite** if its matrix is positive definite.

Key Terms

Conic section
Quadric surface
Real quadratic form
Matrix of a quadratic form
Congruent matrices
Equivalent quadratic forms
Principal axis theorem
Rank
Canonical form
Signature
Positive definite

9.4 Exercises

In Exercises 1 and 2, write each quadratic form as $\mathbf{x}^T A\mathbf{x}$, where A is a symmetric matrix.

1. (a) $-3x^2 + 5xy - 2y^2$

(b) $2x_1^2 + 3x_1x_2 - 5x_1x_3 + 7x_2x_3$

(c) $3x_1^2 + x_2^2 - 2x_3^2 + x_1x_2 - x_1x_3 - 4x_2x_3$

2. (a) $x_1^2 - 3x_2^2 + 4x_3^2 - 4x_1x_2 + 6x_2x_3$

(b) $4x^2 - 6xy - 2y^2$

(c) $-2x_1x_2 + 4x_1x_3 + 6x_2x_3$

In Exercises 3 and 4, for each given symmetric matrix A find a diagonal matrix D that is congruent to A.

3. (a) $A = \begin{bmatrix} -1 & 0 & 0 \\ 0 & 1 & 1 \\ 0 & 1 & 1 \end{bmatrix}$

(b) $A = \begin{bmatrix} 1 & 1 & 1 \\ 1 & 1 & 1 \\ 1 & 1 & 1 \end{bmatrix}$

(c) $A = \begin{bmatrix} 0 & 2 & 2 \\ 2 & 0 & 2 \\ 2 & 2 & 0 \end{bmatrix}$

4. (a) $A = \begin{bmatrix} 3 & 4 & 0 \\ 4 & -3 & 0 \\ 0 & 0 & 5 \end{bmatrix}$

(b) $A = \begin{bmatrix} 2 & 1 & 1 \\ 1 & 2 & 1 \\ 1 & 1 & 2 \end{bmatrix}$

(c) $A = \begin{bmatrix} 0 & 0 & 1 \\ 0 & 1 & 0 \\ 1 & 0 & 0 \end{bmatrix}$

In Exercises 5 through 10, find a quadratic form of the type in Theorem 9.2 that is equivalent to the given quadratic form.

5. $2x^2 - 4xy - y^2$

6. $x_1^2 + x_2^2 + x_3^2 + 2x_2x_3$

7. $2x_1x_3$

8. $2x_2^2 + 2x_3^2 + 4x_2x_3$

9. $-2x_1^2 - 4x_2^2 + 4x_3^2 - 6x_2x_3$

10. $6x_1x_2 + 8x_2x_3$

In Exercises 11 through 16, find a quadratic form of the type in Theorem 9.3 that is equivalent to the given quadratic form.

11. $2x^2 + 4xy + 2y^2$.

12. $x_1^2 + x_2^2 + x_3^2 + 2x_1x_2$

13. $2x_1^2 + 4x_2^2 + 4x_3^2 + 10x_2x_3$

14. $2x_1^2 + 3x_2^2 + 3x_3^2 + 4x_2x_3$

15. $-3x_1^2 + 2x_2^2 + 2x_3^2 + 4x_2x_3$

16. $-3x_1^2 + 5x_2^2 + 3x_3^2 - 8x_1x_3$

17. Let $g(\mathbf{x}) = 4x_2^2 + 4x_3^2 - 10x_2x_3$ be a quadratic form in three variables. Find a quadratic form of the type in Theorem 9.3 that is equivalent to g. What is the rank of g? What is the signature of g?

18. Let $g(\mathbf{x}) = 3x_1^2 - 3x_2^2 - 3x_3^2 + 4x_2x_3$ be a quadratic form in three variables. Find a quadratic form of the type in Theorem 9.3 that is equivalent to g. What is the rank of g? What is the signature of g?

19. Find all quadratic forms $g(\mathbf{x}) = \mathbf{x}^T A\mathbf{x}$ in two variables of the type described in Theorem 9.3. What conics do the equations $\mathbf{x}^T A\mathbf{x} = 1$ represent?

20. Find all quadratic forms $g(\mathbf{x}) = \mathbf{x}^T A\mathbf{x}$ in two variables of rank 1 of the type described in Theorem 9.3. What conics do the equations $\mathbf{x}^T A\mathbf{x} = 1$ represent?

In Exercises 21 and 22, which of the given quadratic forms in three variables are equivalent?

21. $g_1(\mathbf{x}) = x_1^2 + x_2^2 + x_3^2 + 2x_1x_2$
$g_2(\mathbf{x}) = 2x_2^2 + 2x_3^2 + 2x_2x_3$
$g_3(\mathbf{x}) = 3x_2^2 - 3x_3^2 + 8x_2x_3$
$g_4(\mathbf{x}) = 3x_2^2 + 3x_3^2 - 4x_2x_3$

22. $g_1(\mathbf{x}) = x_2^2 + 2x_1x_3$
$g_2(\mathbf{x}) = 2x_1^2 + 2x_2^2 + x_3^2 + 2x_1x_2 + 2x_1x_3 + 2x_2x_3$
$g_3(\mathbf{x}) = 2x_1x_2 + 2x_1x_3 + 2x_2x_3$
$g_4(\mathbf{x}) = 4x_1^2 + 3x_2^2 + 4x_3^2 + 10x_1x_3$

In Exercises 23 and 24, which of the given matrices are positive definite?

23. (a) $\begin{bmatrix} 2 & -1 \\ -1 & 2 \end{bmatrix}$ (b) $\begin{bmatrix} 2 & 1 \\ 1 & 2 \end{bmatrix}$

(c) $\begin{bmatrix} 3 & 1 & 0 \\ 1 & 3 & 0 \\ 0 & 0 & 3 \end{bmatrix}$ (d) $\begin{bmatrix} 1 & 0 & 0 \\ 0 & 2 & 0 \\ 0 & 0 & -3 \end{bmatrix}$

(e) $\begin{bmatrix} 2 & 2 \\ 2 & 2 \end{bmatrix}$

24. (a) $\begin{bmatrix} 0 & -1 \\ -1 & 0 \end{bmatrix}$ (b) $\begin{bmatrix} 1 & 1 \\ 1 & 1 \end{bmatrix}$

(c) $\begin{bmatrix} 0 & 0 & 0 \\ 0 & 1 & 2 \\ 0 & 2 & 1 \end{bmatrix}$ (d) $\begin{bmatrix} 7 & 4 & 4 \\ 4 & 7 & 4 \\ 4 & 4 & 7 \end{bmatrix}$

(e) $\begin{bmatrix} 2 & 0 & 0 & 0 \\ 0 & 1 & 0 & 0 \\ 0 & 0 & 3 & 4 \\ 0 & 0 & 4 & -3 \end{bmatrix}$

Theoretical Exercises

T.1. Show that if A is a symmetric matrix, then $P^T AP$ is also symmetric.

T.2. If A, B, and C are $n \times n$ symmetric matrices, show the following.

(a) A and A are congruent.

(b) If A and B are congruent, then B and A are congruent.

(c) If A and B are congruent and if B and C are congruent, then A and C are congruent.

T.3. Show that if A is symmetric, then A is congruent to a diagonal matrix D.

T.4. Let

$$A = \begin{bmatrix} a & b \\ b & d \end{bmatrix}$$

be a 2×2 symmetric matrix. Show that A is positive definite if and only if $\det(A) > 0$ and $a > 0$.

T.5. Show that a symmetric matrix A is positive definite if and only if $A = P^T P$ for a nonsingular matrix P.

MATLAB Exercises

In MATLAB, eigenvalues of a matrix A can be determined using the command **eig(A)**. *(See the MATLAB Exercises in Section 8.3.) Hence the rank and signature of a quadratic form are easily determined.*

ML.1. Determine the rank and signature of the quadratic form with matrix A.

(a) $A = \begin{bmatrix} -1 & 0 & 0 \\ 0 & 1 & 1 \\ 0 & 1 & 1 \end{bmatrix}$

(b) $A = \begin{bmatrix} 1 & 1 & 1 \\ 1 & 1 & 1 \\ 1 & 1 & 1 \end{bmatrix}$

(c) $A = \begin{bmatrix} 2 & 1 & 0 & -2 \\ 1 & -1 & 1 & 3 \\ 0 & 1 & 2 & -1 \\ -2 & 3 & -1 & 0 \end{bmatrix}$

(d) $A = \begin{bmatrix} 2 & -1 & 0 & 0 \\ -1 & 2 & -1 & 0 \\ 0 & -1 & 2 & -1 \\ 0 & 0 & -1 & 2 \end{bmatrix}$

ML.2. Use **eig** to determine which of the matrices in Exercise ML.1 are positive definite.

9.5 CONIC SECTIONS

Prerequisite. Section 9.4, Quadratic Forms.

In this section we discuss the classification of the conic sections in the plane. A **quadratic equation** in the variables x and y has the form

$$ax^2 + 2bxy + cy^2 + dx + ey + f = 0, \tag{1}$$

where a, b, c, d, e, and f are real numbers. The graph of Equation (1) is a **conic section**, a curve so named because it is obtained by intersecting a plane with a right circular cone that has two nappes. In Figure 9.17 we show that a plane cuts the cone and the resulting intersection is a circle, ellipse, parabola, or hyperbola. Degenerate cases of the conic sections are a point, a line, a pair of lines, or the empty set.

The nondegenerate conic sections are said to be in **standard position** if their graphs and equations are shown in Figure 9.18 and the equation is said to be in **standard form**.

EXAMPLE 1 Identify the graph of the given equation.

(a) $4x^2 + 25y^2 - 100 = 0$

(b) $9y^2 - 4x^2 = -36$

(c) $x^2 + 4y = 0$

(d) $y^2 = 0$

(e) $x^2 + 9y^2 + 9 = 0$

(f) $x^2 + y^2 = 0$

Figure 9.17 ► The nondegenerate conic sections

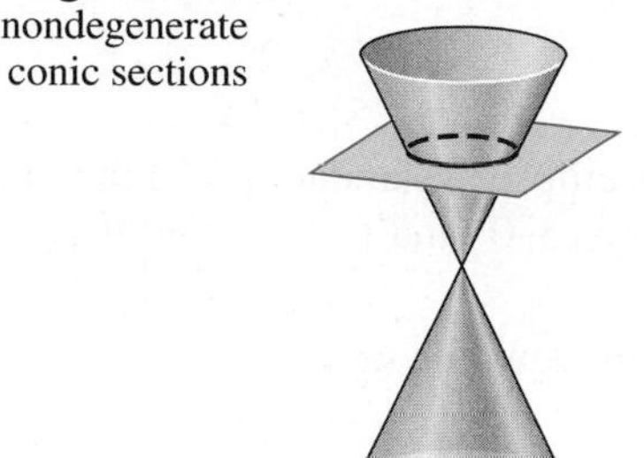
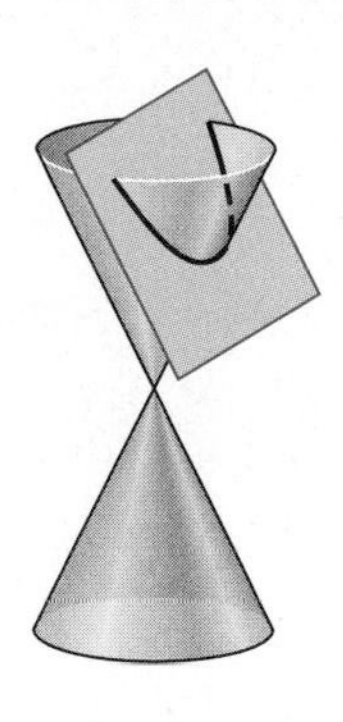
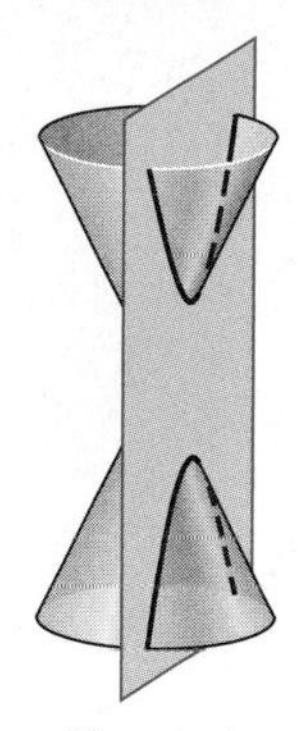

Figure 9.18 ► The conic sections in standard position

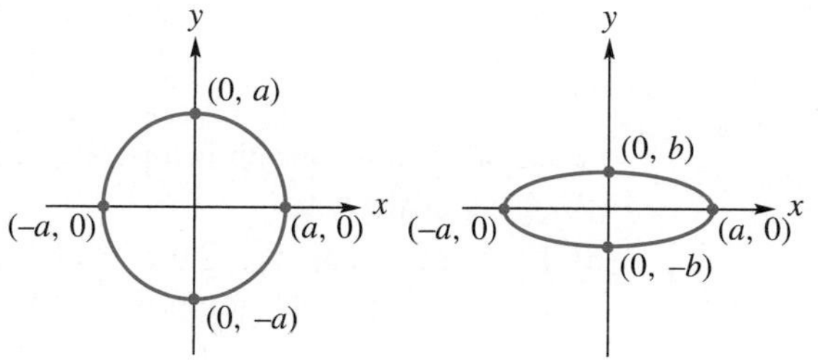

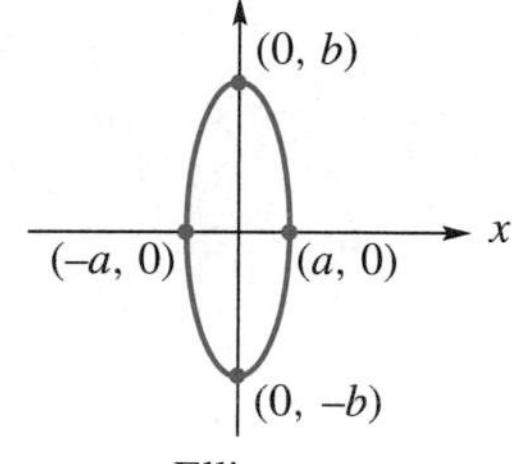

Circle
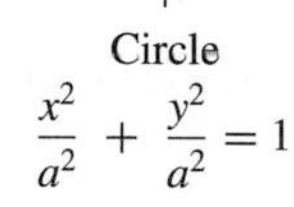
$$\frac{x^2}{a^2} + \frac{y^2}{a^2} = 1$$

Ellipse
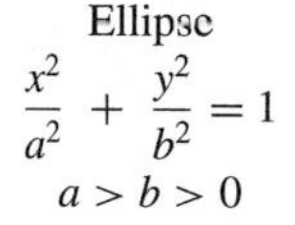
$$\frac{x^2}{a^2} + \frac{y^2}{b^2} = 1$$
$a > b > 0$

Ellipse
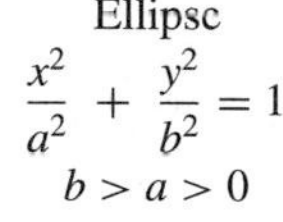
$$\frac{x^2}{a^2} + \frac{y^2}{b^2} = 1$$
$b > a > 0$

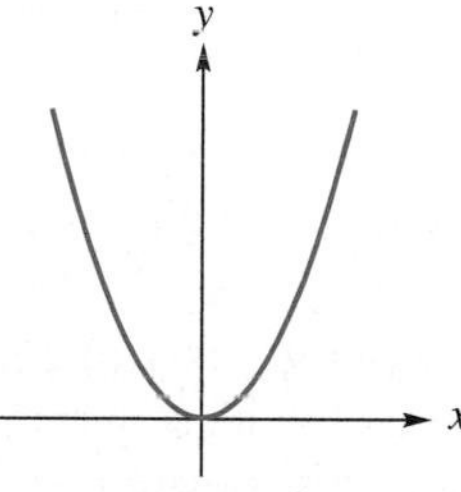

Parabola
$x^2 = ay$
$a > 0$

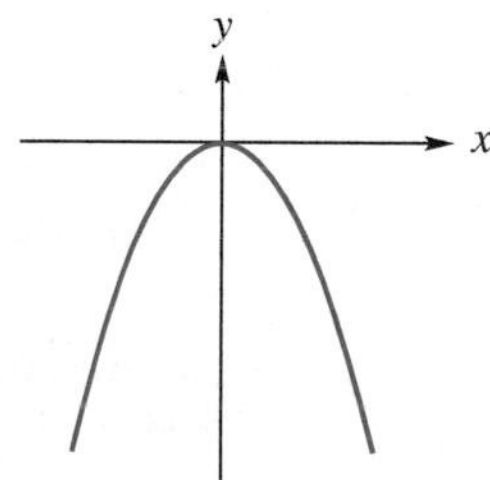

Parabola
$x^2 = ay$
$a < 0$

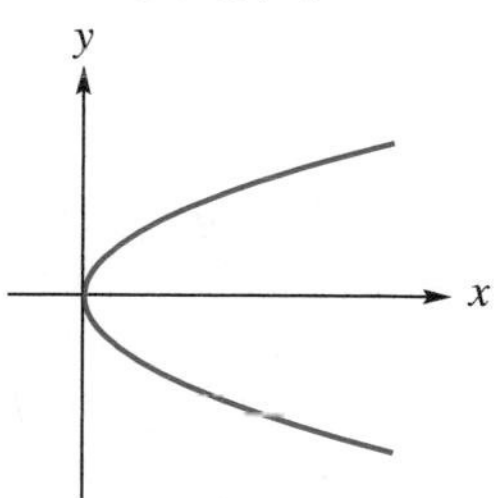

Parabola
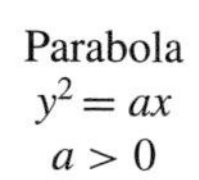
$y^2 = ax$
$a > 0$

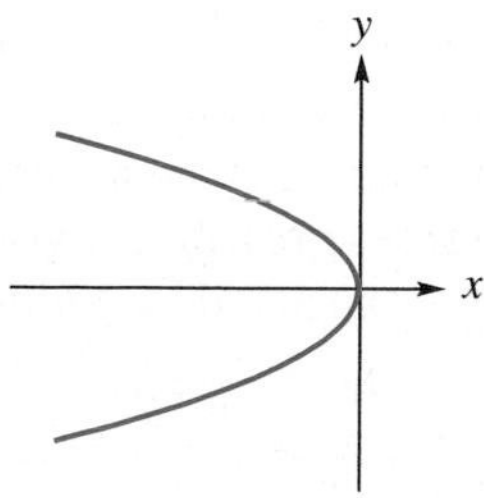

Parabola
$y^2 = ax$
$a < 0$

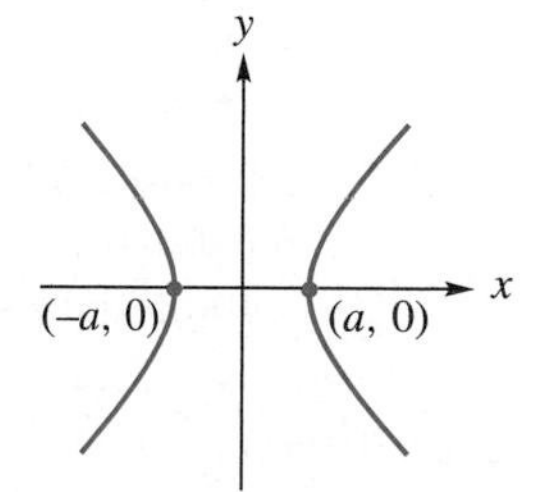

Hyperbola
$$\frac{x^2}{a^2} - \frac{y^2}{b^2} = 1$$
$a > 0, b > 0$

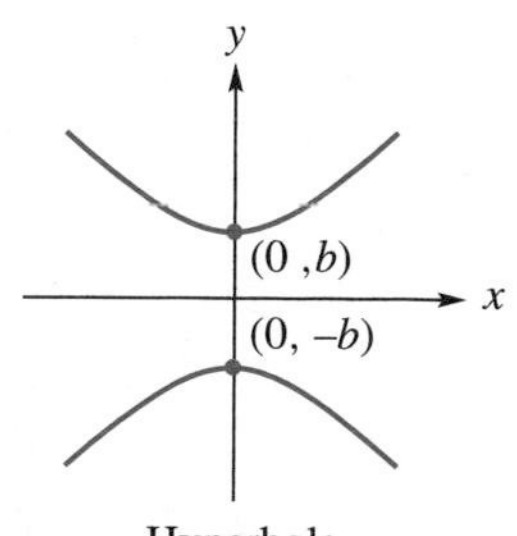

Hyperbola
$$\frac{y^2}{a^2} - \frac{x^2}{b^2} = 1$$
$a > 0, b > 0$

Solution (a) We rewrite the given equation as

$$\frac{4}{100}x^2 + \frac{25}{100}y^2 = \frac{100}{100}$$

or

$$\frac{x^2}{25} + \frac{y^2}{4} = 1,$$

whose graph is an ellipse in standard position with $a = 5$ and $b = 2$. Thus the x-intercepts are $(5, 0)$ and $(-5, 0)$ and the y-intercepts are $(0, 2)$ and $(0, -2)$.

(b) Rewriting the given equation as

$$\frac{x^2}{9} - \frac{y^2}{4} = 1,$$

we see that its graph is a hyperbola in standard position with $a = 3$ and $b = 2$. The x-intercepts are $(3, 0)$ and $(-3, 0)$.

(c) Rewriting the given equation as

$$x^2 = -4y,$$

we see that its graph is a parabola in standard position with $a = -4$, so it opens downward.

(d) Every point satisfying the given equation must have a y-coordinate equal to zero. Thus the graph of this equation consists of all the points on the x-axis.

(e) Rewriting the given equation as

$$x^2 + 9y^2 = -9,$$

we conclude that there are no points in the plane whose coordinates satisfy the given equation.

(f) The only point satisfying the equation is the origin $(0, 0)$, so the graph of this equation is the single point consisting of the origin. ■

We next turn to the study of conic sections whose graphs are not in standard position. First, notice that the equations of the conic sections whose graphs are in standard position do not contain an xy-term (called a **cross-product term**). If a cross-product term appears in the equation, the graph is a conic section that has been rotated from its standard position [see Figure 9.19(a)]. Also notice that none of the equations in Figure 9.18 contain an x^2-term and an x-term or a y^2-term and a y-term. If either of these cases occurs and there is no xy-term in the equation, the graph is a conic section that has been translated from its standard position [see Figure 9.19(b)]. On the other hand, if an xy-term is present, the graph is a conic section that has been rotated and possibly also translated [see Figure 9.19(c)].

Figure 9.19 ►

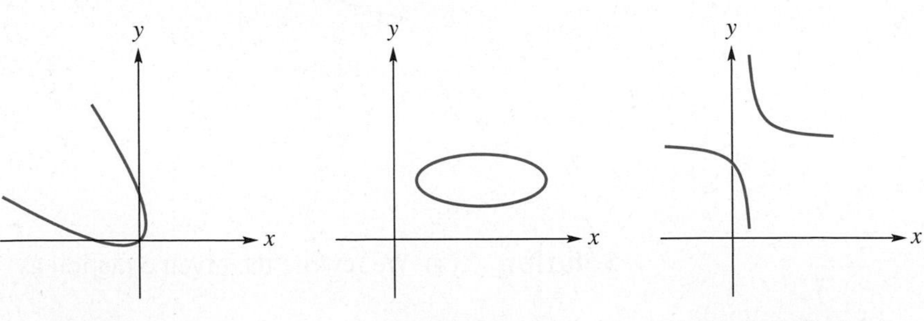

(a) A parabola that has been rotated

(b) An ellipse that has been translated

(c) A hyperbola that has been rotated and translated

> To identify a nondegenerate conic section whose graph is not in standard position, we proceed as follows:
>
> ***Step 1.*** If a cross-product term is present in the given equation, rotate the xy-coordinate axes by means of an orthogonal linear transformation so that in the resulting equation the xy-term no longer appears.
>
> ***Step 2.*** If an xy-term is not present in the given equation, but an x^2-term and an x-term or a y^2-term and a y-term appear, translate the xy-coordinate axes by completing the square so that the graph of the resulting equation will be in standard position with respect to the origin of the new coordinate system.

Thus, if an xy-term appears in a given equation, we first rotate the xy-coordinate axes and then, if necessary, translate the rotated axes. In the next example, we deal with the case requiring only a translation of axes.

EXAMPLE 2 Identify and sketch the graph of the equation

$$x^2 - 4y^2 + 6x + 16y - 23 = 0. \tag{2}$$

Also write its equation in standard form.

Solution Since there is no cross-product term, we only need to translate axes. Completing the squares in the x- and y-terms, we have

$$\begin{aligned} x^2 + 6x + 9 - 4(y^2 - 4y + 4) - 23 &= 9 - 16 \\ (x+3)^2 - 4(y-2)^2 &= 23 + 9 - 16 - 16. \end{aligned} \tag{3}$$

Letting

$$x' = x + 3 \quad \text{and} \quad y' = y - 2,$$

we can rewrite Equation (3) as

$$x'^2 - 4y'^2 = 16$$

or in standard form as

$$\frac{x'^2}{16} - \frac{y'^2}{4} = 1. \tag{4}$$

If we translate the xy-coordinate system to the $x'y'$-coordinate system, whose origin is at $(-3, 2)$, then the graph of Equation (4) is a hyperbola in standard position with respect to the $x'y'$-coordinate system (see Figure 9.20). ■

Figure 9.20 ►

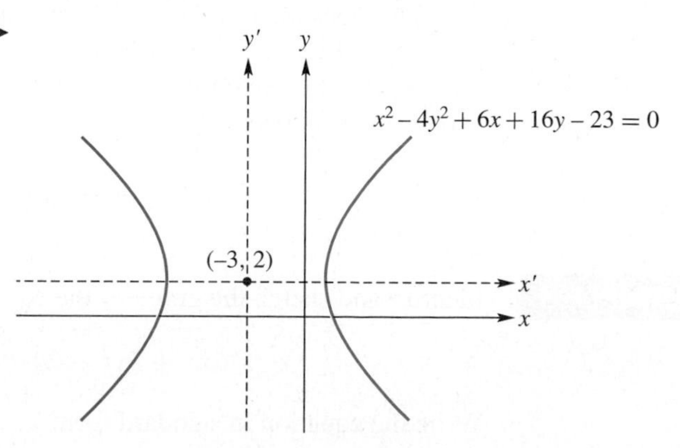

We now turn to the problem of identifying the graph of Equation (1), where we assume that $b \neq 0$, that is, a cross-product term is present. This equation can be written in matrix form as

$$\mathbf{x}^T A\mathbf{x} + B\mathbf{x} + f = 0, \tag{5}$$

where

$$\mathbf{x} = \begin{bmatrix} x \\ y \end{bmatrix}, \quad A = \begin{bmatrix} a & b \\ b & c \end{bmatrix}, \quad \text{and} \quad B = \begin{bmatrix} d & e \end{bmatrix}.$$

Since A is a symmetric matrix, we know from Section 8.3 that it can be diagonalized by an orthogonal matrix P. Thus

$$P^T A P = \begin{bmatrix} \lambda_1 & 0 \\ 0 & \lambda_2 \end{bmatrix},$$

where λ_1 and λ_2 are the eigenvalues of A and the columns of P are $\mathbf{x}_1$ and $\mathbf{x}_2$, orthonormal eigenvectors of A associated with λ_1 and λ_2, respectively.

Letting

$$\mathbf{x} = P\mathbf{y}, \quad \text{where} \quad \mathbf{y} = \begin{bmatrix} x' \\ y' \end{bmatrix},$$

we can rewrite Equation (5) as

$$\begin{aligned} (P\mathbf{y})^T A(P\mathbf{y}) + B(P\mathbf{y}) + f &= 0 \\ \mathbf{y}^T(P^T AP)\mathbf{y} + BP\mathbf{y} + f &= 0 \end{aligned}$$

or

$$\begin{bmatrix} x' & y' \end{bmatrix} \begin{bmatrix} \lambda_1 & 0 \\ 0 & \lambda_2 \end{bmatrix} \begin{bmatrix} x' \\ y' \end{bmatrix} + B(P\mathbf{y}) + f = 0 \tag{6}$$

or

$$\lambda_1 x'^2 + \lambda_2 y'^2 + d'x' + e'y' + f = 0. \tag{7}$$

Equation (7) is the resulting equation for the given conic section and it has no cross-product term.

As discussed in Section 9.4, the x'- and y'-coordinate axes lie along the eigenvectors $\mathbf{x}_1$ and $\mathbf{x}_2$, respectively. Since P is an orthogonal matrix, $\det(P) = \pm 1$ and, if necessary, we can interchange the columns of P (the eigenvectors $\mathbf{x}_1 = \begin{bmatrix} x_{11} \\ x_{21} \end{bmatrix}$ and $\mathbf{x}_2$ of A) or multiply a column of P by -1, so that $\det(P) = 1$. As noted in Section 9.4, it then follows that P is the matrix of a counterclockwise rotation of R^2 through an angle θ that can be determined as follows. First, it is not difficult to show that if $b \neq 0$, then $x_{11} \neq 0$. Since θ is the angle between the directed line segment from the origin to the point (x_{11}, x_{21}), it follows that

$$\theta = \tan^{-1}\left(\frac{x_{21}}{x_{11}}\right).$$

EXAMPLE 3 Identify and sketch the graph of the equation

$$5x^2 - 6xy + 5y^2 - 24\sqrt{2}\,x + 8\sqrt{2}\,y + 56 = 0. \tag{8}$$

Write the equation in standard form.

Solution Rewriting the given equation in matrix form, we obtain

$$\begin{bmatrix} x & y \end{bmatrix}\begin{bmatrix} 5 & -3 \\ -3 & 5 \end{bmatrix}\begin{bmatrix} x \\ y \end{bmatrix} + \begin{bmatrix} -24\sqrt{2} & 8\sqrt{2} \end{bmatrix}\begin{bmatrix} x \\ y \end{bmatrix} + 56 = 0.$$

We now find the eigenvalues of the matrix

$$A = \begin{bmatrix} 5 & -3 \\ -3 & 5 \end{bmatrix}.$$

Thus

$$\begin{aligned} |\lambda I_2 - A| &= \begin{vmatrix} \lambda - 5 & 3 \\ 3 & \lambda - 5 \end{vmatrix} \\ &= (\lambda - 5)(\lambda - 5) - 9 = \lambda^2 - 10\lambda + 16 \\ &= (\lambda - 2)(\lambda - 8), \end{aligned}$$

so the eigenvalues of A are

$$\lambda_1 = 2, \quad \lambda_2 = 8.$$

Associated eigenvectors are obtained by solving the homogeneous system

$$(\lambda I_2 - A)\mathbf{x} = \mathbf{0}.$$

Thus, for $\lambda_1 = 2$, we have

$$\begin{bmatrix} -3 & 3 \\ 3 & -3 \end{bmatrix}\mathbf{x} = \mathbf{0},$$

so an eigenvalue of A associated with $\lambda_1 = 2$ is

$$\begin{bmatrix} 1 \\ 1 \end{bmatrix}.$$

For $\lambda_2 = 8$ we have

$$\begin{bmatrix} 3 & 3 \\ 3 & 3 \end{bmatrix}\mathbf{x} = \mathbf{0},$$

so an eigenvector of A associated with $\lambda_2 = 8$ is

$$\begin{bmatrix} -1 \\ 1 \end{bmatrix}.$$

Normalizing these eigenvectors, we obtain the orthogonal matrix

$$P = \begin{bmatrix} \frac{1}{\sqrt{2}} & -\frac{1}{\sqrt{2}} \\ \frac{1}{\sqrt{2}} & \frac{1}{\sqrt{2}} \end{bmatrix}.$$

Then

$$P^T A P = \begin{bmatrix} 2 & 0 \\ 0 & 8 \end{bmatrix}.$$

Letting $\mathbf{x} = P\mathbf{y}$, we write the transformed equation for the given conic section, Equation (7), as

$$2x'^2 + 8y'^2 - 16x' + 32y' + 56 = 0$$

or

$$x'^2 + 4y'^2 - 8x' + 16y' + 28 = 0.$$

To identify the graph of this equation, we need to translate axes, so we complete the squares, obtaining

$$(x' - 4)^2 + 4(y' + 2)^2 + 28 = 16 + 16$$

$$(x' - 4)^2 + 4(y' + 2)^2 = 4$$

$$\frac{(x' - 4)^2}{4} + \frac{(y' + 2)^2}{1} = 1. \tag{9}$$

Letting $x'' = x' - 4$ and $y'' = y' + 2$, we find that Equation (9) becomes

$$\frac{x''^2}{4} + \frac{y''^2}{1} = 1, \tag{10}$$

whose graph is an ellipse in standard position with respect to the $x''y''$-coordinate axes, as shown in Figure 9.21, where the origin of the $x''y''$-coordinate system is in the $x'y'$-coordinate system at $(4, -2)$, which is $(3\sqrt{2}, \sqrt{2})$ in the xy-coordinate system. Equation (10) is the standard form of the equation of the ellipse. Since

$$\mathbf{x}_1 = \begin{bmatrix} \frac{1}{\sqrt{2}} \\ \frac{1}{\sqrt{2}} \end{bmatrix},$$

the xy-coordinate axes have been rotated through the angle θ, where

$$\theta = \tan^{-1}\left(\frac{\frac{1}{\sqrt{2}}}{\frac{1}{\sqrt{2}}}\right) = \tan^{-1} 1,$$

so $\theta = 45°$. ■

Figure 9.21 ►

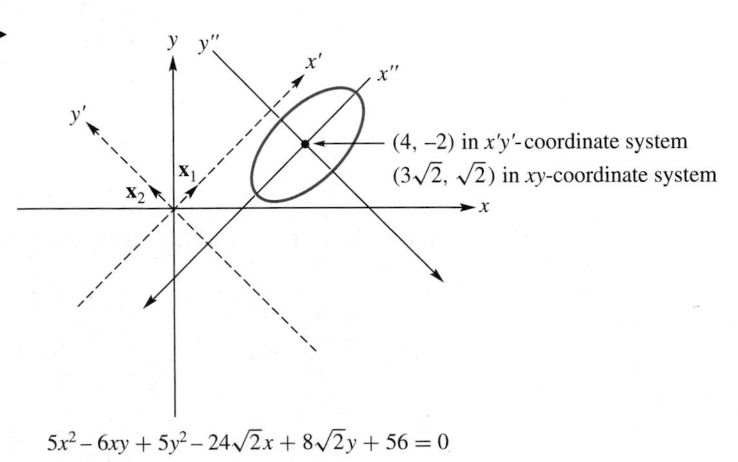

The graph of a given quadratic equation in x and y can be identified from the equation that is obtained after rotating axes, that is, from Equation (6) or (7). The identification of the conic section given by these equations is shown in Table 9.1.

Table 9.1 Identification of the Conic Sections

λ_1, λ_2 *both nonzero*		*Exactly one of* λ_1, λ_2 *is zero*
$\lambda_1\lambda_2 > 0$	$\lambda_1\lambda_2 < 0$	
Ellipse*	Hyperbola**	Parabola†

Key Terms

Quadratic equation
Conic section
Circle
Ellipse
Parabola
Hyperbola
Standard position
Standard form
Cross-product term

9.5 Exercises

In Exercises 1 through 10, identify the graph of the equation.

1. $x^2 + 9y^2 - 9 = 0$

2. $x^2 = 2y$

3. $25y^2 - 4x^2 = 100$

4. $y^2 - 16 = 0$

5. $3x^2 - y^2 = 0$

6. $y = 0$

7. $4x^2 + 4y^2 - 9 = 0$

8. $-25x^2 + 9y^2 + 225 = 0$

9. $4x^2 + y^2 = 0$

10. $9x^2 + 4y^2 + 36 = 0$

In Exercises 11 through 18, translate axes to identify the graph of the equation and write the equation in standard form.

11. $x^2 + 2y^2 - 4x - 4y + 4 = 0$

12. $x^2 - y^2 + 4x - 6y - 9 = 0$

13. $x^2 + y^2 - 8x - 6y = 0$

14. $x^2 - 4x + 4y + 4 = 0$

15. $y^2 - 4y = 0$

16. $4x^2 + 5y^2 - 30y + 25 = 0$

17. $x^2 + y^2 - 2x - 6y + 10 = 0$

18. $2x^2 + y^2 - 12x - 4y + 24 = 0$

In Exercises 19 through 24, rotate axes to identify the graph of the equation and write the equation in standard form.

19. $x^2 + xy + y^2 = 6$

20. $xy = 1$

21. $9x^2 + y^2 + 6xy = 4$

22. $x^2 + y^2 + 4xy = 9$

23. $4x^2 + 4y^2 - 10xy = 0$

24. $9x^2 + 6y^2 + 4xy - 5 = 0$

In Exercises 25 through 30, identify the graph of the equation and write the equation in standard form.

25. $9x^2 + y^2 + 6xy - 10\sqrt{10}\,x + 10\sqrt{10}\,y + 90 = 0$

26. $5x^2 + 5y^2 - 6xy - 30\sqrt{2}\,x + 18\sqrt{2}\,y + 82 = 0$

27. $5x^2 + 12xy - 12\sqrt{13}\,x = 36$

28. $6x^2 + 9y^2 - 4xy - 4\sqrt{5}\,x - 18\sqrt{5}\,y = 5$

29. $x^2 - y^2 + 2\sqrt{3}\,xy + 6x = 0$

30. $8x^2 + 8y^2 - 16xy + 33\sqrt{2}\,x - 31\sqrt{2}\,y + 70 = 0$

9.6 QUADRIC SURFACES

Prerequisite. Section 9.5, Conic Sections.

In Section 9.5 conic sections were used to provide geometric models for quadratic forms in two variables. In this section we investigate quadratic forms in three variables and use particular surfaces called quadric surfaces as geometric

*In this case, the graph of the quadratic equation could be a single point or the empty set (that is, no graph).

**In this case, the graph of the quadratic equation could be a degenerate hyperbola; that is, two intersecting lines.

†In this case, the graph of the quadratic equation could be a degenerate parabola; that is, two parallel lines or a single line.

models. Quadric surfaces are often studied and sketched in analytic geometry and calculus. Here we use Theorems 9.2 and 9.3 to develop a classification scheme for quadric surfaces.

A **second-degree polynomial equation** in three variables x, y, and z has the form

$$ax^2 + by^2 + cz^2 + 2dxy + 2exz + 2fyz + gx + hy + iz = j, \qquad (1)$$

where coefficients a through j are real numbers with $a, b, \dots, f$ not all zero. Equation (1) can be written in matrix form as

$$\mathbf{x}^T A\mathbf{x} + B\mathbf{x} = j, \qquad (2)$$

where

$$A = \begin{bmatrix} a & d & e \\ d & b & f \\ e & f & c \end{bmatrix}, \quad B = \begin{bmatrix} g & h & i \end{bmatrix}, \quad \text{and} \quad \mathbf{x} = \begin{bmatrix} x \\ y \\ z \end{bmatrix}.$$

We call $\mathbf{x}^T A\mathbf{x}$ the **quadratic form (in three variables) associated with the second-degree polynomial** in (1). As in Section 9.4, the symmetric matrix A is called the matrix of the quadratic form.

The graph of (1) in R^3 is called a **quadric surface**. As in the case of the classification of conic sections in Section 9.5, the classification of (1) as to the type of surface represented depends on the matrix A. Using the ideas in Section 9.5, we have the following strategies to determine a simpler equation for a quadric surface.

1. If A is not diagonal, then a rotation of axes is used to eliminate any cross-product terms xy, xz, or yz.
2. If $B = \begin{bmatrix} g & h & i \end{bmatrix} \neq \mathbf{0}$, then a translation of axes is used to eliminate any first-degree terms.

The resulting equation will have the standard form

$$\lambda_1 x''^2 + \lambda_2 y''^2 + \lambda_3 z''^2 = k$$

or, in matrix form,

$$\mathbf{y}^T C\mathbf{y} = k, \qquad (3)$$

where

$$\mathbf{y} = \begin{bmatrix} x'' \\ y'' \\ z'' \end{bmatrix},$$

k is some real constant, and C is a diagonal matrix with diagonal entries λ_1, λ_2, λ_3, which are the eigenvalues of A.

We now turn to the classification of quadric surfaces.

DEFINITION Let A be an $n \times n$ symmetric matrix. The **inertia** of A, denoted $\text{In}(A)$, is an ordered triple of numbers

$$(\text{pos}, \text{neg}, \text{zer}),$$

where pos, neg, and zer are the number of positive, negative, and zero eigenvalues of A, respectively.

EXAMPLE 1 Find the inertia of each of the following matrices:

$$A_1 = \begin{bmatrix} 2 & 2 \\ 2 & 2 \end{bmatrix}, \quad A_2 = \begin{bmatrix} 2 & 1 \\ 1 & 2 \end{bmatrix}, \quad A_3 = \begin{bmatrix} 0 & 2 & 2 \\ 2 & 0 & 2 \\ 2 & 2 & 0 \end{bmatrix}.$$

Solution We determine the eigenvalues of each of the matrices. It follows that (verify)

$\det(\lambda I_2 - A_1) = \lambda(\lambda - 4) = 0,$ so $\lambda_1 = 0$, $\lambda_2 = 4$, and $\text{In}(A_1) = (1, 0, 1)$.

$\det(\lambda I_2 - A_2) = (\lambda - 1)(\lambda - 3) = 0,$ so $\lambda_1 = 1$, $\lambda_2 = 3$, and $\text{In}(A_2) = (2, 0, 0)$.

$\det(\lambda I_3 - A_3) = (\lambda + 2)^2(\lambda - 4) = 0,$ so $\lambda_1 = \lambda_2 = -2$, $\lambda_3 = 4$, and $\text{In}(A_3) = (1, 2, 0)$. ■

From Section 9.4 the signature of a quadratic form $\mathbf{x}^T A\mathbf{x}$ is the difference between the number of positive eigenvalues and the number of negative eigenvalues of A. In terms of inertia, the signature of $\mathbf{x}^T A\mathbf{x}$ is $s = \text{pos} - \text{neg}$.

In order to use inertia for classification of quadric surfaces (or conic sections), we assume that the eigenvalues of an $n \times n$ symmetric matrix A of a quadratic form in n variables are denoted by

$$\begin{aligned} \lambda_1 \geq \cdots \geq \lambda_{\text{pos}} &> 0 \\ \lambda_{\text{pos}+1} \leq \cdots \leq \lambda_{\text{pos}+\text{neg}} &< 0 \\ \lambda_{\text{pos}+\text{neg}+1} = \cdots = \lambda_n &= 0. \end{aligned}$$

The largest positive eigenvalue is denoted by λ_1 and the smallest one by λ_{pos}. We also assume that $\lambda_1 > 0$ and $j \geq 0$ in (2), which eliminates redundant and impossible cases. For example, if

$$A = \begin{bmatrix} -1 & 0 & 0 \\ 0 & -2 & 0 \\ 0 & 0 & -3 \end{bmatrix}, \quad B = \begin{bmatrix} 0 & 0 & 0 \end{bmatrix}, \quad \text{and} \quad j = 5,$$

then the second-degree polynomial is $-x^2 - 2y^2 - 3z^2 = 5$, which has an empty solution set. That is, the surface represented has no points. However, if $j = -5$, then the second-degree polynomial is $-x^2 - 2y^2 - 3z^2 = -5$, which is identical to $x^2 + 2y^2 + 3z^2 = 5$. The assumptions $\lambda_1 > 0$ and $j \geq 0$ avoid such a redundant representation.

EXAMPLE 2 Consider a quadratic form in two variables with matrix A and assume that $\lambda_1 > 0$ and $f \geq 0$ in Equation (1) of Section 9.5. Then there are only three possible cases for the inertia of A, which we summarize as follows.

1. $\text{In}(A) = (2, 0, 0)$; then the quadratic form represents an ellipse.

2. $\text{In}(A) = (1, 1, 0)$; then the quadratic form represents a hyperbola.

3. $\text{In}(A) = (1, 0, 1)$; then the quadratic form represents a parabola.

This classification is identical to that given in Table 9.1, taking the assumptions into account. ■

Note that the classification of the conic sections in Example 2 does not distinguish between special cases within a particular geometric class. For example, both $y = x^2$ and $x = y^2$ have inertia $(1, 0, 1)$.

Before classifying quadric surfaces using inertia, we present the quadric surfaces in the standard forms met in analytic geometry and calculus. (In the following, a, b, and c are positive unless otherwise stated.)

Ellipsoid (See Figure 9.22.)

$$\frac{x^2}{a^2} + \frac{y^2}{b^2} + \frac{z^2}{c^2} = 1$$

The special case $a = b = c$ is a sphere.

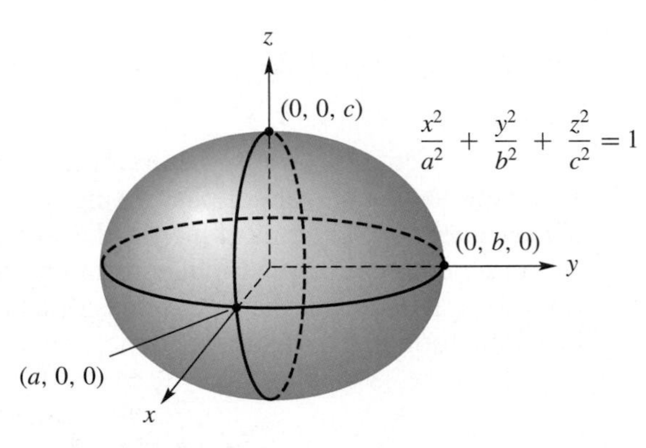

Figure 9.22 ▲
Ellipsoid

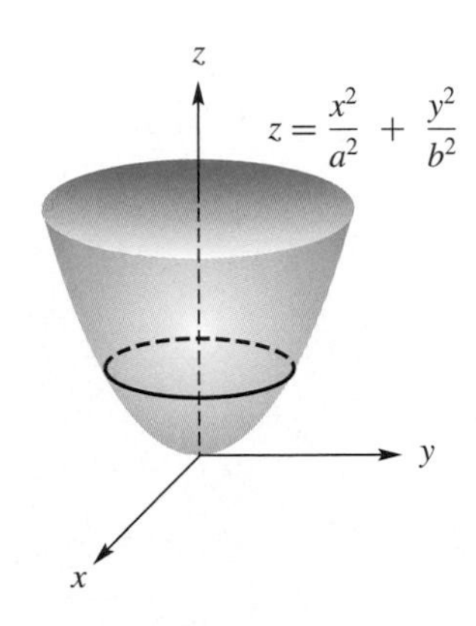

Figure 9.23 ▲
Elliptic paraboloid

Elliptic Paraboloid (See Figure 9.23.)

$$z = \frac{x^2}{a^2} + \frac{y^2}{b^2}, \qquad y = \frac{x^2}{a^2} + \frac{z^2}{c^2}, \qquad x = \frac{y^2}{b^2} + \frac{z^2}{c^2}$$

A degenerate case of a parabola is a line, so a degenerate case of an elliptic paraboloid is an **elliptic cylinder** (see Figure 9.24), which is given by

$$\frac{x^2}{a^2} + \frac{y^2}{b^2} = 1, \qquad \frac{x^2}{a^2} + \frac{z^2}{c^2} = 1, \qquad \frac{y^2}{b^2} + \frac{z^2}{c^2} = 1.$$

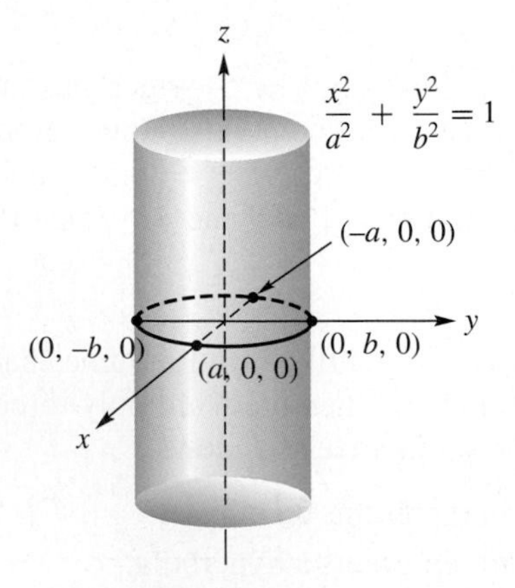

Figure 9.24 ▲
Elliptic cylinder

Hyperboloid of One Sheet (See Figure 9.25.)

$$\frac{x^2}{a^2} + \frac{y^2}{b^2} - \frac{z^2}{c^2} = 1, \qquad \frac{x^2}{a^2} - \frac{y^2}{b^2} + \frac{z^2}{c^2} = 1, \qquad -\frac{x^2}{a^2} + \frac{y^2}{b^2} + \frac{z^2}{c^2} = 1.$$

A degenerate case of a hyperboloid is a pair of lines through the origin; hence a degenerate case of a hyperboloid of one sheet is a **cone** (Figure 9.26), which is given by

$$\frac{x^2}{a^2} + \frac{y^2}{b^2} - \frac{z^2}{c^2} = 0, \qquad \frac{x^2}{a^2} - \frac{y^2}{b^2} + \frac{z^2}{c^2} = 0, \qquad -\frac{x^2}{a^2} + \frac{y^2}{b^2} + \frac{z^2}{c^2} = 0.$$

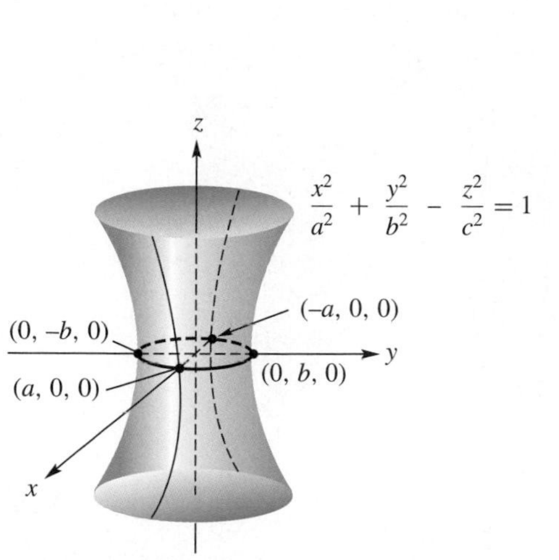

Figure 9.25 ▲
Hyperboloid of one sheet

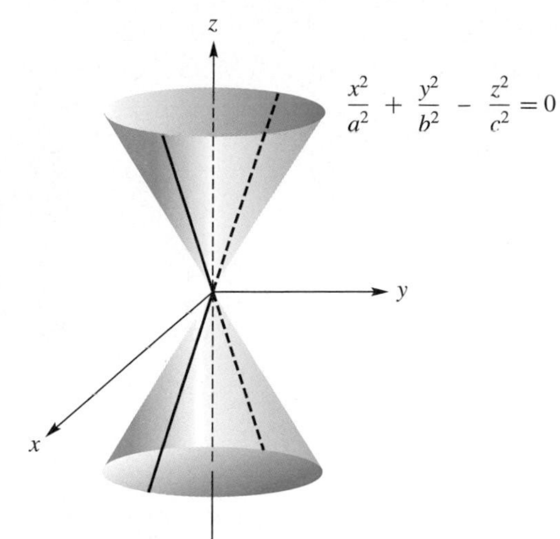

Figure 9.26 ▲
Cone

Hyperboloid of Two Sheets (See Figure 9.27.)

$$\frac{x^2}{a^2} - \frac{y^2}{b^2} - \frac{z^2}{c^2} = 1, \qquad -\frac{x^2}{a^2} - \frac{y^2}{b^2} + \frac{z^2}{c^2} = 1, \qquad -\frac{x^2}{a^2} + \frac{y^2}{b^2} - \frac{z^2}{c^2} = 1$$

Hyperbolic Paraboloid (See Figure 9.28.)

$$\pm z = \frac{x^2}{a^2} - \frac{y^2}{b^2}, \qquad \pm y = \frac{x^2}{a^2} - \frac{z^2}{b^2}, \qquad \pm x = \frac{y^2}{a^2} - \frac{z^2}{b^2}$$

A degenerate case of a parabola is a line, so a degenerate case of a hyperbolic paraboloid is a hyperbolic cylinder (see Figure 9.29), which is given by

$$\frac{x^2}{a^2} - \frac{y^2}{b^2} = \pm 1, \qquad \frac{x^2}{a^2} - \frac{z^2}{b^2} = \pm 1, \qquad \frac{y^2}{a^2} - \frac{z^2}{b^2} = \pm 1.$$

Parabolic Cylinder (See Figure 9.30.) One of a or b is not zero.

$$x^2 = ay + bz, \qquad y^2 = ax + bz, \qquad z^2 = ax + by$$

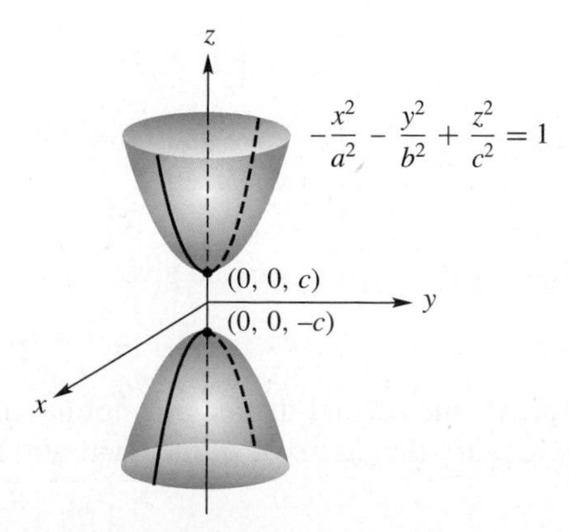

Figure 9.27 ▲
Hyperboloid of two sheets

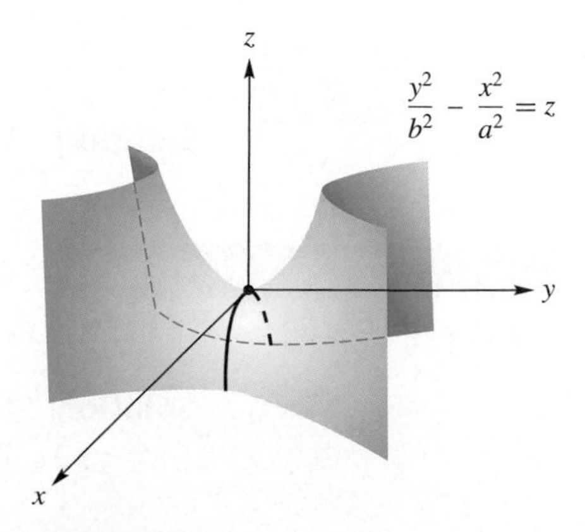

Figure 9.28 ▲
Hyperbolic paraboloid

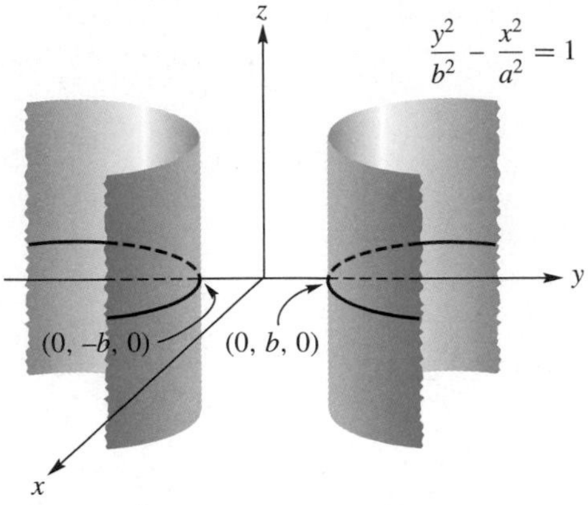

Figure 9.29 ▲
Hyperbolic cylinder

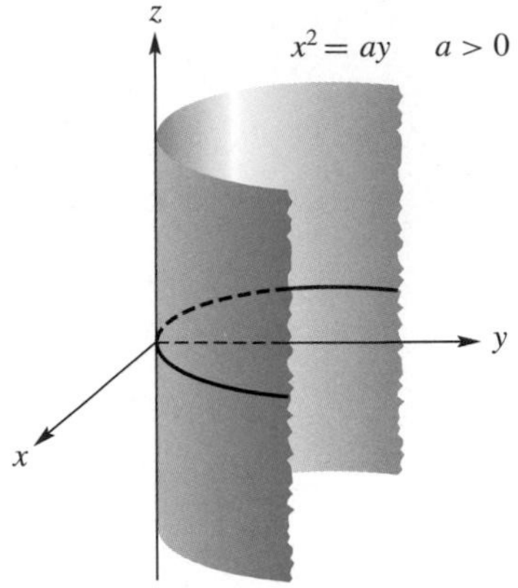

Figure 9.30 ▲
Parabolic cylinder

For a quadratic form in three variables with matrix A, under the assumptions $\lambda_1 > 0$ and $j \geq 0$ in (2), there are exactly six possibilities for the inertia of A. We present these in Table 9.2. As with the conic section classification in Example 2, the classification of quadric surfaces in Table 9.2 does not distinguish between special cases within a particular geometric class.

Table 9.2 Identification of the Quadric Surfaces

$\text{In}(A) = (3, 0, 0)$	Ellipsoid
$\text{In}(A) = (2, 0, 1)$	Elliptic paraboloid
$\text{In}(A) = (2, 1, 0)$	Hyperboloid of one sheet
$\text{In}(A) = (1, 2, 0)$	Hyperboloid of two sheets
$\text{In}(A) = (1, 1, 1)$	Hyperbolic paraboloid
$\text{In}(A) = (1, 0, 2)$	Parabolic cylinder

EXAMPLE 3 Classify the quadric surface represented by the quadratic form $\mathbf{x}^T A\mathbf{x} = 3$, where

$$A = \begin{bmatrix} 0 & 2 & 2 \\ 2 & 0 & 2 \\ 2 & 2 & 0 \end{bmatrix} \quad \text{and} \quad \mathbf{x} = \begin{bmatrix} x \\ y \\ z \end{bmatrix}.$$

Solution From Example 1 we have that $\text{In}(A) = (1, 2, 0)$ and hence the quadric surface is a hyperboloid of two sheets. ■

EXAMPLE 4 Classify the quadric surface given by

$$2x^2 + 4y^2 - 4z^2 + 6yz - 5x + 3y = 2.$$

Solution Rewrite the second-degree polynomial as a quadratic form in three variables to identify the matrix A of the quadratic form. We have

$$A = \begin{bmatrix} 2 & 0 & 0 \\ 0 & 4 & 3 \\ 0 & 3 & -4 \end{bmatrix}.$$

Its eigenvalues are $\lambda_1 = 5$, $\lambda_2 = 2$, and $\lambda_3 = -5$ (verify). Thus $\text{In}(A) = (2, 1, 0)$ and hence the quadric surface is a hyperboloid of one sheet. ■

The classification of a quadric surface is much easier than the problem of transforming it to the standard forms that are used in analytic geometry and calculus. The algebraic steps to obtain an equation in standard form from a second-degree polynomial equation (1) require, in general, a rotation and translation of axes, as mentioned earlier. The rotation requires both the eigenvalues and eigenvectors of the matrix A of the quadratic form. The eigenvectors of A are used to form an orthogonal matrix P so that $\det(P) = 1$, and hence the change of variables $\mathbf{x} = P\mathbf{y}$ represents a rotation. The resulting associated form is that obtained in the principal axes theorem, Theorem 9.2; that is, all cross-product terms are eliminated. We illustrate this with the next example.

EXAMPLE 5 For the quadric surface in Example 4,

$$\mathbf{x}^T A\mathbf{x} + \begin{bmatrix} -5 & 3 & 0 \end{bmatrix} \mathbf{x} = 2,$$

determine the rotation so that all cross-product terms are eliminated.

Solution The eigenvalues and eigenvectors of

$$A = \begin{bmatrix} 2 & 0 & 0 \\ 0 & 4 & 3 \\ 0 & 3 & -4 \end{bmatrix}$$

are (verify), respectively,

$$\lambda_1 = 5, \qquad \lambda_2 = 2, \qquad \lambda_3 = -5$$

and

$$\mathbf{v}_1 = \begin{bmatrix} 0 \\ 3 \\ 1 \end{bmatrix}, \qquad \mathbf{v}_2 = \begin{bmatrix} 1 \\ 0 \\ 0 \end{bmatrix}, \qquad \mathbf{v}_3 = \begin{bmatrix} 0 \\ 1 \\ -3 \end{bmatrix}.$$

The eigenvectors $\mathbf{v}_i$ are mutually orthogonal, since they correspond to distinct eigenvalues of a symmetric matrix (see Theorem 8.7 in Section 8.3). We normalize the eigenvectors as

$$\mathbf{u}_1 = \frac{1}{\sqrt{10}} \begin{bmatrix} 0 \\ 3 \\ 1 \end{bmatrix}, \qquad \mathbf{u}_2 = \mathbf{v}_2, \qquad \mathbf{u}_3 = \frac{1}{\sqrt{10}} \begin{bmatrix} 0 \\ 1 \\ -3 \end{bmatrix}$$

and define $P = \begin{bmatrix} \mathbf{u}_1 & \mathbf{u}_2 & \mathbf{u}_3 \end{bmatrix}$. Then $|P| = 1$ (verify), so we let $\mathbf{x} = P\mathbf{y}$ and obtain the representation

$$(P\mathbf{y})^T A(P\mathbf{y}) + \begin{bmatrix} -5 & 3 & 0 \end{bmatrix} P\mathbf{y} = 2$$
$$\mathbf{y}^T (P^T AP)\mathbf{y} + \begin{bmatrix} -5 & 3 & 0 \end{bmatrix} P\mathbf{y} = 2.$$

Since $P^T AP = D$, and letting $\mathbf{y} = \begin{bmatrix} x' \\ y' \\ z' \end{bmatrix}$, we have

$$\mathbf{y}^T D\mathbf{y} + \begin{bmatrix} -5 & 3 & 0 \end{bmatrix} P\mathbf{y} = 2,$$

$$\mathbf{y}^T \begin{bmatrix} 5 & 0 & 0 \\ 0 & 2 & 0 \\ 0 & 0 & -5 \end{bmatrix} \mathbf{y} + \begin{bmatrix} \frac{9}{\sqrt{10}} & -5 & \frac{3}{\sqrt{10}} \end{bmatrix} \mathbf{y} = 2$$

(if $|P| \neq 1$, we redefine P by reordering its columns until we get its determinant to be 1), or

$$5x'^2 + 2y'^2 - 5z'^2 + \frac{9}{\sqrt{10}}x' - 5y' + \frac{3}{\sqrt{10}}z' = 2. \qquad \blacksquare$$

To complete the transformation to standard form, we introduce a change of variable to perform a translation that eliminates any first-degree terms. Algebraically, we complete the square in each of the three variables.

EXAMPLE 6 Continue with Example 5 to eliminate the first-degree terms.

Solution The last expression for the quadric surface in Example 5 can be written as

$$5x'^2 + \frac{9}{\sqrt{10}}x' + 2y'^2 - 5y' - 5z'^2 + \frac{3}{\sqrt{10}}z' = 2.$$

Completing the square in each variable, we have

$$\begin{aligned} &5\left(x'^2 + \frac{9}{5\sqrt{10}}x' + \frac{81}{1000}\right) + 2\left(y'^2 - \frac{5}{2}y' + \frac{25}{16}\right) \\ &\qquad\qquad - 5\left(z'^2 - \frac{3}{5\sqrt{10}}z' + \frac{9}{1000}\right) \\ &= 5\left(x' + \frac{9}{10\sqrt{10}}\right)^2 + 2\left(y' - \frac{5}{4}\right)^2 - 5\left(z' - \frac{3}{10\sqrt{10}}\right)^2 \\ &= 2 + \frac{405}{1000} + \frac{50}{16} - \frac{45}{1000}. \end{aligned}$$

Letting

$$x'' = x' + \frac{9}{10\sqrt{10}}, \qquad y'' = y' - \frac{5}{4}, \qquad z'' = z' - \frac{3}{10\sqrt{10}},$$

we can write the equation of the quadric surface as

$$5x''^2 + 2y''^2 - 5z''^2 = \frac{5485}{1000} = 5.485.$$

This can be written in standard form as

$$\frac{x''^2}{\frac{5.485}{5}} + \frac{y''^2}{\frac{5.485}{2}} - \frac{z''^2}{\frac{5.485}{5}} = 1. \qquad \blacksquare$$

Key Terms

Second-degree polynomial equation in three variables
Quadratic form
Quadric surface
Inertia
Ellipsoid
Elliptic paraboloid
Elliptic cylinder
Hyperboloid of one sheet
Cone
Hyperboloid of two sheets
Hyperbolic paraboloid
Hyperbolic cylinder
Parabolic cylinder

9.6 Exercises

In Exercises 1 through 14, use inertia to classify the quadric surface given by each equation.

1. $x^2 + y^2 + 2z^2 - 2xy - 4xz - 4yz + 4x = 8$

2. $x^2 + 3y^2 + 2z^2 - 6x - 6y + 4z - 2 = 0$

3. $z = 4xy$

4. $x^2 + y^2 + z^2 + 2xy = 4$

5. $x^2 - y = 0$

6. $2xy + z = 0$

7. $5y^2 + 20y + z - 23 = 0$

8. $x^2 + y^2 + 2z^2 - 2xy + 4xz + 4yz = 16$

9. $4x^2 + 9y^2 + z^2 + 8x - 18y - 4z - 19 = 0$

10. $y^2 - z^2 - 9x - 4y + 8z - 12 = 0$

11. $x^2 + 4y^2 + 4x + 16y - 16z - 4 = 0$

12. $4x^2 - y^2 + z^2 - 16x + 8y - 6z + 5 = 0$

13. $x^2 - 4z^2 - 4x + 8z = 0$

14. $2x^2 + 2y^2 + 4z^2 + 2xy - 2xz - 2yz + 3x - 5y + z = 7$

In Exercises 15 through 28, classify the quadric surface given by each equation and determine its standard form.

15. $x^2 + 2y^2 + 2z^2 + 2yz = 1$

16. $x^2 + y^2 + 2z^2 - 2xy + 4xz + 4yz = 16$

17. $2xz - 2z - 4y - 4z + 8 = 0$

18. $x^2 + 3y^2 + 3z^2 - 4yz = 9$

19. $x^2 + y^2 + z^2 + 2xy = 8$

20. $-x^2 - y^2 - z^2 + 4xy + 4xz + 4yz = 3$

21. $2x^2 + 2y^2 + 4z^2 - 4xy - 8xz - 8yz + 8x = 15$

22. $4x^2 + 4y^2 + 8z^2 + 4xy - 4xz - 4yz + 6x - 10y + 2z = \frac{9}{4}$

23. $2y^2 + 2z^2 + 4yz + \frac{16}{\sqrt{2}}x + 4 = 0$

24. $x^2 + y^2 - 2z^2 + 2xy + 8xz + 8yz + 3x + z = 0$

25. $-x^2 - y^2 - z^2 + 4xy + 4xz + 4yz + \frac{3}{\sqrt{2}}x - \frac{3}{\sqrt{2}}y = 6$

26. $2x^2 + 3y^2 + 3z^2 - 2yz + 2x + \frac{1}{\sqrt{2}}y + \frac{1}{\sqrt{2}}z = \frac{3}{8}$

27. $x^2 + y^2 - z^2 - 2x - 4y - 4z + 1 = 0$

28. $-8x^2 - 8y^2 + 10z^2 + 32xy - 4xz - 4yz = 24$

Key Ideas for Review

- **Fibonacci sequence.**

$$u_n = \frac{1}{\sqrt{5}}\left[\left(\frac{1+\sqrt{5}}{2}\right)^{n+1} - \left(\frac{1-\sqrt{5}}{2}\right)^{n+1}\right]$$

- **Theorem 9.1.** If the $n \times n$ matrix A has n linearly independent eigenvectors $\mathbf{p}_1, \mathbf{p}_2, \dots, \mathbf{p}_n$ associated with the eigenvalues $\lambda_1, \lambda_2, \dots, \lambda_n$, respectively, then the general solution to the system of differential equations

$$\mathbf{x}' = A\mathbf{x}$$

is given by

$$\mathbf{x}(t) = b_1\mathbf{p}_1 e^{\lambda_1 t} + b_2\mathbf{p}_2 e^{\lambda_2 t} + \cdots + b_n\mathbf{p}_n e^{\lambda_n t}.$$

- **Theorem 9.2 (Principal Axes Theorem).** Any quadratic form in n variables $g(\mathbf{x}) = \mathbf{x}^T A\mathbf{x}$ is equivalent to a quadratic form,

$$h(\mathbf{y}) = \lambda_1 y_1^2 + \lambda_2 y_2^2 + \cdots + \lambda_n y_n^2,$$

where

$$\mathbf{y} = \begin{bmatrix} y_1 \\ y_2 \\ \vdots \\ y_n \end{bmatrix}$$

and $\lambda_1, \lambda_2, \dots, \lambda_n$ are the eigenvalues of the matrix A of g.

- **Theorem 9.3.** A quadratic form $g(\mathbf{x}) = \mathbf{x}^T A\mathbf{x}$ in n variables is equivalent to a quadratic form

$$h(\mathbf{y}) = y_1^2 + y_2^2 + \cdots + y_p^2 - y_{p+1}^2 - y_{p+2}^2 - \cdots - y_r^2.$$

- The trajectories of the 2×2 dynamical system of the form

$$\frac{dx}{dt} = ax + by$$

$$\frac{dy}{dt} = cx + dy$$

are completely determined by the eigenvalues and eigenvectors of the matrix $A = \begin{bmatrix} a & b \\ c & d \end{bmatrix}$.

Supplementary Exercises

1. Let $A(t) = \left[a_{ij}(t)\right]$ be an $n \times n$ matrix whose entries are all functions of t; $A(t)$ is called a **matrix function**. The derivative and integral of $A(t)$ is defined componentwise; that is,

$$\frac{d}{dt}[A(t)] = \left[\frac{d}{dt}a_{ij}(t)\right]$$

and

$$\int_a^t A(s)\,ds = \left[\int_a^t a_{ij}(s)\,ds\right].$$

For each of the following matrices $A(t)$, compute $\frac{d}{dt}[A(t)]$ and $\int_0^t A(s)\,ds$.

(a) $A(t) = \begin{bmatrix} t^2 & \frac{1}{t+1} \\ 4 & e^{-t} \end{bmatrix}$

(b) $A(t) = \begin{bmatrix} \sin 2t & 0 & 0 \\ 0 & 1 & -t \\ 0 & te^{t^2} & \frac{t}{t^2+1} \end{bmatrix}$

2. For $\mathbf{x}_0 = \begin{bmatrix} 1 \\ 1 \end{bmatrix}$ and each of the following matrices A, solve the initial value problem defined in Exercise T.6.

(a) $A = \begin{bmatrix} 2 & -1 \\ -1 & 2 \end{bmatrix}$

(b) $A = \begin{bmatrix} -1 & 1 \\ 1 & -1 \end{bmatrix}$

3. For $\mathbf{x}_0 = \begin{bmatrix} 1 \\ 0 \\ 1 \end{bmatrix}$ and each of the following matrices A, solve the initial value problem defined in Exercise T.6.

(a) $A = \begin{bmatrix} -1 & 1 & 0 \\ 0 & 3 & -12 \\ 1 & -1 & 0 \end{bmatrix}$

(b) $A = \begin{bmatrix} 0 & 1 & 0 \\ 0 & 0 & 1 \\ 0 & 8 & -2 \end{bmatrix}$

4. Using a handheld calculator or MATLAB, compute the Fibonacci number u_{25}.

5. Consider the homogeneous linear system of differential equations

$$\begin{bmatrix} x_1' \\ x_2' \end{bmatrix} = \begin{bmatrix} 1 & 1 \\ 3 & -1 \end{bmatrix} \begin{bmatrix} x_1 \\ x_2 \end{bmatrix}.$$

(a) Find the general solution.

(b) Find the solution to the initial value problem determined by the initial value conditions $x_1(0) = 4$, $x_2(0) = 6$.

6. Find a quadratic form of the type in Theorem 9.3 that is equivalent to the quadratic form

$$x^2 + 2y^2 + z^2 - 2xy - 2yz.$$

7. Describe the trajectories of the dynamical system represented by the system given in Exercise 5.

8. For the dynamical system with matrix

$$A = \begin{bmatrix} 1 & k \\ 2 & 1 \end{bmatrix}$$

determine integer values of k so that the trajectories have the following behavior:

(a) The origin is a saddle point.

(b) The origin is an unstable equilibrium point.

Theoretical Exercises

T.1. The usual rules for differentiation and integration of functions introduced in calculus also apply to matrix functions. Let $A(t)$ and $B(t)$ be $n \times n$ matrix functions whose entries are differentiable and let c_1 and c_2 be real numbers. Prove the following properties.

(a) $\frac{d}{dt}[c_1A(t) + c_2B(t)] = c_1\frac{d}{dt}[A(t)] + c_2\frac{d}{dt}[B(t)]$

(b) $\int_a^t (c_1A(s) + c_2B(s))\,ds = c_1\int_a^t A(s)\,ds + c_2\int_a^t B(s)\,ds$

(c) $\frac{d}{dt}[A(t)B(t)] = B(t)\frac{d}{dt}[A(t)] + A(t)\frac{d}{dt}[B(t)]$

T.2. If A is an $n \times n$ matrix, then the matrix function

$$B(t) = I_n + At + A^2\frac{t^2}{2!} + A^3\frac{t^3}{3!} + \cdots$$

is called the **matrix exponential** function and we use the notation $B(t) = e^{At}$.

(a) Prove that $\frac{d}{dt}[e^{At}] = Ae^{At}$.

(b) Let

$$A = \begin{bmatrix} 0 & 1 \\ 0 & 0 \end{bmatrix} \quad \text{and} \quad B = \begin{bmatrix} 0 & 0 \\ 1 & 0 \end{bmatrix}.$$

Prove or disprove that $e^Ae^B = e^{A+B}$.

(c) Prove that $e^{iA} = \cos A + i \sin A$, where $i = \sqrt{-1}$ (see Section A.1).

T.3. Let A and B be $n \times n$ matrices that commute, that is, $AB = BA$. Prove that $e^Ae^B = e^{A+B}$.

T.4. Let $B(t) = \left[b_{ij}(t)\right]$ be a diagonal matrix function with $b_{ii}(t) = e^{\lambda_{ii} t}$, where λ_{ii} is a scalar, $i = 1, 2, \ldots, n$, and $b_{ij}(t) = 0$ if $i \neq j$. Let D be the diagonal matrix with diagonal entries λ_{ii}, $i = 1, 2, \ldots, n$. Prove that $B(t) = e^{Dt}$.

T.5. Let A be an $n \times n$ matrix that is diagonalizable with eigenvalues λ_i and associated eigenvectors $\mathbf{x}_i$, $i = 1, 2, \ldots, n$. Then we can choose the eigenvectors $\mathbf{x}_i$ so that they form a linearly independent set; the matrix P whose jth column is $\mathbf{x}_j$ is nonsingular, and $P^{-1}AP = D$, where D is the diagonal matrix whose diagonal entries are the eigenvalues of A. Prove that Equation (17) in Section 9.2 can be written as

$$\mathbf{x}(t) = Pe^{Dt}B,$$

where

$$B = \begin{bmatrix} b_1 \\ b_2 \\ \vdots \\ b_n \end{bmatrix}.$$

T.6. Let A be an $n \times n$ matrix and

$$\mathbf{x}(t) = \begin{bmatrix} x_1(t) \\ x_2(t) \\ \vdots \\ x_n(t) \end{bmatrix}.$$

Assume that A is diagonalizable, as in Exercise T.5, and prove that the solution to the initial value problem

$$\mathbf{x}' = A\mathbf{x}$$
$$\mathbf{x}(0) = \mathbf{x}_0$$

can be written as

$$\mathbf{x}(t) = Pe^{Dt}P^{-1}\mathbf{x}_0 = e^{At}\mathbf{x}_0.$$

Chapter Test

1. Using a hand calculator or MATLAB, compute the Fibonacci number u_{30}.

2. Find the general solution to the linear system of differential equations:

$$\begin{bmatrix} x_1' \\ x_2' \end{bmatrix} = \begin{bmatrix} 3 & 2 \\ 6 & -1 \end{bmatrix} \begin{bmatrix} x_1 \\ x_2 \end{bmatrix}.$$

3. Let $g(\mathbf{x}) = 2x^2 + 6xy + 2y^2 = 1$ be the equation of a conic section. Identify the conic by finding a quadratic form of the type in Theorem 9.3 that is equivalent to g.

4. Describe the trajectories of the dynamical system

$$\frac{dx}{dt} = 8x + 6y$$

$$\frac{dy}{dt} = -3x - y.$$

5. Determine an integer value for k so that the dynamical system with matrix

$$A = \begin{bmatrix} -5 & -1 \\ k & 1 \end{bmatrix}$$

will have trajectories that all tend toward an equilibrium point at the origin.

CHAPTER 10

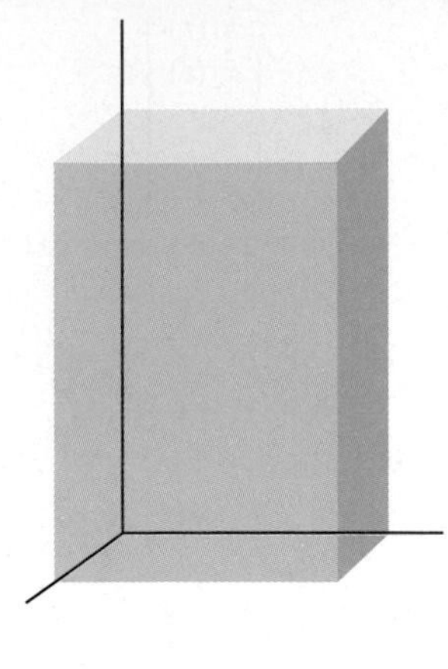

LINEAR TRANSFORMATIONS AND MATRICES

In Section 4.3 we gave the definition, basic properties, and some examples of linear transformations mapping R^n into R^m. In this chapter we consider linear transformations mapping a vector space V into a vector space W.

10.1 DEFINITION AND EXAMPLES

DEFINITION

Let V and W be vector spaces. A **linear transformation** L of V into W is a function assigning a unique vector $L(\mathbf{u})$ in W to each $\mathbf{u}$ in V such that:

(a) $L(\mathbf{u}+\mathbf{v}) = L(\mathbf{u}) + L(\mathbf{v})$, for every $\mathbf{u}$ and $\mathbf{v}$ in V.

(b) $L(k\mathbf{u}) = kL(\mathbf{u})$, for every $\mathbf{u}$ in V and every scalar k.

In the definition above, observe that in (a) the $+$ in $\mathbf{u}+\mathbf{v}$ on the left side of the equation refers to the addition operation in V, whereas the $+$ in $L(\mathbf{u}) + L(\mathbf{v})$ on the right side of the equation refers to the addition operation in W. Similarly, in (b) the scalar product $k\mathbf{u}$ is in V, while the scalar product $kL(\mathbf{u})$ is in W.

As in Section 4.3, we shall write the fact that L maps V into W, even if it is not a linear transformation, as

$$L\colon V \to W.$$

If $V = W$, the linear transformation $L\colon V \to V$ is also called a **linear operator** on V.

In Sections 4.3 and 2.3 we gave a number of examples of linear transformations mapping R^n into R^m. Thus, the following are linear transformations that we have already discussed:

Projection: $L\colon R^3 \to R^2$ defined by $L(x, y, z) = (x, y)$.

Dilation: $L\colon R^3 \to R^3$ defined by $L(\mathbf{u}) = r\mathbf{u}$, $r > 1$.

Contraction: $L\colon R^3 \to R^3$ defined by $L(\mathbf{u}) = r\mathbf{u}$, $0 < r < 1$.

Reflection: $L\colon R^2 \to R^2$ defined by $L(x, y) = (x, -y)$.

Rotation: $L\colon R^2 \to R^2$ defined by $L(\mathbf{u}) = \begin{bmatrix} \cos\phi & -\sin\phi \\ \sin\phi & \cos\phi \end{bmatrix}\mathbf{u}$.

Shear in the x-direction: $L\colon R^2 \to R^2$ defined by $L(\mathbf{u}) = \begin{bmatrix} 1 & k \\ 0 & 1 \end{bmatrix} \mathbf{u}$, where k is a scalar.

Shear in the y-direction: $L\colon R^2 \to R^2$ defined by $L(\mathbf{u}) = \begin{bmatrix} 1 & 0 \\ k & 1 \end{bmatrix} \mathbf{u}$, where k is a scalar.

Recall that P_1 is the vector space of all polynomials of degree ≤ 1; in general, P_n is the vector space of all polynomials of degree $\leq n$, and M_{nn} is the vector space of all $n \times n$ matrices.

As in Section 4.3, to verify that a given function is a linear transformation, we have to check that conditions (a) and (b) in the preceding definition are satisfied.

EXAMPLE 1 Let $L\colon P_1 \to P_2$ be defined by

$$L(at + b) = t(at + b).$$

Show that L is a linear transformation.

Solution Let $at + b$ and $ct + d$ be vectors in P_1 and let k be a scalar. Then

$$\begin{aligned} L[(at + b) + (ct + d)] &= t[(at + b) + (ct + d)] \\ &= t(at + b) + t(ct + d) = L(at + b) + L(ct + d) \end{aligned}$$

and

$$L[k(at + b)] = t[k(at + b)] = k[t(at + b)] = kL(at + b).$$

Hence L is a linear transformation. ■

EXAMPLE 2 Let $L\colon P_1 \to P_2$ be defined by

$$L[p(t)] = tp(t) + t^2.$$

Is L a linear transformation?

Solution Let $p(t)$ and $q(t)$ be vectors in P_1 and let k be a scalar. Then

$$L[p(t) + q(t)] = t[p(t) + q(t)] + t^2 = tp(t) + tq(t) + t^2$$

and

$$L[p(t)] + L[q(t)] = [tp(t) + t^2] + [tq(t) + t^2] = t[p(t) + q(t)] + 2t^2.$$

Since $L[p(t) + q(t)] \neq L[p(t)] + L[q(t)]$, we conclude that L is not a linear transformation. ■

EXAMPLE 3 Let $L\colon M_{mn} \to M_{nm}$ be defined by

$$L(A) = A^T$$

for A in M_{mn}. Is L a linear transformation?

Solution Let A and B be in M_{mn}. Then by Theorem 1.4 in Section 1.4, we have

$$L(A + B) = (A + B)^T = A^T + B^T = L(A) + L(B),$$

and, if k is a scalar,

$$L(kA) = (kA)^T = kA^T = kL(A).$$

Hence L is a linear transformation. ■

EXAMPLE 4 (Calculus Required) Let W be the vector space of all real-valued functions and let V be the subspace of all differentiable functions. Let $L: V \to W$ be defined by

$$L(f) = f',$$

where f' is the derivative of f. It is easy to show (Exercise 13), from the properties of differentiation, that L is a linear transformation. ■

EXAMPLE 5 (Calculus Required) Let $V = C[0, 1]$ denote the vector space of all real-valued continuous functions defined on $[0, 1]$. Let $W = R^1$. Define $L: V \to W$ by

$$L(f) = \int_0^1 f(x)\,dx.$$

It is easy to show (Exercise 14), from the properties of integration, that L is a linear transformation. ■

EXAMPLE 6 Let V be an n-dimensional vector space and $S = \{\mathbf{v}_1, \mathbf{v}_2, \ldots, \mathbf{v}_n\}$ a basis for V. If $\mathbf{v}$ is a vector in V, then

$$\mathbf{v} = c_1\mathbf{v}_1 + c_2\mathbf{v}_2 + \cdots + c_n\mathbf{v}_n,$$

where $c_1, c_2, \ldots, c_n$ are the coordinates of $\mathbf{v}$ with respect to S (see Section 6.7). We define $L: V \to R^n$ by

$$L(\mathbf{v}) = \left[\mathbf{v}\right]_S.$$

It is easy to show (Exercise 15) that L is a linear transformation. ■

EXAMPLE 7 Let A be an $m \times n$ matrix. In Section 4.3 we have observed that if $L: R^n \to R^m$ is defined by

$$L(\mathbf{x}) = A\mathbf{x}$$

for $\mathbf{x}$ in R^n, then L is a linear transformation (see Exercise 16). Specific cases of this type of linear transformation have been seen in Example 3, and Exercise 12 in Section 4.3. ■

The following two theorems give some basic properties of linear transformations.

THEOREM 10.1 *If $L: V \to W$ is a linear transformation, then*

$$L(c_1\mathbf{v}_1 + c_2\mathbf{v}_2 + \cdots + c_k\mathbf{v}_k) = c_1L(\mathbf{v}_1) + c_2L(\mathbf{v}_2) + \cdots + c_kL(\mathbf{v}_k)$$

for any vectors $\mathbf{v}_1, \mathbf{v}_2, \ldots, \mathbf{v}_k$ in V and any scalars $c_1, c_2, \ldots, c_k$.

Proof Exercise T.1. ■

THEOREM 10.2 *Let $L: V \to W$ be a linear transformation. Then:*

(a) $L(\mathbf{0}_V) = \mathbf{0}_W$, *where $\mathbf{0}_V$ and $\mathbf{0}_W$ are the zero vectors in V and W, respectively.*

(b) $L(\mathbf{u} - \mathbf{v}) = L(\mathbf{u}) - L(\mathbf{v})$.

Proof (a) We have

$$\mathbf{0}_V = \mathbf{0}_V + \mathbf{0}_V.$$

Then

$$L(\mathbf{0}_V) = L(\mathbf{0}_V + \mathbf{0}_V) = L(\mathbf{0}_V) + L(\mathbf{0}_V). \tag{1}$$

Adding $-L(\mathbf{0}_V)$ to both sides of Equation (1), we obtain

$$L(\mathbf{0}_V) = \mathbf{0}_W.$$

(b) Exercise T.2. ■

The proof of the following corollary is very similar to the proof of the analogous corollary, Corollary 4.1 in Section 4.3.

COROLLARY 10.1 *Let $T: V \to W$ be a function. If $T(\mathbf{0}_V) \neq \mathbf{0}_W$, then T is not a linear transformation.*

Proof Exercise T.3. ■

Remarks

1. Example 2 could be solved more easily by using Corollary 10.1 as follows. Since $T(0) = t(0)+t^2 = t^2$, it follows that T is not a linear transformation.
2. Let $T: V \to W$ be a function. Note that $T(\mathbf{0}_V) = \mathbf{0}_W$ does *not* imply that T is a linear transformation. For example, consider $T: R^2 \to R^2$ defined by

$$T\left(\begin{bmatrix} a \\ b \end{bmatrix}\right) = \begin{bmatrix} a^2 \\ b^2 \end{bmatrix}.$$

Then

$$T\left(\begin{bmatrix} 0 \\ 0 \end{bmatrix}\right) = \begin{bmatrix} 0 \\ 0 \end{bmatrix},$$

but T is not a linear transformation (verify).

A function f mapping a set V into a set W can be specified by a formula that assigns to every member of V a unique element of W. On the other hand, we can also specify a function by listing next to each member of V its assigned element of W. An example of this would be provided by listing the names of all charge account customers of a department store along with their charge account number. At first glance, it appears impossible to describe a linear transformation $L: V \to W$ of a vector space $V \neq \{\mathbf{0}\}$ into a vector space W in this latter manner, since V has infinitely many members in it. However, the following very useful theorem tells us that once we know what L does to a basis for V, then we have completely determined L. Thus, for a finite-dimensional vector space V, it is possible to describe L by giving only the images of a finite number of vectors in V.

THEOREM 10.3 *Let $L: V \to W$ be a linear transformation of an n-dimensional vector space V into a vector space W. Also, let $S = \{\mathbf{v}_1, \mathbf{v}_2, \ldots, \mathbf{v}_n\}$ be a basis for V. If $\mathbf{u}$ is any vector in V, then $L(\mathbf{u})$ is completely determined by $\{L(\mathbf{v}_1), L(\mathbf{v}_2), \ldots, L(\mathbf{v}_n)\}$.*

Proof Since $\mathbf{u}$ is in V, we can write

$$\mathbf{u} = c_1\mathbf{v}_1 + c_2\mathbf{v}_2 + \cdots + c_n\mathbf{v}_n, \tag{2}$$

where $c_1, c_2, \ldots, c_n$ are uniquely determined scalars. Then

$$L(\mathbf{u}) = L(c_1\mathbf{v}_1 + c_2\mathbf{v}_2 + \cdots + c_n\mathbf{v}_n) = c_1L(\mathbf{v}_1) + c_2L(\mathbf{v}_2) + \cdots + c_nL(\mathbf{v}_n),$$

by Theorem 10.1. Thus $L(\mathbf{u})$ has been completely determined by the elements $L(\mathbf{v}_1), L(\mathbf{v}_2), \ldots, L(\mathbf{v}_n)$. ■

It might be noted that in the proof of Theorem 10.3, the scalars c_i, $i = 1, 2, \ldots, n$ determined in Equation (2) depend on the basis vectors in S. Thus, if we change S, then we may change the c_i's.

EXAMPLE 8 Let $L: P_1 \to P_2$ be a linear transformation for which we know that

$$L(t+1) = t^2 - 1 \quad \text{and} \quad L(t-1) = t^2 + t.$$

(a) What is $L(7t+3)$?

(b) What is $L(at+b)$?

Solution (a) First, note that $\{t+1, t-1\}$ is a basis for P_1 (verify). Next, we find that (verify)

$$7t + 3 = 5(t+1) + 2(t-1).$$

Then

$$\begin{aligned} L(7t+3) &= L(5(t+1) + 2(t-1)) \\ &= 5L(t+1) + 2L(t-1) \\ &= 5(t^2-1) + 2(t^2+t) = 7t^2 + 2t - 5. \end{aligned}$$

(b) Writing $at+b$ as a linear combination of the given basis vectors, we find that (verify)

$$at + b = \left(\frac{a+b}{2}\right)(t+1) + \left(\frac{a-b}{2}\right)(t-1).$$

Then

$$\begin{aligned} L(at+b) &= L\left(\left(\frac{a+b}{2}\right)(t+1) + \left(\frac{a-b}{2}\right)(t-1)\right) \\ &= \left(\frac{a+b}{2}\right)L(t+1) + \left(\frac{a-b}{2}\right)L(t-1) \\ &= \left(\frac{a+b}{2}\right)(t^2-1) + \left(\frac{a-b}{2}\right)(t^2+t) \\ &= at^2 + \left(\frac{a-b}{2}\right)t - \left(\frac{a+b}{2}\right). \end{aligned}$$

■

Key Terms

Linear transformation
Linear operator

10.1 Exercises

1. Which of the following are linear transformations?

(a) $L(x, y) = (x + y, x - y)$

(b) $L\left(\begin{bmatrix} x \\ y \\ z \end{bmatrix}\right) = \begin{bmatrix} x + 1 \\ y - z \end{bmatrix}$

(c) $L\left(\begin{bmatrix} x \\ y \\ z \end{bmatrix}\right) = \begin{bmatrix} 1 & 2 & 3 \\ -1 & 2 & 4 \end{bmatrix}\begin{bmatrix} x \\ y \\ z \end{bmatrix}$

2. Which of the following are linear transformations?

(a) $L(x, y, z) = (0, 0)$

(b) $L(x, y, z) = (1, 2, -1)$

(c) $L(x, y, z) = (x^2 + y, y - z)$

3. Let $L: P_1 \to P_2$ be defined as indicated. Is L a linear transformation? Justify your answer.

(a) $L[p(t)] = tp(t) + p(0)$

(b) $L[p(t)] = tp(t) + t^2 + 1$

(c) $L(at + b) = at^2 + (a - b)t$

4. Let $L: P_2 \to P_1$ be defined as indicated. Is L a linear transformation? Justify your answer.

(a) $L(at^2 + bt + c) = at + b + 1$

(b) $L(at^2 + bt + c) = 2at - b$

(c) $L(at^2 + bt + c) = (a + 2)t + (b - a)$

5. Let $L: P_2 \to P_2$ be defined as indicated. Is L a linear transformation? Justify your answer.

(a) $L(at^2 + bt + c) = (a + 1)t^2 + (b - c)t + (a + c)$

(b) $L(at^2 + bt + c) = at^2 + (b - c)t + (a - b)$

(c) $L(at^2 + bt + c) = 0$

6. Let C be a fixed $n \times n$ matrix and let $L: M_{nn} \to M_{nn}$ be defined by $L(A) = CA$. Show that L is a linear transformation.

7. Let $L: M_{22} \to M_{22}$ be defined by

$$L\left(\begin{bmatrix} a & b \\ c & d \end{bmatrix}\right) = \begin{bmatrix} b & c - d \\ c + d & 2a \end{bmatrix}.$$

Is L a linear transformation?

8. Let $L: M_{22} \to M_{22}$ be defined by

$$L\left(\begin{bmatrix} a & b \\ c & d \end{bmatrix}\right) = \begin{bmatrix} a - 1 & b + 1 \\ 2c & 3d \end{bmatrix}.$$

Is L a linear transformation?

9. Let $L: M_{22} \to R^1$ be defined by

$$L\left(\begin{bmatrix} a & b \\ c & d \end{bmatrix}\right) = a + d.$$

Is L a linear transformation?

10. Let $L: M_{22} \to R^1$ be defined by

$$L\left(\begin{bmatrix} a & b \\ c & d \end{bmatrix}\right) = a + b - c - d + 1.$$

Is L a linear transformation?

11. Consider the function $L: M_{34} \to M_{24}$ defined by

$$L(A) = \begin{bmatrix} 2 & 3 & 1 \\ 1 & 2 & -3 \end{bmatrix} A$$

for A in M_{34}.

(a) Find $L\left(\begin{bmatrix} 1 & 2 & 0 & -1 \\ 3 & 0 & 2 & 3 \\ 4 & 1 & -2 & 1 \end{bmatrix}\right)$.

(b) Show that L is a linear transformation.

12. Let $L: M_{nn} \to R^1$ be defined by $L(A) = a_{11}a_{22} \cdots a_{nn}$, for an $n \times n$ matrix $A = \left[a_{ij}\right]$. Is L a linear transformation?

13. ***(Calculus Required)*** Verify that the function in Example 4 is a linear transformation.

14. ***(Calculus Required)*** Verify that the function in Example 5 is a linear transformation.

15. Verify that the function in Example 6 is a linear transformation.

16. Verify that the function in Example 7 is a linear transformation.

17. Let $L: R^2 \to R^2$ be a linear transformation for which we know that $L(1, 1) = (1, -2)$, $L(-1, 1) = (2, 3)$.

(a) What is $L(-1, 5)$?

(b) What is $L(a_1, a_2)$?

18. Let $L: P_2 \to P_3$ be a linear transformation for which we know that $L(1) = 1$, $L(t) = t^2$, and $L(t^2) = t^3 + t$.

(a) Find $L(2t^2 - 5t + 3)$.

(b) Find $L(at^2 + bt + c)$.

19. Let $L: P_1 \to P_1$ be a linear transformation for which we know that $L(t + 1) = 2t + 3$ and $L(t - 1) = 3t - 2$.

(a) Find $L(6t - 4)$.

(b) Find $L(at + b)$.

Theoretical Exercises

T.1. Prove Theorem 10.1.

T.2. Prove (b) of Theorem 10.2.

T.3. Prove Corollary 10.1.

T.4. Show that $L: V \to W$ is a linear transformation if and only if

$$L(a\mathbf{u} + b\mathbf{v}) = aL(\mathbf{u}) + bL(\mathbf{v}),$$

for any scalars a and b and any vectors $\mathbf{u}$ and $\mathbf{v}$ in V.

T.5. Consider the function $\text{Tr}: M_{nn} \to R^1$ (the **trace**): If $A = \left[a_{ij}\right]$ is in V, then $\text{Tr}(A) = a_{11} + a_{22} + \cdots + a_{nn}$. Show that Tr is a linear transformation (see Supplementary Exercise T.1 in Chapter 1).

T.6. Let $L: M_{nn} \to M_{nn}$ be the function defined by

$$L(A) = \begin{cases} A^{-1} & \text{if } A \text{ is nonsingular} \\ O & \text{if } A \text{ is singular} \end{cases}$$

for A in M_{nn}. Is L a linear transformation?

T.7. Let V and W be vector spaces. Show that the function $O: V \to W$ defined by $O(\mathbf{v}) = \mathbf{0}_W$ is a linear transformation, which is called the **zero linear transformation**.

T.8. Let $I: V \to V$ be defined by $I(\mathbf{v}) = \mathbf{v}$, for $\mathbf{v}$ in V. Show that I is a linear transformation, which is called the **identity operator** on V.

T.9. Let $L: V \to W$ be a linear transformation from a vector space V into a vector space W. The **image** of a subspace V_1 of V is defined as

$$L(V_1) = \{\mathbf{w} \text{ in } W \mid \mathbf{w} = L(\mathbf{v}) \text{ for some } \mathbf{v} \text{ in } V\}.$$

Show that $L(V_1)$ is a subspace of W.

T.10. Let L_1 and L_2 be linear transformations from a vector space V into a vector space W. Let $\{\mathbf{v}_1, \mathbf{v}_2, \ldots, \mathbf{v}_n\}$ be a basis for V. Show that if $L_1(\mathbf{v}_i) = L_2(\mathbf{v}_i)$ for $i = 1, 2, \ldots, n$, then $L_1(\mathbf{v}) = L_2(\mathbf{v})$ for any $\mathbf{v}$ in V.

T.11. Let $L: V \to W$ be a linear transformation from a vector space V into a vector space W. The **preimage** of a subspace W_1 of W is defined as

$$L^{-1}(W_1) = \{\mathbf{v} \text{ in } V \mid L(\mathbf{v}) \text{ is in } W_1\}.$$

Show that $L^{-1}(W_1)$ is a subspace of V.

T.12. Let $T: V \to W$ be the function defined by $T(\mathbf{v}) = \mathbf{v} + \mathbf{b}$, for $\mathbf{v}$ in V, where $\mathbf{b}$ is a fixed nonzero vector in V. T is called a **translation** by vector $\mathbf{v}$. Is T a linear transformation? Explain.

T.13. Let $L: V \to V$ be a linear operator. A nonempty subspace U of V is called **invariant** under L if $L(U)$ is contained in U. Let A be an $n \times n$ matrix and λ an eigenvalue of A. Let $L: R^n \to R^n$ be defined by $L(\mathbf{x}) = A\mathbf{x}$. Show that the eigenspace of A associated with λ (see Exercise T.1 in Section 8.1) is an invariant subspace of R^n.

MATLAB Exercises

MATLAB cannot be used to show that a function between vector spaces is a linear transformation. However, MATLAB can be used to construct an example that shows that a function is not a linear transformation. The following exercises illustrate this point.

ML.1. Let $L: M_{nn} \to R^1$ be defined by $L(A) = \det(A)$.

(a) Find a pair of 2×2 matrices A and B such that

$$L(A + B) \neq L(A) + L(B).$$

Use MATLAB to do the computations. It follows that L is not a linear transformation.

(b) Find a pair of 3×3 matrices A and B such that

$$L(A + B) \neq L(A) + L(B).$$

It follows that L is not a linear transformation. Use MATLAB to do the computations.

ML.2. Let $L: M_{nn} \to R^1$ be defined by $L(A) = \text{rank } A$.

(a) Find a pair of 2×2 matrices A and B such that

$$L(A + B) \neq L(A) + L(B).$$

It follows that L is not a linear transformation. Use MATLAB to do the computations.

(b) Find a pair of 3×3 matrices A and B such that

$$L(A + B) \neq L(A) + L(B).$$

It follows that L is not a linear transformation. Use MATLAB to do the computations.

10.2 THE KERNEL AND RANGE OF A LINEAR TRANSFORMATION

In this section we study special types of linear transformations; we formulate the notions of one-to-one linear transformations and onto linear transformations. We also develop methods for determining when a linear transformation is one-to-one or onto.

DEFINITION A linear transformation $L: V \to W$ is said to be **one-to-one** if for all $\mathbf{v}_1$, $\mathbf{v}_2$ in V, $\mathbf{v}_1 \neq \mathbf{v}_2$ implies that $L(\mathbf{v}_1) \neq L(\mathbf{v}_2)$. An equivalent statement is that L is one-to-one if for all $\mathbf{v}_1$, $\mathbf{v}_2$ in V, $L(\mathbf{v}_1) = L(\mathbf{v}_2)$ implies that $\mathbf{v}_1 = \mathbf{v}_2$.

This definition says that L is one-to-one if $L(\mathbf{v}_1)$ and $L(\mathbf{v}_2)$ are distinct whenever $\mathbf{v}_1$ and $\mathbf{v}_2$ are distinct (Figure 10.1).

Figure 10.1 ►

(a) L is one-to-one. (b) L is not one-to-one.

EXAMPLE 1 Let $L: R^2 \to R^2$ be defined by

$$L(x, y) = (x + y, x - y).$$

To determine whether L is one-to-one, we let

$$\mathbf{v}_1 = (a_1, a_2) \quad \text{and} \quad \mathbf{v}_2 = (b_1, b_2).$$

Then if

$$L(\mathbf{v}_1) = L(\mathbf{v}_2),$$

we have

$$\begin{aligned} a_1 + a_2 &= b_1 + b_2 \\ a_1 - a_2 &= b_1 - b_2. \end{aligned}$$

Adding these equations, we obtain $2a_1 = 2b_1$, or $a_1 = b_1$, which implies that $a_2 = b_2$. Hence, $\mathbf{v}_1 = \mathbf{v}_2$ and L is one-to-one. ■

EXAMPLE 2 Let $L: R^3 \to R^2$ be the linear transformation defined in Example 6 of Section 1.5 (the projection function) by

$$L(x, y, z) = (x, y).$$

Since $(1, 3, 3) \neq (1, 3, -2)$ but

$$L(1, 3, 3) = L(1, 3, -2) = (1, 3),$$

we conclude that L is not one-to-one. ■

We shall now develop some more efficient ways of determining whether or not a linear transformation is one-to-one.

DEFINITION Let $L: V \to W$ be a linear transformation. The **kernel** of L, $\ker L$, is the subset of V consisting of all vectors $\mathbf{v}$ such that $L(\mathbf{v}) = \mathbf{0}_W$.

We observe that property (a) of Theorem 10.2 in Section 10.1 assures us that $\ker L$ is never an empty set, since $\mathbf{0}_V$ is in $\ker L$.

EXAMPLE 3 Let $L: R^3 \to R^2$ be as defined in Example 2. The vector $(0, 0, 2)$ is in $\ker L$, since $L(0, 0, 2) = (0, 0)$. However, the vector $(2, -3, 4)$ is not in $\ker L$, since $L(2, -3, 4) = (2, -3)$. To find $\ker L$, we must determine all $\mathbf{x}$ in R^3 so that $L(\mathbf{x}) = \mathbf{0}$. That is, we seek $\mathbf{x} = (x_1, x_2, x_3)$ so that

$$L(\mathbf{x}) = L(x_1, x_2, x_3) = \mathbf{0} = (0, 0).$$

However, $L(\mathbf{x}) = (x_1, x_2)$. Thus $(x_1, x_2) = (0, 0)$, so $x_1 = 0$, $x_2 = 0$, and x_3 can be any real number. Hence, $\ker L$ consists of all vectors in R^3 of the form $(0, 0, r)$, where r is any real number. It is clear that $\ker L$ consists of the z-axis in three-dimensional space R^3. ■

EXAMPLE 4 If L is as defined in Example 1, then $\ker L$ consists of all vectors $\mathbf{x}$ in R^2 such that $L(\mathbf{x}) = \mathbf{0}$. Thus we must solve the linear system

$$\begin{aligned} x + y &= 0 \\ x - y &= 0 \end{aligned}$$

for x and y. The only solution is $\mathbf{x} = \mathbf{0}$, so $\ker L = \{\mathbf{0}\}$. ■

EXAMPLE 5 If $L: R^4 \to R^2$ is defined by

$$L\left(\begin{bmatrix} x \\ y \\ z \\ w \end{bmatrix}\right) = \begin{bmatrix} x + y \\ z + w \end{bmatrix},$$

then $\ker L$ consists of all vectors $\mathbf{u}$ in R^4 such that $L(\mathbf{u}) = \mathbf{0}$. This leads to the linear system

$$\begin{aligned} x + y &= 0 \\ z + w &= 0. \end{aligned}$$

Thus $\ker L$ consists of all vectors of the form

$$\begin{bmatrix} r \\ -r \\ s \\ -s \end{bmatrix} = r \begin{bmatrix} 1 \\ -1 \\ 0 \\ 0 \end{bmatrix} + s \begin{bmatrix} 0 \\ 0 \\ 1 \\ -1 \end{bmatrix},$$

where r and s are any real numbers. ■

In Example 5, $\ker L$ consists of all linear combinations of

$$\begin{bmatrix} 1 \\ -1 \\ 0 \\ 0 \end{bmatrix} \quad \text{and} \quad \begin{bmatrix} 0 \\ 0 \\ 1 \\ -1 \end{bmatrix},$$

a subspace of R^4. The next theorem generalizes this result.

THEOREM 10.4 *If $L: V \to W$ is a linear transformation, then* $\ker L$ *is a subspace of V.*

Proof First, observe that $\ker L$ is not an empty set, since $\mathbf{0}_V$ is in $\ker L$. Also, let $\mathbf{u}$ and $\mathbf{v}$ be in $\ker L$. Then since L is a linear transformation,

$$L(\mathbf{u} + \mathbf{v}) = L(\mathbf{u}) + L(\mathbf{v}) = \mathbf{0}_W + \mathbf{0}_W = \mathbf{0}_W,$$

so $\mathbf{u}+\mathbf{v}$ is in $\ker L$. Also, if c is a scalar, then since L is a linear transformation,

$$L(c\mathbf{u}) = cL(\mathbf{u}) = c\mathbf{0}_W = \mathbf{0}_W,$$

so $c\mathbf{u}$ is in $\ker L$. Hence $\ker L$ is a subspace of V. ■

EXAMPLE 6 If L is as in Example 1, then $\ker L$ is the subspace $\{\mathbf{0}\}$; its dimension is zero.

EXAMPLE 7 If L is as in Example 2, then a basis for $\ker L$ is

$$\{(0, 0, 1)\}$$

and $\dim(\ker L) = 1$. Thus $\ker L$ consists of the z-axis in three-dimensional space R^3. ■

EXAMPLE 8 If L is as in Example 5, then a basis for $\ker L$ consists of the vectors

$$\begin{bmatrix} 1 \\ -1 \\ 0 \\ 0 \end{bmatrix} \quad \text{and} \quad \begin{bmatrix} 0 \\ 0 \\ 1 \\ -1 \end{bmatrix};$$

thus $\dim(\ker L) = 2$. ■

If $L: R^n \to R^m$ is a linear transformation defined by $L(\mathbf{x}) = A\mathbf{x}$, where A is an $m \times n$ matrix, then the kernel of L is the solution space of the homogeneous system $A\mathbf{x} = \mathbf{0}$.

An examination of the elements in $\ker L$ allows us to decide whether L is or is not one-to-one.

THEOREM 10.5 *A linear transformation $L: V \to W$ is one-to-one if and only if $\ker L = \{\mathbf{0}_V\}$.*

Proof Let L be one-to-one. We show that $\ker L = \{\mathbf{0}_V\}$. Let $\mathbf{x}$ be in $\ker L$. Then $L(\mathbf{x}) = \mathbf{0}_W$. Also, we already know that $L(\mathbf{0}_V) = \mathbf{0}_W$. Thus $L(\mathbf{x}) = L(\mathbf{0}_V)$. Since L is one-to-one, we conclude that $\mathbf{x} = \mathbf{0}_V$. Hence $\ker L = \{\mathbf{0}_V\}$.

Conversely, suppose that $\ker L = \{\mathbf{0}_V\}$. We wish to show that L is one-to-one. Assume that $L(\mathbf{u}) = L(\mathbf{v})$, for $\mathbf{u}$ and $\mathbf{v}$ in V. Then

$$L(\mathbf{u}) - L(\mathbf{v}) = \mathbf{0}_W,$$

so by Theorem 10.2, $L(\mathbf{u} - \mathbf{v}) = \mathbf{0}_W$, which means that $\mathbf{u} - \mathbf{v}$ is in $\ker L$. Therefore, $\mathbf{u} - \mathbf{v} = \mathbf{0}_V$, so $\mathbf{u} = \mathbf{v}$. Thus L is one-to-one. ■

Note that we can also state Theorem 10.5 as: *L is one-to-one if and only if* $\dim(\ker L) = 0$.

The proof of Theorem 10.5 has also established the following result, which we state as Corollary 10.2.

COROLLARY 10.2 *If $L(\mathbf{x}) = \mathbf{b}$ and $L(\mathbf{y}) = \mathbf{b}$, then $\mathbf{x} - \mathbf{y}$ belongs to $\ker L$. In other words, any two solutions to $L(\mathbf{x}) = \mathbf{b}$ differ by an element of the kernel of L.*

Proof Exercise T.1. ■

EXAMPLE 9 The linear transformation in Example 1 is one-to-one; the one in Example 2 is not. ■

In Section 10.3 we shall prove that for every linear transformation $L: R^n \to R^m$, we can find a unique $m \times n$ matrix A so that if $\mathbf{x}$ is in R^n, then $L(\mathbf{x}) = A\mathbf{x}$. It then follows that to find $\ker L$, we need to find the solution space of the homogeneous system $A\mathbf{x} = \mathbf{0}$. Hence to find $\ker L$ we need only use techniques with which we are already familiar.

DEFINITION If $L: V \to W$ is a linear transformation, then the **range** of L, denoted by range L, is the set of all vectors in W that are images, under L, of vectors in V. Thus a vector $\mathbf{w}$ is in range L if there exists some vector $\mathbf{v}$ in V such that $L(\mathbf{v}) = \mathbf{w}$. If range $L = W$, we say that L is **onto**. That is, L is onto if and only if, given any $\mathbf{w}$ in W, there is a $\mathbf{v}$ in V such that $L(\mathbf{v}) = \mathbf{w}$.

THEOREM 10.6 *If $L: V \to W$ is a linear transformation, then* range L *is a subspace of W.*

Proof First, observe that range L is not an empty set, since $\mathbf{0}_W = L(\mathbf{0}_V)$, so $\mathbf{0}_W$ is in range L. Let $\mathbf{w}_1$ and $\mathbf{w}_2$ be in range L. Then $\mathbf{w}_1 = L(\mathbf{v}_1)$ and $\mathbf{w}_2 = L(\mathbf{v}_2)$ for some $\mathbf{v}_1$ and $\mathbf{v}_2$ in V. Now

$$\mathbf{w}_1 + \mathbf{w}_2 = L(\mathbf{v}_1) + L(\mathbf{v}_2) = L(\mathbf{v}_1 + \mathbf{v}_2),$$

which implies that $\mathbf{w}_1 + \mathbf{w}_2$ is in range L. Also, if c is a scalar, then $c\mathbf{w}_1 = cL(\mathbf{v}_1) = L(c\mathbf{v}_1)$, so $c\mathbf{w}_1$ is in range L. Hence range L is a subspace of W. ■

EXAMPLE 10 Let L be the linear transformation defined in Example 2. To find out whether L is onto, we choose any vector $\mathbf{y} = (y_1, y_2)$ in R^2 and seek a vector $\mathbf{x} = (x_1, x_2, x_3)$ in R^3 such that $L(\mathbf{x}) = \mathbf{y}$. Since $L(\mathbf{x}) = (x_1, x_2)$, we find that if $x_1 = y_1$ and $x_2 = y_2$, then $L(\mathbf{x}) = \mathbf{y}$. Therefore, L is onto and the dimension of range L is 2. ■

EXAMPLE 11 Let $L: R^3 \to R^3$ be defined by

$$L\left(\begin{bmatrix} a_1 \\ a_2 \\ a_3 \end{bmatrix}\right) = \begin{bmatrix} 1 & 0 & 1 \\ 1 & 1 & 2 \\ 2 & 1 & 3 \end{bmatrix} \begin{bmatrix} a_1 \\ a_2 \\ a_3 \end{bmatrix}.$$

(a) Is L onto?
(b) Find a basis for range L.
(c) Find ker L.
(d) Is L one-to-one?

Solution (a) Given any

$$\mathbf{w} = \begin{bmatrix} a \\ b \\ c \end{bmatrix}$$

in R^3, where a, b, and c are any real numbers, can we find

$$\mathbf{v} = \begin{bmatrix} a_1 \\ a_2 \\ a_3 \end{bmatrix}$$

so that $L(\mathbf{v}) = \mathbf{w}$? We seek a solution to the linear system

$$\begin{bmatrix} 1 & 0 & 1 \\ 1 & 1 & 2 \\ 2 & 1 & 3 \end{bmatrix} \begin{bmatrix} a_1 \\ a_2 \\ a_3 \end{bmatrix} = \begin{bmatrix} a \\ b \\ c \end{bmatrix}$$

and we find the reduced row echelon form of the augmented matrix to be (verify)

$$\left[\begin{array}{ccc|c} 1 & 0 & 1 & a \\ 0 & 1 & 1 & b - a \\ 0 & 0 & 0 & c - b - a \end{array}\right].$$

Thus a solution exists only when $c - b - a = 0$. Hence, L is not onto; that is, there exist values of a, b, and c for which there is no vector $\mathbf{v}$ in R^3 such that

$$L(\mathbf{v}) = \begin{bmatrix} a \\ b \\ c \end{bmatrix}.$$

(b) To find a basis for range L, we note that

$$L\left(\begin{bmatrix} a_1 \\ a_2 \\ a_3 \end{bmatrix}\right) = \begin{bmatrix} 1 & 0 & 1 \\ 1 & 1 & 2 \\ 2 & 1 & 3 \end{bmatrix} \begin{bmatrix} a_1 \\ a_2 \\ a_3 \end{bmatrix} = \begin{bmatrix} a_1 + a_3 \\ a_1 + a_2 + 2a_3 \\ 2a_1 + a_2 + 3a_3 \end{bmatrix}$$

$$= a_1 \begin{bmatrix} 1 \\ 1 \\ 2 \end{bmatrix} + a_2 \begin{bmatrix} 0 \\ 1 \\ 1 \end{bmatrix} + a_3 \begin{bmatrix} 1 \\ 2 \\ 3 \end{bmatrix}.$$

This means that

$$\left\{ \begin{bmatrix} 1 \\ 1 \\ 2 \end{bmatrix}, \begin{bmatrix} 0 \\ 1 \\ 1 \end{bmatrix}, \begin{bmatrix} 1 \\ 2 \\ 3 \end{bmatrix} \right\}$$

spans range L. That is, range L is the subspace of R^3 spanned by the columns of the matrix defining L.

The first two vectors in this set are linearly independent, since they are not constant multiples of each other. The third vector is the sum of the first two. Therefore, the first two vectors form a basis for range L, and $\dim(\text{range } L) = 2$.

(c) To find $\ker L$, we wish to find all $\mathbf{v}$ in R^3 so that $L(\mathbf{v}) = \mathbf{0}_{R^3}$. Solving the resulting homogeneous system, we find (verify) that $a_1 = -a_3$ and $a_2 = -a_3$. Thus $\ker L$ consists of all vectors of the form

$$\begin{bmatrix} -r \\ -r \\ r \end{bmatrix} = r \begin{bmatrix} -1 \\ -1 \\ 1 \end{bmatrix},$$

where r is any real number. Moreover, $\dim(\ker L) = 1$.

(d) Since $\ker L \neq \{\mathbf{0}_{R^3}\}$, it follows from Theorem 10.5 that L is not one-to-one. ■

> The problem of finding a basis for $\ker L$ always reduces to the problem of finding a basis for the solution space of a homogeneous system; this latter problem has been solved in Example 1 of Section 6.5.
>
> If range L is a subspace of R^m, then a basis for range L can be obtained by the method discussed in the alternative constructive proof of Theorem 6.6 or by the procedure given in Section 6.6. Both approaches are illustrated in the next example.

EXAMPLE 12 Let $L: R^4 \to R^3$ be defined by

$$L(a_1, a_2, a_3, a_4) = (a_1 + a_2, a_3 + a_4, a_1 + a_3).$$

Find a basis for range L.

Solution We have

$$L(a_1, a_2, a_3, a_4) = a_1(1, 0, 1) + a_2(1, 0, 0) + a_3(0, 1, 1) + a_4(0, 1, 0).$$

Thus

$$S = \{(1, 0, 1), (1, 0, 0), (0, 1, 1), (0, 1, 0)\}$$

spans range L. To find a subset of S that is a basis for range L, we proceed as in Theorem 6.6 by first writing

$$a_1(1, 0, 1) + a_2(1, 0, 0) + a_3(0, 1, 1) + a_4(0, 1, 0) = (0, 0, 0).$$

The reduced row echelon form of the augmented matrix of this homogeneous system is (verify)

$$\left[\begin{array}{cccc:c} 1 & 0 & 0 & -1 & 0 \\ 0 & 1 & 0 & 1 & 0 \\ 0 & 0 & 1 & 1 & 0 \end{array}\right].$$

Since the leading 1s appear in columns 1, 2, and 3, we conclude that the first three vectors in S form a basis for range L. Thus

$$\{(1, 0, 1), (1, 0, 0), (0, 1, 1)\}$$

is a basis for range L.

Alternatively, we may proceed as in Section 6.6 to form the matrix whose rows are the given vectors

$$\begin{bmatrix} 1 & 0 & 1 \\ 1 & 0 & 0 \\ 0 & 1 & 1 \\ 0 & 1 & 0 \end{bmatrix}.$$

Transforming this matrix to reduced row echelon form, we obtain (verify)

$$\begin{bmatrix} 1 & 0 & 0 \\ 0 & 1 & 0 \\ 0 & 0 & 1 \\ 0 & 0 & 0 \end{bmatrix}.$$

Hence $\{(1, 0, 0), (0, 1, 0), (0, 0, 1)\}$ is a basis for range L. ■

To determine if a linear transformation is one-to-one or onto, we must solve a linear system. This is one further demonstration of the frequency with which linear systems must be solved to answer many questions in linear algebra. Finally, from Example 11, where $\dim(\ker L) = 1$, $\dim(\text{range } L) = 2$, and $\dim(\text{domain } L) = 3$, we saw that

$$\dim(\ker L) + \dim(\text{range } L) = \dim(\text{domain } L).$$

This very important result is always true, and we now prove it in the following theorem.

THEOREM 10.7 *If $L: V \to W$ is a linear transformation of an n-dimensional vector space V into a vector space W, then*

$$\dim(\ker L) + \dim(\text{range } L) = \dim V. \tag{1}$$

Proof Let $k = \dim(\ker L)$. If $k = n$, then $\ker L = V$ (Exercise T.7, Section 6.4), which implies that $L(\mathbf{v}) = \mathbf{0}_W$ for every $\mathbf{v}$ in V. Hence range $L = \{\mathbf{0}_W\}$, $\dim(\text{range } L) = 0$, and the conclusion holds. Next, suppose that $1 \leq k < n$. We shall prove that $\dim(\text{range } L) = n - k$. Let $\{\mathbf{v}_1, \mathbf{v}_2, \ldots, \mathbf{v}_k\}$ be a basis for $\ker L$. By Theorem 6.8 we can extend this basis to a basis

$$S = \{\mathbf{v}_1, \mathbf{v}_2, \ldots, \mathbf{v}_k, \mathbf{v}_{k+1}, \ldots, \mathbf{v}_n\}$$

for V. We prove that the set

$$T = \{L(\mathbf{v}_{k+1}), L(\mathbf{v}_{k+2}), \ldots, L(\mathbf{v}_n)\}$$

is a basis for range L.

First, we show that T spans range L. Let $\mathbf{w}$ be any vector in range L. Then $\mathbf{w} = L(\mathbf{v})$ for some $\mathbf{v}$ in V. Since S is a basis for V, we can find a unique set of real numbers $a_1, a_2, \ldots, a_n$ such that

$$\mathbf{v} = a_1\mathbf{v}_1 + a_2\mathbf{v}_2 + \cdots + a_n\mathbf{v}_n.$$

Then

$$\begin{aligned}\mathbf{w} &= L(\mathbf{v}) \\ &= L(a_1\mathbf{v}_1 + a_2\mathbf{v}_2 + \cdots + a_k\mathbf{v}_k + a_{k+1}\mathbf{v}_{k+1} + \cdots + a_n\mathbf{v}_n) \\ &= a_1L(\mathbf{v}_1) + a_2L(\mathbf{v}_2) + \cdots + a_kL(\mathbf{v}_k) + a_{k+1}L(\mathbf{v}_{k+1}) + \cdots + a_nL(\mathbf{v}_n) \\ &= a_{k+1}L(\mathbf{v}_{k+1}) + \cdots + a_nL(\mathbf{v}_n)\end{aligned}$$

since $L(\mathbf{v}_1) = L(\mathbf{v}_2) = \cdots = L(\mathbf{v}_k) = \mathbf{0}$, because $\mathbf{v}_1, \mathbf{v}_2, \ldots, \mathbf{v}_k$ are in $\ker L$. Hence T spans range L.

Now we show that T is linearly independent. Suppose that

$$a_{k+1}L(\mathbf{v}_{k+1}) + a_{k+2}L(\mathbf{v}_{k+2}) + \cdots + a_nL(\mathbf{v}_n) = \mathbf{0}_W.$$

Then by (b) of Theorem 10.2,

$$L(a_{k+1}\mathbf{v}_{k+1} + a_{k+2}\mathbf{v}_{k+2} + \cdots + a_n\mathbf{v}_n) = \mathbf{0}_W.$$

Hence the vector $a_{k+1}\mathbf{v}_{k+1} + a_{k+2}\mathbf{v}_{k+2} + \cdots + a_n\mathbf{v}_n$ is in $\ker L$, and we can write

$$a_{k+1}\mathbf{v}_{k+1} + a_{k+2}\mathbf{v}_{k+2} + \cdots + a_n\mathbf{v}_n = b_1\mathbf{v}_1 + b_2\mathbf{v}_2 + \cdots + b_k\mathbf{v}_k,$$

where $b_1, b_2, \ldots, b_k$ are uniquely determined real numbers. We then have

$$b_1\mathbf{v}_1 + b_2\mathbf{v}_2 + \cdots + b_k\mathbf{v}_k - a_{k+1}\mathbf{v}_{k+1} - a_{k+2}\mathbf{v}_{k+2} - \cdots - a_n\mathbf{v}_n = \mathbf{0}_V.$$

Since S is linearly independent, we find that

$$b_1 = b_2 = \cdots = b_k = a_{k+1} = a_{k+2} = \cdots = a_n = 0.$$

Hence T is linearly independent and forms a basis for range L.

If $k = 0$, then $\ker L$ has no basis; we let $\{\mathbf{v}_1, \mathbf{v}_2, \ldots, \mathbf{v}_n\}$ be a basis for V. The proof now proceeds as previously. ■

The dimension of $\ker L$ is also called the **nullity** of L, and the dimension of range L is called the **rank** of L. With this terminology the conclusion of Theorem 10.7 is very similar to that of Theorem 6.12. This is not a coincidence, since in the next section we shall show how to attach a unique $m \times n$ matrix to L, whose properties reflect those of L.

The following example illustrates Theorem 10.7 graphically.

EXAMPLE 13 Let $L\colon R^3 \to R^3$ be the linear transformation defined by

$$L\left(\begin{bmatrix} a_1 \\ a_2 \\ a_3 \end{bmatrix}\right) = \begin{bmatrix} a_1 + a_3 \\ a_1 + a_2 \\ a_2 - a_3 \end{bmatrix}.$$

A vector $\begin{bmatrix} a_1 \\ a_2 \\ a_3 \end{bmatrix}$ is in $\ker L$ if

$$L\left(\begin{bmatrix} a_1 \\ a_2 \\ a_3 \end{bmatrix}\right) = \begin{bmatrix} 0 \\ 0 \\ 0 \end{bmatrix}.$$

We must then find a basis for the solution space of the homogeneous system

$$\begin{aligned} a_1 + \phantom{a_2 - {}} a_3 &= 0 \\ a_1 + a_2 \phantom{{} - a_3} &= 0 \\ a_2 - a_3 &= 0. \end{aligned}$$

We find (verify) that a basis for $\ker L$ is $\left\{\begin{bmatrix} -1 \\ 1 \\ 1 \end{bmatrix}\right\}$, so $\dim(\ker L) = 1$, and $\ker L$ is a line through the origin.

Next, every vector in range L is of the form $\begin{bmatrix} a_1 + a_3 \\ a_1 + a_2 \\ a_2 - a_3 \end{bmatrix}$, which can be written as

$$a_1 \begin{bmatrix} 1 \\ 1 \\ 0 \end{bmatrix} + a_2 \begin{bmatrix} 0 \\ 1 \\ 1 \end{bmatrix} + a_3 \begin{bmatrix} 1 \\ 0 \\ -1 \end{bmatrix}.$$

Then a basis for range L is

$$\left\{\begin{bmatrix} 1 \\ 1 \\ 0 \end{bmatrix}, \begin{bmatrix} 0 \\ 1 \\ 1 \end{bmatrix}\right\}$$

(explain), so $\dim(\text{range } L) = 2$ and range L is a plane passing through the origin. These results are illustrated in Figure 10.2. Moreover,

$$\dim R^3 = 3 = \dim(\ker L) + \dim(\text{range } L) = 1 + 2,$$

verifying Theorem 10.7. ■

We have seen that a linear transformation may be one-to-one and not onto or onto and not one-to-one. However, the following corollary shows that each of these properties implies the other if the vector spaces V and W have the same dimensions.

COROLLARY 10.3 *Let $L\colon V \to W$ be a linear transformation and let* $\dim V = \dim W$.

(a) *If L is one-to-one, then it is onto.*

(b) *If L is onto, then it is one-to-one.*

Proof Exercise T.2. ■

Figure 10.2 ▶

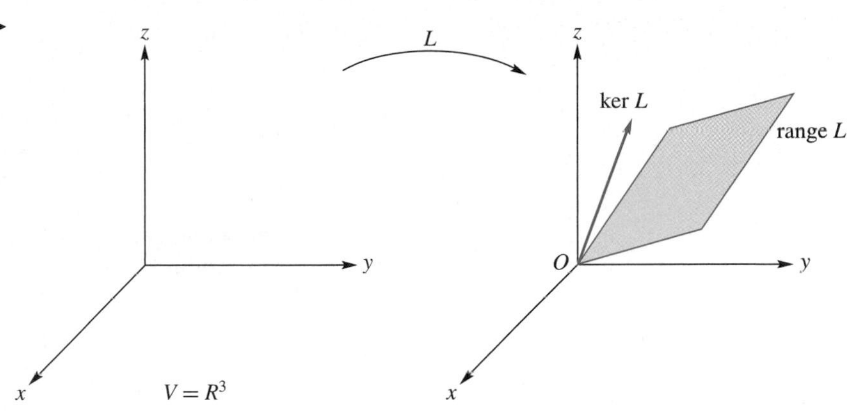

EXAMPLE 14 Let $L: P_2 \to P_2$ be the linear transformation defined by

$$L(at^2 + bt + c) = (a + 2b)t + (b + c).$$

(a) Is $-4t^2 + 2t - 2$ in ker L?
(b) Is $t^2 + 2t + 1$ in range L?
(c) Find a basis for ker L.
(d) Is L one-to-one?
(e) Find a basis for range L.
(f) Is L onto?
(g) Verify Theorem 10.7.

Solution (a) Since

$$L(-4t^2 + 2t - 2) = (-4 + 2 \cdot 2)t + (-2 + 2) = 0,$$

we conclude that $-4t^2 + 2t - 2$ is in ker L.

(b) The vector $t^2 + 2t + 1$ is in range L if we can find a vector $at^2 + bt + c$ in P_2 such that

$$L(at^2 + bt + c) = t^2 + 2t + 1.$$

Since $L(at^2 + bt + c) = (a + 2b)t + (b + c)$, we have

$$(a + 2b)t + (b + c) = t^2 + 2t + 1.$$

The left side of this equation can also be written as $0t^2+(a+2b)t+(b+c)$. Thus

$$0t^2 + (a + 2b)t + (b + c) = t^2 + 2t + 1.$$

We must then have

$$\begin{aligned} 0 &= 1 \\ a + 2b &= 2 \\ b + c &= 1. \end{aligned}$$

Since this linear system has no solution, the given vector is not in range L.

(c) The vector $at^2 + bt + c$ is in ker L if

$$L(at^2 + bt + c) = \mathbf{0},$$

that is, if

$$(a + 2b)t + (b + c) = 0.$$

Then

$$\begin{aligned} a + 2b \quad &= 0 \\ b + c &= 0. \end{aligned}$$

Transforming the augmented matrix of this linear system to reduced row echelon form, we find (verify) that a basis for the solution space is

$$\left\{ \begin{bmatrix} 2 \\ -1 \\ 1 \end{bmatrix} \right\},$$

so a basis for $\ker L$ is $\{2t^2 - t + 1\}$.

(d) Since $\ker L$ does not consist only of the zero vector, L is not one-to-one.

(e) Every vector in range L has the form

$$(a + 2b)t + (b + c),$$

so the vectors t and 1 span range L. Since these vectors are also linearly independent, they form a basis for range L.

(f) The dimension of P_2 is 3, while range L is a subspace of P_2 of dimension 2, so range $L \neq P_2$. Hence, L is not onto.

(g) From (c), $\dim(\ker L) = 1$, and from (e), $\dim(\text{range } L) = 2$, so

$$3 = \dim P_2 = \dim(\ker L) + \dim(\text{range } L). \quad \blacksquare$$

If $L: R^n \to R^n$ is a linear transformation defined by $L(\mathbf{x}) = A\mathbf{x}$, where A is an $n \times n$ matrix, then using Theorem 10.7, Equation (1), and Corollary 6.2, we can show (Exercise T.4) that L is one-to-one if and only if $\det(A) \neq 0$.

We now make one final remark for a linear system $A\mathbf{x} = \mathbf{b}$, where A is $n \times n$. We again consider the linear transformation $L: R^n \to R^n$ defined by $L(\mathbf{x}) = A\mathbf{x}$, for $\mathbf{x}$ in R^n. If A is a nonsingular matrix, then

$$\dim(\text{range } L) = \text{rank } A = n,$$

so

$$\dim(\ker L) = 0.$$

Thus L is one-to-one and hence onto. This means that the given linear system has a unique solution (of course, we already knew this result from other considerations). Now assume that A is singular. Then $\text{rank } A < n$. This means that $\dim(\ker L) = n - \text{rank } A > 0$, so L is not one-to-one and not onto. Therefore, there exists a vector $\mathbf{b}$ in R^n, for which the system $A\mathbf{x} = \mathbf{b}$ has no solution. Moreover, since A is singular, $A\mathbf{x} = \mathbf{0}$ has a nontrivial solution $\mathbf{x}_0$. If $A\mathbf{x} = \mathbf{b}$ has a solution $\mathbf{y}$, then $\mathbf{x}_0 + \mathbf{y}$ is a solution to $A\mathbf{x} = \mathbf{b}$ (verify). Thus, for A singular, if a solution to $A\mathbf{x} = \mathbf{b}$ exists, then it is not unique.

Key Terms

One-to-one
Range
Onto
Nullity
Rank

10.2 Exercises

1. Let $L\colon R^2 \to R^2$ be the linear transformation defined by $L(a_1, a_2) = (a_1, 0)$.

(a) Is $(0, 2)$ in $\ker L$? (b) Is $(2, 2)$ in $\ker L$?

(c) Is $(3, 0)$ in range L? (d) Is $(3, 2)$ in range L?

(e) Find $\ker L$. (f) Find range L.

2. Let $L\colon R^2 \to R^2$ be the linear transformation defined by

$$L\left(\begin{bmatrix} a_1 \\ a_2 \end{bmatrix}\right) = \begin{bmatrix} 1 & 2 \\ 2 & 4 \end{bmatrix}\begin{bmatrix} a_1 \\ a_2 \end{bmatrix}.$$

(a) Is $\begin{bmatrix} 1 \\ 2 \end{bmatrix}$ in $\ker L$? (b) Is $\begin{bmatrix} 2 \\ -1 \end{bmatrix}$ in $\ker L$?

(c) Is $\begin{bmatrix} 3 \\ 6 \end{bmatrix}$ in range L? (d) Is $\begin{bmatrix} 2 \\ 3 \end{bmatrix}$ in range L?

(e) Find $\ker L$.

(f) Find a set of vectors spanning range L.

3. Let $L\colon R^2 \to R^3$ be defined by

$$L(x, y) = (x, x + y, y).$$

(a) Find $\ker L$.

(b) Is L one-to-one?

(c) Is L onto?

4. Let $L\colon R^4 \to R^3$ be defined by

$$L(x, y, z, w) = (x + y, z + w, x + z).$$

(a) Find a basis for $\ker L$.

(b) Find a basis for range L.

(c) Verify Theorem 10.7.

5. Let $L\colon R^5 \to R^4$ be defined by

$$L\left(\begin{bmatrix} x_1 \\ x_2 \\ x_3 \\ x_4 \\ x_5 \end{bmatrix}\right) = \begin{bmatrix} 1 & 0 & -1 & 3 & -1 \\ 1 & 0 & 0 & 2 & -1 \\ 2 & 0 & -1 & 5 & -1 \\ 0 & 0 & -1 & 1 & 0 \end{bmatrix}\begin{bmatrix} x_1 \\ x_2 \\ x_3 \\ x_4 \\ x_5 \end{bmatrix}.$$

(a) Find a basis for $\ker L$.

(b) Find a basis for range L.

(c) Verify Theorem 10.7.

6. Let $L\colon R^3 \to R^3$ be defined by

$$L\left(\begin{bmatrix} x \\ y \\ z \end{bmatrix}\right) = \begin{bmatrix} 4 & 2 & 2 \\ 2 & 3 & -1 \\ -1 & 1 & -2 \end{bmatrix}\begin{bmatrix} x \\ y \\ z \end{bmatrix}.$$

(a) Is L one-to-one?

(b) Find the dimension of range L.

7. Let $L\colon R^4 \to R^3$ be defined by

$$L\left(\begin{bmatrix} x \\ y \\ z \\ w \end{bmatrix}\right) = \begin{bmatrix} x + y \\ y - z \\ z - w \end{bmatrix}.$$

(a) Is L onto?

(b) Find the dimension of $\ker L$.

(c) Verify Theorem 10.7.

8. Let $L\colon R^3 \to R^3$ be defined by

$$L(x, y, z) = (x - y, x + 2y, z).$$

(a) Find a basis for $\ker L$.

(b) Find a basis for range L.

(c) Verify Theorem 10.7.

9. Verify Theorem 10.7 for the following linear transformations.

(a) $L(x, y) = (x + y, y)$.

(b) $L\left(\begin{bmatrix} x \\ y \\ z \end{bmatrix}\right) = \begin{bmatrix} 4 & -1 & -1 \\ 2 & 2 & 3 \\ 2 & -3 & -4 \end{bmatrix}\begin{bmatrix} x \\ y \\ z \end{bmatrix}$.

(c) $L(x, y, z) = (x + y - z, x + y, y + z)$.

10. Let $L\colon R^4 \to R^4$ be defined by

$$L\left(\begin{bmatrix} x \\ y \\ z \\ w \end{bmatrix}\right) = \begin{bmatrix} 1 & 2 & 1 & 3 \\ 2 & 1 & -1 & 2 \\ 1 & 0 & 0 & -1 \\ 4 & 1 & -1 & 0 \end{bmatrix}\begin{bmatrix} x \\ y \\ z \\ w \end{bmatrix}.$$

(a) Find a basis for $\ker L$.

(b) Find a basis for range L.

(c) Verify Theorem 10.7.

11. Let $L\colon P_2 \to P_2$ be the linear transformation defined by $L(at^2 + bt + c) = (a + c)t^2 + (b + c)t$.

(a) Is $t^2 - t - 1$ in $\ker L$?

(b) Is $t^2 + t - 1$ in $\ker L$?

(c) Is $2t^2 - t$ in range L?

(d) Is $t^2 - t + 2$ in range L?

(e) Find a basis for $\ker L$.

(f) Find a basis for range L.

12. Let $L\colon P_3 \to P_3$ be the linear transformation defined by

$$L(at^3 + bt^2 + ct + d) = (a - b)t^3 + (c - d)t.$$

(a) Is $t^3 + t^2 + t - 1$ in $\ker L$?

(b) Is $t^3 - t^2 + t - 1$ in $\ker L$?

(c) Is $3t^3 + t$ in range L?

(d) Is $3t^3 - t^2$ in range L?

(e) Find a basis for $\ker L$.

(f) Find a basis for range L.

13. Let $L\colon M_{22} \to M_{22}$ be the linear transformation defined by

$$L\left(\begin{bmatrix} a & b \\ c & d \end{bmatrix}\right) = \begin{bmatrix} a + b & b + c \\ a + d & b + d \end{bmatrix}.$$

(a) Find a basis for $\ker L$.

(b) Find a basis for range L.

14. Let $L: P_2 \to R^2$ be the linear transformation defined by $L(at^2 + bt + c) = (a, b)$.

(a) Find a basis for $\ker L$.

(b) Find a basis for range L.

15. Let $L: M_{22} \to M_{22}$ be the linear transformation defined by

$$L(\mathbf{v}) = \begin{bmatrix} 1 & 2 \\ 1 & 1 \end{bmatrix} \mathbf{v} - \mathbf{v} \begin{bmatrix} 1 & 2 \\ 1 & 1 \end{bmatrix}.$$

(a) Find a basis for $\ker L$.

(b) Find a basis for range L.

16. Let $L: M_{22} \to M_{22}$ be the linear transformation defined by $L(A) = A^T$.

(a) Find a basis for $\ker L$.

(b) Find a basis for range L.

17. ***(Calculus Required)*** Let $L: P_2 \to P_1$ be the linear transformation defined by

$$L[p(t)] = p'(t).$$

(a) Find a basis for $\ker L$.

(b) Find a basis for range L.

18. ***(Calculus Required)*** Let $L: P_2 \to R^1$ be the linear transformation defined by

$$L[p(t)] = \int_0^1 p(t)\,dt.$$

(a) Find a basis for $\ker L$.

(b) Find a basis for range L.

19. Let $L: R^4 \to R^6$ be a linear transformation.

(a) If $\dim(\ker L) = 2$, what is $\dim(\text{range } L)$?

(b) If $\dim(\text{range } L) = 3$, what is $\dim(\ker L)$?

20. Let $L: V \to R^5$ be a linear transformation.

(a) If L is onto and $\dim(\ker L) = 2$, what is $\dim V$?

(b) If L is one-to-one and onto, what is $\dim V$?

Theoretical Exercises

T.1. Prove Corollary 10.2.

T.2. Prove Corollary 10.3.

T.3. Let A be an $m \times n$ matrix and let $L: R^n \to R^m$ be defined by $L(\mathbf{x}) = A\mathbf{x}$ for $\mathbf{x}$ in R^n. Show that the column space of A is the range of L.

T.4. Let $L: R^n \to R^n$ be a linear transformation defined by $L(\mathbf{x}) = A\mathbf{x}$, where A is an $n \times n$ matrix. Show that L is one-to-one if and only if $\det(A) \neq 0$. [*Hint*: Use Theorem 10.7, Equation (1), and Corollary 6.2.]

T.5. Let $L: V \to W$ be a linear transformation. If $\{\mathbf{v}_1, \mathbf{v}_2, \dots, \mathbf{v}_k\}$ spans V, show that $\{L(\mathbf{v}_1), L(\mathbf{v}_2), \dots, L(\mathbf{v}_k)\}$ spans range L.

T.6. Let $L: V \to W$ be a linear transformation.

(a) Show that $\dim(\text{range } L) \leq \dim V$.

(b) Show that if L is onto, then $\dim W \leq \dim V$.

T.7. Let $L: V \to W$ be a linear transformation, and let $S = \{\mathbf{v}_1, \mathbf{v}_2, \dots, \mathbf{v}_n\}$ be a set of vectors in V. Show that if $T = \{L(\mathbf{v}_1), L(\mathbf{v}_2), \dots, L(\mathbf{v}_n)\}$ is linearly independent, then so is S. (*Hint*: Assume that S is linearly dependent. What can we say about T?)

T.8. Let $L: V \to W$ be a linear transformation. Show that L is one-to-one if and only if $\dim(\text{range } L) = \dim V$.

T.9. Let $L: V \to W$ be a linear transformation. Show that L is one-to-one if and only if the image of every linearly independent set of vectors in V is a linearly independent set of vectors in W.

T.10. Let $L: V \to W$ be a linear transformation, and let $\dim V = \dim W$. Show that L is one-to-one if and only if the image under L of a basis for V is a basis for W.

T.11. Let V be an n-dimensional vector space and $S = \{\mathbf{v}_1, \mathbf{v}_2, \dots, \mathbf{v}_n\}$ be a basis for V. Let $L: V \to R^n$ be defined by $L(\mathbf{v}) = \left[\mathbf{v}\right]_S$. Show that

(a) L is a linear transformation.

(b) L is one-to-one.

(c) L is onto.

MATLAB Exercises

In order to use MATLAB *in this section, you should first read Section 12.8. Find a basis for the kernel and range of the linear transformation* $L(\mathbf{x}) = A\mathbf{x}$ *for each of the following matrices* A.

ML.1. $A = \begin{bmatrix} 1 & 2 & 5 & 5 \\ -2 & -3 & -8 & -7 \end{bmatrix}$

ML.2. $A = \begin{bmatrix} -3 & 2 & -7 \\ 2 & -1 & 4 \\ 2 & -2 & 6 \end{bmatrix}$

ML.3. $A = \begin{bmatrix} 3 & 3 & -3 & 1 & 11 \\ -4 & -4 & 7 & -2 & -19 \\ 2 & 2 & -3 & 1 & 9 \end{bmatrix}$

10.3 THE MATRIX OF A LINEAR TRANSFORMATION

We have shown in Theorem 4.8 that if $L\colon R^n \to R^m$ is a linear transformation, then there is a unique $m \times n$ matrix A so that $L(\mathbf{x}) = A\mathbf{x}$ for $\mathbf{x}$ in R^n. In this section we generalize this result for a linear transformation $L\colon V \to W$ of a finite-dimensional vector space V into a finite-dimensional vector space W.

THE MATRIX OF A LINEAR TRANSFORMATION

THEOREM 10.8 *Let $L\colon V \to W$ be a linear transformation of an n-dimensional vector space V into an m-dimensional vector space W ($n \neq 0$ and $m \neq 0$) and let $S = \{\mathbf{v}_1, \mathbf{v}_2, \dots, \mathbf{v}_n\}$ and $T = \{\mathbf{w}_1, \mathbf{w}_2, \dots, \mathbf{w}_m\}$ be bases for V and W, respectively. Then the $m \times n$ matrix A, whose jth column is the coordinate vector $\left[L(\mathbf{v}_j)\right]_T$ of $L(\mathbf{v}_j)$ with respect to T, is associated with L and has the following property: If $\mathbf{x}$ is in V, then*

$$\left[L(\mathbf{x})\right]_T = A\left[\mathbf{x}\right]_S, \tag{1}$$

where $\left[\mathbf{x}\right]_S$ and $\left[L(\mathbf{x})\right]_T$ are the coordinate vectors of $\mathbf{x}$ and $L(\mathbf{x})$ with respect to the respective bases S and T. Moreover, A is the only matrix with this property.

Proof The proof is a constructive one; that is, we show how to construct the matrix A. It is more complicated than the proof of Theorem 4.8. Consider the vector $\mathbf{v}_j$ in V for $j = 1, 2, \dots, n$. Then $L(\mathbf{v}_j)$ is a vector in W, and since T is a basis for W, we can express this vector as a linear combination of the vectors in T in a unique manner. Thus

$$L(\mathbf{v}_j) = c_{1j}\mathbf{w}_1 + c_{2j}\mathbf{w}_2 + \cdots + c_{mj}\mathbf{w}_m \quad (1 \le j \le n). \tag{2}$$

This means that the coordinate vector of $L(\mathbf{v}_j)$ with respect to T is

$$\left[L(\mathbf{v}_j)\right]_T = \begin{bmatrix} c_{1j} \\ c_{2j} \\ \vdots \\ c_{mj} \end{bmatrix}.$$

We now define the $m \times n$ matrix A by choosing $\left[L(\mathbf{v}_j)\right]_T$ as the jth column of A and show that this matrix satisfies the properties stated in the theorem. We leave the rest of the proof as Exercise T.1 and amply illustrate it in the following examples. ■

DEFINITION The matrix A of Theorem 10.8 is called the **matrix representing L with respect to the bases S and T**, or the **matrix of L with respect to S and T**.

We now summarize the procedure given in Theorem 10.8.

> The procedure for computing the matrix of a linear transformation $L\colon V \to W$ with respect to the bases $S = \{\mathbf{v}_1, \mathbf{v}_2, \dots, \mathbf{v}_n\}$ and $T = \{\mathbf{w}_1, \mathbf{w}_2, \dots, \mathbf{w}_m\}$ for V and W, respectively, is as follows.
>
> ***Step 1.*** Compute $L(\mathbf{v}_j)$ for $j = 1, 2, \dots, n$.
>
> ***Step 2.*** Find the coordinate vector $\left[L(\mathbf{v}_j)\right]_T$ of $L(\mathbf{v}_j)$ with respect to the basis T. This means that we have to express $L(\mathbf{v}_j)$ as a linear combination of the vectors in T [see Equation (2)].
>
> ***Step 3.*** The matrix A of L with respect to S and T is formed by choosing $\left[L(\mathbf{v}_j)\right]_T$ as the jth column of A.

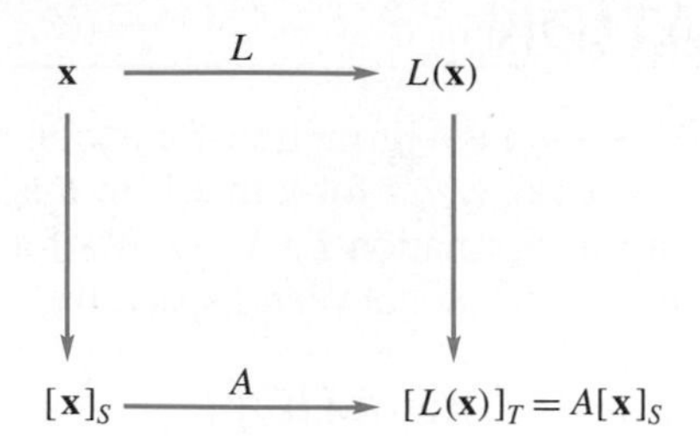

Figure 10.3 ▲

Figure 10.3 gives a graphical interpretation of Equation (1), that is, of Theorem 10.8. The top horizontal arrow represents the linear transformation L from the n-dimensional vector space V into the m-dimensional vector space W and takes the vector $\mathbf{x}$ in V to the vector $L(\mathbf{x})$ in W. The bottom horizontal line represents the matrix A. Then $\left[L(\mathbf{x})\right]_T$, a coordinate vector in R^m, is obtained simply by multiplying $\left[\mathbf{x}\right]_S$, a coordinate vector in R^n, by the matrix A. We can thus always work with matrices rather than with linear transformations.

Physicists and others who deal at great length with linear transformations perform most of their computations with the matrices of the linear transformations.

EXAMPLE 1 Let $L\colon R^3 \to R^2$ be defined by

$$L\left(\begin{bmatrix} x \\ y \\ z \end{bmatrix}\right) = \begin{bmatrix} x + y \\ y - z \end{bmatrix}. \tag{3}$$

Let

$$S = \{\mathbf{v}_1, \mathbf{v}_2, \mathbf{v}_3\} \quad \text{and} \quad T = \{\mathbf{w}_1, \mathbf{w}_2\}$$

be bases for R^3 and R^2, respectively, where

$$\mathbf{v}_1 = \begin{bmatrix} 1 \\ 0 \\ 0 \end{bmatrix}, \quad \mathbf{v}_2 = \begin{bmatrix} 0 \\ 1 \\ 0 \end{bmatrix}, \quad \mathbf{v}_3 = \begin{bmatrix} 0 \\ 0 \\ 1 \end{bmatrix},$$

$$\mathbf{w}_1 = \begin{bmatrix} 1 \\ 0 \end{bmatrix}, \quad \text{and} \quad \mathbf{w}_2 = \begin{bmatrix} 0 \\ 1 \end{bmatrix}.$$

We now find the matrix A of L with respect to S and T. We have

$$L(\mathbf{v}_1) = \begin{bmatrix} 1 + 0 \\ 0 - 0 \end{bmatrix} = \begin{bmatrix} 1 \\ 0 \end{bmatrix},$$

$$L(\mathbf{v}_2) = \begin{bmatrix} 0 + 1 \\ 1 - 0 \end{bmatrix} = \begin{bmatrix} 1 \\ 1 \end{bmatrix},$$

$$L(\mathbf{v}_3) = \begin{bmatrix} 0 + 0 \\ 0 - 1 \end{bmatrix} = \begin{bmatrix} 0 \\ -1 \end{bmatrix}.$$

Since T is the natural basis for R^2, the coordinate vectors of $L(\mathbf{v}_1)$, $L(\mathbf{v}_2)$, and $L(\mathbf{v}_3)$ with respect to T are the same as $L(\mathbf{v}_1)$, $L(\mathbf{v}_2)$, and $L(\mathbf{v}_3)$, respectively. That is,

$$\left[L(\mathbf{v}_1)\right]_T = \begin{bmatrix} 1 \\ 0 \end{bmatrix}, \quad \left[L(\mathbf{v}_2)\right]_T = \begin{bmatrix} 1 \\ 1 \end{bmatrix}, \quad \left[L(\mathbf{v}_3)\right]_T = \begin{bmatrix} 0 \\ -1 \end{bmatrix}.$$

Hence

$$A = \begin{bmatrix} 1 & 1 & 0 \\ 0 & 1 & -1 \end{bmatrix}.$$

■

EXAMPLE 2 Let $L: R^3 \to R^2$ be defined as in Example 1. Now let

$$S = \{\mathbf{v}_1, \mathbf{v}_2, \mathbf{v}_3\} \quad \text{and} \quad T = \{\mathbf{w}_1, \mathbf{w}_2\}$$

be bases for R^3 and R^2, respectively, where

$$\mathbf{v}_1 = \begin{bmatrix} 1 \\ 0 \\ 1 \end{bmatrix}, \quad \mathbf{v}_2 = \begin{bmatrix} 0 \\ 1 \\ 1 \end{bmatrix}, \quad \mathbf{v}_3 = \begin{bmatrix} 1 \\ 1 \\ 1 \end{bmatrix},$$

$$\mathbf{w}_1 = \begin{bmatrix} 1 \\ 2 \end{bmatrix}, \quad \text{and} \quad \mathbf{w}_2 = \begin{bmatrix} -1 \\ 1 \end{bmatrix}.$$

Find the matrix of L with respect to S and T.

Solution We have

$$L(\mathbf{v}_1) = \begin{bmatrix} 1 \\ -1 \end{bmatrix}, \quad L(\mathbf{v}_2) = \begin{bmatrix} 1 \\ 0 \end{bmatrix}, \quad L(\mathbf{v}_3) = \begin{bmatrix} 2 \\ 0 \end{bmatrix}.$$

To find the coordinate vectors $\left[L(\mathbf{v}_1)\right]_T$, $\left[L(\mathbf{v}_2)\right]_T$, and $\left[L(\mathbf{v}_3)\right]_T$, we write

$$L(\mathbf{v}_1) = \begin{bmatrix} 1 \\ -1 \end{bmatrix} = a_1\mathbf{w}_1 + a_2\mathbf{w}_2 = a_1 \begin{bmatrix} 1 \\ 2 \end{bmatrix} + a_2 \begin{bmatrix} -1 \\ 1 \end{bmatrix},$$

$$L(\mathbf{v}_2) = \begin{bmatrix} 1 \\ 0 \end{bmatrix} = b_1\mathbf{w}_1 + b_2\mathbf{w}_2 = b_1 \begin{bmatrix} 1 \\ 2 \end{bmatrix} + b_2 \begin{bmatrix} -1 \\ 1 \end{bmatrix},$$

$$L(\mathbf{v}_3) = \begin{bmatrix} 2 \\ 0 \end{bmatrix} = c_1\mathbf{w}_1 + c_2\mathbf{w}_2 = c_1 \begin{bmatrix} 1 \\ 2 \end{bmatrix} + c_2 \begin{bmatrix} -1 \\ 1 \end{bmatrix}.$$

That is, we must solve three linear systems, each of two equations in two unknowns. Since their coefficient matrix is the same, we solve them all at once, as in Example 4 of Section 6.7. Thus we form the matrix

$$\left[\begin{array}{cc:c:c:c} 1 & -1 & 1 & 1 & 2 \\ 2 & 1 & -1 & 0 & 0 \end{array}\right],$$

which we transform to reduced row echelon form, obtaining (verify)

$$\left[\begin{array}{cc:c:c:c} 1 & 0 & 0 & \frac{1}{3} & \frac{2}{3} \\ 0 & 1 & -1 & -\frac{2}{3} & -\frac{4}{3} \end{array}\right].$$

Hence the matrix A of L with respect to S and T is

$$A = \begin{bmatrix} 0 & \frac{1}{3} & \frac{2}{3} \\ -1 & -\frac{2}{3} & -\frac{4}{3} \end{bmatrix}.$$

Equation (1) is then

$$\left[L(\mathbf{x})\right]_T = \begin{bmatrix} 0 & \frac{1}{3} & \frac{2}{3} \\ -1 & -\frac{2}{3} & -\frac{4}{3} \end{bmatrix} \left[\mathbf{x}\right]_S. \tag{4}$$

To illustrate Equation (4), let

$$\mathbf{x} = \begin{bmatrix} 1 \\ 6 \\ 3 \end{bmatrix}.$$

Then from the definition of L, Equation (3), we have

$$L(\mathbf{x}) = \begin{bmatrix} 1+6 \\ 6-3 \end{bmatrix} = \begin{bmatrix} 7 \\ 3 \end{bmatrix}.$$

Now (verify)

$$\left[\mathbf{x}\right]_S = \begin{bmatrix} -3 \\ 2 \\ 4 \end{bmatrix}.$$

Then from (4),

$$\left[L(\mathbf{x})\right]_T = A\left[\mathbf{x}\right]_S = \begin{bmatrix} \frac{10}{3} \\ -\frac{11}{3} \end{bmatrix}.$$

Hence

$$L(\mathbf{x}) = \frac{10}{3}\begin{bmatrix} 1 \\ 2 \end{bmatrix} - \frac{11}{3}\begin{bmatrix} -1 \\ 1 \end{bmatrix} = \begin{bmatrix} 7 \\ 3 \end{bmatrix},$$

which agrees with the previous value for $L(\mathbf{x})$. ■

Notice that the matrices obtained in Examples 1 and 2 are different even though L is the same in both cases. It can be shown that there is a relationship between these two matrices. The study of this relationship is beyond the scope of this book.

The procedure used in Example 2 can be used to find the matrix representing a linear transformation $L\colon R^n \to R^m$ with respect to given bases S and T for R^n and R^m, respectively.

> The procedure for computing the matrix of a linear transformation $L\colon R^n \to R^m$ with respect to the bases $S = \{\mathbf{v}_1, \mathbf{v}_2, \ldots, \mathbf{v}_n\}$ and $T = \{\mathbf{w}_1, \mathbf{w}_2, \ldots, \mathbf{w}_m\}$ for R^n and R^m, respectively, is as follows.
>
> ***Step 1.*** Compute $L(\mathbf{x}_j)$ for $j = 1, 2, \ldots, n$.
>
> ***Step 2.*** Form the matrix
>
> $$\left[\mathbf{w}_1 \;\; \mathbf{w}_2 \;\; \cdots \;\; \mathbf{w}_m \;\vdots\; L(\mathbf{v}_1) \;\vdots\; L(\mathbf{v}_2) \;\vdots\; \cdots \;\vdots\; L(\mathbf{v}_n)\right],$$
>
> which we transform to reduced row echelon form, obtaining the matrix
>
> $$\left[I_n \;\vdots\; A\right].$$
>
> ***Step 3.*** The matrix A is the matrix representing L with respect to bases S and T.

EXAMPLE 3 Let $L\colon R^3 \to R^2$ be as defined in Example 1. Now let

$$S = \{\mathbf{v}_3, \mathbf{v}_2, \mathbf{v}_1\} \quad \text{and} \quad T = \{\mathbf{w}_1, \mathbf{w}_2\},$$

where $\mathbf{v}_1$, $\mathbf{v}_2$, $\mathbf{v}_3$, $\mathbf{w}_1$, and $\mathbf{w}_2$ are as in Example 2. Then the matrix of L with respect to S and T is

$$A = \begin{bmatrix} \frac{2}{3} & \frac{1}{3} & 0 \\ -\frac{4}{3} & -\frac{2}{3} & -1 \end{bmatrix}.$$

■

Remark Note that if we change the order of the vectors in the bases S and T, then the matrix A of L may change.

EXAMPLE 4 Let $L\colon R^3 \to R^2$ be defined by

$$L\left(\begin{bmatrix} x \\ y \\ z \end{bmatrix}\right) = \begin{bmatrix} 1 & 1 & 1 \\ 1 & 2 & 3 \end{bmatrix}\begin{bmatrix} x \\ y \\ z \end{bmatrix}.$$

Let

$$S = \{\mathbf{v}_1, \mathbf{v}_2, \mathbf{v}_3\} \quad \text{and} \quad T = \{\mathbf{w}_1, \mathbf{w}_2\}$$

be the natural bases for R^3 and R^2, respectively. Find the matrix of L with respect to S and T.

Solution We have

$$L(\mathbf{v}_1) = \begin{bmatrix} 1 & 1 & 1 \\ 1 & 2 & 3 \end{bmatrix}\begin{bmatrix} 1 \\ 0 \\ 0 \end{bmatrix} = \begin{bmatrix} 1 \\ 1 \end{bmatrix} = 1\mathbf{w}_1 + 1\mathbf{w}_2, \quad \text{so} \quad \left[L(\mathbf{v}_1)\right]_T = \begin{bmatrix} 1 \\ 1 \end{bmatrix};$$

$$L(\mathbf{v}_2) = \begin{bmatrix} 1 & 1 & 1 \\ 1 & 2 & 3 \end{bmatrix}\begin{bmatrix} 0 \\ 1 \\ 0 \end{bmatrix} = \begin{bmatrix} 1 \\ 2 \end{bmatrix} = 1\mathbf{w}_1 + 2\mathbf{w}_2, \quad \text{so} \quad \left[L(\mathbf{v}_2)\right]_T = \begin{bmatrix} 1 \\ 2 \end{bmatrix}.$$

Also,

$$\left[L(\mathbf{v}_3)\right]_T = \begin{bmatrix} 1 \\ 3 \end{bmatrix} \quad \text{(verify)}.$$

Then the matrix of L with respect to S and T is

$$A = \begin{bmatrix} 1 & 1 & 1 \\ 1 & 2 & 3 \end{bmatrix}.$$

■

Remark Of course, in Example 4, the reason that A is the same matrix as the one involved in the definition of L is that the natural bases are being used for R^3 and R^2.

EXAMPLE 5 Let $L\colon R^3 \to R^2$ be defined as in Example 4. Now let

$$S = \{\mathbf{v}_1, \mathbf{v}_2, \mathbf{v}_3\} \quad \text{and} \quad T = \{\mathbf{w}_1, \mathbf{w}_2\},$$

where

$$\mathbf{v}_1 = \begin{bmatrix} 1 \\ 1 \\ 0 \end{bmatrix}, \quad \mathbf{v}_2 = \begin{bmatrix} 0 \\ 1 \\ 1 \end{bmatrix}, \quad \mathbf{v}_3 = \begin{bmatrix} 0 \\ 0 \\ 1 \end{bmatrix},$$

$$\mathbf{w}_1 = \begin{bmatrix} 1 \\ 2 \end{bmatrix}, \quad \text{and} \quad \mathbf{w}_2 = \begin{bmatrix} 1 \\ 3 \end{bmatrix}.$$

Find the matrix of L with respect to S and T.

Solution We have

$$L(\mathbf{v}_1) = \begin{bmatrix} 2 \\ 3 \end{bmatrix}, \quad L(\mathbf{v}_2) = \begin{bmatrix} 1 \\ 2 \end{bmatrix}, \quad \text{and} \quad L(\mathbf{v}_3) = \begin{bmatrix} 1 \\ 3 \end{bmatrix}.$$

We now form (verify)

$$\begin{bmatrix} \mathbf{w}_1 & \mathbf{w}_2 \mid L(\mathbf{v}_1) \mid L(\mathbf{v}_2) \mid L(\mathbf{v}_3) \end{bmatrix} = \left[\begin{array}{cc|c|c|c} 1 & 1 & 2 & 1 & 1 \\ 2 & 3 & 3 & 2 & 3 \end{array}\right].$$

Transforming this matrix to reduced row echelon form, we obtain (verify)

$$\left[\begin{array}{cc|c|c|c} 1 & 0 & 3 & 1 & 0 \\ 0 & 1 & -1 & 1 & 1 \end{array}\right],$$

so the matrix of L with respect to S and T is

$$A = \begin{bmatrix} 3 & 1 & 0 \\ -1 & 1 & 1 \end{bmatrix}.$$

This matrix is, of course, far different from the one that defined L. Thus, although a matrix A may be involved in the definition of a linear transformation L, we cannot conclude that it is necessarily the matrix representing L with respect to two given bases S and T. ■

EXAMPLE 6 Let $L\colon P_1 \to P_2$ be defined by $L[p(t)] = tp(t)$.

(a) Find the matrix of L with respect to the bases $S = \{t, 1\}$ and $T = \{t^2, t, 1\}$ for P_1 and P_2, respectively.

(b) If $p(t) = 3t - 2$, compute $L[p(t)]$ directly and using the matrix obtained in (a).

Solution (a) We have

$$L(t) = t \cdot t = t^2 = 1(t^2) + 0(t) + 0(1), \quad \text{so} \quad \begin{bmatrix} L(t) \end{bmatrix}_T = \begin{bmatrix} 1 \\ 0 \\ 0 \end{bmatrix};$$

$$L(1) = t \cdot 1 = t = 0(t^2) + 1(t) + 0(1), \quad \text{so} \quad \begin{bmatrix} L(1) \end{bmatrix}_T = \begin{bmatrix} 0 \\ 1 \\ 0 \end{bmatrix}.$$

Hence the matrix of L with respect to S and T is

$$A = \begin{bmatrix} 1 & 0 \\ 0 & 1 \\ 0 & 0 \end{bmatrix}.$$

(b) Computing $L[p(t)]$ directly, we have

$$L[p(t)] = tp(t) = t(3t - 2) = 3t^2 - 2t.$$

To compute $L[p(t)]$ using A, we first write

$$p(t) = 3 \cdot t + (-2)1, \quad \text{so} \quad \begin{bmatrix} p(t) \end{bmatrix}_S = \begin{bmatrix} 3 \\ -2 \end{bmatrix}.$$

Then

$$\left[L[p(t)]\right]_T = A\left[p(t)\right]_S = \begin{bmatrix} 1 & 0 \\ 0 & 1 \\ 0 & 0 \end{bmatrix} \begin{bmatrix} 3 \\ -2 \end{bmatrix} = \begin{bmatrix} 3 \\ -2 \\ 0 \end{bmatrix}.$$

Hence

$$L[p(t)] = 3t^2 + (-2)t + 0(1) = 3t^2 - 2t. \qquad \blacksquare$$

EXAMPLE 7 Let $L\colon P_1 \to P_2$ be as defined in Example 6.

(a) Find the matrix of L with respect to the bases $S = \{t, 1\}$ and $T = \{t^2, t-1, t+1\}$ for P_1 and P_2, respectively.

(b) If $p(t) = 3t - 2$, compute $L[p(t)]$ using the matrix obtained in (a).

Solution (a) We have (verify)

$$L(t) = t^2 = 1(t^2) + 0(t-1) + 0(t+1), \quad \text{so} \quad \left[L(t)\right]_T = \begin{bmatrix} 1 \\ 0 \\ 0 \end{bmatrix};$$

$$L(1) = t = 0(t^2) + \tfrac{1}{2}(t-1) + \tfrac{1}{2}(t+1), \quad \text{so} \quad \left[L(1)\right]_T = \begin{bmatrix} 0 \\ \frac{1}{2} \\ \frac{1}{2} \end{bmatrix}.$$

Then the matrix of L with respect to S and T is

$$A = \begin{bmatrix} 1 & 0 \\ 0 & \frac{1}{2} \\ 0 & \frac{1}{2} \end{bmatrix}.$$

(b) We have

$$\left[L[p(t)]\right]_T = A\left[p(t)\right]_S = \begin{bmatrix} 1 & 0 \\ 0 & \frac{1}{2} \\ 0 & \frac{1}{2} \end{bmatrix} \begin{bmatrix} 3 \\ -2 \end{bmatrix} = \begin{bmatrix} 3 \\ -1 \\ -1 \end{bmatrix}.$$

Hence

$$L[p(t)] = 3t^2 + (-1)(t-1) + (-1)(t+1) = 3t^2 - 2t. \qquad \blacksquare$$

Suppose that $L\colon V \to W$ is a linear transformation and that A is the matrix of L with respect to bases for V and W. Then the problem of finding $\ker L$ reduces to the problem of finding the solution space of $A\mathbf{x} = \mathbf{0}$. Moreover, the problem of finding range L reduces to the problem of finding the column space of A.

If $L\colon V \to V$ is a linear operator (a linear transformation from V to W, where $V = W$) and V is an n-dimensional vector space, then to obtain a matrix representing L, we fix bases S and T for V and obtain the matrix of L with respect to S and T. However, it is often convenient in this case to choose $S = T$. To avoid redundancy in this case, we refer to A as the **matrix of L with respect to S**. If $L\colon R^n \to R^n$ is a linear operator, then the matrix representing L with respect to the natural basis for R^n has already been discussed in Theorem 4.8 in Section 4.3, where it was called the **standard matrix** representing L.

Let $I: V \to V$ be the identity linear operator on an n-dimensional vector space defined by $I(\mathbf{v}) = \mathbf{v}$ for every $\mathbf{v}$ in V. If S is a basis for V, then the matrix of I with respect to S is I_n (Exercise T.2). Let T be another basis for V. Then the matrix of I with respect to S and T is the transition matrix (see Section 6.7) from the S-basis to the T-basis (Exercise T.5).

If $L: R^n \to R^n$ is a linear operator defined by $L(\mathbf{x}) = A\mathbf{x}$, for $\mathbf{x}$ in R^n, then we can show that L is one-to-one and onto if and only if A is nonsingular.

We can now extend our list of nonsingular equivalences.

List of Nonsingular Equivalences

The following statements are equivalent for an $n \times n$ matrix A.

1. A is nonsingular.
2. $\mathbf{x} = \mathbf{0}$ is the only solution to $A\mathbf{x} = \mathbf{0}$.
3. A is row equivalent to I_n.
4. The linear system $A\mathbf{x} = \mathbf{b}$ has a unique solution for every $n \times 1$ matrix $\mathbf{b}$.
5. $\det(A) \neq 0$.
6. A has rank n.
7. A has nullity 0.
8. The rows of A form a linearly independent set of n vectors in R^n.
9. The columns of A form a linearly independent set of n vectors in R^n.
10. Zero is *not* an eigenvalue of A.
11. The linear operator $L: R^n \to R^n$ defined by $L(\mathbf{x}) = A\mathbf{x}$, for $\mathbf{x}$ in R^n, is one-to-one and onto.

CHANGE OF BASIS LEADS TO A NEW MATRIX REPRESENTING A LINEAR OPERATOR

If $L: V \to V$ is a linear operator and S is a basis for V, then the matrix A representing L with respect to S will change if we use the basis T for V instead of S. The following theorem tells us how to find the matrix of L with respect to T using the matrix A.

THEOREM 10.9 *Let $L: V \to V$ be a linear operator, where V is an n-dimensional vector space. Let $S = \{\mathbf{v}_1, \mathbf{v}_2, \dots, \mathbf{v}_n\}$ and $T = \{\mathbf{w}_1, \mathbf{w}_2, \dots, \mathbf{w}_n\}$ be bases for V and let P be the transition matrix from T to S.* If A is the matrix representing L with respect to S, then $P^{-1}AP$ is the matrix representing L with respect to the basis T.*

Proof If P is the transition matrix from T to S and $\mathbf{x}$ is a vector in V, then by Equation (5) in Section 6.7 we have

$$\left[\mathbf{x}\right]_S = P\left[\mathbf{x}\right]_T, \tag{5}$$

where the jth column of P is the coordinate vector $\left[\mathbf{w}_j\right]_S$ of $\mathbf{w}_j$ with respect to S. From Theorem 6.15 (Section 6.7) we know that P^{-1} is the transition

*In Section 6.7, P was denoted by $P_{S\leftarrow T}$. In this section we denote it by P to simplify the notation.

matrix from S to T, where the jth column of P^{-1} is the coordinate vector $[\mathbf{v}_j]_T$ of $\mathbf{v}_j$ with respect to T. By Equation (5) in Section 6.7 we have

$$[\mathbf{y}]_T = P^{-1}[\mathbf{y}]_S \tag{6}$$

for $\mathbf{y}$ in V. If A is the matrix representing L with respect to S, then

$$[L(\mathbf{x})]_S = A[\mathbf{x}]_S \tag{7}$$

for $\mathbf{x}$ in V. Substituting $\mathbf{y} = L(\mathbf{x})$ in (6), we have

$$[L(\mathbf{x})]_T = P^{-1}[L(\mathbf{x})]_S.$$

Using first (7) and then (5) in this last equation, we obtain

$$[L(\mathbf{x})]_T = P^{-1}[L(\mathbf{x})]_S = P^{-1}A[\mathbf{x}]_S = P^{-1}AP[\mathbf{x}]_T.$$

The equation

$$[L(\mathbf{x})]_T = P^{-1}AP[\mathbf{x}]_T$$

means that $B = P^{-1}AP$ is the matrix representing L with respect to T. ■

Theorem 10.9 can be illustrated by the diagram shown in Figure 10.4. This figure shows that there are two ways of going from $\mathbf{x}$ in V to $L(\mathbf{x})$: directly, using the matrix B; or indirectly, using the matrices P, A, and P^{-1}.

Figure 10.4 ►

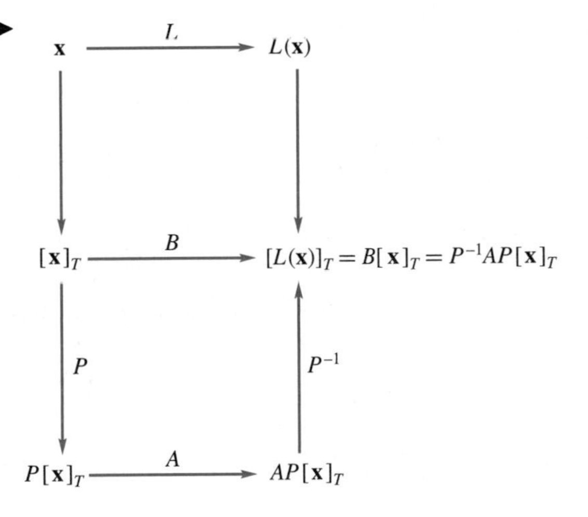

EXAMPLE 8 Let $L: R^2 \to R^2$ be defined by

$$L\left(\begin{bmatrix} a_1 \\ a_2 \end{bmatrix}\right) = \begin{bmatrix} a_1 + a_2 \\ a_1 - 2a_2 \end{bmatrix}.$$

Let

$$S = \left\{\begin{bmatrix} 1 \\ 0 \end{bmatrix}, \begin{bmatrix} 0 \\ 1 \end{bmatrix}\right\} \quad \text{and} \quad T = \left\{\begin{bmatrix} 1 \\ -1 \end{bmatrix}, \begin{bmatrix} 2 \\ 1 \end{bmatrix}\right\}$$

be bases for R^2. We can easily show (verify) that

$$A = \begin{bmatrix} 1 & 1 \\ 1 & -2 \end{bmatrix}$$

is the matrix representing L with respect to S.

The transition matrix P from T to S is the matrix whose jth column is the coordinate vector of the jth vector in the basis T with respect to S. Thus

$$P = \begin{bmatrix} 1 & 2 \\ -1 & 1 \end{bmatrix}.$$

The transition matrix from S to T is (verify)

$$P^{-1} = \frac{1}{3}\begin{bmatrix} 1 & -2 \\ 1 & 1 \end{bmatrix}.$$

Then the matrix representing L with respect to T is (verify)

$$P^{-1}AP = \begin{bmatrix} -2 & 1 \\ 1 & 1 \end{bmatrix}.$$

On the other hand, we can compute the matrix of L with respect to T directly. We have

$$L\left(\begin{bmatrix} 1 \\ -1 \end{bmatrix}\right) = \begin{bmatrix} 0 \\ 3 \end{bmatrix}, \qquad L\left(\begin{bmatrix} 2 \\ 1 \end{bmatrix}\right) = \begin{bmatrix} 3 \\ 0 \end{bmatrix}.$$

We now form the matrix

$$\left[\begin{array}{rr|r|r} 1 & 2 & 0 & 3 \\ -1 & 1 & 3 & 0 \end{array}\right],$$

which we transform to reduced row echelon form, obtaining

$$\left[\begin{array}{rr|r|r} 1 & 0 & -2 & 1 \\ 0 & 1 & 1 & 1 \end{array}\right].$$

Hence the matrix representing L with respect to T is

$$\begin{bmatrix} -2 & 1 \\ 1 & 1 \end{bmatrix},$$

as before. ■

DIAGONALIZABILITY AND SIMILARITY REVISITED

Recall from Section 8.2 that an $n \times n$ matrix B is similar to an $n \times n$ matrix A if there is a nonsingular matrix P such that $B = P^{-1}AP$. It then follows from Theorem 10.9 that any two matrices representing the same linear operator $L: V \to V$ with respect to different bases are similar. Conversely, it can be shown (we omit the proof) that if A and B are similar matrices, then they represent the same linear transformation $L: V \to V$ with respect to bases for the vector space V.

We can now prove the following, which follows from Theorem 8.4.

THEOREM 10.10 *Consider the linear operator $L: R^n \to R^n$ defined by $L(\mathbf{x}) = A\mathbf{x}$ for $\mathbf{x}$ in R^n. Then A is diagonalizable with n linearly independent eigenvectors $\mathbf{x}_1, \mathbf{x}_2, \dots, \mathbf{x}_n$ if and only if the matrix of L with respect to $S = \{\mathbf{x}_1, \mathbf{x}_2, \dots, \mathbf{x}_n\}$ is diagonal.*

Proof Suppose that A is diagonalizable. Then by Theorem 8.4 it has n linearly independent eigenvectors $\mathbf{x}_1, \mathbf{x}_2, \ldots, \mathbf{x}_n$ with corresponding eigenvalues $\lambda_1, \lambda_2, \ldots, \lambda_n$. Since n linearly independent vectors in R^n form a basis (Theorem 6.9 in Section 6.4), we conclude that $S = \{\mathbf{x}_1, \mathbf{x}_2, \ldots, \mathbf{x}_n\}$ is a basis for R^n. Now

$$L(\mathbf{x}_j) = A\mathbf{x}_j = \lambda_j\mathbf{x}_j = 0\mathbf{x}_1 + \cdots + 0\mathbf{x}_{j-1} + \lambda_j\mathbf{x}_j + 0\mathbf{x}_{j+1} + \cdots + 0\mathbf{x}_n,$$

so the coordinate vector $\left[L(\mathbf{x}_j)\right]_S$ of $L(\mathbf{x}_j)$ with respect to S is

$$\begin{bmatrix} 0 \\ \vdots \\ 0 \\ \lambda_j \\ 0 \\ \vdots \\ 0 \end{bmatrix} \longleftarrow j\text{th row.} \tag{8}$$

Hence the matrix of L with respect to S is

$$\begin{bmatrix} \lambda_1 & 0 & \cdots & 0 \\ 0 & \lambda_2 & \cdots & 0 \\ \vdots & & & \vdots \\ 0 & \cdots & 0 & \lambda_n \end{bmatrix}. \tag{9}$$

Conversely, suppose that there is a basis $S = \{\mathbf{x}_1, \mathbf{x}_2, \ldots, \mathbf{x}_n\}$ for R^n with respect to which the matrix of L is diagonal, say, of the form in (9). Then the coordinate vector of $L(\mathbf{x}_j)$ with respect to S is (8), so

$$L(\mathbf{x}_j) = 0\mathbf{x}_1 + \cdots + 0\mathbf{x}_{j-1} + \lambda_j\mathbf{x}_j + 0\mathbf{x}_{j+1} + \cdots + 0\mathbf{x}_n = \lambda_j\mathbf{x}_j.$$

Since $L(\mathbf{x}_j) = A\mathbf{x}_j$, we have

$$A\mathbf{x}_j = \lambda_j\mathbf{x}_j,$$

which means that $\mathbf{x}_1, \mathbf{x}_2, \ldots, \mathbf{x}_n$ are eigenvectors of A. Since they form a basis for R^n, they are linearly independent, and by Theorem 8.4 we conclude that A is diagonalizable. ■

Remark Let $L: R^n \to R^n$ be a linear transformation defined by $L(\mathbf{x}) = A\mathbf{x}$. If A is diagonalizable, then, by Theorem 8.4, R^n has a basis S consisting of eigenvectors of A. Moreover, by Theorem 10.10, the matrix of L with respect to S is the diagonal matrix whose entries on the main diagonal are the eigenvalues of A. Thus the problem of diagonalizing A becomes the problem of finding a basis S for R^n such that the matrix of L with respect to S will be diagonal.

ORTHOGONAL MATRICES REVISITED

We now look at the geometric implications of orthogonal matrices. Let A be an orthogonal $n \times n$ matrix and consider the linear transformation $L: R^n \to R^n$ defined by $L(\mathbf{x}) = A\mathbf{x}$, for $\mathbf{x}$ in R^n. We first compute $L(\mathbf{x}) \cdot L(\mathbf{y})$ for any vectors $\mathbf{x}$ and $\mathbf{y}$ in R^n. We have, using Exercise T.14 in Section 1.3 and the fact that $A^TA = I_n$,

$$\begin{aligned} L(\mathbf{x}) \cdot L(\mathbf{y}) = (A\mathbf{x}) \cdot (A\mathbf{y}) &= (A\mathbf{x})^T(A\mathbf{y}) \\ &= \mathbf{x}^T(A^TA)\mathbf{y} \\ &= \mathbf{x} \cdot (A^TA\mathbf{y}) = \mathbf{x} \cdot (I_n\mathbf{y}) = \mathbf{x} \cdot \mathbf{y}. \end{aligned} \tag{10}$$

This means that L preserves the inner product of two vectors, and consequently L preserves length. It is, of course, clear that if θ is the angle between vectors $\mathbf{x}$ and $\mathbf{y}$ in R^n, then the angle between $L(\mathbf{x})$ and $L(\mathbf{y})$ is also θ (Exercise T.8). A linear transformation satisfying Equation (10) is called an **isometry**. Conversely, let $L\colon R^n \to R^n$ be an isometry, so that

$$L(\mathbf{x}) \cdot L(\mathbf{y}) = \mathbf{x} \cdot \mathbf{y},$$

for any $\mathbf{x}$ and $\mathbf{y}$ in R^n. Let A be the matrix of L with respect to the natural basis for R^n. Then

$$L(\mathbf{x}) = A\mathbf{x}.$$

If $\mathbf{x}$ and $\mathbf{y}$ are any vectors in R^n, we have, using Equation (1) in Section 8.3,

$$\mathbf{x} \cdot \mathbf{y} = L(\mathbf{x}) \cdot L(\mathbf{y}) = (A\mathbf{x}) \cdot (A\mathbf{y}) = \mathbf{x} \cdot (A^T A\mathbf{y}).$$

Since this holds for all $\mathbf{x}$ in R^n, we conclude by Exercise T.8 in Section 4.2 that

$$A^T A\mathbf{y} = \mathbf{y}$$

for any $\mathbf{y}$ in R^n. It then follows from Exercise T.20 in Section 1.4 that $A^T A = I_n$, so A is an orthogonal matrix.

Key Terms

Matrix representing a linear transformation
Isometry

10.3 Exercises

1. Let $L\colon R^2 \to R^2$ be defined by

$$L(x, y) = (x - 2y, x + 2y).$$

Let $S = \{(1, -1), (0, 1)\}$ be a basis for R^2 and let T be the natural basis for R^2. Find the matrix representing L with respect to

(a) S (b) S and T (c) T and S (d) T

(e) Compute $L(2, -1)$ using the definition of L and also using the matrices obtained in (a), (b), (c), and (d).

2. Let $L\colon R^2 \to R^2$ be defined by

$$L\left(\begin{bmatrix} x \\ y \end{bmatrix}\right) = \begin{bmatrix} x + 2y \\ 2x - y \end{bmatrix}.$$

Let S be the natural basis for R^2 and let

$$T = \left\{\begin{bmatrix} -1 \\ 2 \end{bmatrix}, \begin{bmatrix} 2 \\ 0 \end{bmatrix}\right\}$$

be another basis for R^2. Find the matrix representing L with respect to

(a) S (b) S and T (c) T and S (d) T

(e) Compute

$$L\left(\begin{bmatrix} 1 \\ 2 \end{bmatrix}\right)$$

using the definition of L and also using the matrices obtained in (a), (b), (c), and (d).

3. Let $L\colon R^2 \to R^3$ be defined by

$$L\left(\begin{bmatrix} x \\ y \end{bmatrix}\right) = \begin{bmatrix} x - 2y \\ 2x + y \\ x + y \end{bmatrix}.$$

Let S and T be the natural bases for R^2 and R^3, respectively. Also, let

$$S' = \left\{\begin{bmatrix} 1 \\ -1 \end{bmatrix}, \begin{bmatrix} 0 \\ 1 \end{bmatrix}\right\}$$

and

$$T' = \left\{\begin{bmatrix} 1 \\ 1 \\ 0 \end{bmatrix}, \begin{bmatrix} 0 \\ 1 \\ 1 \end{bmatrix}, \begin{bmatrix} 1 \\ -1 \\ 1 \end{bmatrix}\right\}$$

be bases for R^2 and R^3, respectively. Find the matrix representing L with respect to

(a) S and T (b) S' and T'

(c) Compute

$$L\left(\begin{bmatrix} 1 \\ 2 \end{bmatrix}\right)$$

using the definition of L and also using the matrices obtained in (a) and (b).

4. Let $L: R^3 \to R^3$ be defined by

$$L(x, y, z) = (x + 2y + z, 2x - y, 2y + z).$$

Let S be the natural basis for R^3 and let $T = \{(1, 0, 1), (0, 1, 1), (0, 0, 1)\}$ be another basis for R^3. Find the matrix of L with respect to

(a) S (b) S and T (c) T and S (d) T

(e) Compute $L(1, 1, -2)$ using the definition of L and also using the matrices obtained in (a), (b), (c), and (d).

5. Let $L: R^3 \to R^2$ be defined by

$$L\left(\begin{bmatrix} x \\ y \\ z \end{bmatrix}\right) = \begin{bmatrix} x + y \\ y - z \end{bmatrix}.$$

Let S and T be the natural bases for R^3 and R^2, respectively. Also, let

$$S' = \left\{\begin{bmatrix} 1 \\ 1 \\ 0 \end{bmatrix}, \begin{bmatrix} 0 \\ 1 \\ 0 \end{bmatrix}, \begin{bmatrix} -1 \\ 1 \\ 1 \end{bmatrix}\right\}$$

and

$$T' = \left\{\begin{bmatrix} -1 \\ 1 \end{bmatrix}, \begin{bmatrix} 1 \\ 2 \end{bmatrix}\right\}$$

be bases for R^3 and R^2, respectively. Find the matrix of L with respect to

(a) S and T (b) S' and T'

(c) Compute

$$L\left(\begin{bmatrix} 1 \\ 2 \\ 3 \end{bmatrix}\right)$$

using the definition of L and also using the matrices obtained in (a) and (b).

6. Let $L: R^2 \to R^3$ be defined by

$$L\left(\begin{bmatrix} x \\ y \end{bmatrix}\right) = \begin{bmatrix} 1 & 1 \\ 1 & -1 \\ 1 & 2 \end{bmatrix}\begin{bmatrix} x \\ y \end{bmatrix}.$$

(a) Find the matrix of L with respect to the natural bases for R^2 and R^3.

(b) Find the matrix of L with respect to the bases S' and T' of Exercise 3.

(c) Compute

$$L\left(\begin{bmatrix} 2 \\ -3 \end{bmatrix}\right)$$

using the definition of L and also using the matrices obtained in (a) and (b).

7. Let $L: P_1 \to P_3$ be defined by $L[p(t)] = t^2 p(t)$. Let

$$S = \{t, 1\} \quad \text{and} \quad S' = \{t, t + 1\}$$

be bases for P_1. Let

$$T = \{t^3, t^2, t, 1\} \quad \text{and} \quad T' = \{t^3, t^2 - 1, t, t + 1\}$$

be bases for P_3. Find the matrix of L with respect to

(a) S and T (b) S' and T'

8. Let $L: P_1 \to P_2$ be defined by $L[p(t)] = tp(t) + p(0)$. Let

$$S = \{t, 1\} \quad \text{and} \quad S' = \{t + 1, t - 1\}$$

be bases for P_1. Let

$$T = \{t^2, t, 1\} \quad \text{and} \quad T' = \{t^2 + 1, t - 1, t + 1\}$$

be bases for P_2. Find the matrix of L with respect to

(a) S and T (b) S' and T'

(c) Find $L(-3t + 3)$ using the definition of L and also using the matrices obtained in (a) and (b).

9. Let $L: M_{22} \to M_{22}$ be defined by $L(A) = A^T$. Let

$$S = \left\{\begin{bmatrix} 1 & 0 \\ 0 & 0 \end{bmatrix}, \begin{bmatrix} 0 & 1 \\ 0 & 0 \end{bmatrix}, \begin{bmatrix} 0 & 0 \\ 1 & 0 \end{bmatrix}, \begin{bmatrix} 0 & 0 \\ 0 & 1 \end{bmatrix}\right\}$$

and

$$T = \left\{\begin{bmatrix} 1 & 1 \\ 0 & 0 \end{bmatrix}, \begin{bmatrix} 0 & 1 \\ 0 & 0 \end{bmatrix}, \begin{bmatrix} 0 & 0 \\ 1 & 1 \end{bmatrix}, \begin{bmatrix} 1 & 0 \\ 0 & 1 \end{bmatrix}\right\}$$

be bases for M_{22}. Find the matrix of L with respect to

(a) S (b) S and T (c) T and S (d) T

10. Let

$$C = \begin{bmatrix} 1 & 2 \\ 2 & 3 \end{bmatrix}.$$

Let $L: M_{22} \to M_{22}$ be defined by $L(A) = CA$. Let S and T be the bases for M_{22} defined in Exercise 9. Find the matrix of L with respect to

(a) S (b) S and T (c) T and S (d) T

11. Let $L: R^3 \to R^3$ be the linear transformation whose matrix with respect to the natural basis for R^3 is

$$\begin{bmatrix} 1 & 3 & 1 \\ 1 & 2 & 0 \\ 0 & 1 & 1 \end{bmatrix}.$$

Find

(a) $L\left(\begin{bmatrix} 1 \\ 2 \\ 3 \end{bmatrix}\right)$ (b) $L\left(\begin{bmatrix} 0 \\ 1 \\ 1 \end{bmatrix}\right)$

12. Let

$$C = \begin{bmatrix} 1 & 2 \\ 3 & 4 \end{bmatrix},$$

and let $L: M_{22} \to M_{22}$ be the linear transformation defined by $L(A) = AC - CA$ for A in M_{22}. Let S and T be the ordered bases for M_{22} defined in Exercise 9. Find the matrix of L with respect to

(a) S (b) T (c) S and T (d) T and S

13. Let $L: R^2 \to R^2$ be a linear transformation. Suppose that the matrix of L with respect to the basis

$$S = \{\mathbf{v}_1, \mathbf{v}_2\}$$

is

$$A = \begin{bmatrix} 2 & -3 \\ -1 & 4 \end{bmatrix},$$

where

$$\mathbf{v}_1 = \begin{bmatrix} 1 \\ 2 \end{bmatrix} \quad \text{and} \quad \mathbf{v}_2 = \begin{bmatrix} 1 \\ -1 \end{bmatrix}.$$

(a) Compute $\left[L(\mathbf{v}_1)\right]_S$ and $\left[L(\mathbf{v}_2)\right]_S$.

(b) Compute $L(\mathbf{v}_1)$ and $L(\mathbf{v}_2)$.

(c) Compute

$$L\left(\begin{bmatrix}-2\\3\end{bmatrix}\right).$$

14. Let the matrix of $L\colon R^3 \to R^2$ with respect to the bases

$$S = \{\mathbf{v}_1, \mathbf{v}_2, \mathbf{v}_3\} \quad \text{and} \quad T = \{\mathbf{w}_1, \mathbf{w}_2\}$$

be

$$A = \begin{bmatrix}1 & 2 & 1\\-1 & 1 & 0\end{bmatrix},$$

where

$$\mathbf{v}_1 = \begin{bmatrix}-1\\1\\0\end{bmatrix}, \quad \mathbf{v}_2 = \begin{bmatrix}0\\1\\1\end{bmatrix}, \quad \text{and} \quad \mathbf{v}_3 = \begin{bmatrix}1\\0\\0\end{bmatrix}$$

and

$$\mathbf{w}_1 = \begin{bmatrix}1\\2\end{bmatrix}, \quad \mathbf{w}_2 = \begin{bmatrix}1\\-1\end{bmatrix}.$$

(a) Compute $\left[L(\mathbf{v}_1)\right]_T$, $\left[L(\mathbf{v}_2)\right]_T$, and $\left[L(\mathbf{v}_3)\right]_T$.

(b) Compute $L(\mathbf{v}_1)$, $L(\mathbf{v}_2)$, and $L(\mathbf{v}_3)$.

(c) Compute

$$L\left(\begin{bmatrix}2\\1\\-1\end{bmatrix}\right).$$

(d) Compute

$$L\left(\begin{bmatrix}a\\b\\c\end{bmatrix}\right).$$

15. Let $L\colon P_1 \to P_2$ be a linear transformation. Suppose that the matrix of L with respect to the basis $S = \{\mathbf{v}_1, \mathbf{v}_2\}$ and $T = \{\mathbf{w}_1, \mathbf{w}_2, \mathbf{w}_3\}$ is

$$A = \begin{bmatrix}1 & 0\\2 & 1\\-1 & -2\end{bmatrix},$$

where

$$\mathbf{v}_1 = t + 1 \quad \text{and} \quad \mathbf{v}_2 = t - 1;$$

$$\mathbf{w}_1 = t^2 + 1, \quad \mathbf{w}_2 = t, \quad \text{and} \quad \mathbf{w}_3 = t - 1.$$

(a) Compute $\left[L(\mathbf{v}_1)\right]_T$ and $\left[L(\mathbf{v}_2)\right]_T$.

(b) Compute $L(\mathbf{v}_1)$ and $L(\mathbf{v}_2)$.

(c) Compute $L(2t + 1)$.

(d) Compute $L(at + b)$.

16. Let $L\colon R^3 \to R^3$ be defined by

$$L\left(\begin{bmatrix}1\\0\\0\end{bmatrix}\right) = \begin{bmatrix}1\\1\\0\end{bmatrix},$$

$$L\left(\begin{bmatrix}0\\1\\0\end{bmatrix}\right) = \begin{bmatrix}2\\0\\1\end{bmatrix},$$

$$L\left(\begin{bmatrix}0\\0\\1\end{bmatrix}\right) = \begin{bmatrix}1\\0\\1\end{bmatrix}.$$

(a) Find the matrix of L with respect to the natural basis S for R^3.

(b) Find

$$L\left(\begin{bmatrix}1\\2\\3\end{bmatrix}\right)$$

using the definition of L and also using the matrix obtained in (a).

(c) Find $L\left(\begin{bmatrix}a\\b\\c\end{bmatrix}\right)$.

17. Let $L\colon P_1 \to P_1$ be defined by

$$L(t + 1) = t - 1,$$
$$L(t - 1) = 2t + 1.$$

(a) Find the matrix of L with respect to the basis $S = \{t + 1, t - 1\}$ for P_1.

(b) Find $L(2t + 3)$ using the definition of L and also using the matrix obtained in (a).

(c) Find $L(at + b)$.

18. Let the matrix of $L\colon R^2 \to R^2$ with respect to the basis

$$S = \left\{\begin{bmatrix}1\\-1\end{bmatrix}, \begin{bmatrix}0\\1\end{bmatrix}\right\}$$

be

$$\begin{bmatrix}1 & 2\\-2 & 3\end{bmatrix}.$$

Find the matrix of L with respect to the natural basis for R^3.

19. Let the matrix of $L\colon P_1 \to P_1$ with respect to the basis $S = \{t + 1, t - 1\}$ be

$$\begin{bmatrix}2 & 3\\-1 & -2\end{bmatrix}.$$

Find the matrix of L with respect to the basis $\{t, 1\}$ for P_1.

20. Let $L\colon R^3 \to R^3$ be defined by

$$L\left(\begin{bmatrix}a_1\\a_2\\a_3\end{bmatrix}\right) = \begin{bmatrix}a_1 - a_2 + a_3\\a_1 + a_2\\a_2 - a_3\end{bmatrix}.$$

Let S be the natural basis for R^3 and let

$$T = \left\{\begin{bmatrix}1\\0\\1\end{bmatrix}, \begin{bmatrix}0\\1\\-1\end{bmatrix}, \begin{bmatrix}0\\0\\1\end{bmatrix}\right\}$$

be another basis for R^3.

(a) Compute the matrix of L with respect to S.

(b) Compute the matrix of L with respect to T directly.

(c) Compute the matrix of L using Theorem 10.9.

21. (***Calculus Required***) Let $L: P_3 \to P_3$ be defined by

$$L[p(t)] = p''(t) + p(0).$$

Let

$$S = \{1, t, t^2, t^3\} \quad \text{and} \quad T = \{t^3, t^2 - 1, t, 1\}$$

be bases for P_3.

(a) Compute the matrix of L with respect to S.

(b) Compute the matrix of L with respect to T directly.

(c) Compute the matrix of L with respect to T using Theorem 10.9.

22. (***Calculus Required***) Let V be the vector space with basis $S = \{\sin t, \cos t\}$, and let

$$T = \{\sin t - \cos t, \sin t + \cos t\}$$

be another basis for V. Find the matrix of the linear operator $L: V \to V$ defined by $L(f) = f'$ with respect to

(a) S (b) T (c) S and T (d) T and S

23. Let V be the vector space with basis $S = \{e^t, e^{-t}\}$. Find the matrix of the linear operator $L: V \to V$ defined by $L(f) = f'$ with respect to S.

24. For the orthogonal matrix

$$A = \begin{bmatrix} \frac{1}{\sqrt{2}} & -\frac{1}{\sqrt{2}} \\ -\frac{1}{\sqrt{2}} & -\frac{1}{\sqrt{2}} \end{bmatrix}$$

verify that $(A\mathbf{x}) \cdot (A\mathbf{y}) = \mathbf{x} \cdot \mathbf{y}$ for any $\mathbf{x}$ and $\mathbf{y}$ in R^2.

25. Let $L: R^2 \to R^2$ be the linear transformation performing a counterclockwise rotation through 45°; and let A be the matrix of L with respect to the natural basis for R^2. Show that A is orthogonal.

26. Let $L: R^2 \to R^2$ be defined by

$$L\left(\begin{bmatrix} x \\ y \end{bmatrix}\right) = \begin{bmatrix} \frac{1}{\sqrt{2}} & \frac{1}{\sqrt{2}} \\ \frac{1}{\sqrt{2}} & -\frac{1}{\sqrt{2}} \end{bmatrix} \begin{bmatrix} x \\ y \end{bmatrix}.$$

Show that L is an isometry of R^2.

Theoretical Exercises

T.1. Complete the proof of Theorem 10.8.

T.2. Let $I: V \to V$ be the identity linear operator on an n-dimensional vector space V defined by $I(\mathbf{v}) = \mathbf{v}$ for every $\mathbf{v}$ in V. Show that the matrix of I with respect to a basis S for V is I_n.

T.3. Let $O: V \to W$ be the zero linear transformation, defined by $O(\mathbf{x}) = \mathbf{0}_W$, for any $\mathbf{x}$ in V. Show that the matrix of O with respect to any bases for V and W is the $m \times n$ zero matrix (where $n = \dim V$, $m = \dim W$).

T.4. Let $L: V \to V$ be a linear operator defined by $L(\mathbf{v}) = c\mathbf{v}$, where c is a fixed constant. Show that the matrix of L with respect to any basis for V is a scalar matrix (see Section 1.2).

T.5. Let $I: V \to V$ be the identity operator on an n-dimensional vector space V defined by $I(\mathbf{v}) = \mathbf{v}$ for every $\mathbf{v}$ in V. Show that the matrix of I with respect to the bases S and T for V is the transition matrix from the S-basis to the T-basis.

T.6. Let $L: V \to V$ be a linear operator. A nonempty subspace U of V is called **invariant** under L if $L(U)$ is contained in U. Let L be a linear operator with invariant subspace U. Show that if $\dim U = m$, and $\dim V = n$, then the matrix of L with respect to a basis S for V is of the form

$$\begin{bmatrix} A & B \\ O & C \end{bmatrix},$$

where A is $m \times m$, B is $m \times (n - m)$, O is the zero $(n - m) \times m$ matrix, and C is $(n - m) \times (n - m)$.

T.7. Let $L: R^n \to R^n$ be a linear operator defined by $L(\mathbf{x}) = A\mathbf{x}$, for $\mathbf{x}$ in R^n. Show that L is one-to-one and onto if and only if A is nonsingular.

T.8. Let A be an $n \times n$ orthogonal matrix and let $L: R^n \to R^n$ be the linear transformation defined by $L(\mathbf{x}) = A\mathbf{x}$, for $\mathbf{x}$ in R^n. Let θ be the angle between the vectors $\mathbf{x}$ and $\mathbf{y}$ in R^n. Show that if A is orthogonal, then the angle between $L(\mathbf{x})$ and $L(\mathbf{y})$ is also θ.

T.9. Let $L: R^n \to R^n$ be a linear operator and $S = \{\mathbf{v}_1, \mathbf{v}_2, \dots, \mathbf{v}_n\}$ an orthonormal basis for R^n. Show that L is an isometry if and only if $T = \{L(\mathbf{v}_1), L(\mathbf{v}_2), \dots, L(\mathbf{v}_n)\}$ is an orthonormal basis for R^n.

T.10. Show that if a linear transformation $L: R^n \to R^n$ preserves length ($\|L(\mathbf{x})\| = \|\mathbf{x}\|$ for all $\mathbf{x}$ in R^n), then it also preserves the inner product; that is, $L(\mathbf{x}) \cdot L(\mathbf{y}) = \mathbf{x} \cdot \mathbf{y}$ for all $\mathbf{x}$ and $\mathbf{y}$ in R^n. [*Hint*: See Equation (10).]

MATLAB Exercises

In MATLAB, follow the steps given in this section to find the matrix of $L: R^n \to R^m$. The solution technique used in the MATLAB exercises of Section 6.7 will be helpful here.

ML.1. Let $L: R^3 \to R^2$ be given by

$$L\left(\begin{bmatrix} x \\ y \\ z \end{bmatrix}\right) = \begin{bmatrix} 2x - y \\ x + y - 3z \end{bmatrix}.$$

Find the matrix A representing L with respect to the bases

$$S = \{\mathbf{v}_1, \mathbf{v}_2, \mathbf{v}_3\} = \left\{\begin{bmatrix} 1 \\ 1 \\ 1 \end{bmatrix}, \begin{bmatrix} 1 \\ 2 \\ 1 \end{bmatrix}, \begin{bmatrix} 0 \\ 1 \\ -1 \end{bmatrix}\right\}$$

and

$$T = \{\mathbf{w}_1, \mathbf{w}_2\} = \left\{\begin{bmatrix} 1 \\ 2 \end{bmatrix}, \begin{bmatrix} 2 \\ 1 \end{bmatrix}\right\}.$$

ML.2. Let $L: R^3 \to R^4$ be given by $L(\mathbf{v}) = C\mathbf{v}$, where

$$C = \begin{bmatrix} 1 & 2 & 0 \\ 2 & 1 & -1 \\ 3 & 1 & 0 \\ -1 & 0 & 2 \end{bmatrix}.$$

Find the matrix A representing L with respect to the bases

$$S = \{\mathbf{v}_1, \mathbf{v}_2, \mathbf{v}_3\} = \left\{\begin{bmatrix} 1 \\ 0 \\ 1 \end{bmatrix}, \begin{bmatrix} 2 \\ 0 \\ 1 \end{bmatrix}, \begin{bmatrix} 0 \\ 1 \\ 2 \end{bmatrix}\right\}$$

and

$$T = \{\mathbf{w}_1, \mathbf{w}_2, \mathbf{w}_3\} = \left\{\begin{bmatrix} 1 \\ 1 \\ 1 \\ 2 \end{bmatrix}, \begin{bmatrix} 1 \\ 1 \\ 1 \\ 0 \end{bmatrix}, \begin{bmatrix} 0 \\ 1 \\ 1 \\ -1 \end{bmatrix}, \begin{bmatrix} 0 \\ 0 \\ 1 \\ 0 \end{bmatrix}\right\}.$$

ML.3. Let $L: R^2 \to R^2$ be defined by

$$L\left(\begin{bmatrix} x \\ y \end{bmatrix}\right) = \begin{bmatrix} -x + 2y \\ 3x - y \end{bmatrix}$$

and let

$$S = \{\mathbf{v}_1, \mathbf{v}_2\} = \left\{\begin{bmatrix} 1 \\ 2 \end{bmatrix}, \begin{bmatrix} -1 \\ 1 \end{bmatrix}\right\}$$

and

$$T = \{\mathbf{w}_1, \mathbf{w}_2\} = \left\{\begin{bmatrix} -2 \\ 1 \end{bmatrix}, \begin{bmatrix} 1 \\ 1 \end{bmatrix}\right\}$$

be bases for R^2.

(a) Find the matrix A representing L with respect to S.

(b) Find the matrix B representing L with respect to T.

(c) Find the transition matrix P from T to S.

(d) Verify that $B = P^{-1}AP$.

10.4 INTRODUCTION TO FRACTALS (OPTIONAL)

Prerequisites. Sections 10.1–10.3

Using a picture from a world atlas, we measured the western coastline of Baja California (see Figure 10.5), from Tijuana, Mexico, to its southern tip using different scales for measurements. The atlas conveniently showed the scale for 50, 100, 200, and 300 kilometers. We set a standard compass used for drawing circles to the respective lengths. As accurately as possible, we estimated the number of compass steps required to march down the coast from Tijuana to the southern tip, making sure that the compass ends were on the shoreline. Our results are shown in Table 10.1.

Table 10.1

Scale Length (in km)	Compass Steps Required	Estimated Length (in km)
50	31	1550
100	14	1400
200	6.5	1300
300	4.25	1275

Figure 10.5 ►

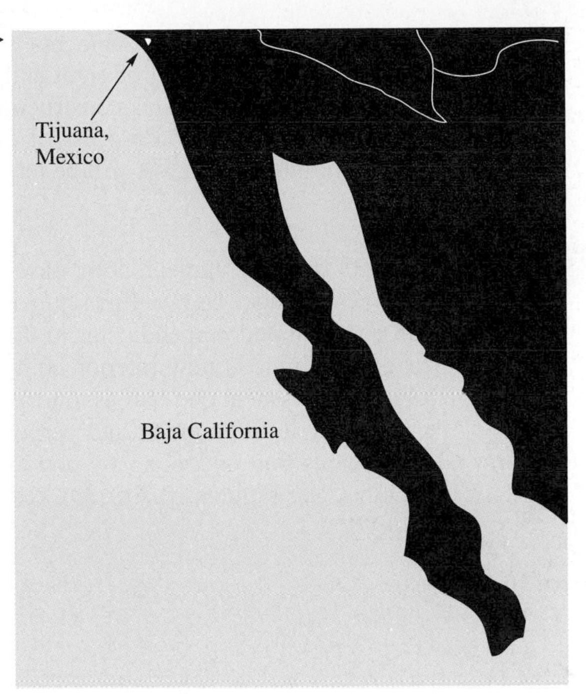

Had we set the scale length to 25 km, 10 km, 5 km, and 1 km, our estimates would have continued to increase because we would be measuring in more detail the harbors, coves, and general changes of the ragged coastline. If the scale length of hypothetical measurements were permitted to become infinitely small, then our estimates of the length of the coastline would grow without bound. The dependency of measured lengths on the measuring scale was a fundamental observation that has led to a new technique that enables us to cope with problems of scale in real-world situations in such diverse fields as biology, medicine, communications, image compression, astronomy, meteorology, economics, ecology, metallurgy, and special effects in cinematography.

Before we provide a general setting for coping with measurements affected by change of measuring scale, let us consider two mathematical examples of a related nature.

EXAMPLE 1

In the late nineteenth century Georg Cantor* constructed a set of points in the interval [0, 1] that is sometimes called **Cantor's point-set** or **Cantor's dust**. This set can be constructed by repeating a very simple principle indefinitely. Starting with the line segment representing the interval [0, 1] (including the endpoints), we omit the middle third, but not the endpoints. The result is two line segments (with a total of four endpoints) that are really copies of one

*Georg Cantor (1845–1918) was a German mathematician at the University of Halle whose initial work involved number theory, but he soon turned to the field of analysis. By 1870 he had solved an open problem on the uniqueness of representation of a function as a trigonometric series. In 1873 Cantor proved that the rational numbers are countable, that is, they may be placed in one-to-one correspondence with the natural numbers. He also showed that the algebraic numbers, that is, the numbers that are roots of polynomial equations with integer coefficients, are countable. However, his attempts to decide whether the real numbers are countable proved harder. He had proved that the real numbers are not countable by December 1873 and published this in a paper in 1874. Cantor continued his work and in 1883 published a paper that included the description of what we now call Cantor's point-set. He made fundamental contributions to the area we call set theory. In his later years Cantor suffered bouts of depression and eventually periods of ill health.

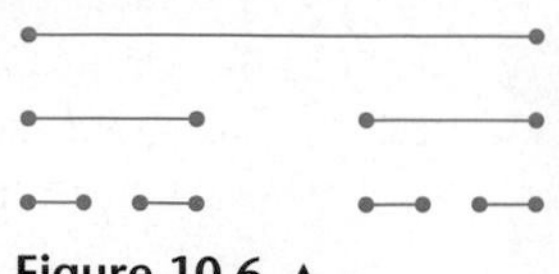

Figure 10.6 ▲

another. From each of these segments we omit the middle third, giving us four line segments (with a total of eight endpoints) and each of these segments is a copy of one another. This construction is illustrated in Figure 10.6. If we continue this construction process of omitting the middle third of each line segment infinitely often, we will be left with a discrete set of points, Cantor's dust. ■

EXAMPLE 2

Start with the line segment along the x-axis from $x = -1$ to $x = 1$. (There are two endpoints.) Draw a line segment one-half as long having its midpoint at an endpoint and perpendicular to the existing line segment. (The original pair of endpoints is now interior points, but there are four new endpoints.) Draw a line segment one-half as long as the previously drawn segment having its midpoint at an endpoint and perpendicular to the existing line segment. The previous four endpoints are now interior points, but there are eight new endpoints. See Figures 10.7(a), (b). [How many endpoints are there in Figure 10.7(c)?]

Figure 10.7 ►

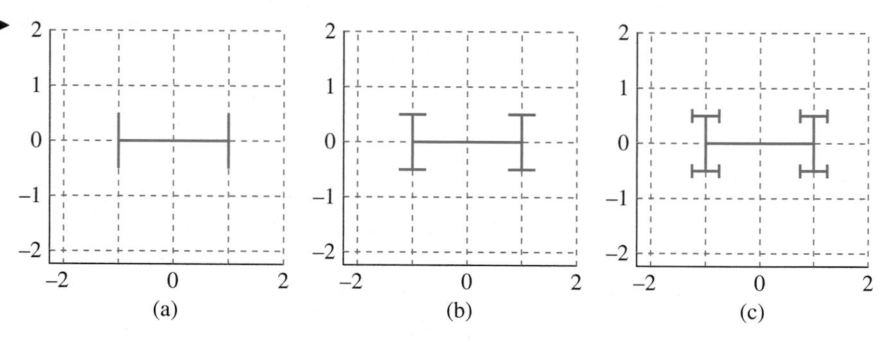

We can continue this process indefinitely. Stopping at any time, we have a portion of a curve known as an **H-curve**. Figure 10.8 shows the H-curve after six repetitions of the process described above.

If we isolate and enlarge the upper left-hand branch of the H-curve, we get the display shown in Figure 10.9. (Many graphics-capable software routines support a zoom feature, so this step is easy to perform.) Figure 10.9 is "similar" to Figure 10.7(b), except for the "supporting center pole." ■

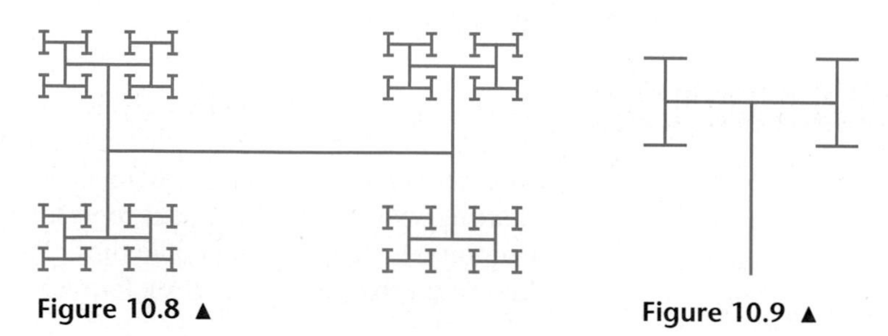

Figure 10.8 ▲

Figure 10.9 ▲

The figures generated in Examples 1 and 2 are really quite complicated if we repeat the construction process many times in spite of the fact that the operations for a single step are very simple. In addition, the line segments of the Cantor construction as well as the H's of Example 2 are copies of one another at each step, and compared to the previous step there is just a change of scale. After many repetitions of the constructive steps for Cantor's dust and the H-curve, the geometric shape is fragmented and possibly quite rough and can be divided into parts that appear to be a reduced copy of the whole shape. Such sets, curves, figures, or surfaces are called **fractals**. The word *fractal* is derived from the Latin word *fractus* that roughly translates as "irregular fragments." The mathematics of fractals was first systematically studied by Benoit B. Mandelbrot.* Hence Cantor's dust is referred to as **Cantor's fractal** and the H-curve as the **H-fractal**.

The mathematical objects in Examples 1 and 2, and others that we will encounter subsequently, are considered *true fractals* in contrast to curves or surfaces in the real world such as coastlines, trees, coral reefs, clouds, or the surface of a metal plate. Real objects can be viewed only over a finite range of scales and hence we do not have infinitely repeated constructive steps like those described in Examples 1 and 2. Thus true fractals are a mathematician's tool that can be used to provide models for physical objects or phenomena. As such they have provided an improved technique for modeling the *geometry of nature* that was not available previously.

We will restrict our investigation to fractals in the plane. That is, we will be concerned with sets of points in R^2 that can be generated by processes such as those described in Examples 1 and 2. To formulate a mathematical description of such processes, we focus on functions from $R^2 \to R^2$, but the general theory goes well beyond R^2. The functions of interest will be able to provide some combination of a change of scale (a contraction), rotation, and translation. (See Section 10.1, where we indicated that a contraction and a rotation are linear transformations, but a translation need not be linear; see Exercise T.12 in Section 10.1.) We make the following definitions.

DEFINITION Let $T: R^2 \to R^2$ be defined by $T(\mathbf{v}) = \mathbf{v} + \mathbf{b}$, where $\mathbf{b}$ is a fixed vector in R^2. We call this a **translation by vector b** (or just a **translation**) and use the notation $\mathbf{tran_b}$; $\mathbf{tran_b}(\mathbf{v}) = \mathbf{v} + \mathbf{b}$.

A translation by vector $\mathbf{b}$, $\mathbf{b} \neq \mathbf{0}$, is not a linear transformation. This follows from noting that $\mathbf{tran_b}(\mathbf{0}) \neq \mathbf{0}$. (See Exercise T.12 in Section 10.1 and Corollary 10.1.)

The composition of a linear transformation from $R^2 \to R^2$ with a translation provides an important class of functions, which we define next.

DEFINITION The function $T: R^2 \to R^2$ defined by $T(\mathbf{v}) = A\mathbf{v} + \mathbf{b}$, where A is a specified 2×2 matrix and $\mathbf{b}$ is a fixed vector in R^2, is called an **affine transformation**.

---

*Benoit Mandelbrot (1924–) was born in Poland into a family with strong academic interests. His father was in the clothing business and his mother was a physician. In 1936 he moved with his family to Paris. As a young boy, Mandelbrot was introduced to mathematics by his two uncles, one of whom was Szolem Mandelbrojt, a distinguished professor of mathematics at the College de France. His years in France during World War II were difficult and dangerous.

In 1947, he graduated from the École Polytechnique, in 1948 he received his M.S. from California Institute of Technology, and in 1952 his Ph.D. from the Université de Paris. In 1958 he joined the IBM Watson Research Center, where he spent many productive years. In 1987 he joined the faculty at Yale University.

All of Benoit Mandelbrot's scientific interests relate to an interdisciplinary field that he originated, fractal geometry. Thus, Mandelbrot is a leader in computer graphics because one aspect of fractal geometry is its extensive reliance on pictures: Some are "forgeries" of reality while others are purely "abstract."

Mandelbrot belongs to the American Academy of Arts and Sciences and the National Academy of Sciences. He has received numerous honorary doctorates and awards; his most recent major award is the 1993 Wolf Prize in Physics.

By our previous discussion, for $\mathbf{b} \neq \mathbf{0}$, an affine transformation is nonlinear. (Verify.) If we let $T_1\colon R^2 \to R^2$ be defined by $T_1(\mathbf{v}) = A\mathbf{v}$ and $T_2\colon R^2 \to R^2$ be defined by $T_2(\mathbf{v}) = \mathbf{tran_b}(\mathbf{v}) = \mathbf{v} + \mathbf{b}$, then the composition of T_1 with T_2, $T_2 \circ T_1$, is equivalent to the affine transformation $T(\mathbf{v}) = A\mathbf{v} + \mathbf{b}$: $(T_2 \circ T_1)(\mathbf{v}) = T_2[T_1(\mathbf{v})] = \mathbf{tran_b}(A\mathbf{v}) = A\mathbf{v} + \mathbf{b}$. The order in which we compose transformations is important because

$$T_2[T_1(\mathbf{v})] = \mathbf{tran_b}(A\mathbf{v}) = A\mathbf{v} + \mathbf{b},$$

but

$$T_1[T_2(\mathbf{v})] = T_1[\mathbf{tran_b}(\mathbf{v})] = T_1(\mathbf{v} + \mathbf{b}) = A(\mathbf{v} + \mathbf{b}) = A\mathbf{v} + A\mathbf{b}.$$

Both $T_2 \circ T_1$ and $T_1 \circ T_2$ are affine transformations, but the translations involved are, in general, different since $A\mathbf{b}$ need not equal $\mathbf{b}$.

We first investigate the behavior of affine transformations on lines in R^2 and then show how they can be used to construct fractals. Recall that a line in R^2 can be determined algebraically by specifying its slope and one of its points. Geometrically, the specification of slope is equivalent to specifying a vector through the origin that is parallel to the line. Let L be a line through the point P_0 and parallel to the vector $\mathbf{u}$. Then L consists of all points $P(x, y)$ such that

$$\mathbf{x} = \begin{bmatrix} x \\ y \end{bmatrix} = \mathbf{w} + t\mathbf{u}, \tag{1}$$

where $\mathbf{w}$ is the vector from the origin to P_0 and t is a real scalar called a *parameter*. We call Equation (1) the *parametric form* for the equation of a line in R^2. Figure 10.10 shows that line L is a translation by the vector $\mathbf{w}$ of the line $t\mathbf{u}$ that goes through the origin.

Figure 10.10 ►

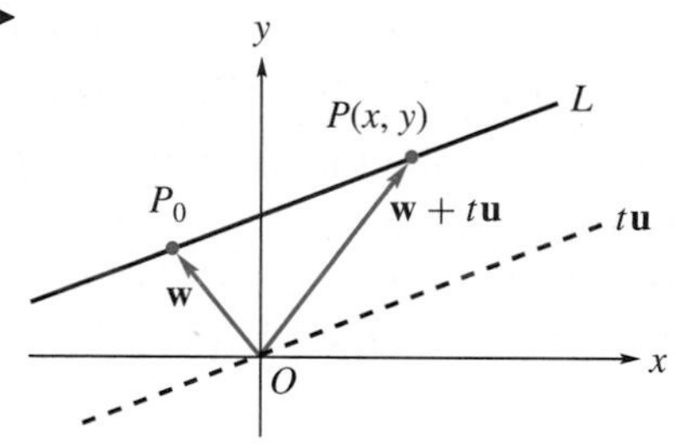

An important property of affine transformations is that they map lines to lines and line segments to line segments. We show this in Examples 3 and 4, respectively.

EXAMPLE 3

Let T be the affine transformation given by $T(\mathbf{v}) = A\mathbf{v} + \mathbf{b}$ and L the line $\mathbf{w} + t\mathbf{u}$. Then the image of L by T is

$$T(\mathbf{w} + t\mathbf{u}) = A(\mathbf{w} + t\mathbf{u}) + \mathbf{b} = (A\mathbf{w} + \mathbf{b}) + t(A\mathbf{u}),$$

which is the translation by vector $A\mathbf{w} + \mathbf{b}$ of the line $tA\mathbf{u}$. Thus the image of a line by an affine transformation is another line. ■

EXAMPLE 4

Let S be the line segment of line L given by $\mathbf{w} + t\mathbf{u}$ determined by the values of the parameter t for $a \leq t \leq c$. Let T be the affine transformation given by $T(\mathbf{v}) = A\mathbf{v} + \mathbf{b}$. Then the image of S by T is

$$T(\mathbf{w} + t\mathbf{u}) = A(\mathbf{w} + t\mathbf{u}) + \mathbf{b} = (A\mathbf{w} + \mathbf{b}) + t(A\mathbf{u}), \quad a \leq t \leq c,$$

which is the translation by vector $A\mathbf{w} + \mathbf{b}$ of the line segment $tA\mathbf{u}$. Thus the image of a line segment by an affine transformation is another line segment. ■

Next we show the result of applying an affine transformation to a figure in R^2 composed of line segments. Each of the line segments of the figure is represented by the coordinates of the endpoints of the segment. From Example 4 we expect that the image will be a plane figure consisting of line segments.

EXAMPLE 5

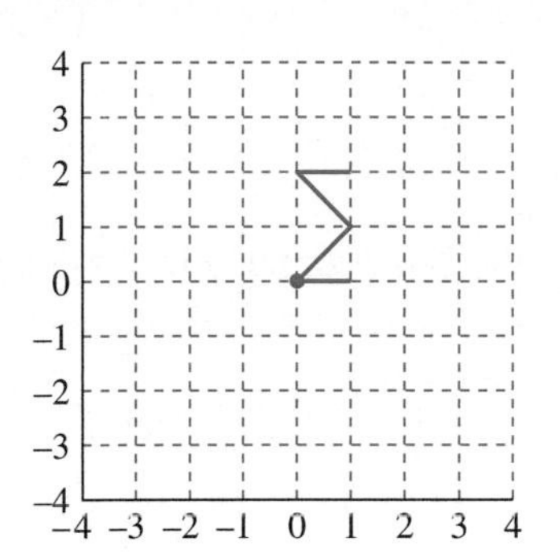

Figure 10.11 ▲

Figure 10.11 displays a summation symbol, a sigma, with a large dot located at the origin. For our purposes we shall consider the sigma as the set of ordered pairs $\{(1, 0), (0, 0), (1, 1), (0, 2), (1, 2)\}$ connected in the order listed by straight line segments. The dot is included so that we can see the image of the origin when transformations are applied to sigma. It is convenient to represent sigma as the matrix

$$S = \begin{bmatrix} 1 & 0 & 1 & 0 & 1 \\ 0 & 0 & 1 & 2 & 2 \end{bmatrix},$$

where the first row is the set of x-coordinates of the points defining sigma and the second row is the corresponding set of y-coordinates. We will assume that the line segments connecting successive points will be drawn as needed.

(a) For ease of manipulation with the vector $\mathbf{b} = \begin{bmatrix} b_1 \\ b_2 \end{bmatrix}$ we assume that a translation $\mathbf{tran_b}(S)$ is computed so that b_1 is added to each of the entries in $\mathbf{row}_1(S)$ and b_2 is added to each of the entries in $\mathbf{row}_2(S)$. Thus

$$\begin{aligned} \mathbf{tran_b}(S) &= \begin{bmatrix} \mathbf{col}_1(S) + \mathbf{b} & \mathbf{col}_2(S) + \mathbf{b} & \mathbf{col}_3(S) + \mathbf{b} & \mathbf{col}_4(S) + \mathbf{b} & \mathbf{col}_5(S) + \mathbf{b} \end{bmatrix} \\ &= \begin{bmatrix} \mathbf{tran_b}(\mathbf{col}_1(S)) & \mathbf{tran_b}(\mathbf{col}_2(S)) & \mathbf{tran_b}(\mathbf{col}_3(S)) & \mathbf{tran_b}(\mathbf{col}_4(S)) & \mathbf{tran_b}(\mathbf{col}_5(S)) \end{bmatrix}, \end{aligned}$$

where $\mathbf{col}_j(S)$ denotes the jth column of the matrix S. For $\mathbf{b} = \begin{bmatrix} 2 \\ -1 \end{bmatrix}$, $\mathbf{tran_b}(S) = \begin{bmatrix} 3 & 2 & 3 & 2 & 3 \\ -1 & -1 & 0 & 1 & 1 \end{bmatrix}$. (Verify.) This image is shown in Figure 10.12 along with the original sigma. We see that the origin has been translated to $(2, -1)$.

Figure 10.12 ►

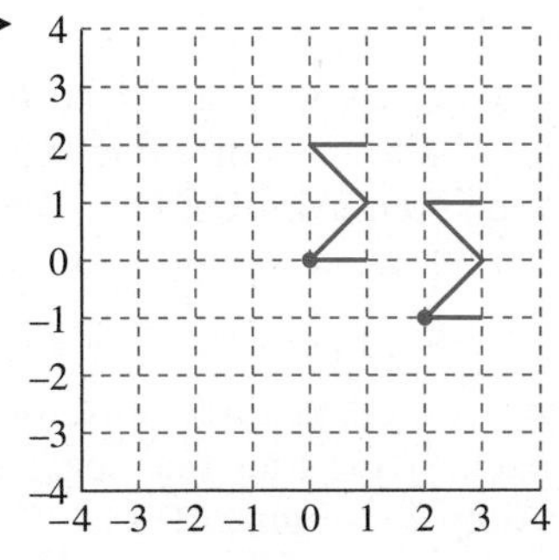

(b) Let T be the affine transformation $T(\mathbf{v}) = A\mathbf{v} + \mathbf{b}$, where $A = \begin{bmatrix} 0 & 1 \\ -1 & 0 \end{bmatrix}$ and $\mathbf{b} = \begin{bmatrix} 2 \\ -1 \end{bmatrix}$. T performs a clockwise rotation of 90° followed by a translation by the vector **b**. The image of sigma by T is shown in Figure 10.13.

Figure 10.13 ►

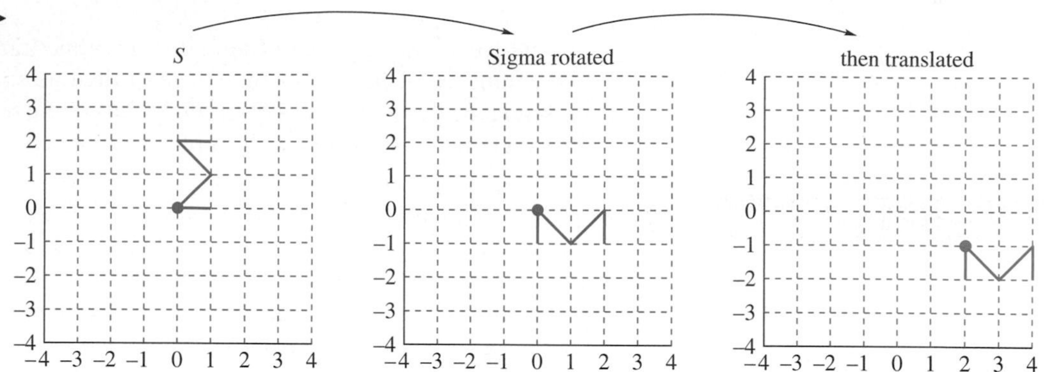

The coordinates of the endpoints of the line segments in the image are given by

$$T(S) = \begin{bmatrix} 2 & 2 & 3 & 4 & 4 \\ -2 & -1 & -2 & -1 & -2 \end{bmatrix} \quad \text{(verify).}$$

■

Affine transformations seem simple enough; they map line segments to line segments. However, the simplicity of the process does not mean that repetitions of it will lead to simple patterns. In fact, it is surprising that a wide variety of complex patterns are the result of repeated applications of affine transformations. It was this type of mathematical observation that laid the foundation for the new areas of mathematics called **fractal geometry** and **chaos theory**. Here we focus on the use of affine transformations to briefly illustrate fractal geometry.

We have described a fractal as a figure that appears self-similar as we make a change of scale. The term **self-similar** is used to describe figures composed of infinite repetitions of the same shape, that is, the figure is composed of smaller objects that are similar to the whole. For example, a tree is made up of branches, limbs, and twigs. Viewed at different scales each of these has a similar shape or appearance. Examples 1 and 2 described two fractals that were easy to visualize. Examples 6 and 7 introduce two more fractals together with the affine transformations that can be used to construct a figure representing the fractal at various scales.

EXAMPLE 6 One of the most familiar fractals is the **Koch curve**. It is constructed by removing the middle third of a line segment, replacing it with an equilateral triangle, and then removing the base. We start with an initial line segment

---

Following the preceding directions, we obtain Step 1 as shown in Figure 10.14. On each of the four line segments in Step 1 we repeat the process. The result is Step 2 in Figure 10.14.

The directions are repeated on succeeding line segments to obtain Step 3 and then Step 4. The self-similarity is built into the construction process. If the initial line segment is 1 unit in length, then the 4 segments in Step 1 are $\frac{1}{3}$ of a unit, the 16 segments in Step 2 are $\frac{1}{9}$ of a unit, and so on. At each step a line segment is scaled down by a factor of 3. Repeating the construction

Figure 10.14 ►

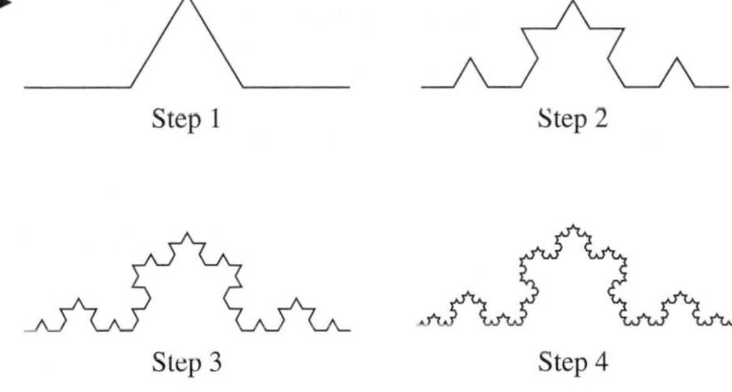

process infinitely often generates a curve that would be extremely rough and in fact does not have a tangent line at any point.

Figure 10.15 ►

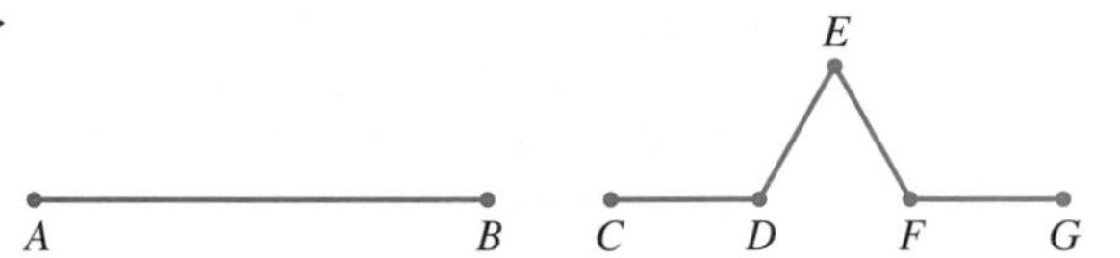

To determine the affine transformations used to obtain the Koch curve we first note that one step generates 4 line segments from each line segment of the current figure. See Figure 10.14. Hence 4 affine transformations, T_1 through T_4, will be used to generate this fractal. Referring to Figure 10.15, $T_1(AB) = CD$, $T_2(AB) = DE$, $T_3(AB) = EF$, and $T_4(AB) = FG$. In each case there will be scaling by a factor of $\frac{1}{3}$. We summarize these as follows:

- T_1 is a contraction by a factor of $\frac{1}{3}$.
- T_2 is a contraction by a factor of $\frac{1}{3}$ followed by a rotation of 60° counter-clockwise and then a translation by the vector $\mathbf{b} = \begin{bmatrix} \frac{1}{3} \\ 0 \end{bmatrix}$.
- T_3 is a contraction by a factor of $\frac{1}{3}$ followed by a rotation of 60° clockwise and then a translation by the vector $\mathbf{b} = \begin{bmatrix} \frac{1}{2} \\ \frac{\sqrt{3}}{6} \end{bmatrix}$.
- T_4 is a contraction by a factor of $\frac{1}{3}$ followed by a translation by the vector $\mathbf{b} = \begin{bmatrix} \frac{2}{3} \\ 0 \end{bmatrix}$.

These four affine transformations are applied to each line segment with adjustments in the translations at each step that take the scale into account. Thus careful structuring of computations is required. ■

Since every affine transformation can be written in the form $T(\mathbf{v}) = A\mathbf{v} + \mathbf{b}$, where $A = \begin{bmatrix} p & r \\ s & t \end{bmatrix}$ and $\mathbf{b} = \begin{bmatrix} b_1 \\ b_2 \end{bmatrix}$, we can specify the transformations that generate a fractal as a table of coefficients for p, r, s, t, b_1, and b_2. The Koch fractal is specified in this manner in Table 10.2.

EXAMPLE 7 Another famous mathematical fractal is the **Sierpinski triangle**. To form this fractal we begin with the equilateral triangle shown in Figure 10.16(a) and

Table 10.2 The Koch Fractal's Affine Transformations

	p	r	s	t	b_1	b_2
T_1	$\frac{1}{3}$	0	0	$\frac{1}{3}$	0	0
T_2	$\frac{1}{6}$	$-\frac{\sqrt{3}}{6}$	$\frac{\sqrt{3}}{6}$	$\frac{1}{6}$	$\frac{1}{3}$	0
T_3	$\frac{1}{6}$	$\frac{\sqrt{3}}{6}$	$-\frac{\sqrt{3}}{6}$	$\frac{1}{6}$	$\frac{1}{2}$	$\frac{\sqrt{3}}{6}$
T_4	$\frac{1}{3}$	0	0	$\frac{1}{3}$	$\frac{2}{3}$	0

remove the triangle formed by the midpoints of the sides. (This is called removing the middle triangle.) The result is Figure 10.16(b). Next we remove the middle triangle from each of the three triangles of Figure 10.16(b), giving Figure 10.16(c). We continually repeat this process to obtain the fractal. Figure 10.16(d) shows the scale changes that result from five repetitions just in the lower left corner.

Figure 10.16 ►

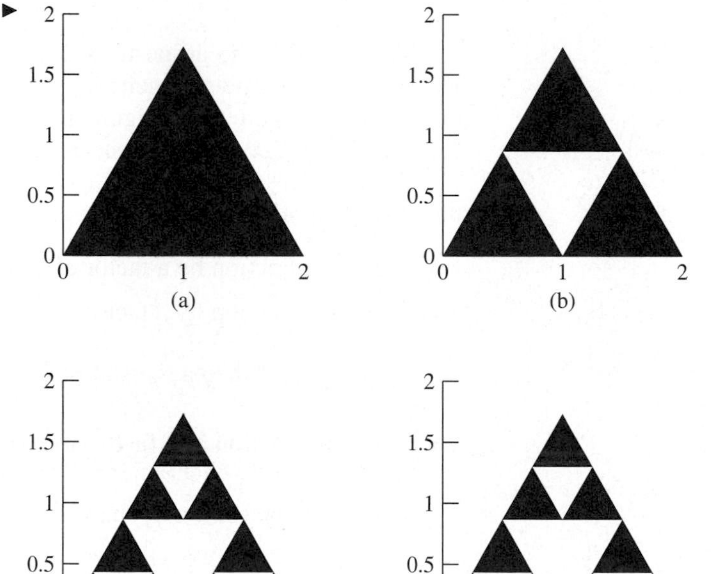

The preceding geometric steps are conceptually easy, but careful thought is needed to determine the affine transformations required to produce Figure 10.16(b) from Figure 10.16(a). Since the middle triangle is formed by connecting the midpoints of the sides of the original equilateral triangle, the sides of the three triangles in Figure 10.16(b) are $\frac{1}{2}$ the size of the sides of the original triangle. Hence there is a contraction by $\frac{1}{2}$, which is given by the affine transformation

$$T_1(\mathbf{v}) = \begin{bmatrix} \frac{1}{2} & 0 \\ 0 & \frac{1}{2} \end{bmatrix} \mathbf{v} + \begin{bmatrix} 0 \\ 0 \end{bmatrix}.$$

The resulting triangle has its lower left corner at the origin. To obtain the second triangle along the horizontal axis, we again use a contraction by $\frac{1}{2}$ but now also translate this triangle so the origin is at the point $(1, 0)$. This requires the affine transformation

$$T_2(\mathbf{v}) = \begin{bmatrix} \frac{1}{2} & 0 \\ 0 & \frac{1}{2} \end{bmatrix} \mathbf{v} + \begin{bmatrix} 1 \\ 0 \end{bmatrix}.$$

To obtain the top triangle in Figure 10.16(b), we again use a contraction by $\frac{1}{2}$ but now translate this triangle so the origin is at the midpoint of the side of the triangle in Figure 10.16(a) with positive slope. That midpoint is $\left(\frac{1}{2}, \frac{\sqrt{3}}{2}\right)$. This requires the affine transformation

$$T_3(\mathbf{v}) = \begin{bmatrix} \frac{1}{2} & 0 \\ 0 & \frac{1}{2} \end{bmatrix} \mathbf{v} + \begin{bmatrix} \frac{1}{2} \\ \frac{\sqrt{3}}{2} \end{bmatrix}.$$

Successive applications require that we apply the affine transformations to each of the existing triangles and "adjust" the translations so thcy go to the midpoints of the sides of each triangle whose image is computed. This approach requires that we carefully structure our computations. The Sierpinski triangle fractal's affine transformations are given in Table 10.3. ■

Table 10.3 The Sierpinski Fractal's Affine Transformations

	p	r	s	t	b_1	b_2
T_1	$\frac{1}{2}$	0	0	$\frac{1}{2}$	0	0
T_2	$\frac{1}{2}$	0	0	$\frac{1}{2}$	1	0
T_3	$\frac{1}{2}$	0	0	$\frac{1}{2}$	$\frac{1}{2}$	$\frac{\sqrt{3}}{2}$

Our description of the fractals in both Examples 6 and 7 indicated that we had to adjust the translations involved as we changed the scale from step to step. This procedure is often called the **deterministic approach**. In summary, this approach, for example, takes any point $\mathbf{x} = \begin{bmatrix} x \\ y \end{bmatrix}$ in the triangle in Figure 10.16(a), computes $T_1(\mathbf{x})$, $T_2(\mathbf{x})$, and $T_3(\mathbf{x})$, and plots these image points. From our discussion in Example 7, each of the images will be in one of thc three triangles of Figure 10.16(b). It follows that if we compute the images of *all* the points from the triangle in Figure 10.16(a), we will obtain Figure 10.16(b). In a corresponding fashion we apply T_1, T_2, and T_3 with translation adjustments to compute the images of all the points in the three triangles in Figure 10.16(b) to generate Figure 10.16(c), and so on to generate the fractal. The adjustments at each stage can become laborious, but there is an alternative approach that will generate the fractals strictly by repeated use of the affine transformations listed in Tables 10.2 and 10.3.

The alternative procedure to the deterministic approach is called the **random iteration approach**. The random iteration approach is simpler to implement in a computer routine than the deterministic approach. We describe this approach next using the Sierpinski triangle. Assign to the affine transformation T_i in Table 10.3 a probability $p_i > 0$, $i = 1, 2, 3$, so that $p_1 + p_2 + p_3 = 1$. Perform thc following steps:

1. Choose an initial point $\mathbf{v}_0 = \begin{bmatrix} x_0 \\ y_0 \end{bmatrix}$ in the triangle in Figure 10.16(a).
2. Compute $T_1(\mathbf{v}_0)$, $T_2(\mathbf{v}_0)$, and $T_3(\mathbf{v}_0)$ and choose a random number r in $[0, 1)$.
3. Define a new point $\mathbf{v}_1$ as follows:

$$\begin{aligned} \mathbf{v}_1 &= T_1(\mathbf{v}_0) \quad \text{if } 0 \leq r < p_1 \\ \mathbf{v}_1 &= T_2(\mathbf{v}_0) \quad \text{if } p_1 \leq r < p_1 + p_2 \\ \mathbf{v}_1 &= T_3(\mathbf{v}_0) \quad \text{if } p_1 + p_2 \leq r < 1. \end{aligned}$$

4. Plot the point $\mathbf{v}_1 = \begin{bmatrix} x_1 \\ y_1 \end{bmatrix}$.
5. Set $\mathbf{v}_0 = \mathbf{v}_1$ and return to Step 2.

The preceding steps contain no mechanism for stopping the computations. An easy modification is to choose a number of points N that we want plotted, keep a running count of the number of times Steps 2 through 4 are executed, and then stop when the running count equals N.

EXAMPLE 8

For the random iteration approach described above for the Sierpinski triangle with

$$p_1 = p_2 = p_3 = \frac{1}{3}, \quad \mathbf{v}_0 = \begin{bmatrix} 1.2 \\ 0.5 \end{bmatrix}, \quad \text{and} \quad N = 20{,}000$$

we obtain Figure 10.17. The pattern displayed in Figure 10.17 is sometimes called **Sierpinski's gasket**. ■

Figure 10.17 ▲

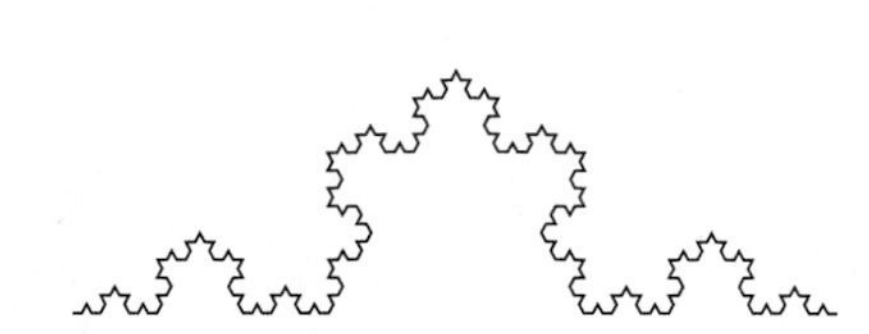

Figure 10.18 ▲

EXAMPLE 9

The random iteration approach applied to the Koch fractal's affine transformations with $p_1 = p_2 = p_3 = p_4 = \frac{1}{4}$, $\mathbf{v}_0 = \begin{bmatrix} 0.5 \\ 0.5 \end{bmatrix}$, and $N = 5{,}000$ yields Figure 10.18. ■

Our description of the random iteration approach is referred to as an **iterated function system**, abbreviated IFS. Michael Barnsley* referred to the random selection of affine transformations from an iterated function system as the **chaos game**. Although the figure generated by an IFS, which is technically called an **attractor**, will be drawn differently every time, there is nothing chaotic about the outcome. However, what is interesting about the chaos game is that the attractor, some particular infinite set of points in R^2, can be generated by a small set of affine transformations, the IFS. This prompted Barnsley

*M. Barnsley, *Fractals Everywhere*, San Diego, California: Academic Press, 1988.

and his colleagues to ask the reverse question. Given an image, a set of points, can we find an iterated function system whose attractor is the specified image? This approach has led to a variety of interesting results, one of which is image compression and reconstruction using iterated function systems. (We earlier mentioned wavelets in this context.)

Multimedia today requires that thousands of images be quickly available. Since graphic images are notoriously large files of digitized information, business and industry are always on the lookout for new ways to reduce storage requirements and increase access speed to such information. Fractals and, in particular, iterated function systems have proven to be a commercially successful technique for image compression. In fact, the thousands of images that appear on a number of popular CD-ROMs have been encoded using a fractal image compression process. Fractal image compression starts with a digitized image and uses image-processing techniques to break it into segments. Each segment is then analyzed to remove as much redundancy as possible and then searched for self-similarity patterns.

The final step of the process determines an iterated function system whose attractor is the self-similar pattern. Hence instead of saving detailed information about each pixel of the segment, only the information about the IFS need be retained to represent the image of the segment. The information about the IFS is just a table of coefficients for the affine transformations like those for the Koch and Sierpinski fractals.

To display the image, the chaos game is played with the IFS. This technique can be quite time-consuming to develop the iterated function systems for the segments of an image, but once completed, the IFS table of coefficients takes up significantly less memory and can generate the image quickly. The entire process has a firm theoretical foundation, which is based on the work of Barnsley and a result known as the "collage theorem."*

Key Terms

Cantor's set (or Cantor's dust)
H-curve
Fractals
Cantor's fractal
H-fractal
Translation
Fractal geometry
Chaos theory
Self-similar
Koch curve
Sierpinski's triangle
Deterministic approach
Sierpinski's gasket
Iterated function system
Chaos game
Attractor
Levy's fractal
Fixed point

10.4 Exercises

1. Let S_0 denote a square with side of length 3 with lower left corner at the origin.

(a) Draw S_0.

(b) Draw the figure S_1 obtained from S_0 by omitting a square with side of length 1 from each corner. (S_1 will consist of five 1×1 squares.)

(c) Draw the figure S_2 obtained from S_1 by omitting a square with side of length $\frac{1}{3}$ from each corner of the five squares composing S_1. How many squares of size $\frac{1}{3} \times \frac{1}{3}$ are there in S_2?

(d) If we construct S_3 from S_2 in a similar manner, how many squares are there in S_3 and what size are they?

(e) Write a short description of the figures S_1, S_2, and S_3.

(f) Compute the areas of S_0, S_1, S_2, and S_3. Next

*For a brief overview of the collage theorem, see "Chaos and Fractals," by R. Burton, *Mathematics Teacher*, Vol. 83, Oct. 1990, pp. 524–529. For a detailed treatment, see "A Better Way to Compress Images," by M. Barnsley and A. Sloan, *BYTE*, Vol. 13, Jan. 1988, pp. 215–223 or "Fractal Image Compression," by M. Barnsley, *Notices of the American Mathematical Society*, Vol. 43, No. 6, 1996, pp. 657–662.

compute the ratios

$$\frac{\text{area}(S_{j+1})}{\text{area}(S_j)}$$

for $j = 0, 1, 2$.

2. Let S_0 denote an isosceles right triangle with hypotenuse along the x-axis from $x = 0$ to $x = \sqrt{2}$.

(a) Draw S_0 and label the lengths of the legs of the triangle.

(b) Explain why S_0 is half-a-square. (Whenever we use the term *half-a-square* we will mean an appropriately sized isosceles right triangle.)

(c) In S_0 omit the hypotenuse and construct half-a-square on each leg, then omit the legs that were in S_0. Call the resulting figure S_1 and draw it. S_1 is composed of 4 line segments of length k. What is k?

(d) Construct half-a-square on the 4 line segments of S_1, then omit the line segments that compose S_1. Call the resulting figure S_2 and draw it. S_2 is composed of 8 line segments of length k. What is k?

(e) Use the procedure of "replacing line segments by half-a-square" to construct S_3 and draw it. How many sides of triangles are in S_3? (Count carefully.)

(f) A portion of the top of S_3 has the form

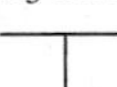

This figure consists of the legs of a pair of isosceles right triangles. Construct half-a-square on each of the 4 sides, then omit the original 4 sides. The result will be a portion of S_4. What feature does S_4 exhibit that does not appear in S_0 through S_3? (*Note*: If we continued the replacement by the half-a-square procedure indefinitely, we would construct a fractal sometimes referred to as **Levy's fractal**. Starting with basic forms other than the isosceles right triangle, S_0 will produce other fractals with varying shapes.)

3. Refer to the H-fractal discussed in Example 2.

(a) How long are the smallest line segments that appear in Figure 10.7(c)?

(b) Using Figure 10.7(c), draw the next step of the fractal construction.

(c) What are the lengths of the smallest line segments that appear in Figure 10.8?

(d) Determine the length of the portions of the H-fractal shown in Figures 10.7(a), 10.7(b), 10.7(c), and 10.8.

4. Refer to the Cantor fractal in Example 1.

(a) Using Figure 10.6, draw the next two portions of the fractal.

(b) Using Figure 10.6 and your results from (a), compute the length of the line segments appearing in these portions of the fractal construction.

(c) Using Figure 10.6 and your results from (a), determine the number of endpoints appearing in these portions of the fractal construction.

(d) How many endpoints will there be at the 10th step of the construction?

5. Let $\mathbf{b} \neq \mathbf{0}$ and $T(\mathbf{v}) = \mathbf{tran_b}(\mathbf{v})$. Compute $T(T(\mathbf{v}))$ and $T(T(T(\mathbf{v})))$. Describe the result of k compositions of T with itself.

6. Let $\mathbf{b} \neq \mathbf{0}$ and T be the affine transformation $T(\mathbf{v}) = A\mathbf{v} + \mathbf{b}$. Compute $T(T(\mathbf{v}))$ and $T(T(T(\mathbf{v})))$. Describe the result of k compositions of T with itself.

7. Let the affine transformation T be defined by

$$T(\mathbf{v}) = \begin{bmatrix} 3v_1 - v_2 + 1 \\ 4v_1 - 5 \end{bmatrix}.$$

Determine the matrix A and the vector $\mathbf{b}$.

8. Let the affine transformation T be defined by

$$T(\mathbf{v}) = \begin{bmatrix} -2v_2 + 4v_1 - 3 \\ 2v_1 + v_2 + 1 \end{bmatrix}.$$

Determine the matrix A and the vector $\mathbf{b}$.

9. The affine transformation T is such that

$$T(\mathbf{0}) = \begin{bmatrix} -2 \\ 1 \end{bmatrix}, \quad T(\mathbf{e}_1) = \begin{bmatrix} 1 \\ -1 \end{bmatrix}, \quad \text{and} \quad T(\mathbf{e}_2) = \begin{bmatrix} 4 \\ 3 \end{bmatrix},$$

where $\mathbf{e}_j = \mathbf{col}_j(I_2)$. Determine the matrix A and the vector $\mathbf{b}$.

10. The affine transformation T is such that

$$T(\mathbf{0}) = \begin{bmatrix} 0 \\ 4 \end{bmatrix}, \quad T(\mathbf{e}_1) = \begin{bmatrix} 2 \\ 2 \end{bmatrix}, \quad \text{and} \quad T(\mathbf{e}_2) = \begin{bmatrix} -1 \\ 1 \end{bmatrix},$$

where $\mathbf{e}_j = \mathbf{col}_j(I_2)$. Determine the matrix A and the vector $\mathbf{b}$.

11. The "house" depicted in Figure 10.19 is made by connecting the set of ordered pairs {(0, 0), (0, 1), (1, 1), (2, 3), (3, 1), (3, 0), (0, 0)} with straight line segments in the order listed. Use a matrix S, as in Example 5, to represent this figure. Compute the affine transformation of this image, $T(S) = AS + \mathbf{b}$, for each of the following pairs A and $\mathbf{b}$ and then sketch the image. (Recall that in this notation we mean add vector $\mathbf{b}$ to each column of the matrix AS.)

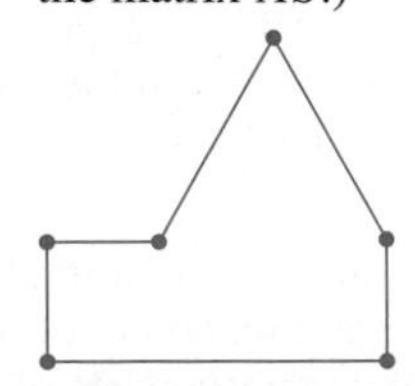

Figure 10.19 ▲

(a) $A = \begin{bmatrix} 2 & -2 \\ 2 & 1 \end{bmatrix}$, $\mathbf{b} = \begin{bmatrix} -2 \\ 1 \end{bmatrix}$

(b) $A = \begin{bmatrix} 2 & -2 \\ 2 & -2 \end{bmatrix}$, $\mathbf{b} = \begin{bmatrix} -2 \\ 1 \end{bmatrix}$

(c) $A = \begin{bmatrix} 2 & 2 \\ -2 & 1 \end{bmatrix}$, $\mathbf{b} = \begin{bmatrix} -2 \\ 1 \end{bmatrix}$

12. The "v-wedge" depicted in Figure 10.20 is made by connecting the set of ordered pairs {(0, 0), (1.5, 1.5), (3, 1), (2.5, 2.5), (4, 4), (4, 0), (0, 0)} with straight line segments in the order listed. Use a matrix S, as in Example 5, to represent this figure. Compute the affine transformation of this image, $T(S) = AS + \mathbf{b}$, for each of the following pairs A and $\mathbf{b}$ and then sketch the image. (Recall that in this notation we mean add vector $\mathbf{b}$ to each column of the matrix AS.)

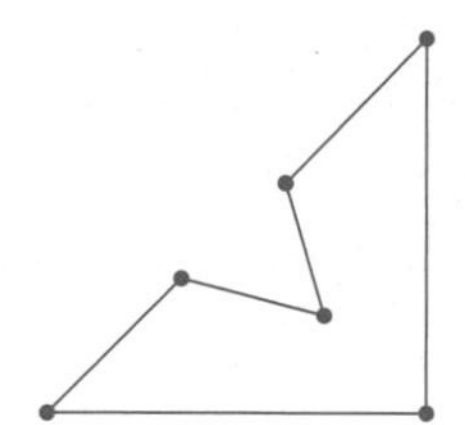

Figure 10.20 ▲

(a) $A = \begin{bmatrix} -1 & 1 \\ -2 & 1 \end{bmatrix}$, $\mathbf{b} = \begin{bmatrix} -2 \\ 1 \end{bmatrix}$

(b) $A = \begin{bmatrix} 2 & 0 \\ -2 & 1 \end{bmatrix}$, $\mathbf{b} = \begin{bmatrix} -1 \\ 0 \end{bmatrix}$

(c) $A = \begin{bmatrix} 1 & -1 \\ -1 & 1 \end{bmatrix}$, $\mathbf{b} = \begin{bmatrix} 1 \\ -2 \end{bmatrix}$

13. An affine transformation T has been applied to the "house" of Exercise 11. The original figure and its image are shown in Figure 10.21. Using just Figure 10.21, determine the translation $\mathbf{b}$ and the matrix A of the affine transformation that has been applied.

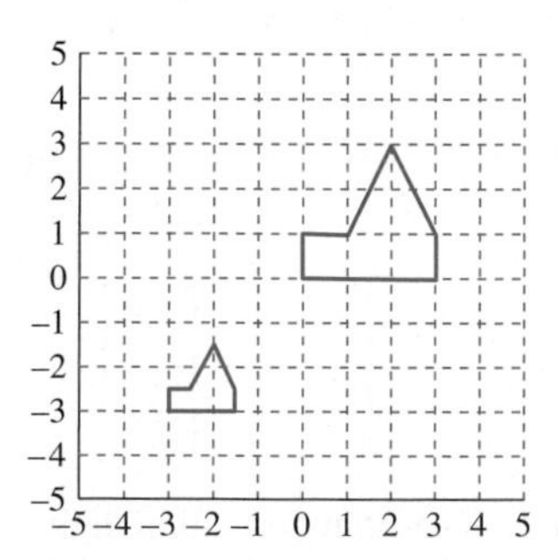

Figure 10.21 ▲

14. An affine transformation T has been applied to the "house" of Exercise 11. The original figure and its image are shown in Figure 10.22. Using just Figure 10.22, determine the translation $\mathbf{b}$ and the matrix A of the affine transformation that has been applied.

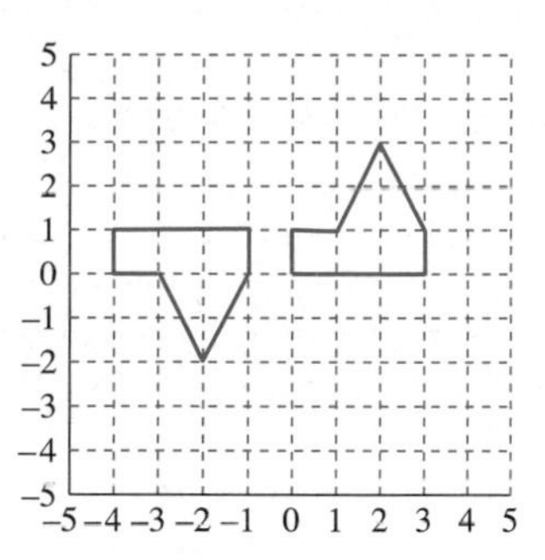

Figure 10.22 ▲

15. A polygon is constructed with vertices {(2, 0), (2, 2), (3, 3), (5, 4), (3, 1), (2, 0)} connected by straight line segments in the order listed. The image of the polygon by the affine transformation T with matrix $A = \begin{bmatrix} p & r \\ s & t \end{bmatrix}$ and vector $\mathbf{b} = \begin{bmatrix} b_1 \\ b_2 \end{bmatrix}$ is given by the corresponding set of vertices {(3, −4), (3, −2), (5, −3), (9, −6), (5, −5), (3, −4)}. Find A and $\mathbf{b}$.

16. A polygon is constructed with vertices {(1, 2), (1, 4), (2, 3), (3, 5), (3, 3), (1, 2)} connected by straight line segments in the order listed. The image of the polygon by the affine transformation T with matrix $A = \begin{bmatrix} p & r \\ s & t \end{bmatrix}$ and vector $\mathbf{b} = \begin{bmatrix} b_1 \\ b_2 \end{bmatrix}$ is given by the corresponding set of vertices {(1, 0), (−1, 0), (1, 1), (0, 2), (2, 2), (1, 0)}. Find A and $\mathbf{b}$.

17. The fractal images shown in Figures 10.17 and 10.18 were drawn with the aid of a computer. In this exercise portions of the fractals can be drawn by hand on graph paper. Let S be the unit square with vertices {(0, 0), (1, 0), (1, 1), (0, 1)}.

(a) Let T be the affine transformation given by $T(\mathbf{v}) = A\mathbf{v} + \mathbf{b}$, where

$$A = \begin{bmatrix} \frac{1}{2} & 0 \\ 0 & \frac{1}{2} \end{bmatrix} \quad \text{and} \quad \mathbf{b} = \begin{bmatrix} 0 \\ 0 \end{bmatrix}.$$

Sketch S and the following images of S on the same sheet of graph paper: $T(S)$, $T(T(S))$, $T(T(T(S)))$. Give a short description of the fractal that would be generated if we were to continue to compute successive images of S. [Recall that in the notation $T(S) = AS + \mathbf{b}$ we mean add vector $\mathbf{b}$ to each column of the matrix AS.]

(b) Follow the directions in (a), but use $\mathbf{b} = \begin{bmatrix} \frac{1}{2} \\ \frac{1}{2} \end{bmatrix}$.

(c) Follow the directions in (a), but use $\mathbf{b} = \begin{bmatrix} \frac{1}{4} \\ \frac{1}{4} \end{bmatrix}$.

(d) Follow the directions in (a), but use $\mathbf{b} = \begin{bmatrix} -\frac{1}{2} \\ -\frac{1}{2} \end{bmatrix}$.

18. The fractal images shown in Figures 10.17 and 10.18 were drawn with the aid of a computer. In this exercise portions of the fractals can be drawn by hand on graph paper. Let S be the triangle with vertices $\{(0, 0), (1, 1), (-1, 1)\}$.

(a) Let T be the affine transformation given by $T(\mathbf{v}) = A\mathbf{v} + \mathbf{b}$, where

$$A = \begin{bmatrix} \frac{1}{2} & 0 \\ 0 & \frac{1}{2} \end{bmatrix} \quad \text{and} \quad \mathbf{b} = \begin{bmatrix} 0 \\ 0 \end{bmatrix}.$$

Sketch S and the following images of S on the same sheet of graph paper: $T(S)$, $T(T(S))$, $T(T(T(S)))$. Give a short description of the fractal that would be generated if we were to continue to compute successive images of S. (Recall that in the notation $T(S) = AS + \mathbf{b}$ we mean add vector $\mathbf{b}$ to each column of the matrix AS.)

(b) Follow the directions in (a), but use $\mathbf{b} = \begin{bmatrix} 0 \\ \frac{1}{2} \end{bmatrix}$.

(c) Follow the directions in (a), but use $\mathbf{b} = \begin{bmatrix} 0 \\ -\frac{1}{2} \end{bmatrix}$.

(d) Follow the directions in (a), but use $\mathbf{b} = \begin{bmatrix} \frac{1}{2} \\ \frac{1}{2} \end{bmatrix}$.

Theoretical Exercises

T.1. Counterclockwise rotations by an angle θ about the origin in R^2 are performed by the linear transformation $T_\theta: R^2 \to R^2$ defined by $T_\theta(\mathbf{v}) = R_\theta \mathbf{v}$, where

$$R_\theta = \begin{bmatrix} \cos(\theta) & -\sin(\theta) \\ \sin(\theta) & \cos(\theta) \end{bmatrix}.$$

(a) Construct an affine transformation that performs a rotation about the y-axis and shifts the figure up two units.

(b) Construct an affine transformation that performs a rotation about the x-axis and shifts the figure down one unit.

T.2. (a) Determine the matrix R_ϕ that performs a rotation in the clockwise direction by an angle ϕ about the origin in R^2.

(b) Construct an affine transformation that performs a rotation about the x-axis and shifts the figure up 1 unit.

T.3. Perform the following constructions of affine transformations to develop the iterated function system for the fractal described in Exercise 1. Figure 10.23 shows S_1 with the 5 squares of size 1×1 numbered for ease of reference.

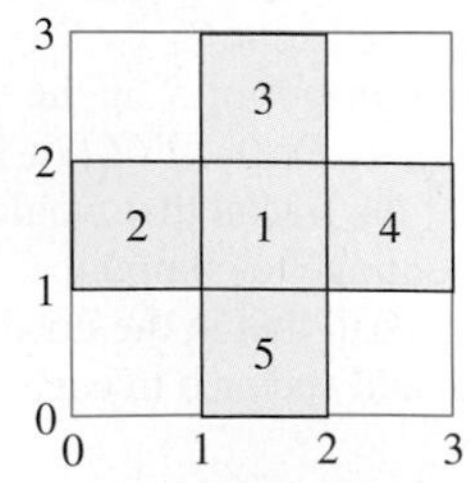

Figure 10.23 ▲

(a) We have applied a contraction to the original figure S_0 to get each of the 5 squares in S_1. Construct the matrix A that performs this contraction.

(b) Each of the 5 squares in S_1 is a translation of the contraction developed in (a). Imagine a 1×1 square with lower left corner at the origin. Now determine the translation of this square to obtain each of the squares labeled 1 through 5. Write the affine transformation $T_j(\mathbf{v}) = A\mathbf{v} + \mathbf{b}_j$, $j = 1, 2, 3, 4, 5$, where A is as in part (a) and $\mathbf{b}_j$ is the translation needed to move the 1×1 square with lower left corner at the origin to the square numbered j. (The IFS here has five transformations.)

T.4. In the theoretical development of fractals the concept of a **fixed point** of an affine transformation plays an important role. We have the following definition:

> $\mathbf{v}$ is a **fixed point** of the affine transformation T provided $T(\mathbf{v}) = \mathbf{v}$.

(a) Let $T: R^2 \to R^2$ be an affine transformation defined by $T(\mathbf{v}) = A\mathbf{v} + \mathbf{b}$, where

$$A = \begin{bmatrix} p & r \\ s & t \end{bmatrix} \quad \text{and} \quad \mathbf{b} = \begin{bmatrix} b_1 \\ b_2 \end{bmatrix}.$$

Show that T has a single fixed point whenever $(p - 1)(t - 1) - rs \neq 0$.

(b) Determine the fixed point of the following affine transformation T, where

$$A = \begin{bmatrix} \frac{1}{6} & \frac{\sqrt{3}}{6} \\ -\frac{\sqrt{3}}{6} & \frac{1}{6} \end{bmatrix}, \quad \mathbf{b} = \begin{bmatrix} \frac{1}{2} \\ \frac{\sqrt{3}}{6} \end{bmatrix}.$$

T.5. Let $T: R^2 \to R^2$ be a linear transformation defined by $T(\mathbf{v}) = A\mathbf{v}$. When does T have a fixed point and how many are there?

T.6. We expressed an affine transformation $T: R^2 \to R^2$ in the form $T(\mathbf{v}) = A\mathbf{v} + \mathbf{b}$, where $A = \begin{bmatrix} p & r \\ s & t \end{bmatrix}$ and $\mathbf{b} = \begin{bmatrix} b_1 \\ b_2 \end{bmatrix}$.

(a) Show that the matrix A can be expressed in terms of sines and cosines in the form

$$A = \begin{bmatrix} r_1\cos(\theta_1) & -r_2\sin(\theta_2) \\ r_1\sin(\theta_1) & r_2\cos(\theta_2) \end{bmatrix},$$

where (r_1, θ_1) are the polar coordinates of (p, s) and $(r_2, \theta_2 + \frac{\pi}{2})$ are the polar coordinates of (r, t).

(b) Use the images of $\mathbf{e}_1 = \mathbf{col}_1(I_2)$ and $\mathbf{e}_2 = \mathbf{col}_2(I_2)$ to determine the role of r_j and θ_j on the result of the affine transformation T.

MATLAB Exercises

ML.1. A well-known fractal is often referred to as the Barnsley fern. Its iterated function system uses different probabilities for the 4 affine transformations involved. In MATLAB type **help fernifs** for directions. Experiment with the routine **fernifs** to generate a fern.

ML.2. To experiment with iterated function systems you can use the MATLAB routine **chaosgame**. In MATLAB type **help chaosgame** for directions. You can enter up to 5 affine transformations and assign probabilities for the use of the transformations. We suggest you use entries between -1 and 1 for the matrix A and entries between -2 and 2 for the entries of the vector $\mathbf{b}$. If you assign unequal probabilities to the affine transformations, make sure the sum of the probabilities is 1.

(a) Use the routine **chaosgame** with the following iterated function system.

	p	r	s	t	b_1	b_2
T_1	0.6	0	0	0.6	0	-0.5
T_2	0.6	0	0	0.6	0	0.5
T_3	$\frac{\sqrt{2}}{4}$	$-\frac{\sqrt{2}}{4}$	$\frac{\sqrt{2}}{4}$	$\frac{\sqrt{2}}{4}$	-0.5	-0.25
T_4	$\frac{\sqrt{2}}{4}$	$\frac{\sqrt{2}}{4}$	$-\frac{\sqrt{2}}{4}$	$\frac{\sqrt{2}}{4}$	0.5	-0.25

Describe the display that is generated.

(b) Use the IFS you constructed in Exercise T.3 in the routine **chaosgame**. Describe the display generated.

Key Ideas for Review

- **Theorem 10.1.** If $L: V \to W$ is a linear transformation, then
$$L(c_1\mathbf{v}_1 + c_2\mathbf{v}_2 + \cdots + c_k\mathbf{v}_k) = c_1L(\mathbf{v}_1) + c_2L(\mathbf{v}_2) + \cdots + c_kL(\mathbf{v}_k).$$
- **Theorem 10.3.** Let $L: V \to W$ be a linear transformation and let $S = \{\mathbf{v}_1, \mathbf{v}_2, \ldots, \mathbf{v}_n\}$ be a basis for V. If $\mathbf{u}$ is any vector in V, then $L(\mathbf{u})$ is completely determined by $\{L(\mathbf{v}_1), L(\mathbf{v}_2), \ldots, L(\mathbf{v}_n)\}$.
- **Theorem 10.4.** If $L: V \to W$ is a linear transformation, then $\ker L$ is a subspace of V.
- **Theorem 10.5.** A linear transformation $L: V \to W$ is one-to-one if and only if $\ker L = \{\mathbf{0}_V\}$.
- **Theorem 10.6.** If $L: V \to W$ is a linear transformation, then range L is a subspace of W.
- **Theorem 10.7.** If $L: V \to W$ is a linear transformation, then
$$\dim(\ker L) + \dim(\text{range } L) = \dim V.$$
- **Matrix of a linear transformation $L: V \to W$.** See Theorem 10.8.
- **Theorem 10.9.** Let $L: V \to V$ be a linear operator where V is an n-dimensional vector space. Let $S = \{\mathbf{v}_1, \mathbf{v}_2, \ldots, \mathbf{v}_n\}$ and $T = \{\mathbf{w}_1, \mathbf{w}_2, \ldots, \mathbf{w}_n\}$ be bases for V and let P be the transition matrix from T to S. If A is the matrix representing L with respect to S, then $P^{-1}AP$ is the matrix representing L with respect to T.
- **Theorem 10.10.** Consider the linear operator $L: R^n \to R^n$ defined by $L(\mathbf{x}) = A\mathbf{x}$ for $\mathbf{x}$ in R^n. Then A is diagonalizable with n linearly independent eigenvectors $\mathbf{x}_1, \mathbf{x}_2, \ldots, \mathbf{x}_n$ if and only if the matrix of L with respect to $S = \{\mathbf{x}_1, \mathbf{x}_2, \ldots, \mathbf{x}_n\}$ is diagonal.
- **List of nonsingular equivalences.** The following statements are equivalent for an $n \times n$ matrix A.
  1. A is nonsingular.
  2. $\mathbf{x} = \mathbf{0}$ is the only solution to $A\mathbf{x} = \mathbf{0}$.
  3. A is row equivalent to I_n.
  4. The linear system $A\mathbf{x} = \mathbf{b}$ has a unique solution for every $n \times 1$ matrix $\mathbf{b}$.
  5. $\det(A) \neq 0$.
  6. A has rank n.
  7. A has nullity 0.
  8. The rows of A form a linearly independent set of n vectors in R^n.
  9. The columns of A form a linearly independent set of n vectors in R^n.
  10. Zero is *not* an eigenvalue of A.
  11. The linear operator $L: R^n \to R^n$ defined by $L(\mathbf{x}) = A\mathbf{x}$, for $\mathbf{x}$ in R^n, is one-to-one and onto.
- A **fractal** is a figure that is self-similar as we make changes of scale. Geometrically, if we "zoom in" or "zoom out" on the figure, we see the same shape.
- An **affine transformation** is nonlinear and consists of a linear transformation followed by a translation.
- An **iterated function system** consists of a random iteration of a set of affine transformations that generates the attractor of the fractal image.

Supplementary Exercises

1. Is $L: R^2 \to R^2$ defined by
$$L(x, y) = (x + y, 2x - y)$$
a linear transformation?

2. Is $L: P_2 \to P_2$ defined by
$$L(at^2 + bt + c) = (a - 1)t^2 + bt - c$$
a linear transformation?

3. Let $L: P_1 \to P_1$ be the linear transformation defined by
$$L(t - 1) = t + 2, \quad L(t + 1) = 2t + 1.$$
Find $L(5t + 1)$. [*Hint*: $\{t - 1, t + 1\}$ is a basis for P_1.]

4. Let $L: P_2 \to P_2$ be defined by
$$L(at^2 + bt + c) = (a - 2c)t^2 + (b - c)t.$$

(a) Find a basis for $\ker L$.

(b) Is L one-to-one?

5. Let $L: R^2 \to R^3$ be defined by
$$L(x, y) = (x + y, x - y, x + 2y).$$

(a) Find a basis for range L.

(b) Is L onto?

6. Find the dimension of the solution space of $A\mathbf{x} = \mathbf{0}$, where
$$A = \begin{bmatrix} 1 & -2 & -1 & 2 & -1 \\ 4 & -7 & 1 & 4 & 0 \\ 1 & -3 & -6 & 4 & -4 \\ 2 & -3 & 3 & 2 & 1 \end{bmatrix}.$$

7. Let $S = \{t^2 - 1, t + 2, t - 1\}$ be a basis for P_2.

(a) Find the coordinate vector of $2t^2 - 2t + 6$ with respect to S.

(b) If the coordinate vector of $p(t)$ with respect to S is
$$\begin{bmatrix} 2 \\ -1 \\ 3 \end{bmatrix},$$
find $p(t)$.

8. Let $L: P_2 \to P_2$ be the linear transformation that is defined by
$$L(at^2 + bt + c) = (a + 2c)t^2 + (b - c)t + (a - c).$$
Let $S = \{t^2, t, 1\}$ and $T = \{t^2 - 1, t, t - 1\}$ be bases for P_2.

(a) Find the matrix of L with respect to S and T.

(b) If $p(t) = 2t^2 - 3t + 1$, compute $L[p(t)]$ using the matrix obtained in part (a).

9. Let $L: P_1 \to P_1$ be a linear transformation. Suppose that the matrix of L with respect to the basis $S = \{p_1(t), p_2(t)\}$ is
$$A = \begin{bmatrix} 2 & -3 \\ 1 & 2 \end{bmatrix},$$
where
$$p_1(t) = t - 2 \quad \text{and} \quad p_2(t) = t + 1.$$

(a) Compute $\left[L[p_1(t)]\right]_S$ and $\left[L[p_2(t)]\right]_S$.

(b) Compute $L[p_1(t)]$ and $L[p_2(t)]$.

(c) Compute $L(t + 2)$.

10. Let $L: P_3 \to P_3$ be defined by
$$L(at^3 + bt^2 + ct + d) = 3at^2 + 2bt + c.$$
Find the matrix of L with respect to the basis $S = \{t^3, t^2, t, 1\}$ for P_3.

11. Let $L: R^3 \to R^3$ be the linear transformation defined by
$$L\left(\begin{bmatrix} x \\ y \\ z \end{bmatrix}\right) = \begin{bmatrix} y + z \\ x + z \\ x + y \end{bmatrix}.$$
Let S be the natural basis for R^3 and let
$$T = \left\{ \begin{bmatrix} 0 \\ 1 \\ 1 \end{bmatrix}, \begin{bmatrix} 1 \\ 0 \\ 1 \end{bmatrix}, \begin{bmatrix} 1 \\ 1 \\ 0 \end{bmatrix} \right\}$$
be another basis for R^3. Find the matrix of L with respect to S and T.

12. Let $L: M_{nn} \to R^1$ be defined by $L(A) = \det(A)$, for A in M_{nn}. Is L a linear transformation? Justify your answer.

13. Let A be a fixed $n \times n$ matrix. Define $L: M_{nn} \to M_{nn}$ by $L(B) = AB - BA$ for B in M_{nn}. Is L a linear transformation? Justify your answer.

14. Let P be a fixed nonsingular $n \times n$ matrix. Define $L: M_{nn} \to M_{nn}$ by $L(A) = P^{-1}AP$ for A in M_{nn}. Is L a linear transformation? Justify your answer.

15. ***(Calculus Required)*** Let $V = C[0, 1]$, the vector space of all real-valued continuous functions that are defined on $[0, 1]$, and let $L: V \to R^1$ be given by $L(f) = f(0)$, for f in V.

(a) Show that L is a linear transformation.

(b) Describe the kernel of L and give examples of polynomials, quotients of polynomials, and trigonometric functions that belong to $\ker L$.

(c) If we redefine L by $L(f) = f\left(\frac{1}{2}\right)$, is it still a linear transformation? Explain.

16. The "jet fighter" shown in Figure 10.24 is constructed by connecting the set of ordered pairs $\{(1, 0), (1, 2), (3, 2), (1, 3), (0, 6), (-1, 3), (-3, 2), (-1, 2), (-1, 0), (1, 0)\}$ with straight lines in the order listed. Use a matrix S, as in Example 5 of Section 10.4, to represent this figure.

(a) Determine the affine transformation that when applied to the "jet fighter" produces the image shown in Figure 10.25.

(b) Determine the affine transformation that when applied to the "jet fighter" produces the image shown in Figure 10.26.

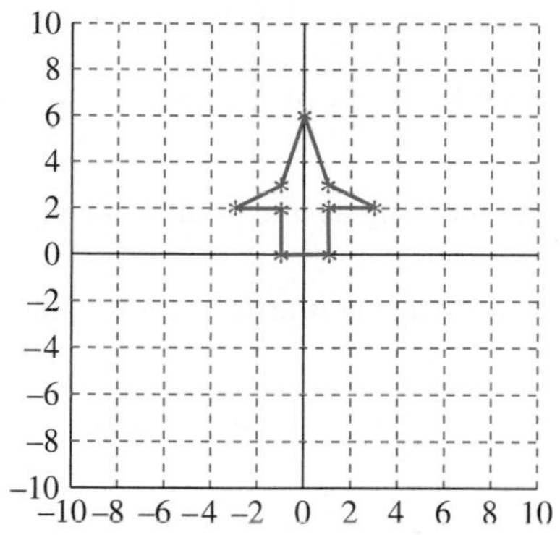

Figure 10.24 ▲

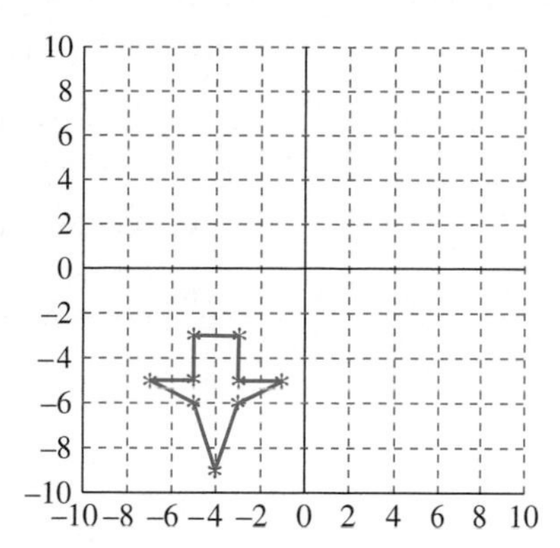

Figure 10.25 ▲

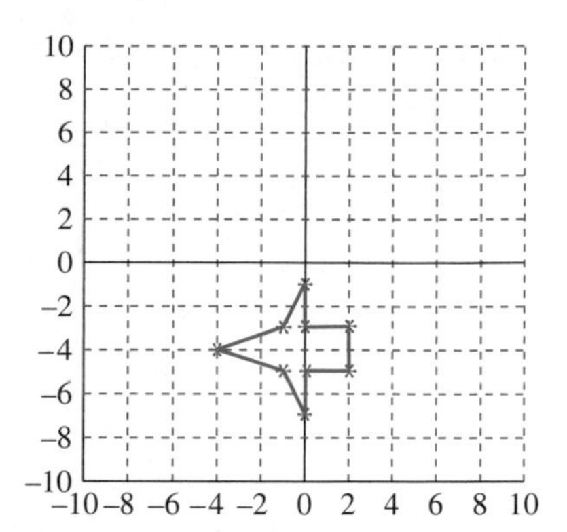

Figure 10.26 ▲

17. A fractal is to be constructed as follows, beginning with the line segment that starts at (0, 0) and ends at (1, 0):

(i) Make the line segment $\frac{1}{2}$ as long.

(ii) Rotate it 90° counterclockwise.

(iii) Draw the result of (i) and (ii) at the end of the previously drawn line segment.

(a) Construct the first four iterations of these steps.

(b) Describe the form of the figure generated after many steps.

Theoretical Exercises

T.1. Let V be an n-dimensional vector space with basis $S = \{\mathbf{v}_1, \mathbf{v}_2, \ldots, \mathbf{v}_n\}$. Show that

$$\{[\mathbf{v}_1]_S, [\mathbf{v}_2]_S, \ldots, [\mathbf{v}_n]_S\}$$

is the natural basis for R^n.

T.2. Let V be an n-dimensional vector space with basis $S = \{\mathbf{v}_1, \mathbf{v}_2, \ldots, \mathbf{v}_n\}$. If $\mathbf{v}$ and $\mathbf{w}$ are vectors in V and c is a scalar, show that

$$[\mathbf{v}+\mathbf{w}]_S = [\mathbf{v}]_S + [\mathbf{w}]_S$$
$$[c\mathbf{v}]_S = c[\mathbf{v}]_S.$$

T.3. Let V and W be two vector spaces of dimensions n and m, respectively. If $L_1: V \to W$ and $L_2: V \to W$ are linear transformations, we define

$$L_1 \boxplus L_2: V \to W$$

by

$$(L_1 \boxplus L_2)(\mathbf{v}) = L_1(\mathbf{v}) + L_2(\mathbf{v}),$$

for $\mathbf{v}$ in V. Also, if $L: V \to W$ is a linear transformation and c is a scalar, we define

$$c \boxdot L: V \to W$$

by

$$(c \boxdot L)(\mathbf{v}) = cL(\mathbf{v}),$$

for $\mathbf{v}$ in V.

(a) Show that $L_1 \boxplus L_2$ is a linear transformation.

(b) Show that $c \boxdot L$ is a linear transformation.

(c) Let $V = R^3$, $W = R^2$, $L_1: V \to W$, and $L_2: V \to W$ be defined by

$$L_1(\mathbf{v}) = L_1(v_1, v_2, v_3) = (v_1 + v_2, v_2 + v_3)$$
$$L_2(\mathbf{v}) = L_2(v_1, v_2, v_3) = (v_1 + v_3, v_2).$$

Compute $(L_1 \boxplus L_2)(\mathbf{v})$ and $(-2 \boxdot L_1)(\mathbf{v})$.

T.4. Show that the set U of all linear transformations of an n-dimensional vector space into an m-dimensional vector space W is a vector space under the operations $\boxplus$ and $\boxdot$ defined in Supplementary Exercise T.3.

T.5. Let $L: R^n \to R^m$ be a linear transformation defined by $L(\mathbf{x}) = A\mathbf{x}$, $\mathbf{x}$ in R^n, where A is an $m \times n$ matrix.

(a) Show that L is one-to-one if and only if rank $A = n$.

(b) Show that L is onto if and only if rank $A = m$.

T.6. Let V be an n-dimensional vector space and $S = \{\mathbf{v}_1, \mathbf{v}_2, \ldots, \mathbf{v}_n\}$ a basis for V. Define $L: R^n \to V$ as follows: If $\mathbf{v} = (a_1, a_2, \ldots, a_n)$ is a vector in R^n, let

$$L(\mathbf{v}) = a_1\mathbf{v}_1 + a_2\mathbf{v}_2 + \cdots + a_n\mathbf{v}_n.$$

Show that:

(a) L is a linear transformation.

(b) L is one-to-one.

(c) L is onto.

Chapter Test

1. Let $L: R^2 \to R^3$ be a linear transformation for which we know that

$$L\left(\begin{bmatrix} 1 \\ -1 \end{bmatrix}\right) = \begin{bmatrix} 1 \\ 2 \\ -1 \end{bmatrix}$$

and

$$L\left(\begin{bmatrix} 2 \\ -1 \end{bmatrix}\right) = \begin{bmatrix} 0 \\ 1 \\ 2 \end{bmatrix}.$$

Find $L\left(\begin{bmatrix} 8 \\ -5 \end{bmatrix}\right)$.

2. Let $L: R^3 \to R^3$ be defined by

$$L(x, y, z) = (x + 2y + z, x + y, 2y + z).$$

(a) Find a basis for $\ker L$.

(b) Is L one-to-one?

3. Let $L: R^2 \to R^3$ be defined by

$$L(x, y) = (x + y, x - y, 2x + y).$$

(a) Find a basis for range L.

(b) Is L onto?

4. Find the dimension of the solution space of $A\mathbf{x} = \mathbf{0}$, where

$$A = \begin{bmatrix} 1 & 0 & -1 & 2 & 3 \\ 2 & 1 & -2 & 2 & 1 \\ 0 & 1 & -6 & 8 & 8 \\ 1 & -1 & -1 & 4 & 8 \end{bmatrix}.$$

5. Let $L: R^2 \to R^2$ be the linear transformation defined by

$$L\left(\begin{bmatrix} x \\ y \end{bmatrix}\right) = \begin{bmatrix} x + 2y \\ x - y \end{bmatrix}.$$

Let

$$S = \left\{\begin{bmatrix} 1 \\ -1 \end{bmatrix}, \begin{bmatrix} 0 \\ 2 \end{bmatrix}\right\}$$

and

$$T = \left\{\begin{bmatrix} 1 \\ 2 \end{bmatrix}, \begin{bmatrix} -1 \\ 2 \end{bmatrix}\right\}$$

be bases for R^2. Find the matrix of L with respect to S and T.

6. Answer each of the following as true or false. Justify your answer.

(a) If $L: P_2 \to P_1$ is the linear transformation defined by

$$L(at^2 + bt + c) = (a - c)t + (b + c),$$

then $t^2 + 2t + 1$ is in $\ker L$.

(b) If $L: R^2 \to R^2$ is the linear transformation defined by

$$L(x, y) = (x - y, x + y),$$

then $(2, 3)$ is in range L.

(c) If A is a 4×7 matrix whose rank is 4, then the homogeneous system $A\mathbf{x} = \mathbf{0}$ has a nontrivial solution.

(d) There are linear transformations $L: R^3 \to R^5$ that are onto.

(e) If $L: R^7 \to R^5$ is a linear transformation such that $\dim(\ker L) = 3$, then $\dim(\text{range } L) = 2$.

7. A circle of radius 2 is drawn with its center at the origin. Let this circle be represented by 100 equispaced points around its circumference that have been stored in a 2×100 matrix S. Determine the affine transformation and the process that you would use to construct a "bulls-eye" of equally spaced rings centered at the origin. (Retain the circular image at each step.)

8. For the circle of Exercise 7, describe the resulting figure when the affine transformation $T(S) = AS + \mathbf{b}$, where

$$A = \begin{bmatrix} 0.5 & 0 \\ 0 & 0.5 \end{bmatrix} \quad \text{and} \quad \mathbf{b} = \begin{bmatrix} 0 \\ 2 \end{bmatrix},$$

is applied as follows and all the images are retained: $T(S)$, $T(T(S))$, $T(T(T(S)))$.

Cumulative Review of Introductory Linear Algebra

Answer each of the following as true (T) or false (F). Justify your answer.

1. If an $n \times n$ matrix A is singular, then A has either a row or a column of zeros.

2. A diagonal matrix is nonsingular if and only if none of the entries on its main diagonal are zero.

3. Two vectors in R^3 always span a two-dimensional subspace.

4. If $|AB| = 12$ and $|A| = 4$, then $|B| = 48$.

5. Let $L: R^6 \to R^{10}$ be a linear transformation defined by $L(\mathbf{x}) = A\mathbf{x}$ for $\mathbf{x}$ in R^6. If $\dim(\text{range } L) = 3$, then $\dim(\ker L) = 7$.

6. If AB is singular, then A is singular or B is singular.

7. Let $L: R^n \to R^n$ be a linear transformation defined by $L(\mathbf{x}) = A\mathbf{x}$. Then L is onto if and only if $\det(A) \neq 0$.

8. The columns of a 5×8 matrix whose rank is 5 form a linearly dependent set.

9. If A is an $m \times n$ matrix with $m < n$, then the linear system $A\mathbf{x} = \mathbf{b}$ has a solution for every $m \times 1$ matrix $\mathbf{b}$.

10. Let $L: R^n \to R^n$ be the linear transformation defined by $L(\mathbf{x}) = A\mathbf{x}$, for $\mathbf{x}$ in R^n. Then L is onto if and only if A is nonsingular.

11. If A is an $n \times n$ matrix, then rank $A < n$ if and only if some eigenvalue of A is zero.

12. Let A be an $n \times n$ matrix. If $A\mathbf{x} = A\mathbf{y}$, then $\mathbf{x} = \mathbf{y}$.

13. Span $\{(1, 1, 0), (0, 1, -1), (1, 0, 1)\} = R^3$.

14. If no row in an $n \times n$ matrix is a multiple of another row of A, then $\det(A) \neq 0$.

15. If $L: V \to W$ is a linear transformation, then $\ker L = V$ if and only if range $L = \{\mathbf{0}_W\}$.

16. The inverse of a nonsingular diagonal matrix is a diagonal matrix.

17. If A and B are $n \times n$ matrices, then $(A + B)^2 = A^2 + 2AB + B^2$.

18. If $\mathbf{u}$ and $\mathbf{v}$ are vectors in R^n, then $\|\mathbf{u} - \mathbf{v}\|^2 = \|\mathbf{u}\|^2 - \|\mathbf{v}\|^2$.

19. There are real vector spaces containing exactly seven vectors.

20. If $\mathbf{u}_1, \mathbf{u}_2, \ldots, \mathbf{u}_k$ are orthogonal to $\mathbf{v}$, then every vector in span $\{\mathbf{u}_1, \mathbf{u}_2, \ldots, \mathbf{u}_k\}$ is orthogonal to $\mathbf{v}$.

21. $\det(A) = 0$ if and only if some eigenvalue of A is zero.

22. If λ is an eigenvalue of A of multiplicity k, then the dimension of the eigenspace associated with λ is k.

23. If A and B are $n \times n$ matrices, then $(A^T B^T)^T = BA$.

24. Let A be an $n \times n$ singular matrix. If the linear system $A\mathbf{x} = \mathbf{b}$ has a solution for $\mathbf{b} \neq 0$, then it has infinitely many solutions.

25. Let $L: R^n \to R^n$ be a linear transformation defined by $L(\mathbf{x}) = A\mathbf{x}$, for $\mathbf{x}$ in R^n. Then A is singular if and only if $\ker L = \{\mathbf{0}_V\}$.

26. The product of two diagonal matrices is always a diagonal matrix.

27. If A is an $n \times n$ matrix that is row equivalent to I_n, then A is singular.

28. If

$$A = \begin{bmatrix} a-3 & 2 & 1 \\ -1 & a & 1 \\ 0 & 0 & 1 \end{bmatrix},$$

then the only value of a for which the linear system $A\mathbf{x} = \mathbf{0}$ has a nontrivial solution is $a = 2$.

29. If A is a 3×3 matrix and $|A| = 3$, then $\left|\frac{1}{2}A^{-1}\right| = \frac{8}{3}$.

30. The linear system $A\mathbf{x} = \mathbf{b}$ has a solution if and only if $\mathbf{b}$ is in the column space of A.

31. Let $L: R^n \to R^n$ be a linear transformation defined by $L(\mathbf{x}) = A\mathbf{x}$, for $\mathbf{x}$ in R^n. Then $\dim(\text{range } L) = n$ if and only if rank $A = n$.

32. If W is a subspace of a finite-dimensional vector space V such that $\dim W = \dim V$, then $W = V$.

33. If A is similar to B, then rank A = rank B.

34. The set of all vectors of the form $(a, b, -a)$ is a subspace of R^3.

35. If the columns of an $n \times n$ matrix span R^n, then the rows are linearly independent.

36. If A is an $n \times n$ matrix, then rank $A = n$.

37. Let λ_1 and λ_2 be eigenvalues of A with associated eigenvectors $\mathbf{x}_1$ and $\mathbf{x}_2$. If $\lambda_1 = \lambda_2$, then $\mathbf{x}_1$ and $\mathbf{x}_2$ are linearly independent.

38. Every orthogonal set of n vectors in R^n is a basis for R^n.

39. The set of all solutions to the linear system $A\mathbf{x} = \mathbf{b}$, where A is $m \times n$ and $\mathbf{b} \neq \mathbf{0}$, is a subspace of R^n.

40. If $\mathbf{u}$ is a vector in R^n such that $\mathbf{u} \cdot \mathbf{v} = 0$ for all $\mathbf{v}$ in R^n, then $\mathbf{u} = \mathbf{0}$.

41. Every set of linearly independent vectors in R^3 contains three vectors.

42. If $\det(A) = 0$, then the linear system $A\mathbf{x} = \mathbf{b}$, $\mathbf{b} \neq \mathbf{0}$, has no solution.

43. Let V be an n-dimensional vector space. If a set of m vectors spans V, then $m = n$.

44. Every set of five orthonormal vectors is a basis for R^5.

45. It is possible for a vector space V to have more than one orthonormal basis.

46. If the set of vectors S spans a vector space V, then every subset of S also spans V.

47. If an $n \times n$ matrix A is diagonalizable, then A^3 is diagonalizable.

48. If $\mathbf{x}$ and $\mathbf{y}$ are eigenvectors of A associated with the eigenvalue λ, then $\mathbf{x} + \mathbf{y}$ is also an eigenvector of A associated with λ.

49. If $\mathbf{x}_0$ is a nontrivial solution to the homogeneous system $A\mathbf{x} = \mathbf{0}$, where A is $n \times n$, then $\mathbf{x}_0$ is an eigenvector of A associated with the eigenvalue 0.

50. If A is similar to B, then A^n is similar to B^n for each positive integer n.

51. If A is an $m \times n$ matrix, then

$$\begin{aligned}&\dim(\text{null space of } A)\\ &\qquad + \dim(\text{column space of } A) = n.\end{aligned}$$

52. The set of solutions to any linear system is always a subspace.

53. The number of linearly independent eigenvectors of a matrix is always greater than or equal to the number of distinct eigenvalues.

54. Every finite set of vectors in a vector space that contains the zero vector is linearly dependent.

55. If the rows of a 4×6 matrix are linearly independent, then the column rank $= 4$.

56. If $\mathbf{v}_1$, $\mathbf{v}_2$, $\mathbf{v}_3$, $\mathbf{v}_4$, and $\mathbf{v}_5$ are linearly dependent vectors in a vector space, then $\mathbf{v}_1$, $\mathbf{v}_2$, and $\mathbf{v}_3$ are linearly dependent.

57. $\det(ABC) = \det(BAC)$.

58. The set of all $n \times n$ symmetric matrices form a subspace of M_{nn}.

59. Every linear system $A\mathbf{x} = \mathbf{0}$, where A is an $m \times n$ matrix, has a solution $\mathbf{x} \neq \mathbf{0}$ if $m < n$.

60. If V is a subspace of R^5, then $1 \leq \dim V < 5$.

61. A diagonalizable $n \times n$ matrix must always have n distinct eigenvalues.

62. Every symmetric matrix is diagonalizable.

63. If c and d are scalars and $\mathbf{u}$ is a vector in R^n such that $c\mathbf{u} = d\mathbf{u}$, then $c = d$.

64. The linear transformation $L\colon P_2 \to P_2$ defined by

$$L(at^2 + bt + c) = 2at + b$$

is one-to-one.

65. If $\mathbf{u}$ and $\mathbf{v}$ are any vectors in R^n, then $-1 \leq \mathbf{u} \cdot \mathbf{v} \leq 1$.

66. If A and B are $n \times n$ matrices, then $(A + B)^{-1} = A^{-1} + B^{-1}$.

67. The rows of an $n \times n$ matrix A always form a basis for the row space of A.

68. If A and B are $n \times n$ matrices with $\det(AB) \neq 0$, then A and B are both nonsingular.

69. If A and B are $n \times n$ orthogonal matrices, then AB and BA are orthogonal matrices.

70. If A is a nonsingular matrix and $A\mathbf{u} = A\mathbf{v}$, then $\mathbf{u} = \mathbf{v}$.

71. If A and B are $n \times n$ nonsingular matrices, then $(3AB)^{-1} = \frac{1}{3}B^{-1}A^{-1}$.

72. If A is a 3×3 matrix and $\det(A) = 2$, then $\det(3A) = 6$.

73. If A is a singular matrix, then A^2 is singular.

74. If λ is an eigenvalue of an $n \times n$ matrix A, then $\lambda I_n - A$ is a singular matrix.

75. If A is an $n \times n$ matrix such that $A^2 = O$, then $A = O$.

76. If A is an $n \times n$ matrix, then $A + A^T$ is symmetric.

77. $A = \begin{bmatrix} 0 & 0 \\ 0 & 1 \end{bmatrix}$ defines a matrix transformation that projects the vector $\begin{bmatrix} x \\ y \end{bmatrix}$ onto the y-axis.

78. Homogeneous systems are always consistent.

79. If a homogeneous system has more equations than unknowns, then it has a nontrivial solution.

80. If A is $n \times n$ and the reduced row echelon form of $\left[A \;\vdots\; I_n\right]$ is $\left[C \;\vdots\; D\right]$, then $C = I_n$ and $D = A^{-1}$.

81. The reduced row echelon form of a singular matrix has a row of zeros.

82. If V is a real vector space, then for every vector $\mathbf{u}$ in V, the scalar 0 times $\mathbf{u}$ gives the zero vector in V.

83. Every subspace of R^3 contains infinitely many vectors.

84. If $\mathbf{v}$ is a multiple of $\mathbf{w}$, then $\mathbf{v}$ and $\mathbf{w}$ are linearly dependent.

85. R^n contains infinitely many vectors, so we say it is infinite dimensional.

86. If A is 4×4 with rank $A = 4$, then $A\mathbf{x} = \mathbf{b}$ has exactly 4 solutions.

87. The cosine of the angle between the vectors $\mathbf{u}$ and $\mathbf{v}$ is given by $\mathbf{u} \cdot \mathbf{v}$.

88. If $W = \operatorname{span}\left\{\begin{bmatrix} 1 \\ 0 \\ 1 \end{bmatrix}\right\}$, then $W^\perp$ is the set of all vectors of the form $\begin{bmatrix} 0 \\ x \\ 0 \end{bmatrix}$, where x is any real number.

89. Given any set of vectors, the Gram–Schmidt process will produce an orthonormal set.

90. If $L: R^2 \to R^2$ is a linear transformation defined by

$$L\left(\begin{bmatrix} u_1 \\ u_2 \end{bmatrix}\right) = \begin{bmatrix} u_1 \\ 0 \end{bmatrix},$$

then L is one-to-one.

91. Let $L: V \to W$ be a linear transformation. If $\mathbf{v}_1$ and $\mathbf{v}_2$ are in $\ker L$, then so is $\operatorname{span}\{\mathbf{v}_1, \mathbf{v}_2\}$.

92. If $L: R^4 \to R^3$ is a linear transformation, then it is possible that $\dim \ker L = 1$ and $\dim \operatorname{range} L = 2$.

93. If $\det(A) = 0$, then A has at least two equal rows.

94. If A is $n \times n$ with $\operatorname{rank} A = n - 1$, then A is singular.

95. If B is the reduced row echelon form of A, then $\det(B) = \det(A)$.

96. If a 3×3 matrix A has eigenvalues $\lambda = 1, -1, 3$, then A is diagonalizable.

97. If $\mathbf{x}$ and $\mathbf{y}$ are eigenvectors of A associated with the eigenvalue λ, then for any nontrivial vector $\mathbf{w}$ in $\operatorname{span}\{\mathbf{x}, \mathbf{y}\}$, $A\mathbf{w} = \lambda\mathbf{w}$.

98. If A is 4×4 and orthogonal, then its columns are a basis for R^4.

99. If A is similar to an upper triangular matrix U, then the eigenvalues of A are the diagonal entries of U.

100. If the general solution of $\mathbf{x}'(t) = A\mathbf{x}(t)$ is given by

$$\mathbf{x}(t) = b_1 \begin{bmatrix} 1 \\ 2 \end{bmatrix} e^{-t} + b_2 \begin{bmatrix} -1 \\ 1 \end{bmatrix} e^{t},$$

then the initial conditions $x_1(0) = 1$ and $x_2(0) = -1$ imply that the solution of the initial value problem is

$$\mathbf{x}(t) = \begin{bmatrix} 1 \\ -1 \end{bmatrix} e^{t}.$$

CHAPTER 11

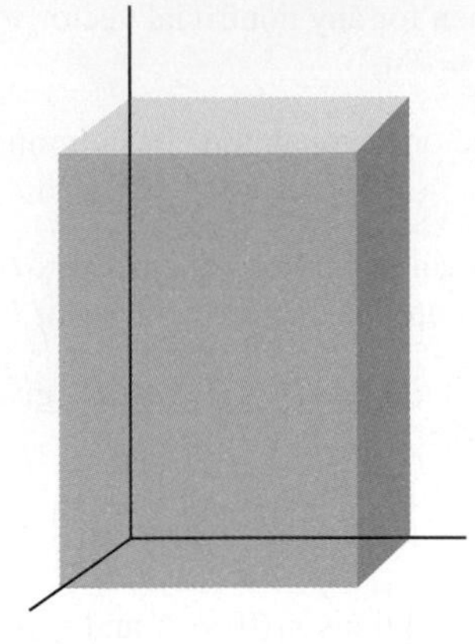

LINEAR PROGRAMMING (OPTIONAL)

PREREQUISITE: Section 1.6, Solutions of Linear Systems of Equations.

In this chapter we provide an introduction to the basic ideas and techniques of linear programming. Linear programming is a recent area of applied mathematics, developed in the late 1940s, to solve a number of problems for the federal government. Since then, it has been applied to an amazing number of problems in many areas. As a vital tool in management science and operations research, it has resulted in enormous savings of money. In the first section we give some examples of linear programming problems, formulate their mathematical models, and describe a geometric solution method. In the second section we present an algebraic method for solving linear programming problems. In the third section, dealing with duality, we discuss several applied interpretations of two related linear programming problems.

11.1 THE LINEAR PROGRAMMING PROBLEM; GEOMETRIC SOLUTION

In many problems in business and industry we are interested in making decisions that will maximize or minimize some quantity. For example, a plant manager may want to determine the most economical way of shipping goods from the factory to the markets, a hospital may want to design a diet satisfying certain nutritional requirements at a minimum cost, an investor may want to select investments that will maximize profits, or a manufacturer may wish to blend ingredients, subject to given specifications, to maximize profit. In this section we give several examples of linear programming problems and show how mathematical models for them can be formulated. Their geometric solution is also considered in this section.

EXAMPLE 1

(A Production Problem) A small manufacturer of photographic products prepares two types of film developers each day, Fine and Extra Fine, using as the raw material solutions A and B. Suppose that each quart of Fine contains 2 ounces of solution A and 1 ounce of solution B, while each quart of Extra Fine contains 1 ounce of solution A and 2 ounces of solution B. Suppose also that the profit on each quart of Fine is 8 cents and that it is 10 cents on each quart of Extra Fine. If the firm has 50 ounces of solution A and 70 ounces of solution B available each day, how many quarts of Fine and how many quarts of Extra Fine should be made each day to maximize profit (assuming that the shop can sell all that is made)?

Mathematical Formulation

Let x be the number of quarts of Fine to be made and let y be the number of quarts of Extra Fine to be made. Since each quart of Fine contains 2 ounces of solution A and each quart of Extra Fine contains 1 ounce of solution A, the total amount of solution A required is

$$2x + y.$$

Similarly, since each quart of Fine contains one ounce of solution B and each quart of Extra Fine contains 2 ounces of solution B, the total amount of solution B required is

$$x + 2y.$$

Since we only have 50 ounces of solution A and 70 ounces of solution B available, we must have

$$\begin{aligned} 2x + y &\le 50 \\ x + 2y &\le 70. \end{aligned}$$

Of course, x and y cannot be negative, so we must also have

$$x \ge 0 \quad \text{and} \quad y \ge 0.$$

Since the profit on each quart of Fine is 8 cents and the profit on each quart of Extra Fine is 10 cents, the total profit (in cents) is

$$z = 8x + 10y.$$

Our problem can be stated in mathematical form as: Find values of x and y that will maximize

$$z = 8x + 10y$$

subject to the following restrictions that must be satisfied by x and y:

$$\begin{aligned} 2x + y &\le 50 \\ x + 2y &\le 70 \\ x &\ge 0 \\ y &\ge 0. \end{aligned}$$

■

EXAMPLE 2

(Pollution) A manufacturer of a certain chemical product has two plants where the product is made. Plant X can make at most 30 tons per week and plant Y can make at most 40 tons per week. The manufacturer wants to make a total of at least 50 tons per week. The amount of particulate matter found weekly in the atmosphere over a nearby town is measured and found to be 20 pounds for each ton of the product made by plant X and 30 pounds for each ton of the product made at plant Y. How many tons should be made weekly at each plant to minimize the total amount of particulate matter in the atmosphere?

Mathematical Formulation

Let x and y denote the number of tons of the product made at plants X and Y, respectively, per week. Then the total amount manufactured per week is

$$x + y.$$

Since we want to make at least 50 tons per week, we must have

$$x + y \geq 50.$$

Since plant X can make at most 30 tons, we must have

$$x \leq 30.$$

Similarly, since plant Y can make at most 40 tons, we must have

$$y \leq 40.$$

Of course, x and y cannot be negative, so we require

$$\begin{aligned} x &\geq 0 \\ y &\geq 0. \end{aligned}$$

The total amount of particulate matter (in pounds) is

$$z = 20x + 30y,$$

which we want to minimize. Thus a mathematical statement of our problem is: Find values of x and y that will minimize

$$z = 20x + 30y$$

subject to the following restrictions that must be satisfied by x and y:

$$\begin{aligned} x + y &\geq 50 \\ x &\leq 30 \\ y &\leq 40 \\ x &\geq 0 \\ y &\geq 0. \end{aligned}$$

■

EXAMPLE 3 **(The Diet Problem)** A nutritionist is planning a menu that includes foods A and B as its main staples. Suppose that each ounce of food A contains 2 units of protein, 1 unit of iron, and 1 unit of thiamine; each ounce of food B contains 1 unit of protein, 1 unit of iron, and 3 units of thiamine. Suppose that each ounce of A costs 30 cents, while each ounce of B costs 40 cents. The nutritionist wants the meal to provide at least 12 units of protein, at least 9 units of iron, and at least 15 units of thiamine. How many ounces of each of the foods should be used to minimize the cost of the meal?

Mathematical Formulation Let x denote the number of ounces of food A to be used and let y be the number of ounces of food B. The number of units of protein supplied by the meal is

$$2x + y,$$

so we must have

$$2x + y \geq 12.$$

The number of units of iron supplied by the meal is

$$x + y,$$

so we must have

$$x + y \geq 9.$$

Since the number of units of thiamine supplied by the meal is

$$x + 3y,$$

we must have

$$x + 3y \geq 15.$$

Of course, we also require

$$x \geq 0, \quad y \geq 0.$$

The cost of the meal (in cents) is

$$z = 30x + 40y,$$

which we want to minimize. Thus a mathematical formulation of our problem is: Find values of x and y that will minimize

$$z = 30x + 40y$$

subject to the restrictions

$$\begin{aligned} 2x + y &\geq 12 \\ x + y &\geq 9 \\ x + 3y &\geq 15 \\ x &\geq 0 \\ y &\geq 0. \end{aligned}$$

■

These examples are typical of linear programming problems, the general form being as follows: Find values of $x_1, x_2, \ldots, x_n$ that will minimize or maximize

$$z = c_1x_1 + c_2x_2 + \cdots + c_nx_n \tag{1}$$

subject to

$$\left.\begin{array}{c} a_{11}x_1 + a_{12}x_2 + \cdots + a_{1n}x_n \ (\leq)\ (\geq)\ (=)\ b_1 \\ a_{21}x_1 + a_{22}x_2 + \cdots + a_{2n}x_n \ (\leq)\ (\geq)\ (=)\ b_2 \\ \vdots \qquad\quad \vdots \qquad\qquad\quad \vdots \qquad\qquad\qquad\quad \vdots \\ a_{m1}x_1 + a_{m2}x_2 + \cdots + a_{mn}x_n \ (\leq)\ (\geq)\ (=)\ b_m \end{array}\right\} \tag{2}$$

$$x_j \geq 0 \qquad (1 \leq j \leq n), \tag{3}$$

where in (2) one and only one of the symbols $\leq$, $\geq$, $=$ occurs in each statement. The linear function in (1) is called the **objective function**. The equalities or inequalities in (2) and (3) are called **constraints**. The term **linear** in linear programming means that the objective function (1) and each of the constraints in (2) are linear functions of the variables $x_1, x_2, \ldots, x_n$. The word *programming*, *not* to be confused with its use in computer programming, refers to the applications in planning or allocation problems.

GEOMETRIC SOLUTION

We now develop a geometric method for solving linear programming problems in two variables. This will enable us to solve the problems posed earlier. Since linear programming deals with systems of linear inequalities, we first consider these from a geometric point of view.

Consider Example 1. Find the set of points that maximize

$$z = 8x + 10y \tag{4}$$

subject to the constraints

$$\begin{aligned} 2x + y &\le 50 \\ x + 2y &\le 70 \\ x &\ge 0 \\ y &\ge 0. \end{aligned} \tag{5}$$

The set of points satisfying the system of four inequalities (5) consists of those points satisfying each of the inequalities. In Figure 11.1(a) we show the set of points satisfying the inequality $x \ge 0$, in Figure 11.1(b) the set of points satisfying the inequality $y \ge 0$. The set of points satisfying both inequalities $x \ge 0$ and $y \ge 0$ is the intersection of the regions in Figures 11.1(a) and (b). This set of points, the set of all points in the first quadrant, is shown in Figure 11.1(c).

Figure 11.1 ►

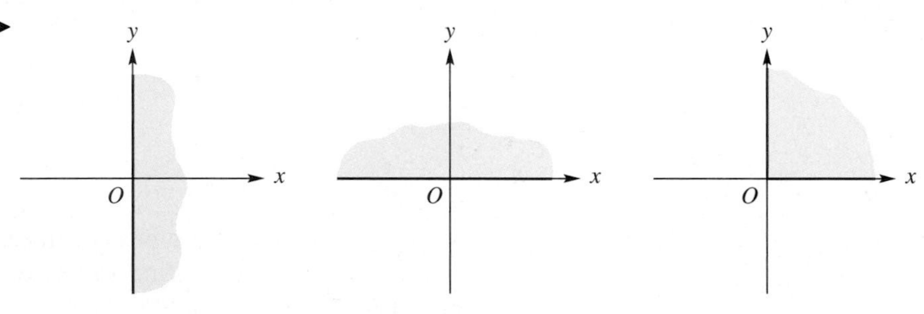

(a) Set of points satisfying $x \ge 0$ (b) Set of points satisfying $y \ge 0$ (c) Set of points satisfying $x \ge 0$ and $y \ge 0$

We must now consider the set of points satisfying the inequality

$$2x + y \le 50. \tag{6}$$

First, consider the set of points satisfying the strict inequality

$$2x + y < 50. \tag{7}$$

The straight line

$$2x + y = 50 \tag{8}$$

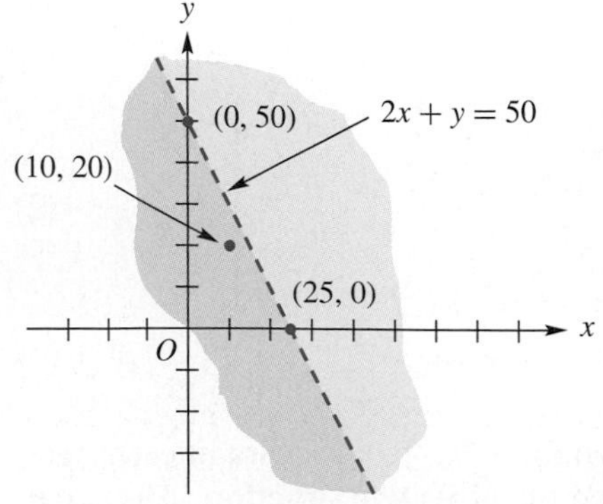

Figure 11.2 ▲

divides the set of points not on the line into two regions (Figure 11.2); one region contains all points satisfying (7) and the other region contains all points not satisfying (7). The line itself, drawn in dashed form, belongs to neither region.

To determine the region determined by inequality (7), we choose any point not on the line as a test point and see on which side of the line it lies. The point (10, 20) does not lie on the line (8) and can thus serve as a test point. Since its coordinates satisfy inequality (7), the correct region determined by (7) is shown in Figure 11.3(a). Another possible test point is (20, 20), since it does not lie on the line. Its coordinates do not satisfy inequality (7), so the test point does not lie in the correct region. Thus the correct region would again be as shown in Figure 11.3(a).

Figure 11.3 ►

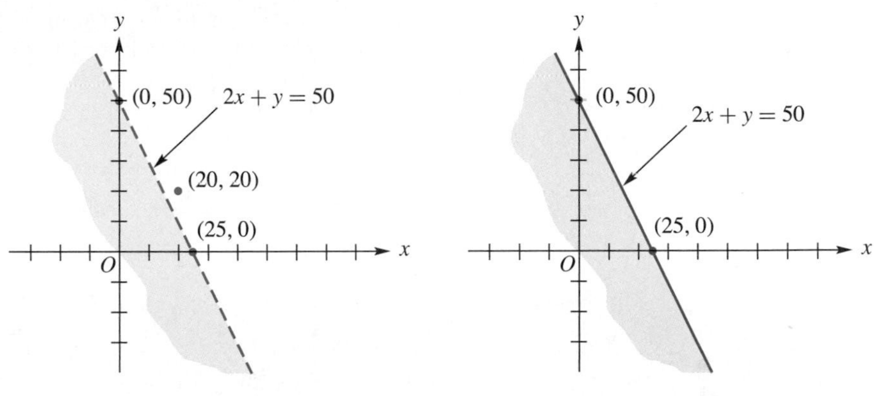

(a) Set of points satisfying $2x + y < 50$ (b) Set of points satisfying $2x + y \leq 50$

The set of points satisfying inequality (6) consists of the set of points satisfying (7) as well as the points on the line (8). Thus the region, shown in Figure 11.3(b), includes the straight line, which is drawn solidly.

Next, the set of points satisfying the inequality

$$x + 2y \leq 70 \tag{9}$$

is shown in Figure 11.4.

Figure 11.4 ►

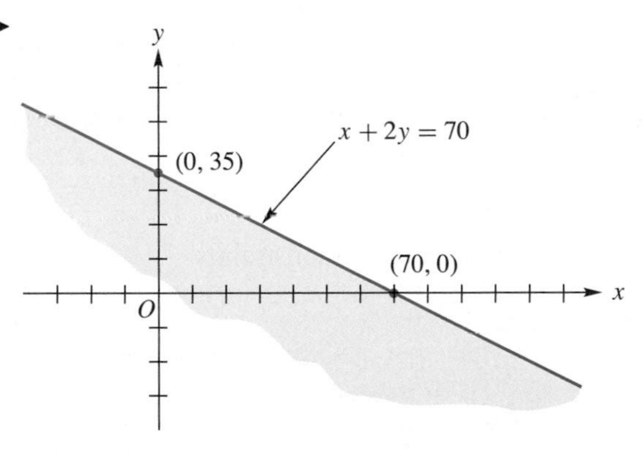

Set of points satisfying $x + 2y \leq 70$

The set of points satisfying inequalities (6) and (9) is the intersection of the regions in Figures 11.3(b) and 11.4; it is the shaded region shown in Figure 11.5.

Finally, the set of points satisfying the four inequalities (5) is the intersection of the regions in Figures 11.1(c) and 11.5. That is, it is the set of points in Figure 11.5 that lie in the first quadrant. This set of points is shown in Figure 11.6. Thus the set of points satisfying a system of inequalities is the intersection of the sets of points satisfying each of the inequalities.

Figure 11.5 ►

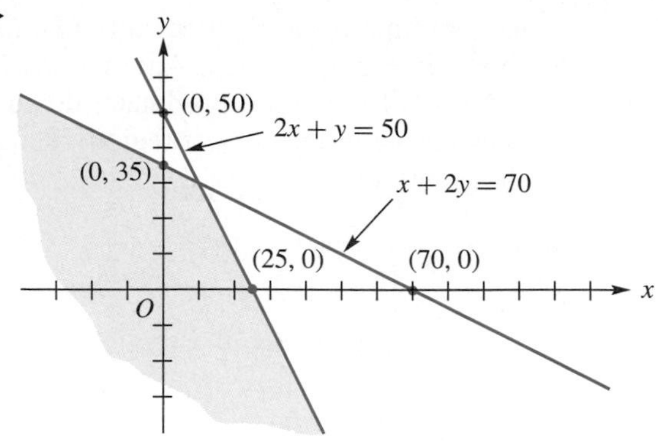

Set of points satisfying $2x + y \leq 50$ and $x + 2y \leq 70$

Figure 11.6 ►

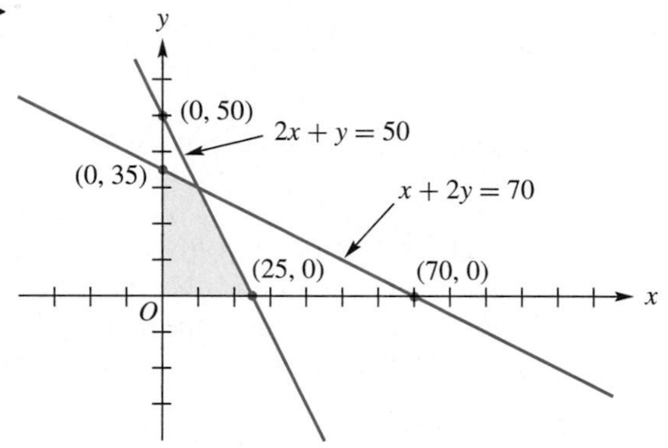

A point satisfying the constraints of a linear programming problem is called a **feasible solution**; the set of all such points is the **feasible region**. It should be noted that not every linear programming problem has a feasible region, as the next example shows.

EXAMPLE 4 Find the set of points that maximize

$$z = 8x + 10y \tag{10}$$

subject to the constraints

$$\begin{aligned} 2x + 3y &\leq 6 \\ x + 2y &\geq 6 \\ x &\geq 0 \\ y &\geq 0. \end{aligned} \tag{11}$$

In Figure 11.7 we have labeled as Region I the set of points satisfying the first, third, and fourth inequalities; that is, Region I is the set of points in the first quadrant satisfying the first inequality in (11). Similarly, Region II is the set of points satisfying the second, third, and fourth inequalities. It is clear that there are no points satisfying all four inequalities. ■

To solve the linear programming problem in Example 1, we must find a feasible solution that makes the objective function (1) as large as possible. Such a solution is called an **optimal solution**. Since there are infinitely many

Figure 11.7 ▶

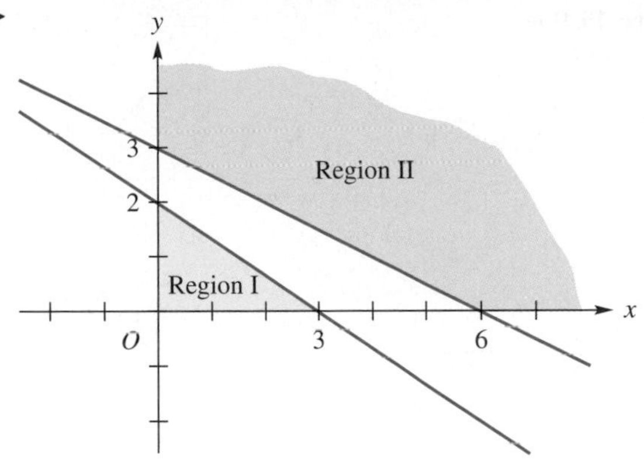

Figure 11.8 ▶

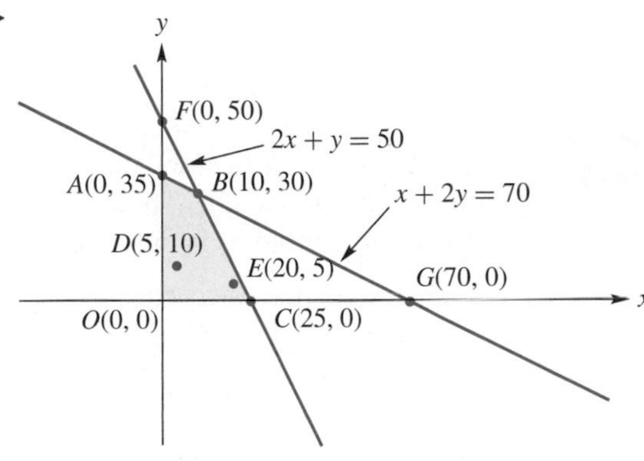

Table 11.1

Point	Value of $z = 8x + 10y$ (cents)
$O(0, 0)$	0
$A(0, 35)$	350
$B(10, 30)$	380
$C(25, 0)$	200
$D(5, 10)$	140
$E(20, 5)$	210

feasible points, it appears, at first glance, quite difficult to tell whether a feasible solution is optimal or not. In Figure 11.8 we have reproduced Figure 11.6 by showing the feasible region for this linear programming problem along with a number of indicated feasible points: O, A, B, C, D, and E. Points F and G are not in the feasible region.

In Table 11.1 we have tabulated the value of the objective function (in cents) for each of the points O, A, B, C, D, and E. Among the feasible solutions O, A, B, C, D, and E we see that the value of the objective function is largest for the point $B(10, 30)$. However, we do not know if there is another feasible solution at which the value of the objective function will be larger than it is at B. Since there are infinitely many feasible solutions in the region, we cannot examine the value of the objective function at every one of these. However, we shall be able to find optimal solutions without examining all feasible solutions, but we first consider some auxiliary notions.

DEFINITION

The **line segment** joining the distinct points $\mathbf{x}_1$ and $\mathbf{x}_2$ of R^n is the set of all points in R^n of the form $\lambda\mathbf{x}_1 + (1 - \lambda)\mathbf{x}_2$, $0 \leq \lambda \leq 1$. Observe that if $\lambda = 0$, we get $\mathbf{x}_2$, and if $\lambda = 1$, we get $\mathbf{x}_1$.

DEFINITION

A nonempty set S of points in R^n is called **convex** if the line segment joining any two points in S also lies entirely in S.

EXAMPLE 5

The shaded sets in Figures 11.9(a), (b), (c), and (d) are convex sets in R^2. The shaded sets in Figures 11.10(a), (b), and (c) are *not* convex sets in R^2, since the line joining the indicated points does not lie entirely in the set. ■

Figure 11.9 ► Convex sets in R^2

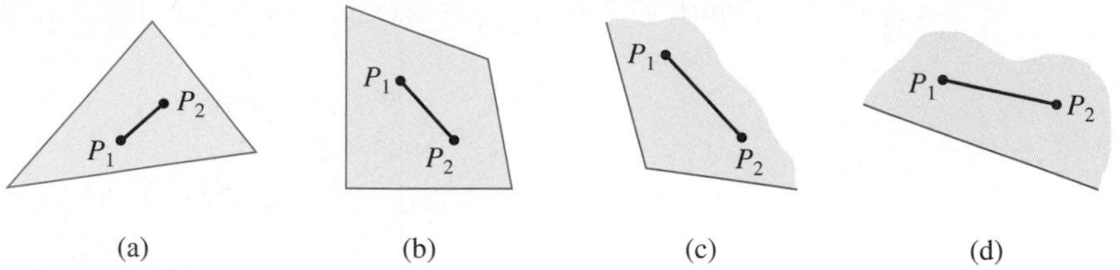

Figure 11.10 ► Sets in R^2 that are not convex

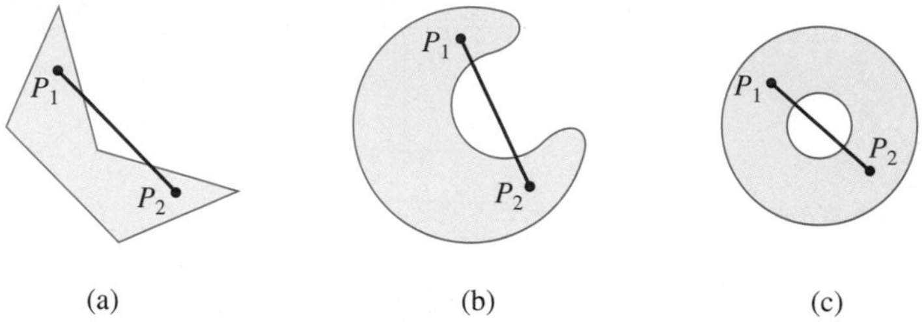

EXAMPLE 6 It is not too difficult to show that the feasible region of a linear programming problem is a convex set. The proof of this result for a broad class of linear programming problems is outlined in Exercise T.1 of Section 11.2. ■

The reader may also check that the feasible regions for the linear programming problems in Examples 2 and 3 are convex.

We shall now limit our further discussion of convex sets to such sets in R^2, although the ideas presented here can be generalized to convex sets in R^n.

Convex sets are either **bounded** or **unbounded**. A bounded convex set is one that can be enclosed by a sufficiently large rectangle; an unbounded convex set is one that cannot be so enclosed. The convex sets in Figures 11.9(a) and (b) are bounded; the convex sets in Figures 11.9(c) and (d) are unbounded.

DEFINITION An **extreme point** in a convex set S is a point *in* S that is not an interior point of any line segment in S.

EXAMPLE 7 In Figure 11.11 we have marked each extreme point of the convex sets in Figures 11.9(a), (b), and (c) with a solid dot. The convex set in Figure 11.9(d) has no extreme points. ■

Figure 11.11 ►

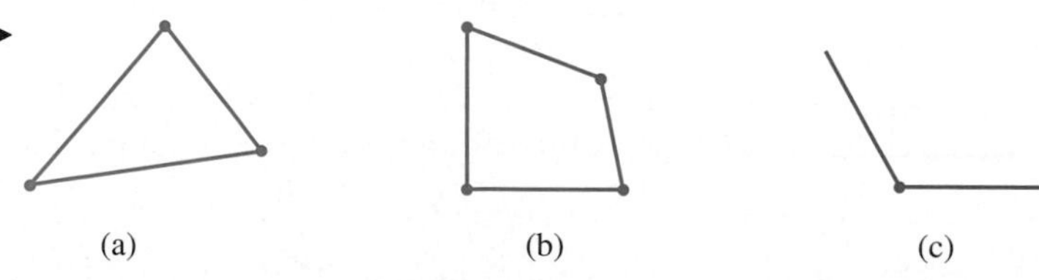

The basic result connecting convex sets, extreme points, and linear programming problems is the following theorem, whose proof we omit.

THEOREM 11.1 *Let S be the feasible region of a linear programming problem.*

(a) *If S is bounded, then the objective function*

$$z = ax + by$$

assumes both a maximum and a minimum value on S; these values occur at extreme points of S.

(b) *If S is unbounded, then there may or may not be a maximum or minimum value on S. If a maximum or minimum value does exist on S, it occurs at an extreme point.* ■

Thus, when the feasible region S of a linear programming problem is bounded, a method for solving the problem consists of finding the extreme points of S and evaluating the objective function $z = ax + by$ at each extreme point. An optimal solution is an extreme point at which the value of z is a maximum or a minimum.

> The procedure for solving a linear programming problem in two variables geometrically is as follows.
>
> ***Step 1.*** Sketch the feasible region S.
>
> ***Step 2.*** Determine all the extreme points of S.
>
> ***Step 3.*** Evaluate the objective function at each extreme point.
>
> ***Step 4.*** Choose an extreme point at which the objective function is largest (smallest) for a maximization (minimization) problem.

EXAMPLE 8

Consider Example 1 again. The feasible region shown in Figure 11.8 is bounded and its extreme points are $O(0, 0)$, $A(0, 35)$, $B(10, 30)$, and $C(25, 0)$. From Table 11.1 we see that the value of z is a maximum at the extreme point $B(10, 30)$. Thus the optimal solution is

$$x = 10, \quad y - 30.$$

This means that the manufacturer should make 10 quarts of Fine and 30 quarts of Extra Fine to maximize its profit. If the firm follows this course of action, its maximum profit will be \$3.80 each day. In this problem we chose small numbers to reduce the computational effort. In a real problem of this type the numbers would be much larger. ■

Referring to Figure 11.8, we note that the points (70, 0) and (0, 50) are points of intersection of boundary lines. However, they are not extreme points in the feasible region, since they are *not* feasible solutions; that is, they do not lie in the feasible region.

EXAMPLE 9

Solve Example 2 geometrically.

Solution

The feasible region S of this linear programming problem is shown in Figure 11.12 (verify). Since S is bounded, we can find the minimum value of z by evaluating z at each of the extreme points. In Table 11.2 we have tabulated the value of the objective function for each of the points $A(10, 40)$, $B(30, 40)$, and $C(30, 20)$. The value of z is a minimum at the extreme point $C(30, 20)$. Thus the optimal solution is

$$x = 30, \quad y = 20,$$

which means that the manufacturer should make 30 tons of the product at plant X and 20 tons at plant Y. If this course of action is followed, the total amount of particulate matter over the town will be 1200 pounds weekly. ■

Figure 11.12 ►

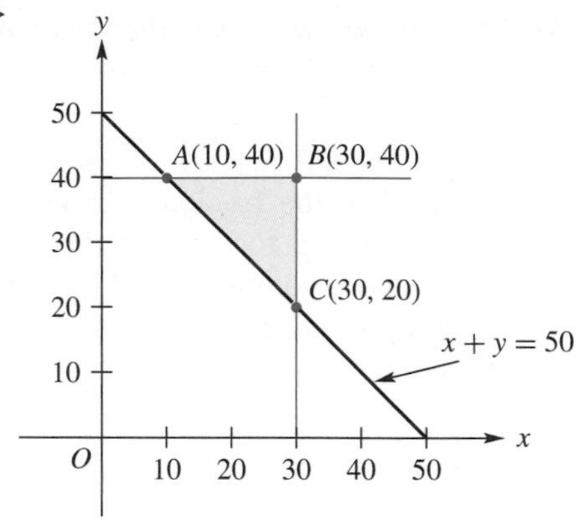

Table 11.2

Point	Value of $z = 20x + 30y$ (pounds)
$A(10, 40)$	1400
$B(30, 40)$	1800
$C(30, 20)$	1200

The feasible region of Example 3 is not bounded (verify). We shall not deal further with such problems in this book.

In general, a linear programming problem may have:

1. No feasible solution; that is, there are no points satisfying all the constraints in Equations (2) and (3).
2. A unique optimal solution.
3. More than one optimal solution (see Exercise 4).
4. No maximum (or minimum) value in the feasible region; that is, it may be possible to choose a point in the feasible region to make the objective function as large (or as small) as we please.

STANDARD LINEAR PROGRAMMING PROBLEMS

We shall now focus our attention on a special class of linear programming problems and will show that every linear programming problem can be transformed into a problem of this special type.

DEFINITION We shall refer to the following linear programming problem as a **standard linear programming problem**. Find values of $x_1, x_2, \ldots, x_n$ that will maximize

$$z = c_1x_1 + c_2x_2 + \cdots + c_nx_n \tag{12}$$

subject to the constraints

$$\left.\begin{array}{l} a_{11}x_1 + a_{12}x_2 + \cdots + a_{1n}x_n \leq b_1 \\ a_{21}x_1 + a_{22}x_2 + \cdots + a_{2n}x_n \leq b_2 \\ \quad\vdots \qquad\quad \vdots \qquad\qquad\quad \vdots \qquad\quad \vdots \\ a_{m1}x_1 + a_{m2}x_2 + \cdots + a_{mn}x_n \leq b_m \end{array}\right\} \tag{13}$$

$$x_j \geq 0 \qquad (1 \leq j \leq n). \tag{14}$$

EXAMPLE 10 Example 1 is a standard linear programming problem. ■

EXAMPLE 11 The linear programming problem

$$\text{Minimize} \quad z = 3x - 4y$$

subject to

$$\begin{aligned} 2x - 3y &\leq 6 \\ x + y &\leq 8 \\ x &\geq 0 \\ y &\geq 0 \end{aligned}$$

is *not* a standard linear programming problem, since the objective function is to be minimized and not maximized. ■

EXAMPLE 12 The linear programming problem

$$\text{Maximize} \quad z = 12x - 15y$$

subject to

$$\begin{aligned} 3x - y &\geq 4 \\ 2x + 3y &\leq 6 \\ x &\geq 0 \\ y &\geq 0 \end{aligned}$$

is *not* a standard linear programming problem, since one of the inequalities is of the form $\geq$; in a standard linear programming problem every inequality of (13) must be of the form $\leq$. ■

EXAMPLE 13 The linear programming problem

$$\text{Maximize} \quad z = 8x + 10y$$

subject to

$$\begin{aligned} 3x + y &= 4 \\ 2x - 3y &\leq 5 \\ x &\geq 0 \\ y &\geq 0 \end{aligned}$$

is *not* a standard linear programming problem, since the first constraint is an equation and not an inequality of the form $\leq$. ■

Every linear programming problem can be transformed into a standard linear programming problem.

MINIMIZATION PROBLEM AS A MAXIMIZATION PROBLEM

Every maximization problem can be viewed as a minimization problem, and conversely. This follows from the observation that

$$\begin{aligned} &\text{minimum of } c_1x_1 + c_2x_2 + \cdots + c_nx_n \\ &\qquad = -\text{ maximum of } \{-(c_1x_1 + c_2x_2 + \cdots + c_nx_n)\}. \end{aligned} \tag{15}$$

REVERSING AN INEQUALITY

Consider the inequality

$$d_1x_1 + d_2x_2 + \cdots + d_nx_n \geq -b.$$

Multiplying both sides of this inequality by -1 reverses the inequality, yielding

$$-d_1x_1 - d_2x_2 - \cdots - d_nx_n \leq b.$$

EXAMPLE 14 Consider the linear programming problem

$$\text{Minimize} \quad w = 5x - 2y$$

subject to

$$\begin{aligned} 2x - 3y &\geq -5 \\ 3x + 2y &\leq 12 \\ x &\geq 0 \\ y &\geq 0. \end{aligned} \tag{16}$$

Using (15) and multiplying the first inequality in (16) by (-1), we then obtain the following standard linear programming problem:

$$\text{Maximize} \quad z = -(5x - 2y) = -5x + 2y$$

subject to

$$\begin{aligned} -2x + 3y &\leq 5 \\ 3x + 2y &\leq 12 \\ x &\geq 0 \\ y &\geq 0. \end{aligned}$$

Once this new maximization problem has been solved, we take the negative of the maximum value of z to obtain the minimum value of w. ■

SLACK VARIABLES

It is not too difficult to change a number of equalities into inequalities of the form $\leq$. Thus Example 13 can be transformed into a standard linear programming problem. In this book we shall not encounter linear programming problems of the type in Example 13, so we shall not continue to examine such problems.

It is not easy to handle systems of linear inequalities algebraically. However, as we have seen in Chapter 1, it is not difficult at all to deal with systems of linear equations. Accordingly, we shall change our given standard linear programming problem into a problem where we must find nonnegative variables maximizing a linear objective function and satisfying a system of linear equations. Every solution to the given problem yields a solution to the new problem, and conversely, every solution to the new problem yields a solution to the given problem.

Consider the constraint

$$d_1x_1 + d_2x_2 + \cdots + d_nx_n \leq b. \tag{17}$$

Since the left side of (17) is not larger than the right side, we can make (17) into an equation by adding the unknown nonnegative quantity u to its left side, to obtain

$$d_1x_1 + d_2x_2 + \cdots + d_nx_n + u = b. \tag{18}$$

The quantity u in (18) is called a **slack variable**, since it takes up the slack between the two sides of the inequality.

We now change each of the constraints in (2) (which we assume represent a standard linear programming problem with only the $\leq$-type inequalities) into an equation by introducing a nonnegative slack variable. Thus the ith inequality

$$a_{i1}x_1 + a_{i2}x_2 + \cdots + a_{in}x_n \leq b_i \qquad (1 \leq i \leq m) \tag{19}$$

is converted into the equation

$$a_{i1}x_1 + a_{i2}x_2 + \cdots + a_{in}x_n + x_{n+i} = b_i \qquad (1 \leq i \leq m),$$

by introducing the nonnegative slack variable x_{n+i}. Our new problem can now be stated as follows.

NEW PROBLEM

Find values of $x_1, x_2, \ldots, x_n, x_{n+1}, \ldots, x_{n+m}$ that will maximize

$$z = c_1x_1 + c_2x_2 + \cdots + c_nx_n \tag{20}$$

subject to

$$\left.\begin{array}{llllll} a_{11}x_1 + a_{12}x_2 + \cdots + a_{1n}x_n & + x_{n+1} & & & & = b_1 \\ a_{21}x_1 + a_{22}x_2 + \cdots + a_{2n}x_n & & + x_{n+2} & & & = b_2 \\ \vdots & & & \ddots & & \vdots \\ a_{m1}x_1 + a_{m2}x_2 + \cdots + a_{mn}x_n & & & & + x_{n+m} & = b_m \end{array}\right\} \tag{21}$$

$$x_1 \geq 0, \ldots, x_n \geq 0, x_{n+1} \geq 0, x_{n+2} \geq 0, \ldots, x_{n+m} \geq 0. \tag{22}$$

Thus the new problem has m equations in $m + n$ unknowns. Solving the original problem is equivalent to solving the new problem in the following sense. If $x_1, x_2, \ldots, x_n$ is a feasible solution to the given problem as defined by (1), (2), and (3), then

$$x_1 \geq 0, \quad x_2 \geq 0, \ldots, \quad x_n \geq 0.$$

Also, $x_1, x_2, \ldots, x_n$ satisfy each of the constraints in (2). Let x_{n+i}, $1 \leq i \leq m$, be defined by

$$x_{n+i} = b_i - a_{i1}x_1 - a_{i2}x_2 - \cdots - a_{in}x_n.$$

That is, x_{n+i} is the difference between the right side of inequality (19) and its left side. Then

$$x_{n+1} \geq 0, \quad x_{n+2} \geq 0, \ldots, \quad x_{n+m} \geq 0$$

so that $x_1, x_2, \ldots, x_n, x_{n+1}, \ldots, x_{n+m}$ satisfy (21) and (22).

Conversely, suppose that $x_1, x_2, \ldots, x_n, x_{n+1}, \ldots, x_{n+m}$ satisfy (21) and (22). It is then clear that $x_1, x_2, \ldots, x_n$ satisfy (2) and (3).

EXAMPLE 15 Consider the problem of Example 1. Introducing the slack variables u and v, we formulate our new problem as follows: Find values of x, y, u, and v that will maximize

$$z = 8x + 10y$$

subject to

$$\begin{aligned} 2x + y + u \phantom{{}+v} &= 50 \\ x + 2y \phantom{{}+u} + v &= 70 \end{aligned}$$
$$x \geq 0, \quad y \geq 0, \quad u \geq 0, \quad v \geq 0.$$

The slack variable u is the difference between the total amount of solution A available, 50 ounces, and the amount $2x + y$ of solution A actually used. The slack variable v is the difference between the total amount of solution B available, 70 ounces, and the amount $x + 2y$ of solution B actually used.

Consider the feasible solution to the given problem

$$x = 5, \qquad y = 10,$$

which represents point D in Figure 11.8. We then obtain the slack variables

$$u = 50 - 2(5) - 10 = 50 - 10 - 10 = 30$$

and

$$v = 70 - 5 - 2(10) = 70 - 5 - 20 = 45$$

so that

$$x = 5, \qquad y = 10, \qquad u = 30, \qquad v = 45$$

is a feasible solution to the new problem. Of course, the solution $x = 5$, $y = 10$ is not an optimal solution, since $z = 8(5) + 10(10) = 140$, and we recall that the maximum value of z is attained for

$$x = 10, \qquad y = 30.$$

In this case, the corresponding optimal solution to the new problem is

$$x = 10, \qquad y = 30, \qquad u = 0, \qquad v = 0.$$ ■

Key Terms

Linear programming problem
Objective function
Constraints
Feasible solution
Feasible region
Optimal solution
Line segment
Convex set
Bounded convex set
Unbounded convex set
Extreme point
Standard linear programming problem
Slack variable

11.1 Exercises

In Exercises 1 through 9, formulate mathematically each linear programming problem.

1. A steel producer makes two types of steel: regular and special. A ton of regular steel requires 2 hours in the open-hearth furnace and 5 hours in the soaking pit; a ton of special steel requires 2 hours in the open-hearth furnace and 3 hours in the soaking pit. The open-hearth furnace is available 8 hours per day and the soaking pit is available 15 hours per day. The profit on a ton of regular steel is $120 and it is $100 on a ton of special steel. Determine how many tons of each type of steel should be made to maximize the profit.

2. A trust fund is planning to invest up to \$6000 in two types of bonds: A and B. Bond A is safer than bond B and carries a dividend of 8%, and bond B carries a dividend of 10%. Suppose that the fund's rules state that no more than \$4000 may be invested in bond B, while at least \$1500 must be invested in bond A. How much should be invested in each type of bond to maximize the fund's return?

3. Solve Exercise 2 if the fund has the following additional rule: "The amount invested in bond B cannot exceed one-half the amount invested in bond A."

4. A trash-removal company carries industrial waste in sealed containers in its fleet of trucks. Suppose that each container from the Smith Corporation weighs 6 pounds and is 3 cubic feet in volume, while each container from the Johnson Corporation weighs 12 pounds and is 1 cubic foot in volume. The company charges the Smith Corporation 30 cents for each container carried on a trip, and 60 cents for each container from the Johnson Corporation. If a truck cannot carry more than 18,000 pounds or more than 1800 cubic feet in volume, how many containers from each customer should the company carry in a truck on each trip to maximize the revenue per truckload?

5. A television producer designs a program based on a comedian and time for commercials. The advertiser insists on at least 2 minutes of advertising time, the station insists on no more than 4 minutes of advertising time, and the comedian insists on at least 24 minutes of the comedy program. Also, the total time allotted for the advertising and comedy portions of the program cannot exceed 30 minutes. If it has been determined that each minute of advertising (very creative) attracts 40,000 viewers and each minute of the comedy program attracts 45,000 viewers, how should the time be divided between advertising and programming to maximize the number of viewers per minute?

6. A small generator burns two types of fuel: low sulfur (L) and high sulfur (H) to produce electricity. For each hour of use, each gallon of L emits 3 units of sulfur dioxide, generates 4 kilowatts, and costs 60 cents, while each gallon of H emits 5 units of sulfur dioxide, generates 4 kilowatts, and costs 50 cents. The environmental protection agency insists that the maximum amount of sulfur dioxide that may be emitted per hour is 15 units. Suppose that at least 16 kilowatts must be generated per hour. How many gallons of L and how many gallons of H should be used hourly to minimize the cost of the fuel used?

7. The Protein Diet Club serves a luncheon consisting of two dishes, A and B. Suppose that each unit of A has 1 gram of fat, 1 gram of carbohydrate, and 4 grams of protein, whereas each unit of B has 2 grams of fat, 1 gram of carbohydrate, and 6 grams of protein. If the dietician planning the luncheon wants to provide no more than 10 grams of fat or more than 7 grams of carbohydrate, how many units of A and how many units of B should be served to maximize the amount of protein consumed?

8. In designing a new airline route, a company is considering two types of planes, types A and B. Each type A plane can carry 40 passengers and requires 2 mechanics for servicing; each type B plane can carry 60 passengers and requires 3 mechanics for servicing. Suppose that the company must transport at least 300 people daily and that insurance rules for the size of the hangar allow no more than 12 mechanics on the payroll. If each type A plane costs \$10,000,000 and each type B plane costs \$15,000,000, how many planes of each type should be bought to minimize the cost?

9. An animal feed producer makes two types of grain: A and B. Each unit of grain A contains 2 grams of fat, 1 gram of protein, and 80 calories. Each unit of grain B contains 3 grams of fat, 3 grams of protein, and 60 calories. Suppose that the producer wants each unit of the final product to yield at least 18 grams of fat, at least 12 grams of protein, and at least 480 calories. If each unit of A costs 10 cents and each unit of B costs 12 cents, how many units of each type of grain should the producer use to minimize the cost?

In Exercises 10 through 13, sketch the set of points satisfying the given system of inequalities.

10.
$$\begin{aligned} x &\le 4 \\ x &\ge 2 \\ y &\le 4 \\ y &\ge 1 \\ x + y &\le 6 \end{aligned}$$

11.
$$\begin{aligned} 2x - y &\le 6 \\ 2x + y &\le 10 \\ x &\ge 0 \\ y &\ge 0 \end{aligned}$$

12.
$$\begin{aligned} x + y &\le 3 \\ 5x + 4y &\ge 20 \\ x &\ge 0 \\ y &\ge 0 \end{aligned}$$

13.
$$\begin{aligned} x + y &\ge 4 \\ x + 4y &\ge 8 \\ x &\ge 0 \\ y &\ge 0 \end{aligned}$$

In Exercises 14 and 15, solve the given linear programming problem geometrically.

14. Maximize $z = 3x + 2y$ subject to

$$\begin{aligned} 2x - 3y &\le 6 \\ x + y &\le 4 \\ x &\ge 0 \\ y &\ge 0. \end{aligned}$$

15. Minimize $z = 3x - y$ subject to

$$\begin{aligned} -3x + 2y &\le 6 \\ 5x + 4y &\ge 20 \\ 8x + 3y &\le 24 \\ x &\ge 0 \\ y &\ge 0. \end{aligned}$$

16. Solve the problem in Exercise 1 geometrically.

17. Solve the problem in Exercise 2 geometrically.

18. Solve the problem in Exercise 3 geometrically.

19. Solve the problem in Exercise 4 geometrically.

20. Solve the problem in Exercise 5 geometrically.

21. Solve the problem in Exercise 6 geometrically.

22. Solve the problem in Exercise 7 geometrically.

23. Solve the problem in Exercise 8 geometrically.

24. Which of the following are standard linear programming problems?

(a) Maximize $z = 2x - 3y$
subject to

$$\begin{aligned} 2x - 3y &\le 4 \\ 3x + 2y &\ge 5. \end{aligned}$$

(b) Minimize $z = 2x + 3y$
subject to

$$\begin{aligned} 2x + 3y &\le 4 \\ 3x + 2y &\le 5 \\ x &\ge 0 \\ y &\ge 0. \end{aligned}$$

(c) Minimize $z = 2x_1 - 3x_2 + x_3$
subject to

$$\begin{aligned} 2x_1 + 3x_2 + 2x_3 &\le 6 \\ 3x_1 \qquad - 2x_3 &\le 4 \\ x_1 &\le 0 \\ x_2 &\ge 0 \\ x_3 &\ge 0. \end{aligned}$$

(d) Maximize $z = 2x + 2y$
subject to

$$\begin{aligned} 2x + 3y &\le 4 \\ 3x &\le 5 \\ x &\ge 0. \end{aligned}$$

25. Which of the following are standard linear programming problems?

(a) Maximize $z = 3x_1 + 2x_2 + x_3$
subject to

$$\begin{aligned} 2x_1 + 3x_2 + x_3 &\le 4 \\ 3x_1 - 2x_2 \qquad &\le 5 \\ x_1 &\ge 0 \\ x_2 &\ge 0 \\ x_3 &\ge 0. \end{aligned}$$

(b) Minimize $z = 2x + 3y$
subject to

$$\begin{aligned} 2x + 4y &\le 2 \\ 3x + 2y &\le 4 \\ x &\ge 0 \\ y &\ge 0. \end{aligned}$$

(c) Maximize $z = 3x_1 + 4x_2 + x_3$
subject to

$$\begin{aligned} 2x_1 + 4x_2 + x_3 &\le 2 \\ 3x_1 - 2x_2 + x_3 &\le 4 \\ 2x_1 \qquad + x_3 &\le 8 \\ x_1 &\ge 0 \\ x_2 &\ge 0. \end{aligned}$$

(d) Maximize $z = 2x_1 + 3x_2 + x_3$
subject to

$$\begin{aligned} 2x_1 + 3x_2 + 5x_3 &\le 8 \\ 3x_1 - 2x_2 + 2x_3 &= 4 \\ 2x_1 + x_2 + 3x_3 &\le 6 \\ x_1 &\ge 0 \\ x_2 &\ge 0 \\ x_3 &\ge 0. \end{aligned}$$

In Exercises 26 and 27, formulate each problem as a standard linear programming problem.

26. Minimize $z = -2x_1 + 3x_2 + 2x_3$
subject to

$$\begin{aligned} 2x_1 + x_2 + 2x_3 &\le 12 \\ x_1 + x_2 - 3x_3 &\le 8 \\ x_1 &\ge 0 \\ x_2 &\ge 0 \\ x_3 &\ge 0. \end{aligned}$$

27. Maximize $z = 3x_1 - x_2 + 6x_3$
subject to

$$\begin{aligned} 2x_1 + 4x_2 + x_3 &\le 4 \\ -3x_1 + 2x_2 - 3x_3 &\ge -4 \\ 2x_1 + x_2 - x_3 &\le 8 \\ x_1 &\ge 0 \\ x_2 &\ge 0 \\ x_3 &\ge 0. \end{aligned}$$

In Exercises 28 and 29, formulate the given linear programming problem as a new problem with slack variables.

28. Maximize $z = 2x + 8y$
subject to

$$\begin{aligned} 2x + 3y &\le 18 \\ 3x - 2y &\le 6 \\ x &\ge 0 \\ y &\ge 0. \end{aligned}$$

29. Maximize $z = 2x_1 + 3x_2 + 7x_3$
subject to

$$\begin{aligned} 3x_1 + x_2 - 4x_3 &\le 3 \\ x_1 - 2x_2 + 6x_3 &\le 21 \\ x_1 - x_2 - x_3 &\le 9 \\ x_1 &\ge 0 \\ x_2 &\ge 0 \\ x_3 &\ge 0. \end{aligned}$$

11.2 THE SIMPLEX METHOD

The simplex method for solving linear programming problems was developed by George B. Dantzig* in 1947 in connection with his work on planning problems for the federal government. In this section we present the essential features of the method, illustrating them with examples. A number of proofs will be omitted, and for further details the interested reader may consult the references given at the end of this chapter. It is convenient to introduce matrix terminology into our further discussion of linear programming.

MATRIX NOTATION

We again restrict our attention to the standard linear programming problem: Maximize

$$c_1x_1 + c_2x_2 + \cdots + c_nx_n \tag{1}$$

subject to

$$\left.\begin{aligned} a_{11}x_1 + a_{12}x_2 + \cdots + a_{1n}x_n &\le b_1 \\ a_{21}x_1 + a_{22}x_2 + \cdots + a_{2n}x_n &\le b_2 \\ \vdots \qquad\quad \vdots \qquad\qquad\quad \vdots \quad &\quad\ \vdots \\ a_{m1}x_1 + a_{m2}x_2 + \cdots + a_{mn}x_n &\le b_m \end{aligned}\right\} \tag{2}$$

$$x_j \ge 0 \quad (1 \le j \le n). \tag{3}$$

If we let

$$A = \begin{bmatrix} a_{11} & a_{12} & \cdots & a_{1n} \\ a_{21} & a_{22} & \cdots & a_{2n} \\ \vdots & \vdots & & \vdots \\ a_{m1} & a_{m2} & \cdots & a_{mn} \end{bmatrix}, \qquad \mathbf{x} = \begin{bmatrix} x_1 \\ x_2 \\ \vdots \\ x_n \end{bmatrix}, \qquad \mathbf{b} = \begin{bmatrix} b_1 \\ b_2 \\ \vdots \\ b_m \end{bmatrix},$$

---

*George B. Dantzig (1914–) was born in Portland, Oregon. He received a bachelor's degree from the University of Maryland, master's degree from the University of Michigan, and a Ph.D. degree from the University of California at Berkeley. He is presently Professor of Operations Research and Computer Science at Stanford University. His simplex method for solving linear programming problems, developed in 1947, was a major breakthrough in the newly developed area of operations research. The rapid growth of computing power coupled with its decreasing cost has produced a great many computer implementations of the simplex method and has resulted in savings of billions of dollars for industry and government. Professor Dantzig has held important positions in government, industry, and universities and is the recipient of numerous honors and awards.

and

$$\mathbf{c} = \begin{bmatrix} c_1 \\ c_2 \\ \vdots \\ c_n \end{bmatrix},$$

then the given problem can be stated as follows: Find a vector $\mathbf{x}$ in R^n that will maximize the objective function

$$z = \mathbf{c}^T\mathbf{x} \tag{4}$$

subject to

$$A\mathbf{x} \leq \mathbf{b} \tag{5}$$

$$\mathbf{x} \geq \mathbf{0}, \tag{6}$$

where $\mathbf{x} \geq \mathbf{0}$ means that each entry of $\mathbf{x}$ is nonnegative and $A\mathbf{x} \leq \mathbf{b}$ means that each entry of $A\mathbf{x}$ is less than or equal to the corresponding entry in $\mathbf{b}$.

A vector $\mathbf{x}$ in R^n satisfying (5) and (6) is called a **feasible solution** to the given problem, and a feasible solution maximizing the objective function (4) is called an **optimal solution**.

EXAMPLE 1 We can write the problem in Example 1 of Section 11.1 in matrix form as follows: Find a vector

$$\mathbf{x} = \begin{bmatrix} x \\ y \end{bmatrix}$$

in R^2 that will maximize

$$z = \begin{bmatrix} 8 & 10 \end{bmatrix} \begin{bmatrix} x \\ y \end{bmatrix}$$

subject to

$$\begin{bmatrix} 2 & 1 \\ 1 & 2 \end{bmatrix} \begin{bmatrix} x \\ y \end{bmatrix} \leq \begin{bmatrix} 50 \\ 70 \end{bmatrix}$$

$$\begin{bmatrix} x \\ y \end{bmatrix} \geq \begin{bmatrix} 0 \\ 0 \end{bmatrix}.$$

Feasible solutions include the vectors

$$\begin{bmatrix} 0 \\ 0 \end{bmatrix}, \quad \begin{bmatrix} 0 \\ 35 \end{bmatrix}, \quad \begin{bmatrix} 10 \\ 30 \end{bmatrix}, \quad \begin{bmatrix} 25 \\ 0 \end{bmatrix}, \quad \begin{bmatrix} 5 \\ 10 \end{bmatrix}, \quad \text{and} \quad \begin{bmatrix} 20 \\ 5 \end{bmatrix}.$$

An optimal solution is the vector

$$\begin{bmatrix} 10 \\ 30 \end{bmatrix}.$$

■

The new problem with slack variables can also be written in matrix form as follows: Find a vector $\mathbf{x}$ that will maximize

$$z = \mathbf{c}^T\mathbf{x} \tag{7}$$

subject to

$$A\mathbf{x} = \mathbf{b} \tag{8}$$

$$\mathbf{x} \geq \mathbf{0}, \tag{9}$$

where now

$$A = \begin{bmatrix} a_{11} & a_{12} & \cdots & a_{1n} & 1 & 0 & \cdots & 0 \\ a_{21} & a_{22} & \cdots & a_{2n} & 0 & 1 & \cdots & 0 \\ \vdots & \vdots & & \vdots & \vdots & \vdots & & \vdots \\ a_{m1} & a_{m2} & \cdots & a_{mn} & 0 & 0 & \cdots & 1 \end{bmatrix}, \qquad \mathbf{x} = \begin{bmatrix} x_1 \\ x_2 \\ \vdots \\ x_n \\ x_{n+1} \\ \vdots \\ x_{n+m} \end{bmatrix},$$

$$\mathbf{b} = \begin{bmatrix} b_1 \\ b_2 \\ \vdots \\ b_m \end{bmatrix}, \quad \text{and} \quad \mathbf{c} = \begin{bmatrix} c_1 \\ c_2 \\ \vdots \\ c_n \\ 0 \\ \vdots \\ 0 \end{bmatrix}.$$

A vector $\mathbf{x}$ satisfying (8) and (9) is called a **feasible solution** to the new problem, and a feasible solution maximizing the objective function (7) is called an **optimal solution**. Throughout this chapter, we now make the additional assumption that in all standard linear programming problems,

$$b_1 \geq 0, \quad b_2 \geq 0, \ldots, \quad b_m \geq 0.$$

We shall use Example 1 of Section 11.1 as our principal illustrative example in this section.

ILLUSTRATIVE PROBLEM

Find values of x and y that will maximize

$$z = 8x + 10y \tag{10}$$

subject to

$$\begin{aligned} 2x + y &\leq 50 \\ x + 2y &\leq 70 \end{aligned} \tag{11}$$

$$x \geq 0, \quad y \geq 0. \tag{12}$$

The new problem with slack variables u and v is: Find values of x, y, u, and v that will maximize

$$z = 8x + 10y \tag{13}$$

subject to

$$\begin{aligned} 2x + y + u \phantom{{}+v} &= 50 \\ x + 2y \phantom{{}+u} + v &= 70 \end{aligned} \tag{14}$$

$$x \geq 0, \quad y \geq 0, \quad u \geq 0, \quad v \geq 0. \tag{15}$$

DEFINITION The vector $\mathbf{x}$ in R^{n+m} is called a **basic solution** to the new problem if it is obtained by setting n of the variables in (8) equal to zero and solving for the remaining m variables. The m variables that we solve for are called **basic variables**, and the n variables set equal to zero are called **nonbasic variables**. The vector $\mathbf{x}$ is called a **basic feasible solution** if it is a basic solution that also satisfies (9).

Basic feasible solutions are important because the following theorem can be established.

THEOREM 11.2 *If a linear programming problem has an optimal solution, then it has a basic feasible solution that is optimal.* ■

Thus to solve a linear programming problem we need only search for basic feasible solutions. In our illustrative example we can select two of the four variables x, y, u, and v as nonbasic variables by setting them equal to zero and solving for the remaining two variables; that is, we solve for the basic variables. Thus, if

$$x = y = 0,$$

then

$$u = 50, \qquad v = 70.$$

The vector

$$\mathbf{x}_1 = \begin{bmatrix} 0 \\ 0 \\ 50 \\ 70 \end{bmatrix}$$

is a basic solution, which gives rise to the feasible solution

$$\begin{bmatrix} 0 \\ 0 \end{bmatrix}$$

to the original problem specified by (10), (11), and (12). The variables x and y are nonbasic, and the variables u and v are basic. The convex region of solutions to the original problem has been sketched in Figure 11.8. Thus the vector $\mathbf{x}_1$ corresponds to the extreme point O.

If we let the variables x and u be nonbasic ($x = u = 0$), then $y = 50$ and $v = -30$. The vector

$$\mathbf{x}_2 = \begin{bmatrix} 0 \\ 50 \\ 0 \\ -30 \end{bmatrix}$$

is a basic solution that is not feasible, since v is negative. It corresponds to the point F in Figure 11.8, which is not a feasible solution to the original problem.

In Table 11.3 we have tabulated all the possible choices for basic solutions. The nonbasic variables are shaded and the corresponding point from Figure 11.8 is indicated in the table. It can be seen in this example, and proved in general, that every basic feasible solution determines an extreme point, and conversely, each extreme point determines a basic feasible solution.

One method of solving our linear programming problem would be to obtain all the basic solutions, discard those which are not feasible, and evaluate the objective function at each basic feasible solution, selecting that one, or

Table 11.3

x	y	u	v	Type of Solution	Corresponding Point in Figure 11.8
0	0	50	70	Basic feasible solution	O
0	50	0	−30	Not a basic feasible solution	F
0	35	15	0	Basic feasible solution	A
25	0	0	45	Basic feasible solution	C
70	0	−90	0	Not a basic feasible solution	G
10	30	0	0	Basic feasible solution	B

ones, for which we get a maximum value of the objective function. The number of possible basic solutions is

$$\binom{n+m}{n} = \frac{(n+m)!}{m!\,n!}.$$

That is, it is the number of ways of selecting n objects out of $m + n$ given objects.

The simplex method is a procedure that enables us to go from a given extreme point (basic feasible solution) to an adjacent extreme point in such a way that the value of the objective function increases as we move from extreme point to extreme point until we either obtain an optimal solution or find that the given problem has no finite optimal solution. The simplex method thus consists of two steps: (1) a way of checking whether a given basic feasible solution is an optimal solution, and (2) a way of obtaining another basic feasible solution with a larger value of the objective function. In most practical problems the simplex method does not consider every basic feasible solution; rather, it works with only a small number of these. However, examples have been given where a large number of basic feasible solutions have been examined by the simplex method. We shall now turn to a detailed discussion of this powerful method, using our illustrative example as a guide.

SELECTING AN INITIAL BASIC FEASIBLE SOLUTION

We can take all the original (nonslack) variables as our nonbasic variables and set them equal to zero. We then solve for the slack variables, our basic variables. In our example, we set

$$x = y = 0$$

and solve for u and v:

$$u = 50, \qquad v = 70.$$

Thus the initial basic feasible solution is the vector

$$\begin{bmatrix} 0 \\ 0 \\ 50 \\ 70 \end{bmatrix},$$

which yields the origin as an extreme point.

It is convenient to develop a tabular method for displaying the given problem and the initial basic feasible solution. First, we write (13) as the equation

$$-8x - 10y + z = 0, \tag{13$'$}$$

with z being considered as another variable. We now form the **initial tableau** (Tableau 1). The variables x, y, u, v, and z are written in the top row as labels on the corresponding columns. Constraints (14) are entered in the top rows followed by Equation (13′) in the bottom row. The bottom row of the tableau is called the **objective row**. Along the left-hand side of the tableau we indicate which variable is a basic variable in the corresponding equation. Thus u is a basic variable in the first equation and v is a basic variable in the second equation. A basic variable can also be described as a variable, other than z, which is present in exactly one equation, and there it appears with a coefficient of $+1$. In the tableau, the value of the basic variable is explicitly given in the rightmost column.

Tableau 1

	x	y	u	v	z	
u	2	1	1	0	0	50
v	1	2	0	1	0	70
	-8	-10	0	0	1	0

The initial tableau (Tableau 1) shows the values of the basic variables u and v, and therefore the nonbasic variables have values

$$x = 0, \qquad y = 0.$$

The value of the objective function for this initial basic feasible solution is

$$c_1x + c_2y + 0(u) + 0(v) = 8(0) + 10(0) + 0(u) + 0(v) = 0,$$

which is the entry in the objective row and rightmost column. It is clear that this solution is not optimal, for using the bottom row of the initial tableau, we can write

$$z = 0 + 8x + 10y - 0u - 0v. \tag{16}$$

Now the value of z can be increased by increasing either x or y, since both of these variables appear in (16) with positive coefficients. Since (16) contains terms with positive coefficients if and only if the objective row of our initial tableau has negative entries under the columns labeled with variables, we see that we can increase z by increasing any variable with a negative entry in the objective row. Thus we obtain the following optimality criterion for determining whether the feasible solution indicated in a tableau is an optimal solution yielding a maximum value for the objective function z.

In general, for the problem given by (7), (8), and (9), the initial tableau is Tableau 2.

Optimality Criterion If the objective row of a tableau has no negative entries in the columns labeled with variables, then the indicated solution is optimal and we can stop our computation.

Tableau 2

	x_1	x_2	$\cdots$	x_n	x_{n+1}	x_{n+2}	$\cdots$	x_{n+m}	z	
x_{n+1}	a_{11}	a_{12}	$\cdots$	a_{1n}	1	0	$\cdots$	0	0	b_1
x_{n+2}	a_{21}	a_{22}	$\cdots$	a_{2n}	0	1	$\cdots$	0	0	b_2
$\vdots$	$\vdots$						$\vdots$			$\vdots$
x_{n+m}	a_{m1}	a_{m2}	$\cdots$	a_{mn}	0	0	$\cdots$	1	0	b_m
	$-c_1$	$-c_2$	$\cdots$	$-c_n$	0	0	$\cdots$	0	1	0

SELECTING THE ENTERING VARIABLE

If the objective row of a tableau has negative entries in the columns labeled with the variables, then the indicated solution is not optimal and we must continue our search for an optimal solution.

The simplex method moves from one extreme point to an adjacent extreme point in such a way that the value of the objective function increases. This is done by increasing *one* variable at a time. The largest increase in z per unit increase in a variable occurs for the variable with the most negative entry in the objective row. In Tableau 1, the most negative entry in the objective row is -10, and since it occurs in the y-column, this is the variable to be increased. The variable to be increased is called the **entering variable**, because in the next iteration it will become a basic variable, thereby *entering* the set of basic variables. If there are several candidates for entering variables, choose one. An increase in y must be accompanied by a decrease in some of the other variables. This can be seen if we solve the equations in (14) for u and v:

$$\begin{aligned} u &= 50 - 2x - y \\ v &= 70 - x - 2y. \end{aligned}$$

Since we only increase y, we keep $x = 0$, and obtain

$$\begin{aligned} u &= 50 - y \\ v &= 70 - 2y \end{aligned} \tag{17}$$

so that as y increases, both u and v decrease. Equations (17) also show by how much we can increase y. That is, since u and v must be nonnegative, we must have

$$\begin{aligned} y &\le \tfrac{50}{1} = 50 \\ y &\le \tfrac{70}{2} = 35. \end{aligned}$$

We now see that the allowable increase in y can be no larger than the smaller of the ratios $\frac{50}{1}$ and $\frac{70}{2}$. Taking y as 35, we obtain the new basic feasible solution:

$$x = 0, \qquad y = 35, \qquad u = 15, \qquad v = 0.$$

The basic variables are y and u; the variables x and v are nonbasic. The objective function for this solution now has the value

$$z = 8(0) + 10(35) + 0(15) + 0(0) = 350,$$

which is much better than the earlier value of 0. This solution yields the extreme point A in Figurc 11.8, which is adjacent to O.

CHOOSING THE DEPARTING VARIABLE

Since the variable $v = 0$, it is not basic and is called the **departing variable**, for it has *departed* from the set of basic variables. The column of the entering variable is called the **pivotal column**; the row that is labeled by the departing variable is called the **pivotal row**.

Let us now look more closely at the selection of the departing variable. The choice of this variable was closely related to the determination of how far we could increase the entering variable (y in our example). To obtain this number, we formed the ratios (called **θ-ratios**) of the entries above the objective row in the rightmost column of the tableau by the corresponding entries in the pivotal column. The smallest of these ratios tells how far the entering variable can be increased. The basic variable labeling the row for which this smallest ratio occurs (the pivotal row) is then the departing variable. In our example, the θ-ratios, formed by using the rightmost column and the y-column, are $\frac{50}{1}$ and $\frac{70}{2}$. The smaller of these ratios, 35, occurs for the second row, which means that the second row is the pivotal row and the basic variable, v, labeling it becomes the departing variable and is no longer basic. If the smallest of the ratios is not selected, then one of the variables in the new solution becomes negative and the new solution is no longer feasible (Exercise T.2). What happens if there are entries in the pivotal column that are either zero or negative? If any entry in the pivotal column is negative, then the corresponding ratio is also negative; in this case the equation associated with the negative entry imposes no restriction on how far the entering variable can be increased. Suppose, for example, that the y-column in our initial tableau is

$$\begin{bmatrix} -3 \\ 2 \end{bmatrix} \quad \text{instead of} \quad \begin{bmatrix} 1 \\ 2 \end{bmatrix}.$$

Then instead of (17) we have

$$\begin{aligned} u &= 50 + 3y \\ v &= 70 - 2y, \end{aligned}$$

and since u must be nonnegative, we have

$$y \geq -\tfrac{50}{3},$$

which puts no limitation at all on how far y can be increased. If an entry in the pivotal column is zero, then the corresponding ratio cannot be formed (we cannot divide by zero), and again the associated equation puts no limitation on how far the entering variable can be increased. Thus, in forming the ratios, we only use the positive entries above the objective row in the pivotal column.

If all the entries above the objective row in the pivotal column are either zero or negative, then the entering variable can be made as large as we please. This means that the problem has no finite optimal solution.

OBTAINING A NEW TABLEAU

We must now obtain a new tableau indicating the new basic variables and the new basic feasible solution. Solving the second equation in (14) for y, we obtain

$$y = 35 - \tfrac{1}{2}x - \tfrac{1}{2}v, \tag{18}$$

and substituting this expression for y in the first equation in (14), we have

$$\tfrac{3}{2}x + u - \tfrac{1}{2}v = 15. \tag{19}$$

The second equation in (14) can be written, upon dividing by 2 (the coefficient of y), as

$$\tfrac{1}{2}x + y + \tfrac{1}{2}v = 35. \tag{20}$$

Substituting (18) for y in (13′), we have

$$-3x + 5v + z = 350. \tag{21}$$

Since $x = 0$, $v = 0$, we obtain the value of z for the current basic feasible solution as

$$z = 350.$$

Equations (19), (20), and (21) yield our new tableau (Tableau 3). Observe in this tableau that we have labeled the basic variables in each row.

Tableau 3

	x	y	u	v	z	
u	$\frac{3}{2}$	0	1	$-\frac{1}{2}$	0	15
y	$\frac{1}{2}$	1	0	$\frac{1}{2}$	0	35
	-3	0	0	5	1	350

Comparing Tableau 1 with Tableau 3, we observe that we can transform the former to the latter by elementary row operations as follows.

Step 1. Locate and circle the entry in the pivotal row and pivotal column. This entry is called the **pivot**. Mark the pivotal column by placing an arrow $\downarrow$ above the entering variable and mark the pivotal row by placing an arrow $\leftarrow$ to the left of the departing variable.

Step 2. If the pivot is k, multiply the pivotal row by $1/k$, making the entry that was the pivot a 1.

Step 3. Add the appropriate multiples of the pivotal row to all other rows (including the objective row) so that all elements in the pivotal column except for the 1 where the pivot was located become zero.

Step 4. In the new tableau replace the label on the pivotal row by the entering variable.

These four steps form a process called **pivotal elimination**. It is one of the iterations of the procedure described in Section 1.6 for transforming a matrix to reduced row echelon form.

We now repeat Tableau 1, with the arrows placed next to the entering and departing variables and with the pivot circled.

Performing pivotal elimination on Tableau 1 yields Tableau 3. We now repeat the entire procedure, using Tableau 3 as our initial tableau. Since the most negative entry in the objective row of Tableau 3, -3, occurs in the first column, x is the entering variable and the first column is the pivotal column. To determine the departing variable, we form the ratios of the entries in the rightmost column (except for the objective row) by the corresponding entries of the pivotal column for those entries in the pivotal column which are positive

Tableau 1

	x	y ↓	u	v	z	
u	2	1	1	0	0	50
$\leftarrow v$	1	②	0	1	0	70
	-8	-10	0	0	1	0

and select the smallest of these ratios. Since both entries in the pivotal column are positive, the ratios are $15/\frac{3}{2}$ and $35/\frac{1}{2}$. The smaller of these, $15/\frac{3}{2} = 10$, occurs for the first row, so the departing variable is u and the pivotal row is the first row. Tableau 3, with the pivotal column, pivotal row, and circled pivot, is shown again here.

Tableau 3

	x ↓	y	u	v	z	
$\leftarrow u$	$\frac{3}{2}$ (circled)	0	1	$-\frac{1}{2}$	0	15
y	$\frac{1}{2}$	1	0	$\frac{1}{2}$	0	35
	-3	0	0	5	1	350

Performing pivotal elimination on Tableau 3 yields Tableau 4.

Tableau 4

	x	y	u	v	z	
x	1	0	$\frac{2}{3}$	$-\frac{1}{3}$	0	10
y	0	1	$-\frac{1}{3}$	$\frac{2}{3}$	0	30
	0	0	2	4	1	380

Since the objective row of Tableau 4 has no negative entries in the columns labeled with variables, we conclude, by the optimality criterion, that we are finished and that the indicated solution is optimal. Thus the optimal solution is

$$x = 10, \qquad y = 30, \qquad u = 0, \qquad v = 0,$$

which corresponds to the extreme point $B(10, 30)$. Thus the simplex method started from the extreme point $O(0, 0)$ moved to the adjacent extreme point $A(0, 35)$ and then to the extreme point $B(10, 30)$, which is adjacent to A. The value of the objective function increased from 0 to 350 to 380, respectively.

We now summarize the simplex method.

The procedure for carrying out the simplex method is as follows.

Step 1. Set up the initial tableau.

Step 2. Apply the optimality test. If the objective row has no negative entries in the columns labeled with variables, then the indicated solution is optimal; we stop our computations.

Step 3. Choose a pivotal column by determining the column with the most negative entry in the objective row. If there are several candidates for a pivotal column, choose any one.

Step 4. Choose a pivotal row. Form the ratios of the entries above the objective row in the rightmost column by the corresponding entries of the pivotal column for those entries in the pivotal column which are positive. The pivotal row is the row for which the smallest of these ratios occurs. If there is a tie, so that the smallest ratio occurs at more than one row, choose any one of the qualifying rows. If none of the entries in the pivotal column above the objective row is positive, the problem has no finite optimum. We stop our computation.

Step 5. Perform pivotal elimination to construct a new tableau and return to Step 2.

Figure 11.13 gives a flow chart for the simplex method.

Figure 11.13 ► The simplex method

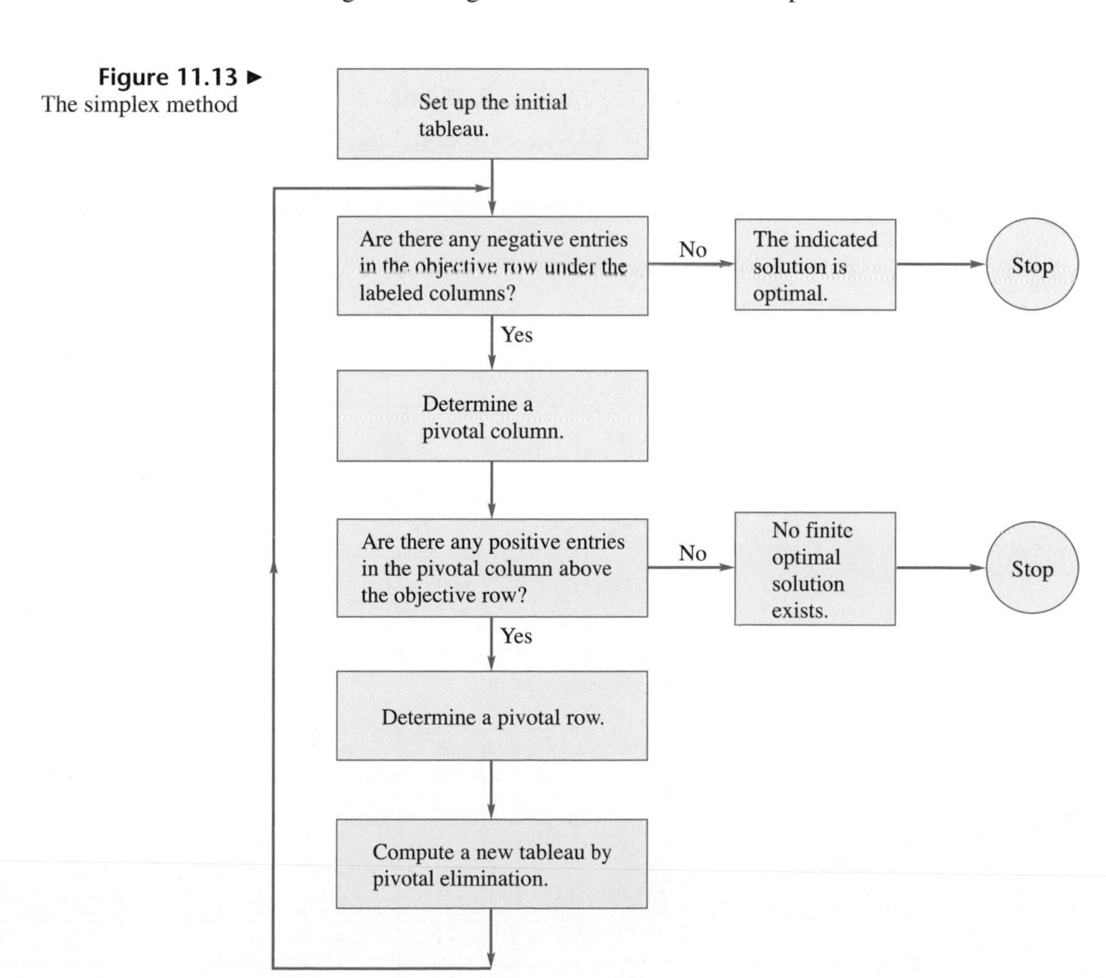

We have restricted our discussion of the simplex method to standard linear programming problems in which all the right-hand entries are nonnegative. It should be noted that the method applies to the general linear programming problem. For additional details we refer the reader to the references at the end of this chapter.

EXAMPLE 2 Maximize

$$z = 4x_1 + 8x_2 + 5x_3$$

subject to

$$\begin{aligned} x_1 + 2x_2 + 3x_3 &\leq 18 \\ x_1 + 4x_2 + x_3 &\leq 6 \\ 2x_1 + 6x_2 + 4x_3 &\leq 15 \end{aligned}$$

$$x_1 \geq 0, \quad x_2 \geq 0, \quad x_3 \geq 0.$$

The new problem with slack variables is

$$\text{Maximize} \quad z = 4x_1 + 8x_2 + 5x_3$$

subject to

$$\begin{aligned} x_1 + 2x_2 + 3x_3 + x_4 \qquad\qquad &= 18 \\ x_1 + 4x_2 + x_3 \qquad + x_5 \qquad &= 6 \\ 2x_1 + 6x_2 + 4x_3 \qquad\qquad + x_6 &= 15 \end{aligned}$$

$$x_1 \geq 0, \quad x_2 \geq 0, \quad x_3 \geq 0, \quad x_4 \geq 0, \quad x_5 \geq 0, \quad x_6 \geq 0.$$

The initial tableau and the succeeding tableaux are as follows.

	x_1	x_2 ↓	x_3	x_4	x_5	x_6	z	
x_4	1	2	3	1	0	0	0	18
← x_5	1	④	1	0	1	0	0	6
x_6	2	6	4	0	0	1	0	15
	−4	−8	−5	0	0	0	1	0

	x_1	x_2	x_3 ↓	x_4	x_5	x_6	z	
x_4	$\frac{1}{2}$	0	$\frac{5}{2}$	1	$-\frac{1}{2}$	0	0	15
x_2	$\frac{1}{4}$	1	$\frac{1}{4}$	0	$\frac{1}{4}$	0	0	$\frac{3}{2}$
← x_6	$\frac{1}{2}$	0	$\textcircled{\frac{5}{2}}$	0	$-\frac{3}{2}$	1	0	6
	−2	0	−3	0	2	0	1	12

	↓ x_1	x_2	x_3	x_4	x_5	x_6	z	
x_4	0	0	0	1	1	-1	0	9
$\leftarrow x_2$	$\textcircled{\frac{1}{5}}$	1	0	0	$\frac{2}{5}$	$-\frac{1}{10}$	0	$\frac{9}{10}$
x_3	$\frac{1}{5}$	0	1	0	$-\frac{3}{5}$	$\frac{2}{5}$	0	$\frac{12}{5}$
	$-\frac{7}{5}$	0	0	0	$\frac{1}{5}$	$\frac{6}{5}$	1	$\frac{96}{5}$

	x_1	x_2	x_3	x_4	x_5	x_6	z	
x_4	0	0	0	1	1	-1	0	9
x_1	1	5	0	0	2	$-\frac{1}{2}$	0	$\frac{9}{2}$
x_3	0	-1	1	0	-1	$\frac{1}{2}$	0	$\frac{3}{2}$
	0	7	0	0	3	$\frac{1}{2}$	1	$\frac{51}{2}$

Hence an optimal solution is

$$x_1 = \tfrac{9}{2}, \qquad x_2 = 0, \qquad x_3 = \tfrac{3}{2}.$$

The slack variables are

$$x_4 = 9, \qquad x_5 = 0, \qquad x_6 = 0,$$

and the optimal value of z is $\frac{51}{2}$. ■

A number of difficulties can arise in using the simplex method. We shall briefly describe one of these, and we again refer the reader to the references at the end of this chapter for an extended discussion of computational considerations.

DEGENERACY

Suppose that a basic variable becomes zero in one of the tableaux in the simplex method. Then one of the ratios used to determine the next pivotal row may be zero, in which case the pivotal row is the one labeled with the basic variable that is zero (in this case, zero is the smallest of the ratios). Recall that the entering variable is increased from zero to the smallest ratio. Since the smallest ratio is zero, the entering variable remains at the value zero. In the new tableau all the old basic variables have the same values that they had in the old tableau; the value of the objective function has not been increased. The new tableau looks like the old one, except that a basic variable with value zero has become nonbasic, and its place has been taken by a nonbasic variable also entering with value zero. A basic feasible solution in which one or more of the basic variables are zero is called a **degenerate basic feasible solution**. It can be shown that when a tie occurs for the smallest ratio in determining the pivotal row, a degenerate solution will arise and there are several optimal solutions. When no degenerate solution occurs, the value of the objective function increases as we move from one basic feasible solution to another. The procedure stops after a finite number of steps, since the number of basic feasible solutions is finite. However, if we have a degenerate basic feasible solution, we may return to a basic feasible solution that had already been found at an earlier iteration and thus enter an infinite cycle. Fortunately, cycling is encountered only occasionally in practical linear programming problems, although several examples have been carefully constructed to show that cycling can occur. In actual practice, when degeneracy occurs, it is handled by merely ignoring it. The value of the objective function may remain constant for a few iterations and will then start to increase. Moreover, a technique called **perturbation** has been developed for handling degeneracy. This technique calls for making slight changes in the rightmost column of the tableau so that the troublesome ties in the ratios will no longer exist.

INTERIOR POINT METHODS

Linear programming is a field that has changed dramatically in the last 15 years. The simplex method, developed by George B. Dantzig in 1947 proceeds from one extreme point to an adjacent extreme point in the convex set of feasible solutions in such a way that the value of the objective function increases, or at worst, remains the same. This method has been used to solve a huge number of linear programming problems for over fifty years. The experience of solving so many real applied problems with the simplex method led to the conjecture that it is a polynomial time algorithm. That is, the running time of the simplex method is proportional to a polynomial in n, where n is the number of variables in the problem. In 1972, Klee and Minty constructed a particular linear programming problem showing that the simplex method is not a polynomial time algorithm. However, this type of behavior has not been seen in the solution of the very large number of practical problems solved with the simplex method.

In the mid-1980s a new approach to solving linear programming problems was proposed. It was based on algorithms that follow paths in the interior of the convex set of feasible solutions. Some of these methods have been shown to be polynomial time algorithms. Instead of following the edges of the convex set from extreme point to extreme point, eventually reaching an optimal solution, the interior point methods burrow through the interior of the set to find an optimal solution. Since they are not constrained to the directions of the edges, nor their length, it is reasonable to assume that the interior point methods might be significantly faster than the edge-following simplex method. However, current computational performance seems to show that interior point methods and the simplex method are pretty comparable; which method is best depends on the particular problem, and interior point methods are probably best the majority of the time, with the advantage of interior point methods becoming slightly more marked as the problem size increases (thousands of variables and/or constraints).

COMPUTER IMPLEMENTATIONS

There are many computer programs implementing the simplex method and other algorithms in the area of mathematical programming (which includes integer programming and nonlinear programming). Some of these programs can do extensive data manipulation to prepare the input for the problem, solve the problem, and then prepare elaborate reports that can be used by management in decision making. Moreover, some programs can handle huge problems, limited only by the available computer resources.

LINDO Systems, Web site: `www.lindo.com` (1415 North Dayton Avenue, Chicago, Illinois, 60622; telephone: 800-441-2378), makes available scaled-down versions of their widely used programs LINDO, LINGO, and WHAT'S BEST, which can be downloaded for free from their Web site. LINDO is a program that solves linear programming problems presented in mathematical form using the simplex and dual simplex methods. LINGO is a program that formulates a mathematical model of a problem and then proceeds to solve it. WHAT'S BEST is a spreadsheet-based program that solves a mathematically presented problem and works with Microsoft's Excel. The free versions of these programs will handle problems with as many as 150 constraints and 300 variables.

AMPL Optimization LLC makes available a scaled down-version of its program AMPL for solving linear programming problems. This program can be downloaded for free from its Web site, `www.ampl.com`. The free version of this program is limited to 300 constraints and 300 variables.

ILOG has developed the program ILOG OPL Studio for developing a mathematical model and solution of a given linear programming problem. A trial version can be downloaded from ILOG's Web site, `www.ILOG.com` (telephone: 800-367-4564).

The professional versions of these programs are used frequently to solve large problems in a wide variety of applications. Moreover, these programs can solve linear programming and related problems and are available for a variety of platforms, including PCs and various workstations.

Key Terms

Feasible solution
Optimal solution
Basic variables
Nonbasic variables
Basic feasible solution
Simplex method
Initial tableau
Objective row
Entering variable
Departing variable
Pivotal column
Pivotal row
θ-ratio
Pivot
Pivotal elimination
Degenerate basic feasible solution
Perturbation

11.2 Exercises

In Exercises 1 through 4, write the initial simplex tableau for each given linear programming problem.

1. Maximize $z = 3x + 7y$
subject to

$$\begin{aligned} 3x - 2y &\leq 7 \\ 2x + 5y &\leq 6 \\ 2x + 3y &\leq 8 \\ x \geq 0, \quad y &\geq 0. \end{aligned}$$

2. Maximize $z = 2x_1 + 3x_2 - 4x_3$
subject to

$$\begin{aligned} 3x_1 - 2x_2 + x_3 &\leq 4 \\ 2x_1 + 4x_2 + 5x_3 &\leq 6 \\ x_1 \geq 0, \quad x_2 \geq 0, \quad x_3 &\geq 0. \end{aligned}$$

3. Maximize $z = 2x_1 + 2x_2 + 3x_3 + x_4$
subject to

$$\begin{aligned} 3x_1 - 2x_2 + x_3 + x_4 &\leq 6 \\ x_1 + x_2 + x_3 + x_4 &\leq 8 \\ 2x_1 - 3x_2 - x_3 + 2x_4 &\leq 10 \\ x_1 \geq 0, \quad x_2 \geq 0, \quad x_3 \geq 0, \quad x_4 &\geq 0. \end{aligned}$$

4. Maximize $z = 2x_1 - 3x_2 + x_3$
subject to

$$\begin{aligned} x_1 - 2x_2 + 4x_3 &\leq 5 \\ 2x_1 + 2x_2 + 4x_3 &\leq 5 \\ 3x_1 + x_2 - x_3 &\leq 7 \\ x_1 \geq 0, \quad x_2 \geq 0, \quad x_3 &\geq 0. \end{aligned}$$

In Exercises 5 through 11, solve each linear programming problem by the simplex method. Some of these problems may have no finite optimal solution.

5. Maximize $z = 2x + 3y$
subject to

$$3x + 5y \leq 6$$
$$2x + 3y \leq 7$$
$$x \geq 0, \quad y \geq 0.$$

6. Maximize $z = 2x + 5y$
subject to

$$3x + 7y \leq 6$$
$$2x + 6y \leq 7$$
$$3x + 2y \leq 5$$
$$x \geq 0, \quad y \geq 0.$$

7. Maximize $z = 2x + 5y$
subject to

$$2x - 3y \leq 4$$
$$x - 2y \leq 6$$
$$x \geq 0, \quad y \geq 0.$$

8. Maximize $z = 3x_1 + 2x_2 + 4x_3$
subject to

$$x_1 - x_2 - x_3 \leq 6$$
$$-2x_1 + x_2 - 2x_3 \leq 7$$
$$3x_1 + x_2 - 4x_3 \leq 8$$
$$x_1 \geq 0, \quad x_2 \geq 0, \quad x_3 \geq 0.$$

9. Maximize $z = 2x_1 - 4x_2 + 5x_3$
subject to

$$3x_1 + 2x_2 + x_3 \leq 6$$
$$3x_1 - 6x_2 + 7x_3 \leq 9$$
$$x_1 \geq 0, \quad x_2 \geq 0, \quad x_3 \geq 0.$$

10. Maximize $z = 2x_1 + 4x_2 - 3x_3$
subject to

$$5x_1 + 2x_2 + x_3 \leq 5$$
$$3x_1 - 2x_2 + 3x_3 \leq 10$$
$$4x_1 + 5x_2 - x_3 \leq 20$$
$$x_1 \geq 0, \quad x_2 \geq 0, \quad x_3 \geq 0.$$

11. Maximize $z = x_1 + 2x_2 - x_3 + 5x_4$
subject to

$$2x_1 + 3x_2 + x_3 - x_4 \leq 8$$
$$3x_1 + x_2 - 4x_3 + 5x_4 \leq 9$$
$$x_1 \geq 0, \quad x_2 \geq 0, \quad x_3 \geq 0, \quad x_4 \geq 0.$$

12. Solve Exercise 1 in Section 11.1 by the simplex method.

13. Solve Exercise 4 in Section 11.1 by the simplex method.

14. Solve Exercise 7 in Section 11.1 by the simplex method.

15. A power plant burns coal, oil, and gas to generate electricity. Suppose that each ton of coal generates 600 kilowatt hours, emits 20 units of sulfur dioxide and 15 units of particulate matter, and costs \$200; each ton of oil generates 550 kilowatt hours, emits 18 units of sulfur dioxide and 12 units of particulate matter, and costs \$220; each ton of gas generates 500 kilowatt hours, emits 15 units of sulfur dioxide and 10 units of particulate matter, and costs \$250. The environmental protection agency restricts the daily emission of sulfur dioxide to no more than 60 units and no more than 75 units of particulate matter. If the power plant wants to spend no more than \$2000 per day on fuel, how much fuel of each type should be bought to maximize the amount of energy generated?

Theoretical Exercises

T.1. Consider the standard linear programming problem:
Maximize $z = \mathbf{c}^T \mathbf{x}$
subject to

$$A\mathbf{x} \leq \mathbf{b}$$
$$\mathbf{x} \geq \mathbf{0},$$

where A is an $m \times n$ matrix. Show that the feasible region (the set of all feasible solutions) of this problem is a convex set. [*Hint*: If $\mathbf{u}$ and $\mathbf{v}$ are in R^n, the line segment joining them is the set of points $\lambda\mathbf{u} + (1 - \lambda)\mathbf{v}$, where $0 \leq \lambda \leq 1$.]

T.2. Suppose that in selecting the departing variable the minimum θ-ratio is not chosen. Show that the resulting solution is not feasible.

MATLAB Exercises

Routine **lpstep** *in* MATLAB *provides a step-by-step procedure for solving linear programming problems using techniques described in this section. Before using* **lpstep**, *you must formulate the initial tableau into a matrix for entry into* MATLAB. *The matrix representing the tableau has the same form as discussed in this section except that there are no variable names for rows and columns. For more information, use* **help lpstep**.

ML.1. For the illustrative problem given in Equations (10) through (12), the initial tableau is displayed in the text as Tableau 1. To use **lpstep**, use the following MATLAB commands and answer the questions posed. You can use the choices for this problem given in the discussion following the steps for pivotal elimination.

A = [**2 1 1 0 0 50**; **1 2 0 1 0 70**;
−8 − 10 0 0 1 0]
lpstep(A)

ML.2. Solve Exercise 5 using **lpstep**.

ML.3. Solve Exercise 6 using **lpstep**.

ML.4. Solve Exercise 8 using **lpstep**.

ML.5. Maximize $z = 8x_1 + 9x_2 + 5x_3$ subject to

$$\begin{aligned} x_1 + x_2 + 2x_3 &\le 2 \\ 2x_1 + 3x_2 + 4x_3 &\le 3 \\ 6x_1 + 6x_2 + 2x_3 &\le 8 \\ x_i \ge 0, \quad i &= 1, 2, 3. \end{aligned}$$

ML.6. Maximize $z = x_1 + 2x_2 + x_3 + x_4$ subject to

$$\begin{aligned} 2x_1 + x_2 + 3x_3 + x_4 &\le 8 \\ 2x_1 + 3x_2 \qquad + 4x_4 &\le 12 \\ 3x_1 + 2x_2 + 2x_3 \qquad &\le 18 \\ x_i \ge 0, \quad i &= 1, 2, 3, 4. \end{aligned}$$

ML.7. Routine **linprog** solves linear programming problems of the form described in this section directly. That is, the steps are performed automatically. Check your solutions to Exercises 10 and 12.

11.3 DUALITY

In this section we show how to associate a minimization problem with each linear programming problem in standard form. There are some interesting interpretations of the associated problem that we also discuss.

Consider the following pair of linear programming problems:

$$\text{Maximize} \quad z = \mathbf{c}^T\mathbf{x}$$

subject to

$$\begin{aligned} A\mathbf{x} &\le \mathbf{b} \\ \mathbf{x} &\ge \mathbf{0} \end{aligned} \tag{1}$$

and

$$\text{Minimize} \quad z' = \mathbf{b}^T\mathbf{y}$$

subject to

$$\begin{aligned} A^T\mathbf{y} &\ge \mathbf{c} \\ \mathbf{y} &\ge \mathbf{0}, \end{aligned} \tag{2}$$

where A is $m \times n$, $\mathbf{b}$ is $m \times 1$, $\mathbf{c}$ is $n \times 1$, $\mathbf{x}$ is $n \times 1$, and $\mathbf{y}$ is $m \times 1$.

These problems are called **dual problems**. The problem given by (1) is called the **primal problem**; the problem given by (2) is called the **dual problem**.

EXAMPLE 1 If the primal problem is

$$\text{Maximize} \quad z = \begin{bmatrix} 3 & 4 \end{bmatrix} \begin{bmatrix} x_1 \\ x_2 \end{bmatrix}$$

subject to

$$\begin{bmatrix} 2 & 3 \\ 3 & -1 \\ 5 & 4 \end{bmatrix} \begin{bmatrix} x_1 \\ x_2 \end{bmatrix} \le \begin{bmatrix} 3 \\ 4 \\ 2 \end{bmatrix}$$

$$x_1 \ge 0, \quad x_2 \ge 0,$$

then the dual problem is

$$\text{Minimize} \quad z' = \begin{bmatrix} 3 & 4 & 2 \end{bmatrix} \begin{bmatrix} y_1 \\ y_2 \\ y_3 \end{bmatrix}$$

subject to

$$\begin{bmatrix} 2 & 3 & 5 \\ 3 & -1 & 4 \end{bmatrix} \begin{bmatrix} y_1 \\ y_2 \\ y_3 \end{bmatrix} \geq \begin{bmatrix} 3 \\ 4 \end{bmatrix}$$

$$y_1 \geq 0, \quad y_2 \geq 0, \quad y_3 \geq 0.$$

■

Observe that in formulating the dual problem, the coefficients of the ith constraint of the primal problem become the coefficients of the variable y_i in the constraints of the dual problem. Conversely, the coefficients of x_j in the primal problem become the coefficients of the jth constraint in the dual problem. Moreover, the coefficients of the objective function of the primal problem become the right-hand sides of the constraints of the dual problem, and conversely.

Since problem (2) can be rewritten as a standard linear programming problem, we can ask for the dual of the dual problem. The answer is provided by the following theorem.

THEOREM 11.3 *Given a primal problem, as in* (1), *the dual of its dual problem is the primal problem.*

Proof Consider the dual problem (2), which we rewrite in standard form as

$$\text{Maximize} \quad z' = -\mathbf{b}^T\mathbf{y}$$

subject to

$$\begin{aligned} -A^T\mathbf{y} &\leq -\mathbf{c} \\ \mathbf{y} &\geq \mathbf{0}. \end{aligned} \tag{3}$$

Now the dual of (3) is

$$\text{Minimize} \quad z'' = -\mathbf{c}^T\mathbf{w}$$

subject to

$$\begin{aligned} (-A^T)^T\mathbf{w} &\geq (-\mathbf{b}^T)^T \\ \mathbf{w} &\geq \mathbf{0} \end{aligned}$$

or

$$\text{Maximize} \quad z'' = \mathbf{c}^T\mathbf{w}$$

subject to

$$\begin{aligned} A\mathbf{w} &\leq \mathbf{b} \\ \mathbf{w} &\geq \mathbf{0}. \end{aligned} \tag{4}$$

Writing $\mathbf{w} = \mathbf{x}$, we see that problem (4) is the primal problem. ■

EXAMPLE 2 Find the dual problem of the linear programming problem

$$\text{Minimize} \quad z' = \begin{bmatrix} 2 & 3 \end{bmatrix} \begin{bmatrix} y_1 \\ y_2 \end{bmatrix}$$

subject to

$$\begin{bmatrix} 3 & 4 \\ 1 & 2 \\ 5 & 3 \end{bmatrix} \begin{bmatrix} y_1 \\ y_2 \end{bmatrix} \geq \begin{bmatrix} 5 \\ 2 \\ 7 \end{bmatrix}$$

$$y_1 \geq 0, \quad y_2 \geq 0.$$

Solution The given problem is the dual of the primal problem that we seek to formulate. This primal problem is obtained by taking the dual of the given problem, obtaining

$$\text{Maximize} \quad z = 5x_1 + 2x_2 + 7x_3$$

subject to

$$\begin{aligned} 3x_1 + x_2 + 5x_3 &\leq 2 \\ 4x_1 + 2x_2 + 3x_3 &\leq 3 \end{aligned}$$

$$x_1 \geq 0, \quad x_2 \geq 0, \quad x_3 \geq 0.$$ ■

The following theorem, whose proof we omit, gives the relations between the optimal solutions to the primal and dual problems.

THEOREM 11.4 **(Duality Theorem)** *If either the primal problem or dual problem has an optimal solution with finite objective value, then the other problem also has an optimal solution. Moreover, the objective values of the two problems are equal.* ■

It can also be shown that when the primal problem is solved by the simplex method, the final tableau contains the optimal solution to the dual problem in the objective row under the columns of the slack variables. That is, y_1, the first dual variable, is found in the objective row under the first slack variable; y_2 is found under the second slack variable, and so on. We can use these facts to solve the diet problem, Example 3 in Section 11.1, as follows.

EXAMPLE 3 Consider the diet problem of Example 3, Section 11.1 (using z' instead of z and y_1, y_2 instead of x, y):

$$\text{Minimize} \quad z' = 30y_1 + 40y_2$$

subject to

$$\begin{aligned} 2y_1 + y_2 &\geq 12 \\ y_1 + y_2 &\geq 9 \\ y_1 + 3y_2 &\geq 15 \end{aligned} \tag{5}$$

$$y_1 \geq 0, \quad y_2 \geq 0.$$

The dual of this problem is

$$\text{Maximize} \quad z = 12x_1 + 9x_2 + 15x_3$$

subject to

$$\begin{aligned} 2x_1 + x_2 + x_3 &\le 30 \\ x_1 + x_2 + 3x_3 &\le 40 \end{aligned} \tag{6}$$

$$x_1 \ge 0, \quad x_2 \ge 0, \quad x_3 \ge 0.$$

This is a standard linear programming problem. Introducing the slack variables x_4 and x_5, we obtain

$$\text{Maximize} \quad z = 12x_1 + 9x_2 + 15x_3$$

subject to

$$\begin{aligned} 2x_1 + x_2 + x_3 + x_4 \phantom{{}+x_5} &= 30 \\ x_1 + x_2 + 3x_3 \phantom{{}+x_4} + x_5 &= 40 \end{aligned}$$

$$x_1 \ge 0, \quad x_2 \ge 0, \quad x_3 \ge 0, \quad x_4 \ge 0, \quad x_5 \ge 0.$$

We now apply the simplex method, obtaining the following tableaux.

	x_1	x_2	x_3 ↓	x_4	x_5	z	
x_4	2	1	1	1	0	0	30
← x_5	1	1	(3)	0	1	0	40
	−12	−9	−15	0	0	1	0

	x_1 ↓	x_2	x_3	x_4	x_5	z	
← x_4	($\frac{5}{3}$)	$\frac{2}{3}$	0	1	$-\frac{1}{3}$	0	$\frac{50}{3}$
x_3	$\frac{1}{3}$	$\frac{1}{3}$	1	0	$\frac{1}{3}$	0	$\frac{40}{3}$
	−7	−4	0	0	5	1	200

	x_1	x_2 ↓	x_3	x_4	x_5	z	
← x_1	1	($\frac{2}{5}$)	0	$\frac{3}{5}$	$-\frac{1}{5}$	0	10
x_3	0	$\frac{1}{5}$	1	$-\frac{1}{5}$	$\frac{2}{5}$	0	10
	0	$-\frac{6}{5}$	0	$\frac{21}{5}$	$\frac{18}{5}$	1	270

	x_1	x_2	x_3	x_4	x_5	z	
x_2	$\frac{5}{2}$	1	0	$\frac{3}{2}$	$-\frac{1}{2}$	0	25
x_3	$-\frac{1}{2}$	0	1	$-\frac{1}{2}$	$\frac{1}{2}$	0	5
	3	0	0	6	3	1	300

The optimal solution to problem (6) is

$$x_1 = 0, \quad x_2 = 25, \quad x_3 = 5,$$

and the value of z is 300.

The optimal solution to the given problem (5), the dual of (6), is found in the objective row under the x_4 and x_5 columns:

$$y_1 = 6, \qquad y_2 = 3.$$

The value of $z' = 30(6) + 40(3) = 300$, as we expect from Theorem 11.4 (the duality theorem). ■

ECONOMIC INTERPRETATION OF THE DUAL PROBLEM

We now give two economic interpretations of the dual linear programming problem. The primal problem (1) can be interpreted as follows. We have m resources, inputs or raw materials, and n activities, each of which produces a product. Let

b_i = amount of resource i that is available $(1 \le i \le m)$;

a_{ij} = amount of resource i that is used by 1 unit of activity j;

c_j = contribution to the total profit z from 1 unit of activity j $(1 \le j \le n)$;

x_j = level of activity j.

We illustrate these interpretations with Example 1 of Section 11.1, the photo shop problem, which we restate for the convenience of the reader (we use x_1 and x_2 instead of x and y as in Section 11.1).

$$\text{Maximize} \quad z = 8x_1 + 10x_2$$

subject to

$$\begin{aligned} 2x_1 + x_2 &\le 50 \\ x_1 + 2x_2 &\le 70 \\ x_1 \ge 0, \quad x_2 &\ge 0. \end{aligned}$$

Here $b_1 = 50$ and $b_2 = 70$ are the amounts of solution A and B, respectively, available each day; activity 1 is the preparation of 1 quart of Fine developer, activity 2 is the preparation of 1 quart of Extra Fine developer; $a_{12} = 1$ is the amount of solution A used in making 1 quart of Extra Fine and so on; $c_1 = 8$ cents is the profit on 1 quart of Fine, $c_2 = 10$ cents is the profit on 1 quart of Extra Fine; x_1 and x_2 are the numbers of quarts of Fine and Extra Fine, respectively, to be made. The dual of the photo shop problem is

$$\text{Minimize} \quad z' = 50y_1 + 70y_2$$

subject to

$$\begin{aligned} 2y_1 + y_2 &\ge 8 \\ y_1 + 2y_2 &\ge 10 \\ y_1 \ge 0, \quad y_2 &\ge 0. \end{aligned}$$

Consider the jth constraint of the dual general problem (2):

$$a_{1j}y_1 + a_{2j}y_2 + \cdots + a_{mj}y_m \ge c_j. \tag{7}$$

Since a_{ij} represents amount of resource i per unit of output j, and c_j is the value per unit of output j, we now show from Equation (7) that the dual variable y_j represents "value per unit of resource i" as follows: The dimensions of c_j/a_{ij} are

$$\frac{\text{value}/\ \cancel{\text{unit of output } j}}{\text{amount of resource } i/\ \cancel{\text{unit of output } j}} = \frac{\text{value}}{\text{amount of resource } i}.$$

Thus the dual variable y_i acts as a price or cost of 1 unit of resource i. The dual variables are called **shadow prices**, **fictitious prices**, or **accounting prices**.

FIRST INTERPRETATION

From (7) we see that the left side of this equation is the contribution to the total profit z of the resources that are used in making 1 unit of the jth product. Since the profit of 1 unit of the jth product is c_j, Equation (7) merely says that the contribution of the resources to the total profit has to be at least c_j, for otherwise, we should be using the resources in some better way. In our example, the optimal solution to the dual problem obtained from Tableau 4 in Section 11.2 is

$$y_1 = 2, \qquad y_2 = 4.$$

This means that a shadow price of 2 cents has been attached to each ounce of solution A, and a shadow price of 4 cents has been attached to each unit of solution B. Thus the contribution of the resources used in making 1 quart of Fine to the profit is (in cents)

$$2y_1 + y_2 = 2(2) + 4 = 8,$$

so that the first constraint of the dual has been satisfied. Of course, the constraints

$$y_1 \geq 0, \quad y_2 \geq 0$$

in the dual merely state that the contribution of each resource to the total profit is nonnegative; if some y_1 were negative, we would be better off not using this resource. Finally, the objective function

$$z' = b_1y_1 + b_2y_2 + \cdots + b_my_m$$

is to be minimized; this can be interpreted to be minimizing the total shadow value of the resources that are being used to make the products.

SECOND INTERPRETATION

If $\mathbf{x}_0$ and $\mathbf{y}_0$ are optimal solutions to the primal and dual problems, respectively, then, by the duality theorem (Theorem 11.4), the maximum profit z_0 satisfies the equation

$$z_0 = \mathbf{b}^T\mathbf{y}_0 = b_1y_1^0 + b_2y_2^0 + \cdots + b_my_m^0, \tag{8}$$

where

$$\mathbf{y}_0 = \begin{bmatrix} y_1^0 \\ \vdots \\ y_m^0 \end{bmatrix}.$$

From Equation (8) we see that the manufacturer can increase the profit by increasing the available amount of at least one of the resources. If b_i is increased by 1 unit, the profit will increase by y_i^0. Thus y_i^0 represents the **marginal value** of the ith resource. Similarly, y_i^0 represents the loss incurred if 1 unit of the ith resource is not used. Thus it can be considered as a replacement value of the ith resource for insurance purposes. Then the objective function of the dual problem

$$z' = b_1y_1 + b_2y_2 + \cdots + b_my_m$$

is the total replacement value. Thus an insurance company would want to minimize this objective function, since it would want to pay out as little as possible to settle a claim.

Applications In the Theory of Games, Section 11.4, the strategies for two competing players are obtained by solving two dual linear programming problems.

Further Readings

CALVERT, JAMES E., and WILLIAM L. VOXMAN. *Linear Programming.* Philadelphia: Harcourt Brace Jovanovich, 1989.

CHVATAL, VASEK. *Linear Programming.* New York: Freeman, 1983.

FOURER, R., D. M. GAY, and B. W. KERNIGHAM. *AMPL: A Modeling Language for Mathematical Programming*, 2nd ed. Pacific Grove, Calif.: Duxbury/Brooks/Cole, 2002.

GASS, SAUL I. *Linear Programming: Methods and Applications*, 5th ed. New York: Dover, 2003.

KOLMAN, BERNARD, and ROBERT E. BECK. *Elementary Linear Programming with Applications*, 2nd ed. Boston: Academic Press, Inc., 1995.

KUESTER, JAMES L., and JOE H. MIZE. *Optimization Techniques with FORTRAN.* New York: McGraw-Hill Book Company, 1974.

MURTY, KATTA G. *Linear and Combinatorial Programming.* New York: John Wiley & Sons, Inc., 1989.

NERING, EVAR D., and ALBERT W. TUCKER. *Linear Programs and Related Problems.* Boston: Academic Press, Inc., 1993.

RARDIN, RONALD L. *Optimization in Operations Research.* Upper Saddle River, N.J.: Prentice Hall, 1997.

SCHRAGE, LINUS. *Optimization Modeling with LINDO*, 5th ed. Pacific Grove, Calif.: Duxbury/Brooks/Cole, 1997.

STRANG, GILBERT. *Introduction to Applied Mathematics.* Wellesley, Mass.: Wellesley-Cambridge Press, 1986.

SULTAN, ALAN. *Linear Programming.* Boston: Academic Press, Inc., 1993.

WALKER, RUSSELL C. *Introduction to Mathematical Programming.* Upper Saddle River, N.J.: Prentice Hall, 1999.

WINSTON, WAYNE L. *Mathematical Programming: Applications and Algorithms*, 4th ed. Pacific Grove, Calif.: Duxbury/Brooks/Cole, 2002.

WRIGHT, STEPHEN J. *Primal-Dual Interior-Point Methods.* Philadelphia: SIAM, 1997.

Key Terms

Dual problems
Primal problem

11.3 Exercises

In Exercises 1 through 4, state the dual of the given linear programming problem.

1. Maximize $z = 3x_1 + 2x_2$
subject to

$$\begin{aligned} 4x_1 + 3x_2 &\leq 7 \\ 5x_1 - 2x_2 &\leq 6 \\ 6x_1 + 8x_2 &\leq 9 \end{aligned}$$

$$x_1 \geq 0, \quad x_2 \geq 0.$$

2. Maximize $z = 10x_1 + 12x_2 + 15x_3$
subject to

$$\begin{aligned} x_1 + 3x_2 + 4x_3 &\leq 5 \\ 2x_1 + 4x_2 - 5x_3 &\leq 6 \end{aligned}$$

$$x_1 \geq 0, \quad x_2 \geq 0, \quad x_3 \geq 0.$$

3. Minimize $z = 3x_1 + 5x_2$
subject to

$$\begin{aligned} 2x_1 + 3x_2 &\geq 7 \\ 8x_1 - 9x_2 &\geq 12 \\ 10x_1 + 15x_2 &\geq 18 \end{aligned}$$

$$x_1 \geq 0, \quad x_2 \geq 0.$$

4. Minimize $z' = 14x_1 + 12x_2 + 18x_3$
subject to

$$\begin{aligned} 3x_1 + 5x_2 - 4x_3 &\geq 9 \\ 5x_1 + 2x_2 + 7x_3 &\geq 12 \end{aligned}$$

$$x_1 \geq 0, \quad x_2 \geq 0, \quad x_3 \geq 0.$$

5. Verify that the dual of the dual of the linear programming problem

$$\text{Maximize} \quad z = 3x_1 + 6x_2 + 9x_3$$

subject to

$$\begin{aligned} 3x_1 + 2x_2 - 3x_3 &\leq 12 \\ 5x_1 + 4x_2 + 7x_3 &\leq 18 \end{aligned}$$

$$x_1 \geq 0, \quad x_2 \geq 0, \quad x_3 \geq 0,$$

is the given problem.

In Exercises 6 through 9, solve the dual of the indicated problem by the method of Example 3.

6. Exercise 5 in Section 11.2

7. Exercise 6 in Section 11.2

8. Exercise 9 in Section 11.2

9. Exercise 10 in Section 11.2

10. A natural cereal consists of dates, nuts, and raisins. Suppose that each ounce of dates contains 24 units of protein, 6 units of iron, and costs 15 cents; each ounce of nuts contains 2 units of protein, 6 units of iron, and costs 18 cents; and each ounce of raisins contains 4 units of protein, 2 units of iron, and costs 12 cents. If each box of cereal is to contain at least 24 units of protein and at least 36 units of iron, how many ounces of each ingredient should be used to minimize the cost of a box of cereal? (*Hint*: Solve the dual problem.)

11.4 THE THEORY OF GAMES

Prerequisites. Sections 11.1–11.3.

There are many problems in economics, politics, warfare, business, and so on, that require decisions to be made in conflicting or competitive situations. The theory of games is a rather new area of applied mathematics that attempts to analyze conflict situations and provides a basis for rational decision making.

The theory of games was developed in the 1920s by John von Neumann* and E. Borel,** but the subject did not come to fruition until the publication, in

*John von Neumann (1903–1957) was born in Hungary and came to the United States in 1930. His talents in mathematics were recognized early, and he is considered one of the greatest mathematicians of the twentieth century. He made fundamental contributions to the foundations of mathematics, quantum mechanics, operator calculus, and the theory of computing machines and automata. He also participated in many defense projects during World War II and helped to develop the atomic bomb. Beginning in 1926 he developed the theory of games, culminating in his joint work with Oskar Morgenstern in 1944, *Theory of Games and Economic Behavior*.

**Emile Borel (1871–1956) was born in Saint Affrique, France. A child prodigy and prolific author of mathematical papers, he married an author, was part of French literary circles, and was active in politics, serving in the National Assembly for a time and joining the Resistance during World War II. In mathematics he contributed to the theory of functions, measure theory, and probability theory. In the 1920s he wrote the first papers in game theory, defining the basic terms, stating the minimax theorem, and discussing applications to war and economics.

1944, of the decisive book *Theory of Games and Economic Behavior* by John von Neumann and Oskar Morgenstern.†

A **game** is a competitive situation in which each of a number of players is pursuing his objective in direct conflict with the other players. Each player is doing everything he can to gain as much as possible for himself. Essentially, games are of two types. First, there are **games of chance**, such as roulette, which require no skill on the part of the players; the outcomes and winnings are determined solely by the laws of probability and can in no way be affected by any actions of the players. Second, there are **games of strategy**, such as chess, checkers, bridge, and poker, which require skill on the part of the players; the outcomes and winnings are determined by the skills of the players. By **game**, we mean a game of strategy. In addition to such parlor games there are many games in economic competition, warfare, geological exploration, farming, administration of justice, and so on, in which each player (competitor) can choose one of a set of possible moves and the outcomes depend on the player's (competitor's) skill. The theory of games attempts to determine the best course of action for each player. The subject is still being developed, and many new theoretical and applied results are needed for handling the complex games that occur in everyday situations.

We shall limit our treatment to games played by two players, usually denoted by R and C. We shall also assume that R has m possible moves (or courses of action) and that C has n moves. We now form an $m \times n$ matrix by labeling its rows, from top to bottom, with the moves of R, and labeling its columns, from left to right, with the moves of C. Entry a_{ij}, in row i and column j, indicates the amount (money or some other valuable item) received by R if R makes his ith move and C makes his jth move. The entry a_{ij} is called a **payoff** and the matrix $A = \left[a_{ij}\right]$ is called the **payoff matrix** (for R). Such games are called **two-person games**. We also refer to them as **matrix games**.

We could also, if we wished, construct a second matrix that represents payoffs for C. However, in a class of games called **constant-sum games**, the sum of the payoff to R and the payoff to C is constant for all mn pairs of RC moves. The more R gains, the less C gains (or the more C loses), and vice versa. A special kind of constant-sum game is the **zero-sum game**, in which the amount won by one player is exactly the amount lost by the other player. Because of this strict interrelation between the payoff matrix for C and that for R, in the case of a constant-sum game it is sufficient to study only the payoff matrix for R. In the following discussion we study only constant-sum games, and the matrix involved is the payoff matrix for R.

In the study of matrix games, it is always assumed that both players are equally capable, that each is playing as well as he can possibly play, and that each player makes his move without knowing what his opponent's move will be.

EXAMPLE 1 Consider the game of matching pennies, consisting of two players, R and C, each of whom has a penny in his hand. Each player shows one side of the coin without knowing his opponent's choice. If both players are showing the same side of the coin, then R wins \$1 from C; otherwise, C wins \$1 from R.

†Oskar Morgenstern (1902–1977) was born in Germany and came to the United States in 1938. He was a professor of economics at Princeton University until 1970 and at New York University until his death. In 1944 he collaborated with von Neumann on the influential book *Theory of Games and Economic Behavior.*

In this two-person, zero-sum game each player has two possible moves: he can show a tail or he can show a head. The payoff matrix is thus

$$R \quad \begin{array}{c} \\ H \\ T \end{array} \begin{array}{c} C \\ \begin{array}{cc} H & T \end{array} \\ \begin{bmatrix} 1 & -1 \\ -1 & 1 \end{bmatrix} \end{array}.$$

■

EXAMPLE 2

There are two suppliers, firms R and C, of a new specialized type of tire that has 100,000 customers. Each company can advertise its product on TV or in the newspapers. A marketing firm determines that if both firms advertise on TV, then firm R gets 40,000 customers (and firm C gets 60,000 customers). If they both use newspapers, then each gets 50,000 customers. If R uses newspapers and C uses TV, then R gets 60,000 customers (and C gets 40,000 customers). If R uses TV and C uses newspapers, they each get 50,000 customers.

We can consider this situation as a game between firms R and C, with the payoff matrix as shown in Figure 11.14. *The entries in the matrix indicate the number of customers secured by firm R.* This is a constant-sum game, for the sum of the R customers and the C customers is always the total population of 100,000 tire customers. ■

Figure 11.14 ►

		Firm C	
		TV	Newspapers
Firm R	TV	40,000	50,000
	Newspapers	60,000	50,000

Consider now a two-person, constant-sum game with the $m \times n$ payoff matrix $A = [a_{ij}]$, so that player R has m moves and player C has n moves. If player R plays his ith move, he is assured of winning at least the smallest entry in the ith row of A, no matter what C does. Thus R's best course of action is to choose that move which will maximize his assured winnings in spite of C's best countermove. Player R will get his largest payoff by maximizing his smallest gain. Player C's goals are in direct conflict with those of player R: he is trying to keep R's winnings to a minimum. If C plays his jth move, he is assured of losing no more than the largest entry in the jth column of A, no matter what R does. Thus C's best course of action is to choose that move which will minimize his assured losses in spite of R's best countermove. Player C will do his best by minimizing his largest loss.

DEFINITION

If the payoff matrix of a matrix game contains an entry a_{rs}, which is at the same time the minimum of row r and the maximum of column s, then a_{rs} is called a **saddle point**. Also, a_{rs} is called the **value** of the game. If the value of a zero-sum game is zero, the game is said to be **fair**.

DEFINITION

A matrix game is said to be **strictly determined** if its payoff matrix has a saddle point.

If a_{rs} is a saddle point for a matrix game, then player R will be assured of winning at least a_{rs} by playing his rth move and player C will be guaranteed that he will lose no more than a_{rs} by playing his sth move. This is the best that each player can do.

EXAMPLE 3 Consider a game with payoff matrix

$$R \overset{\displaystyle C}{\begin{bmatrix} 0 & -3 & -1 & 3 \\ 3 & 2 & 2 & 4 \\ 1 & 4 & 0 & 6 \end{bmatrix}}.$$

To determine whether this game has a saddle point, we write the minimum of each row to the right of the row and the maximum of each column at the bottom of each column. Thus we have

		C				Row minima
R	0	−3	−1	3		−3
	3	2	2	4		2
	1	4	0	6		0
Column maxima	3	4	2	6		

Entry $a_{23} = 2$ is both the least entry in the second row and the largest entry in the third column. Hence it is a saddle point for the game, which is then a strictly determined game. The value of the game is 2 and player R has an advantage. The best course of action for R is to play his second move; he will win at least 2 units from C, no matter what C does. The best course of action for C is to play his third move; he will limit his loss to not more than 2 units, no matter what R does. ■

EXAMPLE 4 Consider the advertising game of Example 2. The payoff matrix is shown in Figure 11.15. Thus entry $a_{22} = 50{,}000$ is a saddle point. The best course of action for both firms is to advertise in newspapers. The game is strictly determined with value 50,000. ■

Figure 11.15 ▶

		Firm C		
		TV	Newspapers	**Row minima**
Firm R	TV	40,000	50,000	40,000
	Newspapers	60,000	50,000	50,000
	Column maxima	60,000	50,000	

There are many games that are not strictly determined.

EXAMPLE 5 Consider the game with payoff matrix

				Row minima
	1	6	−1	−1
	3	−2	4	−2
	4	5	−3	−3
Column maxima	4	6	4	

It is clear that there is no saddle point. ■

On the other hand, a game may have more than one saddle point. However, it can be proved that all saddle points must have the same value.

EXAMPLE 6 Consider the game with payoff matrix

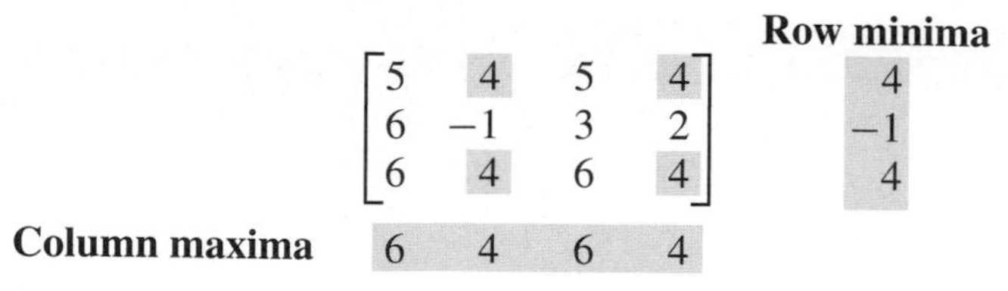

Entries a_{12}, a_{14}, a_{32}, and a_{34} are all saddle points and have the same value, 4. They appear shaded in the payoff matrix. The value of the game is also 4. ■

Consider now the penny-matching game of Example 1 with payoff matrix

$$\begin{array}{cccc} & & \multicolumn{2}{c}{C} & \\ & & H & T & \textbf{Row minima} \\ R & H & 1 & -1 & -1 \\ & T & -1 & 1 & -1 \\ \textbf{Column maxima} & & 1 & 1 & \end{array}$$

It is clear that this game is not strictly determined; that is, it has no saddle point.

To analyze this type of situation, we assume that a game is played repeatedly and that each player is trying to determine his best course of action. Thus player R tries to maximize his winnings while player C tries to minimize his losses. A **strategy** for a player is a decision for choosing his moves.

Consider now the above penny-matching game. Suppose that in the repeated play of the game, player R always chooses the first row (he chooses to show heads), in the hope that player C will always choose the first column (play heads), thereby ensuring a win of \$1 for himself. However, as player C begins to notice that R always chooses his first row, then player C will choose his second column, resulting in a loss of \$1 for R. Similarly, if R always chooses the second row, then C will choose the first column, resulting in a loss of \$1 for R. We can thus conclude that each player must somehow keep the other player from anticipating his choice of moves. This situation is in marked contrast with the case in strictly determined games. In a strictly determined game each player will make the same move whether or not he has advanced knowledge of his opponent's move. Thus, in a nonstrictly determined game, each player will make each move with a certain relative frequency.

DEFINITION Suppose that we have a matrix game with an $m \times n$ payoff matrix A. Let p_i, $1 \leq i \leq m$, be the probability that R chooses the ith row of A (that is, chooses his ith move). Let q_j, $1 \leq j \leq n$, be the probability that C chooses the jth column of A. The vector $\mathbf{p} = \begin{bmatrix} p_1 & p_2 & \cdots & p_m \end{bmatrix}$ is called a **strategy** for player R; the vector

$$\mathbf{q} = \begin{bmatrix} q_1 \\ q_2 \\ \vdots \\ q_n \end{bmatrix}$$

is called a **strategy** for player C.

Of course, the probabilities p_i and q_j in the definition satisfy

$$p_1 + p_2 + \cdots + p_m = 1$$
$$q_1 + q_2 + \cdots + q_n = 1.$$

If a matrix game is strictly determined, then optimal strategies for R and C are strategies having 1 as one component and zero for all other components. Such strategies are called **pure strategies**. A strategy that is not pure is called a **mixed strategy**. Thus, in Example 3, the pure strategy for R is

$$\mathbf{p} = \begin{bmatrix} 0 & 1 & 0 \end{bmatrix},$$

and the pure strategy for C is

$$\mathbf{q} = \begin{bmatrix} 0 \\ 0 \\ 1 \\ 0 \end{bmatrix}.$$

Consider now a matrix game with payoff matrix

$$A = \begin{bmatrix} a_{11} & a_{12} \\ a_{21} & a_{22} \end{bmatrix}. \tag{1}$$

Suppose that

$$\mathbf{p} = \begin{bmatrix} p_1 & p_2 \end{bmatrix} \quad \text{and} \quad \mathbf{q} = \begin{bmatrix} q_1 \\ q_2 \end{bmatrix}$$

are strategies for R and C, respectively. Then if R plays his first row with probability p_1 and if C plays his first column with probability q_1, then R's expected payoff is $p_1 q_1 a_{11}$. Similarly, we can examine the remaining three possibilities, obtaining Table 11.4. The expected payoff $E(\mathbf{p}, \mathbf{q})$ of the game to R is then the sum of the four quantities in the rightmost column. We obtain

$$E(\mathbf{p}, \mathbf{q}) = p_1 q_1 a_{11} + p_1 q_2 a_{12} + p_2 q_1 a_{21} + p_2 q_2 a_{22},$$

which can be written in matrix form (verify) as

$$E(\mathbf{p}, \mathbf{q}) = \mathbf{p}A\mathbf{q}. \tag{2}$$

Table 11.4

Moves				
Player R	Player C	Probability	Payoff to Player R	Expected Payoff to Player R
Row 1	Column 1	$p_1 q_1$	a_{11}	$p_1 q_1 a_{11}$
Row 1	Column 2	$p_1 q_2$	a_{12}	$p_1 q_2 a_{12}$
Row 2	Column 1	$p_2 q_1$	a_{21}	$p_2 q_1 a_{21}$
Row 2	Column 2	$p_2 q_2$	a_{22}	$p_2 q_2 a_{22}$

The same analysis applies to a matrix game with an $m \times n$ payoff matrix A. Thus if

$$\mathbf{p} = \begin{bmatrix} p_1 & p_2 & \cdots & p_m \end{bmatrix} \quad \text{and} \quad \mathbf{q} = \begin{bmatrix} q_1 \\ q_2 \\ \vdots \\ q_n \end{bmatrix}$$

are strategies for R and C, respectively, then the payoff to player R is given by (2).

EXAMPLE 7 Consider a matrix game with payoff matrix

$$A = \begin{bmatrix} 2 & -2 & 3 \\ 4 & 0 & -3 \end{bmatrix}.$$

If

$$\mathbf{p} = \begin{bmatrix} \frac{1}{4} & \frac{3}{4} \end{bmatrix} \quad \text{and} \quad \mathbf{q} = \begin{bmatrix} \frac{1}{3} \\ \frac{1}{3} \\ \frac{1}{3} \end{bmatrix}$$

are strategies for R and C, respectively, then the expected payoff to R is

$$E(\mathbf{p}, \mathbf{q}) = \mathbf{p}A\mathbf{q} = \begin{bmatrix} \frac{1}{4} & \frac{3}{4} \end{bmatrix} \begin{bmatrix} 2 & -2 & 3 \\ 4 & 0 & -3 \end{bmatrix} \begin{bmatrix} \frac{1}{3} \\ \frac{1}{3} \\ \frac{1}{3} \end{bmatrix} = \frac{1}{2}.$$

If

$$\mathbf{p} = \begin{bmatrix} \frac{3}{4} & \frac{1}{4} \end{bmatrix} \quad \text{and} \quad \mathbf{q} = \begin{bmatrix} \frac{1}{3} \\ \frac{2}{3} \\ 0 \end{bmatrix}$$

are strategies for R and C, respectively, then the expected payoff to R is $-\frac{1}{6}$. Thus, in the first case, R gains $\frac{1}{2}$ from C, whereas in the second case R loses $\frac{1}{6}$ to C. ■

A strategy for player R is said to be **optimal** if it guarantees R the largest possible payoff no matter what his opponent may do. Similarly, a strategy for player C is said to be **optimal** if it guarantees the smallest possible payoff to R no matter what R may do.

If $\mathbf{p}$ and $\mathbf{q}$ are optimal strategies for R and C, respectively, then the expected payoff to R, $v = E(\mathbf{p}, \mathbf{q})$, is called the **value** of the game. Although $E(\mathbf{p}, \mathbf{q})$ is a 1×1 matrix, we think of it merely as a number v. If the value of a zero-sum game is zero, the game is said to be **fair**. The principal task of the theory of games is the determination of optimal strategies for each player.

Consider again a matrix game with the 2×2 payoff matrix (1) and suppose that the game is not strictly determined. It can then be shown that

$$a_{11} + a_{22} - a_{12} - a_{21} \neq 0.$$

To determine an optimal strategy for R, we proceed as follows. Suppose that R's strategy is $\begin{bmatrix} p_1 & p_2 \end{bmatrix}$. Then if C plays his first column, the expected payoff to R is

$$a_{11}p_1 + a_{21}p_2. \tag{3}$$

If C plays his second column, the expected payoff to R is

$$a_{12}p_1 + a_{22}p_2. \tag{4}$$

If v is the minimum of the expected payoffs (3) and (4), then R expects to gain at least v units from C no matter what C does. Thus we have

$$a_{11}p_1 + a_{21}p_2 \geq v \tag{5}$$

$$a_{12}p_1 + a_{22}p_2 \geq v. \tag{6}$$

Moreover, player R seeks to make v as large as possible. Thus player R seeks to find p_1, p_2, and v such that

$$v \text{ is a maximum}$$

and

$$\begin{aligned} a_{11}p_1 + a_{21}p_2 - v &\geq 0 \\ a_{12}p_1 + a_{22}p_2 - v &\geq 0 \\ p_1 + p_2 &= 1 \\ p_1 \geq 0, \quad p_2 \geq 0, \quad v &\geq 0. \end{aligned} \tag{7}$$

We shall see later (in a more general situation) that problem (7) is a linear programming problem. It can be shown that a solution to (7), giving an optimal strategy for R, is

$$p_1 = \frac{a_{22} - a_{21}}{a_{11} + a_{22} - a_{12} - a_{21}}, \qquad p_2 = \frac{a_{11} - a_{12}}{a_{11} + a_{22} - a_{12} - a_{21}} \tag{8}$$

and

$$v = \frac{a_{11}a_{22} - a_{12}a_{21}}{a_{11} + a_{22} - a_{12} - a_{21}}. \tag{9}$$

We now find an optimal strategy for C. Suppose that C's strategy is

$$\begin{bmatrix} q_1 \\ q_2 \end{bmatrix}.$$

If R plays the first row, then the expected payoff to R is

$$a_{11}q_1 + a_{12}q_2, \tag{10}$$

while if R plays the second row, the expected payoff to R is

$$a_{21}q_1 + a_{22}q_2. \tag{11}$$

If v' is the maximum of the expected payoffs (10) and (11), then

$$a_{11}q_1 + a_{12}q_2 \leq v'$$

$$a_{21}q_1 + a_{22}q_2 \leq v'.$$

Since player C wishes to lose as little as possible, he seeks to make v' as small as possible. Thus C wants to find q_1, q_2, and v such that

$$v' \text{ is a minimum}$$

and

$$\begin{aligned} a_{11}q_1 + a_{12}q_2 - v' &\leq 0 \\ a_{21}q_1 + a_{22}q_2 - v' &\leq 0 \\ q_1 + q_2 &= 1 \\ q_1 \geq 0, \quad q_2 \geq 0, \quad v' &\geq 0. \end{aligned} \tag{12}$$

Problem (12) is also a linear programming problem. It can be shown that a solution to (12), giving an optimal strategy for C, is

$$q_1 = \frac{a_{22} - a_{12}}{a_{11} + a_{22} - a_{12} - a_{21}}, \qquad q_2 = \frac{a_{11} - a_{21}}{a_{11} + a_{22} - a_{12} - a_{21}} \tag{13}$$

and

$$v' = \frac{a_{11}a_{22} - a_{12}a_{21}}{a_{11} + a_{22} - a_{12} - a_{21}}. \tag{14}$$

Thus $v = v'$ when both players use their optimal strategies.

EXAMPLE 8

For the zero-sum penny-matching game of Example 1, we have upon substituting in (8), (9), and (13),

$$p_1 = p_2 = \tfrac{1}{2} \quad \text{and} \quad q_1 = q_2 = \tfrac{1}{2}, \quad v = 0,$$

so that optimal strategies for R and C are

$$\begin{bmatrix} \frac{1}{2} & \frac{1}{2} \end{bmatrix} \quad \text{and} \quad \begin{bmatrix} \frac{1}{2} \\ \frac{1}{2} \end{bmatrix},$$

respectively. This means that half the time R should show heads and half the time he should show tails; likewise for player C. The value of the game is zero, so the game is fair. ■

EXAMPLE 9

Consider a matrix game with payoff matrix

$$\begin{bmatrix} 2 & -5 \\ 1 & 3 \end{bmatrix}.$$

Again substituting in (8), (9), and (13), we obtain

$$p_1 = \frac{3-1}{2+3-1+5} = \frac{2}{9}, \qquad p_2 = \frac{2+5}{2+3-1+5} = \frac{7}{9},$$

$$q_1 = \frac{3+5}{2+3-1+5} = \frac{8}{9}, \qquad q_2 = \frac{2-1}{2+3-1+5} = \frac{1}{9},$$

$$v = \frac{6+5}{2+3-1+5} = \frac{11}{9}.$$

Thus optimal strategies for R and C are

$$\begin{bmatrix} \frac{2}{9} & \frac{7}{9} \end{bmatrix} \quad \text{and} \quad \begin{bmatrix} \frac{8}{9} \\ \frac{1}{9} \end{bmatrix},$$

respectively; when both players use their optimal strategies, the value of the game (the expected payoff to R) is $\frac{11}{9}$. If this matrix represents a zero-sum game, the game is not fair and in the long run favors player R. ■

We can now generalize our discussion to a game with an $m \times n$ payoff matrix $A = \left[a_{ij}\right]$. First, let us observe that if we add a constant r to every entry of A, then the optimal strategies for R and C do not change, and the value of the new game is r plus the value of the old game (Exercise T.2). Thus we can assume that, by adding a suitable constant to every entry of the payoff matrix, every entry of A is positive.

Player R seeks to find $p_1, p_2, \dots, p_m$, and v such that

$$v \text{ is a maximum}$$

subject to

$$\begin{aligned}
a_{11}p_1 + a_{21}p_2 + \cdots + a_{m1}p_m - v &\geq 0 \\
a_{12}p_1 + a_{22}p_2 + \cdots + a_{m2}p_m - v &\geq 0 \\
\vdots \qquad \vdots \qquad\quad \vdots \quad\; &\vdots \;\; \vdots \\
a_{1n}p_1 + a_{2n}p_2 + \cdots + a_{mn}p_m - v &\geq 0 \\
p_1 + p_2 + \cdots + p_m &= 1 \\
p_1 \geq 0,\ p_2 \geq 0, \dots,\ p_m \geq 0, \quad v &\geq 0.
\end{aligned} \tag{15}$$

Since every entry of A is positive, we may assume that $v > 0$. Now divide each of the constraints in (15) by v, and let

$$y_i = \frac{p_i}{v}.$$

Observe that

$$y_1 + y_2 + \cdots + y_m = \frac{p_1}{v} + \frac{p_2}{v} + \cdots + \frac{p_m}{v} = \frac{1}{v}(p_1 + p_2 + \cdots + p_m) = \frac{1}{v}.$$

Thus v is a maximum if and only if $y_1 + y_2 + \cdots + y_m$ is a minimum. We can now restate problem (15), R's problem, as follows:

$$\text{Minimize} \quad y_1 + y_2 + \cdots + y_m$$

subject to

$$\begin{aligned}
a_{11}y_1 + a_{21}y_2 + \cdots + a_{m1}y_m &\geq 1 \\
a_{12}y_1 + a_{22}y_2 + \cdots + a_{m2}y_m &\geq 1 \\
\vdots \qquad \vdots \qquad\quad \vdots \quad\; &\;\; \vdots \\
a_{1n}y_1 + a_{2n}y_2 + \cdots + a_{mn}y_m &\geq 1 \\
y_1 \geq 0,\ y_2 \geq 0, \dots,\ y_m &\geq 0.
\end{aligned} \tag{16}$$

Observe that (16) is a linear programming problem and that it has one fewer constraint and one fewer variable than (15).

Turning next to C's problem, we note that he seeks to find $q_1, q_2, \ldots, q_n$ and v' such that

$$v' \text{ is a minimum}$$

subject to

$$\begin{aligned}
a_{11}q_1 + a_{12}q_2 + \cdots + a_{1n}q_n - v' &\le 0 \\
a_{21}q_1 + a_{22}q_2 + \cdots + a_{2n}q_n - v' &\le 0 \\
\vdots \qquad \vdots \qquad\qquad \vdots \qquad \vdots \quad &\vdots \\
a_{m1}q_1 + a_{m2}q_2 + \cdots + a_{mn}q_n - v' &\le 0 \\
q_1 + q_2 + \cdots + q_n &= 1 \\
q_1 \ge 0, q_2 \ge 0, \ldots, q_n \ge 0, \quad v' &\ge 0.
\end{aligned} \tag{17}$$

The fundamental theorem of matrix games, which we now state, says that every matrix game has a solution.

THEOREM 11.5 **(Fundamental Theorem of Matrix Games)** *Every matrix game has a solution. That is, there are optimal strategies for R and C. Moreover, $v = v'$.* ■

Since $v = v'$, we can divide each of the constraints in (17) by $v = v'$ and let

$$x_i = \frac{q_i}{v}.$$

Now,

$$x_1 + x_2 + \cdots + x_n = \frac{1}{v},$$

so v is a minimum if and only if $x_1 + x_2 + \cdots + x_n$ is a maximum. We can now restate problem (17), C's problem, as follows:

$$\text{Maximize} \quad x_1 + x_2 + \cdots + x_n$$

subject to

$$\begin{aligned}
a_{11}x_1 + a_{12}x_2 + \cdots + a_{1n}x_n &\le 1 \\
a_{21}x_1 + a_{22}x_2 + \cdots + a_{2n}x_n &\le 1 \\
\vdots \qquad \vdots \qquad\qquad \vdots \quad &\vdots \\
a_{m1}x_1 + a_{m2}x_2 + \cdots + a_{mn}x_n &\le 1 \\
x_1 \ge 0, x_2 \ge 0, \ldots, x_n &\ge 0.
\end{aligned} \tag{18}$$

Observe that (18) is a linear programming problem in standard form, which is the dual of (16). From the results of Section 11.3, it follows that when (18) is solved by the simplex method, the final tableau will contain the optimal strategies for R in the objective row under the columns of the slack variables. That is, y_1 is found in the objective row under the first slack variable, y_2 is found in the objective row under the second slack variable, and so on.

EXAMPLE 10 Consider a game with payoff matrix

$$\begin{bmatrix} 2 & -3 & 0 \\ 3 & 1 & -2 \end{bmatrix}.$$

Adding 4 to each element of the matrix, we obtain a matrix A with positive entries.

$$A = \begin{bmatrix} 6 & 1 & 4 \\ 7 & 5 & 2 \end{bmatrix}$$

We now find optimal strategies for the game with payoff matrix A. Problem (18), C's problem, becomes

$$\text{Maximize} \quad x_1 + x_2 + x_3$$

subject to

$$\begin{aligned} 6x_1 + x_2 + 4x_3 &\le 1 \\ 7x_1 + 5x_2 + 2x_3 &\le 1 \\ x_1 \ge 0, \quad x_2 \ge 0, \quad x_3 &\ge 0. \end{aligned}$$

If we introduce the slack variables x_4 and x_5, our problem becomes

$$\text{Maximize} \quad x_1 + x_2 + x_3$$

subject to

$$\begin{aligned} 6x_1 + x_2 + 4x_3 + x_4 &= 1 \\ 7x_1 + 5x_2 + 2x_3 + x_5 &= 1 \\ x_1 \ge 0,\ x_2 \ge 0,\ x_3 \ge 0,\ x_4 \ge 0,\ x_5 &\ge 0. \end{aligned}$$

Using the simplex method, we obtain the following:

	x_1	x_2	x_3 ↓	x_4	x_5	z	
← x_4	6	1	(4)	1	0	0	1
x_5	7	5	2	0	1	0	1
	-1	-1	-1	0	0	1	0

	x_1	x_2 ↓	x_3	x_4	x_5	z	
x_3	$\frac{3}{2}$	$\frac{1}{4}$	1	$\frac{1}{4}$	0	0	$\frac{1}{4}$
← x_5	4	$(\frac{9}{2})$	0	$-\frac{1}{2}$	1	0	$\frac{1}{2}$
	$\frac{1}{2}$	$-\frac{3}{4}$	0	$\frac{1}{4}$	0	1	$\frac{1}{4}$

	x_1	x_2	x_3	x_4	x_5	z	
x_3	$\frac{23}{18}$	0	1	$\frac{5}{18}$	$-\frac{1}{18}$	0	$\frac{2}{9}$
x_2	$\frac{8}{9}$	1	0	$-\frac{1}{9}$	$\frac{2}{9}$	0	$\frac{1}{9}$
	$\frac{7}{6}$	0	0	$\frac{1}{6}$	$\frac{1}{6}$	1	$\frac{1}{3}$

Thus we have

$$x_1 = 0, \quad x_2 = \tfrac{1}{9}, \quad \text{and} \quad x_3 = \tfrac{2}{9}.$$

The maximum value of $x_1 + x_2 + x_3$ is $\frac{1}{3}$, so the minimum value of v is 3. Hence

$$q_1 = x_1 v = 0, \qquad q_2 = x_2 v = \left(\tfrac{1}{9}\right)(3) = \tfrac{1}{3},$$

and

$$q_3 = x_3 v = \left(\tfrac{2}{9}\right)(3) = \tfrac{2}{3}.$$

Thus an optimal strategy for C is

$$\mathbf{q} = \begin{bmatrix} 0 \\ \frac{1}{3} \\ \frac{2}{3} \end{bmatrix}.$$

An optimal solution to (16), R's problem, is found in the objective row under the columns of slack variables. Thus, under the slack variables x_4 and x_5, we find

$$y_1 = \tfrac{1}{6} \quad \text{and} \quad y_2 = \tfrac{1}{6},$$

respectively. Since $v = 3$,

$$p_1 = y_1 v = \left(\tfrac{1}{6}\right)(3) = \tfrac{1}{2} \quad \text{and} \quad p_2 = y_2 v = \left(\tfrac{1}{6}\right)(3) = \tfrac{1}{2}.$$

Thus an optimal strategy for R is

$$\mathbf{p} = \begin{bmatrix} \frac{1}{2} & \frac{1}{2} \end{bmatrix}.$$

The zero-sum game with payoff matrix A is not fair, since the value is 3, with player R having the advantage. Since matrix A was obtained from the game as initially proposed by adding $r = 4$ to all matrix entries, the value of the *initial* game is $3 - 4 = -1$. That initial game is to C's advantage. ■

Sometimes it is possible to solve a matrix game by reducing the size of the payoff matrix A. If each element of the rth *row* of A is *less than or equal* to the corresponding element of the sth row of A, then the rth row is called **recessive** and the sth row is said to **dominate** the rth row. If each element of the rth *column* of A is *greater than or equal* to the corresponding element in the sth column of A, then the rth column is called **recessive** and the sth column is said to **dominate** the rth column.

EXAMPLE 11 In the payoff matrix

$$\begin{bmatrix} 2 & -1 & 3 \\ 0 & 3 & 4 \\ 3 & 2 & 4 \end{bmatrix},$$

the first row is recessive; the third row dominates the first row. In the payoff matrix

$$\begin{bmatrix} -2 & 4 & 3 \\ 3 & -3 & -3 \\ 5 & 2 & 1 \end{bmatrix},$$

the second column is recessive; the third column dominates the second column. ■

Consider a matrix game in which the rth row is recessive and the sth row dominates the rth row. It is then obvious that player R will always tend to choose the sth row rather than the rth row, since he will be guaranteed a gain equal to or greater than the gain realized by choosing the rth row. Thus, since the rth row will never be chosen, it can be dropped from further consideration. Suppose now that the rth column is recessive and that the sth column dominates the rth column. Since player C wishes to keep his losses to a minimum, by choosing the sth column he will be guaranteed a loss equal to or smaller than the loss incurred by choosing the rth column. Since the rth column will never be chosen, it can be dropped from further consideration. These techniques, when applicable, result in a smaller payoff matrix.

EXAMPLE 12 Consider the matrix game with payoff matrix

$$A = \begin{bmatrix} 2 & -1 & 3 \\ -2 & 2 & 4 \\ 3 & 0 & 4 \end{bmatrix}.$$

Since the third row of A dominates its first row, the latter can be dropped, obtaining

$$A_1 = \begin{bmatrix} -2 & 2 & 4 \\ 3 & 0 & 4 \end{bmatrix}.$$

Since the second column of A_1 dominates its third column, the latter can be dropped, obtaining

$$A_2 = \begin{bmatrix} -2 & 2 \\ 3 & 0 \end{bmatrix},$$

which has no saddle point. The solution to the matrix game with payoff matrix A_2 can be obtained from Equations (8), (9), and (13). We have

$$p_1 = \frac{0-3}{-2+0-2-3} = \frac{-3}{-7} = \frac{3}{7},$$

$$p_2 = \frac{-2-2}{-2+0-2-3} = \frac{-4}{-7} = \frac{4}{7},$$

$$q_1 = \frac{0-2}{-2+0-2-3} = \frac{-2}{-7} = \frac{2}{7},$$

$$q_2 = \frac{-2-3}{-2+0-2-3} = \frac{-5}{-7} = \frac{5}{7},$$

and

$$v = \frac{0-6}{-2+0-2-3} = \frac{-6}{-7} = \frac{6}{7}.$$

Since in A, the original payoff matrix, the first row and third column were dropped, we obtain

$$\mathbf{p} = \begin{bmatrix} 0 & \frac{3}{7} & \frac{4}{7} \end{bmatrix}$$

as an optimal strategy for player R. Similarly,

$$\mathbf{q} = \begin{bmatrix} \frac{2}{7} \\ \frac{5}{7} \\ 0 \end{bmatrix}$$

is an optimal strategy for player C. ■

Further Readings

OWEN, G. *Game Theory*, 3rd ed. Orlando, Fla.: Academic Press, 1995.

STRAFFIN, PHILIP D. *Game Theory and Strategy*. Washington, D.C.: New Mathematical Library, No. 36, 1996.

THIE, PAUL R. *An Introduction to Linear Programming and Game Theory*, 2nd ed. New York: John Wiley & Sons, Inc., 1988.

Key Terms

Game
Games of chance
Games of strategy
Payoff
Payoff matrix
Two-person games
Matrix games
Constant-sum games
Zero-sum games
Saddle point
Value
Fair game
Strictly determined
Strategy
Pure strategy
Mixed strategy
Optimal strategy
Fundamental theorem of matrix games
Recessive row (column)
Dominate

11.4 Exercises

In Exercises 1 through 4, write the payoff matrix for the given game.

1. Each of two players shows two or three fingers. If the sum of the fingers shown is even, then R pays C an amount equal to the sum of the numbers shown; if the sum is odd, then C pays R an amount equal to the sum of the numbers shown.

2. (***Stone, Scissors, Paper***) Each of two players selects one of the words *stone*, *scissors*, *paper*. Stone beats scissors, scissors beats paper, and paper beats stone. In case of a tie, there is no payoff. In case of a win, the winner collects \$1.

3. Firms A and B, both handling specialized sporting equipment, are planning to locate in either Abington or Wyncote. If they both locate in the same town, each will capture 50% of the trade. If A locates in Abington and B locates in Wyncote, then A will capture 60% of the business (and B will keep 40%); if A locates in Wyncote and B locates in Abington, then A will hold on to 25% of the business (and B to 75%).

4. Player R has a nickel and a dime with him. He chooses one of the coins and player C must guess R's choice. If C guesses correctly, he keeps the coin; if he guesses incorrectly, he must give R an amount equal to the coin shown.

5. Find all saddle points for the following matrix games.

(a) $\begin{bmatrix} 5 & 4 \\ 3 & -2 \end{bmatrix}$ (b) $\begin{bmatrix} 2 & 1 & 0 \\ 3 & 1 & -2 \\ 4 & 2 & -4 \end{bmatrix}$

(c) $\begin{bmatrix} 3 & 4 & 5 \\ -2 & 5 & 1 \\ -1 & 0 & 1 \end{bmatrix}$

(d) $\begin{bmatrix} 5 & 2 & 4 & 2 \\ 0 & -1 & 2 & 0 \\ 3 & 2 & 3 & 2 \\ 1 & 0 & -1 & -1 \end{bmatrix}$

6. Find optimal strategies for the following strictly determined games. Give the payoff for R.

(a) $\begin{bmatrix} -3 & 4 \\ 3 & 5 \end{bmatrix}$ (b) $\begin{bmatrix} -1 & -3 & -2 \\ 3 & -1 & 4 \\ -1 & -2 & 5 \end{bmatrix}$

(c) $\begin{bmatrix} -2 & 3 & -2 & 4 \\ -1 & 2 & -2 & 4 \\ -2 & 3 & -3 & 5 \\ -1 & 2 & -3 & 1 \end{bmatrix}$

7. Find optimal strategies for the following strictly determined games. Give the payoff for R.

(a) $\begin{bmatrix} 2 & 1 & 3 \\ -2 & 0 & 2 \end{bmatrix}$

(b) $\begin{bmatrix} -2 & -2 & 4 & 5 \\ -2 & -2 & 1 & 0 \\ 0 & 1 & 1 & 2 \end{bmatrix}$

(c) $\begin{bmatrix} 6 & 4 \\ 7 & 4 \end{bmatrix}$

8. Consider a matrix game with payoff matrix

$$\begin{bmatrix} 2 & -3 & -2 \\ -4 & 5 & 6 \end{bmatrix}.$$

Find $E(\mathbf{p}, \mathbf{q})$, the expected payoff to R, if

(a) $\mathbf{p} = \begin{bmatrix} \frac{1}{4} & \frac{3}{4} \end{bmatrix}$ and $\mathbf{q} = \begin{bmatrix} \frac{1}{3} \\ \frac{1}{6} \\ \frac{1}{2} \end{bmatrix}$

(b) $p_1 = \frac{2}{3}$, $p_2 = \frac{1}{3}$; $q_1 = \frac{1}{2}$, $q_2 = \frac{1}{4}$, and $q_3 = \frac{1}{4}$

9. Consider a matrix game with payoff matrix

$$\begin{bmatrix} 3 & -3 \\ 2 & 5 \\ 1 & 0 \end{bmatrix}.$$

Find $E(\mathbf{p}, \mathbf{q})$, the expected payoff to R, if

(a) $\mathbf{p} = \begin{bmatrix} \frac{1}{2} & \frac{1}{3} & \frac{1}{6} \end{bmatrix}$ and $\mathbf{q} = \begin{bmatrix} \frac{1}{6} \\ \frac{5}{6} \end{bmatrix}$

(b) $p_1 = 0$, $p_2 = 0$, $p_3 = 1$; $q_1 = \frac{1}{7}$, $q_2 = \frac{6}{7}$

In Exercises 10 and 11, solve the given matrix game using (8) and (13). Find the value of the game using (9).

10. $\begin{bmatrix} 4 & 8 \\ 6 & -2 \end{bmatrix}$ **11.** $\begin{bmatrix} -3 & 2 \\ 4 & -5 \end{bmatrix}$

In Exercises 12 and 13, solve the given matrix game by linear programming.

12. $\begin{bmatrix} -2 & 3 \\ 4 & 5 \\ 5 & 2 \end{bmatrix}$ **13.** $\begin{bmatrix} 2 & -3 & 4 \\ 4 & 0 & 1 \\ 3 & 2 & -2 \end{bmatrix}$

In Exercises 14 and 15, solve the given matrix game using the method of Example 11.

14. $\begin{bmatrix} -3 & 1 & 3 \\ 1 & -2 & 2 \\ 2 & -1 & 3 \end{bmatrix}$ **15.** $\begin{bmatrix} 0 & -4 & 3 & 0 \\ 2 & -3 & 4 & 1 \\ -1 & 2 & 2 & 2 \\ 1 & -4 & 3 & 0 \end{bmatrix}$

16. Solve Exercise 1.

17. Solve Exercise 2.

18. Solve Exercise 3.

19. Solve Exercise 4.

20. In a labor–management dispute, labor can make one of three different moves, L_1, L_2, and L_3, while management can make one of two moves, M_1 and M_2. Suppose that the following payoff matrix is obtained (the entries represent millions of dollars). Determine the best courses of action for both labor and management.

$$L \quad \begin{matrix} \\ L_1 \\ L_2 \\ L_3 \end{matrix} \overset{M}{\begin{matrix} M_1 & M_2 \\ \begin{bmatrix} 2 & 4 \\ 3 & 2 \\ 2 & 5 \end{bmatrix} \end{matrix}}$$

Theoretical Exercises

T.1. Consider a matrix game with $m \times n$ payoff matrix A. Verify that if player R uses strategy $\mathbf{p}$ and player C uses strategy $\mathbf{q}$, then the expected payoff to R is $\mathbf{p}A\mathbf{q}$.

T.2. Consider a matrix game with payoff matrix A. Show that if a constant r is added to each entry of A, then we get a new game whose optimal strategies are the same as those for the original game, and the value of the new game is r plus the value of the old game.

Key Ideas for Review

- **Linear programming problem.** See page 561.
- **Theorem 11.1.** Let S be the feasible region of a linear programming problem.
  (a) If S is bounded, then the objective function $z = ax + by$ assumes both a maximum and a minimum value on S; these values occur at extreme points of S.
  (b) If S is unbounded, then there may or may not be a maximum or minimum value on S. If a maximum or minimum value does exist on S, it occurs at an extreme point.
- **Theorem 11.2.** If a linear programming problem has an optimal solution, then it has a basic feasible solution that is optimal.
- **Simplex method.** See page 585.
- **Primal and dual problems.** See page 591.
- **Theorem 11.4 (duality theorem).** If either the primal problem or dual problem has an optimal solution with finite objective value, then the other problem also has an optimal solution. Moreover, the objective values of the two problems are equal.
- If a_{rs} is a saddle point for a matrix game, then the optimal strategy for player R is his rth move, the optimal strategy for player C is his sth move. The value of the game is a_{rs}.
- The optimal strategies

$$\mathbf{p} = \begin{bmatrix} p_1 & p_2 \end{bmatrix} \quad \text{and} \quad \mathbf{q} = \begin{bmatrix} q_1 \\ q_2 \end{bmatrix}$$

for players R and C, respectively, in a 2×2 matrix game are given by

$$p_1 = \frac{a_{22} - a_{21}}{a_{11} + a_{22} - a_{12} - a_{21}}$$

$$p_2 = \frac{a_{11} - a_{12}}{a_{11} + a_{22} - a_{12} - a_{21}}$$

$$q_1 = \frac{a_{22} - a_{12}}{a_{11} + a_{22} - a_{12} - a_{21}}$$

$$q_2 = \frac{a_{11} - a_{21}}{a_{11} + a_{22} - a_{12} - a_{21}}.$$

The value of the game is

$$v = \frac{a_{11}a_{22} - a_{12}a_{21}}{a_{11} + a_{22} - a_{12} - a_{21}} = v'.$$

■ **Theorem 11.5 (fundamental theorem of matrix games).** Every matrix game has a solution. That is, there are optimal strategies for R and C. Moreover, $v = v'$.

Supplementary Exercises

1. Solve the following linear programming problem geometrically.

$$\text{Maximize} \quad z = 2x + 3y$$

subject to

$$\begin{aligned} 3x + y &\leq 6 \\ x + y &\leq 4 \\ x + 2y &\leq 6 \\ x \geq 0, \quad y &\geq 0. \end{aligned}$$

2. Solve the following linear programming problem geometrically.

A microprocessor manufacturer makes two types of microprocessors, model A and model B. The size of the work force limits the total daily production to at most 600 microprocessors. On the other hand, the suppliers of components limit production to at most 400 model A units and 500 model B units. If the net profit on each model A unit is \$80 and it is \$100 on each model B unit, how many microprocessors of each type should the manufacturer produce daily to maximize the profit?

3. Solve the following linear programming problem by the simplex method.

$$\text{Maximize} \quad z = 50x + 100y$$

subject to

$$\begin{aligned} x + 2y &\leq 16 \\ 3x + 2y &\leq 24 \\ 2x + 2y &\leq 18 \\ x \geq 0, \quad y &\geq 0. \end{aligned}$$

4. Determine the dual of the following linear programming problem.

$$\text{Minimize} \quad z = 6x_1 + 5x_2$$

subject to

$$\begin{aligned} 2x_1 + 3x_2 &\geq 6 \\ 5x_1 + 2x_2 &\geq 10 \\ x_1 \geq 0, \quad x_2 &\geq 0. \end{aligned}$$

5. Solve the problem in Exercise 4 by solving its dual.

6. Solve the following matrix game:

$$\begin{bmatrix} 6 & 2 & 3 \\ 3 & 4 & 2 \\ 4 & 1 & 2 \end{bmatrix}.$$

Chapter Test

1. Solve the following linear programming problem geometrically.

A farmer who has a 120-acre farm plants corn and wheat. The expenses are \$12 for each acre of corn planted and \$24 for each acre of wheat planted. Each acre of corn requires 32 bushels of storage and yields a profit of \$40; each acre of wheat requires 8 bushels of storage and yields a profit of \$50. If the total amount of storage available is 160 bushels and the farmer has \$1200 of capital, how many acres of corn and how many acres of wheat should be planted to maximize profit?

2. Solve the following linear programming problem by the simplex method.

$$\text{Maximize} \quad z = 8x_1 + 9x_2 + 5x_3$$

subject to

$$\begin{aligned} x_1 + x_2 + 2x_3 &\leq 2 \\ 2x_1 + 3x_2 + 4x_3 &\leq 3 \\ 3x_1 + 3x_2 + x_3 &\leq 4 \\ x_1 \geq 0, \quad x_2 \geq 0, \quad x_3 &\geq 0. \end{aligned}$$

3. Determine the dual of the following linear programming problem.

$$\text{Minimize} \quad z = 3x_1 + 4x_2$$

subject to

$$\begin{aligned} x_1 + 4x_2 &\geq 8 \\ 2x_1 + 3x_2 &\geq 12 \\ 2x_1 + x_2 &\geq 6 \\ x_1 \geq 0, \quad x_2 &\geq 0. \end{aligned}$$

4. Solve the following matrix game by linear programming:

$$\begin{bmatrix} -3 & 2 & 4 \\ 4 & 1 & 5 \end{bmatrix}.$$

5. Show that the following matrix game is strictly determined regardless of the value of a:

$$\begin{bmatrix} 2 & 3 \\ 1 & a \end{bmatrix}.$$

CHAPTER 12

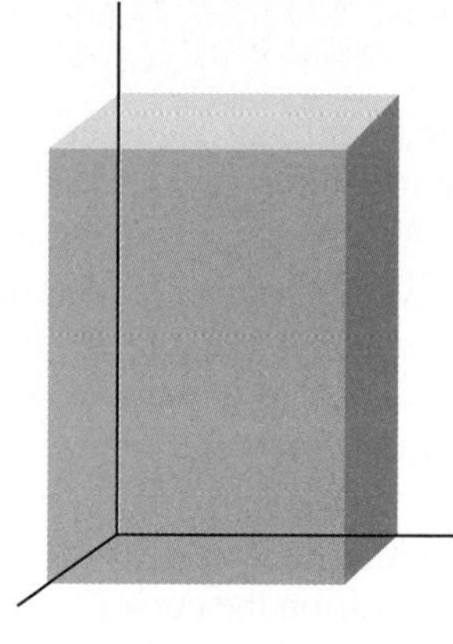

MATLAB FOR LINEAR ALGEBRA

INTRODUCTION*

MATLAB is a versatile piece of computer software with linear algebra capabilities as its core. MATLAB stands for MATrix LABoratory. It incorporates portions of professionally developed projects of quality computer routines for linear algebra computation. The code employed by MATLAB is written in the C language however. Many of the routines/functions are written in the MATLABlanguage, and are upgraded as new versions of MATLAB are released. MATLAB is available for Microsoft Windows, and for Unix and VMS workstations.

MATLAB has a wide range of capabilities. In this book we will use only a small portion of its features. We will find that MATLAB's command structure is very close to the way we write algebraic expressions and linear algebra operations. The names of many MATLAB commands closely parallel those of the operations and concepts of linear algebra. We give descriptions of commands and features of MATLAB that relate directly to this course. A more detailed discussion of MATLAB commands can be found in *The MATLAB User's Guide* that accompanies the software and in the books, *Experiments in Computational Matrix Algebra*, by David R. Hill (New York: Random House, 1988) and *Linear Algebra LABS with MATLAB*, 3rd ed., by David R. Hill and David E. Zitarelli (Upper Saddle River, N.J.: Prentice Hall, Inc., 2004). Alternatively, the MATLAB software provides immediate on-screen descriptions using the **help** command. Typing

help

displays a list of MATLAB subdirectories and alternative directories containing files corresponding to commands and data sets. Typing **help** *name*, where *name* is the name of a command, accesses information on the specific command named. In some cases the description displayed goes much further than we need for this course. Hence you may not fully understand all of the description displayed by **help**. We provide a list of the majority of MATLAB commands we use in this book in Section 12.9.

Once you initiate the MATLAB software, you will see the MATLAB logo appear and the MATLAB prompt ≫. The prompt ≫ indicates that MATLAB is awaiting a command. In Section 12.1 we describe how to enter matrices into MATLAB and give explanations of several commands. However, there

*This material on MATLAB refers to the Microsoft Windows version.

are certain MATLAB features you should be aware of before you begin the material in Section 12.1.

□ *Starting execution of a command.*
After you have typed a command name and any arguments or data required, you must press ENTER before it will begin to execute.

□ *The command stack.*
As you enter commands, MATLAB saves a number of the most recent commands in a stack. Previous commands saved on the stack can be recalled using the **up arrow** key. The number of commands saved on the stack varies depending on the length of the commands and other factors.

□ *Editing commands.*
If you make an error or mistype something in a command, you can use the **left arrow** and **right arrow** keys to position the cursor for corrections. The **home** key moves the cursor to the beginning of a command, and the **end** key moves the cursor to the end. The **backspace** and **delete** keys can be used to remove characters from a command line. The **insert** key is used to initiate the insertion of characters. Pressing the insert key a second time exits the insert mode. If MATLAB recognizes an error after you have pressed ENTER, then MATLAB responds with a beep and a message that helps define the error. You can recall the command line using the up arrow key in order to edit the line.

□ *Continuing commands.*
MATLAB commands that do not fit on a single line can be continued to the next line using an ellipsis, which is three consecutive periods, followed by ENTER.

□ *Stopping a command.*
To stop execution of a MATLAB command, press **Ctrl** and **C** simultaneously, then press ENTER. Sometimes this sequence must be repeated.

□ *Quitting.*
To quit MATLAB, type **exit** or **quit** followed by ENTER.

12.1 INPUT AND OUTPUT IN MATLAB

MATRIX INPUT

To enter a matrix into MATLAB just type the entries enclosed in square brackets [...], with entries separated by a space and rows terminated with a semicolon. Thus, matrix

$$\begin{bmatrix} 9 & -8 & 7 \\ -6 & 5 & -4 \\ 11 & -12 & 0 \end{bmatrix}$$

is entered by typing

[9 −8 7;−6 5 −4;11 −12 0]

and the accompanying display is

```
ans =

     9    -8     7
    -6     5    -4
    11   -12     0
```

Notice that no brackets are displayed and that MATLAB has assigned this matrix the name **ans**. Every matrix in MATLAB must have a name. If you do not assign a matrix a name, then MATLAB assigns it **ans**, which is called the **default variable name**. To assign a matrix name we use the assignment operator =. For example,

A = [4 5 8;0 −1 6]

is displayed as

```
A =

     4     5     8
     0    -1     6
```

Warning

1. All rows must have the same number of entries.
2. MATLAB distinguishes between uppercase and lowercase letters. So matrix B is not the same as matrix **b**.
3. A matrix name can be reused. In such a case the "old" contents are lost.

To assign a matrix but *suppress the display of its entries*, follow the closing square bracket,], with a semicolon.

A = [4 5 8;0 −1 6];

assigns the same matrix to name A as above, but no display appears. To assign a currently defined matrix a new name, use the assignment operator =. Command **Z** = **A** assigns the contents of A to Z. Matrix **A** is still defined.

To determine the matrix names that are in use, use the **who** command. To delete a matrix, use the **clear** command followed by a space and then the matrix name. For example, the command

clear A

deletes name A and its contents from MATLAB. The command **clear** by itself deletes all currently defined matrices.

To determine the number of rows and columns in a matrix, use the **size** command, as in

size(A)

which, assuming that A has not been cleared, displays

```
ans =

     2     3
```

meaning that there are two rows and three columns in matrix A.

SEEING A MATRIX

To see all of the components of a matrix, type its name. If the matrix is large, the display may be broken into subsets of columns that are shown successively. For example, use the command

hilb(9)

which displays the first seven columns followed by columns 8 and 9. (For information on command **hilb**, use **help hilb**.) If the matrix is quite large, the screen display will scroll too fast for you to see the matrix. To see a portion of a matrix, type command **more on** followed by ENTER, then type the matrix name or a command to generate it. Press the Space Bar to reveal more of the matrix. Continue pressing the Space Bar until the "--more--" no longer appears near the bottom of the screen. Try this with **hilb(20)**. To disable this paging feature, type command **more off**. If a scroll bar is available, you can use your mouse to move the scroll bar to reveal previous portions of displays.

We have the following conventions to see a portion of a matrix in MATLAB. For purposes of illustration, suppose matrix A has been entered into MATLAB as a 5×5 matrix.

☐ To see the (2, 3) entry of A, type

A(2,3)

☐ To see the fourth row of A, type

A(4,:)

☐ To see the first column of A, type

A(:,1)

In the preceding situations the : is interpreted to mean "all." The colon can also be used to represent a range of rows or columns. For example, typing

2:8

displays

```
ans =
         2     3     4     5     6     7     8
```

We can use this feature to display a subset of rows or columns of a matrix. As an illustration, to display rows 3 through 5 of matrix A, type

A(3:5,:)

Similarly, columns 1 through 3 are displayed by typing

A(:,1:3)

For more information on the use of the colon operator, type **help colon**. The colon operator in MATLAB is very versatile, but we will not need to use all of its features.

DISPLAY FORMATS

MATLAB stores matrices in decimal form and does its arithmetic computations using a decimal-type arithmetic. This decimal form retains about 16 digits, but not all digits must be shown. Between what goes on in the machine and what is shown on the screen are routines that convert or format the numbers into displays. Here we give an overview of the display formats that we will use. (For more information, see the *MATLAB User's Guide* or type **help format**.)

☐ If the matrix contains *all* integers, then the entire matrix is displayed as integer values; that is, no decimal points appear.

☐ If any entry in the matrix is not exactly represented as an integer, then the entire matrix is displayed in what is known as **format short**. Such a display shows four places behind the decimal point and the last place may have been rounded. The exception to this is zero. If an entry is exactly zero, then it is displayed as an integer zero. Enter the matrix

Q = [5 0 1/3 2/3 7.123456]

into MATLAB. The display is

```
Q =

     5.0000     0    0.3333    0.6667    7.1235
```

Warning If a value is displayed as 0.0000, then it is not identically zero. You should change to **format long**, discussed next, and display the matrix again.

☐ To see more than four places, change the display format. One way to proceed is to use the command

format long

which shows 15 places. The matrix Q in format long is

```
Q =
  Columns 1 through 4
  5.00000000000000 0 0.33333333333333 0.66666666666667
  Column 5
  7.12345600000000
```

Other display formats use an exponent of 10. They are **format short e** and **format long e**. The "e-formats" are often used in numerical analysis. Try these formats with matrix Q.

☐ MATLAB can display values in rational form. The command **format rat**, short for rational display, is used. Inspect the output from the following sequence of MATLAB commands.

format short
V = [1 1/2 1/6 1/12]

displays

```
V =

     1.0000    0.5000    0.1667    0.0833
```

and

format rat
V

displays

```
V =

     1    1/2    1/6    1/12
```

Finally, type **format short** to return to a decimal display form.

Warning Rational output is displayed in what is called "string" form. Strings are not numeric data and hence cannot be used with arithmetic operators. Thus rational output is for "looks" only.

When MATLAB starts, the format in effect is **format short**. If you change the format, it remains in effect until another format command is executed. Some MATLAB routines change the format within the routine.

12.2 MATRIX OPERATIONS IN MATLAB

The operations of addition, subtraction, and multiplication of matrices in MATLAB follow the same definitions as in Sections 1.2 and 1.3. If A and B are $m \times n$ matrices that have been entered into MATLAB, then their sum in MATLAB is computed using command

A+B

and their difference by command

A−B

(spaces can be used on either side of + or −). If A is $m \times n$ and C is $n \times k$, then the product of A and C in MATLAB must be written as

A∗C

In MATLAB, ∗ *must be specifically placed between the names of matrices to be multiplied.* In MATLAB, writing AC does not perform an implied multiplication. In fact, MATLAB considers AC as a new matrix name and, if it has not been previously defined, an error will result. If the matrices involved are not compatible for the operation specified, then an error message will be displayed. Compatibility for addition and subtraction means that the matrices are the same size. Matrices are compatible for multiplication if the number of columns in the first matrix equals the number of rows in the second.

EXAMPLE 1 Enter the matrices

$$A = \begin{bmatrix} 1 & 2 \\ 2 & 4 \end{bmatrix}, \quad \mathbf{b} = \begin{bmatrix} -3 \\ 1 \end{bmatrix}, \quad \text{and} \quad C = \begin{bmatrix} 3 & -5 \\ 5 & 2 \end{bmatrix}$$

into MATLAB and compute the following expressions. We display the results from MATLAB.

Solution (a) **A+C** displays

```
ans =

     4  -3
     7   6
```

(b) **A∗C** displays

```
ans =

    13  -1
    26  -2
```

(c) **b*A** displays

```
??? Error using ==> *
    Inner matrix dimensions must agree.
```

■

Scalar multiplication in MATLAB *requires the use of the multiplication symbol* $*$. For the matrix A in Example 1, $5A$ designates scalar multiplication in the text, while **5*A** is required in MATLAB.

In MATLAB the transpose operator (or symbol) is the single quotation mark or prime, $'$. Using the matrices in Example 1, in MATLAB

Q = C′ displays

```
Q =
     3     5
    -5     2
```

and

p = b′ displays

```
p =
    -3     1
```

As a convenience, we can enter column matrices into MATLAB using $'$. To enter the matrix

$$\mathbf{x} = \begin{bmatrix} 1 \\ 3 \\ -5 \end{bmatrix},$$

we can use either the command

x = [1;3;−5]

or the command

x = [1 3 −5]′

Suppose that we are given the linear system $A\mathbf{x} = \mathbf{b}$, where coefficient matrix A and right-hand side $\mathbf{b}$ have been entered into MATLAB. The augmented matrix $\left[A \;\vdots\; \mathbf{b}\right]$ is formed in MATLAB by typing

[A b]

or, if we want to name it **aug**, by typing

aug = [A b]

No bar will be displayed separating the right-hand side from the coefficient matrix. Using matrices A and $\mathbf{b}$ from Example 1, form the augmented matrix in MATLAB for the system $A\mathbf{x} = \mathbf{b}$.

Forming augmented matrices is a special case of building matrices in MATLAB. Essentially we can "paste together" matrices as long as sizes are appropriate. Using the matrices A, **b**, and C in Example 1, we give some examples:

[A C] displays

```
ans =
     1     2     3    -5
     2     4     5     2
```

[A;C] displays

```
ans =
     1     2
     2     4
     3    -5
     5     2
```

[A b C] displays

```
ans =
     1     2    -3     3    -5
     2     4     1     5     2
```

[C A;A C] displays

```
ans =
     3    -5     1     2
     5     2     2     4
     1     2     3    -5
     2     4     5     2
```

MATLAB has a command to build diagonal matrices by entering only the diagonal entries. The command is **diag**, and

D = diag([1 2 3]) displays

```
D =
     1     0     0
     0     2     0
     0     0     3
```

Command **diag** also works to "extract" a set of diagonal entries. If

$$R = \begin{bmatrix} 5 & 2 & 1 \\ -3 & 7 & 0 \\ 6 & 4 & -8 \end{bmatrix}$$

has been entered into MATLAB, then

diag(R) displays

```
ans =
     5
     7
    -8
```

Note that

diag(diag(R)) displays

```
ans =
     5     0     0
     0     7     0
     0     0    -8
```

For more information on **diag**, use **help**. Commands related to **diag** are **tril** and **triu**.

12.3 MATRIX POWERS AND SOME SPECIAL MATRICES

In MATLAB, to raise a matrix to a power we must use the exponentiation operator ^. If A is square and k is a positive integer, then A^k is denoted in MATLAB by

A^k

which corresponds to a matrix product of A with itself k times. The rules for exponents given in Section 1.4 apply in MATLAB. In particular,

A^0

displays an identity matrix having the same size as A.

EXAMPLE 1 Enter matrices

$$A = \begin{bmatrix} 1 & -1 \\ 1 & 1 \end{bmatrix} \quad \text{and} \quad B = \begin{bmatrix} 1 & -2 \\ 2 & 1 \end{bmatrix}$$

into MATLAB and compute the following expressions. We display the MATLAB results.

(a) **A^2** displays
```
ans =
     0    -2
     2     0
```

(b) **(A∗B)^2** displays
```
ans =
    -8     6
    -6    -8
```

(c) **(B−A)^3** displays
```
ans =
     0     1
    -1     0
```
■

The $n \times n$ identity matrix is denoted by I_n throughout this book. MATLAB has a command to generate I_n when it is needed. The command is **eye**, and it behaves as follows:

eye(2)	displays a 2×2 identity matrix.
eye(5)	displays a 5×5 identity matrix.
t = 10;eye(t)	displays a 10×10 identity matrix.
eye(size(A))	displays an identity matrix the same size as A.

Two other MATLAB commands, **zeros** and **ones**, behave in a similar manner. The command **zeros** produces a matrix of all zeros, while the command **ones** generates a matrix of all ones. Rectangular matrices of size $m \times n$ can be generated using

zeros(m,n), ones(m,n)

where m and n have been previously defined with positive integer values in MATLAB. Using this convention we can generate a column with four zeros using the command

zeros(4,1)

From algebra you are familiar with polynomials in x such as

$$4x^3 - 5x^2 + x - 3 \quad \text{and} \quad x^4 - x - 6.$$

The evaluation of such polynomials at a value of x is easily handled in MATLAB using the command **polyval**. Define the coefficients of the polynomial as a vector (a row or column matrix) with the coefficient of the largest power first, the coefficient of the next largest power second, and so on down to the constant term. If any power is explicitly missing, its coefficient must be set to zero in the corresponding position in the coefficient vector. In MATLAB, for the polynomials above we have coefficient vectors

v = [4 −5 1 −3] and **w = [1 0 0 −1 −6]**

respectively. The command

polyval(v,2)

evaluates the first polynomial at $x = 2$ and displays the computed value of 11. Similarly, the command

t = −1;polyval(w,t)

evaluates the second polynomial at $x = -1$ and displays the value -4.

Polynomials in a square matrix A have the form

$$5A^3 - A^2 + 4A - 7I.$$

Note that the constant term in a matrix polynomial is an identity matrix of the same size as A. This convention is a natural one if we recall that the constant term in an ordinary polynomial is the coefficient of x^0 and that $A^0 = I$. We often meet matrix polynomials when evaluating a standard polynomial such as $p(x) = x^4 - x - 6$ at an $n \times n$ matrix A. The resulting matrix polynomial is

$$p(A) = A^4 - A - 6I_n.$$

Matrix polynomials can be evaluated in MATLAB using the command **polyvalm**. Define the square matrix A and the coefficient vector

w = [1 0 0 −1 −6]

in MATLAB. Then the command

polyvalm(w,A)

produces the value of $p(A)$, which will be a matrix the same size as A.

EXAMPLE 2 Let

$$A = \begin{bmatrix} 1 & -1 & 2 \\ -1 & 0 & 1 \\ 0 & 3 & 1 \end{bmatrix} \quad \text{and} \quad p(x) = 2x^3 - 6x^2 + 2x + 3.$$

To compute $p(A)$ in MATLAB, use the following commands. We show the MATLAB display below the commands.

```
A = [1 −1 2;−1 0 1;0 3 1];
v = [2 −6 2 3];
Q = polyvalm(v,A)

Q =

   −13   −18    10
    −6   −25    10
     6    18   −17
```

■

At times you may want a matrix with integer entries to use in testing some matrix relationship. MATLAB commands can generate such matrices quite easily. Type

C = fix(10∗rand(4))

and you will see displayed a 4×4 matrix C with integer entries. To investigate what this command does, use **help** with the commands **fix** and **rand**.

EXAMPLE 3 In MATLAB, generate several $k \times k$ matrices A for $k = 3, 4, 5$ and display $B = A + A^T$. Look over the matrices displayed and try to determine a property that these matrices share. We show several such matrices below. Your results may not be the same because of the random number generator **rand**.

k = 3;
A = fix(10∗rand(k));
B = A+A′

The display is

```
B =

     4     6    11
     6    18    11
    11    11     0
```

Using the **up arrow** key recall the previous commands one at a time, pressing ENTER after each command. This time the matrix displayed is

```
B =

     0     5    10
     5     6     6
    10     6    10
```

See Exercise T.27 at the end of Section 1.4. ■

12.4 ELEMENTARY ROW OPERATIONS IN MATLAB

The solution of linear systems of equations as discussed in Section 1.6 uses elementary row operations to obtain a sequence of linear systems whose augmented matrices are row equivalent. Row equivalent linear systems have the same solutions, hence we choose elementary row operations to produce row equivalent systems that are easy to solve. It is shown that linear systems in **reduced row echelon form** are easily solved using the Gauss–Jordan procedure and systems in **row echelon form** are easily solved using Gaussian elimination with back substitution. Using either of these procedures requires that we perform row operations that introduce zeros into the augmented matrix of the linear system. We show how to perform such row operations using MATLAB. The arithmetic involved is done by the MATLAB software and we are able to concentrate on the strategy to produce the reduced row echelon form or row echelon form.

Given a linear system $A\mathbf{x} = \mathbf{b}$, we enter the coefficient matrix A and the right-hand side $\mathbf{b}$ into MATLAB. We form the augmented matrix (see Section 12.2) as

C = [A   b]

Now we are ready to begin applying row operations to the augmented matrix C. Each row operation replaces an existing row by a new row. Our strategy is to construct the row operation so that the resulting new row moves us closer to the goal of reduced row echelon form or row echelon form. There are many different choices that can be made for the sequence of row operations to transform $\left[A \mid \mathbf{b}\right]$ to one of these forms. Naturally we try to use the fewest number of row operations, but many times it is convenient to avoid introducing fractions (if possible), especially when doing calculations by hand. Since MATLAB will be doing the arithmetic for us, we need not be concerned about fractions, but it is visually pleasing to avoid them anyway.

As described in Section 1.6, there are three row operations:

- Interchange two rows.
- Multiply a row by a nonzero number.
- Add a multiple of one row to another row.

To perform these operations on an augmented matrix $C = \left[A \mid \mathbf{b}\right]$ in MATLAB, we employ the colon operator, which was discussed in Section 12.1. We illustrate the technique on the linear system in Example 8 of Section 1.6. When the augmented matrix is entered into MATLAB, we have

```
C =

     1     2     3     9
     2    -1     1     8
     3     0    -1     3
```

To produce the reduced row echelon form given in Equation (2) of Section 1.6, we proceed as follows.

Description	**MATLAB Commands and Display**
add (−2) times row 1 to row 2	**C(2,:) = −2 * C(1,:) + C(2,:)**
[Explanation of MATLAB command: Row 2 is replaced by (or set equal to) the sum of −2 times row 1 and row 2.]	C = 1 2 3 9 0 −5 −5 −10 3 0 −1 3
add (−3) times row 1 to row 3	**C(3,:) = −3 * C(1,:) + C(3,:)**
	C = 1 2 3 9 0 −5 −5 −10 0 −6 −10 −24
multiply row 2 by (−1/5)	**C(2,:) = (− 1/5) * C(2,:)**
[Explanation of MATLAB command: Row 2 is replaced by (or set equal to) $\left(-\frac{1}{5}\right)$ times row 2.]	C = 1 2 3 9 0 1 1 2 0 −6 −10 −24

add (−2) times row 2 to row 1 **C(1,:) = −2 * C(2,:) + C(1,:)**

```
C =
     1     0     1     5
     0     1     1     2
     0    -6   -10   -24
```

add 6 times row 2 to row 3 **C(3,:) = 6 * C(3,:) + C(3,:)**

```
C =
     1     0     1     5
     0     1     1     2
     0     0    -4   -12
```

multiply row 3 by (−1/4) **C(3,:) = (− 1/4) * C(3,:)**

```
C =
     1     0     1     5
     0     1     1     2
     0     0     1     3
```

add (−1) times row 3 to row 2 **C(2,:) = −1 * C(3,:) + C(2,:)**

```
C =
     1     0     1     5
     0     1     0    -1
     0     0     1     3
```

add (−1) times row 3 to row 1 **C(1,:) = −1 * C(3,:) + C(1,:)**

```
C =
     1     0     0     2
     0     1     0    -1
     0     0     1     3
```

This last augmented matrix implies that the solution of the linear system is $x = 2$, $y = -1$, $z = 3$.

In the preceding reduction of the augmented matrix to reduced row echelon form, no row interchanges were required. Suppose at some stage we had to interchange rows 2 and 3 of the augmented matrix C. To accomplish this we use a temporary storage area. (We choose to name this area **temp** here.) In MATLAB we proceed as follows.

Description	MATLAB Commands
Assign row 2 to temporary storage.	**temp = C(2,:);**
Assign the contents of row 3 to row 2.	**C(2,:) = C(3,:);**
Assign the contents of row 2 contained in temporary storage to row 3.	**C(3,:) = temp;**

(The semicolons at the end of each command just suppress the display of the contents.)

Using the colon operator and the assignment operator, =, as previously, we can instruct MATLAB to perform row operations to obtain the reduced row echelon form or row echelon form of a matrix. MATLAB does the arithmetic and we concentrate on choosing the row operations to perform the reduction.

We also must enter the appropriate MATLAB command. If we mistype a multiplier or row number, the error can be corrected, but the correction process requires a number of steps. To permit us to concentrate completely on choosing row operations for the reduction process, there is a routine called **reduce** in the set of auxiliary MATLAB routines available to users of this book.* Once you have incorporated these routines into MATLAB, you can type **help reduce** and see the following display.

```
REDUCE  Perform row reduction on matrix A by
        explicitly choosing row operations to use.
        A row operation can be "undone," but this
        feature cannot be used in succession.  This
        routine is for small matrices, real or
        complex.

        Use in the form ===> reduce <=== to select a
        demo or enter your own matrix A

        Use the form    ===> reduce(A) <===.
```

Routine **reduce** alleviates all the command typing and instructs MATLAB to perform the associated arithmetic. To use **reduce**, enter the augmented matrix C of your system as discussed previously and type

reduce(C)

We display the first three steps of **reduce** for Example 8 in Section 1.6. The matrices involved will be the same as those in the first three steps of the reduction process above, where we made direct use of the colon operator to perform the row operations in MATLAB. Screen displays are shown between rows of plus signs, and all input appears in boxes.

```
+++++++++++++++++++++++++++++++++++++++

     ***** "REDUCE" a Matrix by Row Reduction *****

The current matrix is:

A =
     1   2   3   9
     2  -1   1   8
     3   0  -1   3

                  OPTIONS
 <1>   Row(i) <===> Row(j)
 <2>   k * Row(i)     (k not zero)
 <3>   k * Row(i) + Row(j) ===> Row(j)
 <4>   Turn on rational display.
 <5>   Turn off rational display.
<-1>   "Undo" previous row operation.
 <0>   Quit reduce!
       ENTER your choice ===> [3]
```

*Some MATLAB commands that follow require the incorporation of the instructional routines that are available through the Prentice Hall Web site.

```
Enter multiplier. -2

Enter first row number. 1

Enter number of row that changes. 2
```

```
Comment: Option 3 in the above menu means the same
as
```

add a multiple of one row to another row

```
The input above performs the operation in the form

      multiplier * (first row) + (second row)
```

```
+++++++++++++++++++++++++++++++++++++

***** Replacement by Linear Combination Complete *****

The current matrix is:

A =
     1     2     3     9
     0    -5    -5   -10
     3     0    -1     3

                    OPTIONS
 <1>   Row(i) <===> Row(j)
 <2>   k * Row(i)     (k not zero)
 <3>   k * Row(i) + Row(j) ===> Row(j)
 <4>   Turn on rational display.
 <5>   Turn off rational display.
<-1>   "Undo" previous row operation.
 <0>   Quit reduce!
       ENTER your choice ===> 3

Enter multiplier. -3

Enter first row number. 1

Enter number of row that changes. 3

+++++++++++++++++++++++++++++++++++++

***** Replacement by Linear Combination Complete *****

The current matrix is:

A =
     1     2     3     9
     0    -5    -5   -10
     0    -6   -10   -24
```

```
                     OPTIONS
 <1>   Row(i) <===> Row(j)
 <2>   k * Row(i)     (k not zero)
 <3>   k * Row(i) + Row(j) ===> Row(j)
 <4>   Turn on rational display.
 <5>   Turn off rational display.
<-1>   "Undo" previous row operation.
 <0>   Quit reduce!
       ENTER your choice ===> 2

Enter multiplier.  -1/5

Enter row number.  2

+++++++++++++++++++++++++++++++++++++

         ***** Multiplication Complete *****

The current matrix is:

A =
      1     2     3     9
      0     1     1     2
      0    -6   -10   -24

                     OPTIONS
 <1>   Row(i) <===> Row(j)
 <2>   k * Row(i)     (k not zero)
 <3>   k * Row(i) + Row(j) ===> Row(j)
 <4>   Turn on rational display.
 <5>   Turn off rational display.
<-1>   "Undo" previous row operation.
 <0>   Quit reduce!
       ENTER your choice ===>

+++++++++++++++++++++++++++++++++++++
```

At this point you should complete the reduction of this matrix to reduced row echelon form using **reduce**.

Comments

1. Although options 1–3 in **reduce** appear in symbols, they have the same meaning as the phrases used to describe the row operations near the beginning of this section. Option <3> forms a *linear combination* of rows to replace a row. This terminology will be used later in this course and appear in certain displays of **reduce**. (See Sections 12.7 and 1.6.)
2. Within routine **reduce**, the matrix on which the row operations are performed is called A, regardless of the name of your input matrix.

EXAMPLE 1 Solve the following linear system using **reduce**.

$$\tfrac{1}{3}x + \tfrac{1}{4}y = \tfrac{13}{6}$$

$$\tfrac{1}{7}x + \tfrac{1}{9}y = \tfrac{59}{63}$$

Solution Enter the augmented matrix into MATLAB and name it C.

C = [1/3 1/4 13/6;1/7 1/9 59/63]

```
C =
      0.3333  0.2500  2.1667
      0.1429  0.1111  0.9365
```

Then type

reduce(C)

The steps from **reduce** are displayed next. The steps are shown with decimal displays unless you choose the rational display option <4>. The corresponding rational displays are shown in braces in the following examples for illustration purposes. Ordinarily the decimal and rational displays are not shown simultaneously.

```
++++++++++++++++++++++++++++++++++++++
    ***** "REDUCE" a Matrix by Row Reduction *****
The current matrix is:

A =
    0.3333    0.2500    2.1667          {1/3    1/4    13/6 }
    0.1429    0.1111    0.9365          {1/7    1/9    59/63}

                  OPTIONS
 <1>  Row(i) <===> Row(j)
 <2>  k * Row(i)     (k not zero)
 <3>  k * Row(i) + Row(j) ===> Row(j)
 <4>  Turn on rational display.
 <5>  Turn off rational display.
<-1>  "Undo" previous row operation.
 <0>  Quit reduce!
      ENTER your choice ===> 2

Enter multiplier.  1/A(1,1)

Enter row number.  1
++++++++++++++++++++++++++++++++++++++
       ***** Row Multiplication Complete *****
The current matrix is:

A =
    1.0000    0.7500    6.5000          {1      3/4    13/2 }
    0.1429    0.1111    0.9365          {1/7    1/9    59/63}

                  OPTIONS
 <1>  Row(i) <===> Row(j)
 <2>  k * Row(i)     (k not zero)
 <3>  k * Row(i) + Row(j) ===> Row(j)
 <4>  Turn on rational display.
 <5>  Turn off rational display.
<-1>  "Undo" previous row operation.
 <0>  Quit reduce!
      ENTER your choice ===> 3
```

```
Enter multiplier. [-A(2,1)]

Enter first row number. [1]

Enter number of row that changes. [2]
++++++++++++++++++++++++++++++++++++++

***** Replacement by Linear Combination Complete *****

The current matrix is:

A =
    1.0000    0.7500    6.5000         {1    3/4      13/2 }
         0    0.0040    0.0079         {0    1/252    1/126}

                 OPTIONS
 <1>  Row(i) <===> Row(j)
 <2>  k * Row(i)    (k not zero)
 <3>  k * Row(i) + Row(j) ===> Row(j)
 <4>  Turn on rational display.
 <5>  Turn off rational display.
<-1>  "Undo" previous row operation.
 <0>  Quit reduce!
      ENTER your choice ===> [2]

Enter multiplier. [1/A(2,2)]

Enter row number. [2]

++++++++++++++++++++++++++++++++++++++

       ***** Row Multiplication Complete *****

The current matrix is:

A =
    1.0000    0.7500    6.5000         {1    3/4    13/2}
         0    1.0000    2.0000         {0    1         2  }

                 OPTIONS
 <1>  Row(i) <===> Row(j)
 <2>  k * Row(i)    (k not zero)
 <3>  k * Row(i) + Row(j) ===> Row(j)
 <4>  Turn on rational display.
 <5>  Turn off rational display.
<-1>  "Undo" previous row operation.
 <0>  Quit reduce!
      ENTER your choice ===> [3]

Enter multiplier. [-A(1,2)]

Enter first row number. [2]

Enter number of row that changes. [1]
```

```
++++++++++++++++++++++++++++++++++++++++
***** Replacement by Linear Combination Complete *****
The current matrix is:

A =
    1.0000         0    5.0000          {1    0    5}
         0    1.0000    2.0000          {0    1    2}

               OPTIONS
 <1>  Row(i) <===> Row(j)
 <2>  k * Row(i)    (k not zero)
 <3>  k * Row(i) + Row(j) ===> Row(j)
 <4>  Turn on rational display.
 <5>  Turn off rational display.
<-1>  "Undo" previous row operation.
 <0>  Quit reduce!
      ENTER your choice ===> 0

  * * ** ===> REDUCE is over.  Your final matrix is:

  A =

    1.0000         0    5.0000
         0    1.0000    2.0000

++++++++++++++++++++++++++++++++++++++++
```

It follows that the solution of the system is $x = 5$, $y = 2$. ■

The **reduce** routine forces you to concentrate on the strategy of the row reduction process. Once you have used **reduce** on a number of linear systems, the reduction process becomes a fairly systematic computation. The reduced row echelon form of a matrix is used in many places in linear algebra to provide information related to concepts (we study a number of these later). As such, the reduced row echelon form of a matrix becomes one step of more involved computational processes. Hence MATLAB provides an automatic way to obtain the **r**educed **r**ow **e**chelon **f**orm. The command is **rref**. Once you have entered the matrix A under consideration, where A could represent an augmented matrix, just type

rref(A)

and MATLAB responds by displaying the reduced row echelon form of A.

EXAMPLE 2 In Section 1.6, Example 14 asks for the solution of the homogeneous system

$$\begin{aligned} x + y + z + w &= 0 \\ x + w &= 0 \\ x + 2y + z &= 0. \end{aligned}$$

Form the augmented matrix C in MATLAB to obtain

```
C =

     1     1     1     1     0
     1     0     0     1     0
     1     2     1     0     0
```

Next type

rref(C)

and MATLAB displays

```
ans =

     1     0     0     1     0
     0     1     0    −1     0
     0     0     1     1     0
```

It follows that unknown w can be chosen arbitrarily—say, $w = r$, where r is any real number. Hence the solution is

$$x = -r, \qquad y = r, \qquad z = -r, \qquad w = r. \qquad \blacksquare$$

12.5 MATRIX INVERSES IN MATLAB

As discussed in Section 1.7, for a square matrix A to be nonsingular, the reduced row echelon form of A must be the identity matrix. Hence in MATLAB we can determine if A is singular or nonsingular by computing the reduced row echelon form of A using either **reduce** or **rref**. If the result is the identity matrix, then A is nonsingular. Such a computation determines whether or not an inverse exists, but does not explicitly compute the inverse when it exists. To compute the inverse of A we can proceed as in Section 1.7 and find the reduced row echelon form of $\left[A \mid I\right]$. If the resulting matrix is $\left[I \mid Q\right]$, then $Q = A^{-1}$. In MATLAB, once a nonsingular matrix A has been entered, the inverse can be found step by step using

reduce([A eye(size(A))])

or computed immediately using

rref([A eye(size(A))])

For example, if we use the matrix A in Example 5 of Section 1.7, then

$$A = \begin{bmatrix} 1 & 1 & 1 \\ 0 & 2 & 3 \\ 5 & 5 & 1 \end{bmatrix}.$$

Entering matrix A into MATLAB and typing the command

rref([A eye(size(A))])

displays

```
ans =

   1.0000        0        0   1.6250  −0.5000  −0.1250
        0   1.0000        0  −1.8750   0.5000   0.3750
        0        0   1.0000   1.2500        0  −0.2500
```

To extract the inverse matrix we use

Ainv = ans(:,4:6)

and obtain

```
Ainv =

     1.6250   -0.5000   -0.1250
    -1.8750    0.5000    0.3750
     1.2500         0   -0.2500
```

To see the result in rational display, use

format rat
Ainv

which gives

```
Ainv =

     13/8    -1/2    -1/8
    -15/8     1/2     3/8
      5/4       0    -1/4
```

Type command

format short

Thus our previous MATLAB commands can be used in a manner identical to the way the hand computations are described in Section 1.7.

For convenience, there is a routine that computes inverses directly. The command is **invert**. For the preceding matrix A we would type

invert(A)

and the result would be identical to that obtained in **Ainv** by using **rref**. If the matrix is not square or is singular, an error message will appear.

12.6 VECTORS IN MATLAB

An n-vector $\mathbf{x}$ (see Section 4.2) in MATLAB can be represented either as a column matrix with n elements,

$$\mathbf{x} = \begin{bmatrix} x_1 \\ x_2 \\ \vdots \\ x_n \end{bmatrix},$$

or as a row matrix with n elements,

$$\mathbf{x} = \begin{bmatrix} x_1 & x_2 & \cdots & x_n \end{bmatrix}.$$

In a particular problem or exercise, choose one way of representing the n-vectors and stay with that form.

The vector operations of Section 4.2 correspond to operations on $n \times 1$ matrices or columns. If the n-vector is represented by row matrices in MATLAB, then the vector operations correspond to operations on $1 \times n$ matrices. These are just special cases of addition, subtraction, and scalar multiplication of matrices, which were discussed in Section 12.2.

The norm or length of vector $\mathbf{x}$ in MATLAB is obtained by using the command

norm(x)

This command computes the square root of the sum of the squares of the components of $\mathbf{x}$, which is equal to $\|\mathbf{x}\|$, as discussed in Section 4.2.

The distance between vectors $\mathbf{x}$ and $\mathbf{y}$ in R^n in MATLAB is given by

norm(x − y)

EXAMPLE 1 Let

$$\mathbf{u} = \begin{bmatrix} 2 \\ 1 \\ 1 \\ -1 \end{bmatrix} \quad \text{and} \quad \mathbf{v} = \begin{bmatrix} 3 \\ 1 \\ 2 \\ 0 \end{bmatrix}.$$

Enter these vectors in R^4 into MATLAB as columns. Then

norm(u)

displays

```
ans =

    2.6458
```

while

norm(v)

gives

```
ans =

    3.7417
```

and

norm(u − v)

gives

```
ans =
    1.7321
```

■

The dot product of a pair of vectors $\mathbf{u}$ and $\mathbf{v}$ in R^n in MATLAB is computed by the command

dot(u,v)

For the vectors in Example 1, MATLAB gives the dot product as

```
ans =
     9
```

As discussed in Section 4.2, the notion of a dot product is useful to define the angle between n-vectors. Equation (4) in Section 4.2 tells us that the cosine of the angle θ between $\mathbf{u}$ and $\mathbf{v}$ is given by

$$\cos\theta = \frac{\mathbf{u}\cdot\mathbf{v}}{\|\mathbf{u}\|\ \|\mathbf{v}\|}.$$

In MATLAB the cosine of the angle between $\mathbf{u}$ and $\mathbf{v}$ is computed by the command

dot(u,v)/(norm(u) ∗ norm(v))

The angle θ can be computed by taking the arccosine of the value of the previous expression. In MATLAB the arccosine function is denoted by **acos**. The result will be an angle in radians.

EXAMPLE 2 For the vectors **u** and **v** in Example 1, the angle between the vectors is computed as

c = dot(u,v)/(norm(u) ∗ norm(v));
angle = acos(c)

which displays

```
angle =
          0.4296
```

and is approximately 24.61°. ■

12.7 APPLICATIONS OF LINEAR COMBINATIONS IN MATLAB

The notion of a linear combination as discussed in Section 6.2 is fundamental to a wide variety of topics in linear algebra. The ideas of span, linear independence, linear dependence, and basis are based on forming linear combinations of vectors. In addition, the elementary row operations discussed in Sections 1.6 and 12.4 are essentially of the form, "Replace an existing row by a linear combination of rows." This is clearly the case when we add a multiple of one row to another row. (See the menu for the routine **reduce** in Section 12.4.) From this point of view, it follows that the reduced row echelon form and the row echelon form are processes for implementing linear combinations of rows of a matrix. Hence the MATLAB routines **reduce** and **rref** should be useful in solving problems that involve linear combinations.

Here we discuss how to use MATLAB to solve problems dealing with linear combinations, span, linear independence, linear dependence, and basis. The basic strategy is to set up a linear system related to the problem and ask questions such as "Is there a solution?" or "Is the only solution the trivial solution?"

THE LINEAR COMBINATION PROBLEM

Given a vector space V and a set of vectors $S = \{\mathbf{v}_1, \mathbf{v}_2, \ldots, \mathbf{v}_k\}$ in V, determine if $\mathbf{v}$, belonging to V, can be expressed as a linear combination of the members of S. That is, can we find some set of scalars $c_1, c_2, \ldots, c_k$ so that

$$c_1\mathbf{v}_1 + c_2\mathbf{v}_2 + \cdots + c_k\mathbf{v}_k = \mathbf{v}?$$

There are several common situations.

Case 1. If the vectors in S are row matrices, then we construct (as shown in Example 11 of Section 6.2) a linear system whose coefficient matrix A is

$$A = \begin{bmatrix} \mathbf{v}_1 \\ \mathbf{v}_2 \\ \vdots \\ \mathbf{v}_k \end{bmatrix}^T$$

and whose right-hand side is $\mathbf{v}^T$. That is, the columns of A are the row matrices of set S converted to columns. Let $\mathbf{c} = \begin{bmatrix} c_1 & c_2 & \cdots & c_k \end{bmatrix}$ and $\mathbf{b} = \mathbf{v}^T$, then transform the linear system $A\mathbf{c} = \mathbf{b}$ using **reduce** or **rref** in MATLAB. If the system is shown to be consistent, so that no rows of the form

$\begin{bmatrix} 0 & 0 & \cdots & 0 & \vdots & q \end{bmatrix}$, $q \neq 0$, occur, then the vector **v** can be written as a linear combination of the vectors in S. In that case the solution of the system gives the values of the coefficients. *Caution*: Many times we need only determine if the system is consistent to decide whether **v** is a linear combination of the members of S. Read the question carefully.

EXAMPLE 1 To apply MATLAB to Example 11 of Section 6.2, proceed as follows. Define

A = [1 2 1;1 0 2; 1 1 0]′
b = [2 1 5]′

Then use the command

rref([A b])

to give

```
ans =

     1     0     0     1
     0     1     0     2
     0     0     1    −1
```

Recall that this display represents the reduced row echelon form of an augmented matrix. It follows that the system is consistent, with solution

$$c_1 = 1, \qquad c_2 = 2, \qquad c_3 = -1.$$

Hence **v** is a linear combination of $\mathbf{v}_1$, $\mathbf{v}_2$, and $\mathbf{v}_3$. ■

Case 2. If the vectors in S are column matrices, then just lay the columns side by side to form the coefficient matrix

$$A = \begin{bmatrix} \mathbf{v}_1 & \mathbf{v}_2 & \cdots & \mathbf{v}_k \end{bmatrix}$$

and set $\mathbf{b} = \mathbf{v}$. Proceed as described in Case 1.

Case 3. If the vectors in S are polynomials, then associate with each polynomial a column of coefficients. Make sure any missing terms in the polynomial are associated with a zero coefficient. One way to proceed is to use the coefficient of the highest-power term as the first entry of the column, the coefficient of the next-highest-power term as the second entry, and so on. For example,

$$t^2 + 2t + 1 \longrightarrow \begin{bmatrix} 1 \\ 2 \\ 1 \end{bmatrix}, \qquad t^2 + 2 \longrightarrow \begin{bmatrix} 1 \\ 0 \\ 2 \end{bmatrix}, \qquad 3t - 2 \longrightarrow \begin{bmatrix} 0 \\ 3 \\ -2 \end{bmatrix}.$$

The linear combination problem is now solved as in Case 2.

Case 4. If the vectors in S are $m \times n$ matrices, then associate with each such matrix A_j a column $\mathbf{v}_j$ formed by stringing together its columns one after the other. In MATLAB this transformation is done using the **reshape** command. Then we proceed as in Case 2.

EXAMPLE 2 Given matrix

$$P = \begin{bmatrix} 1 & 2 & 3 \\ 4 & 5 & 6 \end{bmatrix}.$$

To associate a column matrix as described above within MATLAB, first enter P into MATLAB, then type the command

v = reshape(P,6,1)

which gives

```
v =

     1
     4
     2
     5
     3
     6
```

For more information, type **help reshape**. ■

THE SPAN PROBLEM

There are two common types of problems related to span. The first is as follows:

> Given the set of vectors $S = \{\mathbf{v}_1, \mathbf{v}_2, \ldots, \mathbf{v}_k\}$ and the vector $\mathbf{v}$ in a vector space V, is $\mathbf{v}$ in span S?

This is identical to the linear combination problem addressed previously because we want to know if $\mathbf{v}$ is a linear combination of the members of S. As shown previously, we can use MATLAB in many cases to solve this problem.

The second type of problem related to span is as follows:

> Given vectors $S = \{\mathbf{v}_1, \mathbf{v}_2, \ldots, \mathbf{v}_k\}$ in a vector space V, does span $S = V$?

Here we are asked if every vector in V can be written as a linear combination of the vectors in S. In this case the linear system constructed has a right-hand side that contains arbitrary values that correspond to an arbitrary vector in V. (See Example 1 in Section 6.3.) Since MATLAB manipulates only numerical values in routines such as **reduce** and **rref**, we cannot use MATLAB here to (fully) answer this question.

For the second type of spanning question there is a special case that arises frequently and can be handled in MATLAB. In Section 6.4 the concept of the dimension of a vector space is discussed. The dimension of a vector space V is the number of vectors in a basis (see Section 6.4), which is the smallest number of vectors that can span V. If we know that V has dimension k and the set S has k vectors, then we can proceed as follows to see if span $S = V$. Develop a linear system $A\mathbf{c} = \mathbf{b}$ associated with the span question. If the reduced row echelon form of the coefficient matrix A has the form

$$\begin{bmatrix} I_k \\ \mathbf{0} \end{bmatrix},$$

where $\mathbf{0}$ is a submatrix of all zeros, then any vector in V is expressible in terms of the members of S. In fact, S is a basis for V. In MATLAB we can use routine **reduce** or **rref** on matrix A. If A is square, we can also use **det**. Try this strategy on Example 1 in Section 6.3.

Another spanning question involves finding a set that spans the set of solutions of a homogeneous system of equations $A\mathbf{x} = \mathbf{0}$. The strategy in MATLAB is to find the reduced row echelon form of $\left[A \mid \mathbf{0}\right]$ using the command

rref(A)

(There is no need to include the augmented column since it is all zeros.) Then form the general solution of the system and express it as a linear combination of columns. The columns form a spanning set for the solution set of the system. See Example 6 in Section 6.3.

THE LINEAR INDEPENDENCE/DEPENDENCE PROBLEM

The linear independence or dependence of a set of vectors $S = \{\mathbf{v}_1, \mathbf{v}_2, \ldots, \mathbf{v}_k\}$ is a linear combination question. Set S is linearly independent if the *only* time the linear combination $c_1\mathbf{v}_1 + c_2\mathbf{v}_2 + \cdots + c_k\mathbf{v}_k$ gives the zero vector is when $c_1 = c_2 = \cdots = c_k = 0$. If we can produce the zero vector with any one of the coefficients $c_j \neq 0$, then S is linearly dependent. Following the discussion on linear combination problems, we produce the associated linear system

$$A\mathbf{c} = \mathbf{0}.$$

Note that the linear system is homogeneous. We have the following result:

S is linearly independent if and only if $A\mathbf{c} = \mathbf{0}$
has only the trivial solution.

Otherwise, S is linearly dependent. See Examples 8 and 9 in Section 6.3. Once we have the homogeneous system $A\mathbf{c} = \mathbf{0}$, we can use MATLAB routine **reduce** or **rref** to analyze whether or not the system has a nontrivial solution.

A special case arises if we have k vectors in a set S in a vector space V whose dimension is k (see Section 6.4). Let the linear system associated with the linear combination problem be $A\mathbf{c} = \mathbf{0}$. It can be shown that

S is linearly independent if and only if
the reduced row echelon form of A is $\begin{bmatrix} I_k \\ \mathbf{0} \end{bmatrix}$,

where $\mathbf{0}$ is a submatrix of all zeros. In fact we can extend this further to say S is a basis for V. (See Theorem 6.9.) In MATLAB we can use **reduce** or **rref** on A to aid in the analysis of such a situation.

12.8 LINEAR TRANSFORMATIONS IN MATLAB

We consider the special case of linear transformations $L\colon R^n \to R^m$. Every such linear transformation can be represented by an $m \times n$ matrix A. (See Section 4.3.) Then, for $\mathbf{x}$ in R^n,

$$L(\mathbf{x}) = A\mathbf{x},$$

which is in R^m. For example, suppose that $L\colon R^4 \to R^3$ is given by $L(\mathbf{x}) = A\mathbf{x}$, where matrix

$$A = \begin{bmatrix} 1 & -1 & -2 & -2 \\ 2 & -3 & -5 & -6 \\ 1 & -2 & -3 & -4 \end{bmatrix}.$$

The image of

$$\mathbf{x} = \begin{bmatrix} 1 \\ 2 \\ -1 \\ 0 \end{bmatrix}$$

under L is

$$L(\mathbf{x}) = A\mathbf{x} = \begin{bmatrix} 1 & -1 & -2 & -2 \\ 2 & -3 & -5 & -6 \\ 1 & -2 & -3 & -4 \end{bmatrix} \begin{bmatrix} 1 \\ 2 \\ -1 \\ 0 \end{bmatrix} = \begin{bmatrix} 1 \\ 1 \\ 0 \end{bmatrix}.$$

The **range of a linear transformation** L is the subspace of R^m consisting of the set of all images of vectors from R^n. It is easily shown that

$$\text{range } L = \text{column space of } A.$$

(See Example 11 in Section 10.2.) It follows that we "know the range of L" when we have a basis for the column space of A. There are two simple ways to find a basis for the column space of A:

1. The transposes of the nonzero rows of **rref(A′)** form a basis for the column space. (See Example 4 in Section 6.6.)
2. If the columns containing the leading 1s of **rref(A)** are $k_1 < k_2 < \cdots < k_r$, then columns $k_1, k_2, \ldots, k_r$ of A are a basis for the column space of A. (See Example 4 in Section 6.6.)

For the matrix A given previously, we have

$$\textbf{rref(A}'\textbf{)} = \begin{bmatrix} 1 & 0 & -1 \\ 0 & 1 & 1 \\ 0 & 0 & 0 \\ 0 & 0 & 0 \end{bmatrix},$$

and hence $\left\{ \begin{bmatrix} 1 \\ 0 \\ -1 \end{bmatrix}, \begin{bmatrix} 0 \\ 1 \\ 1 \end{bmatrix} \right\}$ is a basis for the range of L. Using method 2,

$$\textbf{rref(A)} = \begin{bmatrix} 1 & 0 & -1 & 0 \\ 0 & 1 & 1 & 2 \\ 0 & 0 & 0 & 0 \end{bmatrix}.$$

Thus it follows that columns 1 and 2 of A are a basis for the column space of A and hence a basis for the range of L. In addition, routine **lisub** can be used. Use **help** for directions.

The **kernel of a linear transformation** is the subspace of all vectors in R^n whose image is the zero vector in R^m. This corresponds to the set of all vectors $\mathbf{x}$ satisfying

$$L(\mathbf{x}) = A\mathbf{x} = \mathbf{0}.$$

Hence it follows that the kernel of L, denoted $\ker L$, is the set of all solutions of the homogeneous system

$$A\mathbf{x} = \mathbf{0},$$

which is the null space of A. Thus we "know the kernel of L" when we have a basis for the null space of A. To find a basis for the null space of A, we form the general solution of $A\mathbf{x} = \mathbf{0}$ and "separate it into a linear combination of columns using the arbitrary constants that are present." The columns employed form a basis for the null space of A. This procedure uses **rref(A)**. For the matrix A given previously, we have

$$\textbf{rref(A)} = \begin{bmatrix} 1 & 0 & -1 & 0 \\ 0 & 1 & 1 & 2 \\ 0 & 0 & 0 & 0 \end{bmatrix}.$$

If we choose the variables corresponding to columns without leading 1s to be arbitrary, we have

$$x_3 = r \quad \text{and} \quad x_4 = t.$$

It follows that the general solution to $A\mathbf{x} = \mathbf{0}$ is given by

$$\mathbf{x} = \begin{bmatrix} x_1 \\ x_2 \\ x_3 \\ x_4 \end{bmatrix} = \begin{bmatrix} r \\ -r - 2t \\ r \\ t \end{bmatrix} = r \begin{bmatrix} 1 \\ -1 \\ 1 \\ 0 \end{bmatrix} + t \begin{bmatrix} 0 \\ -2 \\ 0 \\ 1 \end{bmatrix}.$$

Thus columns

$$\begin{bmatrix} 1 \\ -1 \\ 1 \\ 0 \end{bmatrix} \quad \text{and} \quad \begin{bmatrix} 0 \\ -2 \\ 0 \\ 1 \end{bmatrix}$$

form a basis for $\ker L$. See also routine **homsoln**, which will display the general solution of a homogeneous linear system. In addition, the command **null** will produce an orthonormal basis for the null space of a matrix. Use **help** for further information on these commands.

In summary, appropriate use of the **rref** command in MATLAB will give bases for both the kernel and range of the linear transformation $L(\mathbf{x}) = A\mathbf{x}$.

12.9 MATLAB COMMAND SUMMARY

In this section we list the principal MATLAB commands and operators used in this book. The list is divided into two parts: commands that come with the MATLAB software, and special instructional routines that are available to users of this book. Both parts are cross-indexed with sections of Chapter 12 where they are discussed and/or initial sections that have MATLAB exercises specifically referring to them. For ease of reference we have included a brief description of each instructional routine that is available to users of this book. These descriptions are also available using MATLAB's **help** command once the installation procedures are complete. A description of any MATLAB command can be obtained by using **help**. (See the introduction to this chapter.)

Built-in MATLAB Commands

ans ⟨12.1⟩	**inv** ⟨12.5⟩	**rref** ⟨12.4, 1.6⟩
clear ⟨12.1, 1.2⟩	**norm** ⟨12.6⟩	**size** ⟨12.1, 1.7⟩
conj ⟨A.1⟩	**null** ⟨12.8⟩	**sqrt** ⟨A.1⟩
det ⟨12.7, 3.1⟩	**ones** ⟨12.3, 1.3⟩	**sum** ⟨2.3⟩
diag ⟨12.2, 1.3⟩	**poly** ⟨8.1⟩	**tril** ⟨12.2, 1.4⟩
dot ⟨12.6, 1.3, 4.2⟩	**polyval** ⟨12.3⟩	**triu** ⟨12.2, 1.4⟩
eig ⟨8.3⟩	**polyvalm** ⟨12.3, 1.4⟩	**zeros** ⟨12.3⟩
exit ⟨12⟩	**quit** ⟨12⟩	\ ⟨1.6⟩
eye ⟨12.3, 1.6⟩	**rand** ⟨12.3, 1.4⟩	; ⟨12.1⟩
fix ⟨12.3, 1.4⟩	**rank** ⟨6.6⟩	: ⟨12.1⟩
format ⟨12.1, 1.2⟩	**rat** ⟨12.1⟩	′ (prime) ⟨12.2, 1.2⟩
help ⟨12⟩	**real** ⟨A.1⟩	+, −, *, /, ^ ⟨12.2, 12.3⟩
hilb ⟨12.1, 1.2⟩	**reshape** ⟨12.7⟩	
image ⟨A.1⟩	**roots** ⟨8.1⟩	

Supplemental Instructional Commands

adjoint ⟨1.2⟩	**crossprd** ⟨4.5⟩	**lsqline** ⟨7.2⟩
binadd ⟨1.2⟩	**crossdemo** ⟨4.5⟩	**matrixtrans** ⟨1.5, 2.3⟩
bingen ⟨1.2⟩	**gschmidt** ⟨6.8⟩	**planelt** ⟨2.3, 5.1⟩
binprod ⟨1.3⟩	**homsoln** ⟨12.8, 6.5⟩	**project** ⟨2.3⟩
binrand ⟨1.4⟩	**invert** ⟨12.5⟩	**reduce** ⟨12.4, 1.6⟩
binreduce ⟨1.6⟩	**linprog** ⟨11.2⟩	**vec2demo** ⟨4.1⟩
cofactor ⟨3.2⟩	**lpstep** ⟨11.2⟩	**vec3demo** ⟨4.2⟩

Notes: ⟨12⟩ refers to the introduction to Chapter 12. Both **rref** and **reduce** are used in many sections. Several utilities required by the instructional commands are also available to users of this book: **arrowh**, **mat2strh**, and **blkmat**. The description given next is that displayed in response to the **help** command. In the description of several commands, the notation differs slightly from that in the text.

Description of Instructional Commands

ADJOINT

Compute the classical adjoint of a square matrix A. If A is not square an empty matrix is returned.
*** This routine should only be used by students to check adjoint computations and should not be used as part of a routine to compute inverses. See invert or inv.

Use in the form ==> adjoint(A) <==

BINADD

Utility to add two bit vectors using binary arithmetic. Checks for addends of the same size and binary are made.

Use in the form ==> sum = binadd(x,y) or binadd(x,y) <==

BINGEN

Generate a matrix of binary codes for integers from start to fin in steps of 1 as columns of num bits.

Use in the form ==> bingen(start,fin,num) <==

or

==> M = bingen(start,fin,num) <==

where start is a nonnegative integer to begin with
fin is a nonnegative integer greater than or equal to start to stop at
num is the number of bits to use in generating the binary form of the integers

BINPROD

Utility to compute the matrix product A*B of two binary matrices. Checks are made to ensure A and B are binary.

Use in the form ==> C = binprod(A,b) or binprod(A,B) <==

BINRAND

Randomly generate an m by n bit matrix

Use in the form ==> binrand(m,n) or B = binrand(m,n)<==

BINREDUCE

Perform row reduction on binary matrix A by explicitly choosing row operations to use. A row operation can be "undone," but this feature cannot be used in succession. (This routine is only for binary matrices.)

Use in the form ==> binreduce <==

or in the form ==> binreduce(A) <==

BKSUB

Perform back substitution on upper triangular system Ax=b. If A is not square, upper triangular, and nonsingular, an error message is displayed. In case of an error the solution returned is all zeros.

Use in the form ==> bksub(A,b) <==

COFACTOR

Computes the (i,j)-cofactor of matrix A. If A is not square, an error message is displayed.
*** This routine should only be used by students to check cofactor computations.

Use in the form ==> cofactor(i,j,A) <==

CROSSDEMO

Display a pair of three-dimensional vectors and their cross product.

The input vectors X and Y are displayed in a three-dimensional perspective along with their cross product. For visualization purposes a set of coordinate 3-D axes are shown.

Use in the form ==> crossdemo(X,Y) <==

CROSSPRD

Compute the cross product of vectors x and y in 3-space. The output is a vector orthogonal to both of the original vectors x and y. The output is returned as a row matrix with 3 components [v1 v2 v3] which is interpreted as v1*i + v2*j + v3*k where i, j, k are the unit vectors in the x, y, and z directions respectively.

Use in the form ==> v=crossprd(x,y) <==

FORSUB

Perform forward substitution on a lower triangular system Ax=b. If A is not square, lower triangular, and nonsingular, an error message is displayed. In case of an error the solution returned is all zeros.

Use in the form ==> forsub(A,b) <==

GSCHMIDT

The Gram-Schmidt process on the columns in matrix x. The orthonormal basis appears in the columns of y unless there is a second argument, in which case y contains only an orthogonal basis. The second argument can have any value.

Use in the form ==> y=gschmidt(x) <==
or ==> y=gschmidt(x,v) <==

HOMSOLN

Find the general solution of a homogeneous system of equations. The routine returns a set of basis vectors for the null space of Ax=0.

Use in the form ==> ns=homsoln(A) <==

If there is a second argument, the general solution is displayed.

Use in the form ==> homsoln(A,1) <==

This option assumes that the general solution has at most 10 arbitrary constants.

INVERT

Compute the inverse of a matrix A by using the reduced row echelon form applied to [A I]. If A is singular, a warning is given.

Use in the form ==> B = invert(A) <==

LINPROG

Directly solves the standard linear programming problem using slack variables as formulated in Introductory Linear Algebra with Applications by B. Kolman and D. R. Hill. This routine is only designed for small problems.

To form the initial tableau A the coefficients of the constraints are entered into the rows where the equations are of the form:

$$a_1X_1 + a_2X_2 + a_3X_3 + \cdots + a_nX_n = C_m$$

and the bottom row consists of the objective function written in the form:

$$z_1X_1 + z_2X_2 + z_3X_3 + \cdots + z_nX_n = 0.$$

Then, as long as there is at least one negative entry in the last row, linprog will find the optimal solution. If no argument containing the initial tableau is present, the routine prompts for the tableau.

Use in the form ==> linprog(A) or linprog <==

LPSTEP

A step-by-step solver for small standard linear programming problems. At each stage you are asked to determine the pivot. Incorrect responses initiate a set of questions to aid in pivot selection. The screens reflect the problem form developed in Introductory Linear Algebra with Applications by B. Kolman and D. R. Hill.

This routine solves the standard linear programming problem using the Simplex Method with slack variables. To form the initial tableau A the coefficients of the constraints are entered into the rows where the equations are of the form

$$a_1X_1 + a_2X_2 + a_3X_3 + \cdots + a_nX_n = C_m$$

and the bottom row consists of the objective function written in the form

$$z_1X_1 + z_2X_2 + z_3X_3 + \cdots + z_nX_n = 0.$$

Then, as long as there is at least one negative entry in the last row, LPSTEP will find an optimal solution. If the tableau A is not supplied as an argument, the user will be prompted to enter it. <<requires utility mat2strh.m>>

Use in the form ==> lpstep(A) or lpstep <==

LSQLINE

This routine will construct the equation of the least square line to a data set of ordered pairs and then graph the line and the data set. A short menu of options is available, including evaluating the equation of the line at points.

Use in the form ==> c = lsqline(x,y) or lsqline(x,y) <==

Here x is a vector containing the x-coordinates and y is a vector containing the corresponding y-coordinates. On output, c contains the coefficients of the least squares line:

$$y = c(1) * x + c(2)$$

LUPR

Perform LU-factorization on matrix A by explicitly choosing row operations to use. No row interchanges are permitted, hence it is possible that the factorization cannot be found. It is recommended that the multipliers be constructed in terms of the elements of matrix U, like $-U(3,2)/U(2,2)$, since the displays of matrices L and U do not show all the decimal places available. A row operation can be "undone," but this feature cannot be used in succession.

This routine uses the utilities mat2strh and blkmat.

Use in the form ==> [L,U] = lupr(A) <==

MATRIXTRANS

A script to show the images of objects from 2-space when mapped by a 2 by 2 matrix. Several objects are available and a composite mapping option can be selected. The graphics user interface is employed for object selection and the initiation of operations and options.

When a matrix is used to perform a mapping we always name it A.

Use in the form ==> matrixtrans <==

and follow the screen directions.

PLANELT

Demonstration of plane linear transformations:
Rotations, Reflections, Expansions/Compressions, Shears

Or you may specify your own transformation.

Graphical results of successive plane linear transformations can be seen using a multiple window display. Standard figures can be chosen or you may choose to use your own figure.

Use in the form ==> planelt <==

PROJECT

Projecting vector u onto vector w. Vectors u and w can be either a pair of 2-D or 3-D vectors. A sketch showing u being projected onto w is displayed.

Use in the form ==> project(u,w) <==

or ==> project <==

In the latter case a menu of options is presented. One option is a demo which randomly selects 2-D or 3-D.

REDUCE

Perform row reduction on matrix A by explicitly choosing row operations to use. A row operation can be "undone," but this feature cannot be used in succession. This routine is for small matrices, real or complex.

Use in the form ==> reduce <== to select a demo or enter your own matrix A

or in the form ==> reduce(A) <==

VEC2DEMO

A graphical demonstration of vector operations for two-dimensional vectors.

Select vectors X = [x1 x2] and Y = [y1 y2]. They will be displayed graphically along with their sum, difference, and a scalar multiple.

Use in the form ==> vec2demo(X,Y) <==

or ==> vec2demo <==

In the latter case you will be prompted for input.

VEC3DEMO

Display a pair of three-dimensional vectors, their sum, difference and scalar multiples.

The input vectors X and Y are displayed in a 3-dimensional perspective along with their sum, difference and selected scalar multiples. For visualization purposes a set of coordinate 3-D axes are shown.

Use in the form ==> vec3demo(X,Y) <==

APPENDIX A

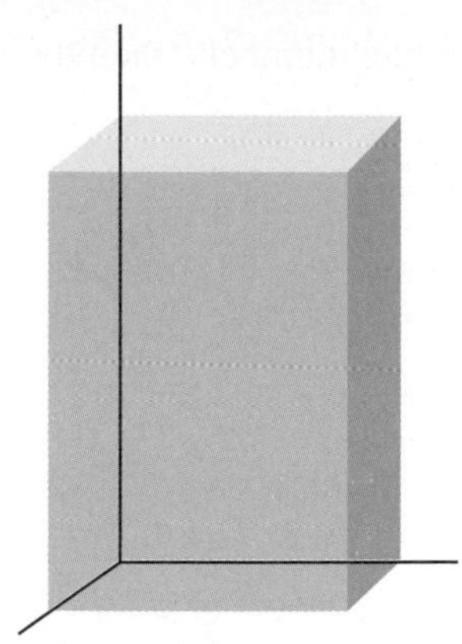

COMPLEX NUMBERS

A.1 COMPLEX NUMBERS

Complex numbers are usually introduced in an algebra course to "complete" the solution to the quadratic equation

$$ax^2 + bx + c = 0, \qquad a \neq 0.$$

In using the quadratic formula

$$x = \frac{-b \pm \sqrt{b^2 - 4ac}}{2a},$$

the case in which $b^2 - 4ac < 0$ is not resolved unless we can cope with the square roots of negative numbers. In the sixteenth century mathematicians and scientists justified this "completion" of the solution of quadratic equations by intuition. Naturally, a controversy arose, with some mathematicians denying the existence of these numbers and others using them along with real numbers. The use of complex numbers did not lead to any contradictions, and the idea proved to be an important milestone in the development of mathematics.

A **complex number** c is of the form $c = a + bi$, where a and b are real numbers and where $i = \sqrt{-1}$; a is called the **real part** of c and b is called the **imaginary part** of c. The term "imaginary part" arose from the mysticism surrounding the beginnings of complex numbers; however, these numbers are as "real" as the real numbers.

EXAMPLE 1

(a) $5 - 3i$ has real part 5 and imaginary part -3.

(b) $-6 + \sqrt{2}\,i$ has real part -6 and imaginary part $\sqrt{2}$. ■

The symbol $i = \sqrt{-1}$ has the property that $i^2 = -1$ and we can deduce the following relationships:

$$i^3 = -i, \quad i^4 = 1, \quad i^5 = i, \quad i^6 = -1, \quad i^7 = -i, \ldots .$$

These results will be handy for simplifying operations involving complex numbers.

We say that two complex numbers $c_1 = a_1 + b_1 i$ and $c_2 = a_2 + b_2 i$ are **equal** if their real and imaginary parts are equal, that is, if $a_1 = a_2$ and $b_1 = b_2$. Of course, every real number a is a complex number with its imaginary part zero: $a = a + 0i$.

OPERATIONS ON COMPLEX NUMBERS

If $c_1 = a_1 + b_1 i$ and $c_2 = a_2 + b_2 i$ are complex numbers, then their **sum** is

$$c_1 + c_2 = (a_1 + a_2) + (b_1 + b_2)i,$$

and their **difference** is

$$c_1 - c_2 = (a_1 - a_2) + (b_1 - b_2)i.$$

In words, to form the sum of two complex numbers, add the real parts and add the imaginary parts. The **product** of c_1 and c_2 is

$$\begin{aligned} c_1 c_2 &= (a_1 + b_1 i) \cdot (a_2 + b_2 i) = a_1 a_2 + (a_1 b_2 + b_1 a_2)i + b_1 b_2 i^2 \\ &= (a_1 a_2 - b_1 b_2) + (a_1 b_2 + b_1 a_2)i. \end{aligned}$$

A special case of multiplication of complex numbers occurs when c_1 is real. In this case we obtain the simple result

$$c_1 c_2 = c_1 \cdot (a_2 + b_2 i) = c_1 a_2 + c_1 b_2 i.$$

If $c = a + bi$ is a complex number, then the **conjugate** of c is the complex number $\overline{c} = a - bi$. It is not difficult to show that if c and d are complex numbers, then the following basic properties of complex arithmetic hold:

1. $\overline{\overline{c}} = c$.
2. $\overline{c + d} = \overline{c} + \overline{d}$.
3. $\overline{cd} = \overline{c}\,\overline{d}$.
4. c is a real number if and only if $c = \overline{c}$.
5. $c\,\overline{c}$ is a nonnegative real number and $c\,\overline{c} = 0$ if and only if $c = 0$.

We prove property 4 here and leave the others as exercises. Let $c = a + bi$ so that $\overline{c} = a - bi$. If $c = \overline{c}$, then $a + bi = a - bi$, so $b = 0$ and c is real. On the other hand, if c is real, then $c = a$ and $\overline{c} = a$, so $c = \overline{c}$.

EXAMPLE 2 Let $c_1 = 5 - 3i$, $c_2 = 4 + 2i$, and $c_3 = -3 + i$.

(a) $c_1 + c_2 = (5 - 3i) + (4 + 2i) = 9 - i$

(b) $c_2 - c_3 = (4 + 2i) - (-3 + i) = (4 - (-3)) + (2 - 1)i = 7 + i$

(c) $c_1 c_2 = (5 - 3i) \cdot (4 + 2i) = 20 + 10i - 12i - 6i^2 = 26 - 2i$

(d) $c_1 \overline{c}_3 = (5 - 3i) \cdot \overline{(-3 + i)} = (5 - 3i) \cdot (-3 - i) = -15 - 5i + 9i + 3i^2$
$= -18 + 4i$

(e) $3c_1 + 2\overline{c}_2 = 3(5 - 3i) + 2\overline{(4 + 2i)} = (15 - 9i) + 2(4 - 2i)$
$= (15 - 9i) + (8 - 4i) = 23 - 13i$

(f) $c_1 \overline{c}_1 = (5 - 3i)\overline{(5 - 3i)} = (5 - 3i)(5 + 3i) = 34$ ■

When we consider systems of linear equations with complex coefficients, we will need to divide complex numbers to complete the solution process and obtain a reasonable form for the solution. Let $c_1 = a_1 + b_1 i$ and $c_2 = a_2 + b_2 i$. If $c_2 \neq 0$, that is, if $a_2 \neq 0$ or $b_2 \neq 0$, then we can **divide** c_1 by c_2:

$$\frac{c_1}{c_2} = \frac{a_1 + b_1 i}{a_2 + b_2 i}.$$

To conform to our practice of expressing a complex number in the form real part + imaginary part $\cdot\, i$, we must simplify the foregoing expression for c_1/c_2.

To simplify this complex fraction, we multiply the numerator and the denominator by the conjugate of the denominator. Thus, dividing c_1 by c_2 gives the complex number

$$\frac{c_1}{c_2} = \frac{a_1 + b_1 i}{a_2 + b_2 i} = \frac{(a_1 + b_1 i)(a_2 - b_2 i)}{(a_2 + b_2 i)(a_2 - b_2 i)} = \frac{a_1 a_2 + b_1 b_2}{a_2^2 + b_2^2} - \frac{a_1 b_2 + a_2 b_1}{a_2^2 + b_2^2} i.$$

EXAMPLE 3 Let $c_1 = 2 - 5i$ and $c_2 = -3 + 4i$. Then

$$\frac{c_1}{c_2} = \frac{2 - 5i}{-3 + 4i} = \frac{(2 - 5i)(-3 - 4i)}{(-3 + 4i)(-3 - 4i)} = \frac{-26 + 7i}{(-3)^2 + (4)^2} = -\frac{26}{25} + \frac{7}{25} i. \quad \blacksquare$$

Finding the reciprocal of a complex number is a special case of division of complex numbers. If $c = a + bi$, $c \neq 0$, then

$$\frac{1}{c} = \frac{1}{a + bi} = \frac{a - bi}{(a + bi)(a - bi)} = \frac{a - bi}{a^2 + b^2}$$
$$= \frac{a}{a^2 + b^2} - \frac{b}{a^2 + b^2} i.$$

EXAMPLE 4

(a) $\dfrac{1}{2 + 3i} = \dfrac{2 - 3i}{(2 + 3i)(2 - 3i)} = \dfrac{2 - 3i}{2^2 + 3^2} = \dfrac{2}{13} - \dfrac{3}{13} i$

(b) $\dfrac{1}{i} = \dfrac{-i}{i(-i)} = \dfrac{-i}{-i^2} = \dfrac{-i}{-(-1)} = -i$ $\blacksquare$

Summarizing, we can say that complex numbers are mathematical objects for which addition, subtraction, multiplication, and division are defined in such a way that these operations on real numbers can be derived as special cases. In fact, it can be shown that complex numbers form a mathematical system that is called a field.

GEOMETRIC REPRESENTATION OF COMPLEX NUMBERS

A complex number $c = a + bi$ may be regarded as an ordered pair (a, b) of real numbers. This ordered pair of real numbers corresponds to a point in the plane. Such a correspondence naturally suggests that we represent $a + bi$ as a point in the **complex plane**, where the horizontal axis is used to represent the real part of c and the vertical axis is used to represent the imaginary part of c. To simplify matters, we call these the **real axis** and **imaginary axis**, respectively (see Figure A.1).

EXAMPLE 5 Plot the complex numbers $c = 2 - 3i$, $d = 1 + 4i$, $e = -3$, and $f = 2i$ in the complex plane.

Solution See Figure A.2. $\blacksquare$

The rules concerning inequality of real numbers, such as less than and greater than, *do not apply to complex numbers.* There is no way to arrange the complex numbers according to size. However, using the geometric representation from the complex plane, we can attach a notion of size to a complex number by measuring its distance from the origin. The distance from the origin to $c = a + bi$ is called the **absolute value** or **modulus** of the complex

Figure A.1 ►
Complex plane

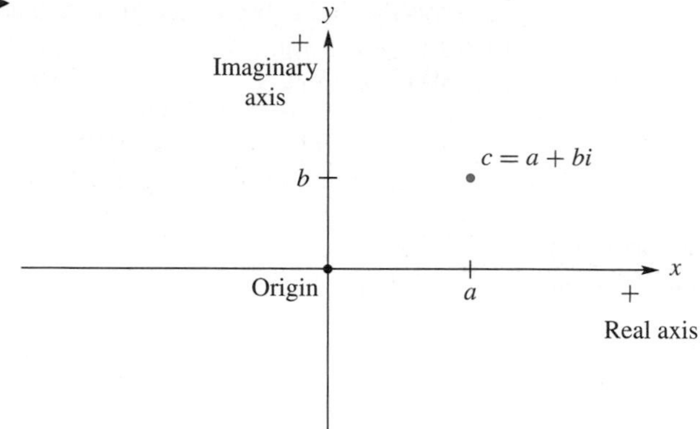

Figure A.2 ►

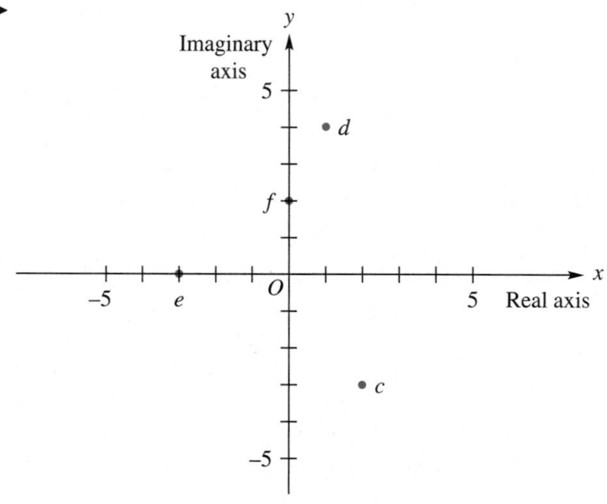

number and is denoted by $|c| = |a + bi|$. Using the formula for the distance between ordered pairs of real numbers, we obtain

$$|c| = |a + bi| = \sqrt{a^2 + b^2}.$$

It follows that $c\,\overline{c} = |c|^2$ (verify).

EXAMPLE 6 Referring to Example 5: $|c| = \sqrt{13}$; $|d| = \sqrt{17}$; $|e| = 3$; $|f| = 2$. ■

A different, but related, interpretation of a complex number is obtained if we associate with $c = a + bi$ the vector OP, where O is the origin $(0, 0)$ and P is the point (a, b). There is an obvious correspondence between this representation and vectors in the plane discussed in calculus, which we reviewed in Section 4.1. Using a vector representation, addition and subtraction of complex numbers can be viewed as the corresponding vector operations. These are represented in Figures 4.14, 4.15, and 4.19. We will not pursue the manipulation of complex numbers by vector operations here, but such a point of view is important for the development and study of complex variables.

MATRICES WITH COMPLEX ENTRIES

If the entries of a matrix are complex numbers, we can perform the matrix operations of addition, subtraction, multiplication, and scalar multiplication in a manner completely analogous to that for real matrices. The validity of these operations can be verified using properties of complex arithmetic, just imitating the proofs for real matrices presented in the text. We illustrate these concepts in the following example.

EXAMPLE 7 Let

$$A = \begin{bmatrix} 4+i & -2+3i \\ 6+4i & -3i \end{bmatrix}, \quad B = \begin{bmatrix} 2-i & 3-4i \\ 5+2i & -7+5i \end{bmatrix},$$

$$C = \begin{bmatrix} 1+2i & i \\ 3-i & 8 \\ 4+2i & 1-i \end{bmatrix}.$$

(a) $$A + B = \begin{bmatrix} (4+i)+(2-i) & (-2+3i)+(3-4i) \\ (6+4i)+(5+2i) & (-3i)+(-7+5i) \end{bmatrix} = \begin{bmatrix} 6 & 1-i \\ 11+6i & -7+2i \end{bmatrix}$$

(b) $$B - A = \begin{bmatrix} (2-i)-(4+i) & (3-4i)-(-2+3i) \\ (5+2i)-(6+4i) & (-7+5i)-(-3i) \end{bmatrix} = \begin{bmatrix} -2-2i & 5-7i \\ -1-2i & -7+8i \end{bmatrix}$$

(c) $$CA = \begin{bmatrix} 1+2i & i \\ 3-i & 8 \\ 4+2i & 1-i \end{bmatrix} \begin{bmatrix} 4+i & -2+3i \\ 6+4i & -3i \end{bmatrix}$$

$$= \begin{bmatrix} (1+2i)(4+i)+(i)(6+4i) & (1+2i)(-2+3i)+(i)(-3i) \\ (3-i)(4+i)+(8)(6+4i) & (3-i)(-2+3i)+(8)(-3i) \\ (4+2i)(4+i)+(1-i)(6+4i) & (4+2i)(-2+3i)+(1-i)(-3i) \end{bmatrix}$$

$$= \begin{bmatrix} -2+15i & -5-i \\ 61+31i & -3-13i \\ 24+10i & -17+5i \end{bmatrix}$$

(d) $$(2+i)B = \begin{bmatrix} (2+i)(2-i) & (2+i)(3-4i) \\ (2+i)(5+2i) & (2+i)(-7+5i) \end{bmatrix} = \begin{bmatrix} 5 & 10-5i \\ 8+9i & -19+3i \end{bmatrix}$$ ■

Just as we can compute the conjugate of a complex number, we can compute the **conjugate of a matrix** by computing the conjugate of each entry of the matrix. We denote the conjugate of a matrix A by $\overline{A}$, and we write

$$\overline{A} = \left[\overline{a_{kj}}\right].$$

EXAMPLE 8 Referring to Example 7, we find that

$$\overline{A} = \begin{bmatrix} 4-i & -2-3i \\ 6-4i & 3i \end{bmatrix} \quad \text{and} \quad \overline{B} = \begin{bmatrix} 2+i & 3+4i \\ 5-2i & -7-5i \end{bmatrix}.$$ ■

The following properties of the conjugate of a matrix hold:

1. $\overline{\overline{A}} = A$.
2. $\overline{A + B} = \overline{A} + \overline{B}$.
3. $\overline{AB} = \overline{A}\,\overline{B}$.
4. For any real number k, $\overline{kA} = k\,\overline{A}$.
5. For any complex number c, $\overline{cA} = \overline{c}\,\overline{A}$.
6. $(\overline{A})^T = \overline{A^T}$.
7. If A is nonsingular, then $(\overline{A})^{-1} = \overline{A^{-1}}$.

We prove properties 5 and 6 here and leave the others as exercises. First property 5: If c is complex, the (k, j) entry of $\overline{cA}$ is

$$\overline{ca_{kj}} = \overline{c}\,\overline{a_{kj}},$$

which is the (k, j) entry of $\overline{c}\,\overline{A}$. Next, property 6: The (k, j) entry of $(\overline{A})^T$ is $\overline{a_{jk}}$, which is the (k, j) entry of $\overline{A^T}$.

SPECIAL TYPES OF COMPLEX MATRICES

As we have already seen, certain types of real matrices satisfy some important properties. The same situation applies to complex matrices and we now discuss several of these types of matrices.

An $n \times n$ complex matrix A is called **Hermitian*** if

$$\overline{A^T} = A.$$

This is equivalent to saying that $\overline{a_{jk}} = a_{kj}$ for all k and j. Every real symmetric matrix is Hermitian [Exercise T.3(c)], so we may consider Hermitian matrices as the analogs of real symmetric matrices.

EXAMPLE 9 The matrix

$$A = \begin{bmatrix} 2 & 3+i \\ 3-i & 5 \end{bmatrix}$$

is Hermitian, since

$$\overline{A^T} = \overline{\begin{bmatrix} 2 & 3-i \\ 3+i & 5 \end{bmatrix}} = \begin{bmatrix} 2 & 3+i \\ 3-i & 5 \end{bmatrix} = A.$$

■

An $n \times n$ complex matrix A is called **unitary** if

$$(\overline{A^T})A = A(\overline{A^T}) = I_n.$$

This is equivalent to saying that $\overline{A^T} = A^{-1}$. Every real orthogonal matrix is unitary [Exercise T.4(a)], so we may consider unitary matrices as the analogs of real orthogonal matrices.

---

*Charles Hermite (1822–1901) was born to a well-to-do merchant family in Lorraine, France, and died in Paris. He was considered one of the greatest algebraists of the nineteenth century. He studied at the École Polytechnique, where he barely passed the entrance examination. His first academic appointment was at the École Polytechnique, then at the École Normale, and then in 1870 he was named to a professorship at the Sorbonne, where he remained for 27 years until his retirement. Two of Hermite's most notable mathematical accomplishments were the proof that the number e is a transcendental number, that is, it is not the root of any polynomial equation with integer coefficients, and a method of solving a fifth-degree polynomial equation.

EXAMPLE 10 The matrix

$$A = \begin{bmatrix} \frac{1}{\sqrt{3}} & \frac{1+i}{\sqrt{3}} \\ \frac{1-i}{\sqrt{3}} & -\frac{1}{\sqrt{3}} \end{bmatrix}$$

is unitary, since (verify)

$$(\overline{A^T})A = \begin{bmatrix} \frac{1}{\sqrt{3}} & \frac{1+i}{\sqrt{3}} \\ \frac{1-i}{\sqrt{3}} & -\frac{1}{\sqrt{3}} \end{bmatrix} \begin{bmatrix} \frac{1}{\sqrt{3}} & \frac{1+i}{\sqrt{3}} \\ \frac{1-i}{\sqrt{3}} & -\frac{1}{\sqrt{3}} \end{bmatrix} = I_2$$

and similarly, $A(\overline{A^T}) = I_2$. ■

There is one more type of complex matrix that is important. An $n \times n$ complex matrix is called **normal** if

$$(\overline{A^T})\,A = A\,(\overline{A^T}).$$

EXAMPLE 11 The matrix

$$A = \begin{bmatrix} 5-i & -1+i \\ -1-i & 3-i \end{bmatrix}$$

is normal, since (verify)

$$(\overline{A^T})\,A = A\,(\overline{A^T}) = \begin{bmatrix} 28 & -8+8i \\ -8-8i & 12 \end{bmatrix}.$$

Moreover, A is not Hermitian, since $\overline{A^T} \neq A$ (verify). ■

COMPLEX NUMBERS AND ROOTS OF POLYNOMIALS

A polynomial of degree n with real coefficients has n complex roots, some, all, or none of which may be real numbers. Thus the polynomial $f_1(x) = x^4 - 1$ has the roots i, $-i$, 1, and -1; the polynomial $f_2(x) = x^2 - 1$ has the roots 1 and -1; and the polynomial $f_3(x) = x^2 + 1$ has the roots i and $-i$.

Key Terms

Complex number
Real part
Imaginary part
Equal complex numbers
Sum of complex numbers
Difference of complex numbers
Product of complex numbers
Conjugate
Division of complex numbers
Complex plane
Real axis
Imaginary axis
Absolute value (or modulus)
Hermitian matrix
Unitary matrix
Normal matrix
Matrix polynomial

A.1 Exercises

1. Let $c_1 = 3 + 4i$, $c_2 = 1 - 2i$, and $c_3 = -1 + i$. Compute each of the following and simplify as much as possible.

(a) $c_1 + c_2$
(b) $c_3 - c_1$
(c) $c_1 c_2$
(d) $c_2 \overline{c_3}$
(e) $4c_3 + \overline{c_2}$
(f) $(-i) \cdot c_2$
(g) $\overline{3c_1 - ic_2}$
(h) $c_1 c_2 c_3$

2. Write in the form $a + bi$.

(a) $\dfrac{1+2i}{3-4i}$ (b) $\dfrac{2-3i}{3-i}$

(c) $\dfrac{(2+i)^2}{i}$ (d) $\dfrac{1}{(3+2i)(1+i)}$

3. Represent each complex number as a point and as a vector in the complex plane.

(a) $4+2i$ (b) $-3+i$

(c) $3-2i$ (d) $i(4+i)$

4. Find the modulus of each complex number in Exercise 3.

5. In the complex plane sketch the vectors corresponding to c and $\overline{c}$ for $c = 2+3i$ and $c = -1+4i$. Geometrically, we can say that $\overline{c}$ is the reflection of c with respect to the real axis. (See also Example 2 in Section 2.3.)

6. Let

$$A = \begin{bmatrix} 2+2i & -1+3i \\ -2 & 1-i \end{bmatrix},$$

$$B = \begin{bmatrix} 2i & 1+2i \\ 0 & 3-i \end{bmatrix}, \qquad C = \begin{bmatrix} 2+i \\ -i \end{bmatrix}.$$

Compute each of the following and simplify each entry as $a + bi$.

(a) $A + B$ (b) $(1-2i)C$ (c) AB

(d) BC (e) $A - 2I_2$ (f) $\overline{B}$

(g) $A\overline{C}$ (h) $(\overline{A+B})C$

7. If

$$A = \begin{bmatrix} 0 & i \\ i & 0 \end{bmatrix},$$

compute A^2, A^3, and A^4. Give a general rule for A^n, n a positive integer.

8. Which of the following matrices are Hermitian, unitary, or normal?

(a) $\begin{bmatrix} 3 & 2+i \\ 2-i & 4 \end{bmatrix}$ (b) $\begin{bmatrix} 2 & 1-i \\ 3+i & -2 \end{bmatrix}$

(c) $\begin{bmatrix} \dfrac{1-i}{2} & \dfrac{1+i}{2} \\ \dfrac{1+i}{2} & \dfrac{1-i}{2} \end{bmatrix}$ (d) $\begin{bmatrix} 1 & -1 \\ 1 & 1 \end{bmatrix}$

(e) $\begin{bmatrix} 1 & 3-i & 4-i \\ 3+i & -2 & 2+i \\ 4+i & 2-i & 3 \end{bmatrix}$

(f) $\begin{bmatrix} 3 & \dfrac{3-i}{2} & \dfrac{4-i}{2} \\ \dfrac{3-i}{2} & -2 & 2+i \\ \dfrac{4-i}{2} & 2-i & 5 \end{bmatrix}$

(g) $\begin{bmatrix} 3+2i & -1 \\ -i & 2+i \end{bmatrix}$ (h) $\begin{bmatrix} i & i \\ -i & 1 \end{bmatrix}$

(i) $\begin{bmatrix} 1 & 0 & 0 \\ 0 & \dfrac{1+i}{\sqrt{3}} & \dfrac{1}{\sqrt{3}} \\ 0 & -\dfrac{1}{\sqrt{3}} & \dfrac{1-i}{\sqrt{3}} \end{bmatrix}$ (j) $\begin{bmatrix} 4+7i & -2-i \\ 1-2i & 3+4i \end{bmatrix}$

9. Find all the roots.

(a) $x^2 + x + 1 = 0$ (b) $x^3 + 2x^2 + x + 2 = 0$

(c) $x^5 + x^4 - x - 1 = 0$

10. Let $p(x)$ denote a polynomial and let A be a square matrix. Then $p(A)$ is called a **matrix polynomial** or a **polynomial in the matrix** A. For $p(x) = 2x^2 + 5x - 3$, compute $p(A) = 2A^2 + 5A - 3I_n$ for each of the following.

(a) $A = \begin{bmatrix} -3 & 0 \\ 0 & -3 \end{bmatrix}$ (b) $A = \begin{bmatrix} 1 & 2 \\ 0 & 1 \end{bmatrix}$

(c) $A = \begin{bmatrix} 0 & i \\ i & 0 \end{bmatrix}$ (d) $A = \begin{bmatrix} 1 & i \\ 0 & 0 \end{bmatrix}$

11. Let $p(x) = x^2 + 1$.

(a) Determine two different 2×2 matrices A of the form kI_2 that satisfy $p(A) = O_2$.

(b) Verify that $p(A) = O_2$, for $A = \begin{bmatrix} 1 & 2 \\ -1 & -1 \end{bmatrix}$.

12. Find all the 2×2 matrices A of the form kI_2 that satisfy $p(A) = O_2$ for $p(x) = x^2 - x - 2$.

13. In Supplementary Exercise 29 in Chapter 1, we introduced the concept of a square root of a matrix with real entries. We can generalize the notion of a square root of a matrix if we permit complex entries.

(a) Compute a complex square root of

$$A = \begin{bmatrix} -1 & 0 \\ 0 & 0 \end{bmatrix}.$$

(b) Compute a complex square root of

$$A = \begin{bmatrix} -2 & 2 \\ 2 & -2 \end{bmatrix}.$$

Theoretical Exercises

T.1. If $c = a + bi$, then we can denote the real part of c by $\text{Re}(c)$ and the imaginary part of c by $\text{Im}(c)$.

(a) For any complex numbers $c_1 = a_1 + b_1 i$ and $c_2 = a_2 + b_2 i$, show that $\text{Re}(c_1 + c_2) = \text{Re}(c_1) + \text{Re}(c_2)$ and $\text{Im}(c_1 + c_2) = \text{Im}(c_1) + \text{Im}(c_2)$.

(b) For any real number k, show that $\text{Re}(kc) = k\,\text{Re}(c)$ and $\text{Im}(kc) = k\,\text{Im}(c)$.

(c) Is part (b) true if k is a complex number?

(d) Prove or disprove:

$$\text{Re}(c_1 c_2) = \text{Re}(c_1) \cdot \text{Re}(c_2).$$

T.2. Let A and B be $m \times n$ complex matrices, and let C be an $n \times n$ nonsingular matrix.

(a) Show that $\overline{A + B} = \overline{A} + \overline{B}$.

(b) Show that for any real number k, $\overline{kA} = k\overline{A}$.

(c) Show that $(\overline{C})^{-1} = \overline{C^{-1}}$.

T.3. (a) Show that the diagonal entries of a Hermitian matrix must be real.

(b) Show that every Hermitian matrix A can be written as $A = B + iC$, where B is real and symmetric and C is real and skew symmetric (see Exercise T.24 in Section 1.4). [*Hint*: Consider $B = (A+\overline{A})/2$ and $C = (A-\overline{A})/2i$.]

(c) Show that every real symmetric matrix is Hermitian.

T.4. (a) Show that every real orthogonal matrix is unitary.

(b) Show that if A is a unitary matrix, then A^T is unitary.

(c) Show that if A is a unitary matrix, then A^{-1} is unitary.

T.5. Let A be an $n \times n$ complex matrix.

(a) Show that A can be written as $B + iC$, where B and C are Hermitian.

(b) Show that A is normal if and only if

$$BC = CB.$$

[*Hint*: Consider $B = (A + \overline{A^T})/2$ and $C = (A - \overline{A^T})/2i$.]

T.6. (a) Show that any Hermitian matrix is normal.

(b) Show that any unitary matrix is normal.

(c) Find a 2×2 normal matrix that is neither Hermitian nor unitary.

T.7. An $n \times n$ complex matrix A is called **skew Hermitian** if

$$\overline{A^T} = -A.$$

Show that a matrix $A = B + iC$, where B and C are real matrices, is skew Hermitian if and only if B is skew symmetric and C is symmetric.

MATLAB Exercises

MATLAB does complex arithmetic automatically. To enter a complex number into MATLAB, first assign the complex unit $\sqrt{-1}$ to a variable name i with command

i = sqrt(−1)

Then to store $3 - 5i$ into variable v, type

v = 3 − 5i

If we enter a second complex number $-2 + 7i$ into w with command

w = − 2 + 7i

we can add, subtract, multiply, and divide v and w using the standard arithmetic symbols. Command

conj(v)

displays the conjugate of v, while **real(v)** *and* **imag(v)** *display the real and imaginary parts of v, respectively. We define a complex matrix just by entering its elements as complex numbers, for example,*

A = [2 − i 3 + 5i;6 − 2i]

Do not leave any extra spaces in complex numbers, otherwise MATLAB takes it as two separate numbers. The commands **conj**, **real**, *and* **imag** *apply to matrices as well.* **A′** *gives the conjugate transpose of matrix A.*

ML.1. For Exercise 1, enter the three values into MATLAB as c1, c2, and c3, no subscripts, and compute the values in parts (a)–(h).

ML.2. Do Exercise 6 in MATLAB.

A.2 COMPLEX NUMBERS IN LINEAR ALGEBRA

The primary goal of this appendix is to provide an easy transition to complex numbers in linear algebra. This is of particular importance in Chapter 8, where complex eigenvalues and eigenvectors arise naturally for matrices with real entries. Hence we only restate the main theorems in the complex case and provide a discussion and examples of the major ideas needed to accomplish this transition. It will soon be evident that the increased computational effort of complex arithmetic becomes quite tedious if done by hand. However, such calculations can be easily done by computers.

SOLVING LINEAR SYSTEMS WITH COMPLEX ENTRIES

The results and techniques dealing with the solution of linear systems that we developed in Chapter 1 carry over directly to linear systems with complex coefficients. We shall illustrate row operations and echelon forms for such systems with Gauss–Jordan reduction using complex arithmetic.

EXAMPLE 1 Solve the following linear system by Gauss–Jordan reduction:

$$\begin{aligned}(1+i)x_1 + (2+i)x_2 &= 5\\ (2-2i)x_1 + \qquad ix_2 &= 1+2i.\end{aligned}$$

Solution We form the augmented matrix and use elementary row operations to transform it to reduced row echelon form. For the augmented matrix $\left[A \mid B\right]$,

$$\left[\begin{array}{cc|c} 1+i & 2+i & 5 \\ 2-2i & i & 1+2i \end{array}\right],$$

multiply the first row by $\dfrac{1}{1+i}$ to obtain

$$\left[\begin{array}{cc|c} 1 & \frac{3}{2}-\frac{1}{2}i & \frac{5}{2}-\frac{5}{2}i \\ 2-2i & i & 1+2i \end{array}\right].$$

We now add $[-(2-2i)]$ times the first row to the second row to get

$$\left[\begin{array}{cc|c} 1 & \frac{3}{2}-\frac{1}{2}i & \frac{5}{2}-\frac{5}{2}i \\ 0 & -2+5i & 1+12i \end{array}\right].$$

Multiply the second row by $\dfrac{1}{-2+5i}$ to obtain

$$\left[\begin{array}{cc|c} 1 & \frac{3}{2}-\frac{1}{2}i & \frac{5}{2}-\frac{5}{2}i \\ 0 & 1 & 2-i \end{array}\right],$$

which is in row echelon form. To get to reduced row echelon form, we add $-\left(\frac{3}{2}-\frac{1}{2}i\right)$ times the second row to the first row to obtain

$$\left[\begin{array}{cc|c} 1 & 0 & 0 \\ 0 & 1 & 2-i \end{array}\right].$$

Hence the solution is $x_1 = 0$ and $x_2 = 2 - i$. ■

If you carry out the arithmetic for the row operations in the preceding example, you will feel the burden of the complex arithmetic even though there were just two equations in two unknowns.

Gaussian elimination with back substitution can also be used on linear systems with complex coefficients.

EXAMPLE 2 Suppose that the augmented matrix of a linear system has been transformed to the following matrix in row echelon form:

$$\left[\begin{array}{ccc|c} 1 & 0 & 1+i & -1 \\ 0 & 1 & 3i & 2+i \\ 0 & 0 & 1 & 2i \end{array}\right].$$

The back-substitution procedure gives us

$$\begin{aligned} x_3 &= 2i \\ x_2 &= 2+i-3i(2i) = 2+i+6 = 8+i \\ x_1 &= -1-(1+i)(2i) = -1-2i+2 = 3-2i. \end{aligned}$$

■

We can alleviate the tediousness of complex arithmetic by using computers to solve linear systems with complex entries. However, we must still pay a high price because the execution time will be approximately twice as long as that for the same size linear system with all real entries. We can illustrate this by showing how to transform an $n \times n$ linear system with complex coefficients to a $2n \times 2n$ linear system with only real coefficients.

EXAMPLE 3 Consider the linear system

$$\begin{aligned} (2+i)x_1 + \ (1+i)x_2 &= 3+6i \\ (3-i)x_1 + (2-2i)x_2 &= 7-i. \end{aligned}$$

If we let $x_1 = a_1 + b_1 i$ and $x_2 = a_2 + b_2 i$, with a_1, b_1, a_2, and b_2 real numbers, then we can write this system in matrix form as

$$\begin{bmatrix} 2+i & 1+i \\ 3-i & 2-2i \end{bmatrix} \begin{bmatrix} a_1 + b_1 i \\ a_2 + b_2 i \end{bmatrix} = \begin{bmatrix} 3+6i \\ 7-i \end{bmatrix}.$$

We first rewrite the given linear system as

$$\left(\begin{bmatrix} 2 & 1 \\ 3 & 2 \end{bmatrix} + i \begin{bmatrix} 1 & 1 \\ -1 & -2 \end{bmatrix} \right) \left(\begin{bmatrix} a_1 \\ a_2 \end{bmatrix} + i \begin{bmatrix} b_1 \\ b_2 \end{bmatrix} \right) = \begin{bmatrix} 3 \\ 7 \end{bmatrix} + i \begin{bmatrix} 6 \\ -1 \end{bmatrix}.$$

Multiplying, we have

$$\left(\begin{bmatrix} 2 & 1 \\ 3 & 2 \end{bmatrix} \begin{bmatrix} a_1 \\ a_2 \end{bmatrix} - \begin{bmatrix} 1 & 1 \\ -1 & -2 \end{bmatrix} \begin{bmatrix} b_1 \\ b_2 \end{bmatrix} \right) + i \left(\begin{bmatrix} 2 & 1 \\ 3 & 2 \end{bmatrix} \begin{bmatrix} b_1 \\ b_2 \end{bmatrix} + \begin{bmatrix} 1 & 1 \\ -1 & -2 \end{bmatrix} \begin{bmatrix} a_1 \\ a_2 \end{bmatrix} \right) = \begin{bmatrix} 3 \\ 7 \end{bmatrix} + i \begin{bmatrix} 6 \\ -1 \end{bmatrix}.$$

The real and imaginary parts on both sides of the equation must agree, respectively, and so we have

$$\begin{bmatrix} 2 & 1 \\ 3 & 2 \end{bmatrix} \begin{bmatrix} a_1 \\ a_2 \end{bmatrix} - \begin{bmatrix} 1 & 1 \\ -1 & -2 \end{bmatrix} \begin{bmatrix} b_1 \\ b_2 \end{bmatrix} = \begin{bmatrix} 3 \\ 7 \end{bmatrix}$$

and

$$\begin{bmatrix} 2 & 1 \\ 3 & 2 \end{bmatrix} \begin{bmatrix} b_1 \\ b_2 \end{bmatrix} + \begin{bmatrix} 1 & 1 \\ -1 & -2 \end{bmatrix} \begin{bmatrix} a_1 \\ a_2 \end{bmatrix} = \begin{bmatrix} 6 \\ -1 \end{bmatrix}.$$

This leads to the linear system

$$\begin{aligned} 2a_1 + a_2 - b_1 - b_2 &= 3 \\ 3a_1 + 2a_2 + b_1 + 2b_2 &= 7 \\ a_1 + a_2 + 2b_1 + b_2 &= 6 \\ -a_1 - 2a_2 + 3b_1 + 2b_2 &= -1, \end{aligned}$$

which can be written as

$$\begin{bmatrix} 2 & 1 & -1 & -1 \\ 3 & 2 & 1 & 2 \\ 1 & 1 & 2 & 1 \\ -1 & -2 & 3 & 2 \end{bmatrix} \begin{bmatrix} a_1 \\ a_2 \\ b_1 \\ b_2 \end{bmatrix} = \begin{bmatrix} 3 \\ 7 \\ 6 \\ -1 \end{bmatrix}.$$

This linear system of four equations in four unknowns is now solved as in Chapter 1. The solution is (verify) $a_1 = 1$, $a_2 = 2$, $b_1 = 2$, and $b_2 = -1$. Thus $x_1 = 1 + 2i$ and $x_2 = 2 - i$ is the solution to the given linear system. ■

DETERMINANTS OF COMPLEX MATRICES

The definition of a determinant and all the properties derived in Chapter 3 apply to matrices with complex entries. The following example is an illustration.

EXAMPLE 4 Let A be the coefficient matrix of Example 3. Compute $|A|$.

Solution

$$\begin{aligned} \begin{vmatrix} 2+i & 1+i \\ 3-i & 2-2i \end{vmatrix} &= (2+i)(2-2i) - (3-i)(1+i) \\ &= (6-2i) - (4+2i) \\ &= 2-4i \end{aligned}$$

■

COMPLEX VECTOR SPACES

A **complex vector space** is defined exactly as was a real vector space in Definition 1 in Section 6.1, except that the scalars in properties (e) through (h) are permitted to be complex numbers. The terms *complex* vector space and *real* vector space emphasize the set from which the scalars are chosen. It happens that in order to satisfy the closure property of scalar multiplication [Definition 1(β) in Section 6.1] in a complex vector space, we must, in most examples, consider vectors that involve complex numbers.

Most of the real vector spaces of Chapter 6 have complex vector space analogs.

EXAMPLE 5 (a) Consider C^n, the set of all $n \times 1$ matrices

$$\begin{bmatrix} a_1 \\ a_2 \\ \vdots \\ a_n \end{bmatrix}$$

with complex entries. Let the operation $\oplus$ be matrix addition and let the operation $\odot$ be multiplication of a matrix by a complex number. We can verify that C^n is a complex vector space by using the properties of matrices established in Section 1.4 and the properties of complex arithmetic established in Section A.1. (Note that if the operation $\odot$ is taken as multiplication of a matrix by a real number, then C^n is a real vector space whose vectors have complex components.)

(b) The set of all $m \times n$ matrices, with complex entries with matrix addition as $\oplus$ and multiplication of a matrix by a complex number as $\odot$, is a complex vector space (verify). We denote this vector space by C_{mn}.

(c) The set of polynomials, with complex coefficients with polynomial addition as $\oplus$ and multiplication of a polynomial by a complex constant as $\odot$, forms a complex vector space. Verification follows the pattern of Example 8 in Section 6.1.

(d) The set of complex-valued functions that are continuous on the interval $[a, b]$ (that is, all functions of the form $f(t) = f_1(t) + if_2(t)$, where f_1 and f_2 are real-valued functions that are continuous on $[a, b]$), with $\oplus$ defined by $(f \oplus g)(t) = f(t) + g(t)$ and $\odot$ defined by $(c \odot f)(t) = cf(t)$ for a complex scalar c, forms a complex vector space. The corresponding real vector space is given in Example 5 in Section 6.1 for the interval $(-\infty, \infty)$. ■

A **complex vector subspace** W of a complex vector space V is defined as in Section 6.1, but with real scalars replaced by complex ones. The analog of Theorem 6.2 can be proved to show that a nonempty subset W of a complex vector space V is a complex vector subspace if and only if the following conditions hold:

(a) If $\mathbf{u}$ and $\mathbf{v}$ are any vectors in W, then $\mathbf{u} \oplus \mathbf{v}$ is in W.

(b) If c is any complex number and $\mathbf{u}$ is any vector in W, then $c \odot \mathbf{u}$ is in W.

EXAMPLE 6

(a) Let W be the set of all vectors in C^3 of the form $(a, 0, b)$, where a and b are complex numbers. It follows that

$$(a, 0, b) \oplus (d, 0, e) = (a + d, 0, b + e)$$

belongs to W and for any complex scalar c,

$$c \odot (a, 0, b) = (ca, 0, cb)$$

belongs to W. Hence W is a complex vector subspace of C^3.

(b) Let W be the set of all vectors in C_{mn} having only real entries. If $A = \left[a_{ij}\right]$ and $B = \left[b_{kj}\right]$ belong to W, then so will $A \oplus B$, because if a_{kj} and b_{kj} are real, then so is their sum. However, if c is any complex scalar and A belongs to W, then $c \odot A = cA$ can have entries ca_{kj} that need not be real numbers. It follows that $c \odot A$ need not belong to W, so W is not a complex vector subspace. ■

LINEAR INDEPENDENCE AND BASIS IN COMPLEX VECTOR SPACES

The notions of linear combinations, spanning sets, linear dependence, linear independence, and basis are unchanged for complex vector spaces, except that we use complex scalars (see Sections 6.2, 6.3, and 6.4).

EXAMPLE 7

Let V be the complex vector space C^3. Let

$$\mathbf{v}_1 = (1, i, 0), \quad \mathbf{v}_2 = (i, 0, 1 + i), \quad \text{and} \quad \mathbf{v}_3 = (1, 1, 1).$$

(a) Determine whether $\mathbf{v} = (-1, -3 + 3i, -4 + i)$ is a linear combination of $\{\mathbf{v}_1, \mathbf{v}_2, \mathbf{v}_3\}$.
(b) Determine whether $\{\mathbf{v}_1, \mathbf{v}_2, \mathbf{v}_3\}$ spans C^3.
(c) Determine whether $\{\mathbf{v}_1, \mathbf{v}_2, \mathbf{v}_3\}$ is a linearly independent subset of C^3.
(d) Is $\{\mathbf{v}_1, \mathbf{v}_2, \mathbf{v}_3\}$ a basis for C^3?

Solution (a) We proceed as in Example 11 of Section 6.2. We form a linear combination of $\mathbf{v}_1$, $\mathbf{v}_2$, and $\mathbf{v}_3$ with unknown coefficients c_1, c_2, and c_3, respectively, and set it equal to $\mathbf{v}$:

$$c_1\mathbf{v}_1 + c_2\mathbf{v}_2 + c_3\mathbf{v}_3 = \mathbf{v}.$$

If we substitute the vectors $\mathbf{v}_1$, $\mathbf{v}_2$, $\mathbf{v}_3$, and $\mathbf{v}$ into this expression, we obtain (verify) the linear system

$$\begin{aligned} c_1 + \quad\quad ic_2 + c_3 &= -1 \\ ic_1 \quad\quad\quad\quad + c_3 &= -3 + 3i \\ (1+i)c_2 + c_3 &= -4 + i. \end{aligned}$$

We next investigate the consistency of this linear system by using elementary row operations to transform its augmented matrix to either row echelon or reduced row echelon form. A row echelon form is (verify)

$$\left[\begin{array}{ccc|c} 1 & i & 1 & -1 \\ 0 & 1 & 1-i & -3+4i \\ 0 & 0 & 1 & -3 \end{array}\right],$$

which implies that the system is consistent, hence $\mathbf{v}$ is a linear combination of $\mathbf{v}_1$, $\mathbf{v}_2$, $\mathbf{v}_3$. In fact, back substitution gives (verify) $c_1 = 3$, $c_2 = i$, and $c_3 = -3$.

(b) Let $\mathbf{v} = (a, b, c)$ be an arbitrary vector of C^3. We form the linear combination

$$c_1\mathbf{v}_1 + c_2\mathbf{v}_2 + c_3\mathbf{v}_3 = \mathbf{v}$$

and solve for c_1, c_2, and c_3. The resulting linear system is

$$\begin{aligned} c_1 + \quad\quad ic_2 + c_3 &= a \\ ic_1 \quad\quad\quad\quad + c_3 &= b \\ (1+i)c_2 + c_3 &= c. \end{aligned}$$

Transforming the augmented matrix to row echelon form, we obtain (verify)

$$\left[\begin{array}{ccc|c} 1 & i & 1 & a \\ 0 & 1 & 1-i & b - ia \\ 0 & 0 & 1 & -c + (1+i)(b - ia) \end{array}\right].$$

Hence we can solve for c_1, c_2, c_3 for any choice of complex numbers a, b, c, which implies that $\{\mathbf{v}_1, \mathbf{v}_2, \mathbf{v}_3\}$ spans C^3.

(c) Proceeding as in Example 8 of Section 6.3, we form the equation

$$c_1\mathbf{v}_1 + c_2\mathbf{v}_2 + c_3\mathbf{v}_3 = \mathbf{0}$$

and solve for c_1, c_2, and c_3. The resulting homogeneous system is

$$\begin{aligned} c_1 + \quad\quad ic_2 + c_3 &= 0 \\ ic_1 \quad\quad\quad\quad + c_3 &= 0 \\ (1+i)c_2 + c_3 &= 0. \end{aligned}$$

Transforming the augmented matrix to row echelon form, we obtain (verify)

$$\left[\begin{array}{ccc|c} 1 & i & 1 & 0 \\ 0 & 1 & 1-i & 0 \\ 0 & 0 & 1 & 0 \end{array}\right],$$

and hence the only solution is $c_1 = c_2 = c_3 = 0$, showing that $\{\mathbf{v}_1, \mathbf{v}_2, \mathbf{v}_3\}$ is linearly independent.

(d) Yes, because $\mathbf{v}_1$, $\mathbf{v}_2$, and $\mathbf{v}_3$ span C^3 [part (b)] and they are linearly independent [part (c)]. ■

Just as for real vector spaces, the questions of spanning sets, linearly independent or linearly dependent sets, and basis in a complex vector space are resolved by using an appropriate linear system. The definition of the dimension of a complex vector space is the same as that given in Section 6.4. In discussing the dimension of a complex vector space like C^n, we must adjust our intuitive picture. For example, C^1 consists of all complex multiples of a single nonzero vector. This collection can be put into one-to-one correspondence with the complex numbers themselves, that is, with all the points in the complex plane (see Figure A.1). Since the elements of a two-dimensional real vector space can be put into a one-to-one correspondence with the points of R^2 (see Section 4.1), we see that a complex vector space of dimension 1 has a geometric model that is in one-to-one correspondence with a geometric model of a two-dimensional real vector space. Similarly, a complex vector space of dimension 2 is the same, geometrically, as a four-dimensional real vector space.

If

$$\mathbf{u} = \begin{bmatrix} u_1 \\ u_2 \\ \vdots \\ u_n \end{bmatrix} \quad \text{and} \quad \mathbf{v} = \begin{bmatrix} v_1 \\ v_2 \\ \vdots \\ v_n \end{bmatrix}$$

are vectors in C^n, we define the dot product $\mathbf{u} \cdot \mathbf{v}$ as

$$\mathbf{u} \cdot \mathbf{v} = u_1\overline{v_1} + u_2\overline{v_2} + \cdots + u_n\overline{v_n},$$

which can also be expressed as

$$\mathbf{u} \cdot \mathbf{v} = \mathbf{u}^T\overline{\mathbf{v}}.$$

EXAMPLE 8 Let

$$\mathbf{u} = \begin{bmatrix} 1-i \\ 2 \\ -3+2i \end{bmatrix} \quad \text{and} \quad \mathbf{v} = \begin{bmatrix} 3+2i \\ 3-4i \\ -3i \end{bmatrix}.$$

Compute $\mathbf{u} \cdot \mathbf{v}$.

Solution We have

$$\begin{aligned} \mathbf{u} \cdot \mathbf{v} &= (1-i)\overline{(3+2i)} + 2\overline{(3-4i)} + (-3+2i)\overline{(-3i)} \\ &= (1-5i) + (6+8i) + (-6-9i) \\ &= 1-6i. \end{aligned}$$

■

We can also define the length of a vector $\mathbf{u}$ in C^n exactly as in the real case:

$$\|\mathbf{u}\| = \sqrt{\mathbf{u} \cdot \mathbf{u}}.$$

Moreover, the vectors $\mathbf{u}$ and $\mathbf{v}$ in C^n are said to be **orthogonal** if $\mathbf{u} \cdot \mathbf{v} = 0$.

EXAMPLE 9 Let

$$\mathbf{u} = \begin{bmatrix} 1+i \\ 2-i \\ 3+i \end{bmatrix} \quad \text{and} \quad \mathbf{v} = \begin{bmatrix} 6-2i \\ 2i \\ -1+i \end{bmatrix}.$$

Then

$$\mathbf{u} \cdot \mathbf{v} = \mathbf{u}^T\overline{\mathbf{v}} = (1+i)(6+2i) + (2-i)(-2i) + (3+i)(-1-i) = 0,$$

so $\mathbf{u}$ and $\mathbf{v}$ are orthogonal. Also,

$$\begin{aligned} \|\mathbf{u}\| &= \sqrt{\mathbf{u} \cdot \mathbf{u}} = \sqrt{\mathbf{u}^T\overline{\mathbf{u}}} \\ &= \sqrt{(1+i)(1-i) + (2-i)(2+i) + (3+i)(3-i)} \\ &= \sqrt{17}. \end{aligned}$$

■

We can prove the following properties for the dot product on C^n (Exercise T.6):

(a) $\mathbf{u} \cdot \mathbf{u} > 0$ for $\mathbf{u} \neq \mathbf{0}$ in C^n; $\mathbf{u} \cdot \mathbf{u} = 0$ if and only if $\mathbf{u} = \mathbf{0}$ in C^n.
(b) $\mathbf{u} \cdot \mathbf{v} = \overline{\mathbf{v} \cdot \mathbf{u}}$ for any $\mathbf{u}$, $\mathbf{v}$ in C^n.
(c) $(\mathbf{u} + \mathbf{v}) \cdot \mathbf{w} = \mathbf{u} \cdot \mathbf{w} + \mathbf{v} \cdot \mathbf{w}$ for any $\mathbf{u}$, $\mathbf{v}$, $\mathbf{w}$ in C^n.
(d) $(c\mathbf{u}) \cdot \mathbf{v} = c(\mathbf{u} \cdot \mathbf{v})$ for any $\mathbf{u}$, $\mathbf{v}$ in C^n and c a complex scalar.

Remark Observe that these properties are somewhat different from those satisfied by the dot product on R^n. (See Theorem 4.3 in Section 4.2.)

COMPLEX EIGENVALUES AND EIGENVECTORS

In the case of complex matrices, we have the following analogs of the theorem, presented in Section 8.3, which show the role played by the special matrices discussed in Section A.1.

THEOREM A.1 *If A is a Hermitian matrix, then the eigenvalues of A are all real. Moreover, eigenvectors belonging to distinct eigenvalues are orthogonal (complex analog of Theorems 8.6 and 8.7).* ■

THEOREM A.2 *If A is a Hermitian matrix, then there exists a unitary matrix U such that $U^{-1}AU = D$, a diagonal matrix. The eigenvalues of A lie on the main diagonal of D (complex analog of Theorem 8.9).* ■

In Section 8.3 we proved that if A is a real symmetric matrix, then there exists an orthogonal matrix P such that $P^{-1}AP = D$, a diagonal matrix, and conversely, if there is an orthogonal matrix P such that $P^{-1}AP$ is a diagonal matrix, then A is a symmetric matrix. For complex matrices, the situation is more complicated. The converse of Theorem A.2 is not true. That is, if A is a matrix for which there exists a unitary matrix U such that $U^{-1}AU = D$, a diagonal matrix, then A need not be a Hermitian matrix. The correct statement involves normal matrices. The following result can be established.

THEOREM A.3 *If A is a normal matrix, then there exists a unitary matrix U such that $U^{-1}AU = D$, a diagonal matrix. Conversely, if A is a matrix for which there exists a unitary matrix U such that $U^{-1}AU = D$, a diagonal matrix, then A is a normal matrix.* ■

Key Terms

Complex vector space
Complex vector subspace
Orthogonal vectors

A.2 Exercises

1. Solve using Gauss–Jordan reduction.

   (a) $\begin{aligned}(1+2i)x_1 + (-2+i)x_2 &= 1-3i\\ (2+i)x_1 + (-1+2i)x_2 &= -1-i\end{aligned}$

   (b) $\begin{aligned}2ix_1 - (1-i)x_2 &= 1+i\\ (1-i)x_1 + x_2 &= 1-i\end{aligned}$

   (c) $\begin{aligned}(1+i)x_1 - x_2 &= -2+i\\ 2ix_1 + (1-i)x_2 &= i\end{aligned}$

2. Transform the given augmented matrix of a linear system to row echelon form and solve by back substitution.

   (a) $\left[\begin{array}{ccc|c} 2 & i & 0 & 1-i\\ 0 & 3i & -2+i & 4\\ 0 & 0 & 2+i & 2-i\end{array}\right]$

   (b) $\left[\begin{array}{ccc|c} i & 2 & 1+i & 3i\\ 0 & 1-i & 0 & 2+i\\ 0 & 0 & 3 & 6-3i\end{array}\right]$

3. Solve by Gaussian elimination with back substitution.

   (a) $\begin{aligned}ix_1 + (1+i)x_2 \qquad &= i\\ (1-i)x_1 + x_2 - ix_3 &= 1\\ ix_2 + x_3 &= 1\end{aligned}$

   (b) $\begin{aligned}x_1 + ix_2 + (1-i)x_3 &= 2+i\\ ix_1 \qquad + (1+i)x_3 &= -1+i\\ 2ix_2 - x_3 &= 2-i\end{aligned}$

4. Compute the determinant and simplify as much as possible.

   (a) $\begin{vmatrix} 1+i & -1\\ 2i & 1+i\end{vmatrix}$

   (b) $\begin{vmatrix} 2-i & 1+i\\ 1+2i & -(1-i)\end{vmatrix}$

   (c) $\begin{vmatrix} 1+i & 2 & 2-i\\ i & 0 & 3+i\\ -2 & 1 & 1+2i\end{vmatrix}$

   (d) $\begin{vmatrix} 2 & 1-i & 0\\ 1+i & -1 & i\\ 0 & -i & 2\end{vmatrix}$

5. Find the inverse of each of the following matrices if possible.

   (a) $\begin{bmatrix} i & 2\\ 1+i & -i\end{bmatrix}$ (b) $\begin{bmatrix} 2 & i & 3\\ 1+i & 0 & 1-i\\ 2 & 1 & 2+i\end{bmatrix}$

6. Determine whether or not the following subsets W of C_{22} are complex vector subspaces.

   (a) W is the set of all 2×2 complex matrices with zeros on the main diagonal.

   (b) W is the set of all 2×2 complex matrices that have diagonal entries with real part equal to zero.

   (c) W is the set of all symmetric 2×2 complex matrices.

7. Let $W = \text{span}\{\mathbf{v}_1, \mathbf{v}_2, \mathbf{v}_3\}$, where

   $$\mathbf{v}_1 = (-1+i, 2, 1), \qquad \mathbf{v}_2 = (1, 1+i, i),$$
   $$\mathbf{v}_3 = (-5+2i, -1-3i, 2-3i).$$

   (a) Does $\mathbf{v} = (i, 0, 0)$ belong to W?

   (b) Is the set $\{\mathbf{v}_1, \mathbf{v}_2, \mathbf{v}_3\}$ linearly independent or linearly dependent?

8. Let $\{\mathbf{v}_1, \mathbf{v}_2, \mathbf{v}_3\}$ be a basis for a complex vector space V. Determine whether or not $\mathbf{w}$ is in span $\{\mathbf{w}_1, \mathbf{w}_2\}$.

   (a) $\begin{aligned}\mathbf{w}_1 &= i\mathbf{v}_1 + (1-i)\mathbf{v}_2 + 2\mathbf{v}_3\\ \mathbf{w}_2 &= (2+i)\mathbf{v}_1 + 2i\mathbf{v}_2 + (3-i)\mathbf{v}_3\\ \mathbf{w} &= (-2-3i)\mathbf{v}_1 + (3-i)\mathbf{v}_2 + (-2-2i)\mathbf{v}_3\end{aligned}$

   (b) $\begin{aligned}\mathbf{w}_1 &= 2i\mathbf{v}_1 + \mathbf{v}_2 + (1-i)\mathbf{v}_3\\ \mathbf{w}_2 &= 3i\mathbf{v}_1 + (1+i)\mathbf{v}_2 + 3\mathbf{v}_3\\ \mathbf{w} &= (2+3i)\mathbf{v}_1 + (2+i)\mathbf{v}_2 + (4-2i)\mathbf{v}_3\end{aligned}$

In Exercises 9 and 10, compute $\mathbf{u} \cdot \mathbf{v}$.

9. (a) $\mathbf{u} = \begin{bmatrix} 1-3i\\ 1+3i\end{bmatrix}$, $\mathbf{v} = \begin{bmatrix} 2i\\ 6\end{bmatrix}$

   (b) $\mathbf{u} = \begin{bmatrix} 2-3i\\ 1+2i\\ 4\end{bmatrix}$, $\mathbf{v} = \begin{bmatrix} 2i\\ 1-i\\ 3+4i\end{bmatrix}$

10. (a) $\mathbf{u} = \begin{bmatrix} 2-i \\ 1+i \\ 3 \end{bmatrix}$, $\mathbf{v} = \begin{bmatrix} -2 \\ -1-2i \\ 3i \end{bmatrix}$

(b) $\mathbf{u} = \begin{bmatrix} 2+2i \\ 3 \\ 1-2i \\ -4i \end{bmatrix}$, $\mathbf{v} = \begin{bmatrix} 2+i \\ i \\ -4 \\ -3-2i \end{bmatrix}$

11. Determine whether $\mathbf{u}$ and $\mathbf{v}$ are orthogonal.

(a) $\mathbf{u} = (3+i, 2-i)$, $\mathbf{v} = (1-i, 2+i)$

(b) $\mathbf{u} = (i, 1+i, 1-i)$, $\mathbf{v} = (3-i, 1+2i, -1+3i)$

(c) $\mathbf{u} = (1+i, 2-i, 3)$, $\mathbf{v} = (i, -2i, 1+i)$

(d) $\mathbf{u} = (1+2i, 2i, 2-i)$, $\mathbf{v} = (-3+5i, 5-5i, 1-i)$

12. Compute $\|\mathbf{u}\|$.

(a) $\mathbf{u} = (1-i, 2, 3+i)$

(b) $\mathbf{u} = (i, -2-3i, 1+i)$

(c) $\mathbf{u} = (i, -i, 1, 0)$

(d) $\mathbf{u} = (1+i, 1-i, 2+i, 3-i)$

13. Find the eigenvalues and associated eigenvectors of the following complex matrices.

(a) $A = \begin{bmatrix} 1 & 1 \\ -1 & 1 \end{bmatrix}$ (b) $A = \begin{bmatrix} 1 & i \\ -i & 1 \end{bmatrix}$

(c) $A = \begin{bmatrix} 2 & 0 & 0 \\ 0 & 2 & i \\ 0 & -i & 2 \end{bmatrix}$

14. For each of the parts in Exercise 13, find a matrix P such that $P^{-1}AP = D$, a diagonal matrix. For part (c), find three different matrices P that diagonalize A.

Theoretical Exercises

T.1. (a) Prove or disprove: The set W of all $n \times n$ Hermitian matrices is a complex vector subspace of C_{nn}.

(b) Prove or disprove: The set W of all $n \times n$ Hermitian matrices is a real vector subspace of the real vector space of all $n \times n$ complex matrices.

T.2. Prove or disprove: The set W of all $n \times n$ unitary matrices is a complex vector subspace of C_{nn}.

T.3. (a) Show that if A is Hermitian, then the eigenvalues of A are real.

(b) Verify that A in Exercise 13(c) is Hermitian.

(c) Are the eigenvectors associated with an eigenvalue of a Hermitian matrix guaranteed to be real vectors? Explain.

T.4. Show that an $n \times n$ complex matrix A is unitary if and only if the columns (and rows) of A form an orthonormal set with respect to the dot product on C^n. (*Hint*: See Theorem 8.8.)

T.5. Show that if A is a skew Hermitian matrix (see Exercise T.7 in Section A.1) and λ is an eigenvalue of A, then the real part of λ is zero.

T.6. Show that the dot product on C^n satisfies the following properties.

(a) $\mathbf{u} \cdot \mathbf{u} > 0$ for $\mathbf{u} \neq \mathbf{0}$ in C^n; $\mathbf{u} \cdot \mathbf{u} = 0$ if and only if $\mathbf{u} = \mathbf{0}$

(b) $\mathbf{u} \cdot \mathbf{v} = \overline{\mathbf{v} \cdot \mathbf{u}}$ for any $\mathbf{u}$, $\mathbf{v}$ in C^n

(c) $(\mathbf{u} + \mathbf{v}) \cdot \mathbf{w} = \mathbf{u} \cdot \mathbf{w} + \mathbf{v} \cdot \mathbf{w}$ for any $\mathbf{u}$, $\mathbf{v}$, $\mathbf{w}$ in C^n

(d) $(c\mathbf{u}) \cdot \mathbf{v} = c(\mathbf{u} \cdot \mathbf{v})$ for any $\mathbf{u}$, $\mathbf{v}$ in C^n and c a complex scalar

MATLAB Exercises

All the routines for solving linear systems, such as **reduce**, **rref**, *and* \, *and the commands* **det**, **invert**, **eig**, **roots**, **poly**, *and so on, apply to complex matrices.*

ML.1. Solve the linear system in Exercise 1 using \.

ML.2. Solve Exercise 4 using **det**.

ML.3. Solve Exercise 5 using **invert**.

ML.4. Solve Exercise 13 using **eig**.

APPENDIX B

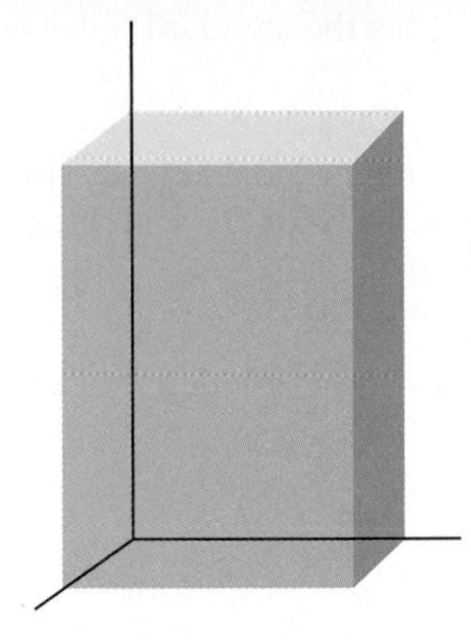

FURTHER DIRECTIONS

B.1 INNER PRODUCT SPACES (CALCULUS REQUIRED)

In this section we use the properties of the standard inner product or dot product on R^3 listed in Theorem 4.1 as our foundation for generalizing the notion of the inner product to any real vector space. Here V is an arbitrary real vector space, not necessarily finite-dimensional.

DEFINITION

Let V be any real vector space. An **inner product** on V is a function that assigns to each ordered pair of vectors $\mathbf{u}$, $\mathbf{v}$ in V a real number denoted by $(\mathbf{u}, \mathbf{v})$ satisfying:

(a) $(\mathbf{u}, \mathbf{u}) \geq 0$; $(\mathbf{u}, \mathbf{u}) = 0$ if and only if $\mathbf{u} = \mathbf{0}_V$, where $\mathbf{0}_V$ is the zero vector in V

(b) $(\mathbf{v}, \mathbf{u}) = (\mathbf{u}, \mathbf{v})$ for any $\mathbf{u}$, $\mathbf{v}$ in V

(c) $(\mathbf{u} + \mathbf{v}, \mathbf{w}) = (\mathbf{u}, \mathbf{w}) + (\mathbf{v}, \mathbf{w})$ for any $\mathbf{u}$, $\mathbf{v}$, $\mathbf{w}$ in V

(d) $(c\mathbf{u}, \mathbf{v}) = c(\mathbf{u}, \mathbf{v})$ for $\mathbf{u}$, $\mathbf{v}$ in V and c a real scalar

From these properties it follows that $(\mathbf{u}, c\mathbf{v}) = c(\mathbf{u}, \mathbf{v})$ because $(\mathbf{u}, c\mathbf{v}) = (c\mathbf{v}, \mathbf{u}) = c(\mathbf{v}, \mathbf{u}) = c(\mathbf{u}, \mathbf{v})$. Also, $(\mathbf{u}, \mathbf{v} + \mathbf{w}) = (\mathbf{u}, \mathbf{v}) + (\mathbf{u}, \mathbf{w})$.

EXAMPLE 1

The dot product on R^n, as defined in Section 1.3, is

$$(\mathbf{u}, \mathbf{v}) = \mathbf{u} \cdot \mathbf{v} = u_1 v_1 + u_2 v_2 + \cdots + u_n v_n,$$

where $\mathbf{u} = (u_1, u_2, \ldots, u_n)$ and $\mathbf{v} = (v_1, v_2, \ldots, v_n)$ is an inner product. This inner product will be called the **standard inner product** on R^n. ■

EXAMPLE 2

Let V be any finite-dimensional vector space and let $S = \{\mathbf{u}_1, \mathbf{u}_2, \ldots, \mathbf{u}_n\}$ be a basis for V. If

$$\mathbf{v} = a_1\mathbf{u}_1 + a_2\mathbf{u}_2 + \cdots + a_n\mathbf{u}_n$$

and

$$\mathbf{w} = b_1\mathbf{u}_1 + b_2\mathbf{u}_2 + \cdots + b_n\mathbf{u}_n,$$

we define

$$(\mathbf{v}, \mathbf{w}) = \left(\left[\mathbf{v}\right]_S, \left[\mathbf{w}\right]_S\right) = a_1 b_1 + a_2 b_2 + \cdots + a_n b_n.$$

It is not difficult to verify that this defines an inner product on V. This definition of $(\mathbf{v}, \mathbf{w})$ as an inner product on V uses the standard inner product on R^n. ■

Example 2 shows that we can define an inner product on any finite-dimensional vector space. Of course, if we change the basis for V in Example 2, we obtain a different inner product.

EXAMPLE 3 Let $\mathbf{u} = (u_1, u_2)$ and $\mathbf{v} = (v_1, v_2)$ be vectors in R^2. We define

$$(\mathbf{u}, \mathbf{v}) = u_1v_1 - u_2v_1 - u_1v_2 + 3u_2v_2.$$

Show that this gives an inner product on R^2.

Solution We have

$$\begin{aligned}(\mathbf{u}, \mathbf{u}) &= u_1^2 - 2u_1u_2 + 3u_2^2 = u_1^2 - 2u_1u_2 + u_2^2 + 2u_2^2 \\ &= (u_1 - u_2)^2 + 2u_2^2 \geq 0.\end{aligned}$$

Moreover, if $(\mathbf{u}, \mathbf{u}) = 0$, then $u_1 = u_2$ and $u_2 = 0$, so $\mathbf{u} = \mathbf{0}$. Conversely, if $\mathbf{u} = \mathbf{0}$, then $(\mathbf{u}, \mathbf{u}) = 0$. We can also verify (see Exercise 1) the remaining three properties of the preceding definition. This inner product is, of course, not the standard inner product on R^2. ■

Example 3 shows that on one vector space we may have more than one inner product, since we also have the standard inner product on R^2.

EXAMPLE 4 **(Calculus Required)** Let V be the vector space $C[0, 1]$ consisting of all real-valued continuous functions that are defined on the unit interval $[0, 1]$. For f and g in V, we let

$$(f, g) = \int_0^1 f(t)g(t)\,dt.$$

We now verify that this is an inner product on V, that is, that the properties of the definition above are satisfied.

Using results from calculus, we have for $f \neq 0$, the zero function,

$$(f, f) = \int_0^1 (f(t))^2\,dt \geq 0.$$

Moreover, if $(f, f) = 0$, then $f = 0$. Conversely, if $f = 0$, then $(f, f) = 0$. Also,

$$(f, g) = \int_0^1 f(t)g(t)\,dt = \int_0^1 g(t)f(t)\,dt = (g, f).$$

Next,

$$\begin{aligned}(f + g, h) &= \int_0^1 (f(t) + g(t))h(t)\,dt = \int_0^1 f(t)h(t)\,dt + \int_0^1 g(t)h(t)\,dt \\ &= (f, h) + (g, h).\end{aligned}$$

Finally,

$$(cf, g) = \int_0^1 (cf(t))g(t)\,dt = c\int_0^1 f(t)g(t)\,dt = c(f, g).$$

Thus, if f and g are the functions defined by $f(t) = t + 1$, $g(t) = 2t + 3$, then

$$(f, g) = \int_0^1 (t + 1)(2t + 3)\,dt = \int_0^1 (2t^2 + 5t + 3)\,dt = \tfrac{37}{6}. \qquad \blacksquare$$

DEFINITION

A real vector space that has an inner product defined on it is called an **inner product space**.

EXAMPLE 5

Let $V = P$, the set of all polynomials. Since P is a subspace of $C[0, 1]$, if we use the inner product defined in Example 4, we see that P is an inner product space. ■

If V is an inner product space, then by the **dimension** of V we mean the dimension of V as a real vector space, and a set S is a **basis** for V if S is a basis for the real vector space V. In an inner product space we define the **length** of a vector $\mathbf{u}$ by

$$\|\mathbf{u}\| = \sqrt{(\mathbf{u}, \mathbf{u})}.$$

This definition of length seems reasonable because at least we have $\|\mathbf{u}\| > 0$ if $\mathbf{u} \neq \mathbf{0}$. We can show [see Exercise T.1(a)] that $\|\mathbf{0}\| = 0$.

Any result in Sections 4.2, 6.8, and 6.9 dealing with R^n is also valid for any *inner product space* if a basis is not involved in the statement; it is true for any finite-dimensional inner product *space* if a basis is involved in the statement. Thus, let V be an inner product space. We define the **distance** between two vectors $\mathbf{u}$ and $\mathbf{v}$ in V as

$$d(\mathbf{u}, \mathbf{v}) = \|\mathbf{u} - \mathbf{v}\|.$$

The vectors $\mathbf{u}$ and $\mathbf{v}$ are defined to be **orthogonal** if $(\mathbf{u}, \mathbf{v}) = 0$. An orthogonal set of vectors $\mathbf{u}_1, \mathbf{u}_2, \ldots, \mathbf{u}_k$ in V is called **orthonormal** if the vectors are all of unit length.

Moreover, the Cauchy–Schwarz inequality (Theorem 4.4), the triangle inequality (Theorem 4.5), Theorems 6.16, 6.18 (the Gram–Schmidt process), 6.20, 6.21, and 6.23 are all valid in any inner product space. Theorem 6.17 holds in a finite-dimensional inner product space. We illustrate these ideas in the following examples.

EXAMPLE 6

(Calculus Required) Let V be the inner product space P_2 with inner product defined as in Example 4. If $p(t) = t + 2$, then the length of $p(t)$ is

$$\|p(t)\| = \sqrt{(p(t), p(t))} = \sqrt{\int_0^1 (t + 2)^2\,dt} = \sqrt{\frac{19}{3}}.$$

If $q(t) = 2t - 3$, then to find the cosine of the angle θ between $p(t)$ and $q(t)$, we proceed as follows. First,

$$\|q(t)\| = \sqrt{\int_0^1 (2t - 3)^2\,dt} = \sqrt{\frac{13}{3}}.$$

Next,

$$(p(t), q(t)) = \int_0^1 (t + 2)(2t - 3)\,dt = \int_0^1 (2t^2 + t - 6)\,dt = -\frac{29}{6}.$$

Then

$$\cos\theta = \frac{(p(t), q(t))}{\|p(t)\| \ \|q(t)\|} = \frac{-\frac{29}{6}}{\sqrt{\frac{19}{3}}\sqrt{\frac{13}{3}}} = \frac{-29}{2\sqrt{(19)(13)}}.$$

■

EXAMPLE 7 Let V be the inner product space P_2 considered in Example 6. The vectors t and $t - \frac{2}{3}$ are orthogonal, since

$$\left(t, t - \frac{2}{3}\right) = \int_0^1 t\left(t - \frac{2}{3}\right) dt = \int_0^1 \left(t^2 - \frac{2t}{3}\right) dt = 0.$$

■

EXAMPLE 8 **(Calculus Required)** Let V be the vector space $C[-\pi, \pi]$ of all real-valued continuous functions that are defined on $[-\pi, \pi]$. For f and g in V, we let $(f, g) = \int_{-\pi}^{\pi} f(t)g(t)\,dt$, which is easily shown to be an inner product on V (see Example 4). Consider the functions

$$1, \quad \cos t, \quad \sin t, \quad \cos 2t, \quad \sin 2t, \ldots, \quad \cos nt, \quad \sin nt, \ldots, \qquad (1)$$

which are clearly in V. The relationships

$$\int_{-\pi}^{\pi} \cos nt\,dt = \int_{-\pi}^{\pi} \sin nt\,dt = \int_{-\pi}^{\pi} \sin nt \cos nt\,dt = 0,$$

$$\int_{-\pi}^{\pi} \cos mt \cos nt\,dt = \int_{-\pi}^{\pi} \sin mt \sin nt\,dt = 0 \quad \text{if } m \neq n$$

demonstrate that $(f, g) = 0$ whenever f and g are distinct functions from (1). Hence every finite subset of functions from (1) is an orthogonal set. Theorem 6.16 generalized to inner product spaces then implies that any finite subset of functions from (1) is linearly independent. The functions in (1) were studied by the French mathematician Jean Baptiste Joseph Fourier. We take a closer look at these functions below. ■

EXAMPLE 9 Let V be the inner product space P_3 with the inner product defined in Example 4. Let W be the subspace of P_3 with basis $\{1, t^2\}$. Find a basis for $W^\perp$.

Solution Let $p(t) = at^3 + bt^2 + ct + d$ be an element of $W^\perp$. Since $p(t)$ must be orthogonal to each of the vectors in the given basis for W, we have

$$(p(t), 1) = \int_0^1 (at^3 + bt^2 + ct + d)\,dt = \frac{a}{4} + \frac{b}{3} + \frac{c}{2} + d = 0$$

$$(p(t), t^2) = \int_0^1 (at^5 + bt^4 + ct^3 + dt^2)\,dt = \frac{a}{6} + \frac{b}{5} + \frac{c}{4} + \frac{d}{3} = 0.$$

Solving the homogeneous system

$$\frac{a}{4} + \frac{b}{3} + \frac{c}{2} + d = 0$$

$$\frac{a}{6} + \frac{b}{5} + \frac{c}{4} + \frac{d}{3} = 0,$$

we obtain (verify)

$$a = 3r + 16s, \qquad b = -\tfrac{15}{4}r - 15s, \qquad c = r, \qquad d = s.$$

Then

$$p(t) = (3r + 16s)t^3 + \left(-\tfrac{15}{4}r - 15s\right)t^2 + rt + s$$
$$= r\left(3t^3 - \tfrac{15}{4}t^2 + t\right) + s(16t^3 - 15t^2 + 1).$$

Hence the vectors $3t^3 - \frac{15}{4}t^2 + t$ and $16t^3 - 15t^2 + 1$ span $W^{\perp}$. Since they are not multiples of each other, they are linearly independent and thus form a basis for $W^{\perp}$. ■

EXAMPLE 10 Let V be the inner product space P_3 with the inner product defined in Example 4. Let W be the subspace of P_3 having $S = \{t^2, t\}$ as a basis. Find an orthonormal basis for W.

Solution First, let $\mathbf{u}_1 = t^2$ and $\mathbf{u}_2 = t$. Now let $\mathbf{v}_1 = \mathbf{u}_1 = t^2$. Then

$$\mathbf{v}_2 = \mathbf{u}_2 - \frac{(\mathbf{u}_2, \mathbf{v}_1)}{(\mathbf{v}_1, \mathbf{v}_1)}\mathbf{v}_1 = t - \frac{\frac{1}{4}}{\frac{1}{5}}t^2 = t - \tfrac{5}{4}t^2,$$

where

$$(\mathbf{v}_1, \mathbf{v}_1) = \int_0^1 t^2t^2\,dt = \int_0^1 t^4\,dt = \frac{1}{5}$$

and

$$(\mathbf{u}_2, \mathbf{v}_1) = \int_0^1 tt^2\,dt = \int_0^1 t^3\,dt = \frac{1}{4}.$$

Then

$$T^* = \left\{t^2, t - \tfrac{5}{4}t^2\right\}$$

is an orthogonal basis for W. We now have to normalize the vectors in T^* to obtain an orthonormal basis T for W. We have already computed

$$(\mathbf{v}_1, \mathbf{v}_1) = \tfrac{1}{5}, \quad \text{so } \|\mathbf{v}_1\| = \sqrt{\tfrac{1}{5}}.$$

We also have

$$(\mathbf{v}_2, \mathbf{v}_2) = \int_0^1 \left(t - \tfrac{5}{4}t^2\right)^2 dt = \tfrac{1}{48}, \quad \text{so } \|\mathbf{v}_2\| = \sqrt{\tfrac{1}{48}}.$$

We now let

$$\mathbf{w}_1 = \frac{1}{\|\mathbf{v}_1\|}\mathbf{v}_1 = \sqrt{5}\,t^2, \qquad \mathbf{w}_2 = \frac{1}{\|\mathbf{v}_2\|}\mathbf{v}_2 = \sqrt{48}\left(t - \tfrac{5}{4}t^2\right).$$

Then $T = \left\{\sqrt{5}\,t^2, \sqrt{48}\left(t - \frac{5}{4}t^2\right)\right\}$ is an orthonormal basis for W. If we choose $\mathbf{u}_1 = t$ and $\mathbf{u}_2 = t^2$, then we obtain (verify) the orthonormal basis

$$\left\{\sqrt{3}\,t, \sqrt{30}\left(t^2 - \tfrac{1}{2}t\right)\right\}$$

for W. ■

FOURIER SERIES (CALCULUS REQUIRED)

In the study of calculus you most likely encountered functions $f(t)$, which had derivatives of all orders at a point $t = t_0$. Associated with $f(t)$ is its Taylor series, defined by

$$\sum_{k=0}^{\infty} \frac{f^{(k)}(t_0)}{k!}(t - t_0)^k. \tag{2}$$

The expression in (2) is called the **Taylor series of f at t_0** (or **about t_0** or **centered at t_0**). When $t_0 = 0$, the Taylor series is called a **Maclaurin series**. The coefficients of Taylor and Maclaurin series expansions involve successive derivatives of the given function evaluated at the center of the expansion. If we take the first $n + 1$ terms of the series in (2), we obtain a Taylor or Maclaurin polynomial of degree n that approximates the given function.

The function $f(t) = |t|$ does not have a Taylor series expansion at $t_0 = 0$ (a Maclaurin series), because f does not have a derivative at $t = 0$. Thus there is no way to compute the coefficients in such an expansion. The expression in (2) is in terms of the functions $1, t, t^2, \dots$. However, it is possible to obtain a series expansion for such a function by using a different type of expansion. One such important expansion involves the set of functions

$$1, \quad \cos t, \quad \sin t, \quad \cos 2t, \quad \sin 2t, \dots, \quad \cos nt, \quad \sin nt, \dots,$$

which we discussed briefly in Example 8. The French mathematician Jean Baptiste Joseph Fourier* showed that every function f (continuous or not) that is defined on $[-\pi, \pi]$, can be represented by a series of the form

$$\begin{aligned} &\tfrac{1}{2}a_0 + a_1 \cos t + a_2 \cos 2t + \cdots + a_n \cos nt \\ &\quad + b_1 \sin t + b_2 \sin 2t + \cdots + b_n \sin nt + \cdots. \end{aligned}$$

It then follows that every function f (continuous or not) that is defined on $[-\pi, \pi]$, can be approximated as closely as we wish by a function of the form

$$\begin{aligned} &\tfrac{1}{2}a_0 + a_1 \cos t + a_2 \cos 2t + \cdots + a_n \cos nt \\ &\quad + b_1 \sin t + b_2 \sin 2t + \cdots + b_n \sin nt \end{aligned} \tag{3}$$

for n sufficiently large. The function in (3) is called a **trigonometric polynomial**, and if a_n and b_n are both nonzero, we say that its **degree** is n. The topic of Fourier series is beyond the scope of this book. We limit ourselves to a brief discussion on how to obtain the best approximation of a function by trigonometric polynomials.

---

*Jean Baptiste Joseph Fourier (1768–1830) was born in Auxere, France. His father was a tailor. Fourier received much of his early education in the local military school, which was run by the Benedictine order, and at the age of 19 he decided to study for the priesthood. His strong interest in mathematics, which started developing at the age of 13, continued while studying for the priesthood. Two years later, he decided not to take his religious vows and became a teacher at the military school where he had studied.

Fourier was active in French politics throughout the French Revolution and the turbulent period that followed. In 1795, he was appointed to a chair at the prestigious École Polytechnique. In 1798, Fourier accompanied Napoleon as a scientific advisor in his invasion of Egypt. Upon returning to France, he served for 12 years as prefect of the department of Isére and lived in Grenoble. During this period he did his pioneering work on the theory of heat. In this work he showed that every function can be represented by a series of trigonometric polynomials. Such a series is now called a Fourier series. He died in Paris in 1830.

It is not difficult to show that

$$\int_{-\pi}^{\pi} 1\,dt = 2\pi, \quad \int_{-\pi}^{\pi} \sin nt\,dt = 0, \quad \int_{-\pi}^{\pi} \cos nt\,dt = 0,$$

$$\int_{-\pi}^{\pi} \sin nt \sin mt\,dt = 0 \quad (n \neq m), \quad \int_{-\pi}^{\pi} \cos nt \cos mt\,dt = 0 \quad (n \neq m),$$

$$\int_{-\pi}^{\pi} \sin nt \cos mt\,dt = 0 \quad (n \neq m), \quad \int_{-\pi}^{\pi} \sin nt \sin nt\,dt = \pi,$$

$$\int_{-\pi}^{\pi} \cos nt \cos nt\,dt = \pi.$$

Let V be the vector space of real-valued continuous functions on $[-\pi, \pi]$. If f and g belong to V, then $(f, g) = \int_{-\pi}^{\pi} f(t)g(t)\,dt$ defines an inner product on V as in Example 8. The preceding relations show that the following set of vectors is an orthonormal set in V:

$$\frac{1}{\sqrt{2\pi}}, \quad \frac{1}{\sqrt{\pi}}\cos t, \quad \frac{1}{\sqrt{\pi}}\sin t, \quad \frac{1}{\sqrt{\pi}}\cos 2t, \quad \frac{1}{\sqrt{\pi}}\sin 2t, \ldots,$$

$$\frac{1}{\sqrt{\pi}}\cos nt, \quad \frac{1}{\sqrt{\pi}}\sin nt, \ldots.$$

Now

$$W = \operatorname{span}\left\{\frac{1}{\sqrt{2\pi}}, \frac{1}{\sqrt{\pi}}\cos t, \frac{1}{\sqrt{\pi}}\sin t, \frac{1}{\sqrt{\pi}}\cos 2t, \frac{1}{\sqrt{\pi}}\sin 2t, \ldots, \right.$$

$$\left. \frac{1}{\sqrt{\pi}}\cos nt, \frac{1}{\sqrt{\pi}}\sin nt\right\}$$

is a finite-dimensional subspace of V. Theorem 6.23 implies that the best approximation to a given function f in V by a trigonometric polynomial of degree n is given by $\operatorname{proj}_W f$, the projection of f onto W. This polynomial is called the **Fourier polynomial of degree *n* for *f***.

EXAMPLE 11 Find Fourier polynomials of degrees one and three for the function $f(t) = |t|$.

Solution First, we compute the Fourier polynomial of degree one. Using a generalization of Equation (1) in Section 6.9, we can compute $\operatorname{proj}_W \mathbf{v}$ for $\mathbf{v} = |t|$ as

$$\operatorname{proj}_W |t| = \left(|t|, \frac{1}{\sqrt{2\pi}}\right)\frac{1}{\sqrt{2\pi}} + \left(|t|, \frac{1}{\sqrt{\pi}}\cos t\right)\frac{1}{\sqrt{\pi}}\cos t$$

$$+ \left(|t|, \frac{1}{\sqrt{\pi}}\sin t\right)\frac{1}{\sqrt{\pi}}\sin t.$$

We have

$$\left(|t|, \frac{1}{\sqrt{2\pi}}\right) = \int_{-\pi}^{\pi} |t| \frac{1}{\sqrt{2\pi}}\, dt$$
$$= \frac{1}{\sqrt{2\pi}} \int_{-\pi}^{0} -t\, dt + \frac{1}{\sqrt{2\pi}} \int_{0}^{\pi} t\, dt = \frac{\pi^2}{\sqrt{2\pi}},$$

$$\left(|t|, \frac{1}{\sqrt{\pi}} \cos t\right) = \int_{-\pi}^{\pi} |t| \frac{1}{\sqrt{\pi}} \cos t\, dt$$
$$= \frac{1}{\sqrt{\pi}} \int_{-\pi}^{0} -t \cos t\, dt + \frac{1}{\sqrt{\pi}} \int_{0}^{\pi} t \cos t\, dt$$
$$= -\frac{2}{\sqrt{\pi}} - \frac{2}{\sqrt{\pi}} = -\frac{4}{\sqrt{\pi}},$$

and

$$\left(|t|, \frac{1}{\sqrt{\pi}} \sin t\right) = \int_{-\pi}^{\pi} |t| \frac{1}{\sqrt{\pi}} \sin t\, dt$$
$$= \frac{1}{\sqrt{\pi}} \int_{-\pi}^{0} -t \sin t\, dt + \frac{1}{\sqrt{\pi}} \int_{0}^{\pi} t \sin t\, dt$$
$$= -\sqrt{\pi} + \sqrt{\pi} = 0.$$

Then

$$\text{proj}_W |t| = \frac{\pi^2}{\sqrt{2\pi}} \frac{1}{\sqrt{2\pi}} - \frac{4}{\sqrt{\pi}} \frac{1}{\sqrt{\pi}} \cos t = \frac{\pi}{2} - \frac{4}{\pi} \cos t.$$

Next, we compute the Fourier polynomial of degree three. By a generalization of Equation (1) in Section 6.9,

$$\text{proj}_W |t| = \left(|t|, \frac{1}{\sqrt{2\pi}}\right) \frac{1}{\sqrt{2\pi}}$$
$$+ \left(|t|, \frac{1}{\sqrt{\pi}} \cos t\right) \frac{1}{\sqrt{\pi}} \cos t + \left(|t|, \frac{1}{\sqrt{\pi}} \sin t\right) \frac{1}{\sqrt{\pi}} \sin t$$
$$+ \left(|t|, \frac{1}{\sqrt{\pi}} \cos 2t\right) \frac{1}{\sqrt{\pi}} \cos 2t + \left(|t|, \frac{1}{\sqrt{\pi}} \sin 2t\right) \frac{1}{\sqrt{\pi}} \sin 2t$$
$$+ \left(|t|, \frac{1}{\sqrt{\pi}} \cos 3t\right) \frac{1}{\sqrt{\pi}} \cos 3t + \left(|t|, \frac{1}{\sqrt{\pi}} \sin 3t\right) \frac{1}{\sqrt{\pi}} \sin 3t.$$

We have

$$\int_{-\pi}^{\pi} |t| \frac{1}{\sqrt{\pi}} \cos 2t\, dt = 0, \qquad \int_{-\pi}^{\pi} |t| \frac{1}{\sqrt{\pi}} \sin 2t\, dt = 0,$$

$$\int_{-\pi}^{\pi} |t| \frac{1}{\sqrt{\pi}} \cos 3t\, dt = -\frac{4}{9\sqrt{\pi}}, \qquad \int_{-\pi}^{\pi} |t| \frac{1}{\sqrt{\pi}} \sin 3t\, dt = 0.$$

Hence

$$\text{proj}_W \mathbf{v} = \frac{\pi}{2} - \frac{4}{\pi} \cos t - \frac{4}{9\pi} \cos 3t.$$

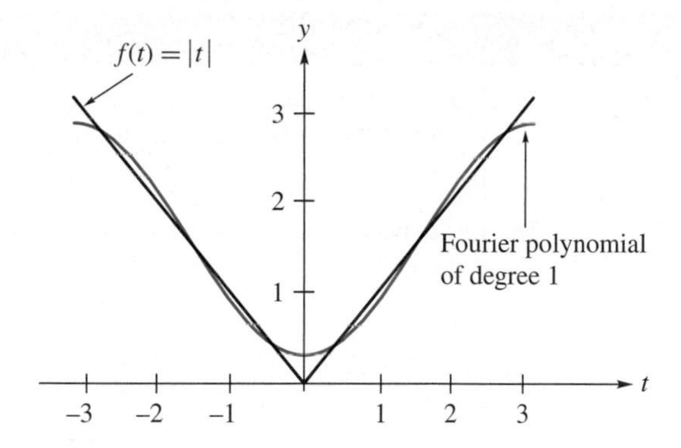

Figure B.1 ▲

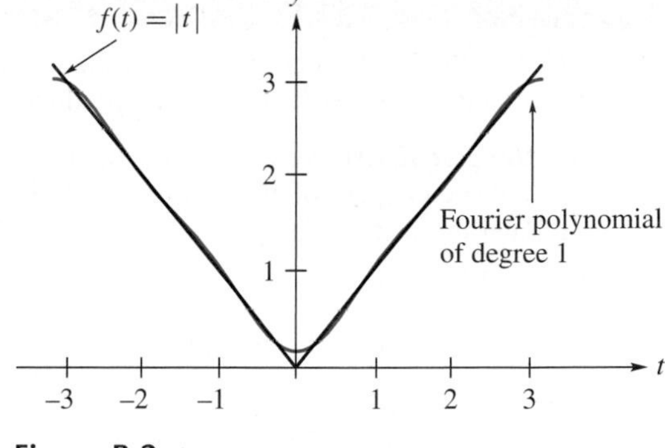

Figure B.2 ▲

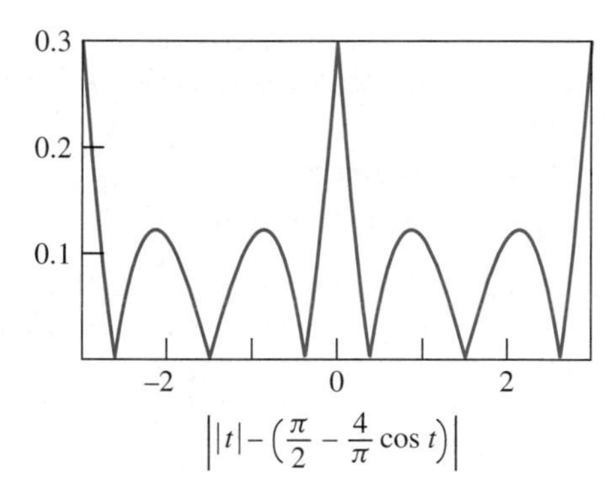

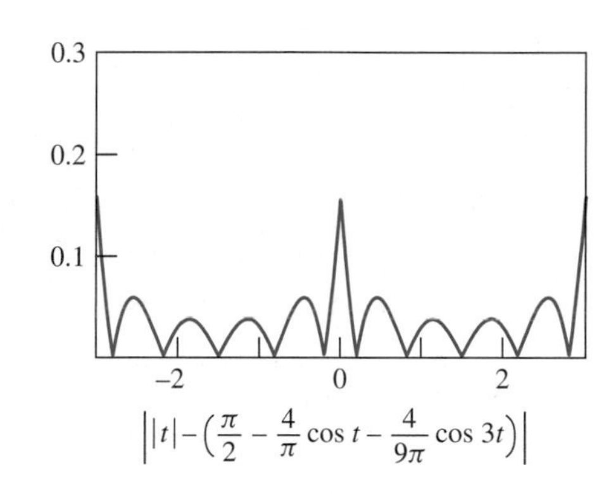

Figure B.3 ▲

Figure B.1 shows the graphs of f and the Fourier polynomial of degree one. Figure B.2 shows the graphs of f and the Fourier polynomial of degree three. Figure B.3 shows the graphs of

$$\left||t|-\left(\tfrac{\pi}{2}-\tfrac{4}{\pi}\cos t\right)\right| \quad \text{and} \quad \left||t|-\left(\tfrac{\pi}{2}-\tfrac{4}{\pi}\cos t-\tfrac{4}{9\pi}\cos 3t\right)\right|.$$

Observe how much better the approximation by a Fourier polynomial of degree three is. ■

Fourier series play an important role in the study of heat distribution and in the analysis of sound waves. The study of projections is important in a number of areas in applied mathematics. We illustrate this in Section 7.2 by considering the topic of least squares, which provides a technique for dealing with inconsistent systems.

Key Terms

Inner product
Standard inner product
Inner product space
Dimension
Basis
Length
Distance
Orthogonal
Orthonormal
Taylor series
Maclaurin series
Trigonometric polynomial
Degree
Fourier polynomial
Parallelogram law
Pythagorean theorem

B.1 Exercises

1. Verify that the function in Example 3 satisfies the remaining three properties of an inner product.

2. ***(Calculus Required)*** Verify that the function defined on P, the vector space of all polynomials, in Example 5 is an inner product.

3. Let $V = R^2$. If
$$\mathbf{u} = (u_1, u_2) \quad \text{and} \quad \mathbf{v} = (v_1, v_2),$$
define
$$(\mathbf{u}, \mathbf{v}) = u_1 v_1 + 5u_2 v_2.$$
Show that this function is an inner product on R^2.

4. Let $V = M_{22}$. If
$$A = \begin{bmatrix} a_{11} & a_{12} \\ a_{21} & a_{22} \end{bmatrix} \quad \text{and} \quad B = \begin{bmatrix} b_{11} & b_{12} \\ b_{21} & b_{22} \end{bmatrix},$$
define
$$(A, B) = a_{11}b_{11} + a_{12}b_{12} + a_{21}b_{21} + a_{22}b_{22}.$$
Show that this function is an inner product on V.

5. Let $V = M_{nn}$ be the real vector space of all $n \times n$ matrices. If A and B are in V, we define $(A, B) = \mathrm{Tr}(B^T A)$, where Tr is the trace function defined in Supplementary Exercise T.1 of Chapter 1. Show that this function is an inner product on V.

6. Let V be the vector space $C[a, b]$ consisting of all real-valued continuous functions that are defined on $[a, b]$. If f and g are in V, let $(f, g) = \int_a^b f(x)g(x)\,dx$. Show that this function is an inner product on V.

In Exercises 7 and 8, use the inner product in Example 3 and compute $(\mathbf{u}, \mathbf{v})$.

7. $\mathbf{u} = (1, 2)$, $\mathbf{v} = (3, -1)$

8. $\mathbf{u} = (0, 1)$, $\mathbf{v} = (-2, 5)$

In Exercises 9 and 10, use the inner product defined in Example 4 and compute (f, g).

9. $f(t) = 1$, $g(t) = 3 + 2t$

10. $f(t) = \sin t$, $g(t) = \cos t$

In Exercises 11 and 12, use the inner product space defined in Exercise 4 and compute (A, B).

11. $A = \begin{bmatrix} 1 & 2 \\ -1 & 3 \end{bmatrix}$, $B = \begin{bmatrix} 1 & 0 \\ 2 & -1 \end{bmatrix}$

12. $A = \begin{bmatrix} 0 & 1 \\ 2 & 3 \end{bmatrix}$, $B = \begin{bmatrix} 1 & 1 \\ 2 & -1 \end{bmatrix}$

13. Let V be the inner product space in Example 3. Compute the length of the given vector.
(a) $(1, 3)$ (b) $(-2, -4)$ (c) $(3, -1)$

14. Let V be the inner product space in Example 4. Compute the length of the given vector.
(a) t^2 (b) e^t

15. Let V be the inner product space in Example 4. Find the distance between $\mathbf{u}$ and $\mathbf{v}$.
(a) $\mathbf{u} = t$, $\mathbf{v} = t^2$ (b) $\mathbf{u} = e^t$, $\mathbf{v} = e^{-t}$

16. Let V be the inner product space in Exercise 3. Find the distance between $\mathbf{u}$ and $\mathbf{v}$.
(a) $\mathbf{u} = (0, 1)$, $\mathbf{v} = (1, -1)$
(b) $\mathbf{u} = (-2, -1)$, $\mathbf{v} = (2, 3)$

17. Let V be the inner product space in Example 4. Find the cosine of the angle between each pair of given vectors in V.
(a) $p(t) = t$, $q(t) = t - 1$
(b) $p(t) = \sin t$, $q(t) = \cos t$

18. Let V be the inner product space in Exercise 3. Find the cosine of the angle between each pair of given vectors in V.
(a) $\mathbf{u} = (2, 1)$, $\mathbf{v} = (3, 2)$
(b) $\mathbf{u} = (1, 1)$, $\mathbf{v} = (-2, -3)$

In Exercises 19 and 20, let V be the inner product space of Example 4.

19. Let $p(t) = 3t + 1$ and $q(t) = at$. For what values of a are $p(t)$ and $q(t)$ orthogonal?

20. Let $p(t) = 3t + 1$ and $q(t) = at + b$. For what values of a and b are $p(t)$ and $q(t)$ orthogonal?

21. Let
$$A = \begin{bmatrix} 1 & 2 \\ 3 & 4 \end{bmatrix}.$$
Find a matrix $B \neq O_2$ such that A and B are orthogonal in the inner product space defined in Exercise 5. Can there be more than one matrix B that is orthogonal to A?

22. Let V be the inner product space in Example 4.
(a) If $p(t) = \sqrt{t}$, find $q(t) = a + bt \neq 0$ such that $p(t)$ and $q(t)$ are orthogonal.
(b) If $p(t) = \sin t$, find $q(t) = a + be^t \neq 0$ such that $p(t)$ and $q(t)$ are orthogonal.

23. Consider the inner product space R^4 with the standard inner product and let
$$\mathbf{u}_1 = (1, 0, 0, 1) \quad \text{and} \quad \mathbf{u}_2 = (0, 1, 0, 1).$$
(a) Show that the set W consisting of all vectors in R^4 that are orthogonal to both $\mathbf{u}_1$ and $\mathbf{u}_2$ is a subspace of R^4.
(b) Find a basis for W.

24. State the Cauchy–Schwarz inequality and the triangle inequality for the inner product space in Example 4.

In Exercises 25 through 28, the inner product on the given vector space is as defined in Example 4.

25. (a) Let $S = \{t, 1\}$ be a basis for a subspace W of the inner product space P_2. Use the Gram–Schmidt process to find an orthonormal basis for W.

(b) Use a generalization of Theorem 6.17 to write $2t - 1$ as a linear combination of the orthonormal basis obtained in part (a).

26. (a) Repeat Exercise 25 with $S = \{t + 1, t - 1\}$.

(b) Use a generalization of Theorem 6.17 to find the coordinate vector of $3t + 2$ with respect to the orthonormal basis found in part (a).

27. Let $S = \{t, \sin 2\pi t\}$ be a basis for a subspace W of the inner product space in Example 9. Use the Gram–Schmidt process to find an orthonormal basis for W.

28. Let $S = \{t, e^t\}$ be a basis for a subspace W of the inner product space in Example 4. Use the Gram–Schmidt process to find an orthonormal basis for W.

29. Let V be the inner product space P_3 with the inner product defined in Example 4. Let W be the subspace of P_3 spanned by $\{t - 1, t^2\}$. Find a basis for $W^\perp$.

30. Let V be the inner product space P_4 with the inner product defined in Example 4. Let W be the subspace of P_4 spanned by $\{1, t\}$. Find a basis for $W^\perp$.

In Exercises 31 and 32, let W be the subspace of continuous functions on $[-\pi, \pi]$ defined in Example 8. Find $\text{proj}_W \mathbf{v}$ *for the given vector* $\mathbf{v}$.

31. $\mathbf{v} = t$

32. $\mathbf{v} = e^t$

In Exercises 33 and 34, let W be the subspace of continuous functions on $[-\pi, \pi]$ defined in Example 8. Write the vector $\mathbf{v}$ *as* $\mathbf{w} + \mathbf{u}$, *with* $\mathbf{w}$ *in W and* $\mathbf{u}$ *in* $W^\perp$.

33. $\mathbf{v} = t - 1$

34. $\mathbf{v} = t^2$

In Exercises 35 and 36, let W be the subspace of continuous functions on $[-\pi, \pi]$ defined in Example 11. Find the distance from $\mathbf{v}$ *to W.*

35. $\mathbf{v} = t$

36. $\mathbf{v} = 1 - \cos t$

In Exercises 37 and 38, find the Fourier polynomial of degree two for f.

37. ***(Calculus required)*** $f(t) = t^2$

38. ***(Calculus required)*** $f(t) = e^t$

Theoretical Exercises

T.1. Let V be an inner product space. Show the following.

(a) $\|\mathbf{0}\| = 0$.

(b) $(\mathbf{u}, \mathbf{0}) = (\mathbf{0}, \mathbf{u}) = 0$ for any $\mathbf{u}$ in V.

(c) If $(\mathbf{u}, \mathbf{v}) = 0$ for all $\mathbf{v}$ in V, then $\mathbf{u} = \mathbf{0}$.

(d) If $(\mathbf{u}, \mathbf{w}) = (\mathbf{v}, \mathbf{w})$ for all $\mathbf{w}$ in V, then $\mathbf{u} = \mathbf{v}$.

(e) If $(\mathbf{w}, \mathbf{u}) = (\mathbf{w}, \mathbf{v})$ for all $\mathbf{w}$ in V, then $\mathbf{u} = \mathbf{v}$.

T.2. Let V be an inner product space. If $\mathbf{u}$ and $\mathbf{v}$ are vectors in V, we define the distance between $\mathbf{u}$ and $\mathbf{v}$ as

$$d(\mathbf{u}, \mathbf{v}) = \|\mathbf{u} - \mathbf{v}\|.$$

Let $\mathbf{u}$, $\mathbf{v}$, and $\mathbf{w}$ be in V. Show that:

(a) $d(\mathbf{u}, \mathbf{v}) \geq 0$

(b) $d(\mathbf{u}, \mathbf{v}) = 0$ if and only if $\mathbf{u} = \mathbf{v}$

(c) $d(\mathbf{u}, \mathbf{v}) = d(\mathbf{v}, \mathbf{u})$

(d) $d(\mathbf{u}, \mathbf{v}) \leq d(\mathbf{u}, \mathbf{w}) + d(\mathbf{w}, \mathbf{v})$

T.3. Show that if T is an orthonormal basis for a finite-dimensional inner product space and

$$\left[\mathbf{v}\right]_T = \begin{bmatrix} a_1 \\ a_2 \\ \vdots \\ a_n \end{bmatrix},$$

then $\|\mathbf{v}\| = \sqrt{a_1^2 + a_2^2 + \cdots + a_n^2}$.

T.4. Let $S = \{\mathbf{v}_1, \mathbf{v}_2, \dots, \mathbf{v}_n\}$ be an orthonormal basis for a finite-dimensional inner product space V and let $\mathbf{v}$ and $\mathbf{w}$ be vectors in V with

$$\left[\mathbf{v}\right]_S = \begin{bmatrix} a_1 \\ a_2 \\ \vdots \\ a_n \end{bmatrix} \quad \text{and} \quad \left[\mathbf{w}\right]_S = \begin{bmatrix} b_1 \\ b_2 \\ \vdots \\ b_n \end{bmatrix}.$$

Show that

$$d(\mathbf{v}, \mathbf{w}) = \sqrt{(a_1 - b_1)^2 + (a_2 - b_2)^2 + \cdots + (a_n - b_n)^2}.$$

T.5. Prove the **parallelogram law** for any two vectors $\mathbf{u}$ and $\mathbf{v}$ in an inner product space:

$$\|\mathbf{u} + \mathbf{v}\|^2 + \|\mathbf{u} - \mathbf{v}\|^2 = 2\|\mathbf{u}\|^2 + 2\|\mathbf{v}\|^2.$$

T.6. Let V be an inner product space. Show that $\|c\mathbf{u}\| = |c|\|\mathbf{u}\|$ for any vector $\mathbf{u}$ and any scalar c.

T.7. Let V be an inner product space. Show that if $\mathbf{u}$ and $\mathbf{v}$ are any vectors in V, then

$$\|\mathbf{u} + \mathbf{v}\|^2 = \|\mathbf{u}\|^2 + \|\mathbf{v}\|^2$$

if and only if $(\mathbf{u}, \mathbf{v}) = 0$, that is, if and only if $\mathbf{u}$ and $\mathbf{v}$ are orthogonal. This result is known as the **Pythagorean theorem**.

T.8. Let $\{\mathbf{u}, \mathbf{v}, \mathbf{w}\}$ be an orthonormal set of vectors in an inner product space V. Compute $\|\mathbf{u} + \mathbf{v} + \mathbf{w}\|^2$.

T.9. Let V be an inner product space. Show that if $\mathbf{v}$ is orthogonal to $\mathbf{w}_1, \mathbf{w}_2, \ldots, \mathbf{w}_k$, then $\mathbf{v}$ is orthogonal to every vector in span $\{\mathbf{w}_1, \mathbf{w}_2, \ldots, \mathbf{w}_k\}$.

B.2 COMPOSITE AND INVERTIBLE LINEAR TRANSFORMATIONS

We have already seen that nonsingular matrices are important and lead to many useful results. In this section we examine the analogous notion for linear transformations.

COMPOSITE LINEAR TRANSFORMATIONS

DEFINITION Let V_1 be an n-dimensional vector space, V_2 an m-dimensional vector space, and V_3 a p-dimensional vector space. Let $L_1: V_1 \to V_2$ and $L_2: V_2 \to V_3$ be linear transformations. The function $L_2 \circ L_1: V_1 \to V_3$ defined by

$$(L_2 \circ L_1)(\mathbf{u}) = L_2(L_1(\mathbf{u}))$$

for $\mathbf{u}$ in V_1 is called the **composite of L_2 with L_1**. See Figure B.4.

Figure B.4 ►

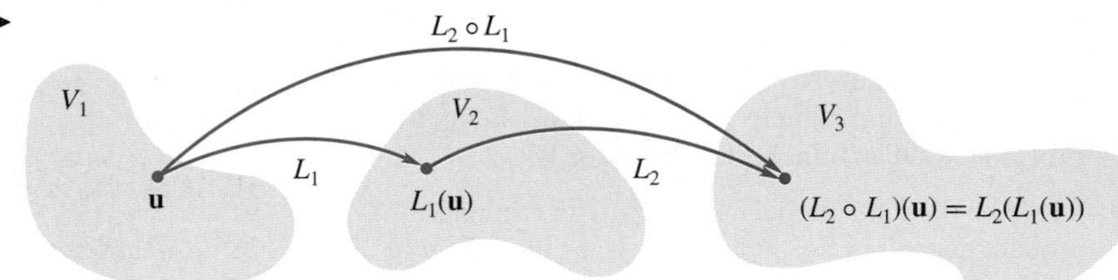

$L_2 \circ L_1$: The composite of L_2 with L_1.

If $V_1 = V_2 = V_3$, and $L_1 = L_2$, we write $L \circ L$ as L^2.

THEOREM B.1 *Let $L_1: V_1 \to V_2$ and $L_2: V_2 \to V_3$ be linear transformations. Then*

$$L_2 \circ L_1: V_1 \to V_3$$

is a linear transformation.

Proof Exercise T.1. ■

EXAMPLE 1 Let $L_1: R^2 \to R^3$ and $L_2: R^3 \to R^4$ be defined by

$$L_1\left(\begin{bmatrix} a_1 \\ a_2 \end{bmatrix}\right) = \begin{bmatrix} a_1 + a_2 \\ a_1 - a_2 \\ a_1 + 2a_2 \end{bmatrix}; \quad L_2\left(\begin{bmatrix} b_1 \\ b_2 \\ b_3 \end{bmatrix}\right) = \begin{bmatrix} b_1 + b_2 \\ b_1 - b_2 \\ b_2 + b_3 \\ 2b_1 + 3b_3 \end{bmatrix}.$$

Then $L_2 \circ L_1: R^2 \to R^4$ is given by

$$(L_2 \circ L_1)\left(\begin{bmatrix} a_1 \\ a_2 \end{bmatrix}\right) = L_2\left(L_1\left(\begin{bmatrix} a_1 \\ a_2 \end{bmatrix}\right)\right) = L_2\left(\begin{bmatrix} a_1 + a_2 \\ a_1 - a_2 \\ a_1 + 2a_2 \end{bmatrix}\right)$$

$$= \begin{bmatrix} (a_1 + a_2) + (a_1 - a_2) \\ (a_1 + a_2) - (a_1 - a_2) \\ (a_1 - a_2) + (a_1 + 2a_2) \\ 2(a_1 + a_2) + 3(a_1 + 2a_2) \end{bmatrix} = \begin{bmatrix} 2a_1 \\ 2a_2 \\ 2a_1 + a_2 \\ 5a_1 + 8a_2 \end{bmatrix}.$$

■

EXAMPLE 2 Let $L_1 \colon P_2 \to P_2$ and $L_2 \colon P_2 \to P_2$ be defined by

$$L_1(at^2 + bt + c) = 2at + b$$
$$L_2(at^2 + bt + c) = 2at^2 + bt.$$

Compute

(a) $L_2 \circ L_1$ (b) $L_1 \circ L_2$

Solution (a) We have

$$\begin{aligned}(L_2 \circ L_1)(at^2 + bt + c) &= L_2(L_1(at^2 + bt + c)) \\ &= L_2(2at + b) = 7at.\end{aligned}$$

(b) We have

$$\begin{aligned}(L_1 \circ L_2)(at^2 + bt + c) &= L_1(L_2(at^2 + bt + c)) \\ &= L_1(7at^2 + bt) = 14at + b.\end{aligned}$$

■

Remark Example 2 shows that, in general, $L_2 \circ L_1 \neq L_1 \circ L_2$.

THEOREM B.2 *Let V_1 be an n-dimensional vector space with basis P, V_2 an m-dimensional vector space with basis S, and V_3 a p-dimensional vector space with basis T. Let $L_1 \colon V_1 \to V_2$ and $L_2 \colon V_2 \to V_3$ be linear transformations. If A_1 represents L_1 with respect to P and S and A_2 represents L_2 with respect to S and T, then A_2A_1 represents $L_2 \circ L_1$ with respect to P and T.*

Proof From Theorem 10.8, it follows that if $\mathbf{x}$ is any vector in V_1 and $\mathbf{y}$ is any vector in V_2, then

$$\begin{aligned}\left[L_1(\mathbf{x})\right]_S &= A_1 \left[\mathbf{x}\right]_P \\ \left[L_2(\mathbf{y})\right]_T &= A_2 \left[\mathbf{y}\right]_S.\end{aligned}$$

Then

$$\begin{aligned}\left[(L_2 \circ L_1)(\mathbf{x})\right]_T &= \left[L_2(L_1(\mathbf{x}))\right]_T \\ &= A_2 \left[L_1(\mathbf{x})\right]_S = A_2 \left(A_1 \left[\mathbf{x}\right]_P\right) = A_2A_1 \left[\mathbf{x}\right]_P.\end{aligned}$$

Since the matrix representing a given linear transformation with respect to two given bases is unique, we conclude that A_2A_1 is the matrix representing $L_2 \circ L_1$ with respect to P and T. ■

Remark Since AB need not equal BA for matrices A and B, it is not surprising that $L_1 \circ L_2$ need not be the same linear transformation as $L_2 \circ L_1$, as we have just seen in Example 2.

EXAMPLE 3 Let $L_1 \colon R^2 \to R^2$ and $L_2 \colon R^2 \to R^3$ be defined by

$$L_1\left(\begin{bmatrix} a_1 \\ a_2 \end{bmatrix}\right) = \begin{bmatrix} a_2 \\ a_1 \end{bmatrix}; \quad L_2\left(\begin{bmatrix} a_1 \\ a_2 \end{bmatrix}\right) = \begin{bmatrix} a_1 + a_2 \\ a_1 - a_2 \\ a_2 \end{bmatrix}.$$

The matrix representing L_1 with respect to the natural basis for R^2 is (verify)

$$A_1 = \begin{bmatrix} 0 & 1 \\ 1 & 0 \end{bmatrix}.$$

The matrix representing L_2 with respect to the natural bases for R^2 and R^3 is (verify)

$$A_2 = \begin{bmatrix} 1 & 1 \\ 1 & -1 \\ 0 & 1 \end{bmatrix}.$$

Then by Theorem B.2, the matrix representing $L_2 \circ L_1 \colon R^2 \to R^3$ with respect to the natural bases for R^2 and R^3 is

$$A_2 A_1 = \begin{bmatrix} 1 & 1 \\ 1 & -1 \\ 0 & 1 \end{bmatrix} \begin{bmatrix} 0 & 1 \\ 1 & 0 \end{bmatrix} = \begin{bmatrix} 1 & 1 \\ -1 & 1 \\ 1 & 0 \end{bmatrix}.$$

Computing $L_2 \circ L_1$, we find that

$$\begin{aligned} (L_2 \circ L_1)\left(\begin{bmatrix} a_1 \\ a_2 \end{bmatrix}\right) &= L_2\left(L_1\left(\begin{bmatrix} a_1 \\ a_2 \end{bmatrix}\right)\right) \\ &= L_2\left(\begin{bmatrix} a_2 \\ a_1 \end{bmatrix}\right) = \begin{bmatrix} a_2 + a_1 \\ a_2 - a_1 \\ a_1 \end{bmatrix} = \begin{bmatrix} a_1 + a_2 \\ -a_1 + a_2 \\ a_1 \end{bmatrix}. \end{aligned}$$

We can then compute the matrix of $L_2 \circ L_1$ directly, obtaining (verify) the same answer as we got from $A_2 A_1$ previously. ■

INVERTIBLE LINEAR TRANSFORMATIONS

DEFINITION A linear transformation $L \colon V \to W$ is called **invertible** if there exists a unique function $L^{-1} \colon W \to V$ such that $L^{-1} \circ L = I_V$, the identity linear operator on V, defined by $I_V(\mathbf{v}) = \mathbf{v}$, and $L \circ L^{-1} = I_W$, the identity linear operator on W, defined by $I_W(\mathbf{w}) = \mathbf{w}$. The function L^{-1} is called the **inverse** of L.

THEOREM B.3 *A linear transformation $L \colon V \to W$ is invertible if and only if L is one-to-one and onto. Moreover, L^{-1} is a linear transformation and $(L^{-1})^{-1} = L$.*

Proof Let L be one-to-one and onto. We define a function $H \colon W \to V$ as follows. If $\mathbf{w}$ is in W, then since L is onto, $\mathbf{w} = L(\mathbf{v})$ for some $\mathbf{v}$ in V, and since L is one-to-one, $\mathbf{v}$ is unique. Let $H(\mathbf{w}) = \mathbf{v}$; H is a function and $L(H(\mathbf{w})) = L(\mathbf{v}) = \mathbf{w}$, so that $L \circ H = I_W$. Also, $H(L(\mathbf{v})) = H(\mathbf{w}) = \mathbf{v}$, so $H \circ L = I_V$. Thus H is an inverse of L. Now H is unique, for if $H_1 \colon W \to V$ is a function such that $L \circ H_1 = I_W$ and $H_1 \circ L = I_V$, then $L(H(\mathbf{w})) = \mathbf{w} = L(H_1(\mathbf{w}))$ for any $\mathbf{w}$ in W. Since L is one-to-one, we conclude that $H(\mathbf{w}) = H_1(\mathbf{w})$. Hence $H = H_1$. Thus $H = L^{-1}$ and L is invertible.

Conversely, let L be invertible; that is, $L \circ L^{-1} = I_W$ and $L^{-1} \circ L = I_V$. We show that L is one-to-one and onto. Suppose that $L(\mathbf{v}_1) = L(\mathbf{v}_2)$ for $\mathbf{v}_1$, $\mathbf{v}_2$ in V. Then $L^{-1}(L(\mathbf{v}_1)) = L^{-1}(L(\mathbf{v}_2))$, so $\mathbf{v}_1 = \mathbf{v}_2$, which means that L is one-to-one. Also, if $\mathbf{w}$ is a vector in W, then $L(L^{-1}(\mathbf{w})) = \mathbf{w}$, so if we let $L^{-1}(\mathbf{w}) = \mathbf{v}$, then $L(\mathbf{v}) = \mathbf{w}$. Thus L is onto.

We now show that L^{-1} is a linear transformation. Let $\mathbf{w}_1$, $\mathbf{w}_2$ be in W, where $L(\mathbf{v}_1) = \mathbf{w}_1$ and $L(\mathbf{v}_2) = \mathbf{w}_2$ for $\mathbf{v}_1$, $\mathbf{v}_2$ in V. Then since

$$L(a\mathbf{v}_1 + b\mathbf{v}_2) = aL(\mathbf{v}_1) + bL(\mathbf{v}_2) = a\mathbf{w}_1 + b\mathbf{w}_2 \quad \text{for } a, b \text{ real numbers,}$$

we have

$$L^{-1}(a\mathbf{w}_1 + b\mathbf{w}_2) = a\mathbf{v}_1 + b\mathbf{v}_2 = aL^{-1}(\mathbf{w}_1) + bL^{-1}(\mathbf{w}_2),$$

which implies (by Exercise T.4 in Section 10.1) that L^{-1} is a linear transformation.

Finally, since $L \circ L^{-1} = I_W$, $L^{-1} \circ L = I_V$, and inverses are unique, we conclude that $(L^{-1})^{-1} = L$. ■

EXAMPLE 4 Let $L\colon R^4 \to R^2$ be the linear transformation defined in Example 5 of Section 10.2:

$$L\left(\begin{bmatrix} x \\ y \\ z \\ w \end{bmatrix}\right) = \begin{bmatrix} x + y \\ z + w \end{bmatrix}.$$

As we have seen in Example 5, Section 10.2, $\ker L$ has dimension 2, so L is not one-to-one and thus is not invertible. ■

EXAMPLE 5 Consider the linear operator $L\colon R^3 \to R^3$ defined by

$$L\left(\begin{bmatrix} a_1 \\ a_2 \\ a_3 \end{bmatrix}\right) = \begin{bmatrix} 1 & 1 & 1 \\ 2 & 2 & 1 \\ 0 & 1 & 1 \end{bmatrix}\begin{bmatrix} a_1 \\ a_2 \\ a_3 \end{bmatrix}.$$

Since $\ker L = \{\mathbf{0}\}$ (verify), L is one-to-one and by Corollary 10.3 it is also onto, so it is invertible. To obtain L^{-1}, we proceed as follows. Since $L^{-1}(\mathbf{w}) = \mathbf{v}$, we must solve $L(\mathbf{v}) = \mathbf{w}$ for $\mathbf{v}$. We have

$$L(\mathbf{v}) = L\left(\begin{bmatrix} a_1 \\ a_2 \\ a_3 \end{bmatrix}\right) = \begin{bmatrix} a_1 + a_2 + a_3 \\ 2a_1 + 2a_2 + a_3 \\ a_2 + a_3 \end{bmatrix} = \mathbf{w} = \begin{bmatrix} b_1 \\ b_2 \\ b_3 \end{bmatrix}.$$

We are then solving the linear system

$$\begin{aligned} a_1 + a_2 + a_3 &= b_1 \\ 2a_1 + 2a_2 + a_3 &= b_2 \\ a_2 + a_3 &= b_3 \end{aligned}$$

for a_1, a_2, and a_3. We find that (verify)

$$\begin{bmatrix} a_1 \\ a_2 \\ a_3 \end{bmatrix} = \mathbf{v} = L^{-1}(\mathbf{w}) = L^{-1}\left(\begin{bmatrix} b_1 \\ b_2 \\ b_3 \end{bmatrix}\right) = \begin{bmatrix} b_1 - b_3 \\ -2b_1 + b_2 + b_3 \\ 2b_1 - b_2 \end{bmatrix}.$$

■

Remark In Example 5 it was rather straightforward to find $L^{-1}(\mathbf{w})$. In general, if $L\colon V \to W$ is an invertible linear transformation, it is not always so easy to find an expression for $L^{-1}(\mathbf{w})$ for $\mathbf{w}$ in W. In Example 6 we shall solve this problem quite readily by using the matrix representing L^{-1} with respect to a basis S for V.

THEOREM B.4 *Let $L\colon V \to V$ be an invertible linear operator and let A be a matrix representing L with respect to a basis S for V. Then A^{-1} is the matrix representing L^{-1} with respect to S.*

Proof Let B be the matrix representing L^{-1} with respect to S. Since $L \circ L^{-1} = I_W$, the identity linear operator on W, the matrix representing $L \circ L^{-1}$ with respect to S is I_n (see Exercise T.2 in Section 10.3). From Theorem B.2, it follows that the matrix representing $L \circ L^{-1}$ with respect to S is AB. Thus

$$AB = I_n,$$

which implies (by Theorem 1.11 in Section 1.7) that $B = A^{-1}$. ■

We can now complete our list of nonsingular equivalences.

List of Nonsingular Equivalences

The following statements are equivalent for an $n \times n$ matrix A.

1. A is nonsingular.
2. $\mathbf{x} = \mathbf{0}$ is the only solution to $A\mathbf{x} = \mathbf{0}$.
3. A is row equivalent to I_n.
4. The linear system $A\mathbf{x} = \mathbf{b}$ has a unique solution for every $n \times 1$ matrix $\mathbf{b}$.
5. $\det(A) \neq 0$.
6. A has rank n.
7. A has nullity 0.
8. The rows of A form a linearly independent set of n vectors in R^n.
9. The columns of A form a linearly independent set of n vectors in R^n.
10. Zero is *not* an eigenvalue of A.
11. The linear operator $L\colon R^n \to R^n$ defined by $L(\mathbf{x}) = A\mathbf{x}$, for $\mathbf{x}$ in R^n, is one-to-one and onto.
12. The linear operator $L\colon R^n \to R^n$ defined by $L(\mathbf{x}) = A\mathbf{x}$, for $\mathbf{x}$ in R^n, is invertible.

EXAMPLE 6 Let $L\colon P_2 \to P_2$ be the linear operator defined by

$$L(at^2 + bt + c) = 2at^2 + bt + c.$$

The matrix representing L with respect to the basis $\{t^2 + 1, t - 1, t\}$ for P_2 is (verify)

$$A = \begin{bmatrix} 2 & 0 & 0 \\ 1 & 1 & 0 \\ -1 & 0 & 1 \end{bmatrix}.$$

We then find (verify) that

$$A^{-1} = \begin{bmatrix} \frac{1}{2} & 0 & 0 \\ -\frac{1}{2} & 1 & 0 \\ \frac{1}{2} & 0 & 1 \end{bmatrix}$$

is the matrix of L^{-1} with respect to S.

We now use A^{-1} to obtain a formula for $L^{-1}(at^2 + bt + c)$ as follows. Since A^{-1} is the matrix of L^{-1} with respect to S, we have

$$\left[L^{-1}(at^2 + bt + c)\right]_S = A^{-1}\left[at^2 + bt + c\right]_S. \tag{1}$$

To compute $\left[at^2+bt+c\right]_S$ we form

$$at^2+bt+c = k_1(t^2+1)+k_2(t-1)+k_3t$$

and solve the resulting linear system for k_1, k_2, and k_3, obtaining (verify)

$$k_1 = a, \quad k_2 = a-c, \quad k_3 = b+c-a.$$

Thus

$$\left[at^2+bt+c\right]_S = \begin{bmatrix} a \\ a-c \\ b+c-a \end{bmatrix}.$$

Substituting this coordinate vector into Equation (1) yields

$$\left[L^{-1}(at^2+bt+c)\right]_S = \begin{bmatrix} \frac{1}{2} & 0 & 0 \\ -\frac{1}{2} & 1 & 0 \\ \frac{1}{2} & 0 & 1 \end{bmatrix} \begin{bmatrix} a \\ a-c \\ b+c-a \end{bmatrix} = \begin{bmatrix} \frac{1}{2}a \\ \frac{1}{2}a-c \\ b+c-\frac{1}{2}a \end{bmatrix}.$$

Then

$$\begin{aligned} L^{-1}(at^2+bt+c) &= \tfrac{1}{2}a(t^2+1)+\left(\tfrac{1}{2}a-c\right)(t-1)+\left(b+c-\tfrac{1}{2}a\right)t \\ &= \tfrac{1}{2}at^2+bt+c. \end{aligned}$$

■

Key Terms

Composite linear transformation
Invertible linear transformation
Inverse linear transformation

B.2 Exercises

1. Let $L_1: R^2 \to R^3$ and $L_2: R^3 \to R^3$ be defined by

$$L_1(x, y) = (x+y, x-y, 2x+y),$$
$$L_2(x, y, z) = (x+y+z, y+z, x+z).$$

Compute

(a) $(L_2 \circ L_1)(-1, 1)$ (b) $(L_2 \circ L_1)(x, y)$

2. Let $L_1: R^2 \to R^2$ and $L_2: R^2 \to R^3$ be defined by

$$L_1\left(\begin{bmatrix} x \\ y \end{bmatrix}\right) = \begin{bmatrix} x \\ 2y-x \end{bmatrix},$$

$$L_2\left(\begin{bmatrix} x \\ y \end{bmatrix}\right) = \begin{bmatrix} 3x-2y \\ x+y \\ x-y \end{bmatrix}.$$

Compute

(a) $(L_2 \circ L_1)\left(\begin{bmatrix} 2 \\ -1 \end{bmatrix}\right)$ (b) $(L_2 \circ L_1)\left(\begin{bmatrix} x \\ y \end{bmatrix}\right)$

3. Let $L_1: P_1 \to P_1$ and $L_2: P_1 \to P_2$ be defined by

$$L_1(at+b) = 2at-b$$

and

$$L_2(at+b) = t(at+b).$$

Compute

(a) $(L_2 \circ L_1)(3t+2)$ (b) $(L_2 \circ L_1)(at+b)$

4. Let $L_1: P_2 \to P_2$ and $L_2: P_2 \to P_2$ be defined by

$$L_1(at^2+bt+c) = 2at+b$$

and

$$L_2(at^2+bt+c) = at+c.$$

Compute

(a) $(L_2 \circ L_1)(2t^2-3t+1)$

(b) $(L_1 \circ L_2)(2t^2-3t+1)$

(c) $(L_2 \circ L_1)(at^2+bt+c)$

(d) $(L_1 \circ L_2)(at^2+bt+c)$

5. Let $L_1: R^2 \to R^2$ and $L_2: R^2 \to R^2$ be defined by

$$L_1(x, y) = (x+y, x-2y)$$

and

$$L_2(x, y) = (y, x-y).$$

Compute

(a) $(L_2 \circ L_1)(1, 2)$ (b) $(L_1 \circ L_2)(1, 2)$
(c) $(L_2 \circ L_1)(x, y)$ (d) $(L_1 \circ L_2)(x, y)$

6. Let L_1 and L_2 be as defined in Exercise 2. Let

$$S = \{(1, 1), (0, 1)\}$$

and

$$T = \{(1, 0, 0), (0, 1, -1), (1, 1, 0)\}$$

be bases for R^2 and R^3, respectively.

(a) Compute the matrix B of $L_2 \circ L_1$ with respect to S and T.

(b) Compute the matrix A_1 of L_1 with respect to S and the matrix A_2 of L_2 with respect to S and T. Verify that $B = A_2A_1$.

7. Repeat Exercise 6 with L_1 and L_2 as defined in Exercise 3, and let

$$S = \{t + 1, t - 1\} \quad \text{and} \quad T = \{t^2 + 1, t, t - 1\}$$

be bases for P_1 and P_2, respectively.

8. Let L_1 and L_2 be as defined in Exercise 5, and let $S = \{(1, -1), (0, 1)\}$ and $T = \{(1, 0), (2, 1)\}$ be bases for R^2. Compute the matrix of

(a) $L_2 \circ L_1$ with respect to S

(b) $L_1 \circ L_2$ with respect to S

(c) $L_2 \circ L_1$ with respect to S and T

(d) $L_1 \circ L_2$ with respect to S and T

9. Let $L_1: R^2 \to R^2$ and $L_2: R^2 \to R^2$ be linear transformations whose matrices with respect to bases S and T for R^2 are

$$A_1 = \begin{bmatrix} 1 & 2 \\ -1 & 3 \end{bmatrix} \quad \text{and} \quad A_2 = \begin{bmatrix} 0 & 1 \\ -2 & 3 \end{bmatrix}.$$

(a) Compute the matrix of $L_2 \circ L_1$ with respect to S and T.

(b) Compute the matrix of $L_1 \circ L_2$ with respect to S and T.

10. Let $L: R^3 \to R^3$ be defined by

$$L\left(\begin{bmatrix} 1 \\ 0 \\ 0 \end{bmatrix}\right) = \begin{bmatrix} 1 \\ 2 \\ 3 \end{bmatrix}, \quad L\left(\begin{bmatrix} 0 \\ 1 \\ 0 \end{bmatrix}\right) = \begin{bmatrix} 0 \\ 1 \\ 1 \end{bmatrix},$$

$$L\left(\begin{bmatrix} 0 \\ 0 \\ 1 \end{bmatrix}\right) = \begin{bmatrix} 1 \\ 1 \\ 0 \end{bmatrix}.$$

(a) Show that L is invertible.

(b) Find $L^{-1}\left(\begin{bmatrix} 2 \\ 3 \\ 4 \end{bmatrix}\right)$.

In Exercises 11 through 18, determine whether the given linear transformation is invertible. If it is, find its inverse.

11. $L: R^2 \to R^3$ defined by
$L(x, y) = (x + y, x - y, x + 2y)$

12. $L: R^2 \to R^2$ defined by $L(x, y) = (x - y, x + 3y)$

13. $L: R^3 \to R^3$ defined by

$$L\left(\begin{bmatrix} x \\ y \\ z \end{bmatrix}\right) = \begin{bmatrix} 1 & 0 & 1 \\ 0 & 1 & 1 \\ 1 & 0 & 2 \end{bmatrix}\begin{bmatrix} x \\ y \\ z \end{bmatrix}$$

14. $L: R^2 \to R^2$ defined by $L(x, y) = (x - y, x - y)$

15. $L: R^3 \to R^3$ defined by

$$L\left(\begin{bmatrix} x \\ y \\ z \end{bmatrix}\right) = \begin{bmatrix} 1 & 1 & 1 \\ 0 & 1 & 2 \\ -2 & -1 & 0 \end{bmatrix}\begin{bmatrix} x \\ y \\ z \end{bmatrix}$$

16. $L: P_1 \to P_1$ defined by $L(at + b) = -bt + a$

17. $L: P_2 \to P_2$ defined by
$L(at^2 + bt + c) = -at^2 + bt - c$

18. $L: P_2 \to P_2$ defined by $L(at^2 + bt + c) = 2at^2 + bt$

In Exercises 19 through 22, determine whether L is invertible from the given information. [*Recall that* nullity $L = \dim(\ker L)$ *and* rank $L = \dim(\text{range } L)$.]

19. $L: R^4 \to R^4$, rank $L = 4$

20. $L: R^4 \to R^4$, nullity $L = 2$

21. $L: P_2 \to P_2$, nullity $L = 1$

22. $L: P_3 \to P_3$, rank $L = 4$

23. Let $L: R^3 \to R^3$ be the linear transformation defined in Exercise 10. Find the matrix representing L^{-1} with respect to the natural basis for R^3.

24. Let $L: R^3 \to R^3$ be the linear transformation defined by $L(\mathbf{x}) = A\mathbf{x}$, where

$$A = \begin{bmatrix} 1 & 1 & 1 \\ 0 & 1 & 2 \\ 1 & 2 & 2 \end{bmatrix}.$$

(a) Show that L is invertible.

(b) Find the matrix representing L^{-1} with respect to the natural basis for R^3.

25. Let $L: R^3 \to R^3$ be the invertible linear transformation represented by

$$A = \begin{bmatrix} 2 & 0 & 4 \\ -1 & 1 & -2 \\ 2 & 3 & 3 \end{bmatrix}$$

with respect to a basis S for R^3. Find the matrix of L^{-1} with respect to S.

26. Let $L: P_1 \to P_1$ be the invertible linear transformation represented by

$$A = \begin{bmatrix} 2 & 3 \\ 1 & 2 \end{bmatrix}$$

with respect to a basis S for P_1. Find the matrix of L^{-1} with respect to S.

Theoretical Exercises

T.1. Prove Theorem B.1.

T.2. Let $L: V \to W$ be a linear transformation and let I_V and I_W be the identity linear transformations on V and W, respectively. Show that

$$L \circ I_V = L$$
$$I_W \circ L = L.$$

T.3. Let $L: V \to V$ be a linear operator and let O_V be the zero linear transformation on V. Show that

$$L \circ O_V = O_V$$
$$O_V \circ L = O_V$$

T.4. Let $L: V \to V$ be a linear operator whose matrix with respect to a basis S for V is A. Show that A^2 is the matrix of $L^2 = L \circ L$ with respect to S. Moreover, show that if k is a positive integer, then A^k is the matrix of $L^k = L \circ L \circ \cdots \circ L$ (k times) with respect to S.

T.5. Let $L_1: V \to V$ and $L_2: V \to V$ be invertible linear operators. Show that $L_2 \circ L_1$ is also invertible and that $(L_2 \circ L_1)^{-1} = L_1^{-1} \circ L_2^{-1}$.

T.6. Let $L: V \to V$ be an invertible linear operator and let c be a nonzero scalar. Show that cL is an invertible linear operator and that $(cL)^{-1} = \dfrac{1}{c}L^{-1}$.

T.7. Let $L: M_{22} \to M_{22}$ be defined by $L(A) = A^T$. Is L invertible? If it is, find L^{-1}.

T.8. Let $L: M_{22} \to M_{22}$ be defined by $L(A) = BA$, where

$$B = \begin{bmatrix} 1 & 2 \\ -2 & -3 \end{bmatrix}.$$

Is L invertible? If it is, find L^{-1}.

T.9. Let $L: V \to V$ be a linear operator, where V is an n-dimensional vector space. Show that the following are equivalent:

(a) L is invertible.

(b) Rank $L = n$.
[Recall that rank L = dim(range L).]

(c) Nullity $L = 0$.
[Recall that nullity L = dim(ker L).]

T.10. Let $L_1: V \to V$ and $L_2: V \to V$ be linear transformations on a vector space V. Show that

$$(L_1 + L_2)^2 = L_1^2 + 2L_1 \circ L_2 + L_2^2$$

if and only if $L_1 \circ L_2 = L_2 \circ L_1$.

T.11. Let V be an inner product space, and let $\mathbf{w}$ be a fixed vector in V. Let $L: V \to V$ be defined by $L(\mathbf{v}) = (\mathbf{v}, \mathbf{w})$ for $\mathbf{v}$ in V. Show that L is a linear transformation.

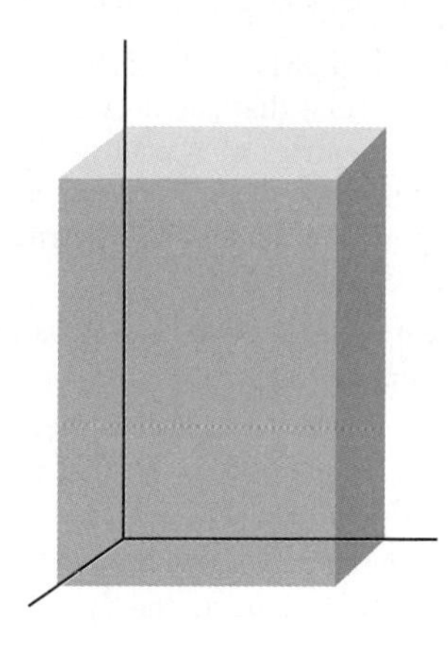

Glossary for Linear Algebra

Additive inverse of a matrix: The additive inverse of an $m \times n$ matrix A is an $m \times n$ matrix B such that $A + B = O$. Such a matrix B is the negative of A, denoted $-A$, which is equal to $(-1)A$.

Adjoint: For an $n \times n$ matrix $A = \left[a_{ij}\right]$ the adjoint of A, denoted adj A is the transpose of the matrix formed by replacing each entry by its cofactor A_{ij}; that is, adj $A = \left[A_{ji}\right]$.

Angle between vectors: For nonzero vectors $\mathbf{u}$ and $\mathbf{v}$ in R^n the angle θ between $\mathbf{u}$ and $\mathbf{v}$ is determined from the expression

$$\cos(\theta) = \frac{\mathbf{u} \cdot \mathbf{v}}{\|\mathbf{u}\| \|\mathbf{v}\|}.$$

Augmented matrix: For the linear system $A\mathbf{x} = \mathbf{b}$, the augmented matrix is formed by adjoining to the coefficient matrix A the right side vector $\mathbf{b}$. We expressed the augmented matrix as $\left[A \,\vdots\, \mathbf{b}\right]$.

Back substitution: If $U = \left[u_{ij}\right]$ is an upper triangular matrix all of whose diagonal entries are not zero, then the linear system $U\mathbf{x} = \mathbf{b}$ can be solved by back substitution. The process starts with the last equation and computes

$$x_n = \frac{b_n}{u_{nn}};$$

we use the next to last equation and compute

$$x_{n-1} = \frac{b_{n-1} - u_{n-1\,n} x_n}{u_{n-1\,n-1}};$$

continuing in this fashion using the jth equation we compute

$$x_j = \frac{b_j - \sum_{k=n}^{j+1} u_{jk} x_k}{u_{jj}}.$$

Basis: A set of vectors $S = \{\mathbf{v}_1, \mathbf{v}_2, \ldots, \mathbf{v}_k\}$ from a vector space V is called a basis for V provided S spans V and S is a linearly independent set.

Cauchy–Schwarz inequality: For vectors $\mathbf{v}$ and $\mathbf{u}$ in R^n, the Cauchy–Schwarz inequality says that the absolute value of the dot product of $\mathbf{v}$ and $\mathbf{u}$ is less than or equal to the product of the lengths of $\mathbf{v}$ and $\mathbf{u}$; that is, $|\mathbf{v} \cdot \mathbf{u}| \leq \|\mathbf{v}\| \|\mathbf{u}\|$.

Characteristic equation: For a square matrix A, its characteristic equation is given by $f(t) = \det(A - tI) = 0$.

Characteristic polynomial: For a square matrix A, its characteristic polynomial is given by $f(t) = \det(A - tI)$.

Closure properties: Let V be a given set, with members that we call vectors, and two operations, one called vector addition, denoted $\oplus$, and the second called scalar multiplication, denoted $\odot$. We say that V is closed under $\oplus$, provided for $\mathbf{u}$ and $\mathbf{v}$ in V, $\mathbf{u} \oplus \mathbf{v}$ is a member of V. We say that V is closed under $\odot$, provided for any real number k, $k \odot \mathbf{u}$ is a member of V.

Coefficient matrix: A linear system of m equations in n unknowns has the form

$$\begin{aligned} a_{11}x_1 + a_{12}x_2 + \cdots + a_{1n}x_n &= b_1 \\ a_{21}x_1 + a_{22}x_2 + \cdots + a_{2n}x_n &= b_2 \\ \vdots \qquad \vdots \qquad\quad & \vdots \qquad \vdots \\ a_{m1}x_1 + a_{m2}x_2 + \cdots + a_{mn}x_n &= b_m. \end{aligned}$$

The matrix

$$A = \begin{bmatrix} a_{11} & a_{12} & \cdots & a_{1n} \\ a_{21} & a_{22} & \cdots & a_{2n} \\ \vdots & \vdots & & \vdots \\ a_{m1} & a_{m2} & \cdots & a_{mn} \end{bmatrix}$$

is called the coefficient matrix of the linear system.

Cofactor: For an $n \times n$ matrix $A = \left[a_{ij}\right]$ the cofactor A_{ij} of a_{ij} is defined as $A_{ij} = (-1)^{i+j} \det(M_{ij})$, where M_{ij} is the ij-minor of A.

Column rank: The column rank of a matrix A is the dimension of the column space of A or equivalently the number of linearly independent columns of A.

Column space: The column space of a real $m \times n$ matrix A is the subspace of R^m spanned by the columns of A.

Complex vector space: A complex vector space V is a set, with members that we call vectors, and two operations: one called vector addition, denoted $\oplus$, and the second called scalar multiplication, denoted $\odot$. We require that V be closed under $\oplus$, that is, for $\mathbf{u}$ and $\mathbf{v}$ in V, $\mathbf{u} \oplus \mathbf{v}$ is a member of V; in addition we require that V be closed under $\odot$, that is, for any complex number k, $k \odot \mathbf{u}$ is a member of V. There are 8 other properties that must be satisfied before V with the two operations $\oplus$ and $\odot$ is called a complex vector space. (See pages A12 and 272 for further details.)

Complex vector subspace: A subset W of a complex vector space V that is closed under addition and scalar multiplication is called a complex subspace of V.

Components of a vector: The components of a vector $\mathbf{v}$ in R^n are its entries;

$$\mathbf{v} = \begin{bmatrix} v_1 \\ v_2 \\ \vdots \\ v_n \end{bmatrix}.$$

Composite linear transformation: Let L_1 and L_2 be linear transformations with $L_1 \colon V \to W$ and $L_2 \colon W \to U$. Then the composition $L_2 \circ L_1 \colon V \to U$ is a linear transformation and for $\mathbf{v}$ in V, we compute $(L_2 \circ L_1)(\mathbf{v}) = L_2(L_1(v))$.

Computation of a determinant via reduction to triangular form: For an $n \times n$ matrix A the determinant of A, denoted $\det(A)$ or $|A|$, can be computed with the aid of elementary row operations as follows. Use elementary row operations on A, keeping track of the operations used, to obtain an upper triangular matrix. Using the changes in the determinant as the result of applying a row operation as discussed in Section 3.1 and the fact that the determinant of an upper triangular matrix is the product of its diagonal entries, we can obtain an appropriate expression for $\det(A)$.

Consistent linear system: A linear system $A\mathbf{x} = \mathbf{b}$ is called consistent if the system has at least one solution.

Coordinates: The coordinates of a vector $\mathbf{v}$ in a vector space V with ordered basis $S = \{\mathbf{v}_1, \mathbf{v}_2, \ldots, \mathbf{v}_n\}$ are the coefficients $c_1, c_2, \ldots, c_n$ such that $\mathbf{v} = c_1\mathbf{v}_1 + c_2\mathbf{v}_2 + \cdots + c_n\mathbf{v}_n$. We denote the coordinates of $\mathbf{v}$ relative to the basis S by $[\mathbf{v}]_S$ and write

$$[\mathbf{v}]_S = \begin{bmatrix} c_1 \\ c_2 \\ \vdots \\ c_n \end{bmatrix}.$$

Cross product: The cross product of a pair of vectors $\mathbf{u}$ and $\mathbf{v}$ from R^3 is denoted $\mathbf{u} \times \mathbf{v}$, and is computed as the determinant

$$\begin{vmatrix} \mathbf{i} & \mathbf{j} & \mathbf{k} \\ u_1 & u_2 & u_3 \\ v_1 & v_2 & v_3 \end{vmatrix},$$

where $\mathbf{i}$, $\mathbf{j}$, and $\mathbf{k}$ are the unit vectors in the x-, y-, and z-directions, respectively.

Defective matrix: A square matrix A is called defective if it has an eigenvalue of multiplicity $m > 1$ for which the associated eigenspace has a basis with fewer than m vectors.

Determinant: For an $n \times n$ matrix A the determinant of A, denoted $\det(A)$ or $|A|$, is a scalar that is computed as the sum of all possible products of n entries of A each with its appropriate sign, with exactly one entry from each row and exactly one entry from each column. (For details and alternative computational procedures see Chapter 3.)

Diagonal matrix: A square matrix $A = [a_{ij}]$ is called diagonal provided $a_{ij} = 0$ whenever $i \neq j$.

Diagonalizable: A square matrix A is called diagonalizable provided it is similar to a diagonal matrix D; that is, there exists a nonsingular matrix P such that $P^{-1}AP = D$.

Difference of matrices: The difference of the $m \times n$ matrices A and B is denoted $A - B$ and is equal to the sum $A + (-1)B$. The difference $A - B$ is the $m \times n$ matrix whose entries are the difference of corresponding entries of A and B.

Difference of vectors: The difference of the vectors $\mathbf{v}$ and $\mathbf{w}$ in a vector space V is denoted $\mathbf{v} - \mathbf{w}$, which is equal to the sum $\mathbf{v} + (-1)\mathbf{w}$. If $V = R^n$, then $\mathbf{v} - \mathbf{w}$ is computed as the difference of corresponding entries.

Dilation: The linear transformation $L \colon R^n \to R^n$ given by $L(\mathbf{v}) = k\mathbf{v}$, for $k > 1$, is called a dilation.

Dimension: The dimension of a nonzero vector space V is the number of vectors in a basis for V. The dimension of the vector space $\{\mathbf{0}\}$ is defined as zero.

Distance between points (or vectors): The distance between the points $(u_1, u_2, \ldots, u_n)$ and $(v_1, v_2, \ldots, v_n)$ is the length of the vector $\mathbf{u} - \mathbf{v}$, where $\mathbf{u} = (u_1, u_2, \ldots, u_n)$ and $\mathbf{v} = (v_1, v_2, \ldots, v_n)$, and is given by

$$\|\mathbf{u} - \mathbf{v}\| = \sqrt{(u_1 - v_1)^2 + (u_2 - v_2)^2 + \cdots + (u_n - v_n)^2}.$$

Thus we see that the distance between vectors in R^n is also $\|\mathbf{u} - \mathbf{v}\|$.

Dot product: For vectors $\mathbf{v}$ and $\mathbf{w}$ in R^n the dot product of $\mathbf{v}$ and $\mathbf{w}$ is also called the standard inner product or just the inner product of $\mathbf{v}$ and $\mathbf{w}$. The dot product of $\mathbf{v}$ and $\mathbf{w}$ in R^n is denoted $\mathbf{v} \cdot \mathbf{w}$ and is computed as

$$\mathbf{v} \cdot \mathbf{w} = v_1 w_1 + v_2 w_2 + \cdots + v_n w_n.$$

Eigenspace: The set of all eigenvectors of a square matrix A associated with a specified eigenvalue λ of A, together with the zero vector, is called the eigenspace associated with the eigenvalue λ.

Eigenvalue: An eigenvalue of an $n \times n$ matrix A is a scalar λ for which there exists a nonzero n-vector $\mathbf{x}$ such that $A\mathbf{x} = \lambda\mathbf{x}$. The vector $\mathbf{x}$ is an eigenvector associated with the eigenvalue λ.

Eigenvector: An eigenvector of an $n \times n$ matrix A is a nonzero n-vector $\mathbf{x}$ such that $A\mathbf{x}$ is a scalar multiple of $\mathbf{x}$; that is, there exists some scalar λ such that $A\mathbf{x} = \lambda\mathbf{x}$. The scalar is an eigenvalue of the matrix A.

Elementary row operations: An elementary row operation on a matrix is any of the following three operations: (1) an interchange of rows, (2) multiplying a row by a nonzero scalar, and (3) replacing a row by adding a scalar multiple of a different row to it.

Equal matrices: The $m \times n$ matrices A and B are equal provided corresponding entries are equal; that is, $A = B$ if $a_{ij} = b_{ij}$, $i = 1, 2, \ldots, m$, $j = 1, 2, \ldots, n$.

Equal vectors: Vectors $\mathbf{v}$ and $\mathbf{w}$ in R^n are equal provided corresponding entries are equal; that is, $\mathbf{v} = \mathbf{w}$ if their corresponding components are equal.

Finite-dimensional vector space: A vector space V that has a basis that is a finite subset of V is said to be finite dimensional.

Forward substitution: If $L = \left[l_{ij}\right]$ is a lower triangular matrix all of whose diagonal entries are not zero, then the linear system $L\mathbf{x} = \mathbf{b}$ can be solved by forward substitution. The process starts with the first equation and computes

$$x_1 = \frac{b_1}{l_{11}};$$

next we use the second equation and compute

$$x_2 = \frac{b_2 - l_{21}x_1}{l_{22}};$$

continuing in this fashion using the jth equation we compute

$$x_j = \frac{b_j - \displaystyle\sum_{k=1}^{j-1} l_{jk}x_k}{l_{jj}}.$$

Fundamental vector spaces associated with a matrix: If A is an $m \times n$ matrix there are four fundamental subspaces associated with A: (1) the null space of A, a subspace of R^n; (2) the row space of A, a subspace of R^n; (3) the null space of A^T, a subspace of R^m; and (4) the column space of A, a subspace of R^m.

Gaussian elimination: For the linear system $A\mathbf{x} = \mathbf{b}$ form the augmented matrix $\left[A \mid \mathbf{b}\right]$. Compute the row echelon form of the augmented matrix; then the solution can be computed using back substitution.

Gauss–Jordan reduction: For the linear system $A\mathbf{x} = \mathbf{b}$ form the augmented matrix $\left[A \mid \mathbf{b}\right]$. Compute the reduced row echelon form of the augmented matrix; then the solution can be computed using back substitution.

General solution: The general solution of a consistent linear system $A\mathbf{x} = \mathbf{b}$ is the set of all solutions to the system. If $\mathbf{b} = \mathbf{0}$, then the general solution is just the set of all solutions to the homogeneous system $A\mathbf{x} = \mathbf{0}$, denoted $\mathbf{x}_h$. If $\mathbf{b} \neq \mathbf{0}$, then the general solution of the nonhomogeneous system consists of a particular solution of $A\mathbf{x} = \mathbf{b}$, denoted $\mathbf{x}_p$, together with $\mathbf{x}_h$; that is, the general solution is expressed as $\mathbf{x}_p + \mathbf{x}_h$.

Gram–Schmidt process: The Gram–Schmidt process converts a basis for a subspace into an orthonormal basis for the same subspace.

Hermitian matrix: An $n \times n$ complex matrix A is called Hermitian provided $\overline{A}^T = A$.

Homogeneous system: A homogeneous system is a linear system in which the right side of each equation is zero. We denote a homogeneous system by $A\mathbf{x} = \mathbf{0}$.

Identity matrix: The $n \times n$ identity matrix, denoted I_n, is a diagonal matrix with diagonal entries of all 1s.

Inconsistent linear system: A linear system $A\mathbf{x} = \mathbf{b}$ that has no solution is called inconsistent.

Infinite-dimensional vector space: A vector space V for which there is no finite subset of vectors that form a basis for V is said to be infinite dimensional.

Inner product: For vectors $\mathbf{v}$ and $\mathbf{w}$ in R^n the inner product of $\mathbf{v}$ and $\mathbf{w}$ is also called the dot product or standard inner product of $\mathbf{v}$ and $\mathbf{w}$. The inner product of $\mathbf{v}$ and $\mathbf{w}$ in R^n is denoted $\mathbf{v} \cdot \mathbf{w}$ and is computed as

$$\mathbf{v} \cdot \mathbf{w} = v_1w_1 + v_2w_2 + \cdots + v_nw_n.$$

Invariant subspace: A subspace W of a vector space V is said to be invariant under the linear transformation $L\colon V \to V$, provided $L(\mathbf{v})$ is in W for all vectors $\mathbf{v}$ in W.

Inverse linear transformation: See invertible linear transformation.

Inverse of a matrix: An $n \times n$ matrix A is said to have an inverse provided there exists an $n \times n$ matrix B such that $AB = BA = I$. We call B the inverse of A and denote it as A^{-1}. In this case, A is also called nonsingular.

Invertible linear transformation: A linear transformation $L\colon V \to W$ is called invertible if there exists a linear transformation, denoted L^{-1}, such that $L^{-1}(L(\mathbf{v})) = \mathbf{v}$, for all vectors $\mathbf{v}$ in V and $L(L^{-1}(\mathbf{w})) = \mathbf{w}$, for all vectors $\mathbf{w}$ in W.

Isometry: An isometry is a linear transformation L that preserves the distance between pairs of vectors; that is, $\|L(\mathbf{v}) - L(\mathbf{u})\| = \|\mathbf{v} - \mathbf{u}\|$, for all vectors $\mathbf{u}$ and $\mathbf{v}$. Since an isometry preserves distances, it also preserves lengths; that is, $\|L(\mathbf{v})\| = \|\mathbf{v}\|$, for all vectors $\mathbf{v}$.

Length (or magnitude) of a vector: The length of a vector $\mathbf{v}$ in R^n is denoted $\|\mathbf{v}\|$ and is computed as the expression

$$\sqrt{v_1^2 + v_2^2 + \cdots + v_n^2}.$$

For a vector $\mathbf{v}$ in a vector space V on which an inner product (dot product) is defined, the length of $\mathbf{v}$ is computed as $\|\mathbf{v}\| = \sqrt{\mathbf{v} \cdot \mathbf{v}}$.

Linear combination: A linear combination of vectors $\mathbf{v}_1, \mathbf{v}_2, \ldots, \mathbf{v}_k$ from a vector space V is an expression of the form $c_1\mathbf{v}_1 + c_2\mathbf{v}_2 + \cdots + c_k\mathbf{v}_k$, where the $c_1, c_2, \ldots, c_k$ are scalars. A linear combination of the $m \times n$ matrices $A_1, A_2, \ldots, A_k$ is given by $c_1A_1 + c_2A_2 + \cdots + c_kA_k$.

Linear operator: A linear operator is a linear transformation L from a vector space to itself; that is, $L\colon V \to V$.

Linear system: A system of m linear equations in n unknowns $x_1, x_2, \ldots, x_n$ is a set of linear equations in the n unknowns. We express a linear system in matrix form as $A\mathbf{x} = \mathbf{b}$, where A is the matrix of coefficients, $\mathbf{x}$ is the vector of unknowns, and $\mathbf{b}$ is the vector of right sides of the linear equations. (See coefficient matrix.)

Linear transformation: A linear transformation $L\colon V \to W$ is a function assigning a unique vector $L(\mathbf{v})$ in W to each vector $\mathbf{v}$ in V such that two properties are satisfied: (1) $L(\mathbf{u}+\mathbf{v}) =$

$L(\mathbf{u}) + L(\mathbf{v})$, for every $\mathbf{u}$ and $\mathbf{v}$ in V, and (2) $L(k\mathbf{v}) = kL(\mathbf{v})$, for every $\mathbf{v}$ in V and every scalar k.

Linearly dependent: A set of vectors $S = \{\mathbf{v}_1, \mathbf{v}_2, \ldots, \mathbf{v}_n\}$ is called linearly dependent provided there exists a linear combination $c_1\mathbf{v}_1 + c_2\mathbf{v}_2 + \cdots + c_n\mathbf{v}_n$ that produces the zero vector when not all the coefficients are zero.

Linearly independent: A set of vectors $S = \{\mathbf{v}_1, \mathbf{v}_2, \ldots, \mathbf{v}_n\}$ is called linearly independent provided the only linear combination $c_1\mathbf{v}_1 + c_2\mathbf{v}_2 + \cdots + c_n\mathbf{v}_n$ that produces the zero vector is when all the coefficients are zero, that is, only when $c_1 = c_2 = \cdots = c_n = 0$.

Lower triangular matrix: A square matrix with zero entries above its diagonal entries is called lower triangular.

LU-factorization (or LU-decomposition): An LU-factorization of a square matrix A expresses A as the product of a lower triangular matrix L and an upper triangular matrix U; that is, $A = LU$.

Main diagonal of a matrix: The main diagonal of an $n \times n$ matrix A is the set of entries $a_{11}, a_{22}, \ldots, a_{nn}$.

Matrix: An $m \times n$ matrix A is a rectangular array of mn entries arranged in m rows and n columns.

Matrix addition: For $m \times n$ matrices $A = [a_{ij}]$ and $B = [b_{ij}]$, the addition of A and B is performed by adding corresponding entries; that is, $A + B = [a_{ij}] + [b_{ij}]$. This is also called the sum of the matrices A and B.

Matrix representing a linear transformation: Let $L\colon V \to W$ be a linear transformation from an n-dimensional space V to an m-dimensional space W. For a basis $S = \{\mathbf{v}_1, \mathbf{v}_2, \ldots, \mathbf{v}_n\}$ in V and a basis $T = \{\mathbf{w}_1, \mathbf{w}_2, \ldots, \mathbf{w}_m\}$ in W there exists an $m \times n$ matrix A, with column j of $A = [L(\mathbf{v}_j)]_T$ such that the coordinates of $L(\mathbf{x})$, for any x in V, with respect to the T basis can be computed as $[L(\mathbf{x})]_T = A[\mathbf{x}]_S$. We say A is the matrix representing the linear transformation L.

Matrix transformation: For an $m \times n$ matrix A the function f defined by $f(\mathbf{u}) = A\mathbf{u}$ for $\mathbf{u}$ in R^n is called the matrix transformation from R^n to R^m defined by the matrix A.

Minor: Let $A = [a_{ij}]$ be an $n \times n$ matrix and M_{ij} the $(n-1) \times (n-1)$ submatrix of A obtained by deleting the ith row and jth column of A. The determinant $\det(M_{ij})$ is called the minor of a_{ij}.

Multiplicity of an eigenvalue: The multiplicity of an eigenvalue λ of a square matrix A is the number of times λ is a root of the characteristic polynomial of A.

Natural (or standard) basis: The natural basis for R^n is the set of vectors $\mathbf{e}_j$ = column j (or, equivalently, row j) of the $n \times n$ identity matrix, $j = 1, 2, \ldots, n$.

Negative of a vector: The negative of a vector $\mathbf{u}$ is a vector $\mathbf{w}$ such that $\mathbf{u} + \mathbf{w} = \mathbf{0}$, the zero vector. We denote the negative of $\mathbf{u}$ as $-\mathbf{u} = (-1)\mathbf{u}$.

Nonhomogeneous system: A linear system $A\mathbf{x} = \mathbf{b}$ is called nonhomogeneous provided the vector $\mathbf{b}$ is not the zero vector.

Nonsingular (or invertible) matrix: An $n \times n$ matrix A is called nonsingular provided there exists an $n \times n$ matrix B such that $AB = BA = I$. We call B the inverse of A and denote it as A^{-1}.

Nontrivial solution: A nontrivial solution of a linear system $A\mathbf{x} = \mathbf{b}$ is any vector $\mathbf{x}$ containing at least one nonzero entry such that $A\mathbf{x} = \mathbf{b}$.

Normal matrix: An $n \times n$ complex matrix A is called normal provided $\left(\overline{A}^T\right) A = A \left(\overline{A}^T\right)$.

***n*-space**: The set of all n-vectors is called n-space. For vectors whose entries are real numbers we denote n-space as R^n. For a special case see 2-space.

Nullity: The nullity of the matrix A is the dimension of the null space of A.

***n*-vector**: A $1 \times n$ or an $n \times 1$ matrix is called an n-vector. When n is understood, we refer to n-vectors merely as vectors.

One-to-one: A function $f\colon S \to T$ is said to be one-to-one provided $f(s_1) \neq f(s_2)$ whenever s_1 and s_2 are distinct elements of S. A linear transformation $L\colon V \to W$ is called one-to-one provided L is a one-to-one function.

Onto: A function $f\colon S \to T$ is said to be onto provided for each member t of T there is some member s in S so that $f(s) = t$. A linear transformation $L\colon V \to W$ is called onto provided range $L = W$.

Ordered basis: A set of vectors $S = \{\mathbf{v}_1, \mathbf{v}_2, \ldots, \mathbf{v}_k\}$ in a vector space V is called an ordered basis for V provided S is a basis for V and if we reorder the vectors in S, this new ordering of the vectors in S is considered a different basis for V.

Orthogonal basis: A basis for a vector space V that is also an orthogonal set is called an orthogonal basis for V.

Orthogonal complement: The orthogonal complement of a set S of vectors in a vector space V is the set of all vectors in V that are orthogonal to all vectors in S.

Orthogonal matrix: A square matrix P is called orthogonal provided $P^{-1} = P^T$.

Orthogonal projection: For a vector $\mathbf{v}$ in a vector space V, the orthogonal projection of $\mathbf{v}$ onto a subspace W of V with orthonormal basis $\{\mathbf{w}_1, \mathbf{w}_2, \ldots, \mathbf{w}_k\}$ is the vector $\mathbf{w}$ in W, where $\mathbf{w} = (\mathbf{v} \cdot \mathbf{w}_1)\mathbf{w}_1 + (\mathbf{v} \cdot \mathbf{w}_2)\mathbf{w}_2 + \cdots + (\mathbf{v} \cdot \mathbf{w}_k)\mathbf{w}_k$. Vector $\mathbf{w}$ is the vector in W that is closest to $\mathbf{v}$.

Orthogonal set: A set of vectors $S = \{\mathbf{w}_1, \mathbf{w}_2, \ldots, \mathbf{w}_k\}$ from a vector space V on which an inner product is defined is an orthogonal set provided none of the vectors is the zero vector and the inner product of any two different vectors is zero.

Orthogonal vectors: A pair of vectors is called orthogonal provided their dot (inner) product is zero.

Orthogonally diagonalizable: A square matrix A is said to be orthogonally diagonalizable provided there exists an orthogonal matrix P such that $P^{-1}AP$ is a diagonal matrix. That is, A is similar to a diagonal matrix using an orthogonal matrix P.

Orthonormal basis: A basis for a vector space V that is also an orthonormal set is called an orthonormal basis for V.

Orthonormal set: A set of vectors $S = \{\mathbf{w}_1, \mathbf{w}_2, \ldots, \mathbf{w}_k\}$ from a vector space V on which an inner product is defined is an orthonormal set provided each vector is a unit vector and the inner product of any two different vectors is zero.

Parallel vectors: Two nonzero vectors are said to be parallel if one is a scalar multiple of the other.

Particular solution: A particular solution of a consistent linear system $A\mathbf{x} = \mathbf{b}$ is a vector $\mathbf{x}_p$ containing no arbitrary constants such that $A\mathbf{x}_p = \mathbf{b}$.

Partitioned matrix: A matrix that has been partitioned into submatrices by drawing horizontal lines between rows and/or vertical lines between columns is called a partitioned matrix. There are many ways to partition a matrix.

Perpendicular (or orthogonal) vectors: A pair of vectors is said to be perpendicular or orthogonal provided their dot product is zero.

Pivot: When using row operations on a matrix A, a pivot is a nonzero entry of a row that is used to zero-out entries in the column in which the pivot resides.

Positive definite: Matrix A is positive definite provided A is symmetric and all of its eigenvalues are positive.

Powers of a matrix: For a square matrix A and nonnegative integer k, the kth power of A, denoted A^k, is the product of A with itself k times; $A^k = A \cdot A \cdot \cdots \cdot A$, where there are k factors.

Projection: The projection of a point P in a plane onto a line L in the same plane is the point Q obtained by intersecting the line L with the line through P that is perpendicular to L. The linear transformation $L\colon R^3 \to R^2$ defined by $L(x, y, z) = (x, y)$ is called a projection of R^3 onto R^2. (See also orthogonal projection.)

Range: The range of a function $f\colon S \to T$ is the set of all members t of T such that there is a member s in S with $f(s) = t$. The range of a linear transformation $L\colon V \to W$ is the set of all vectors in W that are images under L of vectors in V.

Rank: Since row rank A = column rank A, we just refer to the rank of the matrix A as rank A. Equivalently, rank A = the number of linearly independent rows (columns) of A = the number of leading 1s in the reduced row echelon form of A.

Real vector space: A real vector space V is a set, with members that we call vectors and two operations: one is called vector addition, denoted $\oplus$, and the second called scalar multiplication, denoted $\odot$. We require that V be closed under $\oplus$; that is, for $\mathbf{u}$ and $\mathbf{v}$ in V, $\mathbf{u} \oplus \mathbf{v}$ is a member of V. In addition we require that V be closed under $\odot$; that is, for any real number k, $k \odot \mathbf{u}$ is a member of V. There are 8 other properties that must be satisfied before V with the two operations $\oplus$ and $\odot$ is called a vector space. (See page 272 for details.)

Reduced row echelon form: A matrix is said to be in reduced row echelon form provided it satisfies the following properties: (1) All zero rows, if there are any, appear as bottom rows. (2) The first nonzero entry in a nonzero row is a 1; it is called a leading 1. (3) For each nonzero row, the leading 1 appears to the right and below any leading 1s in preceding rows. (4) If a column contains a leading 1, then all other entries in that column are zero.

Reflection: The linear transformation $L\colon R^2 \to R^2$ given by $L(x, y) = (x, -y)$ is called a reflection with respect to the x-axis. Similarly, $L(x, y) = (-x, y)$ is called a reflection with respect to the y-axis.

Roots of the characteristic polynomial: For a square matrix A, the roots of its characteristic polynomial $f(t) = \det(A - tI)$ are the eigenvalues of A.

Rotation: The linear transformation $L\colon R^2 \to R^2$ given by

$$L\left(\begin{bmatrix} x \\ y \end{bmatrix}\right) = \begin{bmatrix} \cos(\theta) & -\sin(\theta) \\ \sin(\theta) & \cos(\theta) \end{bmatrix} \begin{bmatrix} x \\ y \end{bmatrix}$$

is called a counterclockwise rotation in the plane by the angle θ.

Row echelon form: A matrix is said to be in row echelon form provided it satisfies the following properties: (1) All zero rows, if there are any, appear as bottom rows. (2) The first nonzero entry in a nonzero row is a 1; it is called a leading 1. (3) For each nonzero row, the leading 1 appears to the right and below any leading 1s in preceding rows.

Row equivalent: The $m \times n$ matrices A and B are row equivalent provided there exists a set of row operations that when performed on A yield B.

Row rank: The row rank of matrix A is the dimension of the row space of A or, equivalently, the number of linearly independent rows of A.

Row space: The row space of a real $m \times n$ matrix A is the subspace of R^n spanned by the rows of A.

Scalar matrix: Matrix A is a scalar matrix provided A is a diagonal matrix with equal diagonal entries.

Scalar multiple of a matrix: For an $m \times n$ matrix $A = \left[a_{ij}\right]$ and scalar r, the scalar multiple of A by r gives the $m \times n$ matrix $rA = \left[ra_{ij}\right]$.

Scalar multiple of a vector: If $\mathbf{v}$ is in real vector space V, then for any real number k, a scalar, the scalar multiple of $\mathbf{v}$ by k is denoted $k\mathbf{v}$. If $V = R^n$, then $k\mathbf{v} = (kv_1, kv_2, \ldots, kv_n)$.

Scalars: In a real vector space V the scalars are real numbers and are used when we form scalar multiples $k\mathbf{v}$, where $\mathbf{v}$ is in V. Also, when we form linear combinations of vectors the coefficients are scalars.

Shear: A shear in the x-direction is defined by the matrix transformation

$$L(\mathbf{u}) = \begin{bmatrix} 1 & k \\ 0 & 1 \end{bmatrix} \begin{bmatrix} u_1 \\ u_2 \end{bmatrix},$$

where k is a scalar. Similarly, a shear in the y-direction is given by

$$L(\mathbf{u}) = \begin{bmatrix} 1 & 0 \\ k & 1 \end{bmatrix} \begin{bmatrix} u_1 \\ u_2 \end{bmatrix}.$$

Similar matrices: Matrices A and B are similar provided there exists a nonsingular matrix P such that $A = P^{-1}BP$.

Singular (or noninvertible) matrix: A matrix A that has no inverse matrix is said to be singular. Any square matrix whose reduced row echelon form is not the identity matrix is singular.

Skew symmetric matrix: A square real matrix A such that $A = -A^T$ is called skew symmetric.

Solution space: The solution space of an $m \times n$ real homogeneous system $A\mathbf{x} = \mathbf{0}$ is the set W of all n-vectors $\mathbf{x}$ such that A times $\mathbf{x}$ gives the zero vector. W is a subspace of R^n.

Solution to a homogeneous system: A solution to a homogeneous system $A\mathbf{x} = \mathbf{0}$ is a vector $\mathbf{x}$ such that A times $\mathbf{x}$ gives the zero vector.

Solution to a linear system: A solution to a linear system $A\mathbf{x} = \mathbf{b}$ is any vector $\mathbf{x}$ such that A times $\mathbf{x}$ gives the vector $\mathbf{b}$.

Span: The span of a set $W = \{\mathbf{w}_1, \mathbf{w}_2, \dots, \mathbf{w}_k\}$, denoted by span W, from a vector space V is the set of all possible linear combinations of the vectors $\mathbf{w}_1, \mathbf{w}_2, \dots, \mathbf{w}_k$. Span W is a subspace of V.

Square matrix: A matrix with the same number of rows as columns is called a square matrix.

Standard inner product: For vectors $\mathbf{v}$ and $\mathbf{w}$ in R^n the standard inner product of $\mathbf{v}$ and $\mathbf{w}$ is also called the dot product of $\mathbf{v}$ and $\mathbf{w}$, denoted $\mathbf{v} \cdot \mathbf{w} = v_1 w_1 + v_2 w_2 + \cdots + v_n w_n$.

Submatrix: A matrix obtained from a matrix A by deleting rows and/or columns is called a submatrix of A.

Subspace: A subset W of a vector space V that is closed under addition and scalar multiplication is called a subspace of V.

Sum of vectors: The sum of two vectors is also called vector addition. In R^n adding corresponding components of the vectors performs the sum of two vectors. In a vector space V, $\mathbf{u} \oplus \mathbf{v}$ is computed using the definition of the operation $\oplus$.

Summation notation: A compact notation to indicate the sum of a set $\{a_1, a_2, \dots, a_n\}$; the sum of a_1 through a_n is denoted in summation notation as $\sum_{i=1}^{n} a_i$.

Symmetric matrix: A square real matrix A such that $A = A^T$ is called symmetric.

Transition matrix: Let $S = \{\mathbf{v}_1, \mathbf{v}_2, \dots, \mathbf{v}_n\}$ and $T = \{\mathbf{w}_1, \mathbf{w}_2, \dots, \mathbf{w}_n\}$ be bases for an n-dimensional vector space V. The transition matrix from the T-basis to the S-basis is an $n \times n$ matrix, denoted $P_{S \leftarrow T}$, that converts the coordinates of a vector $\mathbf{v}$ relative to the T-basis into the coordinates of $\mathbf{v}$ relative to the S-basis; $\left[\mathbf{v}\right]_S = P_{S \leftarrow T} \left[\mathbf{v}\right]_T$.

Translation: Let $T: V \rightarrow V$ be defined by $T(\mathbf{v}) = \mathbf{v} + \mathbf{b}$ for all $\mathbf{v}$ in V and any fixed vector $\mathbf{b}$ in V. We call this the translation by the vector $\mathbf{b}$.

Transpose of a matrix: The transpose of an $m \times n$ matrix A is the $n \times m$ matrix obtained by forming columns from each row of A. The transpose of A is denoted A^T.

Trivial solution: The trivial solution of a homogeneous system $A\mathbf{x} = \mathbf{0}$ is the zero vector.

2-space: The set of all 2-vectors is called 2-space. For vectors whose entries are real numbers we denote 2-space as R^2.

Unit vector: A vector of length 1 is called a unit vector.

Unitary matrix: An $n \times n$ complex matrix A is called unitary provided $A^{-1} = \overline{A}^T$.

Upper triangular matrix: A square matrix with zero entries below its diagonal entries is called upper triangular.

Vector: The generic name for any member of a vector space. (See also 2-vector and n-vector.)

Vector addition: The sum of two vectors is also called vector addition. In R^n adding corresponding components of the vectors performs vector addition.

Zero matrix: A matrix with all zero entries is called the zero matrix.

Zero polynomial: A polynomial all of whose coefficients are zero is called the zero polynomial.

Zero subspace: The subspace consisting of exactly the zero vector of a vector space is called the zero subspace.

Zero vector: A vector with all zero entries is called the zero vector.

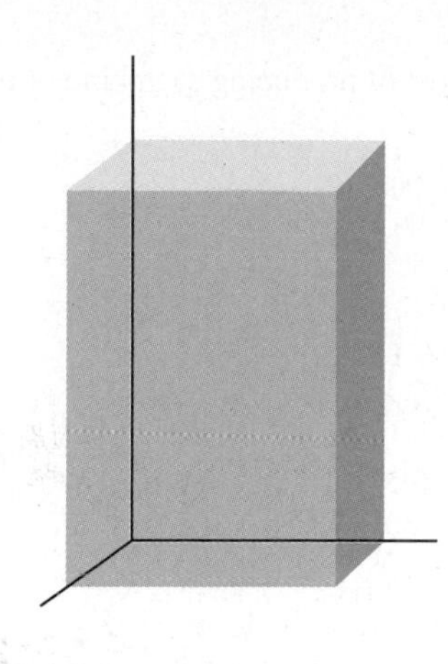

ANSWERS TO ODD-NUMBERED EXERCISES AND CHAPTER TESTS

Chapter 1

Section 1.1, page 8

1. $x = 4$, $y = 2$.

3. $x = -4$, $y = 2$, $z = 10$.

5. $x = 2$, $y = -1$, $z = -2$.

7. $x = -20$, $y = \frac{1}{4}r + 8$, $z = r$, where r is any real number.

9. No solution.

11. $x = 5$, $y = 1$.

13. No solution.

15. (a) $t = 10$. (b) One value is $t = 3$.

(c) The choice $t = 3$ in part (b) was arbitrary. Any choice for t, other than $t = 10$ makes the system inconsistent. Hence there are infinitely many ways to choose a value of t in part (b).

17. $x = 1$, $y = 1$, $z = 4$.

19. $r = -3$.

21. One, zero, infinitely many.

23. 20 tons of each type of fuel.

25. 3.2 ounces of A, 4.2 ounces of B, 2.0 ounces of C.

27. (a)
$$\begin{aligned} a + b + c &= -5 \\ a - b + c &= 1 \\ 4a + 2b + c &= 7. \end{aligned}$$
(b) $a = 5$, $b = -3$, $c = -7$.

Section 1.2, page 19

1. (a) $-3, -5, 4$. (b) $4, 5$.

(c) $2, 6, -1$.

3. $a = 0$, $b = 2$, $c = 1$, $d = 2$.

5. (a) $\begin{bmatrix} 1 & 4 \\ 10 & 18 \end{bmatrix}$.

(b) $3(2A) = 6A = \begin{bmatrix} 6 & 12 & 18 \\ 12 & 6 & 24 \end{bmatrix}$.

(c) $3A + 2A = 5A = \begin{bmatrix} 5 & 10 & 15 \\ 10 & 5 & 20 \end{bmatrix}$.

(d) $2(D + F) = 2D + 2F = \begin{bmatrix} -2 & 6 \\ 8 & 14 \end{bmatrix}$.

(e) $(2 + 3)D = 2D + 3D = \begin{bmatrix} 15 & -10 \\ 10 & 20 \end{bmatrix}$.

(f) Impossible.

7. (a) $\begin{bmatrix} 2 & 4 \\ 4 & 2 \\ 6 & 8 \end{bmatrix}$. (b) Impossible.

(c) $\begin{bmatrix} 1 & -4 \\ 2 & 1 \\ 3 & -2 \end{bmatrix}$.

(d) Impossible.

(e) $(-A)^T = -(A^T) = \begin{bmatrix} -1 & -2 \\ -2 & -1 \\ -3 & -4 \end{bmatrix}$.

(f) Impossible.

9. No.

11. (a) $\begin{bmatrix} 1 & 1 & 0 \\ 0 & 1 & 1 \\ 1 & 0 & 1 \end{bmatrix}$. (b) $\begin{bmatrix} 1 & 0 & 1 \\ 1 & 1 & 0 \\ 0 & 1 & 1 \end{bmatrix}$.

(c) $\begin{bmatrix} 0 & 0 & 0 \\ 0 & 0 & 0 \\ 0 & 0 & 0 \end{bmatrix}$. (d) $\begin{bmatrix} 0 & 0 & 0 \\ 0 & 0 & 0 \\ 0 & 0 & 0 \end{bmatrix}$.

(e) $\begin{bmatrix} 1 & 0 & 1 \\ 1 & 1 & 0 \\ 0 & 1 & 1 \end{bmatrix}$.

13. (a) $B = \begin{bmatrix} 1 & 0 \\ 0 & 0 \end{bmatrix}$. (b) $C = \begin{bmatrix} 0 & 1 \\ 1 & 1 \end{bmatrix}$.

15. $\mathbf{v} = \begin{bmatrix} 1 & 0 & 1 & 0 \end{bmatrix}$.

ML.1. (a) Commands: **A(2,3)**, **B(3,2)**, **B(1,2)**.

(b) For $\text{row}_1(\mathbf{A})$, use command **A(1,:)**.
For $\text{col}_3(\mathbf{A})$, use command **A(:,3)**.
For $\text{row}_2(\mathbf{B})$, use command **B(2,:)**.
(In this context the colon means "all.")

(c) Matrix B in **format long** is

$$\begin{bmatrix} 8.00000000000000 & 0.66666666666667 \\ 0.00497512437811 & -3.20000000000000 \\ 0.00001000000000 & 4.33333333333333 \end{bmatrix}.$$

Section 1.3, page 34

1. (a) 2. (b) 1. (c) 4. (d) 1.

3. ± 2. **5.** $\pm\frac{\sqrt{2}}{2}$.

7. (a) $\begin{bmatrix} 10 & -6 \\ 14 & -6 \end{bmatrix}$. (b) $\begin{bmatrix} 7 & 6 & -11 \\ 18 & 4 & -14 \\ 19 & -2 & -7 \end{bmatrix}$.

(c) Impossible. (d) $\begin{bmatrix} 26 & -9 \\ 4 & -5 \end{bmatrix}$.

(e) Impossible.

9. (a) 4. (b) 13. (c) 3. (d) 12.

11. $AB = \begin{bmatrix} -4 & 7 \\ 0 & 5 \end{bmatrix}$; $BA = \begin{bmatrix} -1 & 2 \\ 9 & 2 \end{bmatrix}$.

13. (a) $\begin{bmatrix} 6 \\ 25 \\ 10 \\ 25 \end{bmatrix}$. (b) $\begin{bmatrix} 12 \\ 11 \\ 17 \\ 20 \end{bmatrix}$.

15. $2\begin{bmatrix} 2 \\ -1 \\ 5 \end{bmatrix} + 1\begin{bmatrix} -3 \\ 2 \\ -1 \end{bmatrix} + 4\begin{bmatrix} 4 \\ 3 \\ -2 \end{bmatrix}$.

19. (a) $\begin{bmatrix} 2 & 0 & 0 & 1 \\ 3 & 2 & 3 & 0 \\ 2 & 3 & -4 & 0 \\ 1 & 0 & 3 & 0 \end{bmatrix}$.

(b) $\begin{bmatrix} 2 & 0 & 0 & 1 \\ 3 & 2 & 3 & 0 \\ 2 & 3 & -4 & 0 \\ 1 & 0 & 3 & 0 \end{bmatrix}\begin{bmatrix} x \\ y \\ z \\ w \end{bmatrix} = \begin{bmatrix} 7 \\ -2 \\ 3 \\ 5 \end{bmatrix}$.

(c) $\left[\begin{array}{cccc|c} 2 & 0 & 0 & 1 & 7 \\ 3 & 2 & 3 & 0 & -2 \\ 2 & 3 & -4 & 0 & 3 \\ 1 & 0 & 3 & 0 & 5 \end{array}\right]$.

21.
$$\begin{aligned} 2x - 4z &= 3 \\ y + 2z &= 5 \\ x + 3y + 4z &= -1. \end{aligned}$$

23. They are equivalent.

25. (a) $x\begin{bmatrix} 1 \\ 2 \end{bmatrix} + y\begin{bmatrix} 2 \\ -1 \end{bmatrix} = \begin{bmatrix} 3 \\ 5 \end{bmatrix}$.

(b) $x\begin{bmatrix} 2 \\ 1 \end{bmatrix} + y\begin{bmatrix} -3 \\ 4 \end{bmatrix} + z\begin{bmatrix} 5 \\ -1 \end{bmatrix} = \begin{bmatrix} -2 \\ 3 \end{bmatrix}$.

27. (a) $r = -5$. (b) BA^T.

29. $A + B = \left[\begin{array}{cc|c} 4 & 5 & 0 \\ 0 & 4 & 1 \\ \hline 6 & -2 & 6 \end{array}\right]$ is one possible answer.

31. AB gives the total cost of producing each kind of product in each city:

	Salt Lake City	Chicago
Chair	38	44
Table	67	78

33. (a) 2800 g. (b) 6000 g.

35. (a) $P = \begin{bmatrix} \mathbf{s}_1 \\ \mathbf{s}_2 \end{bmatrix} = \begin{bmatrix} 18.95 & 14.75 & 8.98 \\ 17.80 & 13.50 & 10.79 \end{bmatrix}$.

(b) $0.80P = \begin{bmatrix} 15.16 & 11.80 & 7.18 \\ 14.24 & 10.80 & 8.63 \end{bmatrix}$.

37. (a) 1. (b) 0.

39. $x = 0$, $y = 1$. **41.** $B = \begin{bmatrix} 1 & 1 \\ 0 & 1 \end{bmatrix}$.

ML.1. (a) $\begin{bmatrix} 4.5000 & 2.2500 & 3.7500 \\ 1.5833 & 0.9167 & 1.5000 \\ 0.9667 & 0.5833 & 0.9500 \end{bmatrix}$.

(b)
```
??? Error using ==> *
Inner matrix dimensions must
agree.
```

(c) $\begin{bmatrix} 5.0000 & 1.5000 \\ 1.5833 & 2.2500 \\ 2.4500 & 3.1667 \end{bmatrix}$.

(d)
```
??? Error using ==> *
Inner matrix dimensions must
agree.
```

(e)
```
??? Error using ==> *
Inner matrix dimensions must
agree.
```

(f)
```
??? Error using ==> -
Inner matrix dimensions must
agree.
```

(g) $\begin{bmatrix} 18.2500 & 7.4583 & 12.2833 \\ 7.4583 & 5.7361 & 8.9208 \\ 12.2833 & 8.9208 & 14.1303 \end{bmatrix}$.

ML.3. $\begin{bmatrix} 4 & -3 & 2 & -1 & -5 \\ 2 & 1 & -3 & 0 & 7 \\ -1 & 4 & 1 & 2 & 8 \end{bmatrix}$.

ML.5. (a) $\begin{bmatrix} 1 & 0 & 0 & 0 \\ 0 & 2 & 0 & 0 \\ 0 & 0 & 3 & 0 \\ 0 & 0 & 0 & 4 \end{bmatrix}$.

(b) $\begin{bmatrix} 0 & 0 & 0 & 0 & 0 \\ 0 & 1.0000 & 0 & 0 & 0 \\ 0 & 0 & 0.5000 & 0 & 0 \\ 0 & 0 & 0 & 0.3333 & 0 \\ 0 & 0 & 0 & 0 & 0.2500 \end{bmatrix}$.

(c) $\begin{bmatrix} 5 & 0 & 0 & 0 & 0 & 0 \\ 0 & 5 & 0 & 0 & 0 & 0 \\ 0 & 0 & 5 & 0 & 0 & 0 \\ 0 & 0 & 0 & 5 & 0 & 0 \\ 0 & 0 & 0 & 0 & 5 & 0 \\ 0 & 0 & 0 & 0 & 0 & 5 \end{bmatrix}$.

ML.9. (a) **bingen(0,7,3)** $= \begin{bmatrix} 0 & 0 & 0 & 0 & 1 & 1 & 1 & 1 \\ 0 & 0 & 1 & 1 & 0 & 0 & 1 & 1 \\ 0 & 1 & 0 & 1 & 0 & 1 & 0 & 1 \end{bmatrix}$.

(b) $AB = \begin{bmatrix} 0 & 1 & 1 & 0 & 1 & 0 & 0 & 1 \\ 0 & 1 & 1 & 0 & 1 & 0 & 0 & 1 \\ 0 & 1 & 1 & 0 & 1 & 0 & 0 & 1 \end{bmatrix}$.

(c) The columns of B that contain an odd number of 1s are dotted with a vector of all 1s (a row of A); hence the result is 1.

ML.11. $n = 2$, $BB = \begin{bmatrix} 0 & 0 \\ 0 & 0 \end{bmatrix}$.

$n = 3$, $BB = \begin{bmatrix} 1 & 1 & 1 \\ 1 & 1 & 1 \\ 1 & 1 & 1 \end{bmatrix}$.

$n = 4$, $BB = O$.

$n = 5$, $BB =$ matrix of all 1s.

$BB = \begin{cases} \text{zero matrix} & \text{if } n \text{ is even} \\ \text{matrix of 1s} & \text{if } n \text{ is odd.} \end{cases}$

Section 1.4, page 49

1. $A + B = \begin{bmatrix} 3 & 2 & -1 \\ 6 & 2 & 10 \end{bmatrix}$.

$A + B + C = \begin{bmatrix} -1 & -4 & 0 \\ 8 & 5 & 10 \end{bmatrix}$.

3. $A(B + C) = \begin{bmatrix} -10 & -8 & 16 \\ 10 & 14 & -28 \end{bmatrix}$.

5. $A(rB) = \begin{bmatrix} -6 & 18 & -42 \\ 9 & -27 & 0 \end{bmatrix}$.

7. $(AB)^T = \begin{bmatrix} 11 & 5 \\ 15 & -4 \end{bmatrix}$.

9. (a) $\begin{bmatrix} 2 & -62 \\ 25 & 33 \\ 30 & 15 \end{bmatrix}$.

(b) $\begin{bmatrix} 3 & -5 \\ 1 & -3 \\ -11 & -3 \end{bmatrix}$.

(c) $\begin{bmatrix} 6 & 10 & 16 \\ -9 & 7 & 18 \end{bmatrix}$.

(d) $\begin{bmatrix} -2 & 30 \\ -6 & 38 \\ -4 & -20 \end{bmatrix}$.

(e) $\begin{bmatrix} 1 & 11 & 28 \\ 7 & 17 & 30 \end{bmatrix}$.

11. $AB = AC = \begin{bmatrix} 8 & -6 \\ -8 & 6 \end{bmatrix}$.

13. (a) $\begin{bmatrix} 30 & 20 \\ 10 & 20 \end{bmatrix}$. (b) $\begin{bmatrix} 247 & 206 \\ 103 & 144 \end{bmatrix}$.

15. $r = 3$.

17. $A^T A = \begin{bmatrix} \mathbf{a}_1^T \\ \mathbf{a}_2^T \\ \mathbf{a}_3^T \end{bmatrix} \begin{bmatrix} \mathbf{a}_1 & \mathbf{a}_2 & \mathbf{a}_3 \end{bmatrix}$

$= \begin{bmatrix} 25 & 14 & -3 \\ 14 & 29 & 2 \\ -3 & 2 & 1 \end{bmatrix}$.

19. (a) $\begin{bmatrix} \frac{4}{9} \\ \frac{5}{9} \end{bmatrix}$. (b) $\begin{bmatrix} \frac{3}{7} \\ \frac{4}{7} \end{bmatrix}$.

21. (a) After 1 year: $\begin{bmatrix} \frac{4}{15} \\ \frac{3}{10} \\ \frac{13}{30} \end{bmatrix} \approx \begin{bmatrix} 0.2666 \\ 0.3000 \\ 0.4333 \end{bmatrix}$.

After 2 years: $\begin{bmatrix} \frac{7}{25} \\ \frac{97}{300} \\ \frac{119}{300} \end{bmatrix} \approx \begin{bmatrix} 0.2800 \\ 0.3233 \\ 0.3967 \end{bmatrix}$.

(c) T. It will gain approximately 7.21% of the market.

25. (a) $\begin{bmatrix} 0 & 0 \\ 0 & 0 \end{bmatrix}$. (b) $\begin{bmatrix} 0 & 0 \\ 1 & 1 \end{bmatrix}$.

ML.1. (a) $k = 3$. (b) $k = 5$.

ML.3. (a) $\begin{bmatrix} 0 & -2 & 4 \\ 4 & 0 & -2 \\ -2 & 4 & 0 \end{bmatrix}$. (b) $\begin{bmatrix} 0 & 0 & 0 \\ 0 & 0 & 0 \\ 0 & 0 & 0 \end{bmatrix}$.

ML.5. The sequence seems to be converging to

$$\begin{bmatrix} 1.0000 & 0.7500 \\ 0 & 0 \end{bmatrix}.$$

ML.7. (a) $A^T A = \begin{bmatrix} 2 & -3 & -1 \\ -3 & 9 & 2 \\ -1 & 2 & 6 \end{bmatrix}$,

$AA^T = \begin{bmatrix} 6 & -1 & -3 \\ -1 & 6 & 4 \\ -3 & 4 & 5 \end{bmatrix}$.

(b) $B = \begin{bmatrix} 2 & -3 & 1 \\ -3 & 2 & 4 \\ 1 & 4 & 2 \end{bmatrix}$,

$C = \begin{bmatrix} 0 & -1 & 1 \\ 1 & 0 & 0 \\ -1 & 0 & 0 \end{bmatrix}$.

(c) $B + C = \begin{bmatrix} 2 & -4 & 2 \\ -2 & 2 & 4 \\ 0 & 4 & 2 \end{bmatrix}$,

$B + C = 2A$.

ML.9. $k = 4$. **ML.11.** $k = 8$.

Section 1.5, page 61

1.

3.

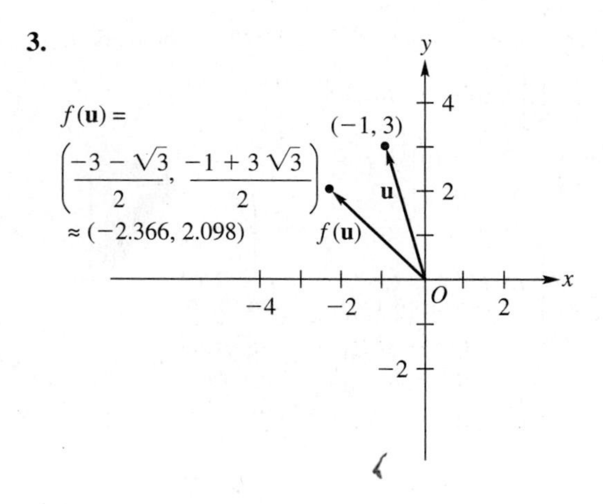

5.

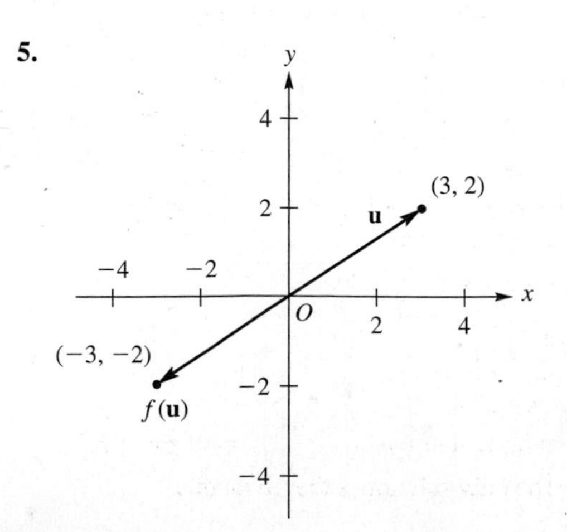

7.

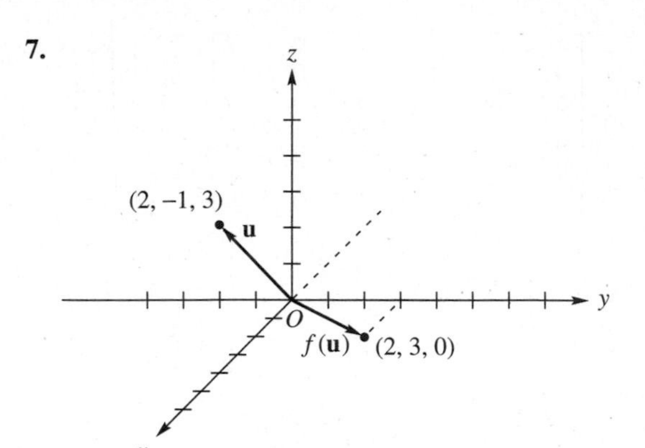

9. Yes. **11.** Yes. **13.** No.

15. (a) Reflection about the y-axis.

(b) Rotate counterclockwise through $\frac{\pi}{2}$.

17. (a) Projection onto the x-axis.

(b) Projection onto the y-axis.

19. (a) Counterclockwise rotation by 60°.

(b) Counterclockwise rotation by 90°.

(c) $k = 12$.

Section 1.6, page 85

1. Reduced row echelon form, row echelon form.

3. Reduced row echelon form, row echelon form.

5. Row echelon form.

7. Neither.

9. (a) $\begin{bmatrix} 1 & 0 & 3 \\ 5 & -1 & 5 \\ 4 & 2 & 2 \\ -3 & 1 & 4 \end{bmatrix}$. (b) $\begin{bmatrix} 1 & 0 & 3 \\ -3 & 1 & 4 \\ 12 & 6 & 6 \\ 5 & -1 & 5 \end{bmatrix}$.

(c) $\begin{bmatrix} 1 & 0 & 3 \\ -3 & 1 & 4 \\ 4 & 2 & 2 \\ 2 & -1 & -4 \end{bmatrix}$.

11. Possible answers:

(a) $\begin{bmatrix} 2 & -1 & 3 & 4 \\ 5 & 2 & -3 & 4 \\ 0 & 1 & 2 & -1 \end{bmatrix}$.

(b) $\begin{bmatrix} 4 & -2 & 6 & 8 \\ 0 & 1 & 2 & -1 \\ 5 & 2 & -3 & 4 \end{bmatrix}$.

(c) $\begin{bmatrix} 2 & -1 & 3 & 4 \\ 0 & 1 & 2 & -1 \\ 7 & 1 & 0 & 8 \end{bmatrix}$.

13. $\begin{bmatrix} 1 & 3 & -1 & 2 \\ 0 & 1 & -2 & -3 \\ 0 & 0 & 1 & \frac{26}{7} \\ 0 & 0 & 0 & 1 \end{bmatrix}$.

15. $\begin{bmatrix} 1 & 2 & -3 & 1 \\ 0 & 1 & 2 & -1 \\ 0 & 0 & 1 & -\frac{7}{4} \\ 0 & 0 & 0 & 1 \end{bmatrix}$.

17. For Exercise 13: $\begin{bmatrix} 1 & 0 & 0 & 0 \\ 0 & 1 & 0 & 0 \\ 0 & 0 & 1 & 0 \\ 0 & 0 & 0 & 1 \end{bmatrix}$.

For Exercise 14: $\begin{bmatrix} 1 & 0 & 0 & 0 \\ 0 & 1 & 0 & 0 \\ 0 & 0 & 1 & 0 \\ 0 & 0 & 0 & 1 \\ 0 & 0 & 0 & 0 \end{bmatrix}$.

For Exercise 15: $\begin{bmatrix} 1 & 0 & 0 & 0 \\ 0 & 1 & 0 & 0 \\ 0 & 0 & 1 & 0 \\ 0 & 0 & 0 & 1 \end{bmatrix}$.

For Exercise 16: $\begin{bmatrix} 1 & 0 & -\frac{1}{3} & -\frac{2}{3} & \frac{11}{3} \\ 0 & 1 & -\frac{2}{3} & -\frac{7}{3} & \frac{10}{3} \\ 0 & 0 & 0 & 0 & 0 \\ 0 & 0 & 0 & 0 & 0 \end{bmatrix}$.

19. (a) Yes. (b) No. (c) Yes. (d) No.

21. (a) $x = -2 + r$, $y = -1$, $z = 8 - 2r$, $w = r$, $r =$ any real number.

(b) $x = 1$, $y = \frac{2}{3}$, $z = -\frac{2}{3}$.

(c) No solution.

(d) $x = -\frac{1}{12} - \frac{7}{12}r$, $y = \frac{23}{12} + \frac{5}{12}r$, $z = -\frac{5}{4} + \frac{1}{4}r$, $r =$ any real number.

23. (a) $a = -2$. (b) $a \neq \pm 2$. (c) $a = 2$.

25. (a) $a = \pm\sqrt{6}$. (b) $a \neq \pm\sqrt{6}$. (c) None.

27. (a) $x = -1$, $y = 4$, $z = -3$.

(b) $x = 0$, $y = 0$, $z = 0$.

29. (a) $x = 1 - r$, $y = 2$, $z = 1$, $w = r$, $r =$ any real number.

(b) No solution.

31. $x = \frac{3}{2} - t$, $y = -2 + t$, $z = t$, $t =$ any real number.

33. $-a + b + c = 0$.

35. $x = -1$, $y = -2$, $x = 2$, $y = -3$.

37. $x = -r$, $y = 0$, $z = r$, $r =$ any real number.

39. $-3a - b + c = 0$.

41. $\mathbf{x} = \begin{bmatrix} r \\ 0 \end{bmatrix}$, where $r \neq 0$.

43. $\mathbf{x} = \begin{bmatrix} -\frac{1}{4}r \\ \frac{1}{4}r \\ r \end{bmatrix}$, where $r \neq 0$.

45. $\mathbf{x} = \begin{bmatrix} \frac{17}{12} \\ \frac{2}{3} \\ \frac{3}{4} \\ 0 \end{bmatrix} + \begin{bmatrix} -\frac{25}{6}r \\ \frac{7}{3}r \\ -\frac{3}{2}r \\ r \end{bmatrix}$.

47. $y = \frac{1}{2}x^2 - \frac{3}{2}x + 3$.

49. $y = \frac{11}{6}x^3 - 2x^2 + \frac{7}{6}x - 1$.

51. 30 chairs, 30 coffee tables, and 20 dining-room tables.

53. $2x^2 + 2x + 1$.

55. $T_1 = 36.25°$, $T_2 = 36.25°$, $T_3 = 28.75°$, $T_4 = 28.75°$.

57. (a) $\begin{bmatrix} 1 \\ 1 \\ 0 \\ 0 \end{bmatrix}$ and $\begin{bmatrix} 0 \\ 0 \\ 1 \\ 0 \end{bmatrix}$. (b) $\begin{bmatrix} 0 \\ 0 \\ 1 \\ 1 \end{bmatrix}$ and $\begin{bmatrix} 1 \\ 1 \\ 1 \\ 1 \end{bmatrix}$.

59. (a) $\begin{bmatrix} 1 \\ 1 \\ 0 \end{bmatrix}$. (b) Inconsistent.

ML.1. (a) $\begin{bmatrix} 1.0000 & 0.5000 & 0.5000 \\ -3.0000 & 1.0000 & 4.0000 \\ 1.0000 & 0 & 3.0000 \\ 5.0000 & -1.0000 & 5.0000 \end{bmatrix}$.

(b) $\begin{bmatrix} 1.0000 & 0.5000 & 0.5000 \\ 0 & 2.5000 & 5.5000 \\ 1.0000 & 0 & 3.0000 \\ 5.0000 & -1.0000 & 5.0000 \end{bmatrix}$.

(c) $\begin{bmatrix} 1.0000 & 0.5000 & 0.5000 \\ 0 & 2.5000 & 5.5000 \\ 0 & -0.5000 & 2.5000 \\ 5.0000 & -1.0000 & 5.0000 \end{bmatrix}$.

(d) $\begin{bmatrix} 1.0000 & 0.5000 & 0.5000 \\ 0 & 2.5000 & 5.5000 \\ 0 & -0.5000 & 2.5000 \\ 0 & -3.5000 & 2.5000 \end{bmatrix}$.

(e) $\begin{bmatrix} 1.0000 & 0.5000 & 0.5000 \\ 0 & -3.5000 & 2.5000 \\ 0 & -0.5000 & 2.5000 \\ 0 & 2.5000 & 5.5000 \end{bmatrix}$.

ML.3. $\begin{bmatrix} 1 & 0 & 0 \\ 0 & 1 & 0 \\ 0 & 0 & 1 \\ 0 & 0 & 0 \end{bmatrix}$.

ML.5. $x = -2 + r$, $y = -1$, $z = 8 - 2r$, $w = r$, $r =$ any real number.

ML.7. Only the trivial solution.

ML.9. $\mathbf{x} = \begin{bmatrix} 0.5r \\ r \end{bmatrix}$.

ML.11. Exercise 27:

(a) Unique solution: $x = -1$, $y = 4$, $z = -3$.

(b) The only solution is the trivial one.

Exercise 28:

(a) $x = r$, $y = -2r$, $z = r$, where r is any real number.

(b) Unique solution: $x = 1$, $y = 2$, $z = 2$.

ML.13. The \ command yields a matrix showing that the system is inconsistent. The **rref** command leads to the display of a warning that the result may contain large roundoff errors.

Section 1.7, page 105

1. $A^{-1} = \begin{bmatrix} \frac{3}{8} & -\frac{1}{8} \\ \frac{1}{4} & \frac{1}{4} \end{bmatrix}$.

3. Nonsingular: $A^{-1} = \begin{bmatrix} 4 & -1 \\ -3 & 1 \end{bmatrix}$.

5. (a) $\begin{bmatrix} \frac{1}{2} & -\frac{1}{4} \\ \frac{1}{6} & \frac{1}{12} \end{bmatrix}$. (b) $\begin{bmatrix} 0 & 1 & -1 \\ 2 & -2 & -1 \\ -1 & 1 & 1 \end{bmatrix}$.

(c) $\begin{bmatrix} \frac{7}{3} & -\frac{1}{3} & -\frac{1}{3} & -\frac{2}{3} \\ \frac{4}{9} & -\frac{1}{9} & -\frac{4}{9} & \frac{1}{9} \\ -\frac{1}{9} & -\frac{2}{9} & \frac{1}{9} & \frac{2}{9} \\ -\frac{5}{3} & \frac{2}{3} & \frac{2}{3} & \frac{1}{3} \end{bmatrix}$.

7. (a) $\begin{bmatrix} -2 & \frac{3}{2} \\ 1 & -\frac{1}{2} \end{bmatrix}$. (b) Singular.

(c) $\begin{bmatrix} \frac{3}{2} & -1 & \frac{1}{2} \\ \frac{1}{2} & 0 & -\frac{1}{2} \\ -\frac{3}{2} & 1 & \frac{1}{2} \end{bmatrix}$.

9. (a) Singular. (b) $\begin{bmatrix} 1 & -1 & 0 \\ 1 & -2 & 1 \\ -\frac{3}{2} & \frac{5}{2} & -\frac{1}{2} \end{bmatrix}$.

(c) $\begin{bmatrix} -1 & \frac{3}{2} & \frac{1}{2} \\ 1 & -\frac{3}{2} & \frac{1}{2} \\ 0 & \frac{1}{2} & -\frac{1}{2} \end{bmatrix}$.

11. (a) and (b). **13.** $\begin{bmatrix} \frac{4}{5} & -\frac{3}{5} \\ -\frac{1}{5} & \frac{2}{5} \end{bmatrix}$.

17. (a) $\begin{bmatrix} -30 \\ 60 \\ 10 \end{bmatrix}$. (b) $\begin{bmatrix} 23 \\ -31 \\ -1 \end{bmatrix}$. **19.** Yes.

21. $\lambda = -1, \lambda = 3$.

23. $\begin{bmatrix} \frac{1}{4} & 0 & 0 \\ 0 & -\frac{1}{2} & 0 \\ 0 & 0 & \frac{1}{3} \end{bmatrix}$. **25.** $\mathbf{x} = \begin{bmatrix} 19 \\ 23 \end{bmatrix}$.

27. $\left[\begin{array}{cc|c} -1 & 2 & 0 \\ 3 & -5 & 0 \\ \hline 0 & 0 & -\frac{1}{4} \end{array}\right]$.

29. (a) $\begin{bmatrix} 0 & 0 & 1 \\ 0 & 1 & 1 \\ 1 & 1 & 0 \end{bmatrix}$. (b) Singular.

(c) $\begin{bmatrix} 1 & 0 & 1 & 1 \\ 0 & 0 & 1 & 1 \\ 1 & 1 & 0 & 0 \\ 1 & 1 & 0 & 1 \end{bmatrix}$.

31. (a) Yes. (b) No.

ML.1. (a) and (c).

ML.3. (a) $\begin{bmatrix} -2 & 3 \\ 1 & -1 \end{bmatrix}$.

(b) $\begin{bmatrix} -\frac{1}{4} & \frac{3}{4} & -\frac{1}{4} \\ -\frac{1}{4} & -\frac{1}{4} & \frac{3}{4} \\ \frac{3}{4} & -\frac{1}{4} & -\frac{1}{4} \end{bmatrix}$.

ML.5. (a) $t = 4$. (b) $t = 3$.

ML.9. $\begin{bmatrix} 1 & 0 & 0 \\ 0 & 1 & 0 \\ 0 & 0 & 1 \end{bmatrix}$ and $\begin{bmatrix} 1 & 1 & 1 \\ 0 & 1 & 1 \\ 0 & 0 & 1 \end{bmatrix}$ have inverses, but there are others.

$\begin{bmatrix} 1 & 0 & 1 \\ 0 & 1 & 1 \\ 0 & 1 & 1 \end{bmatrix}$ and $\begin{bmatrix} 0 & 0 & 0 \\ 0 & 1 & 1 \\ 1 & 0 & 1 \end{bmatrix}$ do not have inverses, but there are others.

Section 1.8, page 113

1. $\mathbf{x} = \begin{bmatrix} 1 \\ 2 \\ 1 \end{bmatrix}$. **3.** $\mathbf{x} = \begin{bmatrix} 1 \\ 0 \\ 2 \\ -4 \end{bmatrix}$.

5. $L = \begin{bmatrix} 1 & 0 & 0 \\ 2 & 1 & 0 \\ 2 & -2 & 1 \end{bmatrix}$, $U = \begin{bmatrix} 2 & 3 & 4 \\ 0 & -1 & 2 \\ 0 & 0 & -2 \end{bmatrix}$,

$\mathbf{x} = \begin{bmatrix} 4 \\ -2 \\ 1 \end{bmatrix}$.

7. $L = \begin{bmatrix} 1 & 0 & 0 \\ 0.5 & 1 & 0 \\ 0.25 & -1.5 & 1 \end{bmatrix}$, $U = \begin{bmatrix} 4 & 2 & 3 \\ 0 & -1 & 3.5 \\ 0 & 0 & 5.5 \end{bmatrix}$,

$\mathbf{x} = \begin{bmatrix} 2 \\ -2 \\ -1 \end{bmatrix}$.

9. $L = \begin{bmatrix} 1 & 0 & 0 & 0 \\ 0.5 & 1 & 0 & 0 \\ -1 & 0.2 & 1 & 0 \\ 2 & 0.4 & 2 & 1 \end{bmatrix}$,

$$U = \begin{bmatrix} 2 & 1 & 0 & -4 \\ 0 & -0.5 & 0.25 & 1 \\ 0 & 0 & 0.2 & 2 \\ 0 & 0 & 0 & 2 \end{bmatrix},$$

$$\mathbf{x} = \begin{bmatrix} 0.5 \\ 2 \\ -2 \\ 1.5 \end{bmatrix}.$$

ML.1. $L = \begin{bmatrix} 1 & 0 & 0 \\ 1 & 1 & 0 \\ 0.5 & 0.3333 & 1 \end{bmatrix},$

$$U = \begin{bmatrix} 2 & 8 & 0 \\ 0 & -6 & -3 \\ 0 & 0 & 8 \end{bmatrix}.$$

ML.3. $L = \begin{bmatrix} 1.0000 & 0 & 0 & 0 \\ 0.5000 & 1.0000 & 0 & 0 \\ -2.0000 & -2.0000 & 1.0000 & 0 \\ -1.0000 & 1.0000 & -2.0000 & 1.0000 \end{bmatrix},$

$$U = \begin{bmatrix} 6 & -2 & -4 & 4 \\ 0 & -2 & -4 & -1 \\ 0 & 0 & 5 & -2 \\ 0 & 0 & 0 & 8 \end{bmatrix},$$

$$\mathbf{z} = \begin{bmatrix} 2 \\ -5 \\ 2 \\ -32 \end{bmatrix}, \mathbf{x} = \begin{bmatrix} 4.5000 \\ 6.9000 \\ -1.2000 \\ -4.0000 \end{bmatrix}.$$

Supplementary Exercises, page 114

1. $\begin{bmatrix} -1 & -3 \\ 26 & 6 \end{bmatrix}.$

3. $\begin{bmatrix} 19 & 10 \\ -6 & 1 \end{bmatrix}.$

5. (a) $\left[\begin{array}{cccc|c} 1 & 2 & -1 & 1 & 7 \\ 2 & -1 & 0 & 2 & -8 \end{array}\right].$

(b) $3x + 2y = -4$
$5x + y = 2$
$3x + 2y = 6.$

7. $k \neq \frac{5}{2}, t \neq 1.$

9. $x = 1, y = 2, z = -2.$

11. (a) $a = -3.$
(b) $a \neq \pm 3.$
(c) $a = 3.$

13. $x = -3r, y = r, z = 0, r =$ any real number.

15. $\begin{bmatrix} -40 & 16 & 9 \\ 13 & -5 & -3 \\ 5 & -2 & -1 \end{bmatrix}.$

17. Yes.

19. $\mathbf{x} = \begin{bmatrix} 4 \\ 1 \\ 4 \end{bmatrix}.$

23. (a) $a \neq 15.$ (b) None. (c) $a = 15.$

25. $a = 1, -1.$

27. (a) $k = 1;\quad B = \begin{bmatrix} b_1 \\ 0 \end{bmatrix}.$

$k = 2;\quad B = \begin{bmatrix} b_{11} & b_{12} \\ 0 & 0 \end{bmatrix}.$

$k = 3;\quad B = \begin{bmatrix} b_{11} & b_{12} & b_{13} \\ 0 & 0 & 0 \end{bmatrix}.$

$k = 4;\quad B = \begin{bmatrix} b_{11} & b_{12} & b_{13} & b_{14} \\ 0 & 0 & 0 & 0 \end{bmatrix}.$

(b) The answers are not unique. The only requirement is that row 2 of B have all zero entries.

29. (a) $\begin{bmatrix} 1 & \frac{1}{2} \\ 0 & 1 \end{bmatrix}.$ (b) $\begin{bmatrix} 1 & 0 & 0 \\ 0 & 0 & 0 \\ 0 & 0 & 0 \end{bmatrix} = B.$

(c) $I_4.$

31. $A^2 = \begin{bmatrix} 1 & \frac{3}{4} \\ 0 & \frac{1}{4} \end{bmatrix}, A^3 = \begin{bmatrix} 1 & \frac{7}{8} \\ 0 & \frac{1}{8} \end{bmatrix},$

$A^4 = \begin{bmatrix} 1 & \frac{15}{16} \\ 0 & \frac{1}{16} \end{bmatrix}, A^5 = \begin{bmatrix} 1 & \frac{31}{32} \\ 0 & \frac{1}{32} \end{bmatrix}.$

It appears that $A^n = \begin{bmatrix} 1 & (2^n - 1)/2^n \\ 0 & 1/2^n \end{bmatrix}.$

33. (a) $\begin{bmatrix} -41 \\ 47 \\ -35 \end{bmatrix}.$ (b) $\begin{bmatrix} 83 \\ -45 \\ -62 \end{bmatrix}.$

35. $L = \begin{bmatrix} 1 & 0 & 0 \\ -3 & 1 & 0 \\ 2 & 4 & 1 \end{bmatrix},$

$U = \begin{bmatrix} -2 & 1 & -2 \\ 0 & 4 & 3 \\ 0 & 0 & -3 \end{bmatrix}, \mathbf{x} = \begin{bmatrix} -1 \\ -2 \\ 3 \end{bmatrix}.$

Chapter Test, page 117

1. All vectors $\mathbf{w}$ such that $d \neq a + 2c.$

2. No solution.

3. (a) $a = 2, 3.$ (b) $a \neq 2, 3.$ (c) None.

4. $\begin{bmatrix} -\frac{1}{2} & 1 & \frac{3}{2} \\ \frac{1}{2} & 0 & -\frac{1}{2} \\ -\frac{1}{2} & 1 & \frac{1}{2} \end{bmatrix}.$

5. $-2, 3.$

6. (a) $\begin{bmatrix} 3 & 6 & 5 \\ -2 & 2 & -8 \\ 0 & 5 & -3 \end{bmatrix}.$ (b) $\mathbf{x} = \begin{bmatrix} -4 \\ 14 \\ 25 \end{bmatrix}.$

(c) $\begin{bmatrix} \frac{15}{14} & \frac{5}{28} & -\frac{9}{28} & -\frac{23}{14} \\ \frac{8}{7} & -\frac{1}{7} & -\frac{1}{7} & -\frac{9}{7} \\ \frac{3}{7} & \frac{1}{14} & \frac{1}{14} & -\frac{6}{7} \\ -\frac{4}{7} & \frac{1}{14} & \frac{1}{14} & \frac{8}{7} \end{bmatrix}.$

7. $L = \begin{bmatrix} 1 & 0 & 0 \\ -4 & 1 & 0 \\ 2 & -3 & 1 \end{bmatrix}$,

$U = \begin{bmatrix} 2 & 2 & -1 \\ 0 & -3 & 1 \\ 0 & 0 & -2 \end{bmatrix}$, $\mathbf{x} = \begin{bmatrix} 2.25 \\ 3.50 \\ 8.50 \end{bmatrix}$.

8. (a) F. (b) T. (c) F. (d) T. (e) F.

Chapter 2

Section 2.1, page 123

1. (a) Yes. (b) $A = \begin{bmatrix} 1 & 0 & 0 \\ 0 & 1 & 0 \\ 0 & 0 & 1 \\ 1 & 0 & 1 \end{bmatrix}$.

3. (a) No, $e(111) = e(110)$.

(b) $A = \begin{bmatrix} 1 & 0 & 0 \\ 0 & 1 & 0 \end{bmatrix}$.

5. (a) 3. (b) 3. (c) 2. (d) 1.

7. (a) Odd. (b) Even. (c) Odd. (d) Even.

9. (a) No. (b) Yes. (c) No. (d) Yes.

11. (a) 000, 011, 101, 110.

(b) (i) No. (ii) Yes. (iii) Yes. (iv) Yes.

Section 2.2, page 134

1. (a)

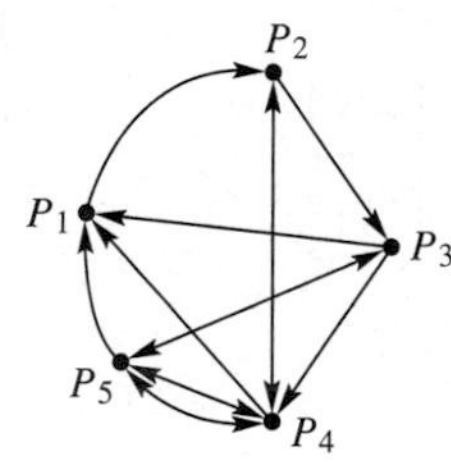

(b)

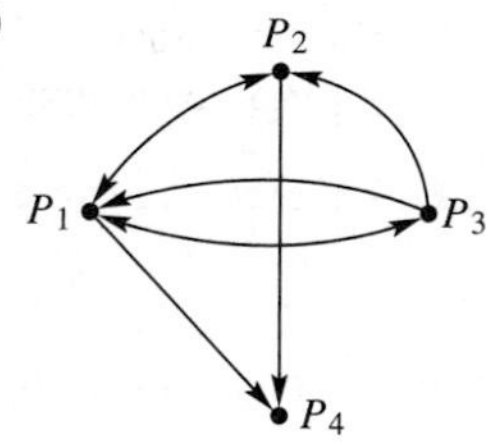

3.

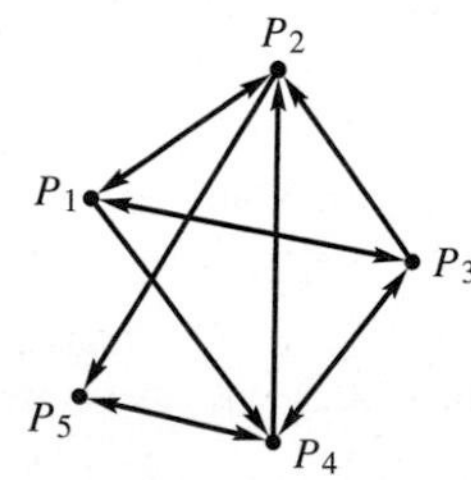

	P_1	P_2	P_3	P_4	P_5
P_1	0	1	1	1	0
P_2	1	0	0	0	1
P_3	1	1	0	1	0
P_4	0	1	1	0	1
P_5	0	0	0	1	0

5. (a) Peters. (b) Russell.

7. (a) No. (b) 3. (c) 5.

9. P_1, P_4, P_5, and P_6.

11. (a) No. (b) 3. (c) 4.

13. (a) Strongly connected. (b) Not strongly connected.

ML.1. P_2, P_3, and P_4 form a clique.

ML.3. (a) Strongly connected.

(b) Not strongly connected.

Section 2.3, page 141

1.

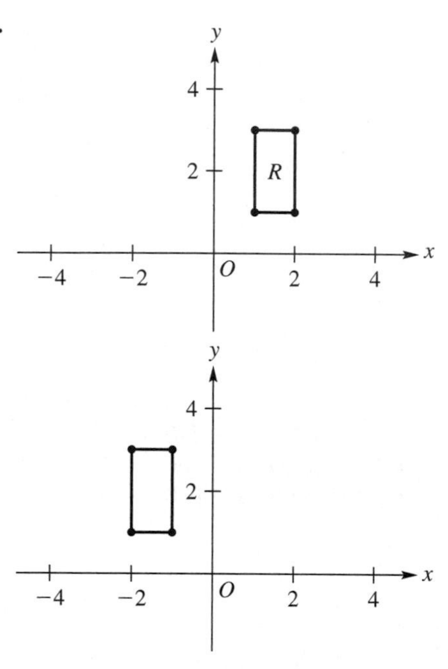

3.

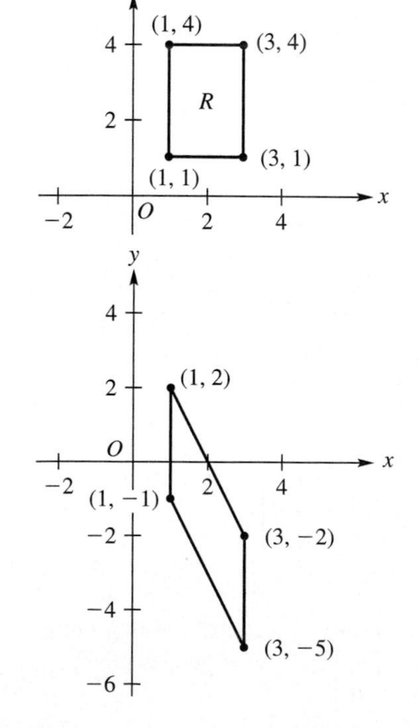

5.

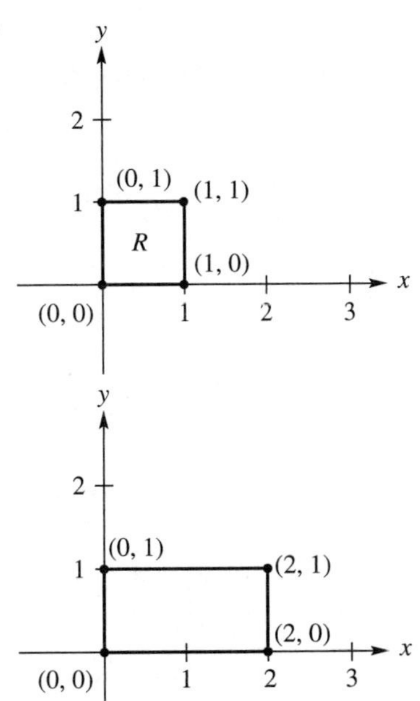

7. $(-10, 15)$, $(3, 12)$, $(-5, 2)$.

9.

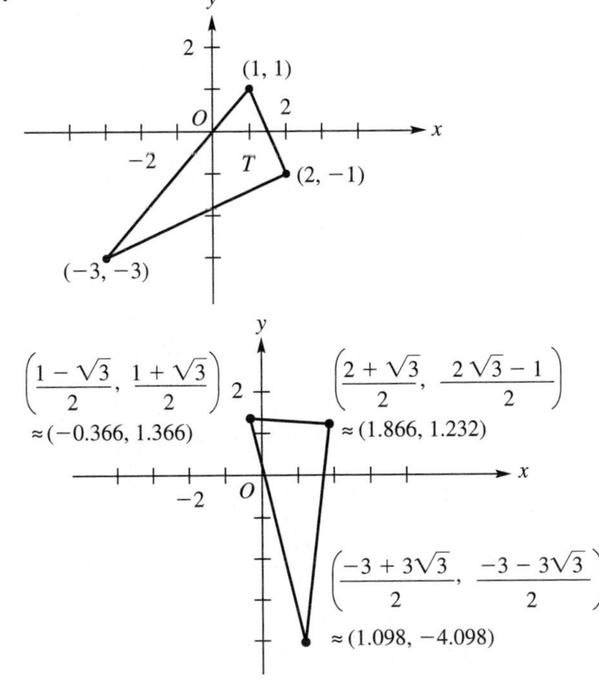

11. The image of the vertices of T under L consists of the points $(-9, -18)$, $(0, 0)$, and $(3, 6)$. Thus the image of T under L is a line segment.

13.

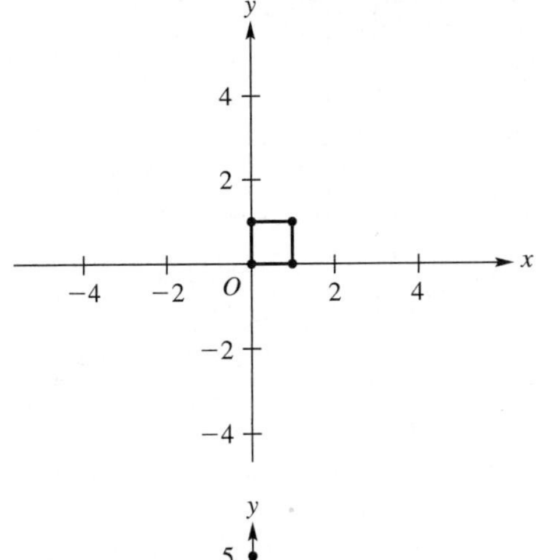

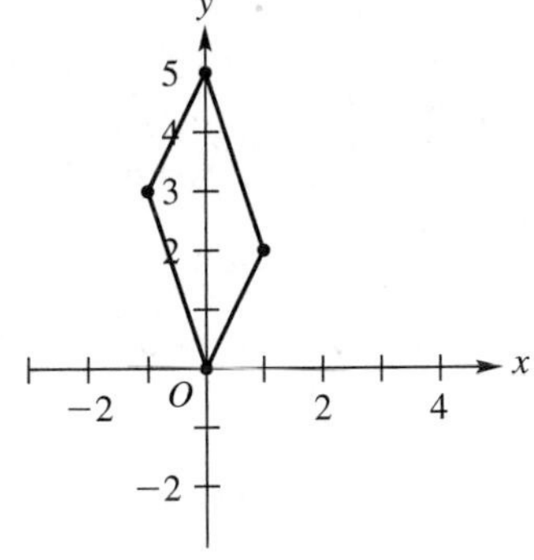

15. (a) Possible answer: First perform f_1 (90° counterclockwise rotation) then f_3.

(b) Possible answer: Perform f_1 (−135° counterclockwise rotation).

ML.1. (c) Part (a) resulted in an ellipse. Part (b) generated another ellipse within that generated in part (a). The two ellipses are nested.

(d) Inside the ellipse generated in part (b).

ML.3. (a) The area of the house is 5 square units. The area of the image is 5 square units. The areas of the original figure and the image are the same.

(b) The area of the image is 5 square units. The areas of the original figure and the image are the same.

(c) The area of the image is 5 square units. The areas of the original figure and the image are the same.

ML.5. (a) Composite transformation

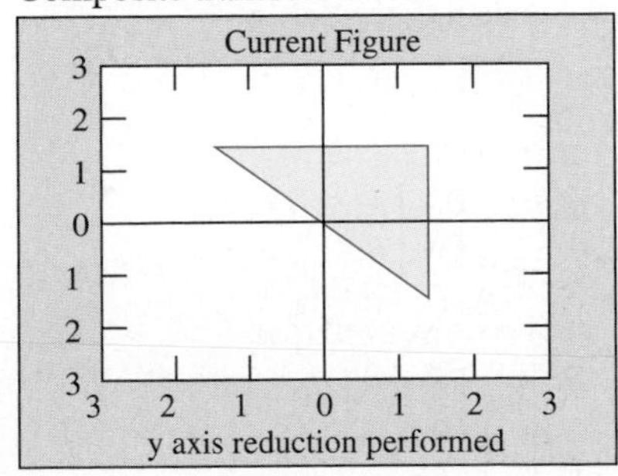

(b)

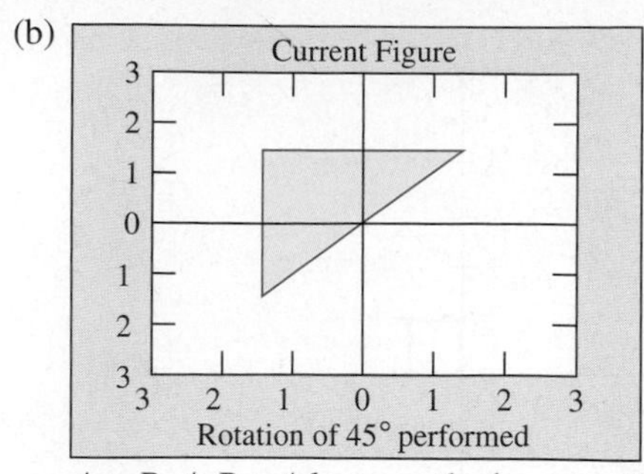

$A * B \neq B * A$ because the images of the composite transformations represented by the matrix products are not equal.

ML.7. (a) The projection is longer than **w** and is in the same direction.

(b) The projection is shorter than **w** and is in the opposite direction.

(c) The projection is shorter than **w** and is in the same direction.

(d) The projection is shorter than **w** and is in the same direction.

Section 2.4, page 148

1. $I_1 = 15$ A from e to a, $I_2 = 8$ A from a to b, $I_3 = 7$ A from a to c, $I_4 = 1$ A from d to c, $I_5 = 16$ A from c to e.

3. $I_1 = 25$ A from f to c, $I_2 = 10$ A from c to b, $I_3 = 15$ A from c to d, $I_4 = 5$ A from f to d, $I_5 = 20$ A from e to a.

5. $I_1 = 5$ A from b to a, $I_2 = 8$ A from c to d, $E = 40$ V.

7. $I_1 = 4$ A from f to a, $I_2 = 14$ A from c to b, $I_3 = 18$ A from b to e, $I_4 = 24$ A from d to e, $R = 1\ \Omega$, $E = 100$ V.

Section 2.5, page 157

1. (b) and (c).

3. Possible answer: $\begin{bmatrix} 0.5 & 0.4 & 0.3 \\ 0.3 & 0.4 & 0.5 \\ 0.2 & 0.2 & 0.2 \end{bmatrix}$.

5. (a) $\mathbf{x}^{(1)} = \begin{bmatrix} 0.7 \\ 0.3 \end{bmatrix}$, $\mathbf{x}^{(2)} = \begin{bmatrix} 0.61 \\ 0.39 \end{bmatrix}$, $\mathbf{x}^{(3)} = \begin{bmatrix} 0.583 \\ 0.417 \end{bmatrix}$.

(b) $\mathbf{T} > \mathbf{0}$, hence it is regular; $\mathbf{u} = \begin{bmatrix} 0.571 \\ 0.429 \end{bmatrix}$.

7. (a) and (d).

11. (a)

$$T = \begin{array}{c} \begin{array}{cc} A & B \end{array} \\ \begin{bmatrix} 0.3 & 0.4 \\ 0.7 & 0.6 \end{bmatrix} \end{array} \begin{array}{c} A \\ B \end{array}$$

(b) 0.364.

(c) $\mathbf{u} = \begin{bmatrix} \frac{4}{11} \\ \frac{7}{11} \end{bmatrix} = \begin{bmatrix} 0.364 \\ 0.636 \end{bmatrix}$.

13. (a) 0.69.

(b) 20.7 percent of the population will be farmers.

15. (a) 35%, 37.5%.

(b) 40%.

ML.3. (a).

Section 2.6, page 165

1. (b) and (d).

3. $\begin{bmatrix} 4 \\ 0 \\ 3 \end{bmatrix}$.

5.

$$\begin{array}{c} \\ \text{Farmer} \\ \text{Carpenter} \\ \text{Tailor} \end{array} \begin{array}{c} \begin{array}{ccc} \text{Farmer} & \text{Carpenter} & \text{Tailor} \end{array} \\ \begin{bmatrix} \frac{2}{5} & \frac{1}{3} & \frac{1}{2} \\ \frac{2}{5} & \frac{1}{3} & \frac{1}{2} \\ \frac{1}{5} & \frac{1}{3} & 0 \end{bmatrix} \end{array};$$

$\mathbf{p} = \begin{bmatrix} 75 \\ 75 \\ 40 \end{bmatrix}$.

7. Not productive.

9. Productive.

11. (a) $\begin{bmatrix} 18 \\ 16 \end{bmatrix}$. (b) $\begin{bmatrix} 12 \\ 8 \end{bmatrix}$.

Section 2.7, page 178

1. Final average: 73.5; detail coefficients: 10.5, 3, −1.

3. Final average: 71.25; detail coefficients: 2.25, 10.5, 8, 3, −1, −1, 7.

5. The averages are computed using a pair of data. At the second stage we will have only three averages and hence cannot use the procedure as discussed. One remedy is to adjoin a pair of zeros after the original six items to give eight items. Then proceed as in the discussion in the text.

7. $A^{-1} = \begin{bmatrix} Q^{-1} & Z & Z & Z \\ Z & I & Z & Z \\ Z & Z & I & Z \\ Z & Z & Z & I \end{bmatrix}$.

Supplementary Exercises, page 179

1. (a) 16.

(b) 10001, 01001, 00101, 00011, 11000, 10100, 10010, 01100, 01010, 00110.

(c) 6.

3. 2.

5. (a) $\begin{bmatrix} 0.2 & 0.6 \\ 0.8 & 0.4 \end{bmatrix}$. (b) 0.4432. (c) $\frac{3}{7}$.

7. (a) Final average: 3; detail coefficients: 3, 0, −1, 2.

(b) Compressed data: 3, 0, 0, 2.
Wavelet y-coordinates: 3, 3, 5, 1.

(c)

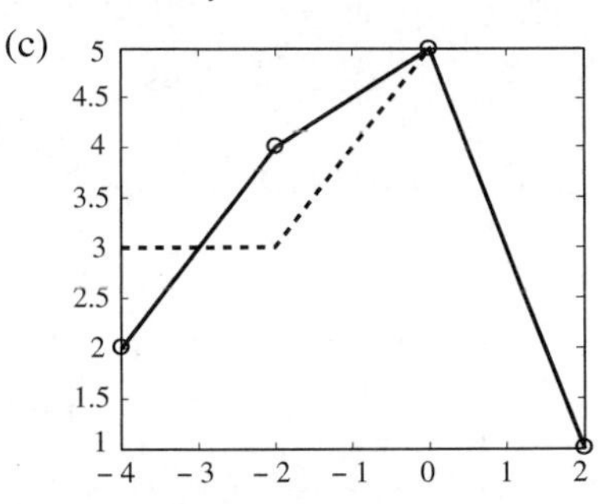

Chapter Test, page 180

1. (a) Yes. (b) $\begin{bmatrix} 1 & 0 \\ 0 & 1 \\ 0 & 1 \\ 1 & 0 \end{bmatrix}$.

(c) $|0000| = 0$, $|0110| = 2$, $|1001| = 2$, $|1111| = 4$.

2. (a) The triangle with vertices (1, 4), (−3, 1), and (−2, 6).

(b) Reflection with respect to the y-axis.

(c) $L(L(T)) = T$.

3. $a = 1, b = 2$.

4. There are two cliques: P_1, P_3, P_5 and P_1, P_3, P_6.

5. $I_1 = 5$ A from a to f, $I_2 = 13$ A from b to a, $R_1 = 3\ \Omega$, $R_2 = 4\ \Omega$, $I_3 = 14$ A from c to d, $I_4 = 22$ A from d to b.

6. (a) $\begin{bmatrix} 0.4 & 0.5 \\ 0.6 & 0.5 \end{bmatrix}$. (b) 0.4546. (c) $\frac{6}{11}$.

7. \$5.65 million of steel, \$5.41 million of coal, and \$2.83 million of transportation.

8. (a) Final average: −0.5; detail coefficients: −1, 1.5, −0.5.

(b) Compressed data: −0.5, 0, 1.5, 0.
Wavelet y-coordinates: 1, −2, −0.5, −0.5.

Chapter 3

Section 3.1, page 192

1. (a) 5. (b) 7. (c) 4. (d) 4. (e) 7. (f) 0.

3. (a) −. (b) +. (c) −. (d) −. (e) +. (f) +.

5. (a) 7. (b) 30. (c) −24. (d) 4.

7. There are 24 terms.

9. $|B| = 3$, $|C| = 9$, $|D| = -3$.

11. (a) $\lambda^2 - 3\lambda - 4$. (b) $\lambda^2 - 5\lambda + 6$.

13. (a) −1, 4. (b) 2, 3.

15. (a) 0. (b) −144. (c) 72.

17. (a) 72. (b) 0. (c) −24.

19. (a) −30. (b) 0. (c) 6.

23. $-\frac{3}{2}$.

25. (a) 1. (b) 1. (c) (1).

27. (a) 0. (b) 1.

ML.1. (a) −18. (b) 5.

ML.3. (a) 4. (b) 0.

ML.5. $t = 3, t = 4$.

Section 3.2, page 207

1. $A_{11} = -11$, $A_{12} = 29$, $A_{13} = 1$,
$A_{21} = -4$, $A_{22} = 7$, $A_{23} = -2$,
$A_{31} = 2$, $A_{32} = -10$, $A_{33} = 1$.

3. (a) −43. (b) 75. (c) 0.

5. (a) 0. (b) −6. (c) −36.

9. (a) $\begin{bmatrix} 24 & -42 & -30 \\ 19 & -2 & -30 \\ -4 & 32 & 30 \end{bmatrix}$. (b) 150.

11. (a) Singular. (b) $\begin{bmatrix} \frac{2}{7} & -\frac{3}{7} \\ \frac{1}{7} & \frac{2}{7} \end{bmatrix}$.

(c) $\begin{bmatrix} \frac{1}{4} & -\frac{1}{20} & \frac{3}{20} \\ 0 & \frac{1}{5} & \frac{2}{5} \\ 0 & \frac{1}{10} & -\frac{3}{10} \end{bmatrix}$.

13. (a) $\begin{bmatrix} 0 & \frac{1}{2} \\ 1 & \frac{3}{2} \end{bmatrix}$.

(b) $\begin{bmatrix} \frac{1}{4} & 0 & 0 \\ 0 & -\frac{1}{3} & 0 \\ 0 & 0 & \frac{1}{2} \end{bmatrix}$.

(c) $\begin{bmatrix} \frac{15}{14} & \frac{5}{28} & -\frac{9}{28} & -\frac{23}{14} \\ \frac{8}{7} & -\frac{1}{7} & -\frac{1}{7} & -\frac{9}{7} \\ \frac{3}{7} & \frac{1}{14} & \frac{1}{14} & -\frac{6}{7} \\ -\frac{4}{7} & \frac{1}{14} & \frac{1}{14} & \frac{8}{7} \end{bmatrix}$.

15. (d) is nonsingular.

17. (a) 1, 4. (b) −5, 0, 3.

19. (a) Has only the trivial solution.

(b) Has nontrivial solutions.

21. $x = 1, y = -1, z = 0, w = 2$.

23. No solution.

25. (a) is nonsingular. (b) is singular.

ML.1. $A_{11} = -11$, $A_{23} = -2$, $A_{31} = 2$.

ML.3. 0.

ML.5. (a) The matrix is singular.

(b) $\begin{bmatrix} \frac{2}{7} & -\frac{3}{7} \\ \frac{1}{7} & \frac{2}{7} \end{bmatrix}$.

(c) $\begin{bmatrix} \frac{1}{4} & -\frac{1}{20} & \frac{3}{20} \\ 0 & \frac{1}{5} & \frac{2}{5} \\ 0 & \frac{1}{10} & -\frac{3}{10} \end{bmatrix}$.

Supplementary Exercises, page 212

1. (a) -24. (b) 24.

3. (a) $\frac{1}{5}$. (b) 80. (c) $\frac{16}{5}$. (d) $\frac{1}{80}$.

5. $0, -1, -4$.

7. 172.

9. -218.

11. $\frac{1}{5}\begin{bmatrix} 0 & 5 & -5 \\ -2 & 7 & -4 \\ 1 & -1 & 2 \end{bmatrix}$.

13. $\lambda \neq -1, 0, 1$.

17. $a \neq 0, a \neq 2$.

Chapter Test, page 213

1. 17.

2. (a) 54. (b) $\frac{27}{2}$. (c) $\frac{1}{54}$.

3. $\frac{20}{3}$.

4. $-3, 0, 3$.

5. $x = 1, y = 0, z = -2$.

6. (a) T. (b) F. (c) F. (d) F. (e) T. (f) T. (g) T. (h) T. (i) F. (j) T.

Chapter 4

Section 4.1, page 227

1.

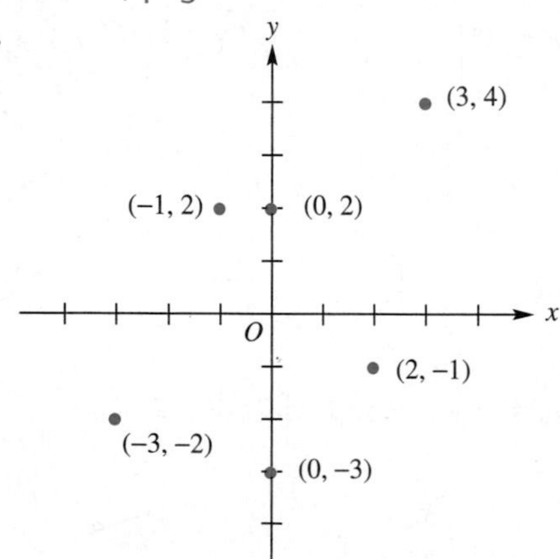

3. (a) $(1, 7)$.

5. (a) $\mathbf{u} + \mathbf{v} = (0, 8)$, $\mathbf{u} - \mathbf{v} = (4, -2)$, $2\mathbf{u} = (4, 6)$, $3\mathbf{u} - 2\mathbf{v} = (10, -1)$.

(b) $\mathbf{u} + \mathbf{v} = (3, 5)$, $\mathbf{u} - \mathbf{v} = (-3, 1)$, $2\mathbf{u} = (0, 6)$, $3\mathbf{u} - 2\mathbf{v} = (-6, 5)$.

(c) $\mathbf{u} + \mathbf{v} = (5, 8)$, $\mathbf{u} - \mathbf{v} = (-1, 4)$, $2\mathbf{u} = (4, 12)$, $3\mathbf{u} - 2\mathbf{v} = (0, 14)$.

7. (a) $\mathbf{w}_1 = 2$. (b) $\mathbf{x}_2 = \frac{8}{3}$. (c) $\mathbf{w}_1 = 3$, $\mathbf{x}_2 = -2$.

9. (a) $\sqrt{5}$. (b) 5. (c) 2. (d) 5.

11. (a) $\sqrt{2}$. (b) 5. (c) $\sqrt{10}$. (d) $\sqrt{13}$.

13. $(-5, 6) = 19(1, 2) - 8(3, 4)$.

15. 6.

17. 6.

19. (a) $\left(\frac{3}{5}, \frac{4}{5}\right)$. (b) $\left(-\frac{2}{\sqrt{13}}, -\frac{3}{\sqrt{13}}\right)$. (c) $(1, 0)$.

21. (a) $\dfrac{-4}{\sqrt{5} \cdot \sqrt{13}}$. (b) 0. (c) 0. (d) -1.

25. $a = \frac{8}{5}$.

27. (a) $\mathbf{i} + 3\mathbf{j}$. (b) $-2\mathbf{i} - 3\mathbf{j}$. (c) $-2\mathbf{i}$. (d) $3\mathbf{j}$.

29.

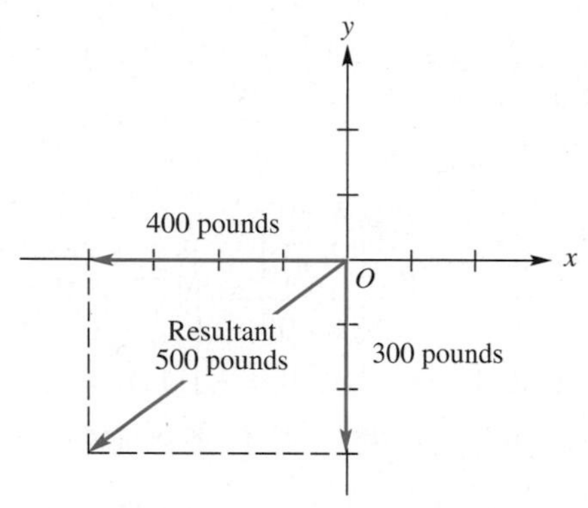

Section 4.2, page 244

1. (a) $\mathbf{u} + \mathbf{v} = (1, 3, -5)$, $\mathbf{u} - \mathbf{v} = (1, 1, -1)$, $2\mathbf{u} = (2, 4, -6)$, $3\mathbf{u} - 2\mathbf{v} = (3, 4, -5)$.

(b) $\mathbf{u} + \mathbf{v} = (3, 0, 6, -1)$, $\mathbf{u} - \mathbf{v} = (5, -4, -4, 7)$, $2\mathbf{u} = (8, -4, 2, 6)$, $3\mathbf{u} - 2\mathbf{v} = (14, -10, -7, 17)$.

3. (a) $a = \frac{1}{2}, b = \frac{3}{2}$. (b) $a = 4, b = 0$. (c) $a = -6, b = 1, c = 0$.

7.

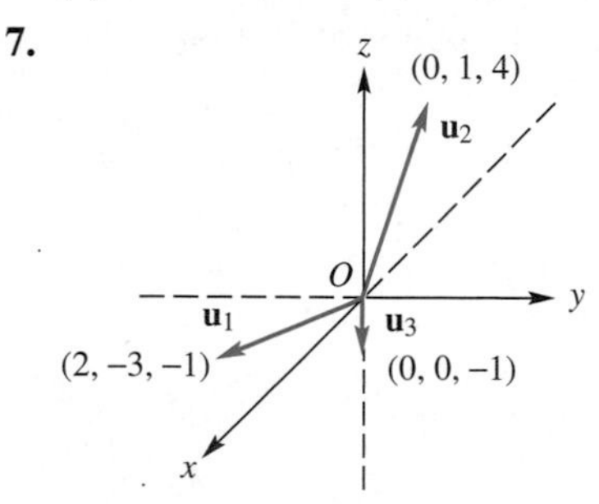

9. $(4, 2, 2)$.

11. (a) $\sqrt{29}$. (b) $\sqrt{14}$. (c) $\sqrt{5}$. (d) $\sqrt{30}$.

13. (a) $\sqrt{18}$. (b) $\sqrt{6}$. (c) $\sqrt{50}$. (d) $\sqrt{10}$.

15. Possible answer: $c_1 = -2, c_2 = -1, c_3 = 1$.

17. $a = 4$ or $a = -1$.

21. (a) 0. (b) $-\dfrac{1}{\sqrt{3}}$. (c) $\dfrac{1}{\sqrt{5}}$. (d) 0.

23. (a) $\mathbf{u}_1$ and $\mathbf{u}_2$, $\mathbf{u}_1$ and $\mathbf{u}_6$, $\mathbf{u}_2$ and $\mathbf{u}_3$, $\mathbf{u}_3$ and $\mathbf{u}_6$, $\mathbf{u}_4$ and $\mathbf{u}_6$.

(b) $\mathbf{u}_1$ and $\mathbf{u}_3$. (c) None.

25. Possible answer: $a = 1, b = 0, c = -1$.

27. (a) $\left(\frac{2}{\sqrt{14}}, -\frac{1}{\sqrt{14}}, \frac{3}{\sqrt{14}}\right)$.

(b) $\left(\frac{1}{\sqrt{30}}, \frac{2}{\sqrt{30}}, \frac{3}{\sqrt{30}}, \frac{4}{\sqrt{30}}\right)$.

(c) $\left(0, \frac{1}{\sqrt{2}}, -\frac{1}{\sqrt{2}}\right)$.

(d) $\left(0, -\frac{1}{\sqrt{6}}, \frac{2}{\sqrt{6}}, -\frac{1}{\sqrt{6}}\right)$.

29. (a) $\mathbf{i} + 2\mathbf{j} - 3\mathbf{k}$. (b) $2\mathbf{i} + 3\mathbf{j} - \mathbf{k}$.
(c) $\mathbf{j} + 2\mathbf{k}$. (d) $-2\mathbf{k}$.

33. $1.08\mathbf{u}$. **35.** $\frac{1}{2}(\mathbf{t} + \mathbf{b})$.

37. $\mathbf{v} = (0, 1, 0, 1)$ is the only such vector, since a vector only has one additive inverse.

39. $(0, 0, 0), (1, 0, 1), (0, 1, 0), (1, 1, 1)$.

ML.3. (a) 2.2361. (b) 5.4772. (c) 3.1623.

ML.5. (a) 19. (b) -11. (c) -55.

ML.9. (a) $\begin{bmatrix} 0.6667 \\ 0.6667 \\ -0.3333 \end{bmatrix}$ or in rational form $\begin{bmatrix} \frac{2}{3} \\ \frac{2}{3} \\ -\frac{1}{3} \end{bmatrix}$.

(b) $\begin{bmatrix} 0 \\ 0.8000 \\ -0.6000 \\ 0 \end{bmatrix}$ or in rational form $\begin{bmatrix} 0 \\ \frac{4}{5} \\ -\frac{3}{5} \\ 0 \end{bmatrix}$.

(c) $\begin{bmatrix} 0.3015 \\ 0 \\ 0.3015 \\ 0 \end{bmatrix}$.

Section 4.3, page 255

1. (b).

3. (a).

5.

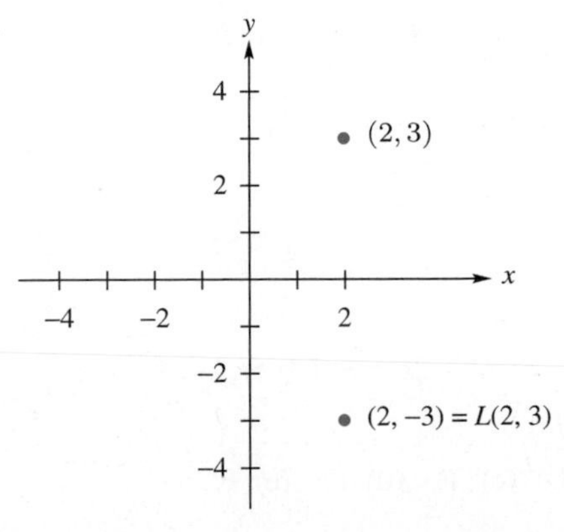

7.

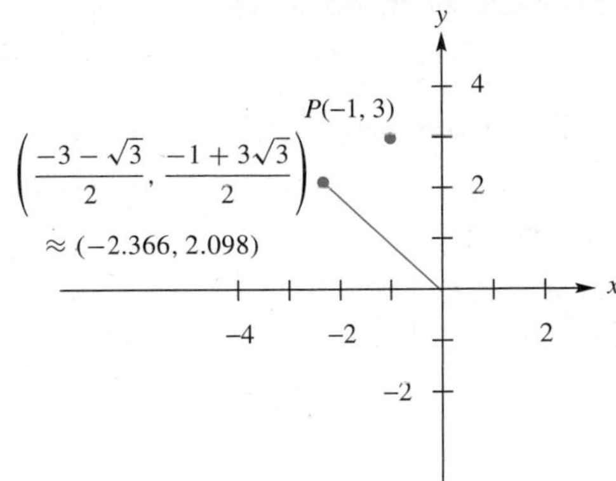

9.

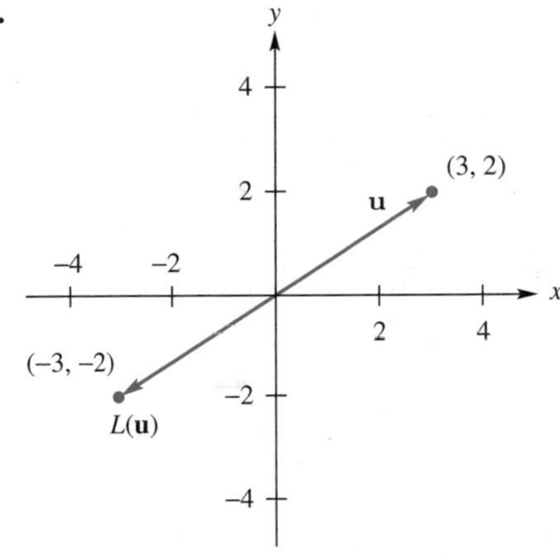

11.

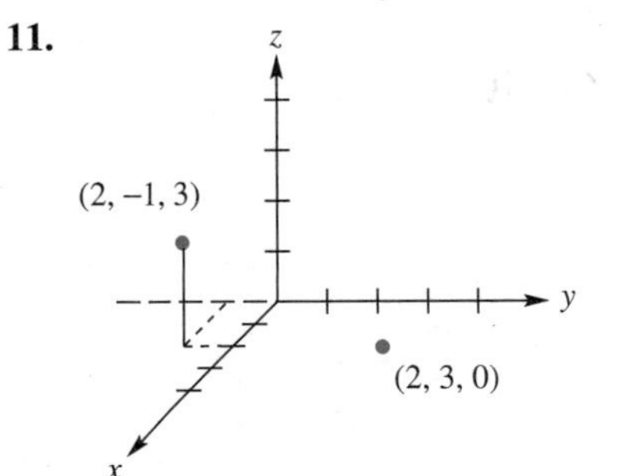

13. (a) Yes. (b) Yes.

15. $c - a + b = 0$. **17.** $\begin{bmatrix} 11 \\ 6 \end{bmatrix}$.

19. $\mathbf{x} = \begin{bmatrix} 0 \\ 0 \\ r \end{bmatrix}$, where r is any real number.

21. (a) Reflection about the y-axis;
(b) Reflection about the origin;
(c) Rotate counterclockwise through $\frac{\pi}{2}$.

23. No.

25. $\begin{bmatrix} -1 & 0 \\ 0 & 1 \end{bmatrix}$. **27.** $\begin{bmatrix} \frac{\sqrt{2}}{2} & -\frac{\sqrt{2}}{2} \\ \frac{\sqrt{2}}{2} & \frac{\sqrt{2}}{2} \end{bmatrix}$.

29. $\begin{bmatrix} 1 & -1 & 0 \\ 1 & 0 & 1 \\ 0 & 1 & -1 \end{bmatrix}$.

31. (a) 71 52 33 47 30 26 84 56 43 99 69 55.
(b) Message: CERTAINLY NOT.

Supplementary Exercises, page 257

1.

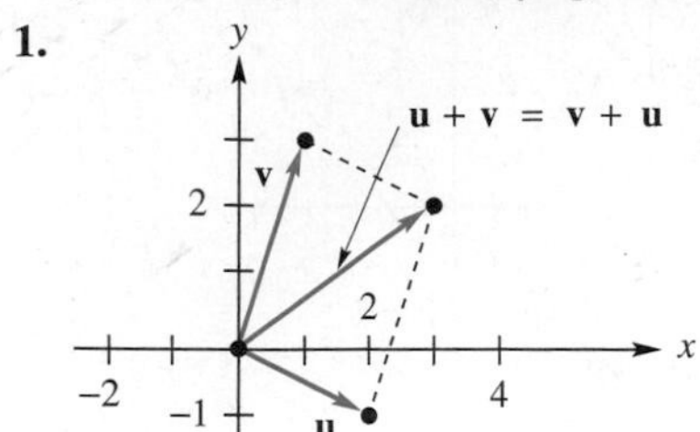

3.

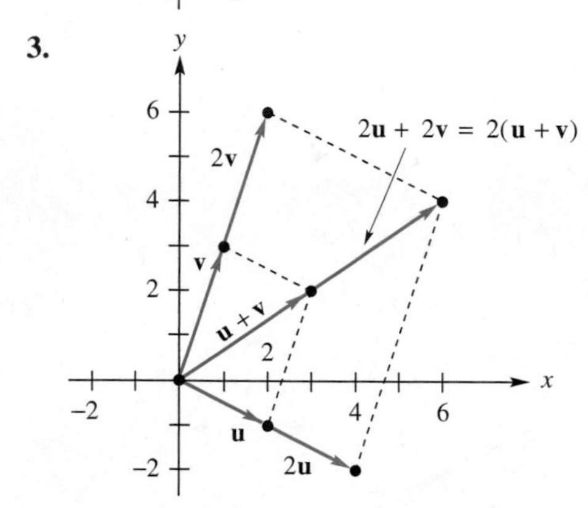

5. $\mathbf{x} = \left(-\frac{1}{2}, -\frac{3}{8}, \frac{7}{8}\right)$.

7. (a) $\sqrt{15}$. (b) $3\sqrt{2}$. (c) $\sqrt{43}$.
(d) -5. (e) $-\frac{1}{3}\sqrt{\frac{5}{6}}$.

9. $c = \pm 3$. **11.** No.

13. $\frac{k}{\sqrt{14}}(-1, 2, 3)$, where $k = \pm 1$.

15. $a = -2, b = 2$. **17.** 26.

19. $\begin{bmatrix} \frac{\sqrt{3}}{2} & \frac{1}{2} \\ -\frac{1}{2} & \frac{\sqrt{3}}{2} \end{bmatrix}$.

21. (a) $(1, 3, -2) = 1(1, 1, 0) + 2(0, 1, 1) - 4(0, 0, 1)$.
(b) $(4, 7)$.

23. $n = 3, m = 5$. **25.** $c + 3a - b = 0$.

Chapter Test, page 258

1. $-\frac{1}{\sqrt{22}\sqrt{30}}$. **2.** $\frac{1}{\sqrt{15}}(2, -1, 1, 3)$.

3. Yes. **4.** Yes. **5.** $\begin{bmatrix} 2 & 3 \\ -2 & 3 \\ 1 & 1 \end{bmatrix}$.

6. (a) F. (b) F. (c) T. (d) F. (e) F.
(f) T. (g) F. (h) T. (i) T. (j) F.

Chapter 5

Section 5.1, page 263

1. (a) $-15\mathbf{i} - 2\mathbf{j} + 9\mathbf{k}$. (b) $-3\mathbf{i} + 3\mathbf{j} + 3\mathbf{k}$.
(c) $7\mathbf{i} + 5\mathbf{j} - \mathbf{k}$. (d) $0\mathbf{i} + 0\mathbf{j} + 0\mathbf{k}$.

9. $\frac{1}{2}\sqrt{478}$. **11.** $\sqrt{150}$. **13.** 39.

ML.1. (a) $\begin{bmatrix} -11 & 2 & 5 \end{bmatrix}$. (b) $\begin{bmatrix} 3 & 1 & -1 \end{bmatrix}$.
(c) $\begin{bmatrix} 1 & -8 & -5 \end{bmatrix}$.

ML.5. 8.

Section 5.2, page 269

1. (a) $-7x + 5y + 1 = 0$. (b) $9x + 5y + 7 = 0$.
(c) $-5x - 3y = 0$. (d) $-7x + 3y - 6 = 0$.

3. (d).

5. (a) $x = 3 + 4t, y = 4 - 5t, z = -2 + 2t, \infty < t < \infty$.
(b) $x = 3 - 2t, y = 2 + 5t, z = 4 + t, -\infty < t < \infty$.
(c) $x = t, y = t, z = t, -\infty < t < \infty$.
(d) $x = -2 + 2t, y = -3 + 3t, z = 1 + 4t$, $-\infty < t < \infty$.

7. (a) $\frac{x-2}{2} = \frac{y+3}{5} = \frac{z-1}{4}$.
(b) $\frac{x+3}{8} = \frac{y+2}{7} = \frac{z+2}{6}$.
(c) $\frac{x+2}{4} = \frac{y-3}{-6} = \frac{z-4}{1}$.
(d) $\frac{x}{4} = \frac{y}{5} = \frac{z}{2}$.

9. (a) $3x - 2y + 4z + 16 = 0$.
(b) $y - 3z + 3 = 0$.
(c) $-z + 4 = 0$.
(d) $-x - 2y + 4z - 3 = 0$.

11. (a) $x = \frac{8}{13} + 23t, y = -\frac{27}{13} + 2t, z = 13t$, $-\infty < t < \infty$.
(b) $x = -\frac{28}{13} + 7t, y = -\frac{16}{13} - 22t, z = 13t$, $-\infty < t < \infty$.
(c) $x = -\frac{16}{3} + 5t, y = -\frac{8}{3} + 4t, z = -3t$, $-\infty < t < \infty$.

13. Yes. **15.** $(5, 1, 2)$.

19. $4x - 4y + z + 16 = 0$.

21. $7x + 2y - 2z - 19 = 0$.

23. $x = -2 + 2t, y = 5 - 3t, z = -3 + 4t$.

Supplementary Exercises, page 271

1. $x = \frac{2}{3}, y = \frac{4}{3}$.

3. (b) and (c).

Chapter Test, page 271

1. $x = 5 + 3t, y = -2 - 2t, z = 1 + 5t, -\infty < t < \infty$.

2. $x - y + 1 = 0$.

3. (a) F. (b) T. (c) T. (d) F. (e) F.

Chapter 6

Section 6.1, page 278

1. Closed under $\oplus$; not closed under $\odot$.

3. Not closed under $\oplus$; not closed under $\odot$.

13. Vector space.

15. Not a vector space; (β) and (d) do not hold.

17. Vector space.

Section 6.2, page 287

1. Yes. Properties (a) and (b) of Theorem 6.2 are satisfied.

3. No. A scalar multiple of a vector in W may not lie in W.

5. (b). **7.** (b) and (c).

9. (a) and (c). **17.** (c).

19. (b). **21.** (a), (b), (c), and (d).

23. (b).

25. (a) No. (b) No. (c) No. (d) No.

27. (a) No. (b) No. (c) No. (d) Yes.

29. Yes.

31. No, since the sum of two vectors from W will have second entry 0.

33. Yes; observe that $\begin{bmatrix}0\\0\\1\end{bmatrix}+\begin{bmatrix}1\\0\\1\end{bmatrix}+\begin{bmatrix}0\\1\\1\end{bmatrix}=\begin{bmatrix}1\\1\\1\end{bmatrix}$.

ML.3. (a) No. (b) Yes.

ML.5. (a) $0\mathbf{v}_1+\mathbf{v}_2-\mathbf{v}_3-\mathbf{v}_4=\mathbf{v}$.
(b) $p_1(t)+2p_2(t)+2p_3(t)=p(t)$.

ML.7. (a) Yes. (b) Yes. (c) Yes.

Section 6.3, page 301

1. (a), (c), and (d). **3.** (a) and (d).

5. No. **7.** $\left\{\begin{bmatrix}1\\0\\0\\1\end{bmatrix},\begin{bmatrix}0\\-2\\1\\0\end{bmatrix}\right\}$.

9. Yes.

11. (a) $(4,6,8,6)=3(1,1,2,1)+(1,0,0,2)+(0,3,2,1)$.
(b) $(-2,4,-6,2)=-2(1,-2,3,-1)$.
(d) $(6,5,-5,1)=2(4,2,-1,3)-(2,-1,3,5)$.

13. (b) and (c) are linearly independent, (a) is linearly dependent.
$\begin{bmatrix}2&6\\4&6\end{bmatrix}=3\begin{bmatrix}1&1\\1&2\end{bmatrix}-\begin{bmatrix}1&0\\0&2\end{bmatrix}+\begin{bmatrix}0&3\\1&2\end{bmatrix}$.

15. $c=1$.

17. Yes. **19.** No. **21.** No.

ML.1. (a) Linearly dependent.
(b) Linearly independent.
(c) Linearly independent.

Section 6.4, page 314

1. (a) and (d). **3.** (a) and (d). **5.** (c).

7. (a) $(2,1,3)=\frac{3}{2}(1,1,1)+\frac{1}{2}(1,2,3)-\frac{3}{2}(0,1,0)$.

9. (a) Forms a basis, $5t^2-3t+8=5(t^2+t)-8(t-1)$.

11. Possible answer: $\{\mathbf{v}_1,\mathbf{v}_2\}$; $\dim W=2$.

13. Possible answer: $\{t^3+t^2-2t+1,t^2+1\}$; $\dim W=2$.

15. $\left\{\begin{bmatrix}1&0&0\\0&0&0\end{bmatrix},\begin{bmatrix}0&1&0\\0&0&0\end{bmatrix},\begin{bmatrix}0&0&1\\0&0&0\end{bmatrix},\begin{bmatrix}0&0&0\\1&0&0\end{bmatrix},\begin{bmatrix}0&0&0\\0&1&0\end{bmatrix},\begin{bmatrix}0&0&0\\0&0&1\end{bmatrix}\right\}$;
$\dim M_{23}=6$; $\dim M_{mn}=mn$.

17. (a) Possible answer: $\{(1,1,0),(0,1,1)\}$.
(b) Possible answer: $\{(1,1,0),(0,0,1)\}$.
(c) Possible answer: $\{(1,0,2),(0,1,1)\}$.

19. (a) 3. (b) 2.

21. $\{t^2-1,t-1\}$.

23. (a) 2. (b) 1. (c) 2. (d) 2.

25. (a) 4. (b) 3. (c) 3. (d) 4.

27. 2.

29. Possible answer:
$\{(1,0,1,0),(0,1,-1,0),(1,0,0,0),(0,0,0,1)\}$.

31. $\left\{\begin{bmatrix}1&0&0\\0&0&0\\0&0&0\end{bmatrix},\begin{bmatrix}0&1&0\\1&0&0\\0&0&0\end{bmatrix},\begin{bmatrix}0&0&1\\0&0&0\\1&0&0\end{bmatrix},\begin{bmatrix}0&0&0\\0&1&0\\0&0&0\end{bmatrix},\begin{bmatrix}0&0&0\\0&0&1\\0&1&0\end{bmatrix},\begin{bmatrix}0&0&0\\0&0&0\\0&0&1\end{bmatrix}\right\}$.

33. The set of all vectors of the form $(a,a+2b,-2a+b,a-2b)$, where a, b, c, and d are real numbers.

35. Possible answer: $\{(3,2,0),(-2,0,1)\}$.

37. Yes. **39.** No.

ML.1. Basis.

ML.3. Basis.

ML.5. Basis.

ML.7. dim (span S) = 3, span $S\neq R^4$.

ML.9. dim (span S) = 3, span $S=P_2$.

ML.11. $\{t^3-t+1,t^3+2,t,1\}$.

Section 6.5, page 327

1. (a) $\mathbf{x}=\begin{bmatrix}\frac{1}{2}r+s\\r\\s\end{bmatrix}$, where r and s are any real numbers.

(b) $\mathbf{x} = r\begin{bmatrix} \frac{1}{2} \\ 1 \\ 0 \end{bmatrix} + s\begin{bmatrix} 1 \\ 0 \\ 1 \end{bmatrix}$.

(c)

z, y, x; $\begin{bmatrix} 1 \\ 0 \\ 1 \end{bmatrix}$, $\begin{bmatrix} \frac{1}{2} \\ 1 \\ 0 \end{bmatrix}$

3. $\left\{\begin{bmatrix} 0 \\ -1 \\ 0 \\ 1 \end{bmatrix}, \begin{bmatrix} 2 \\ -3 \\ 1 \\ 0 \end{bmatrix}\right\}$; dimension $= 2$.

5. $\left\{\begin{bmatrix} 1 \\ -\frac{8}{3} \\ -\frac{4}{3} \\ 1 \end{bmatrix}\right\}$; dimension $= 1$.

7. $\left\{\begin{bmatrix} -2 \\ 1 \\ 0 \\ 0 \\ 0 \end{bmatrix}, \begin{bmatrix} -3 \\ 0 \\ 1 \\ 1 \\ 0 \end{bmatrix}\right\}$; dimension $= 2$.

9. $\left\{\begin{bmatrix} -2 \\ \frac{1}{2} \\ 0 \\ 0 \\ 1 \end{bmatrix}, \begin{bmatrix} 1 \\ 1 \\ 0 \\ 1 \\ 0 \end{bmatrix}\right\}$; dimension $= 2$.

11. $\left\{\begin{bmatrix} -3 \\ 2 \\ 0 \\ 1 \end{bmatrix}, \begin{bmatrix} 5 \\ -4 \\ 1 \\ 0 \end{bmatrix}\right\}$.

13. $\left\{\begin{bmatrix} -1 \\ 1 \end{bmatrix}\right\}$.

15. $\left\{\begin{bmatrix} 1 \\ -2 \\ 1 \end{bmatrix}\right\}$.

17. $\lambda = 3$ or -4.

19. $\lambda = 0$ or 1.

21. $\mathbf{x}_p = \begin{bmatrix} 3 \\ -\frac{5}{7} \\ -\frac{10}{7} \\ 0 \end{bmatrix}$, $\mathbf{x}_h = r\begin{bmatrix} 0 \\ \frac{1}{7} \\ -\frac{5}{7} \\ 1 \end{bmatrix}$, where r is any real number.

23. $\left\{\begin{bmatrix} 0 \\ 1 \\ 1 \\ 1 \end{bmatrix}\right\}$ is a basis. Dimension of solution space $= 1$.

25. $\left\{\begin{bmatrix} 1 \\ 1 \\ 1 \\ 0 \\ 0 \end{bmatrix}, \begin{bmatrix} 1 \\ 0 \\ 0 \\ 0 \\ 1 \end{bmatrix}\right\}$ is a basis. Dimension of solution space $= 2$.

27. $\mathbf{x}_p = \begin{bmatrix} 1 \\ 0 \\ 0 \\ 0 \end{bmatrix}$, $\mathbf{x}_h = b\begin{bmatrix} 0 \\ 1 \\ 1 \\ 1 \end{bmatrix}$, where b is any bit.

ML.1. $\left\{\begin{bmatrix} -2 \\ 0 \\ 1 \\ 0 \\ 0 \end{bmatrix}, \begin{bmatrix} -1 \\ -1 \\ 0 \\ 1 \\ 0 \end{bmatrix}, \begin{bmatrix} -2 \\ 1 \\ 0 \\ 0 \\ 1 \end{bmatrix}\right\}$.

ML.3. $\left\{\begin{bmatrix} 1 \\ -2 \\ 1 \\ 0 \end{bmatrix}, \begin{bmatrix} \frac{4}{3} \\ -\frac{1}{3} \\ 0 \\ 1 \end{bmatrix}\right\}$.

ML.5. $\mathbf{x} = \begin{bmatrix} t \\ t \\ t \end{bmatrix}$, where t is any nonzero real number.

Section 6.6, page 337

1. Possible answer: $\{(1, 0, 0), (0, 1, 0), (0, 0, 1)\}$.

3. Possible answer: $\left\{\begin{bmatrix} 1 \\ 0 \\ 1 \\ 0 \end{bmatrix}, \begin{bmatrix} 0 \\ 1 \\ 0 \\ 1 \end{bmatrix}\right\}$.

5. (a) $\{(1, 0, -1), (0, 1, 0)\}$.
(b) $\{(1, 2, -1), (1, 9, -1)\}$.

7. (a) $\left\{\begin{bmatrix} 1 \\ 0 \\ 0 \\ 0 \end{bmatrix}, \begin{bmatrix} 0 \\ 1 \\ 0 \\ \frac{1}{5} \end{bmatrix}, \begin{bmatrix} 0 \\ 0 \\ 1 \\ \frac{3}{5} \end{bmatrix}\right\}$.

(b) $\left\{\begin{bmatrix} 1 \\ 1 \\ 3 \\ 2 \end{bmatrix}, \begin{bmatrix} -2 \\ -1 \\ 2 \\ 1 \end{bmatrix}, \begin{bmatrix} 0 \\ 0 \\ 5 \\ 3 \end{bmatrix}\right\}$.

9. Basis for row space of $A = \left\{\begin{bmatrix} 1 & 0 & \frac{19}{7} \end{bmatrix}, \begin{bmatrix} 0 & 1 & -\frac{8}{7} \end{bmatrix}\right\}$.

Basis for column space of $A = \left\{\begin{bmatrix} 1 \\ 0 \\ 2 \end{bmatrix}, \begin{bmatrix} 0 \\ 1 \\ -1 \end{bmatrix}\right\}$.

Basis for row space of $A^T = \left\{\begin{bmatrix} 1 & 0 & 2 \end{bmatrix}, \begin{bmatrix} 0 & 1 & -1 \end{bmatrix}\right\}$.

Basis for column space of $A^T = \left\{\begin{bmatrix} 1 \\ 0 \\ \frac{19}{7} \end{bmatrix}, \begin{bmatrix} 0 \\ 1 \\ \frac{8}{7} \end{bmatrix}\right\}$.

A basis for the column space of A^T consists of the transposes of a corresponding basis for the row space of

A. Similarly, a basis for the row space of A^T consists of the transposes of a corresponding basis for the column space of A.

11. Row rank = column rank = 3.

13. Rank $A = 2$, nullity $A = 2$.

15. Rank $A = 3$, nullity $A = 0$.

17. Rank $A = 2$, nullity $A = 1$.

23. Singular.

25. Nonsingular.

27. Has a unique solution.

29. Linearly dependent.

31. Nontrivial solution.

33. Has a solution.

35. Has no solution.

37. 2. **39.** 4.

ML.3. (a) $\left\{\begin{bmatrix}1\\2\\4\\6\end{bmatrix}, \begin{bmatrix}3\\5\\11\\9\end{bmatrix}, \begin{bmatrix}1\\0\\2\\1\end{bmatrix}\right\}$. (b) $\left\{\begin{bmatrix}2\\0\\1\\4\\3\end{bmatrix}, \begin{bmatrix}1\\0\\2\\5\\3\end{bmatrix}\right\}$.

ML.5. (a) Consistent. (b) Inconsistent. (c) Inconsistent.

Section 6.7, page 349

1. $\begin{bmatrix}3\\-2\end{bmatrix}$. **3.** $\begin{bmatrix}2\\-1\end{bmatrix}$. **5.** $\begin{bmatrix}1\\-1\\0\\2\end{bmatrix}$.

7. $\begin{bmatrix}0\\3\end{bmatrix}$. **9.** $4t - 3$. **11.** $\begin{bmatrix}-1 & 0\\9 & 7\end{bmatrix}$.

13. (a) $[\mathbf{v}]_T = \begin{bmatrix}-7\\4\end{bmatrix}$; $[\mathbf{w}]_T = \begin{bmatrix}7\\-1\end{bmatrix}$.

(b) $\begin{bmatrix}1 & 2\\-1 & -1\end{bmatrix}$.

(c) $[\mathbf{v}]_S = \begin{bmatrix}1\\3\end{bmatrix}$; $[\mathbf{w}]_S = \begin{bmatrix}5\\-6\end{bmatrix}$.

(d) Same as (c). (e) $\begin{bmatrix}-1 & -2\\1 & 1\end{bmatrix}$.

(f) Same as (a).

15. (a) $[\mathbf{v}]_T = \begin{bmatrix}3\\2\\-7\end{bmatrix}$; $[\mathbf{w}]_T = \begin{bmatrix}2\\3\\-3\end{bmatrix}$.

(b) $\begin{bmatrix}2 & 1 & 0\\1 & -\frac{2}{5} & \frac{3}{5}\\0 & \frac{2}{5} & \frac{2}{5}\end{bmatrix}$.

(c) $[\mathbf{v}]_S = \begin{bmatrix}8\\-2\\-2\end{bmatrix}$; $[\mathbf{w}]_S = \begin{bmatrix}7\\-1\\0\end{bmatrix}$.

(d) Same as (c).

(e) $\begin{bmatrix}\frac{1}{3} & \frac{1}{3} & -\frac{1}{2}\\\frac{1}{3} & -\frac{2}{3} & 1\\-\frac{1}{3} & \frac{2}{3} & \frac{3}{2}\end{bmatrix}$.

(f) Same as (a).

17. (a) $[\mathbf{v}]_T = \begin{bmatrix}1\\1\\1\\0\end{bmatrix}$; $[\mathbf{w}]_T = \begin{bmatrix}2\\-2\\1\\-1\end{bmatrix}$.

(b) $\begin{bmatrix}1 & 0 & 0 & 1\\\frac{1}{3} & \frac{2}{3} & -\frac{2}{3} & 0\\\frac{1}{3} & -\frac{1}{3} & \frac{1}{3} & 0\\-\frac{1}{3} & \frac{1}{3} & \frac{2}{3} & 0\end{bmatrix}$.

(c) $[\mathbf{v}]_S = \begin{bmatrix}1\\\frac{1}{3}\\\frac{1}{3}\\\frac{2}{3}\end{bmatrix}$; $[\mathbf{w}]_S = \begin{bmatrix}1\\-\frac{4}{3}\\\frac{5}{3}\\-\frac{2}{3}\end{bmatrix}$.

(d) Same as (c).

(e) $\begin{bmatrix}0 & 1 & 2 & 0\\0 & 1 & 0 & 1\\0 & 0 & 1 & 1\\1 & -1 & -2 & 0\end{bmatrix}$.

(f) Same as (a).

19. $\begin{bmatrix}5\\3\end{bmatrix}$. **21.** $\begin{bmatrix}4\\-1\\3\end{bmatrix}$.

23. $\left\{\begin{bmatrix}3\\2\\0\end{bmatrix}, \begin{bmatrix}2\\1\\0\end{bmatrix}, \begin{bmatrix}3\\1\\3\end{bmatrix}\right\}$. **25.** $\left\{\begin{bmatrix}2\\5\end{bmatrix}, \begin{bmatrix}1\\3\end{bmatrix}\right\}$.

ML.1. $\left\{\begin{bmatrix}1\\2\\3\end{bmatrix}, \begin{bmatrix}-1\\2\\-1\end{bmatrix}, \begin{bmatrix}1\\1\\1\end{bmatrix}\right\}$.

ML.3. (a) $\begin{bmatrix}0.5000\\-0.5000\\0\\-0.5000\end{bmatrix}$. (b) $\begin{bmatrix}1.0000\\0.5000\\0.3333\\0\end{bmatrix}$.

(c) $\begin{bmatrix}0.5000\\0.1667\\-0.3333\\-1.5000\end{bmatrix}$.

ML.5. $\begin{bmatrix}-0.5000 & -1.0000 & -0.5000 & 0\\-0.5000 & 0 & 1.5000 & 0\\1.0000 & 0 & -1.0000 & 1.0000\\0 & 0 & 0 & 1.0000\end{bmatrix}$.

ML.7. (a) $\begin{bmatrix} 1.0000 & -1.6667 & 2.3333 \\ 1.0000 & 0.6667 & -1.3333 \\ 0 & 1.3333 & -0.6667 \end{bmatrix}$.

(b) $\begin{bmatrix} 2 & 0 & 1 \\ -1 & 1 & -1 \\ 0 & -1 & 2 \end{bmatrix}$.

(c) $\begin{bmatrix} 2 & -2 & 4 \\ 0 & 1 & -3 \\ -1 & 2 & 0 \end{bmatrix}$. (d) QP.

Section 6.8, page 359

1. (b). **3.** $a = 5$.

5. $\left\{\left(\frac{1}{\sqrt{2}}, -\frac{1}{\sqrt{2}}, 0\right), \left(\frac{1}{\sqrt{3}}, \frac{1}{\sqrt{3}}, \frac{1}{\sqrt{3}}\right)\right\}$.

7. $\left\{\left(\frac{1}{\sqrt{3}}, -\frac{1}{\sqrt{3}}, 0, \frac{1}{\sqrt{3}}\right), \left(\frac{5}{\sqrt{42}}, \frac{1}{\sqrt{42}}, 0, -\frac{4}{\sqrt{42}}\right), (0, 0, 1, 0)\right\}$.

9. (a) $\{(1, 2), (-4, 2)\}$.

(b) $\left\{\left(\frac{1}{\sqrt{5}}, \frac{2}{\sqrt{5}}\right), \left(-\frac{2}{\sqrt{5}}, \frac{1}{\sqrt{5}}\right)\right\}$.

11. $\left\{\left(\frac{2}{3}, -\frac{2}{3}, \frac{1}{3}\right), \left(\frac{2}{3}, \frac{1}{3}, -\frac{2}{3}\right), \left(\frac{1}{3}, \frac{2}{3}, \frac{2}{3}\right)\right\}$.

13. Possible answer: $\left\{\left(\frac{1}{\sqrt{2}}, \frac{1}{\sqrt{2}}, 0, 0\right), \left(\frac{3}{\sqrt{22}}, -\frac{3}{\sqrt{22}}, 0, \frac{2}{\sqrt{22}}\right), \left(\frac{1}{\sqrt{11}}, -\frac{1}{\sqrt{11}}, 0, -\frac{3}{\sqrt{11}}\right)\right\}$.

15. $\left\{\left(\frac{1}{\sqrt{2}}, \frac{1}{\sqrt{2}}, 0, 0\right), \left(-\frac{1}{\sqrt{6}}, \frac{1}{\sqrt{6}}, 0, \frac{2}{\sqrt{6}}\right), \left(\frac{1}{\sqrt{12}}, -\frac{1}{\sqrt{12}}, \frac{3}{\sqrt{12}}, \frac{1}{\sqrt{12}}\right)\right\}$.

17. $\left\{\left(\frac{1}{\sqrt{2}}, \frac{1}{\sqrt{2}}, 0, 0\right), \left(\frac{1}{\sqrt{3}}, -\frac{1}{\sqrt{3}}, \frac{1}{\sqrt{3}}, 0\right), \left(-\frac{1}{\sqrt{42}}, \frac{1}{\sqrt{42}}, \frac{2}{\sqrt{42}}, \frac{6}{\sqrt{42}}\right)\right\}$.

19. $\left\{\frac{1}{\sqrt{42}}\begin{bmatrix} -4 \\ 5 \\ 1 \end{bmatrix}\right\}$.

21. $\frac{4}{\sqrt{5}}\left(\frac{1}{\sqrt{5}}, 0, \frac{2}{\sqrt{5}}\right) - \frac{3}{\sqrt{5}}\left(-\frac{2}{\sqrt{5}}, 0, \frac{1}{\sqrt{5}}\right) - 3(0, 1, 0) = (2, -3, 1)$.

ML.1. $\left\{\begin{bmatrix} 0.7071 \\ 0.7071 \\ 0 \end{bmatrix}, \begin{bmatrix} 0.7071 \\ -0.7071 \\ 0 \end{bmatrix}, \begin{bmatrix} 0 \\ 0 \\ 1.0000 \end{bmatrix}\right\}$

$= \left\{\begin{bmatrix} \frac{\sqrt{2}}{2} \\ \frac{\sqrt{2}}{2} \\ 0 \end{bmatrix}, \begin{bmatrix} \frac{\sqrt{2}}{2} \\ -\frac{\sqrt{2}}{2} \\ 0 \end{bmatrix}, \begin{bmatrix} 0 \\ 0 \\ 1 \end{bmatrix}\right\}$.

ML.3. (a) $\begin{bmatrix} -1.4142 \\ 1.4142 \\ 1.0000 \end{bmatrix}$. (b) $\begin{bmatrix} 0 \\ 1.4142 \\ 1.0000 \end{bmatrix}$.

(c) $\begin{bmatrix} 0.7071 \\ 0.7071 \\ -1.0000 \end{bmatrix}$.

Section 6.9, page 369

1. $\mathbf{v} = (2, 2, 0) = (1, 2, 1) + (1, 0, -1)$, where $(1, 2, 1) = \mathbf{w}$ is in W and $(1, 0, -1) = \mathbf{u}$ is in $W^\perp$.

3. (a) $\left\{\left(\frac{3}{2}, 1, 0\right), \left(-\frac{1}{2}, 0, 1\right)\right\}$.

(b) The set of all points $P(x, y, z)$ such that $2x - 3y + z = 0$. $W^\perp$ is the plane whose normal is $\mathbf{w}$.

5. $\left\{\begin{bmatrix} -\frac{17}{5} \\ \frac{6}{5} \\ 5 \\ 1 \\ 0 \end{bmatrix}, \begin{bmatrix} \frac{8}{5} \\ \frac{1}{5} \\ -3 \\ 0 \\ 1 \end{bmatrix}\right\}$.

7. Basis for null space of A: $\left\{\begin{bmatrix} -\frac{1}{3} \\ \frac{7}{3} \\ 1 \\ 0 \end{bmatrix}, \begin{bmatrix} -\frac{7}{3} \\ -\frac{2}{3} \\ 0 \\ 1 \end{bmatrix}\right\}$.

Basis for row space of A: $\left\{\left(1, 0, \frac{1}{3}, \frac{7}{3}\right), \left(0, 1, -\frac{7}{3}, \frac{2}{3}\right)\right\}$.

Basis for null space of A^T: $\left\{\begin{bmatrix} -\frac{1}{2} \\ \frac{1}{2} \\ 1 \\ 0 \end{bmatrix}, \begin{bmatrix} -\frac{1}{2} \\ \frac{3}{2} \\ 0 \\ 1 \end{bmatrix}\right\}$.

Basis for column space of A: $\left\{\begin{bmatrix} 1 \\ 0 \\ \frac{1}{2} \\ \frac{1}{2} \end{bmatrix}, \begin{bmatrix} 0 \\ 1 \\ -\frac{1}{2} \\ -\frac{3}{2} \end{bmatrix}\right\}$.

9. (a) $\left(\frac{7}{5}, \frac{11}{5}, \frac{9}{5}, -\frac{3}{5}\right)$.

(b) $\left(-\frac{2}{5}, -\frac{1}{5}, \frac{1}{5}, -\frac{2}{5}\right)$.

(c) $\left(\frac{1}{10}, \frac{9}{5}, \frac{1}{5}, \frac{31}{10}\right)$.

11. $\mathbf{w} = (1, 0, 2, 3)$, $\mathbf{u} = (0, 0, 0, 0)$. **13.** 2.

ML.1. (a) $\begin{bmatrix} 0 \\ \frac{5}{6} \\ \frac{5}{3} \\ \frac{5}{6} \end{bmatrix}$. (b) $\begin{bmatrix} \frac{3}{5} \\ \frac{3}{5} \\ \frac{3}{5} \\ \frac{3}{5} \\ \frac{3}{5} \end{bmatrix}$.

ML.3. (a) $\begin{bmatrix} 2.4286 \\ 3.9341 \\ 7.9011 \end{bmatrix}$.

(b) $\sqrt{(2.4286 - 2)^2 + (3.9341 - 4)^2 + (7.9011 - 8)^2} \approx 0.4448$.

ML.5. $\mathbf{p} = \begin{bmatrix} 0.8571 \\ 0.5714 \\ 1.4286 \\ 0.8571 \\ 0.8571 \end{bmatrix}.$

Supplementary Exercises, page 372

1. No. **3.** No.

5. Linearly dependent; possible answer: $-t-3 = (2t^2+3t+1) - 2(t^2+2t+2)$.

7. Possible answer: $\{(1,0,1,0), (1,1,-1,1)\}$; dimension is 2.

9. Possible answer: $\left\{\begin{bmatrix} 0 \\ -1 \\ 0 \\ 0 \\ 1 \end{bmatrix}, \begin{bmatrix} 7 \\ -4 \\ 0 \\ 1 \\ 0 \end{bmatrix}, \begin{bmatrix} -5 \\ 3 \\ 1 \\ 0 \\ 0 \end{bmatrix}\right\}$; dimension is 3.

11. $\lambda \neq \pm 2$. **13.** $a = 1$.

17. (a) m arbitrary and $b = 0$. (b) $r = 0$.

21. (b) $k = 0$. **23.** 3.

27. (a) $\left[\mathbf{v}\right]_T = \begin{bmatrix} -6 \\ 11 \\ 8 \end{bmatrix}$. (b) $\left[\mathbf{v}\right]_S = \begin{bmatrix} 2 \\ 3 \\ 4 \end{bmatrix}$.

(c) $P_{S\leftarrow T} = \begin{bmatrix} 1 & 0 & 1 \\ 0 & 1 & -1 \\ 1 & -2 & 4 \end{bmatrix}$. (d) Same as (b).

(e) $Q_{T\leftarrow S} = \begin{bmatrix} 2 & -2 & -1 \\ -1 & 3 & 1 \\ -1 & 2 & 1 \end{bmatrix}$.

(f) Same as (a).

29. $a = b = 0$.

31. (a) One such basis is

$$\left\{\frac{1}{\sqrt{30}}\begin{bmatrix} -5 \\ 2 \\ 1 \\ 0 \end{bmatrix}, \frac{1}{\sqrt{30}}\begin{bmatrix} 2 \\ 5 \\ 0 \\ 1 \end{bmatrix}\right\}.$$

(b) One such basis is

$$\left\{\frac{1}{\sqrt{30}}\begin{bmatrix} -5 \\ 2 \\ 1 \\ 0 \end{bmatrix}, \frac{1}{\sqrt{255}}\begin{bmatrix} -5 \\ -14 \\ 3 \\ 5 \end{bmatrix}\right\}.$$

33. Possible answer:

$$\left\{\left(\frac{1}{\sqrt{2}}, 0, 0, -\frac{1}{\sqrt{2}}\right), \left(\frac{1}{\sqrt{6}}, -\frac{2}{\sqrt{6}}, 0, \frac{1}{\sqrt{6}}\right), \left(\frac{1}{\sqrt{3}}, \frac{1}{\sqrt{3}}, 0, \frac{1}{\sqrt{3}}\right)\right\}.$$

35. (a) Possible answer: $\{(-1,0,1)\}$.

(c) (i) $\mathbf{w} = \left(\frac{1}{2}, 0, \frac{1}{2}\right)$, $\mathbf{u} = \left(\frac{1}{2}, 0, -\frac{1}{2}\right)$.

(ii) $\mathbf{w} = (2,2,2)$, $\mathbf{u} = (-1,0,1)$.

37. Basis for null space of $A = \left\{\begin{bmatrix} -\frac{37}{11} \\ \frac{20}{11} \\ \frac{8}{11} \\ 1 \end{bmatrix}\right\}$.

Basis for row space of $A = \left\{\left(1,0,0,\frac{37}{11}\right), \left(0,1,0,-\frac{20}{11}\right), \left(0,0,1,-\frac{8}{11}\right)\right\}$.

There is no basis for the null space of A^T, since the null space of $A^T = \{\mathbf{0}\}$.

Basis for the column space of $A = \left\{\begin{bmatrix} 1 \\ 0 \\ 0 \end{bmatrix}, \begin{bmatrix} 0 \\ 1 \\ 0 \end{bmatrix}, \begin{bmatrix} 0 \\ 0 \\ 1 \end{bmatrix}\right\}$.

Chapter Test, page 374

1. Yes.

2. Possible answer: $\left\{\begin{bmatrix} 0 \\ -1 \\ 1 \\ 0 \\ 0 \end{bmatrix}, \begin{bmatrix} 4 \\ -1 \\ 0 \\ 1 \\ 0 \end{bmatrix}, \begin{bmatrix} -2 \\ 0 \\ 0 \\ 0 \\ 1 \end{bmatrix}\right\}$.

3. Yes. **4.** $\lambda = \pm 3$.

5. Possible answer:

$$\left\{\left(\frac{1}{\sqrt{2}}, 0, -\frac{1}{\sqrt{2}}, 0\right), \left(\frac{1}{\sqrt{6}}, -\frac{2}{\sqrt{6}}, \frac{1}{\sqrt{6}}, 0\right), \left(\frac{1}{\sqrt{3}}, \frac{1}{\sqrt{3}}, \frac{1}{\sqrt{3}}, 0\right)\right\}.$$

6. (a) T. (b) F. (c) F. (d) F. (e) T. (f) F. (g) T. (h) F. (i) F. (j) T.

Chapter 7

Section 7.1, page 378

1. $Q = \begin{bmatrix} \frac{1}{\sqrt{2}} & \frac{1}{\sqrt{2}} \\ -\frac{1}{\sqrt{2}} & \frac{1}{\sqrt{2}} \end{bmatrix} \approx \begin{bmatrix} 0.7071 & 0.7071 \\ -0.7071 & 0.7071 \end{bmatrix}$,

$R = \begin{bmatrix} \sqrt{2} & -\frac{1}{\sqrt{2}} \\ 0 & \frac{5}{\sqrt{2}} \end{bmatrix} \approx \begin{bmatrix} 1.4142 & 0.7071 \\ 0 & 3.5355 \end{bmatrix}$.

3. $Q = \begin{bmatrix} \frac{1}{\sqrt{6}} & \frac{4}{\sqrt{21}} & -\frac{1}{\sqrt{14}} \\ \frac{2}{\sqrt{6}} & -\frac{1}{\sqrt{21}} & \frac{2}{\sqrt{14}} \\ -\frac{1}{\sqrt{6}} & \frac{2}{\sqrt{21}} & \frac{3}{\sqrt{14}} \end{bmatrix} \approx \begin{bmatrix} 0.4082 & 0.8729 & -0.2673 \\ 0.8165 & -0.2182 & 0.5345 \\ -0.4082 & 0.4364 & 0.8018 \end{bmatrix}$,

$R = \begin{bmatrix} \frac{6}{\sqrt{6}} & -\frac{8}{\sqrt{6}} & \frac{1}{\sqrt{6}} \\ 0 & \frac{7}{\sqrt{21}} & \frac{1}{\sqrt{21}} \\ 0 & 0 & \frac{19}{\sqrt{14}} \end{bmatrix} \approx \begin{bmatrix} 2.4495 & -3.2660 & 0.4082 \\ 0 & 1.5275 & 0.2182 \\ 0 & 0 & 5.0780 \end{bmatrix}$.

5. $Q = \begin{bmatrix} \frac{1}{\sqrt{3}} & 0 & \frac{2}{\sqrt{6}} \\ -\frac{1}{\sqrt{3}} & \frac{1}{\sqrt{2}} & \frac{1}{\sqrt{6}} \\ -\frac{1}{\sqrt{3}} & -\frac{1}{\sqrt{2}} & \frac{1}{\sqrt{6}} \end{bmatrix} \approx \begin{bmatrix} 0.5774 & 0 & 0.8165 \\ -0.5774 & 0.7071 & 0.4082 \\ -0.5774 & -0.7071 & 0.4082 \end{bmatrix}$,

$R = \begin{bmatrix} \sqrt{3} & 0 & 0 \\ 0 & \sqrt{8} & -\sqrt{2} \\ 0 & 0 & \sqrt{6} \end{bmatrix} \approx \begin{bmatrix} 1.7321 & 0 & 0 \\ 0 & 2.8284 & -1.4142 \\ 0 & 0 & 2.4495 \end{bmatrix}$.

7. $G = \begin{bmatrix} 1 & 1 & 1 & 1 & 0 \\ 1 & 0 & 0 & 0 & 1 \end{bmatrix}$.

9. $\left\{ \begin{bmatrix} 1 \\ 1 \\ 0 \\ 1 \\ 0 \end{bmatrix}, \begin{bmatrix} 1 \\ 0 \\ 1 \\ 0 \\ 1 \end{bmatrix} \right\}$.

11. 3.

13. (a) Yes. (b) No.

15. (a) No error was detected.

(b) A single error was detected in the first bit. The corrected word is

$$\mathbf{x}_t = \begin{bmatrix} 0 \\ 0 \\ 1 \\ 0 \\ 1 \\ 1 \end{bmatrix}.$$

(c) A single error was detected in the fifth bit. The corrected word is

$$\mathbf{x}_t = \begin{bmatrix} 0 \\ 1 \\ 1 \\ 1 \\ 0 \\ 1 \end{bmatrix}.$$

17. $H(6) = \begin{bmatrix} 0 & 0 & 0 & 1 & 1 & 1 \\ 0 & 1 & 1 & 0 & 0 & 1 \\ 1 & 0 & 1 & 0 & 1 & 0 \end{bmatrix}$.

19. $C = \begin{bmatrix} 1 & 1 & 0 \\ 1 & 0 & 1 \\ 1 & 0 & 0 \\ 0 & 1 & 1 \\ 0 & 1 & 0 \\ 0 & 0 & 1 \end{bmatrix}$.

21. (a) No error was detected.

(b) A single error was detected in the first bit. The corrected word is

$$\mathbf{x}_t = \begin{bmatrix} 0 \\ 1 \\ 1 \\ 1 \\ 1 \\ 0 \\ 0 \end{bmatrix}.$$

Section 7.2, page 388

1. $\widehat{\mathbf{x}} = \begin{bmatrix} \frac{24}{17} \\ -\frac{8}{17} \end{bmatrix} \approx \begin{bmatrix} 1.4118 \\ -0.4706 \end{bmatrix}$.

3. $\widehat{\mathbf{x}} \approx \begin{bmatrix} -1.5333 \\ -1.8667 \\ 4.2667 \end{bmatrix}$.

7. $y = 0.4x + 0.6$.

9. $y = 0.086x + 3.114$.

11. $y = 0.5718x^2 - 3.1314x + 3.4627$.

13. (a) $y = 0.426x + 0.827$.
(b) 5.087 hours.

15. (a) $y = 0.974x - 2.657$.
(b) 10.979 millions of dollars.

17. $\widehat{\mathbf{x}} = \begin{bmatrix} -\frac{5}{11} \\ \frac{4}{11} \\ 0 \end{bmatrix}$.

ML.1. $y = 0.08571x + 3.114$.

ML.3. (a) $T = -8.278t + 188.1$, where t = time.
(b) $T(1) = 179.7778°$ F.
$T(6) = 138.3889°$ F.
$T(8) = 121.8333°$ F.
(c) 3.3893 minutes.

Section 7.3, page 404

1. All vectors of the form $\begin{bmatrix} b_1 \\ b_2 \\ \vdots \\ b_m \\ b_1 + b_2 + \cdots + b_m \end{bmatrix}$.

3. The code words are

$$\begin{bmatrix} 0 \\ 0 \\ 0 \\ 0 \end{bmatrix}, \begin{bmatrix} 0 \\ 0 \\ 1 \\ 1 \end{bmatrix}, \begin{bmatrix} 0 \\ 1 \\ 0 \\ 1 \end{bmatrix}, \begin{bmatrix} 0 \\ 1 \\ 1 \\ 0 \end{bmatrix}, \begin{bmatrix} 1 \\ 0 \\ 0 \\ 1 \end{bmatrix}, \begin{bmatrix} 1 \\ 0 \\ 1 \\ 0 \end{bmatrix}, \begin{bmatrix} 1 \\ 1 \\ 0 \\ 0 \end{bmatrix}, \begin{bmatrix} 1 \\ 1 \\ 1 \\ 1 \end{bmatrix}.$$

(c) A single error was detected in the fifth bit. The corrected word is

$$\mathbf{x}_t = \begin{bmatrix} 1 \\ 0 \\ 0 \\ 0 \\ 0 \\ 1 \\ 1 \end{bmatrix}.$$

ML.1. (a) $H(8) = \begin{bmatrix} 0 & 0 & 0 & 0 & 0 & 0 & 0 & 1 \\ 0 & 0 & 0 & 1 & 1 & 1 & 1 & 0 \\ 0 & 1 & 1 & 0 & 0 & 1 & 1 & 0 \\ 1 & 0 & 1 & 0 & 1 & 0 & 1 & 0 \end{bmatrix}.$

(b) $C = \begin{bmatrix} 1 & 1 & 0 & 1 \\ 1 & 0 & 1 & 1 \\ 1 & 0 & 0 & 0 \\ 0 & 1 & 1 & 1 \\ 0 & 1 & 0 & 0 \\ 0 & 0 & 1 & 0 \\ 0 & 0 & 0 & 1 \\ 0 & 0 & 0 & 0 \end{bmatrix}.$

ML.3. (a) $H(15) =$

$$\begin{bmatrix} 0 & 0 & 0 & 0 & 0 & 0 & 0 & 1 & 1 & 1 & 1 & 1 & 1 & 1 & 1 \\ 0 & 0 & 0 & 1 & 1 & 1 & 1 & 0 & 0 & 0 & 0 & 1 & 1 & 1 & 1 \\ 0 & 1 & 1 & 0 & 0 & 1 & 1 & 0 & 0 & 1 & 1 & 0 & 0 & 1 & 1 \\ 1 & 0 & 1 & 0 & 1 & 0 & 1 & 0 & 1 & 0 & 1 & 0 & 1 & 0 & 1 \end{bmatrix}$$

(b) $C = \begin{bmatrix} 1 & 1 & 0 & 1 & 1 & 0 & 1 & 0 & 1 & 0 & 1 \\ 1 & 0 & 1 & 1 & 0 & 1 & 1 & 0 & 0 & 1 & 1 \\ 1 & 0 & 0 & 0 & 0 & 0 & 0 & 0 & 0 & 0 & 0 \\ 0 & 1 & 1 & 1 & 0 & 0 & 0 & 1 & 1 & 1 & 1 \\ 0 & 1 & 0 & 0 & 0 & 0 & 0 & 0 & 0 & 0 & 0 \\ 0 & 0 & 1 & 0 & 0 & 0 & 0 & 0 & 0 & 0 & 0 \\ 0 & 0 & 0 & 1 & 0 & 0 & 0 & 0 & 0 & 0 & 0 \\ 0 & 0 & 0 & 0 & 1 & 1 & 1 & 1 & 1 & 1 & 1 \\ 0 & 0 & 0 & 0 & 1 & 0 & 0 & 0 & 0 & 0 & 0 \\ 0 & 0 & 0 & 0 & 0 & 1 & 0 & 0 & 0 & 0 & 0 \\ 0 & 0 & 0 & 0 & 0 & 0 & 1 & 0 & 0 & 0 & 0 \\ 0 & 0 & 0 & 0 & 0 & 0 & 0 & 1 & 0 & 0 & 0 \\ 0 & 0 & 0 & 0 & 0 & 0 & 0 & 0 & 1 & 0 & 0 \\ 0 & 0 & 0 & 0 & 0 & 0 & 0 & 0 & 0 & 1 & 0 \\ 0 & 0 & 0 & 0 & 0 & 0 & 0 & 0 & 0 & 0 & 1 \end{bmatrix}.$

Supplementary Exercises, page 407

1. (a) $\left\{ \begin{bmatrix} 1 \\ 1 \\ 1 \\ 0 \\ 0 \\ 0 \\ 0 \\ 0 \end{bmatrix}, \begin{bmatrix} 1 \\ 0 \\ 0 \\ 1 \\ 1 \\ 0 \\ 0 \\ 0 \end{bmatrix}, \begin{bmatrix} 0 \\ 1 \\ 0 \\ 1 \\ 0 \\ 1 \\ 0 \\ 0 \end{bmatrix}, \begin{bmatrix} 1 \\ 1 \\ 0 \\ 1 \\ 0 \\ 0 \\ 1 \\ 0 \end{bmatrix} \right\}.$

(b) It is a code word since $H(8)\mathbf{x}_t$ is the zero vector.

3. $y = \frac{19}{17}x - \frac{15}{68}$.

Chapter Test, page 407

1. $\left\{ \begin{bmatrix} 1 \\ 1 \\ 1 \\ 0 \\ 0 \\ 0 \end{bmatrix}, \begin{bmatrix} 1 \\ 0 \\ 0 \\ 1 \\ 1 \\ 0 \end{bmatrix}, \begin{bmatrix} 0 \\ 1 \\ 0 \\ 1 \\ 0 \\ 1 \end{bmatrix} \right\}.$

2. It is not a code word. From Example 10, $G\mathbf{x}_t = \begin{bmatrix} 0 \\ 1 \\ 0 \end{bmatrix}$.

Thus bit number 2 is in error.

3. $Q = \begin{bmatrix} \frac{2}{3} & \frac{5}{\sqrt{90}} \\ -\frac{1}{3} & -\frac{4}{\sqrt{90}} \\ -\frac{2}{3} & \frac{7}{\sqrt{90}} \end{bmatrix} \approx \begin{bmatrix} 0.6667 & 0.5270 \\ -0.3333 & -0.4216 \\ -0.6667 & 0.7379 \end{bmatrix},$

$R = \begin{bmatrix} 3 & -1 \\ 0 & \frac{10}{\sqrt{10}} \end{bmatrix} \approx \begin{bmatrix} 3.0000 & -1.0000 \\ 0 & 3.1623 \end{bmatrix}.$

4. $\widehat{\mathbf{x}} \approx \begin{bmatrix} -0.6284 \\ 1.0183 \end{bmatrix}$.

5. (a) $y = \frac{173}{290}x + \frac{31}{58}$.

(b) Approximately 12.47 calories.

Chapter 8

Section 8.1, page 420

1. (a) $A\mathbf{x}_1 = 1\mathbf{x}_1$. (b) $A\mathbf{x}_2 = 4\mathbf{x}_2$.

3. $\lambda^3 - 4\lambda^2 + 7$.

5. $(\lambda - 4)(\lambda - 2)(\lambda - 3) = \lambda^3 - 9\lambda^2 + 26\lambda - 24$.

7. $p(\lambda) = \lambda^3 - 5\lambda^2 + 2\lambda + 8 = (\lambda + 1)(\lambda - 2)(\lambda - 4)$. The eigenvalues and associated eigenvectors are

$$\lambda_1 = -1; \quad \mathbf{x}_1 = \begin{bmatrix} -8 \\ 10 \\ 7 \end{bmatrix}$$

$$\lambda_2 = 2; \quad \mathbf{x}_2 = \begin{bmatrix} 1 \\ -2 \\ 1 \end{bmatrix}$$

$$\lambda_3 = 4; \quad \mathbf{x}_3 = \begin{bmatrix} 1 \\ 0 \\ 1 \end{bmatrix}.$$

9. $f(\lambda) = (\lambda - 1)(\lambda - 3)(\lambda + 2)$;
$\lambda_1 = 1, \lambda_2 = 3, \lambda_3 = -2$;
$\mathbf{x}_1 = \begin{bmatrix} 6 \\ 3 \\ 8 \end{bmatrix}, \mathbf{x}_2 = \begin{bmatrix} 0 \\ 5 \\ 2 \end{bmatrix}, \mathbf{x}_3 = \begin{bmatrix} 0 \\ 0 \\ 1 \end{bmatrix}.$

11. $f(\lambda) = \lambda^2 - 5\lambda + 6$; $\lambda_1 = 2, \lambda_2 = 3$;
$\mathbf{x}_1 = \begin{bmatrix} 1 \\ -1 \end{bmatrix}, \mathbf{x}_2 = \begin{bmatrix} 1 \\ -2 \end{bmatrix}.$

13. $f(\lambda) = \lambda^3 - 5\lambda^2 + 2\lambda + 8$;
$\lambda_1 = -1, \lambda_2 = 2, \lambda_3 = 4$;
$$\mathbf{x}_1 = \begin{bmatrix} 1 \\ 0 \\ -1 \end{bmatrix}, \mathbf{x}_2 = \begin{bmatrix} -2 \\ -3 \\ 2 \end{bmatrix}, \mathbf{x}_3 = \begin{bmatrix} 8 \\ 5 \\ 2 \end{bmatrix}.$$

15. $f(\lambda) = (\lambda - 1)(\lambda + 1)(\lambda - 3)(\lambda - 2)$;
$\lambda_1 = 1, \lambda_2 = -1, \lambda_3 = 3, \lambda_4 = 2$;
$$\mathbf{x}_1 = \begin{bmatrix} 1 \\ 0 \\ 0 \\ 0 \end{bmatrix}, \mathbf{x}_2 = \begin{bmatrix} 1 \\ -1 \\ 0 \\ 0 \end{bmatrix}, \mathbf{x}_3 = \begin{bmatrix} 9 \\ 3 \\ 4 \\ 0 \end{bmatrix}, \mathbf{x}_4 = \begin{bmatrix} 29 \\ 7 \\ 9 \\ -3 \end{bmatrix}.$$

17. (a) $p(\lambda) = \lambda^2 + \lambda + 1 - i = [\lambda - i][\lambda - (-1 - i)]$. The eigenvalues and associated eigenvectors are
$$\lambda_1 = i; \qquad \mathbf{x}_1 = \begin{bmatrix} i \\ 1 \end{bmatrix}$$
$$\lambda_2 = -1 - i; \qquad \mathbf{x}_2 = \begin{bmatrix} -1 - i \\ 1 \end{bmatrix}.$$

(b) $p(\lambda) = (\lambda - 1)(\lambda^2 - 2i\lambda - 2) = (\lambda - 1)[\lambda - (1 + i)][\lambda - (-1 + i)]$. The eigenvalues and associated eigenvectors are
$$\lambda_1 = 1; \qquad \mathbf{x}_1 = \begin{bmatrix} 0 \\ 0 \\ 1 \end{bmatrix}$$
$$\lambda_2 = 1 + i; \qquad \mathbf{x}_2 = \begin{bmatrix} 1 \\ 1 \\ 0 \end{bmatrix}$$
$$\lambda_3 = -1 + i; \qquad \mathbf{x}_3 = \begin{bmatrix} -1 \\ 1 \\ 0 \end{bmatrix}.$$

(c) $p(\lambda) = \lambda^3 + \lambda = \lambda(\lambda - i)(\lambda + i)$. The eigenvalues and associated eigenvectors are
$$\lambda_1 = 0; \qquad \mathbf{x}_1 = \begin{bmatrix} 0 \\ 0 \\ 1 \end{bmatrix}$$
$$\lambda_2 = i; \qquad \mathbf{x}_2 = \begin{bmatrix} -1 \\ i \\ 1 \end{bmatrix}$$
$$\lambda_3 = -i; \qquad \mathbf{x}_3 = \begin{bmatrix} -1 \\ -i \\ 1 \end{bmatrix}.$$

(d) $p(\lambda) = \lambda^2(\lambda - 1) + 9(\lambda - 1) = (\lambda - 1)(\lambda - 3i)(\lambda + 3i)$.

The eigenvalues and associated eigenvectors are
$$\lambda_1 = 1; \qquad \mathbf{x}_1 = \begin{bmatrix} 0 \\ 1 \\ 0 \end{bmatrix}$$
$$\lambda_2 = 3i; \qquad \mathbf{x}_2 = \begin{bmatrix} 3i \\ 0 \\ 1 \end{bmatrix}$$
$$\lambda_3 = -3i; \qquad \mathbf{x}_3 = \begin{bmatrix} -3i \\ 0 \\ 1 \end{bmatrix}.$$

19. Basis for eigenspace associated with $\lambda_1 = \lambda_2 = 2$ is
$$\left\{ \begin{bmatrix} 1 \\ 0 \\ 0 \\ 0 \end{bmatrix} \right\}.$$
Basis for eigenspace associated with $\lambda_3 = \lambda_4 = 1$ is
$$\left\{ \begin{bmatrix} 3 \\ -3 \\ 1 \\ 0 \end{bmatrix} \right\}.$$

21. $\left\{ \begin{bmatrix} -1 \\ 0 \\ 1 \end{bmatrix} \right\}$. **23.** $\left\{ \begin{bmatrix} 0 \\ 0 \\ 1 \\ 0 \end{bmatrix} \right\}$.

25. (a) $\left\{ \begin{bmatrix} \frac{3}{2} \\ 0 \\ 1 \\ -1 \end{bmatrix}, \begin{bmatrix} 0 \\ 1 \\ 0 \\ 0 \end{bmatrix} \right\}$. (b) $\left\{ \begin{bmatrix} 0 \\ 0 \\ 1 \\ i \end{bmatrix} \right\}$.

27. $\begin{bmatrix} 8 \\ 2 \\ 1 \end{bmatrix}$.

ML.1. (a) $\lambda^2 - 5$. (b) $\lambda^3 - 6\lambda^2 + 4\lambda + 8$.
(c) $\lambda^4 - 3\lambda^3 - 3\lambda^2 + 11\lambda - 6$.

ML.3. (a) $\begin{bmatrix} 1 \\ 1 \end{bmatrix}$. (b) $\begin{bmatrix} 0 \\ 0 \\ 1 \end{bmatrix}$. (c) $\begin{bmatrix} 1 \\ -2 \\ 1 \end{bmatrix}$.

Section 8.2, page 431

1. Diagonalizable. The eigenvalues are $\lambda_1 = -3$ and $\lambda_2 = 2$. The result follows by Theorem 8.5.

3. Diagonalizable. The eigenvalues are $\lambda_1 = 0$, $\lambda_2 = 2$, and $\lambda_3 = 3$. The result follows by Theorem 8.5.

5. Not diagonalizable.

7. Not diagonalizable.

9. $\begin{bmatrix} -\frac{4}{3} & -\frac{5}{3} \\ -\frac{10}{3} & \frac{1}{3} \end{bmatrix}$.

11. Not possible.

13. $P = \begin{bmatrix} 1 & -3 & 1 \\ 0 & 0 & -6 \\ 1 & 2 & 4 \end{bmatrix}$.

15. Not possible.

17. $P = \begin{bmatrix} 1 & 2 & 1 \\ 0 & 1 & 0 \\ 0 & 0 & -3 \end{bmatrix}$.

19. Not possible. **21.** Not possible.

23. $P = \begin{bmatrix} -1 & 2 \\ 1 & 1 \end{bmatrix}$.

25. Possible answers: $\begin{bmatrix} 3 & 0 \\ 0 & 0 \end{bmatrix}, \begin{bmatrix} 0 & 0 \\ 0 & 3 \end{bmatrix}$.

27. Possible answers: $\begin{bmatrix} 1 & 0 & 0 \\ 0 & 1 & 0 \\ 0 & 0 & 2 \end{bmatrix}, \begin{bmatrix} 1 & 0 & 0 \\ 0 & 2 & 0 \\ 0 & 0 & 1 \end{bmatrix}$.

29. Similar to a diagonal matrix.

31. Similar to a diagonal matrix.

33. $D = \begin{bmatrix} 6 & 0 \\ 0 & 1 \end{bmatrix}$. **35.** $D = \begin{bmatrix} 2 & 0 & 0 \\ 0 & 4 & 0 \\ 0 & 0 & 1 \end{bmatrix}$.

37. A is upper triangular with the multiple eigenvalue $\lambda_1 = \lambda_2 = 1$ with associated eigenvector $\begin{bmatrix} 1 \\ 0 \end{bmatrix}$.

39. A has the multiple eigenvalue $\lambda_1 = \lambda_2 = -1$ with associated eigenvector $\begin{bmatrix} -1 \\ 1 \\ 0 \end{bmatrix}$.

41. Defective. **43.** Not defective.

45. $\begin{bmatrix} 2^9 & 0 \\ 0 & (-2)^9 \end{bmatrix} = \begin{bmatrix} 512 & 0 \\ 0 & -512 \end{bmatrix}$.

Section 8.3, page 433

5. $\begin{bmatrix} 0 & 0 \\ 0 & 4 \end{bmatrix}$; $P = \begin{bmatrix} \frac{1}{\sqrt{2}} & \frac{1}{\sqrt{2}} \\ -\frac{1}{\sqrt{2}} & \frac{1}{\sqrt{2}} \end{bmatrix}$.

7. $\begin{bmatrix} 0 & 0 & 0 \\ 0 & 0 & 0 \\ 0 & 0 & 4 \end{bmatrix}$; $P = \begin{bmatrix} 1 & 0 & 0 \\ 0 & -\frac{1}{\sqrt{2}} & \frac{1}{\sqrt{2}} \\ 0 & \frac{1}{\sqrt{2}} & \frac{1}{\sqrt{2}} \end{bmatrix}$.

9. $\begin{bmatrix} -2 & 0 & 0 \\ 0 & 1 & 0 \\ 0 & 0 & 1 \end{bmatrix}$; $P = \begin{bmatrix} \frac{1}{\sqrt{3}} & -\frac{1}{\sqrt{2}} & -\frac{1}{\sqrt{6}} \\ \frac{1}{\sqrt{3}} & \frac{1}{\sqrt{2}} & -\frac{1}{\sqrt{6}} \\ \frac{1}{\sqrt{3}} & 0 & \frac{2}{\sqrt{6}} \end{bmatrix}$.

11. $\begin{bmatrix} 3 & 0 \\ 0 & 1 \end{bmatrix}$.

13. $\begin{bmatrix} 1 & 0 & 0 \\ 0 & 2 & 0 \\ 0 & 0 & 0 \end{bmatrix}$. **15.** $\begin{bmatrix} 1 & 0 & 0 \\ 0 & 0 & 0 \\ 0 & 0 & 2 \end{bmatrix}$.

17. $\begin{bmatrix} 1 & 0 & 0 \\ 0 & 1 & 0 \\ 0 & 0 & 4 \end{bmatrix}$.

ML.1. (a) $\lambda_1 = 0, \lambda_2 = 12$; $P = \begin{bmatrix} 0.7071 & 0.7071 \\ -0.7071 & 0.7071 \end{bmatrix}$.

(b) $\lambda_1 = -1, \lambda_2 = -1, \lambda_3 = 5$;

$$P = \begin{bmatrix} 0.7743 & -0.2590 & 0.5774 \\ -0.6115 & -0.5411 & 0.5774 \\ -0.1629 & 0.8001 & 0.5774 \end{bmatrix}.$$

(c) $\lambda_1 = 5.4142, \lambda_2 = 4.0000, \lambda_3 = 2.5858$.

$$P = \begin{bmatrix} 0.5000 & -0.7071 & -0.5000 \\ 0.7071 & -0.0000 & 0.7071 \\ 0.5000 & 0.7071 & -0.5000 \end{bmatrix}.$$

Supplementary Exercises, page 445

1. $f(\lambda) = (\lambda + 2)(\lambda^2 - 8\lambda + 15)$; $\lambda_1 = -2, \lambda_2 = 3, \lambda_3 = 5$;

$$\mathbf{x}_1 = \begin{bmatrix} -35 \\ 12 \\ 19 \end{bmatrix}, \mathbf{x}_2 = \begin{bmatrix} 0 \\ 3 \\ 1 \end{bmatrix}, \mathbf{x}_3 = \begin{bmatrix} 0 \\ 1 \\ 1 \end{bmatrix}.$$

3. Yes.

5. Not diagonalizable; the roots of the characteristic polynomial are not all real.

7. For $\lambda = 0$, possible answer: $\left\{\begin{bmatrix} 1 \\ 0 \\ 0 \end{bmatrix}\right\}$;

for $\lambda = 2$, possible answer: $\left\{\begin{bmatrix} 1 \\ 0 \\ 2 \end{bmatrix}, \begin{bmatrix} 0 \\ 1 \\ 0 \end{bmatrix}\right\}$.

9. $P = \begin{bmatrix} -\frac{1}{\sqrt{2}} & \frac{1}{\sqrt{6}} & \frac{1}{\sqrt{3}} \\ \frac{1}{\sqrt{2}} & -\frac{1}{\sqrt{6}} & \frac{1}{\sqrt{3}} \\ 0 & \frac{2}{\sqrt{6}} & \frac{1}{\sqrt{3}} \end{bmatrix}$,

$$D = \begin{bmatrix} 0 & 0 & 0 \\ 0 & 0 & 0 \\ 0 & 0 & 3 \end{bmatrix}.$$

11. $\lambda = 0, \lambda = 1$.

Chapter Test, page 446

1. Not diagonalizable; $\lambda_1 = 1, \lambda_2 = \lambda_3 = 2$.

2. Check that $AA^T = I_3$.

3. $P = \begin{bmatrix} -2 & 1 & 1 \\ 1 & 2 & 0 \\ 2 & 0 & 1 \end{bmatrix}$, $D = \begin{bmatrix} 9 & 0 & 0 \\ 0 & -9 & 0 \\ 0 & 0 & -9 \end{bmatrix}$.

4. (a) F. (b) F. (c) T. (d) T. (e) F.

Chapter 9

Section 9.1, page 450

3. (a) $u_8 = 34$. (b) $u_{12} = 233$. (c) $u_{20} = 10{,}946$.

Section 9.2, page 460

1. (a) $\mathbf{x}(t) = \begin{bmatrix} x_1(t) \\ x_2(t) \\ x_3(t) \end{bmatrix} = \begin{bmatrix} b_1 e^{-3t} \\ b_2 e^{4t} \\ b_3 e^{2t} \end{bmatrix}$

$$= b_1 \begin{bmatrix} 1 \\ 0 \\ 0 \end{bmatrix} e^{-3t} + b_2 \begin{bmatrix} 0 \\ 1 \\ 0 \end{bmatrix} e^{4t} + b_3 \begin{bmatrix} 0 \\ 0 \\ 1 \end{bmatrix} e^{2t}.$$

(b) $\begin{bmatrix} 3e^{-3t} \\ 4e^{4t} \\ 5e^{2t} \end{bmatrix} = 3 \begin{bmatrix} 1 \\ 0 \\ 0 \end{bmatrix} e^{-3t} + 4 \begin{bmatrix} 0 \\ 1 \\ 0 \end{bmatrix} e^{4t} + 5 \begin{bmatrix} 0 \\ 0 \\ 1 \end{bmatrix} e^{2t}.$

3. $\mathbf{x}(t) = b_1 \begin{bmatrix} 6 \\ 2 \\ 7 \end{bmatrix} e^{4t} + b_2 \begin{bmatrix} 0 \\ 7 \\ -1 \end{bmatrix} e^{-5t} + b_3 \begin{bmatrix} 0 \\ 0 \\ 1 \end{bmatrix} e^{2t}.$

5. $\mathbf{x}(t) = b_1 \begin{bmatrix} 1 \\ 0 \\ 0 \end{bmatrix} e^{5t} + b_2 \begin{bmatrix} 0 \\ 1 \\ 3 \end{bmatrix} e^{5t} + b_3 \begin{bmatrix} 0 \\ -3 \\ 1 \end{bmatrix} e^{-5t}.$

7. $\mathbf{x}(t) = b_1 \begin{bmatrix} 1 \\ 0 \\ 1 \end{bmatrix} e^{4t} + b_2 \begin{bmatrix} -3 \\ 0 \\ 2 \end{bmatrix} e^{-t} + b_3 \begin{bmatrix} 1 \\ -6 \\ 4 \end{bmatrix} e^{t}.$

9. $\mathbf{x}(t) = 220 \begin{bmatrix} 2 \\ 1 \end{bmatrix} + 20 \begin{bmatrix} 3 \\ -1 \end{bmatrix} e^{-5t} = \begin{bmatrix} 440 + 60e^{-5t} \\ 220 - 20e^{-5t} \end{bmatrix}.$

ML.1. $\mathbf{x}(t) = b_1 \begin{bmatrix} -0.5774 \\ -0.5774 \\ -0.5774 \end{bmatrix} e^{t} +$

$$b_2 \begin{bmatrix} 0.2182 \\ 0.4364 \\ 0.8729 \end{bmatrix} e^{2t} + b_3 \begin{bmatrix} 0.0605 \\ 0.2421 \\ 0.9684 \end{bmatrix} e^{4t}.$$

ML.3. $\mathbf{x}(t) = b_1 \begin{bmatrix} -0.8321 \\ 0 \\ 0.5547 \end{bmatrix} e^{-t} +$

$$b_2 \begin{bmatrix} -0.7071 \\ 0 \\ -0.7071 \end{bmatrix} e^{4t} + b_3 \begin{bmatrix} -0.1374 \\ 0.8242 \\ -0.5494 \end{bmatrix} e^{t}.$$

Section 9.3, page 474

1. The origin is a stable equilibrium point. The phase portrait shows all trajectories tending toward the origin.

3. The origin is a stable equilibrium point. The phase portrait shows all trajectories tending toward the origin with those passing through points not on the eigenvector aligning themselves to be tangent to the eigenvector at the origin.

5. The origin is a saddle point. The phase portrait shows trajectories not in the direction of an eigenvector heading toward the origin, but bending away as $t \to \infty$.

7. The origin is a stable equilibrium point. The phase portrait shows all trajectories tending toward the origin.

9. The origin is called marginally stable.

Section 9.4, page 483

1. (a) $\begin{bmatrix} x & y \end{bmatrix} \begin{bmatrix} -3 & \frac{5}{2} \\ \frac{5}{2} & -2 \end{bmatrix} \begin{bmatrix} x \\ y \end{bmatrix}.$

(b) $\begin{bmatrix} x_1 & x_2 & x_3 \end{bmatrix} \begin{bmatrix} 2 & \frac{3}{2} & -\frac{5}{2} \\ \frac{3}{2} & 0 & \frac{7}{2} \\ -\frac{5}{2} & \frac{7}{2} & 0 \end{bmatrix} \begin{bmatrix} x_1 \\ x_2 \\ x_3 \end{bmatrix}.$

(c) $\begin{bmatrix} x_1 & x_2 & x_3 \end{bmatrix} \begin{bmatrix} 3 & \frac{1}{2} & -1 \\ \frac{1}{2} & 1 & -2 \\ -1 & -2 & -2 \end{bmatrix} \begin{bmatrix} x_1 \\ x_2 \\ x_3 \end{bmatrix}.$

3. (a) $\begin{bmatrix} -1 & 0 & 0 \\ 0 & 2 & 0 \\ 0 & 0 & 0 \end{bmatrix}.$ (b) $\begin{bmatrix} 3 & 0 & 0 \\ 0 & 0 & 0 \\ 0 & 0 & 0 \end{bmatrix}.$

5. $2x'^2 - 3y'^2.$

7. $y_2^2 - y_3^2.$

9. $-2y_1^2 + 5y_2^2 - 5y_3^2.$

11. $y_1'^2.$

13. $y_1^2 + y_2^2 - y_3^2.$

15. $y_1^2 - y_2^2.$

17. $h(\mathbf{y}) = y_1^2 - y_2^2$, rank of g is 2 and signature of g is 0.

19. $y_1^2 + y_2^2 = 1$ is a circle.
$-y_1^2 - y_2^2 = 1$ is empty; it represents no conic.
$y_1^2 - y_2^2 = 1$ is a hyperbola.
$y_1^2 = 1$ is a pair of lines; $y_1 = 1$, $y_1 = -1$.
$-y_1^2 = 1$ is empty; it represents no conic.

21. g_1, g_2, and g_4.

23. (a), (b), and (c).

ML.1. (a) rank = 2, signature = 0.
(b) rank = 1, signature = 1.
(c) rank = 4, signature = 2.
(d) rank = 4, signature = 4.

Section 9.5, page 491

1. Ellipse.

3. Hyperbola.

5. Two intersecting lines.

7. Circle.

9. Point.

11. Ellipse; $\dfrac{x'^2}{2} + y'^2 = 1.$

13. Circle; $\dfrac{x'^2}{5^2} + \dfrac{y'^2}{5^2} = 1.$

15. Pair of parallel lines; $y' = 2$, $y' = -2$; $y'^2 = 4.$

17. Point $(1, 3)$; $x'^2 + y'^2 = 0.$

19. Possible answer: ellipse; $\dfrac{x'^2}{12} + \dfrac{y'^2}{4} = 1.$

21. Possible answer: pair of parallel lines $y' = \frac{2}{\sqrt{10}}$ and $y' = -\frac{2}{\sqrt{10}}$; $y'^2 = \frac{4}{10}.$

23. Possible answer: two intersecting lines $y' = 3x'$ and $y' = -3x'$; $9x'^2 - y'^2 = 0$.

25. Possible answer: parabola; $y''^2 = -4x''$.

27. Possible answer: hyperbola; $\frac{x''^2}{4} - \frac{y''^2}{9} = 1$.

29. Possible answer: hyperbola; $\frac{x''^2}{\frac{9}{8}} - \frac{y''^2}{\frac{9}{8}} = 1$.

Section 9.6, page 499

1. Hyperboloid of one sheet.

3. Hyperbolic paraboloid.

5. Parabolic cylinder.

7. Parabolic cylinder.

9. Ellipsoid.

11. Elliptic paraboloid.

13. Hyperbolic paraboloid.

15. Ellipsoid; $x'^2 + y'^2 + \frac{z'^2}{\frac{1}{3}} = 1$.

17. Hyperbolic paraboloid; $\frac{x''^2}{4} - \frac{y''^2}{4} = z''$.

19. Elliptic paraboloid; $\frac{x'^2}{4} + \frac{y'^2}{8} - 1$.

21. Hyperboloid of one sheet; $\frac{x''^2}{2} + \frac{y''^2}{4} - \frac{z''}{4} = 1$.

23. Parabolic cylinder; $x''^2 = \frac{4}{\sqrt{2}} y''$.

25. Hyperboloid of two sheets;

$$\frac{x''^2}{\frac{7}{4}} - \frac{y''^2}{\frac{7}{4}} - \frac{z''^2}{\frac{7}{4}} = 1.$$

27. Cone; $x''^2 + y''^2 - z''^2 = 0$.

Supplementary Exercises, page 500

1. (a) $\frac{d}{dt}[A(t)] = \begin{bmatrix} 2t & \frac{-1}{(t+1)^2} \\ 0 & -e^{-t} \end{bmatrix}$.

$$\int_0^t A(s)\,ds = \begin{bmatrix} \frac{t^3}{3} & \ln(1+t) \\ 4t & -e^{-t}+1 \end{bmatrix}.$$

(b) $\frac{d}{dt}[A(t)] = \begin{bmatrix} 2\cos 2t & 0 & 0 \\ 0 & 0 & -1 \\ 0 & e^{t^2} + 2t^2 e^{t^2} & \frac{1-t^2}{(t^2+1)^2} \end{bmatrix}$.

$$\int_0^t A(s)\,ds = \begin{bmatrix} -\frac{\cos 2t}{2} + \frac{1}{2} & 0 & 0 \\ 0 & t & -\frac{t^2}{2} \\ 0 & \frac{e^{t^2}}{2} - \frac{1}{2} & \frac{1}{2}\ln(t^2+1) \end{bmatrix}.$$

3. (a) $\mathbf{x}(t) = \frac{2}{5}\begin{bmatrix} 4 \\ 4 \\ 1 \end{bmatrix} + \frac{7}{20}\begin{bmatrix} -1 \\ -6 \\ 1 \end{bmatrix} e^{5t} + \frac{1}{4}\begin{bmatrix} -1 \\ 2 \\ 1 \end{bmatrix} e^{-3t}$.

(b) $\mathbf{x}(t) = \frac{7}{8}\begin{bmatrix} 1 \\ 0 \\ 0 \end{bmatrix} + \frac{1}{12}\begin{bmatrix} 1 \\ 2 \\ 4 \end{bmatrix} e^{2t} + \frac{1}{24}\begin{bmatrix} 1 \\ -4 \\ 16 \end{bmatrix} e^{-4t}$.

5. (a) $\mathbf{x} = b_1 \begin{bmatrix} 1 \\ 1 \end{bmatrix} e^{2t} + b_2 \begin{bmatrix} -1 \\ 3 \end{bmatrix} e^{-2t}$.

(b) $\mathbf{x} = \frac{9}{2}\begin{bmatrix} 1 \\ 1 \end{bmatrix} e^{2t} + \frac{1}{2}\begin{bmatrix} -1 \\ 3 \end{bmatrix} e^{-2t}$.

7. The origin is a saddle point.

Chapter Test, page 501

1. 1,346,269.

2. $\mathbf{x} = b_1 \begin{bmatrix} 1 \\ 1 \end{bmatrix} e^{5t} + b_2 \begin{bmatrix} -1 \\ 3 \end{bmatrix} e^{-3t}$.

3. $y_1^2 - y_2^2$, a hyperbola.

4. The origin is an unstable equilibrium point. The phase portrait shows all trajectories tending to move away from the origin.

5. Possible answer: $k = 2$.

Chapter 10

Section 10.1, page 507

1. (a) and (c).

3. (a) Yes. (b) No. (c) Yes.

5. (a) No. (b) Yes. (c) Yes.

7. Yes.

9. Yes.

11. (a) $\begin{bmatrix} 15 & 5 & 4 & 8 \\ -5 & -1 & 10 & 2 \end{bmatrix}$.

17. (a) $\begin{bmatrix} 8 & 5 \end{bmatrix}$. (b) $\begin{bmatrix} \frac{-a_1+3a_2}{2} & \frac{-5a_1+a_2}{2} \end{bmatrix}$.

19. (a) $17t - 7$. (b) $\left(\frac{5a-b}{2}\right)t + \frac{a+5b}{2}$.

Section 10.2, page 519

1. (a) Yes. (b) No. (c) Yes. (d) No.
(e) $\{(0, r)\}$, r = any real number.
(f) $\{(r, 0)\}$, r = any real number.

3. (a) $\{(0, 0)\}$. (b) Yes. (c) No.

5. (a) Possible answer: $\left\{\begin{bmatrix}-2\\0\\1\\1\\0\end{bmatrix}, \begin{bmatrix}0\\1\\0\\0\\0\end{bmatrix}\right\}$.

(b) Possible answers:

$\left\{\begin{bmatrix}1\\0\\0\\1\end{bmatrix}, \begin{bmatrix}0\\1\\0\\-1\end{bmatrix}, \begin{bmatrix}0\\0\\1\\0\end{bmatrix}\right\}, \left\{\begin{bmatrix}1\\1\\2\\0\end{bmatrix}, \begin{bmatrix}-1\\0\\-1\\-1\end{bmatrix}, \begin{bmatrix}-1\\-1\\-1\\0\end{bmatrix}\right\}$.

7. (a) Yes. (b) 1.

11. (a) No. (b) Yes. (c) Yes. (d) No.
(e) Possible answer: $\{-t^2 - t + 1\}$.
(f) Possible answer: $\{t^2, t\}$.

13. (a) $\ker L = \left\{\begin{bmatrix}0 & 0\\0 & 0\end{bmatrix}\right\}$, so $\ker L$ has no basis.

(b) $\left\{\begin{bmatrix}1 & 0\\1 & 0\end{bmatrix}, \begin{bmatrix}1 & 1\\0 & 1\end{bmatrix}, \begin{bmatrix}0 & 1\\0 & 0\end{bmatrix}, \begin{bmatrix}0 & 0\\1 & 1\end{bmatrix}\right\}$.

15. (a) Possible answer: $\left\{\begin{bmatrix}1 & 0\\0 & 1\end{bmatrix}, \begin{bmatrix}0 & 1\\\frac{1}{2} & 0\end{bmatrix}\right\}$.

(b) Possible answer: $\left\{\begin{bmatrix}0 & -2\\1 & 0\end{bmatrix}, \begin{bmatrix}-1 & 0\\0 & 1\end{bmatrix}\right\}$.

17. (a) Possible answer: $\{1\}$.
(b) Possible answer: $\{t, 1\}$.

19. (a) 2. (b) 1.

ML.1. Basis for $\ker L$: $\left\{\begin{bmatrix}-1\\-2\\1\\0\end{bmatrix}, \begin{bmatrix}1\\-3\\0\\1\end{bmatrix}\right\}$.

Basis for range L: $\left\{\begin{bmatrix}1\\0\end{bmatrix}, \begin{bmatrix}0\\1\end{bmatrix}\right\}$.

ML.3. Basis for $\ker L$: $\left\{\begin{bmatrix}-2\\0\\1\\-2\\1\end{bmatrix}, \begin{bmatrix}-1\\1\\0\\0\\0\end{bmatrix}\right\}$.

Basis for range L: $\left\{\begin{bmatrix}1\\0\\0\end{bmatrix}, \begin{bmatrix}0\\1\\0\end{bmatrix}, \begin{bmatrix}0\\0\\1\end{bmatrix}\right\}$.

Section 10.3, page 532

1. (a) $\begin{bmatrix}3 & -2\\2 & 0\end{bmatrix}$. (b) $\begin{bmatrix}3 & -2\\-1 & 2\end{bmatrix}$.
(c) $\begin{bmatrix}1 & -2\\2 & 0\end{bmatrix}$. (d) $\begin{bmatrix}1 & -2\\1 & 2\end{bmatrix}$. (e) $(4, 0)$.

3. (a) $\begin{bmatrix}1 & -2\\2 & 1\\1 & 1\end{bmatrix}$. (b) $\begin{bmatrix}\frac{7}{3} & -\frac{4}{3}\\-\frac{2}{3} & \frac{5}{3}\\\frac{2}{3} & -\frac{2}{3}\end{bmatrix}$. (c) $\begin{bmatrix}-3\\4\\3\end{bmatrix}$.

5. (a) $\begin{bmatrix}1 & 1 & 0\\0 & 1 & -1\end{bmatrix}$. (b) $\begin{bmatrix}-1 & -\frac{1}{3} & 0\\1 & \frac{2}{3} & 0\end{bmatrix}$.
(c) $\begin{bmatrix}3\\-1\end{bmatrix}$.

7. (a) $\begin{bmatrix}1 & 0\\0 & 1\\0 & 0\\0 & 0\end{bmatrix}$. (b) $\begin{bmatrix}1 & 1\\0 & 1\\0 & -1\\0 & 1\end{bmatrix}$.

9. (a) $\begin{bmatrix}1 & 0 & 0 & 0\\0 & 0 & 1 & 0\\0 & 1 & 0 & 0\\0 & 0 & 0 & 1\end{bmatrix}$.

(b) $\begin{bmatrix}1 & 1 & 0 & -1\\-1 & -1 & 1 & 1\\0 & 1 & 0 & 0\\0 & -1 & 0 & 1\end{bmatrix}$.

(c) $\begin{bmatrix}1 & 0 & 0 & 1\\0 & 0 & 1 & 0\\1 & 1 & 0 & 0\\0 & 0 & 1 & 1\end{bmatrix}$.

(d) $\begin{bmatrix}2 & 1 & -1 & 0\\-2 & -1 & 2 & 0\\1 & 1 & 0 & 0\\-1 & -1 & 1 & 1\end{bmatrix}$.

11. (a) $\begin{bmatrix}10\\5\\5\end{bmatrix}$. (b) $\begin{bmatrix}4\\2\\2\end{bmatrix}$.

13. (a) $\left[L(\mathbf{v}_1)\right]_S = \begin{bmatrix}2\\-1\end{bmatrix}, \left[L(\mathbf{v}_2)\right]_S = \begin{bmatrix}-3\\4\end{bmatrix}$.
(b) $L(\mathbf{v}_1) = \begin{bmatrix}1\\5\end{bmatrix}, L(\mathbf{v}_2) = \begin{bmatrix}1\\-10\end{bmatrix}$.
(c) $\begin{bmatrix}-2\\25\end{bmatrix}$.

15. (a) $\left[L(\mathbf{v}_1)\right]_T = \begin{bmatrix}1\\2\\-1\end{bmatrix}, \left[L(\mathbf{v}_2)\right]_T = \begin{bmatrix}0\\1\\-2\end{bmatrix}$.
(b) $L(\mathbf{v}_1) = t^2 + t + 2$, $L(\mathbf{v}_2) = -t + 2$.
(c) $L(2t+1) = \frac{3}{2}t^2 + t + 4$.
(d) $L(at+b) = \left(\dfrac{a+b}{2}\right)t^2 + bt + 2a$.

17. (a) $\begin{bmatrix}0 & \frac{3}{2}\\1 & \frac{1}{2}\end{bmatrix}$. (b) $\frac{3}{2}t - 3$.
(c) $\left(\dfrac{3a-b}{2}\right)t - b$.

19. $\begin{bmatrix}1 & 0\\4 & -1\end{bmatrix}$.

21. (a) $\begin{bmatrix}1 & 0 & 2 & 0\\0 & 0 & 0 & 6\\0 & 0 & 0 & 0\\0 & 0 & 0 & 0\end{bmatrix}$.

(b) $\begin{bmatrix} 0 & 0 & 0 & 0 \\ 0 & 0 & 0 & 0 \\ 6 & 0 & 0 & 0 \\ 0 & 1 & 0 & 1 \end{bmatrix}$. (c) Same as (b).

23. $\begin{bmatrix} 1 & 0 \\ 0 & -1 \end{bmatrix}$.

ML.1. $A = \begin{bmatrix} -1 & 0 & 3 \\ 1 & 0 & -2 \end{bmatrix}$.

ML.3. (a) $A = \begin{bmatrix} 1.3333 & -0.3333 \\ -1.6667 & -3.3333 \end{bmatrix}$.

(b) $B = \begin{bmatrix} -3.6667 & 0.3333 \\ -3.3333 & 1.6667 \end{bmatrix}$.

(c) $P = \begin{bmatrix} -0.3333 & 0.6667 \\ 1.6667 & -0.3333 \end{bmatrix}$.

Section 10.4, page 547

1. (a) S_0

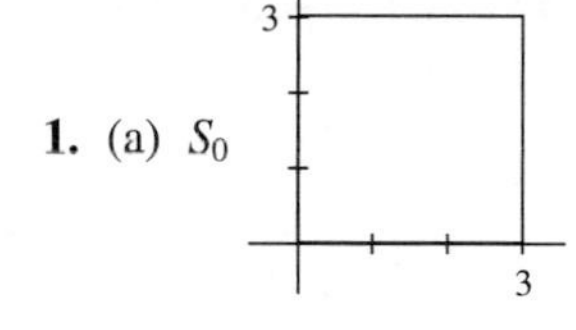

(b) S_1

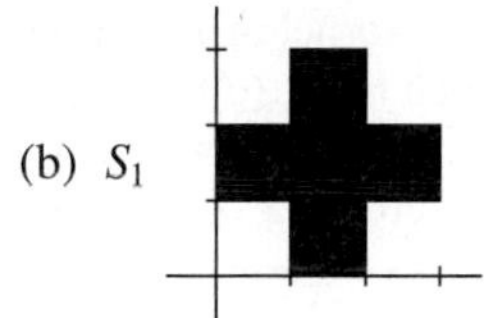

(c) S_2

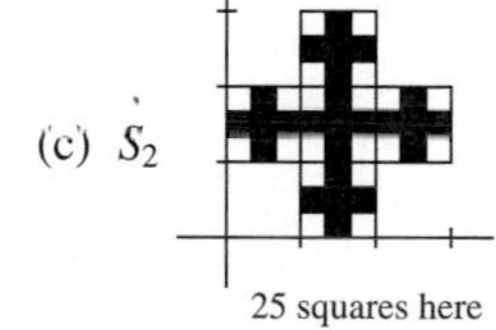

(d) $5 \times (25)$ squares of size $\frac{1}{9} \times \frac{1}{9}$.

(e) S_1, S_2, and S_3 are composed of crosses made up of 5 squares of the same size.

(f) area $(S_0) = 9$, area $(S_1) = 5$, area $(S_2) = \frac{25}{9}$;

area $(S_3) = \dfrac{125}{81}$;

$\dfrac{\text{area}(S_1)}{\text{area}(S_0)} = \dfrac{5}{9}$;

$\dfrac{\text{area}(S_2)}{\text{area}(S_1)} = \dfrac{\frac{25}{9}}{9} = \dfrac{5}{9}$;

$\dfrac{\text{area}(S_3)}{\text{area}(S_2)} = \dfrac{\frac{125}{81}}{\frac{25}{9}} = \dfrac{5}{9}$.

3. (a) $\frac{1}{8}$ (b) (c) 2^{-6}.

(d) For 10.8(a) length $= 2 = 1 + 2(\frac{1}{2})$.

For 10.8(b) length $= 3 = 1 + 2(\frac{1}{2}) + 4(\frac{1}{4})$.

For 10.8(c) length $= 4 = 1 + 2(\frac{1}{2}) + 4(\frac{1}{4}) + 8(\frac{1}{8})$.

For 10.9 length $= 7$.

5. $T(T(\mathbf{v})) = \mathbf{v} + 2\mathbf{b}$, $T(T(T(\mathbf{v}))) = \mathbf{v} + 3\mathbf{b}$, $T^k(\mathbf{v}) = \mathbf{v} + k\mathbf{b}$; the vector $\mathbf{v}$ is translated by $k\mathbf{b}$.

7. $A = \begin{bmatrix} 3 & -1 \\ 4 & 0 \end{bmatrix}$, $\mathbf{b} = \begin{bmatrix} 1 \\ -5 \end{bmatrix}$.

9. $A = \begin{bmatrix} 1 & 4 \\ -1 & 3 \end{bmatrix}$, $\mathbf{b} = \begin{bmatrix} -2 \\ 1 \end{bmatrix}$.

11. $S = \begin{bmatrix} 0 & 0 & 1 & 2 & 3 & 3 & 0 \\ 0 & 1 & 1 & 3 & 1 & 0 & 0 \end{bmatrix}$. Recall from Example 1 that to compute $T(S)$ we compute AS, then add the vector $\mathbf{b}$ to each column of the result of AS.

(a) $T(S) = AS + \mathbf{b}$

$$= \begin{bmatrix} 2 & -2 \\ 1 & 2 \end{bmatrix} \begin{bmatrix} 0 & 0 & 1 & 2 & 3 & 3 & 0 \\ 0 & 1 & 1 & 3 & 1 & 0 & 0 \end{bmatrix} + \begin{bmatrix} -2 & -2 & -2 & -2 & -2 & -2 & -2 \\ 1 & 1 & 1 & 1 & 1 & 1 & 1 \end{bmatrix}$$

$$= \begin{bmatrix} -2 & -4 & -2 & -4 & 2 & 4 & -2 \\ 1 & 2 & 4 & 8 & 8 & 7 & 1 \end{bmatrix}.$$

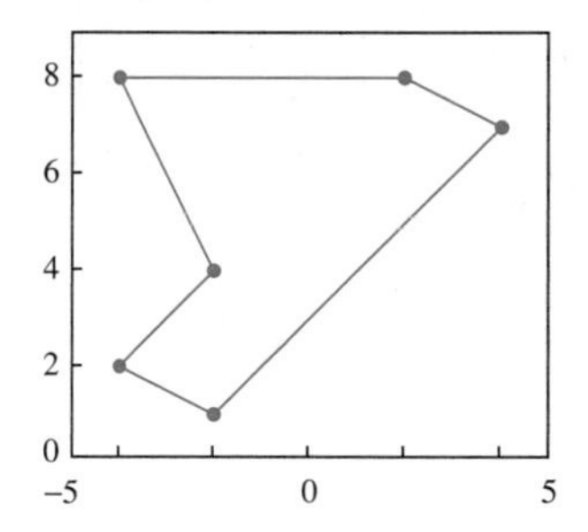

(b) $T(S) = AS + \mathbf{b}$

$$= \begin{bmatrix} 2 & -2 \\ 2 & -2 \end{bmatrix} \begin{bmatrix} 0 & 0 & 1 & 2 & 3 & 3 & 0 \\ 0 & 1 & 1 & 3 & 1 & 0 & 0 \end{bmatrix} + \begin{bmatrix} -2 & -2 & -2 & -2 & -2 & -2 & -2 \\ 1 & 1 & 1 & 1 & 1 & 1 & 1 \end{bmatrix}$$

$$= \begin{bmatrix} -2 & -4 & -2 & -4 & 2 & 4 & -2 \\ 1 & -1 & 1 & -1 & 5 & 7 & 1 \end{bmatrix}.$$

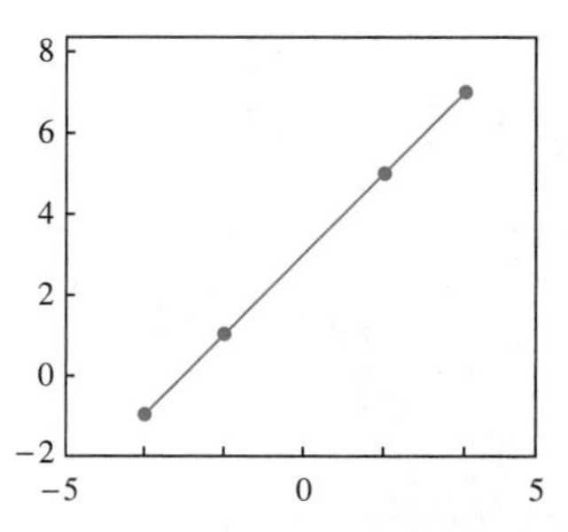

(c) $T(S) = AS + \mathbf{b}$

$$= \begin{bmatrix} 2 & 2 \\ -2 & 1 \end{bmatrix} \begin{bmatrix} 0 & 0 & 1 & 2 & 3 & 3 & 0 \\ 0 & 1 & 1 & 3 & 1 & 0 & 0 \end{bmatrix} + \begin{bmatrix} -2 & -2 & -2 & -2 & -2 & -2 & -2 \\ 1 & 1 & 1 & 1 & 1 & 1 & 1 \end{bmatrix}$$

$$= \begin{bmatrix} -2 & 0 & 2 & 8 & 6 & 4 & -2 \\ 1 & 2 & 0 & 0 & -4 & -5 & 1 \end{bmatrix}.$$

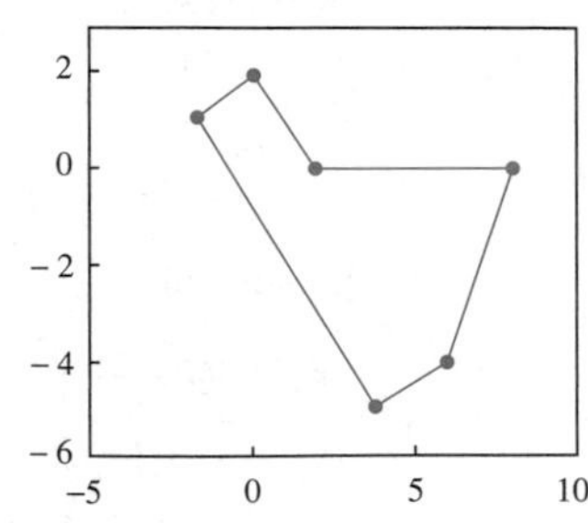

13. $\mathbf{b} = \begin{bmatrix} -3 \\ -3 \end{bmatrix}$, $A = \begin{bmatrix} \frac{1}{2} & 0 \\ 0 & \frac{1}{2} \end{bmatrix}$.

15. $A = \begin{bmatrix} 2 & 0 \\ -2 & 1 \end{bmatrix}$, $\mathbf{b} = \begin{bmatrix} -1 \\ 0 \end{bmatrix}$.

17. (a)

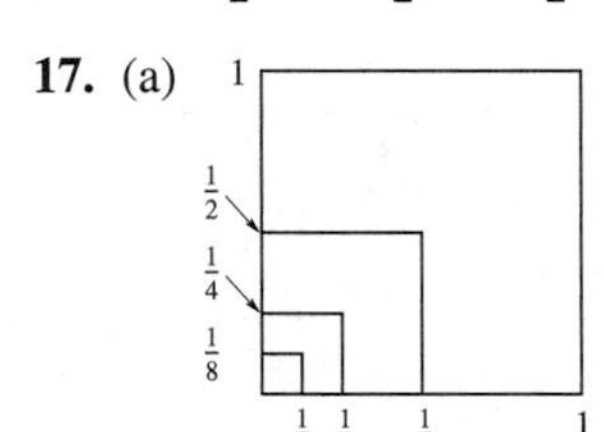

(b)

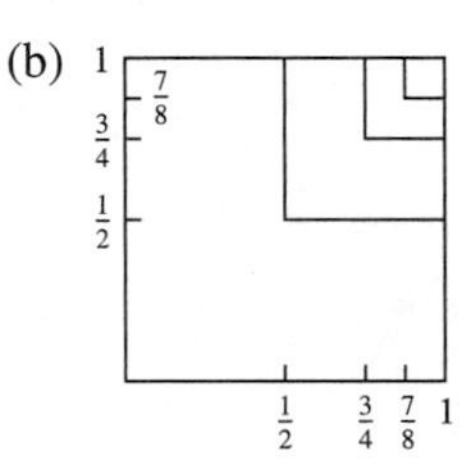

(c)

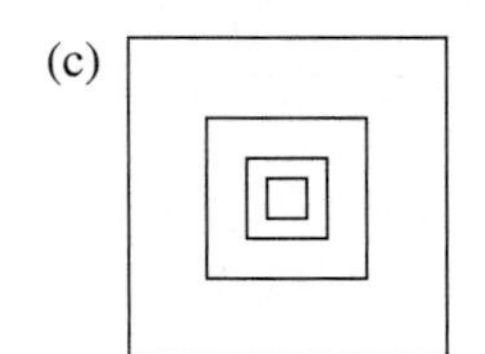

(d)

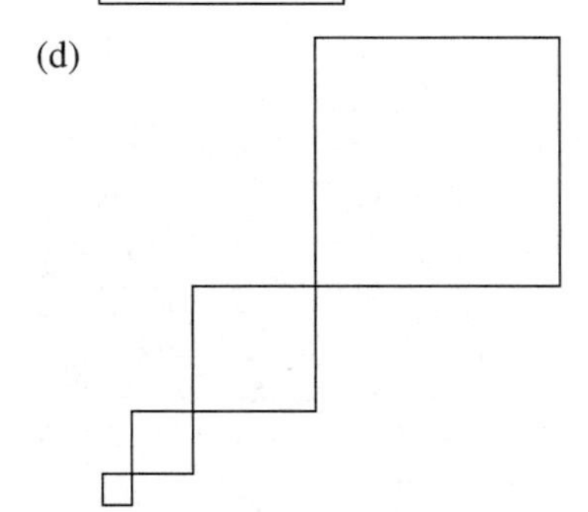

ML.1. Command **fernifs([0 .2],30000)** produces the following figure.

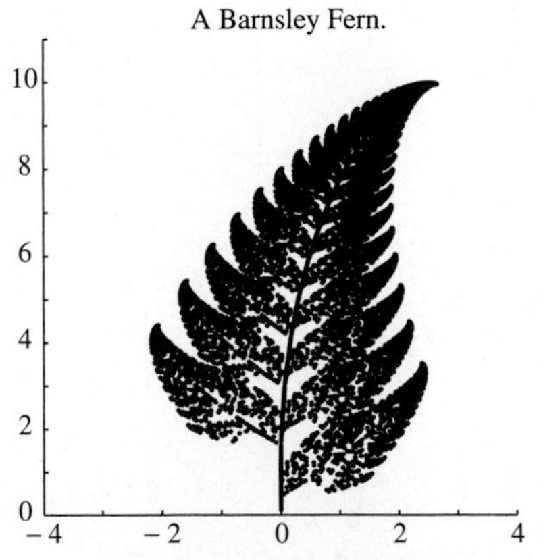

Routine is over. Press ENTER twice.

Supplementary Exercises, page 552

1. Yes.

3. $8t + 7$.

5. (a) Possible answer: $\{(1, 1, 1), (1, -1, 2)\}$.

(b) No.

7. (a) $\begin{bmatrix} 2 \\ 2 \\ -4 \end{bmatrix}$. (b) $2t^2 + 2t - 7$.

9. (a) $\begin{bmatrix} 2 \\ 1 \end{bmatrix}, \begin{bmatrix} -3 \\ 2 \end{bmatrix}$. (b) $3t - 3, -t + 8$.

(c) $-\frac{7}{3}t + \frac{35}{3}$.

11. $\begin{bmatrix} 1 & 0 & 0 \\ 0 & 1 & 0 \\ 0 & 0 & 1 \end{bmatrix}$.

13. Yes.

15. (b) The kernel of L consists of any continuous function f such that $L(f) = f(0) = 0$. That is, f is in $\ker L$ provided the value of f at $x = 0$ is zero.

(c) Yes.

17. (a)

(b) A rectangular spiral.

Chapter Test, page 554

1. $\begin{bmatrix} 2 \\ 7 \\ 4 \end{bmatrix}$.

2. (a) $\ker L = \{(0, 0, 0)\}$, so there is no basis.

(b) Yes.

3. (a) Possible answer: $\{(1, 1, 2), (1, -1, 1)\}$.

(b) No.

4. 2.

5. $\begin{bmatrix} 0 & \frac{3}{2} \\ 1 & -\frac{5}{2} \end{bmatrix}$.

6. (a) F. (b) T. (c) T. (d) F. (e) F.

7. $T(\mathbf{v}) = A\mathbf{v}$, where $A = \begin{bmatrix} \frac{1}{2} & 0 \\ 0 & \frac{1}{2} \end{bmatrix}$.

8.

Cumulative Review of Introductory Linear Algebra, page 555

1. F.	**2.** T.	**3.** F.	**4.** F.	**5.** F.
6. T.	**7.** T.	**8.** T.	**9.** F.	**10.** T.
11. T.	**12.** F.	**13.** F.	**14.** F.	**15.** T.
16. T.	**17.** F.	**18.** F.	**19.** F.	**20.** T.
21. T.	**22.** F.	**23.** T.	**24.** T.	**25.** F.
26. T.	**27.** F.	**28.** F.	**29.** F.	**30.** T.
31. T.	**32.** T.	**33.** T.	**34.** T.	**35.** T.
36. F.	**37.** F.	**38.** F.	**39.** F.	**40.** T.
41. F.	**42.** F.	**43.** F.	**44.** T.	**45.** T.
46. F.	**47.** T.	**48.** T.	**49.** T.	**50.** T.
51. T.	**52.** F.	**53.** T.	**54.** T.	**55.** T.
56. F.	**57.** T.	**58.** T.	**59.** T.	**60.** F.
61. F.	**62.** T.	**63.** F.	**64.** F.	**65.** F.
66. F.	**67.** F.	**68.** T.	**69.** T.	**70.** T.
71. T.	**72.** F.	**73.** T.	**74.** T.	**75.** F.
76. T.	**77.** T.	**78.** T.	**79.** F.	**80.** F.
81. T.	**82.** T.	**83.** F.	**84.** T.	**85.** F.
86. F.	**87.** F.	**88.** F.	**89.** F.	**90.** F.
91. T.	**92.** F.	**93.** F.	**94.** T.	**95.** F.
96. T.	**97.** T.	**98.** T.	**99.** T.	**100.** T.

Chapter 11

Section 11.1, page 572

1. Maximize $z = 120x + 100y$ subject to

$$\begin{aligned} 2x + 2y &\leq 8 \\ 5x + 3y &\leq 15 \\ x \geq 0, \quad y &\geq 0. \end{aligned}$$

3. Maximize $z = 0.08x + 0.10y$ subject to

$$\begin{aligned} x + y &\leq 6000 \\ x &\geq 1500 \\ y &\leq 4000 \\ y &\leq \tfrac{1}{2}x \\ x \geq 0, \quad y &\geq 0. \end{aligned}$$

5. Maximize $z = 40{,}000x + 45{,}000y$ subject to

$$\begin{aligned} x + y &\leq 30 \\ y &\geq 24 \\ x &\geq 2 \\ x &\leq 4 \\ x \geq 0, \quad y &\geq 0. \end{aligned}$$

7. Maximize $z = 4x + 6y$ subject to

$$\begin{aligned} x + 2y &\leq 10 \\ x + y &\leq 7 \\ x \geq 0, \quad y &\geq 0. \end{aligned}$$

9. Minimize $z = 10x + 12y$ subject to

$$\begin{aligned} 2x + 3y &\geq 18 \\ x + 3y &\geq 12 \\ 80x + 60y &\geq 480 \\ x \geq 0, \quad y &\geq 0. \end{aligned}$$

11.

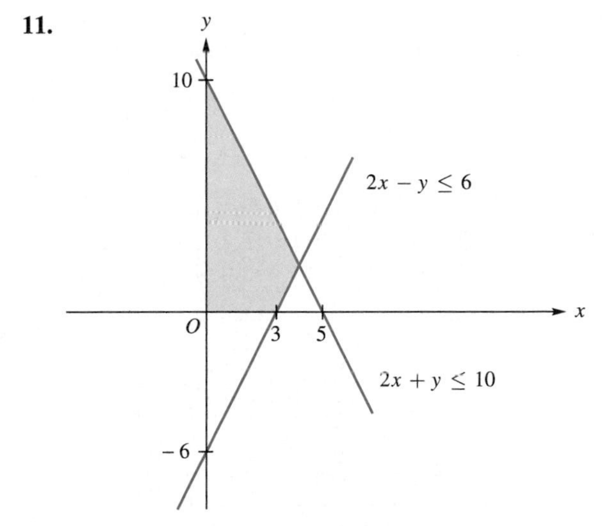

13.

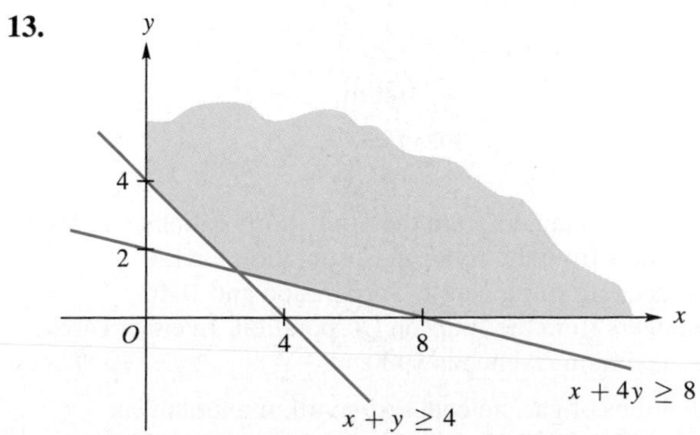

15. $x = \frac{8}{11}$, $y = \frac{45}{11}$, optimal value of $z = -\frac{21}{11}$.

17. Invest \$2000 in bond A and \$4000 in bond B; maximum return is \$560.

19. Carry no containers from the Smith corporation and 1500 containers from the Johnson Corporation, or 120 containers from the Smith Corporation and 1440 containers from the Johnson Corporation. In either case the maximum revenue is \$900.

21. Use $\frac{5}{2}$ gallons of L and $\frac{3}{2}$ gallons of H; minimum cost is \$2.25.

23. No solution. **25.** (a).

27. Maximize $z = 3x_1 - x_2 + 6x_3$
subject to

$$\begin{aligned} 2x_1 + 4x_2 + x_3 &\le 4 \\ 3x_1 - 2x_2 + 3x_3 &\le 4 \\ 2x_1 + x_2 - x_3 &\le 8 \\ x_1 \ge 0, \quad x_2 \ge 0, \quad x_3 &\ge 0. \end{aligned}$$

29. Maximize $z = 2x_1 + 3x_2 + 7x_3$
subject to

$$\begin{aligned} 3x_1 + x_2 - 4x_3 + x_4 \qquad\qquad &= 3 \\ x_1 - 2x_2 + 6x_3 \qquad + x_5 \qquad &= 21 \\ x_1 - x_2 - x_3 \qquad\qquad + x_6 &= 9 \end{aligned}$$
$$x_1 \ge 0,\ x_2 \ge 0,\ x_3 \ge 0,\ x_4 \ge 0,\ x_5 \ge 0,\ x_6 \ge 0.$$

Section 11.2, page 589

1.

	x	y	u	v	w	z	
u	3	−2	1	0	0	0	7
v	2	5	0	1	0	0	6
w	2	3	0	0	1	0	8
	−3	−7	0	0	0	1	0

3.

	x_1	x_2	x_3	x_4	x_5	x_6	x_7	z	
x_5	3	−2	1	1	1	0	0	0	6
x_6	1	1	1	1	0	1	0	0	8
x_7	2	−3	−1	2	0	0	1	0	10
	−2	−2	−3	−1	0	0	0	1	0

5. $x = 2$, $y = 0$, optimal $z = 4$.

7. No finite optimal solution.

9. $x_1 = 0$, $x_2 = \frac{33}{20}$, $x_3 = \frac{27}{10}$, optimal $z = \frac{69}{10}$.

11. $x_1 = 0$, $x_2 = 0$, $x_3 = 49$, $x_4 = 41$, optimal $z = 156$.

13. Carry no containers from the Smith corporation and 1500 containers from the Johnson Corporation, or 120 containers from the Smith Corporation and 1440 containers from the Johnson Corporation. In either case the maximum revenue is \$900.

15. Use 4 tons of gas, no coal and no oil; maximum energy generated is 2000 kilowatt hours.

ML.3. $x = 0$, $y = 0.8571$, optimal $z = 4.286$.

ML.5. $x_1 = 1$, $x_2 = 0.3333$, $x_3 = 0$, optimal $z = 11$.

ML.7. Exercise 10: $x_1 = 0$, $x_2 = 2.5$, $x_3 = 0$, optimal $z = 10$. Exercise 12: 1.5 tons of regular steel and 2.5 tons of special steel; maximum profit is \$430.

Section 11.3, page 598

1. Minimize $z' = 7w_1 + 6w_2 + 9w_3$
subject to

$$\begin{aligned} 4w_1 + 5w_2 + 6w_3 &\ge 3 \\ 3w_1 - 2w_2 + 8w_3 &\ge 2 \\ w_1 \ge 0, \quad w_2 \ge 0, \quad w_3 &\ge 0. \end{aligned}$$

3. Maximize $z' = 7w_1 + 12w_2 + 18w_3$
subject to

$$\begin{aligned} 2w_1 + 8w_2 + 10w_3 &\le 3 \\ 3w_1 - 9w_2 + 15w_3 &\le 5 \\ w_1 \ge 0, \quad w_2 \ge 0, \quad w_3 &\ge 0. \end{aligned}$$

7. $w_1 = \frac{5}{7}$, $w_2 = 0$, $w_3 = 0$, optimal $z' = \frac{30}{7}$.

9. $w_1 = 2$, $w_2 = 0$, $w_3 = 0$, optimal $z' = 10$.

Section 11.4, page 612

1.

		C	
		2 fingers shown	3 fingers shown
R	2 fingers shown	−4	5
	3 fingers shown	5	−6

3.

		Firm B	
		Abington	Wyncote
Firm A	Abington	50	60
	Wyncote	25	50

5. (a) $\begin{bmatrix} 5 & \textcircled{4} \\ 3 & -2 \end{bmatrix}$. (b) $\begin{bmatrix} 2 & 1 & \textcircled{0} \\ 3 & 1 & -2 \\ 4 & 2 & -4 \end{bmatrix}$.

(c) $\begin{bmatrix} \textcircled{3} & 4 & 5 \\ -2 & 5 & 1 \\ -1 & 0 & 1 \end{bmatrix}$.

(d) $\begin{bmatrix} 5 & \textcircled{2} & 4 & \textcircled{2} \\ 0 & -1 & 2 & 0 \\ 3 & \textcircled{2} & 3 & \textcircled{2} \\ 1 & 0 & -1 & -1 \end{bmatrix}$.

7. (a) $\mathbf{p} = \begin{bmatrix} 1 & 0 \end{bmatrix}$, $\mathbf{q} = \begin{bmatrix} 0 \\ 1 \\ 0 \end{bmatrix}$, $v = 1$.

(b) $\mathbf{p} = \begin{bmatrix} 0 & 0 & 1 \end{bmatrix}$, $\mathbf{q} = \begin{bmatrix} 1 \\ 0 \\ 0 \\ 0 \end{bmatrix}$, $v = 0$.

(c) $\mathbf{p} = \begin{bmatrix} 1 & 0 \end{bmatrix}$ or $\begin{bmatrix} 0 & 1 \end{bmatrix}$, $\mathbf{q} = \begin{bmatrix} 0 \\ 1 \end{bmatrix}$, $v = 4$.

9. (a) $\frac{19}{36}$. (b) $\frac{1}{7}$.

11. $p_1 = \frac{9}{14}$, $p_2 = \frac{5}{14}$, $q_1 = \frac{1}{2}$, $q_2 = \frac{1}{2}$, $v = -\frac{1}{2}$.

13. $p_1 = 0$, $p_2 = \frac{4}{5}$, $p_3 = \frac{1}{5}$, $q_1 = 0$, $q_2 = \frac{3}{5}$, $q_3 = \frac{2}{5}$, $v = \frac{22}{5}$.

15. $\mathbf{p} = \begin{bmatrix} 0 & \frac{3}{8} & -\frac{5}{8} & 0 \end{bmatrix}$, $\mathbf{q} = \begin{bmatrix} \frac{5}{8} \\ \frac{3}{8} \\ 0 \\ 0 \end{bmatrix}$, $v = \frac{1}{8}$.

17. $\mathbf{p} = \begin{bmatrix} \frac{1}{3} & \frac{1}{3} & \frac{1}{3} \end{bmatrix}$, $\mathbf{q} = \begin{bmatrix} \frac{1}{3} \\ \frac{1}{3} \\ \frac{1}{3} \end{bmatrix}$, $v = 0$.

19. $\mathbf{p} = \begin{bmatrix} \frac{2}{3} & \frac{1}{3} \end{bmatrix}$, $\mathbf{q} = \begin{bmatrix} \frac{1}{2} \\ \frac{1}{2} \end{bmatrix}$, $v = 0$.

Supplementary Exercises, page 614

1. $x = \frac{6}{5}$, $y = \frac{12}{5}$, optimal $z = \frac{48}{5}$.

3. $x = 0$, $y = 8$, or $x = 2$, $y = 7$, optimal $z = 800$.

5. $x_1 = \frac{18}{11}$, $x_2 = \frac{10}{11}$, optimal $z = \frac{158}{11}$.

Chapter Test, page 614

1. Plant no corn, plant 20 acres of wheat.

2. $x_1 = 1$, $x_2 = \frac{1}{3}$, $x_3 = 0$, optimal $z = 11$.

3. Maximize $z' = 8y_1 + 12y_2 + 6y_3$
subject to

$$\begin{aligned} y_1 + 2y_2 + 2y_3 &\leq 3 \\ 4y_1 + 3y_2 + \ \ y_3 &\leq 4 \\ y_1 \geq 0, \quad y_2 \geq 0, \quad y_3 &\geq 0. \end{aligned}$$

4. $\mathbf{p} = \begin{bmatrix} \frac{3}{8} & \frac{5}{8} \end{bmatrix}$, $\mathbf{q} = \begin{bmatrix} \frac{1}{8} \\ \frac{7}{8} \\ 0 \end{bmatrix}$, $v = \frac{11}{8}$.

5. $a_{11} = 2$ is a saddle point for any value of a.

Appendix A

Section A.1, page A7

1. (a) $4 + 2i$. (b) $-4 - 3i$. (c) $11 - 2i$.
(d) $-3 + i$. (e) $-3 + 6i$. (f) $-2 - i$.
(g) $7 - 11i$. (h) $-9 + 13i$.

3. (a)

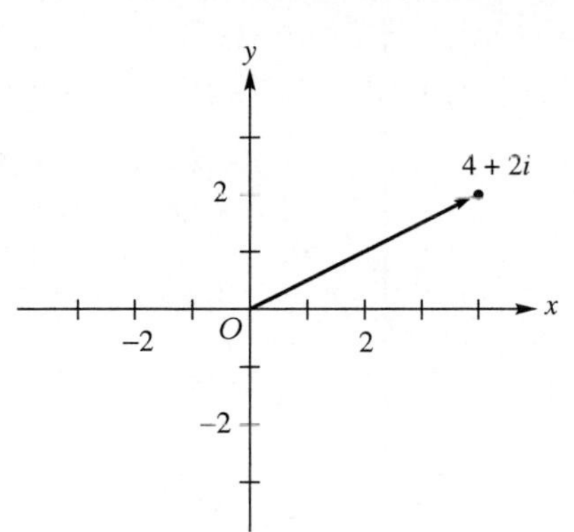

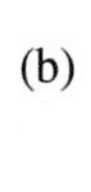
(b)

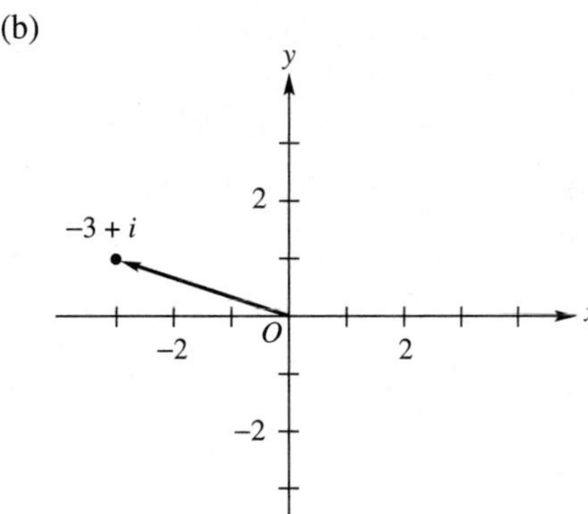

(c)

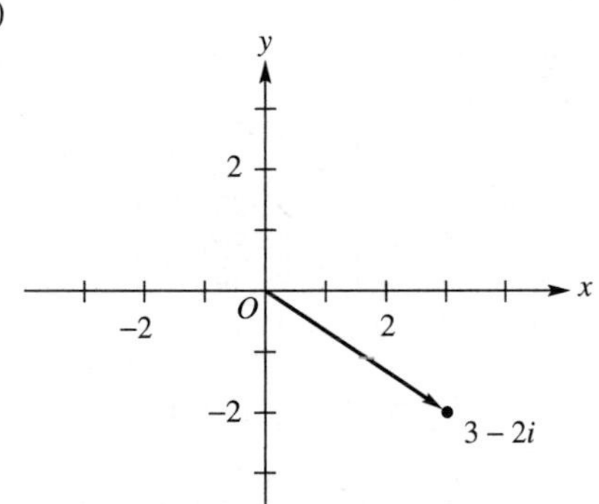

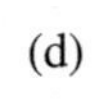
(d)

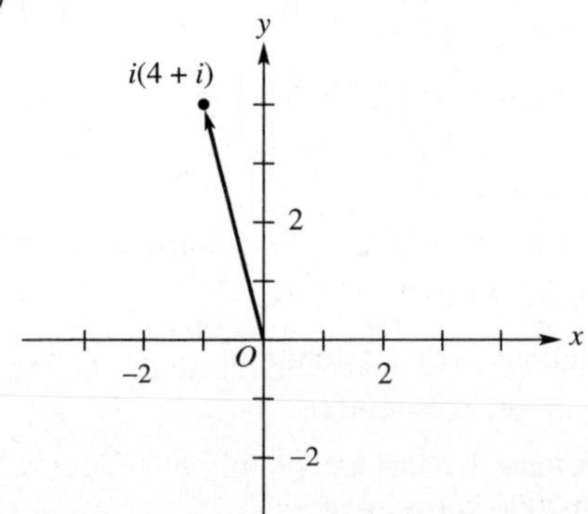

5.

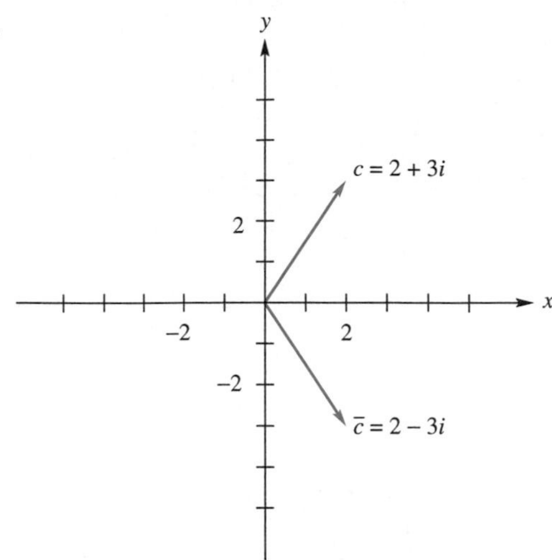

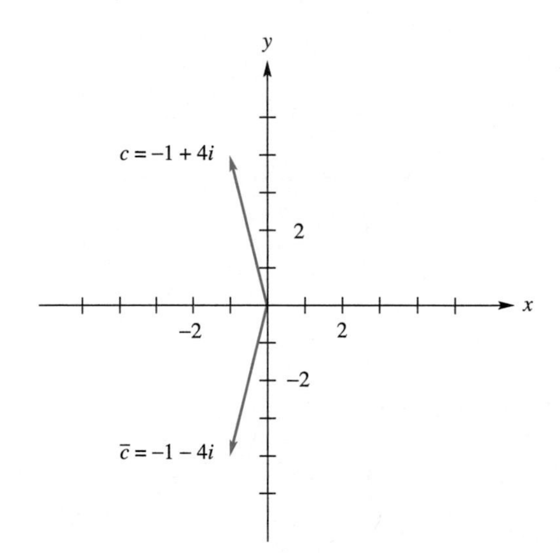

7. $A^2 = \begin{bmatrix} -1 & 0 \\ 0 & -1 \end{bmatrix}$, $A^3 = \begin{bmatrix} 0 & -i \\ -i & 0 \end{bmatrix}$,
$A^4 = \begin{bmatrix} 1 & 0 \\ 0 & 1 \end{bmatrix}$, $A^{4n} = I_2$, $A^{4n+1} = A$,
$A^{4n+2} = A^2 = -I_2$, $A^{4n+3} = A^3 = -A$.

9. (a) $\dfrac{-1 \pm i\sqrt{3}}{2}$. (b) $-2, \pm i$. (c) $\pm 1, \pm i$.

11. (a) Possible answers: $A_1 = \begin{bmatrix} i & 0 \\ 0 & i \end{bmatrix}$, $A_2 = \begin{bmatrix} -i & 0 \\ 0 & -i \end{bmatrix}$.

13. (a) Possible answers: $\begin{bmatrix} i & 0 \\ 0 & 0 \end{bmatrix}, \begin{bmatrix} -i & 0 \\ 0 & 0 \end{bmatrix}$.
(b) Possible answers: $\begin{bmatrix} i & -i \\ -i & i \end{bmatrix}, \begin{bmatrix} -i & i \\ i & -i \end{bmatrix}$.

Section A.2, page A17

1. (a) No solution. (b) No solution.
(c) $x_1 = \frac{3}{4} + \frac{5}{4}i$, $x_2 = \frac{3}{2} + i$.

3. (a) $x_1 = i$, $x_2 = 1$, $x_3 = 1 - i$.
(b) $x_1 = 0$, $x_2 = -i$, $x_3 = i$.

5. (a) $\dfrac{1}{5}\begin{bmatrix} 2+i & 2-4i \\ 3-i & -2-i \end{bmatrix}$.
(b) $\dfrac{1}{6}\begin{bmatrix} i & 1-3i & 1 \\ -2-3i & 2i & 3+2i \\ 1 & 2i & -i \end{bmatrix}$.

7. (a) Yes. (b) Linearly independent.

9. (a) $16i$. (b) $5 - 17i$.

11. (a) No. (b) No. (c) No. (d) Yes.

13. (a) The eigenvalues are $\lambda_1 = 1 + i$, $\lambda_2 = 1 - i$. Associated eigenvectors are
$$\mathbf{x}_1 = \begin{bmatrix} -i \\ 1 \end{bmatrix} \quad \text{and} \quad \mathbf{x}_2 = \begin{bmatrix} i \\ 1 \end{bmatrix}.$$
(b) The eigenvalues are $\lambda_1 = 0$, $\lambda_2 = 2$. Associated eigenvectors are
$$\mathbf{x}_1 = \begin{bmatrix} -i \\ 1 \end{bmatrix} \quad \text{and} \quad \mathbf{x}_2 = \begin{bmatrix} i \\ 1 \end{bmatrix}.$$
(c) The eigenvalues are $\lambda_1 = 1$, $\lambda_2 = 2$, $\lambda_3 = 3$. Associated eigenvectors are
$$\mathbf{x}_1 = \begin{bmatrix} 0 \\ -i \\ 1 \end{bmatrix}, \quad \mathbf{x}_2 = \begin{bmatrix} 1 \\ 0 \\ 0 \end{bmatrix}, \quad \text{and} \quad \mathbf{x}_3 = \begin{bmatrix} 0 \\ i \\ 1 \end{bmatrix}.$$

Appendix B

Section B.1, page A28

7. -8. **9.** 4. **11.** -4.

13. (a) $\sqrt{22}$. (b) 6. (c) $\sqrt{18}$.

15. (a) $\sqrt{\frac{1}{30}}$. (b) $\sqrt{\frac{1}{2}\left(e^2 - e^{-2}\right) - 2}$.

17. (a) $-\frac{1}{2}$. (b) $\dfrac{2\sin^2 1}{\sqrt{4 - \sin^2 2}}$. **19.** $a = 0$.

21. $B = \begin{bmatrix} b_{11} & b_{12} \\ b_{21} & b_{22} \end{bmatrix}$ with $b_{11} + 3b_{21} + 2b_{12} + 4b_{22} = 0$.

23. (b) $\{(0, 0, 1, 0), (-1, 1, 0, 1)\}$.

25. (a) $\{\sqrt{3}\,t, 2 - 3t\}$.
(b) $2t - 1 = \frac{\sqrt{3}}{6}(\sqrt{3}\,t) - \frac{1}{2}(2 - 3t)$.

27. $\left\{\sqrt{3}\,t, \dfrac{\sin 2\pi t + \left(\frac{3}{2\pi}\right)t}{\sqrt{\frac{1}{2} - \frac{3}{4\pi^2}}}\right\}$.

29. $\left\{\frac{45}{14}t^3 - \frac{55}{14}t^2 + t,\ \frac{130}{7}t^3 - \frac{120}{7}t^2 + 1\right\}$.

31. $2\sin t$.

33. $\mathbf{w} = 2\sin t - 1$, $\mathbf{u} = t - 1 - (2\sin t - 1) = t - 2\sin t$.

35. $\sqrt{\frac{2}{3}\pi^3 - 4\pi}$.

37. $\dfrac{\pi^2}{3} - 4\cos t + \cos 2t$.

Section B.2, page A35

1. (a) $(-3, -3, -1)$. (b) $(4x + y, 3x, 3x + 2y)$.

3. (a) $6t^2 - 2t$. (b) $2at^2 - bt$.

5. (a) $(-3, 6)$. (b) $(1, 4)$.
(c) $(x - 2y, 3y)$. (d) $(x, -2x + 3y)$.

7. (a) $B = \begin{bmatrix} 2 & 2 \\ -3 & -1 \\ 2 & 2 \end{bmatrix}$.

(b) $A_1 = \begin{bmatrix} \frac{1}{2} & \frac{3}{2} \\ \frac{3}{2} & \frac{1}{2} \end{bmatrix}$, $A_2 = \begin{bmatrix} 2 & 2 \\ -3 & -1 \\ 2 & 2 \end{bmatrix}$.

9. (a) $\begin{bmatrix} -1 & 3 \\ -5 & 5 \end{bmatrix}$. (b) $\begin{bmatrix} -4 & 7 \\ -6 & 8 \end{bmatrix}$.

11. Not invertible.

13. Invertible. $L^{-1}\left(\begin{bmatrix} b_1 \\ b_2 \\ b_3 \end{bmatrix}\right) = \begin{bmatrix} 2b_1 - b_3 \\ b_1 + b_2 - b_3 \\ -b_1 + b_3 \end{bmatrix}$.

15. Not invertible.

17. Invertible. $L^{-1}(dt^2 + et + f) = -dt^2 + et - f$.

19. Invertible.

21. Not invertible.

23. $\begin{bmatrix} \frac{1}{2} & -\frac{1}{2} & \frac{1}{2} \\ -\frac{3}{2} & \frac{3}{2} & -\frac{1}{2} \\ \frac{1}{2} & \frac{1}{2} & -\frac{1}{2} \end{bmatrix}$.

25. $\begin{bmatrix} -\frac{9}{2} & -6 & 2 \\ \frac{1}{2} & 1 & 0 \\ \frac{5}{2} & 3 & -1 \end{bmatrix}$.

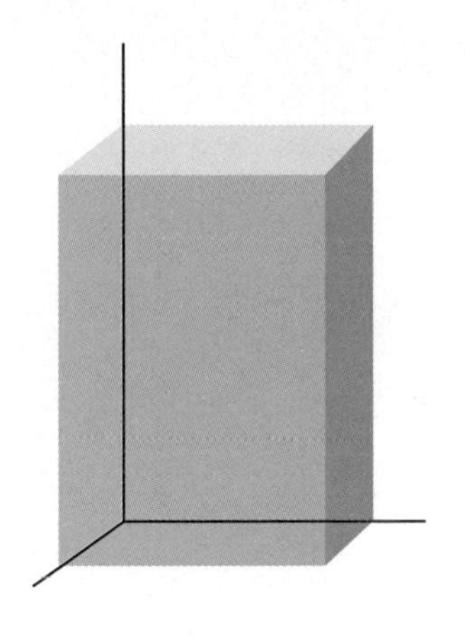

INDEX

D

G

H

I

J

N

O

P

T

LIST OF FREQUENTLY USED SYMBOLS

Symbol	Meaning
$A = [a_{ij}]$	An $m \times n$ matrix, p. 10
A^T	The transpose of the matrix A, p. 16
$[A \mid B]$	An augmented matrix, p. 28
$\sum_{i=1}^{n} a_i$	Summation notation, p. 32
O	The zero $m \times n$ matrix, p. 39
I_n	The $n \times n$ identity matrix, p. 42
A^{-1}	The inverse of the matrix A, p. 91
B^m	The set of all bit m-vectors, p. 119
e	An encoding function, p. 121
$A(G)$	The adjacency matrix of a digraph G, p. 126
$\mathbf{x}^{(k)}$	The state vector of a Markov process at the observation period k, p. 151
$p_j^{(k)}$	The probability that a system is in state j at the observation period k, p. 151
$\mathbf{x}^{(0)}$	The initial state vector of a Markov process, p. 151
$j_1 j_2 \cdots j_n$	A permutation of $S = \{1, 2, \ldots, n\}$, p. 182
$\det(A)$	The determinant of the matrix A, p. 183
$\|A\|$	The determinant of the matrix A, p. 183
$\det(M_{ij})$	The minor of a_{ij}, p. 196
A_{ij}	The cofactor of a_{ij}, p. 196
adj A	The adjoint of the matrix A, p. 200
$\mathbf{u}, \mathbf{v}, \mathbf{w}, \mathbf{x}, \mathbf{y}, \mathbf{z}$	Vectors in a vector space, pp. 214, 272
$\overrightarrow{OP}$	A directed line segment, p. 216
$\\|\mathbf{u}\\|$	Length of the vector $\mathbf{u}$, pp. 219, 235
$\mathbf{0}$	The zero vector, pp. 222, 272
$-\mathbf{u}$	The negative of the vector $\mathbf{u}$, pp. 222, 272
$\mathbf{x} \cdot \mathbf{y}$	Dot product, standard inner product, pp. 22, 235
θ	Angle between two nonzero vectors, pp. 224, 237
R^n	n-space; the vector space of all n-vectors, p. 229
$\\|\mathbf{u} - \mathbf{v}\\|$	Distance between the vectors $\mathbf{u}$ and $\mathbf{v}$ in R^n, p. 235
L	A linear transformation, p. 247, 502
$L(\mathbf{u})$	The image of $\mathbf{u}$, p. 247
range L	The range of the linear transformation L, pp. 247, 512
$\times$	The cross product operator, p. 259
V, W	Vector spaces, p. 272
M_{mn}	The vector space of all $m \times n$ matrices, p. 274
$p(t)$	A polynomial in t, p. 274
P_n	The vector space of all polynomials of degree $\leq n$ and the zero polynomial, p. 274
P	The vector space of all polynomials, p. 276